Journal of Communication Disorders
Journal of Consumer Affairs
Journal of Drug Issues
Journal of Education for Business
Journal of Engineering for Industry
Journal of Environmental Education
Journal of Experimental Education
Journal of Experimental Social Psychology
Journal of Financial Planning
Journal of Gang Research
Journal of Genetic Psychology
Journal of Geography
Journal of Gerontological Nursing
Journal of Hazardous Materials
Journal of Heat Transfer
Journal of Information Systems
Journal of Interdisciplinary History
Journal of the Institute for Socioeconomic Studies
Journal of Literacy Research
Journal of the Market Research Society
Journal of Marriage and the Family
Journal of Modern History
Journal of Moral Education
Journal of the National Cancer Institute
Journal of Nonverbal Behavior
Journal of Nursing Education
Journal of Nutrition
Journal of Occupational and Organizational Psychology
Journal of Parapsychology
Journal of Peace Research
Journal of Performance of Constructed Facilities
Journal of Personal Selling & Sales Management
Journal of Psychology and Aging
Journal of Quality Technology
Journal of Research in Music Education
Journal of Speech, Language, and Hearing Research
Journal of Sport & Social Issues
Journal of Teaching in Physical Education
Journal of Testing and Evaluation
Journal of Transportatio[illegible] Engineering
Journal of Travel Researc[illegible]
Journal of Visual Impairment & Blindness
Journal of Zoology
Kelly Uscategui
Kimberly-Clark Corp.
Lisa Schwartz
Lottery Buster
Major League Baseball
Management Science
Marine Technology
Medical Interfaces
Memory & Cognition
Merck Research Labs
Microelectronics and Reliability
Morbidity and Mortality Weekly Report
National Aeronautic and Space Administration
National Center for Education Statistics
National Center for Health Statistics
National Earthquake Information Center
National Highway Traffic Safety Administration
National Institute of Justice
National Science Foundation
National Underwriters
Nature
National Wildlife
Networks
New Accountant
New England Journal of Medicine
New Jersey Governor's Council
New York Times
Newsday
Newsweek
Nuclear Science and Engineering
Oncology Nursing Forum
Online
Organizational Science
Parade Magazine
Perception & Psychophysics
Prison Journal
Proceedings of the Institute of Civil Engineers
[illegible] National Science Council, Republic of China
Procter & Gamble
Prog. Oceanog.
Psychological Assessment
Psychological Reports
Psychological Science
Psychonomic Bulletin & Review
Public Health Reports
Quality
Quality Engineering
Quarterly Journal of Experimental Psychology
Risk Management
Science
Science Education
Scientific American
Scripps Institution of Oceanography
Self
Shere Hite
Simulation
Skeptical Inquirer
Sleep
Sloan Management Review
Small Group Behavior
Sociology of Sport Journal
Sports Illustrated
Statistical Abstract of the United States
Statistics in Sport
Susan L. Farber
Tampa Tribune
Teaching Psychology
Teaching Statistics
Technometrics
Tennis
The Sporting News
Time
Transportation Journal
University of South Florida
U.S. Army Corps of Engineers
U.S. Department of Agriculture
U.S. Golf Association
U.S. News & World Report
USA Today
Wall Street Journal
Washington Post
Wetlands
Women & Therapy
Zagat Survey

SECOND EDITION

Practical Statistics by Example Using Microsoft® Excel and MINITAB®

TERRY SINCICH
University of South Florida

DAVID M. LEVINE
Bernard M. Baruch College (CUNY)

DAVID STEPHAN
Bernard M. Baruch College (CUNY)

Prentice Hall

PRENTICE HALL
Upper Saddle River, New Jersey 07458

Library of Congress Cataloging-in-Publication Data

Sincich, Terry.
Practical statistics by example using Microsoft Excel and MINITAB / Terry Sincich, David M. Levine, David Stephan.—2nd ed.
p. cm.
Rev. ed. of: Practical statistics by example using Microsoft Excel. c1999.
Includes bibliographical references and index.
ISBN 0-13-041521-9
1. Statistics. 2. Microsoft Excel (Computer file) 3. Electronic spreadsheets. I. Levine, David M., 1946– II. Stephan, David. III. Sincich, Terry. Practical statistics by example using Microsoft Excel. IV. Title.

QA276.12 .S558 2002
519.5—dc21 2001032713

Acquisition Editor: Quincy McDonald
Editor-in-Chief: Sally Yagan
Vice President/Director of Production and Manufacturing: David W. Riccardi
Executive Managing Editor: Kathleen Schiaparelli
Senior Managing Editor: Linda Mihatov Behrens
Project Management: Elm Street Publishing Services, Inc.
Manufacturing Buyer: Alan Fischer
Manufacturing Manager: Trudy Pisciotti
Marketing Manager: Angela Battle
Assistant Editor of Media: Vince Jansen
Editorial Assistant/Supplements Editor: Joanne Wendelken
Art Director: Joseph Sengotta
Assistant to the Art Director: John Christiana
Interior Designer: Studio Montage
Cover Designer: Maureen Eide
Managing Editor, Audio/Video Assets: Grace Hazeldine
Creative Director: Carole Anson
Director of Creative Services: Paul Belfanti
Cover Photo: Diane Fenster "Hold the World"
Art Studio: Elm Street Publishing Services, Inc.

Printed in the United States of America
10 9 8 7 6 5 4 3 2 1

0-13-041521-9

Pearson Education Ltd., *London*
Pearson Education Australia Pty., Limited, *Sydney*
Pearson Education Singapore, Pte. Ltd.
Pearson Education North Asia Ltd., *Hong Kong*
Pearson Education Canada, Ltd., *Toronto*
Pearson Educación de Mexico, S.A. de C.V.
Pearson Education—Japan, *Tokyo*
Pearson Education Malaysia, Pte. Ltd
Pearson Education, *Upper Saddle River, New Jersey*

Contents

CHAPTER 1 Introduction: Statistics and Data 21

CHAPTER 2 Exploring Data with Graphs and Tables 57

CHAPTER 7 Estimation of Population Parameters Using Confidence Intervals: One Sample 337

CHAPTER 8 Testing Hypotheses about Population Parameters: One Sample 388

Preface

Educational Philosophy

In our many years of teaching introductory statistics courses at the University of South Florida and Baruch College, we have continually searched for ways to improve the teaching of these courses. Our vision for teaching these introductory statistics courses has been shaped by active participation in a series of professional conferences as well as the reality of serving a diverse group of students at a large university. Over the years, our vision has come to include these principles:

- Students need a frame of reference when learning about a subject, especially one that is not their major. That frame of reference for introductory statistics students should be the various areas in which statistics can be applied, including business, biology, education, engineering, mathematics, political science, psychology, and sociology. Each statistical topic needs to be related to at least one of these areas of application.
- Virtually all the students taking introductory statistics courses are majoring in areas other than statistics. Introductory courses should focus on underlying principles that are important for non-statistics majors.
- The use of spreadsheet and/or statistical software should be integrated into all aspects of the introductory statistics course. The reality that exists in the workplace is that spreadsheet software (and sometimes statistical software) is most typically available on the desktop. Our teaching approach needs to recognize this reality and make our courses more consistent with the workplace environment.
- Textbooks that use software must provide instructions at a depth that maximizes the student's ability to use the software with a minimum risk of failure.
- The focus in teaching each topic should be on (1) the application of the topic to a specific problem, (2) the interpretation of results, (3) the presentation of assumptions, (4) the evaluation of the assumptions, and (5) the discussion of what should be done if the assumptions are violated. These points are particularly important in regression and forecasting and in hypothesis testing. Although the illustration of some computations is inevitable, the focus on computations should be minimized.
- Both classroom examples and homework exercises should relate to actual or realistic data as much as possible. Students should be encouraged to look beyond the statistical analysis of data to the interpretation of results in an applied context, preferably through the use of case studies.

This philosophy led us to develop *Practical Statistics by Example Using Microsoft® Excel and MINITAB®*. Designed as an introductory text in statistics for students with a background in college algebra, our text contains the following features that distinguish it from the many other statistics texts available.

"By Example" Introduction of Concepts

Each new idea is introduced and illustrated by real data-based examples taken from a wide variety of disciplines and sources. These examples demonstrate how

to solve various types of statistical problems encountered in the real world. We believe that students better understand definitions, generalizations, and concepts *after* seeing a real application. Each example is set off for easy identification and contains a full, detailed solution to the problem.

Microsoft Excel and MINITAB as Tools for Statistical Analysis

The spreadsheet application Microsoft Excel and the statistical software MINITAB are integrated throughout the entire text. Many texts published and revised in the past twenty years have incorporated the use of popular statistical software packages such as SAS, SPSS, and MINITAB. Few, however, have successfully integrated Excel. With the increasing functionality and power of worksheet applications, virtually all kinds of statistical analyses taught in the introductory course can now be supported by Excel and the statistics add-in provided with this text (PHStat2). In addition to its possible use in a statistics course, students are usually exposed to a worksheet application such as Microsoft Excel in an introductory computer course or when they need to analyze data in advanced courses in their major area of study. Even if they are not familiar with Excel, they have almost certainly heard of this software, or Microsoft Office, and its use in the statistics course will give added relevancy to the course.

Emphasis on Critical Thinking and Interpretation of Computer Output

Both Excel- and MINITAB-generated graphs and output accompany every statistical technique presented, allowing instructors to focus on the statistical analysis of data and the interpretation of the results rather than the calculations required to obtain the results. Free from memorizing formulas and performing hand calculations, students are encouraged to develop critical thinking skills that will allow them to realize greater success in the workplace. Examples on hand calculations are provided for those instructors who desire flexibility in teaching the course.

Tutorials on Using Microsoft Excel and MINITAB

For the novice, *The Excel Primer* provides basic instruction on using Windows and Microsoft Excel and *The MINITAB Primer* provides basic instruction on using MINITAB. *Excel* and *MINITAB Tutorials* appear at the end of pertinent chapters and give step-by-step instructions and screen shots for using the applications to perform the statistical techniques presented in the chapter.

All data sets that are stored in the Excel and MINITAB directories on the CD that accompanies the text are identified with a CD-ROM icon and the name is provided.

Statistics Add-In for Microsoft Excel: PHStat

The CD-ROM that accompanies the text also includes PHStat2, the latest version of PHStat, Prentice Hall's statistical add-in for Microsoft Excel for Windows. PHStat2 minimizes the work associated with setting up statistical solutions in Microsoft Excel by automating the creation of worksheets and charts. PHStat2, in combination with Excel's Data Analysis ToolPAK add-in and table and chart wizards, allows users to perform statistical analyses on virtually all topics covered in

an introductory statistics course. (Compared to its predecessor, PHStat2 contains a number of new or enhanced procedures and now includes a full help system for easy reference. For more information about PHStat2, see Appendix E.)

"Statistics in the Real World" Application in Each Chapter

Each chapter opens with a real-world application and data set to motivate the material presented in the chapter and to provide a real-life context for learning statistics. The "Statistics in the Real World" problem is revisited throughout the chapter in relevant sections. At the end of each of these sections, the data set is analyzed using the method presented in the section and relevant conclusions are drawn from the analysis.

Built-In Study Guide

The following features are incorporated throughout the text to help students learn and retain new ideas:

- *Self-test questions* appear immediately after important ideas have been introduced to test the student's comprehension of the concept and to help develop good study habits. Answers given in the Appendix allow students to check their work.
- *Summary boxes* are set off to provide step-by-step instructions for the statistical techniques presented.
- *Side notes* provide additional explanations of key ideas adjacent to where the concept is first referenced.
- Each chapter ends with a list of *Key Terms, Formulas,* and *Symbols* with page references that guide the student back to the text in order to review the element in context.
- Each chapter begins with a set of *Objectives.* Students can determine if the objectives have been met by answering a series of *Checking Your Understanding* questions at the end of the chapter.

Topical Coverage at the Introductory Level

This text includes all the topics covered in a basic introductory statistics course, including data collection (Chapter 1), descriptive statistics (Chapters 2 and 3), probability and probability distributions (Chapters 4 – 6), confidence intervals (Chapters 7 and 9), hypothesis tests (Chapters 8 and 9), regression (Chapter 10), and analysis of variance (Chapter 11). A minimal amount of probability is presented, allowing more time for instructors to teach statistical inference (Chapters 7–11). Unique to this introductory text is a section on proper graphical presentation (Section 2.6), which promotes E. R. Tufte's principles of graphical excellence.

New to This Edition

The second edition of *Practical Statistics by Example Using Microsoft® Excel and MINITAB®* includes a number of additions and enhancements:

- Chapter 1—*Collecting Internet Data.* Section 1.4 now includes a discussion and example on obtaining data available via the Internet.

- Chapter 2—*Dot Plots.* We've added dot plots as another method for graphing quantitative data in Section 2.3.
- Chapter 5—*Binomial Tables.* In addition to obtaining binomial probabilities using Excel and MINITAB, we demonstrate how to use cumulative binomial tables.
- Chapter 6—*Normal Approximation to the Binomial.* A new section (Section 6.4) has been added on using the normal distribution to approximate binomial probabilities.
- Chapter 8—*Testing Category Probabilities for a Qualitative Variable.* The chi-square goodness of fit test for category probabilities of a single qualitative variable is now included in Section 8.9.
- Chapters 9–11—*Integration of Nonparametric Methods.* The following nonparametric tests (in Chapter 12 of the previous edition) are now integrated as optional sections in the chapters with their parametric alternatives: Wilcoxon rank sum test (Section 9.7), Wilcoxon signed ranks test (Section 9.8), Spearman's rank correlation test (Section 10.13), and Kruskal-Wallis one-way ANOVA test (Section 11.8).
- *PHStat2*—The latest version of the PHStat statistics add-in now includes additional features relating to tables and charts, confidence intervals, hypothesis tests, and regression, along with a help system.
- *MINITAB Tutorials*—The second edition of the text now includes MINITAB tutorials in addition to Excel tutorials at the end of each chapter.

Supplements for the Instructor

Each element in the package has been accuracy-checked to ensure clarity, adherence to the approaches presented in the main text, and freedom from computational, typographical, and statistical errors.

Instructor's Solutions Manual

(by Mark Dummeldinger) (ISBN 0-13-041587-1). Complete solutions to all even-numbered problems are provided in this manual. Manual solutions are most frequently provided for the "Using the Tools" problems while a combination of hand and Excel solutions are presented for the "Applying the Concepts" problems. Solutions are also provided for the Statistics in the Real World application that begins each chapter. Solutions to the odd-numbered problems are found in the *Student's Solutions Manual.*

Test Bank

(by Tom Bratcher) (ISBN 0-13-041580-4). The *Test Bank* offers a full complement of more than 1,000 additional problems that correlate to exercises presented in the text. Microsoft Word™ files for this *Test Bank* are available from the publisher.

TestGen EQ Computerized Test Bank

(ISBN 0-13-041586-3)

Data Files

Data files for most problems and for the Statistics in the Real World applications are contained on the CD-ROM that is packaged with each copy of the text. When a given data set is referenced, a disk icon with the file name will appear in the text near the exercise. The data files may also be downloaded from the World Wide Web.

Companion Web Site: http://www.prenhall.com/sincich

The Companion Web site provides self-scoring quizzes, an online syllabus maker, technology projects, and data files for the textbook.

Supplements Available for Purchase by Students

Student's Solutions Manual

(by Mark Dummeldinger) (ISBN 0-13-041592-8). Fully worked out solutions to all of the odd-numbered problems are provided in this manual. Careful attention has been paid to ensure that all methods of solution and notation are consistent with those used in the core text.

Acknowledgments

We are extremely grateful to the following reviewers who provided excellent suggestions for revising the text: Julia Hassett, Oakton Community College; Patty Monroe, Greenville Technical College; Robert L. Raymond, University of St. Thomas; Shannon Schumann, University of Colorado at Colorado Springs; and Judy Eng Woo, Bellevue Community College. In addition we would like to thank the Biometrika Trustees, American Cyanamid Company, Annals of Mathematical Statistics, and the Chemical Rubber Company for their kind permission to publish various tables in Appendix B.

This book reflects the efforts of a great many people over a number of years. Professor Emeritus William Mendenhall (University of Florida) and publisher Don Dellen (now deceased) were instrumental in developing and shaping earlier editions of *Statistics by Example,* upon which this text is partially based. Special thanks are due to our ancillary author, Mark Dummeldinger, and his typist Kelly Barber. Phyllis Barnidge and Lynda Kay Steele of Laurel Technical Services have done an excellent job of accuracy checking the manuscript and have helped us to ensure a clean answer appendix and solutions. The Prentice Hall staff of Quincy McDonald, Joanne Wendelken, Angela Battle, Amy Lysik, Linda Behrens, and Alan Fischer, and Elm Street Publishing Services' Martha Beyerlein helped greatly with all phases of the text development, production, and marketing effort. Pam Johnson did an outstanding job as copy editor. Finally, we would like to thank our wives and children for their patience, understanding, love, and assistance in making this book a reality. It is to them that we dedicate this book.

Correspondence with the Authors

We have gone to great lengths to make this text both pedagogically sound and error-free. If you have any suggestions or material requiring clarification, or should you find potential errors, please contact Terry Sincich at tsincich@coba.usf.edu or David Levine at David_Levine@baruch.cuny.edu. For questions concerning PHStat2, see Appendix E and the PHStat Web site located at www.prenhall.com/phstat.

A Guide to Using This Text

This text was designed with you, the student, in mind. We understand that you bring a variety of backgrounds and interests to this course. As a result, we strive to offer a unique presentation of statistics that reflects these interests. Additionally, we support the integration of technology to exemplify and help students to visualize statistics.

The following pages will demonstrate how to use this text effectively to make studying easier and to understand the connection between statistics, technology, and your world.

Chapter Openers

- Objectives are provided to help students understand chapter goals.
- Chapter contents detail the topics covered in each section.
- The end-of-chapter Excel and MINITAB tutorials are correlated to each chapter section.

CHAPTER 2

Exploring Data with Graphs and Tables

OBJECTIVES

1. To develop tables and graphs that describe a qualitative variable
2. To develop tables and graphs that describe a quantitative variable
3. To develop tables and graphs that describe the relationship between two qualitative variables
4. To develop graphs that describe the relationship between two quantitative variables
5. To demonstrate the principles of proper graphical presentation

CONTENTS

EXCEL TUTORIAL

MINITAB TUTORIAL

Page 57

"Statistics in the Real World" Applications

- Each chapter begins with an engaging, detailed Real World Application.
- Each problem is solved using the methods presented in the chapter, then applied to additional chapter topics.
- The real data presented in the application are provided in Excel and MINITAB workbooks.

Statistics in the Real World

Crash Ratings for New Cars—Stars or Scars?

Each year the National Highway Traffic Safety Administration (NHTSA) crash tests new car models to determine how well they protect the driver and front-seat passenger in a head-on collision. During the NHTSA's crash test procedure, new vehicles are crashed into a fixed barrier at 35 miles per hour (mph), which is equivalent to a head-on collision between two identical vehicles, each moving at 35 mph. Two dummies of average human size—one in the driver's seat and one in the front passenger seat—occupy each tested car. The dummies contain electronic instruments that register the forces and impacts that occur to the head, chest, and legs during the crash.

In response to Congress' request to provide consumers with easily understandable vehicle safety performance information, the NHTSA developed a "star" scoring system for the frontal crash test and publishes the results in the *New Car Assessment Program (NCAP)*. The NCAP crash results are reported in a range of one to five stars, allowing consumers to compare crash ratings for vehicles within the same weight class. The more stars in the rating, the better the level of crash protection in a head-on collision.

★★★★★ = 10% or less chance of serious injury
★★★★ = 11% to 20% chance of serious injury
★★★ = 21% to 35% chance of serious injury
★★ = 36% to 45% chance of serious injury
★ = 46% or greater chance of serious injury

(*Note:* A serious injury is considered to be one requiring immediate hospitalization and may be life threatening.)

CRASH

The NCAP test results for 98 cars (model year 1997) are stored in the Excel workbook **CRASH.XLS** and the MINITAB worksheet **CRASH.MTW**. A description of the variables reported by the NCAP for each car tested is provided below.

CLASS (Compact, Heavy, Light, Light Truck, Medium, Mini, Sport, or Van)
MAKE (e.g., Chrysler, Ford, Honda, ...)
MODEL (e.g., Town & Country, Mustang, Accord, ...)
DOORS (number of doors)
WEIGHT (vehicle weight in pounds)
DRIVSTAR (Driver Overall Star rating—1 to 5 stars)
PASSSTAR (Passenger Overall Star rating—1 to 5 stars)
DRIVAIR (Driver-side air bag—Yes or No)
PASSAIR (Passenger-side air bag—Yes or No)
DRIVHEAD (Driver's severity of head injury score—0 to 1,500 scale)
PASSHEAD (Passenger's severity of head injury score—0 to 1,500 scale)
DRIVCHST (Driver's severity of chest deceleration score—0 to 100 scale)
PASSCHST (Passenger's severity of chest deceleration score—0 to 100 scale)
DRIVFEML (Driver's severity of left femur injury score—0 to 3,000 scale)
DRIVFEMR (Driver's severity of right femur injury score—0 to 3,000 scale)
PASSFEML (Passenger's severity of left femur injury score—0 to 3,000 scale)

Page 58

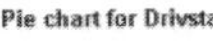

Statistics in the Real World Revisited

Interpreting a Pie Chart

Recall that the National Highway Traffic Safety Administration (NHTSA) crash tests new car models and publishes the results of the tests in the *New Car Assessment Program* (NCAP). (See p. 58.) One of the qualitative variables reported by the NCAP is Driver Star Rating, which ranges from one star (★) to five stars (★★★★★). The more stars in the rating, the better the level of crash protection to the driver in a head-on collision.

We can use either a bar graph or pie chart to summarize the Driver Star Rating data for the 98 cars in the NCAP report. Using Excel, we produced the pie chart displayed in Figure 2.3. Each slice of the Excel pie chart represents one of the categories of Driver Star Rating. Note that the percentage of the 98 cars falling into each category are shown on the pie chart. We see that 18% had the highest level of crash protection (five stars), 61% were rated four stars, 17% were rated three stars, and only 4% were rated two stars. The actual number of cars that fall into each Driver Star Rating category is displayed in the MINITAB frequency table, Figure 2.4. Note that none of the cars crash tested were rated one star (the lowest level of protection).

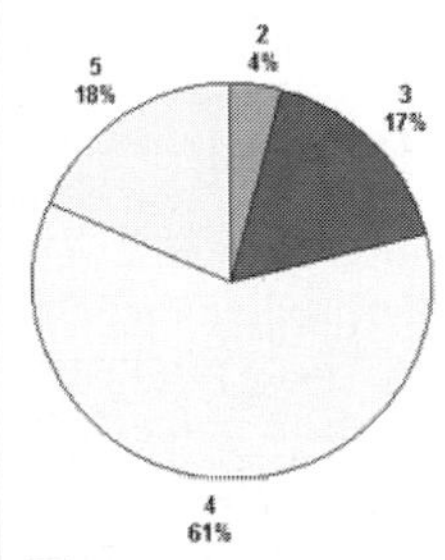

E Figure 2.3 Excel Pie Chart for NCAP Passenger Star Rating

Tally for Discrete Variables: DrivStar

DriverSt	Count	Percent
2	4	1.08
3	17	17.35
4	59	60.20
5	18	18.37
N=	98	

M Figure 2.4 MINITAB Frequency Table for Driver Star Rating

Page 62

Interesting Examples with Complete Solutions

- Examples, with complete solutions and explanations, illustrate every concept.
- Real data integrated into examples.
- Graphical output of data provided in examples.
- All examples are numbered for easy reference and the end of the solution is marked with a ▬ symbol.

EXAMPLE 2.7 **CONSTRUCTING A SCATTERPLOT**

Annually, the Federal Trade Commission (FTC) collects data on domestic cigarette brands. Smoking machines are used to "smoke" cigarettes to a certain length, and the residual "dry" particulate matter is tested for the amounts of various hazardous substances. The variables of interest to the FTC are carbon monoxide (CO) content and amount of nicotine, both measured in milligrams. Data collected on these two variables for 20 cigarette brands are listed in Table 2.12.

SMOKING

TABLE 2.12 Carbon Monoxide–Nicotine Data for Example 2.7

Brand	CO	Nicotine	Brand	CO	Nicotine
1	15	0.9	11	6	0.5
2	6	0.4	12	22	1.2
3	13	1.3	13	15	0.8
4	12	0.8	14	17	1.3
5	12	1.2	15	13	1.0
6	9	0.7	16	11	0.7
7	13	1.0	17	14	1.1
8	16	1.1	18	14	1.4
9	13	1.1	19	12	0.9
10	12	0.9	20	8	0.4

a. Construct a scatterplot for the data. **b.** Interpret the graph.

Solution

a. A **scatterplot** is a two-dimensional graph, with a vertical axis and a horizontal axis as shown in the Excel graph, Figure 2.15. The values of one of the

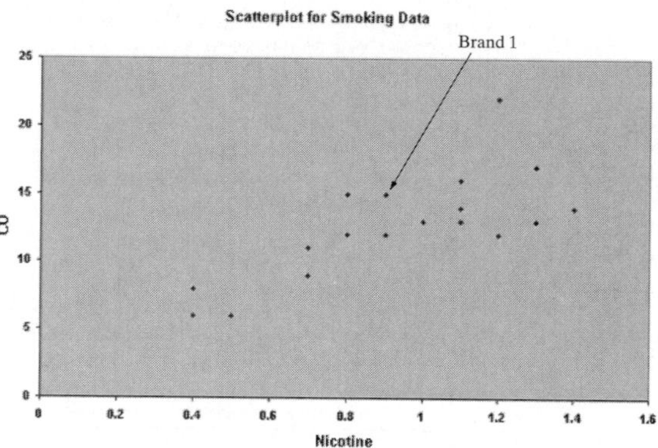

E Figure 2.15 Excel Scatterplot of CO Ranking versus Nicotine Content for Example 2.7

quantitative variables of interest are located on the vertical axis while the other variable's values are located on the horizontal axis of the graph. Note that

Pages 90-91

Self-Test 2.5

Consider the data on two quantitative variables for $n = 12$ experimental units shown in Table 2.13.

TABLE 2.13 Data for Two Quantitative Variables

Experimental Unit	Variable #1	Variable #2
1	−1	7
2	0	8
3	1	5
4	2	2
5	−3	10
6	1	6
7	−2	8
8	4	1
9	5	0
10	3	0
11	0	7
12	2	5

Construct a scatterplot for the data. What type of relationship is revealed?

Page 92

"Self-Test" Boxes

- Most examples are followed by a similar problem in Self-Test boxes. These problems allow students to immediately test their knowledge of the topic just presented.

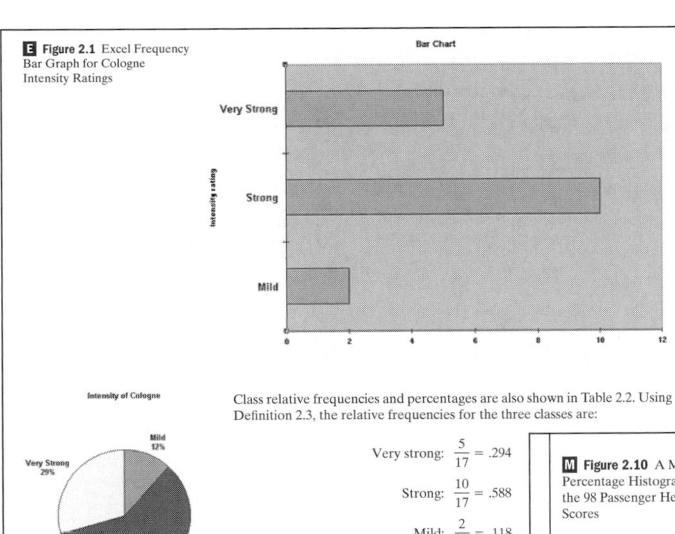

E Figure 2.1 Excel Frequency Bar Graph for Cologne Intensity Ratings

E Figure 2.2 Excel Pie Chart for Cologne Intensity Ratings

Class relative frequencies and percentages are also shown in Table 2.2. Using Definition 2.3, the relative frequencies for the three classes are:

$$\text{Very strong: } \frac{5}{17} = .294$$

$$\text{Strong: } \frac{10}{17} = .588$$

$$\text{Mild: } \frac{2}{17} = .118$$

Class percentages are obtained by multiplying eac by 100.

b. A bar graph displays the class frequencies, relativ ages. Figure 2.1 is a horizontal frequency bar gr created using Microsoft Excel. The figure contain each intensity class; the length of the bar is prop quency. (Optionally, the bar heights can be proport frequencies or percentages.) The bar graph makes jority of the 17 men's cologne brands had a stron Some software packages reverse the axes and disp fashion.]

c. A pie chart conveys the same information as a b relative frequencies are shown on a pie chart. Figur cologne data, also created using Excel. Note that the pie is divided into three slices, one for each of the three classes. The size (a... of each slice is pro- portional to the class relative frequency. For exampl the slice assigned to a strong intensity rating is 59 212°. It is common to show the percentage of meas the pie chart as indicated.

Qualitative variables can be either *nominal* or *ordinal* variables. The values of an ordinal variable can be naturally ordered like "mild," "strong," and "very strong" in Example 2.1. The values of a nominal variable (e.g., "male" and "female" for gender) have no natural ordering. The bars in an ordinal variable bar graph are usually arranged according to their natural ordering as in Figure 2.1.

Page 61

Excel and MINITAB Printouts Integrated

- Almost every statistical technique presented is illustrated with Excel and MINITAB printouts. The solutions teach students how to interpret the printouts and draw conclusions based on the data.
- Allows instructors to focus on analysis and interpretation rather than calculations.
- The data for the example are provided in an Excel Workbook and the file name is given.

Page 73

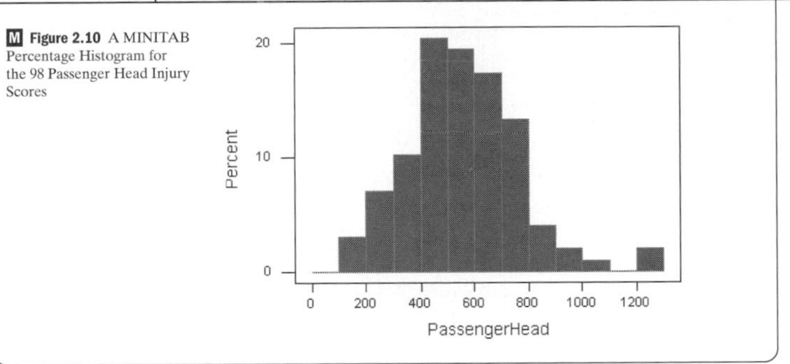

M Figure 2.10 A MINITAB Percentage Histogram for the 98 Passenger Head Injury Scores

PROBLEMS FOR SECTION 2.5

Using the Tools

2.41 Construct a scatterplot for each of the following sets of data. Determine the nature of the relationship between the two variables.

(a)

Variable #1	−1	0	1	2	3	4	5
Variable #2	−1	1	2	4	5	8	9

(b)

Variable #1	−1	0	1	2	3	4	5
Variable #2	4	1	5	2	1	7	3

(c)

Variable #1	−1	0	1	2	3	4	5
Variable #2	11	8	7	3	1	0	−2

2.42 Two quantitative variables are measured for each of 11 experimental units. The data are shown in the accompanying table.

Variable #1	7	5	8	3	6	10	12	4	9	15	18
Variable #2	21	15	24	9	18	30	36	12	27	45	54

a. Graph the data using a scatterplot.
b. Do you detect a relationship between the variables? Explain.

2.43 Two quantitative variables are measured for each of 10 experimental units. The data are shown in the accompanying table.

Variable #1	100	200	250	400	350	300	450	500	100	250
Variable #2	6	20	10	4	12	30	35	15	18	5

a. Graph the data using a scatterplot.
b. Do you detect a relationship between the variables? Explain.

2.44 For each scatterplot, determine the nature of the relationship (positive, negative, or none) between the two variables plotted.

Pages 93-94

Challenging and Engaging Problem Sets

- Problem sets are divided into two parts:

  Using the Tools has straightforward applications to test your mastery of definitions, concepts, and basic computation.

  Applying the Concepts tests your understanding of concepts and requires you to apply statistical techniques in solving real-world problems. These problems help you develop your critical thinking skills.
- Integration of relevant, current real data into problem sets. Source lines are provided.
- Graphical display of data included in problem sets.

Built-In Excel and MINITAB Manuals

Microsoft Excel and MINITAB Primers

- Designed for the newcomer to Excel or MINITAB, these Primers will acquaint you with the software and its basic functionality.
- More experienced users may skip portions of the Primer.

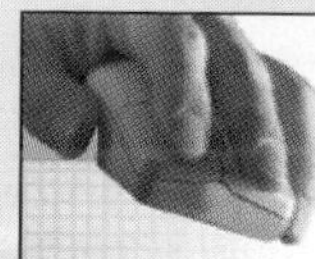

Microsoft® Excel Primer

Overview

This primer reviews the Microsoft Excel operational skills that are necessary for those who are using this software with this textbook. *Working with Windows and Microsoft Excel* introduces the operation of Windows-based programs and is written for novices. *Common Workbook Operations* reviews the basic Excel commands that all readers will find useful. *Basic Worksheet Concepts and Operations* discusses the concepts and procedures required to understand and modify worksheet designs. *Enhancing Worksheet Appearance* presents techniques that will interest readers who seek to format their worksheets in a manner similar to the examples presented in this textbook. *Using Microsoft Excel Wizards* introduces the concept of a wizard and presents a summary of how the Chart Wizard operates.

Working with Windows and Microsoft Excel

EP.

Page 1

Page 19

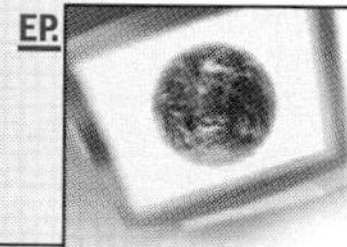

MINITAB® Primer

What Are Statistical Software Application Programs?

Statistical software application programs contain a collection of statistical methods that help provide solutions to problems in many disciplines. These programs allow users who are relatively unskilled in statistics to access a wide variety of statistical methods for their data sets of interest. In this text, the MINITAB statistical software will be illustrated and explanations on the use of MINITAB will be provided in tutorials at the end of the chapters.

MP.1.1 Entering Data Using MINITAB

There are two fundamental methods for obtaining data for use with MINITAB, entering data at a keyboard or importing data from a file. As an introduction to the operation of MINITAB, open MINITAB (typically by double-clicking the MINITAB icon on the desktop or selecting Start, then Programs, and then MINITAB). The window obtained should look similar to Figure MP.1.

Page 115

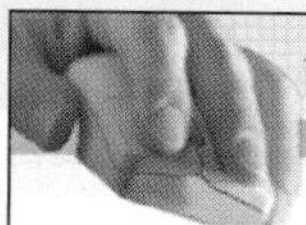

Using Microsoft® Excel

In this chapter a variety of tables and charts have been developed. These tables and charts can be obtained using PHStat and Microsoft Excel. If you have not already read the Microsoft Excel Primer, you should do so now.

2.E.1 Using PHStat to Obtain a One-Way Summary Table, a Bar Chart, and a Pie Chart

Use the PHStat **Descriptive Statistics | One-Way Tables & Charts** procedure to generate a summary table and charts for categorical data on new chart sheets. For example, to generate a summary table similar to Table 2.2 on p. 60 and bar and pie charts similar to the ones shown in Figures 2.1 and 2.2, open the **COLOGNE.XLS** workbook to the Data worksheet and:

1. Select PH
2. In the On

a. Select the

b. Enter B1:

c. Select the

d. Enter a tit

Microsoft Excel and MINITAB Tutorials

- Provided at the end of pertinent chapters for ease of use.
- Includes step-by-step guidelines and screen shots explaining how to perform the statistical techniques presented.

Page 122

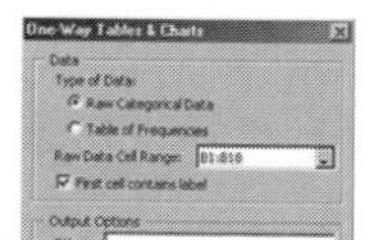

Using MINITAB®

In this chapter a variety of tables and charts have been developed. Many of these tables and charts can be obtained using MINITAB. If you have not already read the MINITAB Primer, you should do so now.

2.M.1 Obtaining a Bar Chart

To obtain a bar chart for a categorical variable, open the **CRASH.MTW** worksheet. Select **Graph | Chart.** In the Chart dialog box (see Figure 2.M.1), in the Graph variables edit box, enter **C6** or **'DriverStar'** in row 1 in the X column. Click the **OK** button.

Figure 2.M.1 MINITAB Chart Dialog Box

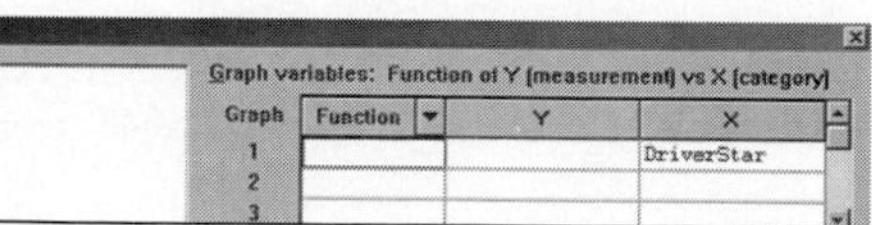

PHStat CD-ROM

- PHStat, a statistics add-in for Excel, provides a custom menu of choices that lead to dialog boxes to help you perform statistical analyses more quickly and easily than "off-the-shelf" Excel permits.
- All of the Excel workbooks used in examples and problems are on the CD.
- See pages 16-17 for more details of the Excel Primer.

2.E.3 Obtaining a Stem-and-Leaf Display

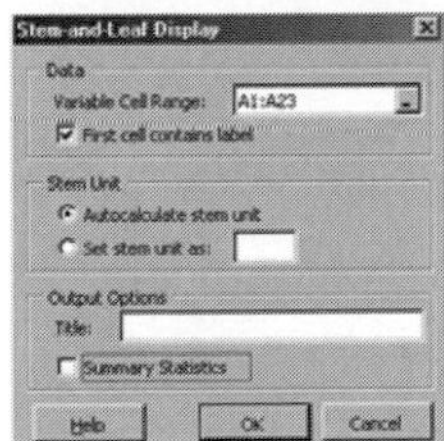

E **Figure 2.E.3** PHStat Stem-and-Leaf Dialog Box

Use the PHStat **Descriptive Statistics | Stem-and-Leaf Display** procedure to generate a stem-and-leaf display that uses the rounding method. For example, to generate the stem-and-leaf display shown in Figure 2.6 on p. 68, open the **PMI.XLS** workbook to the Data worksheet and:

1. Select Descriptive Statistics | Stem-and-Leaf Display.
2. In the Stem-and-Leaf Display dialog box (see Figure 2.E.3):

a. Enter A1:A23 in the Variable Cell Range edit box.
b. Select the First cell contains label check box.
c. Select the Autocalculate stem unit option button.
d. Enter a title in the Output Title edit box.
e. Leave unchecked the Summary Statistics check box.
f. Click the OK button.

2.E.4 Using PHStat to Obtain Frequency Distributions and Histograms

Use the PHStat **Descriptive Statistics | Histogram & Polygons** procedure to generate a frequency distribution and histogram on new sheets. For example, to generate a histogram for the data shown in Table 2.5 and Figure 2.8, open the **SAL50.XLS** workbook to the Data worksheet and:

1. Select PHStat | Descriptive Statistics | Histogram & Polygons. In the Histogram & Polygons dialog box (see Figure 2.E.4):

a. Enter A1:A51 in the Variable Cell Range edit box.
b. Enter B1:B13 in the Bins Cell Range edit box.
c. Enter C1:C12 in the Midpoints Cell Range edit box.
d. Select the First cell in each range contains label check box.
e. Select the Single Group Variable option button.

E **Figure 2.E.4** PHStat Histogram & Polygons Dialog Box

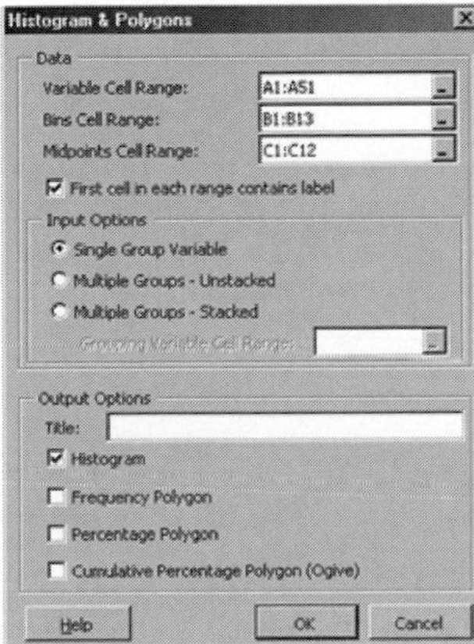

Page 116

Microsoft® Excel Primer

Overview

This primer reviews the Microsoft Excel operational skills that are necessary for those who are using this software with this textbook. *Working with Windows and Microsoft Excel* introduces the operation of Windows-based programs and is written for novices. *Common Workbook Operations* reviews the basic Excel commands that all readers will find useful. *Basic Worksheet Concepts and Operations* discusses the concepts and procedures required to understand and modify worksheet designs. *Enhancing Worksheet Appearance* presents techniques that will interest readers who seek to format their worksheets in a manner similar to the examples presented in this textbook. *Using Microsoft Excel Wizards* introduces the concept of a wizard and presents a summary of how the Chart Wizard operates.

Working with Windows and Microsoft Excel

EP.1.1 The "Point and Click" User Interface

The Windows system, and Microsoft Excel, an application that runs under Windows, share a "point and click" user interface in which a pointing device, such as a mouse, is the primary means by which to make selections and choices.

Moving a pointing device moves the onscreen graphic known as the **mouse pointer,** that in Microsoft Excel commonly takes the form of an arrow (see Figure EP.1a) or an outlined plus sign (see Figure EP.1b). Selections and choices are made by moving this pointer over an object (such as a menu) and pressing one of the pointing device buttons. Four distinct operations commonly used throughout this text are defined as follows:

E Figure EP.1 a and b
Arrow and Outlined Plus Sign Mouse Pointers

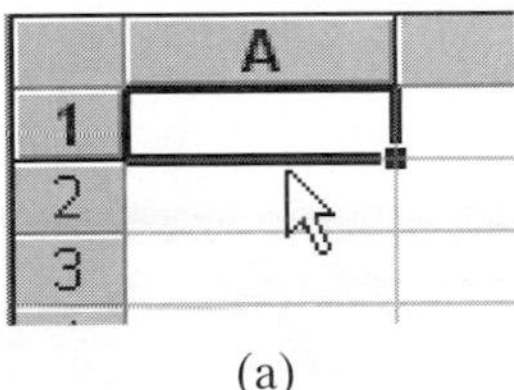

(a)

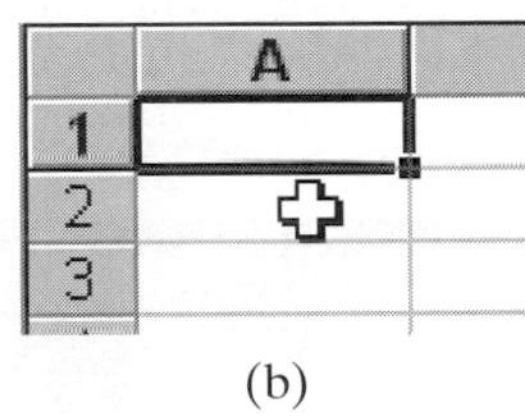

(b)

1. **Selecting** an on-screen object means moving the mouse pointer directly over an object and pressing the left mouse button. Sometimes selecting is described as **clicking,** as in the phrase "click the OK button."
2. **Double-clicking** an object means moving the mouse pointer directly over an object and pressing the left mouse button twice in rapid succession.
3. **Right-clicking** an object means moving the mouse pointer directly over an object and pressing the right mouse button.
4. **Dragging** an object requires first moving the mouse pointer over an object and then, while holding down the left mouse button, moving the pointing device to move the object to a new onscreen position. Releasing the mouse button stops the operation.

Selection typically results in the highlighting of some object. In Figure EP.2a, The Microsoft Excel icon on the Windows desktop of a computer system at fictional Tadashi College has been selected. Right-clicking an object typically displays a shortcut menu of context-sensitive commands as shown in Figure EP.2b. Double-clicking typically selects the default shortcut operation, sometimes shown on the shortcut menu as the boldfaced choice. In this example, double-clicking the Microsoft Excel icon would **open,** or load, Microsoft Excel.

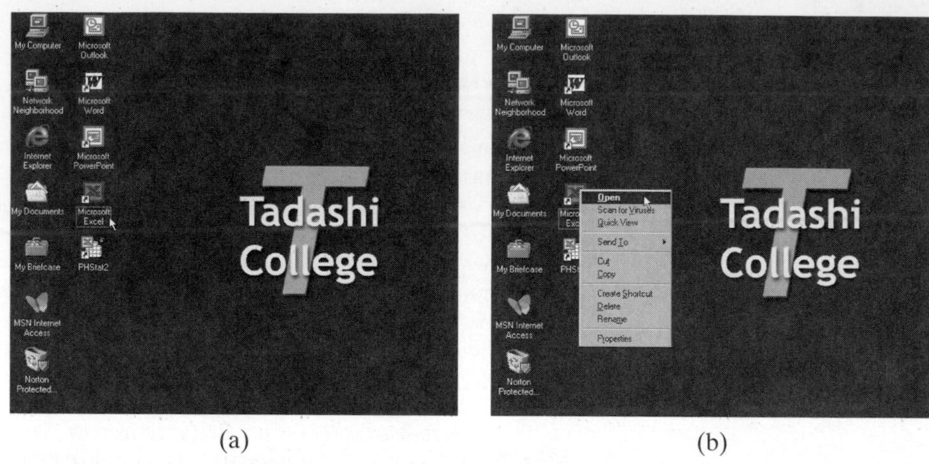

Figure EP.2a and b Selecting and Right-clicking an Icon

The Mousing Practice workbook (MOUSING PRACTICE.XLS), supplied in the Excel folder on the companion CD, allows you to practice these operations.

EP.1.2 Elements of the Microsoft Excel Application Window

Opening Microsoft Excel loads the Excel program and causes an Excel application **window** to appear. This window, a framed portion of the screen, typically will take up the entire display area, but can be **resized** to a smaller rectangular area as well as customized to display only certain elements. Figure EP.3 identifies and explains the elements of the Excel window of most relevance to readers of this text.

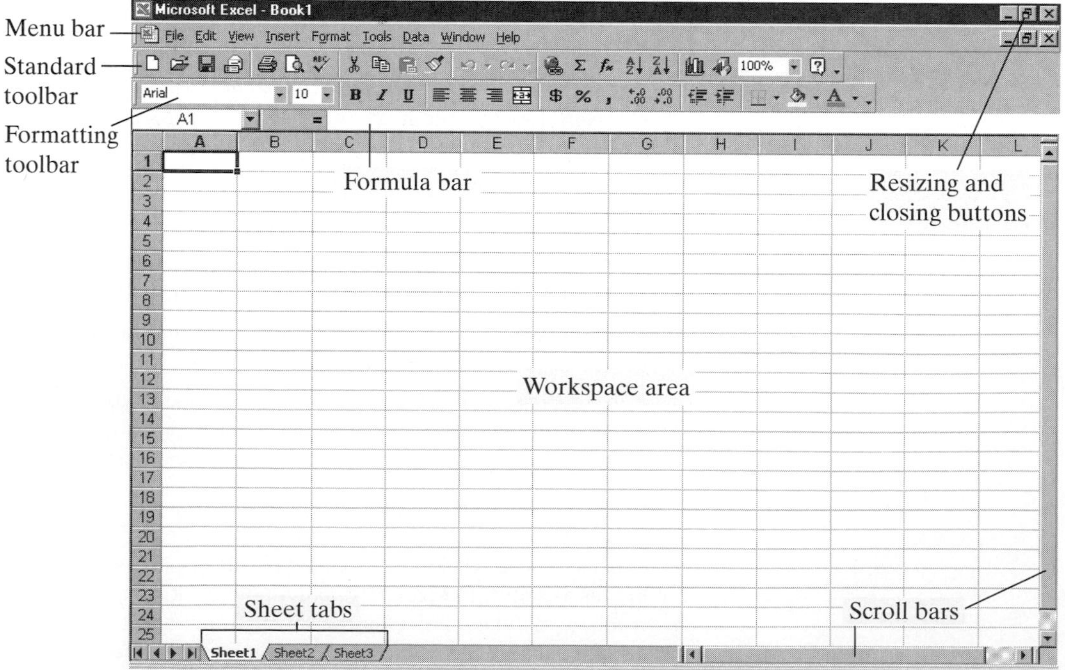

Figure EP.3 The Microsoft Excel Application Window

System minimize, resize, and close buttons respectively shrink (hide) the Excel application window, resize the Excel window, and close Microsoft Excel. The **menu bar** is the horizontal list of words that represent sets of command choices.

The **standard toolbar** contains icons that represent shortcuts to many file-oriented commands, including the common workbook operations discussed in section EP.2.

The **formatting toolbar** contains icons that represent shortcuts to many formatting commands (see section EP.2 for further details).

The **formula bar** indicates the currently selected worksheet cell (see section EP.2) and its contents.

The **workspace area** contains the currently opened workbook or workbooks. (Usually only one workbook is opened, but it is possible to open more than one and switch between them by selecting choices from the Window menu.)

Scroll bars, both horizontal and vertical, allow for the display of parts of the worksheet currently off screen (such as row 100 or column T).

Sheet tabs identify the names of individual sheets. Clicking a sheet tab selects a sheet and makes it the currently **active sheet.**

See Appendix D for the procedure by which you can customize an Excel application window to match the one shown in Figure EP.3.

EP.1.3 Standard Conventions and Features of Microsoft Excel Menus and Dialog Boxes

In Microsoft Excel, menus and **dialog boxes,** special windows that are designed to accept user-supplied information or to report a status message and that appear over the Excel application window, are the primary means of making command selections. These objects contain standard conventions and features that are used in a consistent manner throughout Microsoft Excel. Conventions and features relevant to this text are identified and explained in Figures EP.4, EP.5, and EP.6.

Menu conventions: (see Figure EP.4 on p. 4)

- Underlining **accelerator keys** that can be pressed to select individual choices.
- Using an **ellipsis** to mark menu choices that lead to a dialog box.
- Using a **triangular marker** to denote that the menu choice leads to a second (sub)menu of choices.
- Displaying to the left of a menu choice the **toolbar button** that is the shortcut for that menu choice.
- Listing to the right of a menu choice the **keyboard shortcut** for that choice.

From this point, this text uses the vertical slash character | to indicate a series of menu selections, as in **File|Open,** instead of the stating "select the File menu and then select the Open choice." When directed to make a menu selection in this text, you can, of course, substitute the appropriate toolbar button, keyboard shortcut, or accelerator key, as appropriate.

Figure EP.4 The Microsoft Excel File Menu

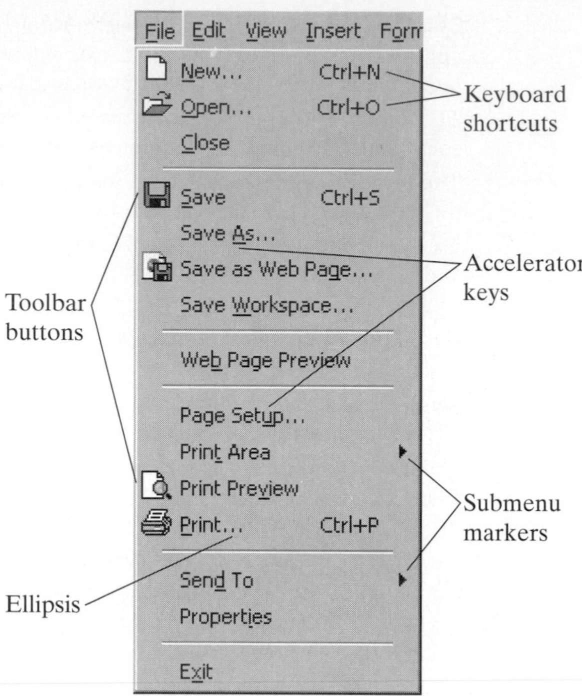

Dialog box standard features: (see Figure EP.5 a and b)

Drop-down list boxes allow the selection of an item from a non-scrollable list that appears when the drop-down button, located on the right edge of the box, is clicked.

List boxes display lists of items, in this case files or folders, for possible selection. When a list is too long to be seen in its entirety in the box, clicking the scroll buttons that automatically appear will reveal the rest of the list.

Edit boxes provide an area into which a value can be typed or edited. Edit boxes are often combined with either a drop-down list box or **spinner buttons** to aid in the entry of a value. (Pressing spinner buttons increases or decreases the numeric value that appears in an edit box.)

Option buttons represent a set of mutually exclusive choices. Only one option can be selected from the set as selecting an option always deselects, or clears, the other option buttons in the set.

Check boxes allow the selection of optional actions. Unlike option buttons, more than one check box in a set can be selected at any given time.

Clicking the **Open** or **OK buttons** causes Microsoft Excel to execute the operation represented by the dialog box with the current values and choices as shown in the dialog box.

Clicking the **Cancel button** closes a dialog box and cancels the operation.

E Figure EP.5a and b Standard Features of Dialog Boxes Found in the File Open (a) and File Print (b) Dialog Boxes

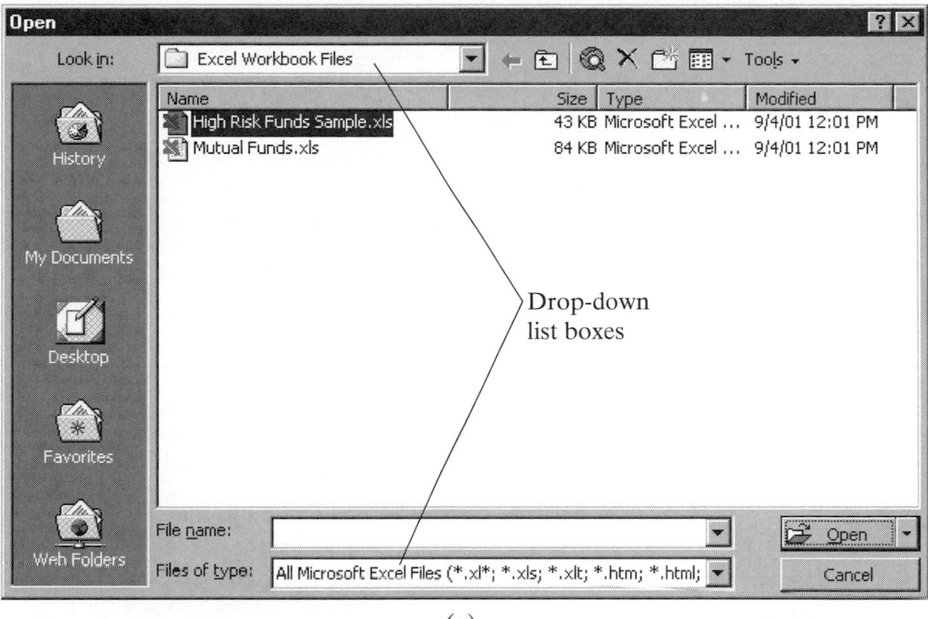

(a)

Option buttons

Edit box with spinner buttons

Check box

(b)

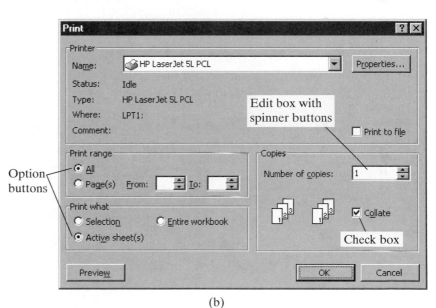

Common Workbook Operations

At some point, most users will want to open, save, and print workbook files and use the context-sensitive Microsoft Excel help system. This section reviews those operations using the dialog boxes and associated displays from Office 2000. (If you are using another version of Microsoft Excel, some of the details of these displays will be different, but the basic operations described will be the same or similar.)

EP.2.1 Opening Workbooks

To open a Microsoft Excel workbook for use, first select **File | Open.** In the Open dialog box that appears (see Figure EP.6):

1. Select the appropriate folder (also know as a directory) from the Look in: drop-down list box.
2. If necessary, change the display of files in the scrollable files list box by selecting the appropriate format button. (Figure EP.6 shows the Preview button selected. Compare to Figure EP.5a in which the List files button has been selected.)
3. Select the proper Files of type: value from the drop-down list. The default Microsoft Excel Files choice will list all Excel workbooks. To list all text files, select Text Files. To list every file in the folder, select All Files.
4. Select the file to be opened from the files list box. If the file does not appear, verify that steps a and c were done correctly.
5. Click the Open button.

After opening a workbook, always verify its contents before proceeding.

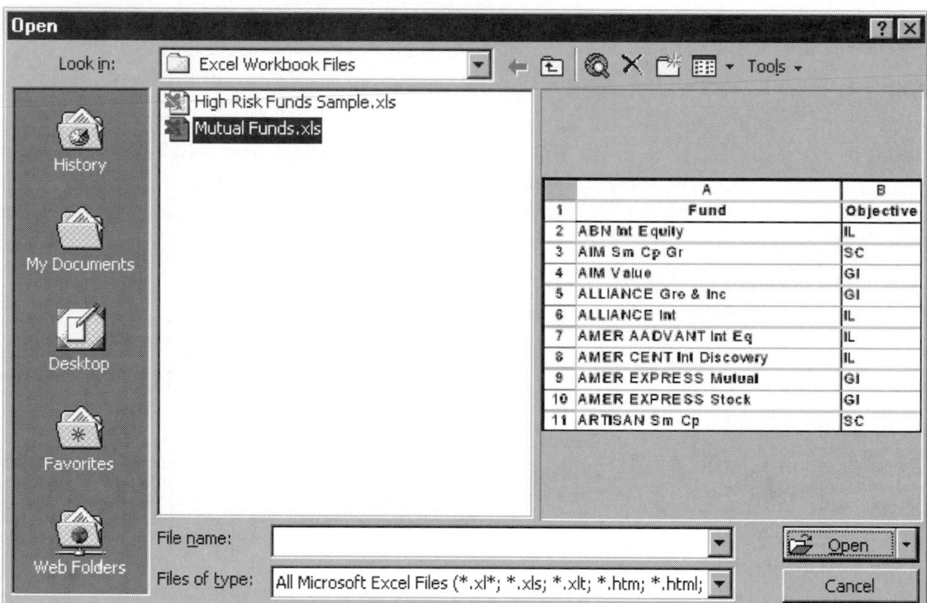

E **Figure EP.6** File Open Dialog Box

EP.2.2 Saving Workbooks

Saving a workbook assures its future availability and is protection against a loss of power or other system failure. To save a workbook, select **File | Save As.** In the Save As dialog box that appears (see Figure EP.7):

1. Select the folder to hold the file (also known as a directory) from the Save in: drop-down list box.
2. Select the appropriate choice from the Save as type: drop-down list. Typically, this choice will be Microsoft Excel Workbook. However, Formatted Text (space delimited), Text (tab delimited), and Microsoft Excel 5.0/95 are useful choices when saving data to be used by earlier versions of Excel or other programs.
3. Enter the value for the name of the file in the File name: edit box
4. Click the Save button.

Saving the workbook under a second name is an easy way to create a backup of the workbook that can be used should some error make the original workbook unusable.

Figure EP.7 File Save As Dialog Box

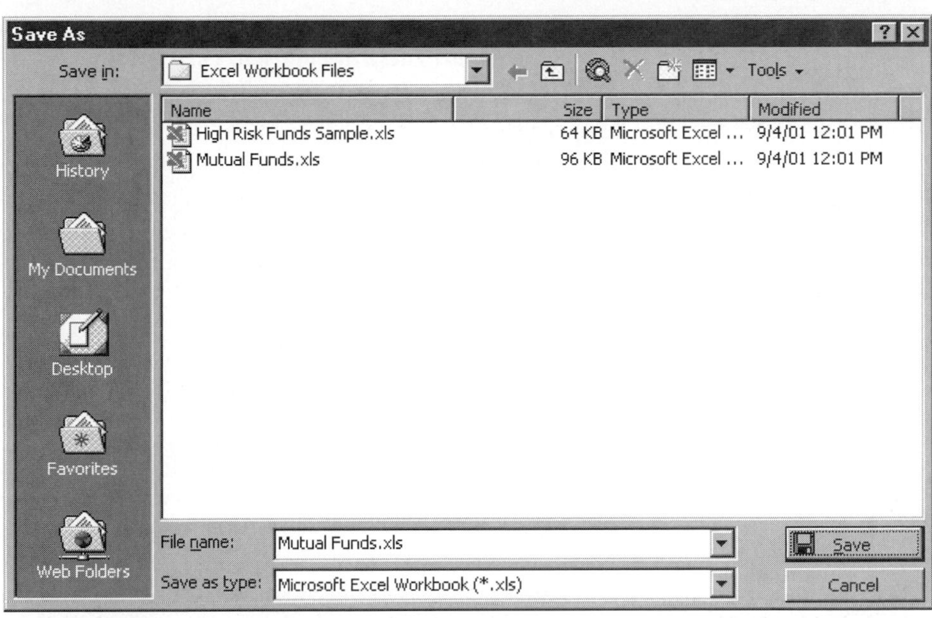

Retrieving a companion CD workbook directly from the CD-ROM will cause Microsoft Excel to mark the workbook as "read-only." Such a workbook *cannot* be saved under its original name, but can be saved under an alternate name using the File | Save As command.

EP.2.3 Printing Worksheets

Printing worksheets gives you the means to study and review results away from the computer screen. To print a specific worksheet, select the worksheet and then select **File | Print Preview** to preview the output (see Figure EP.8). If the preview contains errors or displays the

Figure EP.8 Print Preview Screen

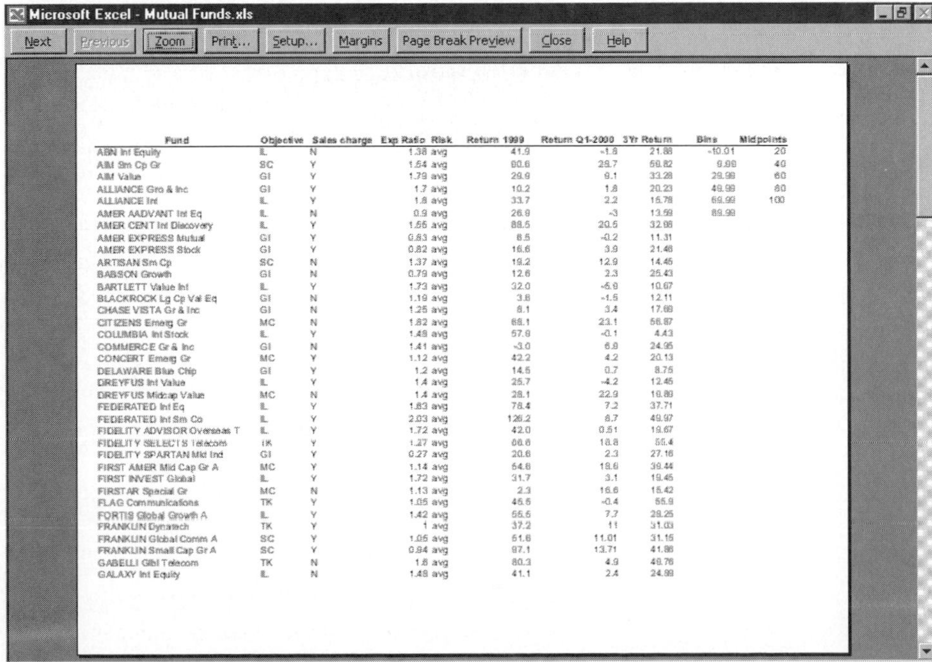

Fund	Objective	Sales charge	Exp Ratio	Risk	Return 1999	Return Q1-2000	3Yr Return	Bins	Midpoints
ABN Int Equity	IL	N	1.38	avg	41.9	-1.6	21.88	-10.01	20
AIM Sm Cp Gr	SC	Y	1.54	avg	90.6	28.7	59.82	9.99	40
AIM Value	GI	Y	1.79	avg	29.9	9.1	33.28	29.99	60
ALLIANCE Gro & Inc	GI	Y	1.7	avg	10.2	1.8	20.23	49.99	80
ALLIANCE Int	IL	Y	1.8	avg	33.7	2.2	15.78	69.99	100
AMER AADVANT Int Eq	IL	N	0.9	avg	26.9	-3	13.59	89.99	
AMER CENT Int Discovery	IL	Y	1.56	avg	88.5	20.5	32.98		
AMER EXPRESS Mutual	GI	Y	0.83	avg	8.5	-0.2	11.31		
AMER EXPRESS Stock	GI	Y	0.82	avg	16.6	3.9	21.48		
ARTISAN Sm Cp	SC	N	1.37	avg	19.2	12.9	14.46		
BABSON Growth	GI	N	0.79	avg	12.6	2.3	25.43		
BARTLETT Value Int	IL	Y	1.73	avg	32.0	-6.9	10.67		
BLACKROCK Lg Cp Val Eq	GI	N	1.19	avg	3.8	-1.5	12.11		
CHASE VISTA Gr & Inc	GI	N	1.25	avg	8.1	3.4	17.69		
CITIZENS Emerg Gr	MC	N	1.82	avg	68.1	23.1	56.87		
COLUMBIA Int Stock	IL	Y	1.49	avg	57.9	-0.1	4.43		
COMMERCE Gr & Inc	GI	N	1.41	avg	-3.0	6.9	24.95		
CONCERT Emerg Gr	MC	Y	1.12	avg	42.2	4.2	20.13		
DELAWARE Blue Chip	GI	Y	1.2	avg	14.5	0.7	8.75		
DREYFUS Int Value	IL	Y	1.4	avg	25.7	-4.2	12.45		
DREYFUS Midcap Value	MC	N	1.4	avg	28.1	22.9	16.89		
FEDERATED Int Eq	IL	Y	1.83	avg	78.4	7.2	37.71		
FEDERATED Int Sm Co	IL	Y	2.03	avg	126.2	8.7	49.97		
FIDELITY ADVISOR Overseas T	IL	Y	1.72	avg	42.0	0.51	19.67		
FIDELITY SELECTS Telecom	TK	Y	1.27	avg	66.6	18.8	55.4		
FIDELITY SPARTAN Mkt Ind	GI	Y	0.27	avg	20.6	2.3	27.16		
FIRST AMER Mid Cap Gr A	MC	Y	1.14	avg	54.8	18.6	39.44		
FIRST INVEST Global	IL	Y	1.72	avg	31.7	3.1	19.45		
FIRSTAR Special Gr	MC	N	1.13	avg	2.3	16.6	15.42		
FLAG Communications	TK	Y	1.05	avg	45.5	-0.4	55.9		
FORTIS Global Growth A	IL	Y	1.42	avg	55.5	7.7	28.25		
FRANKLIN Dynatech	TK	Y	1	avg	37.2	11	31.03		
FRANKLIN Global Comm A	SC	Y	1.05	avg	51.6	11.01	31.15		
FRANKLIN Small Cap Gr A	SC	Y	0.94	avg	97.1	13.71	41.86		
GABELLI Glbl Telecom	TK	N	1.8	avg	80.3	4.9	48.76		
GALAXY Int Equity	IL	N	1.48	avg	41.1	2.4	24.89		

worksheet in a manner not desired, click the Close button, make the changes necessary, and reselect **File | Print Preview.** Then click the Print button in the print preview screen, or, if the preview has been closed, select **File | Print** .

In the Print dialog box that appears (see Figure EP.5b on p. 5):

1. Select the printer to be used from the Name drop-down list box.
2. Select the All option button under the Print range heading.
3. Select the Active sheet option button under the Print what heading. (The authors recommend not selecting the Entire workbook option button.)
4. Set the number of copies to the proper value.
5. Click the OK button.

When printing ends, verify the contents of the printout. Most failures to print will result in the display of an informational dialog box. Click the OK button of that dialog box and correct any problems, if possible, before attempting to print the worksheet a second time. Printouts can be customized by selecting **File | Page Setup** (or clicking the Setup button in the Print Preview screen) to display the Page Setup dialog box. Information for using this dialog box can be found in Appendix D.

EP.2.4 Using the Microsoft Excel Help System

Microsoft Excel includes several methods of getting help including a standard Windows help command **(Help | Microsoft Excel Help)** and the context-sensitive help methods of tool tips, "what's this?" helps, and the Office Assistant.

Use **Help | Microsoft Excel Help** in Excel 2000 to look up an explanation of a specific menu command or a worksheet function (if you see the animated Office Assistant, see p. 9). Selecting this command opens the Microsoft Excel Help dialog box to the index tab. Type the menu command or worksheet function in the Type keywords edit box and click the Search button. Figure EP.9 shows the result of clicking Search after typing the word open in the Type keywords edit box. Note that the first help topic listed is automatically displayed in the pane to the right and other help topics can be displayed by selecting them. (If you use Microsoft Excel 97, select **Help | Contents and Index** for a similar feature.)

E Figure EP.9 Microsoft Excel Help Dialog Box

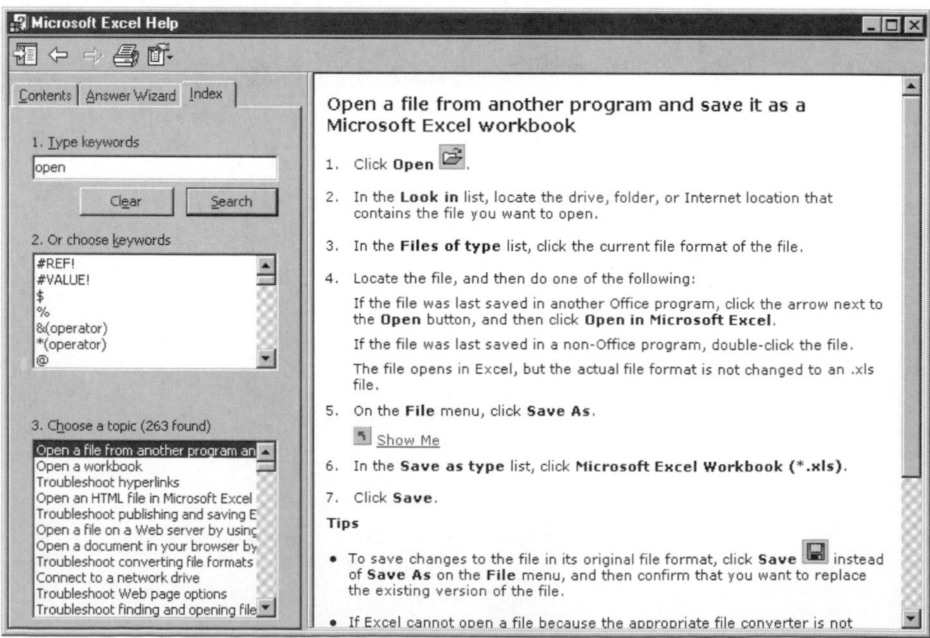

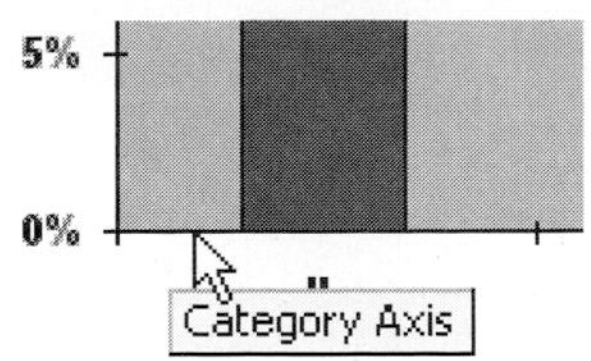

E Figure EP.10 Tool Tip for a Chart Axis

Use **tool tips** to quickly identify elements of the Excel application window such as toolbar buttons or to help select a part of a chart for further editing. To display a tool tip, pause the mouse pointer over an on-screen object. For example, Figure EP.10 shows the tool tip for a chart axis.

Use **"What's This?" Help** to display an informational message about an element of the Excel application window. To use this help, first select **Help | What's This?** to change the normal mouse pointer into the special question-mark pointer (see Figure EP.11a). Then click on the element of interest to display an informational message for that element (see Figure EP.11b). To remove the message from the screen, click anywhere in the application window. (Many dialog boxes contain a **question mark button** on their top borders that can be used to get "what's this?" help for the contents of those dialog boxes.)

E Figure EP.11a "What's This?" Mouse Pointer

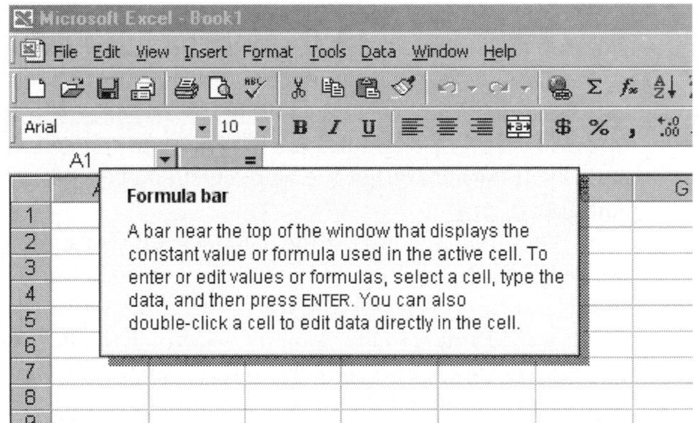

E Figure EP.11b "What's This?" Message for the Formula Bar

Use the **Office Assistant,** an animated cartoon character that appears in the lower right of the worksheet, to get help and advice in comic-like speech balloons and to conduct help searches similar to those in the Microsoft Excel Help dialog box. Although many Microsoft Excel users find the animations of the Office Assistant distracting, this help method can be useful when you are not quite certain what to search for. To display the Assistant, select **Help | Show the Office Assistant** in Excel 2000. (To remove the Assistant from the screen, select **Help | Hide the Office Assistant** in Excel 2000.) Figure EP.12a shows "Clippit," one of the several cartoon identities that the Assistant can assume and Figure EP.12b shows the contents of the balloon after a search was done on the word "help."

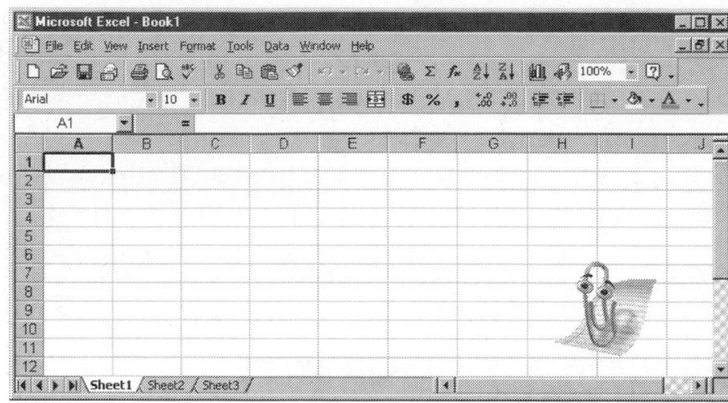

E Figure EP.12a "Clippit" Office Assistant

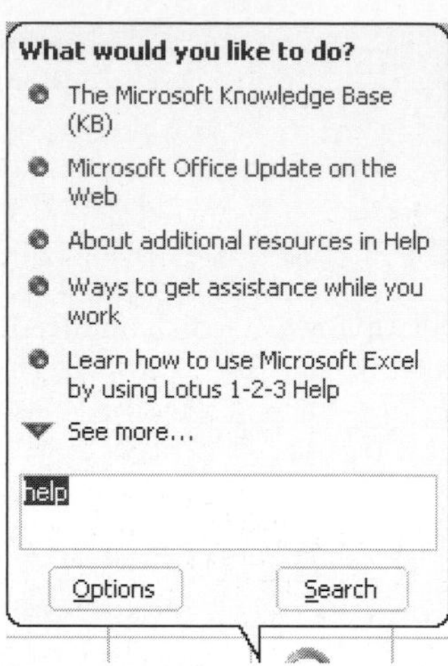

E Figure EP.12b Balloon Help for Topic "Help"

The Office Assistant is an optional component of Microsoft Excel and may need to be installed (by running the Excel or Office setup program) before it can be used. If you are using Microsoft Excel 97, select **Help | Microsoft Excel Help** to display the Assistant. To remove the Asistant from the screen, right-click on the Assistant and select **Hide Assistant** from the shortcut menu that appears.

Basic Worksheet Concepts and Operations

EP.3.1 Specifying Worksheet Locations

Worksheets are comprised of lettered columns and numbered rows, the intersections of which form **cells,** places to make individual entries into the worksheet. Cells are the basic unit for specifying worksheet locations and each cell is identified by its column letter and row number. For example, cell A1 refers to the cell in the first column and first row (the upper left corner cell) of the worksheet and cell B4 refers to the cell in the second column and fourth row.

When you need to distinguish between two similarly positioned cells on two different worksheets, the cell reference must be written in the form *Sheetname!ColumnRow*. For example, Data!A1 refers to the cell in the upper left corner of a sheet named Data, while Calculations!A1 refers to the similarly placed cell on the Calculations sheet. Using the expanded notation is always necessary when referring to cells that are not located on the currently active sheet and may be a requirement when entering cell references in certain Microsoft Excel dialog boxes.

Cell references can also be made to rectangular groups of adjacent cells in a worksheet. Such **cell ranges** are specified using the cell references to the upper leftmost and lower rightmost cells in the block in the form UpperLeft:LowerRight. For example, the cell range A1:B3 refers to the six-cell worksheet block containing the cells A1, B1, A2, B2, A3, and B3, and the range A1:A8 refers to the first eight cells in the first column of the worksheet. Ranges can also refer to an entire column, as in A:A, or an entire row, as in 1:1. Ranges in the form Sheetname!UpperLeft:LowerRight, such as Data!A1:A8, are allowed and refer to cell ranges located on a sheet that is not the currently active one.

> It is also possible to distinguish between two similarly located cells on two similarly named worksheets in two different workbooks using the form '[Workbookname]Sheetname'!ColumnRow. For example, '[ChapterS2]Data'!A1, referring to the upper left corner cell on the Data worksheet in the ChapterS2 workbook.

EP.3.2 Formulas and Functions

Formulas are instructions that manipulate the data of a workbook. Formulas always begin with the = (equals sign) symbol and can contain symbols, or **operators,** that represent arithmetic operations (see Table EP.1), cell references, and **functions,** pre-programmed calculations that perform common arithmetic, business, engineering, and statistical operations that might otherwise be hard or tedious to express using only the operators. For example, the formula =A1−A2 calculates the difference between the values in cells A1 and A2 and the formula =B1+B2+B3+B4+B5+B6 calculates the sum of the values found in the first six cells of the second column of the current worksheet. In the case of the second example, the SUM function could be used to simplify the formula as =SUM(B1:B6).

TABLE EP.1 Microsoft Excel Arithmetic Operators

Operation	Excel Operator
Addition	+
Subtraction	-
Multiplication	*
Division	/
Exponentiation	^

EP.3.3 Making Worksheet Cell Entries

Making an entry into a worksheet cell begins by selecting the cell into which the entry is to be made. The selection frames the cell with a border and causes the cell's column letter and row number to be highlighted along the border of the worksheet and the cell's reference to appear in the name box portion of the formula bar. As you type an entry, it appears in the edit box portion of the formula bar, where you can accept, cancel, or revise the entry (see Figure EP.13). Either pressing the Enter key, which moves the cell highlight down one row, or the Tab key, which moves the cell highlight one column to the right, accepts the entry. Clicking the "X" button on the formula bar or pressing the Escape key cancels the entry. Either retyping the entire entry or editing it using the cursor keys and either the Backspace or Delete keys revises the entry.

E **Figure EP.13** Formula Bar

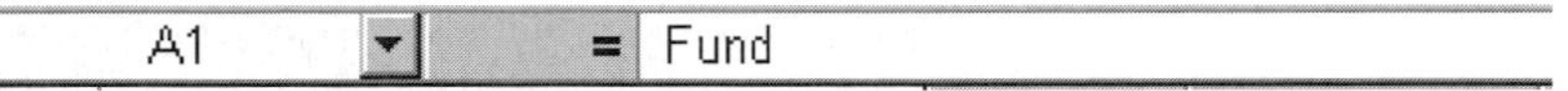

After you have completed all worksheet cell entries, each entry should be reviewed for errors. One technique that makes this task easier is to switch the worksheet to Formulas view. Formulas view causes Microsoft Excel to display the contents of each cell as it would appear in the formula bar edit box if the cell had been selected. For formula entries, this means that the actual formulas, and not their results, are displayed on the screen. To switch to this view, select **Tools | Options.** In the Options dialog box that appears, select the Formulas checkbox under the View tab and click the OK button.

Some formulas, especially long formulas entered into narrow width cells, may appear truncated after switching to Formulas view. To see the entire formula on the screen, select the cell with the truncated formula and then select **Format | Column | AutoFit Selection.**

EP.3.4 Copying Formulas

Generally, copying the contents of a cell (the cell entry) to another cell involves the simple procedure of first selecting the cell, then selecting **Edit | Copy,** then selecting the cell to receive the copy, and then selecting **Edit | Paste.** Copying entries that contain formulas are a special case as exact duplicates may or may not result, depending on how cell references have been entered. If cell references are entered in the form ColumnletterRownumber, such as in A1 or any of the cell references used in Section EP.3.2, the cell references are called **relative** and will be *changed* to reflect their new relative positions. For example, a cell C2 formula =A2+B2 when copied to cell C3 will be changed to =A3+B3 to reflect its new relative position one row down from the original. Likewise, a cell A5 formula =SUM(A1:A4) when copied to cell B5 will be changed to =SUM(B1:B4).

When you do not desire this automatic adjustment, the cell reference is entered as an **absolute** reference using the form $Columnletter$Rownumber. For example, the formula =$A$2+$B$2 will always sum the contents of the row 2 cells in the first and second columns regardless of the cell to which it is copied. Note the dollar sign symbol does not imply currency values; it is used solely to prevent the changing of a cell reference during a copy operation. Formulas that mix relative and absolute cell references are allowed. For example, a cell C2 formula =A2/$B$10 when copied to cell C3 will be changed to =A3/$B$10. This mixing of cell reference types often simplifies the implementation of sets of worksheet formulas and is seen in many of the worksheets used in this text.

COPYING FORMULAS BETWEEN WORKSHEETS

You can copy formulas between cells of different worksheets using **Edit | Copy** and **Edit | Paste,** too. In such cases, you should ensure that all cell references are entered in the form *Sheetname!UpperLeft:LowerRight* as discussed in section EP.3.1 on p. 10. If, in copying the formula, you seek only to transfer results from one sheet to another, consider entering a formula in the receiving cell that refers to the cell containing the formula you otherwise would seek to copy. For example, if you were thinking of copying a formula found in cell B10 of a sheet named Results to cell A5 in a sheet named Summary in order to show a result, consider entering the formula = Results!B10 in cell A5 of the Summary sheet in lieu of copying the actual formula.

Enhancing Worksheet Appearance

EP.4.1 Common Formatting Operations

Any number of formatting selections can be used to enhance the appearance of a worksheet. Many common formatting operations have formatting toolbar icons, although these (and all other) formatting operations can always be selected from the Format menu as well. Formatting toolbar icons that were used to produce the worksheets shown in this text are identified in Figure EP.14 and discussed below.

To display cell values in boldface type, select the cell (or cell range) containing the values to be boldfaced and click the Boldface button on the formatting toolbar.

To display cell values centered in a column, select the cell (or cell range) containing the values to be centered and click the Center button on the formatting toolbar. (The Align Left and Align Right buttons to either side of the Center button similarly left or right justify values in a column.)

E **Figure EP.14** Formatting Toolbar

Arial 10 B I U $ % , .00 .0 A

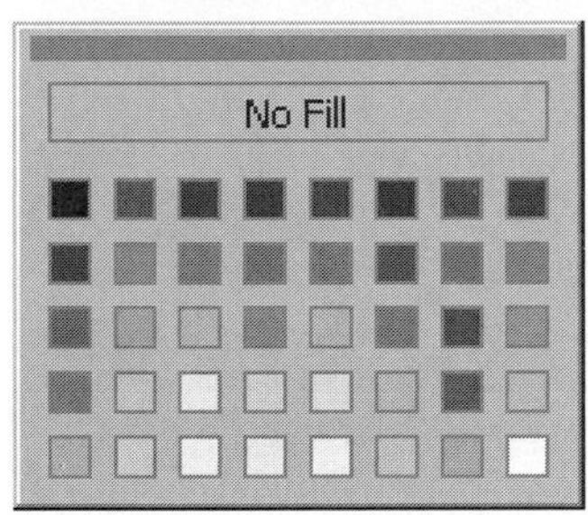

E Figure EP.15 Fill Color Palette Dialog Box

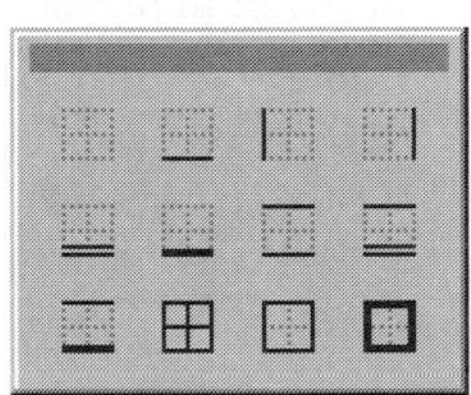

E Figure EP.16 Borders Palette Dialog Box

To center a cell entry representing a title over a range of columns, select the cell row range over which the entry is to be centered (this range must include the cell containing the title) and click the Merge and Center button on the formatting toolbar.

To display numeric values as percentages, select the cell range containing the numeric entries to be displayed as percentages and click the Percent button on the formatting toolbar.

To align the decimal point in a series of numeric entries, select the cell range containing the numeric entries to be aligned and click either the Increase Decimal or Decrease Decimal button on the formatting toolbar until the desired decimal alignment is produced.

To change the background color, select the cell range containing the cells to be changed and click the Fill Color drop-down list button. In the Fill Color palette dialog box that appears (see Figure EP.15), select the new background color. (Many of the worksheets illustrated in this text use the color "Light Turquoise" for "Data" areas and "Light Yellow" for Results areas. These are the fifth and third choices, respectively, on the bottom row of the standard palette.)

To change the cell border effect, select the cell range containing the cells whose borders are to be changed and click the Borders drop-down list button. In the Borders palette dialog box that appears (see Figure EP.16), select the new border effect. (Many of the worksheets illustrated in this text use a variety of border effects including those identified by their tool tips as "All Borders," "Outside Borders," and "Top Border.")

To adjust the width of a column to fully display all displayed cell values in that column, select the column to be formatted by clicking its column heading in the border of the worksheet and then select **Format | Column | AutoFit Selection.**

Additional formatting operations can be executed through the Format | Cells and Format | Autoformat dialog boxes. See the Microsoft Excel Help for further information on these topics.

EP.4.2 Copying and Renaming Worksheets

Copying worksheets can be helpful when a worksheet needs to be formatted in two different ways or when a Formulas view version of a worksheet is required to verify the contents of a worksheet (see Section EP.3.3). To copy a worksheet, first select the worksheet by clicking on its sheet tab. Then, select **Edit | Move or Copy Sheet** and in the Move or Copy dialog box (see Figure EP.17):

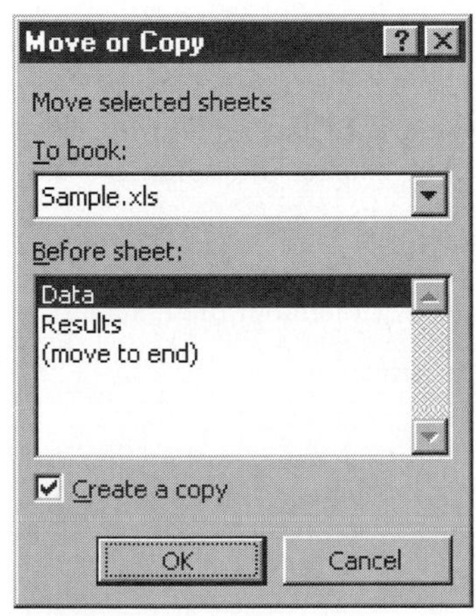

E Figure EP.17 Move or Copy Dialog Box

1. Select the Create a copy check box.
2. Select (new book) from the To book: drop-down list if the worksheet copy is to be placed in a new workbook, or, if the copy is to be placed in the same workbook, specify the position of the copy in the opened workbook by selecting the appropriate choice from the Before sheet: list box.
3. Click the OK button.

Microsoft Excel assigns the copied worksheet the name of the original sheet plus a number in parenthesis. For example, copying a sheet named Calculations will result in a sheet named "Calculations (2)." Renaming the copied worksheet to give it a more self-descriptive name such as "Calculations in Report Format" is a good practice and one that can be extended to the sheets of a new workbook that Microsoft Excel poorly names in the serial form of Sheet1, Sheet2, Sheet3, etc. To give a worksheet a (better) self-descriptive name, double-click the sheet tab of the sheet to be renamed (see Figure EP.3, p. 2), type the new sheet name, and press the Enter key. Chart sheets, introduced in Chapter 2, can also be renamed using this procedure.

Using Microsoft Excel Wizards

EP.5.1 Introduction to Microsoft Excel Wizards

Wizards are sets of linked dialog boxes that guide you through the task of creating workbook objects such as charts and summary tables. You enter information and make selections in each of the linked boxes, advancing through each one by clicking a Next button and, ultimately, a Finish button, to create an object. At any time, you can click a Cancel button to cancel the wizard or you can move backwards in the linked set by clicking Back.

Wizards are generally the easiest way to create objects. For example, the Microsoft Excel **Chart Wizard,** referenced in this text, simplifies the creation of workbook charts from worksheet data. The **Text Import Wizard** aids in the transfer of text file data into a Microsoft Excel worksheet. More information about using the Text Import Wizard can be found on the CD-ROM that accompanies this text.

EP.5.2 Using the Chart Wizard

The Microsoft Excel Chart Wizard asks you to make selections and enter data as it guides you through the process of generating a chart from worksheet data. To use this series of four linked dialog boxes (shown in Figure EP.18a–e as they would appear when completing the scatter diagram shown in Figure 2.5 on p. 62), select **Insert|Chart** from the Excel menu bar to begin the wizard. Then:

1. Select the desired chart type in the Step 1 dialog box (see Figure EP.18a).
2. Enter the cell range for the chart data in the Data range edit box of the Data Range tab of the Step 2 dialog box (see Figure EP.18b). If necessary, select the Series tab and enter the necessary information for the data series by either clicking the Add button or selecting a data series from the Series list box (see Figure EP.18c).
3. Select, and supply values as appropriate, the chart options in the Step 3 dialog box (see Figure EP.18d).
4. Select to place the chart either on a new chart sheet (the preferred style in this text) or as a chart object on a preexisting worksheet in the Step 4 dialog box (see Figure EP.18e).

In most cases, you will not want your chart to contain all of the optional features that Microsoft Excel automatically selects in the Step 3 dialog box (see Figure EP.18d). To eliminate these preselected features, do the following in the Step 3 dialog box:

1. Select the Titles tab and enter values as appropriate.
2. Select the Axes tab and then select both the (X) axis and (Y) axis check boxes. Also select the Automatic option button under the (X) axis check box.
3. Select the Gridlines tab and deselect (uncheck) all the choices under the (X) axis heading and under the (Y) axis heading.
4. Select the Legend tab and deselect (uncheck) the Show legend check box.
5. Select the Data Labels tab and select the None option button under the Data labels heading.

E **Figure EP.18a–e** Chart Wizard Dialog Boxes

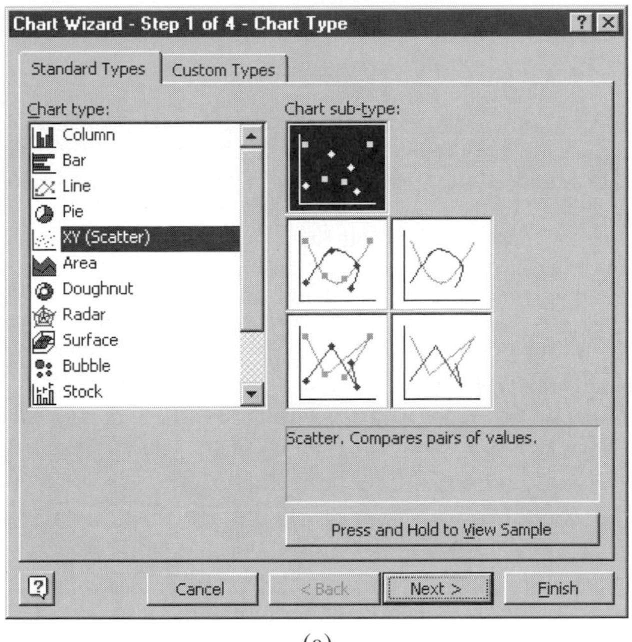

(a)

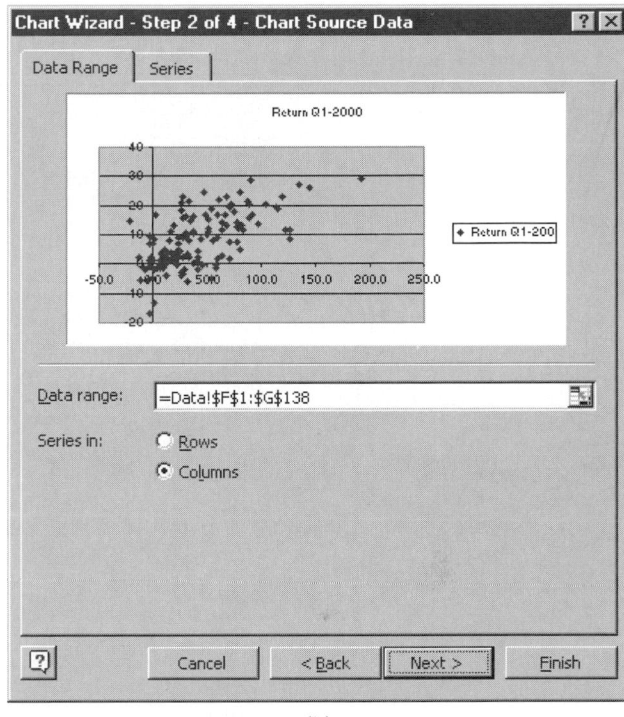

(b)

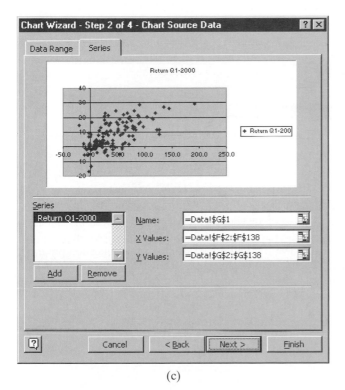

Chart Wizard - Step 2 of 4 - Chart Source Data

Data Range | Series

Return Q1-2000

Series: Return Q1-2000

Name: =Data!$G$1

X Values: =Data!$F$2:$F$138

Y Values: =Data!$G$2:$G$138

Add | Remove

Cancel | < Back | Next > | Finish

(c)

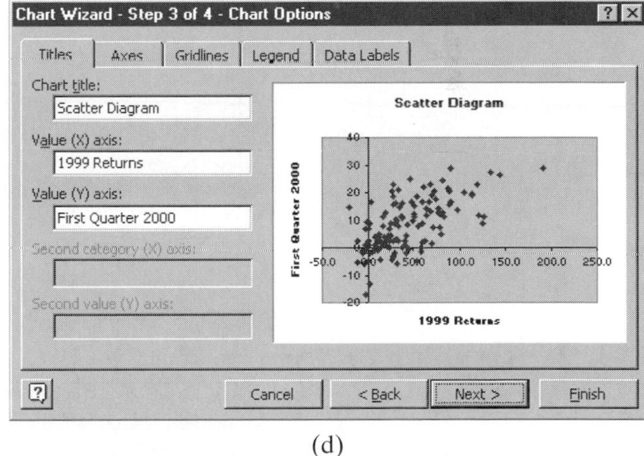

(d)

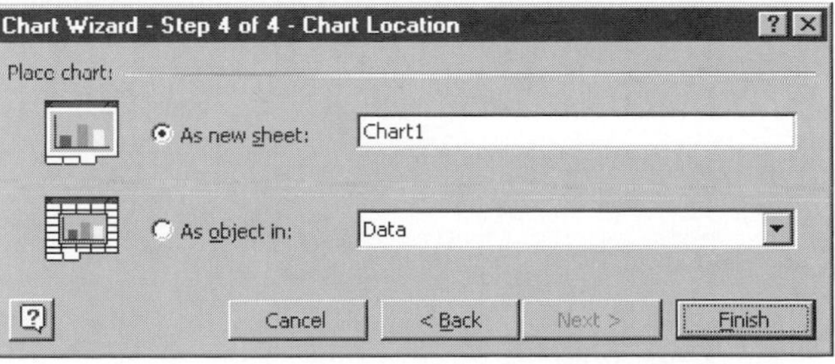

(e)

Using Add-Ins

EP.6.1 Introduction to Microsoft Excel Add-Ins

Add-ins are optional programs that contain procedures that extend the functionality of Microsoft Excel. Microsoft supplies a variety of add-ins, including the Analysis ToolPak add-in, on the Microsoft Excel or Microsoft Office program CD. Other add-ins are available from a variety of third parties.

Add-ins typically modify the menu bar of Microsoft Excel by inserting either a new menu on the menu bar or adding a new menu choice to a pre-existing menu. When selected, some add-ins choices lead to dialog boxes that ask the user to enter information and make selections, while others generate new objects or results directly.

For example, the Analysis ToolPak add-in inserts the choice Data Analysis to the preexisting Tools menu. Selecting **Tools | Data Analysis** causes the ToolPak add-in to display the Data Analysis dialog box (see Figure EP.19) from which a statistical analysis can be selected from the Analysis Tools list box. Clicking the OK button after selection begins the analysis, typically with the display of a dialog box that asks you to make selections and enter data about the analysis.

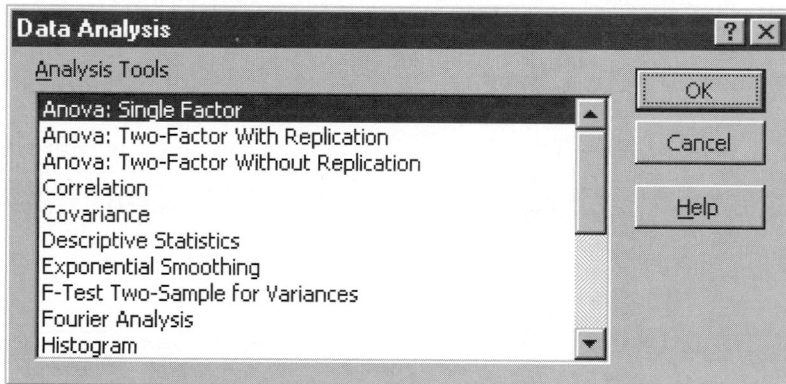

E Figure EP.19 Data Analysis Dialog Box

EP.6.2 Opening Add-Ins

All add-ins must be opened in Microsoft Excel before they can be used. Add-ins that users want opened automatically every time Excel runs can be installed in Microsoft Excel. Installing an add-in, not to be confused with installing or setting up an application program, is accomplished by first selecting **Tools | Add-Ins** and then selecting the check box in the Add-Ins available list of the Add-Ins dialog box that corresponds to the add-in to be installed (see Figure EP.20).

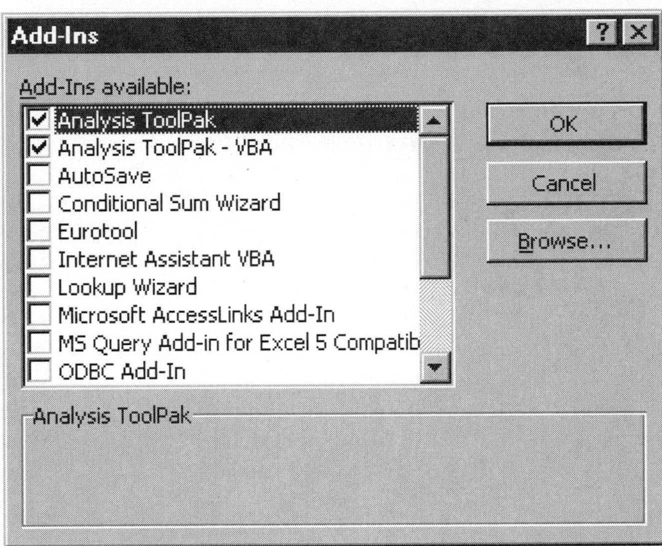

E **Figure EP.20** Add-Ins Dialog Box

If an add-in does not appear in the Add-Ins available list, it can be added to that list by clicking the Browse button and locating and selecting it in the Browse dialog box (which operates similar to the File Open dialog box).

Add-ins that are not "installed" can be opened by using the **File | Open** dialog box that is used to open workbook files. Opening an add-in using this method may trigger a macro virus dialog box that warns of the possibility of the add-in containing a computer virus. Click the Enable Macros button to allow a virus-free add-in to be opened. If you are using Excel 2000 version SR-1 or later, you will need to first select **Tools | Macros | Security** and select the Medium option button in the Security Level tab of the Security dialog box (see Figure EP.21) in order to use PHStat (see p. 51) successfully.

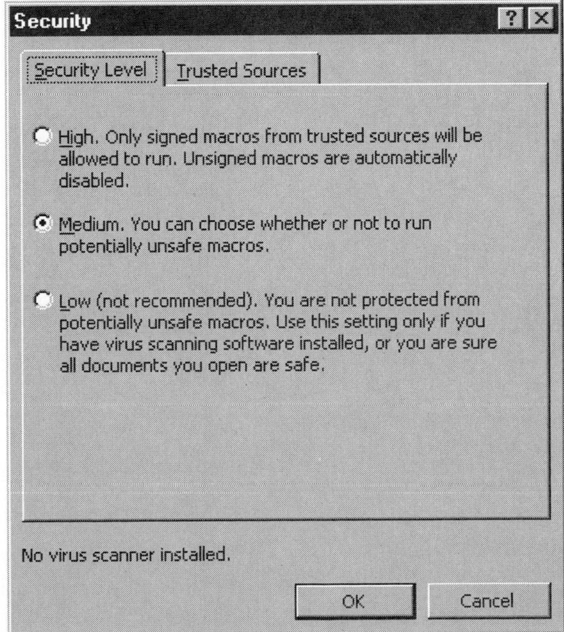

E **Figure EP.21** Security Dialog Box

KEY TERMS

Absolute cell reference 12
Active sheet 3
Add-ins 17
Arithmetic operators 11
Cancel button 4
Cell ranges 10
Cells 10
Chart Wizard 14
Check box 4
Click mouse operation 1
Dialog box 3
Double-click operation 1
Drag operation 1
Drop-down list box 4
Edit box 4
Formatting tool bar 3
Formula 10
Formula bar 3
Function 10
List box 4
Menu bar 2
Mouse pointer 1
Office Assistant 9
Open/OK buttons 4
Operators 11
Option button 4
Question-mark button 9
Relative cell reference 12
Resizing button 3
Right-click operation 1
Scroll bar 2
Select operation 1
Sheet tab 3
Spinner buttons 4
Standard tool bar 3
Text files 6
Text Import Wizard 14
Toolbar button 3
Tool tips 9
Window 2
Wizards 14
Workspace area 3
Worksheets 10

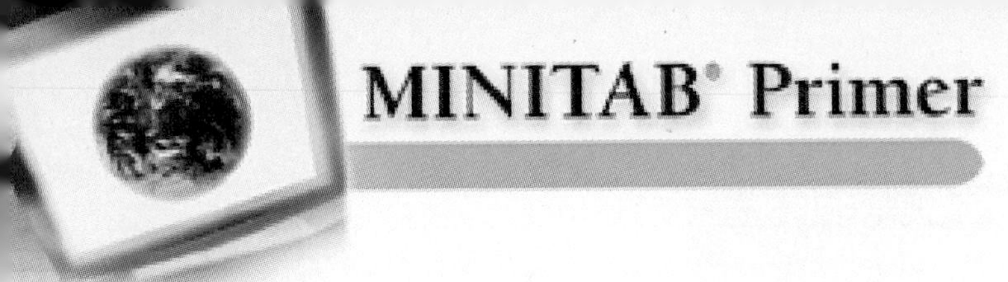

MINITAB® Primer

What Are Statistical Software Application Programs?

Statistical software application programs contain a collection of statistical methods that help provide solutions to problems in many disciplines. These programs allow users who are relatively unskilled in statistics to access a wide variety of statistical methods for their data sets of interest. In this text, the MINITAB statistical software will be illustrated and explanations on the use of MINITAB will be provided in tutorials at the end of the chapters.

MP.1.1 Entering Data Using MINITAB

There are two fundamental methods for obtaining data for use with MINITAB, entering data at a keyboard or importing data from a file. As an introduction to the operation of MINITAB, open MINITAB (typically by double-clicking the MINITAB icon on the desktop or selecting Start, then Programs, and then MINITAB). The window obtained should look similar to Figure MP.1.

M Figure MP.1 The MINITAB Worksheet and Session Window

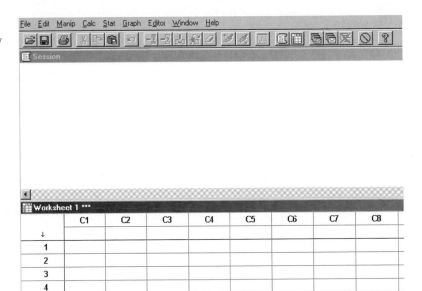

At the top of the window is a menu bar from which selections are made that allow you to obtain statistics, display graphs, store and save data, and perform many other operations. Below the Menu bar is a Session window that displays text output such as tables of statistics. Below the Session window is the Data window, a rectangular array of rows and columns where you enter, edit, and view data. Note that the first row of the Data window lists the columns, which are generically labeled C1, C2, and so on. The following row contains an arrow in the first column and then a set of blank columns. This row serves as a placeholder in which labels for variable names are entered.

Data can be entered via the keyboard in the cells in the Data window. If the arrow points down, pressing Enter moves the active cell (into which numbers are entered) down, while if the arrow points right, pressing Enter moves the active cell to the right. Clicking the arrow will change the direction of the data entry.

As an illustrative example, suppose that data are available concerning the amount of money spent by five customers at a department store. The results are illustrated in Table MP.1.

TABLE MP.1 Amount Spent by a Sample of Five Customers at a Department Store

Name	Amount ($)
Allen	125
Barry	250
Diane	72
Kim	105
Susan	48

Begin in the row above the horizontal line containing the numbered rows. This row is used to provide labels for each variable. In the first column (labeled C1) enter the label of the first variable (Name), and press the Enter key. This moves the cursor to the first row of this column. Enter Allen in row 1, Barry in row 2, Diane in row 3, Kim in row 4, and Susan in row 5. Move the cursor to the heading area at the top of column 2 (under C2), and enter Amount as the label for this column. After pressing Enter to move to row 1 in column C2, enter 125, the amount spent by Allen. Continuing, enter the amount spent for the other four customers in rows 2 to 5, respectively.

When performing statistical analyses throughout this text, data sets with a large number of observations are often encountered. When such sets have been previously entered and stored in a data file, it makes sense to import the contents of the file into a Data sheet to avoid having to reenter each observation one at a time. In this text, each data set has been stored in several different formats including the MINITAB format (.MTW). These data sets are included on the CD that accompanies the text in a directory labeled MINITAB. The name of each data set used in the text (either in the chapter or in the problems) is displayed along with a CD icon. Documentation for the data sets is provided in Appendix C. To import a MINITAB worksheet, select **File|Open Worksheet** on the menu bar and then select the appropriate file from the directory.

In addition to files in MINITAB format (.MTW), files in other formats such as Microsoft Excel (.XLS), data (.DAT), and text (.TXT) can be opened from this dialog box. Making sure that the proper file type appears in the Files of type box, select the file that you wish to open. To see how the file will look on the worksheet, click the Preview button.

CHAPTER 1

Introduction: Statistics and Data

OBJECTIVES

1. To introduce some key ideas and concepts of statistics and demonstrate how they apply to everyday life
2. To identify the different types of data that can be collected and the methods used to collect the data
3. To differentiate between population and sample data
4. To differentiate between descriptive and inferential statistical methods

CONTENTS

EXCEL TUTORIAL

MINITAB TUTORIAL

Statistics in the Real World

Obedience to Authority—The Shocking Truth

A basic principle upon which our society is organized is obedience to authority. Psychologists and sociologists agree that without obedience, our society would soon be very chaotic. But what if a person is asked to obey orders that appear to be evil or malicious? Could we expect that person to disobey? In a classic psychology study conducted in the early 1960s, Stanley Milgram performed a series of experiments designed to isolate the psychological factors that influence a person's behavior in such a setting.

The basic set-up of the experiment was as follows. Two people were brought into a room at an appointed time; one person was assigned to play the role of the "learner" and the other was assigned to play the role of the "teacher." The learner's task was to learn a list of word pairs, for example, box-boat. After reviewing the entire list with the learner, the teacher (the subject of the experiment) went through the list one-by-one, giving the first member of the word pair (e.g., box). The learner was then asked to respond with the corresponding second member of the pair (e.g., boat). If the learner answered correctly, the teacher went on to the next word pair. However, if the learner gave the wrong answer, the teacher was instructed to deliver an electric shock by depressing one of 30 levers, ranging from 15 volts to 450 volts in increments of 15 volts. In addition to the voltage, the first 28 levers were labeled in groups of four as follows: Slight shock, Moderate shock, Strong shock, Very strong shock, Intense shock, Extreme intense shock, and Danger: Severe shock. The final two levers were simply labeled "XXX." Teachers were instructed to start with the lowest-level shock and to increase the level of shock with each error until the entire list was learned correctly. [*Note:* Unknown to the teacher, the learner was an actor who conspired with the experimenter. The shocks were imaginary, but the learner responded in a manner that led the teacher to believe that, in fact, they were real.]

If the teacher raised questions about the experiment and whether it should be continued, the experimenter would respond with one of the following four statements:

Statement 1: Please continue or Please go on.

Statement 2: The experiment requires that you continue.

Statement 3: It is absolutely certain that you continue.

Statement 4: You have no other choice; you must go on.

Only if the teacher refused to continue after all four statements had been made was the experiment stopped. Of interest to Milgram was how far (in the sequence of shocks) the teacher proceeded before the experiment was terminated. For this study, we will refer to this variable as the shock level (measured in volts).

Milgram manipulated the basic experiment in several ways to study the impact of different situational factors on obedience. Teachers (i.e., the subjects) were randomly assigned to one of five different experimental settings:

Setting 1: *Predicted behavior.* Subjects (i.e., teachers) had the experimental set-up described to them and were asked to predict how far they would go (although they did not actually perform the experiment).

Setting 2: *Remote.* The learner could not be heard or seen but pounded on the walls at 300 volts and ceased responding and pounding at 315 volts.

Setting 3: *Voice feedback.* The learner could not be seen but vocal protest could be clearly heard.

Setting 4: *Proximity*. The learner was placed in the same room so that he or she could be both seen and heard.

Setting 5: *Touch proximity.* The learner received a shock only when his hand rested on a shock plate. At 150 volts, the learner demanded to be released from the experiment and refused to put his hand on the shock plate. The teacher was instructed to take the learner's hand and force it onto the plate.

What were the results of Milgram's experiment? Surprisingly, two-thirds of all the teachers in the study continued to obey the experimenter and "applied" shock levels of 400 or more volts despite the pleas of the learners. From this, Milgram concluded that a high level of obedience to authority exists in our society. In *Obedience to Authority: Current Perspectives on the Milgram Paradigm* (2000), Thomas Blass demonstrates that Milgram's obedience theory holds true some 35 years later. Milgram's experiments

continue to be discussed and reproduced in disciplines such as business ethics, philosophy, nursing, and law—in addition to psychology.

In the following Statistics in the Real World Revisited sections, we discuss several key statistical concepts covered in this chapter that are relevant to Milgram's shock experiment.

Statistics in the Real World Revisited

- Identifying the Experimental Units and Variables (p. 22)
- Identifying the Population, Sample, and Inference (p. 33)
- Identifying the Data Collection Method (p. 40)

1.1 What Is Statistics?

Consider the following recent items from the news media:

1. *The Associated Press,* Sept. 19, 2000—Apparently the city that never sleeps is also too busy to wash up. A new survey of public restroom habits in five U.S. cities finds New York commuters are least likely to clean their hands after using the bathroom. . . . Four years ago, 60% of folks using public restrooms at Grand Central and Penn stations washed up afterward. This time, it was just 49%.
2. *Time,* Sept. 25, 2000—So much for man's best friend. Over the two years 1997 and 1998, dogs bit some 9 million Americans—27 of them fatally. The breeds to blame in more than half the fatalities: pit bulls and rottweilers.
3. *The Sporting News,* July 12, 1999—Our statisticians have developed a unique player-performance ratio—The TSNdex—that indicates how a player's overall performance compares to his peers within a given season, and how a player's season then stacks up against players from other times. . . . According to the TSNdex, Babe Ruth (New York Yankees) in 1921 had the best ever single season baseball performance.
4. *USA Today,* Sept. 19, 2000—Long working days with too few hours of sleep can slow responses as much as alcohol, researchers say. After 17 to 19 hours of staying awake—a normal day for many people on shift work, working more than one job, or putting in lots of overtime—reaction times are up to 50% slower than they are after drinking alcohol, [according to] research in the *Occupational and Environmental Medicine Journal.*
5. *Online Magazine,* May/June 1999—World Wide Web search engines are the most heavily used on-line services, with millions of searches performed each day. In a test of the ability of five major search engines to satisfy the search criterion specified, Lycos had the overall best average performance score (based on the top 100 "hits").

Every day we are inundated with bits of information—**data**—like those in the citations above, whether we are in the classroom, on the job, or at home. These data often guide the many decisions that we make during our lifetime. What decision-making tools will enable you to make sense of data? The answer is *statistics.*

A common misconception is that a statistician is simply a "number cruncher" or a person who calculates and summarizes numbers, like baseball batting averages or unemployment rates. Statistics involves numbers, but there is much more to it than that.

According to *The Random House College Dictionary* (2000 ed.), statistics is "the science that deals with the collection, classification, analysis, and interpretation of numerical facts or data." In short, statistics is the *science of data*—a science that will enable you to be proficient data producers and efficient data users.

Definition 1.1

Statistics is the science of data. It involves collecting, classifying, summarizing, organizing, analyzing, and interpreting data.

In this chapter we explore the different types of data and problems that you will encounter in your field of study and introduce you to some ideas on methods for collecting data. Various statistical methods for summarizing, analyzing, and interpreting those data are presented in the chapters that follow.

1.2 Types of Data

Data are obtained by measuring a characteristic or property of the objects (usually people or things) of interest to us. These objects upon which the measurements (or observations) are made are called *experimental units,* and the properties being measured are called *variables* (since, in virtually all studies of interest, the property varies from one observation to another).

A finer breakdown of data types into nominal, ordinal, interval, and ratio data is possible. *Nominal* data are qualitative data with categories that cannot be meaningfully ordered. *Ordinal* data are also qualitative data, but a distinct ranking of the groups from high to low exists. *Interval* and *ratio* data are two different types of quantitative data. For most statistical applications, it is sufficient to classify data as either quantitative or qualitative.

Definition 1.2

An **experimental unit** is an object (person or thing) upon which we collect data.

Definition 1.3

A **variable** is a characteristic (property) that differs or varies from one observation to the next.

All data (and, consequently, the variables we measure) are either *quantitative* or *qualitative* in nature. Quantitative data are data that can be measured on a numerical scale. In general, qualitative data take values that are nonnumerical; they can only be classified. The statistical tools that we use to analyze data depend on whether the data are quantitative or qualitative. Thus, it is important to be able to distinguish between the two types of data.

Definition 1.4

Quantitative data are observations measured on a natural numerical scale.

Definition 1.5

Nonnumerical data that can only be classified into one of a group of categories are **qualitative** (or **categorical**) **data.**

EXAMPLE 1.1 **IDENTIFYING FEATURES OF A DATA SET**

Many universities provide their students with a Career Resource Center. Periodically, the center will mail out questionnaires pertaining to the employment status and starting salary of students who recently graduated. A University of Florida

(UF) questionnaire gathered the following information from each graduate who was contacted:

1. Gender
2. Degree
3. Major
4. Job type
5. Salary
6. Age
7. Overall grade point average

Data for 10 (fictitious) graduates from the data set are shown in Table 1.1.

a. For this data set, identify the experimental units and the variables measured.
b. Determine the type (quantitative or qualitative) of each variable measured.

TABLE 1.1 Data for 10 University of Florida Graduates

Gender	Degree	Major	Job Type	Salary	Age	GPA
M	PhD	Law	Criminal	$72,000	31	3.97
M	Bachelor	Accounting	Auditing	$45,000	22	3.51
M	Bachelor	Ind/Systems	Sales	$43,000	25	2.90
F	Bachelor	Physical Therapy	Therapy	$27,600	34	3.05
M	Masters	Latin Amer. Studies	U.S. Army	$49,000	40	2.88
F	Bachelor	Medical Technology	Therapy	$31,100	24	3.15
F	Bachelor	Political Science	Ins. Claims	$27,500	26	2.74
M	Bachelor	Economics	Bank/Finance	$33,000	21	3.82
F	PhD	Dentistry	Dentist	$65,000	39	4.00
F	Bachelor	Nursing	Reg. Nurse	$30,000	25	4.00

Solution

a. Since the center collects data on each UF graduate surveyed, the UF graduates are the experimental units. The variables (properties) measured on each graduate are the seven items previously listed, i.e., Age, Grade point average, Gender, etc. These are called variables since their values vary from one graduate to another (i.e., they are not constant).

b. The first four variables listed are qualitative since the data they produce are values that are nonnumerical; they can only be classified into categories or groups. For example, Gender is either M (for male) or F (for female). Similarly, values for Major are accounting, nursing, political science, etc. We can classify a graduate according to Gender or Major, but we cannot represent Gender or Major as a meaningful numerical quantity.

The last three variables are quantitative since they are all measured on a natural numerical scale. Salary is measured in dollars, Age in years, and Grade point average on a 4-point scale.

Caution: The values of a qualitative variable may be coded or recorded numerically. For example, Gender in Example 1.1 could have been coded 1 for males and 0 for females. Nevertheless, the data are qualitative since each observation can only be classified into one of a group of categories.

EXAMPLE 1.2

CLASSIFYING VARIABLES

The U.S. Army Corps of Engineers conducted a study of contaminated fish inhabiting the Tennessee River (in Alabama) and its tributaries. A total of 144 fish were captured and the following variables measured for each:

1. Location of capture
2. Species
3. Length (centimeters)
4. Weight (grams)
5. DDT concentration (parts per million)

Classify each of the five variables measured as quantitative or qualitative.

Solution

The variables Length, Weight, and DDT are quantitative because they are all measured on a natural numerical scale: length in centimeters, weight in grams, and DDT in parts per million. In contrast, Location and Species cannot be measured quantitatively; they can only be classified (e.g., channel catfish, largemouth bass, and smallmouth buffalo for species). Consequently, data on location and species are qualitative.

Self-Test 1.1

State whether each of the following variables measured on graduating high school students is quantitative or qualitative.

a. National Honor Society member or not
b. Scholastic Assessment Test (SAT) score
c. Number of colleges applied to
d. Part-time job status

EXAMPLE 1.3

CLASSIFYING VARIABLES

Good Housekeeping (July 1996) conducted an experiment to answer the question: "Do shaving creams and gels really work better than soap when shaving your legs?" Each of 30 women shaved one leg using a cream or gel and the other leg using soap or a beauty bar. At the end of the study, 20 of the 30 women stated their preference for shaving with creams and gels. Identify the experimental unit for this study and describe the variable of interest as quantitative or qualitative.

Solution

Since we are interested in the opinions of the 30 women who participated in the leg-shaving study, the experimental unit is a woman who shaves her legs. The variable measured, Shaving preference, can assume one of two categories in this experiment: (1) prefer a cream or gel, or (2) prefer soap or a beauty bar. Consequently, Shaving preference is a qualitative variable.

To summarize, knowing the type (quantitative or qualitative) of data that you want to analyze is one of the keys to selecting the appropriate statistical method to use. A second key involves the concept of population and samples, as discussed in the next section.

Statistics in the Real World Revisited

Identifying the Experimental Units and Variables

In Milgram's classic shock study introduced on pp. 22–23, several variables were measured for each teacher who participated, including the shock level that the teacher "applied" to the learner and the experimental setting (one of five different settings) that the teacher was assigned to. Thus, the *experimental units for the study are the teachers* (i.e., the subjects). The *Shock level variable is quantitative* since it is measured in volts, a natural numeric scale. The *Setting variable is qualitative* since its values (predicted behavior, remote, voice feedback, proximity, and touch proximity) are categorical in nature.

PROBLEMS FOR SECTION 1.2

Applying the Concepts

1.1 Determine whether each of the following variables is quantitative or qualitative.

a. Number of telephones per household

b. Whether there is a telephone line connected to a computer modem in the household

c. Whether there is a FAX machine in the household

d. Length (in minutes) of longest long-distance call made per month

e. Number of local calls made per month

1.2 Consider the following variables recorded in the Environmental Protection Agency Gas Mileage Guide for new automobiles. Determine whether each variable is quantitative or qualitative.

a. Manufacturer of automobile

b. Model name

c. Transmission type (automatic or manual)

d. Engine size

e. Number of cylinders

f. Estimated city gas miles per gallon

g. Estimated highway gas miles per gallon

1.3 A few years ago, only estates were permitted to own land in Hawaii and homeowners leased the land from the estate (a law that dates back to Hawaii's feudal period). The law now requires estates to sell land to homeowners at a fair market price. As part of a study to estimate the fair market value of its land (called the "leased fee" value), a large Hawaiian estate collected the data shown in the accompanying table for five properties.

Property	Leased Fee Value (thousands of dollars)	Lot Size (thousands of sq. ft.)	Neighborhood	Location of Lot
1	70.7	13.5	Cove	Cul-de-sac
2	52.6	9.6	Highlands	Interior
3	87.1	17.6	Cove	Corner
4	43.2	7.9	Highlands	Interior
5	144.3	13.8	Golf Course	Cul-de-sac

a. Identify the experimental units.

b. State whether each of the variables measured is quantitative or qualitative.

1.4 The Cutter Consortium recently surveyed 154 U.S. companies to determine the extent of their involvement in e-commerce. They found that "a stunning 65% of companies . . . do not have an overall e-commerce strategy." Four of the questions

they asked are listed below. For each question, determine the variable of interest and classify it as quantitative or qualitative. (*Internet Week,* Sept. 6, 1999, www.internetwk.com)

a. Do you have an overall e-commerce strategy?

b. If you don't already have an e-commerce plan, when will you implement one? (Never, Later than 2000, By the Second Half of 2000, By the First Half of 2000, By the End of 1999)?

c. Are you delivering products over the Internet?

d. What was your company's total revenue in the last fiscal year?

1.5 Work conflicts are generally considered counterproductive, but research conducted by Mississippi State professor Allen Amason found that some conflicts can be helpful in the decision-making process (*Academy of Management,* June 1996). Amason studied management teams and classified their conflicts as either cognitive (clarification of issues resulting in better decisions) or affective (personal, finger-pointing, conflicts that can tear a team apart).

a. Each team member was asked to respond "yes" or "no" to each of the following 10 questions: "Do disagreements in your group tend to . . . "

1. become arguments?
2. bring out the worst in us?
3. dissolve into a conflict among personalities?
4. become emotional?
5. cause hurt feelings?
6. build on irrational justifications?
7. sound self-serving?
8. seek political gain?
9. build cliques within the group?
10. result in ineffective decisions?

Consider each response as a variable measured on a management team member. Classify the 10 responses as quantitative or qualitative.

b. Three or more "yes" responses to the questions in part **a** suggest a high degree of dysfunctional, affective conflict within the group. For each team member, consider the total number of "yes" responses as a variable. Is this a quantitative or qualitative variable?

1.6 *Bioethics* describes ethical issues that arise in the life sciences. Examples include environmental pollution, wildlife conservation, animal rights, birth control (contraception/abortion), and herbicide/pesticide use. The *Journal of Moral Education* (Dec. 1997) examined the coverage of bioethics in biology textbooks for grades five to ten. Three questions of interest to the researchers were:

1. How much space in the textbook (in inches of type) is dedicated to bioethics?
2. How often are catchwords such as "danger," "threatened," "polluted," "poisoned," and "protection/rescue" used in the textbook?
3. Does the textbook use pictures in discussion of bioethics?

a. Identify the experimental units in the study.

b. Identify the type of data (quantitative or qualitative) extracted from each question.

1.7 D. Pritchard and K. D. Hughes, mass communication researchers at the University of Wisconsin—Milwaukee, collected and analyzed data on 100 Milwaukee homicides (*Journal of Communication,* Summer 1997). The purpose of the research was to identify important determinants of the "newsworthiness" of a crime such as homicide. Several of the many variables measured for each homicide are listed below. Identify the type (quantitative or qualitative) of each variable.

a. Average length (in words) of news stories published

b. Number of news items published

c. Gender of the victim
d. Gender of the suspect
e. Whether or not the police banned the press from publishing crime-scene photos
f. Per capita income of area in which crime occurred
g. Relationship between the victim and suspect
h. Age of victim

1.8 List five or more variables that your family physician considers while giving you a complete physical examination. State whether each is qualitative or quantitative.

1.9 The Gallup organization presents the results of recent polls at its home address on the World Wide Web: gallup.com. Access this Internet site and click on "Business and the Economy." For the survey indicated,

a. give an example of a qualitative variable measured in this survey.
b. give an example of a quantitative variable measured in this survey.
c. develop three questions that could be used in this survey. Identify the type (quantitative or qualitative) of data that the questions will produce.

1.3 Descriptive vs. Inferential Statistics

When you examine a data set in the course of your studies, you will be doing so because the data characterize some variable (or variables) of interest to you. In statistics, the data set that is the target of your interest is called a *population*. This data set, which is typically large, exists in fact or is part of an ongoing operation and hence is conceptual. Some examples of variables and their corresponding populations are shown in Table 1.2.

Definition 1.6

A **population** is a collection (or set) of data that describe some variable of interest to you.

TABLE 1.2 Some Typical Populations

Phenomenon	Experimental Units	Variable Measured	Type of Data	Population
a. White blood cell count of a hemophiliac	Hemophiliacs	White blood cell count	Quantitative	Set of white blood cell counts of all hemophiliacs
b. Grade point average (GPA) of a college freshman	College freshman	GPA	Quantitative	Set of grade point averages of all college freshman
c. July high temperature of a U.S. state	States	Highest temperature in July	Quantitative	Set of July high temperatures for all 50 U.S. states
d. Quality of items produced on an assembly line	Manufactured items	Defective status	Qualitative	Set of defective/nondefective measurements for all items manufactured over the recent past and future

If every measurement in the population is known and available, then statistical methodology can help you describe this set of data. In future chapters we will find graphical (Chapter 2) and numerical (Chapter 3) ways to make sense out of a large mass of data, i.e., to summarize and draw conclusions from it. The branch of statistics devoted to this application is called *descriptive statistics.*

Definition 1.7

The branch of statistics devoted to the organization, summarization, description, and presentation of data sets is called **descriptive statistics.**

Many populations are too large to measure each observation; others cannot be measured because they are conceptual. For example, population (a) in Table 1.2 cannot be measured since it would be impossible to identify all hemophiliacs. Even if we could identify them, it would be too costly and time-consuming to measure and record their white blood cell counts. Population (d) in Table 1.2 cannot be measured because it is partly conceptual. Even though we may be able to record the quality measurements of all items manufactured over the recent past, we cannot measure quality in the future. Because of this problem, we are required to select a subset of values from a population, called a *sample.*

Definition 1.8

A **sample** is a subset of data selected from a population.

EXAMPLE 1.4 POPULATION AND SAMPLE

Refer to the starting salary data described in Example 1.1 (p. 24). Recall that the Career Resource Center (CRC) surveys all University of Florida (UF) graduates at the time of their graduation. However, the CRC estimates that only about half of the UF graduates return the questionnaire, and of these, only half indicate that they have secured a job as of the date of graduation. Thus, it is important to note that the data set collected by the CRC is not a complete listing of the annual starting salaries of all recent University of Florida graduates.

a. Suppose the target of your interest is the starting salary data collected by the CRC last year. Describe the target population.

b. Suppose the target of your interest is the first-year salaries of all recent University of Florida graduates. Is the data set collected by the CRC a population or a sample?

Solution

a. The measurements of particular interest to us are the starting salaries of UF graduates. Since the data set collected by the CRC includes only the starting salaries of last year's UF graduates who returned the CRC survey and secured a job, the target population consists of the collection of starting salaries of UF graduates *who returned the CRC questionnaire and had secured a job at graduation time last year.*

b. Since we are interested in *all* recent UF graduates, the target population is the set of starting salaries of all recent graduates—not only those who returned the CRC questionnaires but also those who failed to return the survey even though they had secured a job by the date of graduation. Since data collected by the CRC are only a subset of this target population, they represent a sample.

EXAMPLE 1.5

POPULATION AND SAMPLE

Potential advertisers value television's well-known Nielsen ratings as a barometer of a TV show's popularity among viewers. The Nielsen rating of a certain TV program (e.g., NBC's hit medical drama *ER*) is an estimate of the proportion of viewers, expressed as a percentage, who tune their sets to the program on a given night at a given time. A typical Nielsen survey consists of 165 families selected nationwide who regularly watch television. Suppose we are interested in the Nielsen ratings for the latest episode of *ER* (currently shown on Thursday nights).

a. Identify the population of interest.

b. Describe the sample.

Solution

a. We want to know which TV program was watched by a family on Thursday during the *ER* time slot; consequently, the population of interest is the collection of programs watched by all families in the United States during this time slot. In this example, the qualitative measurements might be recorded as {family was tuned to *ER*} and {family was not tuned to *ER*}.

b. The sample—that is, the subset of the population—consists of the set of qualitative measurements defined in part **a** for each of the 165 families in the Nielsen survey.

Self-Test 1.2

To evaluate the current status of the dental health of all U.S. school children, the American Dental Association conducted a survey. One thousand grade school children from across the country were selected and examined by a dentist; the number of cavities was recorded for each.

a. Identify the population of interest to the American Dental Association.

b. Identify the sample.

c. What is the variable of interest? Is it quantitative or qualitative?

In Chapters 2 and 3, we demonstrate several statistical methods for organizing, describing, summarizing, and presenting data sets similar to those of Examples 1.4 and 1.5. However, statistics may involve much more than data description. In succeeding chapters, we will study statistical methods that enable us to infer the nature of the population from the information in the sample. The branch of statistics devoted to this is called *inferential statistics.* In addition, this *statistical methodology provides measures of reliability for each inference obtained from a sample.* This is one of the major contributions of inferential statistics. Anyone can examine a sample and make a "guess" about the nature of the population. For example, we might estimate that the average grade point average of college freshmen is 2.35, or that the starting salary range of a college graduate is between \$20,000 and \$30,000. But statistical methodology enables us to go one step further. When the sample is selected in a specified way from the population, we can also say how accurate our estimate will be, i.e., how close the estimate of 2.35 will be to the population average GPA, or how confident we are that the typical starting salary falls between \$20,000 and \$30,000.

Definition 1.9

The branch of statistics concerned with using sample data to draw conclusions about a population is called **inferential statistics.** When proper sampling techniques are used, this methodology also provides a measure of **reliability** for the inference.

EXAMPLE 1.6 POPULATION, SAMPLE, AND INFERENCE

The American Society for Microbiology sponsored a study to determine how likely people are to wash their hands in public restrooms. Researchers monitored 1,000 New York City commuters in the restrooms at the Grand Central and Penn train stations and observed that only 49% washed their hands before leaving (*Tampa Tribune,* Sept. 19, 2000). *Note:* Four years ago, a similar survey revealed that 60% of New Yorkers washed up in public restrooms. In this study, identify:

a. the population
b. the variable measured
c. the sample
d. the inference made about the population

Solution

a. The researchers are interested in the hygiene of people using public restrooms. Consequently, the experimental units in the target population are all users of public restrooms.

b. The measurement (or variable) of interest is the person's hand-wash status—wash hands before leaving the restroom or not. Note that this is a qualitative variable.

c. The sample consists of the collection of hand-wash status measurements (wash or not) for the 1,000 people observed in the New York City public restrooms.

d. Since 49% of the people in the sample washed their hands, the inference is that 49% of public restroom users in the population also wash up before leaving. This leads to the conclusion reached by the researchers, namely, that fewer New Yorkers wash up in public restrooms than in 1996. Note that the researchers do not provide a measure of reliability for this inference. Enough information is available, however, to calculate such a measure. We will show in Chapter 7 that, with a high degree of "confidence," the estimate of 49% is within 3% of the population percentage. That is, the population percentage is no lower than 46% and no higher than 52%.

The key facts to remember in this section are summarized in the accompanying boxes.

THE OBJECTIVE OF STATISTICS

1. To describe data sets (populations or samples)
2. To use sample data to make inferences about a population

THE MAJOR CONTRIBUTION OF DESCRIPTIVE STATISTICS

Conclusions about data can be made based on patterns revealed by descriptive statistical data.

THE MAJOR CONTRIBUTION OF INFERENTIAL STATISTICS

Statistical methodology allows us to provide a measure of reliability for every statistical inference based on a properly selected sample.

Statistics in the Real World Revisited

Identifying the Population, Sample, and Inference

The *data collected* and analyzed by Milgram in his classic shock study (pp. 22–23) *represent sample data* collected on the "teachers." The *collection of all possible "teachers,"* that is, all possible people who could theoretically participate in a similar experiment, *represents Milgram's target population.* Based on the sample results, Milgram made an inference about the population, namely, that a high level of obedience to authority exists in our society.

PROBLEMS FOR SECTION 1.3

Applying the Concepts

1.10 A sociologist is interested in describing the socioeconomic status levels of all Florida households. Identify each of the following data sets as a population or sample.

a. Set of annual incomes of all Florida households

b. Set of job status measurements (employed or unemployed) for each of 50 households located in the city of Miami

c. Set of poverty levels for each of 1,000 households selected from across the state of Florida

d. Set of job type measurements (white collar or blue collar) for all Florida households

1.11 Identify whether the data sets analyzed in the following studies represent populations or samples.

a. Flavor quality ratings (poor, fair, good, very good, excellent) for the 16 best-selling regular and ice beer brands (*Consumer Reports,* June 1996)

b. Presence or absence of Hessian fly damage for 600 wheat plants selected from a large midwestern wheat field (*Journal of Agricultural, Biological, and Environmental Statistics,* June 1997)

c. Magnitudes (on the Richter scale) of the 16,287 earthquakes that occurred worldwide in 1998 (National Earthquake Information Center)

d. Numbers of whale bones found at 35 prehistoric Thule Eskimo burial sites in the Canadian Arctic (*Journal of Archaeological Sciences,* Oct. 1997)

1.12 Are men or women more adept at remembering where they leave misplaced items (such as car keys)? According to University of Florida psychology professor Robin West, women show greater competence in actually finding these objects (*Explore,* Fall 1998). Approximately 300 men and women from Gainesville, Florida, participated in a study in which each person placed 20 common objects in a 12-room "virtual" house represented on a computer screen. Thirty minutes later, the subjects

were asked to recall where they put each of the objects. For each object, a recall variable was measured as "yes" or "no."

a. Identify the population (or sample) of interest to the psychology professor.

b. Does the study involve descriptive or inferential statistics? Explain.

c. Are the variables measured in the study quantitative or qualitative?

1.13 What percentage of Web users are addicted to the Internet? To find out, a psychologist designed a series of 10 questions based on a widely used set of criteria for gambling addiction and distributed them through the Web site ABCNews.com. (A sample question: "Do you use the Internet to escape problems?") A total of 17,251 Web users responded to the questionnaire. If participants answered "yes" to at least half of the questions, they were viewed as addicted. The findings, released at the 1999 annual meeting of the American Psychological Association, revealed that 990 respondents, or 5.7%, are addicted to the Internet (*Tampa Tribune,* Aug. 23, 1999).

a. Identify the data collection method.

b. Identify the target population.

c. Are the sample data representative of the population?

1.14 *Postpartum depression* is the term used to describe the usually short-lived period of emotional sensitivity that many women suffer following childbirth. Studies show that men, too, can suffer from postpartum depression. A developmental psychologist wants to estimate the proportion of fathers who suffer from postpartum blues. Fifty men who have recently fathered a child are interviewed and observed in the home, and the number experiencing some form of postpartum depression is recorded.

a. What is the population of interest to the developmental psychologist?

b. Describe the sample in this problem.

c. Suppose that 31 of the 50 men are diagnosed as having postpartum blues. The psychologist then estimates that 62% of all fathers experience postpartum depression. Do you believe that this estimate is equal to the proportion for the entire population? Explain.

1.15 As part of an undergraduate course in children's literature, a group of education majors read weekly to individually selected children. During the reading session, readers asked questions in an effort to draw the children into the literature. Three University of Colorado researchers analyzed all of the approximately 2,000 questions that readers asked and published the results in the *Journal of Literacy Research* (Dec. 1996). Each question was classified into one of four types:

1. Known information (e.g., "What's that?")
2. Opinion (e.g., "Why? What do you think about that?")
3. Conditional (e.g., "If you were in the story, what would you do?")
4. Connection (e.g., "Does that remind you of something in your life?")

a. How many variables are measured in this study?

b. What is the experimental unit?

c. Identify the type of data collected as quantitative or qualitative.

d. Would you classify the data as a population or a sample? Explain.

1.16 Tel Aviv University law professor A. Likhovski examined the history of the concept of "tyranny" in nineteenth-century American legal cases (*Journal of Interdisciplinary History,* Autumn 1997). Likhovski used "Lexis," a computer-searchable database containing all cases in which lawyers and/or judges used keywords associated with tyranny (e.g., "tyranny," "despotism," "dictatorship," "autocracy," and "absolutism"). In the decade 1850–1859, there were a total of 3,854 U.S. federal cases. Likhovski's computer search found that 45 of these cases (1.17%) used one or more of these keywords.

a. Identify the experimental unit in this analysis.

b. For the decade 1850–1859, describe the target population.

c. Describe the variable measured on each experimental unit and its type.

d. Does the percentage, 1.17%, describe a population or a sample?

1.17 A panel of tobacco experts convened by the National Cancer Institute recommended that cigarette manufacturers put more descriptive labels on their cigarette packages, including a disclaimer that "light" brands are not really more healthful than "regular" (nonlight) brands (*Tampa Tribune,* Dec. 7, 1994). The panel's recommendations are based, in part, on data collected by the Federal Trade Commission (FTC). Each year, the FTC tests all domestic cigarette brands for carcinogens such as tar and nicotine. Suppose our goal is to compare the average nicotine content of all domestic light cigarette brands to the average nicotine content of all domestic regular cigarette brands. To do this, we record the nicotine contents (in milligrams) of 25 light cigarette brands and 25 regular cigarette brands.

a. Describe the target populations. (Give the precise statistical definitions.)

b. Describe the samples.

1.18 *Euthanasia,* the act of painlessly putting to death a person suffering from an incurable and painful disease or condition, has long been a dilemma of medical ethics. Recently, Dr. Jack Kevorkian's physician-assisted suicide efforts have made national headlines. Suppose you work for a major opinion pollster and you have been assigned the task of conducting a survey. The purpose of the survey is to estimate the proportion of American adults who support Dr. Jack Kevorkian's euthanasia movement.

a. Clearly define the population of interest to the pollster.

b. Do you think it is possible to obtain the entire population? Explain.

c. Why should the sample you select for the survey be representative of the population?

1.19 The slope of a river delta region can sometimes be accurately predicted from knowledge of the typical size of stones found there. With this in mind, a geographer studying South America would like to estimate the average size of stones found in the delta region of the Amazon River. To obtain this estimate, the geographer collects 50 stones and measures the diameter of each.

a. What is the population of interest to the geographer?

b. Describe the sample in this problem.

c. Suppose that the average diameter of the 50 stones is 7.2 inches. Do you believe that this sample average will equal the average for the population? Explain.

1.4 Collecting Data

Once you decide on the type of data—quantitative or qualitative—appropriate for the problem at hand, you will need to collect the data. Generally, you can obtain the data in four different ways:

1. Data from a *published source*
2. Data from a *designed experiment*
3. Data from a *survey*
4. Data from an *observational study*

Sometimes, the data set of interest has already been collected for you and is available in a **published source,** such as a book, journal, newspaper, data base, or the World Wide Web. For example, you may want to examine and summarize the birth rates (i.e., number of births per 1,000 people) in the 50 states of the United States. You can find this data set (as well as numerous other data sets) at your library in the *Statistical Abstract of the United States,* published annually by the U.S. Department of Commerce. Some other examples of published data sources include *The Wall Street Journal* (financial data) and *The Sporting News* (sports information).

EXAMPLE 1.7 OBTAINING PUBLISHED DATA FROM THE WORLD WIDE WEB

To illustrate how data from a published source can be obtained from the World Wide Web, suppose we would like to obtain information concerning the birth rate for each state in the United States for 1997 and 1998. How can these data be obtained using the World Wide Web?

Solution

The United States Federal Government maintains a World Wide Web site (www.fedstats.gov) that provides statistics of interest for more than 70 federal agencies. (If you were not aware that such a site exists, you could reach this site through a World Wide Web search engine such as Yahoo at www.yahoo.com.) To obtain these data we went through the following steps as illustrated in Figure 1.1a–e.*

1. After signing on to the World Wide Web using your Web browser, enter the web address www.fedstats.gov.
2. You will see the Web page illustrated in Figure 1.1a.
3. Click on **Agencies listed alphabetically.**
4. You will see a Web page that contains a list of U.S. federal agencies. Scroll until you reach NCHS (National Center for Health Statistics) as illustrated in Figure 1.1b. Click on **National Center for Health Statistics.**
5. You will see a Web page that lists various statistics provided by the National Center for Health Statistics as illustrated in Figure 1.1c. Click **Births** under **Tabulated State Data.**

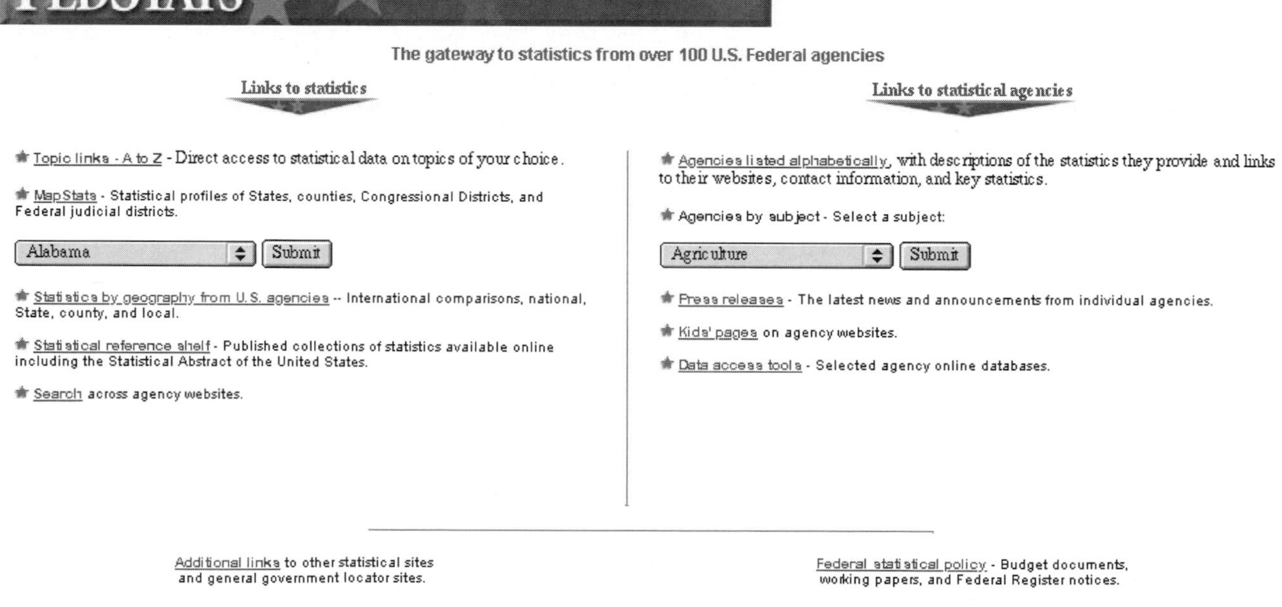

(a)

Figure 1.1a–e Finding State Birth Rates from the World Wide Web

* Web pages often change over time (although governmental sites tend to change less often than commercial sites). This search was conducted in October 2000.

Statistical Agencies FEDSTATS

Back to Fedstats home page | Topic links - A to Z

Federal Agencies with Statistical Programs

A B C D E F G H I J K L M N O P Q R S T U V W X Y Z Principal Statistical Agencies

National Center for Education Statistics collects, analyzes, and publishes statistics on education and public school district finance information in the United States; conducts studies on international comparisons of education statistics; and provides leadership in developing and promoting the use of standardized terminology and definitions for the collection of those statistics. Also supports the Data Analysis System (DAS) and the National Public School Locator.	• Contact Information • Key Statistics
National Center for Health Statistics responsible for the collection, analysis, and dissemination of statistics on the nature and extent of the health, illness, and disability of the U.S. population; the impact of illness and disability on the economy; the effects of environmental, social, and other health hazards; the use of health care services; health resources; family formation, growth, and dissolution; and vital events (i.e., births and deaths). CDC also provides data on morbidity, infectious and chronic diseases, occupational diseases and injuries, vaccine efficacy, and safety studies.	• Contact Information • Key Statistics
National Eye Institute supports more than 80 percent of the vision research conducted in the U.S. at approximately 250 medical centers, hospitals, universities, and other institutions. In addition, the Institute conducts studies in its own facilities to combat the myriad of eye disorders affecting millions of people worldwide.	• Contact Information
National Heart, Lung and Blood Institute plans, conducts, fosters, and supports an integrated and coordinated program of basic research, clinical investigations and trials, observational studies, and demonstration and education projects. Research is related to the causes, prevention, diagnosis, and treatment of heart, blood vessel, lung, and blood diseases; and sleep disorders. It also supports research on clinical use of blood and all aspects of the management of blood resources.	• Contact Information
National Highway Traffic Safety Administration under the U.S. Department of Transportation, is responsible for reducing deaths, injuries and economic losses resulting from motor vehicle crashes. This is accomplished by setting and enforcing safety performance standards for motor vehicles and motor vehicle equipment, and through grants to state and local governments to enable them to conduct effective local highway safety programs.	• Contact Information • Key Statistics
National Institute on Aging leads a broad scientific effort to understand the nature of aging and to extend the healthy, active years of life. In 1974, Congress granted authority to form the NIA to provide leadership in aging research, training, health information dissemination, and other programs relevant to aging and older people. Subsequent amendments to this legislation designated the NIA as the primary federal agency on Alzheimer's disease research.	• Contact Information
National Institute on Alcohol Abuse and Alcoholism supports and conducts biomedical and behavioral research on the causes, consequences, treatment, and prevention of alcoholism and alcohol-related problems. NIAAA also provides leadership in the national effort to reduce the severe and often fatal consequences of these problems.	• Contact Information

(b)

- Births
- Deaths
- Deaths, Infant
- Health, United States
- Healthy People 2000, State Data
- 2000 State Health Profiles
- State Health Statistics by Sex and Race/Ethnicity
- Trends in Health and Aging

(c)

Data Highlights

■ Selected charts from: National Vital Statistics System, Declines in Teenage Birth Rates, 1991-98: Update of National State Trends, (PHS) 2000-1120 *(Click figure to view full size)*

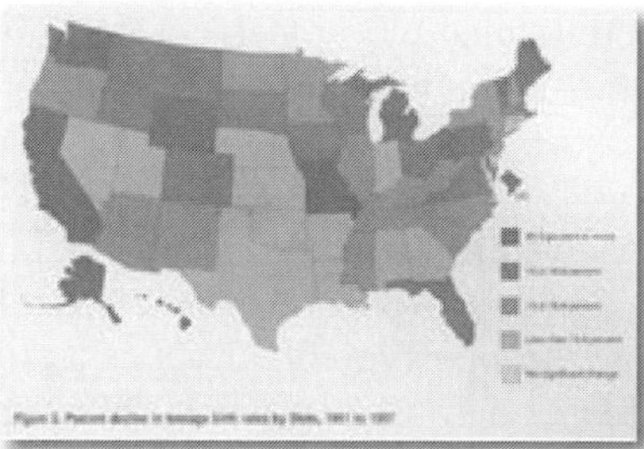

NVSR, Vol. 47, No. 26, Figure 3
View/Download PDF 153KB

■ Teen pregnancy

■ Preliminary number of live-births
Table excerpted from the Births and Deaths: United States, 1998
View/download PDF 11KB

■ Preliminary number of live births: Each division and State
Table excerpted from the Births and Deaths: United States, 1998
View/download PDF 14KB

■ Live Births, Birth rates, and Fertility Rates
Table excerpted from the Report of Final Natality Statistics, 1997
View/download PDF 13KB

(d)

Table 4. Live births by race and Hispanic origin of mother: United States, each State, Puerto Rico, Virgin Islands, and Guam, preliminary 1998, and birth and fertility rates, final 1997 and preliminary 1998

[By place of residence. Data are based on a continuous file of records received from the States. Birth rates are total births per 1,000 total population; fertility rates are total births per 1,000 women aged 15-44 years. Figures for 1998 are based on weighted data rounded to the nearest individual, so categories may not add to totals]

Area	Number							Birth rate		Fertility rate	
	All races	*White, total* [1]	*White, non-Hispanic*	*Black* [1]	*American Indian* [1,2]	*Asian or Pacific Islander* [1]	*Hispanic* [3]	*1998*	*1997*	*1998*	*1997*
United States [4]	3,944,046	3,122,391	2,364,907	610,203	40,167	171,284	735,019	14.6	14.5	65.6	65.0
Alabama	62,126	41,564	40,242	20,041	143	378	1,317	14.3	14.1	63.3	62.1
Alaska	9,935	6,684	6,180	405	2,388	458	595	16.2	16.3	73.2	72.4
Arizona	78,296	68,428	38,729	2,663	5,437	1,768	29,792	16.8	16.6	78.3	78.1
Arkansas	36,891	28,311	26,553	7,990	219	371	1,730	14.5	14.5	67.6	67.3
California	521,801	424,662	177,152	36,738	3,380	57,020	247,668	16.0	16.3	70.7	72.3
Colorado	59,800	54,527	40,086	2,880	654	1,739	14,710	15.1	14.5	67.4	64.2
Connecticut	43,812	36,938	28,957	5,375	110	1,389	6,183	13.4	13.2	61.3	60.3
Delaware	10,583	7,703	6,939	2,623	34	222	748	14.2	14.0	61.2	60.3
District of Columbia	7,694	2,054	1,321	5,467	8	164	734	14.7	15.0	60.9	61.7
Florida	195,636	146,218	107,758	44,386	910	4,121	39,541	13.1	13.1	65.1	64.9

(e)

6. You will see a Web page that lists different birth information as illustrated in Figure 1.1d. Click **View/download.PDF** (.PDF is a type of file format) under **Preliminary number of live births: Each division and state.**
7. You will see a Web page that provides the birth rates for the states along with additional information. The first portion of this Web page is illustrated in Figure 1.1e.
8. Save the data in the format that you select. Different data sets will be able to be saved in alternate formats.

With published data, we often make a distinction between the *primary source* and a *secondary source.* If the publisher is the original collector of the data, the source is primary. Otherwise, the data are secondary source data.

A second method of collecting data involves conducting a **designed experiment,** in which the researcher exerts strict control over the units (people, objects, or events) in the study. For example, a recent medical study investigated the potential of aspirin in preventing heart attacks. Volunteer physicians were divided into two groups—the *treatment* group and the *control* group. In the treatment group, each physician took one aspirin tablet a day for one year, while each physician in the control group took an aspirin-free placebo (a drug with no active ingredient) made to look like an aspirin tablet. The researchers, not the physicians under study, controlled who received the aspirin (the treatment) and who received the placebo. As you will learn in Chapter 11, a properly designed experiment allows you to extract more information from the data than is possible with an uncontrolled study.

Surveys are a third source of data. With a **survey,** the researcher samples a group of people, asks one or more questions, and records the responses. Probably the most familiar type of survey is a public opinion poll conducted by any one of a number of organizations (e.g., Harris, Gallup, Roper, and CNN). Another well-known survey is the Nielsen survey, which provides the major networks with information on the most watched television programs. Surveys can be conducted through the mail, with telephone interviews, or with in-person interviews. Although in-person interviews are more expensive than mail or telephone surveys, they may be necessary when complex information must be collected.

Finally, observational studies can be employed to collect data. In an **observational study,** the researcher observes the experimental units in their natural setting and records the variable(s) of interest. For example, a psychologist might observe and record the level of "Type A" behavior of a sample of male managers. Similarly, a marine biologist may observe and record the length and gender of blue whales

captured in Antarctica. Unlike a designed experiment, an observational study is one in which the researcher makes no attempt to control any aspect of the experimental units.

EXAMPLE 1.8

DATA COLLECTION METHODS

Identify the data collection method in each of the following studies:

a. The U.S. Army Corps of Engineers study of fish inhabiting the Tennessee River, Example 1.2 (p. 26).

b. The *Good Housekeeping* study of leg shaving preferences of women, Example 1.3 (p. 26).

c. The salary study of recent University of Florida graduates, Examples 1.1 and 1.4 (p. 24 and p. 30).

Solution

a. Recall that the U.S. Army Corps of Engineers captured 144 fish in their natural habitat—the Tennessee River—then recorded the values of several variables, e.g., capture location and species type. Consequently, the resulting data are observational in nature and the study is an observational study. The engineers made no attempt to control experimental variables such as weight and length.

b. The researchers in the *Good Housekeeping* study controlled the experimental units (women shavers) by requiring each woman shaver to shave one leg with a cream or gel and the other leg with soap or a beauty bar. Thus, it is a designed experiment. We learn in Chapter 7 that such a design attempts to remove an extraneous source of variation (the different shaving styles of women) in order to obtain a more reliable estimate of the proportion of women who prefer shaving their legs with a cream or gel.

c. The data collection method used by the Career Resource Center is a mail survey since questionnaires were mailed to University of Florida graduates to determine the employment status and salary of each.

Self-Test 1.3

All highway bridges in the United States are inspected for structural deficiency by the Federal Highway Administration (FHA). For each bridge, the variables measured include length of maximum span, number of vehicle lanes, and structural condition of bridge deck. Identify each of these variables as quantitative or qualitative, the data collection method, and whether or not the data represent a population or a sample.

Regardless of the data collection method employed, it is likely that the data will be a sample from a population. And if we wish to apply inferential statistics, we must obtain a *representative sample.*

Definition 1.10

A **representative sample** exhibits characteristics typical of those possessed by the target population.

For example, consider a political poll conducted during a presidential election year. Assume the pollster wants to estimate the percentage of all 120,000,000 registered voters in the United States who favor the incumbent president. The pollster would be unwise to base the estimate on survey data collected for a sample of voters from the incumbent's own state. Such an estimate would almost certainly be *biased* high.

The most common way to satisfy the representative sample requirement is to select a *random sample.* Random samples are the topic of the next section.

Statistics in the Real World Revisited

Identifying the Data Collection Method

Since psychologist Stanley Milgram exerted strict control over the experimental units in his shock study (pp. 22–23)—he randomly assigned teachers to one of the five different experimental settings—his data collection method is a *designed experiment.* The random assignment enabled Milgram to compare the average shock levels administered under each setting and draw valid conclusions about which setting (if any) would likely yield the highest shock levels.

PROBLEMS FOR SECTION 1.4

Applying the Concepts

1.20 The Dow Jones Industrial Average (DJIA) at the end of each of the past 10 years is listed in the table. Suppose you are interested in using these data to forecast the DJIA in the year 2003. Identify the data collection method.

Year	DJIA
1990	2,634
1991	3,169
1992	3,330
1993	3,754
1994	3,834
1995	5,117
1996	6,448
1997	7,908
1998	9,181
1999	11,497

Source: Standard & Poors *NYSE Daily Stock Price Record,* 1990–1999.

Problem 1.20

1.21 Refer to the *Journal of Communication* (Summer 1997) study of 100 Milwaukee homicides, Problem 1.7 on p. 28. Identify the data collection method.

1.22 Refer to the Gallup poll you selected in Problem 1.9 on p. 29. Identify the data collection method.

1.23 Refer to the study of web users and addiction to the Internet, Problem 1.13 on p. 34. Identify the data collection method.

1.24 The *Journal of Experimental Education* (Winter 1997) published a study that examined the benefit of repeated presentations of a lecture on student learning. The study participants were 109 college juniors and seniors enrolled in an educational psychology course where lectures are offered on videotape. The researchers were interested in how two variables, the number of times the videotaped lecture was presented (once, twice, or three times) and the type of organizational material presented before the lecture (conventional, linear, or matrix), impact lecture recall test scores. The students were randomly assigned to one of the $3 \times 3 = 9$ combinations of number of videotaped lectures and organization material; then the test scores of the nine groups were compared.

a. Identify the experimental units for this study.

b. Describe the variables measured and their type.

c. Is this a designed study? Explain why or why not.

d. Are the data collected in the study population data or sample data?

1.25 Can you tell when an unseen person is staring at you? *Skeptical Inquirer* (Sept./Oct. 2000) reports on a series of experiments designed to investigate the "psychic staring" phenomenon. Twelve volunteers who all believed they could detect unseen staring were tested in a controlled setting. Each volunteer sat with his or her back to a one-way mirror and in a series of trials were either stared at or not by a person on the other side of the mirror. (The sequence of staring—stare, stare, no stare, etc.—was randomized.) At the end of each trial, the volunteer was asked if he or she had been stared at. The number of correct responses for each volunteer was

recorded and the data analyzed. The results showed the volunteers had no ability to detect unseen staring.

a. Identify the population and the sample for this study.

b. What data collection method was used by the researchers and why is this key to the conclusions drawn from the study?

1.26 The Data and Story Library (DASL) is an on-line library of data files and stories that illustrate the use of basic statistical methods. Each data set has one or more associated stories. The stories are classified by method and by subject. Access this World Wide Web site at lib.stat.cmu.edu/DASL and, after reading a story, summarize how statistics have been used in one of the subject areas. Have inferential or descriptive statistics been used?

1.5 Random Sampling

A careful analysis of most data sets will reveal that they are samples from larger data sets that are really the object of interest. Consequently, most applications of modern statistics involve sampling and using information in the sample to make inferences about the population. For these applications, it is essential that we obtain a representative sample.

Assume we want to select a sample of size n from a population with N experimental units. The most common way is to select the sample in such a way that every different sample of size n has an equal probability (or chance) of selection from the population.* This procedure is called *random sampling* and the resulting sample is called a *(simple) random sample.* (Throughout this text, a "random sample" refers to a simple random sample as defined below.) Random samples are very likely (although not guaranteed) to be representative of the target population.

*A more formal definition of "probability" will be provided in Chapter 4.

Definition 1.11

A **(simple) random sample** of n experimental units is one selected from the population in such a way that every different sample of size n has an equal chance of selection.

How can a random sample be generated? If the population is not too large, each observation may be recorded on a piece of paper and placed in a suitable container. After the collection of papers is thoroughly mixed, the researcher can remove n pieces of paper from the container; the elements named on these n pieces of paper are the ones to be included in the sample. Lottery officials utilize such a technique in generating the winning numbers for the state of Florida's weekly 6/53 Lotto game. Fifty-three white Ping-Pong balls (the population), each identified with a number from 1 to 53 in black numerals, are placed into a clear plastic drum and mixed by blowing air into the container. The Ping-Pong balls bounce at random until a total of six balls "pop" into a tube attached to the drum. The numbers on the six balls (the random sample) are the winning Lotto numbers.

This method of random sampling is fairly easy to implement if the population is relatively small. It is not feasible, however, when the population consists of a large number of observations. Since it is also very difficult to achieve a thorough mixing, the procedure provides only an approximation to random sampling. Most scientific studies, however, rely on **random number generators** to automatically generate the random sample. Random number generators can be found in a table of random numbers (such as Table B.1 in Appendix B) and in software (such as Microsoft Excel or MINITAB). We illustrate the use of these random number generators in the following example.

EXAMPLE 1.9

SELECTING A RANDOM SAMPLE

Suppose we want to conduct a study of state lottery winners. Assume there were 18,000 people who won at least $1 million in state lotteries in the past decade and we are interested in knowing what proportion of these 18,000 winners quit their jobs within a year of winning. The target population consists of the job status of these 18,000 state lottery winners. Use the random numbers in Table B.1 to generate a random sample of size $n = 5$ from the population of size $N = 18{,}000$.

Solution

Begin by numbering the state lottery winners in the population from 1 to 18,000. A portion of Table B.1 is reproduced in Table 1.3. The steps for obtaining a random sample using Table B.1 are outlined in the box below. The five random numbers are shown in Table 1.4. The state lottery winners assigned these numbers comprise our random sample of size 5.

TABLE 1.3 Reproduction of a Portion of Table B.1

	Column					
	1	2	3	4	5	6
Row 1	10480	15011	01536	02011	81647	91646
2	22368	46573	25595	85393	30995	89198
3	24130	48360	22527	97265	76393	64809
4	42167	93093	06243	61680	07856	16376
5	37570	39975	81837	16656	06121	91782
6	77921	06907	11008	42751	27756	53498
7	99562	72905	56420	69994	98872	31016
8	96301	91977	05463	07972	18876	20922
9	89579	14342	63661	10281	17453	18103
10	85475	36857	53342	53988	53060	59533
11	28918	69578	88231	33276	70997	79936
12	63553	40961	48235	03427	49626	69445
13	09429	93969	52636	92737	88974	33488
14	10365	61129	87529	85689	48237	52267
15	07119	97336	71048	08178	77233	13916

TABLE 1.4 Random Sample of Five Selected from a Population of 18,000

Random Number
10,480
15,011
1,536
2,011
6,243

Using a Table of Random Numbers to Generate a Random Sample of Size *n* from a Population of *N* Elements

Step 1 The elements in the population are numbered from 1 to $N = 18{,}000$. This labeling implies that we will obtain random numbers of five digits from the table, selecting only those numbers with values less than or equal to 18,000. Note that Table 1.3 gives 5-digit random numbers in each column.

Step 2 Arbitrarily, let us begin in row 1, column 1 of the table. The random number entry given there is 10480 (highlighted in Table 1.3). Thus, we'll choose the state lottery winner numbered 10,480 as our first element of the random sample.

Step 3 Proceeding horizontally to the right across the columns (this choice of direction is arbitrary), the next entry in the table is 15011. Therefore, the lottery winner numbered 15,011 represents the second element in the sample. Continuing in this manner, the remaining

elements to be included in the sample are those numbered 01536, 02011, 81647 (skip), 91646 (skip), 22368 (skip), 46573 (skip), . . . , (proceeding to row 4), 42167 (skip), 93093 (skip), and 06243.

Self-Test 1.4

Use the random number table, Table B.1, to select a random sample of size $n = 3$ from a population of size $N = 100$.

[*Note:* In the tutorials at the end of this chapter (pp. 55–56) we show you how to obtain a printout of 50 random numbers using either Excel or MINITAB.]

Although random sampling represents one of the simplest of the multitude of sampling techniques available for research, most of the statistical techniques presented in this introductory text assume that a random sample (or a sample that closely approximates a random sample) has been collected. We consider a few of the other, more sophisticated, sampling methods in optional Section 1.6.

PROBLEMS FOR SECTION 1.5

Using the Tools

1.27 Consider a random sample from a population of size $N = 750$. To obtain the sample, you must assign each experimental unit in the population a code number. What number would you assign to:

a. the first experimental unit on the list?

b. the twentieth experimental unit on the list?

c. the last experimental unit on the list?

1.28 Starting in row 12, column 1 of the random number table, Table B.1, list the first random number for a random sample drawn from a population of size:

a. $N = 300$ **b.** $N = 1{,}000$

c. $N = 500$ **d.** $N = 50$

1.29 Use Table B.1 to draw a random sample of size $n = 5$ from a population of size $N = 5{,}000$.

Applying the Concepts

FISH

1.30 Refer to the U.S. Army Corps of Engineers study of contaminated fish inhabiting the Tennessee River, Example 1.2, p. 26. The data collected for all 144 fish captured are stored in the file named FISH. Select a random sample of size 10 from the population of fish weights.

CHKOUT

1.31 At a supermarket, customer checkout time is measured as the total length of time (in seconds) required for a checker to scan and total the prices of the customer's food items, accept payment, return change, and bag the items. Data collected on customer checkout times for 500 shoppers at a supermarket in Gainesville, Florida, are stored in the CHKOUT file. Randomly select a sample of 5 from the 500 checkout times in the data set.

1.32 A clinical psychologist is asked to view tapes in which each of six experimental subjects is discussing his or her recent dreams. Three of the six subjects have been previously classified as "high-anxiety" individuals, and the other three as "low-anxiety." The psychologist is told only that there are three of each type and is asked to identify the three high-anxiety subjects.

a. List the different samples of three subjects that may be selected by the psychologist.

b. Do you think the sample chosen by the psychologist will be random? Explain.

1.33 A file clerk is assigned the task of selecting a random sample of 26 company accounts (from a total of 5,000) to be audited. The clerk is considering two sampling methods:

Method A: Organize the 5,000 company accounts in alphabetical order (according to the first letter of the client's last name). Then randomly select one account card for each of the 26 letters of the alphabet.

Method B: Assign each company account a 4-digit number from 0001 to 5000. Using a computer random number generator, choose 26 4-digit numbers (from 0001 to 5000) and match the numbers with the corresponding company account.

Which of the two methods would you recommend to the file clerk? Which sampling method could possibly yield a biased sample?

1.6 Other Types of Samples (Optional)

Selecting a random sample can be difficult and costly. For example, suppose we want to sample the opinion of all residents age 18 or over in a certain community on the issue of mandatory drug testing in the workplace. The first obstacle to the sample selection is acquiring a *frame,* a complete listing of all the experimental units in the population. In this example, the frame is a list of all residents age 18 or over in the community. The second obstacle is contacting the residents who were selected in the sample to obtain their opinions.

Definition 1.12

A **frame** is a list of all experimental units in the population.

To reduce the difficulty and costs associated with acquiring a frame and selecting the sample, and to increase the precision of the sample information, experts trained in the art of *survey sampling* have devised some modifications to the simple random sampling procedure. In this section we discuss three alternative methods of sampling: *stratified random sampling, cluster sampling,* and *systematic sampling.*

EXAMPLE 1.10

STRATIFIED RANDOM SAMPLE

Suppose we wish to sample the opinions of all heads of household in a state on an issue such as mandatory drug testing and further suppose that the state contains ten counties (see Figure 1.2). It might be difficult to obtain a frame listing all households within the state, but suppose we know that each county possesses a frame—namely, the households listed on its tax roll. How could we proceed?

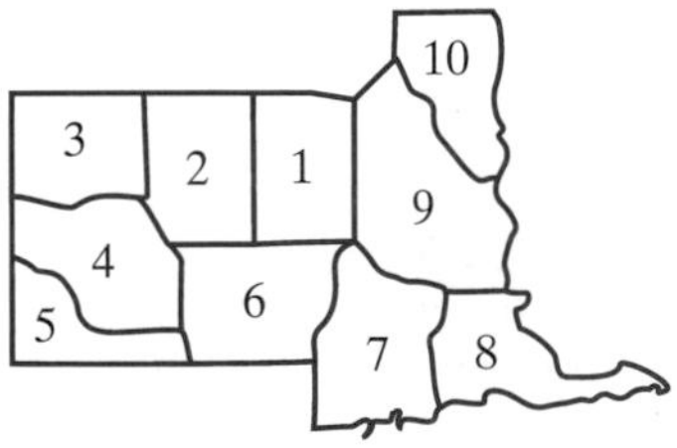

Figure 1.2 Map of a Fictitious State with 10 Counties

Solution

Instead of combining the 10 frames and selecting a random sample from the whole state, it would be easier and less costly to select a random sample of heads of household from each county. This would enable us to obtain sample opinions not only for each county (so that particular counties could be compared) but also for the entire state. This method of sampling is called *stratified random sampling* because the population is partitioned into a number of *strata* (in this example, counties), and random samples are then selected from among the elements in each *stratum.*

Stratification is advantageous because it allows you to acquire sample information on the individual strata (e.g., counties) as well as on the entire population (e.g., state). It may be less costly to select the sample and, because the variability of the responses within a stratum (county) is usually less than the variability in

responses between strata, stratification may provide more accurate estimates of strata and population parameters. However, if the target population is stratified incorrectly, stratified sampling may produce less accurate results than simple random sampling.

Definition 1.13

A **stratified random sample** is obtained by partitioning the sampling units in the population into nonoverlapping subpopulations called *strata*. Random samples are then selected from each stratum.

EXAMPLE 1.11

CLUSTER SAMPLING

Suppose we want to sample the opinions of the heads of household in a city but we know that many households, for one reason or another, are not listed on the tax roll. How could we proceed?

Solution

Since the tax roll cannot be used as a frame, we could construct a frame by numbering each of the city blocks on a map. The list of all blocks in the city would then provide a frame for selecting a random sample of city blocks, each representing a *cluster* of households. After the random sample of clusters is selected, interviewers are sent out to contact and interview all heads of household within each cluster (city block) that appears in the sample. The information contained in the clusters is ultimately combined to make inferences about the population of heads of household within the city. This method of sampling is commonly known as *cluster sampling*. The advantages of cluster sampling are clear: It is often easy to form a frame consisting of clusters, and it is less costly to conduct interviews within clusters than it is to interview individual elements selected at random from the population. Cluster sampling, however, will produce unreliable results if improper clusters are selected.

Definition 1.14

A **cluster sample** is a random sample in which the sampling units consist of clusters of elements.

EXAMPLE 1.12

SYSTEMATIC SAMPLING

Suppose our objective is to sample the opinions of students at your college or university concerning a prospective increase in on-campus parking fees. We have obtained an alphabetical list of all students with campus parking permits from the Parking and Services Division at your college and consider this list to be the frame. However, due to cost considerations, we are unable to produce a random sample from the frame. How could we proceed?

Solution

The sample could be collected by *systematically* selecting every fifth student (or every 10th student or, in general, every kth student) on the alphabetical list and contacting and questioning each. This technique, known as *systematic sampling*, is very valuable when choosing large samples. Instead of generating many random numbers, you can order the sampling frame and select every kth element.

Although this type of sampling is easy and inexpensive, it is difficult to assess the accuracy of estimates based on a systematic sample. Also, systematic sampling produces poor results if a periodic or cyclical pattern exists in the data. (For example, if prices are sampled every 30th day at a retail outlet, the prices will tend to be too low because of end-of-the-month clearance sales.)

Definition 1.15

A **systematic sample** is obtained by systematically selecting every kth element in the population when the elements are ordered from 1 to N.

1.7 Ethical and Other Issues in Statistical Applications

According to H. G. Wells, author of such science-fiction classics as *The War of the Worlds* and *The Time Machine,* "Statistical thinking will one day be as necessary for effective citizenship as the ability to read and write." Written more than a hundred years ago, Wells' prediction is proving true. Every day we read about the results of surveys or opinion polls in our newspaper, or we hear some titillating commentary on our radio or television. Clearly, advances in information technology have led to a proliferation of data-driven studies. Not all this information, unfortunately, is good, meaningful, or important. For example, a recent segment of the popular ABC television program *20/20* exposed the truth behind a "survey" that claimed that today's teachers worry most about being assaulted by their students. The original source of this information was not a survey at all, but a Texas oilman who simply published his opinions in a newsletter.* It is essential that we learn to evaluate critically what we read and hear and discount studies that lack objectivity or credibility.

*"Fact or Fiction? Exposés of So-Called Surveys." Mar. 31, 1995, segment of *20/20,* ABC Television.

When evaluating information that is based on a survey, we should first determine whether or not the survey results were obtained from a random sample. Surveys that employ nonrandom sampling methods are subject to serious, perhaps unintentional biases that may render the results meaningless. Even when surveys employ random sampling, they are subject to potential errors. Three of these problems are due to *selection bias, nonresponse error,* and *measurement error,* as the following examples illustrate.

EXAMPLE 1.13

SELECTION BIAS

One of the most infamous examples of improper sampling was conducted in 1936 by *Literary Digest* magazine to determine the winner of the Landon–Roosevelt presidential election in the United States. The poll, which predicted Landon to be the winner, was conducted by sending ballots to a random sample of persons selected from among the names listed in the telephone directories, magazine subscription lists, club membership lists, and automobile registrations of that year. In the actual election, Landon won in Maine and Vermont but lost in the remaining 46 states. The *Literary Digest*'s erroneous forecast is believed to be the major reason for its eventual failure. What was the cause of the *Literary Digest*'s erroneous forecast?

Solution

In 1936, the United States was still suffering from the Great Depression. During this period of economic hardship, the majority of the voting population could not afford such amenities as telephones, magazine subscriptions, club memberships,

and automobiles. Consequently, the majority of voters in the election were excluded from being selected in the sample because they were not on the lists compiled by *Literary Digest*—only the rich participated in the poll.

Recall, from Definition 1.12, that a frame is a listing of all possible experimental units in the population. *Selection bias* results from the exclusion of certain experimental units from the frame so that they have no chance of being selected in the random sample. The *Literary Digest* poll, then, suffered from selection bias, leading to an erroneous forecast of the presidential winner.

Definition 1.16

Selection bias results when a subset of the experimental units in the population is excluded from the population frame so that they have no chance of being selected in the sample.

EXAMPLE 1.14

NONRESPONSE BIAS

The *Journal of the Institute for Socioeconomic Studies* published a study on state lottery winners and the likelihood that they quit their job within one year of winning a big payoff. The researcher mailed job status questionnaires to over 2,000 lottery winners, and 576 (only 29%) responded. Why might the results of the mail survey be biased?

Solution

In mail surveys, not everyone mailed a questionnaire will be willing to respond. Busy people, a particular and unique social class, are often too preoccupied to complete the survey's questionnaire. Or, as in the study of state lottery winners, those surveyed may be totally disinterested in the survey question (remember, they have just struck it rich in the lottery) and, therefore, less likely to respond. *Nonresponse error* results from the failure to collect data on all experimental units (i.e., state lottery winners) in the sample. This type of error results in nonresponse bias. Methods for coping with nonresponse (based primarily on re-sampling the nonresponders) are available and are discussed in the references.

Definition 1.17

Nonresponse error occurs when those conducting a survey or study are unable to obtain data on all experimental units in the sample.

EXAMPLE 1.15

MISLEADING SURVEY RESULTS

The Holocaust—the Nazi Germany extermination of the Jews during World War II—has been well documented through film, books, and interviews with the concentration camp survivors. But a recent Roper poll found that one in five U.S. residents doubts the Holocaust really occurred. A national sample of over 1,000 adults and high school students were asked, "Does it seem possible or does it seem impossible to you that the Nazi extermination of the Jews never happened?" Twenty-two percent of the adults and 20% of the high school students responded "Yes" (*New York Times,* May 16, 1994). Why might the results of this survey be misleading?

Solution

A good questionnaire is designed with the intent of extracting meaningful and valid data. Do responses to the survey questions actually measure the phenomenon of interest to the researcher? Clearly, the pollsters in this example want to know the proportion of U.S. residents who believe that the Holocaust never occurred. But the question, "Does it seem possible or does it seem impossible to you that the Nazi extermination of the Jews never happened?", is poorly worded. It is unclear whether answering "Yes" to the question means that you believe the Holocaust was real or whether you believe it never happened! This results in *measurement error,* since the pollster is unsure of what the question is really measuring. In addition to ambiguous questions, other common sources of measurement error are "leading questions" (i.e., the questions are not objectively presented in a neutral manner) and the perceived obligation on the part of the respondent to "please" the interviewer by answering questions in a certain manner.

Definition 1.18

Measurement error refers to inaccuracies in the values of the data recorded. In surveys, this error may be due to ambiguous or leading questions and the interviewer's effect on the respondent.

A fourth source of error in survey research is *sampling error.* This error reflects the chance differences that occur from one sample to another when selecting random samples from the target population. For example, surveys or polls often include a statement regarding the margin of error or precision of the results—"the results of this poll are expected to be within ±4 percentage points of the actual value." In all the inferential statistical procedures presented in this text, we show how to measure and interpret sampling error.

Definition 1.19

Sampling error results from the chance differences that occur from sample to sample, when random samples are selected from the population. It measures the margin of error or precision of a sample estimate, i.e., how close the sample estimate is to the true population value.

Self-Test 1.5

An editor of a magazine for Harley-Davidson bikers argued that motorcyclists should not be required by law to wear helmets (*New York Times,* June 17, 1995). The editor cited a survey of 2,500 bikers at a rally that found 98% opposed the law. Identify the potential sources of error in this survey.

We conclude this section with some comments about ethics in surveys and other statistical studies. Eric Miller, editor of the newsletter *Research Alert,* recently stated that "there's been a slow sliding in ethics. The scary part is that people make decisions based on this stuff. It may be an invisible crime, but it's not a victimless one." As we stated earlier, not all surveys are good, meaningful, or important—and not all survey research is ethical. However, there is a distinction between poor survey design and **unethical survey design.** The key is *intent.*

For example, selection bias becomes an ethical issue only if particular groups or individuals are purposely excluded from the population frame so that the survey's results will be more likely to favor the position of the survey's sponsor. Similarly, nonresponse error becomes an ethical issue if there are individuals that are less likely to respond to a particular survey format and the researcher knowingly designs the survey with this in mind. Any time the survey designer purposely chooses loaded, lead-in questions that will guide responses in a particular direction, or advises the interviewer through mannerism and tone to guide the responses, measurement error becomes an ethical issue. Finally, if the researcher purposely presents a survey's results without reference to sampling error in order to promote a certain viewpoint, ethics can be called into question.

The ideas presented in this section will, hopefully, cause you to critically evaluate the results of surveys or other studies presented in the media. Examine the purpose of the study, why it was conducted, and for whom, and then discard the results if the study is found to be lacking in objectivity or credibility, or is unethical.

PROBLEMS FOR SECTION 1.7

Applying the Concepts

1.34 To gauge consumer preferences for a new product, companies sometimes employ *simulated test marketing* over conventional sampling methods. In simulated test marketing, a consumer is recruited at a shopping center and given a free sample of a new product to try at home. Later, the consumer rates the product in a telephone interview. The telephone responses are used by the test-marketing firm to predict potential sales volume. Researchers have found that simulated marketing tests identify potential product failures reasonably well, but not potential successful products. Why might the sampling procedure yield a sample of consumer preferences that underestimates sales volume of a successful new product?

1.35 Researchers who publish in professional business journals are typically university professors. Consequently, the data upon which the research is based are sometimes obtained by using students in an experimental setting. For example, the *Academy of Management Journal* (Oct. 1993) published a study on the relationship between employee commitment and employee turnover at an aerospace firm. A major part of the study involved measuring and analyzing the commitment of employees (and former employees) of the firm. A secondary portion of the study, however, utilized commitment data measured on a sample of students in an undergraduate management class. In general, comment on the use of students as experimental units in business studies.

1.36 The California Department of Health Services, Licensing and Certification (CDHSL&C) recently commissioned a survey of skilled nursing facilities in the state. The survey's purpose was to examine the infection prevention and control programs implemented at the nursing facilities (*Journal of Gerontological Nursing,* Nov. 1997). A survey was mailed to the managers of all 1,454 skilled nursing facilities in the state. (The survey was mailed only once and no follow-up reminders were sent.) A total of 444 surveys were returned, for a response rate of 30.5%. Several of the survey questions are listed below.

1. Is your facility a profit or nonprofit facility?
2. Is your facility a free-standing or hospital-based facility?
3. How many licensed beds are at your facility?
4. How many private rooms are available at your facility?
5. How many hours per week do you spend in infection prevention and control activities?
6. Do you culture new admissions for infectious diseases?
7. Do you admit patients infected with vancomycin-resistant enterococcus (VRE)?

8. On a scale of 1–4 (where 1 = not important and 4 = very important), rate the importance of teaching health care workers, patients, and visitors to wash hands at your facility.

a. Identify the type, quantitative or qualitative, of the data extracted from each survey question.
b. Identify the population of interest to the CDHSL&C.
c. What are the experimental units in this survey?
d. Explain why the survey data are a sample.
e. Is the sample randomly selected from the population?
f. What type of error might this survey suffer from?

1.37 Many opinion surveys are conducted by mail. In such a sampling procedure, a random sample of persons is selected from among a list of people who are supposed to constitute a target population (e.g., purchasers of a product). Each is sent a questionnaire and is requested to complete and return the questionnaire to the pollster. Why might this type of survey yield a sample that would produce biased inferences?

1.38 A popular ratings gimmick for cable television networks (e.g., CNN, MSNBC) is to conduct a telephone survey of its viewers concerning some newsworthy issue. The TV network usually provides two or more "900" telephone numbers (each corresponding to a different opinion). The viewer chooses which number to call and is charged for the call. Because the calls are electronically monitored, the results can be shown "live" almost immediately after the polling period ends. A cable TV network recently polled its viewers on whether or not they agree with the Food and Drug Administration's approval of the French abortion pill, RU-486. Explain why the sample of opinions collected by this TV survey could have produced biased results.

1.39 According to *Nature* (Aug. 27, 1998), a tenured professor at the University of Arizona was fired for scientific misconduct. The professor, who used data collected on mice to conclude that vitamin E supplements delayed the effect of aging on the immune system and the brain, was accused of "dropping, selecting, or manipulating data points for the desired statistical results." Assuming the claim is true, explain why basing the conclusion on the data is unethical.

1.8 What Readers Need to Know about Microsoft Excel, the PHStat Software, and MINITAB for This Textbook

In this text we will illustrate output from an electronic worksheet (or spreadsheet) program, Microsoft Excel, and from a statistical package, MINITAB. In addition, tutorials that follow all chapters of this text contain step-by-step explanations of how to use Microsoft Excel and MINITAB for the topics that have been discussed in the chapter.

Using Microsoft Excel

To use this textbook most effectively, one approach involves working out the statistical problems using Microsoft Excel, the electronic worksheet program marketed separately as well as included as part of the suite of programs known as Microsoft Office. Although not without its flaws (see the Comment box on p. 51), Microsoft Excel is an excellent way of presenting and demonstrating statistical methods to the introductory statistics student and has the added advantage of being commonly available.

The instructions provided for using Microsoft Excel in the chapter tutorials either make use of standard features that are provided with the "off-the-shelf" version of Excel or through the use of the PHStat statistical software for Microsoft Excel. For certain analyses, PHStat preprocesses data in a way that would otherwise

be awkward or error-prone to do manually or combines basic features of Excel in a novel way to produce additional analyses.

Using PHStat requires only minimal competency in Microsoft Excel. However, installing and setting up PHStat on a computer system requires the user to follow a series of steps that are detailed in Appendix E.

Comment: The use of Microsoft Excel in this text does not constitute an endorsement of its use in place of statistical packages in real-world applications. As others have noted (see McCullough and Wilson, 1999), some of the statistical routines of Microsoft Excel contain flaws that can sometimes lead to invalid results, especially when *using data sets that are very large or that have unusual statistical properties.* Some flaws *always* produce invalid results and in those cases, the flaw is noted and an alternate means of producing the result is presented. For all other cases, the data being analyzed has been selected to avoid the effects of the flaws and to allow the reader to draw the correct conclusion from the results produced by Microsoft Excel.

Using MINITAB

A second approach to working out the statistical problems in the text involves using the statistical package MINITAB. Student versions of MINITAB are readily available and can be "bundled" with this textbook for a nominal additional cost.

Although statistical and spreadsheet software have made even the most sophisticated analyses feasible, we need to be aware that problems may arise when statistically unsophisticated users who do not understand the assumptions behind or limitations of the statistical procedures are misled by results obtained from computer software. Therefore, we believe that for pedagogical reasons it is important that the applications of the methods covered in the text be illustrated through worked-out examples.

KEY TERMS

Starred (*) terms are from the optional section of this chapter.

*Cluster sample 45
Data 23
Descriptive statistics 30
Designed experiment 38
Experimental unit 24
*Frame 44
Inferential statistics 32
Measurement error 48
Nonresponse error 47
Observational study 38
Population 29
Published source 35
Qualitative (categorical) data 24
Quantitative data 24
Random number generator 41
Random sample 41
Reliability 32
Representative sample 39
Sample 30
Sampling error 48
Selection bias 47
Statistics 24
*Stratified random sample 45
Survey 38
*Systematic sample 46
Unethical survey design 48
Variable 24

CHECK YOUR UNDERSTANDING

1. What is the difference between quantitative and qualitative data?
2. What is the difference between a population and a sample?
3. What is the difference between descriptive and inferential statistics?
4. List the different methods of data collection.
5. What is the difference between a representative and a biased sample?
6. How do you select a random sample?
7. What potential errors may occur with random sampling?
8. What is the distinction between poor survey design and unethical survey design?

SUPPLEMENTARY PROBLEMS

Using the Tools

1.40 Generate a random sample of 10 observations from a population with 40,000 elements.

1.41 Generate a random sample of 100 observations from a population with 350,000 elements.

Applying the Concepts

1.42 The *Journal of Performance of Constructed Facilities* (Feb. 1990) reported on the performance dimensions of water distribution networks in the Philadelphia area. For one part of the study, the following data were collected for a sample of water pipe sections:

1. Pipe diameter (inches)
2. Pipe material
3. Age (year of installation)
4. Location
5. Pipe length (feet)
6. Stability of surrounding soil (unstable, moderately stable, or stable)
7. Corrosiveness of surrounding soil (corrosive or noncorrosive)
8. Internal pressure (pounds per square inch)
9. Percentage of pipe covered with land cover
10. Breakage rate (number of times pipe had to be repaired due to breakage)

Identify each variable as quantitative or qualitative.

1.43 Procter & Gamble (P&G), which markets several different brands of laundry detergent, is frequently asked to provide information on which detergent cleans best or is most effective on stains. Consequently, P&G has developed a set of guidelines for researchers in designing a study for comparing brands. P&G recommends measuring the following variables in such a study. Identify the type (quantitative or qualitative) of each.

a. amount of staining agent
b. amount of detergent
c. water temperature setting
d. length of agitation
e. length of time stain has aged
f. fabric material
g. soft or hard water
h. fabric color

1.44 Pesticides applied to an extensively grown crop can result in inadvertent air contamination. *Environmental Science & Technology* (Oct. 1993) reported on thion residues of the insecticide chlorpyrifos used on dormant orchards in the San Joaquin Valley, California. Surrounding air specimens were collected daily at an orchard site during an intensive period of spraying a total of 13 days—and the thion level (ng/m^3) measured each day.

a. Identify the population of interest to the researchers.
b. Identify the sample.

1.45 Hundreds of sea turtle hatchlings, instinctively following the bright lights of condominiums, wandered to their deaths across a coastal highway in Florida (*Tampa Tribune,* Sept. 16, 1990). This incident led researchers to begin experimenting with special low-pressure sodium lights. On one night, 60 turtle hatchlings were released on a dark beach and their direction of travel noted. The next night, the special lights were installed and the same 60 hatchlings were released. Finally, on the third night,

tar paper was placed over the sodium lights and the direction of travel was recorded. Consequently, turtle hatchlings were observed under three experimental conditions—darkness, sodium lights, and sodium lights covered with tar paper.

a. Identify the population of interest to the researchers.

b. Identify the sample.

c. What type of data were collected, quantitative or qualitative?

d. What data collection method was employed?

1.46 When Nissan first introduced its new Infiniti luxury cars, its television ad campaign was renowned for a novel gimmick: The automobiles were nowhere in sight. The Infiniti ads, which depicted lushly photographed trees, boulders, lightning bolts, and ocean waves (but no cars) were found by a nationwide Gallup telephone poll of 1,000 consumers to be one of the best-recalled commercials on television (*Time,* Jan. 22, 1990).

a. Identify the experimental units in this study.

b. Describe the population of interest to the pollsters.

c. Identify the sample.

d. What is the inference made by the Gallup poll?

e. Describe the data collection method used by the pollsters.

f. What types of errors might bias the study results? Explain.

1.47 A Brandeis University researcher experimented with giving propranolol, one of the class of heart drugs called beta blockers, to nervous high school students prior to taking their SATs (*Newsweek,* Nov. 16, 1987). In theory, the same calming effect that beta blockers provide heart patients could also be used to reduce anxiety in test takers. To test this theory, the researcher selected 22 high school juniors who had not performed as well on the SAT as they should have based on IQ and other academic evaluations. One hour before the students repeated the test in their senior year, each was administered a dosage of a beta blocker. [*Note:* Typically, students who retake the test without special preparations will increase their scores by an average of 38 points. These 22 students improved their scores by an average of 120 points!]

a. Identify the experimental units in this study.

b. Identify the measured variable.

c. Describe the population of interest to the researcher. (Give the precise statistical definition.)

d. Identify the data collection method used.

e. Describe the sample.

f. Based on the sample results, what inference would you make about the use of beta blockers to increase SAT scores? (In Chapter 9 we show you how to assess the reliability of this type of inference.)

g. The director of research and development for the College Board (sponsors of the SAT) warns that "the findings have to be taken with a great deal of caution" and that they should not be interpreted to mean "that someone has discovered the magic pill that will unlock the SAT for thousands of teenagers who believe they do not do as well as they should because they're nervous." Give several reasons for issuing such a warning.

1.48 In her book, *Women and Love: A Cultural Revolution in Progress* (Knopf Press, 1988), Shere Hite reveals some startling statistics describing how women she surveyed feel about contemporary relationships: 84% of women are not emotionally satisfied with their relationship; 95% of women report "emotional and psychological harassment" from their men; 70% of women married 5 years or more are having extramarital affairs; and only 13% of women married more than 2 years are "in love." Hite conducted the survey by mailing out 100,000 questionnaires to organizations such as church groups, women's voting/political groups, women's rights organizations, and counseling centers for women, and asked the groups to circulate the questionnaires to their members. Included were volunteer respondents who wrote in

for copies of the survey. Each questionnaire consisted of 127 open-ended questions, many with numerous subquestions and follow-ups. Hite's instructions read: "It is not necessary to answer every question! Feel free to skip around and answer those questions you choose." Approximately 4,500 completed questionnaires were returned for a response rate of 4.5%, and they form the data set from which these percentages were determined.

a. Identify the population of interest to Shere Hite. What are the experimental units?

b. Identify the variables of interest to Hite. Are they quantitative or qualitative variables?

c. Describe Hite's data collection method.

d. Hite claims that the 4,500 women surveyed are a representative sample of all women in the United States, and therefore the survey results imply that vast numbers of women are "suffering a lot of pain in their love relationships with men." Do you agree?

e. Discuss the difficulty in obtaining a random sample of women across the United States to take part in a survey similar to the one conducted by Shere Hite.

1.49 Political polling has traditionally been accomplished through telephone interviews. Researchers at Harris Black International Ltd. argue that Internet polling is less expensive, faster, and offers higher response rates than telephone surveys. Critics are concerned about the scientific reliability of this approach (*Wall Street Journal,* April 13, 1999). Even amid this strong criticism, Internet polling is becoming more and more common. What concerns, if any, do you have about data produced from Internet polling?

REFERENCES

Blass, T., ed. *Obedience to Authority: Current Perspectives on the Milgram Paradigm.* Mahwah, N.J.: Lawrence Erlbaum Assoc., 2000.

Careers in Statistics. American Statistical Association, Biometric Society and the Institute of Mathematical Statistics, 1995.

Cochran, W. G. *Sampling Techniques,* 3rd. ed. New York: Wiley, 1977.

Deming, W. E. *Sample Design in Business Research.* New York: Wiley, 1963.

Ethical Guidelines for Statistical Practice. American Statistical Association, 1995.

Hansen, M. H., Hurwitz, W. N., and Madow, W. G. *Sample Survey Methods and Theory,* Vol. 1. New York: Wiley, 1953.

Kirk, R. E., ed. *Statistical Issues: A Reader for the Behavioral Sciences.* Monterey, Calif.: Brooks/Cole, 1972.

Kish, L. *Survey Sampling.* New York: Wiley, 1965.

McCullough, B. D., and Wilson, B. "On the accuracy of statistical procedures in Microsoft Excel 97," *Computational Statistics and Data Analysis,* 31, 1999, pp. 27–37.

Meet MINITAB: Release 13. State College, Pa.: Minitab, Inc., 2000.

Microsoft Excel 2000. Redmond, Wash.: Microsoft Corporation, 1999.

Milgram, S. *Obedience to Authority.* New York: Harper & Row, 1974.

MINITAB User's Guide 1: Data, Graphics, and Macros: Release 13. State College, PA: Minitab, Inc., 2000.

MINITAB User's Guide 2: Data Analysis and Quality Tools: Release 13. State College, PA: Minitab, Inc., 2000.

Scheaffer, R., Mendenhall, W., and Ott, R. L. *Elementary Survey Sampling,* 2nd. ed. Boston: Duxbury, 1979.

Tanur, J. M., Mosteller, F., Kruskal, W. H., Link, R. F., Pieters, R. S., and Rising, G. R., eds. *Statistics: A Guide to the Unknown,* 3rd. ed. San Francisco: Holden-Day, 1989.

What Is a Survey? Section on Survey Research Methods, American Statistical Association, 1995.

Yamane, T. *Elementary Sampling Theory,* 3rd. ed. Englewood Cliffs, N.J.: Prentice-Hall, 1967.

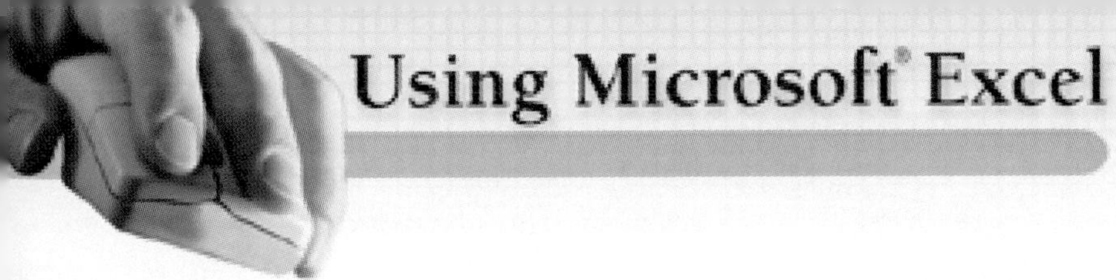

Using Microsoft® Excel

1.E.1 Using Microsoft Excel to Select a Random Sample

Solution Summary:

Use the PHStat add-in to select a random sample.

EXAMPLE Lottery Winners Sample from Example 1.9.

To select a random sample of $n = 50$ from a population of 18,000 state lottery winners, do the following:

1. Select File | New to open a new workbook.
2. Select PHStat | Sampling | Random Sample Generator.
3. In the Random Sample Generator, enter the information and make the selections shown in Figure 1.E.1. Click the OK button.

E **Figure 1.E.1** Random Sample Generator Dialog Box

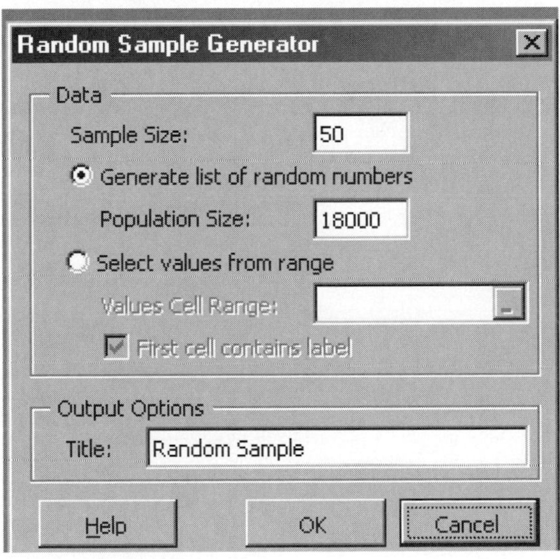

Using the settings of Figure 1.E.1, the add-in produces a randomly generated list of 50 different numbers.

Using MINITAB®

1.M.1 Using MINITAB to Select a Random Sample

Solution Summary:

Use MINITAB to select a random sample.

EXAMPLE Lottery Winners Sample from Example 1.9.

To select a random sample of $n = 50$ from a population of 18,000 state lottery winners, do the following:

1. Open MINITAB to a new worksheet.
2. Select Calc | Random Data | Uniform. (This provides the selection of a simple random sample.)
3. In the Uniform Distribution dialog box, enter the information and make the selections shown in Figure 1.M.1. Click the OK button.

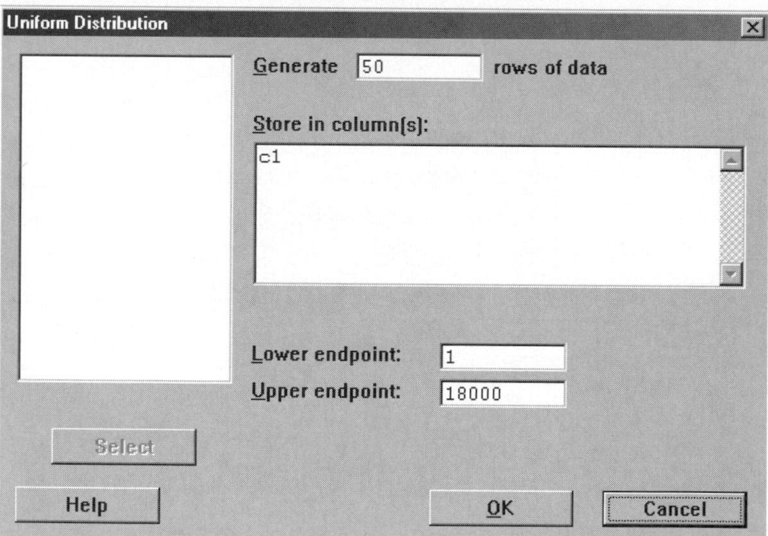

M Figure 1.M.1 Uniform Distribution Dialog Box

Using the settings of Figure 1.M.1, MINITAB produces a randomly generated list of 50 values between the lower endpoint (1 in this case) and the upper endpoint (18,000 in this case). Convert these values to random numbers by rounding each to the nearest integer. If any number is repeated, discard it if you are sampling without replacement, and select additional random numbers until the desired sample size is obtained.

CHAPTER 2

Exploring Data with Graphs and Tables

OBJECTIVES

1. To develop tables and graphs that describe a qualitative variable
2. To develop tables and graphs that describe a quantitative variable
3. To develop tables and graphs that describe the relationship between two qualitative variables
4. To develop graphs that describe the relationship between two quantitative variables
5. To demonstrate the principles of proper graphical presentation

CONTENTS

EXCEL TUTORIAL

MINITAB TUTORIAL

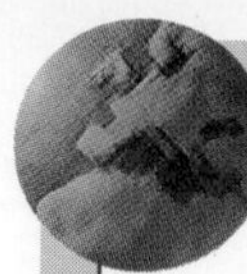

Statistics in the Real World

Crash Ratings for New Cars—Stars or Scars?

Each year the National Highway Traffic Safety Administration (NHTSA) crash tests new car models to determine how well they protect the driver and front-seat passenger in a head-on collision. During the NHTSA's crash test procedure, new vehicles are crashed into a fixed barrier at 35 miles per hour (mph), which is equivalent to a head-on collision between two identical vehicles, each moving at 35 mph. Two dummies of average human size—one in the driver's seat and one in the front passenger seat—occupy each tested car. The dummies contain electronic instruments that register the forces and impacts that occur to the head, chest, and legs during the crash.

In response to Congress' request to provide consumers with easily understandable vehicle safety performance information, the NHTSA developed a "star" scoring system for the frontal crash test and publishes the results in the *New Car Assessment Program (NCAP)*. The NCAP crash results are reported in a range of one to five stars, allowing consumers to compare crash ratings for vehicles within the same weight class. The more stars in the rating, the better the level of crash protection in a head-on collision.

★★★★★ = 10% or less chance of serious injury
★★★★ = 11% to 20% chance of serious injury
★★★ = 21% to 35% chance of serious injury
★★ = 36% to 45% chance of serious injury
★ = 46% or greater chance of serious injury

(*Note:* A serious injury is considered to be one requiring immediate hospitalization and may be life threatening.)

CRASH

The NCAP test results for 98 cars (model year 1997) are stored in the Excel workbook **CRASH.XLS** and the MINITAB worksheet **CRASH.MTW.** A description of the variables reported by the NCAP for each car tested is provided below.

CLASS (Compact, Heavy, Light, Light Truck, Medium, Mini, Sport, or Van)
MAKE (e.g., Chrysler, Ford, Honda, ...)
MODEL (e.g., Town & Country, Mustang, Accord, ...)
DOORS (number of doors)
WEIGHT (vehicle weight in pounds)
DRIVSTAR (Driver Overall Star rating—1 to 5 stars)
PASSSTAR (Passenger Overall Star rating—1 to 5 stars)
DRIVAIR (Driver-side air bag—Yes or No)
PASSAIR (Passenger-side air bag—Yes or No)
DRIVHEAD (Driver's severity of head injury score—0 to 1,500 scale)
PASSHEAD (Passenger's severity of head injury score—0 to 1,500 scale)
DRIVCHST (Driver's severity of chest deceleration score—0 to 100 scale)
PASSCHST (Passenger's severity of chest deceleration score—0 to 100 scale)
DRIVFEML (Driver's severity of left femur injury score—0 to 3,000 scale)
DRIVFEMR (Driver's severity of right femur injury score—0 to 3,000 scale)
PASSFEML (Passenger's severity of left femur injury score—0 to 3,000 scale)
PASSFEMR (Passenger's severity of right femur injury score—0 to 3,000 scale)

In this chapter, several Statistics in the Real World examples demonstrate how to obtain and interpret graphical descriptions of the NCAP data.

Statistics in the Real World Revisited

· Interpreting a Pie Chart (p. 62)
· Interpreting a Histogram (p. 73)
· Interpreting a Side-by-Side Bar Graph (p. 83)
· Interpreting Scatterplots (p. 92)

2.1 The Objective of Data Description

In Chapters 2 and 3 we demonstrate how to explore data using descriptive methods. The objective of data description is to summarize the characteristics of a data set, to identify any patterns in the data, and to present that information in a convenient form. In this chapter we will show you how to construct charts, graphs, and tables that convey the nature of a data set. The procedure that we will use to accomplish this objective in a particular situation depends on the type of data, qualitative or quantitative, that you want to describe, and the number of variables measured. In the next chapter we present numerical descriptive measures of a data set.

2.2 Describing a Single Qualitative Variable: Frequency Tables, Bar Graphs, and Pie Charts

Consider the data collected from measuring one qualitative variable. For example, suppose we observe the rank of a university professor, where rank is recorded as assistant, associate, or full professor. Consequently, each value of rank falls into one of three different categories or *classes*.

Definition 2.1

A **class** is one of the categories into which the qualitative data can be classified.

The summary information we seek about a qualitative variable is either the number of observations in each class (called a *frequency*), the proportion of the total number of observations in each class (called a *relative frequency*), or the *percentage* of the total number of observations in each class.

Definition 2.2

The **class frequency** for a particular class is the number of observations in that class.

Definition 2.3

The **class relative frequency** for a particular class is equal to the class frequency divided by the total number of observations.

$$\text{Class relative frequency} = \frac{\text{Class frequency}}{\text{Total number of observations}}$$

Definition 2.4

A **class percentage** is the relative frequency for the class multiplied by 100.

$$\text{Class percentage} = (\text{Class relative frequency}) \times 100$$

Class frequencies, relative frequencies, or percentages can be displayed in table form or graphically. **Bar graphs** and **pie charts** are two of the most widely used graphical methods for describing qualitative data. We illustrate in Example 2.1.

EXAMPLE 2.1 BAR GRAPHS AND PIE CHARTS

The popular magazine *Consumer Reports* evaluated 17 men's cologne (fragrance) brands. A number of characteristics were rated, including intensity of the cologne. The qualitative variable, intensity, was measured as mild, strong, or very strong. The intensity ratings for the 17 brands are listed in Table 2.1.

a. Construct a frequency table for the data.
b. Portray the data in a bar graph.
c. Portray the data in a pie chart.

COLOGNE

TABLE 2.1 Intensity Ratings of 17 Men's Cologne Brands

Fragrance Brand	Intensity Rating
Aramis	Strong
Brut	Very strong
Drakkar Noir	Very strong
Egoïste	Strong
English Leather	Mild
Escape for Men	Strong
Eternity for Men	Very strong
Gravity	Strong
Lancer	Strong
Obsession for Men	Strong
Old Spice	Strong
Polo	Very strong
Preferred Stock	Strong
Realm for Men	Strong
Safari for Men	Very strong
Stetson	Mild
Tribute	Strong

Data Source: Consumer Reports, Dec. 1993, p. 773.

Solution

a. Examining Table 2.1, we observe that 5 brands are rated very strong, 10 are rated strong, and 2 are rated mild. These numbers—5, 10, and 2—represent the class frequencies for the three classes of intensity rating and are shown in the accompanying frequency table, Table 2.2.

TABLE 2.2 Frequency Table for Cologne Intensity Rating Data

Class	Frequency	Relative Frequency	Percentage
Very strong	5	.294	29.4
Strong	10	.588	58.8
Mild	2	.118	11.8
TOTALS	17	1.000	100.0

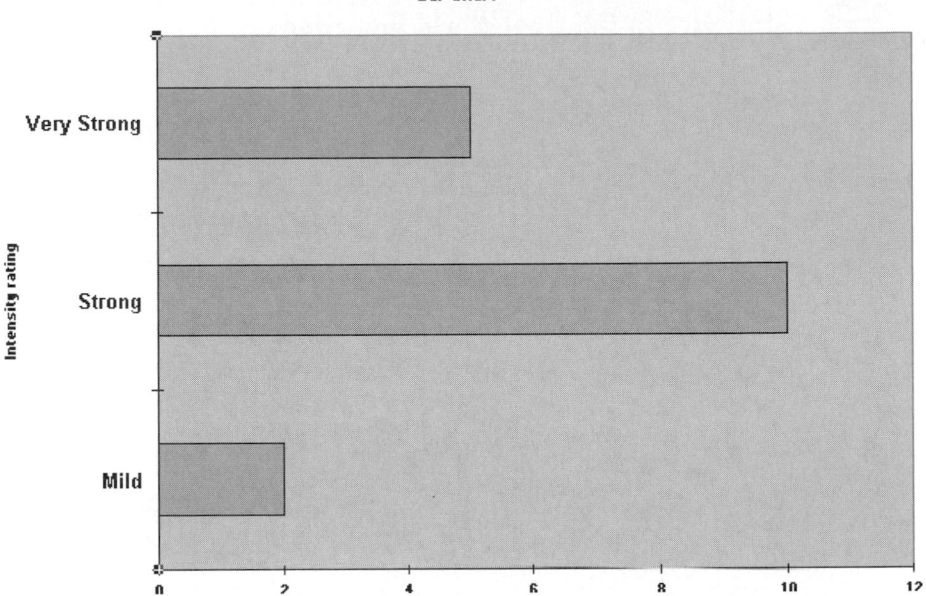

E Figure 2.1 Excel Frequency Bar Graph for Cologne Intensity Ratings

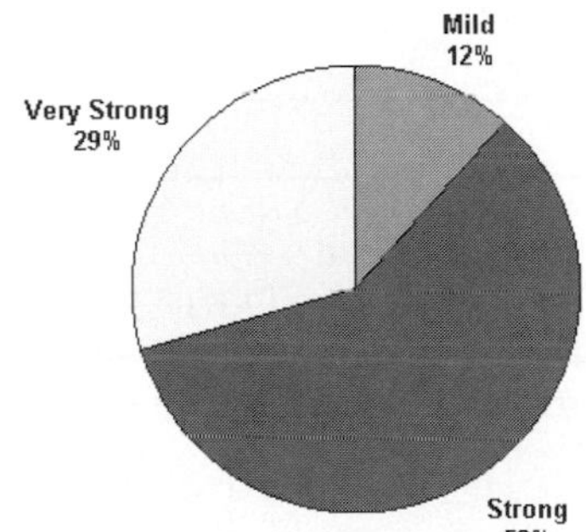

E Figure 2.2 Excel Pie Chart for Cologne Intensity Ratings

Qualitative variables can be either *nominal* or *ordinal* variables. The values of an ordinal variable can be naturally ordered—like "mild," "strong," and "very strong" in Example 2.1. The values of a nominal variable (e.g., "male" and "female" for gender) have no natural ordering. The bars in an ordinal variable bar graph are usually arranged according to their natural ordering as in Figure 2.1.

Class relative frequencies and percentages are also shown in Table 2.2. Using Definition 2.3, the relative frequencies for the three classes are:

$$\text{Very strong: } \frac{5}{17} = .294$$

$$\text{Strong: } \frac{10}{17} = .588$$

$$\text{Mild: } \frac{2}{17} = .118$$

Class percentages are obtained by multiplying each of the class frequencies by 100.

b. A bar graph displays the class frequencies, relative frequencies, or percentages. Figure 2.1 is a horizontal frequency bar graph for the cologne data created using Microsoft Excel. The figure contains a rectangle, or bar, for each intensity class; the length of the bar is proportional to the class frequency. (Optionally, the bar heights can be proportional to the class relative frequencies or percentages.) The bar graph makes it easy to see that the majority of the 17 men's cologne brands had a strong intensity rating. [*Note:* Some software packages reverse the axes and display the bars in a vertical fashion.]

c. A pie chart conveys the same information as a bar chart. Typically, class relative frequencies are shown on a pie chart. Figure 2.2 is a pie chart for the cologne data, also created using Excel. Note that the pie is divided into three slices, one for each of the three classes. The size (angle) of each slice is proportional to the class relative frequency. For example, since a circle spans 360°, the slice assigned to a strong intensity rating is 59% of 360°, or .59(360) = 212°. It is common to show the percentage of measurements in each class on the pie chart as indicated.

Statistics in the Real World Revisited

Interpreting a Pie Chart

Recall that the National Highway Traffic Safety Administration (NHTSA) crash tests new car models and publishes the results of the tests in the *New Car Assessment Program* (NCAP). (See p. 58.) One of the qualitative variables reported by the NCAP is Driver Star Rating, which ranges from one star (★) to five stars (★★★★★). The more stars in the rating, the better the level of crash protection to the driver in a head-on collision.

We can use either a bar graph or pie chart to summarize the Driver Star Rating data for the 98 cars in the NCAP report. Using Excel, we produced the pie chart displayed in Figure 2.3. Each slice of the Excel pie chart represents one of the categories of Driver Star Rating. Note that the percentage of the 98 cars falling into each category are shown on the pie chart. We see that 18% had the highest level of crash protection (five stars), 61% were rated four stars, 17% were rated three stars, and only 4% were rated two stars. The actual number of cars that fall into each Driver Star Rating category is displayed in the MINITAB frequency table, Figure 2.4. Note that none of the cars crash tested were rated one star (the lowest level of protection).

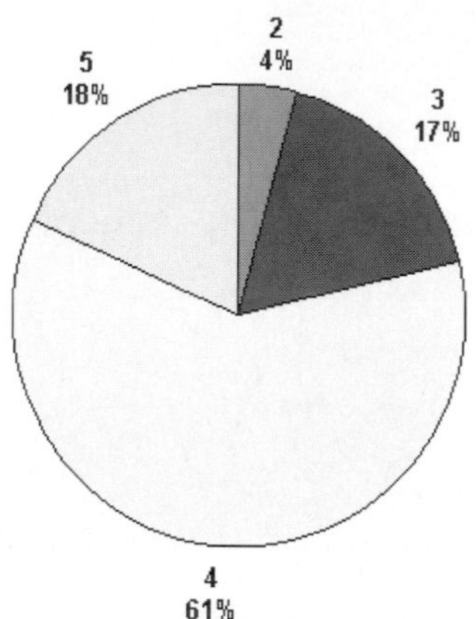

E Figure 2.3 Excel Pie Chart for NCAP Passenger Star Rating

Tally for Discrete Variables: DrivStar

DriverSt	Count	Percent
2	4	4.08
3	17	17.35
4	59	60.20
5	18	18.37
N=	98	

M Figure 2.4 MINITAB Frequency Table for Driver Star Rating

Self-Test 2.1

Fetal alcohol syndrome is a group of abnormalities found in children born to chronic alcoholic mothers. Each of 60 children diagnosed as having the syndrome was examined for the abnormality of the most serious nature, with the results illustrated by the bar graph in Figure 2.5. Discuss the information provided by the graph. Which abnormality occurs most often as the child's most serious problem?

E Figure 2.5 Excel Bar Chart of Fetal Alcohol Syndrome Abnormalities

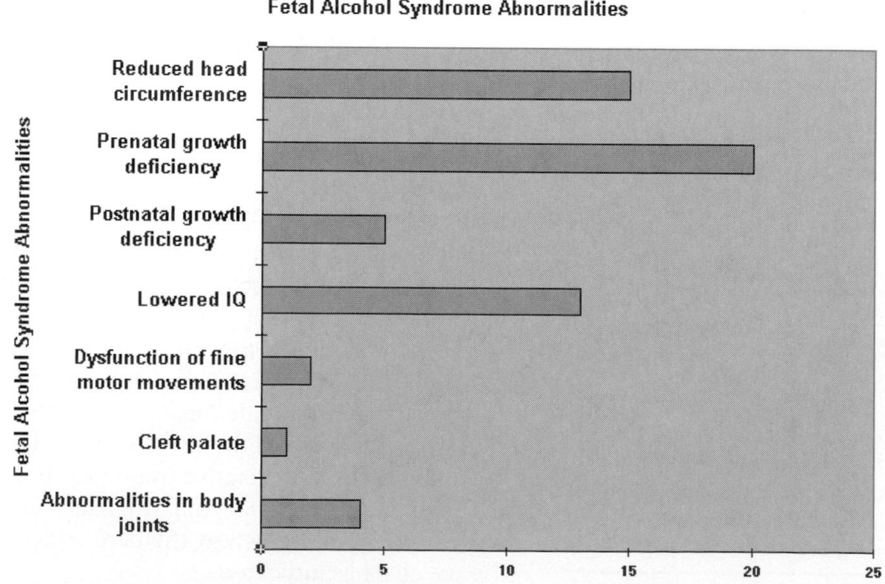

PROBLEMS FOR SECTION 2.2

Using the Tools

Category	Frequency
A	13
B	28
C	9

Problem 2.1

2.1 Suppose that a qualitative variable has three categories with the frequency of occurrence shown here.

a. Compute the percentage of values in each category.

b. Construct a bar graph.

c. Construct a pie chart.

2.2 Complete the following table.

Grade on Statistics Exam	Frequency	Relative Frequency	Percentage
A	16	.08	—
B	36	—	—
C	90	—	—
D	30	—	—
F	28	—	—
TOTALS		1.00	100

Category	Relative Frequency
A	12
B	29
C	35
D	24

Problem 2.3

2.3 Suppose that a qualitative variable has four categories with the relative frequencies shown at left.

a. Construct a bar graph. **b.** Construct a pie chart.

2.4 A qualitative variable with four classes (W, X, Y, and Z) is measured for each of 25 experimental units sampled from the population. The data are listed below.

Y Y W Z X X Y W Z Y X Y X X Z X Y Y Y X W Y W X Y

a. Compute the frequency of each class.

b. Compute the relative frequency of each class.

c. Construct a relative frequency bar graph for the data.

Applying the Concepts

2.5 The accompanying pie chart describes the fate of the (estimated) 240 million automobile tires that are scrapped in the U.S. each year.

a. Interpret the pie chart.

b. Convert the pie chart into a relative frequency bar graph.

c. Convert the pie chart into a frequency bar graph.

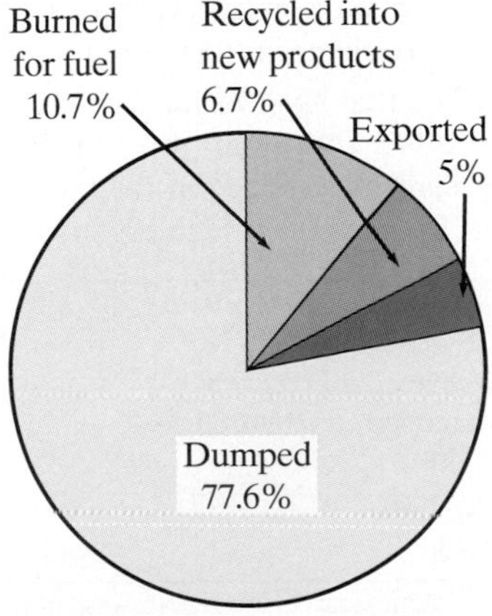

Source: U.S. Environmental Protection Agency and National Solid Waste Management Association.

2.6 The National Science Foundation periodically conducts a survey of graduate students in Science and Engineering. The accompanying pie chart, obtained from Excel, summarizes the results of a Fall 1997 survey in which each of 280,612 full-time graduate students were classified according to source of financial support.

a. What percentage of the 280,612 graduate students are supported by a fellowship?

b. How many of the 280,612 graduate students are supported by a teaching assistantship?

c. Give the proportion of the 280,612 graduate students who are supported by either a research or teaching assistantship.

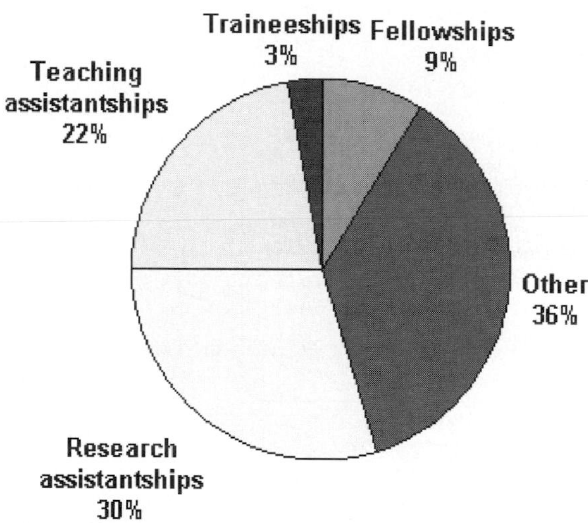

E Problem 2.6

Data Source: National Science Foundation, Division of Science Resources Studies, *Graduate Students and Postdoctorates in Science and Engineering,* Fall 1997.

2.7 According to the American Optometric Association (AOA), about 10% of the U.S. population have normal visual acuity, while 60% suffer from farsightedness and 30% suffer from nearsightedness.

a. Given this information, construct a pie chart to describe the distribution of visual acuity problems in the United States.

b. Of those that have visual acuity problems (nearsighted or farsighted), the AOA has found that many do not wear any sort of corrective lenses. Use the information provided in the table below to construct a relative frequency bar chart for the four categories shown.

Vision Problem Category	Percent
Nearsighted, corrected	27
Nearsighted, uncorrected	7
Farsighted, corrected	27
Farsighted, uncorrected	39
TOTAL	100

Source: American Optometric Association.

c. What percentage of those with vision impairments do not wear any corrective lenses?

2.8 An "armed conflict" is defined as a contested incompatibility between a government of a state and another party in which armed forces are used and at least 25 battle-related deaths have occurred. According to the *Journal of Peace Research* (Aug. 1997), there were 36 active armed conflicts around the world in 1996. Of these, 17 were classified as "minor" (less than 1,000 deaths overall), 13 were classified as "intermediate" (more than 1,000 deaths overall, but less than 1,000 in any year), and 6 were classified as "war" (more than 1,000 deaths in any year). Based on this information, use a graphical method to display the percentages of armed conflicts in 1996 in each category. Interpret the results.

2.9 Archaeologists use the distribution of artifacts found at an archaeological site to make inferences about the economy and lifestyle of ancient inhabitants. Recent archaeological digs at Motupore Island in Papua, New Guinea, have revealed that shell bead manufacturing was a significant part of the island's prehistoric economy (*Archaeol. Oceania,* Apr. 1997). The site was excavated for artifacts related to the manufacture of beads: stone drillpoints, shell beads, and grinding slabs. The distribution of these artifacts at the excavation site is shown in the accompanying table.

Type of Artifact	Number Discovered
Drillpoint	566
Bead	385
Grinding slab	43

Source: Allen, J., et al. "Identifying specialization, production, and exchange in the archaeological record: The case of shell bead manufacture on Motupore Island, Papua." *Archaeol. Oceania,* Vol. 32, No. 1, Apr. 1997, p. 27 (Table 1).

a. Identify the experimental unit and the variable measured for this study.

b. Portray the artifact distribution with a graph.

c. Identify the artifact found most often at the site. What percentage of the time does it occur?

2.10 Researchers at York University and Queen's University in Canada examined the formation of melodic expectancies in a melody completion task (*Perception & Psychophysics,* Oct. 1997). Fifty subjects with a high training in music (i.e., 5 or more years of formal training and currently playing an instrument) and fifty subjects with little musical training participated in the study. After listening to a short (2-note) melody on a piano keyboard, each subject was asked to produce a melody that followed naturally from the two notes. The pitch of the initial note selected by the

Trained Listeners		Untrained Listeners	
Note	Frequency	Note	Frequency
D_3	1	D_3	1
G_3	1	F_3	1
$A_3^{\#}$	1	G_3	2
C_4	12	B_3	1
$C_4^{\#}$	1	C_4	15
D_4	1	D_4	1
E_4	15	$D_4^{\#}$	2
F_4	7	E_4	15
G_4	8	F_4	3
A_4	3	G_4	6
		B_4	2
		C_5	1
TOTALS	50		50

Source: Thompson, W. F., et al. "Expectancies generated by melodic intervals: Evaluation of principles of melodic implication in a melody-completion task." *Perception & Psychophysics,* Vol. 59, No. 7, Oct. 1997, p. 1076.

subject was recorded. The frequencies of the continuation notes selected are shown in the table on p. 65.

a. Use a relative frequency bar graph to describe the notes selected by the trained listeners.

b. Repeat part **a** for the untrained listeners.

c. Compare the two bar graphs. What can you conclude from the data? Do the untrained listeners select certain notes with the same relative frequency as the trained listeners?

2.11 *Quality Engineering* (Vol. 11, 1999) reported on a successful process improvement program implemented at a large injection-molding company that manufactures plastic molded components. The company had been receiving complaints about defective components used in computer keyboards. The accompanying table summarizes the causes of 6,324 defects found in computer keyboards produced during a 3-month period.

Cause	Frequency	Percentage
Black spot	413	6.53
Damage	1039	16.43
Jetting	258	4.08
Pin mark	834	13.19
Scratches	442	6.99
Shot mold	275	4.35
Silver streak	413	6.53
Sink mark	371	5.87
Spray mark	292	4.62
Warpage	1987	31.42
TOTAL	6324	100.00

Source: Acharya, U. H., and Mahesh, C. "Winning back the customer's confidence: A case study on the application of design of experiments to an injection-molding process," *Quality Engineering,* 11, 1999, 357–363.

a. Construct a bar graph to describe the causes of defects in computer keyboards.

b. A tool that is quite useful in process improvement is the *Pareto diagram.* A Pareto diagram is simply a bar graph that has ordered the categories by frequency of occurrence, from largest to smallest, enabling the analyst to separate the "vital few" from the "trivial many." Reorder the categories in the bar graph of part **a** to form a Pareto diagram. Identify the problem areas where the company should concentrate its efforts on reducing defects.

2.12 Refer to the *Journal of Literacy Research* (Dec. 1996) study of teachers who read weekly to children, Problem 1.15, p. 34. The teachers asked approximately 2,000 questions during three different sessions with the children. Each question was classified according to one of four different types: known-information questions, opinion questions, connection questions, and conditional questions. (See Problem 1.15 for an example of each.) The percentage of questions falling into each category during each session is listed in the table.

Question Type	First Session	Second Session	Third Session
Known-information	28%	23%	15%
Opinion	49%	51%	56%
Connection	15%	18%	22%
Conditional	8%	8%	7%
TOTALS	100%	100%	100%

Source: Wolf, S. A., et al. "What's after 'What's that?': Preservice teachers learning to ask literary questions." *Journal of Literacy Research,* Vol. 28, No. 4, Dec. 1996, p. 466 (Figure 2).

a. For each reading session, construct a bar graph to describe the distribution of question types.

b. Do you detect any trends in the bar graphs, part **a**?

2.13 Consider a study of aphasia published in the *Journal of Communication Disorders* (Mar. 1995). Aphasia is the "impairment or loss of the faculty of using or understanding spoken or written language." Three types of aphasia have been identified by researchers: Broca's, conduction, and anomic. They wanted to determine whether one type of aphasia occurs more often than any other, and, if so, how often. Consequently, they measured aphasia type for a sample of 22 adult aphasics. The accompanying table gives the type of aphasia diagnosed for each aphasic in the sample. Analyze the data for researchers and present the results in graphical form.

Type of Aphasia	Number of Subjects	Proportion
Broca's	5	.227
Conduction	7	.318
Anomic	10	.455
TOTALS	22	1.000

2.14 Each year *Forbes* magazine conducts a salary survey of chief executive officers. In addition to salary information, *Forbes* collects and reports personal data on the CEOs, including level of education. Do most CEOs have advanced degrees, such as masters degrees or doctorates? The data in the table represent the highest degree obtained for each of the top 25 best-paid CEOs of 1999. Use a graphical method to summarize the highest degree obtained for these CEOs. What is your opinion about whether most CEOs have advanced degrees?

CEODEGREES

CEO	Company	Degree
1. Michael D. Eisner	Walt Disney	Bachelors
2. Mel Karmazin	CBS	Bachelors
3. Stephen M. Case	America Online	Bachelors
4. Stephen C. Hilbert	Conseco	none
5. Craig R. Barrett	Intel	Doctorate
6. Millard Drexler	GAP	Masters
7. John F. Welch, Jr.	General Electric	Doctorate
8. Thomas G. Stemberg	Staples	Masters
9. Henry R. Silverman	Cendant	JD
10. Reuben Mark	Colgate-Palmolive	Masters
11. Philip J. Purcell	Morgan Stanley Dean Witter	Masters
12. Scott G. McNealy	Sun Microsystems	Masters
13. Margaret C. Whitman	eBay	Masters
14. Louis V. Gerstner, Jr.	IBM	Masters
15. John F. Gifford	Maxim Integrated Products	Bachelors
16. Robert L. Waltrip	Service Corp. International	Bachelors
17. M. Douglas Ivester	Coca-Cola	Bachelors
18. Gordon M. Binder	Amgen	Masters
19. Charles R. Schwab	Charles Schwab	Masters
20. William R. Steere, Jr.	Pfizer	Bachelors
21. Nolan D. Archibald	Black & Decker	Masters
22. Charles A. Heimbold, Jr.	Bristol-Myers Squibb	LLB (law)
23. William L. Larson	Network Association	JD
24. Maurice R. Greenberg	American International Group	LLB (law)
25. Richard Jay Kogan	Schering-Plough	Masters

Source: Forbes, May 17, 1999.

2.3 Describing a Single Quantitative Variable: Frequency Tables, Dot Plots, Stem-and-Leaf Displays, and Histograms

Dot plots, stem-and-leaf displays, and **histograms** are popular graphical methods for describing quantitative data sets. Like the bar graphs and pie charts of Section 2.2, they show class frequencies, class relative frequencies, or class percentages. The difference is that the classes do not represent categories of a qualitative variable; instead, they are formed by grouping the numerical values of the quantitative variable that you want to describe.

For small data sets (say, 30 or fewer observations) with measurements with only a few digits, dot plots and stem-and-leaf displays can be constructed quickly by hand. Histograms, on the other hand, are better suited to the description of large data sets, and they permit greater flexibility in the choice of the classes. Of course, we can generate any of these graphs with computer software.

EXAMPLE 2.2 STEM-AND-LEAF DISPLAY

Postmortem interval (PMI) is defined as the elapsed time between death and an autopsy. Knowledge of PMI is considered essential when conducting research on human cadavers. Table 2.3 gives the PMI (in hours) for a sample of 22 human brain specimens obtained at autopsy.

a. Summarize the quantitative data with a stem-and-leaf display.

b. Interpret the graph.

PMI

TABLE 2.3 Postmortem Intervals for 22 Human Brain Specimens

5.5	14.5	6.0	5.5	5.3	5.8	11.0	6.1	7.0
14.5	10.4	4.6	4.3	7.2	10.5	6.5	3.3	7.0
4.1	6.2	10.4	4.9					

Source: Hayes, T. L., and Lewis, D. A. "Anatomical specialization of the anterior motor speech area: Hemispheric differences in magnopyramidal neurons." *Brain and Language,* Vol. 49, No. 3, June 1995, p. 292 (Table 1).

Solution

a. In a stem-and-leaf display, each quantitative measurement is broken into a *stem* and a *leaf.* One or more of the digits will make up the stem, while the remaining digits (or digit) will be the leaf. The stems represent the classes in a graph; the leaves reflect the number of measurements in each class.

Figure 2.6 is a stem-and-leaf display for the PMI data, produced using PHStat and Excel. The different stems are listed vertically to the left of the line in Figure 2.6. You can see that these stems, 3, 4, 5, 6, . . . , 13, and 14, represent the digit(s) to the left of the decimal point. The leaf of each number in the data set (i.e., the digit to the right of the decimal) is placed in the row of the display corresponding to the number's stem. For example, for the PMI value of 7.0 hours, the leaf 0 is placed in stem row 7. Similarly, for the PMI of 4.9 hours, the leaf 9 is placed in stem row 4. The usual convention is to list the leaves of each stem in increasing order, as shown in Figure 2.6.

	A	B	C
1	Stem-and-Leaf Display for PMI		
2	for Postmortem interval		
3	Stem unit:	1	
4			
5	3	3	
6	4	1 3 6 9	
7	5	3 5 5 8	
8	6	0 1 2 5	
9	7	0 0 2	
10	8		
11	9		
12	10	4 4 5	
13	11	0	
14	12		
15	13		
16	14	5 5	

E **Figure 2.6** Stem-and-Leaf Display for PMI Data Obtained Using PHStat and Excel

b. You can see that the stem-and-leaf display in Figure 2.6 partitions the data set into 12 classes corresponding to the 12 stems listed. The class corresponding to the stem 3 would contain all PMI values from 3.0 to 3.9 hours; the class corresponding to the stem 4 would contain all PMI values from 4.0 to 4.9 hours;

TABLE 2.4 Frequency Table for Stem-and-Leaf Display of PMI Data

PMI Stem Class	Frequency	Relative Frequency
3	1	1/22
4	4	4/22
5	4	4/22
6	4	4/22
7	3	3/22
8	0	0/22
9	0	0/22
10	3	3/22
11	1	1/22
12	0	0/22
13	0	0/22
14	2	2/22
TOTALS	22	1

etc. The number of leaves in each class gives the class frequency. Thus, a stem-and-leaf display makes it easy to calculate class frequencies and relative frequencies.

The frequencies and relative frequencies for the 12 PMI classes (stems) are shown in Table 2.4.

Note that the 22 PMI values are scattered over the interval from 3.0 hours to 14.9 hours, with most falling between stems 4 and 11. Summing the relative frequencies for the stems 4, 5, 6, 7, 8, 9, 10, and 11, we obtain

$$\frac{4}{22} + \frac{4}{22} + \frac{4}{22} + \frac{3}{22} + \frac{0}{22} + \frac{0}{22} + \frac{3}{22} + \frac{1}{22} = \frac{19}{22} = .86$$

Thus, 86% of the 22 postmortem intervals range between 4 hours and 11 hours.

Self-Test 2.2

Consider the sample data shown here.

26	34	21	32	42	36	28	38	17	39	22	12
56	39	25	41	30	23	27	19				

a. Using the first digit as a stem, list the stem possibilities in order.

b. Place the leaf for each observation in the appropriate stem row to form a stem-and-leaf display.

c. Compute the relative frequencies for each stem.

EXAMPLE 2.3

DOT PLOT

Refer to the 22 postmortem interval (PMI) values given in Table 2.3 (p. 68). Construct and interpret a dot plot for the data.

Solution

A dot plot for the 22 PMI values, produced using MINITAB, is shown in Figure 2.7. The horizontal axis of Figure 2.7 is a scale for the quantitative variable, postmortem interval. The numerical value of each PMI measurement in the data set

Dotplot of Postmortem Interval

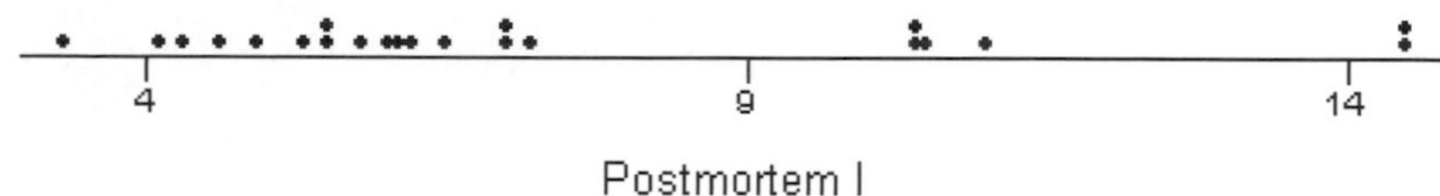

M **Figure 2.7** A MINITAB Dot Plot for the 22 PMI Values

is located on the horizontal scale by a dot. When data values repeat, the dots are placed above one another, forming a "pile" at that particular location. Like the stem-and-leaf display (Figure 2.6), the dot plot shows that most of the postmortem intervals range between 4 and 11 hours, with a "gap" between 8 and 10 hours.

EXAMPLE 2.4

FREQUENCY HISTOGRAM

Table 2.5 gives the starting salaries for 50 graduates of the University of South Florida who returned a mail questionnaire shortly after obtaining a job.

a. Construct a frequency histogram for the data.

b. Interpret the resulting figure.

TABLE 2.5 Starting Salaries for a Sample of College Graduates

$40,000	$30,600	$42,400	$24,200	$25,700
35,200	36,100	40,300	30,400	48,900
32,800	30,400	27,400	40,900	39,400
25,300	34,400	42,700	33,200	33,100
44,600	41,400	32,100	36,700	43,100
36,600	46,700	32,800	31,400	26,800
30,700	53,900	35,200	36,700	50,500
21,000	46,600	38,200	36,300	30,100
31,500	30,500	35,600	34,900	30,000
30,400	47,500	38,700	23,200	38,300

Solution

a. Step 1 The first step in constructing the frequency histogram for this sample is to define the **class intervals** (categories) into which the data will fall. To do this, we need to know the smallest and largest starting salaries in the data set. These salaries are $21,000 and $53,900, respectively. Since we want the smallest salary to fall in the lowest class interval and the largest salary to fall in the highest class interval, the class intervals must span starting salaries ranging from $21,000 to $53,900.

Step 2 The second step is to choose the **class interval width;** this will depend on how many intervals we want to use to span the starting salary range and whether we want to use equal or unequal interval widths. For this example, we will use equal class interval widths (the most popular choice) and 11 class intervals. (See the General Rule in the box on p. 72 on choosing the number of class intervals.)

Note that the starting salary range is equal to

$$\begin{aligned}\text{Range} &= \text{Largest measurement} - \text{Smallest measurement}\\ &= \$53{,}900 - \$21{,}000\\ &= \$32{,}900\end{aligned}$$

Since we chose to use 11 class intervals, the class interval width should approximately equal

$$\begin{aligned}\text{Class interval width} &= \frac{\text{Range}}{\text{Number of class intervals}}\\ &= \frac{32{,}900}{11} = 2{,}990.9\\ &\approx \$3{,}000\end{aligned}$$

We shall start the first class slightly below the smallest observation ($21,000) and choose the starting point so that no observation can fall on a **class boundary.** Since starting salaries are recorded to the nearest hundred dollars, we can do this by choosing the lower class boundary of the first class interval to be $20,950. [*Note:* We could just as easily have chosen $20,955, $20,975, $20,990, or any one of many other points below and near $21,000.] Then the class intervals will be $20,950 to $23,950, $23,950 to $26,950, and so on. The 11 class intervals are shown in the second column of Table 2.6.

Step 3 The third step in constructing a histogram is to obtain each class frequency—i.e., the number of observations falling within each class. This is usually done using the computer. For illustration purposes, we performed this task by examining each starting salary in Table 2.5 and recording by tally (as shown in the third column of Table 2.6) the class in which it falls. The tally for each class gives the class frequencies shown in column 4 of Table 2.6. (Optionally, class relative frequencies

TABLE 2.6 Tabulation of Data for the Starting Salaries of Table 2.5

Class	Class Interval	Tally	Class Frequency	Class Relative Frequency
1	20,950–23,950	II	2	.04
2	23,950–26,950	IIII	4	.08
3	26,950–29,950	I	1	.02
4	29,950–32,950	~~IIII~~ ~~IIII~~ III	13	.26
5	32,950–35,950	~~IIII~~ II	7	.14
6	35,950–38,950	~~IIII~~ III	8	.16
7	38,950–41,950	~~IIII~~	5	.10
8	41,950–44,950	IIII	4	.08
9	44,950–47,950	III	3	.06
10	47,950–50,950	II	2	.04
11	50,950–53,950	I	1	.02
TOTALS			50	1.00

or percentages can also be calculated. Relative frequencies are shown in the fifth column of Table 2.6.)

Step 4 The final step is to produce the graph using computer software. The bars in the histogram will have heights proportional to the class frequency, the class relative frequency, or the class percentage. Since we desire a frequency histogram, the bar heights will be proportional to the class frequencies. The resulting histogram, generated by Excel, is shown in Figure 2.8. Note that Excel displays the midpoints of each class interval (called *bins*) on the histogram.

General Rule for Determining the Number of Classes in a Histogram

Number of Observations in a Data Set	Number of Classes
Less than 25	5 or 6
25–50	7–14
More than 50	15–20

Note: You may want to try several different numbers of classes until you obtain a histogram that presents the clearest picture of the data.

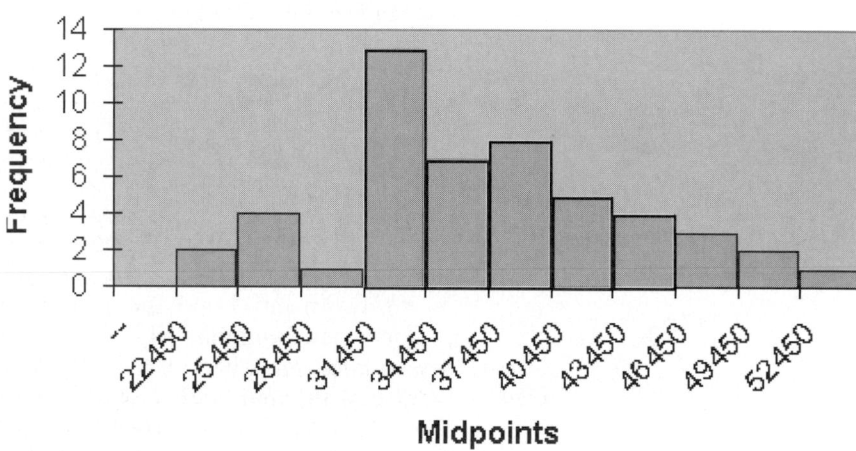

Figure 2.8 Excel Frequency Histogram for the Data of Table 2.5

b. A histogram, like a stem-and-leaf display or dot plot, conveys a visual picture of the quantitative data. In particular, a histogram will identify a range of values where most of the data fall. You can see that most of the 50 graduates in the sample had starting salaries between \$29,950 and \$44,950. The five classes (bars) associated with this interval have frequencies of 13, 7, 8, 5, and 4, respectively. These bars are highlighted on Figure 2.8. The sum of these frequencies is 37. Thus, $37/50 = .74$, or 74%, of the graduates in the sample had starting salaries between \$29,950 and \$44,950.

Note also that none of the starting salaries was less than \$20,950, but several high salaries caused the distribution to have a long tail on the right. We say that such a distribution is **right skewed** (or **positively skewed**). Similarly a **left** (or **negatively**) **skewed** distribution has a histogram that has a long tail on the left because of several unusually small values. Illustrations of right and left skewed distributions, as well as a **nonskewed** (**symmetric**) distribution, are shown in Figure 2.9.

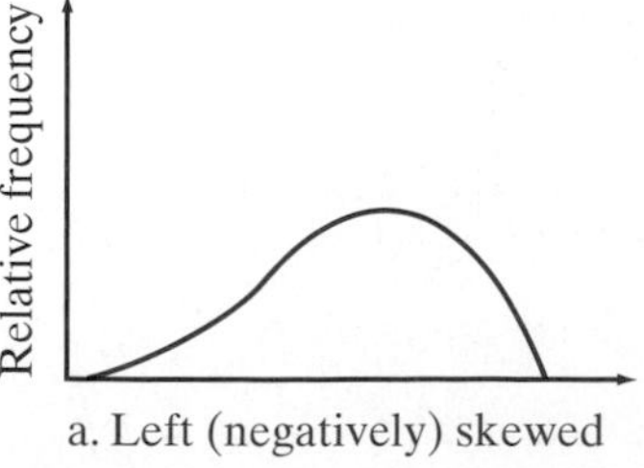

a. Left (negatively) skewed

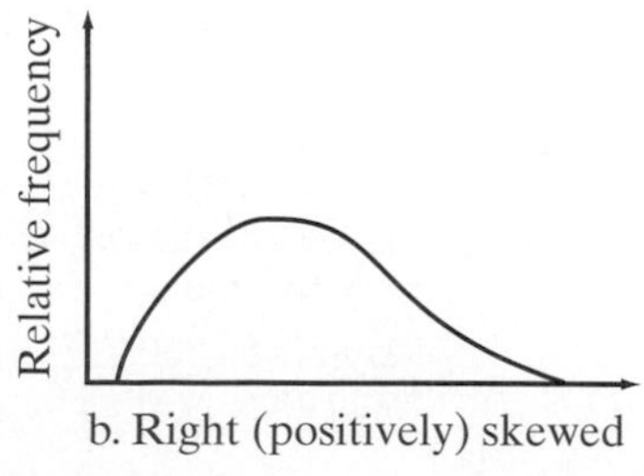

b. Right (positively) skewed

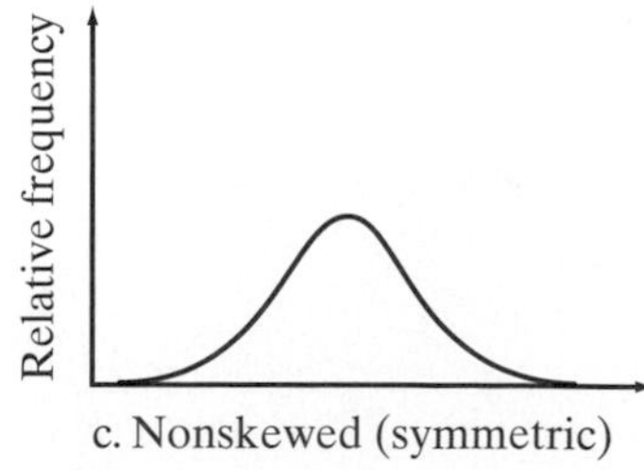

c. Nonskewed (symmetric)

Figure 2.9 Skewed and Symmetric Distributions

INTERPRETING A HISTOGRAM

The percentage of the total number of measurements falling within a particular interval is proportional to the area of the bar that is constructed above the interval. For example, if 30% of the area under the distribution lies over a particular interval, then 30% of the observations fall in that interval.

Statistics in the Real World Revisited

Interpreting a Histogram

As part of its *New Car Assessment Program* (NCAP), the National Highway Traffic Safety Administration (NHTSA) crash tests vehicles into a fixed barrier at 35 mph. Electronic instruments attached to human dummies placed in the driver's and front passenger's seats register the forces and impacts that occur to the head during the crash. (See p. 58.) One of the quantitative variables measured by the NHTSA (on a 0 to 1,500 point scale) is passenger's severity of head injury score. (The higher the score, the more severe the head injury.) We can describe the passenger head injury scores for the 98 cars in the NCAP report using a dot plot, stem-and-leaf display, or histogram.

A MINITAB percentage histogram for the 98 passenger head injury scores is shown in Figure 2.10. This is a relative frequency histogram since the vertical axis is shown as a percentage (rather than a frequency). The values shown on the horizontal axis represent the class boundaries for every other class interval. These values imply that the class intervals are 100-200, 200-300, 300-400, 400-500, . . . , 1100-1200, and 1200-1300. You can see that most of the head injury scores tend to pile up between 400 and 600 points. In fact, the interval from 400 to 500 has the highest relative frequency (approximately 20%) and the interval from 500 to 600 has the next highest relative frequency (approximately 18%). Since the histogram has a slight tail to the right, it is moderately right skewed.

The NHTSA considers a head injury score of 700 points or higher to represent a severe head injury. From the histogram, we can estimate the proportion of passenger head injuries that are "severe" by summing the relative frequencies of the bars corresponding to the classes 700-800, 800-900, 900-1000, 1000-1100, 1100-1200, and 1200-1300. The relative frequencies are (approximately) .13, .04, .02, .01, 0, and .02, respectively. Summing these relative frequencies, we obtain $.13 + .04 + .02 + .01 + 0 + .02 = .22$. Thus, about 22% of the cars had severe passenger head injury scores, i.e., passenger head injury scores of 700 points or higher.

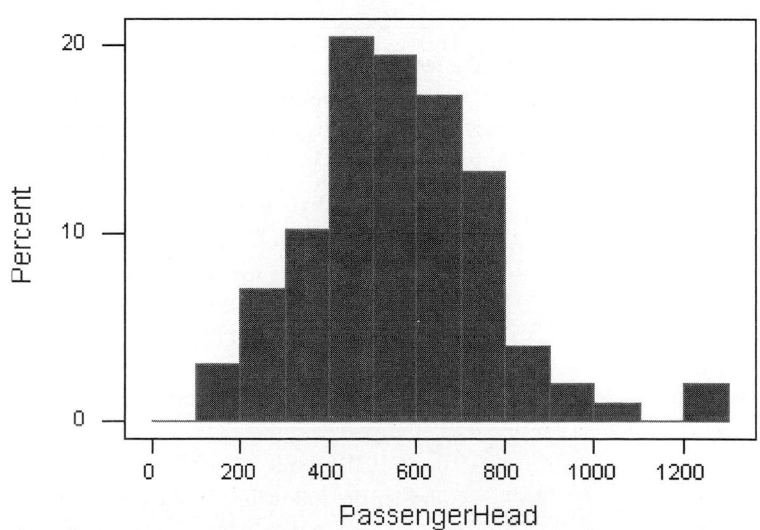

M **Figure 2.10** A MINITAB Percentage Histogram for the 98 Passenger Head Injury Scores

Self-Test 2.3

In a *Developmental Psychology* study on the sexual maturation of college students, 75 male college undergraduates were asked to report the age (in years) when they began to shave regularly. The frequency histogram of age at regular shaving is shown in Figure 2.11.

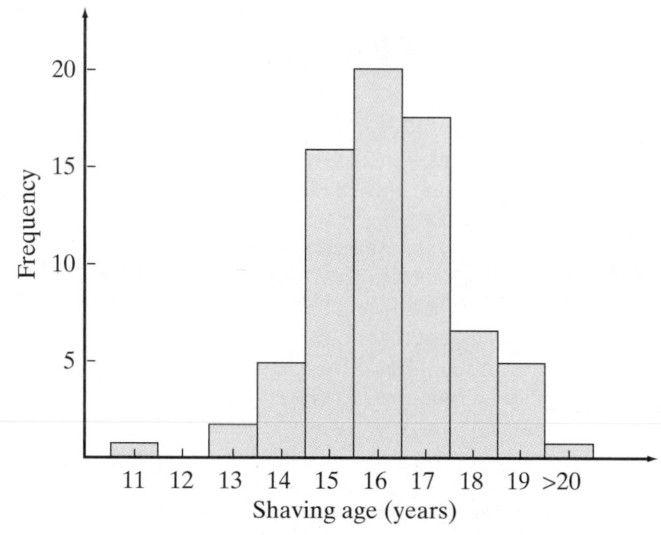

Figure 2.11 Frequency Histogram of Shaving Age

a. Approximately how many of the 75 male students began to shave regularly at age 16?

b. Approximately what proportion of the 75 male students began to shave regularly at age 16?

PROBLEMS FOR SECTION 2.3

Using the Tools

2.15 Consider the following sample of 15 quantitative measurements:

35 40 50 35 40 40 45 35 30 40 30 35
45 50 40

a. Using the digit in the tens place as a stem, list the stems in order.

b. Identify the leaf for each measurement.

c. Place the leaves for each measurement in the appropriate stem row to form a stem-and-leaf display.

d. Construct a dot plot for the data.

2.16 Construct both a stem-and-leaf display and a dot plot for the following sample of 20 quantitative measurements:

8.1 10.5 6.6 6.7 3.3 5.1 8.3 7.5 6.6 4.0 6.9
9.0 10.2 5.0 4.2 6.1 6.3 7.2 5.4 8.7

2.17 Consider the following sample data:

213 228 241 268 234 303 274 316 319 320 227 226
224 267 303 266 265 237 288 291 285 270 254 215

a. Using the first two digits of each number as a stem, list the stems in order.
b. Place the leaf for each observation in the appropriate stem row to form a stem-and-leaf display.

2.18 Consider the sample data shown here.

5.9	5.3	1.6	7.4	8.6	1.2	2.1	4.0	7.3	8.4	8.9	6.7
4.5	6.3	7.6	9.7	3.5	1.1	4.3	3.3	8.4	1.6	8.2	6.5
1.1	5.0	9.4	6.4								

a. Find the difference between the largest and smallest measurements.
b. Divide the difference obtained in part **a** by 5 to determine the approximate class interval width for five class intervals.
c. Specify upper and lower boundaries for each of the five class intervals.
d. Construct a relative frequency distribution for the data.

2.19 A sample of 20 measurements follows.

26	34	21	32	32	22	12	26	39	25	36	28
38	17	39	31	30	23	27	19				

a. Using a class interval width of 5, give the upper and lower boundaries for six class intervals, where the lower boundary of the first class is 10.5.
b. Determine the relative frequency for each of the six classes specified in part **a**.
c. Construct a relative frequency distribution using the results of part **b**.

Applying the Concepts

2.20 The National Earthquake Information Center located 16,287 earthquakes worldwide in 1998. The magnitudes (on the Richter scale) of the earthquakes are summarized in the table. Display the data in a graph. Identify the type of skewness (if any) in the data.

Magnitude	Number of Earthquakes
0	1,774
0.1–0.9	6
1.0–1.9	580
2.0–2.9	3,080
3.0–3.9	4,518
4.0–4.9	5,548
5.0–5.9	674
6.0–6.9	97
7.0–7.9	9
8.0–8.9	1

Source: National Earthquake Information Center, Department of Interior, U.S. Geological Survey, 1999.

BANKCOST

2.21 *Consumer Reports* (June 2000) conducted a study of fees charged by U.S. metropolitan banks. The following data, collected from a sample of 23 banks, represent the bounced check fee ($) for direct-deposit customers who maintain a $100 balance.

26	28	20	20	21	22	25	25	18	25	15	20	18
20	25	25	22	30	30	30	15	20	29			

Source: "The New Face of Banking," *Consumer Reports,* June 2000.

a. Construct a stem-and-leaf display for these data.
b. Construct a dot plot for these data.

c. Which of these two graphs seems to provide more information? Discuss.

d. Around what value, if any, do the bounced check fees seem to be concentrated? Explain.

2.22 The number of calories per 12-ounce serving for each of 22 beer brands is provided in the table.

a. Construct a stem-and-leaf display for the data.

b. Locate the data values for the six light beers on the stem-and-leaf display. Do light beers really have a lower caloric content than regular beers?

BEERCAL

Beer	Calories	Beer	Calories
Old Milwaukee	145	Miller High Life	143
Stroh's	142	Pabst Blue Ribbon	144
Red Dog	147	Milwaukee's Best	133
Budweiser	148	Miller Genuine Draft	143
Icehouse	149	Rolling Rock	143
Molson Ice	155	Michelob Light	134
Michelob	159	Bud Light	110
Bud Ice	148	Natural Light	110
Busch	143	Coors Light	105
Coors Original	137	Miller Light	96
Gennessee Cream Ale	153	Amstel Light	95

Source: Consumer Reports, June 1996, Vol. 61, No. 6, p. 16.

SHIPSANIT

2.23 To minimize the potential for gastrointestinal disease outbreaks, all passenger cruise ships arriving at U.S. ports are subject to unannounced sanitation inspections. Ships are rated on a 100-point scale by the Centers for Disease Control. In general, the lower the score, the lower the level of sanitation. The January 1999 sanitation scores for 121 cruise ships are saved in the file **SHIPSANIT.** (The first 5 observations and the last 5 observations are listed in the accompanying table.)

Ship	Score
Americana	75
Arcadia	93
Arkona	89
Astor	74
Asuka	88
.	.
.	.
.	.
Westerdam	92
Wind Spirit II	93
World Discoverer	89
Yorktown Clipper	91
Zenith	93

Source: Center for Environmental Health and Injury Control. *Tampa Tribune,* Feb. 7, 1999.

a. A MINITAB stem-and-leaf display of the data is shown here. Identify the stems and leaves of the graph.

```
Stem-and-leaf of Score      N  = 121
Leaf Unit = 1.0


    1     3 6
    1     4
    1     4
    1     5
    1     5
    1     6
    1     6
    3     7 04
    7     7 5889
    8     8 1
   40     8 66666666677777788888888899999999
  (56)    9 00000001111111111111222222222333333333333333333444444444
   25     9 5555555555556666667777889
```

M **Problem 2.23**

b. Locate the inspection score of 75 (Americana) on the stem-and-leaf display.

c. A score of 86 or higher at the time of inspection indicates the ship is providing an accepted standard of sanitation. Use the stem-and-leaf display in part **a** to estimate the proportion of ships that have an accepted sanitation standard. Does this answer influence your decision to take a cruise in the future?

2.24 The United States Golf Association (USGA) Handicap System is designed to allow golfers of differing abilities to enjoy fair competition. The handicap index is a

measure of a player's potential scoring ability on an 18-hole golf course of standard difficulty. For example, on a par-72 course, a golfer with a handicap of 7 will typically have a score of 79 (seven strokes over par). Over 4.5 million golfers have an official USGA handicap index. The handicap indexes for both male and female golfers were obtained from the USGA and are summarized in the two MINITAB histograms shown below.

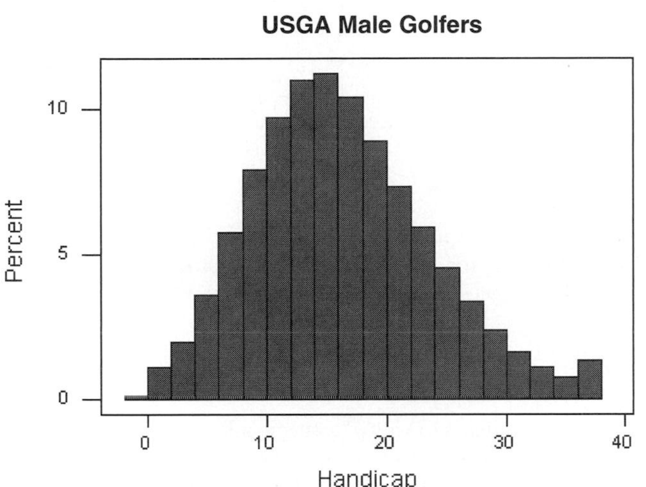

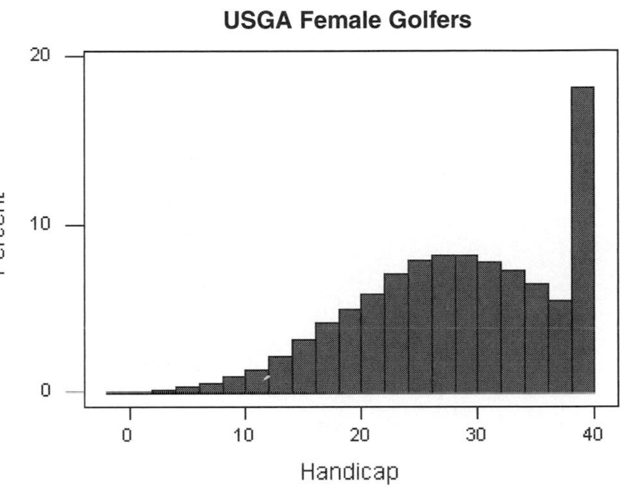

M Problem 2.24

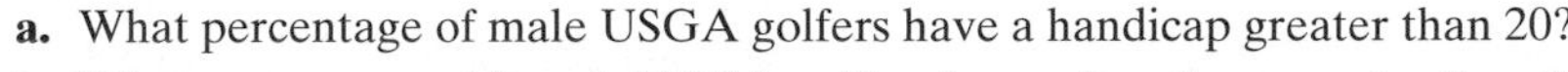

a. What percentage of male USGA golfers have a handicap greater than 20?

b. What percentage of female USGA golfers have a handicap greater than 20?

c. Identify the type of skewness (if any) present in the male handicap distribution.

d. Identify the type of skewness (if any) present in the female handicap distribution.

2.25 Access the **FISH** data set that contains the data on contaminated fish in the Tennessee River (Alabama) collected by the U.S. Army Corps of Engineers. Consider the 50 DDT measurements corresponding to fish specimens identified on the data set by observations numbered 51–100.

a. Use Excel or MINITAB to construct a frequency histogram for the 50 DDT values. Use 10 classes to span the range.

b. Repeat part **a**, but use only three classes to span the range. Compare the result with the frequency histogram you constructed in part **a**. Which is more informative? Why do an inadequate number of classes limit the information conveyed by a frequency histogram?

c. Repeat part **a**, but use 25 classes. Comment on the information provided by this histogram and compare it with the result of part **a**.

2.26 "Deep hole" drilling is a family of drilling processes used when the ratio of hole depth to hole diameter exceeds 10. Successful deep hole drilling depends on the satisfactory discharge of the drill chip. An experiment was conducted to investigate the performance of deep hole drilling when chip congestion exists (*Journal of Engineering for Industry,* May 1993). An analysis of drill chip congestion was performed using data generated via computer simulation. The simulated distribution of the length (in millimeters) of 50 drill chips is displayed on p. 78 in a frequency histogram.

a. Convert the frequency histogram into a relative frequency histogram.

b. Based on the graph in part **a,** would you expect to observe a drill chip with a length of at least 190 mm? Explain.

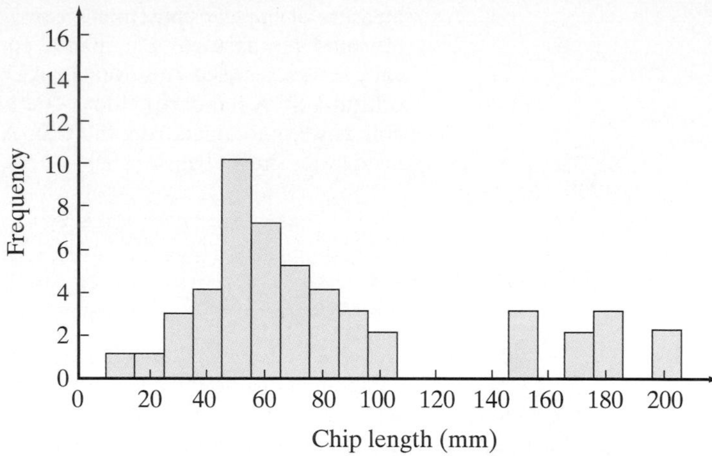

Source: Chin, Jih-Hua, et al. "The computer simulation and experimental analysis of chip monitoring for deep hole drilling." *Journal of Engineering for Industry, Transactions of the ASME,* Vol. 115, May 1993, p. 187 (Figure 12).

Problem 2.26

2.27 Computer software that allows students to view electronic atlases and produce their own interactive maps has become popular in the geography classroom. A study was conducted to determine whether using computer software has a positive influence on learning geography (*Journal of Geography,* May/June 1997). One group of seventh graders (43 students) used computer resources (electronic atlases, encyclopedias, and interactive maps) to research the regions of Africa. Another group of seventh graders (22 students) studied the same African regions using notes from a presentation and teacher-produced worksheets. At the end of the study period, all students were given a test on the African regions. The test scores are listed in the tables below.

Group 1 (Computer Resources) Student Scores

41	53	44	41	66	91	69	44	31	75	66	69
75	53	44	78	91	28	69	78	72	53	72	63
66	75	97	84	63	91	59	84	75	78	88	59
97	69	75	69	91	91	84					

Group 2 (No Computer) Student Scores

56	59	9	59	25	66	31	44	47	50	63	19
66	53	44	66	50	66	19	53	78	78		

Source: Linn, S. E. "The effectiveness of interactive maps in the classroom: A selected example in studying Africa." *Journal of Geography,* Vol. 96, No. 3, May/June 1997, p. 167 (Table 1).

a. Use a graphical technique to describe the test scores for the first group of geography students.

b. Repeat part **a** for the second group of geography students.

c. Compare the graphs. What do you learn from this comparison?

2.28 Many Vietnam veterans have dangerously high levels of the dioxin 2,3,7,8-TCDD in their blood as a result of their exposure to the defoliant Agent Orange. A study published in *Chemosphere* (Vol. 20, 1990) reported on the TCDD levels of

20 Massachusetts Vietnam veterans who were possibly exposed to Agent Orange. The amounts of TCDD (measured in parts per trillion) in blood plasma drawn from each veteran are shown in the table. Use a graphical technique to describe the distribution of TCDD levels. Identify the type of skewness in the data.

2.5	1.8	6.9	1.8	3.5	2.5	1.6	36.0	6.8	3.1
20.0	3.3	4.7	3.1	4.1	7.2	4.6	3.0	2.1	2.0

Source: Schecter, A. et al. "Partitioning of 2,3,7,8-chlorinated dibenzo-p-dioxins and dibenzofurans between adipose tissue and plasma lipid of 20 Massachusetts Vietnam veterans." *Chemosphere,* Vol. 20, Nos. 7–9, 1990, pp. 954–955 (Table I).

2.29 Nondestructive evaluation (NDE) describes the process of evaluating or inspecting components without causing any permanent physical change to the components. NDE was applied in a system for inspecting fluorescent-penetrant specimens that are prone to synthetic cracks (*Technometrics,* May 1996). The crack size (in inches) of each in a sample of 58 fluorescent-penetrant specimens was measured as well as whether or not the specimen was previously determined to be flawed. The crack sizes are reported in the table. [*Note:* Specimens judged to have a flaw are marked with an asterisk (*).]

a. Summarize the data on crack sizes with a stem-and-leaf display.

b. Locate the specimens judged to have a flaw on the graph. Do flawed specimens tend to have larger crack sizes?

Crack Sizes (in inches) of 58 Specimens

.003	.004	.012	.014	.021	.022*	.023	.024
.026*	.026*	.030*	.030	.031*	.034	.034*	.041
.041	.042*	.042	.043	.043*	.044*	.045	.046*
.046*	.052*	.055*	.057	.058*	.060*	.060*	.063
.070*	.071*	.073*	.073*	.074	.076	.078*	.079*
.079*	.083*	.090*	.095*	.095*	.096*	.100*	.102*
.103*	.105*	.114*	.119*	.120*	.130*	.160*	.306*
.328*	.440*						

Source: Olin, B. D. and Meeker, W. Q. "Applications of statistical methods to nondestructive evaluation." *Technometrics,* Vol. 38, No. 2, May 1996, p. 101 (Table 1).

2.4 Exploring the Relationship between Two Qualitative Variables: Cross-Classification Tables and Side-by-Side Bar Graphs

Experimental units in a study are often simultaneously categorized according to two qualitative variables. For example, every 10 years the U.S. Census Bureau collects demographic data on each American citizen. Two of the many variables measured are gender and marital status. In this section, we demonstrate how to describe two categorical variables (e.g., gender and marital status) simultaneously with *cross-classification tables* and *side-by-side bar graphs.* These methods will enable us to explore a possible pattern or relationship between the two qualitative variables.

EXAMPLE 2.5 **CROSS-CLASSIFICATION TABLE**

In group discussions, do men and women interrupt the speaker equally often? This was the research question investigated in the *American Sociological Review* (June 1989). Undergraduate sociology students (half male and half female) were organized into discussion groups and their interactions recorded on video. Each

time the speaker was interrupted, the researchers recorded two variables: (1) the gender of the speaker and (2) the gender of the interrupter. Hypothetical data for a sample of 40 interruptions are listed in Table 2.7. Use a cross-classification table to summarize the data in Table 2.7.

INTERRUPT

TABLE 2.7 Data on Speaker and Interrupter Gender for 40 Interruptions

Interruption	Speaker	Interrupter	Interruption	Speaker	Interrupter
1	M	M	21	F	M
2	M	M	22	F	F
3	F	F	23	F	F
4	M	F	24	M	M
5	F	M	25	F	F
6	M	F	26	M	M
7	M	M	27	F	M
8	F	M	28	F	F
9	F	M	29	M	M
10	F	M	30	M	M
11	F	F	31	F	F
12	M	M	32	M	M
13	F	M	33	F	M
14	F	M	34	M	F
15	M	M	35	M	M
16	M	M	36	M	M
17	F	F	37	F	M
18	M	F	38	F	F
19	M	M	39	F	F
20	M	F	40	M	M

Solution

Each of the two qualitative variables of interest, speaker gender and interrupter gender, has two categories: male (M) and female (F). In order to investigate the relationship between the two variables, we determine the frequency of interruptions for each of the $2 \times 2 = 4$ *combined* categories. That is, we count the number of observations that fall into the four speaker/interrupter categories: M/M, M/F, F/M, and F/F. These four categories and their frequencies are displayed in Table 2.8. The summary table is known as a **cross-classification table,** since the values of one variable are classified across the values of the other variable.

TABLE 2.8 Cross-Classification Table for Data of Table 2.7

		Interrupter		
		Male	Female	Totals
Speaker	Male	15	5	20
	Female	10	10	20
	TOTALS	25	15	40

For example, the number 15 in the upper-left cell of the table is a count of the number of male speakers who were interrupted by a male. Similarly, the number 5 in the upper-right cell is the number of male speakers who were interrupted by a female. You can see that the sum of the counts in the four categories equals 40, the total number of interruptions in the sample.

To explore a possible relationship between speaker gender and interrupter gender, it is useful to compute the relative frequency for each cell of the cross-classification table. One way to obtain these relative frequencies is to divide each cell frequency by the corresponding total for the row in which the cell appears. Alternatively, the relative frequencies can be obtained by dividing the cell frequencies by either their column totals or their overall total (in this case, 40). The choice of divisor will depend on the objective of the analysis, as the next example illustrates.

EXAMPLE 2.6

SIDE-BY-SIDE BAR GRAPHS

Refer to Example 2.5. The researchers want to explore the nature of the relationship between speaker gender and interrupter gender. Specifically, they want to know the gender that is more likely to interrupt a male speaker and the gender that is more likely to interrupt a female speaker. To do this, they require relative frequencies of male interrupters and female interrupters for each speaker gender.

a. Compute these relative frequencies and display them in a table.

b. Portray the results, part **a**, graphically. Interpret the graph.

Solution

a. First, we consider male speakers (i.e., focus on the first row of Table 2.8). Of the 20 male speakers in the study, 15 were interrupted by a male. Thus, the relative frequency for the male speaker/male interrupter cell is 15/20 = .75. Our interpretation is that 75% of the male speakers in the study are interrupted by males. Similarly, 5 of the 20 male speakers were interrupted by females. The associated relative frequency is 5/20 = .25; this implies that 25% of the male speakers were interrupted by a female.

Now, consider the 20 female speakers in the study (i.e., focus on the second row of Table 2.8). Since 10 were interrupted by males and 10 were interrupted by females, the associated relative frequencies are 10/20 = .5 and 10/20 = .5. In other words, half of the female speakers in the study were interrupted by males and half were interrupted by females.

All four of these relative frequencies are shown in Table 2.9. Note that the relative frequencies in each row sum to 1.

TABLE 2.9 Relative Frequencies for Table 2.8 Based on Row Totals

		Gender of Interrupter		
		Male	**Female**	**Totals**
Gender of Speaker	**Male**	.75	.25	1.00
	Female	.50	.50	1.00

b. The relative frequencies of Table 2.9 can be visually displayed using bar graphs. Since the cross-classification table contains two rows (male speaker and female speaker), we form one bar graph for male speakers and one for female speakers, and then place them side-by-side for comparison purposes. A **side-by-side horizontal bar graph,** generated by Excel, is shown in Figure 2.12 on p. 82.

Figure 2.12 clearly shows the dramatic difference in proportions of male and female interrupters for the two speaker genders. Males interrupt three times as much as females (.75 vs. .25) when a male is speaking. However, when a female is speaking, females are just as likely to interrupt as males (.50 vs. .50). Consequently, it appears that males are more likely to interrupt than females, but only when the speaker is a male.

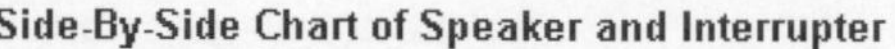

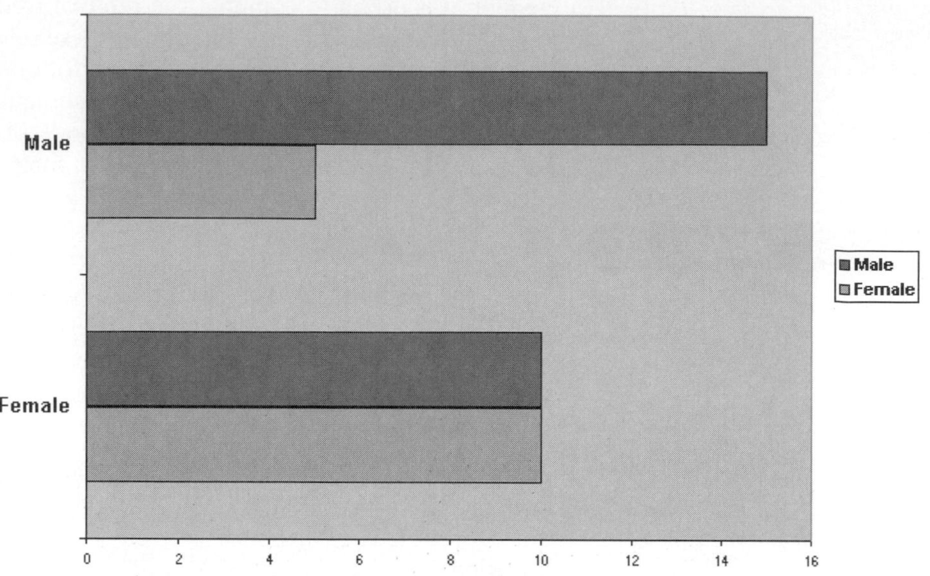

E **Figure 2.12** Excel Side-by-Side Bar Graph for Data Summarized in Table 2.9

Self-Test 2.4

Each of 274 introductory psychology students (80 males and 194 females) was asked to indicate when he/she reached puberty in relation to others of the same gender. One of five response categories were possible: much earlier, earlier, same time, later, and much later. The proportions responding in each of the categories for males and females are reported in Table 2.10.

TABLE 2.10 Summary for *Developmental Psychology* Study

Time Reached Puberty in Relation to Others	Males	Females
Much earlier	.02	.05
Earlier	.22	.20
Same time	.56	.57
Later	.20	.17
Much later	.00	.01
TOTALS	1.00	1.00

Data Source: Sanders, B. and Soares, M. P. "Sexual maturation and spatial ability in college students." *Developmental Psychology,* Vol. 22, No. 2, Mar. 1986, pp. 199–203 (Figure 1).

a. Interpret the value .22 shown in the table.

b. Construct a side-by-side bar chart for the data. Interpret the graph.

Cross-classification tables and side-by-side bar graphs are useful methods for describing the simultaneous occurrence of two qualitative variables, and can be used to make inferences about whether or not the two variables are related. However, no measure of reliability can be attached to such an inference without a more sophisticated numerical analysis. In Chapter 9 we present a formal inferential method for analyzing cross-classification (two-way contingency) tables.

Statistics in the Real World Revisited

Interpreting a Side-by-Side Bar Graph

Let us return to the crash-test data collected by the National Highway Traffic Safety Administration (NHTSA) (p. 58). Recall that the qualitative variable Driver Star Rating gives the level of crash protection for the driver of the car. A rating of five stars implies that the driver has a 10% or less chance of a serious injury in a head-on collision, a 4-star rating implies the driver has a chance of serious injury between 11% and 20%, a 3-star rating implies the driver has a chance of serious injury between 21% and 35%, and a 2-star rating implies the driver has a chance of serious injury between 36% and 45%. (None of the cars tested has a rating of one star.) One question of interest to the NHTSA is: Is the distribution of Driver Star Ratings for 2-door car models the same as the corresponding distribution for car models with three or four doors? To answer this question, we need to examine the relationship between two qualitative variables—Driver Star Rating at four levels (2, 3, 4, and 5 stars) and Door Model at three levels (2, 3, and 4 doors). We do this by constructing a cross-classification table and a side-by-side bar graph for the data.

Both an Excel and a MINITAB cross-classification table for the 98 cars tested are shown in Figure 2.13. Note that the rows of the table represent Door Model, the columns represent Driver Star Rating, and there are 12 cells in the table—one for each of the $3 \times 4 = 12$ combined categories of the

E Figure 2.13a
M Figure 2.13b Cross-Classification Tables for 98 Crash-Tested Cars: Door Model by Driver Star Rating

	A	B	C	D	E	F
1	Cross Tabulation of Doors by DrivStar					
2						
3	Count of Doors	Driver Star				
4	Doors	2	3	4	5	Grand Total
5	2	4	3	14	9	30
6	3	0	4	8	1	13
7	4	0	10	37	8	55
8	Grand Total	4	17	59	18	98

(a) Excel

```
Rows: Doors      Columns: DriverSt

             2         3         4         5       All

2            4         3        14         9        30
         13.33     10.00     46.67     30.00    100.00

3            0         4         8         1        13
            --     30.77     61.54      7.69    100.00

4            0        10        37         8        55
            --     18.18     67.27     14.55    100.00

All          4        17        59        18        98
          4.08     17.35     60.20     18.37    100.00

  Cell Contents --
                      Count
                      % of Row
```

(b) MINITAB

(Continued)

two qualitative variables. In addition to a count of the number of cars in each cell, the MINITAB cross-classification table (Figure 2.13b) gives relative frequencies for the 12 cells. These relative frequencies are computed by dividing each of the cell frequencies by their respective row totals, as shown in Table 2.11. For example, the relative frequency for the 2-door/5-star cell is 9/30 = .300. Similarly, the relative frequency for the 4-door/4-star cell is 37/55 = .673. [Remember, you could also choose to divide the cell counts by column totals or by the overall total. We selected to divide by row (Door Model) totals since the NHTSA wants to examine the Driver Star Ratings for each level of Door Model.]

The frequencies used to compute the percentages in Figure 2.13b are displayed in the Excel side-by-side bar graphs shown in Figure 2.14. From these figures you can see that the 4-star driver rating has the highest percentage of cars for all Door Model levels. However, this percentage is much greater for 4-door models (67.3%) than for 2-door models (46.7%). That is, 67.3% of all 4-door models have a driver rating of four stars while 46.7% of all 2-door models have a driver rating of four stars. Similarly, 4-door models have a higher percentage of 3-star ratings (18.2%) than 2-door models (10%). Interestingly, 2-door models have higher relative frequencies of 2-star and 5-star ratings than 4-door models. From Figure 2.14 it is clear that the distribution of Driver Star Ratings for 2-door car models is *not* the same as the distribution of Driver Star Ratings for car models with three or four doors. It appears that 2-door models tend to have star ratings that cluster at four stars and five stars, while more-than-2-door models tend to have star ratings that cluster at three and four stars.

TABLE 2.11 Calculation of Relative Frequencies Using Row (Door Model) Totals

	Driver Star Rating				
Doors	**2-Stars**	**3-Stars**	**4-Stars**	**5-Stars**	**Row Total**
2	4/30 = .133	3/30 = .100	14/30 = .467	9/30 = .300	1.000
3	0/13 = .000	4/13 = .308	8/13 = .615	1/13 = .077	1.000
4	0/55 = .000	10/55 = .182	37/55 = .673	8/55 = .145	1.000

Side-By-Side Chart of Doors by DrivStar

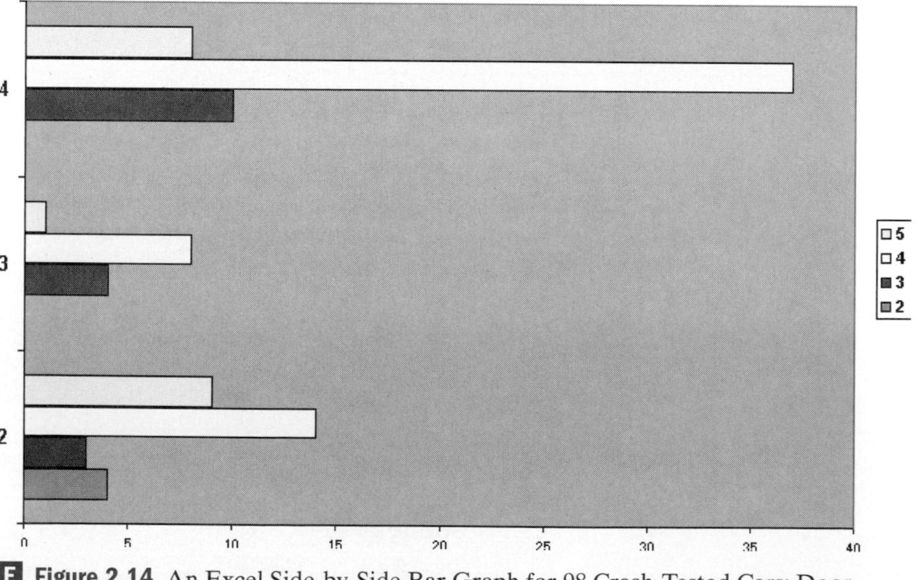

Figure 2.14 An Excel Side-by-Side Bar Graph for 98 Crash-Tested Cars: Door Model by Driver Star Rating

PROBLEMS FOR SECTION 2.4

Using the Tools

2.30 Two qualitative variables were measured for each of 10 experimental units. The data are listed in the accompanying table.

Experimental Unit	Variable 1	Variable 2
1	Yes	A
2	Yes	B
3	No	B
4	Yes	A
5	Yes	B
6	No	B
7	No	A
8	No	A
9	No	B
10	No	A

a. List the possible outcomes of the experiment (i.e., the different combined categories of the two qualitative variables).

b. Find the frequency of each category. Present the results in the form of a cross-classification table.

c. Using the totals for each category of Variable 2, compute the relative frequencies for the different classes of Variable 1.

d. Using the totals for each category of Variable 1, compute the relative frequencies for the different classes of Variable 2.

e. Present the results, part **c**, in the form of a side-by-side bar graph.

f. Present the results, part **d**, in the form of a side-by-side bar graph.

2.31 Two qualitative variables were measured for each of 12 experimental units. The data are listed in the table.

Experimental Unit	Variable 1	Variable 2
1	X	A
2	X	B
3	Z	B
4	Y	A
5	Y	A
6	X	B
7	Z	A
8	Z	B
9	X	B
10	Y	A
11	Y	B
12	Y	B

a. List the possible outcomes of the experiment (i.e., the different combined categories of the two qualitative variables).

b. Find the frequency of each category. Present the results in the form of a cross-classification table.

c. Using the totals for each category of Variable 2, compute the relative frequencies for the different classes of Variable 1.

d. Using the totals for each category of Variable 1, compute the relative frequencies for the different classes of Variable 2.

e. Present the results, part **c**, in the form of a side-by-side bar graph.

f. Present the results, part **d**, in the form of a side-by-side bar graph.

2.32 Consider the cross-classification table shown below.

		Color			
		Red	**White**	**Blue**	**Totals**
Age	**Old**	10	21	3	34
	New	2	14	0	16
TOTALS		12	35	3	50

a. How many experimental units are included in the analysis?

b. How many experimental units are old and white?

c. How many experimental units are new and blue?

d. Use row totals to calculate the relative frequencies for the three colors.

e. Use column totals to calculate the relative frequencies for the two ages.

f. Use the overall total to calculate the relative frequency for each cell of the table.

g. Present the results, part **d**, in a side-by-side bar chart.

h. Present the results, part **e**, in a side-by-side bar chart.

2.33 Consider the cross-classification table shown below.

		Location				
		North	**East**	**South**	**West**	**Totals**
	A	20	40	80	50	190
Type	**B**	50	40	10	25	125
	C	30	20	10	25	85
TOTALS		100	100	100	100	400

a. How many experimental units are included in the analysis?

b. How many experimental units are located in the north?

c. How many experimental units of type B are located in the west?

d. Use row totals to calculate relative frequencies for the four locations.

e. Use column totals to calculate relative frequencies for the three types.

f. Use the overall total to calculate relative frequencies for each cell of the table.

g. Present the results, part **d**, in a side-by-side bar chart.

h. Present the results, part **e**, in a side-by-side bar chart.

Applying the Concepts

2.34 In order to create a behavioral profile of pleasure travelers, M. Bonn (Florida State University), L. Forr (Georgia Southern University), and A. Susskind (Cornell University) interviewed 5,026 pleasure travelers in the Tampa Bay region (*Journal of Travel Research,* May 1999). Two of the characteristics they investigated were the travelers' education level and their use of the Internet to seek travel information. The table on p. 87 summarizes the results of the interviews. The researchers concluded that travelers who use the Internet to search for travel information are likely to be people who are college educated.

		Use of Internet to Seek Travel Information	
		Yes	**No**
Education	**College Degree or More**	1,072	1,287
	Less than a College Degree	640	2,027

Source: Bonn, M., Forr, L., and Susskind, A., "Predicting a Behavioral Profile for Pleasure Travelers on the Basis of Internet Use Segmentation," *Journal of Travel Research,* Vol. 37, May 1999, pp. 333–340.

Problem 2.34

a. Identify the two qualitative variables measured in this study. List the classes for each variable.

b. Construct a table of column percentages.

c. Construct a side-by-side bar chart to visually highlight the results in part **b**.

d. Interpret the bar chart. Does the graph support the researchers' conclusion?

2.35 *Industrial Marketing Management* (Feb. 1993) published a study of humor in trade magazine advertisements. Each in a sample of 665 ads was classified according to nationality (British, German, American) of the trade magazine in which they appear and whether or not they were considered humorous by a panel of judges. The number of ads falling into each of the categories is provided in the accompanying table.

		Humorous	
		Yes	**No**
Nationality	**British**	52	151
	German	44	148
	American	56	214

Source: McCullough, L. S., and Taylor, R. K. "Humor in American, British, and German ads." *Industrial Marketing Management,* Vol. 22, No. 1, Feb. 1993, p. 22 (Table 3).

a. Identify the two qualitative variables measured in this study. List the classes for each variable.

b. Construct a table of row percentages.

c. Construct a side-by-side bar chart to visually highlight the results in part **b**.

d. Interpret the bar chart.

2.36 In professional sports, "stacking" is a term used to describe the practice of African-American players being excluded from certain positions because of race. Contingency tables are often used to demonstrate the stacking phenomenon. As an illustration, the *Sociology of Sport Journal* (Vol. 14, 1997) presented the table shown below. The table summarized the race and positions of 368 National Basketball Association (NBA) players in 1993.

	Position			
	Guard	**Forward**	**Center**	**Totals**
White	26	30	28	84
Black	128	122	34	284
TOTALS	154	152	62	368

Source: Leonard, W. M. and Phillips, J. "The cause and effect rule for percentaging tables; an overdue statistical correction for 'stacking' studies," *Sociology of Sport Journal,* Vol. 14, No. 3, 1997 (Table 2).

a. For each position, compute the proportions of white and black NBA players. (For each position, these two proportions should add to 1.)

b. Form a side-by-side bar graph for the proportions of part **a**. Interpret the graph.

c. For each race, compute the proportions of guards, forwards, and centers. (For each race, these three proportions should add to 1.)

d. Form a side-by-side bar graph for the proportions of part **c**. Interpret the graph.

e. The researchers are interested in whether black and white NBA players are distributed proportionately across positions. They argue that the proportions illustrated in the side-by-side bar graph of part **d** are more useful for this purpose than the side-by-side bar graph of part **b**. Do you agree? Explain.

2.37 A group of cardiac physicians has been studying a new drug designed to reduce blood loss in coronary artery bypass operations. Data were collected for 114 coronary artery bypass patients—half who received a dosage of the new drug and half who received a placebo (i.e., no drug). In addition to monitoring who received the drug, the physicians also recorded the type of complication (if any) experienced by the patient. A summary table for the data on drug administered and complication type is shown below. [*Note:* The data for this study are real. For confidentiality reasons, the drug name and physician group shall remain anonymous.]

		Drug		
		Yes	**No**	**Totals**
Complication Type	**Redo surgery**	7	5	12
	Post-op infection	7	4	11
	Both	3	1	4
	None	40	47	87
	TOTALS	57	57	114

a. Identify the two qualitative variables measured in this study. List the classes for each variable.

b. Construct a table of column percentages.

c. Display the results, part **b**, in a side-by-side bar graph.

d. Do you believe that patients who are administered the drug have more complications than those who receive a placebo? Explain.

2.38 Distance education at colleges or universities refers to education or training courses delivered to remote (off-campus) locations via audio, video (live or prerecorded), or computer technologies. The National Center for Education Statistics used the side-by-side bar graph shown on p. 89 to compare the percentage of institutions that offered distance education courses in Fall 1995 and in 1997–1998. Write several sentences that describe the changes in distance education offerings at colleges and universities over this time period.

2.39 Since 1948, research psychologists have used the "water-level task" to test basic perceptual and conceptual skills. Subjects are shown a drawing of a glass tilted at a 45° angle and asked to assume the glass is filled with water. The task is to draw a line representing the surface of the water. *Psychological Science* (Mar. 1995) reported on the results of the water-level task given to 120 subjects. Each subject was classified by group and by performance on the test. A summary of the results is provided in the table on p. 89.

a. Use a graphical method to describe the overall results of the study (i.e., the performance responses for all 120 subjects).

b. Construct graphs that the researchers could use to explore for group differences on the water-level task.

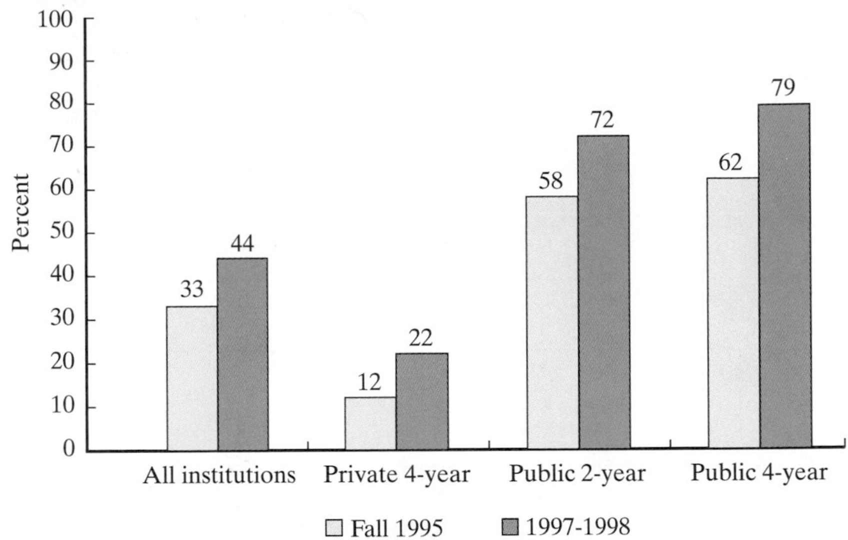

Note: Percentages for 1995 are based on an estimated 3,460 higher education institutions, and for 1997–1998 are based on an estimated 3,580 higher education institutions.

Source: U.S. Department of Education. National Center for Education Statistics. *Distance Education at Postsecondary Education Institutions: 1997–98* (NCES 2000-013), table 21.

Problem 2.38

		Group						
		Females			Males			
		Students	Waitresses	Housewives	Students	Bartenders	Bus Drivers	Totals
Judged Line	**More than 5° below surface**	0	0	1	1	1	1	4
	More than 5° above surface	7	15	13	3	11	4	53
	Within 5° of surface	13	5	6	16	8	15	63
	TOTALS	20	20	20	20	20	20	120

Source: Hecht, H., and Proffitt, D. R. "The price of experience: Effects of experience on the water-level task." *Psychological Science,* Vol. 6, No. 2, Mar. 1995, p. 93 (Table 1).

Problem 2.39

c. Psychologists theorize that males do better than females, that younger adults do better than older adults, and that those experienced in handling liquid-filled containers do better than those who are not. Are these theories supported by the data?

2.40 The National Gang Crime Research Center (NGCRC) has developed a 6-level gang classification system for both adults and juveniles. The six categories are shown in the table on p. 90. The classification system was developed as a potential predictor of a gang member's propensity for violence when in prison, jail, or a correctional facility. To test this theory, the NGCRC collected data on approximately 10,000 confined offenders and assigned each a score using the gang classification system (*Journal of Gang Research,* Winter 1997). One of several other variables measured by the NGCRC was whether or not the offender has ever carried a homemade weapon (e.g., knife) while in custody. The data on gang score and homemade weapon are summarized in the table. In your opinion, is gang classification score related to whether or not an offender in custody will carry a homemade weapon? Support your conclusion with a graph.

		Homemade Weapon Carried	
		Yes	No
Gang Classification Score	0 (Never joined a gang, no close friends in a gang)	255	2,551
	1 (Never joined a gang, 1–4 close friends in a gang)	110	560
	2 (Never joined a gang, 5 or more friends in a gang)	151	636
	3 (Inactive gang member)	271	959
	4 (Active gang member, no position of rank)	175	513
	5 (Active gang member, holds position of rank)	476	831

Source: Knox, G.W., et al. "A gang classification system for corrections." *Journal of Gang Research,* Vol. 4, No. 2, Winter 1997, p. 54 (Table 4).

Problem 2.40

2.5 Exploring the Relationship between Two Quantitative Variables: Scatterplots

In situations where two quantitative variables are measured on the experimental unit, it may be important to examine the nature of the relationship between the two variables. For example, a realtor may be interested in the relationship between the asking price and sale price of a home; an educator may be interested in relating a student's SAT score and high school GPA; a psychologist might want to know whether or not a person's annual salary is related to his or her IQ. In this section, we show how to graphically describe the relationship between two quantitative variables using a simple two-dimensional graph, called a *scatterplot.*

EXAMPLE 2.7 **CONSTRUCTING A SCATTERPLOT**

Annually, the Federal Trade Commission (FTC) collects data on domestic cigarette brands. Smoking machines are used to "smoke" cigarettes to a certain length, and the residual "dry" particulate matter is tested for the amounts of various hazardous substances. The variables of interest to the FTC are carbon monoxide (CO) content and amount of nicotine, both measured in milligrams. Data collected on these two variables for 20 cigarette brands are listed in Table 2.12.

SMOKING

TABLE 2.12 Carbon Monoxide—Nicotine Data for Example 2.7

Brand	CO	Nicotine	Brand	CO	Nicotine
1	15	0.9	11	6	0.5
2	6	0.4	12	22	1.2
3	13	1.3	13	15	0.8
4	12	0.8	14	17	1.3
5	12	1.2	15	13	1.0
6	9	0.7	16	11	0.7
7	13	1.0	17	14	1.1
8	16	1.1	18	14	1.4
9	13	1.1	19	12	0.9
10	12	0.9	20	8	0.4

a. Construct a scatterplot for the data. **b.** Interpret the graph.

Solution

a. A **scatterplot** is a two-dimensional graph, with a vertical axis and a horizontal axis as shown in the Excel graph, Figure 2.15. The values of one of the

E **Figure 2.15** Excel Scatterplot of CO Ranking versus Nicotine Content for Example 2.7 (*Note:* Two points are hidden beneath others on the graph.)

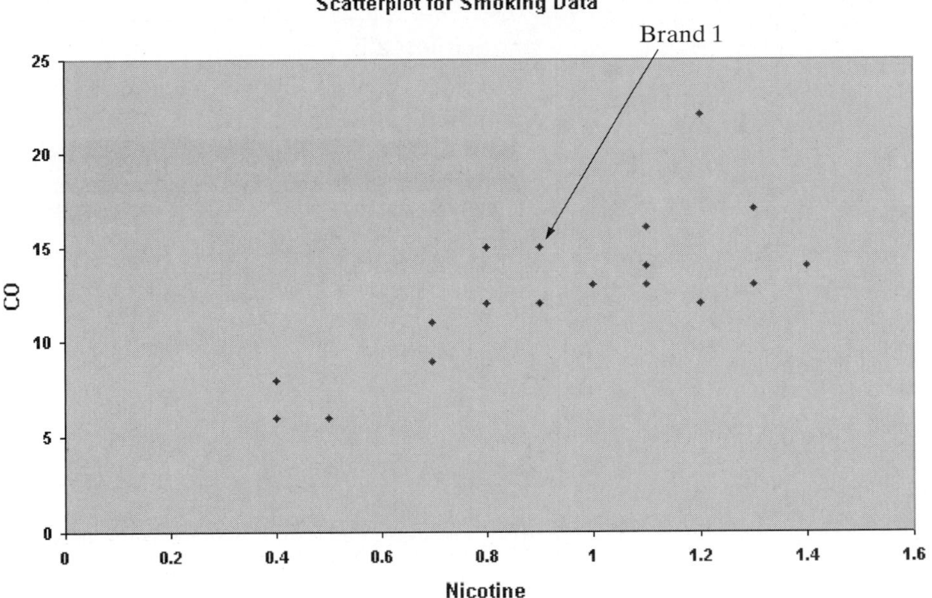

quantitative variables of interest are located on the vertical axis while the other variable's values are located on the horizontal axis of the graph. Note that we have selected (arbitrarily) to locate CO content on the vertical axis and nicotine amount on the horizontal axis. Each of the 20 pairs of observations on CO and nicotine listed in Table 2.12 are in Figure 2.15. (For example, the point corresponding to a CO content of 15 mg and a nicotine amount of .9 mg for brand 1 is identified on the scatterplot.)

b. Scatterplots often reveal a pattern or trend that clearly indicates how the quantitative variables are related. From Figure 2.15 you can see that small values of CO content tend to be associated with small nicotine amounts, and large values of CO tend to be associated with large nicotine amounts. Stated another way, CO content tends to *increase* as nicotine amount *increases*. Since the two variables tend to move in the same direction, we say that there is a **positive association** between CO content and nicotine amount.

In contrast, a **negative association** occurs between two variables when one tends to *decrease* as the other *increases.* If no distinct pattern is revealed by the scatterplot, we say there is *little or no association* between the two variables. That is, as one variable increases (or decreases), there is no definite trend in the values of the other variable. Graphs showing each of these three types of association are displayed in Figure 2.16.

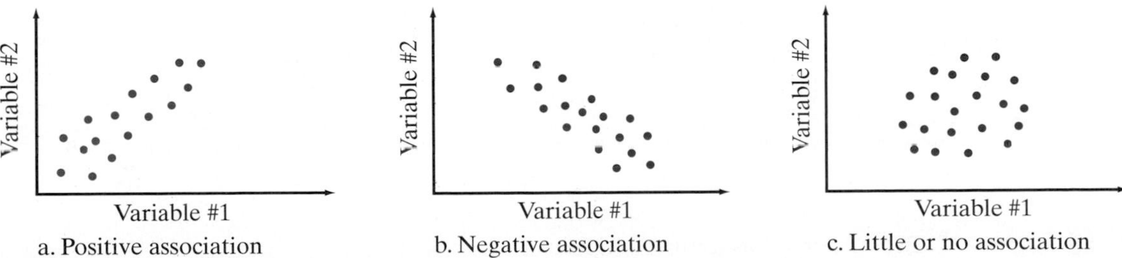

Figure 2.16 Scatterplots for Three Data Sets

Self-Test 2.5

Consider the data on two quantitative variables for $n = 12$ experimental units shown in Table 2.13.

TABLE 2.13 Data for Two Quantitative Variables

Experimental Unit	Variable #1	Variable #2
1	−1	7
2	0	8
3	1	5
4	2	2
5	−3	10
6	1	6
7	−2	8
8	4	1
9	5	0
10	3	0
11	0	7
12	2	5

Construct a scatterplot for the data. What type of relationship is revealed?

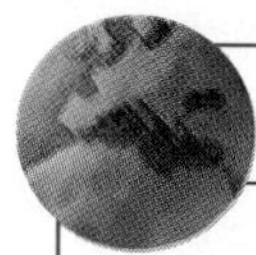

Statistics in the Real World Revisited

Interpreting Scatterplots

Consider the data on 98 automobiles crash tested for the National Highway Traffic Safety Administration's (NHTSA) New Car Assessment Program. Several quantitative variables measured and published in the report are: (1) driver's severity of head injury score (DRIVHEAD), (2) car weight in pounds (WEIGHT), (3) driver's rate of chest deceleration score (DRIVCHEST), and (4) overall driver injury rating (DRIVSTAR). [*Note:* For this example, we treat driver star rating as a quantitative variable.] The NHTSA wants to examine the relationship that driver's severity of head injury score has with each of the other three quantitative variables. We can accomplish this by using scatterplots. Figure 2.17 shows three Excel scatterplots for the NCAP data: (a) DRIVHEAD vs. WEIGHT, (b) DRIVHEAD vs. DRIVCHEST, and (c) DRIVHEAD vs. DRIVSTAR.

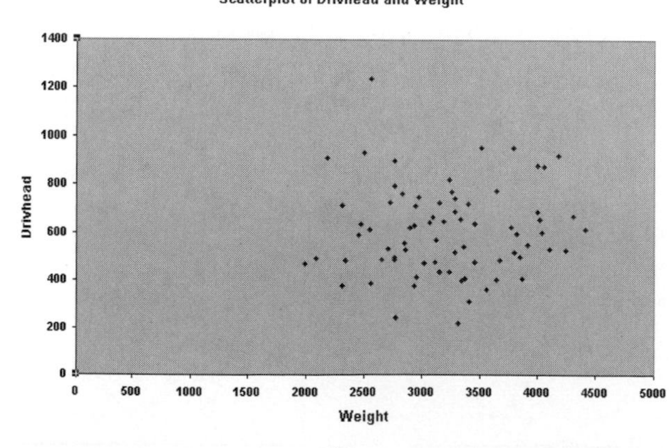

(a)

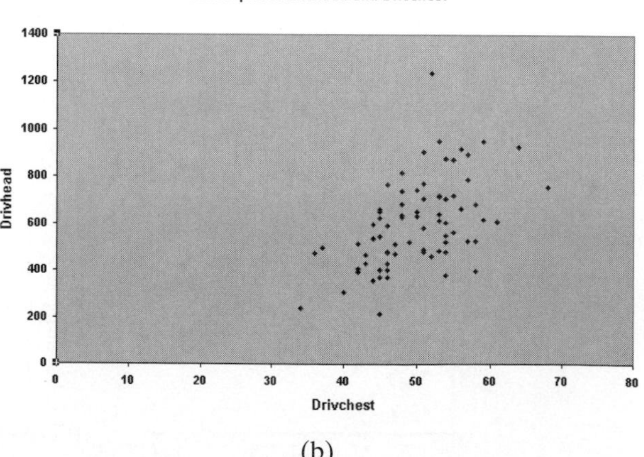

(b)

Figure 2.17 Excel Scatterplots for the NCAP Data

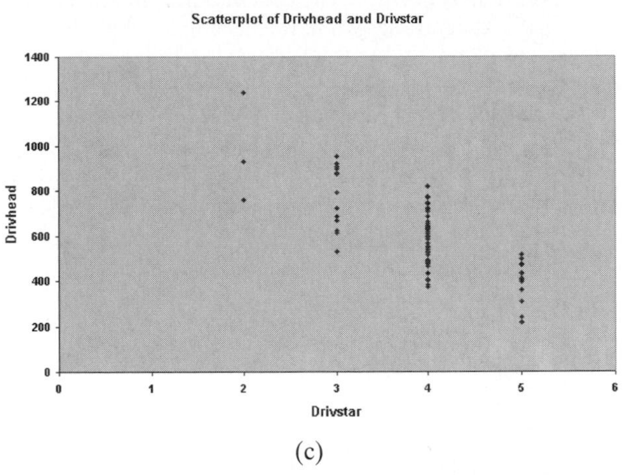

(c)

Figure 2.17a shows little or no association between driver's head injury score and car weight. Figure 2.17b reveals a fairly strong positive association between severity of driver's head injury and chest deceleration. In contrast, Figure 2.17c shows a negative relationship between the driver's head injury score and overall injury rating; the higher the head injury severity score, the lower the overall injury rating.

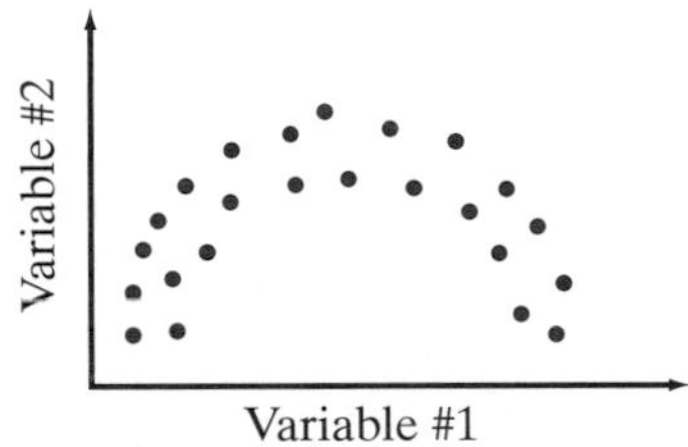

Figure 2.18 A Curvilinear Relationship

Comment: The positive and negative relationships displayed in Figures 2.17b and c show a general straight-line (or *linear*) relationship between the quantitative variables plotted. A scatterplot may reveal a nonlinear (or *curvilinear*) relationship similar to the one shown in Figure 2.18. In this case, we cannot label the relationship as positive or as negative. Later in this text (Chapter 10), we develop mathematical "models" for the different types of relationships that may exist between two quantitative variables.

PROBLEMS FOR SECTION 2.5

Using the Tools

2.41 Construct a scatterplot for each of the following sets of data. Determine the nature of the relationship between the two variables.

(a)

Variable #1	−1	0	1	2	3	4	5
Variable #2	−1	1	2	4	5	8	9

(b)

Variable #1	−1	0	1	2	3	4	5
Variable #2	4	1	5	2	1	7	3

(c)

Variable #1	−1	0	1	2	3	4	5
Variable #2	11	8	7	3	1	0	−2

2.42 Two quantitative variables are measured for each of 11 experimental units. The data are shown in the accompanying table.

Variable #1	7	5	8	3	6	10	12	4	9	15	18
Variable #2	21	15	24	9	18	30	36	12	27	45	54

a. Graph the data using a scatterplot.

b. Do you detect a relationship between the variables? Explain.

2.43 Two quantitative variables are measured for each of 10 experimental units. The data are shown in the accompanying table.

Variable #1	100	200	250	400	350	300	450	500	100	250
Variable #2	6	20	10	4	12	30	35	15	18	5

a. Graph the data using a scatterplot.

b. Do you detect a relationship between the variables? Explain.

2.44 For each scatterplot, determine the nature of the relationship (positive, negative, or none) between the two variables plotted.

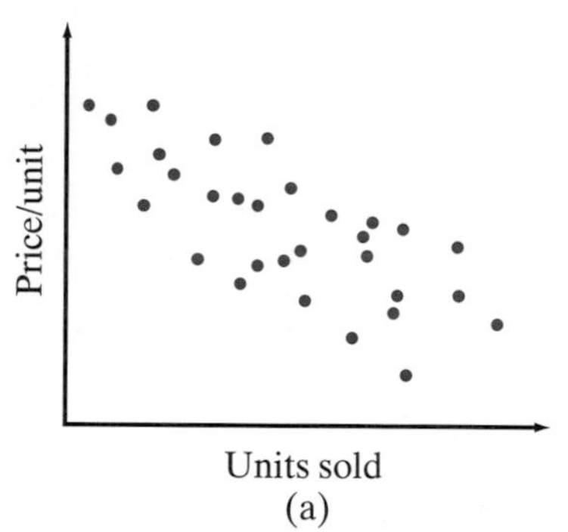

(a)

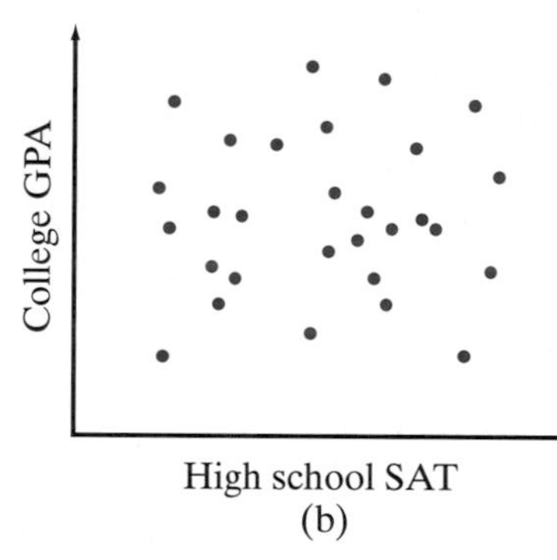

(b)

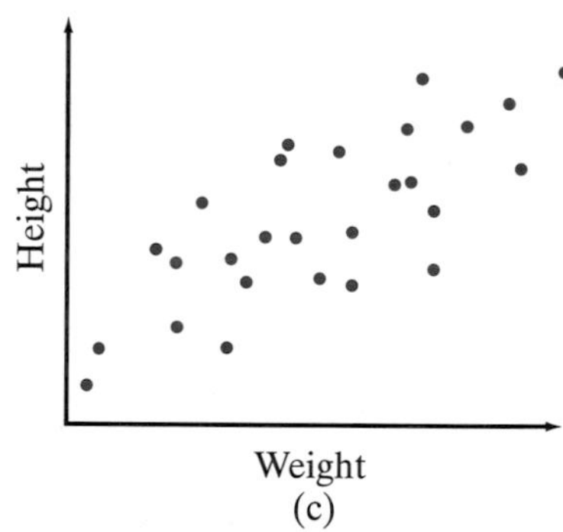

(c)

Applying the Concepts

2.45 The following data represent the approximate retail price (in \$) and the energy cost per year (in \$) of 10 medium-size top-freezer refrigerators, as determined by *Consumer Reports* (Feb. 1999).

Brand	Price ($)	Energy Cost per Year ($)
Maytag MTB2156B	850	48
Amana TR21V2	760	54
Kenmore(Sears) 7820	900	58
Whirlpool Gold GT22DXXG	870	66
GEProfile TBX22PRY	1100	77
KitchenAid Prestige KTRP22KG	800	66
Whirlpool ET21PKXG	650	70
GE TBX22ZIB	750	81
FrigidaireGallery FRT20NGC	750	72
Hotpoint CTX21DIB	570	78

Source: "Cold storage," *Consumer Reports,* Feb. 1999, p. 49.

a. With energy cost on the horizontal axis and price on the vertical axis, set up a scatterplot for the data.

b. Does there appear to be a relationship between price and energy cost? If so, is the relationship positive or negative?

c. Would you expect the higher-priced refrigerators to have greater energy efficiency? Is this borne out by the data?

2.46 Two processes for hydraulic drilling of rock are dry drilling and wet drilling. In a dry hole, compressed air is forced down the drill rods in order to flush the cuttings and drive the hammer; in a wet hole, water is forced down. An experiment was conducted to determine whether the time it takes to dry drill a distance of five feet in rock increases with depth (*The American Statistician,* Feb. 1991). The results for one portion of the experiment are shown in the accompanying table.

DRILLROCK

Depth at Which Drilling Begins (Feet)	Time to Drill 5 Feet (Minutes)
0	4.90
25	7.41
50	6.19
75	5.57
100	5.17
125	6.89
150	7.05
175	7.11
200	6.19
225	8.28
250	4.84
275	8.29
300	8.91
325	8.54
350	11.79
375	12.12
395	11.02

Source: Penner, R., and Watts, D. G. "Mining information." *The American Statistician,* Vol. 45, No. 1, Feb. 1991, p. 6 (Table 1).

Number of Calories	Fat (%)
270	18
150	8
170	10
140	8
160	9
150	9
160	9
290	17
170	9
190	10
190	10
160	8
170	9
150	8
150	8
110	3
180	9

Source: "Cold storage" *Consumer Reports,* Feb. 1999, p. 49.

Problem 2.47

a. With depth on the horizontal axis and drill time on the vertical axis, construct a scatterplot for the data.

b. Identify the nature of the relationship between the two variables.

2.47 The data at left represent the number of calories and the percentage of fat per half-cup serving of 17 chocolate ice cream brands, as determined by *Consumer Reports* (Feb. 1999).

ICECREAM

a. With fat content on the horizontal axis and calories on the vertical axis, construct a scatterplot for the data.

b. Does there appear to be a relationship between calories and fat content? If so, is the relationship positive or negative?

c. Would you expect the ice cream brands with higher fat content to have more calories? Is this borne out by the data?

2.48 Neurologists have found that the hippocampus, a structure of the brain, plays an important role in short-term memory. Research published in the *American Journal of Psychiatry* (July 1995) attempted to establish a link between hippocampal volume and short-term verbal memory of patients with combat-related posttraumatic stress disorder (PTSD). A sample of 21 Vietnam veterans with a history of

combat-related PTSD participated in the study. Magnetic resonance imaging was used to measure the volume of the right hippocampus (in cubic millimeters) of each subject. The verbal memory retention of each subject was also measured by the percent retention subscale of the Wechsler Memory Scale. The data for the 21 patients is plotted in the scatterplot. The researchers "hypothesized that smaller hippocampal volume would be associated with deficits in short-term verbal memory in patients with PTSD." Does the scatterplot provide visual evidence to support this theory?

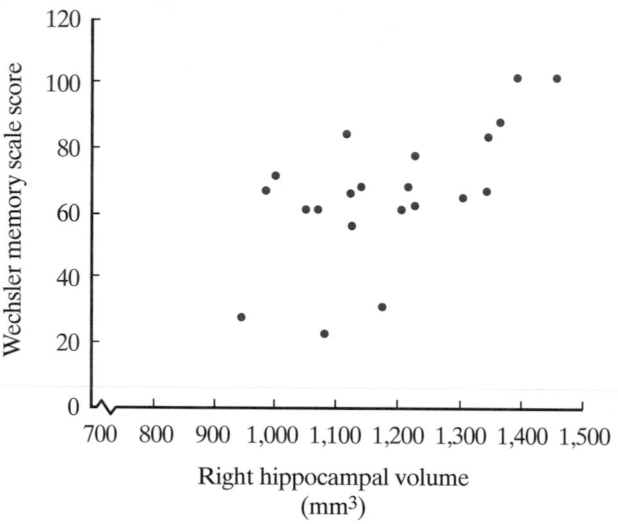

Problem 2.48

2.49 Since 1958, air samples have been collected hourly at Mauna Loa Observatory, Hawaii, for the purpose of determining the carbon dioxide (CO_2) concentration. The Mauna Loa atmospheric data constitute the longest continuous record of atmospheric CO_2 concentrations in the world. Since local influences of vegetation or human activity are minimal at Mauna Loa, these data are considered by experts to be a reliable indicator of the trend in atmospheric CO_2 concentrations in the middle layers of the troposphere (Keeling, C. D., and Whorf, T. P., Scripps Institution of Oceanography, Aug. 2000). The data in the table are the average annual CO_2 measurements (in parts per million) during April at Mauna Loa from 1981 to 1999.

Year	CO_2	Year	CO_2
1981	340.6	1991	358.5
1982	342.3	1992	359.1
1983	343.4	1993	359.4
1984	344.8	1994	361.2
1985	348.2	1995	363.5
1986	349.4	1996	364.8
1987	350.8	1997	366.4
1988	353.4	1998	368.7
1989	355.3	1999	371.2
1990	356.0		

Source: Keeling, C. D., and Whorf, T. P. "Atmospheric CO_2 records from sites in the SIO air sampling network." In *Trends: A Compendium of Data on Global Change.* Carbon Dioxide Information Analysis Center, Oak Ridge National Laboratory, U.S. Department of Energy, Oak Ridge, Tenn., Aug. 2000.

a. Construct a scatterplot for the data, with year on the horizontal axis and CO_2 concentration on the vertical axis.

b. Do you detect a trend in atmospheric CO_2 concentrations at Mauna Loa? If so, describe the nature of the relationship.

2.50 On January 28, 1986, the space shuttle *Challenger* exploded and seven American astronauts died. Experts agreed that the disaster was caused by two leaky rubber O-rings. Prior to the flight, some rocket engineers had theorized that the O-rings would not seal properly due to cold weather. (The predicted temperature for the launch was 26°F to 29°F.) The engineers based their theory on data collected on O-ring damage and temperature for all 23 previous launches of the space shuttle. These data are given in the accompanying table. [*Note:* O-ring damage index is a measure of the total number of incidents of O-ring erosion, heating, and blow-by.]

a. Graph the data in a scatterplot. Do you detect a trend?

b. Based on the graph, would you have permitted *Challenger* to launch on the date of the explosion? Explain.

ORING

Flight Number	Temperature (°F)	O-Ring Damage Index
1	66	0
2	70	4
3	69	0
5	68	0
6	67	0
7	72	0
8	73	0
9	70	0
41-B	57	4
41-C	63	2
41-D	70	4
41-G	78	0
51-A	67	0
51-B	75	0
51-C	53	11
51-D	67	0
51-F	81	0
51-G	70	0
51-I	67	0
51-J	79	0
61-A	75	4
61-B	76	0
61-C	58	4

Note: Data for flight number 4 is omitted due to an unknown O-ring condition.

Primary Sources: Report of the Presidential Commission on the Space Shuttle Challenger Accident, Washington, D.C., 1986, Vol. II (pp. H1–H3) and Volume IV (p. 664). *Post-Challenger Evaluation of Space Shuttle Risk Assessment and Management,* Washington, D.C., 1988, pp. 135–136.

Secondary Source: Tufte, E. R., *Visual and Statistical Thinking: Displays of Evidence for Making Decisions.* Graphics Press, 1997.

2.51 Before a drug may be released for use, a pharmaceutical company is required to conduct a chemical assay to determine if the drug's potency conforms to specifications. An assay method called High Performance Liquid Chromatography (HPLC) was investigated for accuracy in *Statistical Case Studies: A Collaboration between Academe and Industry* (SIAM, ASA, 1998). Ten specimens of a drug compound designed to treat organ rejection after transplants were prepared independently. The potency

level added to the compound—measured as a percentage of label strength (LS)—was varied for each specimen. The HPLC method was then applied to each specimen to determine the estimated (found) potency level. The assay data are listed below.

HPLC

Specimen	% LS Added	% LS Found	% Recovery
1	70	69.80	99.72
2	70	70.62	100.88
3	85	85.70	100.82
4	85	85.43	100.50
5	100	100.47	100.47
6	100	101.05	101.05
7	115	117.13	101.85
8	115	116.66	101.44
9	130	131.59	101.22
10	130	132.51	101.93

a. Construct a scatterplot for the variables % LS Found and % LS Added. What type of relationship do you detect? Use the scatterplot to make an inference about the accuracy of the HPLC method.

b. A variable found to be more useful in practice than percentage of label strength found (% LS Found) is the amount of the drug recovered as a percentage of the amount added to the compound (% Recovery). If the method is accurate, then % Recovery will be 100% regardless of the amount added. Construct a scatterplot for the variables % Recovery and % LS Added. Do you detect a trend? Explain why the scatterplot supports the inference of part **a**.

2.6 Proper Graphical Presentation

According to the Chinese sage Confucius, "a picture is worth a thousand words." The previous sections demonstrated that well-designed graphical displays are powerful tools for presenting data. However, these pictures (e.g., bar graphs, pie charts, histograms, scatterplots) are susceptible to distortion, whether unintentional or as a result of unethical practices. The proliferation of graphics software and spreadsheet applications in recent years has compounded the problem. With a simple "point and click," almost anyone can generate a graph. But if the graph is not presented clearly and carefully, the true message given by the data can become clouded and distorted.

In this section, we present some of the ideas of Professor Edward R. Tufte, who has written a series of books devoted to proper methods of graphical design. We start with Tufte's five principles of **graphical excellence,*** listed in the box.

*Tufte, E. R. *Visual Display of Quantitative Information.* Cheshire, Conn.: Graphics Press, 1983.

PRINCIPLES OF GRAPHICAL EXCELLENCE

1. Graphical excellence is a well-designed presentation of data that provides substance, statistics, and design.
2. Graphical excellence communicates complex ideas with clarity, precision, and efficiency.
3. Graphical excellence gives the viewer the largest number of ideas in the shortest time with the least ink.
4. Graphical excellence almost always involves several dimensions.
5. Graphical excellence requires telling the truth about the data.

The excellence of a graph is usually compromised when the data are not presented clearly or when the data's message is distorted. To present clear graphs, Tufte suggests maximizing *data-ink* and minimizing *chart junk.* The following examples illustrate these ideas.

Definition 2.5

Data-ink is the amount of ink used in a graph that is devoted to non-redundant display of the data.

Definition 2.6

Chart junk is decoration in a graph that is non-data-ink.

EXAMPLE 2.8

CHART JUNK

The graph shown in Figure 2.19 was published in *Newsday,* a New York daily newspaper. Figure 2.19 employs a form of scatterplot to show the trend in annual milk prices in the New York City metropolitan area from 1987 to 1997. Comment on the graphical excellence of Figure 2.19.

Figure 2.19 Plot of New York City Milk Prices. *Source:* "Pricing Milk" by Steve Madden, *Newsday,* Feb. 8, 1998. Reprinted with permission © Newsday, Inc., 1998.

Solution

In this graph, milk prices per gallon are plotted on the vertical axis and years (1987 through 1997) are plotted on the horizontal axis. Note that the years are not clearly marked. (Each vertical grid line represents a different year.) Also, two different milk prices are being plotted. The points at the top of the graph (connected by lines) represent annual price per gallon at New York City supermarkets.

The point of the graph is to compare the selling prices of milk at supermarkets to the selling prices established by dairy farmers. The graph, however, suffers from an overabundance of chart junk. The amount of ink used to display the decorative cow is too large relative to the data-ink. (Tufte refers to this type of graph as "The Duck." The chart junk shown in Figure 2.19 is, of course, a variation of "The Duck"—we would call it "A Cow.")

Avoiding distortions of data is a central feature of graphical excellence. A graph does not distort if its visual representation is consistent with its numerical representation. One way to measure the amount of distortion in a graph is with the *lie factor*.

Definition 2.7

The **lie factor** in a graph is the ratio of the size of the effect shown in the graph to the actual size of the effect in the data.

$$\text{Lie factor} = \frac{\text{Size of effect shown in graph}}{\text{Size of effect in the data}}$$

EXAMPLE 2.9

LIE FACTOR

Consider Figure 2.20, a graph extracted from the *New York Times.* The figure is a horizontal bar graph showing the oyster catch (in millions of bushels) in Chesapeake Bay for six time periods: 1890–1899, 1930–1939, 1962, 1972, 1982, and 1992. Comment on the graphical excellence of Figure 2.20. Compute the lie factor to support your observation.

Solution

Figure 2.20 uses oyster icons rather than rectangles to represent the "bars" in the bar graph. According to the numerical values shown at the end of the icon, 20 million bushels of oysters were caught during the 1890s and 4 million bushels were caught in 1962. Thus, according to the data, five times as many oysters were caught in Chesapeake Bay during the 1890s than during 1962 (i.e., an actual effect size of 5). The oyster icons, however, do not appear to be sized proportionally. In fact the 1890s icon, measuring lengthwise, is only about twice the size of the 1962 icon (i.e., an effect size of 2). Consequently, the lie factor is

$$\text{Lie factor} = \frac{\text{Visual ratio of 1890s catch to 1962 catch}}{\text{Actual ratio of 1890s catch to 1962 catch}} = \frac{2}{5} = .4$$

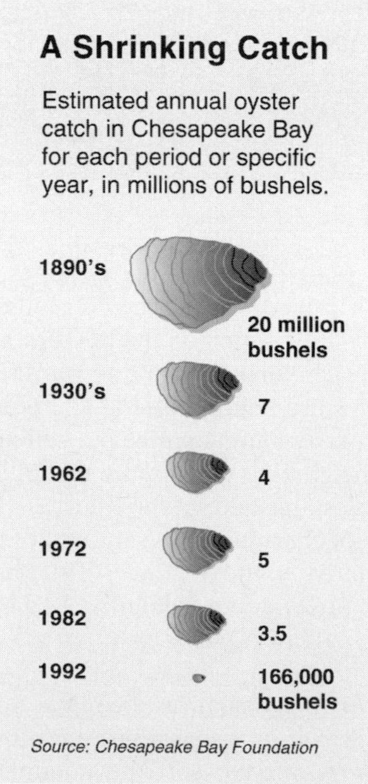

Figure 2.20 Oyster Catch in Chesapeake Bay for Six Time Periods. *Source: The New York Times,* October 17, 1993, p. 26. Reprinted with permission.

In other words, the decrease in the oyster catch from the 1890s to 1962 shown on the graph is only about 40% of the actual decrease conveyed by the data. The oyster icons shown in Figure 2.20 could also be considered chart junk, resulting in distortion of the visual impact of the declining oyster catch in Chesapeake Bay.

Self-Test 2.6

Figure 2.21 is a relative frequency bar graph for data collected on book reviews in American history. Identify one improper or misleading feature of the graph.

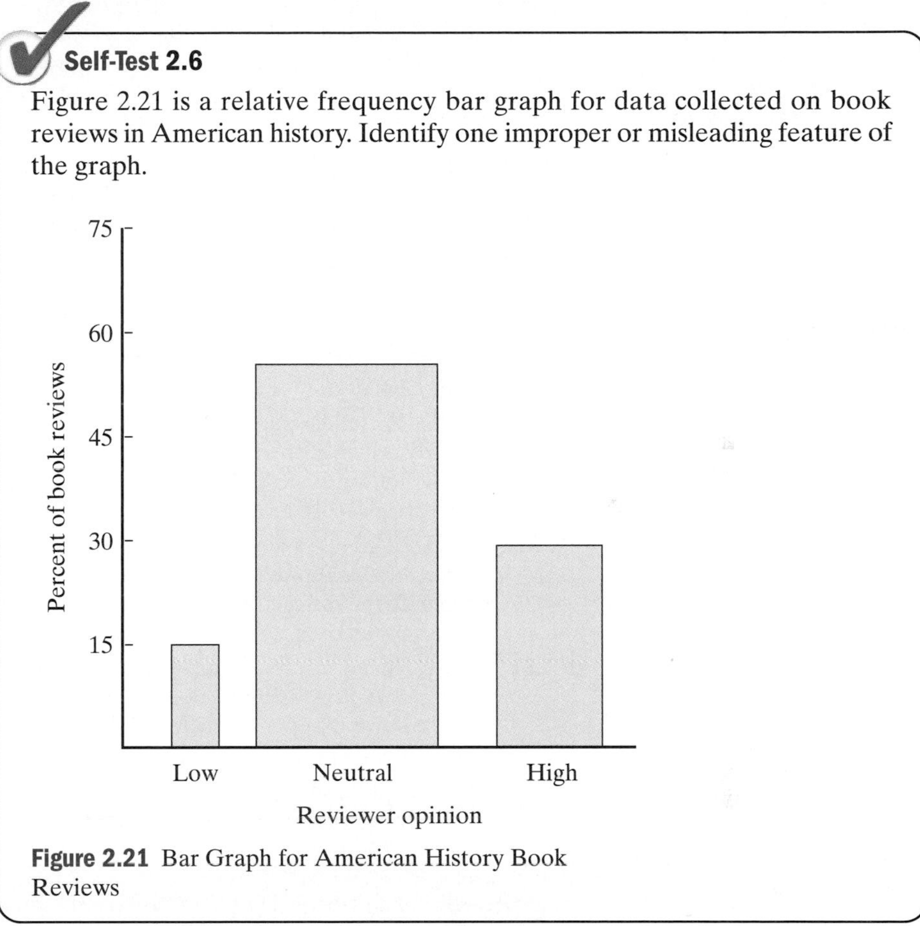

Figure 2.21 Bar Graph for American History Book Reviews

In addition to chart junk, another common way to distort a graph is to change the scale on the vertical axis, the horizontal axis, or both. This occurs most often when the zero point is not displayed or when a broken axis line—called a **scale break**—is used.

EXAMPLE 2.10

SCALE BREAK

Refer to the Chesapeake Bay oyster catch data, Example 2.9. Figure 2.22 is a bar graph comparing the 1890s catch to the 1962 catch. Identify the distortion in this graph.

Solution

Figure 2.22 uses a scale break on the vertical axis, effectively diminishing the perceived decline in number of oysters caught. We know from the data (displayed in Figure 2.20) that 20 million bushels were caught during the 1890s and 4 million were caught in 1962, a difference of 16 million bushels of oysters. Due to the scale break, however, the bar height for the 1890s does not appear to be 16 million bushels more than the bar height for 1962.

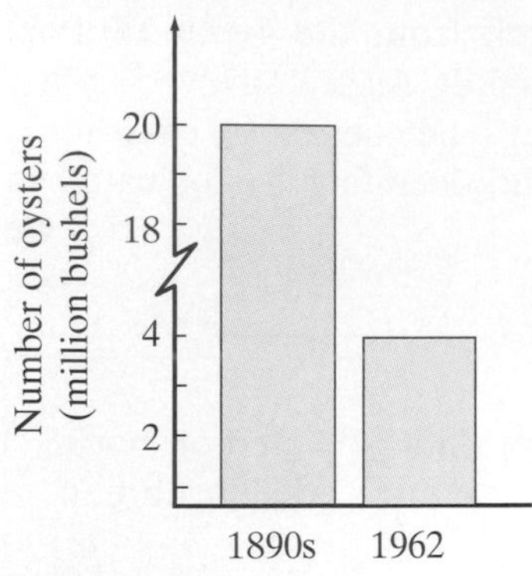

Figure 2.22 Bar Graph for Chesapeake Bay Oyster Catch: 1890s and 1962

According to Tufte, for many people the first word that comes to mind when they think about statistical charts is "lie." Too many graphics distort the underlying data, making it hard for the reader to learn the truth. Ethical considerations arise when we are deciding what data to present in tabular and chart format and what not to present. It is vitally important when presenting data to document both good and bad results. When making oral presentations and presenting written reports, it is essential that the results be given in a fair, objective, and neutral manner. Thus, we must try to distinguish between poor data presentation and unethical presentation. Again, as in our discussion of ethical considerations in data collection (Section 1.7), the key is *intent.* Often, when fancy tables and chart junk are presented or pertinent information is omitted, it is simply done out of ignorance. However, unethical behavior occurs when an individual willfully hides the facts by distorting a table or chart or by failing to report pertinent findings.

PROBLEMS FOR SECTION 2.6

Applying the Concepts

2.52 Bring to class a graphic from a newspaper or magazine that you believe to be a poorly drawn representation of some *quantitative* variable. Be prepared to submit the graph to the instructor with comments as to why you feel it is inappropriate. Also, be prepared to present and comment on this in class.

2.53 Bring to class a graphic from a newspaper or magazine that you believe to be a poorly drawn representation of some *qualitative* variable. Be prepared to submit the graph to the instructor with comments as to why you feel it is inappropriate. Also, be prepared to present and comment on this in class.

2.54 Bring to class a graphic from a newspaper or magazine that you believe contains too much *chart junk* that may cloud the message given by the data. Be prepared to submit the graph to the instructor with comments as to why you feel it is inappropriate. Also, be prepared to present and comment on this in class.

2.55 Comment on the graphical excellence of the pie charts illustrated in the following graphic extracted from the newspaper.

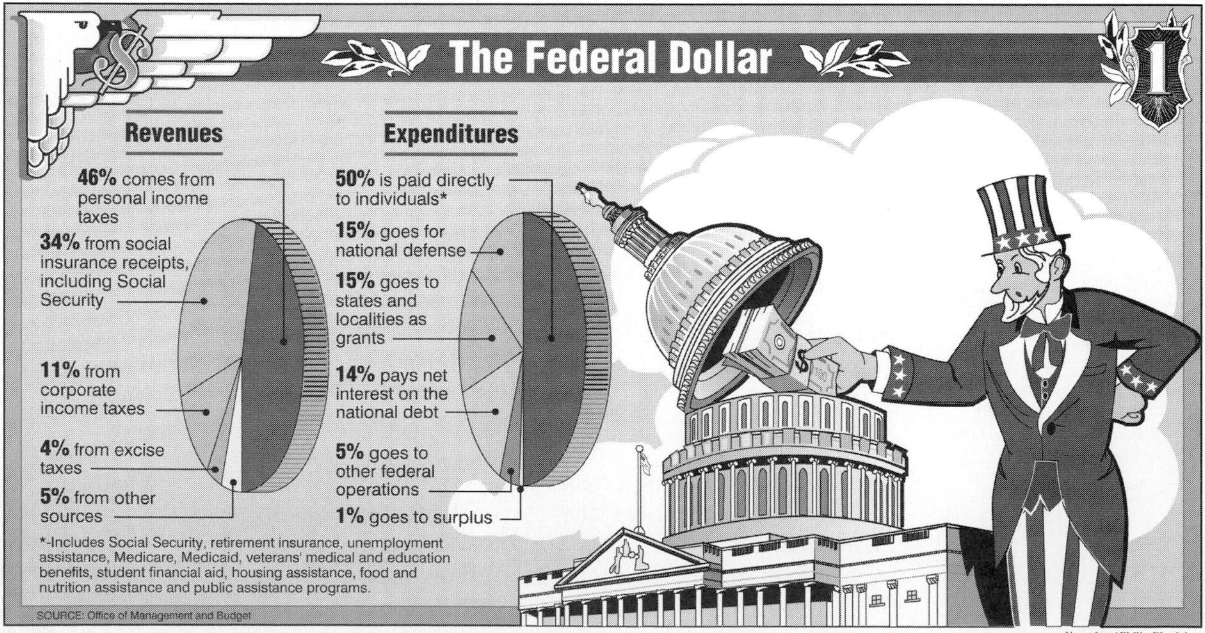

Source: "The Federal Dollar" by Philip Dionisio, *Newsday,* Feb. 3, 1998. Reprinted with permission © Newsday, Inc., 1998.

2.56 The visual display below is based on one that appeared in *USA Today* (Feb. 2000) and permits a comparison of the relative size of police departments in major U.S. cities.

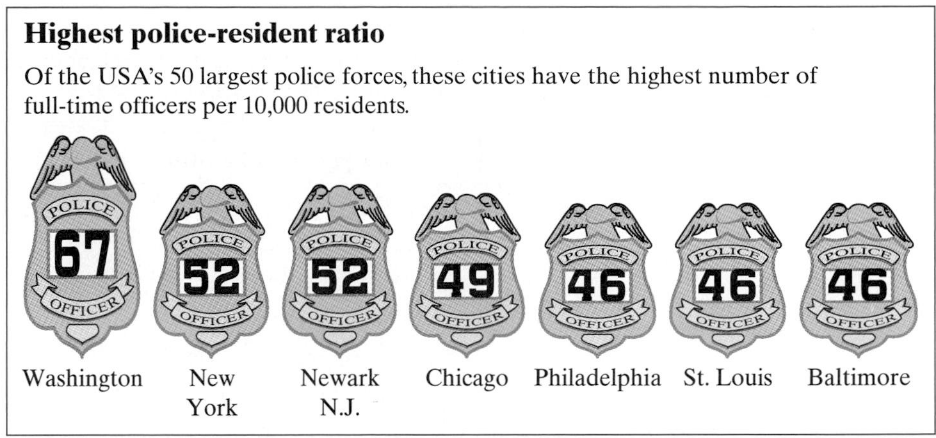

Source: Adapted from *USA Today,* Feb. 29, 2000.

a. Indicate a feature of this chart that violates Tufte's principles of graphical excellence.

b. Set up an alternative graph for the data provided in this figure. Interpret the results.

2.57 The following visual display contains an overembellished chart that is based on one that appeared in *USA Today* (April 6, 1999) dealing with what Americans do with produce that gets overly ripe before they are ready to eat it.

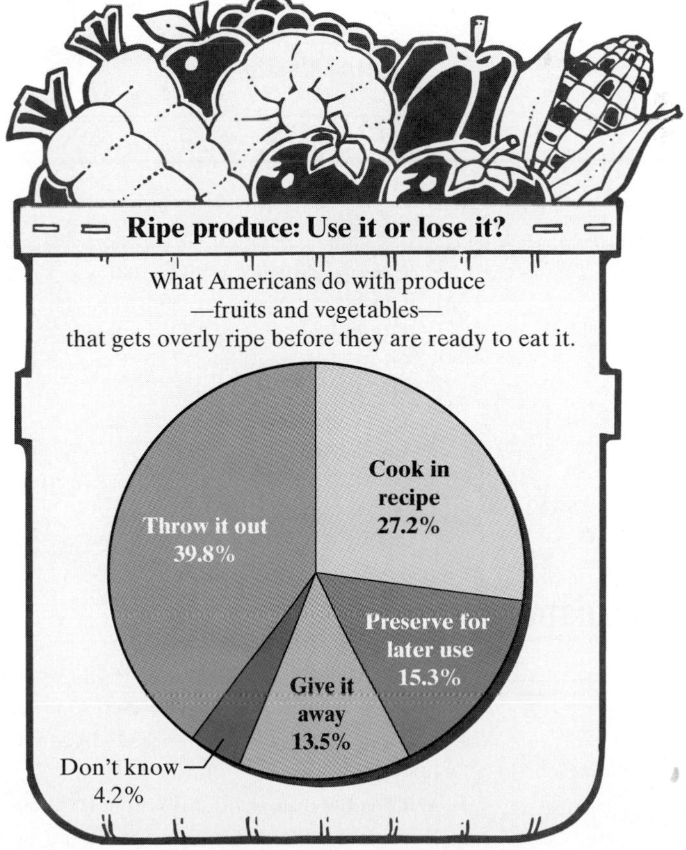

Source: Adapted from *USA Today,* April 6, 1999.

a. Describe at least one good feature of this visual display.
b. Describe at least one bad feature of this visual display.
c. Redraw the graph using the principles of graphical excellence. Interpret the results.

2.58 Comment on the graphical excellence of the bar graph illustrated in the following graphic extracted from *Self* (Oct. 1992).

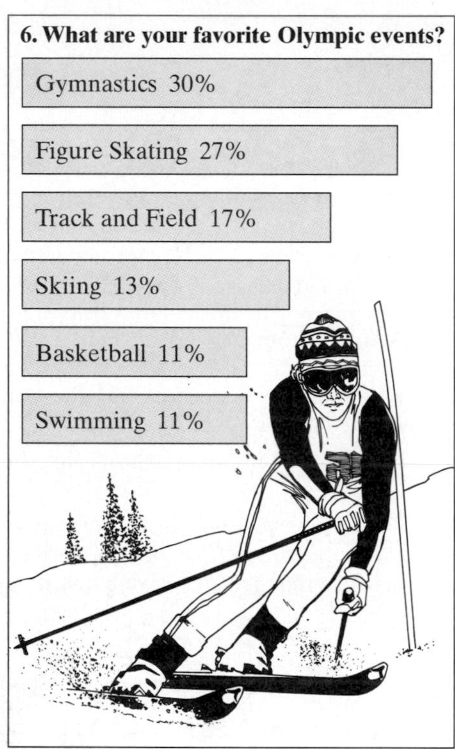

Data Source: Self, October 1992.

KEY TERMS

Bar graph 60
Chart junk 99
Class 59
Class boundary 71
Class frequency 59
Class interval 70
Class interval width 71
Class percentage 59
Class relative frequency 59
Cross-classification table 80
Data-ink 99
Dot plot 68
Graphical excellence 98
Histogram 68
Left (negatively) skewed 72
Lie factor 100
Negative association 91
Nonskewed (symmetric) 72
Pie chart 60
Positive association 91
Right (positively) skewed 72
Scale break 101
Scatterplot 90
Side-by-side horizontal bar chart 81
Stem-and-leaf display 68
Symmetric 72

CHECK YOUR UNDERSTANDING

1. What graphical methods are used to describe a single qualitative variable?
2. What graphical methods are used to describe a single quantitative variable?
3. What are the main differences between a dot plot, a stem-and-leaf display, and a histogram?
4. What is the difference between a frequency and a relative frequency table?
5. What is the difference between a bar graph and a pie chart?

6. Compare and contrast the bar chart for qualitative data with the histogram for quantitative data.
7. What graphical method is used to describe the relationship between two qualitative variables?
8. What graphical method is used to describe the relationship between two quantitative variables?
9. What is a cross-classification table?
10. What are some of the ethical issues to be concerned with when presenting data in tables or graphs?

SUPPLEMENTARY PROBLEMS

Applying the Concepts

2.59 The International Rhino Federation estimates that there are 13,585 rhinoceroses living in the wild in Africa and Asia. A breakdown of the number of rhinos of each species is reported in the accompanying table.

a. Construct a relative frequency table for the data.

b. Display the relative frequencies in a bar graph.

c. What proportion of the 13,585 rhinos are African rhinos? Asian?

Rhino Species	Population Estimate
African Black	2,600
African White	8,465
(Asian) Sumatran	400
(Asian) Javan	70
(Asian) Indian	2,050
TOTAL	13,585

Source: International Rhino Federation, July 1998.

2.60 The data on p. 106 indicate fat and cholesterol information concerning popular protein foods (fresh red meats, poultry, and fish). For the data relating to the amount of calories, protein, the percentage of calories from fat and saturated fat, and the cholesterol for the popular protein foods:

a. Construct the stem-and-leaf display.

b. Construct the frequency distribution and the percentage distribution.

c. Construct scatterplots for each pair of variables.

d. What conclusions can you reach concerning the amount of calories of these foods?

e. What conclusions can you reach concerning the amount of protein of these foods?

f. What conclusions can you reach concerning the percentage of calories from fat and saturated fat of these foods?

g. Are there any foods that seem different from the others in terms of these variables? Explain.

h. Based on part **c**, what conclusions can you reach about the relationships among these five variables?

2.61 At a supermarket, customer checkout time is defined as the total length of time required for a checker to scan and total the prices of the customer's food items, accept payment, return change, and bag the items. The data file **CHKOUT** contains the checkout times (in seconds) for 500 customers who recently shopped at a Gainesville, Florida, supermarket. (The first five checkout times and last five

PROTEIN

Food	Calories	Protein (g)	Fat (%)	Saturated Fat (%)	Cholesterol (mg)
Beef, ground, extra lean	250	25	58	23	82
Beef, ground, regular	287	23	66	26	87
Beef, round	184	28	24	12	82
Brisket	263	28	54	21	91
Flank steak	244	28	51	22	71
Lamb leg roast	191	28	38	16	89
Lamb loin chop, broiled	215	30	42	17	94
Liver, fried	217	27	36	12	482
Pork loin roast	240	27	52	18	90
Sirloin	208	30	37	15	89
Spareribs	397	29	67	27	121
Veal cutlet, fried	183	33	42	20	127
Veal rib roast	175	26	37	15	131
Chicken, with skin, roasted	239	27	51	14	88
Chicken, no skin, roasted	190	29	37	10	89
Turkey, light meat, no skin	157	30	18	6	69
Clams	98	16	6	0	39
Cod	98	22	8	1	74
Flounder	99	21	12	2	54
Mackerel	199	27	77	20	100
Ocean perch	110	23	13	3	53
Salmon	182	27	24	5	93
Scallops	112	23	8	1	56
Shrimp	116	24	15	2	156
Tuna	181	32	41	10	48

Source: United States Department of Agriculture.

Problem 2.60

CHKOUT

checkout times are shown in the table.) An Excel frequency histogram for the 500 customer checkout times is shown below.

18	37	63	6	116	...	85	10	27	20	25

a. Interpret the graph.

b. Estimate the proportion of supermarket customers that had checkout times of one minute or less.

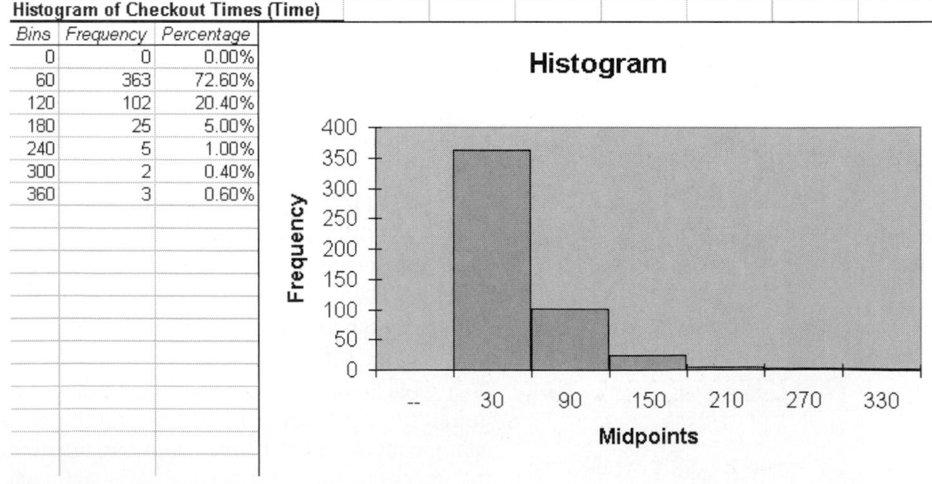

Histogram of Checkout Times (Time)

Bins	Frequency	Percentage
0	0	0.00%
60	363	72.60%
120	102	20.40%
180	25	5.00%
240	5	1.00%
300	2	0.40%
360	3	0.60%

E **Problem 2.61**

2.62 The *Journal of Agricultural, Biological, and Environmental Statistics* (Sep. 2000) published a study on the impact of the *Exxon Valdez* tanker oil spill on the seabird population in Prince William Sound, Alaska. A subset of the data analyzed is stored in the file called **EVOS.** Data were collected on 96 shoreline locations (called transects) of constant width but variable length. For each transect, the number of seabirds found is recorded as well as the length (in kilometers) of the transect and whether or not the transect was in an oiled area. (The first five and last five observations in the **EVOS** file are listed in the table.)

EVOS

Transect	Seabirds	Length	Oil
1	0	4.06	No
2	0	6.51	No
3	54	6.76	No
4	0	4.26	No
5	14	3.59	No
.	.	.	.
.	.	.	.
.	.	.	.
92	7	3.40	Yes
93	4	6.67	Yes
94	0	3.29	Yes
95	0	6.22	Yes
96	27	8.94	Yes

Source: McDonald, T. L., Erickson, W. P., and McDonald, L. L. "Analysis of count data from before–after control-impact studies." *Journal of Agricultural, Biological, and Environmental Statistics*, Vol. 5, No. 3, Sep. 2000, pp. 277–8 (Table A.1).

a. Identify the variables measured as quantitative or qualitative.

b. Identify the experimental unit.

c. Use a pie chart to describe the percentage of transects in oiled and unoiled areas.

d. Use a graphical method to examine the relationship between observed number of seabirds and transect length.

e. The researchers are interested in studying the observed seabird densities of the transects. Observed density is defined as the observed count divided by the length of the transect. Use Excel or MINITAB to calculate the observed seabird densities.

f. Use the methods of this chapter to assess whether the distribution of seabird densities differs for transects in oiled and unoiled areas.

2.63 Refer to the *Journal of Peace Research* (Aug. 1997) study of 36 active armed conflicts, Problem 2.8 (p. 65). Recall that an armed conflict is classified as "minor" (less

	Conflict Type		
	Minor	Intermediate	War
Europe	0	0	1
Middle East	1	3	1
Asia	7	5	2
Africa	9	3	2
America	0	2	0
TOTALS	17	13	6

Source: Wallensteen, P., and Sollenberg, M. "Armed conflicts, conflict termination, and peace agreements, 1989–96." *Journal of Peace Research,* Vol. 34, No. 3, Aug. 1997, p. 341 (Table II).

than 1,000 deaths overall), "intermediate" (more than 1,000 deaths overall, but less than 1,000 in any year), or "war" (more than 1,000 deaths in any year). In addition, the researchers categorized the conflicts according to location. The 36 armed conflicts are summarized in the cross-classification table on p. 107. Construct a side-by-side bar chart for the data and interpret the graph.

2.64 *Sports Illustrated* (Dec. 8, 1997) polled 1,835 middle school and high school students from across the U.S. on how race impacts sports participation. Students were asked to agree or disagree with the statement: "African-American players have become so dominant in sports such as football and basketball that many whites feel they can't compete at the same level as blacks." The results are summarized in the accompanying Excel pie charts. Interpret the results.

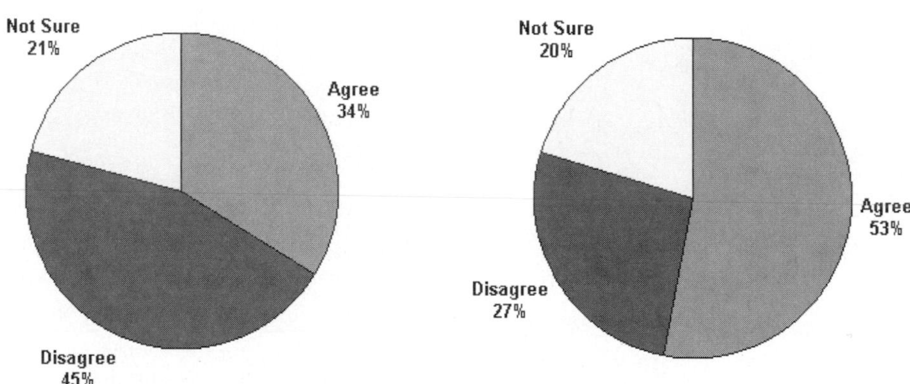

E Problem 2.64

Large Litter Type	Percentage
Paper	30
Plastic	27
Aluminum	13
Glass	6
Other	24
TOTAL	100

Problem 2.65

2.65 Researchers at the University of Florida Center for Solid and Hazardous Waste Management inspected roadside trash at four sites in each of 67 Florida counties (*Focus,* Spring 1995). Each piece of trash was classified according to type. A percentage breakdown of the large litter pieces found is provided in the table at left. Portray the data graphically with a pie chart.

2.66 Age-related macular degeneration (AMD), a progressive disease that deprives a person of central vision, is the leading cause of blindness in older adults. In a study published in the *Journal of the American Geriatrics Society* (Jan. 1998), researchers investigated the link between AMD incidence and alcohol consumption. Each individual in a national sample of 3,072 adults was asked about the most frequent

Most Frequent Alcohol Type Consumed	Total Number	Number Diagnosed with AMD*
None	965	87
Beer	430	30
Wine	1,284	51
Liquor	393	20
TOTALS	3,072	188

**Numbers based on percentages given in Figure 1.*

Source: Obisean, T. O., et al. "Moderate wine consumption is associated with decreased odds of developing age-related macular degeneration in NHANES-1." *Journal of the American Geriatrics Society,* Vol. 46, No. 1, Jan. 1998, p. 3 (Figure 1).

alcohol type (beer, wine, liquor, or none) consumed in the past year. In addition, each adult was clinically tested for AMD. The results are displayed in the table on p. 108.

a. Identify the two qualitative variables measured in this study.

b. Summarize the data on the two variables using a cross-classification table.

c. Construct a bar graph to describe the most frequent alcohol type consumed by all 3,072 adults in the study.

d. Construct a pie chart to describe the incidence of AMD among all 3,072 adults in the study.

e. Construct a side-by-side bar graph to compare the incidence of AMD among the four alcohol type response groups. Interpret the graph.

2.67 The physical layout of a warehouse must be carefully designed to prevent vehicle congestion and optimize response time. Optimal design of an automated warehouse was studied in the *Journal of Engineering for Industry* (Aug. 1993). The layout assumes that vehicles do not block each other when they travel within the warehouse, i.e., that there is no congestion. The validity of this assumption was checked by simulating warehouse operations. In each simulation, the number of vehicles was varied and the congestion time (total time one vehicle blocked another) was recorded. The data are shown in the accompanying table. Of interest to the researchers is the relationship between congestion time and number of vehicles.

WAREHOUS

Number of Vehicles	Congestion Time (Minutes)	Number of Vehicles	Congestion Time (Minutes)
1	0	9	.02
2	0	10	.04
3	.02	11	.04
4	.01	12	.04
5	.01	13	.03
6	.01	14	.04
7	.03	15	.05
8	.03		

Source: Pandit, R., and Palekar, U. S. "Response time considerations for optimal warehouse layout design." *Journal of Engineering for Industry,* Transactions of the ASME, Vol. 115, Aug. 1993, p. 326 (Table 2).

a. Construct a scatterplot for the data.

b. Identify the nature of the relationship between the two variables.

2.68 *Child Welfare* (Jan./Feb. 1992) reported on a study of 124 adoptees who recently reunited with their biological mothers and relatives. One aspect of the study dealt with how often the adoptees experienced "quest" feelings (i.e., the desire to meet or learn about their biological parents) before their reunion. The accompanying table reports

		Age Category		
		Before 10	10–17	18 or Older
Frequency of Quest Feelings	Very often, or somewhat often	19.8%	47.4%	87.5%
	Occasionally	16.5	27.8	12.5
	Almost never	47.2	22.7	0
	Uncertain	16.5	2.1	0
	TOTALS	100.0%	100.0%	100.0%

Source: Sachdev, P. "Adoption reunion and after: A study of the search process and experience of adoptees." *Child Welfare,* Vol. 71, No. 1, Jan./Feb. 1992, p. 59 (Table 2).

the frequency of quest feelings experienced by the 124 adoptees at three different ages: before age 10, between 10 and 17, and 18 or older.

a. For each category, summarize the data on frequency of quest feelings with a bar graph.

b. Compare and contrast the three graphs, part **a.** What inference can you make?

2.69 Is our behavior influenced by the different phases of the moon? Although scientific studies have found no actual link between human behavior and the phases of the moon, other studies have indicated that a significant percentage of people *believe* in lunar effects. To quantify the degree to which people believe in this phenomenon, psychologists J. Rotton and I. W. Kelly (*Psychological Reports,* Vol. 57, 1985) developed a Belief in Lunar Effects (BILE) Scale. The BILE score measures the degree to which a person "believes in lunar effects on behavior," with a maximum value of 81 indicating strong agreement and a minimum value of 9 indicating strong disagreement. The BILE scores for a sample of 157 undergraduate students are saved in the file **BILE.** (The first five scores and last five scores are listed in the table.) Use the methods of this chapter to summarize and describe the distribution of BILE scores. Does a large proportion of respondents have either an overall strong disagreement or strong agreement with the notion that the moon affects human behavior?

BILE

21	43	72	27	50	...	30	17	39	20	49

Summary of Customer Interest Level

Category	Percentage
Watchers	21
Impatients	45
Neutrals	34

Source: Katz, K. L., Larson, B. M., and Larson, R. C. "Prescription for the waiting-in-line blues: Entertain, enlighten, and engage." *Sloan Management Review,* Winter 1991, p. 49 (Fig. 5).

Problem 2.70

2.70 An investigation of the waiting time experienced by customers at a large Boston bank was conducted (*Sloan Management Review,* Winter 1991). With the aid of video cameras, the researchers recorded the actual time each of 277 customers waited in line for a bank teller. As customers finished their transactions, they were interviewed by the researchers and asked about perceived waiting time. In addition to perceived and actual waiting time, the researchers asked customers to describe their wait in line. Customers generally fell into three categories: "watchers" who enjoy observing people and events at the bank, "impatients" who could think of nothing more boring than waiting in line, and "neutrals" who fell somewhere in the middle. The table at left gives the breakdown, in percentages, of the 277 customers falling into the three categories of watchers, impatients, and neutrals.

a. Are the data quantitative or qualitative?

b. Describe the data summarized in the table graphically. Interpret the graph.

2.71 Refer to the *Sloan Management Review* (Winter 1991) study in Problem 2.70.

a. Consider the data set consisting of the differences between the perceived and actual waiting times of the 277 customers. Are the data qualitative or quantitative?

b. A graph describing the data of part **a** is shown on p. 111. What type of graph is displayed? Interpret the graph. In particular, obtain the approximate percentage of customers who overestimated their waiting time in line at the Bank of Boston.

2.72 The *Journal of Modern History* (Dec. 1997) published a study of radio broadcasting in Germany during the 1920s. The aim of the research was to determine whether the introduction of the modern mass medium of radio broadcasting impacted German cultural distinctions. In 1929, the German radio station *Deutsche Stunde in Bayern* broadcast 3,578 hours of programming. The journal article classified each hour of programming into one of several categories. A summary of the number of broadcasting hours in each category is displayed in the first table on p. 111.

a. Summarize the data using a graphical method.

b. Identify the broadcast category that occurred most often.

c. What proportion of programming time was dedicated to lectures?

2.73 Chemical researchers experimented with a new copper-selenium metal compound and reported the results in *Dalton Transactions* (Dec. 1997). The chemists varied the ratio of copper to selenium in each metal cluster, then used x-ray crystallography

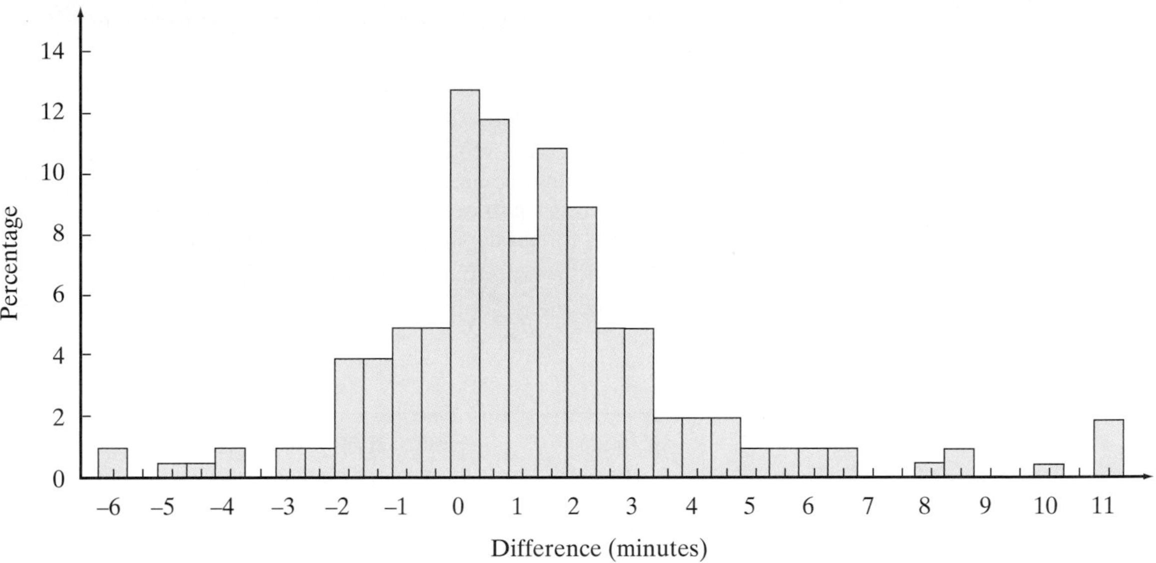

Source: Katz, K. L., Larson, B. M., and Larson, R. C. "Prescription for the waiting-in-line blues: Entertain, enlighten, and engage." *Sloan Management Review,* Winter 1991, p. 48 (Fig. 3).

Problem 2.71

Broadcast Category	Number of Programming Hours
Music	1,943
Lectures	576
News	494
Plays/Literature	279
Programming for women	82
Programming for children	61
General reports	25
Sports reports	21
Miscellaneous	97
TOTAL	3,578

Source: Führer, K. C. "A medium of modernity? Broadcasting in Weimer Germany, 1923–1932." *The Journal of Modern History,* Vol. 69, No. 4, Dec. 1997, p. 743 (adapted from Table 3).

Problem 2.72

to measure the bond length (pm) of each cluster. The bond lengths for 43 metal clusters are listed in the following table. [*Note:* Bond lengths marked with an asterisk (*) represent copper-phosphorus metal compounds.]

256.1	259.0	256.0	264.7	223.8*	262.5
239.0	238.9	237.4	222.7*	265.9	240.2
239.7	238.3	223.3*	273.8	246.4	246.0
247.4	290.6	260.9	255.6	253.1	283.7
261.4	271.9	267.9	238.7	237.6	240.0
226.1*	241.1	240.3	240.8	226.3*	246.5
248.3	246.0	226.0*	250.3	250.1	254.6
225.0*					

Source: Deveson, A., et al. "Syntheses and structures of four new copper (I)-selenium clusters: Size dependence of the cluster on the reaction conditions." *Dalton Transactions,* No. 23, Dec. 1997, p. 4492 (Table 1).

Problem 2.73

a. Graphically summarize the data for all 43 metal clusters.

b. What type of skewness exists in the data?

c. If possible, locate the bond lengths for the copper-phosphorus metal compounds on the graph, part **a**. What do you observe?

2.74 A study was conducted to determine whether a student's final grade in an introductory sociology course was linearly related to his or her performance on the verbal ability test administered before college entrance. The verbal test scores and final grades for a random sample of 10 students are shown in the table.

a. Construct a scatterplot for the data.

b. Describe the relationship between verbal ability test score and final sociology grade.

SOCIOLOGY

Student	Verbal Ability Test Score	Final Sociology Grade
1	39	65
2	43	78
3	21	52
4	64	82
5	57	92
6	47	89
7	28	73
8	75	98
9	34	56
10	52	75

2.75 The *American Journal on Mental Retardation* (Jan. 1992) published a study of the social interactions of two groups of children. Independent random samples of 15 children who did and 15 children who did not display developmental delays (i.e., mild mental retardation) were taken in the experiment. After observing the children during "freeplay," the number of children who exhibited disruptive behavior (e.g., ignoring or rejecting other children, taking toys from another child) was recorded for each group. The data are summarized in the following table. Graphically portray the data with a side-by-side bar chart. Interpret the graph.

	Disruptive Behavior	Nondisruptive Behavior	Totals
With Developmental Delays	12	3	15
Without Developmental Delays	5	10	15
TOTALS	17	13	30

Source: Kopp, C. B., Baker, B., and Brown K. W. "Social skills and their correlates: Preschoolers with developmental delays." *American Journal on Mental Retardation,* Vol. 96, No. 4, Jan. 1992.

2.76 In the book *Identical Twins Reared Apart,* author Susan Farber offers a survey and reanalysis of all published cases of identical (monozygotic) twins separated and reared apart. One chapter deals extensively with the evaluation and analysis of IQ tests conducted on the twins. A portion of the data for 19 pairs of twins is reproduced in the table on p. 113. (The complete data set is described and analyzed in Chapter 9.)

a. Classify the variable of interest, IQ score, as quantitative or qualitative.

b. What type of graphical method is appropriate for summarizing the data in the table?

IDTWIN

Pair	IQ Twin A	IQ Twin B	Pair	IQ Twin A	IQ Twin B
1	99	101	11	115	105
2	85	84	12	90	88
3	92	116	13	91	90
4	94	95	14	66	78
5	105	106	15	85	97
6	92	77	16	102	94
7	122	127	17	89	106
8	96	77	18	102	96
9	89	93	19	88	79
10	116	109			

Data Source: Adapted from *Identical Twins Reared Apart,* by Susan L. Farber. Basic Books, Inc., 1981.

c. Construct an appropriate graph to summarize the data for the 38 twins in the sample.

d. Use the graph to find the percentage of twins in the sample with IQ scores above 100.

e. What would be the advantages and disadvantages of using a stem-and-leaf display, rather than a relative frequency distribution, to describe the IQ data?

2.77 The average New York state household produces about 44,600 pounds of carbon dioxide each year. The graphic shown, extracted from *The New York Times,* is intended to describe the amount of carbon dioxide emissions attributed to each household item. Comment on the excellence of the graph.

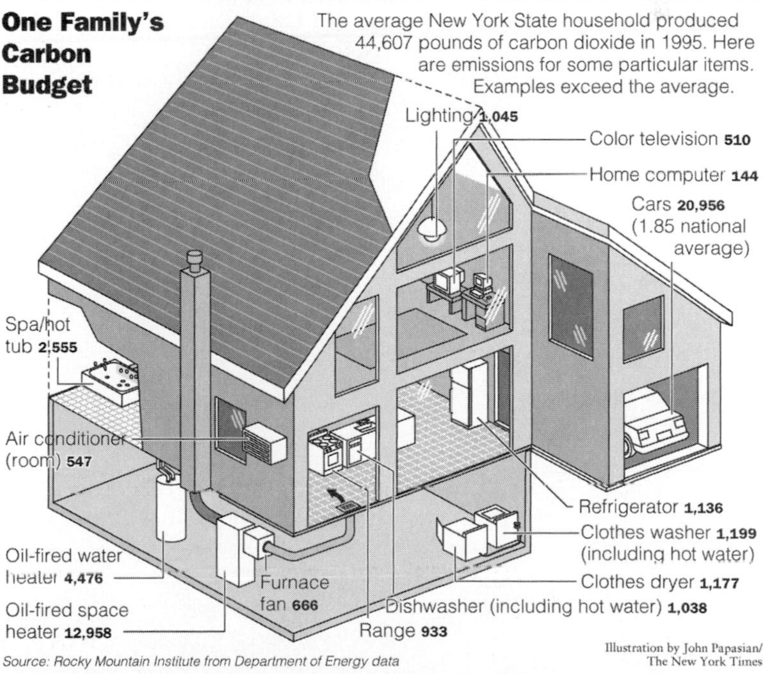

Source: John Papasian/*The New York Times.* Reprinted with permission.

2.78 The following figure appeared in an advertisement for satellite television in a recent issue of *Sports Illustrated.* Comment on the graphical excellence of the figure.

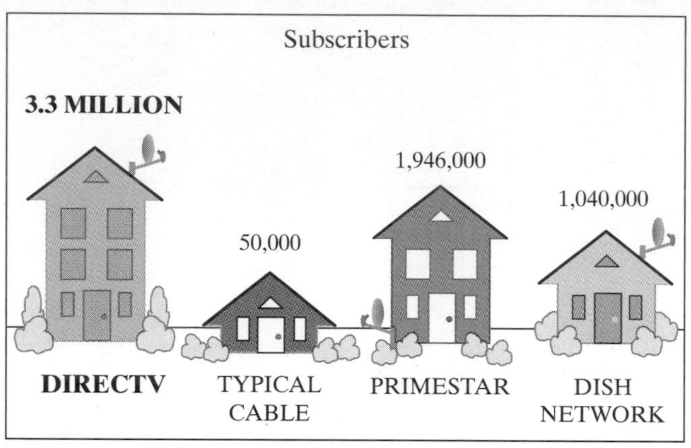

Data Source: Media Business Corp., 12/31/97.

REFERENCES

Huff, D. *How to Lie with Statistics.* New York: Norton, 1954.

Mendenhall, W. *Introduction to Probability and Statistics,* 9th ed. North Scituate, Mass.: Duxbury, 1994.

Microsoft Excel 2000. Redmond, Wash.: Microsoft Corporation, 1999.

MINITAB User's Guide 1: Data, Graphics, and Macros: Release 13. State College, Pa.: Minitab, Inc., 2000.

MINITAB User's Guide 2: Data Analysis and Quality Tools: Release 13. State College, Pa.: Minitab, Inc., 2000.

Tanur, J. M., Mosteller, F., Kruskal, W. H., Link, R. F., Pieters, R. S., and Rising, G. R., eds. *Statistics: A Guide to the Unknown,* 2nd ed. San Francisco: Holden-Day, 1978.

Tufte, E. R. *Envisioning Information.* Cheshire, Conn.: Graphics Press, 1990.

Tufte, E. R. *Visual Display of Quantitative Information.* Cheshire, Conn.: Graphics Press, 1983.

Tufte, E. R. *Visual Explanations.* Cheshire, Conn.: Graphics Press, 1997.

Tukey, J. *Exploratory Data Analysis.* Reading, Mass.: Addison-Wesley, 1977.

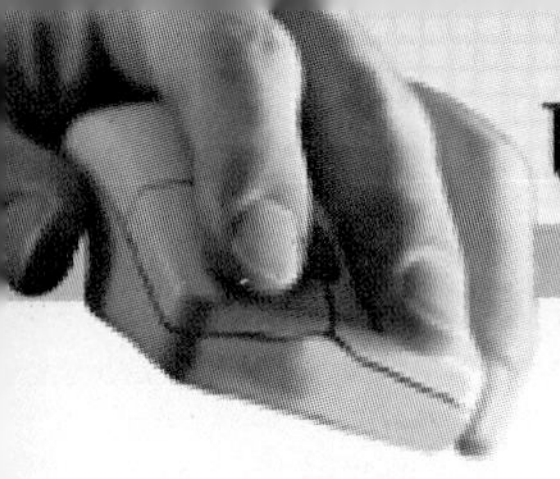

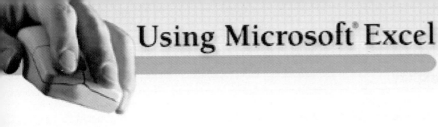

Using Microsoft® Excel

In this chapter a variety of tables and charts have been developed. These tables and charts can be obtained using PHStat and Microsoft Excel. If you have not already read the Microsoft Excel Primer, you should do so now.

2.E.1 Using PHStat to Obtain a One-Way Summary Table, a Bar Chart, and a Pie Chart

Use the PHStat **Descriptive Statistics | One-Way Tables & Charts** procedure to generate a summary table and charts for categorical data on new chart sheets. For example, to generate a summary table similar to Table 2.2 on p. 60 and bar and pie charts similar to the ones shown in Figures 2.1 and 2.2, open the **COLOGNE.XLS** workbook to the Data worksheet and:

1. Select PHStat | Descriptive Statistics | One-Way Tables & Charts.
2. In the One-Way Tables & Charts dialog box (see Figure 2.E.1):
   a. Select the Raw Categorical Data option button.
   b. Enter B1:B18 in the Raw Data Cell Range edit box.
   c. Select the First cell contains label check box.
   d. Enter a title in the Title edit box
   e. Select (check) the Bar Chart and Pie Chart check boxes.
   f. Click the OK button.

E Figure 2.E.1 PHStat One-Way Tables & Charts Dialog Box

This procedure places the summary table and each chart on separate sheets. The One-Way Tables & Charts procedure can also produce charts from frequency tables such as the one given in Table 2.2 on p. 60. To use the procedure with such a table, first enter the table (with column headings) into a worksheet. Then select the Table of Frequencies option button in the Type of Data dialog box and enter the range of the table in the cell range edit box. (When you select the Table of Frequencies option, the label for the cell range edit box changes from "Raw Data Cell Range:" to "Freq. Table Cell Range:".)

2.E.2 Ordering a Set of Data

Use the **Data | Sort** command to sort the contents of a worksheet or cell range to generate an ordered array. For example, to generate an ordered array of the **CRASH.XLS** data by the number of doors, open the **CRASH.XLS** workbook to the Data worksheet and:

1. Select Data | Sort.
2. In the Sort dialog box (see Figure 2.E.2):
   a. Select Doors from the Sort by drop-down list.
   b. Leave selected the first Ascending option button and the Header row option button.
   c. Click the OK button.

The entire table of automobiles is resorted so that the automobiles with the fewest doors appear first and the automobiles with the largest number of doors appear last in the table.

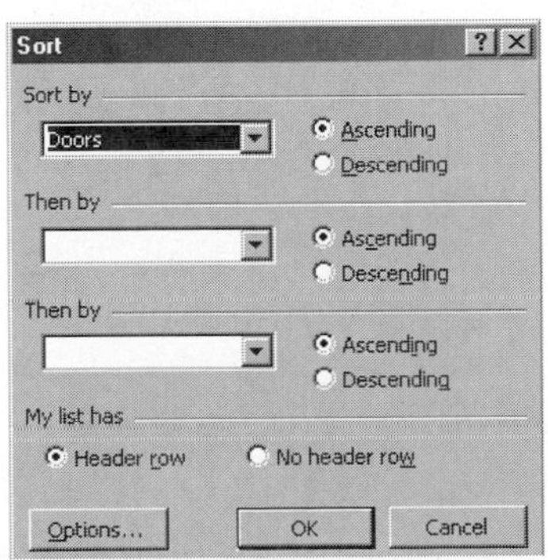

E Figure 2.E.2 Sort Dialog Box

2.E.3 Obtaining a Stem-and-Leaf Display

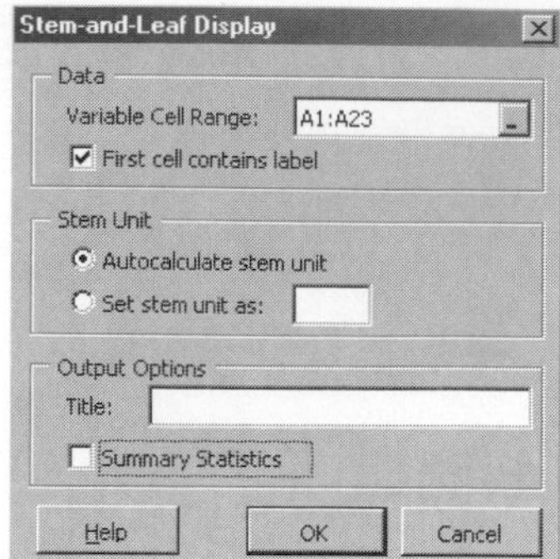

E Figure 2.E.3 PHStat Stem-and-Leaf Dialog Box

Use the PHStat **Descriptive Statistics | Stem-and-Leaf Display** procedure to generate a stem-and-leaf display that uses the rounding method. For example, to generate the stem-and-leaf display shown in Figure 2.6 on p. 68, open the **PMI.XLS** workbook to the Data worksheet and:

1. Select Descriptive Statistics | Stem-and-Leaf Display.

2. In the Stem-and-Leaf Display dialog box (see Figure 2.E.3):

a. Enter A1:A23 in the Variable Cell Range edit box.
b. Select the First cell contains label check box.
c. Select the Autocalculate stem unit option button.
d. Enter a title in the Title edit box.
e. Leave unchecked the Summary Statistics check box.
f. Click the OK button.

2.E.4 Using PHStat to Obtain Frequency Distributions and Histograms

Use the PHStat **Descriptive Statistics | Histogram & Polygons** procedure to generate a frequency distribution and histogram on new sheets. For example, to generate a histogram for the data shown in Table 2.5 and Figure 2.8, open the **SAL50.XLS** workbook to the Data worksheet and:

1. Select PHStat | Descriptive Statistics | Histogram & Polygons. In the Histogram & Polygons dialog box (see Figure 2.E.4):

a. Enter A1:A51 in the Variable Cell Range edit box.
b. Enter B1:B13 in the Bins Cell Range edit box.
c. Enter C1:C12 in the Midpoints Cell Range edit box.
d. Select the First cell in each range contains label check box.
e. Select the Single Group Variable option button.

E Figure 2.E.4 PHStat Histogram & Polygons Dialog Box

f. Enter a title in the Title edit box.

g. Leave the Histogram check box selected.

h. Click the OK button.

Because Microsoft Excel uses an ordered list of maximum class values, and not interval ranges, to specify class intervals (which are called "bins" in the dialog box), the Bins Cell Range B1:B13 contains values that approximate the maximum value of each class including 20,949 for the first class and 53,949 for the last class.

2.E.5 Using the Data Analysis Tool to Obtain Frequency Distributions and Histograms

To generate a frequency distribution and histogram, the **Analysis ToolPak Histogram** procedure can be used as an alternative to PHStat. Select **Tools | Data Analysis.** Select Histogram from the list box in the Data Analysis dialog box and click the OK button. In the Histogram dialog box, enter the cell range for the Input Range (what PHStat calls the Variable Cell Range) and Bin Range. (There is no provision for entering a midpoints range.) Select the Label check box if the first cell in these ranges contains a label. Select the New Worksheet Ply option button and enter a new sheet name of Histogram and select the Chart Output check box and click OK.

The Histogram procedure incorrectly adds an additional class "More" to the frequency distribution. To eliminate this class, first manually add the More frequency count to the count of the preceding class and set the cumulative percentage of the preceding class to 100%. (Note that because the More class has a frequency of zero for these data, these steps would not change the entries for the preceding 20,949 class in this example.) Then, select the cells containing the More row (A14:B14) and select **Edit | Delete.** In the Delete dialog box, select the Shift cells up option button and click OK.

Note that your histogram contains errors: there are gaps between the bars that correspond to the class intervals in the histogram. In addition, you may want to change the X axis legend "Bins" to some other value. To accomplish these tasks use the following procedures:

To eliminate the gaps between bars

Right-click on one of the histogram bars. (The tool tip that begins with the phrase "Series 'Frequency'" will appear when the mouse pointer is properly positioned over a bar.) Select **Format Data Series** from the shortcut menu. In the Format Data Series dialog box, click the Options tab and change the value in the Gap width edit box to 0. Click OK.

To change the X axis legend

Click on the legend Bins and type another value such as "Classes".

2.E.6 Using PHStat to Obtain a Two-Way Summary Table and Side-by-Side Bar Chart

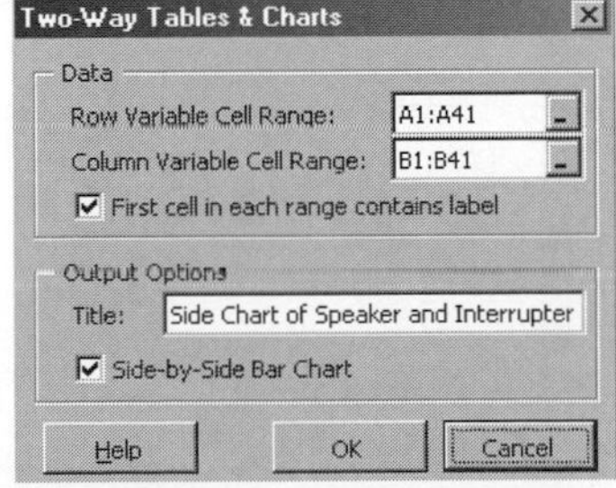

E **Figure 2.E.5** PHStat Two-Way Tables & Charts Dialog Box

Use the PHStat **Descriptive Statistics | Two-Way Tables & Charts** procedure to generate a contingency table and side-by-side bar chart for bivariate categorical data on a new worksheet. For example, to generate a contingency table similar to the one shown in Table 2.8 on p. 80 and a side-by-side bar chart similar to the one shown in Figure 2.12 on p. 82, open the **INTERRUPT.XLS** workbook to the Data worksheet and:

1. Select PHStat | Descriptive Statistics | Two-Way Tables & Charts.

2. In the Two-Way Tables & Charts dialog box (see Figure 2.E.5):

a. Enter A1:A41 in the Row Variable Cell Range edit box.

b. Enter B1:B41 in the Column Variable Cell Range edit box.

c. Select the First cell in each range contains label check box.

d. Enter a title in the Title edit box.

e. Select the Side-by-Side Bar Chart check box.

f. Click the OK button.

Note that this procedure will use the row variable values to form categories.

2.E.7 Generating Scatter Diagrams

Use the Microsoft Excel Chart Wizard (introduced on p. 14) to generate a scatter diagram on a new chart sheet. For example, to generate the scatter diagram shown in Figure 2.15 on p. 91, open the **SMOKING.XLS** workbook to the Data worksheet, select **Insert | Chart** from the Excel menu bar and then:

1. In the Step 1 dialog box (see Figure 2.E.6):

a. Select the Standard Types tab and then select XY (Scatter) from the Chart type list box.

b. Select the first (top) Chart sub-type choice, described as "Scatter. Compares pairs of values." when selected, and click the Next button.

2. In the Step 2 dialog box:

a. Select the Data Range tab. Enter Data!B2:B21 in the Data range edit box.

b. Select the Columns option button in the Series in group and click the Series tab (see Figure 2.E.7).

c. When generating scatter diagrams, the Chart Wizard always assumes that the first column (or row) of data in the data range entered in the Step 2 dialog box contains values for the X variable. Since, in the case of Figure 2.15, nicotine (the second column of data) is the X variable, change the cell ranges in the X Values edit box to =Data!C2:C21 and the Y Values edit box to Data!B2:B21. Note the revised cell ranges must be entered as formulas that include the sheet name. Click Next.

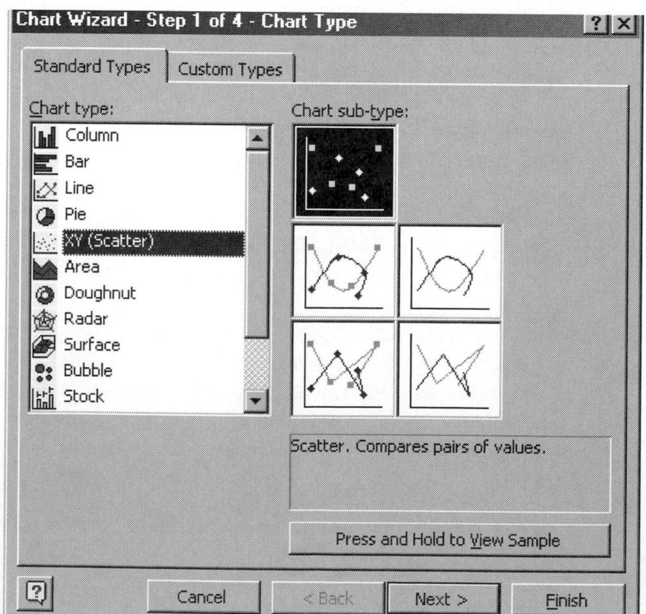

E Figure 2.E.6 Chart Wizard Step 1 Dialog Box

E **Figure 2.E.7** Chart Wizard Step 2 Series Tab Dialog Box

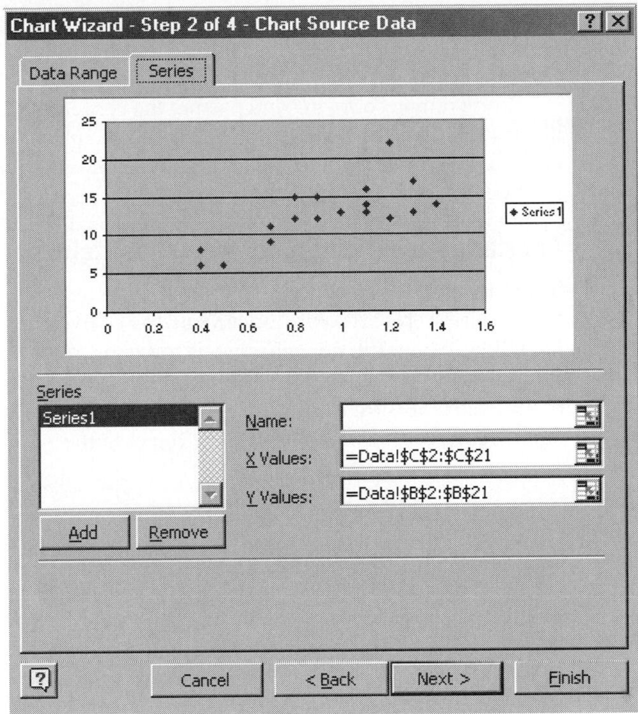

3. In the Step 3 dialog box:

a. Select the Titles tab. Enter Scatter Diagram as the Chart title, Nicotine in the Value (X) axis edit box, and CO in the Value (Y) axis edit box.

b. Select, in turn, the Axes, Gridlines, Legend, and Data Table tabs and adjust settings as discussed in section EP.5.2 on p. 14.

c. Click Next.

4. In the Step 4 dialog box, select the As new sheet option button, enter a name for the new sheet, and click the Finish button to generate the chart.

2.E.8 Using Microsoft Office: Sharing Information between Microsoft Excel and Microsoft Word

In many situations, the worksheet tables and charts generated in Microsoft Excel need to be included as part of a report. If you use Microsoft Word to prepare such a report, select the **Edit | Copy** and either **Edit | Paste** or **Edit | Paste Special** commands to share such information between Microsoft Excel and Word. Variations of the Paste Special command allow for simple copying and pasting as well as creating a link between the original information in Microsoft Excel and its pasted copy. Such a link automatically updates the pasted copy in Microsoft Word when changes are made to the original information in Microsoft Excel and can save time when the original information is undergoing constant revision.

Simple copy and paste for worksheet tables

Select the cell range containing the information to be copied and then select **Edit | Copy.** Open the Microsoft Word document to receive the information, click the point in the document where the copied range is to be inserted, and select **Edit | Paste.** This inserts a

Microsoft Word table that contains all of the formatted text and numbers of the cell range. During the paste operation, any formulas in the cell range are replaced by their current values.

For example, copying and pasting the contents of a cell range A3:D9 in the worksheet illustrated in Figure 2.E.8a creates the Word table shown in Figure 2.E.8b. During this operation the formula in cell B9 and the formulas in columns C and D are replaced by their then-current values.

Copy and preserve link for worksheet tables

Select the cell range containing the information to be copied and then select **Edit | Copy.** Open the Microsoft Word document to receive the information, click the point in the document where the copied range is to be inserted, and select **Edit | Paste Special.** In the Paste Special dialog box (see Figure 2.E.9 for the Excel 2000 version of this dialog box; the Excel 97 version of this dialog box is similar), select the Paste link option button, the default Microsoft Excel Worksheet Object in the As list box, and click OK.

E Figure 2.E.8a Cell range to be copied

	A	B	C	D
1	Sample Table			
2				
3	**Objective**	**Total**	**Percentage**	**Cumulative %**
4	IL	42	30.66%	30.66%
5	SC	37	27.01%	57.66%
6	GI	26	18.98%	76.64%
7	MC	20	14.60%	91.24%
8	TK	12	8.76%	100.00%
9	Grand Total	137		
10				
11				

E Figure 2.E.8b Word document after paste operation

Pasted copy of worksheet table:

Objective	Total	Percentage	Cumulative %
IL	42	30.66%	30.66%
SC	37	27.01%	57.66%
GI	26	18.98%	76.64%
MC	20	14.60%	91.24%
TK	12	8.76%	100.00%
Grand Total	137		

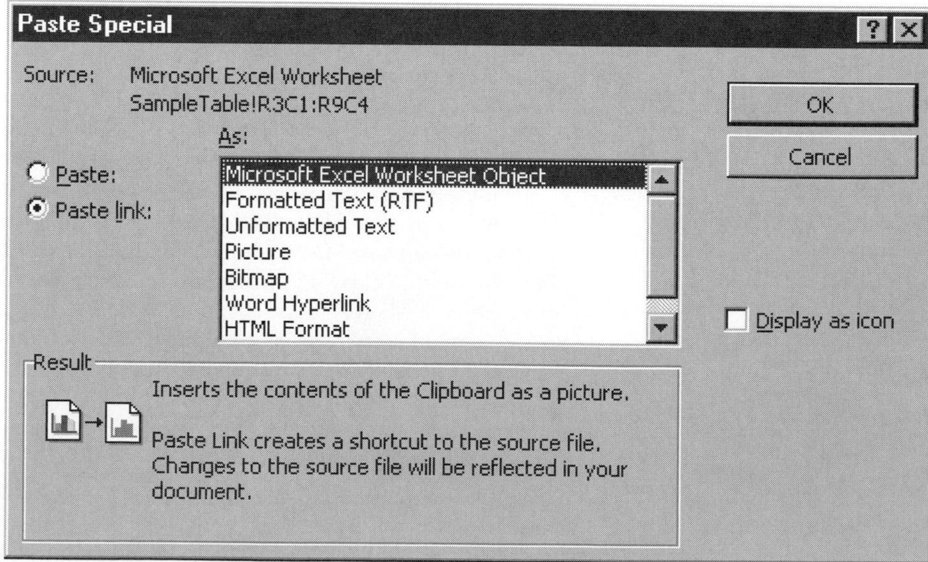

E Figure 2.E.9 Excel 2000 Paste Special Dialog Box for Worksheet Tables

Other choices in the As list box can be selected and may affect the display, formatting, and printing of the worksheet data in minor ways. (See the Result caption for each and the Microsoft Excel help system for further information.) If you are using a system with an older, slower processor and small memory size, you may want to avoid selecting the Microsoft Excel Worksheet Object because this choice, which allows you to directly edit the contents of the original worksheet without first separately opening Microsoft Excel, may cause performance problems. (Note that in the version of Excel 2000 that was used to generate Figure 2.E.9, the Result caption for the Picture choice is incorrectly displayed for the Microsoft Excel Worksheet Object choice.)

Copying charts

Select the cell range containing the information to be copied and then select **Edit | Copy.** For best performance, use **Edit | Paste Special** for all copying operations involving a Microsoft Excel chart. In the Paste Special dialog box for charts (see Figure 2.E.10), select either the Paste or Paste link option button, depending on whether or not you wish to preserve a link between the original data and its pasted copy. Then select the default Microsoft Excel Chart Object in the As list box and click OK. As with copying cell ranges, other choices in the As list box can be selected and may affect the display, formatting, and printing of the chart in minor ways.

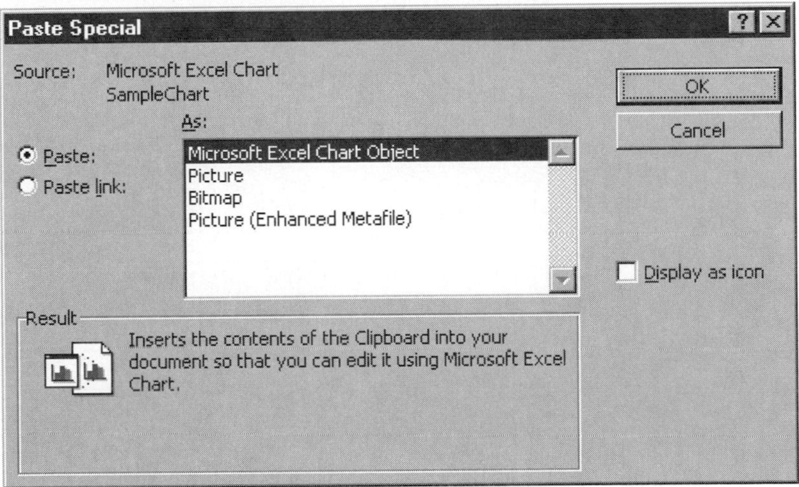

E Figure 2.E.10 Excel 2000 Paste Special Dialog Box for Chart

Using MINITAB®

In this chapter a variety of tables and charts have been developed. Many of these tables and charts can be obtained using MINITAB. If you have not already read the MINITAB Primer, you should do so now.

2.M.1 Obtaining a Bar Chart

To obtain a bar chart for a categorical variable, open the **CRASH.MTW** worksheet. Select **Graph | Chart.** In the Chart dialog box (see Figure 2.M.1), in the Graph variables edit box, enter **C6** or **'DriverStar'** in row 1 in the X column. Click the **OK** button.

Figure 2.M.1 MINITAB Chart Dialog Box

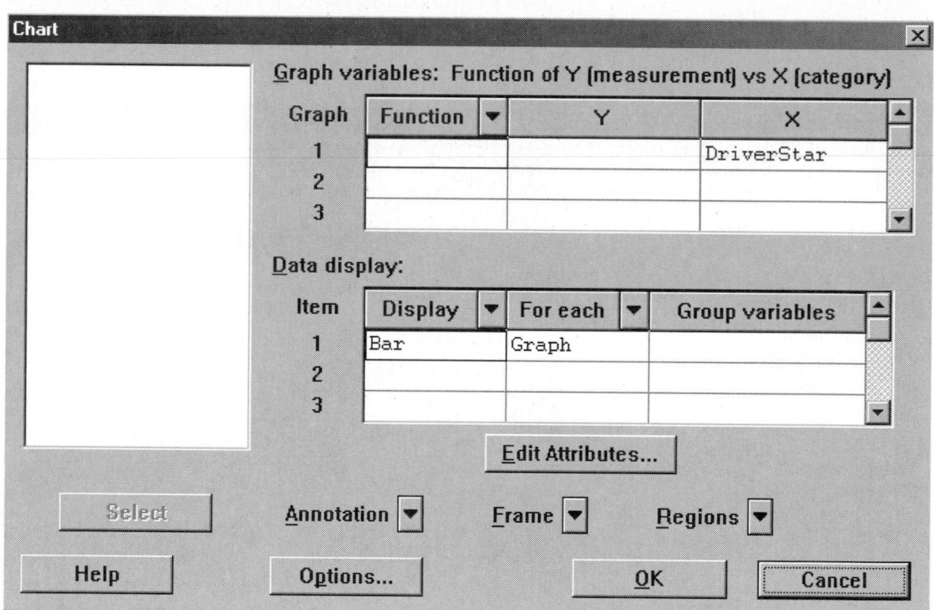

2.M.2 Obtaining a Pie Chart

To obtain a pie chart for a categorical variable, open the **CRASH.MTW** worksheet. Select **Graph | Pie Chart.** In the Pie Chart dialog box (see Figure 2.M.2), in the Chart data in edit box, enter **C6** or **'DriverStar'.** Enter a title in the Title edit box. Click the **OK** button.

2.M.3 Obtaining a Frequency Table

To obtain a frequency table for a categorical variable, open the **CRASH.MTW** worksheet. Select **Stat | Tables | Tally.** In the Tally dialog box (see Figure 2.M.3), in the Variables edit box, enter **C6** or **'DriverStar'.** Select the **Counts** and **Percents** check boxes. Click the **OK** button. The results will be similar to those of Table 2.4 on p. 69.

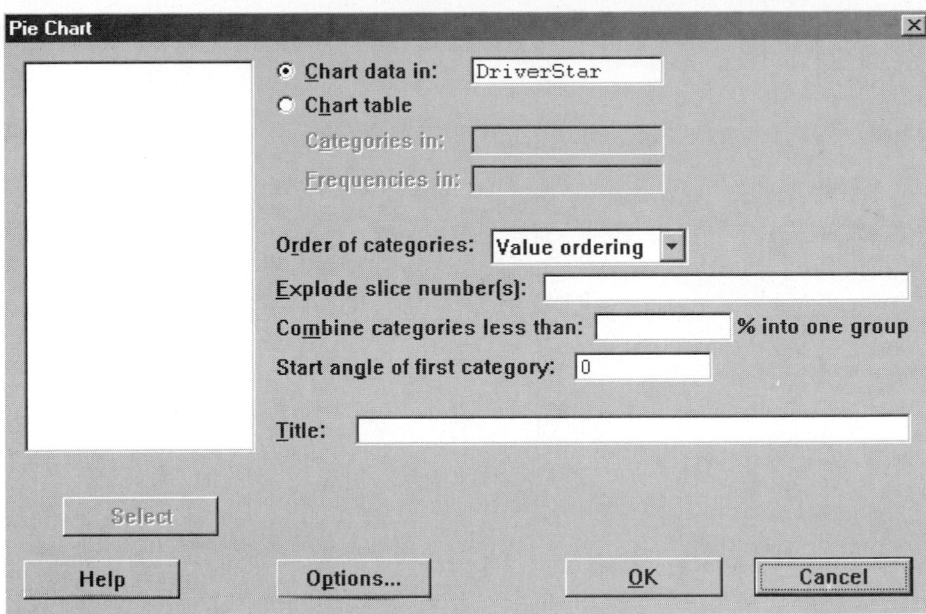

M **Figure 2.M.2** MINITAB Pie Chart Dialog Box

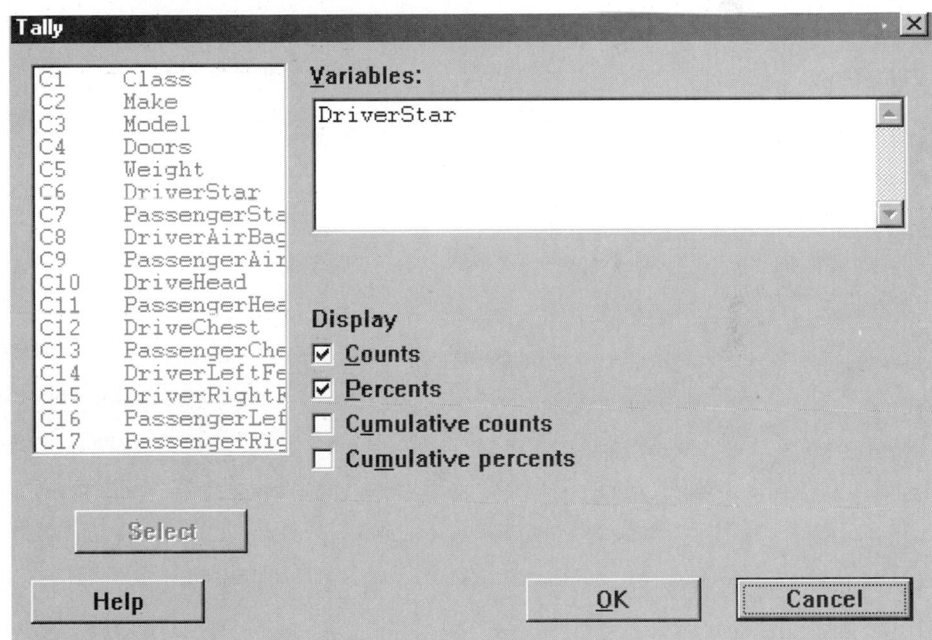

M **Figure 2.M.3** MINITAB Tally Dialog Box

2.M.4 Obtaining a Stem-and-Leaf Display

To obtain a stem-and-leaf display similar to Figure 2.6 on p. 68, open the **PMI.MTW** worksheet. Select **Stat | EDA | Stem-and-Leaf.** In the Stem-and-Leaf dialog box (see Figure 2.M.4 on p. 124), enter **C1** or **'Postmortem Interval'** in the Variables edit box. Click the **OK** button.

2.M.5 Obtaining a Dot Plot

To obtain the dot plot of Figure 2.7 on p. 70, open the **PMI.MTW** worksheet. Select **Graph | Dotplot.** In the Dotplot dialog box (see Figure 2.M.5 on p. 124), enter **C1** or **'Postmortem Interval'** in the Variables edit box. Click the **OK** button.

Figure 2.M.4 MINITAB Stem-and-Leaf Dialog Box

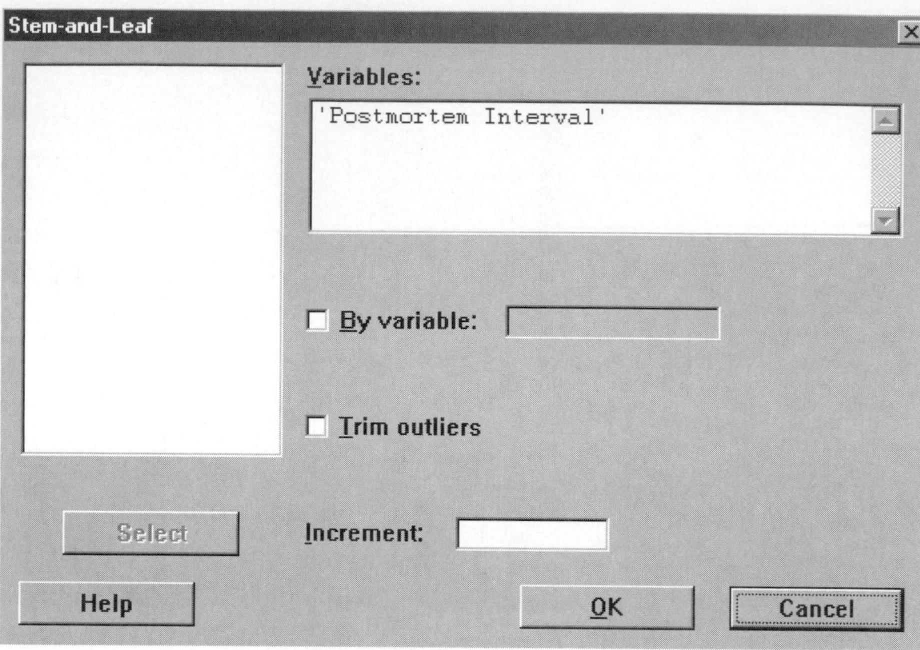

Figure 2.M.5 MINITAB Dotplot Dialog Box

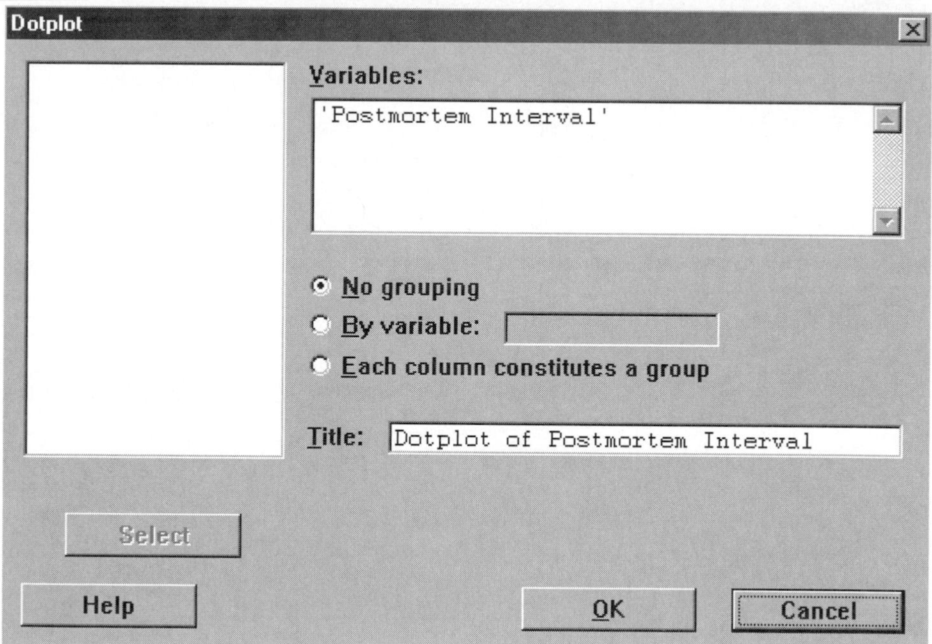

2.M.6 Obtaining a Histogram

To obtain the histogram in Figure 2.10 on p. 73, open the **CRASH.MTW** worksheet and select **Graph | Histogram.** In the Histogram dialog box (see Figure 2.M.6), enter **C11** or **'PassengerHead'** in the Graph variables edit box. Click the **Options** button. In the Histogram Options dialog box (see Figure 2.M.7), under Type of Histogram, select the **Percent** option button. Under Type of Intervals, select the **CutPoint** option button. Under Definition of Intervals, select the **Midpoint/cutpoint positions** option button, and enter the values **0 100 200 300 400 500 600 700 800 900 1000 1100 1200 1300.** Click the **OK** button to return to the Histogram dialog box. Click the **OK** button to obtain the histogram.

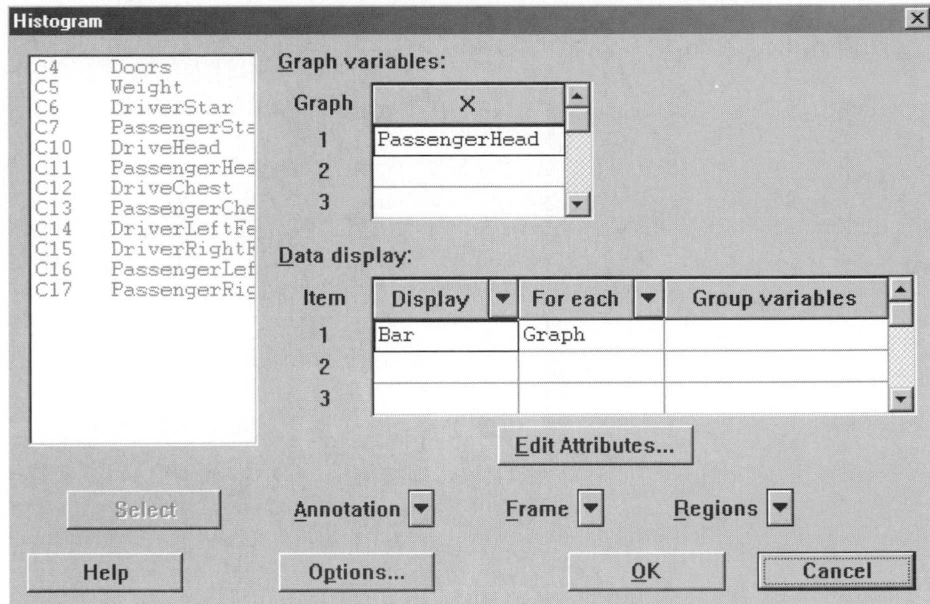

M **Figure 2.M.6** MINITAB Histogram Dialog Box

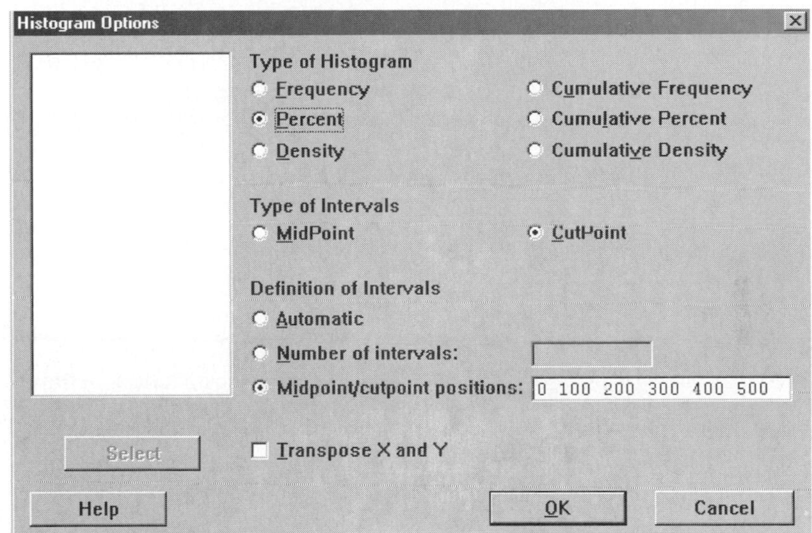

M **Figure 2.M.7** MINITAB Histogram Options Dialog Box

2.M.7 Obtaining a Contingency Table

To obtain a two-way contingency table, open the **CRASH.MTW** worksheet. Select **Stat | Tables | Cross Tabulation.** In the Cross Tabulation dialog box (see Figure 2.M.8 on p. 126), in the Classification variables edit box, enter **C4** or **'Doors'** and **C6** or **'DriverStar'.** Under Display, select the Counts, Row percents, Column percents, and Total percents check boxes. Click the **OK** button. The results are similar to those of Figure 2.13 on p. 83.

2.M.8 Obtaining a Side-by-Side Bar Chart

To obtain a side-by-side bar chart, open the **CRASH.MTW** worksheet. Select **Graph | Chart.** Enter **C4** or **'Doors'** in the Graph variables edit box in row 1 in the X column. In the Data display edit box, be sure that in row 1 the Display column indicates Bar and the For each

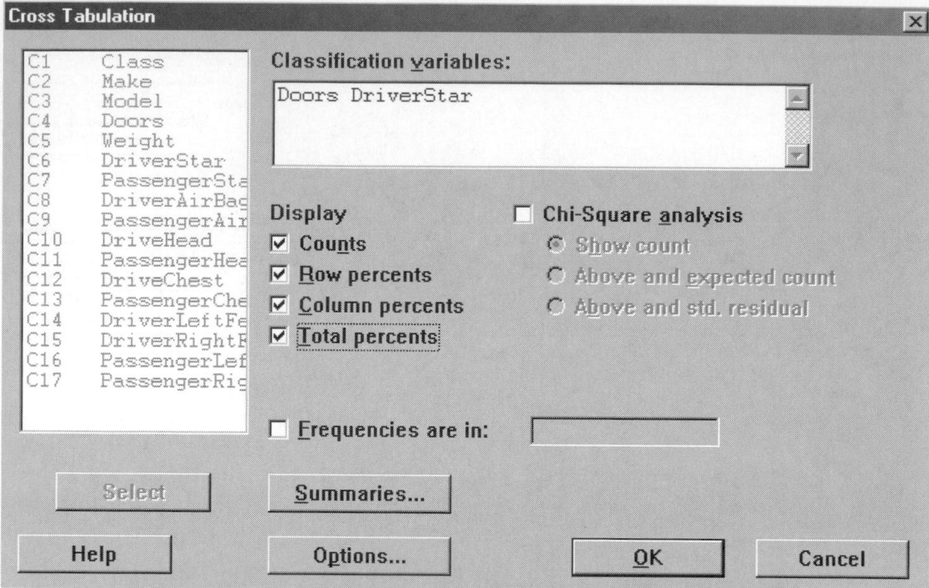

Figure 2.M.8 MINITAB Cross Tabulation Dialog Box

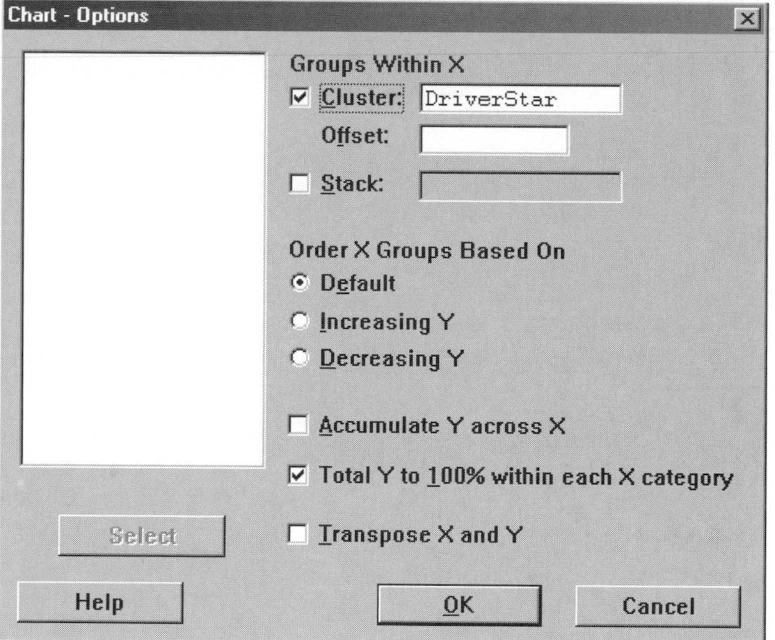

Figure 2.M.9 MINITAB Chart–Options Dialog Box

column indicates Graph. In the Group variables column enter **C6** or **'DriverStar'.** Click the **Options** button. In the Chart–Options dialog box (see Figure 2.M.9), in the Groups Within X edit box, select Cluster and enter **C6** or **'DriverStar'.** Under Order X Groups Based on, select the **Total Y to 100% within each X category** check box to obtain percentages for each group. Click the **OK** button to return to the Chart dialog box. Click the **OK** button again to obtain the side-by-side bar chart.

2.M.9 Obtaining a Scatterplot

To obtain a scatterplot such as in Figure 2.17a on p. 92, open the **CRASH.MTW** worksheet. Select **Graph | Plot.** In the Plot dialog box (see Figure 2.M.10), in the Graph variables edit box, in row 1, enter **C11** or **'PassengerHead'** in the Y column and **C5** or **'Weight'** in the X column. Be sure the Data display edit box appears as in Figure 2.M.10. Click the **OK** button.

M **Figure 2.M.10** MINITAB Plot Dialog Box

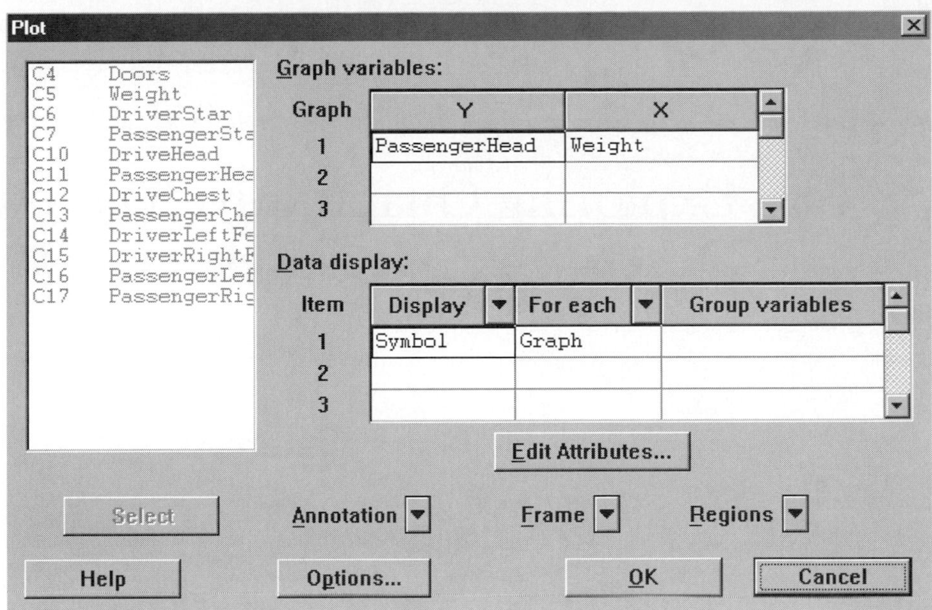

CHAPTER 3

Exploring Quantitative Data with Numerical Descriptive Measures

OBJECTIVES

1. To describe the central tendency of quantitative data
2. To describe the variation in quantitative data
3. To describe the shape of a distribution of quantitative data
4. To describe measures of relative standing for quantitative data
5. To identify outliers in a quantitative data set
6. To describe the level of correlation between two quantitative variables
7. To introduce descriptive measures for populations

CONTENTS

EXCEL TUTORIAL

MINITAB TUTORIAL

Statistics in the Real World

Bid-Collusion in the Highway Contracting Industry

Many products and services are purchased by governments, cities, states, and businesses on the basis of competitive, sealed bids, with the contract awarded to the lowest bidder. Unfortunately, this process is prone to "price-fixing" (also known as "bid-rigging" or "bid-collusion") by the bidders. Price-fixing involves conspiring to set the bid price above the fair, or competitive, price in order to predetermine the winner as well as to increase profit margin. Some recent examples of bid-rigging publicized in the media include the following:

- *Glass & Law in the News* (Sept. 29, 1999). Three Virginia glass companies plead guilty to rigging bids and raising prices on commercial glass installations.
- *The South Florida Business Journal* (June 19, 2000). The state of Florida rescinded a $24 million health care contract after allegations of bid-rigging and improper procedures.
- *Industry Canada* (April 25, 2000). A Calgary manufacturer of pine shakes has pleaded guilty and been convicted of rigging the bids on the purchase of commercial timber permits in Alberta, Canada.
- *Channelnewsasia.com* (Sept. 27, 2000). More than 30 Japanese construction firms are under investigation for bid-collusion on public works projects.
- *Engineering News-Record* (Sept. 8, 2000). Two German construction firms pleaded guilty to federal bid-rigging charges on wastewater treatment projects in Egypt.

In this real-world application, we consider a data set involving road construction contracts in the state of Florida. During a recent 10-year period, the Office of the Florida Attorney General suspected numerous contractors of practicing bid-collusion. The investigations led to an admission of guilt by several of the contractors. Although these contractors were heavily fined, they avoided harsher punishment by identifying which road construction contracts were competitively bid and which involved fixed bids. By comparing the bid prices (and other important bid variables) of the fixed contracts to the competitive contracts, the Florida Attorney General (FLAG) and Florida Department of Transportation (DOT) was able to establish invaluable benchmarks for detecting bid-rigging in the future. (In fact, the benchmarks led to a virtual elimination of bid-rigging in road construction in the state.)

The data set contains 194 competitively bid and 85 fixed contracts–a total of 279 contracts. These data are stored in the **BIDRIG** file. A description of the variables measured for each road contract is provided below.

BIDRIG

LOWBID (Price bid by lowest bidder–thousands of dollars)
DOTEST (DOT engineer's estimate of fair price–thousands of dollars)
LBERAT (Ratio of low bid price to DOT estimate)
STATUS (Collusion status–Fixed or Competitive)
DISTRICT (Florida district in which contract was let–A, B, or C)
BIDS (Number of bidders on the contract)
DAYS (Estimated number of days to complete construction)
LENGTH (Length of road built–miles)
PCTASPH (Percent of contract costs allocated to liquid asphalt–0 to 100%)
SUB (Use of subcontractor–Yes or No)

In this chapter, several Statistics in the Real World examples demonstrate how numerical descriptive measures can be used to investigate bid-collusion in the highway contract industry in Florida.

Statistics in the Real World Revisited

- Interpreting Measures of Central Tendency (p. 137)
- Interpreting the Standard Deviation (p. 150)
- Detecting Outliers (p. 167)
- Interpreting the Correlation Coefficient (p. 174)

3.1 Types of Numerical Descriptive Measures

It is often said that a picture is worth a thousand words. This is certainly true when the goal is to describe a data set. But sometimes you need to discuss the major features of a data set and it may not be sufficient to produce a graph (e.g., bar chart, stem-and-leaf display, or histogram) for the data. When this situation occurs, we seek characteristics called **numerical descriptive measures** (or *descriptive statistics*) that describe the distribution of data. We have already encountered numerical descriptive measures for a *qualitative* data set. Recall in Section 2.2 that we can compute either a frequency or a relative frequency for each value of the qualitative variable of interest. In this chapter, we present numerical descriptive measures for a *quantitative* data set.

Consider a data set containing the scores of all students who have taken a statistics exam. Two numbers that will help you construct a mental image of the distribution of exam scores are:

1. A number that is located near the *center* of the distribution (see Figure 3.1a)
2. A number that measures the *spread* of the distribution (see Figure 3.1b)

Figure 3.1 Numerical Descriptive Measures

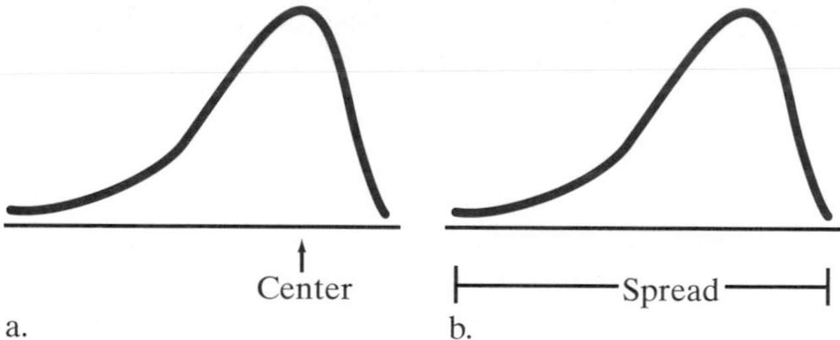

A number that describes the center of the distribution is located near the spot where most of the exam scores are concentrated. Consequently, numbers that fulfill this role are called **measures of central tendency.** We will define and describe several measures of central tendency in Section 3.3.

The amount of spread in the exam scores is a measure of the variation in the data. Consequently, numerical descriptive measures that perform this function are called **measures of variation** or **measures of dispersion.** As you will subsequently see (Section 3.4), there are several ways to measure the variation in a data set.

Measures of central tendency and data variation are not the only types of numerical measures for describing data sets. Some are constructed to measure the **skewness** of a distribution. Recall from Section 2.3 that skewness describes the tendency of the distribution to have long tails out to the right or left. (For example, the distributions in Figure 3.1 are skewed to the left—or negatively skewed.) Although numerical descriptive measures of skewness are beyond the scope of this text, knowing whether a distribution of data is skewed or symmetric is important when describing the data with measures of central tendency and variation (Section 3.5).

Two other important numerical descriptive measures covered in this chapter are **measures of relative standing** (Section 3.5) and **measures of association** (Section 3.8). A measure of relative standing is used to determine where a particular measurement lies *relative* to all other measurements in the data set. A measure of association is used to help determine the nature of the relationship between two different variables in a data set.

As you read this chapter, keep in mind our goal of using a few summarizing numbers to create a mental image of a distribution. Relate each numerical de-

scriptive measure to this objective, and verify that it fulfills the role it is intended to play.

TYPES OF NUMERICAL DESCRIPTIVE MEASURES FOR QUALITATIVE DATA

1. A *category frequency* measures the number of occurrences of a particular category. (Section 2.2)
2. A *category relative frequency* measures the number of occurrences of a particular category as a proportion of the total number of measurements in the data set. (Section 2.2)

TYPES OF NUMERICAL DESCRIPTIVE MEASURES FOR QUANTITATIVE DATA

1. Measures of *central tendency* locate the center of the distribution. (Section 3.3)
2. Measures of *variation* measure the spread of the distribution. (Section 3.4)
3. Measures of *skewness* measure the lack of symmetry in the distribution. (Section 3.6)
4. Measures of *relative standing* represent the location of any one observation relative to the others in the distribution. (Section 3.5)
5. Measures of *association* measure the nature of the relationship between two variables. (Section 3.8)

3.2 Summation Notation

Suppose a data set was obtained by observing a quantitative variable, x. For example, x may represent the grade point average (GPA) of a college student. If we want to represent a particular observation in a data set—say, thc 35th—we represent it by the symbol x with a subscript of 35. If Linda Marcus, the 35th student on the list, has a GPA of 3.27, then we write $x_{35} = 3.27$. If we are interested in the GPAs of a group of 180 students, then the complete data set would be represented by the symbols $x_1, x_2, x_3, \ldots, x_{180}$.

Most of the formulas that we use require the summation of numbers. For example, we may want to sum the observations in a data set, or we may want to square each observation and then sum the squares of all the observations. The sum of the observations in a data set will be represented by the symbol

$$\sum x$$

This symbol is read "summation x." The symbol Σ (sigma) gives an instruction: it is telling you to sum a set of numbers. The variable to be summed, x, is shown to the right of the Σ symbol.

EXAMPLE 3.1 **FINDING SUMS**

Suppose that the variable x is used to represent the length of time (in years) for a college student to obtain his or her bachelor's degree. Five graduates are selected and the value of x is recorded for each. The observations are 2, 4, 3, 5, 4.

a. Find Σx

b. Find Σx^2

Solution

a. The symbol $\sum x$ tells you to sum the x values in the data set. Therefore,

$$\sum x = 2 + 4 + 3 + 5 + 4 = 18$$

b. The symbol $\sum x^2$ tells you to sum the squares of the x values in the data set. Therefore,

$$\sum x^2 = (2)^2 + (4)^2 + (3)^2 + (5)^2 + (4)^2$$
$$= 4 + 16 + 9 + 25 + 16 = 70$$

EXAMPLE 3.2 **FINDING SUMS**

Refer to Example 3.1.

a. Find $\sum(x - 4)$. **b.** Find $\sum(x - 4)^2$. **c.** Find $\sum x^2 - 4$.

Solution

a. The symbol $\sum(x - 4)$ tells you to subtract 4 from each x value and then sum. Therefore,

$$\sum (x - 4) = (2 - 4) + (4 - 4) + (3 - 4) + (5 - 4) + (4 - 4)$$
$$= (-2) + 0 + (-1) + 1 + 0 = -2$$

b. The symbol $\sum(x - 4)^2$ tells you to subtract 4 from each x value in the data set, square these differences, and then sum them as follows:

$$\sum (x - 4)^2 = (2 - 4)^2 + (4 - 4)^2 + (3 - 4)^2 + (5 - 4)^2 + (4 - 4)^2$$
$$= (-2)^2 + (0)^2 + (-1)^2 + (1)^2 + (0)^2$$
$$= 4 + 0 + 1 + 1 + 0 = 6$$

The Meaning of Summation Notation $\sum x$
Sum observations on the variable that appears to the right of the summation symbol.

c. The symbol $\sum x^2 - 4$ tells you to first sum the squares of the x values, and then subtract 4 from this sum:

$$\sum x^2 - 4 = (2)^2 + (4)^2 + (3)^2 + (5)^2 + (4)^2 - 4$$
$$= 4 + 16 + 9 + 25 + 16 - 4 = 66$$

Self-Test 3.1

A data set contains the observations 5, 1, 3, 2, 1. Find:

a. $\sum x$ **b.** $\sum x^2$
c. $\sum(x - 1)$ **d.** $\sum(x - 1)^2$

PROBLEMS FOR SECTION 3.2

Using the Tools

3.1 A data set contains the observations 10, 15, 10, 11, 12. Find:

a. $\sum x$ **b.** $\sum x^2$ **c.** $(\sum x)^2$
d. $\sum(x - 13)$ **e.** $\sum(x - 13)^2$

3.2 A data set contains the observations 3, 8, 4, 5, 3, 4, 6. Find:

a. Σx **b.** Σx^2 **c.** $\Sigma(x - 5)^2$

d. $\Sigma(x - 2)^2$ **e.** $(\Sigma x)^2$

3.3 Refer to Problem 3.1. Find:

a. $\Sigma x^2 - \frac{(\Sigma x)^2}{5}$ **b.** $\Sigma(x - 10)^2$ **c.** $\Sigma x^2 - 10$

3.4 Refer to Problem 3.2. Find:

a. $\Sigma x^2 - \frac{(\Sigma x)^2}{7}$ **b.** $\Sigma(x - 4)^2$ **c.** $\Sigma x^2 - 15$

3.5 A data set contains the observations 6, 0, −2, −1, 3. Find:

a. Σx **b.** Σx^2 **c.** $\Sigma x^2 - \frac{(\Sigma x)^2}{5}$

3.6 A data set contains the observations .7, 1.1, .5, .3. Find:

a. Σx **b.** Σx^2 **c.** $\Sigma x^2 - \frac{(\Sigma x)^2}{4}$

3.3 Measures of Central Tendency: Mean, Median, and Mode

The word *center,* as applied to a distribution of data, is not a well-defined term. In our minds, we know vaguely what we mean: a number somewhere near the "middle" of the distribution, a single number that tends to "typify" the data set. The measures of central tendency that we define often generate different numbers for the same data set but all satisfy the general objective of a number somewhere near the middle of the distribution.

The most common measure of the central tendency of a data set is called the *arithmetic mean* of the data. The arithmetic mean, or *average,* is defined in the box.

Definition 3.1

The **arithmetic mean** $\bar{x}$ (read "x-bar") of a sample of n observations, $x_1, x_2, \ldots, x_n$, is the sum of the x values divided by n:

$$\bar{x} = \frac{\text{Sum of the } x \text{ values}}{\text{Number of observations}} = \frac{\Sigma x}{n} \tag{3.1}$$

[*Note:* From now on, we will refer to an arithmetic mean simply as a **mean.**]

EXAMPLE 3.3 **COMPUTING A MEAN**

Find the mean for the data set consisting of the observations 5, 1, 6, 2, 4.

Solution

The data set contains $n = 5$ observations. Using equation 3.1, we have

$$\bar{x} = \frac{\Sigma x}{n} = \frac{5 + 1 + 6 + 2 + 4}{5} = \frac{18}{5} = 3.6$$

EXAMPLE 3.4 INTERPRETING A MEAN

In Example 1.2 (p. 26) we introduced a U.S. Army Corps of Engineers study of 144 contaminated fish captured in the Tennessee River (Alabama). The data are saved in the file named **FISH.** One of the variables measured for each fish is weight (in grams). Find the mean for the 144 fish weights. Does the mean fall near the center of the distribution?

Solution

FISH

With such a large data set, it is impractical to calculate numerical descriptive measures by hand or calculator. In practice, we rely on an available software package such as Excel or MINITAB. Figure 3.2 is an Excel printout giving descriptive statistics for the fish weight data. The mean of the 144 weights, highlighted on the printout, is (rounded) $\bar{x} = 1{,}049.7$ grams. This mean, or average, weight should be located near the center of the histogram for the 144 fish weights. Figure 3.3, produced with MINITAB, is a frequency histogram for the data. Note that the mean $\bar{x}$ does indeed fall near the center of the bell-shaped portion of the distribution. If we did not have Figure 3.3 available, we could reconstruct the distribution in our minds as a bell-shaped figure centered in the vicinity of $\bar{x} = 1{,}049.7$ grams.

	A	B
1	Weight	
2		
3	Mean	1049.715
4	Standard Error	31.37884
5	Median	1000
6	Mode	1186
7	Standard Deviation	376.5461
8	Sample Variance	141787
9	Kurtosis	0.368447
10	Skewness	0.500552
11	Range	2129
12	Minimum	173
13	Maximum	2302
14	Sum	151159
15	Count	144
16	Largest(1)	2302
17	Smallest(1)	173

Figure 3.2 Excel Descriptive Measures for the Fish Weight Data

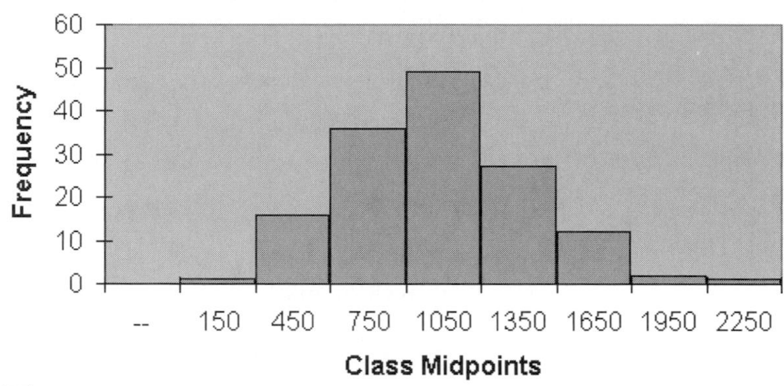

Figure 3.3 MINITAB Histogram for the Fish Weight Data

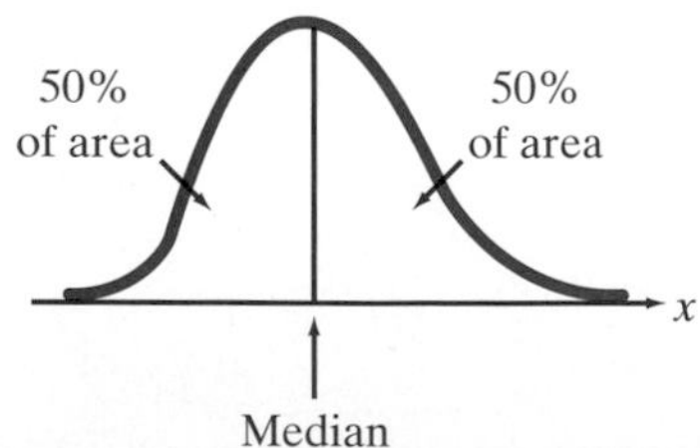

Figure 3.4 The Median Divides the Area of a Relative Frequency Distribution into Two Equal Portions

A second measure of central tendency for a data set is the *median.* For large data sets, the median M is a number chosen so that half the observations are less than the median and half are larger. Since the areas of the bars used to construct the relative frequency histogram are proportional to the numbers of observations falling within the classes, it follows that the median is a value of x that divides the area of the histogram into two equal portions. Half the area will lie to the left of the median (see Figure 3.4) and half will lie to the right. For example, the median for the 144 fish weights (highlighted in Figure 3.2) is 1,000 grams. This weight divides the data into two sets of equal size. Half the 144 weights are less than 1,000 grams; half are larger. The next two examples illustrate how to find the median for small data sets.

EXAMPLE 3.5

COMPUTING A MEDIAN: *n* ODD

Find the median for the data set consisting of the observations 7, 4, 3, 5, 3.

Solution

We first arrange the data in increasing order:

3 3 4 5 7

With an *odd number of measurements,* the median is the middle ordered value. Thus, the median is 4; half the remaining measurements are less than 4 and half are greater than 4.

EXAMPLE 3.6

COMPUTING A MEDIAN: *n* EVEN

Find the median for the data set consisting of the observations 5, 7, 3, 1, 4, 6.

Solution

If we arrange the data in increasing order, we obtain

1 3 4 5 6 7

You can see that there is no single middle value. Any number between 4 and 5 will divide the data set into two groups of three each. There are many ways to choose this number, but the simplest is to choose the median as the point halfway between the two middle numbers when the data are arranged in order. Thus, the median is

$$M = \frac{4 + 5}{2} = 4.5$$

Definition 3.2

The **median *M*** of a sample of n observations, $x_1, x_2, \ldots, x_n$, is defined as follows:

If n is odd: The middle observation when the data are arranged in order.

If n is even: The number halfway between the two middle observations—that is, the mean of the two middle observations—when the data are arranged in order.

$$M = \begin{cases} \left(\dfrac{n+1}{2}\right) \text{ ranked observation} & \text{if } n \text{ is odd} \\[2ex] \dfrac{\left[\left(\dfrac{n}{2}\right) \text{ ranked observation}\right] + \left[\left(\dfrac{n+2}{2}\right) \text{ ranked observation}\right]}{2} & \text{if } n \text{ is even} \end{cases} \quad (3.2)$$

A third measure of central tendency for a data set is the *mode*. The mode is the value of x that occurs with greatest frequency (see Figure 3.5). If the data have been grouped into classes, we can define the modal class as the class with the largest class frequency (or relative frequency). For example, you can see from the Excel printout, Figure 3.2 (p. 134), that the modal fish weight is 1,186 grams. The modal class, however, is 900–1,200 grams. The midpoint of this modal class, 1,050, is shown in Figure 3.3.

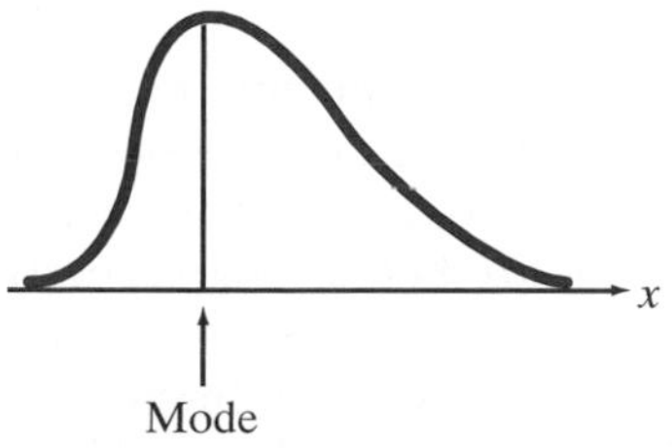

Mode

x

Figure 3.5 The Mode Is the Value of x that Occurs with Greatest Frequency

Definition 3.3

The **mode** of a data set is the value of x that occurs with greatest frequency. The **modal class** is the class with the largest class frequency.

Self-Test 3.2

Find the mean, median, and mode for the following sample of ten measurements: 6, 10, 3, 4, 4, 5, 8, 11, 4, 2.

The mean, median, and mode are shown (Figure 3.6) on the MINITAB histogram for the 144 fish weights. Which is the best measure of central tendency for this data set? The answer is that it depends on the type of descriptive information you want. If you want the "average" weight or "center of gravity" for the weight distribution, then the mean is the desired measure. If your notion of a typical or "central" weight is one that is less than half the fish weights and greater than the remainder, then you will prefer the median. The mode is rarely the measure of central tendency chosen because it does not necessarily lie in the "center" of the distribution. There are situations, however, where the mode is preferred. For example, a retailer of women's shoes would be interested in the modal shoe size of potential customers.

M Figure 3.6 MINITAB Histogram of Fish Weights: Mean, Median, and Mode Located

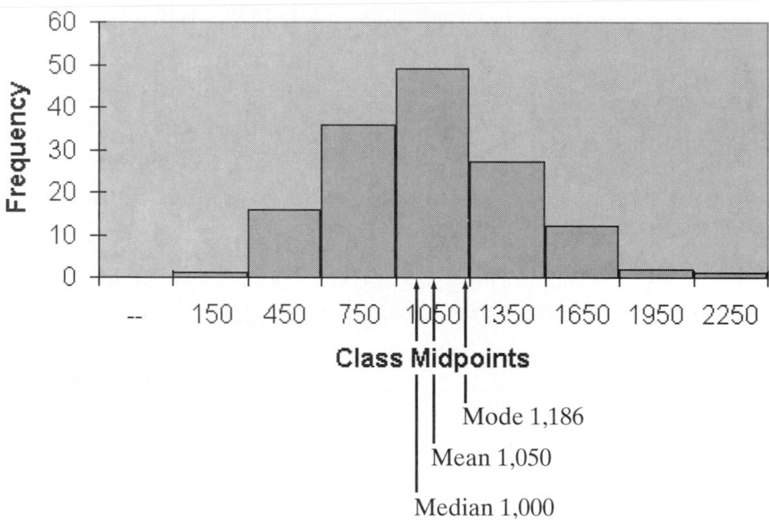

Caution

For data sets that are extremely skewed, be wary of using the mean as a measure of the "center" of the distribution. In this situation, a more meaningful measure of central tendency may be the median, which is more resistant to the influence of extreme measurements.

In making your decision, you should know *the mean is sensitive to very large or very small measurements*. Consequently, the mean will shift toward the direction of skewness and may be a misleading measure of central tendency in some situations. You can see from Figure 3.6 that the mean is slightly larger than the median and that the fish weights are slightly skewed to the right. The relatively few captured fish that have a very high weight will influence the mean more than the median. For this reason, the median is sometimes called a *resistant* measure of central tendency since it, unlike the mean, is resistant to the influence of extreme measurements. For data sets that are extremely skewed, the median would better represent the "center" of the distribution.

Most of the inferential statistical methods discussed in this text are based, theoretically, on bell-shaped distributions of data with little or no skewness. For these situations, the mean and the median will be the same, for all practical purposes. Since the mean has better mathematical properties than the median, it is the preferred measure of central tendency for these inferential techniques.

Statistics in the Real World Revisited

Interpreting Measures of Central Tendency

Consider the highway construction bid data (**BIDRIG** file) collected by the Florida Department of Transportation (p. 129). Two of the many variables measured for each of the 279 road contracts in the data set are (1) STATUS—a qualitative variable representing the bid status (fixed or competitive) of the contract, and (2) LBERAT—a quantitative variable representing the ratio of the winning (low) bid to the DOT engineer's estimate of the "fair" bid price. Theoretically, the ratio will be near 1 for competitive bids. In fact, the ratio is likely to be smaller than 1 since most winning bidders will slightly undervalue the cost of the construction in order to win the bid. The larger the ratio, the more evidence that an unusually high bid (and possibly price-fixing) has occurred. Consequently, the distribution of ratios for competitive winning bids should have a "center" around 1 (or slightly less), while the "center" of the distribution of ratios for fixed winning bids should exceed 1. A MINITAB analysis of the 279 contracts revealed the information shown in the printouts, Figures 3.7 and 3.8.

Descriptive Statistics

Variable	STATUS	N	Mean	Median	TrMean	StDev
LBERAT	Compet	194	0.90747	0.91000	0.90776	0.13699
	Fixed	85	1.1124	1.0900	1.1082	0.1481

Variable	STATUS	SE Mean	Minimum	Maximum	Q1	Q3
LBERAT	Compet	0.00984	0.53000	1.35000	0.82000	0.99250
	Fixed	0.0161	0.8100	1.5600	1.0150	1.2000

M **Figure 3.7** MINITAB Analysis of Florida DOT Data

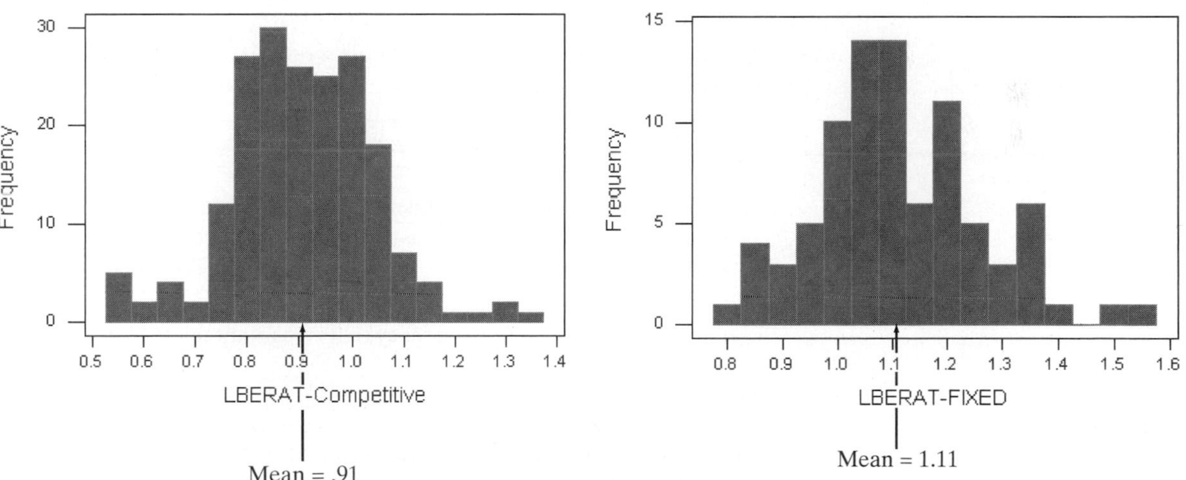

M **Figure 3.8** MINITAB Histograms for Competitive and Fixed Contracts

Figure 3.7 provides numerical descriptive measures for LBERAT for both competitive and fixed contracts. The means and medians for the two distributions are highlighted on the printout. The mean (or average) ratio for competitive winning bids is .907 while the mean (or average) for fixed bids is 1.112. (These means are also shown in the annotated MINITAB histograms in Figure 3.8.) Similarly, the median for competitive bids is .91 while the median for fixed bids is 1.09. This implies that half of the competitive bids have ratios below .91 and half of the fixed-bid ratios fall below 1.09. Clearly, the distribution of winning bid ratios for competitively bid contracts is centered below 1, while the distribution of winning bid ratios for fixed bids is centered above 1. The analysis indicates that, "on average," competitive contracts have winning bid prices below the DOT estimate of the fair price, and fixed contracts have winning bid prices above the DOT estimate.

PROBLEMS FOR SECTION 3.3

Using the Tools

3.7 Calculate the mean for samples with the following characteristics:

a. $n = 5$ and $\Sigma x = 30$

b. $n = 100$ and $\Sigma x = 1{,}220$

c. $n = 800$ and $\Sigma x = 200$

3.8 Calculate the mean for samples with the following characteristics:

a. $n = 10$ and $\Sigma x = 500$

b. $n = 20$ and $\Sigma x = 400$

c. $n = 500$ and $\Sigma x = 100$

3.9 Find the mean and median for a sample consisting of these five measurements: 3, 9, 0, 7, 4.

3.10 Find the mean and median for the following sample of $n = 6$ measurements: 7, 3, 4, 1, 5, 6.

3.11 Calculate the mean, median, and mode for each of the following samples:

a. 3, 4, 4, 5, 5, 5, 6, 6, 7

b. 3, 4, 4, 5, 5, 5, 6, 6, 70

c. −50, −49, 0, 0, 49, 50

d. −50, −49, 0, 9, 9, 81

3.12 Calculate the mean, median, and mode for each of the following samples:

a. 1.3, 1.2, 1.2, 1.1, 1.3, 1.7, 1.5, 1.2, 1.8, 1.0

b. 1.3, 1.2, 1.2, 1.1, 1.3, 1.7, 1.5, 1.2, 1.8, 10.0

c. .3, 1.2, 1.2, 1.1, 1.3, 1.7, 1.5, 1.2, 1.8, 1.0

Applying the Concepts

3.13 The accompanying data set contains quiz scores for 15 students in an introductory statistics class:

QUIZ

8	7	9	6	8	10	9	9	5	7	9	1	7	9	10

a. Find a measure of central tendency that represents the quiz score that occurs most often.

b. Find a measure of central tendency that represents the average of the 15 quiz scores.

c. Find a measure of central tendency that is resistant to the influence of extremely small quiz scores.

3.14 Three-way catalytic converters have been installed in new vehicles in order to reduce pollutants from motor vehicle exhaust emissions. However, these converters unintentionally increase the level of ammonia in the air. *Environmental Science & Technology* (Sept. 1, 2000) published a study on the ammonia levels near the exit ramp of a San Francisco highway tunnel. The data in the table represent daily ammonia concentrations (parts per million) on eight randomly selected days during afternoon drive-time in the summer of 1999.

AMMONIA

1.53	1.50	1.37	1.51	1.55	1.42	1.41	1.48

a. What is the mean daily ammonia level in the air in the tunnel? The median ammonia level?

b. Interpret the values obtained in part **a**.

ICECREAM

3.15 The data file **ICECREAM** contains the number of calories and the percentage of fat per half-cup serving of 17 chocolate ice cream brands, as determined by *Consumer*

Number of Calories	270	150	170	140	160	150	160	290	170	190	190	160	170	150	150	110	180
Fat (%)	18	8	10	8	9	9	9	17	9	10	10	8	9	8	8	3	9

Source: "Cold storage," *Consumer Reports,* February 1999, 49.

Reports (Feb. 1999). The data, first presented in Problem 2.47 (p. 95), are listed in the table above.

a. Find the mean number of calories per half-cup serving for the 17 ice cream brands. Interpret the result.

b. Find the median number of calories per half-cup serving for the 17 ice cream brands. Interpret the result.

c. Find the mode of the number of calories per half-cup serving for the 17 ice cream brands. Interpret the result.

d. Which of the three measures of central tendency best describes the center of the distribution of number of calories?

e. Repeat parts **a–d** for the percentage of fat per half-cup serving.

3.16 The *Journal of Applied Behavior Analysis* (Summer 1997) published a study of a treatment designed to reduce destructive behavior. A 16-year-old boy diagnosed with severe mental retardation was observed playing with toys for 10 sessions, each ten minutes in length. The number of destructive responses (e.g., head banging, pulling hair, pinching, etc.) per minute was recorded for each session in order to establish a baseline behavior pattern. The data for the 10 sessions are listed here:

BOY16

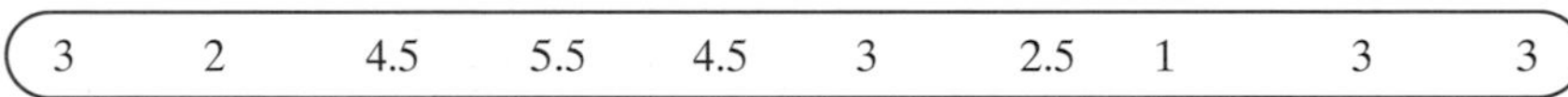

3	2	4.5	5.5	4.5	3	2.5	1	3	3

a. Find and interpret the mean for the data set.

b. Find and interpret the median for the data set.

c. Find and interpret the mode for the data set.

SHIPSANIT

3.17 Refer to the Centers for Disease Control study of sanitation levels for 121 international cruise ships, Problem 2.23, p. 76. An Excel printout of the descriptive statistics for the data is shown at left below. (Recall that sanitation scores range from 0 to 100.) Interpret the numerical descriptive measures of central tendency displayed on the printout.

3.18 Computer scientists at the University of Virginia developed a computer algorithm for quickly solving problems that involve multiple input terminals on a workstation (*Networks,* Mar. 1995). The running time (in seconds) for each of 100 runs of the algorithm was recorded for different workstation configurations. The mean and median running times for workstations with 10, 20, and 30 terminals are given in the table. For each of the three workstation configurations, comment on the type of skewness present in the distribution of algorithm running times.

Number of Workstations	Mean	Median
10	0.3	0.3
20	4.0	2.5
30	2,210.0	171.0

	A	B
1	*Sanitation*	
2		
3	Mean	90.33884
4	Standard Error	0.631535
5	Median	92
6	Mode	93
7	Standard Deviation	6.946886
8	Sample Variance	48.25923
9	Kurtosis	31.25116
10	Skewness	-4.50124
11	Range	63
12	Minimum	36
13	Maximum	99
14	Sum	10931
15	Count	121
16	Largest(1)	99
17	Smallest(1)	36

E Problem 3.17

3.19 More than 4,000 members of the American Institute of Certified Public Accountants (CPAs) responded to a survey on their use of software tools (*Journal of Accountancy,* Nov. 1997). Responses to the question, "How satisfied are you with the tax software program you use?" were recorded on a scale of 1 (worst) to 5 (best). The satisfaction scores for 13 tax software programs are listed in the table on p. 140. Compute the mean, median, and mode of the 13 satisfaction scores. Interpret their values.

TAXRATE

Software Program	Satisfaction Score
AM Tax Pro	5.0
Tax $imple	4.5
Ultra Tax	4.5
TaxWorks	4.4
ProSystem fx	4.4
Lacerte	4.4
Tax Machine	4.2
Professional Tax System	4.2
Turbo Tax Pro	4.1
GoSystem	4.0
A-Plus-Tax	3.9
Tax Relief	3.8
Pencil Pushers	3.8

Problem 3.19

3.20 Spinifex pigeons are one of the few bird species that inhabit the desert of Western Australia. The pigeons rely almost entirely on seeds for food. The *Australian Journal of Zoology* (Vol. 43, 1995) reported on a study of the diets and water requirements of these spinifex pigeons. Sixteen pigeons were captured in the desert and the crop (i.e., stomach) contents of each examined. The accompanying table reports the weight (in grams) of dry seed in the crop of each pigeon. Calculate three different measures of central tendency for the 16 crop weights. Interpret the value of each.

CROPWT

.457	3.751	.238	2.967	2.509	1.384	1.454	.818
.335	1.436	1.603	1.309	.201	.530	2.144	.834

Source: Excerpted from Williams, J. B., Bradshaw, D., and Schmidt, L. "Field metabolism and water requirements of spinifex pigeons (*Geophaps plumifera*) in Western Australia." *Australian Journal of Zoology,* Vol. 43, No. 1, 1995, p. 7 (Table 2).

3.21 Patients with normal hearing in one ear and unaidable sensorineural hearing loss in the other are characterized as suffering from unilateral hearing loss. In a study reported in the *American Journal of Audiology* (Mar. 1995), eight patients with unilateral hearing loss were fitted with a special wireless hearing aid in the "bad" ear. The absolute sound pressure level (SPL) was then measured near the eardrum of the ear when noise was produced at a frequency of 500 hertz. The SPLs of the eight patients, recorded in decibels, are listed below.

SPL

73.0	80.1	82.8	76.8	73.5	74.3	76.0	68.1

Calculate three different measures of central tendency for the eight SPLs. Interpret the value of each.

3.22 Social psychologists label the process by which decision makers escalate their commitment to an ineffective course of action as "entrapment." Fifty-two introductory psychology students took part in a laboratory experiment designed to explore whether individuals' tendencies to view prior outcomes as revealing of their self-identity would heighten entrapment (*Administrative Science Quarterly,* Mar. 1986). The experiment consisted of 30 trials in which points were "awarded" based on the accuracy of students' judgments of geometric patterns of various shapes. The total points awarded on each trial are listed in the table on p. 141.

ENTRAP

5	5	4	7	24	6
10	12	11	15	11	10
3	23	4	20	5	4
7	5	6	6	15	5
15	10	13	9	4	6

Source: Brockner, J., et al. "Escalation of commitment to an ineffective course of action: The effect of feedback having negative implications for self-identity." *Administrative Science Quarterly,* Vol. 31, No. 1, Mar. 1986, p. 115.

a. Construct a stem-and-leaf display for the data.

b. Compute the mean, median, and mode for the data set and locate them on the graph, part **a**.

c. Which of these measures of central tendency appears to best locate the center of the distribution of data?

3.23 Refer to the *Journal of Geography* (May/June 1997) study of the effectiveness of computer software in the geography classroom, Problem 2.27 (p. 78). Recall that two groups of seventh graders participated in the study. Group 1 (43 students) used computer resources to conduct research on Africa. Group 2 (22 students) used traditional methods to conduct the same research. All students were then given a test on the African regions studied. Test scores for the two groups of seventh graders are reproduced in the table.

GEO

Group 1 (Computer Resources) Student Scores											
41	53	44	41	66	91	69	44	31	75	66	69
75	53	44	78	91	28	69	78	72	53	72	63
66	75	97	84	63	91	59	84	75	78	88	59
97	69	75	69	91	91	84					
Group 2 (No Computer) Student Scores											
56	59	9	59	25	66	31	44	47	50	63	19
66	53	44	66	50	66	19	53	78	78		

Source: Linn, S. E. "The effectiveness of interactive maps in the classroom: A selected example in studying Africa." *Journal of Geography,* Vol. 96, No. 3, May/June 1997, p. 167 (Table 1).

a. Find the mean and median test scores for the Group 1 students.

b. Find the mean and median test scores for the Group 2 students.

c. Based on the results, parts **a** and **b**, determine the type of skewness (if any) present in the two test score distributions.

d. Do you think using computer resources had a positive impact on student test scores? Explain.

3.4 Measures of Data Variation: Range, Variance, and Standard Deviation

Just as measures of central tendency locate the center of a relative frequency distribution, *measures of variation* measure its spread. The simplest measure of spread for a data set is the range.

Definition 3.4

The **range** of a quantitative data set is equal to the difference between the largest and the smallest measurements in the set.

$$\text{Range} = (\text{Largest measurement} - \text{Smallest measurement}) \qquad \textbf{(3.3)}$$

EXAMPLE 3.7

COMPUTING A RANGE

Find the range for the data set consisting of the observations 3, 7, 2, 1, 8.

Solution

The smallest and largest members of the data set are 1 and 8, respectively. Using equation 3.3, we find

$$\text{Range} = \text{Largest measurement} - \text{Smallest measurement} = 8 - 1 = 7$$

The range of a data set is easy to find, but it is an insensitive measure of variation and is not very informative. For an example of its insensitivity, consider Figure 3.9. Both distributions have the *same range,* but it is clear that the distribution in Figure 3.9b indicates much less data variation than the distribution in Figure 3.9a. Most of the observations in Figure 3.9b lie close to the mean. In contrast, most of the observations in Figure 3.9a deviate substantially from the center of the distribution. Since the ranges for the two distributions are equal, it is clear that the range is a fairly insensitive measure of data variation. It was unable to detect the differences in data variation for the data sets represented in Figure 3.9.

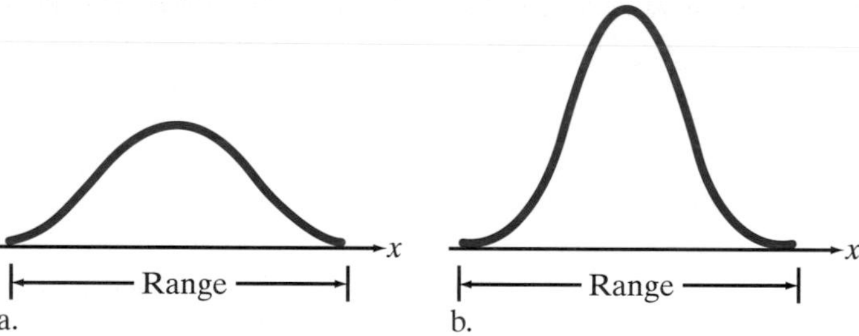

Figure 3.9 Two Distributions that Have Equal Ranges, but Show Differing Amounts of Data Variation

A more useful measure of spread is the *variance.* The variance of a data set is based on how much the measurements "deviate" from their mean. The *deviation* between an observation x and the mean $\bar{x}$ of a sample is the difference

$$x - \bar{x}$$

If a sample data set contains n observations, the **sample variance** is equal to the "average" of the squared deviations of all n observations. The formula for computing the sample variance, denoted by the symbol s^2, is given in the box.

We use $(n - 1)$ rather than n in the denominator of s^2 in order to obtain an estimator of the true population variance with desirable mathematical properties. When n is used in the denominator, the value of s^2 tends to underestimate the population variance. Dividing by $(n - 1)$ adjusts for the underestimation problem.

Definition 3.5

The **variance, s^2,** of a set of n sample measurements is equal to the sum of squares of deviations of the measurements about their mean, divided by $(n - 1)$:

$$s^2 = \frac{\Sigma(x - \bar{x})^2}{n - 1} \quad \textbf{(3.4)}$$

The larger the value of s^2, the more spread out (i.e., the more variable) the sample data.

EXAMPLE 3.8

COMPUTING A VARIANCE

Find the variance for the sample measurements: 3, 7, 2, 1, 8.

Solution

The five observations are listed in the first column of Table 3.1. You can see that $\Sigma x = 21$. Applying equation 3.1, the sample mean is

$$\bar{x} = \frac{\Sigma x}{n} = \frac{21}{5} = 4.2$$

This value of $\bar{x}$, 4.2, is subtracted from each observation to determine how much each observation deviates from the mean. These deviations are shown in the second column of Table 3.1. A *negative deviation* means that the observation fell *below* the mean; a *positive deviation* indicates that the observation fell *above* the mean. **Notice that the sum of the deviations around the mean equals 0. This will be true for all data sets.**

TABLE 3.1 Data and Computation Table

	Observation x	$x - \bar{x}$	$(x - \bar{x})^2$
	3	−1.2	1.44
	7	2.8	7.84
	2	−2.2	4.84
	1	−3.2	10.24
	8	3.8	14.44
TOTALS	21	0	38.8

The squares of the deviations are shown in the third column of Table 3.1. The total at the bottom of the column gives the sum of squares of the deviations,

$$\sum (x - \bar{x})^2 = 38.8$$

Applying equation 3.4, the sample variance is

$$s^2 = \frac{\Sigma(x - \bar{x})^2}{n - 1} = \frac{38.8}{4} = 9.7$$

How can we interpret the value of the sample variance calculated in Example 3.8? We know that data sets with large variances are more variable (i.e., more spread out) than data sets with smaller variances. But what information can we obtain from the number $s^2 = 9.7$? One interpretation is that the average squared deviation of the sample measurements from their mean is 9.7. However, a more practical interpretation can be obtained by calculating the square root of the variance.

A third measure of data variation, the *standard deviation,* is obtained by taking the square root of the variance. This results in a number that has the same units of measurement of the original data. That is, if the units of measurement for the sample observations are feet, dollars, or hours, the standard deviation of the sample is also measured in feet, dollars, or hours (instead of feet^2, dollars^2, or hours^2).

Definition 3.6

The **standard deviation, s,** of a set of n sample measurements is equal to the square root of the variance:

$$s = \sqrt{s^2} = \sqrt{\frac{\Sigma(x - \bar{x})^2}{n - 1}} \qquad \textbf{(3.5)}$$

The standard deviation of the five sample measurements in Example 3.8 is

$$s = \sqrt{s^2} = \sqrt{9.7} = 3.1$$

Self-Test 3.3

Find the range, variance, and standard deviation of the following sample of ten measurements: 6, 10, 3, 4, 4, 5, 8, 11, 4, 5.

Now, we give two rules for interpreting the standard deviation. Both rules use the mean and standard deviation of a data set to determine an interval of values within which most of the measurements fall. For samples, the intervals take the form

$$\bar{x} \pm (k)s$$

where k is any positive constant (usually 1, 2, or 3). The particular rule applied will depend on the shape of the relative frequency histogram for the data set, as the following examples illustrate.

EXAMPLE 3.9 FORMING INTERVALS AROUND THE MEAN

Periodically, the Federal Trade Commission (FTC) ranks domestic cigarette brands according to hazardous substances such as tar, nicotine, and the amount of carbon monoxide (CO) in smoke residue. The CO amounts (measured in milligrams) of a sample of 500 cigarette brands for a recent year are available in the **FTC** data file. Suppose we want to describe the distribution of CO measurements for this sample. To do so, we require the mean and standard deviation of the measurements.

a. Calculate $\bar{x}$ and s for the data set.

b. Form an interval by measuring 1 standard deviation on each side of the mean, i.e., $\bar{x} \pm s$. Also, form the intervals $\bar{x} \pm 2s$ and $\bar{x} \pm 3s$.

c. Find the proportions of the total number (500) of CO measurements falling within these intervals.

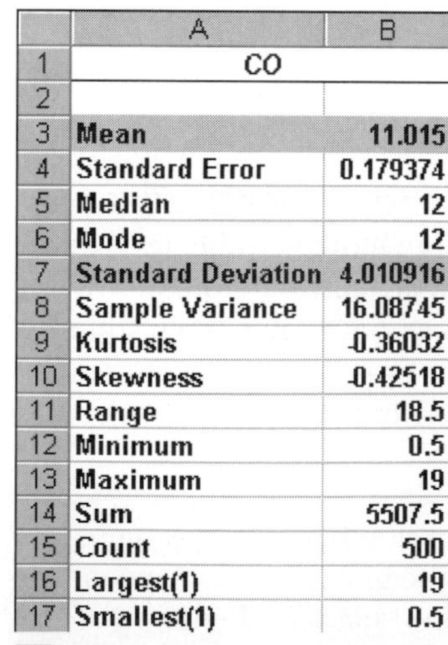

	A	B
1	CO	
2		
3	Mean	11.015
4	Standard Error	0.179374
5	Median	12
6	Mode	12
7	Standard Deviation	4.010916
8	Sample Variance	16.08745
9	Kurtosis	-0.36032
10	Skewness	-0.42518
11	Range	18.5
12	Minimum	0.5
13	Maximum	19
14	Sum	5507.5
15	Count	500
16	Largest(1)	19
17	Smallest(1)	0.5

E **Figure 3.10** Excel Printout: Descriptive Statistics for 500 CO Measurements

Solution

a. An Excel printout giving numerical descriptive measures for the data set is shown in Figure 3.10. The mean and standard deviation, highlighted on the printout, are (rounded)

$$\bar{x} = 11 \quad \text{and} \quad s = 4$$

b. The intervals $\bar{x} \pm s$, $\bar{x} \pm 2s$, and $\bar{x} \pm 3s$, are formed as follows:

$$\bar{x} \pm s = 11 \pm 4 = (11 - 4, 11 + 4) = (7, 15)$$

$$\bar{x} \pm 2s = 11 \pm 2(4) = (11 - 8, 11 + 8) = (3, 19)$$

$$\bar{x} \pm 3s = 11 \pm 3(4) = (11 - 12, 11 + 12) = (-1, 23)$$

c. We used Excel to determine the total number of CO measurements falling within the three intervals. The proportions of the total number of CO measurements falling within the three intervals are shown in Table 3.2. The three intervals—$\bar{x} \pm s$, $\bar{x} \pm 2s$, and $\bar{x} \pm 3s$—are also shown on an Excel frequency histogram for the CO data displayed in Figure 3.11. If you estimate the proportions of the total area under the histogram that lie over the three intervals, you will obtain proportions approximately equal to those given in Table 3.2.

TABLE 3.2 Proportions of the Total Number of CO Measurements in Intervals $\bar{x} \pm s$, $\bar{x} \pm 2s$, and $\bar{x} \pm 3s$

Interval	Number of Observations in Interval	Proportion in Interval
$\bar{x} \pm s$ or (7, 15)	324	.648
$\bar{x} \pm 2s$ or (3, 19)	485	.970
$\bar{x} \pm 3s$ or (−1, 23)	500	1.000

E Figure 3.11 Excel Histogram for Carbon Monoxide Measurements

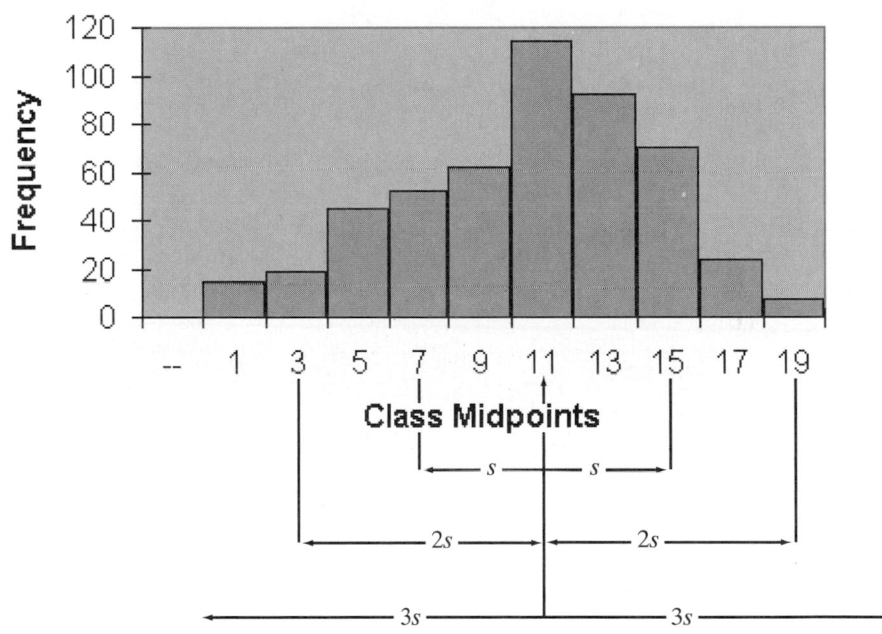

Will the proportions of the total number of observations falling within the intervals $\bar{x} \pm s$, $\bar{x} \pm 2s$, and $\bar{x} \pm 3s$ remain fairly stable for most distributions of data? To examine this possibility, consider the next example.

EXAMPLE 3.10

DESCRIPTIVE STATISTICS

Calculate the mean and standard deviation of each of the following data sets.

a. The 98 driver head-injury ratings for crash-tested cars in the **CRASH** data file
b. The 144 weights of fish specimens in the **FISH** data file
c. The 500 tar contents of cigarettes in the **FTC** data file
d. The 500 checkout times of supermarket customers in the **CHKOUT** data file

Solution

We used Excel to compute the means and standard deviations of each of the four data sets. They are shown in Table 3.3. The accompanying Excel frequency histograms are shown in Figures 3.12–3.15.

TABLE 3.3 Means and Standard Deviations for Four Data Sets

Data Set	Mean	Standard Deviation
a. Driver head-injury ratings (impact points)	604	185
b. Fish weights (grams)	1,050	377
c. Cigarette tar contents (milligrams)	11.1	4.9
d. Supermarket checkout times (seconds)	50	49

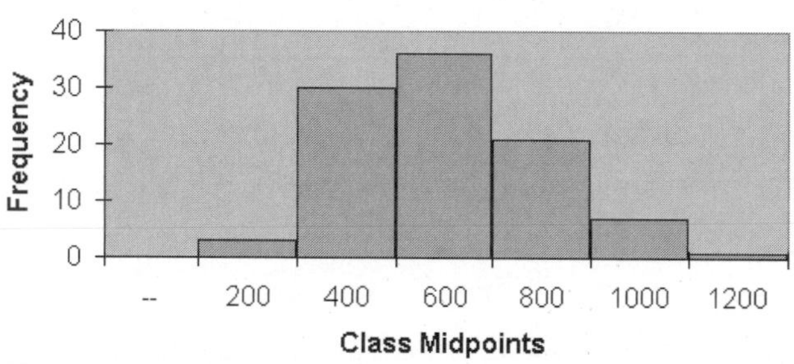

Figure 3.12 Excel Frequency Histogram for the 98 Driver Head-Injury Ratings

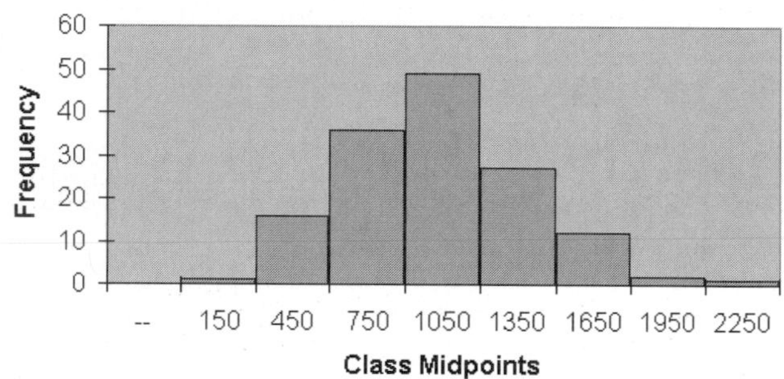

Figure 3.13 Excel Frequency Histogram for the 144 Fish Weights

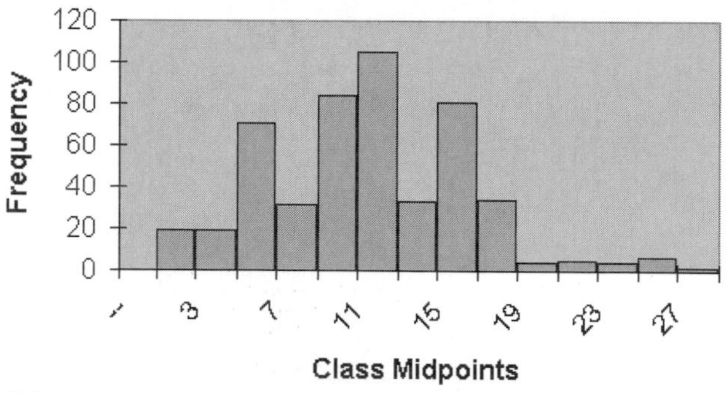

Figure 3.14 Excel Frequency Histogram for the 500 Tar Contents of Cigarettes

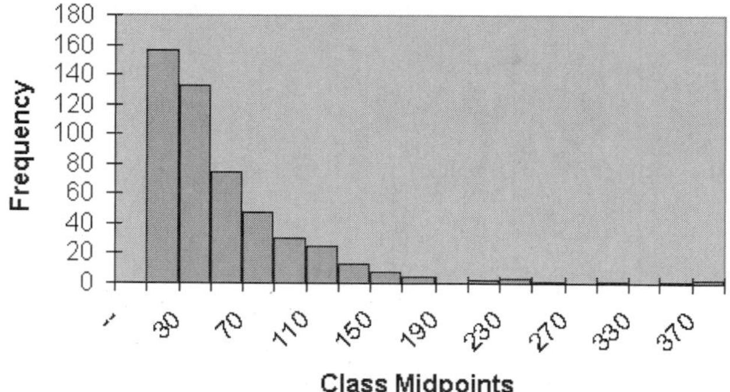

Figure 3.15 Excel Frequency Histogram for the 500 Supermarket Checkout Times

The means and standard deviations of Table 3.3 were used to calculate the intervals $\bar{x} \pm s$, $\bar{x} \pm 2s$, and $\bar{x} \pm 3s$ for each data set. From Excel, we obtained the number and proportion of the total number of observations falling within each interval. These proportions are presented in Table 3.4a–d (p. 147).

Tables 3.2 and 3.4 demonstrate a property that is common to many data sets. The percentage of observations that lie within one standard deviation of the mean $\bar{x}$, i.e., in the interval $\bar{x} \pm s$, is fairly large and variable, usually from 60% to 80% of the total number, but the percentage can reach 90% or more for highly skewed distributions of data. The percentage within two standard deviations of $\bar{x}$, i.e., in

TABLE 3.4 Proportions of the Total Number of Observations Falling within $\bar{x} \pm s$, $\bar{x} \pm 2s$, and $\bar{x} \pm 3s$

a. Driver Head-Injury Ratings	
Interval (impact points)	**Proportion in Interval**
$\bar{x} \pm s$ or (419, 789)	.694
$\bar{x} \pm 2s$ or (234, 974)	.980
$\bar{x} \pm 3s$ or (49, 1,159)	.990
b. Fish Weights	
Interval (grams)	**Proportion in Interval**
$\bar{x} \pm s$ or (673, 1,427)	.681
$\bar{x} \pm 2s$ or (296, 1,804)	.972
$\bar{x} \pm 3s$ or (−81, 2,181)	.986
c. Tar Contents of Cigarettes	
Interval (milligrams)	**Proportion in Interval**
$\bar{x} \pm s$ or (6.2, 16.0)	.582
$\bar{x} \pm 2s$ or (1.3, 20.9)	.940
$\bar{x} \pm 3s$ or (−3.6, 25.8)	.999
d. Supermarket Checkout Times	
Interval (seconds)	**Proportion in Interval**
$\bar{x} \pm s$ or (1, 99)	.860
$\bar{x} \pm 2s$ or (−48, 148)	.962
$\bar{x} \pm 3s$ or (−97, 197)	.980

the interval $\bar{x} \pm 2s$, is close to 95% but, again, the percentage will be larger for highly skewed sets of data. Finally, the percentage of observations within three standard deviations of $\bar{x}$, i.e., in the interval $\bar{x} \pm 3s$, is almost 100%, meaning that almost all of the observations in a data set will fall within this interval. This property, which seems to hold for most data sets that contain at least 20 observations and are *bell-shaped,* is called the **Empirical Rule.** The Empirical Rule provides a very good general rule for forming a mental image of a distribution of data when you know the mean and standard deviation of the data set. Calculate the intervals $\bar{x} \pm s$, $\bar{x} \pm 2s$, and $\bar{x} \pm 3s$ and then picture the observations grouped as shown in Figure 3.16 and described in the box (p. 148).

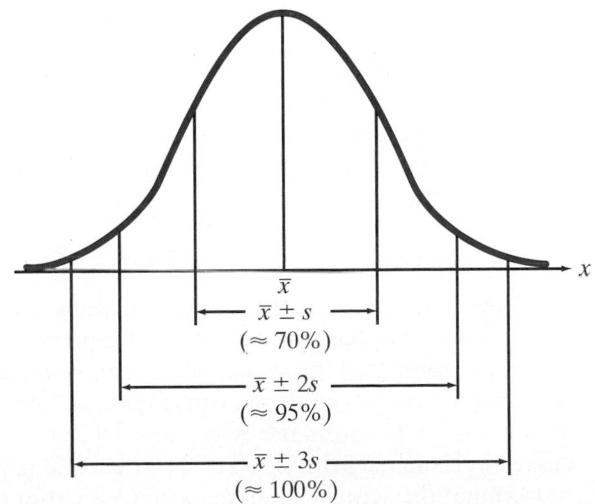

Figure 3.16 Illustration of the Empirical Rule

THE EMPIRICAL RULE

If a distribution of sample data is bell-shaped with mean $\bar{x}$ and standard deviation s, then the proportions of the total number of observations falling within the intervals $\bar{x} \pm s$, $\bar{x} \pm 2s$, and $\bar{x} \pm 3s$ are as follows:

$\bar{x} \pm s$: Usually between 60% and 80%. The percentage will be approximately 70% for distributions that are nearly symmetric, but larger (near 90%) for highly skewed distributions.

$\bar{x} \pm 2s$: Close to 95% for symmetric distributions. The percentage will be larger (near 100%) for highly skewed distributions.

$\bar{x} \pm 3s$: Near 100%.

Note that the frequency histogram of supermarket checkout times (Figure 3.15) is highly skewed. Consequently, actual percentages of observations falling within the intervals $\bar{x} \pm s$, $\bar{x} \pm 2s$, and $\bar{x} \pm 3s$ for this data set will tend to be on the high side of the range of values given by the Empirical Rule. On the other hand, the bell-shaped distributions of head-injury ratings (Figure 3.12), fish weights (Figure 3.13), and tar contents (Figure 3.14) are nearly symmetric; consequently, the percentages falling within the three intervals are very close to the values given by the Empirical Rule.

Self-Test 3.4

Suppose a set of sample data has a bell-shaped, symmetric distribution. Make a statement about the percentage of measurements contained in each of the following intervals:

a. $\bar{x} \pm s$ **b.** $\bar{x} \pm 2s$ **c.** $\bar{x} \pm 3s$

Can the Empirical Rule be applied to data sets with non-bell-shaped histograms or histograms of unknown shape? The answer, unfortunately, is no. However, in these situations we can apply a more conservative rule, called **Tchebysheff's Theorem.**

For $k = 1$,

$$1 - \frac{1}{k^2} = 1 - \frac{1}{(1)^2} = 0.$$

Thus, Tchebysheff's Theorem states that at least 0% of the observations fall within $\bar{x} \pm s$. Consequently, no useful information is provided about the interval.

TCHEBYSHEFF'S THEOREM

For any set of sample measurements with mean $\bar{x}$ and standard deviation s, at least $(1 - 1/k^2)$ of the total number of observations in a sample data set will fall within the interval $\bar{x} \pm ks$, where k is a constant. For two useful values of k, $k = 2$ and $k = 3$, the theorem states that the proportions of the total number of observations in the sample falling within the intervals $\bar{x} \pm 2s$ and $\bar{x} \pm 3s$ are:

$\bar{x} \pm 2s$: At least 75% $\bar{x} \pm 3s$: At least 89%

Note that Tchebysheff's Theorem applies to any set of sample measurements, regardless of the shape of the relative frequency histogram. The rule is conservative in the sense that the specified percentage for any interval is a lower bound on the actual percentage of measurements falling in that interval. For example, Tchebysheff's Theorem states that at least 75% of the 500 CO measurements discussed in Example 3.9 will fall in the interval $\bar{x} \pm 2s$. We know (from Table 3.2 on p. 145) that the actual percentage (97%) is closer to the Empirical Rule's value of

95%. Consequently, whenever you know that a relative frequency histogram for a data set is bell-shaped, the Empirical Rule will give more precise estimates of the true percentages falling within the intervals $\bar{x} \pm s$, $\bar{x} \pm 2s$, and $\bar{x} \pm 3s$.

The last example in this section demonstrates the use of these rules in statistical inference.

EXAMPLE 3.11

APPLICATION OF THE EMPIRICAL RULE

Travelers who have no intention of showing up often fail to cancel their hotel reservations in a timely manner. These travelers are known, in the parlance of the hospitality trade, as "no-shows." To protect against no-shows and late cancellations, hotels invariably overbook rooms. A study reported in the *Journal of Travel Research* examined the problems of overbooking rooms in the hotel industry. The data in Table 3.5, extracted from the study, represent daily numbers of late cancellations and no-shows for a random sample of 30 days at a large (500-room) hotel. Based on this sample, how many rooms, at minimum, should the large hotel overbook each day?

NOSHOW

TABLE 3.5 Hotel No-Shows for a Sample of 30 Days

18	16	16	16	14	18	16	18	14	19
15	19	9	20	10	10	12	14	18	12
14	14	17	12	18	13	15	13	15	19

Source: Toh, R. S. "An inventory depletion overbooking model for the hotel industry." *Journal of Travel Research,* Vol. 23, No. 4, Spring 1985, p. 27. The *Journal of Travel Research* is published by the Travel and Tourism Research Association (TTRA) and the Business Research Division, University of Colorado at Boulder.

Solution

To answer this question, we need to know a range of values within which most of the daily numbers of no-shows fall. This requires that we compute $\bar{x}$ and s, and examine the shape of the relative frequency distribution for the data.

Figures 3.17a and b (shown on p. 150) are a histogram and descriptive statistics for the sample data obtained using MINITAB. Notice from the histogram that the distribution of daily no-shows is bell-shaped, and only slightly skewed on the low (left) side of Figure 3.17a. Thus, the Empirical Rule should give a good estimate of the percentage of days that fall within one, two, and three standard deviations of the mean.

The mean and standard deviation of the sample data, highlighted on Figure 3.17b, are $\bar{x} = 15.133$ and $s = 2.945$. From the Empirical Rule, we know that about 95% of the daily number of no-shows fall within two standard deviations of the mean, i.e., within the interval

$$\bar{x} \pm 2s = 15.133 \pm 2(2.945)$$
$$= 15.133 \pm 5.890$$

or between 9.243 no-shows and 21.023 no-shows. (If we count the number of measurements in this data set, we find that actually 29 out of 30, or 96.7%, fall in this interval).

From this result, the large hotel can infer that there will be at least 9.243 (or, rounding up, 10) no-shows per day. Consequently, the hotel can overbook at least 10 rooms per day and still be highly confident that all reservations can be honored.

Figure 3.17 MINITAB Printout Describing the No-Show Data, Example 3.11

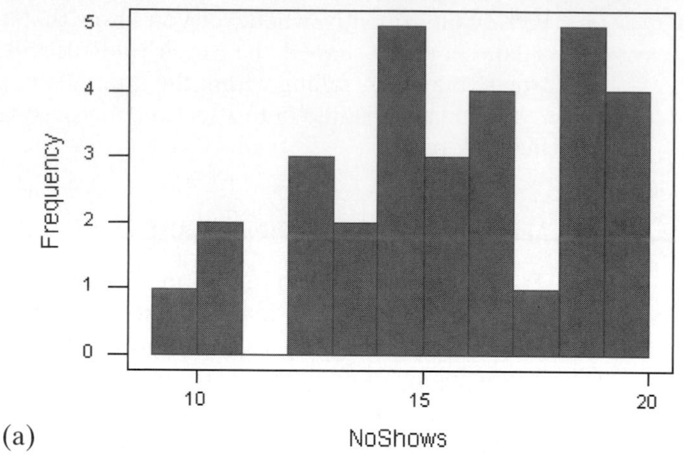

(a)

(b) **Descriptive Statistics**

Variable	N	Mean	Median	TrMean	StDev	SE Mean
NoShows	30	15.133	15.000	15.231	2.945	0.538

Variable	Minimum	Maximum	Q1	Q3
NoShows	9.000	20.000	13.000	18.000

Statistics in the Real World Revisited

Interpreting the Standard Deviation

We return to the highway construction bid data (**BIDRIG** file) collected by the Florida Department of Transportation (p. 129) and the comparison of the ratios of winning bid and DOT estimate for fixed and competitively bid road contracts. The MINITAB descriptive statistics printout for the 279 contracts is reproduced in Figure 3.18, with the means and standard deviations highlighted.

Figure 3.18 MINITAB Analysis of Florida DOT Data

Descriptive Statistics

Variable	STATUS	N	Mean	Median	TrMean	StDev
LBERAT	Compet	194	0.90747	0.91000	0.90776	0.13699
	Fixed	85	1.1124	1.0900	1.1082	0.1481

Variable	STATUS	SE Mean	Minimum	Maximum	Q1	Q3
LBERAT	Compet	0.00984	0.53000	1.35000	0.82000	0.99250
	Fixed	0.0161	0.8100	1.5600	1.0150	1.2000

We substitute these values into the formula $\bar{x} \pm 2s$ to obtain the intervals:

Competitive: $.907 \pm 2(.137) = .907 \pm .274 = (.633, 1.181)$

Fixed: $1.112 \pm 2(.148) = 1.112 \pm .296 = (.816, 1.408)$

From the Empirical Rule, we know that around 95% of the competitive winning bids will have bid-to-estimate ratios between .633 and 1.181. Similarly, we know that about 95% of the fixed contracts will have bid-to-estimate ratios between .816 and 1.408. FLAG used this type of information to identify future rigged bids. For example, if a road contract has a winning bid/DOT estimate ratio of 1.3, it is very unlikely to come from the distribution for competitively bid contracts since 1.3 is above the $\bar{x} \pm 2s$ interval for competitive contracts. Rather, the ratio of 1.3 is much more likely to come from the distribution of rigged contracts.

PROBLEMS FOR SECTION 3.4

Using the Tools

3.24 Calculate the variance and standard deviation of samples for which

a. $n = 10$ $\Sigma x^2 = 331$ $\Sigma x = 50$

b. $n = 25$ $\Sigma x^2 = 163{,}456$ $\Sigma x = 2{,}000$

c. $n = 5$ $\Sigma x^2 = 26.46$ $\Sigma x = 11.5$

3.25 Calculate the range, variance, and standard deviation of each of the following examples:

a. 0, 2, 4, 6, 8, 10 **b.** 0, 4, 5, 5, 6, 10 **c.** 4, 4, 4, 4, 4, 4

3.26 Find the range, variance, and standard deviation for the following data set: 3, 9, 0, 7, 4.

3.27 Find the mean and standard deviation for the following $n = 25$ measurements: 2, 1, 7, 6, 5, 3, 8, 5, 2, 4, 5, 6, 3, 4, 4, 6, 9, 4, 3, 4, 5, 5, 7, 3, 5.

3.28 Refer to the data given in Problem 3.27.

a. Construct the intervals $\bar{x} \pm s$, $\bar{x} \pm 2s$, and $\bar{x} \pm 3s$.

b. Count the number of observations falling within each interval and find the corresponding proportions. Compare your results to the Empirical Rule and Tchebysheff's Theorem.

3.29 Find the mean and standard deviation for the following $n = 20$ measurements: 11, 16, 12, 12, 21, 4, 13, 10, 17, 12, 15, 12, 9, 11, 14, 13, 12, 15, 10, 16.

3.30 Refer to the data given in Problem 3.29.

a. Construct the intervals $\bar{x} \pm s$, $\bar{x} \pm 2s$, and $\bar{x} \pm 3s$.

b. Count the number of observations falling within each interval and find the corresponding proportions. Compare your results to the Empirical Rule and Tchebysheff's Theorem.

Applying the Concepts

3.31 The *Consumer Reports* (June 2000) data on bounced-check fees charged by each in a sample of 23 U.S. metropolitan banks, Problem 2.21 (p. 75), are repeated below.

BANKCOST

26 28 20 20 21 22 25 25 18 25 15 20 18 20 25 25
22 30 30 30 15 20 29

Source: "The new face of banking," *Consumer Reports,* June 2000.

a. Find the range for the data set.

b. Find the variance for the data set.

c. Find the standard deviation for the data set.

d. Find the mean for the data set.

e. Construct an interval that will include about 95% of the 23 bank fees.

3.32 The data on calories per 12-ounce serving for 22 beer brands, Problem 2.22 (p. 76), are repeated in the table on p. 152.

a. Find the range for the data set.

b. Find the variance for the data set.

c. Find the standard deviation for the data set.

d. Find the mean for the data set.

e. Construct an interval that will include about 95% of the 22 caloric measurements.

3.33 Refer to the *Journal of Applied Behavior Analysis* (Summer 1997) study of a 16-year-old boy's destructive behavior, Problem 3.16, p. 139. The data on the number of destructive responses per minute observed during each of 10 sessions is reproduced below.

BOY16

3 2 4.5 5.5 4.5 3 2.5 1 3 3

BEERCAL

Beer	Calories	Beer	Calories
Old Milwaukee	145	Miller High Life	143
Stroh's	142	Pabst Blue Ribbon	144
Red Dog	147	Milwaukee's Best	133
Budweiser	148	Miller Genuine Draft	143
Icehouse	149	Rolling Rock	143
Molson Ice	155	Michelob Light	134
Michelob	159	Bud Light	110
Bud Ice	148	Natural Light	110
Busch	143	Coors Light	105
Coors Original	137	Miller Light	96
Gennessee Cream Ale	153	Amstel Light	95

Source: Consumer Reports, June 1996, Vol. 61, No. 6, p. 16.

Problem 3.32

a. Find the range for the data set.

b. Find the standard deviation for the data set.

c. Use the mean from Problem 3.16 and the standard deviation from part **b** to construct an interval that will be likely to include the number of destructive responses per minute in any given 10-minute session.

3.34 Refer to the *American Journal of Audiology* (Mar. 1995) study of patients with unilateral hearing loss, Problem 3.21, p. 140. The absolute sound pressure levels (SPLs) for the eight patients (recorded in decibels) is reproduced below.

SPL

73.0	80.1	82.8	76.8	73.5	74.3	76.0	68.1

a. Find the range for the data set.

b. Find the standard deviation for the data set.

c. Use the mean from Problem 3.21 and the standard deviation from part **b** to construct an interval that will be likely to include the SPL of any given patient.

EVOS

3.35 Refer to the *Journal of Agricultural, Biological, and Environmental Statistics* (Sept. 2000) study on the impact of the *Exxon Valdez* tanker oil spill on the Alaska seabird population, Problem 2.62 (p. 107). The **EVOS** file contains data on the number of seabirds found at each of 96 transects and whether or not the transect was in an oiled area. Recall that the researchers are interested in studying the observed seabird densities of the transects (where density is defined as the observed count divided by the length of the transect). MINITAB descriptive statistics for seabird densities in unoiled (No) and oiled (Yes) transects are displayed below, followed by histograms for each type of transect on p. 153.

Descriptive Statistics: DENSITY by OIL

```
Variable   Oil          N      Mean    Median    TrMean     StDev
Density    no          36      3.27      0.89      2.05      6.70
           yes         60     3.495     0.700     2.542     5.968

Variable   Oil    SE Mean   Minimum   Maximum        Q1        Q3
Density    no        1.12      0.00     36.23      0.00      3.87
           yes      0.770     0.000    32.836     0.000     5.233
```

M **Problem 3.35**

a. For unoiled transects, give an interval of values that is likely to contain at least 75% of the seabird densities.

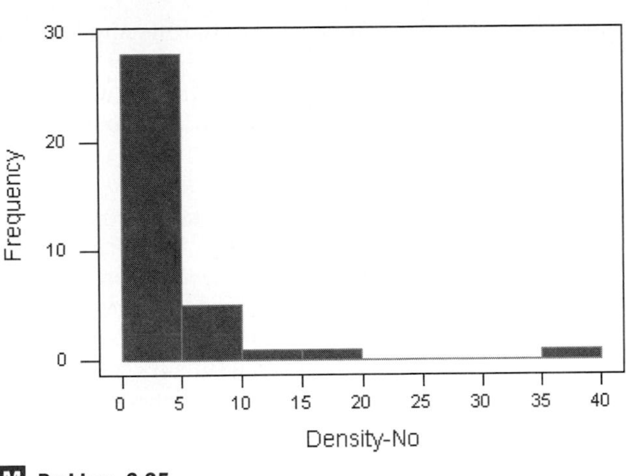

M Problem 3.35

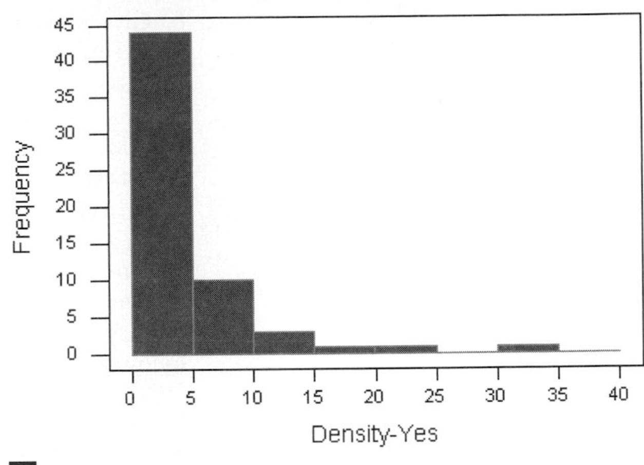

M Problem 3.35

b. For oiled transects, give an interval of values that is likely to contain at least 75% of the seabird densities.

c. Which type of transect, an oiled or unoiled one, is more likely to have a seabird density of 16? Explain.

3.36 Refer to the *Psychological Science* (Mar. 1995) experiment in which 120 subjects were given the "water-level task," Problem 2.39 (p. 88). Recall that the subjects are shown a drawing of a glass tilted at a 45° angle and asked to draw a line representing the surface of the water. The researchers recorded the deviation (in angle degrees) of the judged line from the true line. The deviations for the 120 subjects are contained in the data file **WATERTASK.** (The first 5 and last 5 observations in the data set are listed below.)

WATERTASK

10	20	9	10	6	...	3	0	−4	5	3

a. Use Excel or MINITAB to produce numerical descriptive measures for the 120 deviations in the data set.

b. Use the mean, standard deviation, and the Empirical Rule to form an interval that should contain about 95% of the deviations in the data set.

c. Perform an actual count of the number of the 120 deviations that actually fall within the interval of part **b**. Is the actual percentage approximately equal to 95%?

d. If one subject (drawn from the same population as the 120 subjects in the study) performs the water-level task, is it likely he or she will draw a line with a deviation of 25°? Explain.

e. If one subject (drawn from the same population as the 120 subjects in the study) performs the water-level task, would it be unusual to observe a line with a deviation of −10°? Explain.

3.37 In a study of how external clues influence performance, psychology professors at the University of Alberta and Pennsylvania State University gave two different forms of a midterm examination to a large group of introductory psychology students. The questions on the exam were identical and in the same order, but one exam was printed on blue paper and the other on red paper. (*Teaching Psychology,* May 1998.) Grading only the difficult questions on the exam, the researchers found that scores on the blue exam had a distribution with a mean of 53% and a standard deviation of 15%, while scores on the red exam had a distribution with a mean of 39% and a standard deviation of 12%. (Assume that both distributions are approximately bell-shaped and symmetric.)

a. Give an interpretation of the standard deviation for the students who took the blue exam.

b. Give an interpretation of the standard deviation for the students who took the red exam.

c. Suppose a student is selected at random from the group of students who participated in the study and the student's score on the difficult questions is 20%. Which exam form is the student more likely to have taken, the blue or the red exam? Explain.

3.38 A Harris Corporation/University of Florida study was undertaken to compare the voltage readings of a manufacturing process performed at two locations—an old, remote location and a new location closer to the plant. Test devices were set up at both locations and voltage readings on the process were obtained. [*Note:* A "good process" was considered to be one with voltage readings of at least 9.2 volts.] The following table contains voltage readings for 30 production runs at each location. Descriptive statistics for both sample data sets are provided in the accompanying Excel printout. Use the Empirical Rule to compare the voltage reading distributions for the two locations.

Old Location			New Location		
9.98	10.12	9.84	9.19	10.01	8.82
10.26	10.05	10.15	9.63	8.82	8.65
10.05	9.80	10.02	10.10	9.43	8.51
10.29	10.15	9.80	9.70	10.03	9.14
10.03	10.00	9.73	10.09	9.85	9.75
8.08	9.87	10.01	9.60	9.27	8.78
10.55	9.55	9.98	10.05	8.83	9.35
10.26	9.95	8.72	10.12	9.39	9.54
9.97	9.70	8.80	9.49	9.48	9.36
9.87	8.72	9.84	9.37	9.64	8.68

Source: Harris Corporation, Melbourne, Fla.

	A	B	C
1	*Old Location*		*New*
2			
3	Mean	9.803667	9.422333
4	Standard Error	0.098757	0.08743
5	Median	9.975	9.455
6	Mode	9.98	8.82
7	Standard Deviation	0.540915	0.478876
8	Sample Variance	0.29259	0.229322
9	Kurtosis	3.473528	-0.89134
10	Skewness	-1.87787	-0.26699
11	Range	2.5	1.61
12	Minimum	8.05	8.51
13	Maximum	10.55	10.12
14	Sum	294.11	282.67
15	Count	30	30
16	Largest(1)	10.55	10.12
17	Smallest(1)	8.05	8.51

E Problem 3.38

3.39 Marine scientists at the University of South Florida investigated the feeding habits of midwater fish inhabiting the eastern Gulf of Mexico (*Prog. Oceanog.,* Vol. 38, 1996). Most of the fish captured fed heavily on copepods, a type of zooplankton. The table on p. 155 gives the percent of shallow-living copepods in the diets of each of 35 species of myctophid fish.

COPEPOD

79	100	100	98	95	96	56	51	90	95	94	93
92	71	100	100	99	100	100	100	99	80	54	56
59	92	88	81	88	59	100	85	82	74	66	

Source: Hopkins, T. L., Sutton, T. T., and Lancraft, T. M. "The tropic structure and predation impact of low latitude midwater fish assemblage." *Prog. Oceanog.,* Vol. 38, No. 3, 1996, p. 223 (Table 6).

a. Find the mean, variance, and standard deviation of the data set.

b. Form the interval $\bar{x} \pm 2s$. Make a statement about the proportion of measurements that will fall within this interval.

c. Count the number of measurements that fall within the interval $\bar{x} \pm 2s$. Compare the result to your answer in part **b**.

3.40 A study published in *Applied Psycholinguistics* (June 1998) compared the language skills of young children (16–30 months old) from low income and middle income families. A total of 260 children—65 in the low income group and 195 in the middle income group—completed the Communicative Development Inventory (CDI) exam. One of the variables measured on each child was sentence complexity score. Descriptive statistics for the scores of the two groups are reproduced in the table.

	Low Income	Middle Income
Sample size	65	195
Mean	7.62	15.55
Median	4	14
Standard deviation	8.91	12.24
Minimum	0	0
Maximum	36	37

Source: Arriaga, R. I., et al. "Scores on the MacArthur Communicative Development Inventory of children from low-income and middle-income families." *Applied Psycholinguistics,* Vol. 19, No. 2, June 1998, p. 217 (Table 7).

a. Assuming the distribution of scores is bell-shaped and symmetric, use this information to give an interval which contains about 95% of the sentence complexity scores for each income group.

b. Which income group is more likely to have a child with a high sentence complexity score (e.g., a score of 30)? Explain.

c. For which of the two income groups is the assumption of a bell-shaped, symmetric distribution unlikely to be true? Explain.

3.41 Refer to the *Chemosphere* study of Vietnam veterans possibly exposed to Agent Orange, Problem 2.28 (p. 78). The data on TCDD levels (in parts per trillion) in the plasma of 20 veterans is reproduced in the table.

a. Compute $\bar{x}$, s^2, and s for this data set.

b. What percentage of measurements would you expect to find in the interval $\bar{x} \pm 2s$?

TCDD

TCDD Levels in Plasma

2.5	1.8	6.9	1.8	3.5	2.5	1.6	36.0	6.8	3.1
20.0	3.3	4.7	3.1	4.1	7.2	4.6	3.0	2.1	2.0

Source: Schecter, A., et al. "Partitioning of 2,3,7,8-chlorinated dibenzo-p-dioxins and dibenzofurans between adipose tissue and plasma lipid of 20 Massachusetts Vietnam veterans." *Chemosphere,* Vol. 20, Nos. 7–9, 1990, pp. 954–955 (Table I).

c. Count the number of measurements that actually fall within the interval of part **b** and express the interval count as a percentage of the total number of measurements. Compare this result with the answer to part **b**.

d. Suppose the veteran that had the highest TCDD level (36) was omitted from the analysis. Would you expect $\bar{x}$ to increase or decrease? Would you expect s to increase or decrease? Explain.

3.5 Measures of Relative Standing: Percentiles and *z* Scores

In some situations, you may want to describe the relative position of a particular measurement in a data set. For example, suppose a recent graduate of your college secured a job with a starting salary of $46,000. You might want to know whether this is a relatively low or high starting salary, etc. What percentage of the starting salaries of graduates of your college were less than $46,000; what percentage were larger? Descriptive measures that locate the relative position of a measurement—in relation to the other measurements—are called *measures of relative standing.* One measure that expresses this position in terms of a percentage is called a *percentile* for the data set.

Definition 3.7

Let $x_1, x_2, \ldots, x_n$ be a set of n measurements arranged in increasing (or decreasing) order. The pth **percentile** is a number x such that $p\%$ of the measurements fall below the pth percentile and $(100 - p)\%$ fall above it.

EXAMPLE 3.12 **INTERPRETING A PERCENTILE**

Suppose a starting salary of $46,000 falls at the 95th percentile of the distribution of starting salaries for recent graduates of your college. What does this imply?

Solution

Using $p = 95$ in Definition 3.7, 95% of the numbers in a data set fall below the 95th percentile and $(100 - 95)\% = 5\%$ fall above it. Thus, 95% of the starting salaries are less than $46,000 and 5% are greater.

The median, by definition, is the 50th percentile. The 25th percentile, the median, and the 75th percentiles are often used to describe a data set because they divide the data set into four groups, with each group containing one-fourth (25%) of the observations. They also divide the relative frequency histogram for a data set into four parts, each containing the same area (.25), as shown in Figure 3.19. Consequently, the 25th percentile, the median, and the 75th percentile are called the *1st quartile,* the *mid-quartile,* and the *3rd quartile,* respectively, for a data set.

Definition 3.8

The **1st quartile,** Q_1, for a data set is the 25th percentile.

Definition 3.9

The **mid-quartile** (median), M, for a data set is the 50th percentile.

Definition 3.10

The **3rd quartile,** Q_3, for a data set is the 75th percentile.

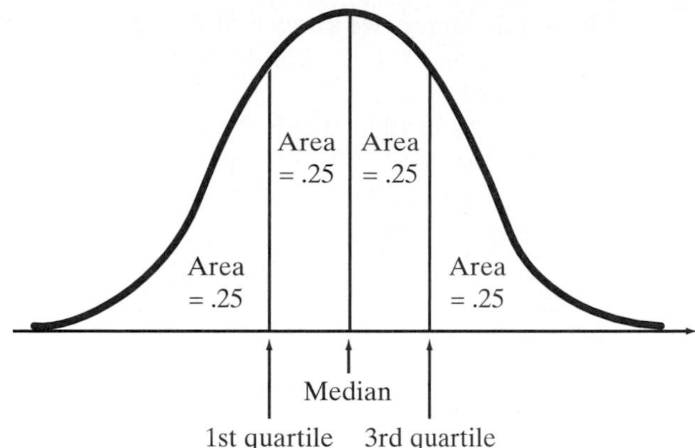

Figure 3.19 Locations of the Lower and Upper Quartiles

For large data sets, percentiles can be found by locating the corresponding areas under the relative frequency distribution. They also can be found by ranking the data and determining the percentiles of interest. Such computations can be performed using Excel or MINITAB.

EXAMPLE 3.13

FINDING PERCENTILES

Refer to the NCAP crash test data in the **CRASH** file. Suppose we are interested in the driver head-injury ratings for the 98 crash-tested cars.

a. Find and interpret the values of Q_1, M, and Q_3.

b. Find and interpret the 90th percentile for the data set.

Solution

a. Descriptive statistics for the driver head-injury ratings are displayed in the Excel printouts, Figure 3.20. The values of Q_1, M, and Q_3, highlighted on Figure 3.20a, are 475, 605, and 724, respectively. From these values, we know that 25% of the 98 head-injury ratings fall below the 1st quartile, 475; 50% fall below the median, 605; and 75% fall below the 3rd quartile, 724.

E **Figure 3.20** PHStat and Excel Descriptive Statistics for Driver Head-Injury Ratings

	A	B	C
1	Five-Number Summary for DriveHead		
2			
3	Five-number Summary		
4	Minimum	216	
5	First Quartile	475	
6	Median	605	
7	Third Quartile	724	
8	Maximum	1240	

a.

	A	B
1	*Drive head*	
2		
3	Mean	603.7449
4	Standard Error	18.72428
5	Median	605
6	Mode	528
7	Standard Deviation	185.3609
8	Sample Variance	34358.67
9	Kurtosis	0.473236
10	Skewness	0.473483
11	Range	1024
12	Minimum	216
13	Maximum	1240
14	Sum	59167
15	Count	98
16	Largest(1)	1240
17	Smallest(88)	874

b.

b. By definition, we know that 90% of the head-injury ratings in the data set fall below the 90th percentile. If we rank the 98 head-injury ratings in increasing order, where the rating with rank 1 is the smallest and the rating with rank 98 is the largest, then the 90th percentile is the value with rank (.90)(98) = 88.2, or 88. This value, labeled **Smallest (88)** on the Excel printout, Figure 3.20b, is 874. Thus, 90% of the head-injury ratings in the data set fall below 874.

Self-Test 3.5

State the percentage of measurements that are above and below each of the following percentiles:

a. 20th percentile
b. Median
c. 86th percentile
d. Lower quartile
e. Upper quartile
f. 10th percentile

When the sample data set is small, it may be impossible to find a measurement in the data set that exceeds, for example, *exactly* 25% of the remaining measurements. Consequently, percentiles for small data sets are not well defined. The box describes a procedure for finding quartiles and other percentiles with small data sets.

Finding Quartiles (and Percentiles) with Small Data Sets

Step 1 Rank the n measurements in the data set in increasing order of magnitude.

Step 2 Calculate the quantity $\frac{1}{4}(n + 1)$ and round to the nearest integer. The measurement with this rank represents the lower quartile or 25th percentile.

Step 3 Calculate the quantity $\frac{3}{4}(n + 1)$ and round to the nearest integer. The measurement with this rank represents the upper quartile or 75th percentile.

General To find the pth percentile, calculate the quantity

$$\frac{p(n + 1)}{100} \qquad \textbf{(3.6)}$$

and round to the nearest integer. The measurement with this rank is the pth percentile.

Another measure of relative standing is the *z score* for a measurement. For example, suppose you were told that $42,000 lies 1.44 standard deviations above the mean of the distribution of starting salaries of recent graduates of your college. Knowing that most of the starting salaries will be less than two standard deviations from the mean and almost all will be within three, you would have a good idea of the relative standing of the $42,000 starting salary. The distance that a measurement x lies above or below the mean $\bar{x}$ of a data set, measured in units of the standard deviation s, is called the *z score* for the measurement. A negative z score indicates that the observation lies to the left of the mean; a positive z score indicates that the observation lies to the right of the mean.

Definition 3.11

The sample ***z* score** for the measurement x is the difference between x and the sample mean, divided by the sample standard deviation:

$$z = \frac{x - \bar{x}}{s} \qquad (3.7)$$

A negative z score indicates that the observation lies to the left of the mean; a positive z score indicates that the observation lies to the right of the mean.

EXAMPLE 3.14

COMPUTING A *z* SCORE

In Example 3.10, we noted that the mean and standard deviation for the 98 driver head-injury ratings in the **CRASH** data file are $\bar{x} = 604$ and $s = 185$. Use these values to find the z score for a driver head-injury rating of 408. Interpret the result.

Solution

Substituting the values of $x, \bar{x}$, and s into the formula for z (equation 3.7), we obtain

$$z = \frac{x - \bar{x}}{s} = \frac{408 - 604}{185} = -1.06$$

Since the z score is negative, we conclude that the head-injury rating of 408 lies a distance of 1.06 standard deviations below (to the left of) the mean of 604.

In the following sections, we discuss how percentiles and z scores can be used to detect skewness and unusual observations in a data set.

PROBLEMS FOR SECTION 3.5

Using the Tools

3.42 Compute the z score corresponding to each x value, assuming $\bar{x} = .7$ and $s = .2$.

a. $x = .9$ **b.** $x = .4$ **c.** $x = 1.3$

3.43 Compute the z score corresponding to each x value, assuming $\bar{x} = 20$ and $s = 5$.

a. $x = 12$ **b.** $x = 23$ **c.** $x = 28$

3.44 The 24 sample measurements of Problem 2.17 (p. 74) are reproduced here.

213	228	241	268	234	303	274	316	319	320	227	226
224	267	303	266	265	237	288	291	285	270	254	215

a. Use the stem-and-leaf display you constructed in Problem 2.17 to find Q_1, M, and Q_3.

b. Find the 90th percentile of the data set.

3.45 The 28 sample measurements of Problem 2.18 (p. 75) are reproduced here.

5.9	5.3	1.6	7.4	8.6	1.2	2.1	4.0	7.3	8.4	8.9	6.7
4.5	6.3	7.6	9.7	3.5	1.1	4.3	3.3	8.4	1.6	8.2	6.5
1.1	5.0	9.4	6.4								

a. Use the stem-and-leaf display you constructed in Problem 2.18 to find Q_1, M, and Q_3.

b. Find the 10th percentile of the data set.

3.46 Compute z scores for each of the following situations. Then determine which x value lies the greatest distance above the mean; the greatest distance below the mean.

a. $x = 77, \bar{x} = 58, s = 8$ **b.** $x = 8.8, \bar{x} = 11, s = 2$

c. $x = 0, \bar{x} = -5, s = 1.5$ **d.** $x = 2.9, \bar{x} = 3, s = .1$

3.47 State the percentage of measurements that are larger than each of the following percentiles:

a. 60th percentile **b.** 95th percentile
c. 33rd percentile **d.** 15th percentile

Applying the Concepts

3.48 The mean and standard deviation of the 50 starting salaries listed in Table 2.5 (p. 70) are $\bar{x}$ = \$35,708 and s = \$9,338. Find and interpret the z score for a college graduate with a starting salary of:

a. \$25,300 **b.** \$40,900

3.49 According to the National Center for Education Statistics (2000), scores on a mathematics assessment test for United States eighth-graders have a mean of 500, a 5th percentile of 356, a 25th percentile of 435, a 75th percentile of 563, and a 95th percentile of 653. Interpret each of these numerical descriptive measures.

3.50 Suicide is the leading cause of death of Americans incarcerated in correctional facilities. To determine what factors increase the risk of suicide in urban jails, a group of researchers collected data on all 37 suicides that occurred from 1967 to 1992 in the Wayne County Jail, Detroit, Michigan (*American Journal of Psychiatry,* July 1995). One of the variables measured was the number of days the inmate spent in jail before committing suicide. The data are listed in the accompanying table.

SUICIDE

3	4	5	7	10	15	15	19	22	29	31	85	126	221	14	22	42	122	309
69	143	6	1	1	1	2	3	4	4	4	6	6	10	18	26	41	86	

a. Construct a stem-and-leaf display for the data.
b. Compute and interpret the values of Q_1, M, and Q_3.
c. Compute and interpret the 90th percentile for the data.

	A	B
1	Five-Number Summary of	
2	Ship Sanitation Scores	
3		
4	Minimum	36
5	First Quartile	88
6	Median	92
7	Third Quartile	94
8	Maximum	99

E **Problem 3.51**

3.51 Refer to the CDC study of sanitation levels for 121 international cruise ships, Problem 3.17 (p. 139). An Excel printout of descriptive statistics for the data is reproduced at left. Locate Q_1, M, and Q_3 on the printout and interpret these values.

3.52 A biologist investigating the parent-young conflict in herring gulls recorded the feeding rates (the number of feedings per hour) for two groups of gulls: those parents with only one chick to feed and those with two or three chicks to feed. The results are summarized in the following table.

	Brood size (Number of Chicks)	
	1	**2 or 3**
Mean feeding rate	.18	.31
Standard deviation	.15	.13

a. A particular pair of parent gulls fed the three chicks in their brood at a rate of .44 times per hour. Find the z score for this feeding rate and interpret its value.
b. Would you expect to observe a feeding rate of .44 for parent gulls with a brood size of only one chick? Explain.

3.53 Refer to the *Administrative Science Quarterly* entrapment experiment, Problem 3.22 (p. 140). The data (i.e., total points awarded) for the 30 trials are reproduced in the table on p. 161.

a. Calculate and interpret a measure of relative standing for a trial in which five points were awarded.
b. Calculate and interpret a measure of relative standing for a trial in which 20 points were awarded.

ENTRAP

5	5	4	7	24	6
10	12	11	15	11	10
3	23	4	20	5	4
7	5	6	6	15	5
15	10	13	9	4	6

Source: Brockner, J., et al. "Escalation of commitment to an ineffective course of action: The effect of feedback having negative implications for self-identity." *Administrative Science Quarterly,* Vol. 31, No. 1, Mar. 1986, p. 115.

Problem 3.53

3.54 Refer to the *Journal of Geography* study comparing test scores of geography students who used computer resources and those who did not, Problem 3.23 (p. 141). Excel descriptive statistics for the test scores of the two groups of students are shown here.

	A	B	C
1		*Computer Resources*	*No Computer Resources*
2			
3	Mean	68.86046512	50.04545455
4	Standard Error	2.693164596	4.068396449
5	Median	69	53
6	Mode	91	66
7	Standard Deviation	17.66026128	19.08247082
8	Sample Variance	311.8848283	364.1406926
9	Kurtosis	-0.352248464	-0.255178597
10	Skewness	-0.482559525	-0.665443331
11	Range	69	69
12	Minimum	28	9
13	Maximum	97	78
14	Sum	2961	1101
15	Count	43	22
16	Largest(1)	97	78
17	Smallest(1)	28	9

E **Problem 3.54**

a. A student in group 1 (computer resources group) scored a 72 on the test. Calculate the z score for this student.

b. A student in group 2 (no computer) scored a 66 on the test. Calculate the z score for this student.

c. One student scored a 30 on the test. (This test score was previously unrecorded.) What group is this student more likely to come from? Explain.

3.55 Refer to the *Prog. Oceanog.* study of the diets of myctophid fish living in the Gulf of Mexico, Problem 3.39 (p. 154). The data on percent of shallow-living copepods in the diets of the 35 fish species are reproduced below.

COPEPOD

79	100	100	98	95	96	56	51	90	95	94	93
92	71	100	100	99	100	100	100	99	80	54	56
59	92	88	81	88	59	100	85	82	74	66	

Source: Hopkins, T. L., Sutton, T. T., and Lancraft, T. M. "The tropic structure and predation impact of low latitude midwater fish assemblage." *Prog. Oceanog.,* Vol. 38, No. 3, 1996, p. 223 (Table 6).

a. Find Q_1, M, and Q_3 for this data set. Interpret these values.

b. Find the 90th percentile for the data set. Interpret this value.

c. Find the z score for the fish species with 81% of copepods in its diet.

3.56 The data on TCDD levels in the plasma of 20 Vietnam veterans from Problem 3.41 (p. 155) is reproduced here. In addition, the TCDD levels in fat tissue are also recorded in the table.

TCDDFAT

TCDD Levels in Plasma				TCDD Levels in Fat Tissue			
2.5	1.8	6.9	1.8	4.9	1.1	7.0	4.2
3.5	2.5	1.6	36.0	6.9	2.3	1.4	41.0
6.8	3.1	20.0	3.3	10.0	5.9	11.0	2.9
4.7	3.1	4.1	7.2	4.4	7.0	2.5	7.7
4.6	3.0	2.1	2.0	4.6	5.5	4.4	2.5

Source: Schecter, A., et al. "Partitioning of 2,3,7,8-chlorinated dibenzo-p-dioxins and dibenzofurans between adipose tissue and plasma lipid of 20 Massachusetts Vietnam veterans." *Chemosphere,* Vol. 20, Nos. 7–9, 1990, pp. 954–955 (Tables I and II).

a. Calculate $\bar{x}$ and s for the TCDD levels in plasma.

b. Calculate $\bar{x}$ and s for the TCDD levels in fat tissue.

c. Calculate the z score for a TCDD level in plasma of 20.

d. Calculate the z score for a TCDD level in fat tissue of 20.

e. Based on the results of parts **c** and **d**, is a TCDD level of 20 or more likely to occur in plasma or fat tissue?

3.6 Box-and-Whisker Plots

To this point, we have presented numerical descriptive measures of central tendency, variation, and relative standing for a quantitative data set. When conducting a preliminary analysis of the sample data—often called an *exploratory data analysis*—it is helpful to present these descriptive measures in a summarized format.

One obvious approach is to give a **two-number summary** consisting of the mean, $\bar{x}$, and standard deviation, s. As demonstrated in Section 3.4, these two numbers and the Empirical Rule provide an interval that captures most of the values in the data set. Two other useful summary techniques are a *five-number summary* and a *box-and-whisker plot.*

A **five-number summary** for a set of n sample observations, $x_1, x_2, \ldots, x_n$, consists of the smallest measurement, the 1st quartile, the median, the 3rd quartile, and the largest measurement:

$$x_{\text{smallest}} \quad Q_1 \quad M \quad Q_3 \quad x_{\text{largest}}$$

Note that this summary consists of one measure of central tendency (the median), two measures of relative standing (Q_1 and Q_3), and the two extreme values that make up the range.

From the five-number summary, several other measures of variation can be computed. An important one is the *interquartile range.*

Definition 3.12

The **interquartile range, IQR**, is the distance between the 1st and 3rd quartiles:

$$\text{IQR} = Q_3 - Q_1 \tag{3.8}$$

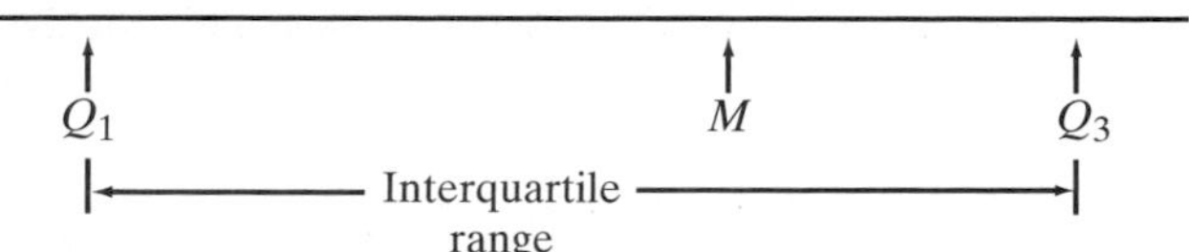

Figure 3.21 The Interquartile Range

You can see from Figure 3.21 that the interquartile range is a measure of data variation. The larger the interquartile range, the more variable the data tend to be. Also, since Q_3 is the 75th percentile and Q_1 is the 25th percentile, we know that (75% − 25%) = 50% of the data lie within the interquartile range.

EXAMPLE 3.15

FIVE-NUMBER SUMMARY

The monthly rental prices for a random sample of 25 two-bedroom apartments are listed in Table 3.6.

TABLE 3.6 Monthly Rents for 25 Two-Bedroom Apartments

Monthly Rents				
$ 660	$ 595	$1,060	$ 500	$ 630
899	1,295	749	820	843
710	950	720	575	760
1,090	770	682	1,016	650
425	367	1,480	945	1,120

a. Give a five-number summary for the data.
b. Find IQR.

	A	B
1	Five-Number Summary	
2	of Monthly Rents	
3		
4	Minimum	367
5	First Quartile	640
6	Median	760
7	Third Quartile	983
8	Maximum	1480

E Figure 3.22 Descriptive Statistics for the Rental Price Data of Table 3.6

Solution

a. We used PHStat and Excel to produce descriptive statistics for the 25 monthly rental prices. These values are displayed in Figure 3.22. From Figure 3.22, we find $x_{\text{smallest}} = 367$, $Q_1 = 640$, $M = 760$, $Q_3 = 983$, and $x_{\text{largest}} = 1{,}480$. Thus, the five-number summary is:

367 640 760 983 1,480

b. Using equation 3.8, the interquartile range is

$$\text{IQR} = Q_3 - Q_1 = 983 - 640 = 343$$

A **box-and-whisker plot** is a graphical representation of the five-number summary. The next example illustrates how to construct such a plot.

EXAMPLE 3.16

CONSTRUCTING A BOX-AND-WHISKER PLOT

Refer to the monthly rental prices for a random sample of 25 two-bedroom apartments as listed in Table 3.6. Construct a box-and-whisker plot for the data.

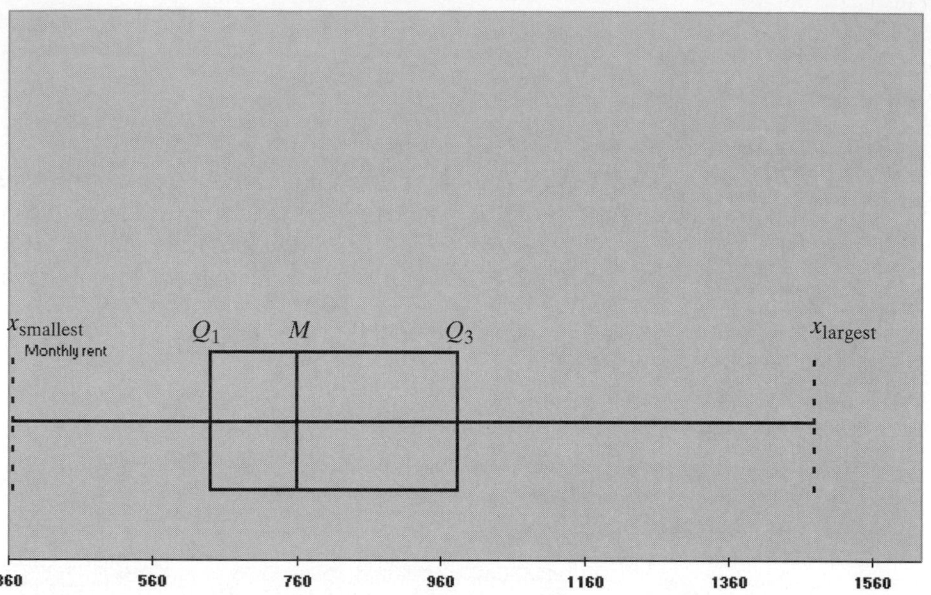

E Figure 3.23 Box-and-Whisker Plot for the Rent Data Using PHStat and Excel

Solution

A horizontal box-and-whisker plot for the 25 monthly rental prices, generated using PHStat and Excel, is shown in Figure 3.23. Recall from Example 3.15 that $Q_1 = 640$, $M = 760$, and $Q_3 = 983$. You can see that the box-and-whisker plot is a box with Q_1 and Q_3 located at the left and right corners, respectively. The length of the box is equal to the interquartile range, IQR; consequently, the box contains the middle 50% of the rental prices in the data set. The vertical line within the box locates the median, M.

The lower 25% of the data are represented by a line (i.e., a *whisker*) connecting the left edge of the box to the location of the smallest rental price, $x_{smallest}$. Similarly, the upper 25% of the data are represented by a line connecting the right edge of the box to $x_{largest}$.

Self-Test 3.6

A sample of measurements has the following five-number summary:

60 70 72 80 100

Calculate IQR and construct a box-and-whisker plot for the data.

A box-and-whisker plot helps us detect possible skewness in the distribution for a data set. For example, if Q_1 is farther away from the median than Q_3, then the distribution is likely to be skewed to the left, as shown in Figure 3.24b. If Q_3 is farther away from the median than Q_1, then the distribution is likely to be skewed to the right, as in Figure 3.24c. Symmetry or lack of skewness is suggested when Q_1 and Q_3 are approximately equidistant from the median and when the whiskers are of approximately equal length, as depicted in Figure 3.24a.

The key ideas of this section are: (1) A two-number summary (i.e., $\bar{x}$ and s) in combination with the Empirical Rule provides us with an interval that contains most of the measurements in a data set; and (2) A five-number summary (i.e., $x_{smallest}$, Q_1, M, Q_3, and $x_{largest}$) illustrated with a box-and-whisker plot allows us to study the shape of the distribution of data.

Figure 3.24 Horizontal Box-and-Whisker Plots for Three Types of Distributions

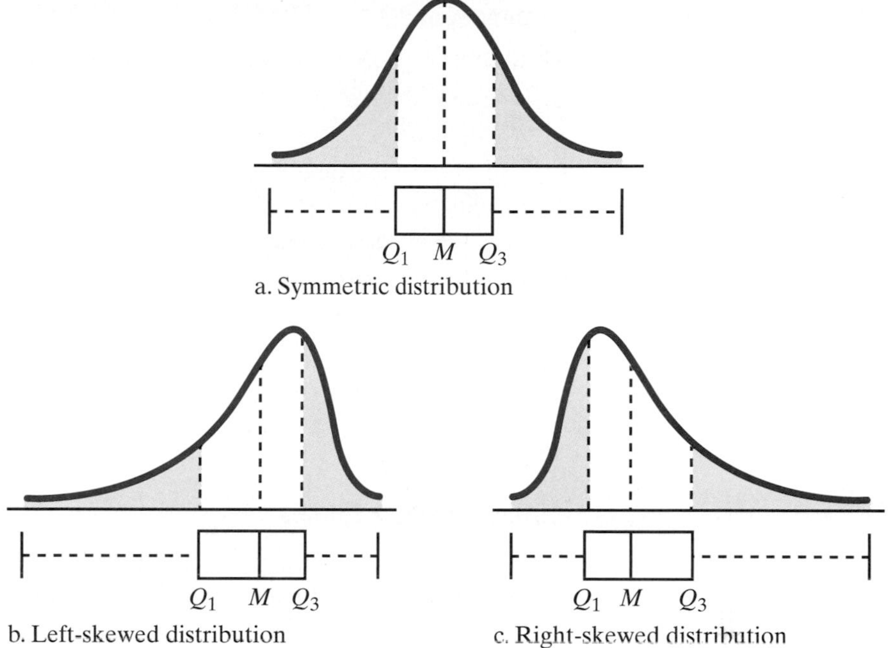

In the next section, we demonstrate how to use both approaches to detect highly unusual observations in the data set.

3.7 Methods for Detecting Outliers

Sometimes observations are included in a data set that are very different from the others. For example, when we discuss starting salaries for college graduates with bachelor's degrees, we generally think of traditional college graduates—those near 22 years of age with four years of college education. But suppose one of the graduates is a basketball player drafted in the first round by a National Basketball Association (NBA) team. Clearly, the starting salary for this graduate will be much larger than the other starting salaries because of the graduate's basketball skill and experience, and we probably would not want to include it in the data set. An unusual observation that lies outside the range of the data values that we want to describe is called an *outlier*.

Outliers are often attributable to one of several causes. First, the measurement associated with the outlier may be invalid. For example, the experimental procedure used to generate the measurement may have malfunctioned, the experimenter may have misrecorded the measurement, or the data might have been coded incorrectly when entered in the computer. Second, the outlier may be the result of a misclassified measurement. That is, the measurement belongs to a population different from that from which the rest of the sample was drawn, as in the case of the basketball player's salary described in the preceding paragraph. Finally, the measurement associated with the outlier may be recorded correctly and from the same population as the rest of the sample, but represents a rare (chance) event. Such outliers occur most often when the relative frequency histogram of the sample data is extremely skewed, because such a distribution has a tendency to include extremely large or small observations relative to the others in the data set.

Definition 3.13

An observation (or measurement) that is unusually large or small relative to the other values in a data set is called an **outlier.** Outliers typically are attributable to one of the following causes:

1. The measurement is observed, recorded, or entered into the computer incorrectly.
2. The measurement comes from a different population.
3. The measurement is correct, but represents a rare (chance) event.

The most obvious method for determining whether an observation is an outlier is to calculate its z score (Section 3.5), as the following example illustrates.

EXAMPLE 3.17 OUTLIERS USING z SCORES

Consider the DDT measurements of 144 fish specimens in the **FISH** data file. One of the fish specimens has a DDT measurement of 1,100 ppm. Is this observation an outlier?

	A	B
1	*DDT*	
2		
3	Mean	24.355
4	Standard Error	8.198215
5	Median	7.15
6	Mode	12
7	Standard Deviation	98.37859
8	Sample Variance	9678.346
9	Kurtosis	102.2388
10	Skewness	9.61166
11	Range	1099.89
12	Minimum	0.11
13	Maximum	1100
14	Sum	3507.12
15	Count	144
16	Largest(1)	1100
17	Smallest(1)	0.11

E **Figure 3.25** Excel Descriptive Statistics for the 144 Fish Specimen DDT Measurements

Solution

To obtain a z score, we need a two-number summary ($\bar{x}$ and s) for the data set. An Excel descriptive statistics printout for the 144 DDT measurements is displayed in Figure 3.25.

Note that $\bar{x} = 24.36$ ppm and $s = 98.38$ ppm. Therefore, the z score for the DDT measurement 1,100 ppm is

$$z = \frac{x - \bar{x}}{s} = \frac{1{,}100 - 24.36}{98.38} = 10.93$$

Both the Empirical Rule and Tchebysheff's Theorem (Section 3.4) tell us that almost all the observations in a data set will have z scores less than 3 in absolute value. Since a z score as large as 10.93 is highly improbable, the DDT measurement of 1,100 ppm is called an outlier. Some research by the Army Corps of Engineers revealed that the DDT value for this fish specimen was correctly recorded, but that the fish was one of the few found at the exact location where the manufacturing plant was discharging its toxic waste materials into the river.

Self-Test 3.7

A sample of measurements has a mean of $\bar{x} = 100$ and a standard deviation of $s = 15$. For each of the following values of x, calculate the z score and determine if the measurement is an outlier.

a. $x = 90$ **b.** $x = 190$

Another procedure for detecting outliers utilizes the information in a five-number summary (Section 3.6). With this method, we construct intervals similar to the intervals of the Empirical Rule; however, the intervals are based on the 1st and 3rd quartiles and the interquartile range.

EXAMPLE 3.18

OUTLIERS USING QUARTILES

Refer to the monthly rental prices for 25 two-bedroom apartments, Examples 3.15 and 3.16 (pp. 163–164). The five-number summary for the data set is:

367 640 760 983 1,480

Use this information to check for outliers in the rental price data set.

Solution

Recall that 50% of the data lie between Q_1 and Q_3. Consequently, observations that fall too far to the left of Q_1 or too far to the right of Q_3 are deemed to be *outliers*. Specifically, observations that fall a distance of (1.5)IQR below Q_1 or a distance of (1.5)IQR above Q_3 are suspect. Thus, any observation in a data set that falls outside the interval $[Q_1 - (1.5)\text{IQR}, Q_3 + (1.5)\text{IQR}]$ is considered an outlier.

For the rental price data, $Q_1 = 640$, $Q_3 = 983$, and $\text{IQR} = Q_3 - Q_1 = 983 - 640 = 343$. Thus,

$$Q_1 - (1.5)\text{IQR} = 640 - (1.5)(343) = 125.5$$
$$Q_3 + (1.5)\text{IQR} = 983 + (1.5)(343) = 1{,}497.5$$

Checking the data set in Table 3.6 (p. 163), you can see that none of the rental prices fall outside the interval; consequently, there are no outliers in the data set.

The z score and quartiles methods both establish limits outside of which a measurement is deemed to be an outlier (see accompanying box). Usually, the two methods produce similar results. However, the presence of one or more outliers in a data set can inflate the computed value of s. Consequently, it will be less likely that an outlier would have a z score larger than 3 in absolute value. In contrast, the values of the quartiles used to calculate the intervals are not as affected by the presence of outliers.

GENERAL RULE FOR DETECTING OUTLIERS

1. *z scores:* Observations with z scores greater than 3 in absolute value are considered outliers. (For some highly skewed data sets, observations with z scores greater than 2 in absolute value may be outliers.)
2. *Quartiles:* Observations falling outside the interval $[Q_1 - (1.5)\text{IQR}, Q_3 + (1.5)\text{IQR}]$ are deemed outliers.

Statistics in the Real World Revisited

Detecting Outliers

Recall that during the early 1980s, the Office of the Florida Attorney General suspected numerous Florida road construction contractors of practicing bid-collusion (p. 129). Subsequent investigations allowed the Florida Department of Transportation to compile highway construction bid data for a sample of 279 contracts (**BIDRIG** file). One of the first analyses conducted by the Attorney General was to identify any unusual winning bid prices relative to the DOT estimate of the "fair" price. If the ratio of the winning bid to the DOT estimate was unusually high, the contractor was suspected of rigging the bid. We can employ both the *z* score method and quartiles method to help identify any outliers in the ratios of low (winning) bid price to DOT estimate (LBERAT).

Descriptive statistics for the variable LBERAT for all 279 contracts are shown in the MINITAB printout, Figure 3.26.

(Continued)

Descriptive Statistics

Variable	N	Mean	Median	TrMean	StDev	SE Mean
LBERAT	279	0.9699	0.9700	0.9671	0.1691	0.0101

Variable	Minimum	Maximum	Q1	Q3
LBERAT	0.5300	1.5600	0.8600	1.0600

M **Figure 3.26** MINITAB Descriptive Statistics for LBERAT for All 279 Contracts

The mean and standard deviation of the low bid/DOT estimate ratios, highlighted on the printout, are $\bar{x} = .97$ and $s = .17$. To check for outliers using the z score method, we use these values to calculate z scores for all $n = 279$ ratios in the data set using the formula

$$z = (\text{LBERAT} - .97)/.17$$

For the purposes of this application, we will focus on only the largest ratios in the sample. The three largest ratios, given on a portion of the Excel spreadsheet shown in Figure 3.27, are 1.56, 1.52, and 1.42. The z scores for these three ratios, also shown on the printout, are 3.49, 3.25, and 2.66, respectively. The two largest ratios both have z scores that exceed 3; consequently, these ratios are outliers. Further investigation of the contractors associated with these winning bids revealed that they did, in fact, engage in bid collusion. (Note that the status of each outlier contract is labeled as "Fixed" by the Attorney General.)

LBERAT	STATUS	ZRATIO
1.56	Fixed	3.49041
1.52	Fixed	3.25382
1.42	Fixed	2.66233

E **Figure 3.27** Excel Spreadsheet Listing of Three Largest Low Bid/Estimate Ratios

To check for outliers using the quartiles method, we require the 1st and 3rd quartiles of the data. These values, $Q_1 = .86$ and $Q_3 = 1.06$, are highlighted on the MINITAB descriptive statistics printout, Figure 3.26. Then,

$$\text{IQR} = 1.06 - .86 = .2$$

$$Q_1 - (1.5)\text{IQR} = .86 - (1.5)(.2) = .56$$

$$Q_3 + (1.5)\text{IQR} = 1.06 + (1.5)(.2) = 1.36$$

Any ratios that fall below .56 or above 1.36 are considered outliers. Note that all three values of LBERAT shown in Figure 3.27 fall above 1.36; consequently, all three are considered outliers based on the quartiles method. This outlier analysis (and other statistical analyses) led to an admission of guilt by these particular contractors. They, in turn, identified other contracts that involved bid collusion, enabling the Florida Attorney General to establish the status (fixed or competitive) of each of the 279 contracts in the **BIDRIG** file.

PROBLEMS FOR SECTION 3.7

Using the Tools

3.57 Consider a sample data set with $n = 50$ observations and the following five-number summary: 75 98 110 122 143.

a. Find Q_1. **b.** Find Q_3. **c.** Find IQR.

d. Find x_{smallest}. **e.** Find x_{largest}.

f. Construct a box-and-whisker plot for the data.

g. What type of skewness, if any, is present in the data?

3.58 Consider a sample data set with $n = 30$ observations and the following five-number summary: 1.7 3.5 5.2 7.8 8.5.

a. Find Q_1. **b.** Find Q_3. **c.** Find IQR.

d. Find x_{smallest}. **e.** Find x_{largest}.

f. Construct a box-and-whisker plot for the data.

g. What type of skewness, if any, is present in the data?

3.59 Construct a box-and-whisker plot for the 24 sample measurements reproduced in Problem 3.44, p. 159.

3.60 Construct a box-and-whisker plot for the 28 sample measurements reproduced in Problem 3.45, p. 159.

3.61 Compute the z score for the value of x in each of the following:

a. $x = 75, \bar{x} = 70, s = 10$
b. $x = .28, \bar{x} = .13, s = .02$
c. $x = 158, \bar{x} = 275, s = 30$
d. $x = 3, \bar{x} = 4, s = .5$

Applying the Concepts

3.62 *Medical Interface* (Oct. 1992) published a study on the cost effectiveness of a sample of 186 physicians in a managed-care HMO. One variable of interest was the monthly cost accrued by each physician divided by the number of patients treated—called the per-patient, per-month cost. The mean and standard deviation of the per-patient, per-month costs were $\bar{x} = \$102.96$ and $s = \$366.31$, respectively. One of the physicians in the data set had a per-patient, per-month cost of \$4,725.10.

a. Calculate and interpret the z score for this physician's cost. Is the observation an outlier?

b. A careful examination of this physician's data revealed that this physician treated only a single patient during a single month of the year at a cost of \$4,725.10. Based on this information, how would you classify this outlier?

3.63 Refer to the **SUICIDE** data file and Problem 3.50 (p. 160).

a. Construct a box-and-whisker plot for the data.

b. Identify the type of skewness present in the data.

c. Use the quartiles method to identify outliers in the data.

d. Use z scores to detect outliers in the data. Do your results agree with part **c**? Explain.

3.64 Refer to the *Psychological Reports* study of the degree to which people believe the moon affects behavior, Problem 2.69 (p. 110). The data file **BILE** contains the Belief in Lunar Effects (BILE) Scale scores for a sample of 157 students. (Higher scores indicate a greater degree of belief in lunar effects.) Numerical descriptive measures for the 157 BILE scores are displayed in the MINITAB printout below. Examine the data set for outliers. Use both the z score and quartiles method.

Descriptive Statistics : BILE

Variable	N	Mean	Median	TrMean	StDev	SE Mean
BILE Sco	157	37.92	39.00	37.77	14.40	1.15

Variable	Minimum	Maximum	Q1	Q3
BILE Sco	11.00	72.00	26.50	48.00

M Problem 3.64

FTC

3.65 Refer to the cigarette nicotine measurements in the **FTC** data file. An Excel box-and-whisker plot and descriptive statistics for the data are shown on p. 170.

a. Identify the type of skewness in the data.

b. Do you detect any outliers? If so, identify them.

3.66 Refer to the passenger head-injury ratings in the **CRASH** data file. An Excel box-and-whisker plot and descriptive statistics for the data are shown on p. 170.

CRASH

a. Identify the type of skewness in the data.

b. Do you detect any outliers? If so, identify them.

3.67 The data in the table at the bottom of p. 170 represent sales, in hundreds of thousands of dollars per week, for a random sample of 24 fast-food outlets located in four cities.

a. Generate a five-number summary for the data.

b. Using quartiles, do you detect any outliers in the data? If so, identify the city and the weekly sales measurement associated with an outlier.

	A	B
1	*Nicotine*	
2		
3	Mean	0.8425
4	Standard Error	0.015452
5	Median	0.9
6	Mode	0.8
7	Standard Deviation	0.345525
8	Sample Variance	0.119388
9	Kurtosis	0.162504
10	Skewness	-0.17036
11	Range	1.85
12	Minimum	0.05
13	Maximum	1.9
14	Sum	421.25
15	Count	500
16	Largest(1)	1.9
17	Smallest(1)	0.05

E Problem 3.65

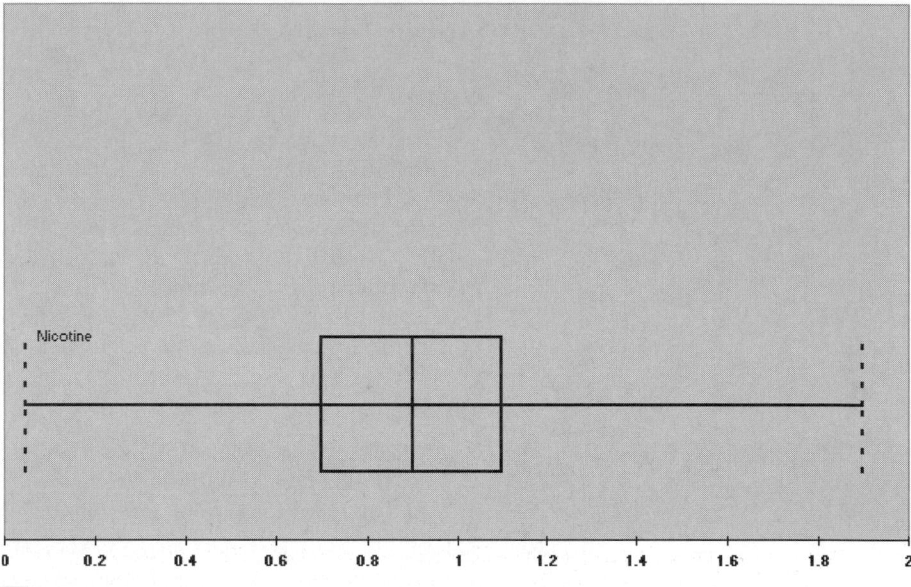

E Problem 3.65

	A	B
1	*Passenger head*	
2		
3	Mean	549.4592
4	Standard Error	21.55076
5	Median	531.5
6	Mode	419
7	Standard Deviation	213.3417
8	Sample Variance	45514.66
9	Kurtosis	0.602666
10	Skewness	0.568088
11	Range	1077
12	Minimum	128
13	Maximum	1205
14	Sum	53847
15	Count	98
16	Largest(1)	1205
17	Smallest(1)	128

E Problem 3.66

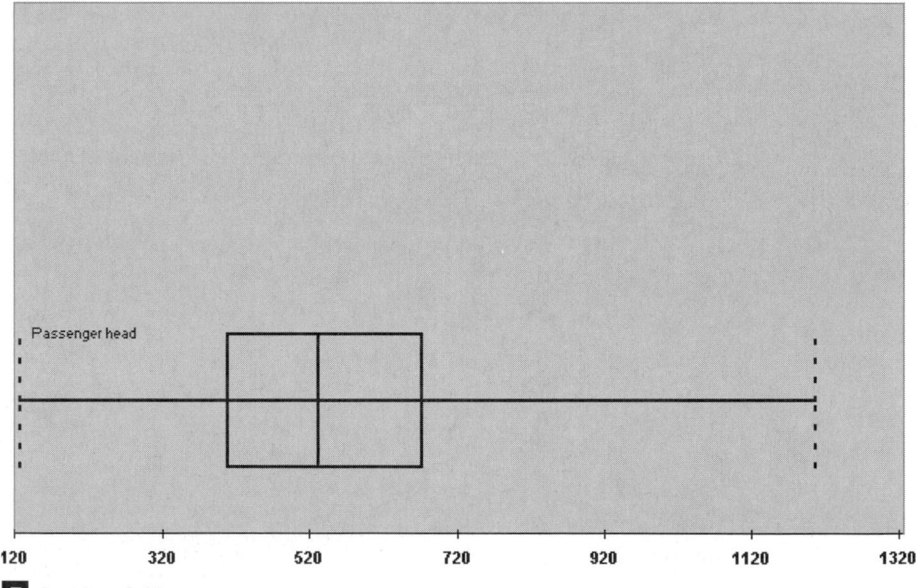

E Problem 3.66

FASTFOOD

City	Weekly Sales (hundred thousand $)
A	6.3, 6.6, 7.6, 3.0, 9.5, 5.9, 6.1, 5.0, 3.6
B	2.8, 6.7, 5.2
C	82.0, 5.0, 3.9, 5.4, 4.1, 3.1, 5.4
D	8.4, 9.5, 8.7, 10.6, 3.3

Problem 3.67

c. Calculate $\bar{x}$ and s for the sample data. Use this information to compute the z score for the outlier(s) identified in part **b**.

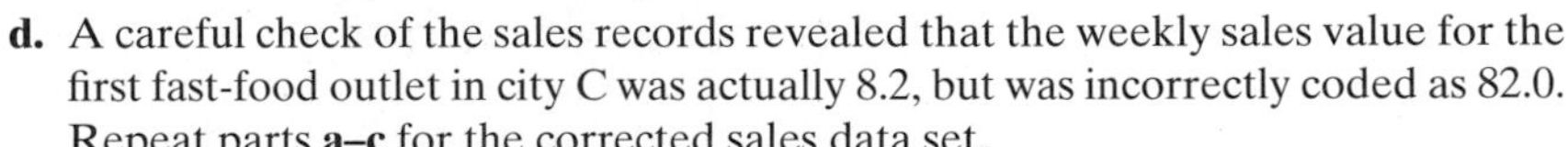

d. A careful check of the sales records revealed that the weekly sales value for the first fast-food outlet in city C was actually 8.2, but was incorrectly coded as 82.0. Repeat parts **a–c** for the corrected sales data set.

3.68 Refer to the *Dalton Transactions* (Dec. 1997) study of the bond strength of a new copper-selenium metal compound, Problem 2.73 (p. 110). The bond lengths (pm) of 36 copper-selenium metal clusters are reproduced in the table.

256.1	259.0	256.0	264.7	262.5	239.0
238.9	237.4	265.9	240.2	239.7	238.3
273.8	246.4	246.0	247.4	290.6	260.9
255.6	253.1	283.7	261.4	271.9	267.9
238.7	237.6	240.0	241.1	240.3	240.8
246.5	248.3	246.0	250.3	250.1	254.6

Source: Deveson, A., et al. "Syntheses and structures of four new copper (I)-selenium clusters: Size dependence of the cluster on the reaction conditions." *Dalton Transactions,* No. 23, Dec. 1997, p. 4492 (Table 1).

a. Find the measures of central tendency and interpret their values.

b. Find the measures of variation and interpret their values.

c. The bond lengths for 7 copper-phosphorus metal clusters are listed below. Comment on whether or not these bond lengths are from the same population as the 36 copper-selenium bond lengths. Explain.

223.8 222.7 223.3 226.1 226.3 226.0 225.0

3.8 A Measure of Association: Correlation

In Section 2.5 we used a scatterplot to investigate the relationship between two quantitative variables, x and y. Depending on the pattern of the points plotted, the two variables may be positively associated, negatively associated, or have little or no association. The strength of the association can be expressed numerically with the sample coefficient of correlation, denoted by the symbol r.

Definition 3.14

The **coefficient of correlation, r,** is computed as follows for a sample of n measurements x and y:

$$r = \frac{SSxy}{\sqrt{(SSxx)(SSyy)}} \tag{3.9}$$

where

$$SSxy = \sum(x - \bar{x})(y - \bar{y}) \tag{3.10}$$

$$SSxx = \sum(x - \bar{x})^2 \tag{3.11}$$

$$SSyy = \sum(y - \bar{y})^2 \tag{3.12}$$

It is a measure of the strength of the *linear* relationship between two quantitative variables x and y. Values of r near $+1$ imply a strong positive linear association between the variables; values of r near -1 imply a strong negative linear association; and values of r near 0 imply little or no linear association.

We illustrate how to calculate r in the following example.

EXAMPLE 3.19

COMPUTING *r*

Suppose a psychologist wants to investigate the relationship between the creativity score y and the flexibility score x of a mentally retarded child. The creativity and flexibility scores (both measured on a scale of 1 to 20) for each in a random sample of five mentally retarded children are listed in Table 3.7. Calculate the coefficient of correlation r between flexibility score and creativity score.

TABLE 3.7 Creativity-Flexibility Scores for Example 3.19

Child	Flexibility Score, x	Creativity Score, y
1	2	2
2	3	8
3	4	7
4	5	10
5	6	8

Solution

Step 1 To calculate r, we need to find the following sums: SSxx, SSyy, and SSxy. As an aid in finding these quantities, we obtain a *sum of squares table* of the type shown in Table 3.8 using Excel. Notice that the quantities needed to compute r appear in the bottom row of the table.

TABLE 3.8 Sum of Squares Obtained from Excel for the Data of Table 3.7

	A	B	C	D	E	F	G	H	I
1		X	Y	(X-XBAR)^2	(Y-YBAR)^2	(X-XBAR)(Y-YBAR)		Summary	
2		2	2	4	25	10		XBAR	4
3		3	8	1	1	-1		YBAR	7
4		4	7	0	0	0		SSX	10
5		5	10	1	9	3		SSY	36
6		6	8	4	1	2		SSXY	14
7	Totals	20	35	10	36	14		r	0.737865

Step 2 Locate the quantities SSxy, SSxx, and SSyy on Table 3.8:

$$\text{SS}xy = \sum(x - \bar{x})(y - \bar{y}) = 14$$

$$\text{SS}xx = \sum(x - \bar{x})^2 = 10$$

$$\text{SS}yy = \sum(y - \bar{y})^2 = 36$$

Step 3 Compute the *coefficient of correlation* r as follows:

$$r = \frac{\text{SS}xy}{\sqrt{(\text{SS}xx)(\text{SS}yy)}} = \frac{14}{\sqrt{(10)(36)}} = .74$$

This value of r is also shown on the Excel spreadsheet, Table 3.8.

The formal name given to r is the sample *Pearson product moment coefficient of correlation,* named for Karl Pearson (1857–1936).

The coefficient of correlation r is *scaleless* (it is not measured in dollars, pounds, etc.); values of r will always be between -1 and $+1$, regardless of the units of measurement of the variables x and y. More importantly, r is a measure of the strength of the *linear* (i.e., straight-line) *association* between x and y in the sample. We can gain insight into this interpretation by examining the scatterplots presented in Figure 3.28.

Figure 3.28 Values of r and Their Implications

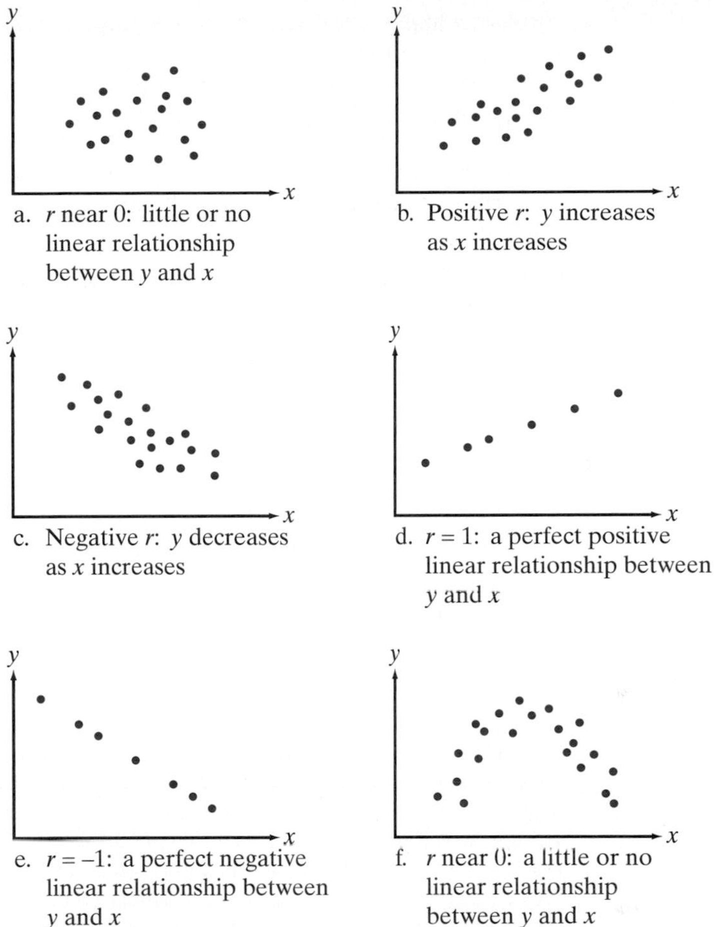
y
x
a. r near 0: little or no linear relationship between y and x
b. Positive r: y increases as x increases
c. Negative r: y decreases as x increases
d. r = 1: a perfect positive linear relationship between y and x
e. r = −1: a perfect negative linear relationship between y and x
f. r near 0: a little or no linear relationship between y and x

Caution

High correlation does not imply causality. If a large positive or negative value of the sample correlation coefficient is observed, it is incorrect to conclude that a change in x causes a change in y. The only valid conclusion is that a linear trend *may* exist between x and y.

Consider first the scatterplot in Figure 3.28a. The correlation coefficient r for this set of points is near 0, implying little or no linear relationship between x and y. In contrast, positive values of r imply a positive linear association between y and x. Consequently, the value of r for the data illustrated in Figure 3.28b is positive. Similarly, a negative value of r implies a negative linear association between y and x (see Figure 3.28c). A perfect linear relationship exists when all the (x, y) points fall exactly along a straight line. A value of $r = +1$ implies a perfect positive linear relationship between y and x (see Figure 3.28d), and a value of $r = -1$ implies a perfect negative linear relationship between y and x (see Figure 3.28e). Finally, consider the points plotted in Figure 3.28f. Although a strong relationship exists between y and x, it is not linear. Hence, the value of r is near 0.

EXAMPLE 3.20

INTERPRETING *r*

Interpret the value of r calculated in Example 3.19.

Solution

The value $r = .74$ implies that creativity score and flexibility score are positively correlated—at least for this sample of five mentally retarded children. The implication is that a positive linear association exists between these variables. We must be careful, however, not to jump to any unwarranted conclusions. For instance, the psychologist may be tempted to conclude that a high flexibility score will *always*

result in a higher creativity score. The implication of such a conclusion is that there is a *causal* relationship between the two variables. However, *high correlation does not imply causality*. Many other factors, such as severity of illness, parental neglect, and IQ, may contribute to the change in creativity score.

Self-Test 3.8

Find and interpret the coefficient of correlation r for the values of x and y listed in the table.

x	−1	2	1	0	3
y	8	4	6	5	1

Statistics in the Real World Revisited

Interpreting the Correlation Coefficient

Consider, once more, the highway construction bid data (**BIDRIG** file) collected for 279 road contracts by the Florida Attorney General's Office (p. 129). After establishing which of the 279 contracts were competitively bid and which had rigged bids, FLAG used the data to help identify variables that have the potential to predict the fixed or competitive status of a future highway construction contract. One approach is to examine the correlation r between low bid/DOT estimate ratio (LBERAT) and certain key bid variables. In addition to the winning bid price (LOWBID) and the DOT engineer's estimated price (DOTEST), two variables of interest to FLAG are number of bidders on the contract (BIDS) and the percentage of the total bid price allocated to supplying liquid asphalt (PCTASPH). Since the correlation between these variables may depend on whether the bids are rigged, FLAG examined these correlations separately for fixed and competitive contracts. The correlation coefficients for competitive contracts are shown in the MINITAB printout, Figure 3.29a, while the correlations for fixed contracts are shown in Figure 3.29b.

(a) STATUS = Compet

	LBERAT	LOWBID	DOTEST	BIDS
LOWBID	0.120			
DOTEST	0.045	0.988		
BIDS	-0.216	0.286	0.308	
PCTASPH	-0.020	-0.329	-0.324	-0.274

Cell Contents: Pearson correlation

(b) STATUS = Fixed

	LBERAT	LOWBID	DOTEST	BIDS
LOWBID	0.037			
DOTEST	-0.035	0.994		
BIDS	0.024	0.203	0.192	
PCTASPH	-0.140	-0.306	-0.291	-0.307

Cell Contents: Pearson correlation

M **Figure 3.29** MINITAB Correlations for **BIDRIG** Data

First, consider the correlation coefficient between low bid price and DOT estimated price. FLAG expects these two variables to be strongly positively correlated regardless of the bid status of the contract—that is, the winning price should increase as the DOT estimate increases. Examining Figure 3.29a and Figure 3.29b, we find the expected result—the correlation between LOWBID and DOTEST is near 1 for both competitive ($r = .988$) and fixed contracts ($r = .994$).

Next, consider the correlation coefficient between low bid/estimate ratio and number of bidders. In a competitive environment, FLAG expects more bidders to lead to a decrease in the ratio, i.e., FLAG expects a negative relationship to exist between these two variables. From Figure 3.29a, we find that the correlation between LBERAT and BIDS is $r = -.216$—a weak to moderate negative relationship. However, for fixed bids (Figure 3.29b), the correlation between LBERAT and BIDS is approximately 0 ($r = .024$). In a collusive environment, it appears that number of bidders has very little impact on the low bid/estimate ratio.

Finally, we examine the correlation between low bid/estimate ratio and liquid asphalt percentage. For competitive contracts (Figure 3.29a), the correlation between LBERAT and PCTASPH is $r = -.020$, while for fixed contracts (Figure 3.29b) the correlation is $r = -.14$. Although both values are negative and small (near 0), there appears to be a stronger negative relationship between the variables for fixed bids. FLAG later discovered that contractors who rigged bids typically did so by overpricing the liquid asphalt required for the job.

PROBLEMS FOR SECTION 3.8

Using the Tools

3.69 Calculate the coefficient of correlation r for a sample data set with the following characteristics:

a. $SSxy = 10$, $SSxx = 100$, $SSyy = 25$

b. $SSxy = -35$, $SSxx = 100$, $SSyy = 25$

c. $SSxy = 125$, $SSxx = 400$, $SSyy = 100$

d. $SSxy = -200$, $SSxx = 400$, $SSyy = 100$

3.70 For each of the three scatterplots **a–c**, decide whether the coefficient of correlation r is positive or negative.

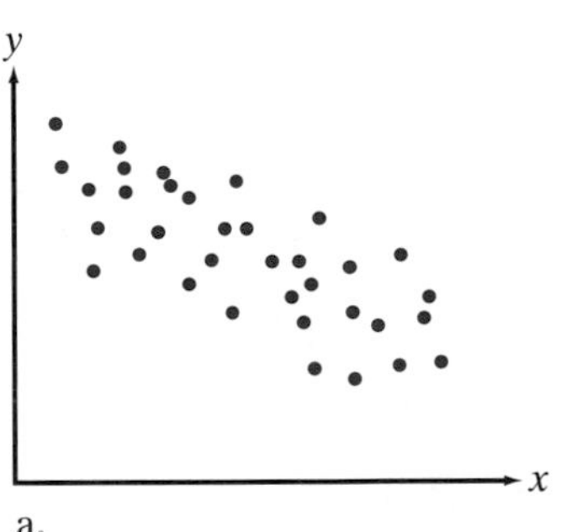

a.

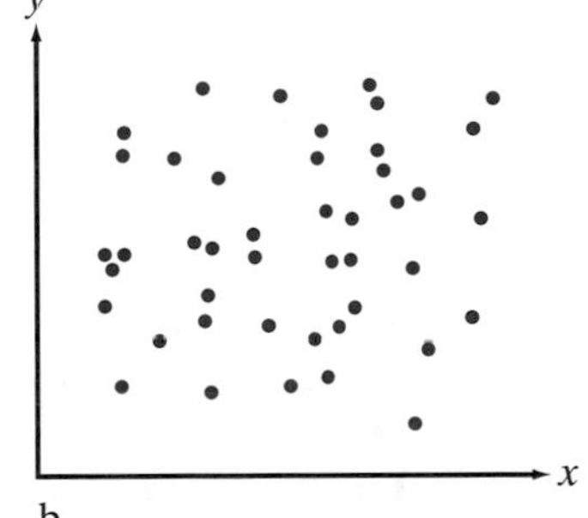

b.

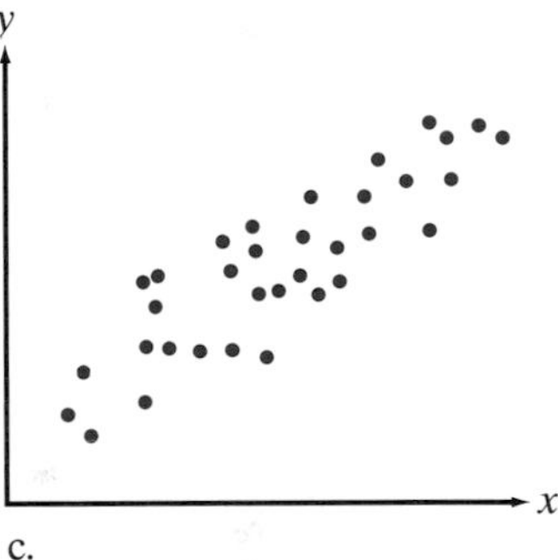

c.

3.71 Calculate the coefficient of correlation r for each of the sample data sets **a–c**:

a.

x	y
−1	−1
0	1
1	2
2	4
3	5
4	8
5	9

b.

x	y
−1	4
0	1
1	5
2	2
3	1
4	7
5	3

c.

x	y
−1	11
0	8
1	7
2	3
3	1
4	0
5	−2

3.72 Calculate the coefficient of correlation r for the following sample values of x and y:

y	7	5	8	3	6	10	12	4	9	15	18
x	21	15	24	9	18	30	36	12	27	45	54

3.73 Calculate the coefficient of correlation r for the following sample values of x and y:

y	100	200	250	400	350	300	450	500	100	250
x	6	20	10	4	12	30	35	15	18	5

Applying the Concepts

3.74 Do you believe the grade point average of a college student is correlated with the student's intelligence quotient (IQ)? If so, will the correlation be positive or negative? Explain.

3.75 Research by law enforcement agencies has shown that the crime rate is correlated with the U.S. population. Would you expect the correlation to be positive or negative? Explain.

3.76 Refer to the *Consumer Reports* (Feb. 1999) data on the retail price and energy cost per year of ten medium-size top-freezer refrigerators, Problem 2.45 (p. 94). The data are repeated in the table below.

REFRIGERATOR

Brand	Price ($)	Energy Cost per Year ($)
Maytag MTB2156B	850	48
Amana TR21V2	760	54
Kenmore(Sears) 7820	900	58
Whirlpool Gold GT22DXXG	870	66
GE Profile TBX22PRY	1100	77
KitchenAid Prestige KTRP22KG	800	66
Whirlpool ET21PKXG	650	70
GE TBX22ZIB	750	81
Frigidaire Gallery FRT20NGC	750	72
Hotpoint CTX21DIB	570	78

Source: "Cold storage," *Consumer Reports,* February 1999, 49.

a. Compute the coefficient of correlation r between price and energy cost.

b. How strong do you think the relationship is between price and energy cost of refrigerators? Explain.

3.77 Refer to the *American Statistician* (Feb. 1991) study of rock drilling, Problem 2.46 (p. 95). The data on depth at which drilling begins and length of time to drill 5 feet are reproduced in the table. Find the coefficient of correlation r and interpret the result.

DRILLROCK

Depth at which drilling begins (feet)	Time to drill 5 feet (minutes)
0	4.90
25	7.41
50	6.19
75	5.57
100	5.17
125	6.89
150	7.05
175	7.11
200	6.19
225	8.28
250	4.84
275	8.29
300	8.91
325	8.54
350	11.79
375	12.12
395	11.02

Source: Penner, R., and Watts, D. G. "Mining information." *The American Statistician,* Vol. 45, No. 1, Feb. 1991, p. 6 (Table 1).

3.78 A math and computer science researcher at Duquesne University investigated the relationship between a student's confidence in math and interest in computers (*Educational Technology,* May–June 1995). A sample of 1,730 high school students—902 boys and 828 girls—from public schools in Pittsburgh, Pennsylvania, participated in the study. Using 5-point Likert scales, where 1 = "strongly disagree" and 5 = "strongly agree," the researcher measured a student's confidence in mathematics (x) and interest in computers (y).

a. For boys, math confidence (x) and computer interest (y) were correlated at $r = .14$. Fully interpret this result.

b. For girls, math confidence (x) and computer interest (y) were correlated at $r = .33$. Fully interpret this result.

Object	Correlation r
Piano	.447
Bench	−.057
Motorbike	.619
Armchair	.294
Teapot	.949

Problem 3.79

3.79 *Perception & Psychophysics* (July 1998) reported on a study of how people view 3-dimensional objects projected onto a rotating 2-dimensional image. Each in a sample of 25 university students viewed a depth-rotated image of a piano until they recognized the object as a piano. The recognition exposure time—that is, the minimum time (in milliseconds) required for the subject to recognize the object—was recorded. In addition, each subject rated the "goodness of view" of the object on a numerical scale, where lower scale values correspond to better views. This experiment was repeated for several other rotated objects, including a bench, motorbike, armchair, and teapot. The table at left gives the correlation coefficient, r, between recognition exposure time and goodness of view for each of these objects. Interpret the value of r for each object.

BRAIN

3.80 Refer to the *American Journal of Psychiatry* (July 1995) study relating verbal memory retention (y) and right hippocampus volume (x) of 21 Vietnam veterans, Problem 2.48 (p. 95). The data for the study are provided in the following table. Find and interpret the coefficient of correlation r for the two variables.

Veteran	Verbal Memory Retention, y	Right Hippocampus Volume, x	Veteran	Verbal Memory Retention, y	Right Hippocampus Volume, x
1	26	960	12	60	1,210
2	22	1,090	13	62	1,220
3	30	1,180	14	66	1,215
4	65	1,000	15	76	1,220
5	70	1,010	16	65	1,300
6	60	1,070	17	66	1,350
7	60	1,080	18	84	1,350
8	58	1,040	19	90	1,370
9	65	1,040	20	102	1,400
10	83	1,030	21	102	1,460
11	69	1,045			

FERTIL

3.81 The fertility rate of a country is defined as the number of children a woman citizen bears, on average, in her lifetime. *Scientific American* (Dec. 1993) reported on the declining fertility rate in developing countries. The researchers found that family planning can have a great effect on fertility rate. The table on p. 178 gives the fertility rate, y, and contraceptive prevalence, x (measured as the percentage of married women who use contraception), for each of 27 developing countries. Find the coefficient of correlation between fertility rate and contraceptive prevalence. Interpret the result.

ORING

3.82 The data on temperature at flight time and O-ring damage index for 23 launches of the space shuttle *Challenger,* first presented in Problem 2.50 (p. 97), are reproduced on p. 178. Find the coefficient of correlation between temperature and damage index. Interpret the result.

Country	Contraceptive Prevalence, x	Fertility Rate, y	Country	Contraceptive Prevalence, x	Fertility Rate, y
Mauritius	76	2.2	Egypt	40	4.5
Thailand	69	2.3	Bangladesh	40	5.5
Colombia	66	2.9	Botswana	35	4.8
Costa Rica	71	3.5	Jordan	35	5.5
Sri Lanka	63	2.7	Kenya	28	6.5
Turkey	62	3.4	Guatemala	24	5.5
Peru	60	3.5	Cameroon	16	5.8
Mexico	55	4.0	Ghana	14	6.0
Jamaica	55	2.9	Pakistan	13	5.0
Indonesia	50	3.1	Senegal	13	6.5
Tunisia	51	4.3	Sudan	10	4.8
El Salvador	48	4.5	Yemen	9	7.0
Morocco	42	4.0	Nigeria	7	5.7
Zimbabwe	46	5.4			

Source: Robey, B., et al. "The fertility decline in developing countries." *Scientific American,* Dec. 1993, p. 62. [*Note:* The data values are estimated from a scatterplot.]

Problem 3.81

Flight Number	Temperature (°F)	O-Ring Damage Index	Flight Number	Temperature (°F)	O-Ring Damage Index
1	66	0	51-A	67	0
2	70	4	51-B	75	0
3	69	0	51-C	53	11
5	68	0	51-D	67	0
6	67	0	51-F	81	0
7	72	0	51-G	70	0
8	73	0	51-I	67	0
9	70	0	51-J	79	0
41-B	57	4	61-A	75	4
41-C	63	2	61-B	76	0
41-D	70	4	61-C	58	4
41-G	78	0			

Note: Data for flight number 4 are omitted due to an unknown O-ring condition.

Primary Sources: Report of the Presidential Commission on the Space Shuttle Challenger Accident, Washington, D.C., 1986, Vol. II (pp. H1–H3) and Volume IV (p. 664). *Post-Challenger Evaluation of Space Shuttle Risk Assessment and Management,* Washington, D.C., 1988, pp. 135–136.

Secondary Source: Tufte, E. R., *Visual and Statistical Thinking: Displays of Evidence for Making Decisions.* Graphics Press, 1997.

Problem 3.82

3.9 Numerical Descriptive Measures for Populations

When you analyze a sample, you are doing so for a reason. Most likely, you want to use the information in the sample to infer the nature of some larger set of data—the population. Numerical descriptive measures that characterize the distribution for a population are called *parameters*. Since we will often use numerical descriptive measures of a sample to estimate the corresponding unknown descriptive measures of the population, we need to make a distinction between the numerical descriptive measure symbols for the population and for the sample.

Definition 3.15

Numerical descriptive measures (e.g., mean, median, standard deviation) of a population are called **parameters.**

SAMPLE AND POPULATION NUMERICAL DESCRIPTIVE MEASURES

Sample	**Population**
mean: $\bar{x}$	mean: μ
variance: s^2	variance: σ^2
standard deviation: s	standard deviation: σ
z score: $z = \frac{x - \bar{x}}{s}$	z score: $z = \frac{x - \mu}{\sigma}$
correlation coefficient: r	correlation coefficient: ρ

In our previous discussion, we used the symbols $\bar{x}$ and s to denote the mean and standard deviation, respectively, of a sample of n observations. Similarly, we will use the symbol μ (mu) to denote the mean of a population and the symbol σ (sigma) to denote the standard deviation of a population. As you will subsequently see, we will use the sample mean $\bar{x}$ to estimate the population mean μ, and the sample standard deviation s to estimate the population standard deviation σ. In doing so, we will be using the sample to help us infer the nature of the population relative frequency distribution.

KEY TERMS

Arithmetic mean 133
Box-and-whisker plot 163
Coefficient of correlation 171
Empirical Rule 147
Five-number summary 162
Interquartile range 163
Mean 133
Measures of association 130
Measures of central tendency 130
Measures of relative standing 130
Measures of variation 130
Median 135
Modal class 135
Mode 135
Numerical descriptive measures 130
Outlier 166
Parameters 178
Percentile 156
Quartiles 156
Range 141
Sample variance 142
Skewness 130
Standard deviation 144
Tchebysheff's Theorem 148
Two-number summary 162
Variance 142
z score 159

KEY FORMULAS

Sample mean:	$\bar{x} = \frac{\Sigma x}{n}$	**(3.1)**, 133
Sample variance:	$s^2 = \frac{\Sigma(x - \bar{x})^2}{n - 1}$	**(3.4)**, 142
Sample standard deviation:	$s = \sqrt{s^2}$	**(3.5)**, 144
Rank of pth percentile:	$\frac{p(n + 1)}{100}$	**(3.6)**, 158
Sample z score:	$z = \frac{x - \bar{x}}{s}$	**(3.7)**, 159
Interquartile range:	$\text{IQR} = Q_3 - Q_1$	**(3.8)**, 163
Sample coefficient of correlation:	$r = \frac{\text{SS}xy}{\sqrt{(\text{SS}xx)(\text{SS}yy)}}$	**(3.9)**, 171

(see Definition 3.14, p. 171, for SSxy, SSxx, and SSyy formulas)

Population z score:	$z = \frac{x - \mu}{\sigma}$	179

KEY SYMBOLS

SYMBOL	DEFINITION
$\overline{x}$	Sample mean
s^2	Sample variance
s	Sample standard deviation
r	Sample coefficient of correlation
μ	Population mean
σ^2	Population variance
σ	Population standard deviation
ρ	Population coefficient of correlation
M	Median
Q_1	1st quartile
Q_3	3rd quartile

CHECKING YOUR UNDERSTANDING

1. What are the types of numerical descriptive measures for qualitative data?
2. What are the types of numerical descriptive measures for quantitative data?
3. Describe the properties of a measure of central tendency.
4. Give three different measures of central tendency.
5. How do the mean, median, and mode differ, and what are the advantages and disadvantages of each?
6. Describe the properties of a measure of variation.
7. Give three different measures of variation.
8. How do the range, variance, and standard deviation differ and what are the advantages and disadvantages of each?
9. When is it appropriate to apply the Empirical Rule for describing the distribution of quantitative data?
10. When is it appropriate to apply Tchebysheff's Theorem for describing the distribution of quantitative data?
11. What are measures of relative standing?
12. What is the difference between the first quartile, median, and third quartile?
13. What numbers make up a two-number summary of quantitative data?
14. What numbers make up a five-number summary of quantitative data?
15. What is a box-and-whisker plot?
16. Give two rules for detecting outliers. What are the advantages and disadvantages of each?
17. What is the purpose of the coefficient of correlation?
18. What are numerical descriptive measures of the population called?

SUPPLEMENTARY PROBLEMS

Applying the Concepts

3.83 The table on p. 181 describes the sale prices for properties sold in six residential neighborhoods located in Tampa, Florida. Use this information to comment on the skewness of the relative frequency distributions for the six data sets.

3.84 According to one study, "The majority of people who die from fire and smoke in compartmented fire-resistive buildings—the type used for hotels, motels, apartments, and other health care facilities—die in the attempt to evacuate" (*Risk*

Neighborhood	Mean Sale Price	Median Sale Price
Avila	$297,004	$192,000
Carrollwood Village	137,492	125,000
Northdale	94,391	91,000
Tampa Palms	198,428	170,000
Town & Country	70,477	69,000
Ybor City	25,764	23,400

Source: Hillsborough County (Florida) property appraiser's office.

Problem 3.83

Management, Feb. 1986). The following data represent the numbers of victims who attempted to evacuate for a sample of 14 fires at compartmented fire-resistive buildings reported in the study.

FIRE

Fire	Number of Victims
Las Vegas Hilton (Las Vegas)	5
Inn on the Park (Toronto)	5
Westchase Hilton (Houston)	8
Holiday Inn (Cambridge, Ohio)	10
Conrad Hilton (Chicago)	4
Providence College (Providence)	8
Baptist Towers (Atlanta)	7
Howard Johnson (New Orleans)	5
Cornell University (Ithaca, New York)	9
Westport Central Apartments (Kansas City, Missouri)	4
Orrington Hotel (Evanston, Illinois)	0
Hartford Hospital (Hartford, Connecticut)	16
Milford Plaza (New York)	0
MGM Grand (Las Vegas)	36

Source: Macdonald, J. N. "Is evacuation a fatal flaw in fire fighting philosophy?" *Risk Management,* Vol. 33, No. 2, Feb. 1986, p. 37.

a. Construct a stem-and-leaf display for the data.

b. Compute the mean, median, and mode for the data set. Which measure of central tendency appears to best describe the center of the distribution of data?

c. The MGM Grand fire in Las Vegas was treated separately in the *Risk Management* analysis because of the size of the high-rise hotel and other unique factors. Do the data support treating the MGM Grand fire deaths differently than the other measurements in the sample? Explain.

RESTRATE

3.85 Zagat's publishes restaurant ratings for various locations in the United States. The data file **RESTRATE** (*Source:* Extracted from *Zagat Survey 2000 New York City Restaurants* and *Zagat Survey 2000 Long Island Restaurants.*) contains the Zagat rating for food, decor, and service, and the price per person for a sample of 50 restaurants located in New York City, and 50 restaurants located on Long Island, a suburb of New York City. Select one of the four quantitative variables in the data set and using Excel or MINITAB:

a. Find and interpret the mean and median for the 50 New York City restaurants.

b. Find and interpret the 1st quartile and 3rd quartile for the 50 New York City restaurants.

c. Find and interpret the range, interquartile range, variance, and standard deviation for the 50 New York City restaurants.

d. Repeat parts **a–c** for the 50 Long Island restaurants.

e. Construct box-and-whisker plots for both the New York City and Long Island restaurants.

f. Are the data for the variable you selected skewed? If so, how?

g. What conclusions can you reach concerning differences between New York City and Long Island restaurants with respect to the variable you selected?

AUTO2000

3.86 Suppose that you wish to study characteristics of the model year 2000 automobiles in terms of the following quantitative variables: miles per gallon, fuel tank capacity, length, wheelbase, width, turning circle requirement, weight, luggage capacity, front shoulder room, front leg room, front head room, rear shoulder room, rear leg room, and rear head room. These data for 107 cars were obtained from *Consumer Reports* (April 2000) and are contained in the data file **AUTO2000** (*Source:* "The 2000 cars," *Consumer Reports,* April 2000, 66–71.)

a. Select one of the variables in the data set and use MINITAB or Excel to obtain descriptive statistics (e.g., mean, standard deviation, quartiles) for the variable.

b. Interpret the values of the descriptive statistics.

c. Obtain a box-and-whisker plot for the data.

d. Are the data skewed? If so, how?

e. What conclusions can you draw concerning the distribution of the variable you selected for the model 2000 automobiles?

f. Suppose you want to compare front-wheel drive cars with rear-wheel drive cars. Perform **a–e** for each of these groups. What conclusions can you reach concerning differences between front-wheel drive cars and rear-wheel drive cars?

PROTEIN

3.87 The sample data on fat and cholesterol levels for popular protein foods, **PROTEIN,** was first presented in Problem 2.60 (p. 105). For each food type, five quantitative variables are measured: amount of calories, amount of protein, percentage of calories from fat, percentage of calories from saturated fat, and amount of cholesterol.

a. For each of these five variables, find the following descriptive statistics: mean, median, mode, range, variance, standard deviation, lower quartile, upper quartile, interquartile range.

b. Construct a box-and-whisker plot for each variable.

c. Identify the type of skewness in the distribution of each variable.

d. For each variable, compute the intervals $\bar{x} \pm s$, $\bar{x} \pm 2s$, and $\bar{x} \pm 3s$. Determine the percentage of measurements that fall within each interval. Compare your results to the Empirical Rule.

e. Identify the outliers for each variable.

f. For each pair of variables, find and interpret the coefficient of correlation, r.

3.88 Passive exposure to environmental tobacco has been associated with growth suppression and an increased frequency of respiratory tract infections in normal children. Is this association more pronounced in children with cystic fibrosis? To answer this question, 43 children (18 girls and 25 boys) attending a 2-week summer camp for cystic fibrosis patients were studied (*New England Journal of Medicine,* Sept. 20, 1990). Among several variables measured were the child's weight percentile (y) and the number of cigarettes smoked per day in the child's home (x).

a. For the 18 girls, the coefficient of correlation between y and x was reported as $r = -.50$. Interpret this result.

b. For the 25 boys, the coefficient of correlation between y and x was reported as $r = -.12$. Interpret this result.

CRASH

3.89 Refer to the NCAP crash test data in the **CRASH** file. In Chapter 2 (p. 92) we used scatterplots to examine the relationship between three pairs of quantitative variables: (a) car weight (WEIGHT) vs. driver's head-injury rating (DRIVHEAD), (b) DRIVHEAD vs. driver's rate of chest deceleration (DRIVCHST), and (c) DRIVHEAD vs. overall driver injury rating (DRIVSTAR). Find the correlation coefficient r relating these three pairs of variables. Interpret the results. Do your conclusions agree with those derived from the scatterplots?

3.90 Nevada continues to be the leading gold producer in the U.S., and according to the U.S. Bureau of Mines, it ranks among the top four regional producers worldwide

GOLD

1,467.8	228.0	111.3	76.0	55.1	40.0
318.0	222.6	89.1	72.5	54.1	32.4
296.9	214.6	82.0	66.0	50.0	30.9
256.0	207.3	81.5	60.4	50.0	30.3
254.5	120.7	78.8	60.0	44.5	30.0

Source: Engineering & Mining Journal, June 1990, p. 38.

(trailing South Africa, Russia, and Australia). The data in the table above represent the production (in thousands of ounces) for the top 30 gold mines in the state.

a. Summarize the data with a graphical technique.

b. Calculate the mean, median, and standard deviation of the data.

c. What proportion of Nevada mines have production values that lie within two standard deviations of the mean?

d. Note the extremely large production value, 1,467.8, for the first mine listed in the table. Recalculate the mean, median, and standard deviation with the production measurement for this mine deleted.

e. Explain how the three numerical descriptive measures (mean, median, and standard deviation) are affected by the deletion of the measurement 1,467.8.

3.91 To investigate the phenomenon of ants protecting plants, a naturalist sampled 50 flower heads of a certain sunflower plant that attracts ants. (Each flower head was on a different sunflower plant.) The flower heads were divided into two groups of 25 each. Ants were prevented from reaching the flower heads of one group (this was accomplished by painting ant repellent around the flower stalks of 25 plants). After a specified period of time, the number of insect seed predators on each flower head was counted. The results are summarized in the table.

	Plants with Ants	Plants without Ants
Mean number of predators per flower head	2.9	7.6
Standard deviation	2.4	4.4

a. For each flower head group, compute the interval $\bar{x} \pm 2s$.

b. Estimate the percentage of measurements that fall within each of the intervals, part **a**.

c. Does it appear that ants protect sunflower plants from insect seed predators?

3.92 Refer to problem 2.74 (p. 112) and the study to determine whether a student's final grade in an introductory sociology course was linearly related to his or her performance on the verbal ability test administered before entrance to college. The verbal test scores and final grades for a random sample of 10 students are repeated in the table. Calculate and interpret a measure of association between verbal ability test score and final sociology grade.

SOCIOLOGY

Student	Verbal Ability Test Score	Final Sociology Grade
1	39	65
2	43	78
3	21	52
4	64	82
5	57	92
6	47	89
7	28	73
8	75	98
9	34	56
10	52	75

3.93 Industrial engineers periodically conduct "work measurement" analyses to determine the time used to produce a single unit of output. At a large processing plant, the total number of man-hours required per day to perform a certain task was recorded for 50 days. This information will be used in a work measurement analysis. The total number of man-hours required for each of the 50 days is listed here.

MANHRS

128	119	95	97	124	113	109	124	132	97
146	128	103	135	114	124	131	133	131	88
100	112	111	150	117	128	142	98	108	120
138	133	136	120	112	109	100	111	131	113
118	116	98	112	138	122	97	116	92	122

a. Find the mean, median, and mode of the data set and interpret their values.

b. Find the range, variance, and standard deviation of the data set and interpret their values.

c. Construct the intervals $\bar{x} \pm s$, $\bar{x} \pm 2s$, and $\bar{x} \pm 3s$. Count the number of observations that fall within each interval and find the corresponding proportions. Compare the results to the Empirical Rule. Do you detect any outliers?

3.94 Neuropsychologists often measure the impact of brain damage on a patient's verbal skills by means of a verbal fluency test. One such test requires subjects to produce as many words as they can that begin with a particular letter. The production figures for one such study are summarized in the accompanying table.

	Mean Productivity	**Standard Deviation**
Normal patients	19.30	5.73
Brain-damaged patients	9.20	2.50

a. A particular brain-damaged patient produced eight words. Find the z score for this observation and interpret the value.

b. Would you expect a normal patient to produce a total of eight words? Explain.

3.95 *Muck* is a rich, highly organic type of soil that serves as a growth medium for most vegetation in the Florida Everglades. Because of its high concentration of organic material, muck can be destroyed by drought, fire, and windstorms. During a recent drought in south Florida, the Everglades lost a considerable amount of muck. To assess the loss, members of the Florida Fish and Game Commission marked 40 plots with stakes at various locations in the Everglades, and measured the depth of the muck (in inches) at each stake. The data are given in the accompanying table.

MUCK

27	30	40	21	41	23	24	30	26	40
35	32	35	7	37	19	30	28	39	33
33	23	35	22	15	32	26	38	26	30
45	15	29	57	27	27	18	16	31	36

a. Give a two-number summary of the data set. Interpret these numbers.

b. Give a five-number summary of the data set. Interpret these numbers.

c. Construct the intervals $\bar{x} \pm s$, $\bar{x} \pm 2s$, and $\bar{x} \pm 3s$. Count the number of observations that fall within each interval and find the corresponding proportions. Compare the results to the Empirical Rule.

d. Find the 10th percentile for the 40 muck depths.

e. Construct a box-and-whisker plot for the data. Use the plot to detect any skewness in the data.

f. Identify any outliers in the data.

3.96 Behavior geneticists are scientists who study the ways in which genetic factors influence behavior. One characteristic that has been studied is the "emotional" behavior of different strains of rats. Emotional behavior is sometimes defined as the tendency to "freeze"—i.e., not move—when presented with a new situation. For rats from different strains that are raised under identical circumstances, differences in emotional behavior may suggest a genetic base. Summary statistics on the number of meters traversed by 500 rats of a particular strain when put in a box with a bright light and white noise are listed here.

$$\bar{x} = 7.65 \qquad s = .88 \qquad M = 7.3$$

a. Find the 50th percentile of the distribution of the number of meters traversed by this sample of rats.

b. Compute the interval $\bar{x} \pm 2s$. Use the Empirical Rule to approximate the percentage of rats that had distance measurements within this interval.

c. Would you expect to observe a rat of this particular strain traverse a distance of 9.50 meters?

REFERENCES

Mendenhall, W. *Introduction to Probability and Statistics,* 9th ed. North Scituate, Mass.: Duxbury, 1994.

Microsoft Excel 2000. Redmond, Wash.: Microsoft Corporation, 1999.

MINITAB User's Guide 1: Data, Graphics, and Macros: Release 13. State College, Pa.: Minitab, Inc., 2000.

MINITAB User's Guide 2: Data Analysis and Quality Tools: Release 13. State College, Pa.: Minitab, Inc., 2000.

Tukey, J. *Exploratory Data Analysis.* Reading, Mass.: Addison-Wesley, 1977.

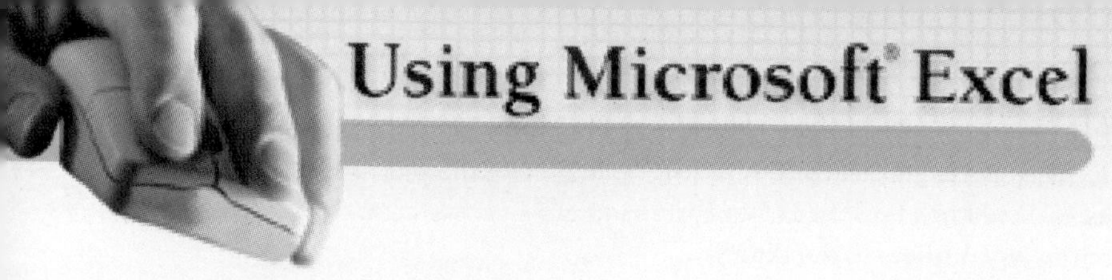

Using Microsoft® Excel

3.E.1 Obtaining Dot Scale Diagrams (Dot Plots)

Use the PHStat **Descriptive Statistics | Dot Scale Diagram** procedure to generate a dot plot on a new worksheet. For example, to generate the dot plot for the weight of fish, open the **FISH.XLS** workbook to the Data worksheet. Then

1. Select PHStat | Descriptive Statistics | Dot Scale Diagram.
2. In the Dot Scale dialog box (see Figure 3.E.1):
   a. Enter D1:D145 in the Variable Cell Range edit box.
   b. Select the First cell contains label check box.
   c. Enter a title in the Title edit box.
   d. Click the OK button.

This procedure places the dot scale diagram as a chart object on a new worksheet that also contains a summary table of statistics below the chart and an ordered array of the values being plotted in column A.

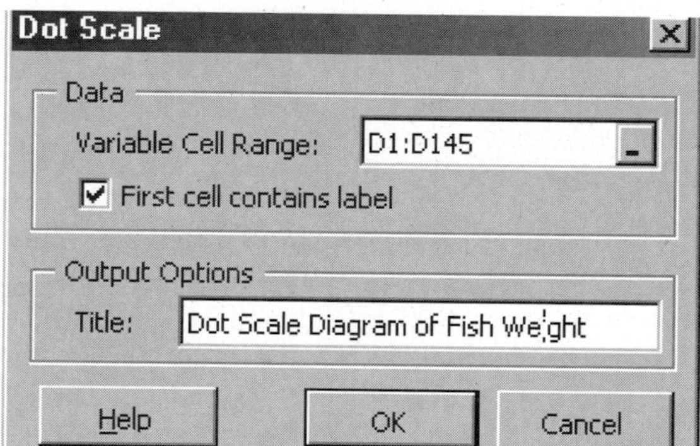

E **Figure 3.E.1** Dot Scale Diagram Dialog Box

3.E.2 Using the Data Analysis Tool to Obtain Descriptive Statistics

Although various Microsoft Excel functions such as AVERAGE, MEDIAN, and STDEV can be used to compute individual statistics, the Data Analysis tool can simultaneously obtain a set of descriptive statistics. For example, to generate the table of descriptive statistics for the weight of fish that is shown in Figure 3.2 on p. 134, open the **FISH.XLS** workbook to the Data worksheet. Then

1. Select Tools | Data Analysis and then select Descriptive Statistics from the Analysis Tools list and click OK.
2. In the Descriptive Statistics dialog box (see Figure 3.E.2):
   a. Enter D1:D145 in the Input Range edit box.
   b. Select the Grouped By Columns option button.
   c. Select the Labels in First Row check box.
   d. Select the New Worksheet Ply option button and enter Descriptive Statistics as the name of the new sheet.
   e. Select the Summary statistics check box.

f. Select the Kth Largest and the Kth Smallest check boxes, leaving the default values (1) in their edit boxes unchanged to obtain the minimum and maximum values.

g. Deselect (uncheck) the Confidence Level for Mean check box. (Chapter 6 discusses confidence levels.)

h. Click the OK button.

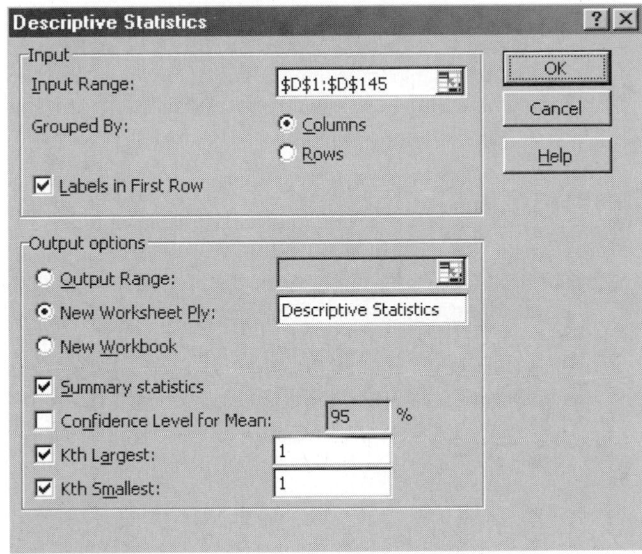

E Figure 3.E.2 Descriptive Statistics Dialog Box

3.E.3 Using PHStat to Obtain a Box-and-Whisker Plot

Use the PHStat **Descriptive Statistics | Box-and-Whisker Plot** procedure to generate a box-and-whisker plot on a new chart sheet and, optionally, a five-number summary table on a new worksheet. For example, to generate the box-and-whisker plot shown in Figure 3.23 on p. 164, open the **MONRENTS.XLS** workbook to the Data worksheet and:

1. Select PHStat | Descriptive Statistics | Box-and-Whisker Plot.

2. In the Box-and-Whisker dialog box (see Figure 3.E.3):

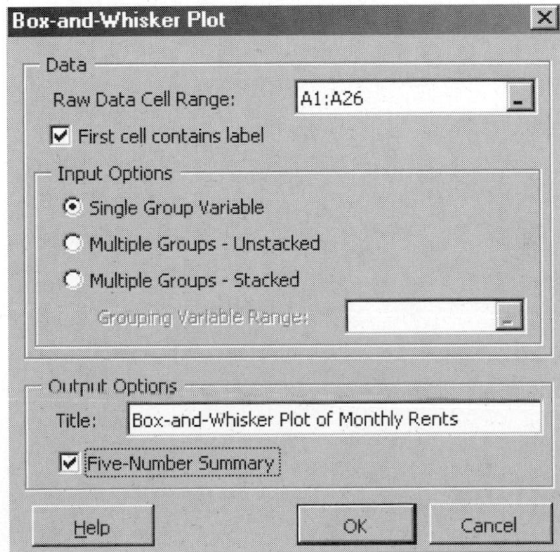

E Figure 3.E.3 Box-and-Whisker Plot Dialog Box

a. Enter A1:A26 in the Raw Data Cell Range edit box.
b. Select the Single Group Variable option button.
c. Enter a title in the Title edit box.
d. Select the Five-Number Summary check box.
e. Click the OK button.

If you wish to obtain a side-by-side box-and-whisker plot, select the Multiple Groups—Unstacked option button if each group is located in a separate column. If the data is stacked with the variable to be analyzed in one column and the grouping variable in a different column, select the Multiple Groups—stacked option button and enter the cell range for the grouping variable in the edit box.

3.E.4 Obtaining a Correlation Coefficient

Use the CORREL Excel worksheet function to calculate the coefficient of correlation. The format of this function is CORREL(*cell range of X*, *cell range of Y*). For example, to obtain the correlation between the length and weight of the fish, open the **FISH.XLS** workbook. The formula =CORREL(Data!C1:C145,Data!D1:D145) entered into any cell of any worksheet would calculate the coefficient of correlation.

Using MINITAB®

3.M.1 Obtaining Descriptive Statistics

To obtain the descriptive statistics for the bid-rigging data, open the **BIDRIG.MTW** file. Select **Stat | Basic Statistics | Display Descriptive Statistics.** In the Display Descriptive Statistics dialog box (see Figure 3.M.1), enter **C3** or **'LBERAT'** in the Variables edit box. Select the **By variable** check box, and enter **C4** or **'STATUS'** in the By variable edit box. Click the **OK** button. The session window should now contain the descriptive summary measures as illustrated in Figure 3.7 on p. 137.

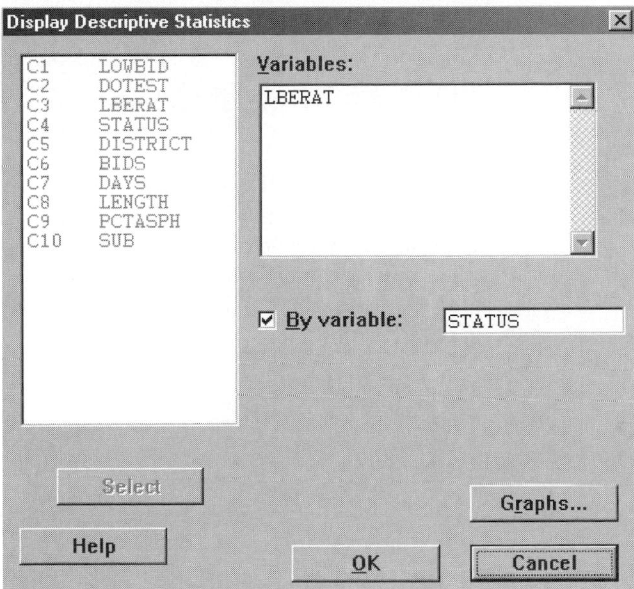

M **Figure 3.M.1** Display Descriptive Statistics Dialog Box

3.M.2 Obtaining Box-and-Whisker Plots (Boxplots)

MINITAB refers to box-and-whisker plots as boxplots. To obtain side-by-side boxplots for the two bid status categories select **Graph | Boxplot.** In the Boxplot dialog box (see Figure 3.M.2 on p. 190), enter **C3** or **'LBERAT'** for Y, and **C4** or **'STATUS'** for X, in the Graph variables edit box. Click the **OK** button. A graphics window containing a side-by-side boxplot should now appear.

3.M.3 Obtaining a Correlation Coefficient

To obtain the coefficient of correlation between the variables LOWBID, DOTEST, LBERAT, BIDS, and PCTASPH in the **BIDRIG.MTW** worksheet, you must first unstack the data.

Select **Manip | Stack/Unstack | Unstack Blocks of Columns.** In the Unstack Blocks of Columns dialog box, enter **C1** or **'LOWBID', C2** or **'DOTEST', C3** or **'LBERAT', C6** or **'BIDS',** and **C9** or **'PCTASPH'** in the Unstack the following columns edit box. In the Using

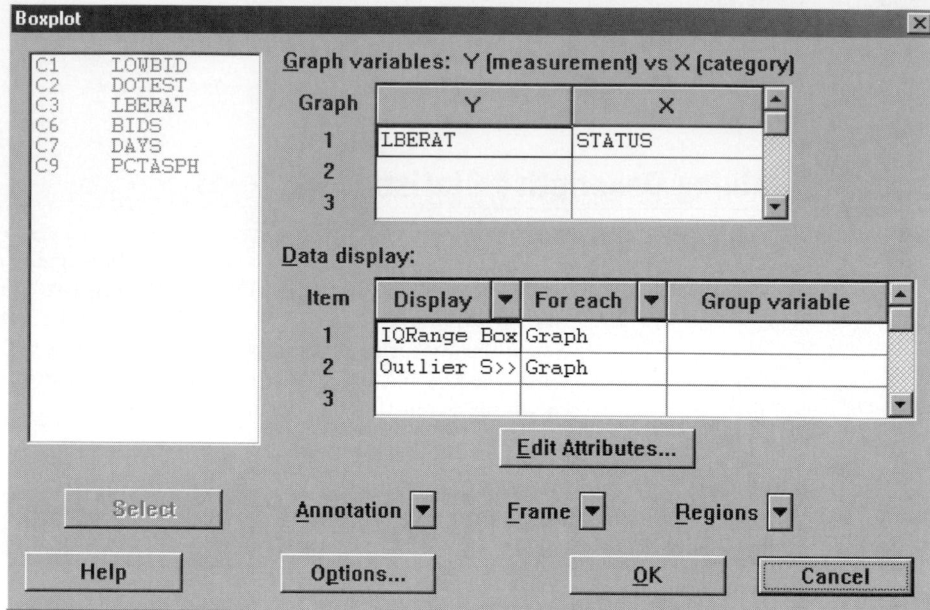

Figure 3.M.2 Boxplot Dialog Box

subscripts in edit box, enter **C4** or **'STATUS'.** In the Store the Unstacked data in blocks edit box, enter **C11–C15** in the first row and **C16–C20** in the second row. Click the **OK** button. Enter labels in **C11–C20.** Select **Stat | Basic Statistics | Correlation.** To obtain the correlations for the competitive bids, in the Correlation dialog box, enter **C11–C15** in the Variables edit box. Click the **OK** button. Repeat the analysis for the fixed bids by selecting **C16–C20** in the Variables edit box.

CHAPTER 4

Probability: Basic Concepts

OBJECTIVES

1. To develop an intuitive understanding of probability as a measure of the uncertainty of an event
2. To define different types of event outcomes
3. To introduce basic rules for finding probabilities of events
4. To introduce conditional probabilities

CONTENTS

EXCEL TUTORIAL

Statistics in the Real World

Lotto Buster!

"Welcome to the Wonderful World of Lottery Bu$ters." So began the premier issue of *Lottery Buster*, a monthly publication for players of the state lottery games. *Lottery Buster* provides interesting facts and figures on the 37 state lotteries currently operating in the United States and, more importantly, tips on how to increase a player's odds of winning the lottery.

New Hampshire, in 1963, was the first state in modern times to authorize a state lottery as an alternative to increasing taxes. (Prior to this time, beginning in 1895, lotteries were banned in America because of corruption.) Since then, lotteries have become immensely popular for two reasons. First, they lure you with the opportunity to win millions of dollars with a $1 investment, and second, when you lose, at least you believe your money is going to a good cause. Many state lotteries, like Florida, designate a high percentage of lottery revenues to fund state education.

The popularity of the state lottery has brought with it an avalanche of "experts" and "mathematical wizards" (such as the editors of *Lottery Buster*) who provide advice on how to win the lottery—for a fee, of course! Many offer guaranteed "systems" of winning through computer software products with catchy names such as Lotto Wizard, Lottorobics, Win4D, and Loto-Luck.

For example, most knowledgeable lottery players would agree that the "golden rule" or "first rule" in winning lotteries is *game selection*. State lotteries generally offer three types of games: Instant (scratch-off tickets or on-line), Daily Numbers (Pick-3 or Pick-4), and the weekly Pick-6 Lotto.

One version of the Instant game involves scratching off the thin opaque covering on a ticket with the edge of a coin to determine whether you have won or lost. The cost of a ticket ranges from 50¢ to $1, and the amount won ranges from $1 to $100,000 in most states, and to as much as $1 million in others. *Lottery Buster* advises against playing the Instant game because it is "a pure chance play, and you can win only by dumb luck. No skill can be applied to this game."

The Daily Numbers game permits you to choose either a three-digit (Pick-3) or four-digit (Pick-4) number at a cost of $1 per ticket. Each night, the winning number is drawn. If your number matches the winning number, you win a large sum of money, usually $100,000. You do have some control over the Daily Numbers game (since you pick the numbers that you play) and, consequently, there are strategies available to increase your chances of winning. However, the Daily Numbers game, like the Instant game, is not available for out-of-state play.

To play Pick-6 Lotto, you select six numbers of your choice from a field of numbers ranging from 1 to *N*, where *N* depends on which state's game you are playing. For example, Florida's current Lotto game involves picking six numbers ranging from 1 to 53. The cost of a ticket is $1 and the payoff, if your six numbers match the winning numbers drawn, is $7 million or more, depending on the number of tickets purchased. (To date, Florida has had the largest state weekly payoff of over $200 million.) In addition to the grand prize, you can win second-, third-, and fourth-prize payoffs by matching five, four, and three of the six numbers drawn, respectively. And you don't have to be a resident of the state to play the state's Lotto game.

In the following Statistics in the Real World Revisited sections, we demonstrate how to use the basic concepts of probability to compute the odds of winning a state lottery game and to assess the validity of the strategies suggested by lottery "experts."

Statistics in the Real World Revisited

- Understanding the Probability of Winning Lotto (p. 196)
- The Probability of Winning Lotto with a Wheel System (p. 203)
- Computing the Probability of Winning 6/53 Lotto (p. 210)
- The Probability of Winning Daily Cash 3 or Play 4 (p. 216)

4.1 The Role of Probability in Statistics

If you play blackjack, a popular casino table game, you know that whether you win in any one game is an outcome that is very uncertain. Similarly, investing in bonds, stock, or a new business is a venture whose success is subject to uncertainty. (In fact, some would argue that investing is a form of educated gambling—one in which knowledge, experience, and good judgment can improve the odds of winning.)

Much like playing blackjack and investing, making inferences based on sample data is also subject to uncertainty. A sample rarely tells a perfectly accurate story about the population from which it was selected. There is always a margin of error (as the pollsters tell us) when sample data are used to estimate the proportion of people in favor of a particular political candidate, some consumer product, or some political or social issue. There is always uncertainty about how far the sample estimate will depart from the true population proportion of affirmative answers that you are attempting to estimate. Consequently, a measure of the amount of uncertainty associated with an estimate (which we called the *reliability of an inference* in Chapter 1) plays a major role in statistical inference.

How do we measure the uncertainty associated with events? Anyone who has observed a daily newscast can answer that question. The answer is *probability*. For example, it may be reported that the probability of rain on a given day is 20%. Such a statement acknowledges that it is uncertain whether it will rain on the given day and indicates that the forecaster measures the likelihood of its occurrence as 20%.

Probability also plays an important role in decision making. To illustrate, suppose you have an opportunity to invest in an oil exploration company. Past records show that for 10 previous oil drillings (a sample of the company's experiences), all 10 resulted in dry wells. What do you conclude? Do you think the chances are better than 50-50 that the company will hit a producing well? Should you invest in this company? We think your answer to these questions will be an emphatic "no." If the company's exploratory prowess is sufficient to hit a producing well 50% of the time, a record of 10 dry wells out of 10 drilled is an event that is just too improbable. Do you agree?

In this chapter we will examine the meaning of probability and develop some properties of probability that will be useful in our study of statistics.

4.2 Experiments, Events, and the Probability of an Event

In the language employed in a study of probability, the word *experiment* has a very broad meaning. In this language, an experiment is a process of making an observation or taking a measurement on the experimental unit. For example, suppose you are dealt a single card from a standard 52-card deck. Observing the outcome (i.e., the number and suit of the card) could be viewed as an experiment. Counting the number of U.S. citizens who live in a particular state or county is an experiment. Similarly, recording a voter's opinion on an important political issue is an experiment. Observing the fraction of insects killed by a new insecticide is an experiment. Note that most experiments result in outcomes (or measurements) that cannot be predicted with certainty in advance.

Definition 4.1

The process of making an observation or taking a measurement on one or more experimental units is called an **experiment.**

EXAMPLE 4.1

LISTING OUTCOMES

Consider the following experiment. You are dealt one card from a standard 52-card deck.* List some possible outcomes of this experiment that cannot be predicted with certainty in advance.

*A standard 52-card deck consists of four types (or suits) of cards. Hearts and diamonds are red; spades and clubs are black. Each suit contains 13 cards, numbered 2, 3, 4, . . . , 10, Jack, Queen, King, and Ace.

Solution

Some possible outcomes of this experiment that cannot be predicted with certainty in advance are as follows (see Figure 4.1):

a. You draw an ace of hearts. **b.** You draw an eight of diamonds.
c. You draw a spade. **d.** You do not draw a spade.

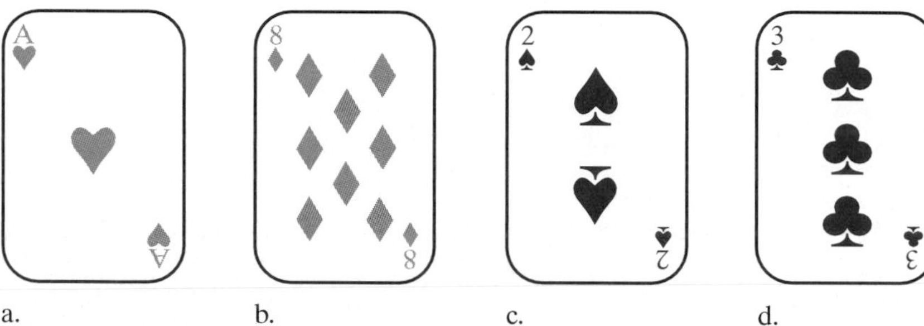

Figure 4.1 Possible Outcomes of Card-Drawing Experiment

EXAMPLE 4.2

LISTING OUTCOMES

During the 2000 U.S. Presidential Election, the outcome was in serious doubt until many weeks after the November 7th election day due to the closeness of the race in Florida between Republican candidate (and current President) George W. Bush and Democratic candidate Al Gore. Consider the following experiment. Five hundred Palm Beach County, Florida, voters are randomly selected from all those who voted in the 2000 presidential election and each is asked who he or she voted for on Nov. 7, 2000. List some possible outcomes of this experiment that cannot be predicted with certainty in advance.

Solution

Since we are observing the choices of 500 Palm Beach County voters, this experiment can result in a very large number of outcomes. Three of the many possible outcomes are listed below.

a. Exactly 287 of the 500 voters voted for Gore.
b. Exactly 302 of the 500 voters voted for Gore.
c. A particular voter, Jessica Jones, voted for Gore.

Clearly, we could define many other outcomes of this experiment that cannot be predicted in advance.

In the language of probability theory, outcomes of experiments are called *events*.

Definition 4.2

Outcomes of experiments are called **events.** [*Note:* To simplify our discussion, we will use italic capital letters $A, B, C, \ldots,$ to denote specific events.]

The outcome for each of the experiments described in Examples 4.1 and 4.2 is shrouded in uncertainty; that is, prior to conducting the experiment, we could not be certain whether a particular event would occur. This uncertainty is measured by the *probability* of the event.

EXAMPLE 4.3

INTERPRETING A PROBABILITY

Suppose we perform the following experiment: Toss a coin and observe whether it results in a head or a tail. Define the event H by H: Observe a head.

What do we mean when we say that the probability of H, denoted by $P(H)$, is equal to $\frac{1}{2}$?

Solution

Stating that the probability of observing a head is $P(H) = \frac{1}{2}$ does *not* mean that exactly half of a number of tosses will result in heads. (For example, we do not expect to observe exactly 1 head in 2 tosses of a coin or exactly 5 heads in 10 tosses of a coin.) Rather, it means that, in a very long series of tosses, we believe that approximately half would result in a head. Therefore, the number $\frac{1}{2}$ measures the likelihood of observing a head on a single toss of the coin.

The "relative frequency" concept of probability discussed in Example 4.3 is illustrated in Figure 4.2. The graph shows the proportion of heads observed after $n = 25, 50, 75, 100, 125, \ldots, 1{,}450, 1{,}475,$ and $1{,}500$ computer-simulated repetitions of a coin-tossing experiment. The number of tosses is marked along the horizontal axis of the graph, and the corresponding proportions of heads are plotted on the vertical axis above the values of n. We have connected the points by line segments to emphasize the fact that the proportion of heads moves closer and closer to .5 as n gets larger (as you move to the right on the graph).

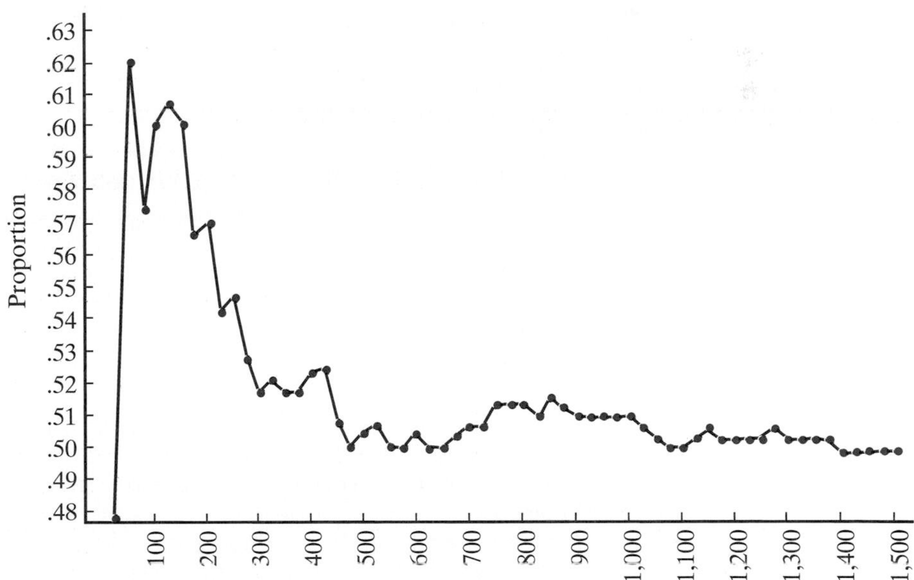

Figure 4.2 The Proportion of Heads in n Tosses of a Coin

Although most people think of the probability of an event as the proportion of times the event occurs in a very long series of trials, some experiments can never be repeated. For example, if you invest \$50,000 in starting a new business, the probability that your business will survive 5 years has some unknown value that you will

never be able to evaluate by repetitive experiments. The probability of this event occurring is a number that has some value, but it is unknown to us. The best that we could do, in estimating its value, would be to attempt to determine the proportion of similar businesses that survived 5 years and take this as an approximation to the desired probability. In spite of the fact that we may not be able to conduct repetitive experiments, the relative frequency definition for probability appeals to our intuition.

Definition 4.3

The **probability of event** A, denoted by $P(A)$, is a number between 0 and 1 that measures the likelihood that A will occur when the experiment is performed. $P(A)$ can be approximated by the proportion of times that A is observed when the experiment is repeated a very large number of times.

Self-Test 4.1

Consider the following experiment: A family with three children whose gender is unknown is selected and the objective is to count the number of boys in the family.

a. List some possible outcomes of this experiment that cannot be predicted in advance.

b. We will learn that the probability of observing three boys is $\frac{1}{8}$. Explain what this probability represents.

In the next section we give some rules for calculating the exact probability of an event.

Statistics in the Real World Revisited

Understanding the Probability of Winning Lotto

In Florida's biweekly lottery game, called Pick-6 Lotto, you select six numbers of your choice from a set of numbers ranging from 1 to 53. The odds of winning the game with a single ticket are 1 in 22,957,480, or, 1 in approximately 23 million. (We show how to calculate this probability in optional Section 4.4.) This probability can be written:

$$P(\text{win Lotto}) = \frac{1}{22{,}957{,}480} = .00000004356.$$

For all practical purposes, this probability is 0, implying that you have almost no chance of winning the lottery with a single ticket. Yet, each week there is almost always a winner in the Florida Lotto. This apparent contradiction can be explained with the following analogy.

Suppose there is a line of minivans, front-to-back, from New York City to Los Angeles, California. Based on the distance between the two cities and the length of a standard minivan, there would be approximately 23 million minivans in line. Lottery officials will select, at random, one of the minivans and put a check for $10 million dollars in the glove compartment. For a cost of $1, you may roam the country and select one (and only one) minivan and check the glove compartment. Do you think you will find $10 million in the minivan you choose? You can be almost certain that you won't. But now permit anyone to enter the lottery for $1 and suppose that 50 million people do so. With such a large number of participants, it is very likely that someone will find the minivan with the $10 million—but it almost certainly won't be you! (This example illustrates an axiom in statistics called the "Law of Large Numbers.")

PROBLEMS FOR SECTION 4.2

Using the Tools

4.1 Consider the following experiment: Toss a fair six-sided die and observe the number of dots showing on the upper face.

a. If this experiment were to be repeated over and over again in a very long series of trials, what proportion of the experimental outcomes do you think would result in a 5?

b. What does it mean to say, "The probability that the outcome is a 5 is $\frac{1}{6}$"?

c. Perform the experiment a large number of times and calculate the proportion of outcomes that result in a 5. Note that as the number of repetitions becomes larger and larger, this proportion moves closer and closer to $\frac{1}{6}$.

Applying the Concepts

4.2 The following facts are extracted from *True Odds: How Risk Affects Your Everyday Life* (Walsh, 1997). Convert each to a probability statement.

a. The prospect of dying from flesh-eating bacteria are one in a million.

b. The chance of being the victim of a violent crime is one in 135.

c. Nine of ten premature deaths are linked to smoking, overeating, alcohol abuse, high blood pressure, not exercising, or not wearing seat belts.

4.3 The United States Department of Agriculture (USDA) reports that, under its standard inspection system, one in every 100 slaughtered chickens pass inspection with fecal contamination (*Tampa Tribune,* Mar. 31, 2000).

a. If a slaughtered chicken is selected at random, what is the probability that it passes inspection with fecal contamination?

b. The probability of part **a** was based on a USDA study that found that 306 of 32,075 chicken carcasses passed inspection with fecal contamination. Do you agree with the USDA's statement about the likelihood of a slaughtered chicken passing inspection with fecal contamination?

4.4 During the 1989 U.S. Open, four professional golfers (Doug Weaver, Jerry Pate, Nick Price, and Mark Wiebe) made holes in one (aces) on the sixth hole at Oak Hill Country Club—all on the same day! How unlikely is such a feat? According to *Golf Digest* (Mar. 1990), the probability of a Professional Golf Association (PGA) tour pro making an ace on a given hole is approximately $\frac{1}{3,000}$. The estimate is based on the ratio of the number of aces made on the PGA tour to the total number of rounds played.

a. Interpret the probability of $\frac{1}{3,000}$.

b. *Golf Digest* also estimates that the probability of any four players getting aces on the same hole on the same day during the next U.S. Open as $\frac{1}{150,000}$. Interpret this probability.

4.5 Each in a random sample of 287 female military veterans was asked: "Imagine that we flip a fair coin 1,000 times. What is your best guess about how many times the coin would come up heads in 1,000 flips?" Surprisingly, only 155 of the women answered correctly (*Annals of Internal Medicine,* Dec. 1, 1997). Use the fact that $P(H) = \frac{1}{2}$ in a single coin toss to estimate the number of heads in 1,000 flips.

4.6 The *San Francisco Examiner* (Oct. 15, 1992) reported on a large-scale study of expectant parents in order to assess the risk of their children being born with cystic fibrosis. The study used genetic DNA analysis of the parents to determine whether the unborn child is likely to have the disease. Based on the study, the risk assessment table shown on p. 198 was compiled.

a. If both parents test positive, what is the probability of the child developing cystic fibrosis?

b. If both parents test negative, what is the probability of the child developing cystic fibrosis?

		Father		
		Untested	Test Positive	Test Negative
Mother	Untested	Unknown	1 in 100	1 in 16,500
	Test Positive	1 in 100	1 in 4	1 in 660
	Test Negative	1 in 16,500	1 in 660	1 in 106,000

Problem 4.6

4.3 Probability Rules for Mutually Exclusive Events

One special property of events can be seen in the examples of the preceding section. Two events are said to be mutually exclusive if, when one occurs, the other cannot occur. Consider the experiment of Example 4.1 (p. 194), where we draw one card at random from a 52-card deck. The events listed under parts **c** and **d** of Example 4.1 are mutually exclusive. You cannot conduct an experiment and "draw a spade" (the event listed under part **c**) and at the same time "not draw a spade" (the event listed under part **d**). If one of these two events occurs when an experiment is conducted, the other event cannot have occurred. Therefore, we say that they are mutually exclusive events.

Definition 4.4

Two events are said to be **mutually exclusive** if, when one of the two events occurs in an experiment, the other cannot occur.

EXAMPLE 4.4

MUTUALLY EXCLUSIVE EVENTS

Refer to Example 4.2 (p. 194) where 500 Palm Beach County, Florida, voters are surveyed about who they voted for in the 2000 Presidential Election. Consider the following events:

A: Exactly 287 of the 500 voters voted for Gore.

B: Exactly 302 of the 500 voters voted for Gore.

C: A particular voter, Jessica Jones, voted for Gore.

State whether the following pairs of events are mutually exclusive:

a. *A* and *B* **b.** *A* and *C* **c.** *B* and *C*

Solution

a. Events *A* and *B* are mutually exclusive. If you have observed exactly 287 Gore voters, then you could not, at the same time, have observed exactly 302 Gore voters.

b. Events *A* and *C* are *not* mutually exclusive. Jessica Jones may be one of the 287 voters in event *A* who voted for Gore. Therefore, it is possible for both events *A* and *C* to occur simultaneously.

c. Events *B* and *C* are not mutually exclusive for the same reason given in part **b**.

EXAMPLE 4.5

MUTUALLY EXCLUSIVE EVENTS

Suppose an experiment consists of selecting two electric light switches from an assembly line for inspection. Define the following events:

A: The first switch is defective.

B: The second switch is defective.

Are A and B mutually exclusive events?

Solution

Events A and B are not mutually exclusive because both the first and the second switch could be defective when the inspection is made. That is, both A and B can occur together.

Self-Test 4.2

An experiment consists of observing whether or not each of the next two people who check out at a supermarket use coupons. List two events that are mutually exclusive.

If two events A and B are mutually exclusive, then the probability that *either A or B* occurs is equal to the sum of their probabilities. We will illustrate with an example.

EXAMPLE 4.6

SUMMING PROBABILITIES

Consider the following experiment: You roll a single six-sided die and observe the number of dots on the upper face of the die. What is the probability that you observe an even number?

Solution

The experiment can result in one of six mutually exclusive events: 1, 2, 3, 4, 5, or 6 dots showing on the die. These six events represent the most basic outcomes of the experiment and are called *simple events*. The collection of all possible simple events in an experiment is called the *sample space*. The sample space for this die-rolling experiment is shown diagrammatically in Figure 4.3.

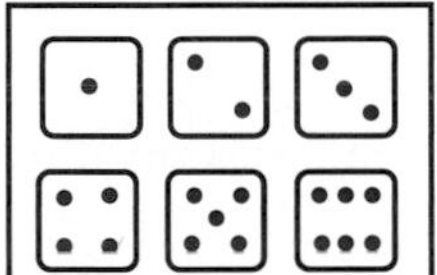

Figure 4.3 Six Mutually Exclusive Outcomes (Simple Events) when Rolling a Die

With a "fair" die, we expect each of the six simple events of Figure 4.3 to occur with approximately equal relative frequency ($\frac{1}{6}$) if the die-rolling experiment were repeated a large number of times. Since you will observe an even number only if a 2 or a 4 or a 6 occurs, and these simple events are mutually exclusive, either one or the other of these events will occur $(\frac{1}{6} + \frac{1}{6} + \frac{1}{6}) = \frac{1}{2}$ of the time. Therefore, the probability of observing an even number in the roll of a die is equal to the probability of observing *either a 2 or a 4 or a 6,* which is $\frac{1}{2}$. You can verify this result experimentally, using the procedure employed in Section 4.2.

Definition 4.5

Simple events are mutually exclusive events that represent the most basic outcomes of an experiment.

Definition 4.6

The collection of all possible simple events in an experiment is called the **sample space**.

Although Example 4.6 utilized the concept of simple events, the additive probability rule shown in the next box applies to any two mutually exclusive events. You can now see why the concept of mutually exclusive events is important. We will illustrate with several more examples.

PROBABILITY RULE #1

The Additive Rule for Mutually Exclusive Events If two events A and B are mutually exclusive, then *the probability that either A or B occurs* is equal to the sum of their respective probabilities:

$$P(A \text{ or } B) = P(A) + P(B) \tag{4.1}$$

EXAMPLE 4.7

APPLYING PROBABILITY RULE #1

A local weather reporter forecasts the following three mutually exclusive weather conditions, with their associated probabilities:

Overcast with some rain: 40%

No rain, but partially or completely overcast: 30%

No rain and a clear day: 30%

What is the probability that the day will be overcast?

Solution

The forecaster suggests that the experiment of observing the day's weather can result in one (and only one) of three mutually exclusive events:

A: Overcast with some rain

B: No rain, but partially or completely overcast

C: No rain and a clear day

where

$$P(A) = .4 \qquad P(B) = .3 \qquad P(C) = .3$$

The event of interest, that the day will be overcast, will occur if either event A or event B occurs. Therefore, using Probability Rule #1, we have:

$$P(A \text{ or } B) = P(A) + P(B) = .4 + .3 = .7$$

The implication is that if we were to forecast the weather for a large number of days with similar weather conditions, 70% of the days would be overcast.

EXAMPLE 4.8

APPLYING PROBABILITY RULE #1

Consider the following experiment: Roll two dice and observe the sum of the dots showing on the two dice. Find the probability that the sum is equal to 7 (an important number in the casino game of craps).

Solution

Mark the dice so that they are identified as die #1 and die #2. Then there are $6 \times 6 = 36$ distinctly different ways that the dice could fall. You could observe a 1 on die #1 and a 1 on die #2; a 1 on die #1 and a 2 on die #2; a 1 on die #1 and a 3 on die #2, etc. In other words, you can pair the six values (shown on the six sides) of die #1 with the six values of die #2 in $6 \times 6 = 36$ mutually exclusive ways. These 36 possibilities represent the simple events for the experiment. The sums associated with the simple events are shown in Figure 4.4 (p. 201).

Since there are 36 possible ways the dice could fall and since these ways should occur with equal frequency, the probability of observing any one of the 36 events shown in the figure is $\frac{1}{36}$. Then to find the probability of rolling a 7, we need only to add the probabilities of those events corresponding to a sum on the dice equal to 7.

Figure 4.4 The Sum of the Dots for the 36 Mutually Exclusive Outcomes in the Tossing of a Pair of Dice

Die #2 \ Die #1	1	2	3	4	5	6
1	2	3	4	5	6	7
2	3	4	5	6	7	8
3	4	5	6	7	8	9
4	5	6	7	8	9	10
5	6	7	8	9	10	11
6	7	8	9	10	11	12

If we denote the simple event that you observe a 6 on die #1 and a 1 on die #2 as (6, 1), etc., then as shown in Figure 4.4, you will toss a 7 if you observe a (6, 1), (5, 2), (4, 3), (3, 4), (2, 5), or (1, 6). Therefore, the probability of tossing a 7 is

$$\begin{aligned} P(7) &= P[(6,1) \text{ or } (5,2) \text{ or } (4,3) \text{ or } (3,4) \text{ or } (2,5) \text{ or } (1,6)] \\ &= P(6,1) + P(5,2) + P(4,3) + P(3,4) + P(2,5) + P(1,6) \\ &= \frac{1}{36} + \frac{1}{36} + \frac{1}{36} + \frac{1}{36} + \frac{1}{36} + \frac{1}{36} = \frac{6}{36} = \frac{1}{6} \end{aligned}$$

Self-Test 4.3

Two marbles are selected at random and without replacement from a box containing two blue marbles and three red marbles. Determine the probability of observing each of the following events:

A: Two blue marbles are selected.
B: A red and a blue marble are selected.
C: Two red marbles are selected.

Example 4.8 suggests a modification of Probability Rule #1 when an experiment can result in one and only one of a number of equally likely (equiprobable) mutually exclusive events.

PROBABILITY RULE #2

The Probability Rule for an Experiment That Results in One of a Number of Equally Likely Mutually Exclusive Events Suppose an experiment can result in one and only one of M equally likely mutually exclusive events and that m of these events result in event A. Then the probability of event A is

$$P(A) = \frac{m}{M} \tag{4.2}$$

EXAMPLE 4.9 **APPLYING PROBABILITY RULE #2**

As a contestant on the popular television game show "Let's Make A Deal," you have selected, by chance, two doors from among four doors offered. Unknown to you, there are prizes behind only two of the four doors. What is the probability that you win at least one of the two prizes?

Solution

Identify the four doors as D_1, D_2, D_3, and D_4, and let D_3 and D_4 be the two doors that hide the two prizes. Then the six distinctly different and mutually exclusive ways that the two doors may be selected from the four are:

$$(D_1, D_2) \quad (D_1, D_3) \quad (D_1, D_4) \quad (D_2, D_3) \quad (D_2, D_4) \quad (D_3, D_4)$$

Step 1 Since the doors were selected by chance, we would expect the likelihood that any one pair would be chosen to be the same as for any other pair. Also, since the experiment can result in only one of these pairs, $M = 6$.

Step 2 You can see from the listed outcomes that the number of pairs that result in a choice of D_3 or D_4 is $m = 5$. (These are the last five pairs listed.)

Step 3 Using Probability Rule #2, the probability of selecting at least one of the two doors with prizes is:

$$P(\text{at least one of } D_3 \text{ or } D_4) = \frac{m}{M} = \frac{5}{6}$$

Examples 4.6–4.9 identify the properties of the probabilities of all events, as summarized in the box.

PROPERTIES OF PROBABILITIES

1. The probability of an event always takes on a value between 0 and 1.
2. If two events A and B are mutually exclusive, then the probability that either A or B occurs is equal to $P(A) + P(B)$.
3. If we list all possible simple events associated with an experiment, then the sum of their probabilities will always equal 1.

Before concluding this section, we will comment on two important mutually exclusive events and their probabilities. Consider again the dice-tossing experiment (Example 4.8, p. 200) and define the following two events:

A: The sum of the dots on the two dice is 7.

A': The sum of the dots on the two dice is not 7.

Thus, A' is the event that *A does not occur*. You can see that A and A' are mutually exclusive events and, further, that if A occurs $\frac{1}{6}$ of the time in a long series of trials (see Example 4.8), then A' will occur $\frac{5}{6}$ of the time. In other words,

$$P(A) + P(A') = 1$$

Definition 4.7

The Rule of Complements The **complement** of an event A, denoted by the symbol A', is the event that A does not occur (see Figure 4.5).

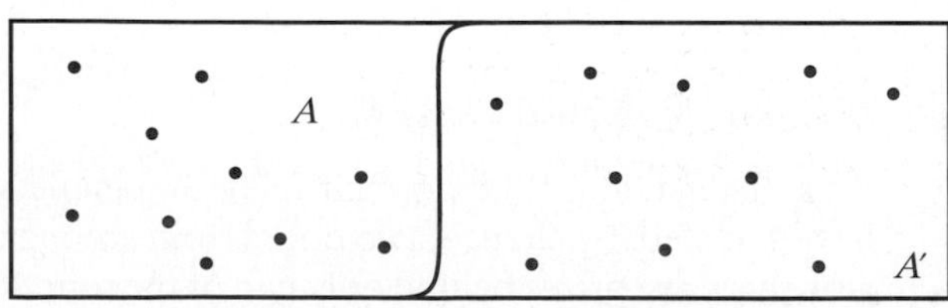

Figure 4.5
Complementary Events A and A' in a Sample Space

PROBABILITY RULE #3

Probability Relationship for Complementary Events

$$P(A) = 1 - P(A') \tag{4.3}$$

Complementary events are important because sometimes it is difficult to find the probability of an event A, but easy to find the probability of its complement A'. In this case, we can find $P(A)$ using the relationship stated in Probability Rule #3.

Statistics in the Real World Revisited

The Probability of Winning Lotto with a Wheel System

Refer to Florida's Pick-6 Lotto game in which you select six numbers of your choice from a field of numbers ranging from 1 to 53. In Section 4.2 (p. 196), we learned that the probability of winning Lotto on a single ticket is only 1 in approximately 23 million. The "experts" at Lotto Buster recommend many strategies for increasing the odds of winning the lottery. One strategy is to employ a *wheeling system*. In a complete wheeling system, you select more than six numbers, say, seven, and play every combination of six of those seven numbers.

Suppose you choose to "wheel" the following seven numbers: 2, 7, 18, 23, 30, 32, and 51. Every combination of six of these seven numbers is listed in Table 4.1. You can see that there are seven different possibilities. Thus, we would purchase seven tickets (at a cost of $7) corresponding to these different combinations in a complete wheeling system.

TABLE 4.1 Wheeling the Six Numbers 2, 7, 18, 23, 30, 32, and 51

Ticket #1	2	7	18	23	30	32
Ticket #2	2	7	18	23	30	51
Ticket #3	2	7	18	23	32	51
Ticket #4	2	7	18	30	32	51
Ticket #5	2	7	23	30	32	51
Ticket #6	2	18	23	30	32	51
Ticket #7	7	18	23	30	32	51

To determine if this strategy does, in fact, increase our odds of winning, we need to find the probability that one of these seven combinations occurs during the 6/53 Lotto draw. In optional Section 4.4, we show you that there are a total of 22,957,480 possible combinations of six numbers that can be selected from the 53. Since these outcomes are mutually exclusive and are equally likely to occur, then $M = 22{,}957{,}480$ in Probability Rule #2. Let event A represent the event that one of our purchased Lotto tickets wins (i.e., that one of the seven combinations shown in Table 4.1 occurs). Then $m = 7$ in Probability Rule #2. Consequently, the probability that we win Lotto with the seven tickets is:

$$P(\text{win Lotto with seven wheeled numbers}) = \frac{m}{M} = \frac{7}{22{,}957{,}480} = .0000003$$

In terms of odds, we now have 3 chances in 10 million of winning the Lotto with the complete wheeling system. The "experts" are correct—our odds of winning Lotto have increased (from 1 in 23 million). However, the probability of winning is so close to 0 we question whether the $7 spent on lottery tickets is worth the negligible increase in odds. In fact, it can be shown that to increase your chance of winning the 6/53 Lotto to 1 chance in 100 (i.e., .01) using a complete wheeling system, you would have to wheel 26 of your favorite numbers—a total of 230,230 combinations at a cost of $230,230!

Self-Test 4.4

Refer to Self-Test 4.3 (p. 201). Recall that two marbles are selected from a box containing two blue and three red marbles. Use the Rule of Complements to find the probability that at least one red marble is selected.

PROBLEMS FOR SECTION 4.3

Using the Tools

4.7 An experiment has five possible outcomes (simple events) with the probabilities shown in the table at left:

Simple Event	Probability
S_1	.15
S_2	.20
S_3	.20
S_4	.25
S_5	.20

Problem 4.7

a. Find the probability of each of the following events:

A: Outcome S_1, S_2, or S_4 occurs.

B: Outcome S_2, S_3, or S_5 occurs.

C: Outcome S_4 does not occur.

b. List the simple events in the complements of events A, B, and C.

c. Find $P(A')$, $P(B')$, and $P(C')$.

4.8 In the roll of an unbalanced ("loaded") die, the outcomes (simple events) occur with the probabilities in the table at left. Define the following events:

Simple Event	Probability
1	$^1/_{12}$
2	$^1/_6$
3	$^1/_6$
4	$^1/_6$
5	$^1/_6$
6	$^3/_{12}$

Problem 4.8

A: Observe a number less than 4.

B: Observe an odd number.

C: Observe an even number.

a. Find $P(A)$, $P(B)$, and $P(C)$.

b. List the outcomes (simple events) in the complements of events A, B, and C.

c. Find $P(A')$, $P(B')$, and $P(C')$.

d. Are any of the pairs of events, A and B, A and C, or B and C, complementary? Explain.

4.9 Suppose an experiment involves tossing two coins and observing the faces of the coins. Find the probabilities of:

a. *A:* Observing exactly two heads.

b. *B:* Observing exactly two tails.

c. *C:* Observing at least one head.

d. Describe the complement of event A and find its probability.

4.10 Two dice are tossed. Use Figure 4.4 on p. 201 to find the probability that the sum of the dots showing on the two dice is equal to:

a. 12 **b.** 5 **c.** 11

4.11 Refer to the dice-tossing experiment in Problem 4.10. Find the approximate probability of tossing a sum of 7 by conducting an experiment similar to the experiments illustrated in Figure 4.2. Toss a pair of dice a large number of times and record the proportion of times a sum of 7 is observed. Compare your value with the exact probability, $\frac{1}{6}$.

4.12 Refer to Examples 4.6 and 4.7 (pp. 199–200). Verify that property 3 (shown in the box titled Properties of Probabilities, p. 202) holds for each of these examples.

Applying the Concepts

4.13 According to *USA Today* (Sept. 19, 2000), there are 650 members of the International Nanny Association (INA). Of these, only three are men. Find the probability that a randomly selected member of the INA is a man.

4.14 *The Wall Street Journal* (Sept. 1, 2000) reported on an independent study of postal workers and violence at post offices. In a sample of 12,000 postal workers, 600 of them were physically assaulted on the job in the past year. Use this information to

estimate the probability that a randomly selected postal worker will be physically assaulted on the job during the year.

4.15 The Excel bar graph below summarizes the market shares of diaper brands in the United States. Suppose we randomly select a recently purchased package of diapers and observe its brand.

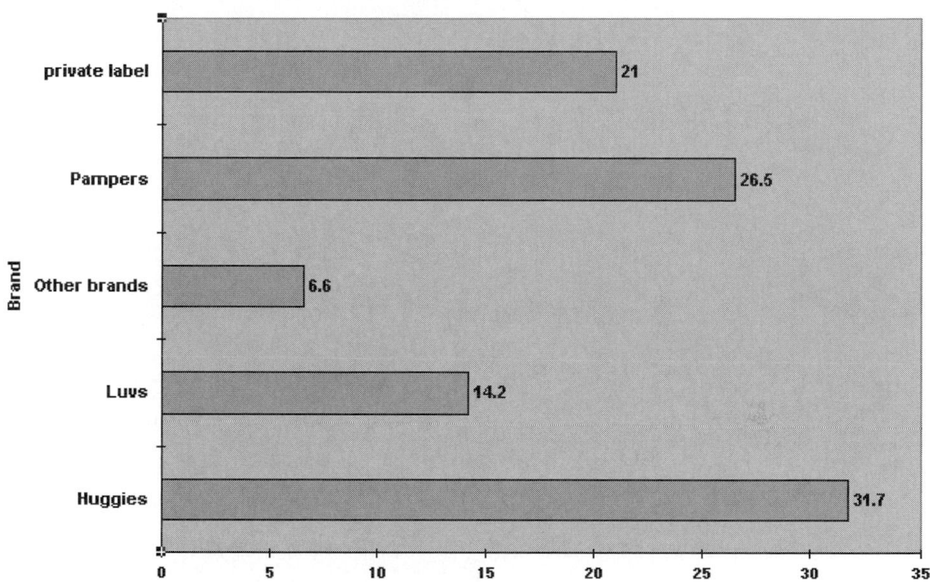

Data Source: A. C. Nielsen.

a. List the simple events for this experiment.

b. Assign probabilities to the simple events.

c. What is the probability the brand selected is Pampers or Luvs?

d. What is the probability the brand selected is not Huggies?

4.16 The National Highway Traffic Safety Administration maintains the Fatal Accident Report System (FARS) file. The FARS file documents every motor vehicle crash (since Jan. 1, 1975) on a U.S. public road in which a fatality occurred. Based on over 800,000 deaths recorded in the FARS file, the types of crashes that account for all car occupant fatalities is distributed as follows:

Type of Crash	Proportion
Single-car crash	.45
Two-car crash	.22
Car crashing with another noncar vehicle (e.g., a truck)	.25
Crash involving 3 or more vehicles	.08
TOTAL	1.00

Source: Evans, L. "Small cars, big cars: What is the safety difference?" *Chance,* Vol. 7, No. 3, Summer 1994, p. 9.

Consider a recent car crash that resulted in a fatality.

a. What is the probability that a single-car crash occurred?

b. What is the probability that a two-car crash occurred?

c. What is the probability that two or more vehicles were involved in the crash?

4.17 A national study was commissioned to gauge the performance of property managers who are employed by nonprofit organizations. The study sample was comprised of 23 rental properties located in six cities—Boston, Chicago, Miami, Minneapolis/St. Paul, New York, and Oakland—all managed by nonprofit organizations (*Journal of the American Planning Association,* Winter 1998). The percentage of total rent collected by each property manager was reported and is summarized in the table. Suppose we randomly select one of the 23 rental properties and observe the percentage of total rent collected by the property manager.

Percentage of Rent Collected	Number of Properties
Less than 80	4
80–84	2
85–89	0
90–94	2
95–99	2
100	13
TOTAL	23

Source: Bratt, R. G., et al. "The status of nonprofit-owned affordable housing." *Journal of the American Planning Association,* Vol. 64, No. 1, Winter 1998, p. 41 (Table 1).

a. Let A be the event that the percentage of total rent collected is less than 80%. Find $P(A)$.

b. Let B be the event that between 85% and 89% of the total rent is collected by the property manager. Find $P(B)$.

c. Are events A and B mutually exclusive? Explain.

d. A widely accepted standard in the rental industry is that management should collect at least 95% of the total rent due. Find the probability that the property manager is meeting this standard.

4.18 Local area merchants often use "scratch off" tickets with promises of grand prize giveaways to entice customers to visit their showrooms. One such game, called Jackpot, was recently used in Tampa, Florida. Residents were mailed game tickets with ten "play squares" and the instructions: "Scratch off your choice of ONLY ONE Jackpot Play Square. If you reveal any 3-OF-A-KIND combination, you WIN. Call NOW to make an appointment to claim your prize." Unknown to the customer, all ten play squares have *winning* 3-of-a-kind combinations. On one such ticket, three cherries (worth up to $1,000 in cash or prizes) appeared in play squares numbered 1, 2, 3, 4, 5, 6, 7, and 9; three lemons (worth a 10-piece microwave cookware set) appeared in play square number 8; and three sevens (worth a whirlpool spa) appeared in play square 10.

a. If you play Jackpot, what is the probability that you win a whirlpool spa?

b. If you play Jackpot, what is the probability that you win $1,000 in cash or prizes?

4.19 In a study conducted by a marketing honors class at the University of South Florida, 257 randomly selected Tampa Bay area residents were asked to select their favorite local attraction. The results are shown in the table on p. 207. Suppose one of the Tampa Bay area residents surveyed is selected at random. Find the probability of each of the following events.

a. The resident's favorite local attraction is Busch Gardens.

b. The resident's favorite local attraction is Sports.

c. The resident's favorite local attraction is not any of the Disney Theme Parks.

Favorite Attraction	Number Surveyed
Adventure Island Water Park	10
Beaches	21
Busch Gardens	69
Disney Theme Parks (Orlando)	31
Museum of Science and Industry	3
Parks	8
Sea World (Orlando)	8
Sports	29
Other	78
TOTAL	257

Source: Department of Marketing, University of South Florida.

Problem 4.19

4.20 The American Association for Marriage and Family Therapy (AAMFT) is a group of professional therapists and family practitioners who treat many of the nation's couples and families. The AAMFT released the findings of a study that tracked the postdivorce history of 98 pairs of former spouses with children. Each divorced couple was classified into one of four groups; the proportions classified into each group are given in the table.

Group	Proportion
"Perfect pals": (Joint-custody parents who get along well, do not remarry)	.12
"Cooperative colleagues": (Occasional conflict, likely to be remarried)	.38
"Angry associates": (Cooperate on issues related to children only, conflicting otherwise)	.25
"Fiery foes": (Communicate only through children, hostile toward each other)	.25

Suppose one of the 98 couples is selected at random.

a. What is the probability that the former spouses are "cooperative colleagues"?

b. What is the probability that the former spouses are "angry associates"?

c. What is the probability that the former spouses are not "perfect pals"?

4.21 Refer to the *Archaeol. Oceania* (Apr. 1997) study of archaeological digs in Papua, New Guinea, Problem 2.9 (p. 65). The table describes the distribution of artifacts (stone drillpoints, shell beads, and grinding slabs) discovered on Motupore Island. If an artifact is found at the site, what is the probability that it is:

Type of Artifact	Number Discovered
Drillpoint	566
Shell bead	385
Grinding slab	43

Source: Allen, J., et al. "Identifying specialization, production, and exchange in the archaeological record: The case of shell bead manufacture on Motupore Island, Papua." *Archaeol. Oceania*, Vol. 32, No. 1, Apr. 1997, p. 27 (Table 1).

a. a stone drillpoint?

b. a shell bead?

c. a grinding slab?

d. a stone drillpoint or grinding slab?

e. not a shell bead?

4.22 Each year the Florida Game and Fresh Water Fish Commission issues permits for trappers to hunt and kill alligators. One year there were 12,685 applications for permits. The Commission issued 678 permits using a computerized random selection system. If you applied for a permit to hunt alligators in Florida during this particular year, what was the probability that you were issued a permit?

4.23 Refer to the population estimates of rhinoceroses living in the wild in Africa and Asia, Problem 2.59 (p. 105). A breakdown of the number of rhinos of each species is reproduced in the table below. Suppose a rhino is captured at random.

a. What is the probability that it is an African Black rhino?

b. What is the probability that it is an Asian rhino?

c. What is the probability that it is an African rhino?

Rhino Species	Population Estimate
African Black	2,600
African White	8,465
(Asian) Sumatran	400
(Asian) Javan	70
(Asian) Indian	2,050
TOTAL	13,585

Source: International Rhino Federation, July 1998.

4.24 Ohio's Learning, Earning, and Parenting (LEAP) program is designed to encourage school attendance among pregnant and parenting teens on welfare. Eligible teens who provide evidence of school enrollment receive a bonus payment, while those who do not attend school have money deducted from their grant (i.e., the teens are "sanctioned"). In a survey of LEAP teens, published in *Children and Youth Services Review* (Vol. 17, 1995), the bonus and/or sanction requests were recorded for each teen. The results are summarized in the accompanying table.

Request	Percentage
No bonuses or sanction (N)	7
Only bonuses (OB)	37
Only sanction (OS)	18
Both, more bonuses than sanctions (BB)	14
Both, more sanctions than bonuses (BS)	18
Both, equal number of bonuses and sanctions (BE)	6
TOTAL	100%

Source: Reprinted from *Children and Youth Services Review,* Vol. 17, Nos. 1/2, 1995, Wood, R. G., et al. "Encouraging school enrollment and attendance among teenage parents on welfare: Early impacts of Ohio's LEAP program," p. 302, Figure 1, Elsevier Science Ltd., The Boulevard, Langford Lane, Kidlington OX5 1GB, UK.

a. Define the experiment that generated the data in the table, and list the simple events.

b. Assign probabilities to the simple events.

c. What is the probability that both bonuses and sanctions are requested for a LEAP teen?

d. What is the probability that only one type of request (either a bonus or a sanction, but not both) is made?

4.25 Entomologists are often interested in studying the effect of chemical attractants (*pheromones*) on insects. One common technique is to release several insects equidistant from the pheromone being studied and from a control substance. If the pheromone has an effect, more insects will travel toward it than toward the control.

Otherwise, the insects are equally likely to travel in either direction. Suppose five insects are released.

a. If we are interested in which insects travel toward the pheromone, how many outcomes (simple events) are possible? [*Hint:* If the five insects are denoted as A, B, C, D, and E, one possible outcome is that insects A and B travel toward the pheromone. Another possible outcome is that insects A, B, C, and D travel toward the pheromone.]

b. Suppose the pheromone under study has no effect and, therefore, it is equally likely that an insect will move toward the pheromone or toward the control—i.e., the possible outcomes are equiprobable. Find the probability that both insects A and E travel toward the pheromone.

4.4 The Combinatorial Rule for Counting Simple Events (Optional)

Consider an experiment that results in M equally likely mutually exclusive outcomes (i.e., M simple events). To utilize Probability Rule #2, we need to determine M. When M is small (as in Example 4.9 on p. 201), we can list the simple events. However, when M is large, it may be inconvenient or practically impossible to list all the different outcomes of the experiment. In this situation, we can rely on a rule for counting the number M of different outcomes in the experiment. The next two examples demonstrate how to use this counting rule.

EXAMPLE 4.10

COUNTING SIMPLE EVENTS

Refer to Example 4.9 (p. 201) and the "Let's Make a Deal" problem of determining the number of ways in which you can select two doors from among four doors offered. Use a counting rule to determine M, the number of mutually exclusive outcomes.

Solution

In this problem, we want to know how many different ways we can select two of the four doors. In other words, we want to find M, the number of distinctly different combinations of two doors that can be selected from the four doors offered. The box below gives a formula—called the *Combinatorial Rule*—for counting the number of ways of selecting n objects from a total of N.

COMBINATORIAL RULE FOR DETERMINING THE NUMBER OF DIFFERENT SAMPLES THAT CAN BE SELECTED FROM A POPULATION

The number of different samples of n objects that can be selected from among a total of N is denoted $\binom{N}{n}$ and is equal to

$$\binom{N}{n} = \frac{N!}{n!(N-n)!} \qquad \textbf{(4.4)}$$

where

$N! = N(N-1)(N-2)(N-3)\cdots(1)$ and is read N factorial

$n! = n(n-1)(n-2)(n-3)\cdots(1)$

$0! = 1$

In this example, $n = 2$ and $N = 4$. Applying the formula, we have

$$M = \binom{N}{n} = \frac{N!}{n!(N-n)!} = \frac{4!}{2!(4-2)!} = \frac{4!}{2!2!} = \frac{(4)(3)(2)(1)}{[(2)(1)][(2)(1)]}$$

Self-Test 4.6

Consider the following experiment: Toss a single, fair die and observe the number of dots on the upper face. Given that the number is even, what is the probability of observing a 6?

Example 4.13 illustrates an important relationship that exists between some pairs of events. If the probability of one event does not depend on whether a second event has occurred, then the events are said to be *independent*.

The notion of independence is particularly important when we want to find the probability that *both* of two events will occur. When the events are independent, the probability that both events will occur is equal to the product of their unconditional probabilities.

Definition 4.9

Two events A and B are said to **independent** if

$$P(A \mid B) = P(A) \qquad \textbf{(4.5)}$$

or if

$$P(B \mid A) = P(B) \qquad \textbf{(4.6)}$$

[*Note:* If one of these equalities is true, then the other will also be true.]

PROBABILITY RULE #4

The Probability that Both of Two Independent Events A and B Occur If two events A and B are independent, then the *probability that both A and B occur* is equal to the product of their respective unconditional probabilities:

$$P(A \text{ and } B) = P(A)P(B) \qquad \textbf{(4.7)}$$

Probability Rule #4 can be extended to apply to any number of independent events. For example, if A, B, and C are independent events, then

$$P(\text{all of the events } A, B, \text{ and } C \text{ occur}) = P(A \text{ and } B)P(C) = P(A)P(B)P(C)$$

EXAMPLE 4.14 **APPLYING PROBABILITY RULE #4**

Find the probability of observing two heads in two tosses of a fair coin.

Solution

Define the following events:

A: Observe a head on the first toss.

B: Observe a head on the second toss.

Since we know that events A and B are independent and that $P(A) = P(B) = \frac{1}{2}$, the probability that we observe two heads, i.e., both events A and B, is

$$\begin{aligned} P(\text{observe two heads}) &= P(A)P(B) \\ &= \left(\frac{1}{2}\right)\left(\frac{1}{2}\right) = \frac{1}{4} \end{aligned}$$

Self-Test 4.7

Consider the following experiment: Roll two dice and observe the number of dots showing on each. Find the probability of observing two 1's (called "snake-eyes" in the game of craps).

We now consider a problem in statistical inference.

EXAMPLE 4.15

USING PROBABILITY TO MAKE AN INFERENCE

Experience has shown that a manufacturing operation produces, on the average, 5% defective units. These are removed from the production line, repaired, and returned to the warehouse. Suppose that during a given period of time you observe five defective units emerging in sequence from the production line.

a. If experience has shown that the defective units usually emerge randomly from the production line, what is the probability of observing a sequence of five consecutive defective units?

b. If the event in part **a** really occurred, what would you conclude about the process?

Solution

a. If the defectives really occur randomly, then whether any one unit is defective should be independent of whether the others are defective. Second, the unconditional probability that any one unit is defective is known to be .05. We will define the following events:

D_1: The first unit is defective.
D_2: The second unit is defective.
⋮
D_5: The fifth unit is defective.

Then

$$P(D_1) = P(D_2) = P(D_3) = P(D_4) = P(D_5) = .05$$

and the probability that all five are defective is

$$\begin{aligned} P(\text{all five are defective}) &= P(D_1)P(D_2)P(D_3)P(D_4)P(D_5) \\ &= (.05)(.05)(.05)(.05)(.05) \\ &= .0000003 \end{aligned}$$

b. We do not need a knowledge of probability to know that something must be wrong with the production line. Intuition would tell us that observing five defectives in sequence is highly improbable (given the past), and we would immediately infer that past experience no longer describes the condition of the process. In fact, we would infer that something is disturbing the stability of the process.

Example 4.15 illustrates how you can use your knowledge of probability and the probability of an event to make an inference about some population. The technique, called the **rare event approach,** is summarized in the box.

RARE EVENT APPROACH TO MAKING STATISTICAL INFERENCES

Suppose we calculate the probability of event A *based on certain assumptions about the sampled population.* If $P(A) = p$ is small (say, p less than .05) and we observe that A occurs, then we can reach one of two conclusions:

1. Our original assumption about the sampled population is correct, and we have observed a rare event, i.e., an event that is highly improbable. (For example, we would conclude that the production line of Example 4.15, in fact, produces only 5% defectives. The fact that we observed five defectives in a sample of five was an unlucky and rare event.)
2. Our original assumption about the sampled population is incorrect. (In Example 4.15, we would conclude that the line is producing more than 5% defectives, a situation that makes the observed sample—five defectives in a sample of five—more probable.)

Using the rare event approach, we prefer conclusion 2. The fact that event A did occur in Example 4.15 leads us to believe that $P(A)$ is much higher than p, and that our original assumption about the population is incorrect.

Statistics in the Real World Revisited

The Probability of Winning Daily Cash 3 or Play 4

In addition to bi-weekly Lotto 6/53, the Florida Lottery runs several other games. Two popular daily games are "Cash 3" and "Play 4." In Cash 3, players pay $1 to select three numbers in sequential order, where each number ranges from 0 to 9. If the three numbers selected (e.g., 2-8-4) match exactly the order of the three numbers drawn, the player wins $500. Play 4 is similar to Cash 3, but players must match four numbers (each number ranging from 0 to 9). For a $1 Play 4 ticket (e.g., 3-8-3-0), the player will win $5,000 if the numbers match the order of the four numbers drawn.

During the official drawing for Cash 3, ten ping pong balls numbered 0, 1, 2, 3, 4, 5, 6, 7, 8, and 9 are placed into each of three chambers. The balls in the 1st chamber are colored pink, the balls in the 2nd chamber are blue, and the balls in the 3rd chamber are yellow. One ball of each color is randomly drawn, with the official order as pink-blue-yellow. In Play 4, a 4th chamber with orange balls is added and the official order is pink-blue-yellow-orange. Since the draws of the colored balls are random and independent, we can apply Probability Rule #4 to find the odds of winning Cash 3 and Play 4. The probability of matching a numbered ball being drawn from a chamber is $\frac{1}{10}$; therefore,

$$\begin{aligned}
P(\text{win Cash 3}) &= P(\text{match pink AND match blue AND match yellow}) \\
&= P(\text{match pink}) \times P(\text{match blue}) \times P(\text{match yellow}) \\
&= \left(\frac{1}{10}\right)\left(\frac{1}{10}\right)\left(\frac{1}{10}\right) = \frac{1}{1000} = .001 \\
P(\text{win Play 4}) &= P(\text{match pink AND match blue AND match yellow AND} \\
&\qquad \text{match orange}) \\
&= P(\text{match pink}) \times P(\text{match blue}) \times P(\text{match yellow}) \\
&\qquad \times P(\text{match orange}) \\
&= \left(\frac{1}{10}\right)\left(\frac{1}{10}\right)\left(\frac{1}{10}\right)\left(\frac{1}{10}\right) = \frac{1}{10{,}000} = .0001
\end{aligned}$$

Although the odds of winning one of these daily games is much better than the odds of winning Lotto 6/53, there is still only a 1 in 1000 chance (for Cash 3) or 1 in 10,000 chance (for Play 4) of winning the daily game. And the payoffs ($500 or $5,000) are much smaller. In fact, it can be shown that you will lose an average of 50¢ every time you play either Cash 3 or Play 4!

PROBLEMS FOR SECTION 4.5

Using the Tools

4.36 Assume that $P(A) = .6$ and $P(B) = .3$. If A and B are independent, find $P(A \text{ and } B)$.

4.37 Assume that $P(A) = .2$ and $P(B) = .9$. If A and B are independent, find $P(A \text{ and } B)$.

4.38 If $P(A) = .5$, $P(B) = .7$, and $P(B \mid A) = .5$, are A and B independent?

4.39 Assume that $P(A) = .6$, $P(B) = .4$, $P(C) = .5$, $P(A \mid B) = .15$, $P(A \mid C) = .5$, and $P(B \mid C) = .3$.

a. Are events A and B independent?

b. Are events A and C independent?

c. Are events B and C independent?

4.40 Consider an experiment that consists of two trials and has the nine possible outcomes (simple events) listed here:

AA *AB* *AC* *BA* *BB* *BC* *CA* *CB* *CC*

where *AC* indicates that *A* occurs on the first trial and *C* occurs on the second trial. Suppose the following events are defined:

D: Observe an *A* on the first trial.

E: Observe a *B* on the second trial.

a. List the simple events associated with event D and find $P(D)$.

b. List the simple events associated with event E and find $P(E)$.

c. Find $P(E \mid D)$.

d. Are D and E independent?

e. Find $P(\text{both } D \text{ and } E \text{ occur})$.

Applying the Concepts

4.41 According to the Newspaper Advertising Bureau, 40% of all those primarily responsible for car maintenance are women. Consider the population consisting of all primary car maintainers.

a. What is the probability that a primary car maintainer, selected from the population, is a woman? A man?

b. What is the probability that both primary car maintainers in a sample of two selected independently from the population are women?

4.42 To develop programs for business travelers staying at convention hotels, Hyatt Hotels Corp. commissioned a study of executives who play golf (*Tampa Tribune,* July 10, 1993). The research revealed two surprising results: (1) 55% of the respondents admitted they had cheated at golf; (2) more than one-third of the respondents that admitted cheating at golf said they have also lied in business. Let A represent the event that an executive has cheated at golf and let B represent the event that the executive has lied in business. Convert the research results into probability statements involving events A and B.

4.43 The *Journal of the National Cancer Institute* (Feb. 16, 2000) published the results of a study that investigated the association between cigar smoking and death from tobacco-related cancers. Data were obtained for a national sample of 137,243 American men. The results are summarized in the table on p. 218. Each male in the study was classified according to his cigar-smoking status and whether or not he died from a tobacco-related cancer.

a. Find the probability that a randomly selected man never smoked cigars and died from cancer.

b. Find the probability that a randomly selected man was a former cigar smoker and died from cancer.

c. Find the probability that a randomly selected man was a current cigar smoker and died from cancer.

	Died from Cancer	Did not Die from Cancer	Totals
Never Smoked Cigars	782	120,747	121,529
Former Cigar Smoker	91	7,757	7,848
Current Cigar Smoker	141	7,725	7,866
TOTALS	1,014	136,229	137,243

Source: Shapiro, J. A., Jacobs, E. J., and Thun, M. J. "Cigar smoking in men and risk of death from tobacco-related cancers." *Journal of the National Cancer Institute,* Vol. 92, No. 4, February 16, 2000 (Table 2).

Problem 4.43

d. Given that a male was a current cigar smoker, find the probability that he died from cancer.

e. Given that a male never smoked cigars, find the probability that he died from cancer.

4.44 A Victoria University psychologist investigated whether Asian immigrants to Australia differed from Anglo-Australians in their attitudes toward mental illness (*Community Mental Health Journal,* Feb. 1999). Each in a sample of 139 Australian students was classified according to ethnic group and degree of contact with mentally ill people. The number in each category is displayed in the following table. Suppose we randomly select one of the 139 students.

		Ethnic Group				
		Anglo-Australian	Short-Term Asian	Long-Term Asian	Other	Totals
Mental Illness	Little or no contact	30	17	20	15	82
	Some contact	17	5	3	11	36
	Close contact	16	0	1	4	21
	TOTALS	63	22	24	30	139

Source: Fan, C. "A comparison of attitudes towards mental illness and knowledge of mental health services between Australian and Asian students." *Community Mental Health Journal,* Vol. 35, No. 1, Feb. 1999, p. 54 (Table 2).

a. Find the probability that the student is an Anglo-Australian who has had little or no contact with mentally ill people.

b. Find the probability that the student has had close contact with mentally ill people.

c. Find the probability that the student is not an Anglo-Australian.

d. Find the probability that the student has had close contact with mentally ill people or is an Asian immigrant.

e. Given the student is an Asian immigrant, what is the probability that he or she has had little or no contact with mentally ill patients?

f. Given the student is Anglo-Australian, what is the probability that he or she has had little or no contact with mentally ill patients?

g. Which ethnic group, Anglo-Australians or Asian immigrants, is most likely to have had little or no contact with mentally ill patients?

4.45 A survey of 12th-graders revealed the following information on youth smoking: 34% of the 12th-graders were classified as smokers; 32% own merchandise promoting cigarette brand names and trademarks; of those 12th-graders who own a promotional item, 58% smoke; and of those 12th-graders who do not own a promotional item, 23% are smokers (*Tampa Tribune,* Dec. 15, 1997).

a. Convert each of these percentages into a probability.

b. Consider the events S = {a 12th-grader smokes} and I = {a 12th-grader owns merchandise promoting cigarettes}. Are S and I independent? Explain.

4.46 Nightmares about college exams appear to be common among college graduates. In a survey of 30- to 45-year-old graduates from Transylvania University (Kentucky), 50 of 188 respondents admitted they had recurring dreams about college exams (*Tampa Tribune,* Dec. 12, 1988). Of these 50, 47 felt distress, anguish, fear, or terror in their dreams. (For example, some dreamers "couldn't find the building or they walked in and all the students were different." Other dreamers either overslept or didn't realize they were enrolled in the class.)

a. Calculate the approximate probability that a 30- to 45-year-old graduate of Transylvania University has recurring dreams about college exams. Why is this probability approximate?

b. Refer to part **a**. Given that the graduate has recurring dreams, what is the approximate probability that the dreams are unpleasant (i.e., that the graduate feels distress, anguish, fear, or terror in the dreams)?

c. Are the events {graduate has recurring dreams} and {dreams are unpleasant} independent?

4.47 In *Parade Magazine*'s (Nov. 26, 2000) column, "Ask Marilyn," the following question was posed: "I have just tossed a [balanced] coin 10 times, and I ask you to guess which of the following three sequences was the result. One (and only one) of the sequences is genuine."

(1) *HHHHHHHHHH*
(2) *HHTTHTTHHH*
(3) *TTTTTTTTTT*

a. Demonstrate that prior to actually tossing the coins, the three sequences are equally likely to occur.

b. Find the probability that the 10 coin tosses result in all heads or all tails.

c. Find the probability that the 10 coin tosses result in a mix of heads and tails.

d. Marilyn's answer to the question posed was: "Though the chances of the three specific sequences occurring randomly are equal . . . it's reasonable for us to choose sequence (2) as the most likely genuine result." If you know that only one of the three sequences actually occurred, explain why Marilyn's answer is correct. [*Hint:* Compare the probabilities in parts **b** and **c**.]

4.48 According to NASA, each space shuttle in the U.S. fleet has 1,500 "critical items" that could lead to catastrophic failure if rendered inoperable during flight. NASA estimates that the chance of at least one critical-item failure within the shuttle's main engines is about 1 in 63 for each mission (*Tampa Tribune,* Dec. 3, 1993). To build the space station *Freedom,* NASA plans to fly eight shuttle missions a year.

a. Find the probability that none of the eight shuttle flights scheduled next year results in a critical-item failure.

b. Use the Rule of Complements and the result, part **a**, to find the probability that at least one of the eight shuttle flights scheduled next year results in a critical-item failure.

4.49 In professional sports, "stacking" is a term used to describe the practice of African-American players being excluded from certain positions because of race. Refer to the *Sociology of Sport Journal* (Vol. 14, 1997) study of "stacking" in the National Basketball Association (NBA), Problem 2.36 (p. 87).

Reconsider the table on p. 220, which summarizes the race and positions of 368 NBA players in 1993. Suppose an NBA player is selected at random from that year's player pool.

a. What is the probability that the player is white?

b. What is the probability that the player is a center?

c. What is the probability that the player is African-American and plays guard?

d. Given the player is white, what is the probability that he is a center?

e. Given the player is African-American, what is the probability that he is a center?

			Position			
			Guard	Forward	Center	Totals
Race	White		26	30	28	84
	Black		128	122	34	284
		TOTALS	154	152	62	368

Problem 4.49

4.50 The *Journal of Teaching in Physical Education* (Apr. 1997) published a research note on sequential analysis of behavioral data collected in physical education classes. An example was given on teacher instruction (event A) and student behavior response (event B). During one 4-minute session with an expert teacher, the following chain of events occurred.

Expert: *ABABAABABABABAABABBABABABABABA*

During another 4-minute session with a novice teacher, the following chain of events occurred.

Novice: *ABAABABBABBAAABABBABAAABBAAABB*

a. For the session with an expert teacher, find the probability of the student behavior response (event B) occurring.

b. Repeat part **a** for the session with a novice teacher.

c. Given that the expert teacher provides an instruction (event A), find the probability that the student behavior (event B) occurs immediately after.

d. Repeat part **c** for the novice teacher.

4.6 The Additive and Multiplicative Laws of Probability (Optional)

In this optional section, we define some standard probability notation and give two laws for finding probabilities. Although these laws are not required for a study of the remaining material in the text, they are needed to complete an introductory coverage of probability.

When both of two events A and B occur (Section 4.5), this is called the *intersection of A and B* and is denoted $A \cap B$. When either A or B occurs (Section 4.3), this is called the *union of A and B* and is denoted $A \cup B$.

Definition 4.10

The **intersection of A and B**, denoted by $A \cap B$, is the event that both A and B occur. (See Figure 4.6.)

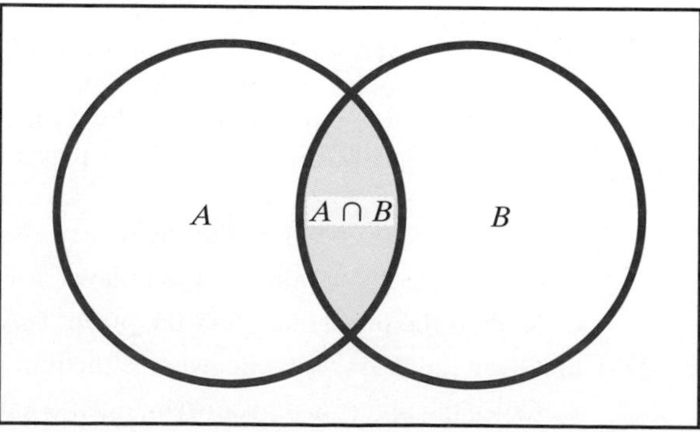

Figure 4.6 Illustration of the Intersection of Two Events, $A \cap B$

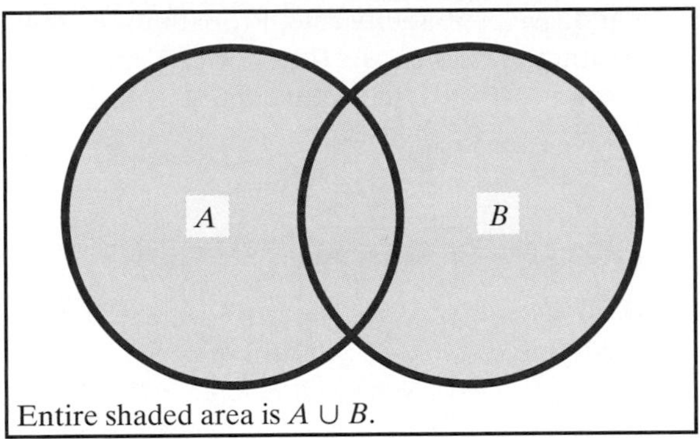

Figure 4.7 Illustration of the Union of Two Events, $A \cup B$

Definition 4.11

The **union of A and B**, denoted by $A \cup B$, is the event that either A or B occurs. (See Figure 4.7.)

The graphical representation that shows the relationships among events in an experiment (as in Figures 4.6 and 4.7) is called a **Venn diagram.**

Probability Rule #4 (Section 4.5) provided a formula for finding the probability that both events A and B occur (i.e., $A \cap B$) for the special case where A and B are independent events. We now give a formula, called the *Multiplicative Law of Probability,* that applies regardless of whether A and B are independent events.

PROBABILITY RULE #5

The Multiplicative Law of Probability The probability that *both* of two events A and B occur is

$$P(A \text{ and } B) = P(A \cap B) = P(A)P(B \mid A) = P(B)P(A \mid B) \quad \textbf{(4.8)}$$

EXAMPLE 4.16

APPLYING PROBABILITY RULE #5

Refer to Example 4.12 (pp. 212–213), where we selected two fuses from a box that contained three, two of which were defective. Use the Multiplicative Law of Probability to find the probability that you first select defective fuse D_1 and then select D_2.

Solution

Define the following events:

A: The second fuse selected is D_2.
B: The first fuse selected is D_1.

The probability of event B is $P(B) = \frac{1}{3}$. Also, from Example 4.12, the conditional probability of A given B is $P(A \mid B) = \frac{1}{2}$. Then the probability that both events A and B occur is

$$P(A \text{ and } B) = P(B)P(A \mid B) = \left(\frac{1}{3}\right)\left(\frac{1}{2}\right) = \frac{1}{6}$$

You can verify this result by rereading Example 4.12.

An additive probability rule, Probability Rule #1 (p. 200) was given for the event that either A or B occurs (i.e., $A \cup B$), but it applies only to the case where A and B are mutually exclusive events. A rule that applies in general is given by the *Additive Law of Probability*.

PROBABILITY RULE #6

The Additive Law of Probability The probability that *either* event A *or* event B *or both* occur is

$$P(A \text{ or } B) = P(A \cup B) = P(A) + P(B) - P(A \cap B) \tag{4.9}$$

EXAMPLE 4.17

APPLYING PROBABILITY RULE #6

An experiment consists of tossing two coins with the sample space shown in Figure 4.8. Define the following events:

A: Observe at least one head.
B: Observe at least one tail.

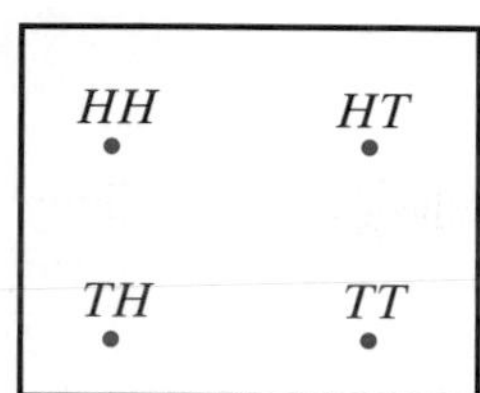

Figure 4.8 Sample Space for Example 4.17

Use the Additive Law of Probability to find the probability of observing either A or B or both.

Solution

We know the answer to this question before we start because the probability of observing at least one head or at least one tail is 1—i.e., the event is a certainty. To obtain this answer using the Additive Law of Probability, we first note that the four simple events in Figure 4.8 are equally likely. Then we find

$$P(A) = P(\text{at least one head}) = P(HH) + P(HT) + P(TH) = \frac{1}{4} + \frac{1}{4} + \frac{1}{4} = \frac{3}{4}$$

$$P(B) = P(\text{at least one tail}) = P(TT) + P(HT) + P(TH) = \frac{1}{4} + \frac{1}{4} + \frac{1}{4} = \frac{3}{4}$$

The event that both A and B occur—observing at least one head and at least one tail—is the event that you observe exactly one head and exactly one tail. The probability of this event is

$$P(\text{both } A \text{ and } B \text{ occur}) = P(HT) + P(TH) = \frac{1}{4} + \frac{1}{4} = \frac{1}{2}.$$

Now, we apply equation 4.9 to obtain

$$P(\text{either } A \text{ or } B \text{ or both occur}) = P(A) + P(B) - P(\text{both } A \text{ and } B \text{ occur})$$

$$= \frac{3}{4} + \frac{3}{4} - \frac{1}{2} = 1$$

Self-Test 4.8

For two events A and B, suppose $P(A) = .8$, $P(B) = .5$, and $P(A \cap B) = .6$.

a. Find $P(A \cup B)$. **b.** $P(B \mid A)$.

In Examples 4.16 and 4.17, two key words helped us identify which probability law to employ. In Example 4.16, the key word was *and,* as in "find the probability that both *A and B* occur." The word *and* implies intersection; therefore, we use the Multiplicative Law of Probability. Alternatively, the key word was *or* in Example 4.17, as in "find the probability that either *A or B or both* occur." The word *or* implies union; therefore, we use the Additive Law of Probability.

EXAMPLE 4.18 APPLYING SEVERAL PROBABILITY RULES

Psychologists believe that there is a relationship between aggressiveness and order of birth. To test this belief, a psychologist randomly chose 1,000 elementary school children and administered each a test designed to measure the student's aggressiveness. Each student was then classified according to aggressiveness (aggressive or unaggressive) and order of birth (firstborn, secondborn, or other). The number of students in the six categories are shown in the cross-classification table, Table 4.2.

TABLE 4.2 Results of Aggressiveness Test

		Order of Birth			
		Firstborn	Secondborn	Other	Totals
Aggressiveness	Aggressive	60	100	80	240
	Unaggressive	120	330	310	760
	TOTALS	180	430	390	1,000

Suppose a single elementary school student is randomly selected from the 1,000 in the study. Define the following events:

A: The student is aggressive.

B: The student was firstborn.

a. Find the probability that both *A* and *B* occur.

b. Find the conditional probability that *A* will occur given that *B* has occurred.

c. Find the probability that *A* will not occur.

d. Find the probability that either *A* or *B* or both occur.

Solution

a. We can see from the table that 60 of the 1,000 students were classified as both aggressive (A) and firstborn (B). Therefore,

$$P(A \text{ and } B) = P(A \cap B) = \frac{60}{1{,}000} = .06$$

b. Intuitively, this conditional probability can be found by focusing on the "firstborn" column of Table 4.2, i.e., the column corresponding to the conditional event B. Now, of the 180 firstborn children, 60 are "aggressive." Hence,

$$P(A \mid B) = \frac{60}{180} = .333$$

Alternatively, we can use the Multiplicative Law of Probability to find $P(A \mid B)$. Examining Table 4.2, we find that 240 of all the students were classified as aggressive and 180 were the firstborn in their family. Therefore,

$$P(A) = \frac{240}{1{,}000} = .24 \quad \text{and} \quad P(B) = \frac{180}{1{,}000} = .18$$

To find $P(A \mid B)$, we substitute the answer to part **a** and the value of $P(B)$ into the formula for the Multiplicative Law of Probability. Thus,

$$P(A \text{ and } B) = P(A \cap B) = P(B)P(A \mid B)$$

or

$$.06 = (.18)P(A \mid B)$$

Solving for $P(A \mid B)$ yields

$$P(A \mid B) = \frac{.06}{.18} = .333$$

This answer agrees with our intuitive approach.

c. The event that A does not occur is the complement of A, denoted by the symbol A'. Since A is the event that a student is aggressive, A' is the event that a student is unaggressive. Recall that $P(A)$ and $P(A')$ bear a special relationship to each other:

$$P(A) + P(A') = 1$$

From part **b**, we have $P(A) = .24$. Therefore,

$$P(A') = 1 - .24 = .76$$

which can be verified by examining Table 4.2.

d. The probability that either A or B or both occur is given by the Additive Law of Probability. From parts **a** and **b** we know that

$$P(A) = .24 \qquad P(B) = .18 \qquad P(A \cap B) = .06$$

Then,

$$P(A \text{ or } B) = P(A) + P(B) - P(A \cap B)$$
$$= .24 + .18 - .06 = .36$$

PROBLEMS FOR SECTION 4.6

Using the Tools

4.51 For two events A and B, $P(A) = .3$, $P(B) = .5$, and $P(A \cap B) = .2$. Find $P(A \cup B)$.

4.52 For two events A and B, $P(A) = .5$, $P(B) = .6$, and $P(A \cap B) = .4$. Find $P(A \cup B)$.

4.53 For two events A and B, $P(A) = .7$, $P(B) = .2$, and $P(A \mid B) = .3$. Find $P(A \cap B)$.

4.54 For two events A and B, $P(A) = .1$, $P(B) = .3$, and $P(B \mid A) = .8$. Find $P(A \cap B)$.

4.55 Consider the experiment of rolling a pair of dice. Define events A and B as follows:

A: Observe a sum of 7.

B: Observe a 4 on at least one die.

a. List the possible outcomes in event A and find $P(A)$.

b. List the possible outcomes in event B and find $P(B)$.

c. Find $P(A \cap B)$ using the Multiplicative Law of Probability. Then list the possible outcomes associated with this event, and find $P(A \cap B)$ by summing the probabilities of these outcomes.

d. Find $P(A \cup B)$ using the Additive Law of Probability. Then list the possible outcomes associated with this event and find $P(A \cup B)$ by summing the probabilities of these outcomes.

Applying the Concepts

4.56 Deluxe Corporation, a company that makes checks, sends all new checking account holders a box of starter checks as well as a brochure meant to entice customers to call the sales team to receive a free gift. Deluxe has found that the probability of a new customer calling for the free gift is .55. If a customer calls, the probability of that customer placing an order for another product is .09 (*Continental Magazine,* April 2000). Use this information to find the probability that a new customer calls for the free gift and places an order for a Deluxe product. [*Note:* The answer to this question led Deluxe Corp. to develop a new marketing plan.]

4.57 Dust mites have been linked to allergies and asthma. A dust mite allergen level that exceeds 2 micrograms per gram (μg/g) of dust has been associated with the development of allergies, while a level that exceeds 10 μg/g has been associated with the trigger of asthma attacks. According to a May 2000 study by the National Institute of Environmental Health Sciences (NIEHS), 45% of U.S. homes have bedding with a dust mite level that exceeds 2 μg/g and 23% have a dust mite level in bedding that exceeds 10 μg/g. Given that a U.S. home has a dust mite level in bedding that exceeds 2 μg/g, find the probability that the level also exceeds 10 μg/g. [*Hint:* Use the fact that if a home has a dust mite level that exceeds 10, it also has a level that exceeds 2.]

4.58 The National Fire Incident Reporting System (NFIRS) is an accumulation of fire reports submitted by fire departments across the country. According to the NFIRS, 5% of all fires in the U.S. are started by children. Of the fires started by children, 25% occur inside the home. (*Tampa Tribune,* Nov. 9, 1996.)

a. Based on this information, a newspaper reporter wrote that "25 percent of all fires in the U.S. are started by children inside the home." Do you agree with this statement?

b. Find the probability that a fire is started by a child inside the home.

4.59 A certain stretch of highway on the New Jersey Turnpike has a posted speed limit of 60 mph. A survey of drivers on this portion of the turnpike revealed the following facts (reported in *The Washington Post,* Aug. 16, 1998):

- 14% of the drivers on this stretch of highway were African-Americans
- 98% of the drivers were exceeding the speed limit by at least 5 mph (and, thus, subject to being stopped by the state police)
- Of these violators, 15% of the drivers were African-American.
- Of the drivers stopped for speeding by the New Jersey state police, 35% were African-American

a. Convert each of the reported percentages above into a probability statement.

b. Find the probability that a driver on this stretch of highway is African-American and exceeding the speed limit.

c. Given a driver on this stretch of highway is exceeding the speed limit, what is the probability the driver is not African-American?

d. Given that a driver on this stretch of highway is stopped by the New Jersey state police, what is the probability that the driver is not African-American?

e. For this stretch of highway, are the events {African-American driver} and {Driver exceeding the speed limit} independent (to a close approximation)? Explain.

f. Use the probabilities in *The Washington Post* report to make a statement about whether African-Americans are stopped more often for speeding than expected on this stretch of turnpike.

4.60 As part of a study of phonological dyslexia in a highly literate individual, researchers asked 22 undergraduate students to spell a list of familiar but commonly misspelled words (*Quarterly Journal of Experimental Psychology,* Aug. 1985). The purpose of the study was to compare the misspellings of the control group (i.e., the 22 undergraduates) with the misspellings of the dyslexic subject. Several words on the list, along with the number of students in the control group who misspelled the words, are given in the table at left. Suppose one of the 22 students in the control group is selected at random.

Word	Number Misspelled
sacrilegious	21
Gandhi	19
professor	8
khaki	5

Problem 4.60

a. What is the probability that the student misspelled the word *sacrilegious*?

b. What is the probability that the student spelled the word *professor* correctly?

c. Suppose that 14% of the students in the control group misspelled both the words *Gandhi* and *khaki*. What is the probability that the selected student misspelled either *Gandhi* or *khaki* or both?

4.61 Many Internet users shop online. In a survey of 1,500 customers who did their 1999 holiday shopping online, 18% indicated that they were not satisfied with their experience. Of the customers that were not satisfied, 53% indicated that they did not receive their product in time for the holidays (*The Wall Street Journal,* April 17, 2000). Suppose the following complete set of results were reported.

			Received Products in Time for Holidays		
			Yes	**No**	**Total**
Satisfied with Experience	**Yes**		1,197	33	1,230
	No		127	143	270
		TOTAL	1,324	176	1,500

If a customer is selected at random, what is the probability that he or she

a. is satisfied with the experience?

b. received the product in time for the holidays?

c. is satisfied with the experience *and* received the product in time for the holidays?

d. is satisfied with the experience *and* did not receive the product in time for the holidays?

e. is satisfied with the experience *or* received the product in time for the holidays?

f. is not satisfied with the experience *or* did not receive the product in time for the holidays?

g. is satisfied with the experience, given he or she received the product in time for the holidays?

4.62 Refer to the *American Journal on Mental Retardation* study of social interactions of children, Problem 2.75 (p. 112). The 30 children in the study were classified according to type of behavior (disruptive or not) and whether or not they displayed developmental delays. The data are summarized in the cross-classification table on p. 227. Consider a randomly selected child from the 30 studied.

a. What is the probability that the child exhibits disruptive behavior and developmental delays?

b. What is the probability that the child exhibits disruptive behavior, given that the child does not display developmental delays?

	Disruptive Behavior	Nondisruptive Behavior	Totals
With developmental delays	12	3	15
Without developmental delays	5	10	15
TOTALS	17	13	30

Source: Kopp, C. B., Baker, B., and Brown, K. W. "Social skills and their correlates: Preschoolers with developmental delays." *American Journal on Mental Retardation,* Vol. 96, No. 4, Jan. 1992.

4.63 A study conducted by the National Opinion Research Center found that Hispanics who are proficient in English do better in school than those who communicate primarily in Spanish. Of those students who speak only Spanish at home, 41% earn a high school grade average of B or better. In contrast, 53% of all students earn a grade average of B or better. It is also known that 5% of all students speak only Spanish at home, while 89% speak only English at home. Consider the following events:

A: A student earns a grade average of B or better.

B: A student speaks only Spanish at home.

C: A student speaks only English at home.

a. Find $P(A)$. **b.** Find $P(B)$. **c.** Find $P(C)$.

d. Find $P(A \mid B)$. **e.** Find $P(A \text{ and } B)$.

f. Which (if any) of the pairs of events, A and B, A and C, or B and C, are mutually exclusive?

4.64 A survey of 300 children in the San Francisco Bay area showed that 90% owned pets. Of these, 30% owned dogs only, 18% owned dogs and other pets, 22% owned cats only, 22% owned cats and other pets, and 8% owned only pets other than dogs and cats (*Psychological Reports*). For a randomly selected child, find the following probabilities:

a. The child owns a pet. **b.** The child owns a dog only.

c. The child owns a cat only. **d.** The child owns a dog.

4.65 In 1941, Ted Williams of the Boston Red Sox became the last major league baseball player to record a batting average over .400, and Joe DiMaggio of the New York Yankees had the longest hitting streak in major league baseball history—56 games. *Chance* (Fall 1991) assessed the probabilities of the events occurring within the next 50 years as $P(\text{bat over } .400) = .035$ and $P(\text{56-game hitting streak}) = .016$ under specific conditions (i.e., a league batting average of .270).

a. Given a current major league baseball star (e.g., Ken Griffey, Jr., of the Cincinnati Reds) has just completed a 56-game hitting streak, explain why it is not possible to calculate the probability that he will end the season with a batting average of over .400. Assume the two events, {bat over .400} and {56-game hitting streak}, are dependent.

b. Calculate the conditional probability of part **a**, assuming the two events are independent. (If you have knowledge of baseball, batting averages, and hitting streaks, do you believe the two events are independent?)

4.66 Refer to the *Psychological Science* experiment in which 120 subjects participated in the "water-level task," Problem 2.39 (p. 88). Recall that the subjects were shown a drawing of a water glass tilted at a 45° angle and asked to draw a line representing the surface of the water. A summary of the results is reproduced in the table on p. 228.

Suppose that we select one subject at random from the 120 subjects in the study.

a. What is the probability that the subject is a female?

b. What is the probability that the subject's line is within 5° of the water surface?

	Group						
	Females			Males			
Judged Line	**Students**	**Waitresses**	**Housewives**	**Students**	**Bartenders**	**Bus Drivers**	**Totals**
More than 5° below surface	0	0	1	1	1	1	4
More than 5° above surface	7	15	13	3	11	4	53
Within 5° of surface	13	5	6	16	8	15	63
TOTALS	20	20	20	20	20	20	120

Source: Hecht, H., and Proffitt, D. R. "The price of experience: Effects of experience on the water-level task." *Psychological Science,* Vol. 6, No. 2, Mar. 1995, p. 93 (Table 1).

Problem 4.66

c. What is the probability that the subject is a waitress or bartender?

d. What is the probability that the subject is a student and judges the line to be within 5° of the water surface?

e. Given the subject is a waitress or bartender, what is the probability that the subject judges the line to be within 5° of the water surface?

f. Given the subject is a male, what is the probability that the subject judges the line to be within 5° of the water surface?

g. Given the subject judges the line to be within 5° of the water surface, is the subject more likely to be a bus driver or a housewife?

4.67 A *USA Today* (May 15, 2000) survey explored things that workers say are "extremely important" aspects of a job. Suppose the survey was based on the responses of 500 men and 500 women. The job aspect "being able to work at home" was considered extremely important by 105 of the men and 170 of the women. Suppose that a respondent is chosen at random.

a. What is the probability that he or she feels that being able to work at home is an important aspect of the job?

b. What is the probability that the person is a female and feels that being able to work at home is an important aspect of the job?

c. Given that the person feels that being able to work at home is an important aspect of the job, what is the probability the person is a male?

d. Is feeling that working at home is an extremely important aspect of a job statistically independent of the gender of the respondent? Explain.

4.68 Children who develop unexpected difficulties with acquisition of spoken language are often diagnosed with specific language impairment (SLI). A study published in the *Journal of Speech, Language, and Hearing Research* (Dec. 1997) investigated the incidence of SLI in kindergarten children. As an initial screen, each in a national sample of over 7,000 children was given a test for language performance. The percentages of children who passed and failed the screen were 73.8% and 26.2%, respectively. All children who failed the screen were tested clinically for SLI. About one-third of those who passed the screen were randomly selected and also tested for SLI. The percentage of children diagnosed with SLI in the "failed screen" group was 20.5%; the percentage diagnosed with SLI in the "pass screen" group was 2.8%.

a. For this problem, let "pass" represent a child who passed the language performance screen, "fail" represent a child who failed the screen, and "SLI" represent a child diagnosed with SLI. Now find each of the following probabilities: $P(\text{Pass})$, $P(\text{Fail})$, $P(\text{SLI} \mid \text{Pass})$, and $P(\text{SLI} \mid \text{Fail})$.

b. Use the probabilities, part **a**, to find and $P(\text{Pass} \cap \text{SLI})$ and $P(\text{Fail} \cap \text{SLI})$. What probability law did you use to calculate these probabilities?

c. Use the probabilities, part **b**, to find $P(\text{SLI})$. What probability law did you use to calculate this probability?

KEY TERMS

Starred (*) terms are from the optional sections of this chapter.

Complementary events 202
Conditional probability 212
Event 194
Experiment 193
Independent events 214
*Intersection 220
Mutually exclusive events 198
Probability 196
Rare event approach 215
Sample space 199
Simple events 199
*Union 221
*Venn diagram 221

KEY FORMULAS AND RULES

Starred (*) formulas refer to the optional sections of this chapter.

1. *Additive Rule:* If two events A and B are mutually exclusive, then

$$P(A \text{ or } B) = P(A \cup B) = P(A) + P(B) \qquad \textbf{(4.1)}, 200$$

*In general,

$$P(A \text{ or } B) = P(A \cup B) = P(A) + P(B) - P(A \cap B) \qquad \textbf{(4.9)}, 222$$

2. *Modified Additive Rule for Equally Likely Mutually Exclusive Events:* If an experiment results in one and only one of M equally likely mutually exclusive (simple) events, of which m of these result in an event A, then

$$P(A) = \frac{m}{M} \qquad \textbf{(4.2)}, 201$$

3. *Multiplicative Rule:* If two events A and B are independent, then

$$P(A \text{ and } B) = P(A \cap B) = P(A)P(B) \qquad \textbf{(4.7)}, 203$$

*In general,

$$P(A \text{ and } B) = P(A \cap B) = P(A)P(B \mid A) = P(B)P(A \mid B) \qquad \textbf{(4.8)}, 221$$

4. *Rule of Complements:*

$$P(A) = 1 - P(A') \qquad \textbf{(4.3)}, 203$$

5. **Combinatorial Rule:* The number of different ways to select n objects from a total of N is

$$\binom{N}{n} = \frac{N!}{n!(N-n)!} \quad \text{where} \quad N! = N(N-1)(N-2)\cdots(2)(1) \qquad \textbf{(4.4)}, 209$$

KEY SYMBOLS

SYMBOL	DEFINITION
$P(A)$	Probability of an event A
A'	Complement of an event A
$\binom{N}{n}$	Number of ways to choose n objects from N objects
$n!$	n factorial
$P(A \mid B)$	Probability of A given B has occurred
$A \cup B$	A union (or) B
$A \cap B$	A intersect (and) B

CHECKING YOUR UNDERSTANDING

1. What is the outcome of an experiment called?
2. What does the probability of an event measure?
3. What are mutually exclusive events?
4. What are the most basic outcomes of an experiment called?

5. What is the sample space?
6. Give three numerical properties of the probability of an event.
7. When is the Combinatorial Rule useful?
8. What is a conditional probability?
9. How does conditional probability relate to the concept of independent events?
10. What events comprise the union of two events A and B?
11. What events comprise the intersection of two events A and B?
12. When is it appropriate to apply the additive law of probability?
13. When is it appropriate to apply the multiplicative law of probability?
14. How does the multiplicative law of probability differ for events that are and are not independent?
15. How does the additive law of probability differ for events that are and are not mutually exclusive?

SUPPLEMENTARY PROBLEMS

Starred (*) problems refer of the optional sections of this chapter.

EVOS

4.69 Refer to the *Journal of Agricultural, Biological, and Environmental Statistics* (Sept. 2000) study of the impact of the *Exxon Valdez* tanker oil spill on the seabird population in Prince William Sound, Alaska, Problem 2.62 (p. 107). Recall that data were collected on 96 shoreline locations (called transects) and it was determined whether or not the transect was in an oiled area. The data are stored in the **EVOS** file. From the data, estimate the probability that a randomly selected transect in Prince William Sound is contaminated with oil.

4.70 Refer to the *Journal of Peace Research* (Aug. 1997) study of the 36 active armed conflicts around the world in 1996, Problem 2.8 (p. 65). Of these 36 armed conflicts, 6 were classified as "war" (more than 1,000 deaths in any year). Provide an estimate of the probability that an armed conflict in any given year would be classified as a war. Comment on the reliability of your probability estimate.

4.71 Certain magazine publishers promote their products by mailing sweepstakes packets to consumers. These packets offer the chance to win a grand prize of $1 million or more, with no obligation to purchase any of the advertised products. Three popular sweepstakes are conducted by Publishers Clearing House, American Family Publishers, and Reader's Digest. On a nationwide basis, the odds of winning the grand prize are 1 in 181,795,000 for the Publishers Clearing House sweepstakes, 1 in 200,000,000 for the American Family Publishers sweepstakes, and 1 in 84,000,000 for the Reader's Digest sweepstakes.

a. Calculate the probability of winning the grand prize for each of the three sweepstakes.

b. Suppose you enter the sweepstakes contests of all three companies. What is the probability that you win the grand prize in all three contests?

c. What is the probability that you do not win any of the three grand prizes?

d. Use the probability computed in part **c** to calculate the probability of winning at least one of the three grand prizes. [*Hint:* The complement of "at least one" is "none."]

4.72 Video card games (e.g., video poker), once only legal in casinos, are now available for play in laundromats and grocery stores of states that employ a lottery. Because these video games are less intimidating than the traditional casino table games (e.g., blackjack, roulette, and craps), experts speculate that more women are becoming addicted to them. *The Wall Street Journal* (July 14, 1992) reported on the gambling habits of 52 women in Gamblers Anonymous.

a. Of the 52 women, 47 were video poker players. Use this information to assess the likelihood that a female member of Gamblers Anonymous (GA) is or was addicted to video poker.

b. Refer to part **a**. Of the 47 video poker players, 35 gambled until they exhausted their family savings. Express this information in a probability statement.

***c.** Use the probabilities, parts **a** and **b**, to find the probability that a female member of GA was addicted to video poker and gambled until she exhausted her family savings.

4.73 A child psychologist is interested in the ability of 5-year-old children to distinguish between imaginative and nonfiction stories. Two 5-year-old children are selected from a kindergarten class and are given a test to determine if they can distinguish imagination from reality. Consider the following events:

A: The first child can distinguish imagination from reality.

B: Neither child can distinguish imagination from reality.

C: The second child can distinguish imagination from reality, but the first child cannot.

Explain whether the pairs of events, A and B, A and C, and B and C, are mutually exclusive.

4.74 There are very few (if any) tests for pregnancy that are 100% accurate. Sometimes a test may indicate that a woman is pregnant even though she really is not. This is known as a *false positive* test. Similarly, a *false negative* result occurs when the test indicates that the woman is not pregnant even though she really is. Suppose a woman takes a certain pregnancy test. Define the events A, B, and C as follows:

A: The woman is really pregnant.

B: The test gives a false positive result.

C: The test gives a false negative result.

Which pair(s) of events are mutually exclusive?

4.75 A nationwide telephone survey of 1,200 people, aged 22 to 61, was conducted to gauge Americans' plans for retirement (*Tampa Tribune,* May 20, 1997). Some of the results are listed below.

- 46% have saved less than $10,000 for their retirement.
- 68% acknowledge they could save more if they made the effort.
- Of those who acknowledge they could save more, 20% say they will make the effort to do so.

a. What is the probability that an American aged 22 to 61 has saved less than $10,000 for retirement?

b. What is the probability that an American aged 22 to 61 can save more for retirement with more effort?

c. Given an American aged 22 to 61 can save more for retirement, what is the probability that he or she will do so?

4.76 To find the probability that a randomly selected United States citizen was born in a given state—Virginia, for example—an introductory statistics student divides the number of favorable outcomes, one, by the total number of states, 50. Thus, the student reports that the probability that John Q. Citizen was born in Virginia is $\frac{1}{50}$. Explain why this probability is incorrect. If you had access to the complete 2000 census, how could the correct probability be obtained?

***4.77** In Florida's "Fantasy 5" lottery game, players select five different whole numbers ranging from 1 to 30. To win, all five numbers must match the five numbers randomly selected each day.

a. How many different combinations of five numbers from 30 are possible?

b. Based on your answer to part **a**, what is the probability of winning the "Fantasy 5" lottery?

4.78 On-line securities trading began just a few years ago, but already accounts for a significant share of the brokerage business. The table on p. 232 reports the number of on-line and traditional accounts (in millions) for five leading brokerages. Suppose

a customer account is to be selected at random from the population of accounts described in the table. Consider the following events:

A: The account is with Merrill Lynch.

B: The account is an on-line account.

C: The account is with E*Trade and is an on-line account.

D: The account is with E*Trade.

Brokerage Firms	On-Line Accounts	Traditional Accounts	Total Accounts
Fidelity Investments	2.80	8.00	10.80
Merrill Lynch & Co.	0.00	8.00	8.00
Charles Schwab & Co.	2.80	3.50	6.30
TD Waterhouse Group	1.00	1.10	2.10
E* Trade Group Inc.	1.24	0.00	1.24
TOTALS	7.84	20.60	28.44

Source: Business Week, Oct. 18, 1999, pp. 185–186.

Problem 4.78

a. Find the probability of each of the above events.

b. Find $P(A \cap B)$.

c. Find $P(A \cup B)$.

d. Find $P(B \cap D)$.

e. Find $P(A \cup D)$.

4.79 Refer to the *Journal of the American Geriatric Society* (Jan. 1998) study of age-related macular degeneration (AMD), Problem 2.66 (p. 108). To investigate the link between AMD incidence and alcohol consumption, 3,072 adults were tested for AMD and classified according to type of alcohol most frequently consumed. The cross-tabulation table for the study is given below.

			Clinical Diagnosis		
			AMD	No AMD	Totals
Most Frequent Alcohol Type Consumed	None		87	878	965
	Beer		30	400	430
	Wine		51	1,233	1,284
	Liquor		20	373	393
		TOTALS	188	2,884	3,072

Consider an adult randomly selected from the 3,072 adults in the study.

a. What is the probability that the adult is diagnosed with AMD?

b. What is the probability that the adult consumes beer most frequently?

c. What is the probability that the adult is diagnosed with AMD and consumes liquor most frequently?

d. Given that the adult does not consume alcohol, what is the probability that he or she is diagnosed with AMD?

e. Given that the adult consumes wine most frequently, what is the probability that he or she is diagnosed with AMD?

f. What is the probability that the adult is not diagnosed with AMD or does not consume alcohol?

4.80 An article in *IEEE Computer Applications in Power* (Apr. 1990) describes "an unmanned watching system to detect intruders in real time without spurious detections, both indoors and outdoors, using video cameras and microprocessors." The system was tested outdoors under various weather conditions in Tokyo, Japan. The numbers of intruders detected and missed under each condition are provided in the table.

	Weather Condition				
	Clear	**Cloudy**	**Rainy**	**Snowy**	**Windy**
Intruders detected	21	228	226	7	185
Intruders missed	0	6	6	3	10
TOTALS	21	234	232	10	195

Source: Kaneda, K., et al. "An unmanned watching system using video cameras." *IEEE Computer Applications in Power,* Apr. 1990, p. 24.

a. Under cloudy conditions, what is the probability that the unmanned system detects an intruder?

b. Given that the unmanned system missed detecting an intruder, what is the probability that the weather condition was snowy?

4.81 The transport of neutral particles in an evacuated duct is an important aspect of nuclear fusion reactor design. In one experiment, particles entering through the duct ends streamed unimpeded until they collided with the inner duct wall. Upon colliding, they are either scattered (reflected) or absorbed by the wall (*Nuclear Science and Engineering,* May 1986). The reflection probability (i.e., the probability a particle is reflected off the wall) for one type of duct was found to be .16.

a. If two particles are released into the duct, find the probability that both will be reflected.

b. If five particles are released into the duct, find the probability that all five will be absorbed.

c. What assumption about the simple events in parts **a** and **b** is required to calculate the probabilities?

***4.82** Despite penicillin and other antibiotics, bacterial pneumonia still kills thousands of Americans every year. An antipneumonia vaccine, called Pneumovax, is designed especially for elderly or debilitated patients, who are usually the most vulnerable to bacterial pneumonia. Suppose the probability that an elderly or debilitated person will be exposed to these bacteria is .45. If exposed, the probability that an elderly or debilitated person inoculated with the vaccine acquires pneumonia is only .10. What is the probability that an elderly or debilitated person inoculated with the vaccine does acquire bacterial pneumonia?

***4.83** One of the most popular card games in the world is the game of bridge. Each player in the game is dealt 13 cards from a standard 52-card deck. Determine the number of different 13-card bridge hands that can be dealt from a 52-card deck.

***4.84** According to the National Association of State Boards of Accountancy, the probability is .20 that a first-time candidate passes all subjects in the Uniform CPA Examination (*New Accountant,* Sept. 1993). Given that a first-time candidate fails, the probability that such a candidate passes all subjects increases to .30 on repeat examinations. Use this information to find the probability that a candidate passes the Uniform CPA Examination on his or her second attempt.

4.85 *Science Digest* magazine features an interesting and informative columnist, known only by the eerie name of Dr. Crypton. Each month Dr. Crypton presents mind-twisters, riddles, puzzles, and enigmas that very often can be solved using the laws of probability. In one issue, Dr. Crypton offered this view of what is commonly called the "birthday problem":

Surely you have been in a situation in which a small group of people compared birthdays and found, to their surprise, that at least two of them were born on the same day of the same month. Suppose there are 10 people in the group. Intuition may suggest that the odds of two sharing a birthday are quite poor. Probability theory, however, shows the odds are better than one in nine.

If you like to win bets, you should keep in mind that for a group of 23 people the odds are in favor of at least two of them sharing a birthday. (For 22 people, the odds are slightly against this.)*

*Excerpt from column by Dr. Krypton, *Science Digest*, January, 1982.

a. Find the probability that no two of a group of 23 people share the same birthday. [*Hint:* Since there are 365 days in a year, any of which may be the birthday of one of the 23 people, the probability that none of the people in the group share a birthday is

$$\underset{\text{Person \#1}}{\left(\frac{365}{365}\right)} \cdot \underset{\text{Person \#2}}{\left(\frac{364}{365}\right)} \cdot \underset{\text{Person \#3}}{\left(\frac{363}{365}\right)} \cdots \underset{\text{Person \#23}}{\left(\frac{343}{365}\right)}$$

Compute this probability.]

b. Using the rule of complements (Probability Rule #3), find the probability that at least two of 23 people will share the same birthday.

c. Use the steps outlined in parts **a** and **b** to find the probability that at least two of 10 people will share the same birthday.

4.86 Marilyn vos Savant, who is listed in the *Guinness Book of World Records Hall of Fame* for "Highest IQ," writes a weekly column in the Sunday newspaper supplement *Parade Magazine*. Her column, "Ask Marilyn," is devoted to games of skill, puzzles, and mind-bending riddles. In one issue, vos Savant posed the following question:* Suppose you're on a game show, and you're given a choice of three doors. Behind one door is a car; behind the others, goats. You pick a door—say, #1—and the host, who knows what's behind the doors, opens another door—say, #3—which has a goat. He then says to you, "Do you want to pick door #2?" Is it to your advantage to switch your choice? By answering the following series of questions, you will arrive at the correct solution.

**Parade Magazine*, Feb. 17, 1991. Reprinted with permission from *Parade*, Copyright © 1991.

a. Before taping of the game show, the host randomly decides behind which of the three doors to put the car; the goats will go behind the remaining two doors. List the simple events for this experiment. [*Hint:* One simple event is $C_1G_2G_3$, i.e., car behind door #1 and goats behind door #2 and door #3.]

b. Randomly choose one of the three doors, as the contestant does in the game show. Now, for each simple event in part **a**, circle the selected door and put an X through one of the remaining two doors that hides a goat. (This is the door that the host shows the contestant—always a goat.)

c. Refer to the altered simple events in part **b**. Assume your strategy is to keep the door originally selected. Count the number of simple events for which this is a "winning" strategy (i.e., you win the car). Assuming equally likely simple events, what is the probability that you win the car?

d. Repeat part **c**, but assume your strategy is to always switch doors.

e. Based on the probabilities of parts **c** and **d**, "Is it to your advantage to switch your choice?"

4.87 Consider another problem posed by Marilyn vos Savant in her weekly column "Ask Marilyn": "A woman and a man (unrelated) each have two children. At least one of the woman's children is a boy, and the man's older child is a boy. Do the chances that the woman has two boys equal the chances that the man has two boys?"*

**Parade Magazine*, Oct. 17, 1997. Reprinted with permission from *Parade*, Copyright © 1997.

a. Answer the question based on your intuition.

b. Use simple events and the rules of probability to solve the problem. Does your answer agree with part **a**?

4.88 Refer to Problem 4.87. A letter from one of vos Savant's readers reported the following statistics from the U.S. Census Bureau's National Health Interview Survey: Of all the households interviewed from 1987 to 1993, 42,888 families had exactly two children. Of these, 9,523 had two girls and 33,365 had at least one boy. Use these figures to support your answer to Problem 4.87.

REFERENCES

Bernstein, P. *Against the Gods: The Remarkable Story of Risk.* New York: John Wiley, 1996.

Epstein, R. A. *The Theory of Gambling and Statistical Logic,* revised ed. New York: Academic Press, 1977.

Feller, W. *An Introduction to Probability Theory and Its Applications,* 3rd ed. Vol. I. New York: Wiley, 1968.

Mendenhall, W., Wackerly, D., and Scheaffer, R. *Mathematical Statistics with Applications,* 4th ed. Boston: PWS-Kent, 1990.

Mosteller, F., Rourke, R., and Thomas, G. *Probability with Statistical Applications,* 2nd ed. Reading, Mass.: Addison-Wesley, 1970.

Parzen, E. *Modern Probability Theory and Its Applications.* New York: Wiley, 1960.

Paulos, J. *Innumeracy.* New York: Hill and Wang, 1988.

Wright, G. and Ayton, P., eds. *Subjective Probability.* New York: John Wiley, 1994.

Using Microsoft® Excel

4.E.1 Calculating Simple Probabilities

Use the PHStat **Probability & Prob. Distributions | Simple & Joint Probabilities** procedure to calculate basic probabilities for a two-way table with two rows and two columns. For example, to calculate the basic probabilities for the on-line holiday shopping data of Problem 4.61 on p. 226, open to an empty worksheet and do the following:

1. Select PHStat | Probability & Prob. Distributions | Simple & Joint Probabilities.
2. In the new Probabilities worksheet, enter these replacement labels for the Sample Space events:
   a. Enter Satisfied with Experience (caption for event A) in cell A5
   b. Enter Satisfied Yes (event A) in cell B5.
   c. Enter Satisfied No (event A') in cell B6.
   d. Enter Received Products in Time for Holidays (event B caption) in cell C3
   e. Enter Received Yes (event B) in cell C4.
   f. Enter Received No (event B') in cell D4.
3. In the same worksheet:
   a. Enter 1197 (the event A and B value) in cell C5.
   b. Enter 127 (the event A' and B value) in cell C6.
   c. Enter 33 (the event A and B' value) in cell D5.
   d. Enter 143 (the event A' and B' value) in cell D6.

4.E.2 Calculating the Number of Combinations

The COMBIN function of Microsoft Excel can be used to obtain the number of combinations. The format of this function is COMBIN(N, n), where N is the total number of objects and n is the number of objects to be selected.

To illustrate the COMBIN function, return to the example on p. 210 concerning the selection of 6 numbers from 53 numbers in the Florida Lotto game. To obtain the number of combinations (equal to 22,957,480) from Microsoft Excel, enter the formula =COMBIN(53,6) in a cell and press the Enter key. The result obtained in that cell will be 22,957,480.

Chapter 5 Discrete Probability Distributions

OBJECTIVES

1. To understand the difference between a discrete random variable and a continuous random variable
2. To present the properties of a probability distribution for a discrete random variable
3. To develop the concept of expected value of a discrete random variable
4. To present applications of the binomial probability distribution
5. To present applications of the Poisson probability distribution
6. To present applications of the hypergeometric probability distribution

CONTENTS

EXCEL TUTORIAL

MINITAB TUTORIAL

Statistics in the Real World

Probability in a Reverse Cocaine Sting

The American Statistician (May 1991) described an interesting application of a discrete probability distribution in a case involving illegal drugs. It all started with a "bust" in a mid-sized Florida city. During the bust, police seized approximately 500 foil packets of a white, powdery substance, presumably cocaine. Since it is not a crime to buy or sell non-narcotic cocaine look-alikes (e.g., inert powders), detectives had to prove that the packets contained genuine cocaine in order to convict their suspects of drug trafficking. When the police laboratory randomly selected and chemically tested four of the packets, all four tested positive for cocaine. This finding led to the conviction of the traffickers.

After the conviction, the police decided to use the remaining foil packets (i.e., those not tested) in reverse sting operations. Two of these packets were randomly selected and sold by undercover officers to a buyer. Between the sale and the arrest, however, the buyer disposed of the evidence. The key question is: Beyond a reasonable doubt, did the defendant really purchase cocaine?

In court, the defendant's attorney argued that his client should not be convicted because the police could not prove that the missing foil packets contained cocaine. The police contended, however, that since four of the original packets tested positive for cocaine, the two packets sold in the reverse sting were also highly likely to contain cocaine. In the following Statistics in the Real World Revisited sections, we demonstrate how to use probability models to solve the dilemma. (The case represented Florida's first cocaine-possession conviction without the actual physical evidence.)

Statistics in the Real World Revisited

- Using the Binomial Model to Solve the Cocaine Sting Case (p. 254)
- Using the Hypergeometric Model to Solve the Cocaine Sting Case (p. 267)

5.1 Random Variables

In a practical setting, an experiment (as defined in Chapter 4) involves selecting a sample of data consisting of one or more observations on some variable. For example, we might survey 1,000 physicians concerning their preferences for aspirin and record x, the number who prefer a particular brand. Or, we might randomly select a single supermarket customer from the **CHKOUT** data file and observe his or her total checkout time, x. Since we can never know with certainty the exact value that we will observe when we record x for a single performance of the experiment, we call x a *random variable.*

Definition 5.1

A **random variable** is a variable that has numerical values associated with events of an experiment.

The random variables described above are examples of two different types of random variables—*discrete* and *continuous.* The number x of physicians in a sample of 1,000 who prefer a particular brand of aspirin is a discrete random variable because it can take on only a *countable* number of values—typically whole numbers such as 0, 1, 2, 3, . . . , 999, 1,000. In contrast, the checkout time of a supermarket customer is a continuous random variable because it could theoretically have any one of an *infinite* number of values—namely, any value greater than or equal to 0 seconds. Of course, in practice we record checkout time to the nearest second but, in *theory,* the checkout time of a supermarket customer could be any value—say 137.21471 seconds—making it impossible to count or list its values.

A good way to distinguish between discrete and continuous random variables is to imagine the values that they have. Discrete random variables may take on any one of a countable number of values (say, 10, 21, or 100). In contrast, a continuous random variable can theoretically take on *any* possible value.

Definition 5.2

A **discrete random variable** is one that can take on only a countable number of values.

Definition 5.3

A **continuous random variable** can take on an infinite number of (non-countable) values.

EXAMPLE 5.1

TYPES OF RANDOM VARIABLES

Suppose you randomly select a student attending your college or university. Classify each of the following random variables as discrete or continuous:

a. Number of credit hours taken by the student this semester
b. Current grade point average of the student

Solution

a. The number of credit hours taken by the student this semester is a discrete random variable because it can take on only a countable number of values (for example, 15, 16, 17, and so on). It is not continuous since the number of credit hours must be an integer; the variable can never have values such as 15.2062, 16.1134, or 17.0398 hours.

b. The grade point average for the student is a continuous random variable since it could theoretically take on any value (for example, 2.87355) from 0 to 4.

Self-Test 5.1

Determine whether the following random variables are discrete or continuous:

a. The time it takes a student to complete an examination
b. The number of registered voters in a national election
c. The number of winners each week in a state lottery
d. The weight of a professional wrestler

The focus of this chapter is on discrete random variables. Continuous random variables are the topic of Chapter 6.

5.2 Probability Models for Discrete Random Variables

We learned in Chapter 4 that we make inferences based on the probability of observing a particular sample outcome. Since we never know the *exact* probability of some events, we must construct probability models for the values assumed by random variables.

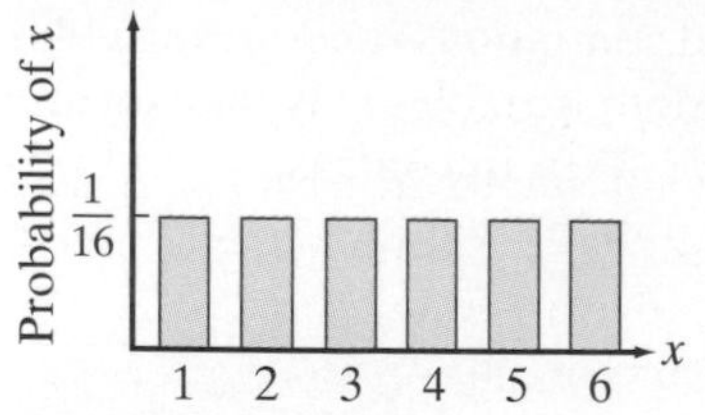

Figure 5.1 The Probability Distribution for x, the Number of Dots Observed on a Balanced Die

For example, if we toss a balanced die, we assume that the values 1, 2, 3, 4, 5, and 6 represent equally likely events, i.e., $P(x = 1) = P(x = 2) = \cdots = P(x = 6) = \frac{1}{6}$. In doing so, we have constructed a probabilistic model for a theoretical population of x values, where x is the number of dots showing on the upper face of the die. The population is "theoretical" in the sense that we would observe an x for an infinite number of die tosses. A bar graph for the theoretical population is shown in Figure 5.1.

Figure 5.1, which gives the relative frequency for each value of x in a very large number of tosses of a die, is called the *probability distribution for the discrete random variable, x.*

Definition 5.4

The **probability distribution for the discrete random variable x** is a table, graph, or formula that gives the probability of observing each value of x. If we denote the probability of x by the symbol $p(x)$, the probability distribution has the following properties:

1. Each probability is between 0 and 1, i.e.,

$$0 \leq p(x) \leq 1 \text{ for all values of } x$$

2. The sum of the probabilities over all possible values of x is equal to 1, i.e.,

$$\sum_{\text{all } x} p(x) = 1$$

EXAMPLE 5.2 PROPERTIES OF A PROBABILITY DISTRIBUTION

Consider the following sampling situation: Suppose we select a random sample of $n = 5$ physicians from a very large number—say, 10,000—and record the number x of physicians who recommend aspirin brand A. Suppose also that 2,000 of the physicians (i.e., 20%) actually prefer brand A. Now suppose we replace the five physicians in the population, randomly select a new sample of $n = 5$ physicians, and record the value of x again. If we repeat this process over and over again 100,000 times, a hypothetical data set containing the 100,000 values of x can be obtained. Table 5.1 is a summary of one such data set (obtained by computer simulation).

TABLE 5.1 Relative Frequencies for 100,000 Hypothetical Observations on x, the Number of Physicians in a Sample of $n = 5$ Who Prefer Aspirin Brand A

x	Frequency	Relative Frequency
0	32,807	.32807
1	40,949	.40949
2	20,473	.20473
3	5,122	.05122
4	645	.00645
5	4	.00004
TOTALS	100,000	1.00000

a. Construct a relative frequency histogram for the 100,000 values of x summarized in Table 5.1.

b. Assuming that the relative frequencies of Table 5.1 are good approximations to the probabilities of x, show that the properties of a probability distribution are satisfied.

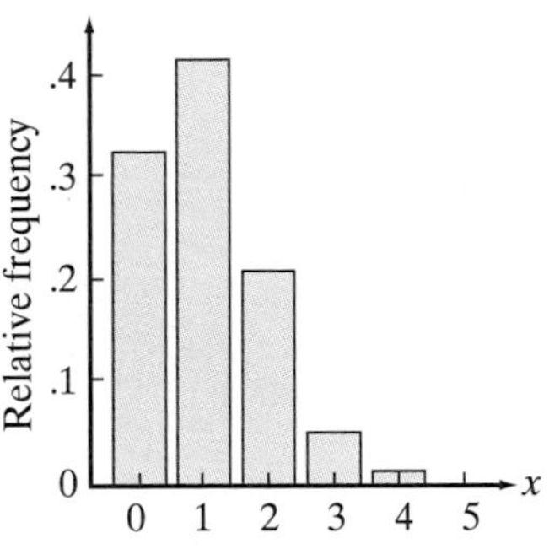

Figure 5.2 Relative Frequency Histogram for Example 5.2

Solution

a. The relative frequency histogram for the 100,000 values of x is shown in Figure 5.2. This figure provides a very good approximation to the probability distribution for x, the number of physicians in a sample of $n = 5$ who prefer brand A (assuming that 20% of the physicians in the population prefer brand A).

b. We use the relative frequencies of Table 5.1 as approximations to the probabilities of $x = 0, x = 1, \ldots, x = 5$. For example, $P(x = 0) = p(0) \approx .32807$. Similarly, $P(x = 1) = p(1) \approx .40949$. Note that each probability (relative frequency) is between 0 and 1. Summing these probabilities, we obtain

$$.32807 + .40949 + .20473 + .05122 + .00645 + .00004 = 1$$

Thus, the properties of a discrete probability distribution are satisfied.

Often, the sample collected in an experiment (such as the physician survey of Example 5.2) is quite large. For example, the Gallup and Harris survey results reported in the news media are usually based on sample sizes from $n = 1{,}000$ to $n = 2{,}000$ people. Since we would not want to calculate $p(x)$ for values of n this large, we need an easy way to describe the probability distribution for x. To do this, we need to know the mean and standard deviation for the distribution. That is, we must know μ and σ for the theoretical population of x values modeled by the probability distribution. Then we can describe it using either the Empirical Rule or Tchebysheff's Theorem from Chapter 3.

Definition 5.5

The **mean μ** (or **expected value**) of a discrete random variable x is equal to the sum of the products of each value of x and the corresponding value of $p(x)$:

$$\mu = \sum xp(x) \tag{5.1}$$

Definition 5.6

The **variance σ^2** of a discrete random variable x is equal to the sum of the products of $(x - \mu)^2$ and the corresponding value of $p(x)$:

$$\sigma^2 = \sum (x - \mu)^2 p(x) \tag{5.2}$$

Definition 5.7

The **standard deviation σ** of a random variable x is equal to the positive square root of the variance:

$$\sigma = \sqrt{\sigma^2} \tag{5.3}$$

EXAMPLE 5.3

FINDING μ AND σ

Consider the sample survey of $n = 5$ physicians, described in Example 5.2. The graph of the probability distribution for x, the number of physicians in the sample that favors aspirin brand A, is reproduced in Figure 5.3 (p. 542).

a. Find the mean μ for this distribution. That is, find the expected value of x.

b. Interpret the value of μ.

c. Find the standard deviation of x.

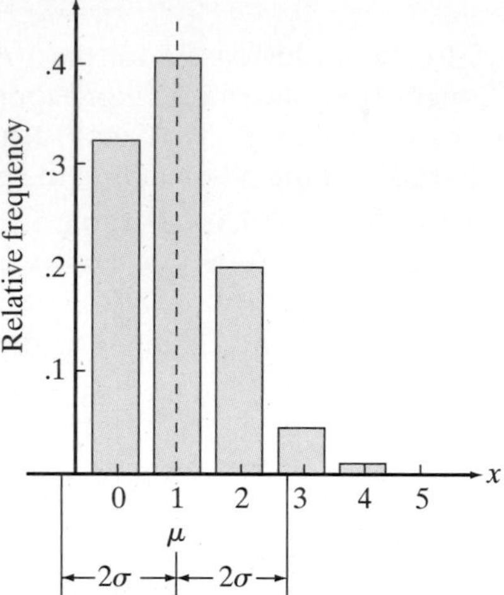

Figure 5.3 Probability Distribution for x, the Number of Physicians in a Sample of $n = 5$ Who Prefer Aspirin Brand A

Solution

a. According to Definition 5.5 (and equation 5.1), the mean μ is given by

$$\mu = \sum xp(x)$$

where $p(x)$ is given in Table 5.1. Since x can take values $x = 0, 1, 2, \ldots, 5$, we have

$$\begin{aligned}\mu &= 0p(0) + 1p(1) + 2p(2) + 3p(3) + 4p(4) + 5p(5) \\ &= 0(.32807) + 1(.40949) + 2(.20473) + 3(.05122) + 4(.00645) + 5(.00004) \\ &= 1.0\end{aligned}$$

b. As we learned in Chapter 3, the estimated value of μ can also be obtained by adding the 100,000 values of x shown in Table 5.1 and dividing the sum by $n = 100{,}000$. Consequently, the value $\mu = 1.0$ implies that *over a long series of surveys* similar to the one described in Example 5.2, the average number of physicians in the sample of 5 who favor aspirin brand A will equal 1. The key to the interpretation of μ is to think in terms of repeating the experiment over a long series of trials (i.e., in the *long run*). Then μ represents the average x value of this large number of trials.

c. By Definition 5.6 (and equation 5.2), we obtain

$$\begin{aligned}\sigma^2 &= \sum (x - \mu)^2 p(x) \\ &= (0 - 1)^2 p(0) + (1 - 1)^2 p(1) + (2 - 1)^2 p(2) + (3 - 1)^2 p(3) \\ &\quad + (4 - 1)^2 p(4) + (5 - 1)^2 p(5) \\ &= (1)(.32807) + (0)(.40949) + (1)(.20473) + (4)(.05122) \\ &\quad + (9)(.00645) + (16)(.00004) \\ &= .8\end{aligned}$$

Then by Definition 5.7 (and equation 5.3), the standard deviation σ is given by

$$\sigma = \sqrt{\sigma^2} = \sqrt{.79} = .8$$

Self-Test 5.2

Consider the probability distribution for a discrete random variable x, shown below:

x	0	1	2	3	4
$p(x)$	.4	.2	.2	—	.1

a. Find $p(3)$. **b.** Find μ. **c.** Find σ.

EXAMPLE 5.4

EMPIRICAL RULE APPLICATION

Refer to Example 5.3. Locate the interval $\mu \pm 2\sigma$ on the graph of the probability distribution for x. Confirm that most of the (theoretical) population falls within this interval.

Solution

Recall that $\mu = 1.0$ and $\sigma = .89$ for this distribution. Then

$$\mu - 2\sigma = 1.0 - 2(.89) = -.78$$

$$\mu + 2\sigma = 1.0 + 2(.89) = 2.78$$

The interval from $-.78$ to 2.78, shown in Figure 5.3, includes the values of $x = 0$, $x = 1$, and $x = 2$. Thus, the probability (relative frequency) that a population value falls within this interval is

$$p(0) + p(1) + p(2) = .32807 + .40949 + .20473 = .94229$$

This certainly agrees with the Empirical Rule, which states that approximately 95% of the data will lie within 2σ of the mean μ.

In the next section, we present a special discrete probability distribution that provides a useful model for many types of data encountered in the real world.

PROBLEMS FOR SECTION 5.2

Using the Tools

5.1 Consider the probability distribution shown in the table.

x	−5	0	2	5
$p(x)$	.2	.3	.4	.1

a. List the values that x may assume.
b. What value of x is most probable?
c. Find the probability that x is greater than 0.
d. What is the probability that $x = -5$?
e. Verify that the sum of the probabilities equals 1.
f. Find μ and σ.

5.2 Suppose a discrete random variable can have five possible values with the probability distribution shown in the table.

x	1	2	3	4	5
$p(x)$	.20	.25	—	.30	.10

a. Find the value for $p(3)$.

b. Find the probability that $x = 2$ or $x = 4$.

c. Find the probability that x is less than or equal to 4.

d. Find μ and σ.

5.3 The probability distribution for a discrete random variable x is given by the formula,

$$p(x) = (.8)(.2)^{x-1}, x = 1, 2, 3, \ldots$$

a. Calculate $p(x)$ for $x = 1, x = 2, x = 3, x = 4$, and $x = 5$.

b. Sum the probabilities of part **a**. Is it likely to observe a value of x greater than 5?

c. Find $P(x = 1 \text{ or } x = 2)$.

5.4 Given the following probability distributions:

Distribution A		Distribution B	
x	$p(x)$	x	$p(x)$
0	.50	0	.05
1	.20	1	.10
2	.15	2	.15
3	.10	3	.20
4	.05	4	.50

a. Compute the mean for each distribution.

b. Compute the standard deviation for each distribution.

c. For each distribution, compute an interval that captures most of the values of x.

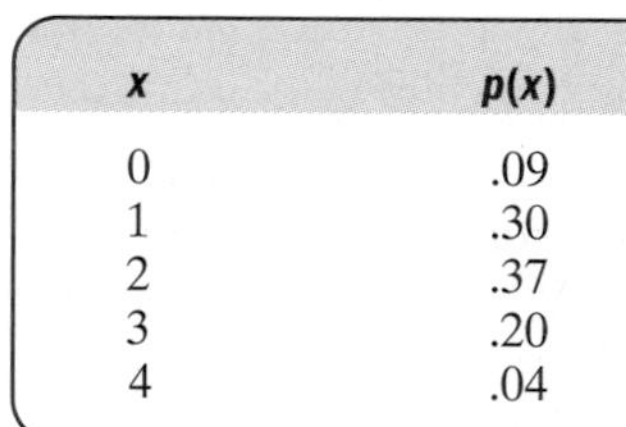

x	$p(x)$
0	.09
1	.30
2	.37
3	.20
4	.04

Problem 5.5

Applying the Concepts

5.5 Refer to the May 2000 study of dust mites by the National Institute of Environmental Health Sciences, Problem 4.57 (p. 225). Recall that 45% of all U.S. homes have bedding with a dust mite level that exceeds 2 micrograms per gram (μg/g)—a level associated with the development of allergies. Consider a random sample of four U.S. homes and let x be the number of homes with a dust mite level that exceeds 2 μg/g. The probability distribution for x is shown in the table at left.

a. Verify that the sum of the probabilities for x in the table is 1.

b. Find the probability that three or four of the homes in the sample have a dust mite level that exceeds 2 μg/g.

c. Find the probability that fewer than two homes in the sample have a dust mite level that exceeds 2 μg/g.

d. Find the expected value of x. Interpret this value.

Yards, x	Probability
−4	.020
−2	.060
−1	.070
0	.150
1	.130
2	.110
3	.090
4	.070
6	.090
8	.060
10	.050
15	.085
30	.010
50	.004
99	.001

Problem 5.6

5.6 In his article "American Football" (*Statistics in Sport,* 1998), Iowa State University statistician Hal Stern evaluates winning strategies of teams in the National Football League (NFL). In a section on estimating the probability of winning a game, Stern used actual NFL play-by-play data to approximate the probabilities associated with certain outcomes (e.g., running plays, short pass plays, and long pass plays). The table at left gives the probability distribution for the yardage gained, x, on a running play. (A negative gain represents a loss of yards on the play.)

a. Find the probability of gaining 10 yards or more on a running play.

Age, x, at First Arrest	$p(x)$
6	.0250
7	.0250
8	.0125
9	.0625
10	.0625
11	.0625
12	.1625
13	.2375
14	.1625
15	.1875

Source: Patterson, G. R., Crosby, L., and Vuchinich, S. "Predicting risk for early police arrest." *Journal of Quantitative Criminology,* Vol. 8, No. 4, 1992, p. 345 (adapted from Table II).

Problem 5.7

Trained Listeners	
Note	**Frequency**
D_3	1
G_3	1
$A_3^{\#}$	1
C_4	12
$C_4^{\#}$	1
D_4	1
E_4	15
F_4	7
G_4	8
A_4	3
TOTALS	50

Source: Thompson, W. F., et al. "Expectancies generated by melodic intervals: Evaluation of principles of melodic implication in a melody-completion task." *Perception & Psychophysics,* Vol. 59, No. 7, Oct. 1997, p. 1076.

Problem 5.8

b. Find the probability of losing yardage on a running play.

c. Find μ. Interpret this value.

d. Find the variance and standard deviation of x.

e. Give a range of values which will likely include the yardage gained, x, on about 95% of running plays.

5.7 Researchers at Oregon State University attempted to develop a model for predicting the risk of early police arrest (*Journal of Quantitative Criminology,* Vol. 8, 1992). The model was based, in part, on data collected for 80 boys arrested for crimes prior to the age of 16. The accompanying table gives the probability distribution for x, the age at first arrest, for the 80 boys.

a. Verify that the probabilities in the table sum to 1.

b. Find $P(x \leq 10)$.

c. Find $P(x > 12)$.

d. Find and interpret μ.

e. Find σ.

f. Compute the interval $\mu \pm 2\sigma$.

g. Find the probability that x falls in the interval of part **f**. Compare your results with the Empirical Rule.

5.8 Refer to the *Perception & Psychophysics* (Oct. 1997) melody-completion task, Problem 2.10 (p. 65). After listening to a short (2-note) melody, each of 50 people trained in music completed the melody on a piano keyboard. The frequency distribution of the initial note selected by the subjects is reproduced in the table at left. Consider a sample of two subjects randomly selected from the 50. Let x be the number of subjects who select the note E_4.

a. Find the probability distribution for x.

b. Compute μ and σ.

c. Interpret the value of μ.

5.9 According to a survey of businesses that use the Internet and its World Wide Web, 70% use Internet Explorer as a Web browser. If two surveyed businesses are selected, find the probability distribution of x, the number of businesses that use Internet Explorer as a Web browser. [*Hint:* List the possible outcomes of the experiment. For each outcome, determine the value of x and then calculate its probability using Probability Rule #4 (p. 214).]

5.10 In a carnival game called "Under-or-over 7," a pair of fair dice are rolled once. A player bets $1 that the sum showing on the dice is either under 7, over 7, or equal to 7.

a. If the player bets that the sum is *under* 7 and wins, he wins $1. Let x be the outcome (win $1 or lose $1) for this bet. Find the probability distribution for x.

b. If the player bets that the sum is *over* 7 and wins, he wins $1. Let y be the outcome (win $1 or lose $1) for this bet. Find the probability distribution for y.

c. If the player bets that the sum *equals* 7 and wins, he wins $4. Let z be the outcome (win $4 or lose $1) for this bet. Find the probability distribution for z.

d. For each of the three bets described in parts **a**, **b**, and **c**, find the expected long-run profit (or loss).

e. For each of the three bets described in parts **a**, **b**, and **c**, find the standard deviation of the profit (or loss).

f. Based on the results, parts **d** and **e**, which bet would you prefer to make in "Under-or-over 7"? Explain.

5.11 The cornrake is a European species of bird in danger of worldwide extinction. A census of singing cornrakes on agricultural land in Great Britain and Ireland was recently conducted (*Journal of Applied Ecology,* Vol. 30, 1993). The census revealed that 12 cornrakes inhabit Mainland Scotland. Suppose that two of these Scottish cornrakes are captured for mating purposes. Let x be the number of these captured

cornrakes that are capable of mating. Suppose exactly 8 of the original 12 cornrakes inhabiting Scotland are capable of mating and 4 are infertile.

a. Find the probability distribution for x. [*Hint:* List the simple events and determine the value of x for each.]

b. Find μ and σ.

c. Interpret the value of μ.

5.12 A survey of American workers commissioned by Northwestern National Life Insurance Company found that only 1% of all workers do not experience stress on the job. Let x be the number of workers that must be sampled until the first worker with no stress is found.

a. List the possible values of x.

b. Refer to part **a**. If you sum the probabilities for these x values, what result will you obtain?

c. Find $p(1)$.

d. Find $p(2)$. [*Hint:* Use Probability Rule #4 for two independent events (p. 214).]

e. It can be shown (proof omitted) that the probability distribution of x is given by the formula $p(x) = p(1 - p)^{x-1}$, where p is the probability that a single worker has no on-the-job stress. Use this formula to find $p(12)$.

5.3 The Binomial Probability Distribution

Opinion polls (i.e., sample surveys) are conducted frequently in politics, psychology, sociology, medicine, and business. Some recent examples include polls to determine (1) the number of voters in favor of legalizing casino gambling in a particular state, (2) the number of Americans who feel the president is a moral leader, and (3) the number of baseball fans who support league realignment of Major League Baseball franchises based on geography. Consequently, it is useful to know the probability distribution of the number x in a random sample of n experimental units (people) who exhibit some characteristic or prefer some specific proposition. This probability distribution can be approximated by a **binomial probability distribution** when the sample size n is small relative to the total number N of experimental units in the population.

Strictly speaking, the binomial probability distribution applies only to sampling that satisfies the conditions of a **binomial experiment,** as listed in the box.

CONDITIONS REQUIRED FOR A BINOMIAL EXPERIMENT

1. A sample of n experimental units is selected from the population *with replacement* (called **sampling with replacement**).
2. Each experimental unit possesses one of two mutually exclusive characteristics. By convention, we call the characteristic of interest a *success* and the other a *failure.*
3. Prior to observing the characteristic (outcome) for an experimental unit, the probability that a single experimental unit is a "success" is equal to p. This probability is the same for all experimental units.
4. The outcome for any one experimental unit is independent of the outcome for any other experimental unit (i.e., the draws are independent).
5. The **binomial random variable** x is the number of "successes" in n experimental units.

In real life, there are probably few experiments that satisfy exactly the conditions for a binomial experiment. However, there are many that approximately satisfy—at least for all practical purposes—these conditions. Consider, for example, a sample survey. When the number N of elements in the population is large and the sample size n is small relative to N, the sampling approximately satisfies the conditions of a binomial experiment. The next two examples illustrate the conditions required for a binomial experiment.

EXAMPLE 5.5

CHECKING BINOMIAL CONDITIONS

Suppose that a sample of $n = 2$ elements is randomly selected from a population containing $N = 10$ elements, three of which are designated as successes and seven as failures. Explain why this sampling procedure violates the conditions of a binomial experiment.

Solution

The probability of selecting a success on the first draw is equal to $\frac{3}{10}$—that is, the number of successes in the population (3) divided by the total number of elements in the population (10). In contrast, the probability of a success on the second draw is either $\frac{2}{9}$ or $\frac{3}{9}$, depending on whether a success was or was not selected on the first draw. In other words, the probability of selecting a success on the second draw is dependent on the outcome of the first draw. This is a violation of condition 4 required for a binomial experiment.

EXAMPLE 5.6

CHECKING BINOMIAL CONDITIONS

Suppose that a sample of size $n = 2$ is randomly selected from a population containing $N = 1{,}000$ elements, 300 of which are successes and 700 of which are failures. Explain why this sampling procedure satisfies, approximately, the conditions required for a binomial experiment.

Solution

The probability of success on the first draw is the same as in Example 5.5—namely, $\frac{300}{1{,}000} = \frac{3}{10}$. The probability of a success on the second draw is either $\frac{299}{999} = .2993$ or $\frac{300}{999} = .3003$, depending on whether the first draw resulted in a success or a failure. But since these conditional probabilities are approximately equal to the unconditional probability ($\frac{3}{10}$) of drawing a success on the second draw, we can say that, *for all practical purposes,* the sampling satisfies the conditions of a binomial experiment. Thus, when N is large and n is small relative to N (say, n/N less than .05), we can use the binomial probability distribution to calculate the probability of observing x successes in a survey sample.

The binomial probability distributions for a sample of $n = 10$ and $p = .1$, $p = .3$, $p = .5$, $p = .7$, and $p = .9$ are shown in Figure 5.4. Note that the probability distribution is skewed to the right for small values of p, skewed to the left for large values of p, and symmetric for $p = .5$.

The formula used for calculating probabilities of the binomial probability distribution is shown in the box.

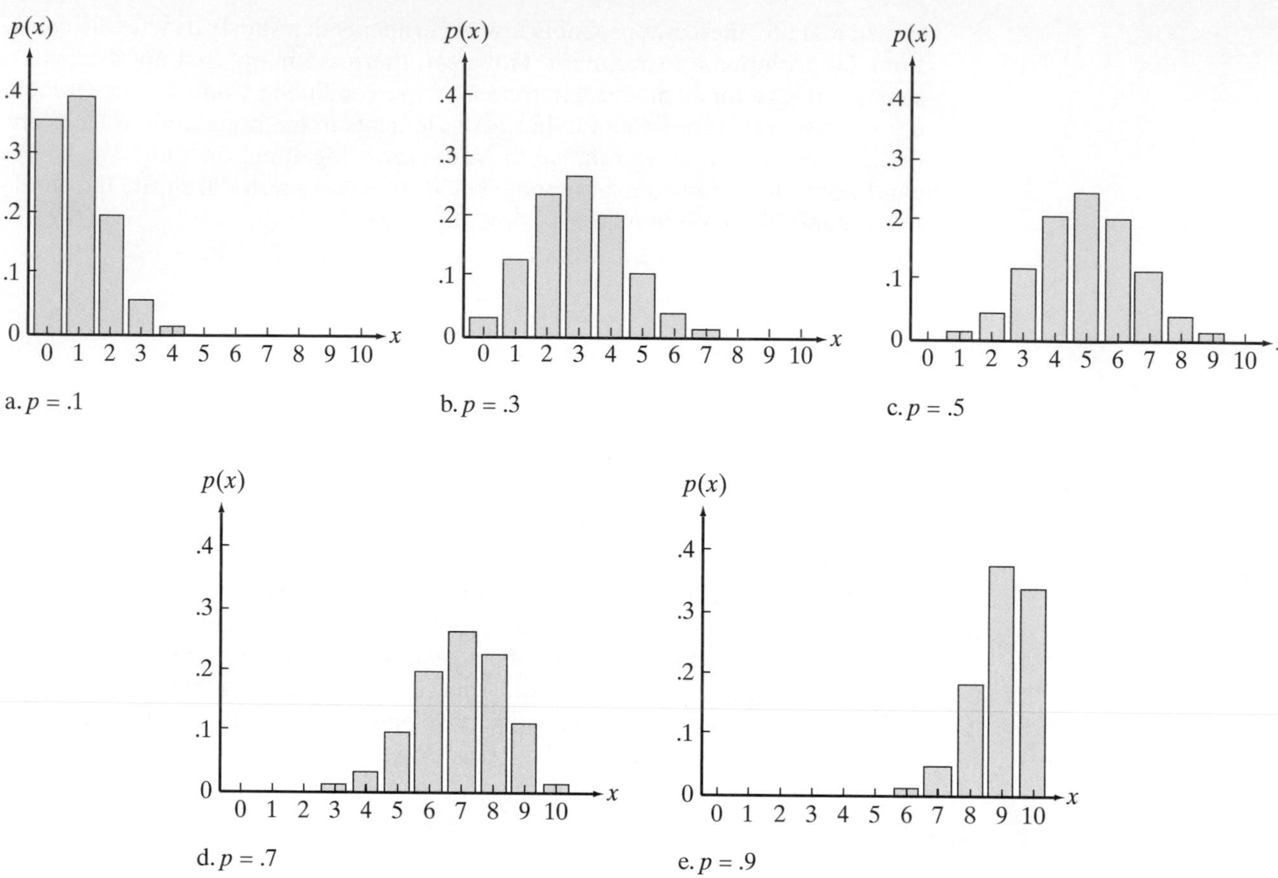

Figure 5.4 Binomial Probability Distributions for $n = 10, p = .1, .3, .5, .7, .9$

THE BINOMIAL PROBABILITY DISTRIBUTION

$$p(x) = \binom{n}{x} p^x(1-p)^{n-x} \qquad x = 0, 1, 2, \dots, n \qquad \textbf{(5.4)}$$

where

n = sample size (number of experimental units)

x = number of successes in n experimental units

p = probability of success on a single experimental unit

$$\binom{n}{x} = \frac{n!}{x!(n-x)!}$$

Assumption: The sample size n is small relative to the number N of elements in the population (say, n/N smaller than 1/20).

[*Note:* $x! = x(x-1)(x-2)(x-3)\cdots(1)$, and $0! = 1$.]

EXAMPLE 5.7 **COMPUTING BINOMIAL PROBABILITIES**

Refer to the physician survey, Example 5.2 (p. 240). Recall that a random sample of 5 physicians is selected from a population of 10,000 physicians. The number x of physicians in the sample who prefer aspirin brand A is recorded. It is known that 20% of the physicians in the population prefer brand A.

a. Verify that the physician survey is a binomial experiment.

b. Use the binomial probability distribution to calculate the probabilities of $x = 0$, $x = 1, \ldots, x = 5$.

Solution

a. The sample survey satisfies the five requirements for a binomial experiment given in the box:

1. A sample of $n = 5$ physicians (experimental units) is selected from a population.
2. Each physician surveyed possesses one of two mutually exclusive characteristics: favors aspirin brand A (a *success*) or does not favor aspirin brand A (a *failure*).
3. The proportion of physicians in the population who prefer aspirin brand A is .2; thus, the probability of a success is $p = .2$. Since the sample size $n = 5$ is small relative to the population size $N = 10{,}000$, this probability remains approximately the same for all trials.
4. The response for any one physician is independent of the response for any other physician.
5. We are counting $x =$ the number of physicians in the sample who favor aspirin brand A.

b. To calculate the binomial probabilities, we will substitute the values of $n = 5$ and $p = .2$ and each value of x into the formula for $p(x)$ given in equation 5.4:

$$p(x) = \binom{n}{x} p^x(1 - p)^{n-x} = \frac{n!}{x!(n - x)!} p^x(1 - p)^{n-x}$$

Thus, remembering that $0! = 1$, we have

$$P(x = 0) = p(0) = \binom{5}{0}(.2)^0(.8)^5$$

$$= \frac{5!}{0!5!}(.2)^0(.8)^5 = (1)(1)(.32768)$$

$$= .32768$$

Similarly,

$$P(x = 1) = p(1) = \binom{5}{1}(.2)^1(.8)^4 = \frac{5!}{1!4!}(.2)^1(.8)^4 = .40960$$

$$P(x = 2) = p(2) = \binom{5}{2}(.2)^2(.8)^3 = \frac{5!}{2!3!}(.2)^2(.8)^3 = .20480$$

$$P(x = 3) = p(3) = \binom{5}{3}(.2)^3(.8)^2 = \frac{5!}{3!2!}(.2)^3(.8)^2 = .05120$$

$$P(x = 4) = p(4) = \binom{5}{4}(.2)^4(.8)^1 = \frac{5!}{4!1!}(.2)^4(.8)^1 = .00640$$

$$P(x = 5) = p(5) = \binom{5}{5}(.2)^5(.8)^0 = \frac{5!}{5!0!}(.2)^5(.8)^0 = .00032$$

(Note that these probabilities are approximately equal to the relative frequencies reported in Table 5.1.) Binomial probabilities can be obtained using PHStat and Excel (or by using MINITAB). Figure 5.5 is an Excel printout that lists the probabilities for the physician survey.

E Figure 5.5 Binomial Probabilities for Example 5.7 Generated Using PHStat and Excel

	A	B	C	D	E	F	G
1	**Binomial Probabilities**						
2							
3	**Data**						
4	**Sample size**	5					
5	**Probability of success**	0.2					
6							
7	**Statistics**						
8	**Mean**	1					
9	**Variance**	0.8					
10	**Standard deviation**	0.894427					
11							
12	**Binomial Probabilities Table**						
13		X	P(X)	P(<=X)	P(<X)	P(>X)	P(>=X)
14		0	0.32768	0.32768	0	0.67232	1
15		1	0.4096	0.73728	0.32768	0.26272	0.67232
16		2	0.2048	0.94208	0.73728	0.05792	0.26272
17		3	0.0512	0.99328	0.94208	0.00672	0.05792
18		4	0.0064	0.99968	0.99328	0.00032	0.00672
19		5	0.00032	1	0.99968	0	0.00032

Self-Test 5.3

Consider a binomial distribution with $n = 10$.

a. Find $p(3)$ when $p = .7$.
b. Find $p(5)$ when $p = .1$.
c. Find $p(8)$ when $p = .3$.

EXAMPLE 5.8

SUMMING BINOMIAL PROBABILITIES

Refer to Example 5.7. Find the probability that three or more physicians in the sample prefer brand A.

Solution

The values that a random variable x can have are always mutually exclusive events—i.e., you could not observe $x = 2$ and, at the same time, observe $x = 3$. Therefore, the event "x is 3 or more" (the event that $x = 3$ or $x = 4$ or $x = 5$) can be found using Probability Rule #1 (p. 200). Thus,

$$P(x = 3 \text{ or } x = 4 \text{ or } x = 5) = P(x = 3) + P(x = 4) + P(x = 5)$$
$$= p(3) + p(4) + p(5)$$

Substituting the probabilities found in Example 5.7, we obtain

$$P(x = 3 \text{ or } x = 4 \text{ or } x = 5) = .05120 + .00640 + .00032 = .05792$$

This sum can also be obtained from one of the many tables that give partial sums of the values of $p(x)$, called **cumulative binomial probabilities**, for a wide range of values of n and p. Table B.2 of Appendix B includes cumulative binomial tables for $n = 5, 6, 7, 8, 9, 10, 15, 20$, and 25. A reproduction of the cumulative binomial probability table for $n = 5$ is shown in Table 5.2. In general, the entry in the table in row k under the appropriate column for p gives the sum

$$P(x \leq k) = p(0) + p(1) + p(2) + p(3) + \cdots + p(k)$$

TABLE 5.2 Reproduction of Table B.2, Appendix B: Cumulative Binomial Probabilities for $n = 5$

k \ p	.01	.05	.10	.20	.30	.40	.50	.60	.70	.80	.90	.95	.99
0	.9510	.7738	.5905	.3277	.1681	.0778	.0313	.0102	.0024	.0003	.0000	.0000	.0000
1	.9990	.9774	.9185	.7373	.5782	.3370	.1875	.0870	.0308	.0067	.0005	.0000	.0000
2	1.0000	.9988	.9914	.9421	.8369	.6826	.5000	.3174	.1631	.0579	.0086	.0012	.0000
3	1.0000	1.0000	.9999	.9933	.9692	.9130	.8125	.6630	.4718	.2627	.0815	.0226	.0010
4	1.0000	1.0000	1.0000	.9997	.9976	.9898	.9687	.9222	.8319	.6723	.4095	.2262	.0490

Looking in Table 5.2 in the row $k = 2$ and the column $p = .2$, we find $P(x \leq 2) = p(0) + p(1) + p(2) = .9421$. Note that the event $\{x \leq 2\}$ is the complement of the event $\{x \geq 3\}$. Consequently, we can find $P(x \geq 3)$ using the Rule of Complements:

$$P(x \geq 3) = 1 - P(x \leq 2) = 1 - .9421 = .0579$$

Except for rounding, this value agrees with our hand-calculated answer above.

In some situations, we will want to compare an observed value of x obtained from a binomial experiment with some theory or claim associated with the sampled population. In particular, we want to see if the observed value of x represents a **rare event,** assuming that the claim is true.

EXAMPLE 5.9

STATISTICAL INFERENCE

A manufacturer of O-ring seals used to prevent hot gases from leaking through the joints of rocket boosters claims that 95% of all seals that it produces will function properly. Suppose you randomly select 10 of these O-ring seals, test them, and find that only six prevent gas from leaking. Is this sample outcome highly improbable (that is, does it represent a *rare event*) if in fact the manufacturer's claim is true?

Solution

If p is in fact equal to .95 (or some larger value), then observing a small number x of O-ring seals that function properly would represent a rare event. Since we observed $x = 6$, we want to know the probability of observing a value of $x = 6$ or some other value of x even more contradictory to the manufacturer's claim, i.e., we want to find the probability that $x = 0$ or $x = 1$ or $x = 2 \ldots$ or $x = 6$. Using the additive rule for values of $p(x)$, we have

$$\begin{aligned} &P(x = 0 \text{ or } x = 1 \text{ or } x = 2 \ldots \text{ or } x = 6) \\ &\quad = p(0) + p(1) + p(2) + p(3) + p(4) + p(5) + p(6) \\ &\quad = P(x \leq 6) \end{aligned}$$

We can find this sum using a table of cumulative binomial probabilities or with computer software. Figure 5.6 is a printout of cumulative probabilities for $p = .95$ and $n = 10$ generated by PHStat and Excel. To find $P(x \leq 6)$ on Figure 5.6, locate the probability in the row corresponding to $x = 6$ and the column labeled **P(<= X)**. This value, highlighted on Figure 5.6, is .001028.

This small probability tells us that observing as few as six good O-ring seals out of 10 is indeed a rare event, if in fact the manufacturer's claim is true. Such a sample result suggests either that the manufacturer's claim is false or that the 10 O-ring seals tested do not represent a random sample from the manufacturer's total production.

E Figure 5.6 Binomial Probabilities for $p = .95$, $n = 10$ Obtained from PHStat and Excel

	A	B	C	D	E	F	G
1	Binomial Probabilities						
2							
3	Data						
4	Sample size	10					
5	Probability of success	0.95					
6							
7	Statistics						
8	Mean	9.5					
9	Variance	0.475					
10	Standard deviation	0.689202					
11							
12	Binomial Probabilities Table						
13		X	P(X)	P(<=X)	P(<X)	P(>X)	P(>=X)
14		0	9.77E-14	9.77E-14	0	1	1
15		1	1.86E-11	1.87E-11	9.77E-14	1	1
16		2	1.59E-09	1.61E-09	1.87E-11	1	1
17		3	8.04E-08	8.2E-08	1.61E-09	1	1
18		4	2.67E-06	2.75E-06	8.2E-08	0.999997	1
19		5	6.09E-05	6.37E-05	2.75E-06	0.999936	0.999997
20		6	0.000965	0.001028	6.37E-05	0.998972	0.999936
21		7	0.010475	0.011504	0.001028	0.988496	0.998972
22		8	0.074635	0.086138	0.011504	0.913862	0.988496
23		9	0.315125	0.401263	0.086138	0.598737	0.913862
24		10	0.598737	1	0.401263	0	0.598737

Self-Test 5.4

A recent survey of American hotels found that 90% offer complimentary shampoo in their guest rooms. In a random sample of 20 hotels, less than 14 offered complimentary shampoo. Does this sample result contradict the survey percentage of 90%? Explain.

Using the formulas in Definitions 5.5, 5.6, and 5.7, you can show (proof omitted) that the mean, variance, and standard deviation for a binomial probability distribution are as listed in the following box.

MEAN, VARIANCE, AND STANDARD DEVIATION FOR A BINOMIAL PROBABILITY DISTRIBUTION

$$\mu = np \tag{5.5}$$

$$\sigma^2 = np(1 - p) \tag{5.6}$$

$$\sigma = \sqrt{np(1 - p)} \tag{5.7}$$

where

n = sample size

p = probability of success on a single trial or the proportion of experimental units in a large population that are successes

For *large* samples, the mean and standard deviation of the binomial probability distribution (in conjunction with the Empirical Rule or Tchebysheff's Theorem) can be used to describe and make inferences about the sampled population.

EXAMPLE 5.10

COMPUTING μ AND σ

People who work at high-stress jobs frequently develop stress-related physical problems (for example, high blood pressure, ulcers, and irritability). In a recent study it was found that 40% of a large number of business executives surveyed

have symptoms of stress-induced problems. Consider a group of 1,500 randomly selected business executives and assume that the probability of an executive with stress-induced problems is $p = .40$. Let x be the number of business executives in the sample of 1,500 who develop stress-related problems.

a. What are the mean and standard deviation of x?

b. Based on the Empirical Rule, within what limits would you expect x to fall?

c. Suppose you observe $x = 800$ executives with symptoms of stress-induced problems. What can you infer about the value of p?

Solution

a. Since $n = 1{,}500$ executives in the sample is small relative to the large number of business executives in the population, the number x of executives with stress-induced problems is a binomial random variable with $p = .4$. We use the formulas for μ and σ given in equations 5.5 and 5.7 to obtain the mean and standard deviation of this binomial distribution:

$$\mu = np = (1{,}500)(.4) = 600$$

$$\sigma = \sqrt{np(1-p)} = \sqrt{(1{,}500)(.4)(.6)} = \sqrt{360} = 18.97$$

b. According to the Empirical Rule, most (about 95%) of the x values will fall within two standard deviations of the mean. Thus, we would expect the number of sampled executives with stress-induced problems to fall in the interval from

$$\mu - 2\sigma = 600 - 2(18.97) = 562.06$$

to

$$\mu + 2\sigma = 600 + 2(18.97) = 637.94$$

c. Using the rare event approach of Chapter 4, we want to determine whether observing $x = 800$ executives with stress-induced problems is unusual, assuming $p = .40$. You can see that this value of x is highly improbable when $p = .40$, since it lies a long way outside the interval $\mu \pm 2\sigma$. The z score for this value of x is

$$z = \frac{x - \mu}{\sigma} = \frac{800 - 600}{18.97} = 10.54$$

Clearly, if $p = .40$, the probability that the number x of executives with stress-induced problems in the sample is 800 or larger is almost 0. Therefore, we are inclined to believe (based on the sample value of $x = 800$) that the proportion of executives with stress-induced problems is much higher than $p = .40$.

Self-Test 5.5

Suppose that 80% of all American hotels provide complimentary shampoo in their guest rooms. In a random sample of 250 hotels, find the mean and standard deviation of x, the number of hotels that actually offer complimentary shampoo.

Statistics in the Real World Revisited

Using the Binomial Model to Solve the Cocaine Sting Case

Refer to the reverse cocaine sting case described on p. 238. During a drug bust, police seized 496 foil packets of a white, powdery substance that appeared to be cocaine. The police laboratory randomly selected four packets and found that all four tested positive for cocaine. This finding led to the conviction of the drug traffickers. Following the conviction, the police used two of the remaining 492 foil packets (i.e., those not tested) in a reverse sting operation. The two randomly selected packets were sold by undercover officers to a buyer who disposed of the evidence before being arrested. Is there a reasonable doubt that the two packets contained cocaine?

To solve the dilemma, we will assume that of the 496 original packets confiscated, 331 contained genuine cocaine and 165 contained an inert (legal) powder. (A statistician, hired as an expert witness on the case, showed that the chance of the defendant being found not guilty is maximized when 331 packets contain cocaine and 165 do not.) First, we'll find the probability that four packets randomly selected from the original 496 will test positive for cocaine. Then, we'll find the probability that the two packets sold in the reverse sting did not contain cocaine. Finally, we'll find the probability that both events occur, i.e., that the four original packets test positive for cocaine but that the two packets used in the reverse sting operation do not. In each of these probability calculations, we will apply the binomial probability distribution to approximate the probabilities.

Let x be the number of packets that contain cocaine in a sample of n selected from the 496 packets. Here, we are defining a "success" as a packet that contains cocaine. If n is small, say $n = 2$ or $n = 4$, then x has an approximate binomial distribution with probability of success $p = 331/496 \approx .67$.

The probability that the first four packets selected contain cocaine, i.e., $P(x = 4)$, is obtained using the binomial formula with $n = 4$ and $p = .67$:

$$P(x = 4) = p(4) = \binom{4}{4} p^4(1 - p)^0 = \frac{4!(.67)^4(.33)^0}{4!0!} = (.67)^4 = .201$$

Thus, there is about a 20% chance that all four of the randomly selected packets contain cocaine.

Given that four of the original packets tested positive for cocaine, the probability that the two packets randomly selected and sold in the reverse sting *do not* contain cocaine is approximated using a binomial distribution with $n = 2$ and $p = .67$. Since a "success" is a packet with cocaine, we find $P(x = 0)$:

$$P(x = 0) = p(0) = \binom{2}{0} p^0(1 - p)^2 = \frac{2!(.67)^0(.33)^2}{0!2!} = (.33)^2 = .109$$

Finally, to compute the probability of interest, we employ the Multiplicative Law of Probability. Let A be the event that the four original packets test positive. Let B be the event that the two reverse sting packets test negative. We want to find the probability of both events occurring, i.e., $P(A \text{ and } B) = P(A \cap B) = P(B|A)P(A)$. Note that this probability is the product of the two previously calculated probabilities:

$$P(A \text{ and } B) = (.109)(.201) = .022$$

Consequently, there is only a .022 probability (i.e., about 2 chances in a hundred) that the four original packets will test positive for cocaine and the two packets used in the reverse sting will test negative for cocaine. A reasonable jury would likely believe that an event with such a small probability is unlikely to occur and conclude that the two "lost" packets contained cocaine. In other words, most of us would infer that the defendant in the reverse cocaine sting was guilty of drug trafficking.

[*Epilogue:* Several of the defendant's lawyers believed that the .022 probability was too high for jurors to conclude guilt "beyond a reasonable doubt." The argument was made moot, however, when, to the surprise of the defense, the prosecution revealed that the remaining 490 packets had not been used in any other reverse sting operations and offered to test a sample of them. On the advice of the statistician, the defense requested that an additional 20 packets be tested. All 20 tested positive for cocaine! As a consequence of this new evidence, the defendant was convicted by the jury.]

PROBLEMS FOR SECTION 5.3

Using the Tools

5.13 Compute each of the following:

a. $4!$ **b.** $\frac{4!}{1!3!}$ **c.** $\binom{5}{3}$

d. $(.4)^3$ **e.** $\binom{5}{3}(.4)^3(.6)^2$

5.14 A coin is tossed ten times and the number of heads is recorded. To a reasonable degree of approximation, is this a binomial experiment? Check to determine whether each of the five conditions required for a binomial experiment is satisfied.

5.15 Four coins are selected without replacement from a group of five pennies and five dimes. Let x equal the number of pennies in the sample of four coins. To a reasonable degree of approximation, is this a binomial experiment? Determine whether each of the five conditions required for a binomial experiment is satisfied.

5.16 Consider a binomial experiment with $n = 4$ trials and probability of success $p = .5$.

a. Find the probabilities for $x = 0, 1, 2, 3,$ and 4.

b. Find the probability that x is less than 2.

c. Find the probability that x is less than or equal to 2.

d. Verify that (except for rounding) the sum of the probabilities for $x = 0, 1, 2, 3,$ and 4 equals 1.

e. Assume $p = .3$ and repeat parts **a–d**.

f. Assume $p = .8$ and repeat parts **a–d**.

5.17 Let x be a binomial random variable with parameters $n = 4$ and $p = .2$.

a. Find the probability that x is less than 2.

b. Find the probability that x is equal to 2 or more.

c. How are the events in parts **a** and **b** related?

d. What relationship must the probabilities of the events in parts **a** and **b** satisfy?

5.18 For a binomial probability distribution with $n = 5$, find:

a. $\sum_{x=0}^{2} p(x) = p(0) + p(1) + p(2)$ when $p = .3$

b. $\sum_{x=0}^{4} p(x)$ when $p = .3$

c. $\sum_{x=4}^{5} p(x)$ when $p = .3$

d. $p(2)$ when $p = .3$

5.19 For a binomial probability distribution with $n = 10$ and $p = .4$, find:

a. the probability that x is less than or equal to 8

b. the probability that x is less than 8

c. the probability that x is greater than 8

5.20 Calculate μ and σ for a binomial probability distribution with

a. $n = 15$ and $p = .1$

b. $n = 15$ and $p = .5$

5.21 Find the mean and standard deviation for each of the following binomial probability distributions:

a. $n = 300$ and $p = .99$

b. $n = 200$ and $p = .8$

c. $n = 100$ and $p = .5$

d. $n = 100$ and $p = .2$

e. $n = 1,000$ and $p = .01$

Applying the Concepts

5.22 According to a nationwide survey, 60% of parents with young children condone spanking their child as a regular form of punishment (*Tampa Tribune,* Oct. 5, 2000). Consider a random sample of three people, each of whom is a parent with young children. Assume that x, the number in the sample that condone spanking, is a binomial random variable.

a. What is the probability that none of the three parents condone spanking as a regular form of punishment for their children?

b. What is the probability that at least one condones spanking as a regular form of punishment?

c. Give the mean and standard deviation of x. Interpret the results.

5.23 Periodically, the Federal Trade Commission (FTC) monitors the pricing accuracy of electronic checkout scanners at stores to ensure consumers are charged the correct price at checkout. A study of over 100,000 items found that the probability that an item is priced incorrectly by the scanner is .03. (*Price Check II: A Follow-Up Report on the Accuracy of Checkout Scanner Prices,* Dec. 16, 1998). Suppose the FTC randomly selects five items at a retail store and checks the accuracy of the scanner price of each. Let x represent the number of the five items that is priced incorrectly.

a. Is x (approximately) a binomial random variable?

b. Use the information in the FTC study to estimate p for the binomial experiment.

c. What is the probability that exactly one of the five items is priced incorrectly by the scanner?

d. What is the probability that at least one of the five items is priced incorrectly by the scanner?

5.24 The warning signs of a stroke are a sudden severe headache, unexplained dizziness, unsteadiness, loss of vision, difficulty speaking, and numbness on one side of the body. According to the *Journal of the American Medical Association* (Apr. 22, 1998), only 57% of Americans who are susceptible to stroke recognize even one warning sign. Consider a random sample of five Americans susceptible to stroke and similar in age, gender, and race. Let x be the number who know at least one warning sign of a stroke.

a. To a reasonable degree of approximation, is this a binomial experiment?

b. What is a "success" in the context of this experiment?

c. What is the value of p?

d. Find the probability that $x = 4$.

e. Find the probability that $x \geq 4$.

5.25 A recent poll by the Gallup Organization, sponsored by Philadelphia-based CIGNA Integrated Care, found that about 40% of employees have missed work due to a musculoskeletal (back) injury of some kind (*National Underwriter,* Apr. 5, 1999). Let x be the number of sampled workers who have missed work due to a back injury.

a. Explain why x is approximately a binomial random variable.

b. Use the Gallup poll data to estimate p for the binomial random variable.

c. A random sample of 10 workers is to be drawn from a particular manufacturing plant. Use the p from part **b** to find the mean and standard deviation of x, the number of workers that missed work due to back injuries.

d. For the sample in part **c**, find the probability that more than one worker missed work due to a back injury.

5.26 *Consumer Reports* (Feb. 1992) found widespread contamination and mislabeling of seafood in supermarkets in New York City and Chicago. One alarming statistic: 40% of the swordfish pieces available for sale had a level of mercury above the Food and Drug Administration (FDA) minimum amount. For a random sample of three swordfish pieces, find the probability that:

a. All three swordfish pieces have mercury levels above the FDA minimum.

b. Exactly one swordfish piece has a mercury level above the FDA minimum.

c. At most one swordfish piece has a mercury level above the FDA minimum.

5.27 The *Journal of Business & Economic Statistics* (July 2000) published a study of an employment discrimination case. At issue was whether an employer's hiring practices were gender neutral. (The employer was accused of hiring a disproportionate number of males.) At the time the charge was filed, females formed 3.4% of the qualified labor force. In a single year, the employer hired 429 people.

a. If the employer's hiring practices were, in fact, gender neutral, how many of the 429 hires would you expect to be female?

b. Refer to part **a**. Find the standard deviation of the number of females hired in the year.

c. Of the 429 people hired, 2 were female. Use the results of parts **a** and **b** to make an inference about whether the employer's hiring practices were gender neutral.

5.28 Do you have the basic skills necessary to succeed in college? Most likely, your college professor does not think so. According to a survey conducted by the Carnegie Foundation for the Advancement of Teaching, over 70% of college professors consider their students "seriously unprepared in basic skills" that should have been learned in high school. In a sample of 25 college faculty at your institution, let x represent the number who agree that students lack basic skills. Assume that $p = .70$ at your institution.

a. What is the probability that x is less than 20?

b. What is the probability that x is less than nine?

c. What is the probability that x is greater than nine?

d. If, in fact, x is less than nine, make an inference about the value of p at your institution.

5.29 Two University of Illinois researchers applied the binomial distribution to the prescription fitting of hearing aids (*American Journal of Audiology,* Mar. 1995). The experiment was designed as follows. Each subject (a hearing aid wearer) listened alternately to speech samples recorded on tape at two different frequency-gain settings—call these two frequencies F1 and F2. The listener chose the frequency-gain setting that best amplified the recorded sounds. As an illustration, the researchers considered a trial with $n = 3$ subjects and counted x, the number of subjects who chose frequency-gain setting F1. The researchers also assumed that there was no real preference for the two settings, i.e., $p = .5$.

a. Find the probability distribution for x. Give the results in table form.

b. Find $P(x \leq 2)$.

c. Suppose two of the three subjects chose frequency-gain setting F1 over F2. Would the result lead you to believe that p exceeds .5, i.e., that hearing aid wearers, in general, prefer F1 over F2? Explain.

5.30 Refer to the *IEEE Computer Applications in Power* study of an outdoor unmanned watching system designed to detect trespassers, Problem 4.80 (p. 233). In snowy weather conditions, the system detected seven out of 10 intruders; thus, the researchers estimated the system's probability of intruder detection in snowy conditions at .70.

a. Assuming the probability of intruder detection in snowy conditions is only .50, find the probability that the unmanned system detects at least seven of the 10 intruders.

b. Based on the result, part **a**, comment on the reliability of the researcher's estimate of the system's detection probability in snowy conditions.

5.31 *Chance* (Summer 1993) reported on a study of the success rate of National Football League (NFL) field goal kickers. Data collected over three NFL seasons revealed that NFL kickers, in aggregate, made 75% of all field goals attempted. However, when the kicks were segmented by distance (yards), the rate of successful kicks varied widely. For example, 95% of all short field goals (under 30 yards) were made, while only 58% of all long field goals (over 40 yards) were successful. Consider x, the number of field goals made in a random sample of 20 kicks attempted last year in an NFL game. Is x (approximately) a binomial random variable? Explain.

5.32 Zoologists have discovered that animals spend a great deal of time resting, although this rest time can have functional importance (e.g., predators lying in wait for their prey). Discounting time spent in deep sleep, a University of Vermont researcher estimated the percentage of time various species spend at rest (*National Wildlife,* Aug.–Sept. 1993). For example, the probability that a female fence lizard will be resting at any given point in time is .95.

a. In a random sample of 20 female fence lizards, what is the probability that at least 15 will be resting at any given point in time?

b. In a random sample of 20 female fence lizards, what is the probability that fewer than 10 will be resting at any given point in time?

c. In a random sample of 200 female fence lizards, would you expect to observe fewer than 190 at rest at any given point in time? Explain.

5.33 Patients on dialysis are susceptible to heart problems. Doctors can often attribute these problems to a condition called mitral annular calcification (MAC). In one study, medical researchers have found that 30% of patients on dialysis suffer moderate or severe cases of MAC (*Collegium Antropologicum,* Vol. 21, 1997). Consider a random sample of 80 dialysis patients.

a. On average, how many of the 80 sampled patients would you expect to suffer a moderate or severe case of MAC? Assume the percentage reported in the study applies to all dialysis patients.

b. Suppose 20 of the 80 dialysis patients are diagnosed with a moderate or severe case of MAC. Does this result tend to support or refute the percentage reported in the study? Explain.

5.34 The degree to which democratic and non-democratic countries attempt to control the news media was examined in the *Journal of Peace Research* (Nov. 1997). Between 1948 and 1996, 80% of all democratic regimes allowed a free press. In contrast, over the same time period, 10% of all non-democratic regimes allowed a free press.

a. In a random sample of 50 democratic regimes, how many would you expect to allow a free press? Give a range that is highly likely to include the number of democratic regimes with a free press.

b. In a random sample of 50 non-democratic regimes, how many would you expect to allow a free press? Give a range that is highly likely to include the number of non-democratic regimes with a free press.

5.35 An exchange rate specifies the price of one nation's currency in terms of another nation's currency. Periodically, the United States enters the foreign exchange market in an attempt to stabilize exchange rates. Are these attempts—called *interventions*—successful? Economic advisor O. W. Humpage found that, if we assume that U.S. interventions are not effective, the probability that an intervention will result in a stable exchange rate by chance is .65 (*Economic Commentary,* Federal Reserve Bank of Cleveland, Mar. 1, 1996). During a recent 3-year period, the U.S. intervened 64 times in an attempt to stabilize the exchange rate of the Japanese yen. Let x be the number of successful interventions of the 64. Humpage showed that x is approximately a binomial random variable with $p = .65$, if, in fact, U.S. interventions are not successful (i.e., no better than chance) in stabilizing exchange rates.

a. Find $P(x = 30)$.

b. Find $P(x \geq 50)$.

c. Find μ and interpret its value.

d. Find σ and interpret its value.

e. Humpage determined that 50 of the interventions were successful in stabilizing the exchange rate of the Japanese yen. Use this result and your answer to part **b** to make an inference about the true value of p. Do you believe the U.S. intervention policy was effective or that the stabilized exchange rates can be attributed to chance?

5.4 The Poisson Probability Distribution

Many real-world discrete random variables can be modeled using the **Poisson probability distribution.** Named for the French mathematician Siméon D. Poisson (1781–1840), the Poisson distribution provides a good model for the probability distribution of the number of "rare events" that occur randomly in time, distance, or space. Some random variables that may possess (approximately) a Poisson probability distribution are the following:

1. The number of car accidents that occur per month (or week, day, etc.) at a busy intersection
2. The number of arrivals per minute (or hour, etc.) at a medical clinic or other servicing facility
3. The number of fleas found on a certain breed of dog
4. The parts per million of a toxic substance found in polluted water
5. The number of diseased Dutch elm trees per acre of a certain woodland

The characteristics of a Poisson random variable are listed in the box, followed by a formula for its probability distribution.

CHARACTERISTICS OF A POISSON RANDOM VARIABLE

1. The experiment consists of counting the number x of times a particular event occurs during a given unit of time, or a given area or volume (or weight, or distance, or any other unit of measurement).
2. The probability that an event occurs in a given unit of time, area, or volume is the same for all the units.
3. The number of events that occur in one unit of time, area, or volume is independent of the number that occur in other units.

THE POISSON PROBABILITY DISTRIBUTION

$$p(x) = \frac{\mu^x e^{-\mu}}{x!}, x = 0, 1, 2, \ldots \qquad \textbf{(5.8)}$$

where

x = number of rare events per unit of time, distance, or space
μ = mean value of x
$e = 2.71828\ldots$

Graphs of $p(x)$ for $\mu = 1, 2, 3$, and 4 are illustrated in Figure 5.7 on p. 260. The figure shows how the shape of the Poisson distribution changes as its mean μ changes.

Poisson probabilities can also be found using software such as Excel or MINITAB.

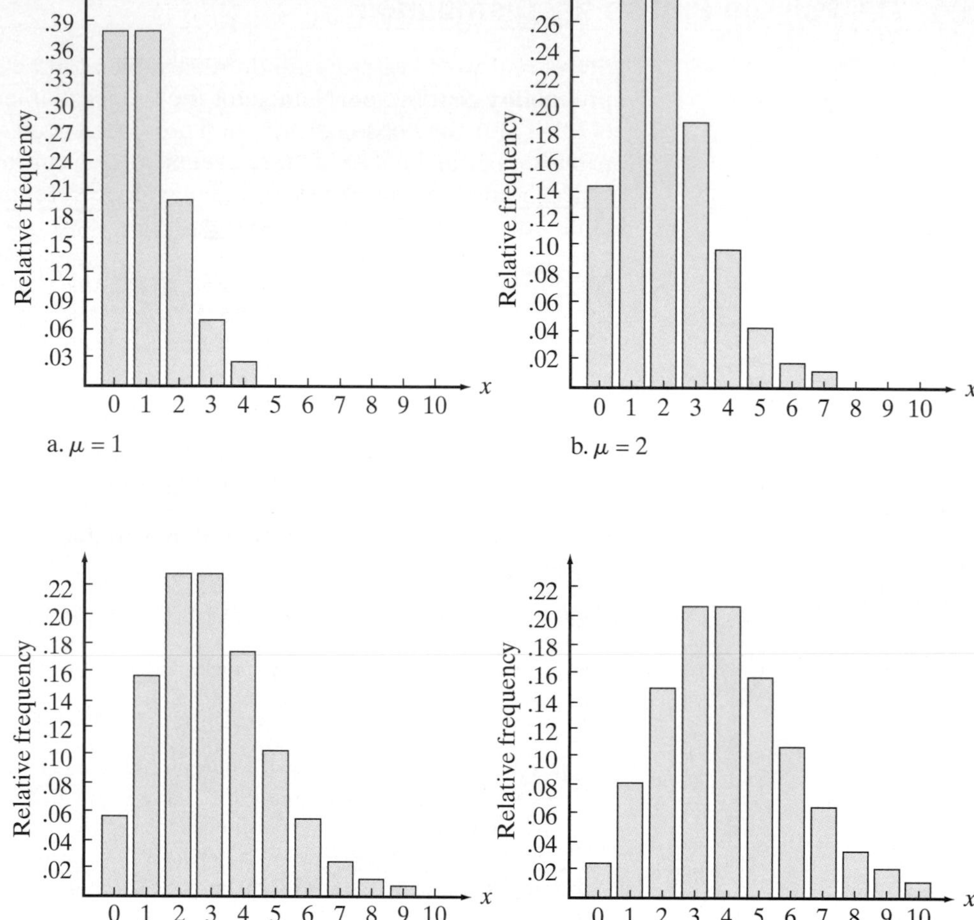

Figure 5.7 Bar Graphs of the Poisson Distribution for $\mu = 1, 2, 3$, and 4

EXAMPLE 5.11

CUMULATIVE POISSON PROBABILITY

The quality control inspector in a diamond-cutting operation has found that the mean number of defects per diamond (defects discernible to the eye of a jeweler) is $\mu = 1.5$. What is the probability that a randomly selected diamond contains more than one such defect?

Solution

The probability distribution of x, the number of discernible defects per diamond, can be approximated by a Poisson probability distribution with $\mu = 1.5$. The probability that a diamond contains more than one defect is

$$\begin{aligned} P(x > 1) &= p(2) + p(3) + p(4) + \cdots \\ &= 1 - [p(0) + p(1)] \\ &= 1 - P(x \leq 1) \end{aligned}$$

To find $p(0)$ and $p(1)$, we can substitute into the formula for $p(x)$. Or we can find the probability using Excel or MINITAB. Figure 5.8 is a PHStat and Excel printout of Poisson probabilities for $\mu = 1.5$. The value of $P(x \leq 1)$, highlighted on the

E **Figure 5.8** Poisson Probabilities Generated Using PHStat and Excel

	A	B	C	D	E	F	G
1	**Poisson Probabilities for Defects per Diamond**						
2							
3			**Data**				
4	**Average/Expected number of successes:**				**1.5**		
5							
6	**Poisson Probabilities Table**						
7		X	P(X)	P(<=X)	P(<X)	P(>X)	P(>=X)
8		0	0.223130	0.223130	0.000000	0.776870	1.000000
9		1	0.334695	0.557825	0.223130	0.442175	0.776870
10		2	0.251021	0.808847	0.557825	0.191153	0.442175
11		3	0.125511	0.934358	0.808847	0.065642	0.191153
12		4	0.047067	0.981424	0.934358	0.018576	0.065642
13		5	0.014120	0.995544	0.981424	0.004456	0.018576
14		6	0.003530	0.999074	0.995544	0.000926	0.004456
15		7	0.000756	0.999830	0.999074	0.000170	0.000926
16		8	0.000142	0.999972	0.999830	0.000028	0.000170
17		9	0.000024	0.999996	0.999972	0.000004	0.000028
18		10	0.000004	0.999999	0.999996	0.000001	0.000004
19		11	0.000000	1.000000	0.999999	0.000000	0.000001
20		12	0.000000	1.000000	1.000000	0.000000	0.000000
21		13	0.000000	1.000000	1.000000	0.000000	0.000000
22		14	0.000000	1.000000	1.000000	0.000000	0.000000
23		15	0.000000	1.000000	1.000000	0.000000	0.000000

printout, is .557825. Therefore, the probability that the number x of defects per diamond is more than one is

$$P(x > 1) = 1 - P(x \leq 1)$$
$$= 1 - .557825 = .442175$$

The mean μ of the Poisson probability distribution appears in the formula for $p(x)$. It is given, with the variance and standard deviation of x, in the accompanying box.

MEAN, VARIANCE, AND STANDARD DEVIATION OF A POISSON PROBABILITY DISTRIBUTION

$$\text{Mean} = \mu \qquad \sigma^2 = \mu \qquad \sigma = \sqrt{\mu}$$

EXAMPLE 5.12

COMPUTING POISSON μ AND σ

Suppose the mean number of employee accidents per month at a particular manufacturing plant is 3.4. Find the mean, variance, and standard deviation of x, the number of accidents in a randomly selected month. Is it likely that x will be as large as 12?

Solution

We are given the fact that $\mu = 3.4$. Therefore,

$$\sigma^2 = 3.4 \qquad \text{and} \qquad \sigma = \sqrt{3.4} = 1.84$$

Then the z score corresponding to $x = 12$ is

$$z = \frac{x - \mu}{\sigma} = \frac{12 - 3.4}{1.84} = 4.67$$

Since $x = 12$ lies 4.67 standard deviations away from its mean, we know (from the Empirical Rule) that this is a highly improbable event.

Self-Test 5.6

A can company reports that the number of breakdowns per 8-hour shift on its machine-operated assembly line follows a Poisson distribution, with a mean of 1.5.

a. What is the probability of exactly two breakdowns during the midnight shift?

b. What is the probability of fewer than two breakdowns during the afternoon shift?

c. What is the variance of the number of breakdowns per 8-hour shift?

PROBLEMS FOR SECTION 5.4

Using the Tools

5.36 Assume that x has a Poisson probability distribution. Find each of the following probabilities:

a. $P(x = 5)$ when $\mu = 3.0$

b. $P(x \geq 4)$ when $\mu = 7.5$

c. $P(x \leq 2)$ when $\mu = 5.5$

d. $P(x = 8)$ when $\mu = 4.0$

5.37 Suppose x has a Poisson probability distribution with $\mu = 2.5$.

a. Find and graph $p(x)$ for $x = 0, 1, 2, \ldots, 10$.

b. Find μ and σ.

c. Locate $\mu \pm 2\sigma$ on the graph of part **a**. What is the probability that x falls within $\mu \pm 2\sigma$?

5.38 Suppose x has a Poisson probability distribution with $\mu = 6.0$.

a. Find and graph $p(x)$ for $x = 0, 1, 2, \ldots, 12$.

b. Find μ and σ.

c. Locate $\mu \pm 2\sigma$ on the graph of part **a**. What is the probability that x falls within $\mu \pm 2\sigma$?

5.39 Assume x has a Poisson probability distribution. Find $P(x \leq 3)$ when

a. $\mu = 1$ **b.** $\mu = 2$ **c.** $\mu = 3.5$

d. $\mu = 5$ **e.** $\mu = 7.5$

Applying the Concepts

5.40 Ecologists often use the number of reported sightings of a rare species of animal to estimate the remaining population size. For example, suppose the number x of reported sightings per week of blue whales is recorded. Assume that x has (approximately) a Poisson probability distribution. Furthermore, assume that the average number of weekly sightings is 2.6.

a. Find the mean and standard deviation of x, the number of blue whale sightings per week.

b. Find the probability that fewer than two sightings are made during a given week.

c. Find the probability that more than five sightings are made during a given week.

d. Find the probability that exactly five sightings are made during a given week.

5.41 Cognitive scientists at the University of Massachusetts designed an experiment to explore how certain word patterns affect a reader's eye movement (*Memory & Cognition,* Sept. 1997). One of the variables measured was x, the number of times a reader's eye fixated on a single word before moving past that word. For this

experiment, x was found to have a mean of 1. Suppose one of the readers in the experiment is randomly selected and assume that x has a Poisson distribution.

a. Find $P(x = 0)$.

b. Find $P(x > 1)$.

c. Find $P(x \leq 2)$.

5.42 Economists at the University of New Mexico studied the effectiveness of the Maritime Safety Program of the U.S. Coast Guard by examining the records of 951 deep-draft U.S. flag vessels. They modeled the number of casualties experienced by a vessel over a 3-year period as a Poisson random variable, x. Casualties were defined as the number of deaths or missing persons in a 3-year interval. Using the data on the 951 vessels, they estimated μ to be .03 (*Management Science,* Jan. 1999).

a. Find the variance of x.

b. Discuss the conditions that would make the researchers' Poisson assumption plausible.

c. What is the probability that a deep-draft U.S. flag vessel will have no casualties in a 3-year time period?

5.43 Researchers at the Massachusetts Institute of Technology (MIT) studied the spectroscopic properties of main-belt asteroids having diameters smaller than 10 kilometers (*Science,* Apr. 3, 1993). Research revealed that, on average, 2.5 independent spectral image exposures are observed per asteroid.

a. Assuming a Poisson distribution, find the probability that exactly one independent spectral image exposure is observed during a main-belt asteroid sighting.

b. Assuming a Poisson distribution, find the probability that at most two independent spectral image exposures are observed during a main-belt asteroid sighting.

c. Would you expect to observe seven or more independent spectral image exposures during a main-belt asteroid sighting? Explain.

5.44 The number of cars that arrive at an intersection during a specified period of time often possesses (approximately) a Poisson probability distribution. When the mean arrival rate μ is known, the Poisson probability distribution can be used to aid a traffic engineer in the design of a traffic control system. Suppose you estimate that the mean number of arrivals per minute at the intersection is one car per minute.

a. What is the probability that in a given minute, the number of arrivals will equal three or more?

b. Can you assure the engineer that the number of arrivals will rarely exceed three per minute?

5.45 The Environmental Protection Agency (EPA) has established national ambient air quality standards in an effort to control air pollution. Currently, The EPA limit on ozone levels in air is set at 12 parts per hundred million (pphm). A 1990 study examined the long-term trend in daily ozone levels in Houston, Texas.* One of the variables of interest is x, the number of days in a year on which the ozone level exceeds the EPA threshold of 12 pphm. The mean number of exceedences in a year is estimated to be 18. Assume that the probability distribution for x can be modeled with the Poisson distribution.

a. Compute $P(x \geq 20)$.

b. Compute $P(5 \leq x \leq 10)$.

c. Estimate the standard deviation of x. Within what range would you expect x to fall in a given year?

d. The study revealed a decreasing trend in the number of exceedences of the EPA threshold over the past several years. The observed values of x for the past six years were 24, 22, 20, 15, 14, and 16. Explain why this trend casts doubt on the validity of the Poisson distribution as a model for x. [*Hint:* Consider characteristic #3 of the Poisson random variable.]

*Shively, Thomas S. "An analysis of the trend in ozone using nonhomogeneous Poisson processes." Paper presented at the annual meeting of the American Statistical Association, Anaheim, Calif., Aug. 1990.

5.5 The Hypergeometric Probability Distribution (Optional)

We noted in Section 5.3 that one of the applications of the binomial probability distribution is its use as the probability distribution for the number x of favorable (or unfavorable) responses in a public opinion or market survey. Ideally, it is applicable when sampling is *with replacement*—that is, each item drawn from the population is observed and returned to the population before the next item is drawn. Practically speaking, sampling in surveys is rarely conducted with replacement. Nevertheless, the binomial probability distribution is still appropriate if the number N of elements in the population is large and the sample size n is small relative to N.

When **sampling is without replacement,** and the number of elements N in the population is small (or when the sample size n is large relative to N), the number of "successes" in a random sample of n items has a **hypergeometric probability distribution.** The defining characteristics and probability for a hypergeometric random variable are stated in the boxes.

CHARACTERISTICS THAT DEFINE A HYPERGEOMETRIC RANDOM VARIABLE

1. The experiment consists of randomly drawing n elements without replacement from a set of N elements, S of which are "successes" and $N - S$ of which are "failures."
2. The hypergeometric random variable x is the number of S's in the draw of n elements.

THE HYPERGEOMETRIC PROBABILITY DISTRIBUTION

$$p(x) = \frac{\binom{S}{x}\binom{N-S}{n-x}}{\binom{N}{n}} \qquad \textbf{(5.9)}$$

where

N = number of elements in the population

S = number of "successes" in the population

n = sample size

x = number of "successes" in the sample

$$\binom{S}{x} = \frac{S!}{x!(S-x)!}$$

$$\binom{N-S}{n-x} = \frac{(N-S)!}{(n-x)!(N-S-n+x)!}$$

$$\binom{N}{n} = \frac{N!}{n!(N-n)!}$$

Assumptions:

1. The sample of n elements is randomly selected from the N elements of the population.
2. The value of x is restricted so that $x \leq n$ and $x \leq S$.

EXAMPLE 5.13

COMPUTING HYPERGEOMETRIC PROBABILITIES

You are given a list of 10 new "dot-com" Internet stocks to invest in. Unknown to you, seven of the 10 will go bankrupt within 2 years. Suppose you randomly select three of the 10 stocks to invest in.

a. What is the probability that all three of the dot-com stocks go bankrupt?

b. What is the probability that at least one of the three stocks goes bankrupt?

Solution

For this problem, define a "success" as a stock that goes bankrupt. Then, the number of elements (stocks) in the population of interest is $N = 10$, the number of successes is $S = 7$, and the sample size (i.e., the number of stocks selected) is $n = 3$.

a. Using equation 5.9, the probability of observing exactly $x = 3$ successes in the sample of $n = 3$ is:

$$p(3) = \frac{\binom{S}{3}\binom{N-S}{n-3}}{\binom{N}{n}} = \frac{\binom{7}{3}\binom{3}{0}}{\binom{10}{3}} = \frac{\left(\frac{7!}{3!4!}\right)\left(\frac{3!}{0!3!}\right)}{\left(\frac{10!}{3!7!}\right)}$$

$$= \frac{\left(\frac{7 \cdot 6 \cdot 5}{3 \cdot 2}\right)(1)}{\left(\frac{10 \cdot 9 \cdot 8}{3 \cdot 2}\right)} = \frac{35}{120} = .292$$

Hypergeometric probabilities can also be obtained using PHStat and Excel. The probabilities for $x = 0, 1, 2, 3$ are shown in the Excel printout, Figure 5.9.

E Figure 5.9 Hypergeometric Probabilities for Example 5.13 Obtained from PHStat and Excel

	A	B	C
1	Hypergeometric Probabilities		
2			
3	Data		
4	Sample size	3	
5	No. of successes in population	7	
6	Population size	10	
7			
8	Hypergeometric Probabilities Table		
9		X	P(X)
10		0	0.008333
11		1	0.175
12		2	0.525
13		3	0.291667

b. The probability of investing in at least one dot-com stock that will go bankrupt can be found by using the Rule of Complements (p. 202).

$$P(x \geq 1) = 1 - P(x = 0)$$

Checking the Excel printout, Figure 5.9, we find $P(x = 0) = .008333$. Therefore,

$$P(x \geq 1) = 1 - P(x = 0) = 1 - .008333 = .991667$$

Consequently, the chance of at least one of the three dot-com stocks going bankrupt is extremely high.

Self-Test 5.7

Suppose you are purchasing small lots of hard disk drives for laptop computers. Assume that each lot contains seven drives. You decide to sample three disk drives per lot and to reject the lot if you observe one or more defectives in the sample.

a. If the lot contains one defective disk drive, what is the probability that you will accept the lot?

b. What is the probability that you will accept the lot if it contains three defective disk drives?

The mean, variance, and standard deviation for a hypergeometric probability distribution are shown in the box.

MEAN, VARIANCE, AND STANDARD DEVIATION FOR A HYPERGEOMETRIC PROBABILITY DISTRIBUTION

$$\mu = \frac{nS}{N} \quad \textbf{(5.10)}$$

$$\sigma^2 = n\left(\frac{S}{N}\right)\left(\frac{N-S}{N}\right)\left(\frac{N-n}{N-1}\right) \quad \textbf{(5.11)}$$

$$\sigma = \sqrt{n\left(\frac{S}{N}\right)\left(\frac{N-S}{N}\right)\left(\frac{N-n}{N-1}\right)} \quad \textbf{(5.12)}$$

EXAMPLE 5.14

FINDING μ AND σ

Refer to Example 5.13.

a. Find μ and σ for the number x of dot-com stocks that will go bankrupt.

b. Interpret μ.

Solution

a. From Example 5.13, we have $N = 10$, $S = 7$, and $n = 3$. Therefore,

$$\mu = \frac{nS}{N} = \frac{(3)(7)}{10} = 2.1$$

$$\sigma = \sqrt{n\left(\frac{S}{N}\right)\left(\frac{N-S}{N}\right)\left(\frac{N-n}{N-1}\right)} = \sqrt{3\left(\frac{7}{10}\right)\left(\frac{3}{10}\right)\left(\frac{7}{9}\right)}$$

$$= .7$$

b. The value $\mu = 2.1$ implies that, on average, we expect to observe 2.1 stocks of the three sampled that go bankrupt. To interpret $\sigma = .7$, we form the interval

$$\mu \pm 2\sigma = 2.1 \pm 2(.7) = (.7, 3.5)$$

From the Empirical Rule, we know that about 95% of the values of x, the number of dot-com stocks that go bankrupt selected from the three, will fall between .7 and 3.5 stocks. Since x must be an integer, this interval includes the values $x = 1$, $x = 2$, and $x = 3$. Consequently, in samples of three dot-com stocks, the number that will go bankrupt will be at least one about 95% of the time.

Statistics in the Real World Revisited

Using the Hypergeometric Model to Solve the Cocaine Sting Case

The reverse cocaine sting case described on p. 238 and solved in Section 5.3 can also be solved by applying the hypergeometric distribution. In fact, the probabilities obtained using the hypergeometric distribution are *exact* probabilities, as compared to the approximate probabilities obtained using the binomial distribution.

Our objective, you will recall, is to find the probability that four packets (randomly selected from 496 packets confiscated in a drug bust) will contain cocaine and two packets (randomly selected from the remaining 492) will not contain cocaine. We assumed that of the 496 original packets, 331 contained genuine cocaine and 165 contained an inert (legal) powder. Since we are sampling *without replacement* from the 496 packets, the probability of a "success" (i.e., the probability of a packet containing cocaine) does not remain *exactly* the same from trial to trial. For example, for the first randomly selected packet, the probability of a "success" is $331/496 = .66734$. If we find that the first three packets selected contain cocaine, the probability of "success" for the fourth packet selected is now $328/493 = .66531$. You can see that these probabilities are not exactly the same. Hence, the binomial distribution will only approximate the distribution of x, the number of packets that contain cocaine in a sample of size n.

To find the probability that four packets randomly selected from the original 496 will test positive for cocaine using the hypergeometric distribution, we first identify the parameters of the distribution:

$N = 496$ is the total number of packets in the population

$S = 331$ is the number of "successes" (cocaine packets) in the population

$n = 4$ is the sample size

$x = 4$ is the number of "successes" (cocaine packets) in the sample

Substituting into equation 5.9 (p. 264), we obtain

$$P(x = 4) = p(4) = \frac{\binom{331}{4}\binom{165}{0}}{\binom{496}{4}} = \frac{\left(\frac{331!}{4!327!}\right)\left(\frac{165!}{0!165!}\right)}{\left(\frac{496!}{4!492!}\right)} = .197$$

To find the probability that two packets randomly selected from the remaining 492 will test negative for cocaine *assuming that the first four packets tested positive,* we identify the parameters of the relevant hypergeometric distribution:

$N = 492$ is the total number of packets in the population

$S = 327$ is the number of "successes" (cocaine packets) in the population

$n = 2$ is the sample size

$x = 0$ is the number of "successes" (cocaine packets) in the sample

Again, we substitute into equation 5.9 (p. 264) to obtain

$$P(x = 0) = p(0) = \frac{\binom{327}{0}\binom{165}{2}}{\binom{492}{2}} = \frac{\left(\frac{327!}{0!327!}\right)\left(\frac{165!}{2!163!}\right)}{\left(\frac{492!}{2!490!}\right)} = .112$$

Using the Multiplicative Law of Probability, the probability that four packets in the first sample test positive for cocaine and the two packets in the second sample test negative is the product of the two probabilities above:

$$P(x = 4 \text{ and } x = 0) = P(x = 4)\ P(x = 0) = (.112)(.197) = .0221$$

Note that this exact probability is almost identical to the approximate probability computed using the binomial distribution in Section 5.3.

PROBLEMS FOR SECTION 5.5

Using the Tools

5.46 Suppose x has a hypergeometric probability distribution with $N = 6$, $n = 4$, and $S = 2$.

a. Compute $p(x)$ for $x = 0, 1, 2$.

b. Graph the probability distribution $p(x)$.

c. Compute μ and σ.

5.47 Suppose x has a hypergeometric probability distribution with $N = 12$, $n = 7$, and $S = 5$. Compute each of the following:

a. $P(x = 3)$ **b.** $P(x \leq 2)$ **c.** $P(x = 5)$

d. $P(x > 3)$ **e.** μ **f.** σ

5.48 Suppose x has a hypergeometric probability distribution with $N = 8$, $n = 5$, and $S = 3$. Compute each of the following:

a. $P(x = 1)$ **b.** $P(x \leq 1)$ **c.** $P(x \geq 2)$

d. $P(x = 4)$ **e.** μ **f.** σ

5.49 Suppose x is a hypergeometric random variable. Compute $p(x)$ for each of the following cases:

a. $N = 5, n = 3, S = 4, x = 1$ **b.** $N = 10, n = 5, S = 3, x = 3$

c. $N = 3, n = 2, S = 2, x = 2$ **d.** $N = 4, n = 2, S = 2, x = 0$

Applying the Concepts

5.50 The dean of a liberal arts school wishes to form an executive committee of five from among the 40 tenured faculty members at the school. The selection is to be random, and at the school there are eight tenured faculty members in sociology.

a. What is the probability that the committee will contain

(1) none of them?

(2) at least one of them?

(3) not more than one of them?

b. What would be your answers to **a** if the committee consisted of seven members?

5.51 A state lottery is conducted in which six winning numbers are selected from a total of 54 numbers. What is the probability that if six numbers are randomly selected

a. All six numbers will be winning numbers?

b. Five numbers will be winning numbers?

c. Four numbers will be winning numbers?

d. Three numbers will be winning numbers?

e. None of the numbers will be winning numbers?

f. What would be your answers to parts **a–e** if the six winning numbers were selected from a total of 40 numbers?

5.52 Based on data provided by the U.S. Department of Health and Human Services, *U.S. News & World Report* (Sept. 28, 1992) estimates that one out of every five kidney transplants fails within a year. Suppose that exactly three of the next 15 kidney transplants will fail within a year. Consider a random sample of three of these 15 patients.

a. Find the probability that all three sampled transplants fail within a year.

b. Find the probability that at least one of the three sampled transplants fails within a year.

5.53 "Hotspots" are species-rich geographical areas. A *Nature* (Sept. 1993) study estimated the probability of a bird species in Great Britain inhabiting a butterfly hotspot at .70. Consider a random sample of four British bird species selected from a total of 10 tagged species. Assume that seven of the 10 tagged species inhabit a butterfly hotspot.

a. What is the probability that exactly two of the bird species sampled inhabit a butterfly hotspot?

b. What is the probability that at least one of the bird species sampled inhabits a butterfly hotspot?

5.54 According to the Centers for Disease Control (CDC), a recent increase in outbreaks of salmonella poisoning can be attributed to the consumption of raw or undercooked eggs. One such outbreak occurred at a cookout in Jacksonville, Florida. Among the people who attended the cookout, 12 cases of salmonella poisoning were identified. Eleven of the 12 ill persons had eaten homemade ice cream served at the cookout. The ice cream, made with raw eggs, tested positive for salmonella contamination (*Morbidity and Mortality Weekly Report,* Sept. 16, 1994). Suppose the CDC randomly selects four of the 12 persons with salmonella poisoning and interviews each.

a. What is the probability that all four of the people ate homemade ice cream at the cookout?

b. What is the probability that at least one person ate homemade ice cream at the cookout?

5.55 Refer to the *Journal of Business & Economic Statistics* (July 2000) study of employment discrimination, Problem 5.27 (p. 257). In a 1980s case in which a charge of gender discrimination was filed against the U.S. Postal Service, there were 302 employees (229 men and 73 women) who applied for promotion. A total of 72 of these employees were awarded promotion. Let x be the number of the promoted employees who are female.

a. Find μ and interpret this value.

b. Find σ.

c. Of the 72 employees promoted, 5 were female. Use the results, parts **a** and **b**, to make an inference about whether or not females at the U.S. Postal Service were promoted fairly.

KEY TERMS *Starred (*) terms are from the optional section of this chapter.*

Binomial experiment 246
Binomial probability distribution 246
Binomial random variable 246
Continuous random variable 239
Cumulative binomial probabilities 250
Discrete random variable 239
*Hypergeometric probability distribution 264
Poisson probability distribution 259
Probability distribution for the discrete random variable x 240
Random variable 238
Rare event 251
Sampling with replacement 246
*Sampling without replacement 264

KEY FORMULAS *Starred (*) formulas are from the optional section of this chapter.*

PROBABILITY DISTRIBUTION		MEAN (μ)		VARIANCE (σ^2)	
General discrete random variable $p(x)$		$\sum_{\text{all } x} xp(x)$	**(5.1)**, 241	$\sum_{\text{all } x} (x-\mu)^2 p(x)$	**(5.2)**, 241
Binomial random variable $\binom{n}{x} p^x(1-p)^{n-x}$	**(5.4)**, 248	np	**(5.5)**, 252	$np(1-p)$	**(5.6)**, 252
Poisson random variable $\frac{\mu^x e^{-\mu}}{x!}$	**(5.8)**, 259	μ		μ	
*Hypergeometric random variable $\frac{\binom{S}{x}\binom{N-S}{n-x}}{\binom{N}{n}}$	**(5.9)**, 264	$\frac{nS}{N}$	**(5.10)**, 266	$n\left(\frac{S}{N}\right)\left(\frac{N-S}{N}\right)\left(\frac{N-n}{N-1}\right)$	**(5.11)**, 266

KEY SYMBOLS

Starred (*) symbols are from the optional section of this chapter.

SYMBOL	DEFINITION
$p(x)$	Probability distribution of the discrete random variable x
μ	Mean of a discrete random variable x
σ	Standard deviation of a discrete random variable x
p	Probability of success in a binomial trial
e	Euler's constant (approximately 2.718) used with a Poisson probability distribution
S	*Number of successes in a hypergeometric probability distribution
N	*Total number of elements in a hypergeometric probability distribution

CHECKING YOUR UNDERSTANDING

1. What is the difference between a discrete random variable and a continuous random variable?
2. Give the properties of a probability distribution for a discrete random variable.
3. Define the expected value of a discrete random variable.
4. What are the properties of a binomial random variable?
5. Give the mean and variance of a binomial probability distribution.
6. What are the properties of a Poisson random variable?
7. Give the mean and variance of a Poisson probability distribution.
8. What are the properties of a hypergeometric random variable?
9. What is the difference between the binomial and hypergeometric probability distributions?

SUPPLEMENTARY PROBLEMS

Starred (*) problems refer to the optional section of this chapter.

5.56 *Developmental Psychology* (Mar. 1986) published a study on the sexual maturation of college students. The study included a probability distribution for the age x (in years) when males begin to shave regularly, shown in the accompanying figure.

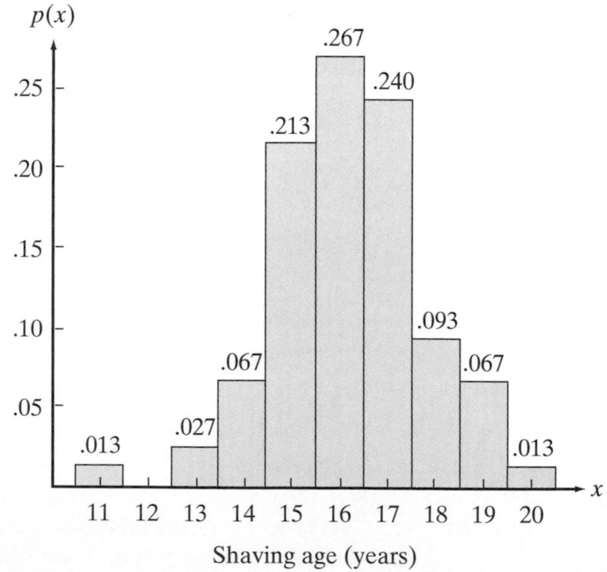

a. Is this a valid probability distribution? Explain.

b. Display the probability distribution in tabular form.

c. What is the probability that a randomly selected male college student began shaving at age 16?

d. What is the probability that a randomly selected male college student began shaving before age 15?

e. Calculate μ and σ.

f. Interpret the value of μ.

g. Compute the interval $(\mu - 2\sigma, \mu + 2\sigma)$.

h. Find the probability that x falls within the interval of part **g**. Compare your results to the Empirical Rule.

5.57 A panel of meteorological and civil engineers studying emergency evacuation plans for Florida's Gulf Coast in the event of a hurricane has estimated that it would take between 13 and 18 hours to evacuate people living in low-lying land, with the probabilities shown in the table at left.

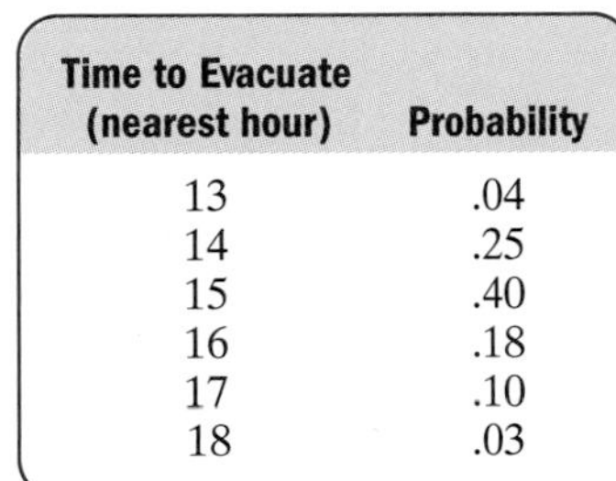

Time to Evacuate (nearest hour)	Probability
13	.04
14	.25
15	.40
16	.18
17	.10
18	.03

Problem 5.57

a. Calculate the mean and standard deviation of the probability distribution of the evacuation times.

b. Within what range would you expect the time to evacuate to fall?

c. Weather forecasters say they cannot accurately predict a hurricane landfall more than 14 hours in advance. If the Gulf Coast Civil Engineering Department waits until the 14-hour warning before beginning evacuation, what is the probability that all residents of low-lying areas are evacuated safely (i.e., before the hurricane hits the Gulf Coast)?

5.58 A random sample of 100 persons was selected to take the "Pepsi Challenge" taste test. Each person tasted one cup of Pepsi and one cup of Coke in random order, then identified the cup of soda that "tastes best." (The cups were disguised so that the taster did not know, a priori, which cup was which.) Let x represent the number of persons preferring Pepsi.

a. Is x a binomial random variable?

b. Suppose that the majority of tasters in fact prefer Pepsi. What does this imply about the value of p?

c. If $p = .5$, what is the implication?

5.59 According to a survey sponsored by the American Society for Microbiology, only 50% of people using the restrooms in New York City's Grand Central and Penn Stations washed their hands after going to the bathroom (*Tampa Tribune,* Sept. 19, 2000). Suppose you randomly observe people using the Penn Station restrooms and record x, the number of people that must be sampled until the first person who does wash up is found. The distribution of x, where x is the number of trials until the first "success" is observed, is called a *geometric distribution.* It can be shown that $P(x = k) = p(1 - p)^{k-1}$, where p is the probability of a "success."

a. Find $P(x = 1)$. b. Find $P(x = 2)$.

c. Find $P(x = 3)$. d. Find $P(x = 4)$.

5.60 Electrical engineers recognize that high neutral current in computer power systems is a potential problem. A recent survey of computer power system load currents at U.S. sites found that 10% of the sites had high neutral current (*IEEE Transactions on Industry Applications,* July/Aug. 1990). In a sample of 20 computer power systems selected from the large number of sites in the country, let x be the number with high neutral current.

a. Find and interpret the mean of x.

b. Find and interpret the standard deviation of x.

5.61 As a college student, are you frequently depressed and overwhelmed by the pressure to succeed? According to the American Council on Education (ACE), the level of stress among college freshmen is rising rapidly. In a survey of 380,000 full-time college freshmen conducted for the ACE by UCLA's Higher Education Research

Institute, over 10% reported frequently "feeling depressed" (*Tampa Tribune*, Jan. 9, 1989). Assume that 10% of all college freshmen frequently feel depressed.

a. On average, how many of the 380,000 college freshmen surveyed would you expect to report frequent feelings of depression if the true percentage in the population is 10%?

b. Find the standard deviation of the number of the 380,000 freshmen surveyed who frequently feel depressed if the true percentage is 10%.

c. What can you infer about the true percentage if 35,000 of the college freshmen surveyed reported frequent feelings of depression? Explain.

5.62 The quality-control manager of Marilyn's Cookies is inspecting a batch of chocolate-chip cookies that has just been baked. If the production process is in control, the average number of chip parts per cookie is 6.0. What is the probability that in any particular cookie being inspected

a. Fewer than five chip parts will be found?

b. Exactly five chip parts will be found?

c. Five or more chip parts will be found?

d. Four or five chip parts will be found?

e. What would be your answers to parts **a–d** if the average number of chip parts per cookie is 5.0?

***5.63** A commercial for a brand of sugarless chewing gum claims that "three out of four dentists who recommend sugarless gum to their patients recommend our brand." Suppose this claim was established following a survey of four dentists randomly selected from a group of 20 dentists who were known to recommend sugarless gum to their patients. What is the probability that at least three of the four dentists surveyed would recommend the advertised brand if in fact only half of the original group of 20 dentists favor that brand? Does your probability calculation strengthen or weaken the gum manufacturer's claim? Explain.

***5.64** Refer to Problem 4.80 (p. 233). As reported in *IEEE Computer Applications in Power* (Apr. 1990), an outdoor, unmanned computerized video monitoring system detected seven out of 10 intruders in snowy conditions. Suppose that two of the intruders had criminal intentions. What is the probability that both of these intruders were detected by the system?

5.65 A group of 20 college graduates contains 10 highly motivated persons, as determined by a company psychologist. Suppose a personnel director selects 10 persons from this group of 20 for employment. Let x be the number of highly motivated persons included in the personnel director's selection. Is this a binomial experiment? Explain.

5.66 According to a Federal Bureau of Investigation (FBI) report, 40% of all assaults on federal workers are directed at IRS employees. For a random sample of 50 federal workers who were assaulted, suppose the FBI records the number x of workers employed by the IRS. Is x a binomial random variable? If so, what are the values of n and p?

5.67 A National Basketball Association (NBA) player makes 80% of his free throws. Near the conclusion of a playoff game, the player is fouled and is rewarded two free throws. If he makes at least one free throw, his team wins the game.

a. Do you think the probability of the NBA player making his first free throw is equal to .8?

b. Is the outcome of the second free throw likely to be independent of the outcome of the first free throw? Why or why not?

c. Considering your answers to parts **a** and **b**, is it likely that the two free throws constitute a binomial experiment? Explain.

5.68 On the average, only 3.5 out of every 1,000 childbirths result in identical twins. That is, the probability that a childbirth will result in identical twins is $p = .0035$. If a random sample of 20 childbirths is selected, what is the probability of observing at least one childbirth in the sample that results in identical twins?

5.69 An international survey involving thousands of past Olympic participants indicated that an overwhelming majority—70%—believe that athletes should be barred from Olympic competition if they have been treated with anabolic steroids during the month prior to competition. In a random sample of 10 former Olympic participants, let x be the number who believe that athletes should be barred from Olympic competition if they have been treated with anabolic steroids during the month prior to competition. What is the probability that:

a. x is equal to 10? **b.** x is at least 3? **c.** x is at most 1?

COMMIT

5.70 Researchers have found a causal link between an employee's commitment to the firm and his or her tendency to leave the firm voluntarily. The *Academy of Management Journal* (Oct. 1993) published one such study involving employees of an aerospace firm. Researchers at the University of South Florida designed and administered a questionnaire to a sample of 270 of the firm's employees. The data for several of the numerous variables measured in the study are stored in the file **COMMIT.** One variable, called LEAVE, classifies employees as either "stayers" or "leavers." For this application, we are interested in the random variable x, where x is the number of the 270 employees who are "leavers." Assume that the $n = 270$ employees represent a random sample from the population of employees at all similar aerospace firms in the U.S.

a. Explain why the binomial probability distribution is the best approximation for the distribution of x.

b. A claim is made that the true proportion of "leavers" at all U.S. aerospace firms is $p = .10$. Find the expected value and variance of x. Interpret these values.

c. Use Excel or MINITAB to generate a relative frequency table for the variable LEAVE. Based on the relative frequency of "leavers" in the data, make an inference about the claim in part **b**.

d. Use Excel or MINITAB to generate a cross-tabulation table for the variables LEAVE and GENDER. Make an inference about whether p, the probability of leaving the firm, depends on gender.

REFERENCES

Feller, W. *An Introduction to Probability Theory and Its Applications,* Vol. I, 3rd ed. New York: Wiley, 1968.

Hogg, R. V., and Craig, A. T. *Introduction to Mathematical Statistics,* 5th ed. Upper Saddle River, N.J.: Prentice Hall, 1995.

Mendenhall, W., Wackerly, D., and Scheaffer, R. L. *Mathematical Statistics with Applications,* 4th ed. Boston: PWS-Kent, 1990.

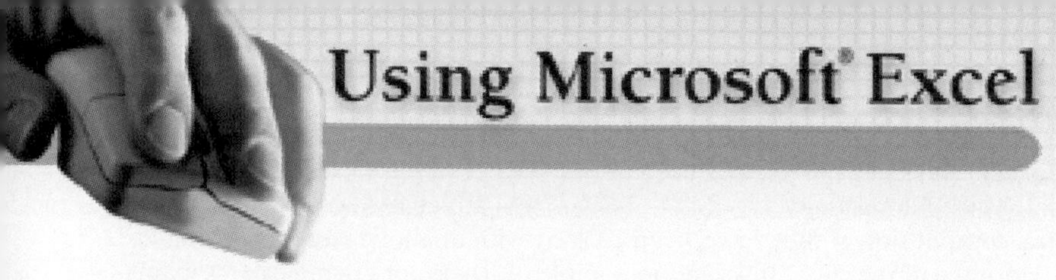

5.E.1 Calculating the Expected Value of a Discrete Random Variable

Implement a simple worksheet to calculate the expected value, variance, and standard deviation of a discrete random variable. The worksheet design below uses the data of Table 5.1 on p. 240, the probability distribution for the example concerning physicians who prefer aspirin brand A. You can adjust this design for other problems by changing the values for $P(X)$ and by either deleting table rows (for problems that have fewer than six outcomes) or adding rows and copying the formula in cell C9 down in the new table rows (for problems with more than six outcomes). To implement this design, open to an empty worksheet in any workbook and enter the values, labels, and formulas shown in the design. Note that the formulas in cells H4 and H5 use the SUM function to calculate the expected value and variance and that the formula in cell H6 uses the SQRT (square root) function to calculate the standard deviation.

Discrete Random Variable Probability Distribution Worksheet Design

	A	B	C	D	E	F	G	H
1	Discrete Random Variable Probability Distribution							
2								
3	X	P(X)	X*P(X)	[X-E(X)]^2	[X-E(X)]^2*P(X)		Statistics	
4	0	0.32807	=A4*B4	=(A4-$H$4)^2	=D4*B4		Expected value	=SUM(C:C)
5	1	0.40949	=A5*B5	=(A5-$H$4)^2	=D5*B5		Variance	=SUM(E:E)
6	2	0.20473	=A6*B6	=(A6-$H$4)^2	=D6*B6		Standard deviation	=SQRT(H5)
7	3	0.05122	=A7*B7	=(A7-$H$4)^2	=D7*B7			
8	4	0.00645	=A8*B8	=(A8-$H$4)^2	=D8*B8			
9	5	0.00004	=A9*B9	=(A9-$H$4)^2	=D9*B9			

5.E.2 Calculating Binomial Probabilities

Use the PHStat **Probability & Prob. Distributions | Binomial** procedure to calculate binomial probabilities on a new worksheet. For example, to calculate the binomial probabilities shown in Figure 5.5 on p. 250, open to an empty worksheet and

1. Select PHStat | Probability & Prob. Distributions | Binomial.
2. In the Binomial Probability Distribution dialog box (see Figure 5.E.1):
   a. Enter 5 in the Sample Size edit box.
   b. Enter 0.2 in the Probability of Success edit box.
   c. Enter 0 (zero) in the Outcomes From edit box and enter 5 in the Outcomes To edit box.
   d. Enter a title in the Title edit box.
   e. Select the Cumulative Probabilities check box and leave the Histogram check box unselected (unchecked).
   f. Click the OK button.

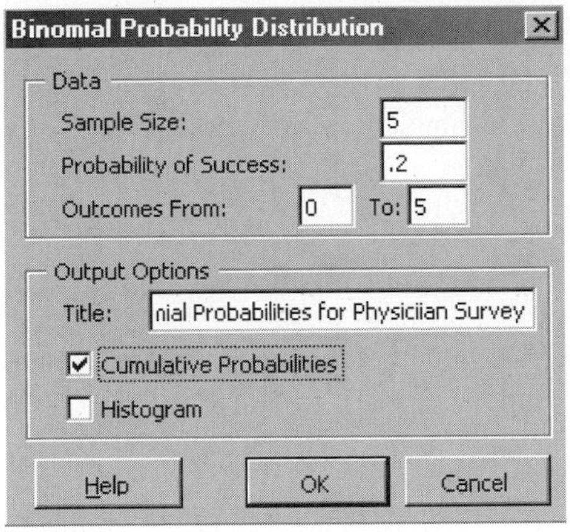

E Figure 5.E.1 PHStat Binomial Probability Distribution Dialog Box

Binomial Probability Distribution
Data
Sample Size: 5
Probability of Success: .2
Outcomes From: 0 To: 5
Output Options
Title: nial Probabilities for Physician Survey
Cumulative Probabilities
Histogram
Help OK Cancel

PHStat uses the BINOMDIST worksheet function of Microsoft Excel. The format of this function is

$$\text{BINOMDIST}(x, n, p, \textit{cumulative})$$

where

x = the number of successes
n = the sample size
p = the probability of success
cumulative = a True or False value that determines whether the function computes the probability of x or fewer successes (True) or computes the probability of exactly x successes (False)

For example, BINOMDIST(2, 5, .20, False) would calculate the probability of obtaining exactly two physicians who prefer aspirin brand A from a sample of five physicians. Changing the False value to True would calculate the probability of two or fewer physicians who prefer aspirin brand A.

5.E.3 Calculating Poisson Probabilities

Use the PHStat **Probability & Prob. Distributions | Poisson** procedure to calculate Poisson probabilities on a new worksheet. For example, to calculate the Poisson probabilities shown in Figure 5.8 on p. 261, open to an empty worksheet and

1. Select PHStat | Probability & Prob. Distributions | Poisson.
2. In the Poisson Probability Distribution dialog box (see Figure 5.E.2):

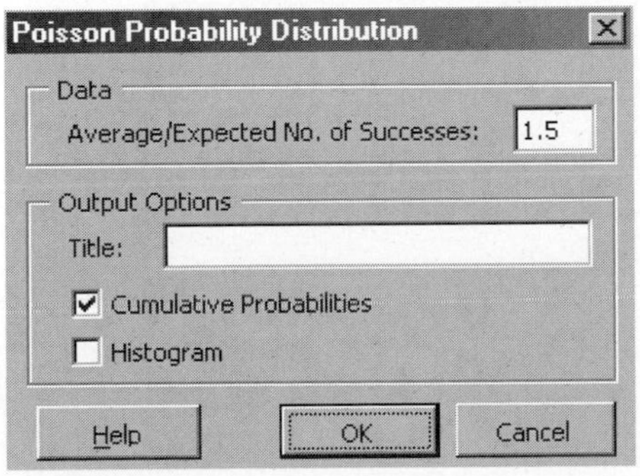

E Figure 5.E.2 PHStat Poisson Probability Distribution Dialog Box

Poisson Probability Distribution
Data
Average/Expected No. of Successes: 1.5
Output Options
Title:
Cumulative Probabilities
Histogram
Help OK Cancel

a. Enter 1.5 in the Average/Expected No. of Successes edit box.

b. Enter a title in the Title edit box.

c. Select the Cumulative Probabilities check box and leave the Histogram check box unselected.

d. Click the OK button.

PHStat uses the POISSON worksheet function of Microsoft Excel. The format of this function is

POISSON(x, *mu*, *cumulative*)

where

x = number of successes

mu = the average or expected number of successes

cumulative = a True or False value that determines whether the function computes the probability of x or fewer successes (True) or computes the probability of exactly x successes (False)

For example, POISSON(1, 1.5, False) would calculate the probability of exactly one discernible defect per diamond when the average number of discernible defects per diamond is 1.5. Changing the False value to True would calculate the probability of 1 or fewer discernible defects per diamond.

5.E.4 Calculating Hypergeometric Probabilities

Use the PHStat **Probability & Prob. Distributions | Hypergeometric** procedure to calculate hypergeometric probabilities on a new worksheet. For example, to calculate the hypergeometric probabilities of Figure 5.9 on p. 265, open to an empty worksheet and

1. Select PHStat | Probability & Prob. Distributions | Hypergeometric.

2. In the Hypergeometric Probability Distribution dialog box (see Figure 5.E.3):

a. Enter 3 in the Sample Size edit box.

b. Enter 7 in the No. of Successes in Population edit box.

c. Enter 10 in the Population Size edit box.

d. Enter a title in the Title edit box.

e. Leave the Histogram check box unselected (unchecked).

f. Click the OK button.

E Figure 5.E.3 PHStat Hypergeometric Probability Distribution Dialog box

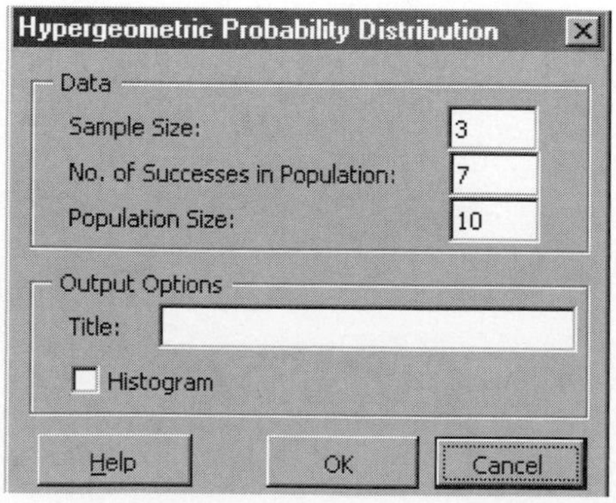

PHStat uses the HYPGEOMDIST worksheet function of Microsoft Excel. The format of this function is

$$\text{HYPGEOMDIST}(x, n, S, N)$$

where

x = number of successes in the sample
n = the sample size
S = the number of successes in the population
N = the population size

Using MINITAB®

In this chapter you studied the probability distribution, and such probability distributions as the binomial, hypergeometric, and Poisson distributions. MINITAB can be used to obtain results for each of these distributions.

5.M.1 Obtaining Binomial Probabilities

To illustrate the use of MINITAB, consider the example concerning physicians who prefer aspirin brand A discussed in Section 5.3. To obtain the binomial probabilities using MINITAB, enter the values **0**, **1**, **2**, **3**, **4**, and **5** in rows 1–6 of column C1. **Select Calc | Probability Distributions | Binomial** to compute binomial probabilities. In the Binomial Distribution dialog box (see Figure 5.M.1), select the **Probability** option button to obtain the exact probabilities of x successes for all values of x. In the Number of trials edit box enter the sample size of **5**. In the Probability of success edit box enter **.20**. Select the **Input column** option button and enter **C1** in its edit box. Click the **OK** button.

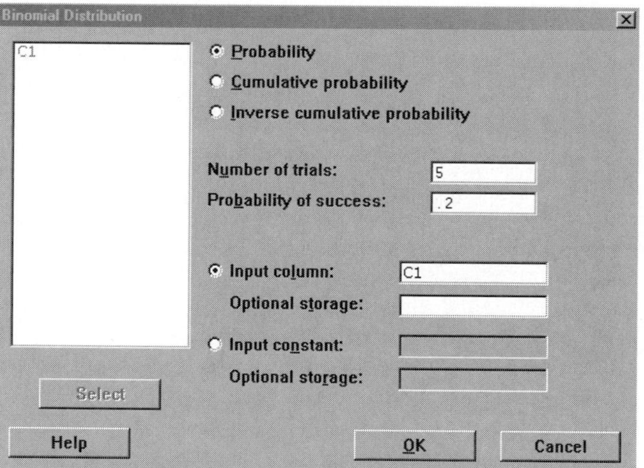

Figure 5.M.1 MINITAB Binomial Distribution Dialog Box

5.M.2 Obtaining Poisson Probabilities

To illustrate how to obtain Poisson probabilities using MINITAB, return to Example 5.11 on p. 260. To obtain Poisson probabilities using MINITAB, enter the values 0 through 12 in rows 1–13 of column C1. Select **Calc | Probability Distributions | Poisson** to compute Poisson probabilities. In the Poisson Distribution dialog box (see Figure 5.M.2), select the **Probability** option button to obtain the exact probabilities of x successes for all values of x. In the Mean edit box enter the μ value of **1.5**. Select the **Input column** option button and enter **C1** in its edit box. Click the **OK** button.

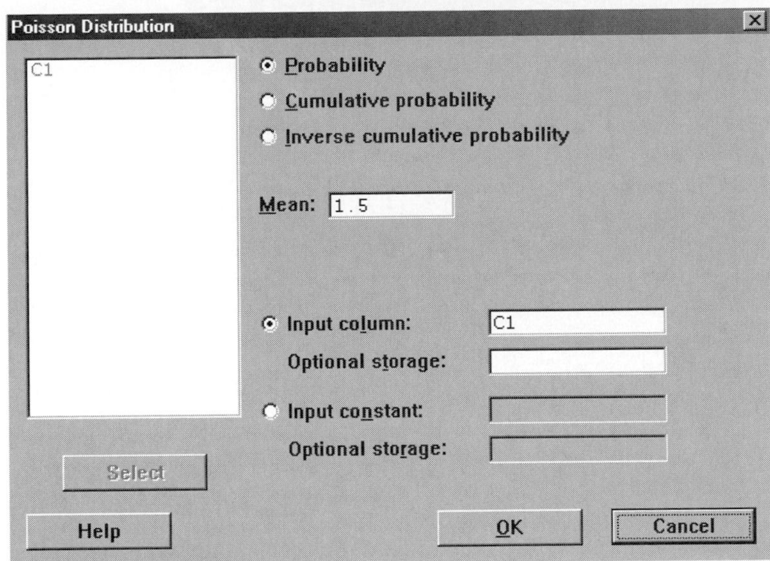

M **Figure 5.M.2** MINITAB Poisson Distribution Dialog Box

5.M.3 Obtaining Hypergeometric Probabilities

To illustrate how to obtain hypergeometric* probabilities, return to Example 5.13 on p. 265. You computed the probability that exactly three dot-com stocks in a sample of three would go bankrupt within 2 years, from a population of 10 dot-com stocks where seven go bankrupt within 2 years, to be .292. To obtain this result, enter the values **0**, **1**, **2**, and **3** in rows 1–4 of column C1. Select **Calc | Probability Distributions | Hypergeometric** to compute hypergeometric probabilities. In the Hypergeometric Distribution dialog box, select the **Probability** option button. In the Population size (N) edit box enter the population size of **10**. In the Successes in population (X) edit box enter the value of **7**. In the Sample size (n) edit box, enter the sample size of **3**. Select the **Input column** option button, and enter **C1** in its edit box. Click the **OK** button.

*Hypergeometric probabilities are not available in versions of MINITAB prior to Version 13.

CHAPTER 6

Normal Probability Distributions

OBJECTIVES

1. To show that the normal probability distribution can be used to represent many types of continuous random variables
2. To present the properties of a normal probability distribution
3. To demonstrate how to find probabilities from a normal distribution
4. To present techniques for assessing whether data are from a normal probability distribution
5. To show how the normal distribution can be used to approximate binomial probabilities
6. To develop the concept of a sampling distribution
7. To present the Central Limit Theorem and its applications to the sampling distribution of the sample mean

CONTENTS

EXCEL TUTORIAL

MINITAB TUTORIAL

Statistics in the Real World

Insurance Fraud in a Furniture Warehouse Fire

A wholesale furniture retailer stores in-stock items at a large warehouse located in Florida. Several years ago, a fire destroyed the warehouse and all the furniture in it. After determining the fire was an accident, the retailer sought to recover costs by submitting a claim to its insurance company.

As is typical in a fire insurance policy of this type, the furniture retailer must provide the insurance company with an estimate of "lost" profit for the destroyed items. Retailers calculate profit margin in percentage form using the Gross Profit Factor (GPF). By definition, the GPF for a single sold item is the ratio of the profit to the item's selling price measured as a percentage, i.e.,

$$\text{Item GPF} = (\text{Profit/Sales price}) \times 100$$

Of interest to both the retailer and the insurance company is the average GPF for all of the items in the warehouse. Since these furniture pieces were all destroyed, their eventual selling prices and profit values are obviously unknown. Consequently, the average GPF for all the warehouse items is unknown.

One way to estimate the mean GPF of the destroyed items is to use the mean GPF of similar, recently sold items. The retailer sold 3,005 furniture items in the year prior to the fire and kept paper invoices on all sales. Rather than calculate the mean GPF for all 3,005 items (the data were not computerized), the retailer sampled a total of 253 of the invoices and computed the mean GPF for these items as 50.8%. The retailer applied this average GPF to the costs of the furniture items destroyed in the fire to obtain an estimate of the "lost" profit.

According to experienced claims adjusters at the insurance company, the GPF for sale items of the type destroyed in the fire rarely exceeds 48%. Consequently, the estimate of 50.8% appeared to be unusually high. (A 1% increase in GPF for items of this type equates to, approximately, an additional $16,000 in profit.) Consequently, a dispute arose between the furniture retailer and the insurance company, and a lawsuit was filed. In one portion of the suit, the insurance company accused the retailer of fraudulently representing its sampling methodology. Rather than selecting the sample randomly, the retailer was accused of selecting an unusual number of "high profit" items from the population in order to increase the average GPF of the sample.

FIREFRAUD

To support its claim of fraud, the insurance company hired a CPA firm to independently assess the retailer's true Gross Profit Factor. Through the discovery process, the CPA firm legally obtained the paper invoices for the entire population of 3,005 items sold the year before the fire and input the information into a computer. The selling price, profit, profit margin, and month sold for these 3,005 furniture items are available in the file **FIREFRAUD.** In the following Statistics in the Real World Revisited sections, we demonstrate how to use normal probability models to analyze the data and determine the likelihood of fraud in the furniture fire case.

Statistics in the Real World Revisited

- Applying the Normal Distribution to the Fire Fraud Data (p. 293)
- Assessing Whether the Fire Fraud Data Is Normal (p. 301)
- Applying the Central Limit Theorem to the Fire Fraud Data (p. 321)

6.1 Probability Models for Continuous Random Variables

Suppose you want to predict your waiting time in a dentist's office, the sale price of a home, the annual amount of rainfall in a city, or your weekly winnings playing the lottery. If you do, you will need to know something about continuous random variables.

Recall that continuous random variables are those that can take on (at least theoretically) any of the infinitely large number of values contained in an interval. Thus, we might envision a population of patient waiting times in a dentist's office, the sale prices of houses, the annual amounts of rainfall in a city since 1900, or the

gains (or losses) of many weeks of playing the lottery. Since our ultimate goal is to make inferences about a population based on the measurements contained in a sample, we need to know the probability that the sample observations (or sample statistics) have specific values.

For example, suppose we are interested in the intelligence quotient (IQ) of a college student. Then the target population is the set of IQs for all students who attend college. What is the probability that a student's IQ will exceed 115? This probability can be obtained only if we know the relative frequency distribution of the population of IQs. For example, let Figure 6.1 represent the *hypothetical* relative frequency distribution for the population, called a **continuous probability distribution.** Then, the probability that a randomly selected college student has an IQ exceeding 115 is .3, the area shaded under the curve of Figure 6.1.

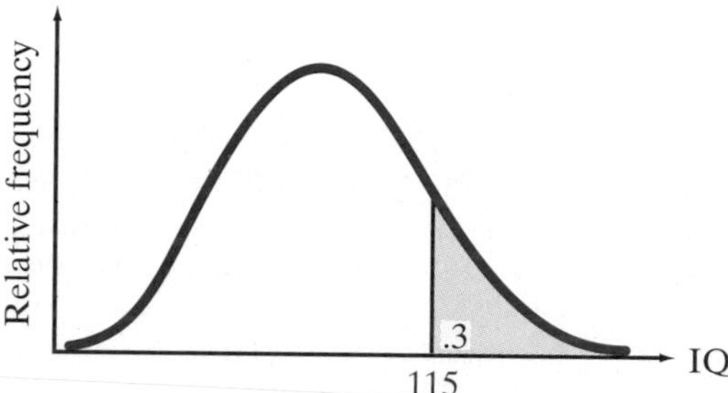

Figure 6.1 Hypothetical Relative Frequency Distribution of College Student IQs

The problem, of course, is that we do not know the IQs of all college students. Hence, we do not know the exact shape of the population relative frequency distribution sketched in Figure 6.1. Rather, we select a smooth curve (similar to the one shown in Figure 6.1) as a **probability model** for the population relative frequency distribution. To find the probability that a particular observation (say, an IQ) will fall in a particular interval, we use the model and find the area under the curve that falls in that interval (as shown in the box). Of course, in order for this approximate probability to be realistic, we need to be fairly certain that the model and the population relative frequency distribution are very similar.

AREAS AND PROBABILITIES FOR A CONTINUOUS RANDOM VARIABLE

The probability that a continuous random variable x falls between two points a and b equals the area under the relative frequency distribution curve that falls over the interval (a, b).

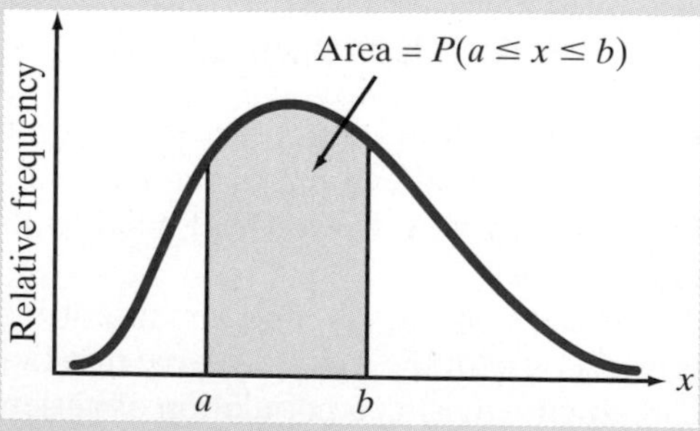

In the next section, we introduce one of the most important and useful probability models for a continuous random variable and show how it can be used to find probabilities associated with specific values of the random variable.

6.2 The Normal Probability Distribution

The normal distribution was proposed by C. F. Gauss (1777–1855) as a model for the relative frequency distribution of *errors,* such as errors of measurement. Amazingly, this curve provides an adequate model for the relative frequency of data collected from many different disciplines.

One of the most useful probability models for population relative frequency distributions is known as the **normal distribution.** A graph of the normal distribution (often called the **normal** or **bell curve**) is shown in Figure 6.2.

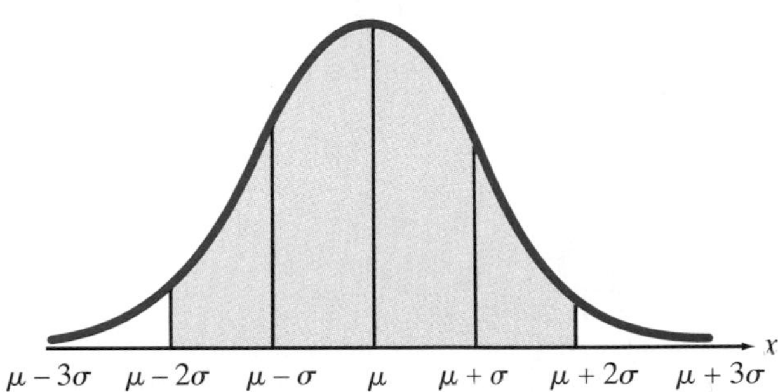

Figure 6.2 The Normal Curve

Definition 6.1

The mathematical model for the normal curve, denoted $f(x)$ and called the ***normal probability density function,*** is

$$f(x) = \frac{1}{\sigma\sqrt{2\pi}} e^{-(1/2)[(x-\mu)/\sigma]^2} \tag{6.1}$$

where

e = the mathematical constant approximated by 2.71828
π = the mathematical constant approximated by 3.14159
μ = the population mean
σ = the population standard deviation
x = any value of the normal random variable, $-\infty < x < \infty$

You can see from Figure 6.2 that the bell-shaped normal curve is symmetric about its mean μ. Furthermore, approximately 68% of the area under a normal curve lies within the interval $\mu \pm \sigma$. Approximately 95% of the area lies within the interval $\mu \pm 2\sigma$ (shaded in Figure 6.2), and almost all (99.7%) lies within the interval $\mu \pm 3\sigma$. Note that these percentages agree with the Empirical Rule of Section 3.5. (This is because the Empirical Rule is based on data that can be modeled by a normal distribution.)

Remember that areas under the normal curve are probabilities. Thus, if a population of measurements has approximately a normal distribution, then the probability that a randomly selected observation falls within the interval $\mu \pm 2\sigma$ is approximately .95.

PROPERTIES OF THE NORMAL CURVE

1. Bell-shaped
2. Symmetric about μ
3. $P(\mu - \sigma < x < \mu + \sigma) \approx .68$
4. $P(\mu - 2\sigma < x < \mu + 2\sigma) \approx .95$
5. $P(\mu - 3\sigma < x < \mu + 3\sigma) \approx .997$

Although always bell-shaped and symmetric, the exact shape of the normal curve will depend on the specific values of μ and σ. Several different normal curves are shown in Figure 6.3. You can see from Figure 6.3 that the mean μ measures the location of the distribution and the standard deviation σ measures its spread.

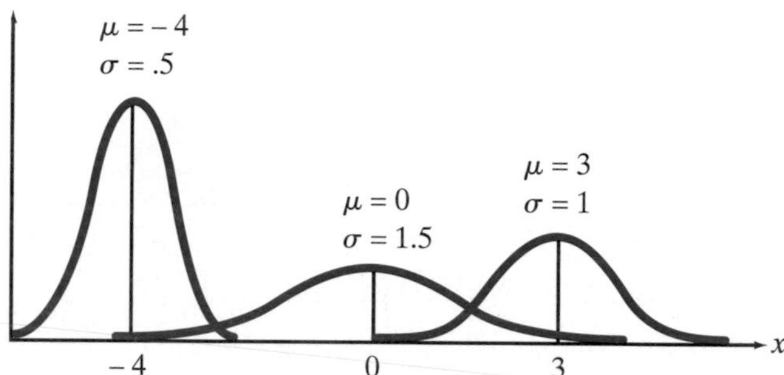

Figure 6.3 Three Normal Distributions with Different Means and Standard Deviations

The areas under the normal curve have been computed and are given in Table B.3 of Appendix B. Since the normal curve is symmetric, we need to give areas on only one side of the mean. Consequently, the entries in Table B.3 are areas between the mean and a value x to the right of the mean.

Since the values of μ and σ vary from one normal distribution to another, the easiest way to express a distance from the mean is in terms of a z score. Recall (Section 3.6) that a z score *measures the number of standard deviations that a value x lies from its mean μ.*

Thus,

$$z = \frac{x - \mu}{\sigma} \tag{6.2}$$

is the distance between x and μ, expressed in units of σ.

EXAMPLE 6.1

COMPUTING A *z* VALUE

Suppose a population relative frequency distribution has mean $\mu = 500$ and standard deviation $\sigma = 100$. Give the z score corresponding to $x = 626$.

Solution

The value $x = 626$ lies 126 units above $\mu = 500$. This distance, expressed in units of σ ($\sigma = 100$), is 1.26. We can get this answer directly by substituting x, μ, and σ into equation 6.1:

$$z = \frac{x - \mu}{\sigma} = \frac{626 - 500}{100} = \frac{126}{100} = 1.26$$

EXAMPLE 6.2

USING THE NORMAL TABLE

Find the area under a normal curve between the mean and a point $z = 1.26$ standard deviations to the right of the mean. This area represents $P(0 \leq z \leq 1.26)$.

Solution

A partial reproduction of Table B.3 is shown in Table 6.1. The entries in the complete table give the areas to the right of the mean for distances from $z = 0.00$ to $z = 3.09$.

TABLE 6.1 Reproduction of Part of Table B.3 of Appendix B

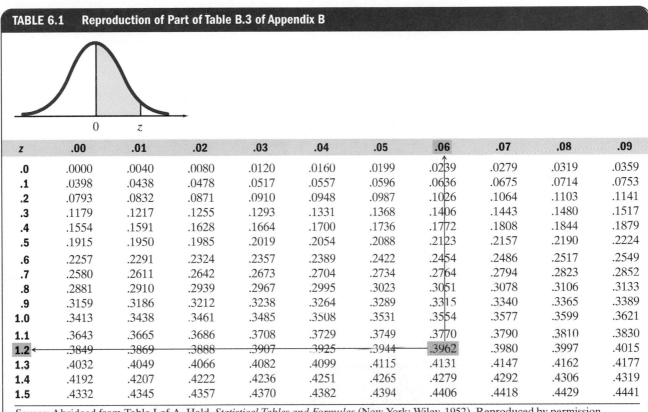

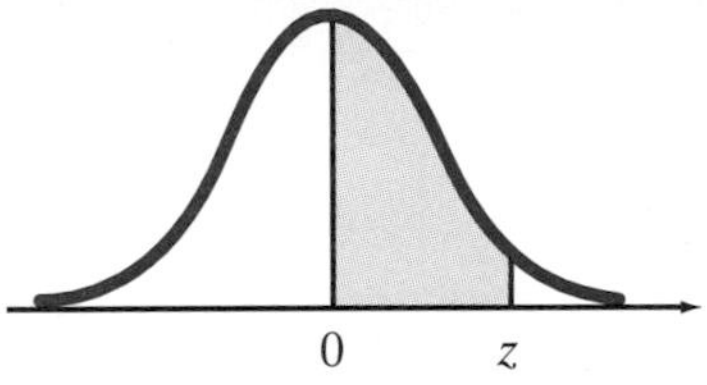

z	.00	.01	.02	.03	.04	.05	.06	.07	.08	.09
.0	.0000	.0040	.0080	.0120	.0160	.0199	.0239	.0279	.0319	.0359
.1	.0398	.0438	.0478	.0517	.0557	.0596	.0636	.0675	.0714	.0753
.2	.0793	.0832	.0871	.0910	.0948	.0987	.1026	.1064	.1103	.1141
.3	.1179	.1217	.1255	.1293	.1331	.1368	.1406	.1443	.1480	.1517
.4	.1554	.1591	.1628	.1664	.1700	.1736	.1772	.1808	.1844	.1879
.5	.1915	.1950	.1985	.2019	.2054	.2088	.2123	.2157	.2190	.2224
.6	.2257	.2291	.2324	.2357	.2389	.2422	.2454	.2486	.2517	.2549
.7	.2580	.2611	.2642	.2673	.2704	.2734	.2764	.2794	.2823	.2852
.8	.2881	.2910	.2939	.2967	.2995	.3023	.3051	.3078	.3106	.3133
.9	.3159	.3186	.3212	.3238	.3264	.3289	.3315	.3340	.3365	.3389
1.0	.3413	.3438	.3461	.3485	.3508	.3531	.3554	.3577	.3599	.3621
1.1	.3643	.3665	.3686	.3708	.3729	.3749	.3770	.3790	.3810	.3830
1.2	.3849	.3869	.3888	.3907	.3925	.3944	.3962	.3980	.3997	.4015
1.3	.4032	.4049	.4066	.4082	.4099	.4115	.4131	.4147	.4162	.4177
1.4	.4192	.4207	.4222	.4236	.4251	.4265	.4279	.4292	.4306	.4319
1.5	.4332	.4345	.4357	.4370	.4382	.4394	.4406	.4418	.4429	.4441

Source: Abridged from Table I of A. Hald, *Statistical Tables and Formulas* (New York: Wiley, 1952). Reproduced by permission of A. Hald.

To locate the proper entry, proceed down the left (z) column of the table to the row corresponding to $z = 1.2$. Then move across the top of the table to the column headed .06. The intersection of the .06 column and the 1.2 row contains the desired area, .3962 (highlighted in Table 6.1), as shown in Figure 6.4.

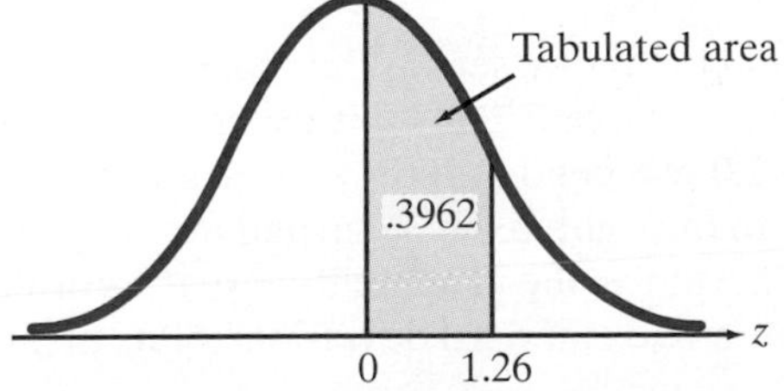

Figure 6.4 The Tabulated Area in Table B.3 Corresponding to $z = 1.26$

The normal distribution associated with the z statistic (as shown in Figure 6.4) is called the *standard normal distribution.* The mean of a standard normal distribution is always equal to 0 (since $z = 0$ when $x = \mu$); the standard deviation is always equal to 1. Since the mean is 0, z values to the right of the mean are positive; those to the left are negative.

Definition 6.2

A **standard normal distribution** is a normal distribution with $\mu = 0$ and $\sigma = 1$. A **standard normal random variable,** z, is a random variable with a standard normal distribution.

EXAMPLE 6.3

AREA BETWEEN 0 AND NEGATIVE *z*

Find the area under the standard normal curve between the mean $z = 0$ and the point $z = -1.26$. This area represents $P(-1.26 \leq z \leq 0)$.

Solution

The best way to solve a problem of this type is to draw a sketch of the distribution (see Figure 6.5). Since $z = -1.26$ is negative, we know that it lies to the left of the mean, and the area that we seek is the shaded area shown.

Since the normal curve is symmetric, the area between the mean 0 and $z = -1.26$ is exactly the same as the area between the mean 0 and $z = +1.26$. We found this area in Example 6.2 to be .3962. Therefore, the area between $z = -1.26$ and $z = 0$ is .3962.

Figure 6.5 Standard Normal Distribution for Example 6.3

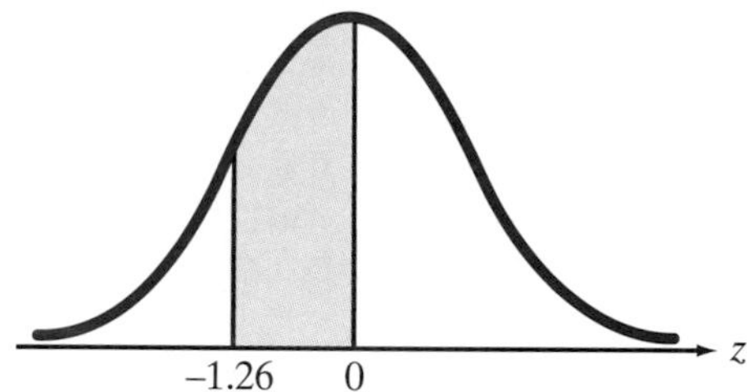

EXAMPLE 6.4

AREA BETWEEN $-z$ AND z

Find the probability that a normally distributed random variable will lie within $z = 2$ standard deviations of its mean; that is, find the probability $P(-2 \leq z \leq 2)$.

Solution

The probability that we seek is the shaded area shown in Figure 6.6. Since the area between the mean and $z = 2.0$ is exactly the same as the area between the mean and $z = -2.0$, we need find only the area between the mean and $z = 2$ standard deviations to the right of the mean and multiply by 2. This area is given in Table B.3 as .4772. Therefore, the probability P that a normally distributed random variable will lie within two standard deviations of its mean is

$$P = 2(.4772) = .9544$$

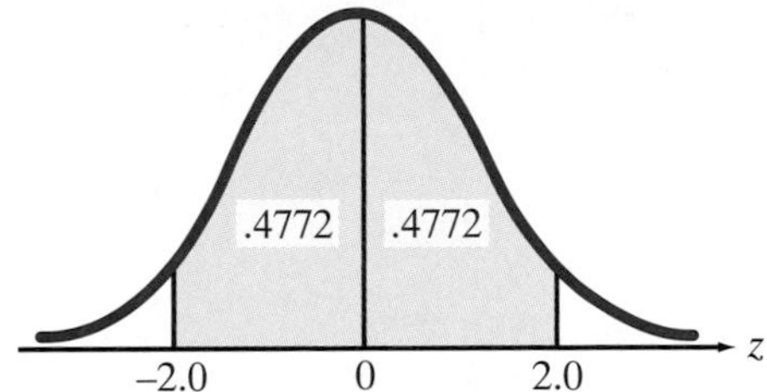

Figure 6.6 Standard Normal Distribution for Example 6.4

EXAMPLE 6.5

AREA ABOVE *z*

Find the probability that a normally distributed random variable x will lie more than $z = 2$ standard deviations above its mean; that is, find $P(z > 2)$.

Solution

The probability we seek is the darker shaded area shown in Figure 6.7. The total area under a standard normal curve is 1; half this area lies to the left of the mean, half to the right. Consequently, the probability P that x will lie more than two standard deviations above the mean is equal to .5 less the area A:

$$P = .5 - A$$

The area A corresponding to $z = 2.0$ is .4772. Therefore,

$$P = .5 - .4772 = .0228$$

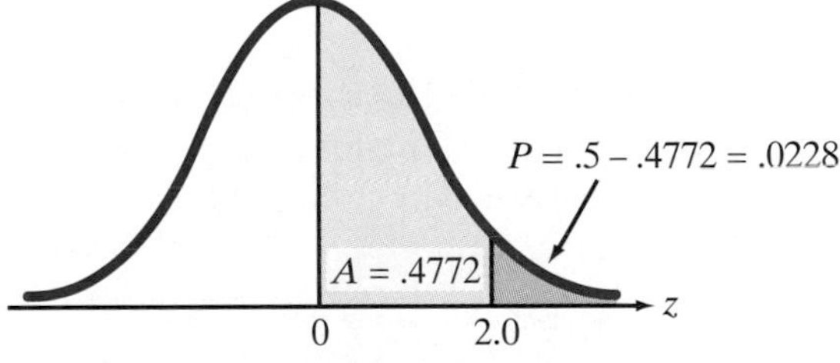

Figure 6.7 Standard Normal Distribution for Example 6.5

EXAMPLE 6.6

AREA BETWEEN TWO POSITIVE *z* VALUES

Find the area under the normal curve between $z = 1.2$ and $z = 1.6$; that is, find $P(1.2 \leq z \leq 1.6)$.

Solution

The area A that we seek lies to the right of the mean because both z values are positive. It will appear as the shaded area shown in Figure 6.8. Let A_1 represent the area between $z = 0$ and $z = 1.2$, and A_2 represent the area between $z = 0$ and $z = 1.6$. Then the area A that we desire is $A = A_2 - A_1$. From Table B.3, we obtain:

$$A_1 = .3849 \quad \text{and} \quad A_2 = .4452$$

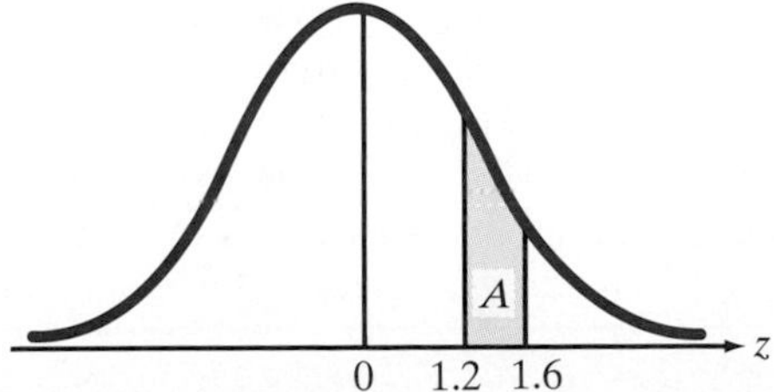

Figure 6.8 Standard Normal Distribution for Example 6.6

Then

$$A = A_2 - A_1$$
$$= .4452 - .3849 = .0603$$

EXAMPLE 6.7

FIND A *z* VALUE FOR A GIVEN AREA

Find a value of z—call it z_0—such that the area to the right of z_0 is equal to .1.

Solution

The z value that we seek appears as shown in Figure 6.9. Note that we show an area to the right of z equal to .1. Since the total area to the right of the mean $z = 0$ is equal to .5, the area between the mean 0 and the unknown z value is $.5 - .1 = .4$ (as shown in the figure).

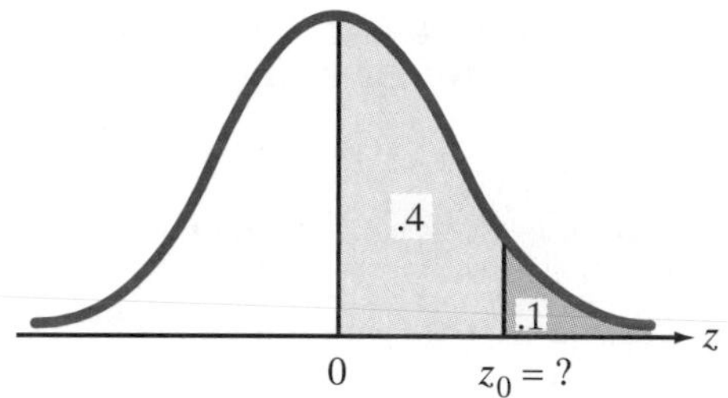

Figure 6.9 Standard Normal Distribution for Example 6.7

Consequently, to find z_0, we must look in the standard normal table (Table B.3) for the z value that corresponds to an area equal to .4. A reproduction of Table B.3 is shown in Table 6.2. The area .4000 does not appear in Table 6.2. The closest values are .3997, corresponding to $z = 1.28$, and .4015, corresponding to $z = 1.29$.

TABLE 6.2 Reproduction of Part of Table B.3 of Appendix B

z	.00	.01	.02	.03	.04	.05	.06	.07	.08	.09
.0	.0000	.0040	.0080	.0120	.0160	.0199	.0239	.0279	.0319	.0359
.1	.0398	.0438	.0478	.0517	.0557	.0596	.0636	.0675	.0714	.0753
.2	.0793	.0832	.0871	.0910	.0948	.0987	.1026	.1064	.1103	.1141
.3	.1179	.1217	.1255	.1293	.1331	.1368	.1406	.1443	.1480	.1517
.4	.1554	.1591	.1628	.1664	.1700	.1736	.1772	.1808	.1844	.1879
.5	.1915	.1950	.1985	.2019	.2054	.2088	.2123	.2157	.2190	.2224
.6	.2257	.2291	.2324	.2357	.2389	.2422	.2454	.2486	.2517	.2549
.7	.2580	.2611	.2642	.2673	.2704	.2734	.2764	.2794	.2823	.2852
.8	.2881	.2910	.2939	.2967	.2995	.3023	.3051	.3078	.3106	.3133
.9	.3159	.3186	.3212	.3238	.3264	.3289	.3315	.3340	.3365	.3389
1.0	.3413	.3438	.3461	.3485	.3508	.3531	.3554	.3577	.3599	.3621
1.1	.3643	.3665	.3686	.3708	.3729	.3749	.3770	.3790	.3810	.3830
1.2	.3849	.3869	.3888	.3907	.3925	.3944	.3962	.3980	.3997	.4015
1.3	.4032	.4049	.4066	.4082	.4099	.4115	.4131	.4147	.4162	.4177
1.4	.4192	.4207	.4222	.4236	.4251	.4265	.4279	.4292	.4306	.4319
1.5	.4332	.4345	.4357	.4370	.4382	.4394	.4406	.4418	.4429	.4441

Source: Abridged from Table I of A. Hald, *Statistical Tables and Formulas* (New York: Wiley, 1952). Reproduced by permission of A. Hald.

Since the area .3997 is closer to .4000 than is .4015, we will choose $z = 1.28$ as our answer. That is, $z_0 = 1.28$.

Examples 6.2–6.7 demonstrate how to solve the following two types of normal probability problems:

1. Examples 6.2–6.6 use Table B.3 to *find areas under the standard normal curve.* These problems may be further classified into one of three types:
   - **a.** Finding the area between the mean $\mu = 0$, and some value of z that is located above or below $\mu = 0$ (Examples 6.2 and 6.3)
   - **b.** Finding the area between the values z_1 and z_2, where neither z_1 nor z_2 is equal to 0 (Examples 6.4 and 6.6)
   - **c.** Finding the area in either the upper or the lower tail of the standard normal z distribution (Example 6.5)
2. Example 6.7 uses Table B.3 to *find the* z *value, denoted* z_0, *corresponding to some area* in the upper tail of the standard normal z distribution. A similar procedure may be used to find a z value corresponding to a lower-tail area under the curve.

Many distributions of data that occur in the real world are approximately normal, but few are *standard* normal. However, Examples 6.8–6.10 use what you have learned about the standard normal curve to solve the same two types of problems involving *any* normal distribution:

1. Finding the probability that a normal random variable x falls between the values x_1 and x_2, or the probability that it falls in either the upper or the lower tail of the normal distribution (Examples 6.8 and 6.9)
2. Finding the value of x that places a probability P in the upper (or lower) tail of a normal distribution (Example 6.10)

Self-Test 6.1

Consider a standard normal random variable z.

a. Find $P(0 \leq z \leq .75)$ **b.** Find $P(-.28 \leq z \leq 1.16)$

c. Find $P(z \leq -1.45)$ **d.** Find z_0 such that $P(z \geq z_0) = .2$

EXAMPLE 6.8

APPLICATION: FINDING A NORMAL PROBABILITY

Salt is America's second leading food additive (after sugar), both in factory-processed foods and home cooking. The average amount of salt consumed per day by an American is 15 grams (15,000 milligrams), although the actual physiological minimum daily requirement for salt is only 220 milligrams. Suppose that the amount of salt intake per day is approximately normally distributed with a standard deviation of 5 grams. What proportion of all Americans consume between 14 and 22 grams of salt per day?

Solution

The proportion P of Americans who consume between $x = 14$ grams and $x = 22$ grams of salt is the total shaded area in Figure 6.10a (p. 290). Before we can

compute this area, we need to determine the z values that correspond to $x = 14$ and $x = 22$. Substituting $\mu = 15$ and $\sigma = 5$ into equation 6.1, we compute the z value for $x = 14$ as

$$z_1 = \frac{x - \mu}{\sigma} = \frac{14 - 15}{5} = \frac{-1}{5} = -.20$$

The corresponding z value for $x = 22$ is

$$z_2 = \frac{x - \mu}{\sigma} = \frac{22 - 15}{5} = \frac{7}{5} = 1.40$$

Figure 6.10 Normal Curve Sketches for Example 6.8

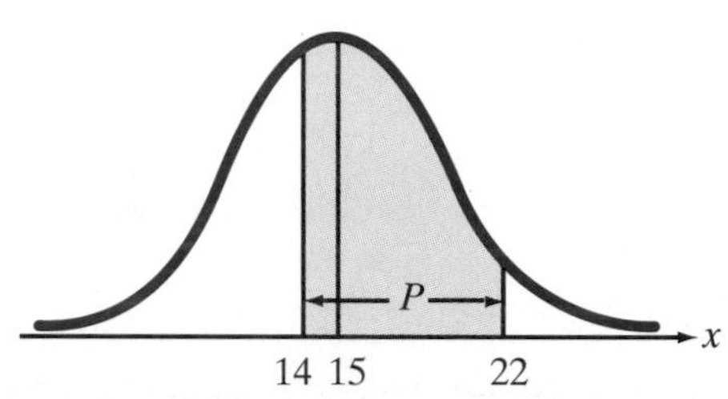

a. Amount of salt intake (in grams) per day

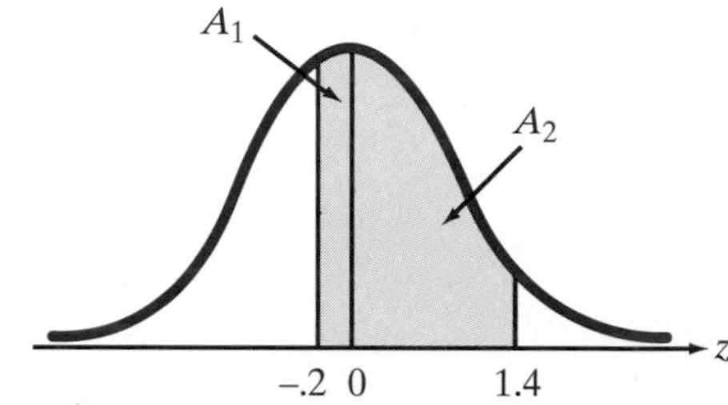

b. Standard normal

Figure 6.10b* shows these z values along with P. From this figure, we see that $P = A_1 + A_2$, where A_1 is the area corresponding to $z_1 = -.20$, and A_2 is the area corresponding to $z_2 = 1.4$. These values, given in Table B.3, are $A_1 = .0793$ and $A_2 = .4192$. Thus,

$$P = A_1 + A_2 = .0793 + .4192 = .4985$$

So 49.85% of Americans consume between 14 and 22 grams of salt per day.

*Actually, the shape of the standard normal distribution shown in Figure 6.10b differs from the shape of the normal distribution in Figure 6.10a because the distributions have different standard deviations. In fact, the normal curve associated with Figure 6.10b ($\sigma = 1$) will be more peaked and narrower than that for Figure 6.10a ($\sigma = 5$) since the standard deviation is smaller. Nevertheless, the tail area P shaded on the two distributions is identical. For pedagogical reasons, we show the distributions with similar shapes.

The technique for finding the probability that a normal random variable falls between two values is summarized in the next box.

FINDING THE PROBABILITY THAT A NORMAL RANDOM VARIABLE FALLS BETWEEN TWO VALUES, x_1 AND x_2

1. Make two sketches of the normal curve, one representing the normal distribution of x and the other, the standard normal z distribution.
2. Show the approximate locations of x_1 and x_2 on the sketch of the x distribution. Be sure to locate x_1 and x_2 correctly relative to the mean μ. For example, if x_1 is less than μ, then it should be located to the left of μ; if x_2 is greater than μ, it should be located to the right of μ.

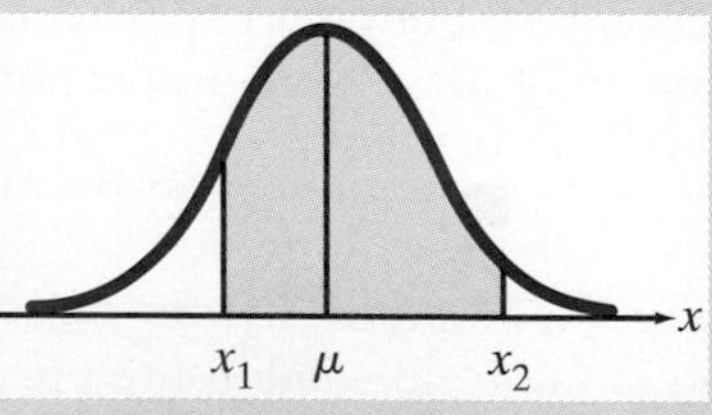

3. Find the values of z corresponding to x_1 and x_2:

$$z_1 = \frac{x_1 - \mu}{\sigma} \quad \text{and} \quad z_2 = \frac{x_2 - \mu}{\sigma}$$

4. Locate the values of z_1 and z_2 on your sketch of the z distribution.

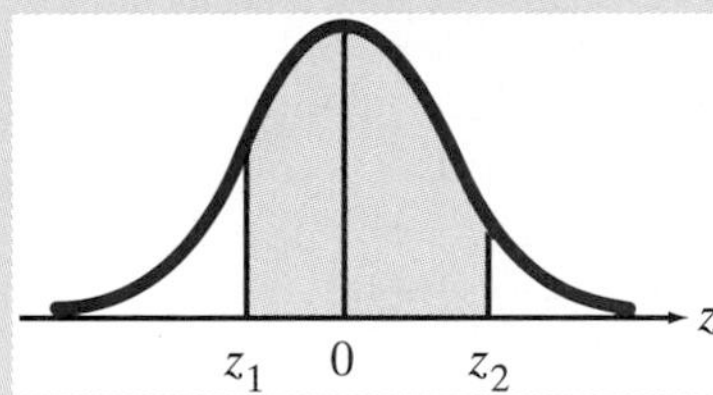

5. Use the table of areas under the standard normal curve, given in Table B.3, to find the area between z_1 and z_2. This will be the probability that x falls between x_1 and x_2.

EXAMPLE 6.9

APPLICATION: FINDING A NORMAL PROBABILITY

Refer to Example 6.8. Medical research has linked excessive consumption of salt to hypertension (high blood pressure). Physicians recommend that those Americans who want to reach a level of salt intake at which hypertension is less likely to occur should consume less than 1 gram of salt per day. What is the probability that a randomly selected American consumes less than 1 gram of salt per day?

Solution

The probability P that a randomly selected American consumes less than 1 gram of salt per day is represented by the shaded area in Figure 6.11a. The z value corresponding to $x = 1$ (shown in Figure 6.11b) is

$$z = \frac{x - \mu}{\sigma} = \frac{1 - 15}{5} = \frac{-14}{5} = -2.80$$

Figure 6.11 Normal Curve Sketches for Example 6.9

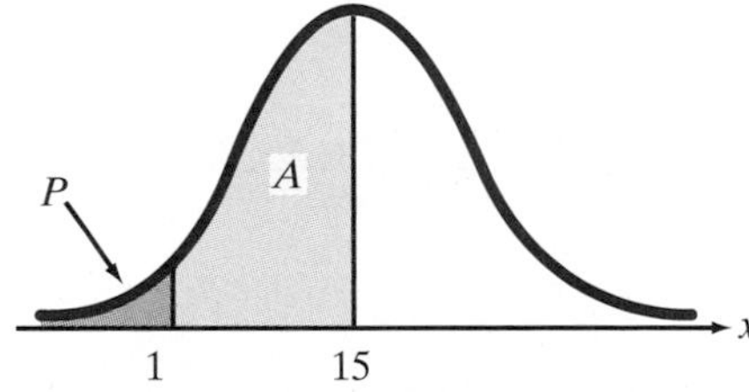

a. Amount of salt intake (in grams) per day

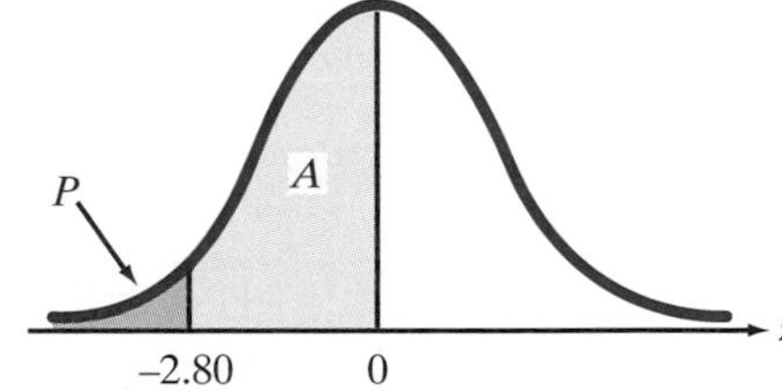

b. Standard normal

Since the area to the left of $z = 0$ is equal to .5, the probability that x is less than or equal to 1 is $P = .5 - A$, where A is the tabulated area corresponding to $z = -2.80$. This value, given in the standard normal table, is $A = .4974$. Then, the probability that a randomly selected American consumes no more than 1 gram of salt per day is

$$P = .5 - A$$
$$= .5 - .4974 = .0026$$

Self-Test 6.2

Steel used for water pipelines is often coated on the inside with cement mortar to prevent corrosion. In a study of the mortar coatings of the pipeline used in a water transmission project in California (*Transportation Engineering Journal*), researchers noted that the mortar thickness was specified to be .4375 inch. A very large sample of thickness measurements produced a mean equal to .635 inch and a standard deviation equal to .082 inch. If the thickness measurements were normally distributed, approximately what proportion were less than .4375 inch?

EXAMPLE 6.10

APPLICATION: FINDING A VALUE OF THE NORMAL RANDOM VARIABLE

Psychologists traditionally treat IQ as a random variable having a normal distribution with a mean of 100 and a standard deviation of 15. Find the 10th percentile of the IQ distribution.

Solution

Let x represent the IQ of a randomly selected person. We know that x is normally distributed with $\mu = 100$ and $\sigma = 15$. We want to find the 10th percentile of the IQ distribution (call this value x_0). By definition, x_0 exceeds exactly 10 percent of the IQs in the distribution. Thus,

$$P(x < x_0) = .10$$

Unlike the two previous examples, where we were given a value of x and asked to find a corresponding probability, here we are given a probability $P = .10$ and asked to locate the corresponding value of x. We call P the **tail probability associated with x_0.** The value x_0, along with its tail probability, is shown in Figure 6.12a.

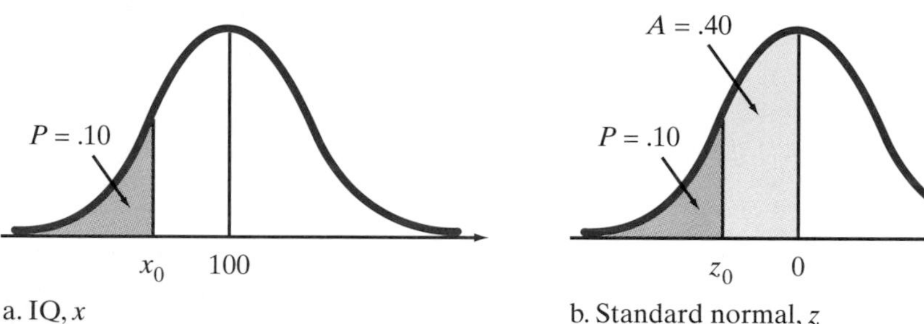

Figure 6.12 Normal Curve Areas for Example 6.10

Now, the 10th percentile, x_0, is located to the left of (i.e., below) the mean of 100. The location of x_0 is not arbitrary, but depends on the value of P. To see this more clearly, try placing x_0 above (to the right of) the mean of 100. Now (mentally) shade in the corresponding tail probability P, i.e., the area to the left of x_0. You can see that this area cannot possibly equal .10—it is too large (.50 or greater). Thus, x_0 must lie to the left of the IQ mean of 100.

The first step, then, in solving problems of this type is to determine the location of x_0 relative to the mean μ. The second step is to find the corresponding z value for x_0—call it z_0—where

$$z_0 = \frac{x_0 - \mu}{\sigma} = \frac{x_0 - (100)}{15}$$

Once we locate z_0, we can use the equivalent relation (equation 6.3 in the next box)

$$x_0 = \mu + z_0(\sigma) = 100 + z_0(15)$$

to find x_0.

You can see from Figure 6.12b that z_0 is the z value that corresponds to the area $A = .40$ in Table B.3. In the body of the table, we see the value closest to .40 is .3997. The z value corresponding to this area is $z_0 = -1.28$. (Note that z_0 is negative because it lies to the left of 0.) Substituting into the formula above, we have

$$x_0 = 100 + z_0(15) = 100 + (-1.28)(15) = 80.8$$

Thus, 10% of the IQs fall below 80.8. In other words, 80.8 represents the 10th percentile of the IQ distribution.

To help in solving problems of this type throughout the remainder of the chapter, we outline the steps leading to the solutions in the next box.

FINDING A PARTICULAR VALUE OF THE NORMAL RANDOM VARIABLE GIVEN AN ASSOCIATED TAIL PROBABILITY

1. Make a sketch of the relative frequency distribution of the normal random variable x. Shade the tail probability P and locate the corresponding value x_0 on the sketch. Remember that x_0 will be to the right or left of the mean μ depending on the value of P.
2. Make a sketch of the corresponding relative frequency distribution of the standard normal random variable z. Locate z_0, the z value corresponding to x_0, and shade the area corresponding to P.
3. Compute the area A associated with z_0 as follows:

$$A = .5 - P$$

4. Use the area A, i.e., the area between 0 and z_0, to find z_0 in Table B.3. (If you cannot find the exact value of A in the table, use the closest value.) Note that z_0 will be negative if you place x_0 to the left of the mean in step 1.
5. Compute x_0 as follows:

$$x_0 = \mu + z_0(\sigma) \qquad \textbf{(6.3)}$$

The preceding examples should help you to understand the use of the table of areas under the normal curve. The practical applications of this information to inference-making will become apparent in the following chapters.

Statistics in the Real World Revisited

Applying the Normal Distribution to the Fire Fraud Data

Refer to furniture fire fraud case described on p. 281. Recall that a fire destroyed a warehouse and all the furniture in it. In an attempt to estimate the "lost" profit, the retailer claimed it randomly sampled invoices for 253 of 3,005 furniture items sold the previous year. The mean gross profit factor (GPF) of the 253 sampled invoices was 50.8%—a value deemed unusually high by experienced claims adjusters at the insurance company. Thus, the insurance company accused the retailer of fraudulently representing its sampling methodology.

(Continued)

For this application, suppose we want to know how likely it is to obtain a GPF value that exceeds the estimated mean GPF of 50.8%. Since the data for all 3,005 items are available in the **FIREFRAUD** file, we can find the actual mean GPF. The mean and standard deviation for the 3,005 gross profit margins are highlighted in the MINITAB printout, Figure 6.13. Treating this data set as a population, we have $\mu = 48.9\%$ and $\sigma = 13.8\%$. Using these parameters, we can find the probability that a randomly selected item will have a GPF that exceeds 50.8%. Assuming that the distribution of gross profit margins is approximately normal, we find (using Table B.3):

$$P(x > 50.8) = P\left(z > \frac{50.8 - \mu}{\sigma}\right) = P\left(z > \frac{50.8 - 48.9}{13.8}\right)$$

$$= P(z > .14) = .5 - .0557 = .4443$$

Consequently, less than half (about 44%) of the gross profit margins in the population of 3,005 furniture items are larger than 50.8%—the mean GPF claimed by the retailer. [*Note:* In Section 6.6, we compute a probability that will allow us to determine the likelihood that the sample was not randomly drawn from the population.]

M **Figure 6.13** MINITAB Descriptive Statistics for 3,005 Gross Profit Margins

Descriptive Statistics

Variable	N	Mean	Median	TrMean	StDev	SE Mean
GPF	3005	48.901	50.210	49.719	13.829	0.252

Variable	Minimum	Maximum	Q1	Q3
GPF	-202.510	148.000	44.380	56.060

PROBLEMS FOR SECTION 6.2

Using the Tools

6.1 Find the area under the standard normal curve:

a. Between $z = 0$ and $z = 1.2$

b. Between $z = 0$ and $z = 1.49$

c. Between $z = -.48$ and $z = 0$

d. Between $z = -1.37$ and $z = 0$

e. For values of z larger than 1.33

Show the z values and the corresponding area of interest on a sketch of the normal curve for each part of the problem.

6.2 Find the area under the standard normal curve:

a. Between $z = 1.21$ and $z = 1.94$

b. For values of z larger than 2.33

c. For values of z less than -2.33

d. Between $z = -1.50$ and $z = 1.79$

Show the z values and the corresponding area of interest on a sketch of the normal curve for each part of the problem.

6.3 Find the z value (to two decimal places) that corresponds to an area in the standard normal table (Table B.3) equal to:

a. .1000 **b.** .3200 **c.** .4000 **d.** .4500 **e.** .4750

Show the area and corresponding value of z on a sketch of the normal curve for each part of the problem.

6.4 Find the value of z (to two decimal places) that cuts off an area in the upper tail of the standard normal curve equal to:

a. .025 **b.** .05 **c.** .005 **d.** .01 **e.** .10

Show the area and corresponding value of z on a sketch of the normal curve for each part of the problem.

6.5 Suppose that a normal random variable x has a mean $\mu = 20.0$ and a standard deviation $\sigma = 4.0$. Find the z score corresponding to:

a. $x = 23.0$ **b.** $x = 16.0$ **c.** $x = 13.5$

d. $x = 28.0$ **e.** $x = 12.0$

For each part of the problem, locate x and μ on a sketch of the normal curve. Check to make sure that the sign and magnitude of your z score agree with your sketch.

6.6 Find the approximate value for z_0 such that the probability that z is larger than z_0 is:

a. $P = .10$ **b.** $P = .15$ **c.** $P = .20$ **d.** $P = .25$

Locate z_0 and the corresponding probability P on a sketch of the normal curve for each part of the problem.

6.7 Find the approximate value for z_0 such that the probability that z is less than z_0 is:

a. $P = .10$ **b.** $P = .15$ **c.** $P = .30$ **d.** $P = .50$

Locate z_0 and the corresponding probability P on a sketch of the normal curve for each part of the problem.

Applying the Concepts

6.8 A set of final examination grades in an introductory statistics course was found to be normally distributed with a mean of 73 and a standard deviation of 8.

a. What is the probability of getting a grade of 91 or lower on this exam?

b. What percentage of students scored between 65 and 89?

c. What percentage of students scored between 81 and 89?

d. What is the final exam grade if only 5% of the students taking the test scored higher?

e. If the professor "curves" (gives A's to the top 10% of the class regardless of the score), are you better off with a grade of 81 on this exam or a grade of 68 on a different exam where the mean is 62 and the standard deviation is 3? Show statistically and explain.

6.9 The *Journal of Visual Impairment & Blindness* (May–June 1997) published a study of the lifestyles of visually impaired students. Using diaries, the students kept track of several variables, including number of hours of sleep they get in a typical day. These visually impaired students had a mean of 9.06 hours and a standard deviation of 2.11 hours. Assume that the distribution of the number of hours of sleep for this group of students is approximately normal.

a. Find the probability that a visually impaired student sleeps less than 6 hours on a typical day.

b. Find the probability that a visually impaired student sleeps between 8 and 10 hours on a typical day.

c. Twenty percent of all visually impaired students get less than how many hours of sleep on a typical day?

6.10 The metropolitan airport commission is considering establishing limits on noise pollution around a local airport. At the present time the noise level per jet takeoff in one neighborhood near the airport is approximately normally distributed with a mean of 100 decibels and a standard deviation of 6 decibels.

a. What is the probability that a randomly selected jet will generate a noise level greater than 108 decibels in this neighborhood?

b. What is the probability that a randomly selected jet will generate a noise level of at least 100 decibels?

c. Suppose a regulation is passed that requires jet noises in this neighborhood to be lower than 105 decibels 95% of the time. Assuming the standard deviation of the noise distribution remains the same, how much will the mean noise level have to be lowered to comply with the regulation?

d. Repeat parts **a–c**, but assume that the standard deviation of the noise levels is 10 decibels.

6.11 Foresters "cruising" British Columbia's boreal forest have determined that the diameter at breast height of white spruce trees in a particular community is approximately normal, with $\mu = 17$ meters and $\sigma = 6$ meters.*

a. Find the probability that the breast height diameter of a randomly selected white spruce in the forest community is less than 12 meters.

b. Suppose you observe a white spruce with a breast height diameter of 12 meters. Is this an unusual event? Explain.

c. Find the probability that the breast height diameter of a randomly selected white spruce in the forest community will exceed 37 meters.

d. Suppose you observe a tree in the community forest with a breast height diameter of 38 meters. Is this tree likely to be a white spruce? Explain.

e. Fifteen percent of white spruce trees in the forest will have a breast height diameter greater than what value?

*Scholz, H. "Fish Creek Community Forest: Exploratory statistical analysis of selected data." Working paper, Northern Lights College, British Columbia, Canada.

6.12 The *Journal of Communication* (Summer 1997) published a study on U.S. television news coverage of major foreign earthquakes. One of the variables measured was the distance (in thousands of miles) the earthquake was from New York City—from where the ABC, NBC, and CBS nightly news broadcasts originate. The distance distribution was found to have a mean of 4.6 thousand miles and a standard deviation of 1.9 thousand miles. Assume the distribution is approximately normal. Find the probability that a randomly selected major foreign earthquake is:

a. Between 2.5 and 3.0 thousand miles from New York City

b. Between 4.0 and 5.0 thousand miles from New York City

c. Less than 6.0 thousand miles from New York City

d. Based on the mean and standard deviation reported, is it reasonable to assume that the distance distribution is approximately normal? Explain.

6.13 Competitive cyclists are thought to have a very low percentage of body fat due to intensive training methods. A group of competitive cyclists were found to have a mean percent body fat of 9% with a standard deviation of 3% (*International Journal of Sport Nutrition,* June 1995). Assume the distribution of body fat measurements for competitive cyclists is normally distributed.

a. What is the median of the percent body fat distribution?

b. What is the 3rd quartile of the percent body fat distribution?

c. What is the 95th percentile of the percent body fat distribution?

6.14 A group of Florida State University psychologists examined the effects of alcohol on the reactions of people to a threat (*Journal of Abnormal Psychology,* Vol. 107, 1998). After obtaining a specified blood alcohol level, experimental subjects were placed in a room and threatened with electric shocks. Using sophisticated equipment to monitor the subjects' eye movements, the startle response (measured in milliseconds) was recorded for each subject. The mean and standard deviation of the startle responses were 37.9 and 12.4, respectively. Assume that the startle response x for a person with the specified blood alcohol level is approximately normally distributed.

a. Find the probability that x is between 40 and 50 milliseconds.

b. Find the probability that x is less than 30 milliseconds.

c. Give an interval for x, centered around 37.9 milliseconds, so that the probability that x falls in the interval is .95.

6.15 The Trail Making Test is frequently used by clinical psychologists to test for brain damage. Patients are required to connect consecutively numbered circles on a sheet of paper. It has been determined that the mean length of time required for a patient to perform this task is 32 seconds and the standard deviation is 4 seconds. Assume that the distribution of the lengths of time required to connect the circles is normal.

a. Find the probability that a randomly selected patient will take longer than 40 seconds to perform the task.

b. Find the probability that a randomly selected patient will take between 24 and 40 seconds to complete the task.

c. A psychologist would like to retest those persons with completion times in the highest 5% of the distribution of times required. What time would a person need to exceed on the Trail Making Test to be considered for retesting?

6.16 In a laboratory experiment, researchers at Barry University (Miami Shores, Florida) studied the rate at which sea urchins ingested turtle grass (*Florida Scientist*, Summer/Autumn 1991). The urchins, without food for 48 hours, were fed 5-cm. blades of green turtle grass. The mean ingestion time was found to be 2.83 hours and the standard deviation was .79 hour. Assume that green turtle grass ingestion time for the sea urchins has an approximate normal distribution.

a. Find the probability that a sea urchin will require 4 or more hours to ingest a 5-cm. blade of green turtle grass.

b. Find the probability that a sea urchin will require between 2 and 3 hours to ingest a 5-cm. blade of green turtle grass.

c. Ninety-nine percent of the urchins require more than how many hours to ingest the grass?

6.17 Researchers have developed sophisticated intrusion-detection algorithms to protect the security of computer-based systems. These algorithms use principles of statistics to identify unusual or unexpected data, i.e., "intruders." One popular intrusion-detection system assumes the data being monitored are normally distributed (*Journal of Information Systems,* Spring 1992). As an example, the researcher considered system data with a mean of .27, a standard deviation of 1.473, and an intrusion-detection algorithm that assumes normal data.

a. Find the probability that a data value observed by the system will fall between $-.5$ and .5.

b. Find the probability that a data value observed by the system exceeds 3.5.

c. Comment on whether a data value of 4 observed by the system should be considered an "intruder."

6.18 In baseball, a "no-hitter" is a regulation 9-inning game in which the pitcher yields no hits to the opposing batters. *Chance* (Summer 1994) reported on a study of no-hitters in Major League Baseball. The initial analysis focused on the total number of hits yielded per game per team for all 9-inning games played between 1989 and 1993. The distribution of hits/9-innings is approximately normal with mean 8.72 and standard deviation 1.10.

a. What percentage of 9-inning games results in fewer than 6 hits?

b. What percentage of 9-inning games results in between 6 and 12 hits?

c. Demonstrate, statistically, why a no-hitter is considered an extremely rare occurrence.

6.19 Pacemakers are used to control the heartbeat of cardiac patients. A single pacemaker is made up of several biomedical components that must be of a high quality for the pacemaker to work. One particular plastic part, called a connector module, mounts on the top of the pacemaker. Connector modules are specified to have a length between .304 inch and .322 inch to work properly. Any module with length outside these limits is *out-of-spec. Quality* (Aug. 1989) reported on one supplier of connector modules that had been shipping out-of-spec parts to the manufacturer for 12 months.

a. The lengths of the connector modules produced by the supplier were found to follow an approximate normal distribution with mean $\mu = .3015$ inch and standard deviation $\sigma = .0016$ inch. Use this information to find the probability that the supplier produces an out-of-spec part.

b. Once the problem was detected, the supplier's inspection crew began to employ an automated data-collection system designed to improve product quality. After two months, the process was producing connector modules with mean $\mu = .3146$ inch and standard deviation $\sigma = .0030$ inch. Find the probability that an out-of-spec part will be produced. Compare your answer to part **a**.

6.20 The physical fitness of a patient is often measured by the patient's maximum oxygen uptake (recorded in milliliters per kilogram, ml/kg). The mean maximum oxygen

uptake for cardiac patients who regularly participate in sports or exercise programs was found to be 24.1 with a standard deviation of 6.30 (*Adapted Physical Activity Quarterly,* Oct. 1997). Assume this distribution is approximately normal.

a. What is the probability that a cardiac patient who regularly participates in sports has a maximum oxygen uptake of at least 20 ml/kg?

b. What is the probability that a cardiac patient who regularly exercises has a maximum oxygen uptake of 10.5 ml/kg or lower?

c. Consider a cardiac patient with a maximum oxygen uptake of 10.5. Is it likely that this patient participates regularly in sports or exercise programs? Explain.

6.3 Descriptive Methods for Assessing Normality

In the chapters that follow, we learn how to make inferences about the population based on information in the sample. Several of these techniques are based on the assumption that the population is approximately normally distributed. However, not all continuous random variables are normally distributed! Consequently, it will be important to determine whether the sample data come from a normal population before we can properly apply these techniques.

Several descriptive methods can be used to check for normality. In this section, we consider the four methods summarized in the box.

DETERMINING WHETHER THE DATA ARE FROM AN APPROXIMATELY NORMAL DISTRIBUTION

1. Construct either a histogram, stem-and-leaf display, or a box-and-whisker plot for the data and note the shape of the graph. If the data are approximately normal, the shape of the histogram or stem-and-leaf display will be similar to the normal curve, Figure 6.2 (i.e., bell-shaped and symmetric about the mean) and the shape of the box-and-whisker plot will be similar to Figure 3.24a (p. 165).
2. Compute the intervals $\bar{x} \pm s$, $\bar{x} \pm 2s$, and $\bar{x} \pm 3s$, and determine the percentage of measurements falling in each. If the data are approximately normal, the percentages will be approximately equal to 68%, 95%, and 100%, respectively.
3. Find the interquartile range, IQR, and standard deviation, s, for the sample, then calculate the ratio IQR/s. If the data are approximately normal, then IQR/$s \approx 1.3$.
4. Construct a *normal probability plot* for the data. If the data are approximately normal, the points will fall (approximately) on a straight line.

WHY IQR/$s \approx$ 1.3 FOR NORMAL DATA

For normal distributions, the z values (obtained from Table B.3) corresponding to the 75th and 25th percentiles are .67 and $-.67$, respectively. Since $\sigma = 1$ for a standard normal (z) distribution,

$$\text{IQR}/\sigma = [.67 - (-.67)]/1 = 1.34$$

EXAMPLE 6.11 **DETECTING NORMALITY**

FTC

Consider the Federal Trade Commission data on 500 cigarette brands, stored in the file **FTC**. Numerical and graphical descriptive measures for the tar content data are shown on the Excel printouts, Figures 6.14a–d. Determine whether the tar contents have an approximate normal distribution.

E **Figure 6.14** Excel Printouts for Example 6.11

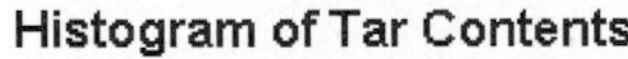

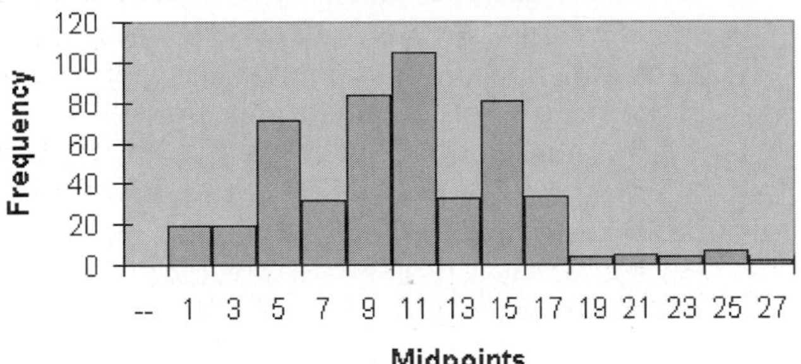

a.

	A	B
1	*Tar*	
2		
3	Mean	11.059
4	Standard Error	0.22053
5	Median	11
6	Mode	11
7	Standard Deviation	4.931192
8	Sample Variance	24.31665
9	Kurtosis	0.37881
10	Skewness	0.360265
11	Range	26.5
12	Minimum	0.5
13	Maximum	27
14	Sum	5529.5
15	Count	500
16	Largest(1)	27
17	Smallest(1)	0.5

b.

	A	B
1	FTC Tar Content Data	
2		
3	Five-number Summary	
4	Minimum	0.5
5	First Quartile	8
6	Median	11
7	Third Quartile	15
8	Maximum	27

c.

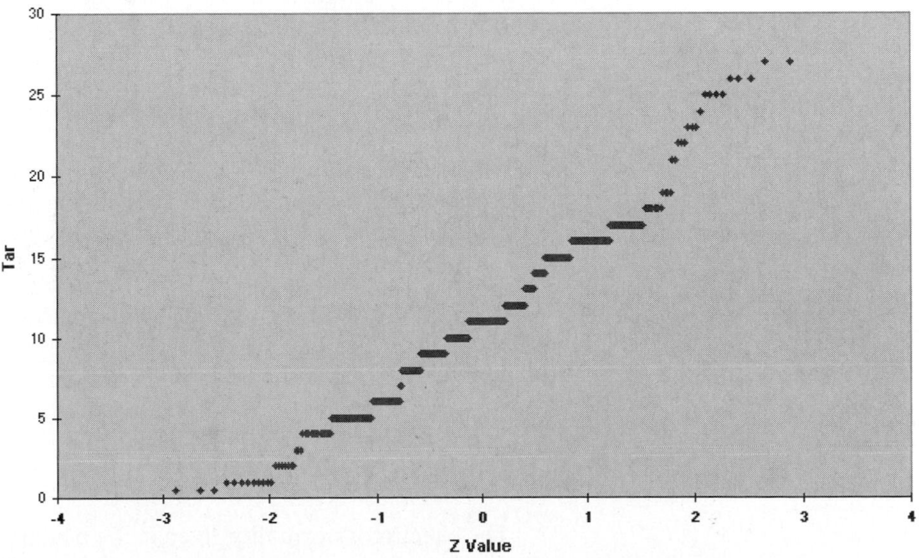

d.

Solution

As a first check, we examine the frequency histogram of the 500 tar contents shown in Figure 6.14a. Clearly, the tar contents fall in an approximately bell-shaped, symmetric distribution centered around the mean of 11.1 milligrams. Thus, from check #1 in the box, the data appear to be approximately normal.

To apply check #2, we obtain $\bar{x} = 11.1$ and $s = 4.93$ from the Excel descriptive statistics printout, Figure 6.14b. The intervals $\bar{x} \pm s$, $\bar{x} \pm 2s$, and $\bar{x} \pm 3s$ are shown in Table 6.3, as well as the percentage of tar measurements that fall in each interval. Except for the one-standard-deviation interval, these percentages agree almost exactly with those from a normal distribution.

TABLE 6.3 Describing the 500 Cigarette Tar Contents

Interval	Percentage in Interval
$\bar{x} \pm s = (6.2, 16.0)$	58.2
$\bar{x} \pm 2s = (1.3, 20.9)$	94.0
$\bar{x} \pm 3s = (-3.6, 25.8)$	99.9

Check #3 in the box requires that we find the interquartile range (i.e., the difference between the 75th and 25th percentiles) and the standard deviation s, and compute the ratio of these two numbers. The ratio IQR/s for a sample from a normal distribution will approximately equal 1.3. The value of s, 4.93, is highlighted in Figure 6.14b. The values of Q_1 and Q_3, 8 and 15 respectively, are highlighted in Figure 6.14c. Then $\text{IQR} = Q_3 - Q_1 = 7$ and the ratio is

$$\frac{\text{IQR}}{s} = \frac{7}{4.93} = 1.42$$

Since this value is approximately equal to 1.3, we have further confirmation that the data are approximately normal.

A fourth descriptive technique for checking normality is a *normal probability plot*. In a normal probability plot, the observations in the data set are ordered from smallest to largest and then plotted against the expected z scores of the observations calculated under the assumption that the data are from a normal distribution. When the data are, in fact, normally distributed, a linear (straight-line) trend will result. A nonlinear trend in the normal probability plot suggests that the data are nonnormal.

Definition 6.3

A **normal probability plot** for a data set is a scatterplot with the ranked data values on the vertical axis and their corresponding z values from a standard normal distribution on the horizontal axis.

Computation of the expected standard normal z scores for a normal probability plot are beyond the scope of this text. A normal probability plot for the 500 tar measurements obtained from PHStat and Excel is shown in Figure 6.14d. Notice that the ordered measurements fall reasonably close to a straight line. Thus, check #4 also suggests that the data are likely to be approximately normally distributed.

The checks for normality given in the box are simple, yet powerful, techniques to apply, but they are only descriptive in nature. It is possible (although unlikely)

that the data are nonnormal even when the checks are reasonably satisfied. Thus, we should be careful not to claim that the 500 tar measurements in the **FTC** file are, in fact, normally distributed. We can only state that it is reasonable to believe that the data are from a normal distribution.*

As we will learn in the next chapter, several inferential methods of analysis require the data to be approximately normal. If the data are clearly nonnormal, inferences derived from the method may be invalid. Therefore, it is advisable to check the normality of the data prior to conducting the analysis.

*Statistical tests of normality that provide a measure of reliability for the inference are available. However, these tests tend to be very sensitive to slight departures from normality, i.e., they tend to reject the hypothesis of normality for any distribution that is not perfectly symmetrical and bell-shaped. Consult the references (see Ramsey & Ramsey, 1990) if you want to learn more about these tests.

Self-Test 6.3

Which of the following suggest that the data set is approximately normal?

a. A data set with $Q_1 = 10$, $Q_3 = 50$, and $s = 10$.

b. A data set with the following stem-and-leaf display:

Stem	*Leaves*
0	1 1 2
1	1 2 5 6 7
2	0 1 1 1 3 8 8
3	4 5 6 6 6 6 7 7
4	1 1 1 2 2 3 3 3 3 4 8 9 9
5	0 0 1 2 5 5 6 8 9
6	0 1 1 1 2 3
7	1 2 5 8
8	0 7
9	2

c. A data set with 93% of the measurements within $\bar{x} \pm s$.

d. A data set with the following normal probability plot.

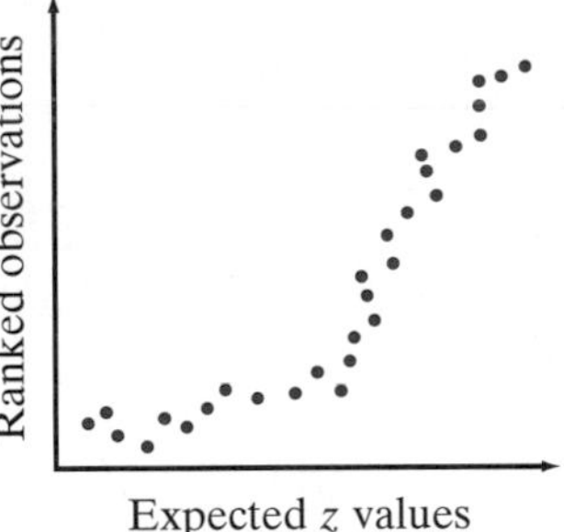

Statistics in the Real World Revisited

Assessing Whether the Fire Fraud Data Is Normal

Refer, once more, to the furniture fire fraud case described on p. 281. In Section 6.2 (p. 293), we applied the normal distribution to calculate the probability that a randomly selected item will have a gross profit margin exceeding 50.8%. Is it reasonable to assume that the GPF is normally distributed? A MINITAB histogram for the gross profit margins of all 3,005 items in the **FIREFRAUD** file is shown in Figure 6.15a, followed by an Excel normal probability plot in Figure 6.15b. The histogram indicates a heavy concentration of values in the center of the distribution, with some extremely small GPFs in the lower tail. This leftward skew to the GPF distribution is also shown in the normal probability plot. Note the deviation of the ordered GPF values from the linear trend in the middle of the plot, Figure 6.15b.

(Continued)

M Figure 6.15a MINITAB Histogram of 3,005 GPF Values

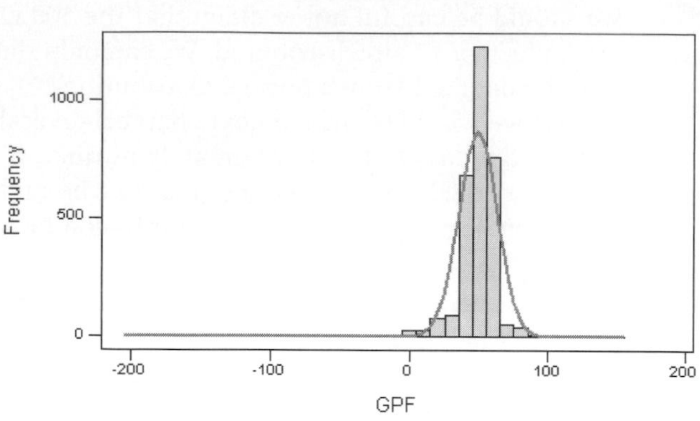

E Figure 6.15b PHStat and Excel Normal Probability Plot for 3,005 GPF Values

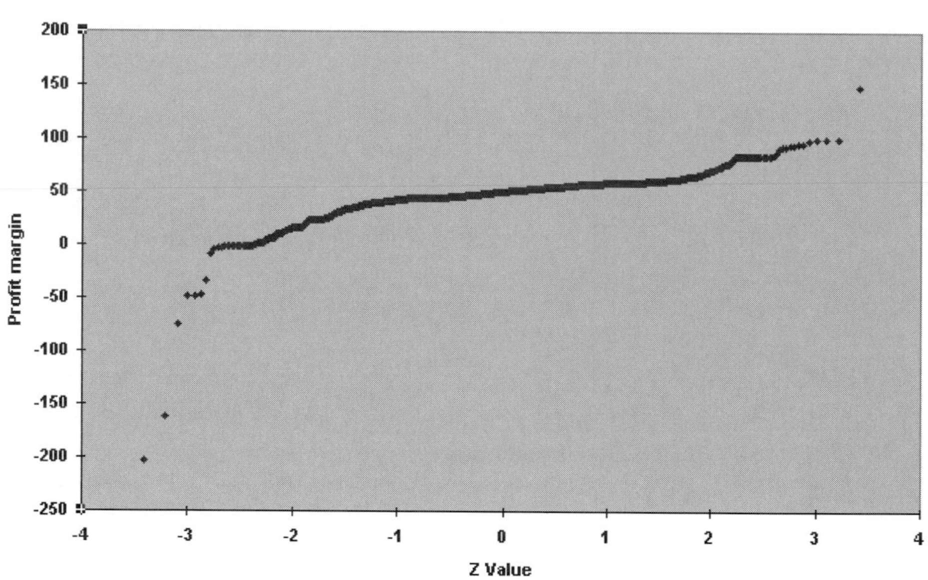

Another way to confirm that the data are not normally distributed is to compare the interquartile range to the standard deviation. From Figure 6.13 (p. 294), $Q_3 = 56.06$, $Q_1 = 44.38$, and $\sigma = 13.83$. Therefore,

$$\text{IQR}/\sigma = (56.06 - 44.38)/13.83 = .84$$

This value is less than the normal distribution value of 1.34.

Since the GPF values do not have an exact normal distribution, the probability calculated in Section 6.2 is only an approximation to the true probability.

PROBLEMS FOR SECTION 6.3

Using the Tools

6.21 If a population data set is normally distributed, what is the proportion of measurements you would expect to fall within the following intervals?

a. $\mu \pm \sigma$ **b.** $\mu \pm 2\sigma$ **c.** $\mu \pm 3\sigma$

6.22 Consider a sample data set with the following summary statistics: $s = 95$, $Q_1 = 72$, $Q_3 = 195$.

a. Calculate IQR. **b.** Calculate IQR/s.

c. Is the value of IQR/s approximately equal to 1.3? What does this imply?

6.23 Normal probability plots for three data sets are shown below. Which plot indicates that the data are approximately normally distributed?

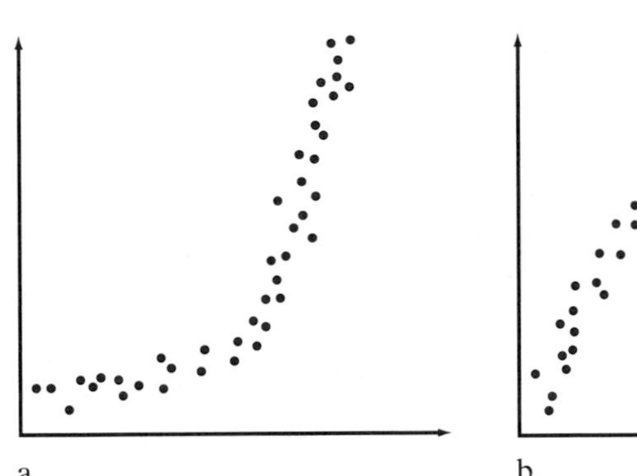

a.

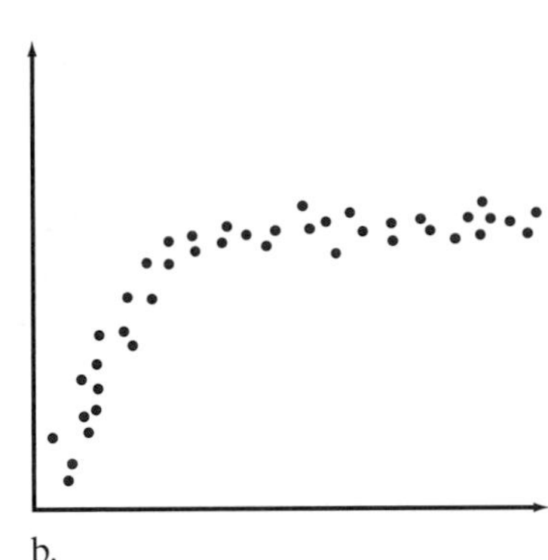

b.

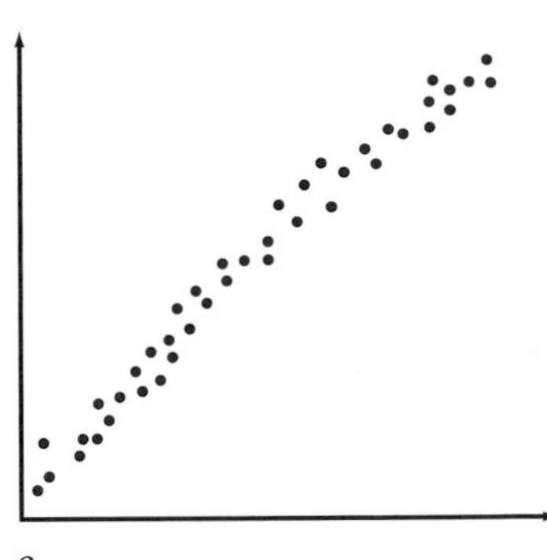

c.

6.24 The sample data for Problems 2.18 (p. 75) and 3.45 (p. 159) are reproduced below.

5.9	5.3	1.6	7.4	8.6	1.2	2.1
4.0	7.3	8.4	8.9	6.7	4.5	6.3
7.6	9.7	3.5	1.1	4.3	3.3	8.4
1.6	8.2	6.5	1.1	5.0	9.4	6.4

a. Use the stem-and-leaf plot from Problem 2.18 to assess whether the data are from an approximately normal distribution.

b. Compute s for the sample data.

c. Use the values of Q_1 and Q_3 from Problem 3.45 and the value of s from part **b** to assess whether the data come from an approximately normal distribution.

d. Generate a normal probability plot for the data and use it to assess whether the data are approximately normal.

Applying the Concepts

6.25 An article published in *Applied Psycholinguistics* (June 1998) studied the language skills of young children (16–30 months old) from low income families. Sixty-five children completed the Communicative Development Inventory (CDI) exam. One of the variables measured on each child was sentence complexity score. The mean sentence complexity score (measured on a 0 to 48 point scale) was 7.62 with a standard deviation of 8.91. Demonstrate why the distribution of sentence complexity scores for low income children is unlikely to be normally distributed.

6.26 How accurate are you at the game of darts? Researchers at Iowa State University attempted to develop a probability model for dart throws (*Chance,* Summer 1997). For each of 590 throws made at certain targets on a dart board, the distance from the dart to the target point was measured (to the nearest millimeter). The error distribution for the dart throws is described by the frequency table on p. 304.

a. Construct a histogram for the data. Is the error distribution for the dart throws approximately normal?

b. Descriptive statistics for the distances from the target for the 590 throws are given below. Use this information to decide whether the error distribution is approximately normal.

$\bar{x} = 24.4$ mm $\quad s = 12.8$ mm $\quad Q_1 = 14$ mm $\quad Q_3 = 34$ mm

c. Construct a normal probability plot for the data and use it to assess whether the error distribution is approximately normal.

Distance from Target (mm)	Frequency	Distance from Target (mm)	Frequency	Distance from Target (mm)	Frequency
0	3	21	15	41	7
2	2	22	19	42	7
3	5	23	13	43	11
4	10	24	11	44	7
5	7	25	9	45	5
6	3	26	21	46	3
7	12	27	16	47	1
8	14	28	18	48	3
9	9	29	9	49	1
10	11	30	14	50	4
11	14	31	13	51	4
12	13	32	11	52	3
13	32	33	11	53	4
14	19	34	11	54	1
15	20	35	16	55	1
16	9	36	13	56	4
17	18	37	5	57	3
18	23	38	9	58	2
19	25	39	6	61	1
20	21	40	5	62	2
				66	1

Source: Stern, H. S., and Wilcox, W. "Shooting darts." *Chance,* Vol. 10, No. 3, Summer 1997, p. 17 (adapted from Figure 2).

Problem 6.26

6.27 Refer to the study of British Columbia's boreal forest, Problem 6.11 (p. 296). The diameters at breast height (in meters) for a sample of 28 trembling aspen trees are listed in the table. Determine whether the sample data are from an approximately normal distribution.

12.4	17.3	27.3	19.1	16.9	16.2	20.0
16.6	16.3	16.3	21.4	25.7	15.0	19.3
12.9	18.6	12.4	15.9	18.8	14.9	12.8
24.8	26.9	13.5	17.9	13.2	23.2	12.7

Source: Schotz, H. "Fish Creek Community Forest: Exploratory statistical analysis of selected data." Working paper, Northern Lights College, British Columbia, Canada.

6.28 Refer to the *Journal of Information Systems* study of intrusion-detection systems, Problem 6.17 (p. 297). According to the study, many intrusion-detection systems assume that the data being monitored are normally distributed when such data are clearly nonnormal. Consequently, the intrusion-detection system may lead to inappropriate conclusions. The researcher considered the following data on input-output (I/O) units utilized by a sample of 44 users of a system. Assess whether the data are normally distributed.

15	5	2	17	4	3	1	1	0	0	0	0
0	0	0	20	9	0	0	0	1	6	1	3
1	0	6	0	0	0	0	1	14	0	7	0
2	9	4	0	0	0	9	10				

Source: O'Leary, D. E. "Intrusion-detection systems." *Journal of Information Systems,* Spring 1992, p. 68 (Table 2).

6.29 Behaviorists have developed a 10-item questionnaire to measure the maturity of small groups. In a study published in *Small Group Behavior* (May 1988), a class of undergraduate college students was divided into two groups, immature and mature, based on their answers to the 10-item questionnaire. A final project was then assigned and, at the end of the semester, student performances were evaluated. The grades for a sample of 25 immature-group students and 25 mature-group students are listed in the table, followed by an Excel stem-and-leaf display and normal probability plot for each of the two data sets. Based on these graphs, assess whether the grade distributions are approximately normal.

SMALLGRP

Immature Group					Mature Group				
74	75	75	75	78	73	75	78	84	85
79	79	81	81	84	85	85	87	91	91
84	85	85	86	86	92	92	93	94	95
87	88	88	89	90	95	98	100	100	100
91	92	93	95	96	100	100	100	100	100

	A	B	C	D
1	Stem-and-Leaf Displays			
2	Stem unit:	10		
3				
4	for Immature			
5	7	4 5 5 5 8 9 9		
6	8	1 1 4 4 5 5 6 6 7 8 8 9		
7	9	0 1 2 3 5 6		
8	10			
9				
10	for Mature			
11	7	3 5 8		
12	8	4 5 5 5 7		
13	9	1 1 2 2 3 4 5 5 8		
14	10	0 0 0 0 0 0 0 0		
15				

E **Problem 6.29**

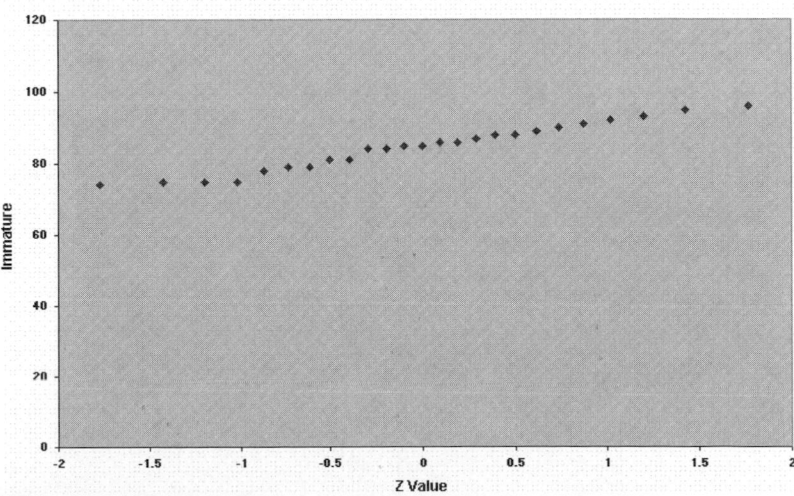

E **Problem 6.29**

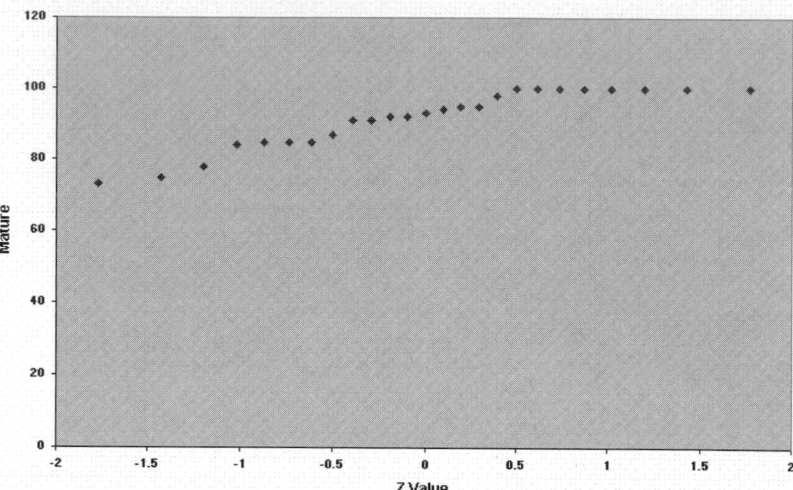

E Problem 6.29

6.30 A chemical analysis of metamorphic rock in western Turkey was reported in *Geological Magazine* (May 1995). The trace amount (in parts per million) of the element rubidium was measured for 20 rock specimens. The data are listed below. Assess whether the sample data come from a normal population.

RUBIDIUM

164	286	355	308	277	330	323	370	241	402
301	200	202	341	327	285	277	247	213	424

Source: Bozkurt, E., et al. "Geochemistry and tectonic significance of augen gneisses from the southern Menderes Massif (West Turkey)." *Geological Magazine,* Vol. 132, No. 3, May 1995, p. 291 (Table 1).

6.31 *Meta-analysis* is a procedure to summarize and compare the findings of independent studies on similar scientific topics. *Chance* (Fall 1997) describes a meta-analysis on the relationship between social behavior and judgments based on first impressions. A literature search revealed 38 independent studies with usable results. The correlation coefficient (r) between the behavioral variables of interest was recorded for each study. A stem-and-leaf display for the 38 correlations is shown at left. Comment on whether the distribution of correlations is approximately normal.

Stem	*Leaf*
.1	0 0 4 5 6 6
.2	1 1 1 2 3 3 4 5 6 6 7 8 9
.3	1 3 5
.4	0 0 0 1 7
.5	0 2 2 3 4 4
.6	3 8
.7	3 4
.8	7
.9	
1.0	

Source: Ambady, N., and Rosenthal, R. "Judging social behavior using 'Thin slices'." *Chance,* Vol. 10, No. 4, Fall 1997, p. 17 (Table 4).

Problem 6.31

6.32 The Harris Corporation data on voltage readings at two locations, Problem 3.38 (p. 154), are reproduced here. Determine whether the voltage readings at each location are approximately normal.

VOLTAGE

Old Location			New Location		
9.98	10.12	9.84	9.19	10.01	8.82
10.26	10.05	10.15	9.63	8.82	8.65
10.05	9.80	10.02	10.10	9.43	8.51
10.29	10.15	9.80	9.70	10.03	9.14
10.03	10.00	9.73	10.09	9.85	9.75
8.05	9.87	10.01	9.60	9.27	8.78
10.55	9.55	9.98	10.05	8.83	9.35
10.26	9.95	8.72	10.12	9.39	9.54
9.97	9.70	8.80	9.49	9.48	9.36
9.87	8.72	9.84	9.37	9.64	8.68

Source: Harris Corporation, Melbourne, Fla.

6.4 The Normal Approximation to the Binomial Distribution (Optional)

It can be shown (proof omitted) that the binomial probability distribution of Chapter 5 approaches a normal distribution as the sample size n becomes larger. That is, the histogram for a binomial random variable is shaped nearly like a normal distribution when n is large. Consequently, we can use the normal probability distribution to approximate binomial probabilities using the technique of Section 6.2.

The normal approximation for a binomial random variable x is reasonably good even for small samples—say, n as small as 10—when $p = .5$ is the probability of success and the distribution of x is symmetric about its mean, $\mu = np$. When p is near 0 (or 1), the binomial probability distribution will tend to be skewed to the right (or left), but this skewness will diminish as n becomes large. In general, the approximation can be used when n is large enough so that both np and $n(1 - p)$ are greater than or equal to 5.

CONDITIONS REQUIRED TO USE THE NORMAL APPROXIMATION TO THE BINOMIAL DISTRIBUTION

$$\text{Both} \quad np \geq 5 \quad \text{and} \quad n(1 - p) \geq 5$$

EXAMPLE 6.12

NORMAL APPROXIMATION TO THE BINOMIAL DISTRIBUTION

Suppose x has a binomial probability distribution with $n = 10$ and $p = .5$.

a. Graph $p(x)$ and superimpose a normal distribution with $\mu = np$ and $\sigma = \sqrt{np(1 - p)}$ on the graph.

b. Use Table B.2 to find $P(x \leq 4)$.

c. Use the normal approximation to the binomial probability distribution to find an approximation to $P(x \leq 4)$.

Solution

a. The graphs of $p(x)$ and a normal distribution with $\mu = np = 10(.5) = 5$ and $\sigma = \sqrt{np(1 - p)} = \sqrt{10(.5)(.5)} = 1.58$ are shown in Figure 6.16. Note that both np and $n(1 - p)$ are equal to 5; thus, the conditions for a good normal approximation are satisfied.

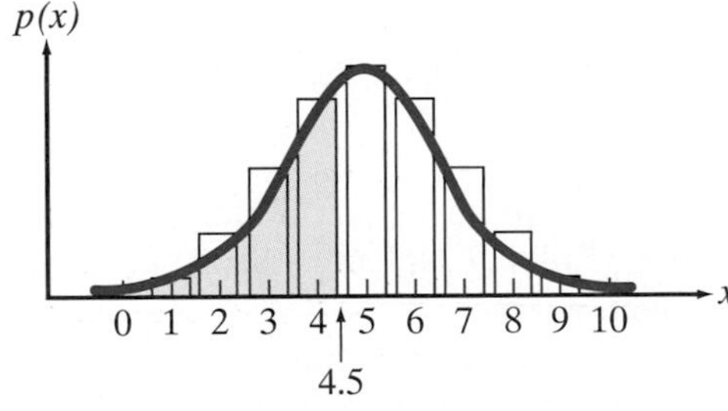

Figure 6.16 Binomial Distribution for $n = 10$ and $p = .5$

b. Using Table B.2 in the Appendix, or MINITAB, or PHStat and Excel, we obtain:

$$P(x \leq 4) = p(0) + p(1) + p(2) + p(3) + p(4) = .3770$$

c. By examining Figure 6.16, you can see that $P(x \leq 4)$ is the area under the normal curve to the left of $x = 4.5$. (Note that the area to the left of $x = 4$ would *not*

be appropriate because it would omit half the probability rectangle corresponding to $x = 4$.) We need to add .5 to 4 before calculating the probability in order to correct for the fact that we are using a continuous probability distribution to approximate a discrete probability distribution. The value .5 is called the **continuity correction factor** for the normal approximation (see the box).

The z value corresponding to the corrected value of $x = 4.5$ is:

$$\begin{aligned} z &= (x - \mu)/\sigma \\ &= (4.5 - 5)/1.58 \\ &= -.32 \end{aligned}$$

Now we use Table B.3 (the standard normal table) to find:

$$P(x \leq 4) \approx P(z \leq -.32) = .5 - .1255 = .3745$$

Even though the sample size is small ($n = 10$ in this case), the normal approximation of .3745 is close to the true probability of .3770.

CONTINUITY CORRECTION FOR THE NORMAL APPROXIMATION TO A BINOMIAL DISTRIBUTION

Let x be a binomial random variable with parameters n and p so that $\mu = np$ and $\sigma = \sqrt{np(1 - p)}$; and, let z be a standard normal random variable. Then

$$P(x \leq a) = P\left[z < \frac{(a + .5) - \mu}{\sigma}\right] \quad \text{(See Figure 6.17a)}$$

$$P(x \geq a) = P\left[z > \frac{(a - .5) - \mu}{\sigma}\right] \quad \text{(See Figure 6.17b)}$$

$$P(a \leq x \leq b) = P\left[\frac{(a - .5) - \mu}{\sigma} < z < \frac{(a + .5) - \mu}{\sigma}\right] \quad \text{(See Figure 6.17c)}$$

[*Note:* $P(x < a) = P(x \leq a - 1)$ and $P(x > a) = P(x \geq a + 1)$]

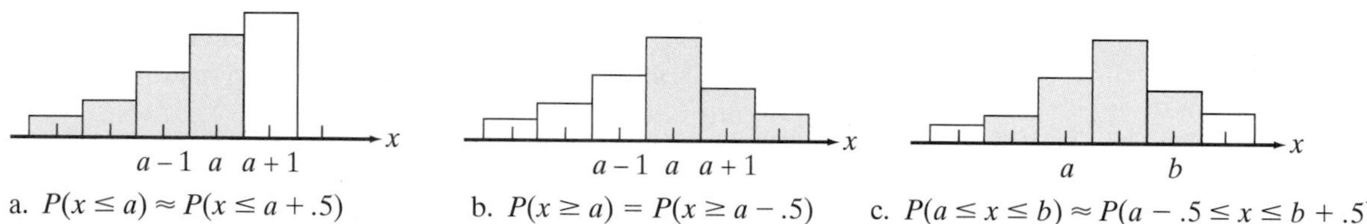

a. $P(x \leq a) \approx P(x \leq a + .5)$ b. $P(x \geq a) = P(x \geq a - .5)$ c. $P(a \leq x \leq b) \approx P(a - .5 \leq x \leq b + .5)$

Figure 6.17 Illustration of Continuity Correction for the Normal Approximation to the Binomial Distribution

In sample surveys and opinion polls conducted by professional pollsters such as Harris, Gallup, CNN, etc., the sample size n is typically large (usually $n \geq 500$). When such a large n is used, the continuity correction can be conveniently omitted from normal approximation calculations without jeopardizing the validity of the results. This point is illustrated in the next example.

EXAMPLE 6.13 **CONTINUITY CORRECTION WITH LARGE *n***

Consider an opinion poll in which we will ask 2,000 randomly selected people whether they prefer Coke or Pepsi. Let x be the number in the sample that prefer

Coke. Then x is a binomial random variable with $n = 2{,}000$. Assume that the probability of success (prefer Coke) in the population is $p = .5$.

a. Use the normal approximation to find $P(x \geq 1{,}049)$.

b. Repeat part **a**, but do not use the continuity correction factor.

Solution

a. First, we find the mean and standard deviation of x, the number of people in the sample of 2,000 that prefer Coke:

$$\mu = np = (2{,}000)(.5) = 1{,}000$$

$$\sigma = \sqrt{np(1-p)} = \sqrt{(2{,}000)(.5)(.5)} = 22.36$$

Using the continuity correction factor, $P(x \geq 1{,}049)$ corresponds to $P(x \geq 1{,}048.5)$. The normal approximation proceeds as follows:

$$P(x \geq 1{,}048.5) = P\left[z > \frac{(1{,}048.5 - \mu)}{\sigma}\right]$$

$$= P\left[z > \frac{(1{,}048.5 - 1{,}000)}{22.36}\right] = P(z > 2.17)$$

$$= .5 - .4850 \quad \text{(from Table B.3)}$$

$$= .0150$$

b. Ignoring the continuity correction factor, we obtain:

$$P(x \geq 1{,}049) = P\left[z > \frac{(1{,}049 - \mu)}{\sigma}\right]$$

$$= P\left[z > \frac{(1{,}049 - 1{,}000)}{22.36}\right] = P(z > 2.19)$$

$$= .5 - .4857 \quad \text{(from Table B.3)}$$

$$= .0143$$

You can see that the two probabilities, parts **a** and **b**, are nearly identical. For practical reasons, then, the continuity correction can be ignored when using the normal approximation to the binomial distribution with a large n.

PROBLEMS FOR SECTION 6.4

Using the Tools

6.33 Assume that x is a binomial random variable with n and p as specified in parts **a–f**. For which cases would it be appropriate to use a normal distribution to approximate the binomial distribution?

a. $n = 100, p = .01$

b. $n = 20, p = .6$

c. $n = 10, p = .4$

d. $n = 1{,}000, p = .05$

e. $n = 100, p = .8$

f. $n = 35, p = .7$

6.34 Suppose x is a binomial random variable with $p = .3$ and $n = 15$.

a. Would it be appropriate to approximate the probability distribution of x with a normal distribution? Explain.

b. Assuming that a normal distribution provides an adequate approximation to the distribution of x, what are the mean and variance of the approximating normal distribution?

c. Use Table B.2 (or MINITAB, or PHStat and Excel) to find the exact value of $P(x \leq 8)$.

d. Use the normal approximation to find $P(x \leq 8)$.

e. Compare the results obtained in parts **c** and **d**.

6.35 Assume that x is a binomial random variable with $n = 20$ and $p = .6$. Use Tables B.2 and B.3 (or MINITAB, or PHStat and Excel) and the normal approximation to find the exact and approximate values, respectively, for the following probabilities:

a. $P(x > 13)$ **b.** $P(x \leq 6)$ **c.** $P(8 < x < 16)$

6.36 Assume that x is a binomial random variable with $n = 1{,}000$ and $p = .40$. Use a normal approximation to find the following:

a. $P(x \leq 440)$ **b.** $P(410 \leq x \leq 415)$ **c.** $P(x \geq 420)$

Applying the Concepts

6.37 In Problem 5.22 (p. 256) you learned that 60% of parents with young children condone spanking as a regular form of punishment (*Tampa Tribune,* Oct. 5, 2000).

a. For a random sample of 150 parents with children, what is the approximate probability that more than half condone spanking?

b. For a random sample of 150 parents with children, what is the approximate probability that fewer than 50 condone spanking?

c. Would you expect to observe fewer than 30 parents who condone spanking in a sample of 150? Explain.

6.38 Refer to the FTC study of the pricing accuracy of supermarket electronic scanners, Problem 5.23 (p. 256). Recall that the probability that a scanned item is priced incorrectly is .03.

a. Suppose 10,000 supermarket items are scanned. What is the approximate probability that you observe at least 100 items with incorrect prices?

b. Suppose 100 items are scanned and you are interested in the probability that fewer than 5 are incorrectly priced. Explain why the approximate method of part **a** may not yield an accurate estimate of the probability.

6.39 The *Statistical Abstract of the United States* reports that 25% of the country's households are composed of one person. If 1,000 randomly selected homes are to participate in a Nielsen survey to determine television ratings, find the approximate probability that no more than 240 of these homes are one-person households.

6.40 A recent poll by the Gallup Organization found that 40% of employees miss work due to a back injury of some kind (*National Underwriter,* Apr. 5, 1999). What is the probability that more than 60 in a random sample of 400 employees will miss work due to a back injury?

6.5 Sampling Distributions

In Chapter 5 and the previous sections of this chapter, we assumed that we knew the probability distribution of a random variable, and using this knowledge we were able to find the mean, variance, and probabilities associated with the random variable. However, in most practical applications, the true mean and standard deviation are unknown quantities that have to be estimated. Thus, our objective is to estimate a numerical characteristic of the population. These population

characteristics are called **parameters** (see Definition 3.15, p. 178). Thus, p, the probability of success in a binomial experiment, and μ and σ, the mean and standard deviation of a normal distribution, are examples of parameters.

Recall (from Chapter 1) that inferential statistics involves using information in a sample to make inferences about unknown population parameters. Thus, we might use the sample mean $\bar{x}$ to estimate μ, or the sample standard deviation s to estimate σ. Quantities like $\bar{x}$ and s that are computed from the sample are called *sample statistics*.

Definition 6.4

A summary measure computed from the observations in a sample is called a **sample statistic**.

Although it is usually unknown to us, the value of a population parameter (e.g., μ) is constant; its value does not vary from sample to sample. However, the value of a sample statistic (e.g., $\bar{x}$) is highly dependent on the particular sample that is selected. Since statistics vary from sample to sample, any inferences based on them will necessarily be subject to some uncertainty. How, then, do we judge the reliability of a sample statistic as a tool in making an inference about the corresponding population parameter? Fortunately, the uncertainty of a statistic generally has characteristic properties that are known to us, and that are reflected in its *sampling distribution*.

Definition 6.5

The **sampling distribution** of a sample statistic (based on n observations) is the distribution of the values of the statistic theoretically generated by taking repeated random samples of size n and computing the value of the statistic for each sample.

The notion of a sampling distribution is illustrated in the next three examples.

EXAMPLE 6.14

CONCEPT OF A SAMPLING DISTRIBUTION

CHKOUT

Consider the 500 supermarket customer checkout times in the file **CHKOUT.** Assume this data set is the target population and we want to estimate μ, the mean checkout time of these 500 observations. Demonstrate how to physically generate the sampling distribution of $\bar{x}$, the mean of a random sample of $n = 5$ observations, from the population of 500 checkout times in **CHKOUT.**

Solution

The sampling distribution for the statistic $\bar{x}$, based on a random sample of $n = 5$ measurements, would be generated in this manner: (1) Select a random sample of five measurements from the population of 500 checkout times; (2) Compute and record the value of $\bar{x}$ for this sample; (3) Return these five measurements to the population and repeat the procedure, i.e., draw another random sample of $n = 5$ measurements and record the value of $\bar{x}$ for this sample; (4) Return these measurements and continue repeating the process.

If this sampling procedure could be repeated an infinite number of times, as shown in Figure 6.18 (p. 312), the infinite number of values of $\bar{x}$ obtained would form a new population. The data in this population could be summarized in a relative frequency distribution, called the **sampling distribution of the sample mean $\bar{x}$.**

Figure 6.18 Generating the Theoretical Sampling Distribution of the Sample Mean $\bar{x}$

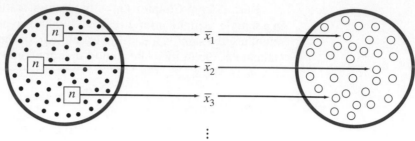

Select sample of size $n = 5$ checkout times from target population

Calculate $\bar{x}$

$\bar{x}_1$

$\bar{x}_2$

$\bar{x}_3$

$\vdots$

Population consisting of 500 checkout times

Repeat this process an infinite number of times

Sampling distribution of $\bar{x}$ (i.e., theoretical population of $\bar{x}$'s)

The task described in Example 6.14, which may seem impractical if not impossible, is not performed in actual practice. Instead, the sampling distribution of a statistic is obtained by applying mathematical theory or computer simulation, as illustrated in the next example.

EXAMPLE 6.15 APPROXIMATING A SAMPLING DISTRIBUTION

CHKOUT

Use computer simulation to find the approximate sampling distribution of $\bar{x}$, the mean of a random sample of $n = 5$ observations from the population of 500 checkout times in **CHKOUT.**

Solution

We obtained a large number—namely, 100—of computer-generated random samples of size $n = 5$ from the target population. The first 10 samples are presented in Table 6.4.

TABLE 6.4 First 10 Samples of $n = 5$ Checkout Times from CHKOUT File

Sample	Customer Checkout Time (seconds)				
1	8	15	28	155	35
2	16	66	45	50	35
3	50	10	50	35	15
4	31	180	53	30	353
5	150	25	15	10	20
6	53	10	113	81	97
7	93	237	6	40	23
8	12	16	35	49	100
9	27	80	50	20	20
10	66	63	50	40	50

For example, the first computer-generated sample contained the following measurements: 8, 15, 28, 155, 35. The corresponding value of the sample mean is

$$\bar{x} = \frac{\Sigma x}{n} = \frac{8 + 15 + 28 + 155 + 35}{5} = 48.2 \text{ seconds}$$

For each sample of five observations, the sample mean $\bar{x}$ was computed. The 100 values of $\bar{x}$ are summarized in the Excel histogram shown in Figure 6.19. This distribution approximates the sampling distribution of $\bar{x}$ for a sample of size $n = 5$.

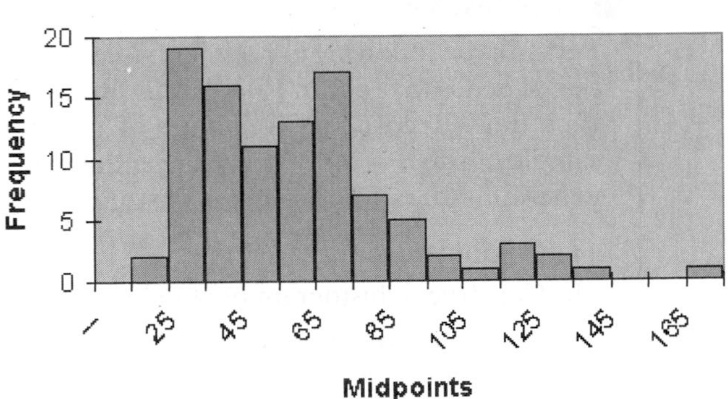

E Figure 6.19 Sampling Distribution of $\bar{x}$: 100 Random Samples of $n = 5$ Checkout Times

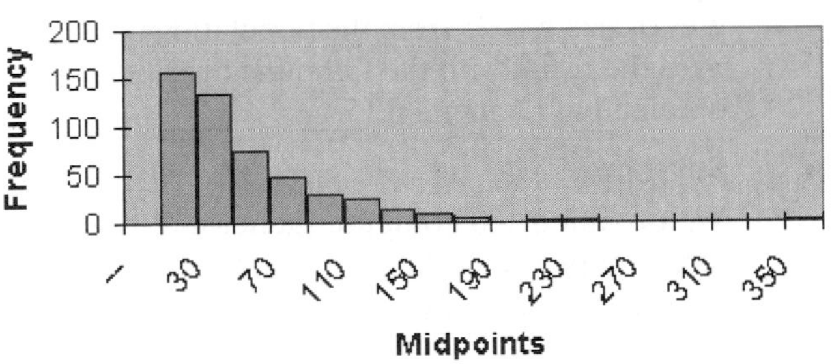

a.

E Figure 6.20 Histogram of the Population of 500 Customer Checkout Times

	A	B
1	Checkout Times	
2		
3	Mean	50.118
4	Standard Error	2.194225
5	Median	35
6	Mode	30
7	Standard Deviation	49.06435
8	Sample Variance	2407.311
9	Kurtosis	9.731435
10	Skewness	2.541635
11	Range	351
12	Minimum	2
13	Maximum	353
14	Sum	25059
15	Count	500
16	Largest(1)	353
17	Smallest(1)	2

b.

A histogram for the population of 500 checkout times in **CHKOUT,** generated using Excel, is displayed in Figure 6.20a above, along with descriptive statistics for the data set (Figure 6.20b at left). From the printout, the population of interest in Example 6.15 has mean $\mu = 50.1$ seconds and standard deviation $\sigma = 49.1$ seconds. Let us compare the histogram for the sampling distribution of $\bar{x}$ (Figure 6.19) with the histogram for the population shown in Figure 6.20a. Note that the values of $\bar{x}$ in Figure 6.19 tend to cluster around the population mean, $\mu = 50.1$. Also, the values of the sample mean are less spread out (that is, they have less variation) than the population values shown in Figure 6.20a. These two observations are borne out by comparing the means and standard deviations of the two sets of observations, as shown in Table 6.5.

TABLE 6.5 Comparison of the Population Distribution and the Approximate Sampling Distribution of $\bar{x}$, Based on 100 Samples of Size $n = 5$ and Size $n = 25$

	Mean	Standard Deviation
Population of 500 checkout times (Figure 6.20)	$\mu = 50.1$	$\sigma = 49.1$
100 values of $\bar{x}$ based on samples of size $n = 5$ (Figure 6.19)	55.6	28.5
100 values of $\bar{x}$ based on samples of size $n = 25$ (Figure 6.21)	50.3	9.4

a. Calculate $\bar{x}$ for each of the 50 samples.

b. Construct a histogram for the 50 sample means. This figure represents an approximation to the sampling distribution of $\bar{x}$ based on samples of size $n = 5$.

c. Compute the mean and standard deviation for the 50 sample means. Locate these values on the histogram of part **b**. Note how the sample means cluster about $\mu = 4.5$.

6.44 Refer to Problem 6.43. Combine pairs of samples (moving down the columns of the table) to obtain 25 samples of $n = 10$ measurements.

a. Calculate $\bar{x}$ for each of the 25 samples.

b. Construct a histogram for the 25 sample means. This figure represents an approximation to the sampling distribution of $\bar{x}$ based on samples of size $n = 10$. Compare with the figure constructed in Problem 6.43.

c. Compute the mean and standard deviation for the 25 sample means, and locate them on the histogram. Note how the sample means cluster about $\mu = 4.5$.

d. Compare the standard deviations of the two sampling distributions in Problems 6.43 and 6.44. Which sampling distribution has less variation?

Applying the Concepts

6.45 Consider a random sample of 25 measurements selected from each of the following populations. Describe the sampling distribution of $\bar{x}$ for each.

a. Percent body fat measurements for all recent Olympic athletes

b. Noise levels (measured in decibels) for all passenger flights taking off at Tampa International airport

c. Task completion times (in minutes) for all patients who submit to a psychological exam requiring consecutively numbered circles to be connected

6.46 The following data represent the number of days absent per year in a population of six employees of a small company.

DAYSABSENT

1	3	6	7	7	12

a. Assuming that you sample *without* replacement, select all possible samples of size 2 and set up the sampling distribution of the mean.

b. Compute the mean of all the sample means and also compute the population mean. Are they equal?

c. Repeat parts **a** and **b** for all possible samples of size 3.

d. Compare the shape of the sampling distribution of the mean obtained in parts **a** and **c**. Which sampling distribution seems to have the least variability? Why?

6.47 Audiologists at Baylor College of Medicine fitted individuals suffering from sensorineural hearing loss with various devices (e.g., hearing aids, cochlear implants, etc.) designed to amplify sounds (*Ear & Hearing,* Apr. 1995). One subject was tested under six different experimental conditions. Under each condition, the signal intensity necessary for a 50% improvement in hearing was recorded. The six intensity scores (measured in db HL) were: 80, 85, 75, 65, 60, and 55. To test the effectiveness of the hearing device used, the audiologists were interested in the sampling distribution of $\bar{x}$, the mean intensity score for a random sample of three scores selected from the six scores recorded for the subject.

DBHL

a. List the different random samples of size $n = 3$ that can be selected from the six intensity scores.

b. For each sample, compute $\bar{x}$.

c. Use a stem-and-leaf display to portray the sampling distribution of $\bar{x}$.

d. Find the mean and the standard deviation of the sampling distribution of $\bar{x}$.

6.48 Use Excel, MINITAB, or Table B.1 to obtain 30 random samples of size $n = 5$ from the "population" of 500 nicotine contents of cigarettes in the **FTC** file. (Alternatively, each class member may generate several random samples, and the results can be pooled.)

FTC

a. Calculate $\bar{x}$ for each of the 30 samples. Construct a histogram for the 30 sample means.

b. Compute the average of the 30 sample means.

c. Compute the standard deviation of the 30 sample means.

d. Locate the average of the 30 sample means, computed in part **b**, on the histogram. This value could be used as an estimate for μ, the mean of the entire population of 500 nicotine measurements.

6.49 Repeat parts **a**, **b**, **c**, and **d** of Problem 6.48, using random samples of size $n = 10$. Compare the relative frequency distribution with that of Problem 6.48**a**. Do the values of $\bar{x}$ generated from samples of size $n = 10$ cluster more closely about μ?

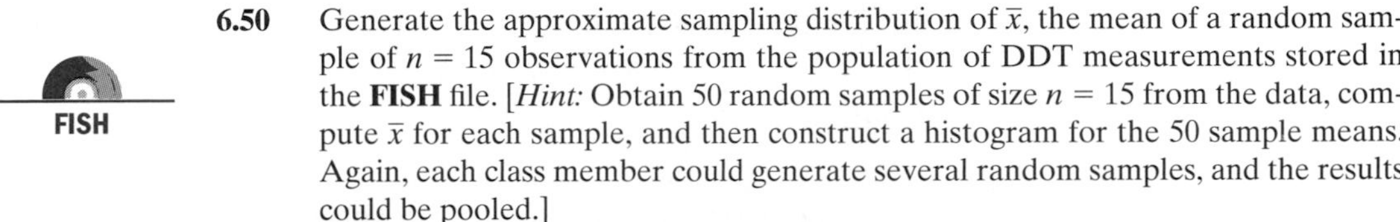

6.50 Generate the approximate sampling distribution of $\bar{x}$, the mean of a random sample of $n = 15$ observations from the population of DDT measurements stored in the **FISH** file. [*Hint:* Obtain 50 random samples of size $n = 15$ from the data, compute $\bar{x}$ for each sample, and then construct a histogram for the 50 sample means. Again, each class member could generate several random samples, and the results could be pooled.]

6.6 The Sampling Distribution of the Mean and the Central Limit Theorem

Estimating the mean checkout time for all customers of a certain supermarket, or the average increase in heart rate for all human subjects after 15 minutes of vigorous exercise, or the mean yield per acre of farmland, are all examples of practical problems in which the goal is to make an inference about the mean μ of some target population. In the previous section, we have indicated that the sample mean $\bar{x}$ is often used as a tool for making an inference about the corresponding population parameter μ, and we have shown how to *approximate* its sampling distribution. The **Central Limit Theorem**, of fundamental importance in statistics, provides information about the *actual* sampling distribution of $\bar{x}$.

The sampling distribution of $\bar{x}$, in addition to being approximately normal, has other known characteristics, which are summarized in the next box.

PROPERTIES OF THE SAMPLING DISTRIBUTION OF $\bar{x}$

If $\bar{x}$ is the mean of a random sample of size n from a population with mean μ and standard deviation σ, then:

1. The sampling distribution of $\bar{x}$ has a mean equal to the mean of the population from which the sample was selected. That is, if we let $\mu_{\bar{x}}$ denote the mean of the sampling distribution of $\bar{x}$, then

$$\mu_{\bar{x}} = \mu$$

2. The sampling distribution of $\bar{x}$ has a standard deviation equal to the standard deviation of the population from which the sample was selected, divided by the square root of the sample size. That is, if we let $\sigma_{\bar{x}}$ denote the standard deviation of the sampling distribution of $\bar{x}$ (also called the **standard error** of $\bar{x}$), then

$$\sigma_{\bar{x}} = \frac{\sigma}{\sqrt{n}} \tag{6.4}$$

THE CENTRAL LIMIT THEOREM

If the sample size is *sufficiently large,* then the mean $\bar{x}$ of a random sample from a population has a sampling distribution that is approximately normal, *regardless of the shape of the relative frequency distribution of the target population.* As the sample size n increases, the sampling distribution of $\bar{x}$ moves closer and closer to the normal distribution.

EXAMPLE 6.17 PROPERTIES OF THE SAMPLING DISTRIBUTION OF $\bar{x}$

Refer to Examples 6.15 and 6.16, where we obtained repeated random samples of sizes $n = 5$ and $n = 25$ from the population of customer checkout times stored in **CHKOUT.** Recall that for this target population, we know the values of the parameters μ and σ:

Population mean: $\mu = 50.1$ seconds

Population standard deviation: $\sigma = 49.1$ seconds

Show that the empirical evidence obtained in Examples 6.15 and 6.16 supports the Central Limit Theorem and the two properties of the sampling distribution of $\bar{x}$.

Solution

In Figures 6.19 and 6.21 (pp. 313 and 314), we noted that the values of $\bar{x}$ cluster about the population mean, $\mu \approx 50$. This is guaranteed by property 1, which implies that, in the long run, the average of *all* values of $\bar{x}$ that would be generated in infinite repeated sampling would be equal to μ.

We also observed, from Table 6.5 (p. 313), that the standard deviation of the sampling distribution of $\bar{x}$, called the *standard error of* $\bar{x}$, decreases as the sample size increases from $n = 5$ to $n = 25$. Property 2 quantifies the decrease and relates it to the sample size. For our approximate (simulated) sampling distribution based on samples of size $n = 5$, we obtained a standard deviation of 28.5, whereas property 2 tells us that, for the actual sampling distribution of $\bar{x}$, the standard deviation is equal to

$$\sigma_{\bar{x}} = \frac{\sigma}{\sqrt{n}} = \frac{49.1}{\sqrt{5}} = 22.0$$

Similarly, for samples of size $n = 25$, the sampling distribution of $\bar{x}$ actually has a standard deviation of

$$\sigma_{\bar{x}} = \frac{\sigma}{\sqrt{n}} = \frac{49.1}{\sqrt{25}} = 9.8$$

The value we obtained by simulation was 9.4.

Finally, for sufficiently large samples, the Central Limit Theorem states that $\bar{x}$ will be approximately normally distributed, regardless of the shape of the original population. In our examples, the population from which the samples were selected is seen in Figure 6.20 (p. 313) to be highly skewed to the right. Note from Figures 6.19 and 6.21 that, although the sampling distribution of $\bar{x}$ tends to be bell-shaped in each case, the normal approximation improves when the sample size is increased from $n = 5$ (Figure 6.19) to $n = 25$ (Figure 6.21).

Self-Test 6.5

Let $\bar{x}$ represent the sample mean for a random sample of size $n = 100$ selected from a population with mean $\mu = 5$ and standard deviation $\sigma = 1$.

a. Find and interpret the value of $\mu_{\bar{x}}$.

b. Find and interpret the value of $\sigma_{\bar{x}}$.

How large must the sample size n be so that the normal distribution provides a good approximation for the sampling distribution of $\bar{x}$? The answer depends on the shape of the distribution of the sampled population, as shown by Figure 6.22. Generally speaking, the greater the skewness of the sampled population distribution, the larger the sample size must be before the normal distribution is an adequate approximation for the sampling distribution of $\bar{x}$. For most sampled populations, sample sizes of $n \geq 30$ will suffice for the normal approximation to be reasonable. We will use the normal approximation for the sampling distribution of $\bar{x}$ when the sample size is at least 30.

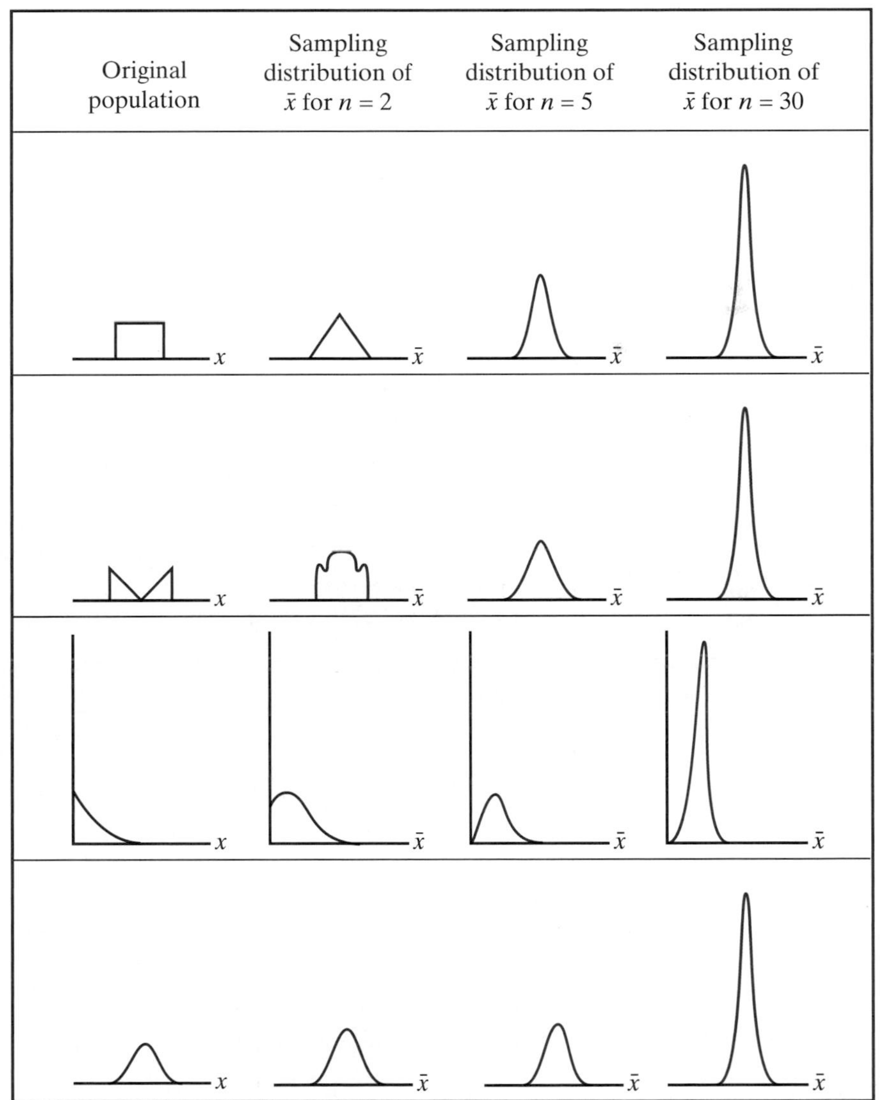

Figure 6.22 Sampling Distributions of $\bar{x}$ for Different Populations and Different Sample Sizes

EXAMPLE 6.18

CENTRAL LIMIT THEOREM APPLICATION

A study was conducted by engineers at Luton Airport (United Kingdom) to assess the suitability of concrete blocks as a surface for aircraft pavements. The original pavement at one end of the runway was overlaid with 80-mm-thick concrete blocks. A series of tests was carried out to determine the load classification number (LCN)—a measure of breaking strength—of the new surface. Let $\bar{x}$ represent the

mean LCN of a random sample of 40 concrete block sections on the repaved end of the runway.

a. Prior to resurfacing, the mean LCN of the original pavement was known to be $\mu = 60$ and the standard deviation was $\sigma = 10$. If the mean strength of the new concrete block surface is no different from that of the original surface, give the characteristics of the sampling distribution of $\bar{x}$.

b. If the mean strength of the new concrete block surface is no different from that of the original surface, find the probability that $\bar{x}$, the sample mean LCN of the 40 concrete block sections, exceeds 64.

c. The tests on the new concrete block surface actually resulted in $\bar{x} = 64.03$. Based on this result, what can you infer about the true mean LCN of the new surface?

Solution

a. Although we have no information about the shape of the distribution of the breaking strengths (LCNs) for sections of the new surface, we can apply the Central Limit Theorem to conclude that the sampling distribution of $\bar{x}$, the mean LCN of the sample, is approximately normally distributed. In addition, if $\mu = 60$ and $\sigma = 10$, the mean, $\mu_{\bar{x}}$, and the standard deviation, $\sigma_{\bar{x}}$, of the sampling distribution are given by

$$\mu_{\bar{x}} = \mu = 60$$

and

$$\sigma_{\bar{x}} = \frac{\sigma}{\sqrt{n}} = \frac{10}{\sqrt{40}} = 1.58$$

b. If the two surfaces are of equal strength, then $P(\bar{x} \geq 64)$, the probability of observing a mean LCN of 64 or more in the sample of 40 concrete block sections, is equal to the darker shaded area shown in Figure 6.23. Since the sampling distribution is approximately normal, with mean and standard deviation as obtained in part **a**, we can compute the desired area by obtaining the z score for $\bar{x} = 64$:

$$z = \frac{\bar{x} - \mu_{\bar{x}}}{\sigma_{\bar{x}}} = \frac{64 - 60}{1.58} = 2.53$$

Thus, $P(\bar{x} \geq 64) = P(z \geq 2.53)$. This probability (area) is found using Table B.3:

$$\begin{aligned} P(\bar{x} \geq 64) &= P(z \geq 2.53) \\ &= .5 - A \qquad \text{(see Figure 6.23)} \\ &= .5 - .4943 = .0057 \end{aligned}$$

If there is no difference between the true mean strengths of the new and original surfaces (i.e., $\mu = 60$ for both surfaces), the probability that we would obtain a sample mean LCN for concrete block of 64 or greater is only .0057.

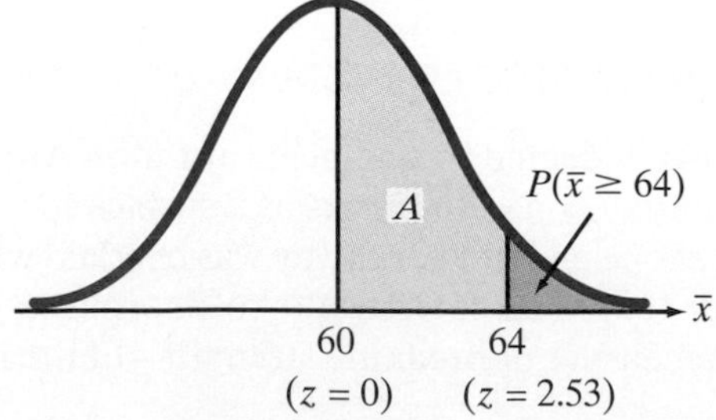

Figure 6.23 Sampling Distribution of $\bar{x}$ in Example 6.18

c. Since $\bar{x} = 64.03$ exceeds 64, this provides strong evidence that the true mean breaking strength of the new surface exceeds $\mu = 60$. Our reasoning stems from the rare event philosophy of Chapter 4, which states that such a large sample mean ($\bar{x} = 64.03$) is very unlikely to occur if $\mu = 60$.

Self-Test 6.6

For a random sample of size $n = 49$ selected from a population with $\mu = 17$ and $\sigma = 14$, find $P(\bar{x} < 15.5)$.

In practical terms, the Central Limit Theorem and the two properties of the sampling distribution of $\bar{x}$ assure us that the sample mean $\bar{x}$ is a reasonable statistic to use in making inferences about the population mean μ, and they allow us to compute a measure of the reliability of inferences made about μ. (This topic will be treated more thoroughly in Chapter 7.)

As we noted earlier, we will not be required to obtain sampling distributions by simulation or by mathematical arguments. Rather, for all the statistics to be used in this text, the sampling distribution and its properties will be presented as the need arises.

Statistics in the Real World Revisited

Applying the Central Limit Theorem to the Fire Fraud Data

In the furniture fire fraud case (first presented on p. 281), a fire destroyed a warehouse and all the furniture in it. In an attempt to estimate the "lost" profit—called gross profit margin (GPF)—the retailer sampled paper invoices for 253 of 3,005 furniture items sold the previous year. The retailer claimed that the 253 items were obtained by selecting an initial random sample of 134 items, and then augmenting this sample with a second random sample of 119 items. The mean GPFs for the two subsamples were calculated to be 50.6% and 51.0%, respectively, yielding an overall average GPF of 50.8%—a value deemed unusually high by experienced claims adjusters at the insurance company. Consequently, the retailer was accused of fraudulently representing its sampling methodology, i.e., of purposely selecting an unusual number of "high profit" items from the population in order to increase the average GPF of the sample.

Is it likely that two independent, random samples of size 134 and 119 will yield mean GPFs of at least 50.6% and 51.0%, respectively? (This was the question posed to a statistician retained by the CPA firm.) That is, we want to find the probability, $P(\bar{x}_1 \geq 50.6 \text{ and } \bar{x}_2 \geq 51.0)$, where $\bar{x}_1$ is the sample mean for the first sample of 134 items and $\bar{x}_2$ is the sample mean for the second sample of 119 items. First note that if, in fact, the two samples are independently selected, then (applying Probability Rule #4 on p. 214)

$$P(\bar{x}_1 \geq 50.6 \text{ and } \bar{x}_2 \geq 51.0) = P(\bar{x}_1 \geq 50.6)P(\bar{x}_2 \geq 51.0)$$

Since both samples are large, the Central Limit Theorem states that the sampling distribution of $\bar{x}$ for either sample mean is approximately normal. The mean and standard deviation of the sampling distribution are also known since we know the mean μ and standard deviation σ of the population of 3,005 GPFs available in the **FIREFRAUD** file. From the MINITAB printout, Figure 6.13 (p. 294), $\mu = 48.9\%$ and $\sigma = 13.8\%$. Consequently, the sampling distributions are given as follows:

Sampling distribution of $\bar{x}_1$: Mean $\mu = 48.9$
Standard error $\sigma/\sqrt{n_1} = 13.8/\sqrt{134} = 1.19$

Sampling distribution of $\bar{x}_2$: Mean $\mu = 48.9$
Standard error $\sigma/\sqrt{n_2} = 13.8/\sqrt{119} = 1.27$

(Continued)

We compute the probability of interest (using Table B.3) as follows:

$$P(\bar{x}_1 \geq 50.6) = P\left(z > \frac{50.6 - 48.9}{1.19}\right)$$
$$= P(z > 1.43) = .5 - .4236 = .0764$$

$$P(\bar{x}_2 \geq 51.0) = P\left(z > \frac{51.0 - 48.9}{1.27}\right)$$
$$= P(z > 1.66) = .5 - .4515 = .0485$$

Then

$$P(\bar{x}_1 \geq 50.6 \text{ and } \bar{x}_2 \geq 51.0) = P(\bar{x}_1 \geq 50.6)P(\bar{x}_2 \geq 51.0)$$
$$= (.0764)(.0485) = .0037$$

Thus, the probability of obtaining independent random samples with mean GPFs at least as large as the values obtained by the retailer (50.6% and 51.0%) is only .0037—less than 4 chances in 1,000. This extremely small probability casts doubt on the believability of the retailer's claim about its sampling methodology.

[*Epilogue:* As a result of this analysis, the retailer admitted fraud and settled the case with the insurance company out of court.]

PROBLEMS FOR SECTION 6.6

Using the Tools

6.51 Suppose a random sample of n measurements is selected from a population with mean $\mu = 60$ and variance $\sigma^2 = 10$. For each of the following values of n, give the mean and standard deviation of the sampling distribution of the sample mean, $\bar{x}$.

a. $n = 10$ **b.** $n = 25$ **c.** $n = 50$ **d.** $n = 75$
e. $n = 100$ **f.** $n = 500$ **g.** $n = 1{,}000$

6.52 Suppose a random sample of $n = 100$ measurements is selected from a population with mean μ and standard deviation σ. For each of the following values of μ and σ, give the values of $\mu_{\bar{x}}$ and $\sigma_{\bar{x}}$:

a. $\mu = 10, \sigma = 20$ **b.** $\mu = 20, \sigma = 10$
c. $\mu = 50, \sigma = 300$ **d.** $\mu = 100, \sigma = 200$

6.53 A random sample of $n = 50$ observations is selected from a population with $\mu = 21$ and $\sigma = 6$. Calculate each of the following probabilities:

a. $P(\bar{x} < 23.1)$ **b.** $P(\bar{x} > 21.7)$ **c.** $P(22.8 < \bar{x} < 23.6)$

6.54 A random sample of $n = 225$ observations is selected from a population with $\mu = 70$ and $\sigma = 30$. Calculate each of the following probabilities:

a. $P(\bar{x} > 72.5)$ **b.** $P(\bar{x} < 73.6)$
c. $P(69.1 < \bar{x} < 74.0)$ **d.** $P(\bar{x} < 65.5)$

Applying the Concepts

6.55 Researchers at the Terry College of Business at the University of Georgia randomly sampled 344 students and asked them this question: "Over the course of your lifetime, what is the maximum number of years you expect to work for any one employer?" The mean and standard deviation of the sample were 19.1 years and 6 years, respectively (*Georgia Trend,* Dec. 1999).

a. Describe the sampling distribution of $\bar{x}$, the sample mean.
b. Assume the population mean is $\mu = 18.5$ years. Find $P(\bar{x} \geq 19.1)$ if $\sigma = 6$.
c. Assume the population mean is $\mu = 19.5$ years. Find $P(\bar{x} \geq 19.1)$ if $\sigma = 6$.

d. If $P(\bar{x} \geq 19.1) = .5$, what is the population mean?

e. If $P(\bar{x} \geq 19.1) = .2$, is the population mean greater or less than 19.1?

6.56 The National Institute for Occupational Safety and Health (NIOSH) evaluated the level of exposure of workers to the chemical dioxin, 2,3,7,8-TCDD. The distribution of TCDD levels in parts per trillion (ppt) of production workers at a Newark, New Jersey, chemical plant had a mean of 293 ppt and a standard deviation of 847 ppt (*Chemosphere,* Vol. 20, 1990). A graph of the distribution is shown here. In a random sample of $n = 50$ workers selected at the New Jersey plant, let $\bar{x}$ represent the sample mean TCDD level.

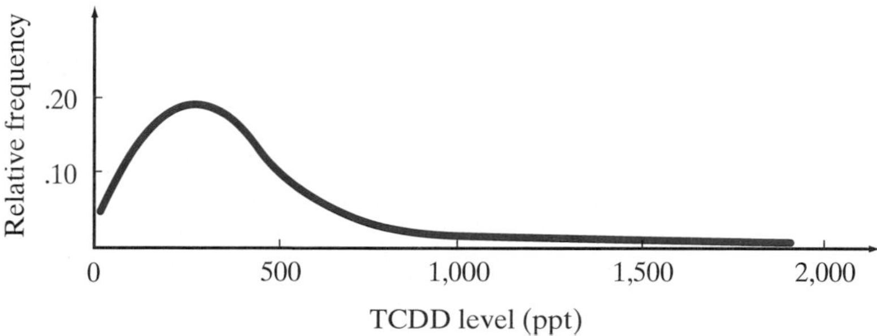

a. Find the mean and standard deviation of the sampling distribution of $\bar{x}$.

b. Draw a sketch of the sampling distribution of $\bar{x}$. Locate the mean on the graph.

c. Find the probability that $\bar{x}$ exceeds 550 ppt.

6.57 A telephone survey of adult literacy was administered to a large number of San Diego (CA) residents (*Journal of Literacy Research,* Dec. 1996). One question asked was: "How many times during a typical week do you read a book for pleasure?" The respondents read a book for pleasure, on average, 3.10 times per week, with a standard deviation of 2.78 times per week. Assume $\mu = 3.10$ and $\sigma = 2.78$ for all American adults.

a. Explain why the distribution of responses (i.e., number of times per week you read a book for pleasure) for the population is unlikely to be normal.

b. In a random sample of 500 American adults, let x represent the mean number of times per week the sampled adults read a book for pleasure. Explain why the sampling distribution of $\bar{x}$ is normal.

c. Refer to part **b**. Find $P(\bar{x} < 3)$.

6.58 Studies by neuroscientists at the Massachusetts Institute of Technology (MIT) reveal that melatonin, which is secreted by the pineal gland in the brain, functions naturally as a sleep-inducing hormone (*Tampa Tribune,* Mar. 1, 1994). Male volunteers were given various doses of melatonin or placebos (i.e., no hormones) and then placed in a dark room at midday and told to close their eyes and fall asleep on demand. Of interest to the MIT researchers is the time x (in minutes) required for each volunteer to fall asleep. With the placebo, the researchers found that the mean time to fall asleep was 15 minutes. Assume that with the placebo treatment $\mu = 15$ and $\sigma = 5$.

a. Consider a random sample of $n = 20$ men who are given the sleep-inducing hormone, melatonin. Let $\bar{x}$ represent the mean time to fall asleep for this sample. If the hormone is *not* effective in inducing sleep, describe the sampling distribution of $\bar{x}$.

b. Refer to part **a**. Find $P(\bar{x} \leq 5)$.

c. In the actual study, the mean time to fall asleep for the 20 volunteers was $\bar{x} = 5$. Use this result to make an inference about the true value of μ for those taking the melatonin.

6.59 Researchers at Marquette University presented MBA students (believed to be representative of entry-level managers) with decision-making situations that were clearly unethical in nature (*Journal of Business Ethics,* Vol. 6, 1987). The subjects' decisions were then rated on a scale of 1 ("definitely unethical") to 5 ("definitely

ethical"). When no references to ethical concern by the "company" were explicitly stated, the ratings had a mean of 3.00 and a standard deviation of 1.03. Assume that these values represent the population mean and standard deviation, respectively, under the condition "no reference to ethical concern."

a. Suppose we present a random sample of 30 entry-level managers with a similar situation and record the ratings of each. Find the probability that $\bar{x}$, the sample mean rating, is greater than 3.40.

b. Refer to part **a**. Prior to making their decisions, the 30 entry-level managers were all read a statement from the president of the "company" concerning the company's code of business ethics. The code advocates socially responsible behavior by all employees. The researchers theorize that the population mean rating of the managers under this condition will be larger than for the "no reference to ethical concern" condition. (A higher mean indicates a more ethical response.) If the sample mean $\bar{x}$ is 3.55, what can you infer about the population mean under the "stated concern" condition?

6.60 One measure of elevator performance is cycle time. Elevator cycle time is the time between successive elevator starts, which includes the time when the car is moving and the time when it is standing at a floor. *Simulation* (Oct. 1993) published a study on the use of a microcomputer-based simulator for elevator cycle time. The simulator produced an average cycle time μ of 26 seconds when traffic intensity was set at 50 persons every five minutes. Consider a sample of 200 simulated elevator runs and let $\bar{x}$ represent the mean cycle time of this sample.

a. What do you know about the distribution of x, the time between successive elevator starts? (Give the value of the mean and standard deviation of x and the shape of the distribution, if possible.)

b. What do you know about the distribution of $\bar{x}$? (Give the value of the mean and standard deviation of $\bar{x}$ and the shape of the distribution, if possible.)

c. Assume σ, the standard deviation of cycle time x, is 20 seconds. Use this information to calculate $P(\bar{x} \geq 26.8)$.

d. Repeat part **c** but assume $\sigma = 10$.

6.61 Refer to Problem 6.60. Cycle time is related to the distance (measured by number of floors) the elevator covers on a particular run, called *running distance*. The simulated distribution of running distance, x, during a down-peak period in elevator

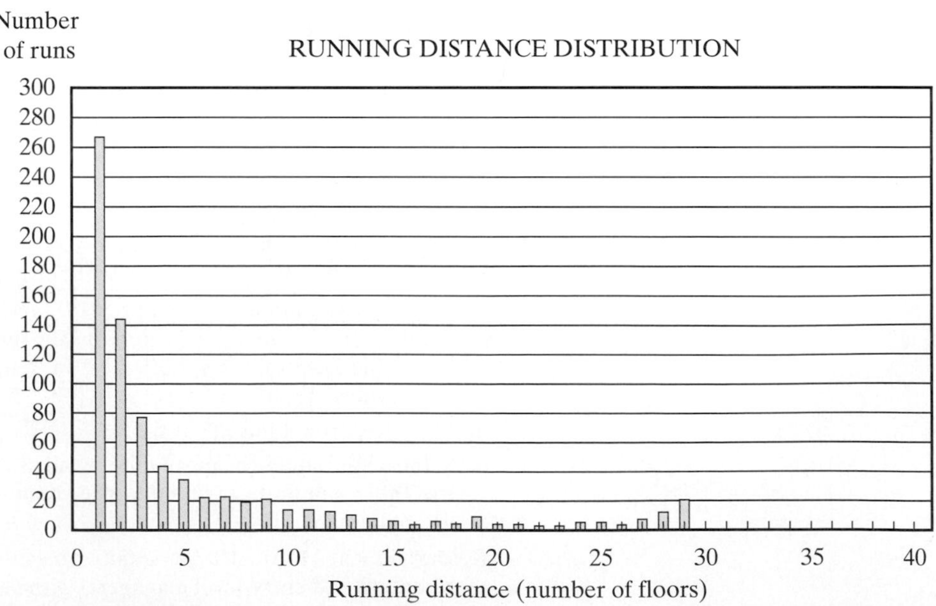

Source: Siikonen, M. L. "Elevator traffic simulation." *Simulation,* Vol. 61, No. 4, Oct. 1993, p. 266 (Figure 8).

traffic intensity is shown in the figure on p. 324. The distribution has mean $\mu = 5.5$ floors and standard deviation $\sigma = 7$ floors. Consider a random sample of 80 simulated elevator runs during a down-peak in traffic intensity. Of interest is the sample mean running distance, $\bar{x}$.

a. Find $\mu_{\bar{x}}$ and $\sigma_{\bar{x}}$.

b. Is the shape of the distribution of $\bar{x}$ similar to the figure? If not, sketch the distribution.

c. During a down-peak in traffic intensity, is it likely to observe a sample mean running distance of less than $\bar{x} = 5.3$ floors? Explain.

KEY TERMS *Starred (*) terms are from the optional section of this chapter.*

Central Limit Theorem 317
*Continuity correction factor 308
Continuous probability distribution 282
*Normal approximation to the binomial distribution 307
Normal (bell) curve 283
Normal distribution 283
Normal probability density function 283
Normal probability plot 300
Parameter 311
Probability model 282
Sample statistic 311
Sampling distribution 311
Sampling distribution of the sample mean 311
Standard error of the mean 317
Standard normal distribution 286
Standard normal random variable 286
Tail probability 292

KEY FORMULAS

Normal distribution:	$z = \dfrac{x - \mu}{\sigma}$	**(6.2)**, 284	$x_0 = \mu + z_0\sigma$	**(6.3)**, 293
Sampling distribution of $\bar{x}$:	$\mu_{\bar{x}} = \mu$		$\sigma_{\bar{x}} = \dfrac{\sigma}{\sqrt{n}}$	**(6.4)**, 317

KEY SYMBOLS

SYMBOL	DEFINITION
z	Standard normal random variable
z_0	Observed value of standard normal random variable
x	Random variable
x_0	Observed value of random variable
$\mu_{\bar{x}}$	Mean of the sampling distribution of $\bar{x}$
$\sigma_{\bar{x}}$	Standard deviation of the sampling distribution of $\bar{x}$ (also called standard error of $\bar{x}$)

CHECKING YOUR UNDERSTANDING

1. How do you find the probability that a continuous random variable falls between two points, a and b?
2. What are the properties of a normal distribution?
3. What is a standard normal distribution?
4. Give four methods for determining whether data are from an approximately normal distribution.
5. What is a normal probability plot?
6. When will the normal distribution provide a good approximation to the binomial distribution?
7. What is a continuity correction factor?
8. What is a sampling distribution of a sample statistic?
9. Give the properties of the sampling distribution of $\bar{x}$.
10. Under what conditions is the sampling distribution of $\bar{x}$ approximately normal?

SUPPLEMENTARY PROBLEMS *Starred (*) problems refer to the optional section of this chapter.*

6.62 In order to graduate, high school students in Florida must demonstrate their competence in mathematics by scoring at least 70 (out of a possible total of 100) on a mathematics achievement test. From published accounts, it is known that the mean score on this test is 78 and the standard deviation is 7.5. Assume that the test scores are approximately normally distributed.

a. What percentage of all high school students in Florida score at least 70 on the mathematics achievement test?

b. What percentage of all high school students in Florida score below 60 on the mathematics achievement test? Locate μ and the desired probability on a sketch of the normal curve.

6.63 A *chewing cycle* is defined as an upward movement followed by a downward movement of the chin. Clinicians have found that the chewing cycles of "normal" children differ from the chewing cycles of children with eating difficulties. *The American Journal of Occupational Therapy* showed that the number of chewing cycles required for a "normal" preschool child to swallow a bite of graham cracker has a mean of 15.09 and a standard deviation of 3.4. Suppose a "normal" preschool child is fed a bite of graham cracker, and assume that x, the number of chewing cycles required to swallow a bite of graham cracker, follows an approximate normal distribution.

a. Find the probability that x is less than 10.

b. Find the probability that x is between 8 and 16.

c. Find the probability that x exceeds 22.

d. One preschool child, thought to have eating difficulties, required $x = 22$ cycles to chew and swallow the bite of graham cracker. Is it likely that this child is from the group of "normal" children? Explain.

e. Find the 90th percentile of the number of chewing cycles distribution.

6.64 The Taylor Manifest Anxiety Scale (TMAS) is a widely used test in psychological research. The TMAS scores of mental patients suffering from manic depression are approximately normally distributed with a mean of 47.6 and a standard deviation of 10.3.

a. Fifty percent of the mental patients have TMAS scores below what value?

b. Ninety-five percent of the mental patients have TMAS scores above what value?

c. In a sample of 50 mental patients, what is the probability that their mean TMAS score is less than 45?

6.65 *New England Journal of Medicine* researchers theorize that the low death rate from coronary heart disease among the Greenland Eskimos is due to their high fish consumption. The average amount of fish consumed by the Eskimos is 400 grams per day. Assume the standard deviation is 50 grams per day.

a. Describe the sampling distribution of $\bar{x}$, the mean amount of fish consumed per day for a sample of 25 Greenland Eskimos.

b. Find the probability that $\bar{x}$ is less than 390 grams.

c. Find the probability that $\bar{x}$ falls between 405 and 425 grams.

6.66 In a particular population of sedentary young adults (i.e., those adults aged 20–40 who spend less than 5 hours per week in active exercise), the distribution of systolic blood pressure is approximately normal with $\mu = 125$ and $\sigma = 15$. Suppose a sedentary young adult is randomly selected from this population.

a. Find the probability that this person will have a systolic blood pressure of 90 or less.

b. Find the probability that this person will have a systolic blood pressure between 140 and 155.

c. Would you expect this person to have a systolic blood pressure below 80? Explain.

6.67 When artificial hearts are implanted there is a chance of internal infection. Experiments show that calves implanted with artificial hearts can live an average of

80 days. Suppose the distribution of the number of days that a calf implanted with an artificial heart can live is approximately normally distributed with a standard deviation of 25 days.

a. What is the probability that a randomly selected calf implanted with an artificial heart will live longer than 120 days?

b. Twenty-five percent of all calves implanted with an artificial heart live longer than how many days? Show the pertinent quantities on a sketch of the normal curve for each part of the exercise.

6.68 The length of time required for a rodent to escape a noxious experimental situation was found to have a normal distribution with a mean of 6.3 minutes and a standard deviation of 1.5 minutes.

a. What is the probability that a randomly selected rodent will take longer than seven minutes to escape the noxious experimental situation?

b. Some rodents may become so disoriented in the experimental situation that they will not escape in a reasonable amount of time. Thus, the experimenter will be forced to terminate the experiment before the rodent escapes. When should each trial be terminated if the experimenter wants to allow sufficient time for 90% of the rodents to escape?

6.69 The average life of a certain steel-belted radial tire is advertised as 60,000 miles. Assume that the life of the tires is normally distributed with a standard deviation of 2,500 miles. (The life of a tire is defined as the number of miles the tire is driven before blowing out.)

a. Find the probability that a randomly selected steel-belted radial tire will have a life of 61,800 miles or less.

b. Find the probability that a randomly selected steel-belted radial tire will have a life between 62,000 miles and 66,000 miles.

c. In order to avoid a tire blowout, the company manufacturing the tires will warn purchasers to replace each tire after it has been used for a given number of miles. What should the replacement time (in miles) be so that only 1% of the tires will blow out?

d. In a sample of 100 steel-belted radial tires, what is the probability that the mean life is less than 61,800 miles? Compare your answer to part **a.**

***6.70** The computer chips in notebook and laptop computers are produced from semiconductor wafers. Certain semiconductor wafers are exposed to an environment that generates up to 100 possible defects per wafer. The number of defects per wafer, x, was found to follow a binomial distribution if the manufacturing process is stable and generates defects that are randomly distributed on the wafers (*IEEE Transactions on Semiconductor Manufacturing,* May 1995). Let p represent the probability that a defect occurs at any one of the 100 points of the wafer. For each of the following cases, determine whether the normal approximation can be used to characterize x.

a. $p = .01$ **b.** $p = .50$ **c.** $p = .90$

***6.71** According to the *American Cancer Society,* melanoma, a form of skin cancer, kills 60% of Americans who suffer from the disease each year. Consider a sample of 10,000 melanoma patients.

a. What are the expected value and variance of x, the number of the 10,000 melanoma patients who die of the affliction this year?

b. Find the probability that x will exceed 6,100 patients per year.

c. Would you expect x, the number of patients dying of melanoma, to exceed 6,500 in any single year? Explain.

6.72 Many species of terrestrial frogs that hibernate at or near the ground surface can survive prolonged exposure to low winter temperatures. In freezing conditions, the frog's body temperature, called its *supercooling temperature,* remains relatively higher because of an accumulation of glycerol in its body fluids. Studies have shown that the supercooling temperature of terrestrial frogs frozen at $-6°C$ has a relative

frequency distribution with a mean of −2.18°C and a standard deviation of .32°C (*Science,* May 1983). Consider the mean supercooling temperature, $\bar{x}$, of a random sample of $n = 42$ terrestrial frogs frozen at −6°C.

a. Find the probability that $\bar{x}$ exceeds −2.05°C.

b. Find the probability that $\bar{x}$ falls between −2.20°C and −2.10°C.

6.73 Refer to the data on 98 crash-tested cars stored in the file **CRASH**. (The data set is described in Appendix C.) For each quantitative variable in the data set, determine whether the data are approximately normal.

6.74 This past year, an elementary school began using a new method to teach arithmetic to first graders. A standardized test, administered at the end of the year, was used to measure the effectiveness of the new method. The distribution of test scores in past years (before implementation of the new teaching method) had a mean of 75 and a standard deviation of 10. Consider the standardized test scores for a random sample of 36 first graders taught by the new method.

a. If the distribution of test scores for first graders taught by the new method is no different from that of the old method, describe the sampling distribution of $\bar{x}$, the mean test score for a random sample of 36 first graders.

b. If the sample mean test score was computed to be $\bar{x} = 79$, what would you conclude about the effectiveness of the new method of teaching arithmetic? Explain.

6.75 Birdwatchers visiting a National Wildlife Refuge located near Atlantic City, New Jersey, report a high degree of satisfaction with their visits (*Leisure Sciences,* Vol. 9, 1987). On a standard satisfaction scale that ranged from 1 to 6 (where 1 = poor rating and 6 = perfect rating), the birdwatchers have a mean score of 5.05 and a standard deviation of .98. Assume the population of satisfaction scores of birdwatchers visiting the National Wildlife Refuge is approximately normal.

a. Find the probability that a randomly selected birdwatcher visiting the refuge will have a satisfaction score of at least 5.

b. One-fourth of the birdwatchers have satisfaction scores below what value?

6.76 *Human Development* (1985) reported on a meta-analysis of the relation between race and subjective well-being based on results obtained from 54 sources. For each source, a measure of the relationship between race and well-being, called "effect size," was calculated. (Effect sizes ranged from −1 to 1, with positive values implying that whites had a higher subjective well-being than minorities and negative values implying the reverse.) A stem-and-leaf display of the 54 effect sizes is reproduced here. The mean and standard deviation of the data were .10 and .09, respectively.

Stem	Leaf
−.2	4
−.1	6
−.1	2 2 1 1 1
−.0	9 9 8
−.0	1 0 0 0
.0	2
.0	5 6 6 6 6 7 7 8 8 8 9
.1	0 0 0 1 1 2 2 2 3 4 4 4
.1	5 5 6 9 9 9
.2	0 0 0 2 4
.2	5 6 7
.3	0 1

Source: Stock, W. A., Okun, M. A., Haring, M. J., and Witter, R. A. "Race and subjective well-being in adulthood: A black-white research synthesis." *Human Development,* Vol. 28, 1985, p. 195.

Problem 6.76

a. Examine the stem-and-leaf display. Is the distribution of effect sizes approximately normal?

b. Assuming normality, calculate the probability that a randomly selected study on the relationship between race and subjective well-being will have a negative effect size. (Assume $\mu = .10$ and $\sigma = .09$.)

c. Use the stem-and-leaf display to calculate the actual percentage of the 54 effect sizes that were negative. Compare the result to your answer to part **b**.

6.77 Suppose you are in charge of student ticket sales for a major college football team. From past experience, you know that the number of tickets purchased by a student standing in line at the ticket window has a distribution with a mean of 2.4 and a standard deviation of 2.0. For today's game, there are 100 eager students standing in line to purchase tickets. If only 250 tickets remain, what is the probability that all 100 students will be able to purchase the tickets they desire?

6.78 It is well-known that children with developmental delays (i.e., mild mental retardation) are slower, cognitively, than normally developing children. Are their social skills also lacking? A study compared the social interactions of the two groups of children in a controlled playground environment (*American Journal on Mental Retardation,* Jan. 1992). One variable of interest was the number of intervals of "no play" by each child. Children with developmental delays had a mean

of 2.73 intervals of "no play" and a standard deviation of 2.58 intervals. Based on this information, is it possible for the variable of interest to be normally distributed? Explain.

6.79 One of the monitoring methods used by the Environmental Protection Agency (EPA) to determine whether sewage treatment plants are conforming to standards is to take 36 1-liter specimens from the plant's discharge during the period of investigation. Chemical methods are applied to determine the percentage of sewage in each specimen. If the sample data provide evidence to indicate that the true mean percentage of sewage exceeds a limit set by the EPA, the treatment plant must undergo mandatory repair and retooling. One particular plant, at which the mean sewage discharge limit has been set at 15%, is suspected of being in violation of the EPA standard.

a. Unknown to the EPA, the relative frequency distribution of sewage percentages in 1-liter specimens at the plant in question has a mean μ of 15.7% and a standard deviation σ of 2.0%. Thus, the plant is in violation of the EPA standard. What is the probability that the EPA will obtain a sample of 36 1-liter specimens with a mean sewage percentage less than 15% even though the plant is violating the sewage discharge limit?

b. Suppose the EPA computes $\bar{x} = 14.95\%$. Does this result lead you to believe that the sample of 36 1-liter specimens obtained by the EPA was not random, but biased in favor of the sewage treatment plant? Explain.

6.80 *Cost estimation* is the term used to describe the process by which engineers estimate the cost of work contracts (e.g., road construction, building construction) that are to be awarded to the lowest bidder. A study investigated the factors that affect the accuracy of engineers' estimates (*Cost Engineering,* Oct. 1988), where accuracy is measured as the percentage difference between the low (winning) bid and the engineers' estimate. One of the most important factors is number of bidders—the more bidders on the contract, the more likely the engineers are to overestimate the cost. For building contracts with five bidders, the mean percentage error was -7.02 and the standard deviation was 24.66. Consider a sample of 50 building contracts, each with five bidders.

a. Describe the sampling distribution of $\bar{x}$, the mean percentage difference between the low bid and the engineers' estimate, for the 50 contracts.

b. Find $P(\bar{x} < 0)$. (This is the probability of an overestimate.)

c. Suppose you observe $\bar{x} = -17.83$ for a sample of 50 building contracts. Based on the information above, are all these contracts likely to have five bidders? Explain.

6.81 The determination of the percent canopy closure of a forest is essential for wildlife habitat assessment, watershed runoff estimation, erosion control, and other forest management activities. One way in which geoscientists estimate the percent forest canopy closure is through the use of a satellite sensor called the Landsat Thematic Mapper. A study of the percent canopy closure in the San Juan National Forest (Colorado) was conducted by examining Thematic Mapper Simulator (TMS) data collected by aircraft at various forest sites (*IEEE Transactions on Geoscience and Remote Sensing,* Jan. 1986). The mean and standard deviation of the readings obtained from TMS Channel 5 were found to be 121.74 and 27.52, respectively.

a. Let $\bar{x}$ be the mean TMS reading for a sample of 32 forest sites. Assuming the figures given are population values, describe the sampling distribution of $\bar{x}$.

b. Use the sampling distribution of part **a** to find the probability that $\bar{x}$ falls between 118 and 130.

***6.82** *Time* (Oct. 11, 1999) reported that a half million Americans underwent laser surgery to correct their vision in 1999. However, 5% of the patients had serious post-laser vision problems. Consider a sample of 400 patients who have undergone laser surgery. What is the approximate probability that 20 or more of these patients suffer serious post-laser vision problems?

REFERENCES

Hogg, R. V., and Craig, A. T. *Introduction to Mathematical Statistics,* 5th ed. Upper Saddle River, N.J.: Prentice Hall, 1995.

Mendenhall, W., and Sincich, T. *Statistics for Engineering and the Sciences,* 4th ed. Upper Saddle River, N.J.: Prentice Hall, 1995.

Mendenhall, W., Wackerly, D., and Scheaffer, R. *Mathematical Statistics with Applications,* 4th ed. Boston: PWS-Kent, 1990.

Ramsey, P. P., and Ramsey, P. H. "Simple tests of normality in small samples." *Journal of Quality Technology,* Vol. 22, 1990, pp. 299–309.

Snedecor, G. W., and Cochran, W. G. *Statistical Methods,* 7th ed. Ames, Iowa: Iowa State University Press, 1980.

E Figure 6.E.2 Normal Probabilities Worksheet for the Salt Consumption Data of Examples 6.8, 6.9, and 6.10

	A	B	C	D	E
1	**Normal Probabilities**				
2					
3	**Common Data**				
4	**Mean**	15			
5	**Standard Deviation**	5			
6				**Probability for a Range**	
7	**Probability for X <=**			**From X Value**	14
8	**X Value**	1		**To X Value**	22
9	**Z Value**	-2.8		**Z Value for 14**	-0.2
10	**P(X<=1)**	0.0025552		**Z Value for 22**	1.4
11				**P(X<=14)**	0.4207
12	**Probability for X >**			**P(X<=22)**	0.9192
13	**X Value**			**P(14<=X<=22)**	0.4985
14	**Z Value**				
15	**P(X>)**			**Find X and Z Given Cum. Pctage.**	
16				**Cumulative Percentage**	10.00%
17		**Z Value**	-1.2816		
18		**X Value**	8.59225		

where $P < x$ is the area under the curve that is less than x.

$$\text{NORMINV}(P < x, \textit{mean}, \textit{standard deviation})$$

where $P < x$, *mean*, and *standard deviation* are as defined for the other functions.

The STANDARDIZE function returns the z value for a particular x value, mean, and standard deviation. The NORMDIST returns the area or probability of less than a given x value. The NORMSINV function returns the z value corresponding to the probability of less than a given x. The NORMINV function returns the x value for a given probability, mean, and standard deviation.

6.E.2 Generating Normal Probability Plots

Use the PHStat **Probability & Prob. Distributions | Normal Probability Plot** procedure to generate a normal probability plot from a set of data. For example, to generate a normal probability plot for the tar measurements for the sample of 500 cigarette brands of Example 6.11, open the **FTC.XLS** workbook to the Data worksheet and

1. Select PHStat | Probability & Prob. Distributions | Normal Probability Plot.
2. In the Normal Probability Plot dialog box (see Figure 6.E.3):

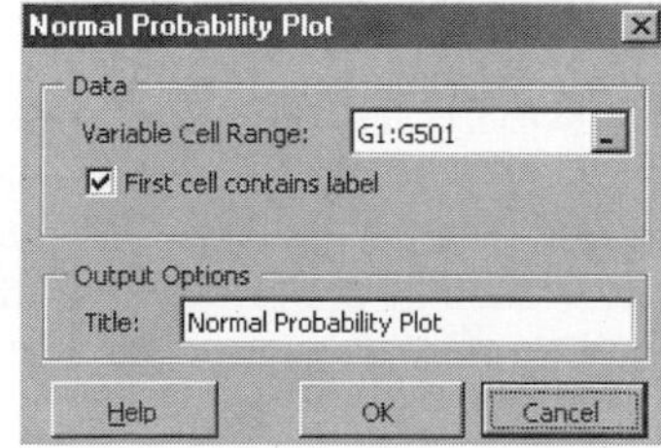

E Figure 6.E.3 Normal Probability Plot Dialog Box

a. Enter G1:G501 in the Variable Cell Range edit box.

b. Select the First cell contains label check box.

c. Enter a title in the Title edit box.

d. Click the OK button.

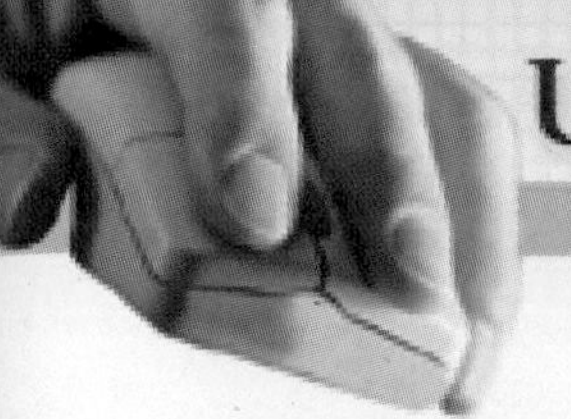

Using Microsoft® Excel

6.E.1 Calculating Normal Probabilities

Use the PHStat **Probability & Prob. Distributions | Normal** procedure to calculate norn probabilities on a new worksheet. For example, to calculate the probabilities for Exan ples 6.8, 6.9, and 6.10, open to an empty worksheet and

1. Select PHStat | Probability & Prob. Distributions | Normal.
2. In the Normal Probability Distribution dialog box (see Figure 6.E.1):

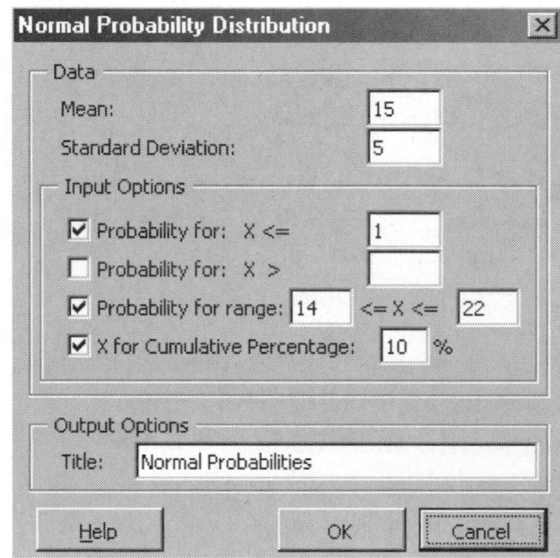

E **Figure 6.E.1** Normal Probability Distribution Dialog Box

a. Enter 15 in the Mean edit box.
b. Enter 5 in the Standard Deviation edit box.
c. Select the Probability for: X <= check box and enter 1 in its edit box (for Example 6.9).
d. Select the Probability for range check box and enter 14 and 22, respectively, in the from and to edit boxes (for Example 6.8).
e. Select the X for Cumulative Percentage check box and enter 10 in its edit box (for Example 6.10).
f. Enter a title in the Title edit box.
g. Click the OK button.

PHStat and Excel insert a worksheet that reports the probabilities, x, and z values for these questions as shown in Figure 6.E.2. You can change the values for the mean and standard deviation in this worksheet to see the effects of these values on all probabilities.

The PHStat normal probabilities option is based on the STANDARDIZE, NORMDIST, NORMSINV, and NORMINV worksheet functions. The formats of these functions are:

STANDARDIZE(x, *mean*, *standard deviation*)

NORMDIST(x, *mean*, *standard deviation*, *True*)

where x is the x value of interest, and *mean* and *standard deviation* are the mean and standard deviation for a normal probability problem.

NORMSINV($P < x$)

This procedure creates two sheets, a worksheet containing a table of ranks, cumulative proportions, z values, and the original tar measurements data and a chart sheet containing the normal probability plot, similar to the one shown in Figure 6.14d on p. 299.

6.E.3 Simulating Sampling Distributions

Use the PHStat **Sampling | Sampling Distributions Simulation** procedure to generate a simulated sampling distribution from either a uniformly distributed, standard normally distributed, or discrete population. For example, to generate a simulated sampling distribution from a uniformly distributed population, using 100 samples of sample size 30, open to an empty worksheet and

1. Select PHStat | Sampling | Sampling Distributions Simulation.
2. In the Sampling Distributions Simulation dialog box (see Figure 6.E.4):

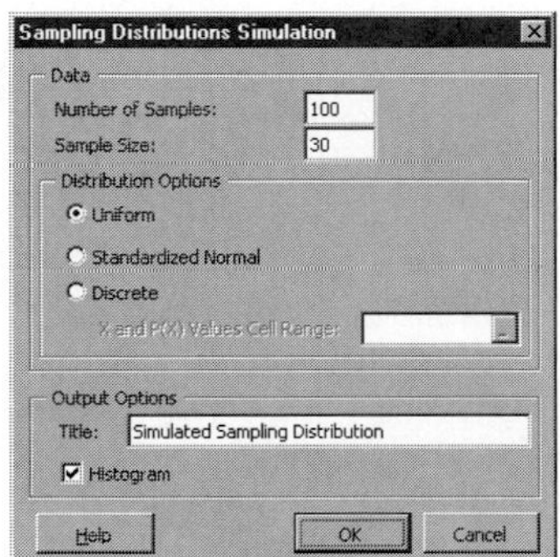

E **Figure 6.E.4** Sampling Distributions Simulation Dialog Box

a. Enter 100 in the Number of Samples edit box.
b. Enter 30 in the Sample Size edit box.
c. Select the Uniform option button.
d. Enter a title in the Title edit box.
e. Select the Histogram check box.
f. Click the OK button.

Select the Standardized Normal option button in step 2c to generate a simulated sampling distribution from a standardized normal population, or begin with a worksheet that contains a table of x and $P(x)$ values and select the Discrete option button to generate a simulated distribution from a discrete population.

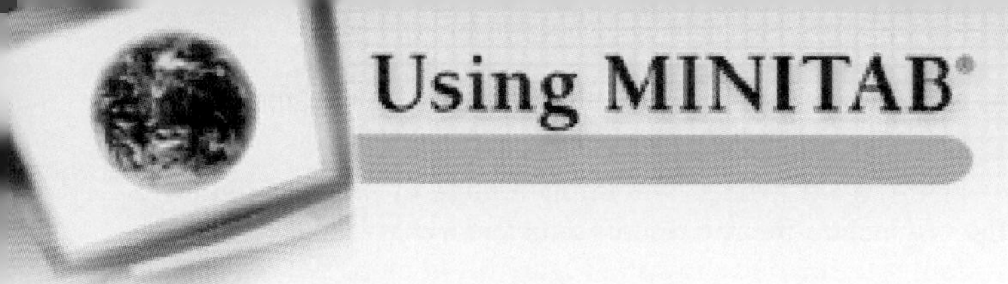

Using MINITAB®

6.M.1 Obtaining Normal Probabilities

MINITAB can be used rather than equations 6.2 and 6.3 and Table B.3 to compute normal probabilities. Referring to Examples 6.8 and 6.10 in which $\mu = 15$ and $\sigma = 5$, to find the area less than one gram (see Figure 6.M.1):

1. Enter **1** in the first row of column C1.
2. Select **Calc | Probability Distributions | Normal.**
3. Select **Cumulative probability.** Enter **15** in the Mean edit box and **5** in the Standard deviation edit box. Enter **C1** in the Input column edit box. Click the **OK** button.

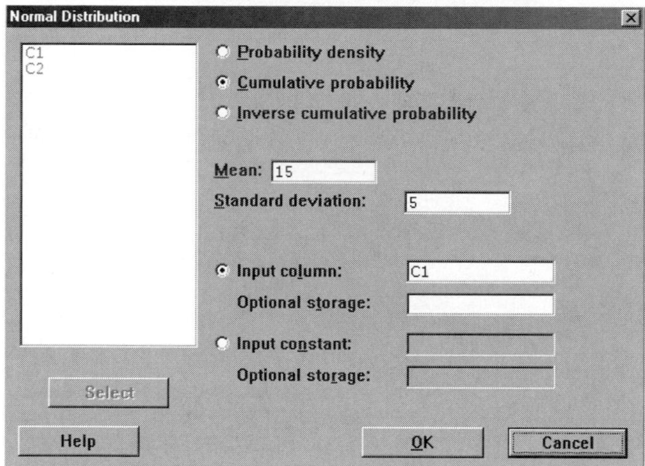

Figure 6.M.1 MINITAB Normal Distribution Dialog Box

To obtain the z value corresponding to a cumulative area of 0.10,

1. Enter **.10** in row 1 of column C2.
2. Select **Calc | Probability Distributions | Normal.**
3. Select **Inverse cumulative probability.** Enter **15** in the Mean edit box and **5** in the Standard deviation edit box. Enter **C2** in the Input column edit box. Click the **OK** button.

6.M.2 Obtaining a Normal Probability Plot

To obtain a normal probability plot in MINITAB for the tar measurements for the sample of 500 cigarette brands of Example 6.11 on p. 298, open the **FTC.MTW** worksheet. Then (see Figure 6.M.2)

1. Select **Graph | Probability Plot.**
2. In the Distribution drop-down list box, select **Normal.** In the Variables edit box, enter **'Tar'** or **C7.** Click the **OK** button.

Note that the normal probability plot is different from the one described in Section 6.3. MINITAB provides a plot in which the variable is plotted on the x axis and the cumulative percentage is plotted on the y axis on a special scale so that if the variable is normally distributed the data will plot along a straight line.

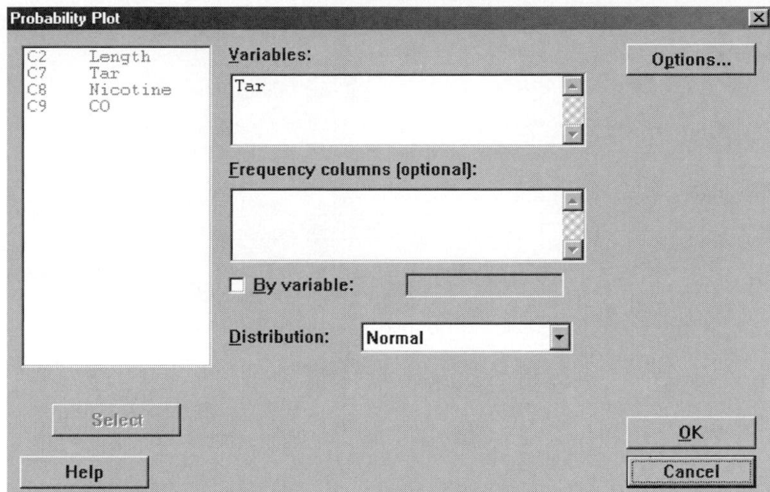

M **Figure 6.M.2** MINITAB Probability Plot Dialog Box

6.M.3 Simulating Sampling Distributions

To develop a simulation of the sampling distribution of the mean from a uniformly distributed population with 100 samples of $n = 30$, select **Calc | Random Data | Uniform.** Then do the following (see Figure 6.M.3).

1. Enter **100** in the Generate edit box.
2. Enter **C1–C30** in the Store in column(s) edit box.
3. Enter **0.0** in the Lower endpoint edit box and **1.0** in the Upper endpoint edit box. Click the **OK** button.

You will observe that 100 rows of values are entered in columns C1–C30.

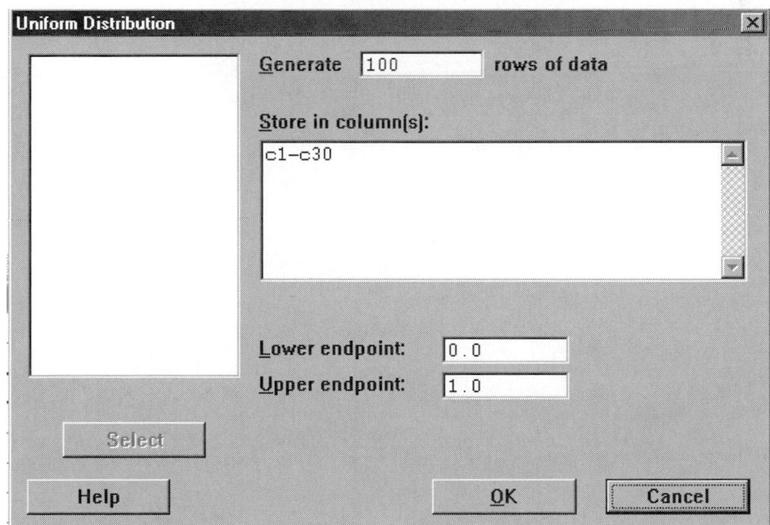

M **Figure 6.M.3** MINITAB Uniform Distribution Dialog Box

To calculate row statistics for each of the 100 samples, do the following (see Figure 6.M.4):

1. Select **Calc | Row Statistics.**
2. Select the **Mean** option button.
3. Enter **C1–C30** in the Input variables edit box. Enter **C31** in the Store result in edit box. Click the **OK** button.

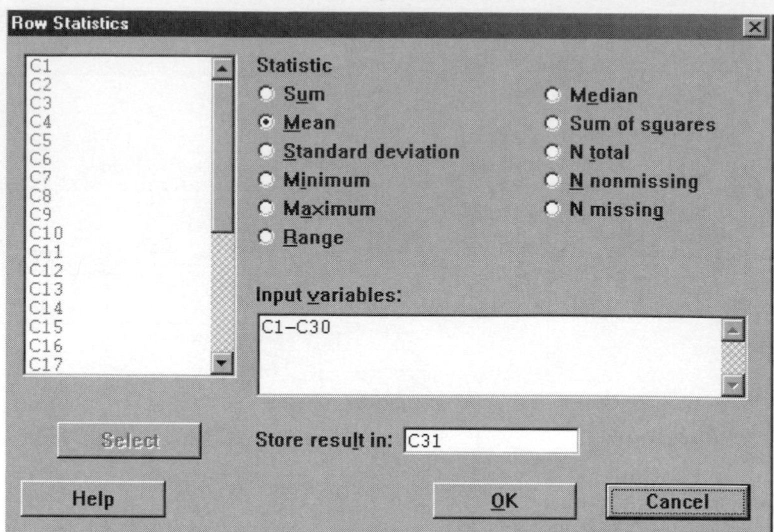

M **Figure 6.M.4** MINITAB Row Statistics Dialog Box

The mean for each of the 100 samples is stored in column C31. To compute statistics for the set of 100 sample means, select **Stat | Basic Statistics | Display Descriptive Statistics.** Enter **C31** in the Variables edit box. Click the **OK** button.

To obtain a histogram of the 100 sample means, select **Graph | Histogram** and do the following:

1. Enter **C31** in row 1 of the Graph Variables edit box. Click the **Options** button.
2. Click **Density** for type of histogram. Click the **OK** button to return to the Histogram dialog box. Click the **OK** button again.

To obtain a simulation of the sampling distribution of the mean for a normal population, select **Calc | Random Data | Normal.** Enter a value for μ in the Mean edit box and for σ in the Standard deviation edit box. Follow the remainder of the instructions given for the Uniform population.

CHAPTER 7

Estimation of Population Parameters Using Confidence Intervals: One Sample

OBJECTIVES

1. To identify point estimates of common population parameters
2. To demonstrate how a confidence interval provides a measure of reliability for an estimate of a population parameter
3. To develop confidence intervals for a population mean, proportion, and variance
4. To show how to select the proper sample size for estimating a population parameter

CONTENTS

EXCEL TUTORIAL

MINITAB TUTORIAL

Statistics in the Real World

Scallops, Sampling, and the Law

Arnold Bennett, a Sloan School of Management professor at the Massachusetts Institute of Technology (MIT), described a recent legal case in *Interfaces* (Mar.-Apr. 1995) in which he served as a statistical "expert." The case involved a ship that fishes for scallops off the coast of New England. In order to protect baby scallops from being harvested, the U.S. Fisheries and Wildlife Service requires that *the average meat per scallop weigh at least 1/36 of a pound.* The ship was accused of violating this weight standard. Bennett lays out the scenario:

> The vessel arrived at a Massachusetts port with 11,000 bags of scallops, from which the harbormaster randomly selected 18 bags for weighing. From each such bag, his agents took a large scoop full of scallops; then, to estimate the bag's average meat per scallop, they divided the total weight of meat in the scoop by the number of scallops it contained. Based on the 18 [numbers] thus generated, the harbormaster estimated that each of the ship's scallops possessed an average of 1/39 of a pound of meat (that is, they were about seven percent lighter than the minimum requirement). Viewing this outcome as conclusive evidence that the weight standard had been violated, federal authorities at once confiscated *95 percent* of the catch (which they then sold at auction). The fishing voyage was thus transformed into a financial catastrophe for its participants.

SCALLOPS

Bennett provided the estimated scallop weight measurements (in pounds) for each of the 18 sampled bags in the article. These data are available in the CD-ROM file called **SCALLOPS.** [*Note:* 1/36 of a pound, the minimum permissible average weight per scallop, is equivalent to .0278 pound. Consequently, weights below .0278 indicate individual bags that do not meet the standard.] The ship's owner filed a lawsuit against the federal government, declaring that his vessel had fully complied with the weight standard. A Boston law firm was hired to represent the owner in legal proceedings and Bennett was retained by the firm to provide statistical litigation support and, if necessary, expert witness testimony.

In the following Statistics in the Real World Revisited sections, we demonstrate how confidence intervals can be used to support the ship's owner in the lawsuit.

Statistics in the Real World Revisited

- Estimating the Mean Weight per Scallop (p. 356)
- Estimating the Proportion of Underweight Scallop Bags (p. 362)
- Determining the Number of Bags of Scallops to Sample (p. 369)
- Estimating the Variance of the Scallop Weights (p. 374)

7.1 Point Estimators

In preceding chapters we learned that populations are characterized by descriptive measures (parameters), and that inferences about parameter values are based on statistics computed from the information in a sample selected from the population of interest. In this chapter, we will demonstrate how to estimate population parameters and assess the reliability of our estimates, based on knowledge of the sampling distributions of the statistics being used.

EXAMPLE 7.1 **SELECTING A POINT ESTIMATE**

Suppose you are interested in estimating the mean grade point average (GPA) of all students who attend your college or university.

a. Identify the target population and the parameter of interest.

b. How could one estimate the parameter of interest in this situation?

Solution

a. The target population consists of the GPAs of all students at your college or university. Since we want to estimate the mean of this population, the parameter of interest is μ, the mean GPA of all students at your institution.

b. An estimate of a population mean μ is the sample mean $\bar{x}$, which is computed from a random sample of n observations from the target population. Assume, for example, that we obtain a random sample of size $n = 100$ students, and then compute the value of the sample mean to be $\bar{x} = 2.63$. This value of $\bar{x}$ provides a *point estimate* of the population mean GPA.

Definition 7.1

A **point estimate** of a parameter is a statistic, a single value computed from the observations in a sample, that is used to estimate the value of the target parameter.

How reliable is a point estimate of a parameter? To be truly practical and meaningful, an inference concerning a parameter (in this case, estimation of the value of μ) must consist not only of a point estimate, but also must be accompanied by a measure of the reliability of the estimate. We need to be able to state how close our estimate is likely to be to the true value of the population parameter. This can be done by forming an **interval estimate** of the parameter using the characteristics of the sampling distribution of the statistic that was used to obtain the point estimate. The procedure will be illustrated in the next section.

7.2 Estimation of a Population Mean: Normal (z) Statistic

Recall from Section 6.6 that, for sufficiently large sample sizes, the sampling distribution of the sample mean $\bar{x}$ is approximately normal, as indicated in Figure 7.1.

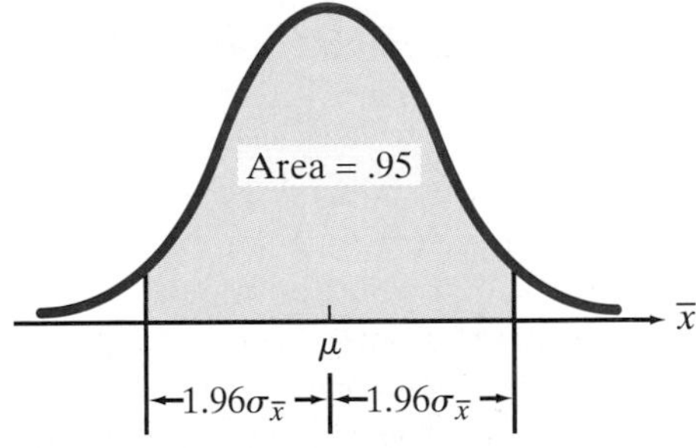

Figure 7.1 Sampling Distribution of $\bar{x}$

EXAMPLE 7.2

RELIABILITY OF INTERVAL ESTIMATE

FTC

Refer to the Federal Trade Commission's cigarette data stored in the **FTC** file. Suppose that our target population is the set of nicotine measurements for the 500 cigarette brands, and we are interested in estimating μ, the mean nicotine content of the cigarettes. We plan to take a sample of $n = 30$ measurements from the population of nicotine contents in the **FTC** file and construct the interval

$$\bar{x} \pm 1.96\sigma_{\bar{x}} = \bar{x} \pm 1.96\left(\frac{\sigma}{\sqrt{n}}\right)$$

where σ is the population standard deviation of the 500 nicotine measurements and $\sigma_{\bar{x}} = \sigma/\sqrt{n}$ is the **standard error** of $\bar{x}$. In other words, we will construct an interval 1.96 standard deviations around the sample mean, $\bar{x}$. How likely is it that this interval will contain the true value of the population mean μ?

Solution

We arrive at a solution by the following three-step process:

Step 1 First note that the area in the sampling distribution of $\bar{x}$ between $\mu - 1.96\sigma_{\bar{x}}$ and $\mu + 1.96\sigma_{\bar{x}}$ is approximately .95. (This area, highlighted in Figure 7.1, is obtained from the standard normal table, Table B.3 in Appendix B.) This implies that before the sample of measurements is drawn, the probability that $\bar{x}$ will fall within the interval $\mu \pm 1.96\sigma_{\bar{x}}$ is .95.

Step 2 If, in fact, the sample yields a value of $\bar{x}$ that falls within the interval $\mu \pm 1.96\sigma_{\bar{x}}$, then it is also true that the interval $\bar{x} \pm 1.96\sigma_{\bar{x}}$ will contain μ. This concept is illustrated in Figure 7.2. For a particular value of $\bar{x}$ (shown with a vertical arrow) that falls within the interval $\mu \pm 1.96\sigma_{\bar{x}}$, a distance of $1.96\sigma_{\bar{x}}$ is marked off both to the left and to the right of $\bar{x}$. You can see that the value of μ must fall within $\bar{x} \pm 1.96\sigma_{\bar{x}}$.

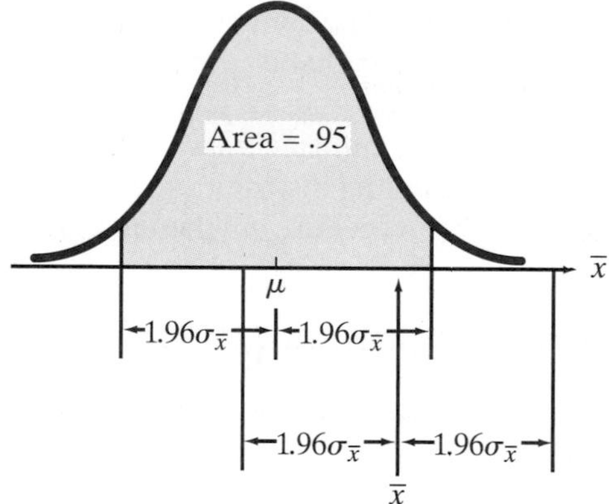

Figure 7.2 Sampling Distribution of $\bar{x}$ in Example 7.2

Step 3 Steps 1 and 2 combined imply that, before the sample is drawn, the probability that the interval $\bar{x} \pm 1.96\sigma_{\bar{x}}$ will enclose μ is approximately .95.

The interval $\bar{x} \pm 1.96\sigma_{\bar{x}}$ in Example 7.2 is called a large-sample 95% confidence interval for the population mean μ. The term *large-sample* refers to the sample being of a sufficiently large size that we can apply the Central Limit Theorem to determine the form of the sampling distribution of $\bar{x}$. Empirical research has found that a sample size n exceeding a value between 20 and 30 will usually yield a sampling distribution of $\bar{x}$ that is approximately normal. This result led many practitioners to adopt the conventional rule that a sample size of $n \geq 30$ is required to use large-sample confidence interval procedures. Keep in mind, though, that 30 is not a magical number, and is, in fact, quite arbitrary.

Definition 7.2

A **confidence interval** for a parameter is an interval within which we expect the true value of the population parameter to be contained. The endpoints of the interval are computed based on sample information.

EXAMPLE 7.3

FTC

95% CONFIDENCE INTERVAL

Refer to Example 7.2. Suppose a random sample of 30 observations from the population of nicotine measurements yielded a sample mean of $\bar{x} = .79$ mg. Construct a 95% confidence interval for μ, the population mean nicotine content of the 500 cigarettes, based on this sample information.

Solution

A 95% confidence interval for μ, based on a sample of size $n = 30$, is given by

$$\bar{x} \pm 1.96\sigma_{\bar{x}} = \bar{x} \pm 1.96\left(\frac{\sigma}{\sqrt{n}}\right) = .79 \pm 1.96\left(\frac{\sigma}{\sqrt{30}}\right)$$

To complete the calculation we need the value of σ, the standard deviation of the population of nicotine measurements. Using Excel, we obtain $\sigma = .3452$ mg. Then, our interval is

$$.79 \pm 1.96\left(\frac{.3452}{\sqrt{30}}\right) = (.666, .914)$$

We estimate that the population mean nicotine content falls within the interval from .666 mg to .914 mg.

How much confidence do we have that μ, the true population nicotine content, lies within the interval (.666, .914)? Although we cannot be certain whether the sample interval contains μ (unless we calculate the true value of μ for all 500 observations), we can be reasonably sure that it does. This confidence is based on the interpretation of the confidence interval procedure: If we were to select repeated random samples of size $n = 30$ nicotine measurements and form a 1.96 standard deviation interval around $\bar{x}$ for each sample, then approximately 95% of the intervals constructed in this manner would contain μ. We say that we are "95% confident" that the particular interval (.666, .914) contains μ; this is our measure of the reliability of the point estimate $\bar{x}$.

EXAMPLE 7.4

FTC

THEORETICAL INTERPRETATION

To illustrate the classical interpretation of a confidence interval, we generated 100 random samples, each of size $n = 30$, from the population of nicotine measurements in **FTC.** For each sample, the sample mean was calculated and used to construct a 95% confidence interval for μ as in Example 7.3. Interpret the results, which are shown in Table 7.1.

Solution

Using Excel, we found the population mean value to be $\mu = .8425$ mg. In the 100 repetitions of the confidence interval procedure described here, note that only five of the intervals (those highlighted in Table 7.1) do not contain the value of μ, whereas the remaining 95 of the 100 intervals (or 95%) do contain the true value of μ. The proportion .95 is called the *confidence coefficient* for the interval.

Keep in mind that, in actual practice, you would not know the true value of μ and you would not perform this repeated sampling; rather you would select a single random sample and construct the associated 95% confidence interval. The one confidence interval you form may or may not contain μ, but you can be fairly sure it does because of your *confidence in the statistical procedure,* the basis for which was illustrated in this example.

TABLE 7.1 Confidence Intervals for μ for 100 Random Samples of 30 Nicotine Measurements Extracted from the FTC File

Sample	$\bar{x}$	95% Confidence Interval Lower Limit	95% Confidence Interval Upper Limit	Sample	$\bar{x}$	95% Confidence Interval Lower Limit	95% Confidence Interval Upper Limit
1	0.8655	0.7419	0.9891	51	0.9967	0.8731	1.1203
2	0.8517	0.7281	0.9753	52	0.8033	0.6797	0.9269
3	0.8600	0.7364	0.9836	53	0.9333	0.8097	1.0569
4	0.8150	0.6914	0.9386	54	0.7967	0.6731	0.9203
5	0.8600	0.7364	0.9836	55	0.7133	0.5897	0.8369
6	0.8400	0.7164	0.9636	56	0.8667	0.7431	0.9903
7	0.8250	0.7014	0.9486	57	0.8233	0.6997	0.9469
8	0.8967	0.7731	1.0203	58	0.8567	0.7331	0.9803
9	0.8103	0.6867	0.9339	59	0.7400	0.6164	0.8636
10	0.8800	0.7564	1.0036	60	0.8333	0.7097	0.9569
11	0.9233	0.7997	1.0469	61	0.8567	0.7331	0.9803
12	0.9117	0.7881	1.0353	62	0.8600	0.7364	0.9836
13	0.8333	0.7097	0.9569	63	0.8117	0.6881	0.9353
14	0.8700	0.7464	0.9936	64	0.6867	0.5631	0.8103
15	0.9133	0.7897	1.0369	65	0.8167	0.6931	0.9403
16	0.7867	0.6631	0.9103	66	0.8400	0.7164	0.9636
17	0.8800	0.7564	1.0036	67	0.9033	0.7797	1.0269
18	0.7100	0.5864	0.8336	68	0.8133	0.6897	0.9369
19	0.9067	0.7831	1.0303	69	0.8833	0.7597	1.0069
20	0.9167	0.7931	1.0403	70	0.7667	0.6431	0.8903
21	0.8717	0.7481	0.9953	71	0.8333	0.7097	0.9569
22	0.9067	0.7831	1.0303	72	0.8700	0.7464	0.9936
23	0.8350	0.7114	0.9586	73	0.8667	0.7431	0.9903
24	0.8233	0.6997	0.9469	74	0.8633	0.7397	0.9869
25	0.8467	0.7231	0.9703	75	0.7900	0.6664	0.9136
26	0.9433	0.8197	1.0669	76	0.8900	0.7664	1.0136
27	0.7233	0.5997	0.8469	77	0.7967	0.6731	0.9203
28	0.8200	0.6964	0.9436	78	0.8433	0.7197	0.9669
29	0.9100	0.7864	1.0336	79	0.8333	0.7097	0.9569
30	0.9067	0.7831	1.0303	80	0.8417	0.7181	0.9653
31	0.8433	0.7197	0.9669	81	0.7883	0.6647	0.9119
32	0.8300	0.7064	0.9536	82	0.8267	0.7031	0.9503
33	0.7933	0.6697	0.9169	83	0.8150	0.6914	0.9386
34	0.7300	0.6064	0.8536	84	0.9500	0.8264	1.0736
35	0.8100	0.6864	0.9336	85	0.8233	0.6997	0.9469
36	0.7567	0.6331	0.8803	86	0.8967	0.7731	1.0203
37	0.8100	0.6864	0.9336	87	0.7567	0.6331	0.8803
38	0.7759	0.6523	0.8995	88	0.7333	0.6097	0.8569
39	0.8833	0.7597	1.0069	89	0.8967	0.7731	1.0203
40	0.7533	0.6297	0.8769	90	0.8100	0.6864	0.9336
41	0.9467	0.8231	1.0703	91	0.9600	0.8364	1.0836
42	0.8267	0.7031	0.9503	92	0.8800	0.7564	1.0036
43	0.8867	0.7631	1.0103	93	0.8650	0.7414	0.9886
44	0.8700	0.7464	0.9936	94	1.0333	0.9097	1.1569
45	0.8467	0.7231	0.9703	95	0.7200	0.5964	0.8436
46	0.8167	0.6931	0.9403	96	0.8733	0.7497	0.9969
47	0.7833	0.6597	0.9069	97	0.8433	0.7197	0.9669
48	0.7700	0.6464	0.8936	98	0.7750	0.6514	0.8986
49	0.8233	0.6997	0.9469	99	0.7533	0.6297	0.8769
50	0.8733	0.7497	0.9969	100	0.8655	0.7419	0.9891

Definition 7.3

The **confidence coefficient** is the proportion of times that a confidence interval encloses the true value of the population parameter if the confidence interval procedure is used repeatedly a very large number of times.

What "95% Confident" Means
A confidence coefficient of .95 implies that, of *all* possible 95% confidence intervals that can be formed from random samples selected from the target population, we are guaranteed 95% of them will contain μ and 5% will not.

Note that for a 95% confidence interval, the confidence coefficient of .95 is equal to the total area under the sampling distribution (1.00) less .05 of the area, which is divided equally between the two tails of the distribution (see Figure 7.1). Thus, each tail has an area of .025 and the tabulated value of z (from Table B.3 in Appendix B) that cuts off an area of .025 in the right tail of the standard normal distribution is 1.96 (see Figure 7.3). The value $z = 1.96$ is also the distance, in terms of standard deviations, that $\bar{x}$ is from each endpoint of the 95% confidence interval.

Now, suppose we want to assign a confidence coefficient other than .95 to a confidence interval. To do this, we change the area under the sampling distribution between the endpoints of the interval, which in turn changes the tail area associated with z. Thus, this z value provides the key to constructing a confidence interval with any desired confidence coefficient. In our subsequent discussion, we will use the notation defined in Definition 7.4.

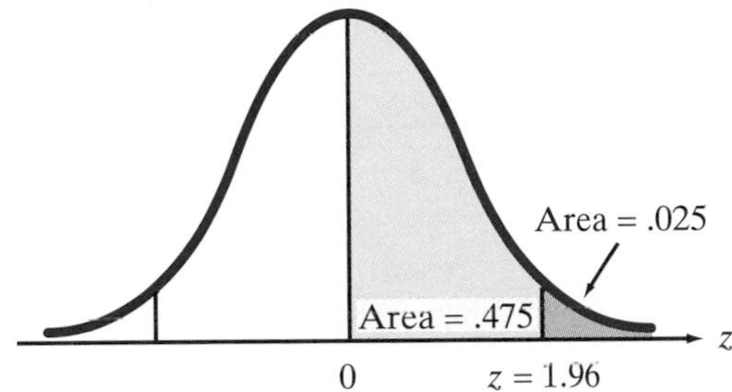

Figure 7.3 Tabulated z Value Corresponding to a Tail Area of .025

Definition 7.4

We define $z_{\alpha/2}$ to be the z value such that an area of $\alpha/2$ lies to its right (see Figure 7.4).

Now, if an area of $\alpha/2$ lies beyond $z_{\alpha/2}$ in the right tail of the standard normal (z) distribution, then an area of $\alpha/2$ lies to the left of $-z_{\alpha/2}$ in the left tail (Figure 7.4) because of the symmetry of the distribution. The remaining area, $(1 - \alpha)$, is equal to the confidence coefficient—that is, the probability that $\bar{x}$ falls within $z_{\alpha/2}$ standard deviations of μ is $(1 - \alpha)$. Thus, a large-sample confidence interval for μ, with confidence coefficient equal to $(1 - \alpha)$, is given by

$$\bar{x} \pm z_{\alpha/2}\sigma_{\bar{x}}$$

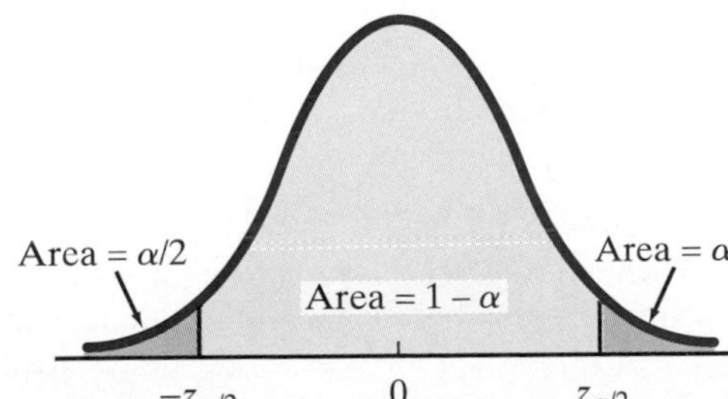

Figure 7.4 Locating $z_{\alpha/2}$ on the Standard Normal Curve

EXAMPLE 7.5

FINDING z FOR A 90% CONFIDENCE INTERVAL

In statistical problems using confidence interval techniques, a common confidence coefficient is .90. Determine the value of $z_{\alpha/2}$ that would be used in constructing a 90% confidence interval for a population mean based on a large sample.

Solution

For a confidence coefficient of .90, we have

$$1 - \alpha = .90 \qquad \alpha = .10 \qquad \alpha/2 = .05$$

and we need to obtain the value of $z_{\alpha/2} = z_{.05}$ that locates an area of .05 in the upper tail of the standard normal distribution. Since the total area to the right of $z = 0$ is .50, $z_{.05}$ is the value such that the area between 0 and $z_{.05}$ is $(.50 - .05) = .45$. From Table B.3, we find $z_{.05} = 1.645$ (see Figure 7.5). We conclude that a large-sample 90% confidence interval for a population mean is given by $\bar{x} \pm 1.645\sigma_{\bar{x}}$.

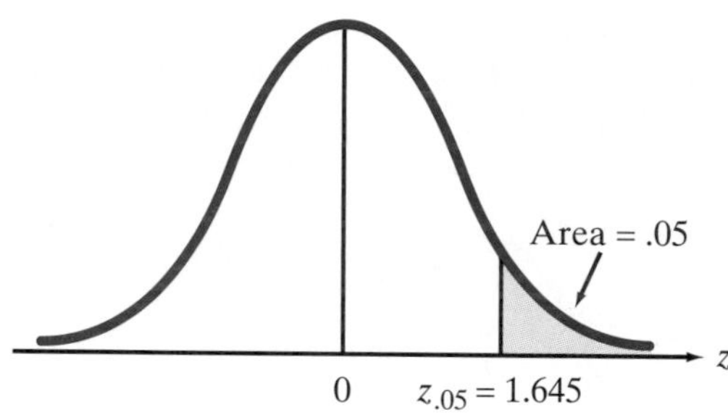

Figure 7.5 Location of $z_{\alpha/2}$ for Example 7.5

TABLE 7.2 Commonly Used Confidence Coefficients and Their Corresponding z Values

Confidence Coefficient $1 - \alpha$	$\alpha/2$	$z_{\alpha/2}$
.90	.05	1.645
.95	.025	1.96
.98	.01	2.33
.99	.005	2.575

In Table 7.2 we present the values of $z_{\alpha/2}$ for the most commonly used confidence coefficients.

Self-Test 7.1

Consider a 99% confidence interval for μ.

a. What is the confidence coefficient? Interpret its value.

b. What is the value of $z_{\alpha/2}$ used to form the interval estimate?

A summary of the large-sample confidence interval procedure for estimating a population mean appears in the following box.

LARGE-SAMPLE $(1 - \alpha)100\%$ CONFIDENCE INTERVAL FOR A POPULATION MEAN μ

$$\bar{x} \pm z_{\alpha/2}\sigma_{\bar{x}} = \bar{x} \pm z_{\alpha/2}\left(\frac{\sigma}{\sqrt{n}}\right) \tag{7.1}$$

where $z_{\alpha/2}$ is the z value that locates an area of $\alpha/2$ to its right, σ is the standard deviation of the population from which the sample was selected, n is the sample size, and $\bar{x}$ is the value of the sample mean.

Assumption: A large random sample (usually, $n \geq 30$) is selected.

[*Note:* When the value of σ is unknown (as will usually be the case), the sample standard deviation s may be used to approximate σ in the formula for the confidence interval. The approximation is generally quite satisfactory for large samples.]

EXAMPLE 7.6 APPLICATION OF A 99% CONFIDENCE INTERVAL

Twins, in their early years, tend to have lower IQs and pick up language more slowly than nontwins. Psychologists have speculated that the slower intellectual growth of twins may be caused by benign parental neglect. Suppose we want to investigate this phenomenon. A random sample of $n = 50$ sets of $2\frac{1}{2}$-year-old twin boys is selected, and the total parental attention time given to each pair during one week is recorded. The data (in hours) are listed in Table 7.3. Estimate μ, the mean attention time given to all $2\frac{1}{2}$-year-old twin boys by their parents, using a 99% confidence interval. Interpret the interval in terms of the problem.

ATTENT

TABLE 7.3 Attention Time for Random Sample of $n = 50$ Sets of Twins

20.7	14.0	16.7	20.7	22.5	48.2	12.1	7.7	2.9	22.2
23.5	20.3	6.4	34.0	1.3	44.5	39.6	23.8	35.6	20.0
10.9	43.1	7.1	14.3	46.0	21.9	23.4	17.5	29.4	9.6
44.1	36.4	13.8	0.8	24.3	1.1	9.3	19.3	3.4	14.6
15.7	32.5	46.6	19.1	10.6	36.9	6.7	27.9	5.4	14.0

Solution

Applying equation 7.1 and Table 7.2, the 99% confidence interval for μ is

$$\bar{x} \pm (2.575)\frac{\sigma}{\sqrt{n}}$$

To compute the interval, we require the sample mean $\bar{x}$ and the population standard deviation σ. In most practical applications, however, the value of σ will be unknown. For large samples, the fact that σ is unknown poses only a minor problem since the sample standard deviation s provides a good approximation to σ. Consequently, we substitute s for σ in the confidence interval formula given in the box.*

*We discuss an alternative form of the confidence interval when σ is unknown in Section 7.3.

An Excel printout showing descriptive statistics for the sample of $n = 50$ attention times is displayed in Figure 7.6. The values of $\bar{x}$ and s, highlighted on the printout, are (rounded) $\bar{x} = 20.85$ and $s = 13.41$. Substituting these values into the formula above, we obtain

$$20.85 \pm 2.575\left(\frac{13.41}{\sqrt{50}}\right)$$

or (15.97, 25.73). We are 99% confident that the interval (15.97, 25.73) encloses the true mean weekly attention time given to $2\frac{1}{2}$-year-old twin boys by their parents. Since all the values in the interval fall below 28 hours, we conclude that there is a general tendency for $2\frac{1}{2}$-year-old twin boys to receive less than 4 hours of parental attention time per day, on the average. Further investigation is required to relate this phenomenon to the intellectual growth of the twins.

	A	B
1	*Attention time*	
2		
3	Mean	20.848
4	Standard Error	1.897001
5	Median	19.65
6	Mode	20.7
7	Standard Deviation	13.41383
8	Sample Variance	179.9307
9	Kurtosis	-0.67289
10	Skewness	0.49963
11	Range	47.4
12	Minimum	0.8
13	Maximum	48.2
14	Sum	1042.4
15	Count	50
16	Largest(1)	48.2
17	Smallest(1)	0.8

E **Figure 7.6** Excel Descriptive Statistics for $n = 50$ Sample Attention Times

Self-Test 7.2

Based on the following summary statistics, construct a 95% confidence interval for μ: $n = 700, \bar{x} = 485, s = 210$.

EXAMPLE 7.7 **WIDTH OF CONFIDENCE INTERVAL VERSUS $(1-\alpha)$**

Refer to Example 7.6.

a. Find a 95% confidence interval for the mean weekly attention time given to all $2\frac{1}{2}$-year-old twin boys by their parents.

b. For a fixed sample size, how is the width of the confidence interval related to the confidence coefficient?

Solution

a. We used PHStat and Excel to generate the large-sample 95% confidence interval for μ. The interval, highlighted in the Excel printout, Figure 7.7, is (17.13, 24.57).

Figure 7.7 95% Confidence Interval for Mean Attention Time Obtained Using PHStat and Excel

	A	B
1	**Confidence Interval Estimate for the Mean**	
2		
3	**Data**	
4	**Population Standard Deviation**	**13.41**
5	**Sample Mean**	**20.85**
6	**Sample Size**	**50**
7	**Confidence Level**	**95%**
8		
9	Intermediate Calculations	
10	Standard Error of the Mean	1.896460387
11	Z Value	-1.95996108
12	Interval Half Width	3.716988553
13		
14	**Confidence Interval**	
15	**Interval Lower Limit**	**17.13301145**
16	**Interval Upper Limit**	**24.56698855**

b. The 99% confidence interval for μ was determined in Example 7.6 to be (15.97, 25.73). The 95% confidence interval in part **a** is narrower than the 99% confidence interval. This relationship holds in general, as noted in the following box.

RELATIONSHIP BETWEEN WIDTH OF CONFIDENCE INTERVAL AND CONFIDENCE COEFFICIENT

For a given sample size, the width of the confidence interval for a parameter increases as the confidence coefficient increases. Intuitively, the interval must become wider for us to have greater confidence that it contains the true parameter value.

EXAMPLE 7.8 **WIDTH OF A CONFIDENCE INTERVAL VERSUS SAMPLE SIZE n**

a. Assume that the given values of the statistics $\bar{x}$ and s were based on a sample of size $n = 100$ instead of a sample of size $n = 50$. Find a 99% confidence interval for μ, the population mean weekly attention time given to $2\frac{1}{2}$-year-old twin boys by their parents.

b. For a fixed confidence coefficient, how is the width of the confidence interval related to the sample size?

Solution

a. A 99% confidence interval for μ, generated using PHStat and Excel, is shown and highlighted in Figure 7.8. The interval is (17.40, 24.30).

E **Figure 7.8** 99% Confidence Interval for Mean Attention Time Using PHStat and Excel

	A	B
1	**Confidence Interval Estimate for the Mean**	
2		
3	**Data**	
4	**Population Standard Deviation**	**13.41**
5	**Sample Mean**	**20.85**
6	**Sample Size**	**100**
7	**Confidence Level**	**99%**
8		
9	Intermediate Calculations	
10	Standard Error of the Mean	1.341
11	Z Value	-2.57583451
12	Interval Half Width	3.454194084
13		
14	**Confidence Interval**	
15	**Interval Lower Limit**	**17.39580592**
16	**Interval Upper Limit**	**24.30419408**

b. The 99% confidence interval based on a sample of size $n = 100$, constructed in part **a**, is narrower than the 99% confidence interval based on a sample of size $n = 50$, constructed in Example 7.6. This will also hold true in general, as noted in the box.

RELATIONSHIP BETWEEN WIDTH OF CONFIDENCE INTERVAL AND SAMPLE SIZE

For a fixed confidence coefficient, the width of the confidence interval decreases as the sample size increases. In other words, larger samples generally provide more information about the target population than do smaller samples.

In this section, we introduced the concepts of point and interval estimation of the population mean μ, based on large samples. The general theory appropriate for the estimation of μ also carries over to the estimation of other population parameters. Hence, in subsequent sections we will present only the point estimate, its sampling distribution, the general form of a confidence interval for the parameter of interest, and any assumptions required for the validity of the procedure.

PROBLEMS FOR SECTION 7.2

Using the Tools

7.1 In a large-sample confidence interval for a population mean, what does the confidence coefficient represent?

7.2 Use Table B.3 in Appendix B to determine the value of $z_{\alpha/2}$ needed to construct a large-sample confidence interval for μ, for each of the following confidence coefficients:

a. .85 **b.** .95 **c.** .975

7.3 Suppose a random sample of size $n = 100$ produces a mean of $\bar{x} = 81$ and a standard deviation of $s = 12$.

a. Construct a 90% confidence interval for μ.

b. Construct a 95% confidence interval for μ.

c. Construct a 99% confidence interval for μ.

7.4 A random sample of size n is selected from a population with unknown mean μ. Calculate a 95% confidence interval for μ for each of the following situations:

a. $n = 35, \bar{x} = 26, \sigma = 15$ **b.** $n = 70, \bar{x} = 26, \sigma = 15$

c. $n = 70, \bar{x} = 26, \sigma = 11$

7.5 A random sample of size 400 is taken from an unknown population with mean μ and standard deviation σ. The following values are computed:

$$\Sigma x = 2{,}280 \qquad \Sigma(x - \bar{x})^2 = 25{,}536$$

a. Find a 90% confidence interval for μ.

b. Find a 99% confidence interval for μ.

7.6 The mean and standard deviation of a random sample of n measurements are equal to 22 and 16, respectively.

a. Construct a 95% confidence interval for μ if $n = 100$.

b. Construct a 95% confidence interval for μ if $n = 500$.

7.7 Give a precise interpretation of the statement, "We are 95% confident that the interval estimate contains μ."

Applying the Concepts

7.8 *Organizational Science* (Mar.–Apr. 2000) published a study on the use of communication media by mid-level managers. One question posed to each in a sample of 426 managers was: "How many electronic mail (e-mail) messages do you send during a typical week?" The sample had a mean of 9.81 messages and a standard deviation of 14.41 messages.

a. Use this information to find a point estimate for the mean number of e-mail messages sent by all mid-level managers during a typical week.

b. Find a 95% confidence interval for the mean number of e-mail messages sent by all mid-level managers during a typical week.

c. Interpret the interval, part **b**.

d. How could you reduce the width of the confidence interval, part **b**? Are there any drawbacks to reducing the interval width? Explain.

7.9 Consider a study designed to investigate the average desired family size among low-income Hispanic women. A sample of 432 Hispanic women aged 18–50 were interviewed at an obstetrics/gynecology clinic of a large Los Angeles public hospital (*Family Planning Perspectives,* Nov./Dec. 1997).

a. Each woman was asked, "How many sons do you desire?" The responses yielded the following summary statistics: $\bar{x} = 2.8$, $s = 1.4$. State the parameter of interest and construct a 90% confidence interval for its value. Interpret the results.

b. Responses to the question, "How many daughters do you desire?" yielded the following summary statistics: $\bar{x} = .1$, $s = .5$. State the parameter of interest and construct a 90% confidence interval for its value. Interpret the results.

c. Compare the results, parts **a** and **b**, and make a statement about the gender mix of the desired family size of low-income Hispanic women.

7.10 A study published in *Applied Psycholinguistics* (June 1998) examined the language skills of low-income children. Each child in a sample of 65 low-income children was administered the Communicative Development Inventory (CDI) exam. One of the

variables measured on each child was sentence complexity score. The sentence complexity scores had a mean of 7.62 and a standard deviation of 8.91.

a. Construct a 95% confidence interval for the mean sentence complexity score of the population of all low-income children.

b. Interpret the interval, part **a**, in the words of the problem.

c. Suppose we know that the true mean sentence complexity score of middle-income children is 15.55. Is there evidence that the true mean for low-income children differs from 15.55? Explain.

SHIPSANIT

7.11 Refer to the Centers for Disease Control study of sanitation levels of cruise ships, Problem 2.23 (p. 76). The inspection scores for a sample of 121 international cruise ships are saved in the **SHIPSANIT** file.

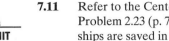

a. Construct a 90% confidence interval for μ, the true mean sanitation inspection score for all international ships.

b. Interpret the interval, part **a**.

c. What assumptions, if any, about the population of inspection scores are required for the confidence interval to be valid?

7.12 The U.S. has 1.4 million tax-exempt organizations including most schools and universities, foundations, and social service organizations such as the Red Cross, the Salvation Army, the YMCA, and the American Cancer Society. Donations to these organizations not only go to the stated charitable purpose, but also are used to cover fundraising expenses and overhead. For a sample of 30 charities, the table lists their charitable commitment (i.e., the percentage of their expenses that go toward the stated charitable purpose).

CHARITY

Organization	Charitable Commitment (%)
American Cancer Society	62
American National Red Cross	91
Big Brothers Big Sisters of America	77
Boy Scouts of America National Council	81
Boys & Girls Clubs of America	81
CARE	91
Covenant House	15
Disabled American Veterans	65
Ducks Unlimited	78
Feed the Children	90
Girl Scouts of the USA	83
Goodwill Industries International	89
Habitat for Humanity International	81
Mayo Foundation	26
Mothers Against Drunk Drivers	71
Multiple Sclerosis Association of America	56
Museum of Modern Art	79
Nature Conservancy	77
Paralyzed Veterans of America	50
Planned Parenthood Federation	81
Salvation Army	84
Shriners Hospital for Children	95
Smithsonian Institution	87
Special Olympics	72
Trust for Public Land	88
United Jewish Appeal/Federation—NY	75
United States Olympic Committee	78
United Way of New York City	85
WGBH Educational Foundation	81
YMCA of the USA	80

Source: "Look before you give," *Forbes,* Dec. 27, 1999, pp. 206–216.

a. Give a point estimate for the mean charitable commitment of all tax-exempt organizations in the U.S.

b. Construct a 98% confidence interval for the mean charitable commitment of all tax-exempt organizations in the U.S.

c. Why is the confidence interval of part **b** a better estimator of the mean charitable commitment than the point estimator of part **a**?

7.13 Animal behaviorists have discovered that the more domestic chickens peck at objects placed in their environment, the healthier the chickens seem to be. White-colored string has been found to be a particularly attractive pecking stimulus. In one experiment, 72 chickens were exposed to a string stimulus. Instead of white string, blue-colored string was used. The number of pecks each chicken took at the blue string over a specified time interval was recorded. Summary statistics for the 72 chickens were $\bar{x} = 1.13$ pecks, $s = 2.21$ pecks (*Applied Animal Behaviour Science,* Oct. 2000).

a. Estimate the population mean number of pecks made by chickens pecking at blue string using a 99% confidence interval. Interpret the result.

b. Previous research has shown that $\mu = 7.5$ pecks if chickens are exposed to white string. Based on the results, part **a**, is there evidence that chickens are more apt to peck at white string than blue string? Explain.

7.14 The velocity of light emitted from a galaxy provides astronomers with valuable information on how the galaxy was formed. The *Astronomical Journal* (July 1995) reported that in a sample of 103 galaxies located in close proximity (called a galaxy cluster), the mean velocity was $\bar{x} = 27{,}117$ kilometers per second and the standard deviation was $s = 1{,}280$ kilometers per second.

a. Estimate the population mean velocity of light emitted from all galaxies in the galaxy cluster using a 95% confidence interval. Interpret the result.

b. How could you reduce the width of the confidence interval from part **a**?

7.15 Unusual rocks at "The Seven Islands," located along the lower St. Lawrence River in Canada, have attracted geologists to the area for over a century. A major geological survey of "The Seven Islands" was completed for the purpose of obtaining an accurate estimate of the rock densities in the area (*Canadian Journal of Earth Sciences,* Vol. 27, 1990). Based on samples of several varieties of rock, the following information on rock density (grams per cubic centimeter) was obtained.

Type of Rock	Sample Size	Mean Density	Standard Deviation
Late gabbro	36	3.04	.13
Massive gabbro	148	2.83	.11
Cumberlandite	135	3.05	.31

Source: Loncarevic, B. D., Feninger, T., and Lefebvre, D. "The Sept-Iles layered mafic intrusion: Geo-physical expression." *Canadian Journal of Earth Sciences,* Vol. 27, Aug. 1990, p. 505.

a. For each rock type, estimate the population mean density with a 90% confidence interval.

b. Interpret the intervals, part **a**.

7.3 Estimation of a Population Mean: Student's *t* Statistic

In the previous section, we discussed the estimation of a population mean based on large random samples and known population standard deviation σ. However, time or cost limitations often restrict the number of sample observations that may be obtained, so that the estimation procedures of Section 7.2 will not be applicable.

With small samples, the following two problems arise:

Problem 1 Since the Central Limit Theorem applies only to large samples, we are not able to assume that the sampling distribution of $\bar{x}$ is approximately normal. For small samples, the sampling distribution of $\bar{x}$ depends on the particular shape of the distribution of the population being sampled.

Problem 2 The sample standard deviation s may not be a satisfactory approximation to the population standard deviation σ if the sample size is small. Thus, replacing σ with s in the large-sample formula given in Section 7.2 is not appropriate.

Fortunately, we may proceed with estimation techniques based on small samples if we can make the following assumption:

ASSUMPTION REQUIRED FOR ESTIMATING μ BASED ON SMALL SAMPLES

The population from which the random sample is selected has an approximately normal distribution.

Confidence Interval for a Normal Population
When sampling is from a normal population with σ known, the appropriate confidence interval is $\bar{x} \pm z_{\alpha/2}(\sigma/\sqrt{n})$ regardless of the size of the sample. This results from the fact that the sampling distribution of $\bar{x}$ is normal whenever the population is normally distributed. (See Section 6.6.)

If this assumption is valid, then we may again use $\bar{x}$ as a point estimate for μ, and the general form of a small-sample confidence interval for μ is shown in the next box.

SMALL-SAMPLE CONFIDENCE INTERVAL FOR μ

$$\bar{x} \pm t_{\alpha/2}\left(\frac{s}{\sqrt{n}}\right) \qquad \textbf{(7.2)}$$

where the distribution of t is based on $(n - 1)$ degrees of freedom.

Now compare the large-sample confidence interval for μ (equation 7.1) to the small-sample confidence interval for μ (equation 7.2). Note that the sample standard deviation s replaces the population standard deviation σ in the formula. Also, the sampling distribution upon which the confidence interval is based is known as a **Student's *t* distribution.** Consequently, we must replace the value of $z_{\alpha/2}$ used in a large-sample confidence interval by a value obtained from the t distribution.

Student's *t* Distribution
The derivation of the t distribution was first published in 1908 by W. S. Gosset, who wrote under the pen name of Student. Thereafter, the distribution became known as Student's t.

The t distribution is very much like the z distribution. In particular, both are symmetric and bell-shaped and have a mean of 0. However, the t distribution is flatter, that is, it has less area in the center and more area in the tails than the z distribution has (see Figure 7.9).

Also, the distribution of t depends on a quantity called its **degrees of freedom** (df), which is equal to $(n - 1)$ when estimating a population mean based on a small sample of size n. Intuitively, we can think of the number of degrees of freedom as

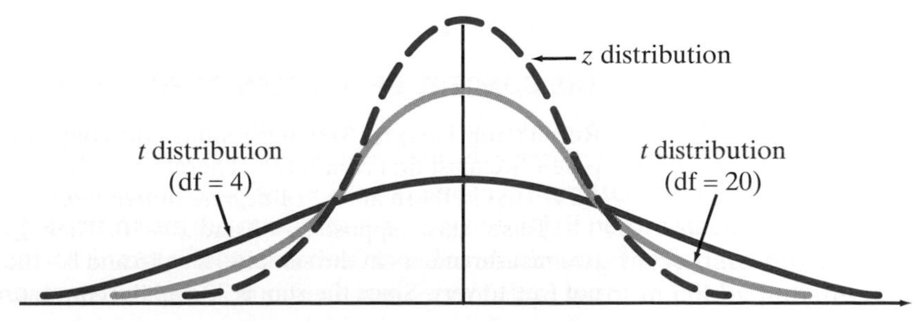

Figure 7.9 Comparison of z Distribution to the t Distribution

SUICIDE14

TABLE 7.5 Days in Jail Before Suicide for 14 Inmates Who Committed Suicide

10	15	3	4	7	19	15
5	22	221	85	126	29	31

	A	B
1	*Days*	
2		
3	Mean	42.28571
4	Standard Error	16.60876
5	Median	17
6	Mode	15
7	Standard Deviation	62.14428
8	Sample Variance	3861.912
9	Kurtosis	5.047523
10	Skewness	2.266428
11	Range	218
12	Minimum	3
13	Maximum	221
14	Sum	592
15	Count	14
16	Largest(1)	221
17	Smallest(1)	3

E **Figure 7.10** Excel Descriptive Statistics for Example 7.11

Solution

The first step in constructing the confidence interval is to compute the sample mean $\bar{x}$ and sample standard deviation s of the data in Table 7.5. These values (rounded), $\bar{x} = 42.28$ and $s = 62.14$, are highlighted in the Excel printout, Figure 7.10.

For a confidence coefficient of $1 - \alpha = .90$, we have $\alpha = .10$ and $\alpha/2 = .05$. Since the sample size is small ($n = 14$), our estimation technique requires the assumption that the distribution of days in jail before suicide for all inmates who eventually commit suicide is approximately normal.

Substituting the values for $\bar{x}$, s, and n into the formula for a small-sample confidence interval for μ (equation 7.2), we obtain

$$\bar{x} \pm t_{\alpha/2}\left(\frac{s}{\sqrt{n}}\right) = \bar{x} \pm t_{.05}\left(\frac{s}{\sqrt{n}}\right)$$

$$= 42.28 \pm t_{.05}\left(\frac{62.14}{\sqrt{14}}\right)$$

where $t_{.05}$ is the value corresponding to an upper-tail area of .05 in the Student's t distribution based on $(n - 1) = 13$ degrees of freedom. From Table B.4, the required t value is $t_{.05} = 1.771$. Substituting this value yields

$$42.28 \pm (1.771)\left(\frac{62.14}{\sqrt{14}}\right) = 42.28 \pm 29.41$$

or 12.87 to 71.69 days. Thus, if the distribution of days spent in jail is approximately normal, then we can be 90% confident that the interval (12.87, 71.69) encloses μ, the true mean number of days spent in jail before suicide for inmates who commit suicide. If, however, the sampled population is nonnormal, then any inferences derived from this small-sample confidence interval may be invalid.

Self-Test 7.4

Based on the following summary statistics, construct a 95% confidence interval for μ: $n = 10, \bar{x} = 485, s = 210$.

EXAMPLE 7.12

COMPUTER-GENERATED CONFIDENCE INTERVAL

Refer to Example 7.6 (p. 345) and the problem of estimating μ, the mean attention time given to $2\frac{1}{2}$-year-old twin boys by their parents. Recall that a 99% confidence interval was computed based on data collected for a random sample of $n = 50$ sets of twins. An Excel printout showing a 99% confidence interval for

E Figure 7.11 Confidence Interval for Example 7.12 Obtained Using PHStat and Excel

	A	B
1	**Confidence Interval Estimate for the Mean**	
2		
3	**Data**	
4	**Sample Standard Deviation**	**13.41382534**
5	**Sample Mean**	**20.848**
6	**Sample Size**	**50**
7	**Confidence Level**	**99%**
8		
9	Intermediate Calculations	
10	Standard Error of the Mean	1.897001372
11	Degrees of Freedom	49
12	t Value	2.679953468
13	Interval Half Width	5.083875405
14		
15	**Confidence Interval**	
16	**Interval Lower Limit**	**15.76**
17	**Interval Upper Limit**	**25.93**

μ is displayed in Figure 7.11. Compare the results to the interval calculated in Example 7.6.

Solution

The 99% confidence interval, highlighted on the Excel printout, is (15.76, 25.93). The interval calculated in Example 7.6 is (15.97, 25.73). The differences in the end points of the interval, although relatively minor, are because σ is unknown for the target population. As do most software packages, PHStat with Excel and MINITAB compute the confidence interval using the t statistic, i.e.,

$$\bar{x} \pm t_{.005}\left(\frac{s}{\sqrt{n}}\right)$$

where $t_{.005} \approx 2.68$ (based on $n - 1 = 49$ df). The confidence interval in Example 7.6, you will recall, was calculated using the z statistic, i.e.,

$$\bar{x} \pm z_{.005}\left(\frac{s}{\sqrt{n}}\right)$$

where $z_{.005} = 2.575$. Theoretically, the Excel confidence interval is the correct one, since $\bar{x}$ has a t distribution when σ is unknown. The confidence interval in Example 7.6 is approximate. But you can see the approximation is good when the sample size n is large.

Before concluding this section, we will comment on the assumption that the sampled population is normally distributed. In the real world, we rarely know whether a sampled population has an exactly normal distribution. However, empirical studies indicate that moderate departures from this assumption do not seriously affect the confidence coefficients for small-sample confidence intervals. For example, if the population of days spent in jail for inmates who commit suicide of Example 7.11 has a distribution that is bell-shaped but nonnormal, it is likely that the actual confidence coefficient for the 90% confidence interval will be close to .90—at least close enough to be of practical use. As a consequence, the small-sample confidence interval given in equation 7.2 is frequently used by experimenters when estimating the population mean of a nonnormal distribution as long as the distribution is bell-shaped and only moderately skewed.

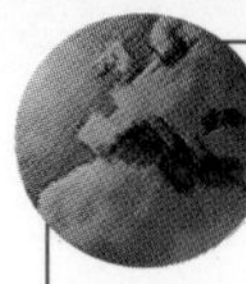

Statistics in the Real World Revisited

Estimating the Mean Weight per Scallop

Refer to the scallop lawsuit described on p. 338. Recall that a ship, returning from a fishing expedition with 11,000 bags of scallops, was accused of violating the U.S. Fisheries and Wildlife Service weight standard for baby scallops. The law requires that the average meat per scallop weigh at least 1/36 of a pound. The harbormaster randomly selected 18 bags for weighing and estimated that the ship caught scallops that averaged only 1/39 of a pound of meat. Consequently, federal authorities confiscated the catch. Was the ship really in violation of the scallop weight standard?

One way to answer this question is to form a confidence interval for the true mean weight per scallop for all 11,000 bags caught by the fishing vessel. The data for the 18 sampled bags in the **SCALLOPS** file are shown in Table 7.6. (Each measurement in the table is the estimated average weight per scallop of the bag. A measurement less than 1/36 = .0278 pound indicates an individual bag that does not meet the standard.)

SCALLOPS

TABLE 7.6 Scallop Weight Measurements (in Pounds) for 18 Bags Sampled

.0258	.0244	.0236	.0253	.0253	.0233	.0250	.0272	.0244
.0247	.0272	.0242	.0253	.0256	.0275	.0317	.0294	.0258

Source: Adapted from Bennett, A. "Misapplications review: Jail terms." *Interfaces,* Vol. 25, No. 2, March–April 1995, p. 20.

The data are stored as a MINITAB worksheet and the software is used to find a 95% confidence interval for the population mean weight per scallop. The MINITAB printout is displayed in Figure 7.12. The interval, highlighted on the printout, is (0.02483, 0.02691). Note that the upper endpoint of the interval falls below .0278 (or 1/36 of a pound). Consequently, we are 95% confident that the true mean weight per scallop for the 11,000 bags caught by the fishing vessel falls below the 1/36 pound minimum set by the U.S. Fisheries and Wildlife Service. It appears that the ship was in violation of the government minimum weight requirement for baby scallops.

```
Variable      N       Mean     StDev   SE Mean        95.0 % CI
Weight       18    0.02587   0.00210   0.00049   ( 0.02483,  0.02691)
```

Figure 7.12 MINITAB Confidence Interval for Scallop Weight Data

[*Note:* Because the ship violated the baby scallop weight rule, federal authorities confiscated 95% of the ship's catch. The ship's owner complained that this was not fair, since not all (or even 95%) of the bags of scallops had an average weight per scallop under 1/36 of a pound. We address this issue in the next Statistics in the Real World Revisited section (p. 362).]

PROBLEMS FOR SECTION 7.3

Using the Tools

7.16 Use Table B.4 in Appendix B to determine the values of $t_{\alpha/2}$ needed to construct a confidence interval for a population mean for each of the following combinations of confidence coefficient and sample size:

a. Confidence coefficient .99, $n = 18$

b. Confidence coefficient .95, $n = 10$

c. Confidence coefficient .90, $n = 15$

7.17 What assumptions are required for the interval estimation procedure of this section to be valid when the sample size is small, i.e., when $n < 30$?

7.18 A random sample of $n = 10$ measurements from a normally distributed population yielded $\bar{x} = 9.4$ and $s = 1.8$.

a. Calculate a 90% confidence interval for μ.

b. Calculate a 95% confidence interval for μ.

c. Calculate a 99% confidence interval for μ.

7.19 The following data represent a random sample of five measurements from a normally distributed population: 7, 4, 2, 5, 7.

a. Find a 90% confidence interval for μ.

b. Find a 99% confidence interval for μ.

7.20 The mean and standard deviation of n measurements randomly sampled from a normally distributed population are 33 and 4, respectively. Construct a 95% confidence interval for μ when:

a. $n = 5$ **b.** $n = 15$ **c.** $n = 25$

7.21 How are the t distribution and the z distribution similar? How are they different?

Applying the Concepts

7.22 A group of Harvard University School of Public Health researchers studied the impact of cooking on the size of indoor air particles (*Environmental Science & Technology,* Sept. 1, 2000). The decay rate (measured as μm/hour) for fine particles produced from oven cooking or toasting was recorded on six randomly selected days. These six measurements are

DECAY

.95	.83	1.20	.89	1.45	1.12

Source: Abt, E. et al. "Relative contribution of outdoor and indoor particle sources to indoor concentrations," *Environmental Science & Technology,* Vol. 34, No. 17, Sept. 1, 2000 (Table 3).

a. Find and interpret a 95% confidence interval for the true average decay rate of fine particles produced from oven cooking or toasting.

b. Explain what the phrase "95% confident" implies in the interpretation of part **a**.

c. What must be true about the distribution of the population of decay rates for the inference to be valid?

7.23 Scientists have discovered increased levels of the hormone adrenocorticotropin in people just before they awake from sleeping (*Nature,* Jan. 7, 1999). In the study, 15 subjects were monitored during their sleep after being told they would be wakened at a particular time. One hour prior to the designated wake-up time, the adrenocorticotropin level (pg/ml) was measured in each, with the following results: $\bar{x} = 37.3, s = 13.9$.

a. Estimate the true mean adrenocorticotropin level of sleepers one hour prior to waking using a 95% confidence interval.

b. Interpret the interval, part **a**, in the words of the problem.

c. The researchers also found that if the subjects were wakened three hours earlier than they anticipated, the average adrenocorticotropin level was 25.5 pg/ml. Assume that $\mu = 25.5$ for all sleepers who are wakened three hours earlier than expected. Use the interval, part **a**, to make an inference about the mean adrenocorticotropin level of sleepers under two conditions: one hour before anticipated wake-up time and three hours before anticipated wake-up time.

7.24 The *Journal of Communication* (Summer 1997) published a study on U.S. media coverage of major earthquakes that occurred in foreign countries. The number of seconds devoted to the earthquake on the nightly news broadcasts of ABC, CBS, and NBC was determined for each in a sample of 22 foreign earthquakes that caused 10 or more deaths. The mean and standard deviation were 1,780 seconds and 2,797 seconds, respectively.

a. Construct a 99% confidence interval for the average amount of time devoted to foreign earthquakes on the nightly news.

b. Interpret the interval, part **a**.

7.25 The *Journal of Applied Behavior Analysis* (Summer 1997) published a study of a treatment designed to reduce destructive behavior. A 16-year-old boy diagnosed with severe mental retardation was observed playing with toys for 10 sessions, each ten minutes in length. The number of destructive responses (e.g., head banging, pulling hair, pinching, etc.) per minute was recorded for each session in order to establish a baseline behavior pattern. The data for the 10 sessions are listed here:

DESTRUCT

3	2	4.5	5.5	4.5	3	2.5	1	3	3

Construct a 95% confidence interval for the true mean number of destructive responses per minute for the 16-year-old boy. Interpret the result.

7.26 Refer to the *Risk Management* study on fires in compartmented fire-resistive buildings, Problem 3.84 (p. 180). The data shown in the table give the number of victims who died attempting to evacuate for a sample of 14 fires.

FIRE

Fire	Number of Victims
Las Vegas Hilton (Las Vegas)	5
Inn on the Park (Toronto)	5
Westchase Hilton (Houston)	8
Holiday Inn (Cambridge, Ohio)	10
Conrad Hilton (Chicago)	4
Providence College (Providence)	8
Baptist Towers (Atlanta)	7
Howard Johnson (New Orleans)	5
Cornell University (Ithaca, New York)	9
Westport Central Apartments (Kansas City, Missouri)	4
Orrington Hotel (Evanston, Illinois)	0
Hartford Hospital (Hartford, Connecticut)	16
Milford Plaza (New York)	0
MGM Grand (Las Vegas)	36

Source: Macdonald, J. N. "Is evacuation a fatal flaw in fire fighting philosophy?" *Risk Management,* Vol. 33, No. 2, Feb. 1986, p. 37.

a. State the assumption, in terms of the problem, that is required for a small-sample confidence interval technique to be valid.

b. Construct a 98% confidence interval for the true mean number of victims per fire who die attempting to evacuate compartmented fire-resistive buildings.

c. Interpret the interval constructed in part **b**.

7.27 The number of calories per 12-ounce serving for a sample of 22 beer brands, first presented in Problem 2.22 (p. 76), is reproduced in the table on p. 359.

a. Find a 90% confidence interval for μ, the true mean number of calories per 12-ounce serving for all beer brands.

b. Interpret the interval, part **a**.

c. What assumption about the population data is required for the confidence interval to be valid?

d. Use the stem-and-leaf display you generated in Problem 2.22 to help you decide whether the assumption of part **c** is reasonably satisfied.

BEERCAL

Beer	Calories	Beer	Calories
Old Milwaukee	145	Miller High Life	143
Stroh's	142	Pabst Blue Ribbon	144
Red Dog	147	Milwaukee's Best	133
Budweiser	148	Miller Genuine Draft	143
Icehouse	149	Rolling Rock	143
Molson Ice	155	Michelob Light	134
Michelob	159	Bud Light	110
Bud Ice	148	Natural Light	110
Busch	143	Coors Light	105
Coors Original	137	Miller Light	96
Gennessee Cream Ale	153	Amstel Light	95

Source: Consumer Reports, June 1996, Vol. 61, No. 6, p. 16.

Problem 7.27

7.28 Refer to the *Chemosphere* (Vol. 20, 1990) study on the TCDD levels of 20 Massachusetts Vietnam veterans who were possibly exposed to the defoliant Agent Orange, Problem 2.28 (p. 78). The amounts of TCDD (measured in parts per trillion) in blood plasma drawn from each veteran are shown in the table.

TCDD

Veteran	TCDD Levels in Plasma	Veteran	TCDD Levels in Plasma
1	2.5	11	6.9
2	3.1	12	3.3
3	2.1	13	4.6
4	3.5	14	1.6
5	3.1	15	7.2
6	1.8	16	1.8
7	6.0	17	20.0
8	3.0	18	2.0
9	36.0	19	2.5
10	4.7	20	4.1

Source: Schecter, A., et al. "Partitioning of 2,3,7,8-chlorinated dibenzo-*p*-dioxins and dibenzofurans between adipose tissue and plasma lipid of 20 Massachusetts Vietnam veterans." *Chemosphere,* Vol. 20, Nos. 7–9, 1990, pp. 954–955 (Table 1).

a. Construct a 90% confidence interval for the true mean TCDD level in plasma of all Vietnam veterans exposed to Agent Orange.

b. Interpret the interval, part **a**.

c. What assumption is required for the interval estimation procedure to be valid?

d. Use one of the methods of Section 6.3 to determine whether the assumption, part **c**, is approximately satisfied.

7.4 Estimation of a Population Proportion

We now consider the method for estimating the binomial proportion of successes—that is, the **proportion** of elements in a population that have a certain characteristic. For example, a sociologist may be interested in the proportion of urban New York City residents who are minorities; a pollster may be interested in the proportion of Americans who favor the president's new economic policy; or a television executive may be interested in the percentage of viewers tuned to a particular TV program on a given night. How would you estimate a binomial proportion p based on information contained in a sample from the population? We illustrate in the next example.

EXAMPLE 7.13 **SELECTING A POINT ESTIMATE**

The United States Commission on Crime is interested in estimating the proportion of crimes related to firearms in an area with one of the highest crime rates in the country. The commission selects a random sample of 300 files of recently committed crimes in the area and determines that a firearm was reportedly used in 180 of them. Estimate the true proportion p of all crimes committed in the area in which some type of firearm was reportedly used.

Solution

A logical candidate for a point estimate of the population proportion p is the proportion of observations in the sample that have the characteristic of interest (called a "success"); we will call this sample proportion $\hat{p}$ (read p-hat). In this example, the sample proportion of crimes related to firearms is given by

$$\hat{p} = \frac{\text{Number of crimes in sample in which a firearm was reportedly used}}{\text{Total number of crimes in sample}}$$

$$= \frac{180}{300} = .60$$

That is, 60% of the crimes in the sample were related to firearms; the value $\hat{p} = .60$ serves as our point estimate of the population proportion p.

To assess the reliability of the point estimate $\hat{p}$, we need to know its sampling distribution. This information may be derived by an application of the Central Limit Theorem (details are omitted here). Properties of the sampling distribution of $\hat{p}$ (illustrated in Figure 7.13) are given in the next box.

Figure 7.13 Sampling Distribution of $\hat{p}$

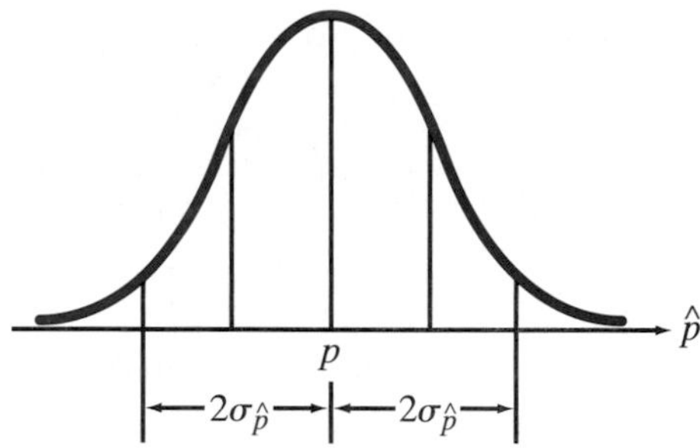

SAMPLING DISTRIBUTION OF THE SAMPLE PROPORTION $\hat{p}$

For sufficiently large samples, the sampling distribution of $\hat{p}$ is approximately normal, with

$$\text{Mean } \mu_{\hat{p}} = p$$

and

$$\text{Standard deviation } \sigma_{\hat{p}} = \sqrt{\frac{p(1-p)}{n}} \quad \textbf{(7.3)}$$

where

p = true population proportion of successes
x = number of successes in the sample
n = sample size
$\hat{p} = x/n$

A large-sample confidence interval for p is constructed by using a procedure analogous to that used for estimating a population mean. We begin with the point estimator $\hat{p}$, then add and subtract a certain number of standard deviations of $\hat{p}$ to obtain the desired level of confidence. The details are given in the next box.

LARGE-SAMPLE $(1 - \alpha)100\%$ CONFIDENCE INTERVAL FOR A POPULATION PROPORTION p

$$\hat{p} \pm z_{\alpha/2}\sigma_{\hat{p}} \approx \hat{p} \pm z_{\alpha/2}\sqrt{\frac{\hat{p}(1-\hat{p})}{n}} \qquad \textbf{(7.4)}$$

where $\hat{p} = x/n$ is the sample proportion of observations with the characteristic of interest.

[*Note:* The interval is approximate since we have substituted the sample value $\hat{p}$ for the corresponding population value p required for $\sigma_{\hat{p}}$.]

Assumption: A sufficiently large random sample of size n is selected from the population.

Note that we must substitute $\hat{p}$ into the formula for

$$\sigma_{\hat{p}} = \sqrt{p(1-p)/n}$$

here. This approximation will be valid as long as the sample size n is sufficiently large. Usually, "sufficiently large" will be satisfied if both $n\hat{p} \geq 5$ and $n(1 - \hat{p}) \geq 5$ (see Section 6.4).

EXAMPLE 7.14

95% CONFIDENCE INTERVAL FOR p

Refer to Example 7.13. Construct and interpret a 95% confidence interval for p, the population proportion of crimes committed in the area in which some type of firearm is reportedly used.

Solution

For a confidence coefficient of .95, we have $1 - \alpha = .95$; $\alpha = .05$; $\alpha/2 = .025$; and the required z value is $z_{.025} = 1.96$. In Example 7.13, we obtained $\hat{p} = 180/300 = .60$. Substituting these values into the formula for an approximate confidence interval for p (equation 7.4) yields

$$\hat{p} \pm z_{\alpha/2}\sqrt{\frac{\hat{p}(1-\hat{p})}{n}} = .60 \pm 1.96\sqrt{\frac{(.60)(.40)}{300}}$$
$$= .60 \pm .06$$

or (.54, .66). We are 95% confident that the true proportion of crimes committed in the area in which firearms were used falls between .54 and .66.

EXAMPLE 7.15

90% CONFIDENCE INTERVAL FOR p

Potential advertisers value television's well-known Nielsen ratings as a barometer of a TV show's popularity among viewers. The Nielsen rating of a certain TV program is an estimate of the proportion of viewers, expressed as a percentage, who tune their sets to the program on a given night. Suppose that in a random sample of 165 families who regularly watch television, a Nielsen survey indicated that 63 of the families were tuned to NBC's *ER* on the night of its season finale. Estimate p, the true proportion of all TV-viewing families who watched the season finale of *ER*, using a 90% confidence interval. Interpret the interval.

7.32 The proportion of successes in a random sample of size n is $\hat{p} = .98$.

a. Find a 90% confidence interval for p if $n = 10{,}000$.

b. Find a 90% confidence interval for p if $n = 1{,}000$.

c. Find a 90% confidence interval for p if $n = 30$.

Applying the Concepts

7.33 Did the movie "Jaws" scare you out of the water or the movie "Psycho" change your shower habits? According to a University of Michigan study, many adults have experienced lingering "fright" effects from a scary movie or TV show they saw as a teenager (*Tampa Tribune,* Mar. 10, 1999). In a survey of 150 college students, 39 said they still experience "residual anxiety" from a scary TV show or movie.

a. Give a point estimate, p, for the true proportion of college students who experience "residual anxiety" from a scary TV show or movie.

b. Find a 95% confidence interval for p.

c. Interpret the interval, part **b**.

7.34 An American Housing Survey (AHS) conducted by the U.S. Department of Commerce revealed that 705 of 1,500 sampled homeowners are "do-it-yourselfers"—they did most of the work themselves on at least one of their home improvements or repairs (Bureau of the Census, *Statistical Brief,* May 1992). Using a 95% confidence interval, estimate the true proportion of American homeowners who do most of their home improvement/repair work themselves. Interpret the result.

7.35 The Gallup Organization recently surveyed 1,252 debit cardholders in the U.S. and found that 180 had used the debit card to purchase a product or service on the Internet (*Card Fax,* Nov. 12, 1999).

a. Describe the population of interest to the Gallup Organization.

b. If you personally were charged with drawing a random sample from this population, what difficulties would you encounter? (Assume in the remainder of the exercise that the 1,252 debit cardholders were randomly selected.)

c. Is the sample size large enough to construct a valid confidence interval for the proportion of debit cardholders who have used their card in making purchases over the Internet?

d. Estimate the proportion referred to in part **c** using a 90% confidence interval. Interpret your result in the context of the problem.

e. If you had constructed a 99% confidence interval instead, would it be wider or narrower?

7.36 In a study reported in *The Wall Street Journal* (April 4, 1999), the Tupperware Corporation surveyed 1,007 U.S. workers. Of the people surveyed, 665 indicated that they take their lunch to work with them. Of these 665 taking their lunch, 200 reported that they take them in brown bags.

a. Find a 95% confidence interval estimate of the population proportion of U.S. workers who take their lunch to work with them.

b. Consider the population of U.S. workers who take their lunch to work with them. Find a 95% confidence interval estimate of the population proportion who take brown-bag lunches.

c. How is this information useful to the Tupperware Corporation?

7.37 Psychologists J. Gottman (University of Washington) and R. Levenson (University of California) recently completed a long-term study of married couples designed to predict when (or if) they will divorce. Of the 79 married couples in their study, 22 had divorced before their 20th year of marriage (*USA Today,* Sept. 13, 2000). Use a 90% confidence interval to estimate the fraction of all married couples that divorce before their 20th year of marriage. Interpret the result.

7.38 Refer to the *Annals of Internal Medicine* (Dec. 1, 1997) study of whether women understand basic probability, Problem 4.5 (p. 197). Recall that a random sample of

287 female military veterans were asked: "How many times will a coin come up heads in 1,000 flips?" Only 155 of the females answered correctly (about 500 heads).

a. Construct a 95% confidence interval for the true proportion of female veterans who answered "about 500 heads" to the coin flip question. Interpret the result.

b. Why is it important to restrict any inferences derived from the interval to female military veterans and not to women in general?

7.39 ABC News and *The Washington Post* conducted a national random-digit-dial telephone survey of 1,500 adults. One question asked: "Have you ever seen anything that you believe was a spacecraft from another planet?" Ten percent (150) of the respondents answered in the affirmative. These 150 people were then asked, "Have you personally ever been in contact with aliens from another planet?" and nine people responded "Yes." (*Chance,* Summer 1997).

a. Estimate the population proportion of adults who believe they have seen spacecraft from another planet with a 90% confidence interval.

b. Estimate the population proportion of adults who believe they have been in contact with aliens from another planet with a 90% confidence interval.

c. It is highly likely that the true value of p estimated in parts **a** and **b** is equal to 0. Given this is true, how do you explain the survey results?

7.40 The concept of tyranny in 19th-century American legal cases was investigated in the *Journal of Interdisciplinary History* (Autumn 1997). In a sample of 3,038 U.S. federal court cases that occurred in the decade 1850–1859, 27 of them used keywords associated with tyranny (e.g., tyranny, despotism, dictatorship, autocracy, and absolutism).

a. Use a 99% confidence interval to estimate the true percentage of federal court cases in the decade 1850–1859 that used keywords associated with tyranny.

b. Do you believe that 2% or more of all the U.S. federal court cases between 1850 and 1859 used keywords associated with tyranny? Explain.

7.41 A University of South Florida research team performed a series of experiments to investigate the effects of stress on memory (*USF Magazine,* Winter 1998). In one experiment, 30 rats were trained to escape from a small pool of water by climbing onto a fixed platform hidden just below the water's surface. After learning the platform's location, all 30 rats were placed in a room with a cat. Although the cat did not touch the rats, the rats exhibited behavioral and physiological signs of stress. The rats were then immediately returned to the water tank and the researchers counted the number that forgot the location of the platform.

a. Suppose that 25 of the 30 rats forgot how to escape from the water tank. Estimate the population proportion of rats that will forget how to escape when placed in a similar experimental setting.

b. Use the result, part **a**, to form a 95% confidence interval for the population proportion of rats that will forget how to escape. Give a practical interpretation of the interval.

7.5 Choosing the Sample Size

In the preceding sections we have overlooked a problem that usually must be faced in the initial stages of an experiment. Before constructing a confidence interval for a parameter of interest, we will have to decide on the number of observations n to be included in a sample. Should we sample $n = 10$ observations, $n = 20$, or $n = 100$? To answer this question we need to decide how wide a confidence interval we are willing to tolerate and the measure of confidence—that is, the confidence coefficient—that we wish to have in the results. The following example will illustrate the method for determining the appropriate sample size for estimating a population mean.

EXAMPLE 7.16 **SAMPLE SIZE FOR ESTIMATING μ**

A mail-order house wants to estimate the mean length of time between shipment of an order and receipt by the customer. The management plans to randomly sample n orders and determine, by telephone, the number of days between shipment and receipt for each order. If the management wants to estimate the mean shipping time correct to within one day with confidence coefficient equal to .95, how many orders should be sampled?

Solution

We will use $\bar{x}$, the sample mean of the n measurements, to estimate μ, the mean shipping time. Its sampling distribution will be approximately normal and the probability that $\bar{x}$ will lie within

$$1.96\sigma_{\bar{x}} = 1.96\left(\frac{\sigma}{\sqrt{n}}\right)$$

of the mean shipping time μ is approximately .95 (see Figure 7.15). Therefore, we want to choose the sample size n so that $1.96\sigma/\sqrt{n}$ equals one day:

$$1.96\left(\frac{\sigma}{\sqrt{n}}\right) = 1 \qquad \textbf{(7.5)}$$

Figure 7.15 Sampling Distribution of the Sample Mean, $\bar{x}$

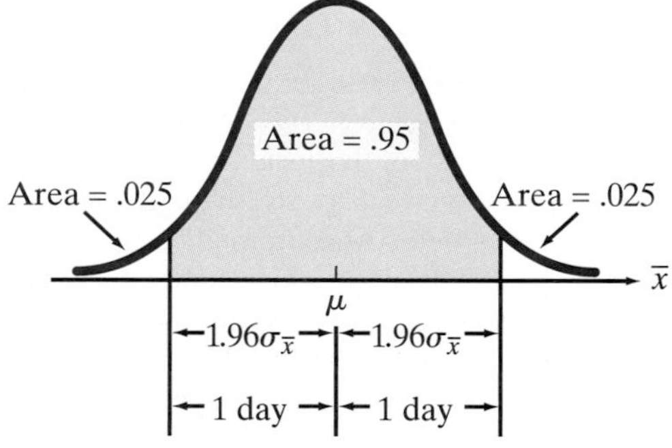

To solve the equation $1.96\sigma/\sqrt{n} = 1$, we need to know the value of σ, a measure of variation of the population of all shipping times. Since σ is unknown (as will usually be the case in practical applications), we must approximate its value using the standard deviation of some previous sample data or deduce an approximate value from other knowledge about the population. Suppose, for example, that we know almost all shipments will be delivered within 21 days. Then the population of shipping times might appear as shown in Figure 7.16.

Figure 7.16 Hypothetical Distribution of Population of Shipping Times for Example 7.16

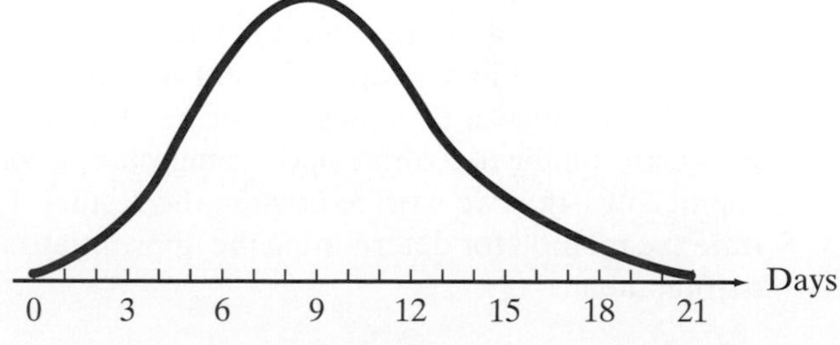

Figure 7.16 provides the information we need to find an approximation for σ. Since the Empirical Rule tells us that almost all the observations in a data set will fall within the interval $\mu \pm 3\sigma$, it follows that the range of a population is approximately 6σ. If the range of the population of shipping times is 21 days, then

$$6\sigma = 21 \text{ days}$$

and σ is approximately equal to 21/6 or 3.5 days.

The final step in determining the sample size is to substitute this approximate value of σ into equation 7.5 and solve for n. Thus, we have

$$1.96\left(\frac{3.5}{\sqrt{n}}\right) = 1$$

or

$$\sqrt{n} = \frac{1.96(3.5)}{1} = 6.86$$

Squaring both sides of this equation yields $n = 47.06$. We will follow the usual convention of rounding the calculated sample size upward. Therefore, the mail-order house needs to sample approximately $n = 48$ shipping times to estimate the mean shipping time correct to within one day with 95% confidence.

In Example 7.16, we wanted our sample estimate to lie within one day of the true mean shipping time μ with 95% confidence. The value 1 (one day) represents the sampling error we desire in the confidence interval. It is the difference between $\bar{x}$, the center of the interval, and the upper (or lower) endpoint of the confidence interval.

Definition 7.5

The **sampling error,** E, in a confidence interval is the difference between the point estimator and either the upper or lower endpoint in the interval.

We could calculate the sample size for some specific sampling error E for a confidence coefficient other than .95 by changing the z value in the equation. In general, if we are willing to accept a sampling error E in estimating μ with confidence coefficient $(1 - \alpha)$, we would solve for n in the equation

$$z_{\alpha/2}\left(\frac{\sigma}{\sqrt{n}}\right) = \text{E}$$

where the value of $z_{\alpha/2}$ is obtained from Table B.3 in Appendix B. The solution is given by

$$n = \left(\frac{z_{\alpha/2}\sigma}{\text{E}}\right)^2 = \frac{(z_{\alpha/2})^2\sigma^2}{\text{E}^2}$$

For example, for a confidence coefficient of .90, we would require a sample size of

$$n = \frac{(1.645)^2\sigma^2}{\text{E}^2}$$

CHOOSING THE SAMPLE SIZE FOR ESTIMATING A POPULATION MEAN μ TO WITHIN E UNITS WITH CONFIDENCE COEFFICIENT $(1 - \alpha)$

$$n = \left(\frac{z_{\alpha/2}\sigma}{\mathrm{E}}\right)^2 = \frac{(z_{\alpha/2})^2\sigma^2}{\mathrm{E}^2} \tag{7.6}$$

[*Note:* The population standard deviation σ will usually have to be approximated.]

Self-Test 7.6

How large a sample is required to estimate μ to within 10 units with 90% confidence? Assume $\sigma = 50$.

The procedures for determining the sample size needed to estimate a population proportion is analogous to the procedure for determining the sample size for estimating a population mean. We present the appropriate formula and illustrate its use with an example.

CHOOSING THE SAMPLE SIZE FOR ESTIMATING A POPULATION PROPORTION p TO WITHIN E UNITS WITH CONFIDENCE COEFFICIENT $(1 - \alpha)$

$$n = \frac{(z_{\alpha/2})^2 p(1-p)}{\mathrm{E}^2} \tag{7.7}$$

where p is the value of the population proportion that you are attempting to estimate. [*Note:* This technique requires a previous estimate of p. If none is available, use $p = .5$ for a conservative choice of n.]

EXAMPLE 7.17

SAMPLE SIZE FOR ESTIMATING p

In Example 7.13, the United States Commission on Crime sampled recently committed crimes to estimate the proportion in which firearms were used. Suppose the Commission wants to obtain an estimate of p that is correct to within .02 with 90% confidence. How many cases would have to be included in the Commission's sample?

Caution
The formulas given in this section are appropriate when the sample size n is small relative to the population size N. For situations in which n may be large relative to N, adjustments to these formulas must be made. Sample size determination for this special case (called *survey sampling*) is beyond the scope of this text. Consult the references (e.g., see Kish, 1965) if you want to learn more about this particular application.

Solution

Here, we desire our estimate to be within .02 of the unknown population proportion p. Thus, our sampling error is $\mathrm{E} = .02$. For a confidence coefficient of $(1 - \alpha) = .90$, we have $\alpha = .10$. Therefore, to calculate n using equation 7.7, we must find the value of $z_{\alpha/2} = z_{.05}$ and find an approximation for p.

From Table B.3, the z value corresponding to an area of $\alpha/2 = .05$ in the upper tail of the standard normal distribution is $z_{.05} = 1.645$. As an approximation to p, we will use the sample estimate, $\hat{p} = .60$, obtained for the sample of 300 cases in Example 7.13 (p. 360).

Substituting the value of .6 for p, $z_{.05} = 1.645$, and $\mathrm{E} = .02$ into equation 7.7, we have

$$n = \frac{(z_{\alpha/2})^2 p(1-p)}{\mathrm{E}^2} = \frac{(1.645)^2(.6)(.4)}{(.02)^2} = 1{,}623.6$$

Therefore, in order to estimate p to within .02 with 90% confidence, the Commission will have to sample approximately $n = 1{,}624$ cases.

USE $p = .5$ WHEN PRIOR ESTIMATES ARE UNAVAILABLE

In Example 7.17, we used a prior estimate of p in computing the required sample size. If such prior information were not available, we could approximate p in the sample size equation using $p = .5$. The nearer the substituted value of p is to .5, the larger will be the sample size obtained from the formula. Hence, if you take $p = .5$ as the approximation to p, you will always obtain a sample size that is at least as large as required.

Statistics in the Real World Revisited

Determining the Number of Bags of Scallops to Sample

In the previous Statistics in the Real World Revisited sections in this chapter, we used confidence intervals to (1) estimate μ, the population mean weight per scallop (p. 356), and (2) estimate p, the population proportion of scallop bags with average weight less than 1/36 pound (p. 362). The confidence interval for p was fairly wide due to the small number of bags sampled ($n = 18$) from the over 11,000 bags in the ship's catch. Also, with such a small sample, the use of the normal (z) statistic to form the confidence interval may be invalid. In order to find a valid, narrower confidence interval for the true proportion, the harbormaster would need to take a larger sample of scallop bags.

How many bags should be sampled from the ship's catch to estimate p to within .03 with 95% confidence? Here, we have $E = .03$ and $z_{.025} = 1.96$ (since, for a 95% confidence interval, $\alpha = .05$ and $\alpha/2 = .025$). From our analysis on p. 362, the estimated proportion is $\hat{p} = .89$. Substituting these values into equation 7.7, we obtain

$$n = \frac{(z_{.025})^2(p)(1-p)}{E^2} = \frac{(1.96)^2(.89)(.11)}{(.03)^2} = 417.88$$

Consequently, the harbormaster would have to sample 418 of the ship's 11,000 bags of scallops in order to find a valid 95% confidence interval for p.

PROBLEMS FOR SECTION 7.6

Using the Tools

7.42 Determine the sample size needed to estimate μ for each of the following situations:

a. $E = 3, \sigma = 40, (1-\alpha) = .95$ **b.** $E = 5, \sigma = 40, (1-\alpha) = .95$

c. $E = 5, \sigma = 40, (1-\alpha) = .99$

7.43 Find the sample size needed to estimate p for each of the following situations:

a. $E = .04, p \approx .9, (1-\alpha) = .90$ **b.** $E = .04, p \approx .5, (1-\alpha) = .90$

c. $E = .01, p \approx .5, (1-\alpha) = .90$

7.44 Find the sample size n needed to estimate:

a. μ to within 5 units with 99% confidence when $\sigma \approx 20$

b. p to within .1 with 90% confidence

7.45 Find the sample size n needed to estimate:

a. p to within .04 with 95% confidence when $p \approx .8$

b. μ to within 70 units with 90% confidence when $\sigma \approx 500$

Applying the Concepts

7.46 Refer to the *Environmental Science & Technology* (Sept. 1, 2000) study of the impact of cooking on the size of indoor air particles, Problem 7.22 (p. 357). How large of a sample of days is required to estimate the true average decay rate of fine particles to within .07 μm/h using a 95% confidence interval?

7.47 A child psychologist wants to estimate the mean age at which a child learns to walk. How many children must be sampled if the psychologist desires an estimate that is correct to within 1 month of the true mean with 99% confidence? Assume the psychologist knows only that the age at which a child begins to walk ranges from 8 to 26 months.

7.48 The chemical benzalkonium chloride (BAC) is an antibacterial agent that is added to some asthma medications to prevent contamination. Researchers at the University of Florida College of Pharmacy have discovered that adding BAC to asthma drugs can cause airway constriction in patients. In a sample of 18 asthmatic patients, each of whom received a heavy dose of BAC, 10 experienced a significant drop in breathing capacity (*Journal of Allergy and Clinical Immunology,* Jan. 2001). Based on this information, a 95% confidence interval for the population percentage of asthmatic patients who experience breathing difficulties after taking BAC is (.326, .785).

a. Why might the confidence interval lead to an erroneous inference?

b. How many asthma patients must be included in the study to estimate the population percentage who experience a significant drop in breathing capacity to within 4% using a 95% confidence interval?

7.49 *Good Housekeeping* (July 1996) reported on a study of women who shave their legs. Each woman was required to shave one leg using a cream or gel and the other leg using soap or a beauty bar. Of interest was the proportion of women who prefer shaving their legs using a cream or gel. How many women must participate in the study to estimate the proportion to within .03 with 95% confidence? Use .5 as a conservative estimate of the true proportion.

7.50 Refer to the *American Journal of Psychiatry* study of jail suicides, Example 7.11 (p. 353). How many jail suicide cases must be sampled to estimate the true mean number of days the suicidal inmate spent in jail to within 10 days with 90% confidence? Use the sample standard deviation calculated in Example 7.11 as an estimate of σ.

7.51 Is the bottled water you drink safe? According to an article in *U.S. News & World Report* (April 12, 1999), the Natural Resources Defense Council warns that the bottled water you are drinking may contain more bacteria and other potentially carcinogenic chemicals than allowed by state and federal regulations. Of the more than 1,000 bottles studied, nearly one-third exceeded government levels. Suppose that the Natural Resources Defense Council wants an updated estimate of the population proportion of bottled water that violates at least one government standard. Determine the sample size (number of bottles) needed to estimate this proportion to within ± 0.01 with 99% confidence.

7.6 Estimation of a Population Variance (Optional)

In the previous sections, we considered interval estimation for population means or proportions. In this optional section, we discuss a confidence interval for a population variance, σ^2.

EXAMPLE 7.18

ESTIMATING σ^2

FISH

Refer to the U.S. Army Corps of Engineers study of contaminated fish in the Tennessee River, Alabama. (Recall that the data are stored in the **FISH** data file.) It is important for the Corps of Engineers to know how stable the weights of the contaminated fish are. That is, how large is the variation in the fish weights?

a. Identify the parameter of interest to the U.S. Army Corps of Engineers.

b. Explain how to build a confidence interval for the parameter, part **a**.

Solution

a. The Corps of Engineers is interested in the *variation* of the fish weights. Consequently, the target population parameter is σ^2, the variance of the weights of all contaminated fish inhabiting the Tennessee River.

b. Intuitively, it seems reasonable to use the sample variance s^2 to estimate σ^2 and to construct our confidence interval around this value. However, unlike sample means and sample proportions, the sampling distribution of the sample variance s^2 does not follow a normal (z) distribution or a t distribution.

Rather, when certain assumptions are satisfied (we discuss these later), the sampling distribution of s^2 possesses approximately a **chi-square (χ^2) distribution.*** The chi-square probability distribution, like the t distribution, is characterized by a quantity called the *degrees of freedom* associated with the distribution. Several chi-square probability distributions with different degrees of freedom are shown in Figure 7.17. Unlike z and t distributions, the chi-square distribution is not symmetric about 0.

*Throughout this section (and this text), we will use the words *chi-square* and the Greek symbol χ^2 interchangeably.

Figure 7.17 Several Chi-Square Probability Distributions

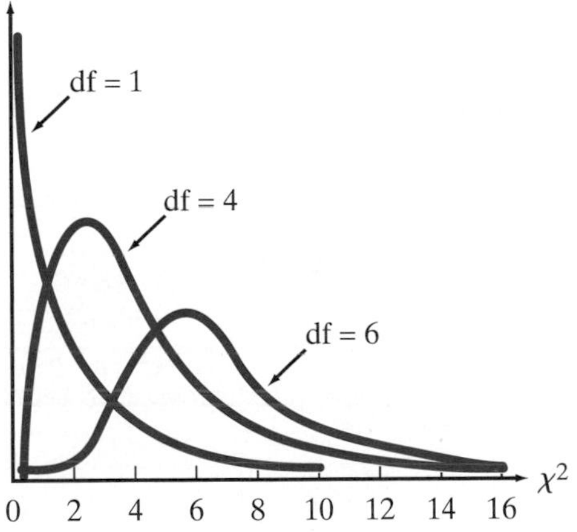

EXAMPLE 7.19 FINDING VALUES OF χ^2

Tabulated values of the χ^2 distribution are given in Table B.5 of Appendix B; a partial reproduction of this table is shown in Table 7.8 (p. 372). Entries in the table give an upper-tail value of χ^2, call it χ^2_α, such that $P(\chi^2 > \chi^2_\alpha) = \alpha$. Find the tabulated value of χ^2 corresponding to 9 degrees of freedom that cuts off an upper-tail area of .05.

Solution

The value of χ^2 that we seek appears (highlighted) in the partial reproduction of Table B.5 given in Table 7.8. The columns of the table identify the value of α associated with the tabulated value of χ^2_α and the rows correspond to the degrees of freedom. For this example, we have df $= 9$ and $\alpha = .05$. Thus, the tabulated value of χ^2 corresponding to 9 degrees of freedom is

$$\chi^2_{.05} = 16.9190$$

TABLE 7.8 Reproduction of Part of Table B.5

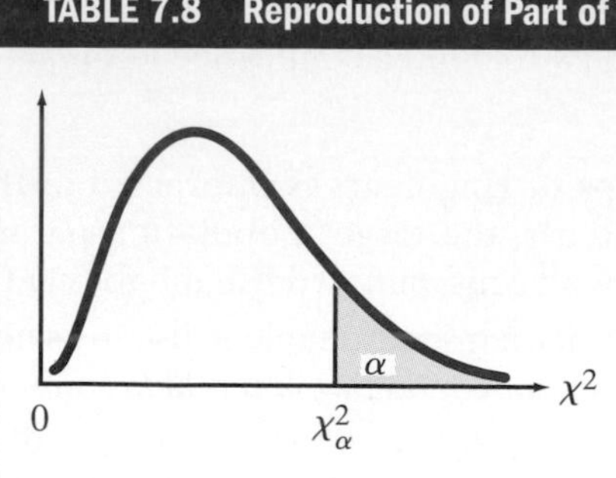

Degrees of Freedom	$\chi^2_{.100}$	$\chi^2_{.050}$	$\chi^2_{.025}$	$\chi^2_{.010}$	$\chi^2_{.005}$
1	2.70554	3.84146	5.02389	6.63490	7.87944
2	4.60517	5.99147	7.37776	9.21034	10.5966
3	6.25139	7.81473	9.34840	11.3449	12.8381
4	7.77944	9.48773	11.1433	13.2767	14.8602
5	9.23635	11.0705	12.8325	15.0863	16.7496
6	10.6446	12.5916	14.4494	16.8119	18.5476
7	12.0170	14.0671	16.0128	18.4753	20.2777
8	13.3616	15.5073	17.5346	20.0902	21.9550
9	14.6837	16.9190	19.0228	21.6660	23.5893
10	15.9871	18.3070	20.4831	23.2093	25.1882
11	17.2750	19.6751	21.9200	24.7250	26.7569
12	18.5494	21.0261	23.3367	26.2170	28.2995
13	19.8119	22.3621	24.7356	27.6883	29.8194
14	21.0642	23.6848	26.1190	29.1413	31.3193
15	22.3072	24.9958	27.4884	30.5779	32.8013
16	23.5418	26.2962	28.8454	31.9999	34.2672
17	24.7690	27.5871	30.1910	33.4087	35.7185
18	25.9894	28.8693	31.5264	34.8053	37.1564
19	27.2036	30.1435	32.8523	36.1908	38.5822

Source: From C. M. Thompson, "Tables of the Percentage Points of the χ^2-Distribution," *Biometrika,* 1941, 32, 188–189. Reproduced by permission of the *Biometrika* Trustees and Oxford University Press.

Self-Test 7.7

Find the value of $\chi^2_{.025}$ when the sample size is $n = 15$.

We use the tabulated values of χ^2 to construct a confidence interval for σ^2, as the next example illustrates.

EXAMPLE 7.20

95% CONFIDENCE INTERVAL FOR σ^2

Refer to Example 7.18. The 144 fish specimens in the U.S. Army Corps of Engineers study produced the following summary statistics: $\bar{x} = 1{,}049.7$ grams, $s = 376.6$ grams. Use this information to construct a 95% confidence interval for the true variation in weights of contaminated fish in the Tennessee River.

Solution

A $(1 - \alpha)100\%$ confidence interval for σ^2 depends on the quantities s^2, $(n - 1)$, and critical values of χ^2 as shown in the box.

A $(1-\alpha)100\%$ CONFIDENCE INTERVAL FOR A POPULATION VARIANCE σ^2

$$\frac{(n-1)s^2}{\chi^2_{\alpha/2}} \leq \sigma^2 \leq \frac{(n-1)s^2}{\chi^2_{(1-\alpha/2)}} \qquad (7.8)$$

where $\chi^2_{\alpha/2}$ and $\chi^2_{(1-\alpha/2)}$ are values of χ^2 that locate an area of $\alpha/2$ to the right and $\alpha/2$ to the left, respectively, of a chi-square distribution based on $(n-1)$ degrees of freedom.

Assumption: The population from which the random sample is selected has an approximate normal distribution.

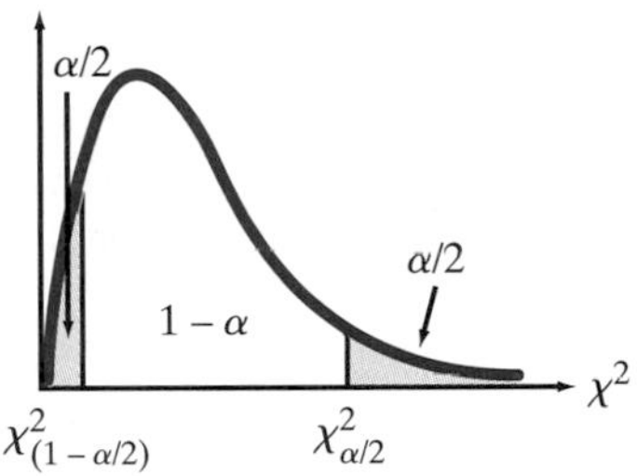

Figure 7.18 The Location of $\chi^2_{(1-\alpha/2)}$ and $\chi^2_{\alpha/2}$ for a Chi-Square Distribution

Note that $(n-1)$ represents the degrees of freedom associated with the χ^2 distribution. To construct the interval, we first locate the critical values $\chi^2_{(1-\alpha/2)}$ and $\chi^2_{\alpha/2}$. These are the values of χ^2 that cut off an area of $\alpha/2$ in the lower and upper tails, respectively, of the chi-square distribution (see Figure 7.18).

For a 95% confidence interval, $(1-\alpha) = .95$ and $\alpha/2 = .025$. Therefore, we need the tabulated values $\chi^2_{.025}$ and $\chi^2_{.975}$ for $(n-1) = 143$ degrees of freedom. Looking in the df = 150 row of Table B.5 (the row with the df value closest to 143), we find $\chi^2_{.025} = 185.800$ and $\chi^2_{.975} = 117.985$. Substituting into equation 7.8, we obtain

$$\frac{(144-1)(376.6)^2}{185.800} \leq \sigma^2 \leq \frac{(144-1)(376.6)^2}{117.985}$$

$$109{,}156.8 \leq \sigma^2 \leq 171{,}897.6$$

We are 95% confident that the true variance in weights of contaminated fish in the Tennessee River falls between 109,156.8 and 171,897.6. The Army Corps of Engineers could use this interval to determine whether the weights of the fish are stable enough to allow further testing for DDT contamination.

Self-Test 7.8

Find a 95% confidence interval for σ^2 based on the following summary statistics: $n = 15$, $\bar{x} = 72$, $s = 11$.

EXAMPLE 7.21

ESTIMATING σ

Refer to Example 7.20. Find a 95% confidence interval for σ, the true standard deviation of the fish weights.

Solution

A confidence interval for σ is obtained by taking the square roots of the lower and upper endpoints of a confidence interval for σ^2. Thus, the 95% confidence interval is

$$\sqrt{109{,}156.8} \leq \sigma \leq \sqrt{171{,}897.6}$$

$$330.4 \leq \sigma \leq 414.6$$

Thus, we are 95% confident that the true standard deviation of the fish weights is between 330.4 grams and 414.6 grams.

CAUTION

The procedure for calculating a confidence interval for σ^2 in the above examples requires an assumption regardless of whether the sample size n is large or small (see previous box). We must assume that the population from which the sample is selected has an approximate normal distribution. Unlike small sample confidence intervals for μ based on the t statistic, *slight to moderate departures from normality will render the χ^2 confidence interval invalid.* It is reasonable to expect this assumption to be satisfied in Examples 7.20 and 7.21 since the histogram of the 144 fish weights in the sample, shown in Figure 3.11 (p. 145), is approximately normal.

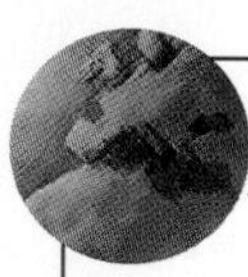

Statistics in the Real World Revisited

Estimating the Variance of Scallop Weights

Refer to the data on weight per scallop for 18 bags of scallops given in Table 7.6 on p. 356. In addition to estimating the true mean weight per scallop for all 11,000 bags caught by the fishing vessel, suppose the harbormaster wants to estimate the variation in weight per scallop for all 11,000 bags. Specifically, the harbormaster wants to find a 90% confidence interval for σ, the true standard deviation of the weights.

The data in the **SCALLOPS** file were analyzed using Excel and the resulting printout shown in Figure 7.19. The lower and upper endpoints of the 90% confidence interval for σ, highlighted on the printout, are (.0016, .0029). Consequently, the harbormaster can be 90% confident that the true standard deviation of the weights per scallop for the 11,000 bags is between .0016 and .0029 pound.

E Figure 7.19 Excel Confidence Interval for Variance of Scallop Weights

	A	B
1	Confidence Interval Estimate of the Population Variance	
2		
3	Sample standard deviation	0.002096
4	sample size	18
5	Confidence level	0.9
6	Degrees of Freedom	17
7	Sum of Squares	7.46847E-05
8	Lower Chi-Square value	8.671753605
9	Upper chi-Square value	27.58710028
10	Lower Confidence Limit for Variance	2.70723E-06
11	Upper Confidence Limit for variance	8.61241E-06
12	Lower Confidence Limit for Standard deviation	0.001645367
13	Upper Confidence Limit for standard deviation	0.00293469

PROBLEMS FOR SECTION 7.6

Using the Tools

7.52 For each of the following combinations of α and degrees of freedom (df), use Table B.5 in Appendix B to find the values of $\chi^2_{\alpha/2}$ and $\chi^2_{(1-\alpha/2)}$ that would be used to form a confidence interval for σ^2.

a. $\alpha = .05$, df $= 7$

b. $\alpha = .10$, df $= 16$

c. $\alpha = .01$, df $= 10$

d. $\alpha = .05$, df $= 20$

7.53 Given the following values of $\bar{x}$, s, and n, form a 90% confidence interval for σ^2.

a. $\bar{x} = 21, s = 2.5, n = 50$ **b.** $\bar{x} = 1.3, s = .02, n = 15$

c. $\bar{x} = 167, s = 31.6, n = 22$ **d.** $\bar{x} = 9.4, s = 1.5, n = 5$

What assumption about the population must be satisfied for the confidence interval to be valid?

7.54 Refer to Problem 7.53. For each part **a–d**, form a 90% confidence interval for σ.

7.55 A random sample of $n = 6$ observations from a normal distribution resulted in the following measurements: 8, 2, 3, 7, 11, 6. Form a 95% confidence interval for σ^2.

Applying the Concepts

7.56 Refer to the *Organizational Science* (Mar.–Apr. 2000) study on the use of e-mail by mid-level managers, Problem 7.8 (p. 348). A sample of 426 managers sent, on average, 9.81 messages with a standard deviation of 14.41 messages. Use this information to find a 90% confidence interval for the population standard deviation of the number of e-mail messages sent by mid-level managers during a typical week. Interpret the interval.

7.57 A machine used to fill beer cans must operate so that the amount of beer actually dispensed varies very little. If too much beer is released, the cans will overflow, causing waste. If too little beer is released, the cans will not contain enough beer, causing complaints from customers. A random sample of the fills for 20 cans yielded a standard deviation of .07 ounce. Estimate the true variance of the fills using a 95% confidence interval.

7.58 Refer to the *Journal of Applied Behavior Analysis* study of a 16-year-old boy's destructive behavior, Problem 7.25 (p. 358). The data on the boy's number of destructive responses for 10 play sessions are reproduced below.

DESTRUCT

3	2	4.5	5.5	4.5	3	2.5	1	3	3

a. Estimate the variation in the boy's number of destructive responses per session using a 99% confidence interval.
b. What assumptions are required for the interval estimate to be valid?

7.59 In an *Exceptional Children* study, a sample of 16 developmentally delayed children had a mean IQ score of 80.2 and a standard deviation of 20.7. Find an estimate of the variance in IQ scores of developmentally delayed children using a 99% confidence interval.

7.60 *IEEE Transactions* (June 1990) presented a computer algorithm for solving a difficult mathematical programming problem. Fifty-two random problems were solved using the algorithm; the times to solution (CPU time in seconds) are listed in the accompanying table.

MATHCPU

.045	3.985	.506	.145	1.267	.049	.333	.379	.091	.036	.336	.219	.209
1.070	.130	.579	.045	.118	1.894	.136	1.639	.064	.258	.412	.209	.070
8.788	3.046	.179	.136	3.888	.242	.227	.182	.136	.600	.394	.258	.327
.445	1.055	.670	.088	4.170	.567	.079	.554	.912	.194	.182	.361	.258

Source: Snyder, W. S., and Chrissis, J. W. "A hybrid algorithm for solving zero–one mathematical programming problems." *IEEE Transactions,* Vol. 22, No. 2, June 1990, p. 166 (Table 1).

	A	B
1	*Time to solution*	
2		
3	Mean	0.810808
4	Standard Error	0.208745
5	Median	0.258
6	Mode	0.136
7	Standard Deviation	1.505282
8	Sample Variance	2.265875
9	Kurtosis	15.72116
10	Skewness	3.652228
11	Range	8.752
12	Minimum	0.036
13	Maximum	8.788
14	Sum	42.162
15	Count	52
16	Largest(1)	8.788
17	Smallest(1)	0.036

E Problem 7.60

An Excel printout giving descriptive statistics for the sample of 52 solution times is provided. Use this information to compute a 95% confidence interval for the variance of the solution times. Interpret the result.

7.61 *Jitter* is a term used to describe the variation in conduction time of a water power system. Low throughput jitter is critical to successful waterline technology. An investigation of throughput jitter in the opening switch of a prototype system (*Journal of Applied Physics,* Sept. 1993) yielded the following descriptive statistics on conduction time for $n = 18$ trials.

$$\bar{x} = 334.8 \text{ nanoseconds} \qquad s = 6.3 \text{ nanoseconds}$$

(Conduction time is defined as the length of time required for the downstream current to equal 10% of the upstream current.)

a. Construct a 95% confidence interval for the true standard deviation of conduction times of the prototype system.

b. A system is considered to have low throughput jitter if the true conduction time standard deviation is less than 7 nanoseconds. Does the prototype system satisfy this requirement? Explain.

KEY TERMS *Starred (*) terms are from the optional section of this chapter.*

*Chi-square (χ^2) distribution 371
Confidence coefficient 343
Confidence interval 340
Degrees of freedom 351
Interval estimate 339
Point estimate 339
Proportion 359
Sampling error 367
Standard error 340
Student's *t* distribution 351

KEY FORMULAS *Starred (*) formulas are from the optional section of this chapter.*

Large-sample confidence interval for means or proportions:

Point estimator $\pm$ $(z_{\alpha/2})$(Standard error) **(7.1)**, 344, **(7.4)**, 361

Small-sample confidence interval for means:

Point estimator $\pm$ $(t_{\alpha/2})$(Standard error) **(7.2)**, 351

[*Note:* The respective point estimator and standard error for each parameter discussed in this chapter are provided in Table 7.9.]

TABLE 7.9 Summary of Estimation Procedures

Parameter	Point Estimate	Standard Error	Estimated Standard Error
μ	$\bar{x}$	$\sigma/\sqrt{n}$	$s/\sqrt{n}$
p	$\hat{p}$	$\sqrt{\frac{p(1-p)}{n}}$	$\sqrt{\frac{\hat{p}(1-\hat{p})}{n}}$

Confidence interval for variances:* $\frac{(n-1)s^2}{\chi^2_{\alpha/2}} \le \sigma^2 \le \frac{(n-1)s^2}{\chi^2_{(1-\alpha/2)}}$ **(7.8)**, 373

KEY SYMBOLS

Symbol	Definition
μ	Population mean
p	Population proportion
σ^2	Population variance
$z_{\alpha/2}$	z value used in a $(1-\alpha)100\%$ large-sample confidence interval
$t_{\alpha/2}$	t value used in a $(1-\alpha)100\%$ small-sample confidence interval
$\chi^2_{\alpha/2}$ and $\chi^2_{(1-\alpha/2)}$	χ^2 values used in a $(1-\alpha)100\%$ confidence interval for a variance
$\bar{x}$	Sample mean (estimates μ)
$\hat{p}$	Sample proportion (estimates p)
s^2	Sample variance (estimates σ^2)
E	Distance to within what we wish to estimate a population parameter

CHECKING YOUR UNDERSTANDING

1. What is the difference between a point estimate and an interval estimate?
2. Explain why the phrase "95% confident" is used as a measure of reliability for a 95% confidence interval for a population parameter.
3. When is it appropriate to apply the t distribution for a confidence interval for the population mean?
4. What key words help you identify whether the parameter of interest is a mean, proportion, or variance?
5. What happens to the width of a confidence interval when the sample size n is increased?
6. What happens to the width of a confidence interval when the confidence level $(1 - \alpha)$ is increased?

SUPPLEMENTARY PROBLEMS

Starred (*) problems refer to the optional section of this chapter.

7.62 Operation Kidsafe is a nationwide safety program designed for children between the ages of 3 and 7. A study commissioned by the sponsor of Operation Kidsafe found that 340 of 500 children between the ages of 3 and 7 do not know their home phone number.

a. Estimate p, the true percentage of children between the ages of 3 and 7 who do not know their home phone number. Use a 95% confidence interval.

b. Give a precise interpretation of the phrase, "We are 95% confident that the interval encloses the true value of p."

c. How would the width of the confidence interval in part **a** change if the confidence coefficient were increased from .95 to .99?

d. How many children must be surveyed to estimate the true proportion who know their phone number to within .02 of its true value, with 95% confidence? Use the sample proportion calculated in part **a** as an estimate of p.

7.63 An experiment was conducted to estimate the mean time needed to transfer heat through sand (*Journal of Heat Transfer,* Aug. 1990). A large-sample 95% confidence interval for the mean time was found to be 20.0 ± 6.4 seconds.

a. Give a practical interpretation of the 95% confidence interval.

b. Give a theoretical interpretation of the 95% confidence interval.

7.64 Substance abuse problems are widespread at New Jersey businesses, according to the *Governor's Council for a Drug Free Workplace Report* (Spring/Summer 1995). A questionnaire on the issue was mailed to all New Jersey businesses that were members of the Governor's Council. Of the 72 companies that responded to the survey, 50 admitted that they had employees whose performance was affected by drugs or alcohol.

a. Use a 95% confidence interval to estimate the proportion of all New Jersey companies with substance abuse problems.

b. What assumptions are necessary to assure the validity of the confidence interval?

c. Interpret the interval in the context of the problem.

d. In interpreting the confidence interval, what does it mean to say you are "95% confident"?

e. Would you use the interval of part **a** to estimate the proportion of all U.S. companies with substance abuse problems? Why or why not?

7.65 Many food products in the typical supermarket contain ingredients from genetically modified corn, soybeans, potatoes, or tomatoes. A poll conducted by the Gallup Organization found that 707 of 1,039 adults want labels disclosing genetically modified ingredients even if it means a higher price (*The Wall Street Journal,* April 4, 2000).

a. Construct a 95% confidence interval for the proportion of U.S. adults that want labels disclosing genetically modified ingredients even if it means a higher price.

b. If you were to conduct a follow-up study that would provide 95% confidence that the point estimate is correct to within ± 0.02 of the population proportion, how large of a sample size would be required?

c. If you were to conduct a follow-up study that would provide 95% confidence that the point estimate is correct to within ± 0.01 of the population proportion, how large of a sample size would be required?

7.66 Obstructive sleep apnea is a sleep disorder that causes a person to stop breathing momentarily and then awaken briefly. These sleep interruptions, which may occur hundreds of times in a night, can drastically reduce the quality of rest and cause fatigue during waking hours. Researchers at Stanford University studied 159 commercial truck drivers and found that 124 of them suffered from obstructive sleep apnea (*Chest,* May 1995).

a. Use the study results to estimate, with 90% confidence, the fraction of truck drivers who suffer from the sleep disorder.

b. Sleep researchers believe that about 25% of the general population suffer from obstructive sleep apnea. Comment on whether or not this value represents the true percentage of truck drivers who suffer from the sleep disorder.

7.67 Research indicates that bicycle helmets save lives. A study reported in *Public Health Reports* (May–June 1992) was intended to identify ways of encouraging helmet use in children. One of the variables measured was the children's perception of the risk involved in bicycling. A four-point scale was used, with scores ranging from 1 (no risk) to 4 (very high risk). A sample of 797 children in grades 4–6 yielded the following results on the perception of risk variable: $\bar{x} = 3.39$, $s = .80$.

a. Calculate a 90% confidence interval for the average perception of risk for all students in grades 4–6. What assumptions did you make to assure the validity of the confidence interval?

b. If the population mean perception of risk exceeds 2.50, the researchers will conclude that students in these grades exhibit an awareness of the risk involved with bicycling. Interpret the confidence interval constructed in part **a** in this context.

7.68 *The Journal of Literacy Research* (Dec. 1996) published a study on the number of times per week a typical adult read a book for pleasure. Suppose you want to estimate the true mean number of times per week an adult reads a book for pleasure. How many adults must be sampled to estimate with 90% confidence the mean to within .5 of its true value? Assume that the true standard deviation for the variable of interest is approximately equal to 3.

7.69 "Are 1990s undergraduates more willing to cheat in order to get good grades than those students in the 1970s?" This was the question posed to a national sample of 5,000 college professors by the Carnegie Foundation for the Advancement of Teaching (*Tampa Tribune,* Mar. 7, 1990). Forty-three percent of the professors responded "yes." Based on this survey, estimate the proportion of all college professors who feel undergraduate students in the 1990s are more willing to cheat to get good grades than undergraduate students in the 1970s. Use a confidence coefficient of .90.

7.70 The *Journal of the American Medical Association* (Apr. 21, 1993) reported on the results of a National Health Interview Survey designed to determine the prevalence of smoking among U.S. adults. Current smokers (over 11,000 adults in the survey) were asked: "On the average, how many cigarettes do you now smoke a day?" The results yielded a mean of 20.0 cigarettes per day with an associated 95% confidence interval of (19.7, 20.3).

a. Interpret the 95% confidence interval.

b. State any assumptions about the target population of current cigarette smokers that must be satisfied for inferences derived from the interval to be valid.

c. A tobacco industry researcher claims that the mean number of cigarettes smoked per day by regular cigarette smokers is less than 15. Comment on this claim.

7.71 Researchers have experimented with generating electricity from the hot, highly saline water of the Salton Sea in southern California. Operating experience has shown that these brines leave silica scale deposits on metallic plant piping, causing excessive plant outages. In one experiment, an antiscalant was added to each of five aliquots of brine, and the solutions were filtered. A silica determination (parts per million of silicon dioxide) was made on each filtered sample after a holding time of 24 hours, with the following results (*Journal of Testing and Evaluation,* Mar. 1981): 229, 255, 280, 203, 229.

a. Estimate the mean amount of silicon dioxide present in the antiscalant solutions with a 99% confidence interval. Interpret the result.

***b.** Estimate the variance of the amount of silicon dioxide present in the antiscalant solutions with a 99% confidence interval. Interpret the result.

7.72 A University of Minnesota survey of brand name (e.g., Levi's, Lee, and Calvin Klein) and private label jeans found a high percentage of jeans with incorrect waist and/or inseam measurements on the label. The study found that only 18 of 240 pairs of men's five-pocket, prewashed jeans sold in Minneapolis stores came within a half inch of all their label measurements (*Tampa Tribune,* May 20, 1991). Let p represent the true proportion of men's five-pocket, prewashed jeans sold in Minneapolis that have inseam and waist measurements that fall within .5 inch of the labeled measurements.

a. Find a point estimate of p.

b. Find an interval estimate of p. Use a confidence coefficient of .90.

c. Interpret the interval, part **b**.

d. How large a sample size is necessary to estimate the proportion of interest to within 2% with 95% confidence?

7.73 Tropical swarm-founding wasps, like ants and bees, rely on workers to raise their offspring. One possible explanation for this behavior is inbreeding, which increases relatedness among the wasps and makes it easier for the workers to pick out and aid their closest relatives. To test this theory, 197 swarm-founding wasps were captured in Venezuela, frozen at $-70°C$, and then subjected to a series of genetic tests (*Science,* Nov. 1988). The data were used to generate an inbreeding coefficient x for each wasp specimen, with the following results: $\bar{x} = .044$ and $s = .884$.

a. Construct a 90% confidence interval for the population mean inbreeding coefficient of this species of wasp.

b. A coefficient of 0 implies that the wasp has no tendency to inbreed. Use the confidence interval, part **a**, to make an inference about the tendency for this species of wasp to inbreed.

7.74 *Human Factors* (Dec. 1988) published a study on the use of color brightness as a body orientation clue. In an experiment in which human subjects were disoriented, 60% used a bright color as a cue to being right-side up. How many subjects are required for a similar experiment to estimate the true proportion who use a bright color level as a cue to being right-side up to within .05 with 95% confidence?

***7.75** Geologists use laser spectroscopy to measure the compositions of fluids present in crystallized rocks. An experiment was conducted to estimate the precision of this laser technique (*Applied Spectroscopy,* Feb. 1986). A chip of natural Brazilian quartz with an artificially produced fluid inclusion was subjected to laser spectroscopy. The amount of liquid carbon dioxide present in the inclusion was recorded on four different days. The data (in mole percentage) are: 86.6, 84.6, 85.5, 85.9.

***a.** Obtain an estimate of the precision of the laser technique by constructing a 99% confidence interval for the variation in the carbon dioxide concentration measurements.

b. Estimate the mean amount of liquid carbon dioxide present in the inclusion with a 95% confidence interval.

***7.76** An experiment was conducted to investigate the precision of measurements of a saturated solution of iodine. The data shown in the table represent $n = 10$ iodine

Run	Concentration
1	5.507
2	5.506
3	5.500
4	5.497
5	5.506
6	5.527
7	5.504
8	5.490
9	5.500
10	5.497

Problen 7.76

concentration measurements on the same solution. The population variance σ^2 measures the variability—i.e., the precision—of a measurement. Use the data to find a 95% confidence interval for σ^2.

7.77 Most conventional opinion polls (e.g., Gallup, Roper, and CNN surveys) utilize a sample size of $n = 1{,}500$. The results of these polls are usually accompanied by the statement: "The estimated percentage is within ±3% of the true percentage." Consider estimating a population proportion p with a 95% confidence interval and sample size $n = 1{,}500$.

a. Calculate $\sigma_{\hat{p}}$, the standard error of the estimate, assuming (conservatively) that $p = .5$.

b. Does your answer to part **a** agree with the ±3% margin of error stated in most polls?

REFERENCES

Agresti, A., and Coull, B. A. "Approximate is better than 'exact' for interval estimation of binomial proportions." *The American Statistician*, Vol. 52, No. 2, May 1998, pp. 119–126.

Cochran, W. G. *Sampling Techniques,* 3rd ed. New York: Wiley, 1977.

Freedman, D., Pisani, R., and Purves, R. *Statistics*. New York: Norton, 1978.

Kish, L. *Survey Sampling*. New York: Wiley, 1965.

Wackerly, D., Mendenhall, W., and Scheaffer, R. *Mathematical Statistics with Applications,* 5th ed. Belmont, Calif.: Duxbury Press, 1996.

Snedecor, G. W., and Cochran, W. G. *Statistical Methods,* 7th ed. Ames, Iowa: Iowa State University Press, 1980.

Wilson, E. B. "Probable inference, the law of succession, and statistical inference." *Journal of the American Statistical Association,* Vol. 22, 1927.

Using Microsoft® Excel

7.E.1 Obtaining the Confidence Interval Estimate for the Mean (σ Known)

Use the PHStat **Confidence Intervals | Estimate for the Mean, sigma known** procedure to calculate the confidence interval estimate for the mean (σ known). For example, to calculate the confidence interval estimate for the mean attention time for Example 7.6 on p. 345, open to an empty worksheet and

1. Select PHStat | Confidence Intervals | Estimate for the Mean, sigma known.
2. In the Estimate for the Mean, sigma known dialog box (see Figure 7.E.1):

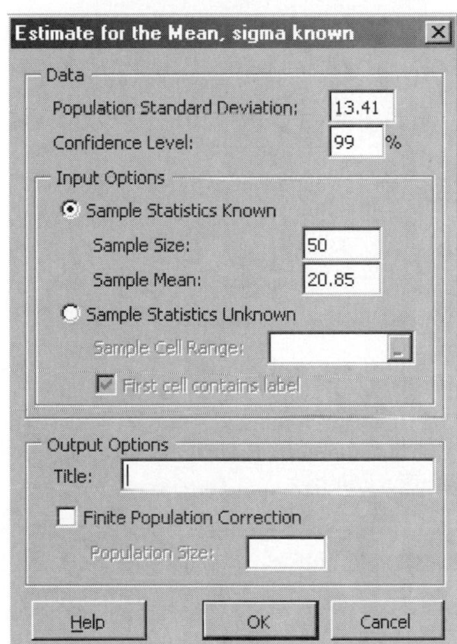

E **Figure 7.E.1** PHStat Estimate for the Mean, Sigma Known Dialog Box

a. Enter 13.41 in the Population Standard Deviation edit box.
b. Enter 99 in the Confidence Level edit box.
c. Select the Sample Statistics Known option button and enter 50 in the Sample Size edit box and 20.85 in the Sample Mean edit box.
d. Enter a title in the Title edit box.
e. Click the OK button.

For similar problems in which the sample mean is not known and needs to be calculated, select the Sample Statistics Unknown option button in step 2c and enter the cell range of the sample data in the Sample Cell Range edit box, shown dimmed in Figure 7.E.1.

7.E.2 Obtaining the Confidence Interval Estimate for the Mean (σ Unknown)

Use the PHStat **Confidence Intervals | Estimate for the Mean, sigma unknown** procedure to calculate the confidence interval estimate for the mean (σ unknown). For example, to

calculate the confidence interval estimate in Example 7.11 on p. 353 for the mean number of days in jail, open to an empty worksheet and

1. Select PHStat | Confidence Intervals | Estimate for the Mean, sigma unknown.
2. In the Estimate for the Mean, sigma unknown dialog box (see Figure 7.E.2):

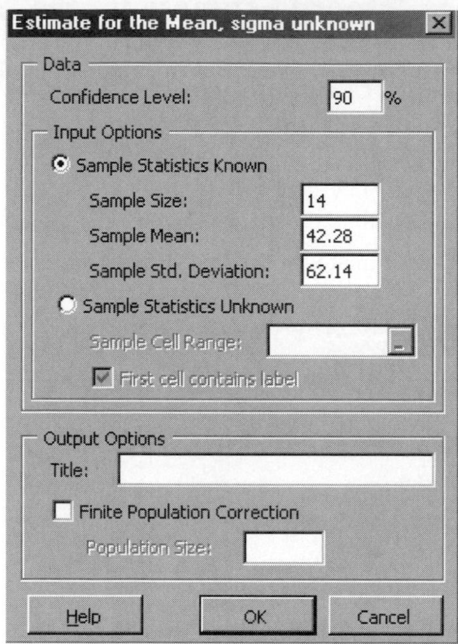

E **Figure 7.E.2** PHStat Estimate for the Mean, Sigma Unknown Dialog Box

a. Enter 90 in the Confidence Level edit box.
b. Select the Sample Statistics Known option button and enter 14 in the Sample Size edit box, 42.28 in the Sample Mean edit box, and 62.14 in the Sample Std. Deviation edit box.
c. Enter a title in the Title edit box.
d. Click the OK button.

For similar problems in which the sample mean and sample standard deviation are not known and need to be calculated, select the Sample Statistics Unknown option button in step 2b and enter the cell range of the sample data in the Sample Cell Range edit box, shown dimmed in Figure 7.E.2.

7.E.3 Obtaining the Confidence Interval Estimate for the Proportion

Use the PHStat **Confidence Intervals | Estimate for the Proportion** procedure to calculate the confidence interval estimate for the proportion. For example, to calculate the confidence interval estimate in Example 7.13 on p. 360 concerning the proportion of crimes committed with firearms, open to an empty worksheet and

1. Select PHStat | Confidence Intervals | Estimate for the Proportion.
2. In the Estimate for the Proportion dialog box (see Figure 7.E.3):

a. Enter 300 in the Sample Size edit box.
b. Enter 180 in the Number of Successes edit box.
c. Enter 95 in the Confidence Level edit box.
d. Enter a title in the Title edit box.
e. Click the OK button.

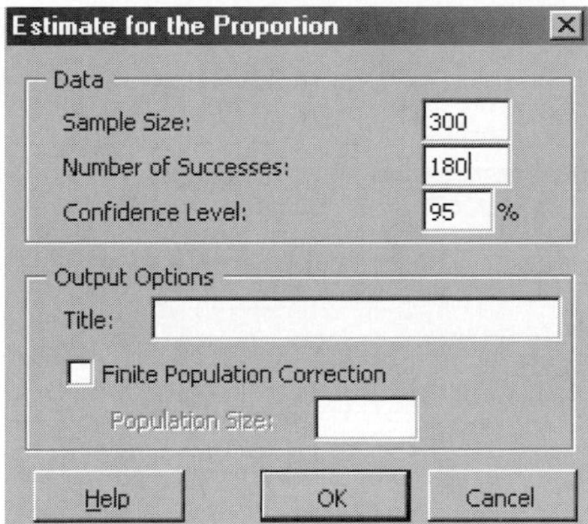

E **Figure 7.E.3** PHStat Estimate for the Proportion Dialog Box

7.E.4 Determining the Sample Size for Estimating the Mean

Use the PHStat **Sample Size | Determination for the Mean** procedure to determine the sample size needed for estimating the mean. For example, to generate the sample size determination calculations for estimating the mean shipping time in Example 7.16 on p. 366, open to an empty worksheet and

1. Select PHStat | Sample Size | Determination for the Mean.
2. In the Sample Size Determination for the Mean dialog box (see Figure 7.E.4):

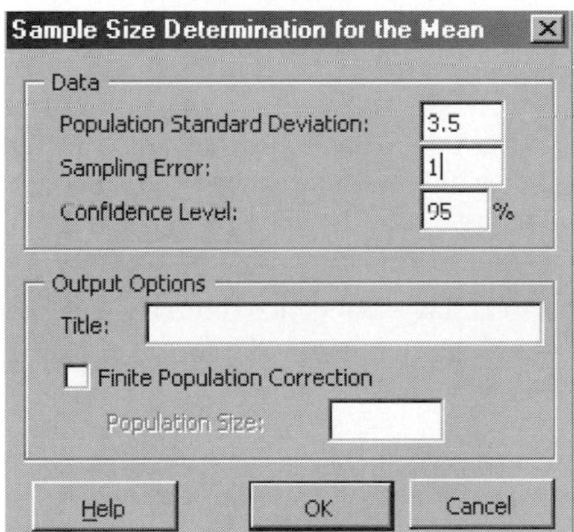

E **Figure 7.E.4** PHStat Sample Size Determination for the Mean Dialog Box

a. Enter 3.5 in the Population Standard Deviation edit box.
b. Enter 1 in the Sampling Error edit box.
c. Enter 95 in the Confidence Level edit box.
d. Enter a title in the Title edit box.
e. Click the OK button.

7.E.5 Determining the Sample Size for Estimating the Proportion

Use the PHStat **Sample Size | Determination for the Proportion** procedure to determine the sample size needed for estimating the proportion. For example, to generate the sample size determination calculations for estimating the proportion of crimes in which a firearm was used in Example 7.17 on p. 370, open to an empty worksheet and

1. Select PHStat | Sample Size | Determination for the Proportion.
2. In the Sample Size Determination for the Proportion dialog box (see Figure 7.E.5):

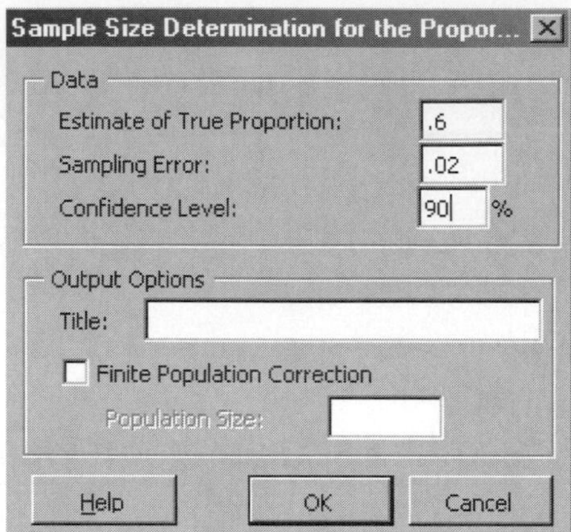

E Figure 7.E.5 PHStat Sample Size Determination for the Proportion Dialog Box

a. Enter .6 in the Estimate of True Proportion edit box.
b. Enter .02 in the Sampling Error edit box.
c. Enter 90 in the Confidence Level edit box.
d. Enter a title in the Title edit box.
e. Click the OK button.

7.E.6 Obtaining the Confidence Interval Estimate for the Variance

Use the PHStat **Confidence Intervals | Estimate for the Population Variance** procedure to calculate the confidence interval estimate for the variance. For example, to calculate the confidence interval estimate in Example 7.20 on p. 372 for the variance in fish weights, open to an empty worksheet and

1. Select PHStat | Confidence Intervals | Estimate for the Population Variance.
2. In the Estimate for the Population Variance dialog box (see Figure 7.E.6):

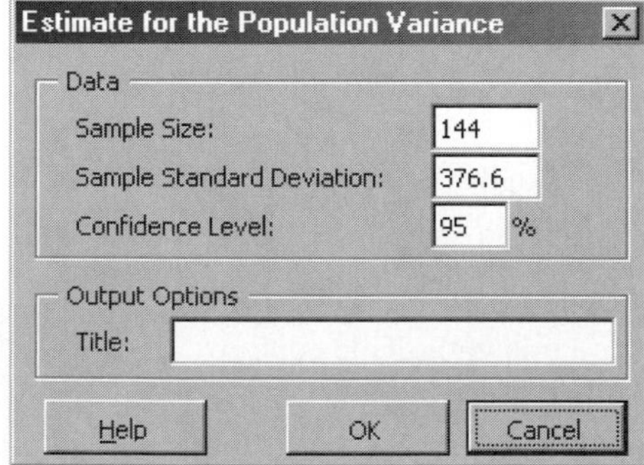

E Figure 7.E.6 PHStat Estimate for the Population Variance Dialog Box

a. Enter 144 in the Sample Size edit box and 376.6 in the Sample Standard Deviation edit box.
b. Enter 95 in the Confidence Level edit box.
c. Enter a title in the Title edit box.
d. Click the OK button.

7.M.1 Obtaining the Confidence Interval Estimation for the Mean (σ Known)

MINITAB can be used to obtain a confidence interval estimate for the mean when σ is known. For example, to calculate the confidence interval estimate for the mean attention time for Example 7.6 on p. 345, open the **ATTENT.MTW** worksheet and do the following.

1. Select **Stat | Basic Statistics | 1-Sample Z.**
2. In the 1-Sample Z dialog box (see Figure 7.M.1), enter **C1** or **'Attention time'** in the Variables edit box.
3. Select the **Confidence interval** button and enter **99** as the level of confidence in the Level edit box. Enter **13.41** in the Sigma edit box. Click the **OK** button.

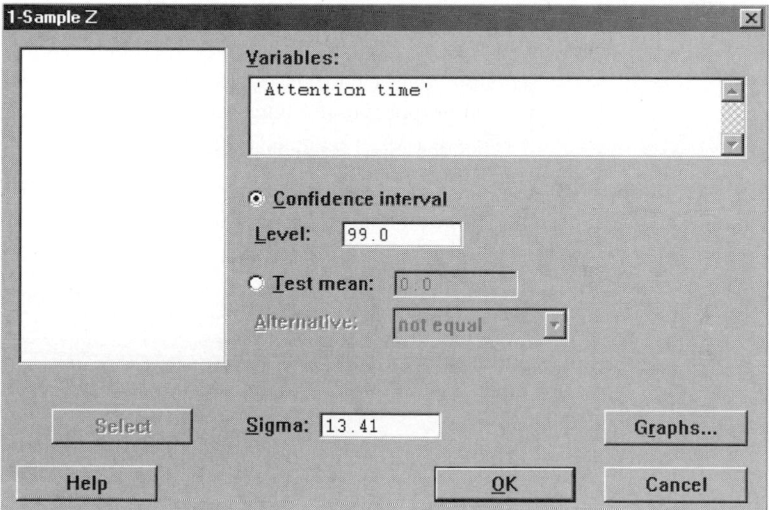

Figure 7.M.1 MINITAB 1-Sample Z Dialog Box

7.M.2 Obtaining the Confidence Interval Estimation for the Mean (σ Unknown)

MINITAB can be used to obtain a confidence interval estimate for the mean when σ is unknown by selecting Stat | Basic Statistics | 1-Sample t. As an example, to calculate the confidence interval estimate in Example 7.11 on p. 353 for the mean number of days in jail, open the **SUICIDE.MTW** worksheet and do the following.

1. Select **Stat | Basic Statistics | 1-Sample t.**
2. In the 1-Sample t dialog box (see Figure 7.M.2 on p. 386) enter **C1** or **'Days'** in the Variables edit box.
3. Select the **Confidence interval** button and enter **90.0** in the Level edit box. Click the **OK** button.

M Figure 7.M.2 MINITAB 1-Sample t Dialog Box

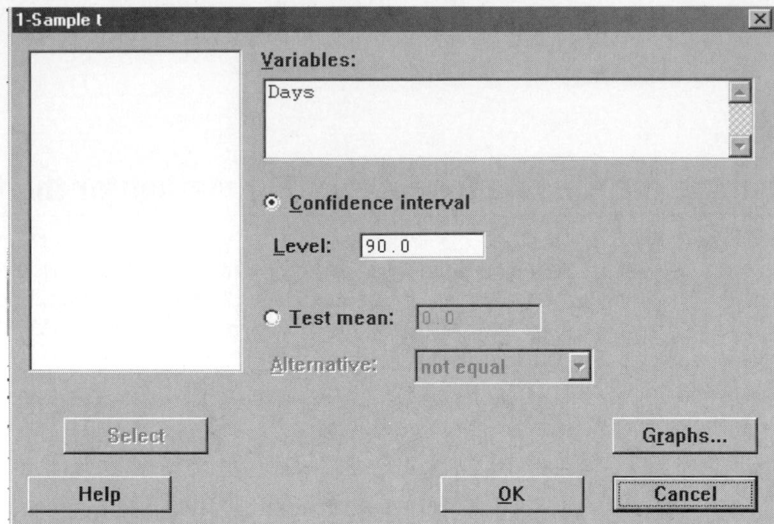

7.M.3 Obtaining the Confidence Interval Estimation for the Proportion

MINITAB can be used to obtain a confidence interval estimate for the proportion. For example, to calculate the confidence interval estimate in Example 7.13 on p. 360 concerning the proportion of crimes committed with firearms,

1. Select **Stat | Basic Statistics | 1 Proportion.**
2. In the 1 Proportion dialog box (see Figure 7.M.3), Select the **Summarized data** option button. Enter **300** in the Number of trials edit box. Enter **180** in the Number of successes edit box.

M Figure 7.M.3 MINITAB 1 Proportion Dialog Box

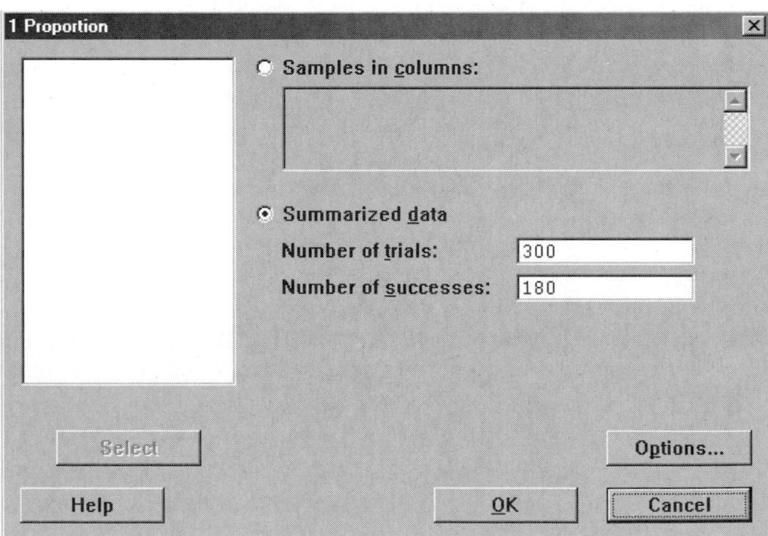

3. Click the **Options** button. In the 1 Proportion—Options dialog box (see Figure 7 M.4), enter **95** in the Confidence level edit box. Select the **Use test and interval based on the normal distribution** check box. Click the **OK** button to return to the 1 Proportion dialog box.

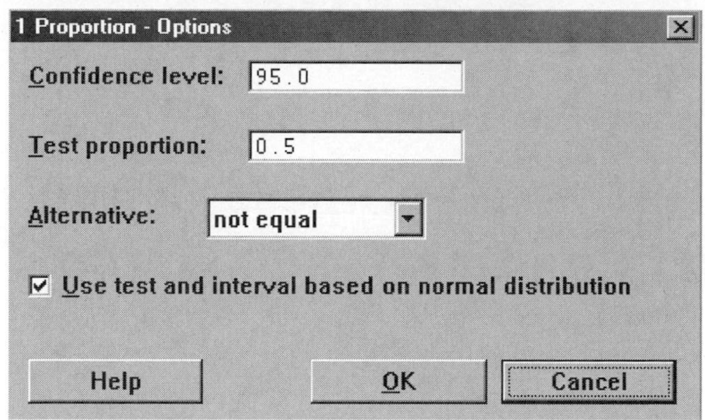

M Figure 7.M.4 MINITAB 1 Proportion—Options Dialog Box

4. Click the **OK** button.

If raw data rather than summarized data is available, in step 2 above, select the Samples in columns option button and enter the name of the variable in the Samples in columns edit box.

CHAPTER 8

Testing Hypotheses about Population Parameters: One Sample

OBJECTIVES

1. To develop hypothesis-testing methodology as a technique for making decisions about a population parameter
2. To understand the relationship between a hypothesis test and a confidence interval
3. To understand the risks involved with making decisions about a population parameter
4. To show how to conduct tests of hypotheses for population parameters based on a single sample
5. To understand the conceptual differences between the traditional critical value approach to hypothesis testing and the *p*-value approach

CONTENTS

EXCEL TUTORIAL

MINITAB TUTORIAL

Statistics in the Real World

Diary of a KLEENEX® User

In 1924, Kimberly-Clark Corporation invented a facial tissue for removing cold cream and began marketing it as KLEENEX® brand tissues. Today, KLEENEX® is recognized as the top-selling brand of tissue in the world. A wide variety of KLEENEX® products are available, ranging from extra-large tissues to tissues with lotion. Over the past 80 years, Kimberly-Clark Corporation has packaged the tissues in boxes of different sizes and shapes, and varied the number of tissues packaged in each box. For example, currently a family-size box contains 144 two-ply tissues, a cold-care box contains 70 tissues (coated with lotion), and a convenience pocket pack contains 15 miniature tissues.

How does Kimberly-Clark Corporation decide how many tissues to put in each box? According to *The Wall Street Journal,* marketing experts at the company use the results of a survey of KLEENEX® customers to help determine how many tissues are packed in a box. In the mid-1980s, when Kimberly-Clark Corporation developed the cold-care box designed especially for people who have a cold, the company conducted their initial survey of customers for this purpose. Hundreds of customers were asked to keep count of their KLEENEX® use in diaries. According to *The Wall Street Journal* report, the survey results left "little doubt that the company should put 60 tissues in each box." The number 60 was "the average number of times people blow their nose during a cold." (Note: In 2000, the company increased the number of tissues packaged in a cold-care box to 70.)

From summary information provided in *The Wall Street Journal* (Sept. 21, 1984) article, we constructed a data set that represents the results of a survey similar to the one described above. In the CD-ROM data file named **TISSUES,** we recorded the number of tissues used by each of 250 consumers during a period when they had a cold. In the following Statistics in the Real World Revisited sections, we apply the hypothesis testing methodology presented in this chapter to this data set.

Statistics in the Real World Revisited

- Identifying the Parameter of Interest for the KLEENEX® Survey (p. 406)
- Testing a Population Mean in the KLEENEX® Survey (p. 418)
- Testing a Population Proportion in the KLEENEX® Survey (p. 424)

8.1 The Relationship between Hypothesis Tests and Confidence Intervals

There are two general methods available for making inferences about population parameters: We can estimate their values using confidence intervals (the subject of Chapter 7) or we can make decisions about them. Making decisions about specific values of the population parameters—*testing hypotheses* about these values—is the topic of this chapter.

It is important to know that confidence intervals and hypothesis tests are related and that either can be used to make decisions about parameters.

EXAMPLE 8.1

INFERENCE ABOUT μ FROM A CONFIDENCE INTERVAL

Suppose an investigator for the Environmental Protection Agency (EPA) wants to determine whether the mean level μ of a certain type of pollutant released into the atmosphere by a chemical company meets the EPA guidelines. If 3 parts per million is the upper limit allowed by the EPA, the investigator would want to use sample data (daily pollution measurements) to decide whether the company is violating the law, i.e., to decide whether $\mu > 3$. Suppose a 99% confidence interval for μ is (3.2, 4.1). What should the EPA conclude?

Solution

The EPA is 99% confident that the mean level μ of the pollutant is between 3.2 ppm and 4.1 ppm. Since both the lower and upper limits given in the interval exceed 3, the EPA is confident that μ exceeds the established limit.

EXAMPLE 8.2 **INFERENCE ABOUT p FROM A CONFIDENCE INTERVAL**

Consider a software manufacturer who purchases 3.5" computer disks in lots of 10,000. Suppose that the supplier of the disks guarantees that no more than 1% of the disks in any given lot are defective. Since the manufacturer cannot test each of the 10,000 disks in a lot, he or she must decide whether to accept or reject a lot based on an examination of a sample of disks selected from the lot. If the number x of defective disks in a sample of, say, $n = 100$ is large, the lot will be rejected and sent back to the supplier. Thus, the manufacturer wants to decide whether the proportion p of defectives in the lot exceeds .01, based on information contained in a sample. Suppose a 95% confidence interval for p is (.005, .008). What decision should the software manufacturer make?

Solution

The manufacturer is 95% confident that the proportion of defectives falls between .005 and .008. Consequently, since both the lower and upper limits fall below .01, the manufacturer will accept the lot and be confident that the proportion of defectives is less than 1%.

The preceding examples illustrate how a confidence interval can be used to make a decision about a parameter. Note that both applications are one-directional: In Example 8.1 the EPA wants to determine whether $\mu > 3$; and in Example 8.2 the manufacturer wants to know if $p > .01$. (In contrast, if the manufacturer is interested in determining whether $p > .01$ or $p < .01$, the inference would be two-directional.)

Recall from Chapter 7 that to find the value of z (or t) used in a $(1 - \alpha)100\%$ confidence interval, the value of α is divided in half and $\alpha/2$ is placed in both the upper and lower tails of the z (or t) distribution. Consequently, confidence intervals are designed to be two-directional. Use of a two-directional technique in a situation where a one-directional method is desired will lead the analyst (e.g., the EPA or the software manufacturer) to understate the level of confidence associated with the method. As we will explain in this chapter, hypothesis tests are appropriate for either one- or two-directional decisions about a population parameter.

The general concepts involved in hypothesis testing are outlined in the next two sections.

8.2 Hypothesis-Testing Methodology: Formulating Hypotheses

Every statistical test of hypothesis consists of five key elements: (1) null hypothesis, (2) alternative hypothesis, (3) test statistic, (4) rejection region, and (5) conclusion. In this section we focus on formulating the null and alternative hypotheses and the consequences of our decision.

Null and Alternative Hypotheses

When a researcher in any field sets out to test a new theory, he or she first formulates a hypothesis, or claim, that he or she believes to be true. In statistical terms, the hypothesis that the researcher tries to establish is called the *alternative hypothesis*, or *research hypothesis*. To be paired with the alternative hypothesis is

the *null hypothesis,* which is the "opposite" of the alternative hypothesis. In this way, the null and alternative hypotheses, both stated in terms of the appropriate population parameters, describe two possible states of nature that are mutually exclusive (i.e., they cannot simultaneously be true). When the researcher begins to collect information about the variable of interest, he or she generally tries to present evidence that lends support to the alternative hypothesis. As you will subsequently learn, we take an indirect approach to obtaining support for the alternative hypothesis: Instead of trying to show that the alternative hypothesis is true, we attempt to produce evidence to show that the null hypothesis (which may often be interpreted as "no change from the status quo") is false.

Definition 8.1

A **statistical hypothesis** is a statement about the numerical value of a population parameter.

Definition 8.2

The **alternative** (or research) **hypothesis,** denoted by H_a, is usually the hypothesis for which the researcher wants to gather supporting evidence.

Definition 8.3

The **null hypothesis,** denoted H_0, is usually the hypothesis that the researcher wants to gather evidence against (i.e., the hypothesis to be "tested").

EXAMPLE 8.3

CHOOSING H_0 AND H_a

A metal lathe is checked periodically by quality control inspectors to determine whether it is producing machine bearings with a mean diameter of 1 centimeter (cm). If the mean diameter of the bearings is larger or smaller than 1 cm, then the process is out of control and needs to be adjusted. Formulate the null and alternative hypotheses that could be used to test whether the bearing production process is out of control.

Solution

The hypotheses must be stated in terms of a population parameter. Thus, we define

μ = True mean diameter (in centimeters) of all bearings produced by the lathe

If either $\mu > 1$ or $\mu < 1$, then the metal lathe's production process is out of control. Since we wish to be able to detect either possibility, the null and alternative hypotheses are

$$H_0\text{: } \mu = 1 \text{ (the process is in control)}$$
$$H_a\text{: } \mu \neq 1 \text{ (the process is out of control)}$$

EXAMPLE 8.4

CHOOSING H_0 AND H_a

Printed cigarette advertisements are required by law to carry the following statement: "Warning: The surgeon general has determined that cigarette smoking is dangerous to your health." However, this warning is often located in inconspicuous corners of the advertisements and printed in small type. Consequently, a spokesperson for the Federal Trade Commission (FTC) believes that over 80% of those who read cigarette advertisements fail to see the warning.

Specify the null and alternative hypotheses that would be used in testing the spokesperson's theory.

Solution

The FTC spokesperson wants to make an inference about p, the true proportion of all readers of cigarette advertisements who fail to see the surgeon general's warning. In particular, the FTC spokesperson wishes to collect evidence to support the claim that p is greater than .80; thus, the null and alternative hypotheses are

$$H_0\text{: } p = .80$$

$$H_a\text{: } p > .80$$

WHY AN EQUALS SIGN IN H_0?

An accepted convention in hypothesis testing is to always write H_0 with an equality sign (=). In Example 8.4, we could have formulated the null hypothesis as H_0: $p \leq .80$ to cover all situations for which H_a: $p > .80$ *does not* occur. However, any evidence that would cause you to reject the null hypothesis H_0: $p = .80$ in favor of H_a: $p > .80$ would also cause you to reject H_0: $p \leq .80$. In other words, H_0: $p = .80$ represents the worst possible case, from the researcher's point of view, if the alternative hypothesis is *not* correct. Thus, for mathematical ease, we combine all possible situations for describing the opposite of H_a into one statement involving an equality.

An alternative hypothesis may hypothesize a change from H_0 in a particular direction, or it may merely hypothesize a change without specifying a direction. In Example 8.4, the researcher is interested in detecting a departure from H_0 in a particular direction; interest focuses on whether the proportion of cigarette advertisement readers who fail to see the surgeon general's warning is *greater than* .80. This test is called a *one-tailed* (or *one-sided*) *test*. In contrast, Example 8.3 illustrates a *two-tailed* (or *two-sided*) *test* in which we are interested in whether the mean diameter of the machine bearings differs in either direction from 1 cm, i.e., whether the process is out of control.

Definition 8.4

A **one-tailed test** of hypothesis is one in which the alternative hypothesis is directional, and includes either the "<" symbol or the ">" symbol.

Definition 8.5

A **two-tailed test** of hypothesis is one in which the alternative hypothesis does not specify a departure from H_0 in a particular direction; such an alternative will be written with the "≠" symbol.

Self-Test 8.1

A recent Food and Drug Administration (FDA) report states that 65% of all domestically produced food products are pesticide-free. Suppose an FDA spokesperson believes that less than 65% of all foreign produced food products are pesticide-free. Specify the null and alternative hypotheses that would be used to test the spokesperson's belief.

Type I and Type II Errors

The goal of any hypothesis test is to make a decision; in particular, we will decide whether to reject the null hypothesis H_0 in favor of the alternative hypothesis H_a. Although we would like to be able to always make a correct decision, we must remember that the decision will be based on sample information. Thus we can make one of two types of errors, as shown in Table 8.1.

TABLE 8.1 Conclusions and Consequences for Testing a Hypothesis

		True State of Nature	
		H_0 True (H_a False)	**H_0 False (H_a True)**
Decision	**Accept H_0**	Correct decision	Type II error
	Reject H_0	Type I error	Correct decision

If we reject H_0 when H_0 is true, we make a *Type I error*. The probability of a Type I error is represented by the Greek letter α. If we accept H_0 when H_0 is false, we make a *Type II error*. The probability of a Type II error is represented by the Greek letter β. There is an intuitively appealing relationship between the probabilities for the two types of error: *As α increases, β decreases; similarly, as β increases, α decreases. The only way to reduce α and β simultaneously is to increase the amount of information available in the sample, i.e., to increase the sample size.*

Definition 8.6

A **Type I error** occurs if we reject a null hypothesis when it is true. The probability of committing a Type I error is denoted by α.

$$\alpha = P(\text{Type I error}) = P(\text{Reject } H_0 \text{ when } H_0 \text{ is true})$$

Definition 8.7

A **Type II error** occurs if we accept a null hypothesis when it is false. The probability of making a Type II error is denoted by β.

$$\beta = P(\text{Type II error}) = P(\text{Accept } H_0 \text{ when } H_0 \text{ is false})$$

EXAMPLE 8.5

TYPE I AND TYPE II ERRORS

Refer to Example 8.3. The hypotheses to be tested are:

$$H_0\colon \mu = 1 \text{ (process is in control)}$$

$$H_a\colon \mu \neq 1 \text{ (process is out of control)}$$

Specify the Type I and Type II errors for the problem.

Solution

From Definition 8.6, a Type I error is that of incorrectly rejecting H_0. In our example, this would occur if we conclude that the process is out of control when, in fact, the process is in control, i.e., if we conclude that the mean bearing diameter is different from 1 cm, when the mean is equal to 1 cm. The consequence of making

such an error would be that unnecessary time and effort would be expended to repair a metal lathe that is operating properly.

From Definition 8.7, a Type II error results from incorrectly accepting H_0. This occurs if we conclude that the mean bearing diameter is equal to 1 cm when, in fact, the mean differs from 1 cm. The practical significance of making a Type II error is that the metal lathe would not be repaired when, in fact, the process is out of control.

Subsequently, we will see that the probability of making a Type I error is controlled by the researcher (Section 8.3); thus, it is often used as a measure of the reliability of the conclusion and is called the *significance level* of the test.

Definition 8.8

The probability, α, of making a Type I error is called the **level of significance** (or **significance level**) for a hypothesis test.

In practice, we will carefully avoid stating a decision in terms of "accept the null hypothesis H_0." Instead, if the sample does not provide enough evidence to support the alternative hypothesis H_a, we prefer to state a decision as "fail to reject H_0," or "insufficient evidence to reject H_0." This is because, if we were to "accept H_0," the reliability of the conclusion would be measured by β, the probability of a Type II error. Unfortunately, the value of β is not constant, but depends on the specific alternative value of the parameter and is difficult to compute in most testing situations.

In summary, we recommend the following procedure for formulating hypotheses and stating conclusions.

FORMULATING HYPOTHESES AND STATING CONCLUSIONS

1. State the hypothesis you want to support as the alternative hypothesis H_a.
2. The null hypothesis H_0 will be the opposite of H_a and will contain an equality sign.
3. If the sample evidence supports the alternative hypothesis, you will reject the null hypothesis and will know that the probability of having made an incorrect decision (when H_0 is true) is α, a quantity that you select (prior to collecting the sample) to be as small as you wish.
4. If the sample does not provide sufficient evidence to support the alternative hypothesis, then you conclude that the null hypothesis cannot be rejected on the basis of your sample. In this situation, you may wish to obtain a larger sample to collect more information about the variable under study.

EXAMPLE 8.6 **HYPOTHESIS-TESTING LOGIC**

The logic used in hypothesis testing has often been likened to that used in the courtroom in which a defendant is on trial for committing a crime. Assume that the judge has issued the standard instruction to the jury: The defendant should be acquitted unless evidence of guilt is beyond a "reasonable doubt."

a. Formulate appropriate null and alternative hypotheses for judging the guilt or innocence of the defendant.

b. Interpret the Type I and Type II errors in this context.

c. If you were the defendant, would you want α to be small or large? Explain.

Solution

a. Under the American judicial system, a defendant is "innocent until proven guilty." That is, the burden of proof is *not* on the defendant to prove his or her innocence; rather, the court must collect sufficient evidence to support the claim that the defendant is guilty "beyond a reasonable doubt." Thus, the null and alternative hypotheses are

$$H_0\text{: Defendant is innocent}$$

$$H_a\text{: Defendant is guilty}$$

b. The four possible outcomes are shown in Table 8.2. A Type I error would occur if the court concludes that the defendant is guilty when, in fact, he or she is innocent; a Type II error would occur if the court concludes that the defendant is innocent when he or she is guilty.

TABLE 8.2 Conclusions and Consequences in Example 8.6

		True State of Nature	
		Defendant Is Innocent	**Defendant Is Guilty**
Decision of Court	**Defendant Is Innocent**	Correct decision	Type II error
	Defendant Is Guilty	Type I error	Correct decision

c. Most would probably agree that the Type I error in this situation is by far the more serious, especially the defendant. The defendant wants α, the probability of committing a Type I error, to be very small so that he or she has little or no chance of erroneously being found guilty.

In the next section we provide details on how to use the sample information to decide whether or not to reject the null hypothesis.

Self-Test 8.2

Last month, a large supermarket chain received many consumer complaints about the quantity of chips in 16-ounce bags of a particular brand of potato chips. The chain decided to test the following hypotheses concerning μ, the mean weight (in ounces) of a bag of potato chips in the next shipment of chips received from their largest supplier:

$$H_0\text{: } \mu = 16$$

$$H_a\text{: } \mu < 16$$

If there is evidence that $\mu < 16$, then the shipment would be refused and a complaint registered with the supplier.

a. What is a Type I error, in terms of the problem?

b. What is a Type II error, in terms of the problem?

c. Which type of error would the chain's customers view as more serious?

d. Which type of error would the chain's supplier view as more serious?

PROBLEMS FOR SECTION 8.2

Using the Tools

8.1 Explain the difference between an alternative hypothesis and a null hypothesis.

8.2 Explain why each of the following statements is incorrect:

a. The probability that the null hypothesis is correct is equal to α.

b. If the null hypothesis is rejected, then the test proves that the alternative hypothesis is correct.

c. In all statistical tests of hypothesis, $\alpha + \beta = 1$.

8.3 Why do we avoid stating a decision in terms of "accept the null hypothesis H_0"?

Applying the Concepts

8.4 According to the *Journal of Advanced Nursing* (Jan. 2001), 45% of senior women (i.e., women over the age of 65) use herbal therapies to prevent or treat health problems. Also, of senior women who use herbal therapies, they use an average of 2.5 herbal products in a year.

a. Give the null hypothesis for testing the first claim by the journal.

b. Give the null hypothesis for testing the second claim by the journal.

8.5 *Science* (Jan. 1, 1999) reported that the mean listening time of 7-month-old infants exposed to a three-syllable sentence (e.g., "ga ti ti") is 9 seconds. Set up the null and alternative hypotheses for testing the claim.

8.6 A herpetologist wants to determine whether the egg hatching rate for a certain species of frog exceeds .5 when the eggs are exposed to ultraviolet radiation.

a. Formulate the appropriate null and alternative hypotheses.

b. Describe a Type I error.

c. Describe a Type II error.

8.7 A manufacturer of fishing line wants to show that the mean breaking strength of a competitor's 22-pound line is really less than 22 pounds.

a. Formulate the appropriate null and alternative hypotheses.

b. Describe a Type I error.

c. Describe a Type II error.

8.8 An individual who has experienced a long run of bad luck playing the game of craps at a casino wants to test whether the casino dice are "loaded," i.e., whether the proportion of "sevens" occurring in many tosses of the two dice is different from $\frac{1}{6}$ (if the dice are fair, the probability of tossing a "seven" is $\frac{1}{6}$).

a. Formulate the appropriate null and alternative hypotheses.

b. Describe a Type I error.

c. Describe a Type II error.

8.9 The Environmental Protection Agency wishes to test whether the mean amount of radium-226 in soil in a Florida county exceeds the maximum allowable amount, 4 pCi/L.

a. Formulate the appropriate null and alternative hypotheses.

b. Describe a Type I error.

c. Describe a Type II error.

8.10 Testing the thousands of compounds of a new drug to find the few that might be effective is known in the pharmaceutical industry as *drug screening*. In its preliminary stage, drug screening can be viewed in terms of a statistical decision problem. Two actions are possible: (1) conclude that the tested drug has little or no effect, in which case it will be set aside and a new drug selected for screening; and (2) conclude, provisionally, that the drug is effective, in which case it will be subjected to further, more refined experimentation. Since it is the goal of the researcher to find

a drug that effects a cure, the null and alternative hypotheses in a statistical test would take the following form:

H_0: Drug is ineffective in treating a particular disease

H_a: Drug is effective in treating a particular disease

a. To abandon a drug when in fact it is a useful one is called a *false negative.* A false negative corresponds to which type of error, Type I or Type II?

b. To proceed with more expensive testing of a drug that is in fact useless is called a *false positive.* A false positive corresponds to which type of error, Type I or Type II?

c. Which of the two errors is more serious? Explain.

8.3 Hypothesis-Testing Methodology: Test Statistics and Rejection Regions

Once we have formulated the null and alternative hypotheses and selected the significance level α, we are ready to carry out the test. For example, suppose we want to test the null hypothesis

$$H_0\colon \mu = \mu_0$$

where μ_0 is some fixed value of the population mean. The next step is to obtain a random sample from the population of interest. The information provided by this sample, in the form of a sample statistic, will help us decide whether to reject the null hypothesis. The sample statistic upon which we base our decision is called the *test statistic.* For this example, we are hypothesizing about the value of the population mean μ. Since our best guess about the value of μ is the sample mean $\overline{x}$ (see Section 7.2), it seems reasonable to use $\overline{x}$ as a test statistic.

In general, when the hypothesis test involves a specific population parameter, the test statistic to be used is the conventional point estimate of that parameter in standardized form. For example, standardizing $\overline{x}$ (i.e., converting $\overline{x}$ to its z score) leads to the test statistic:

$$z = \frac{\overline{x} - \mu_0}{\sigma_{\overline{x}}} = \frac{\overline{x} - \mu_0}{\sigma/\sqrt{n}} \qquad \textbf{(8.1)}$$

Definition 8.9

The **test statistic** is a sample statistic, computed from the information provided by the sample, upon which the decision concerning the null and alternative hypotheses is based.

Now we need to specify the range of possible computed values of the test statistic for which the null hypothesis will be rejected. That is, what specific values of the test statistic will lead us to reject the null hypothesis in favor of the alternative hypothesis? These specific values are known collectively as the *rejection region* for the test. We learn (in the next example) that the rejection region depends on the value of α selected by the researcher.

Definition 8.10

The **rejection region** is the set of possible computed values of the test statistic for which the null hypothesis will be rejected.

a drug that effects a cure, the null and alternative hypotheses in a statistical test would take the following form:

H_0: Drug is ineffective in treating a particular disease

H_a: Drug is effective in treating a particular disease

a. To abandon a drug when in fact it is a useful one is called a *false negative*. A false negative corresponds to which type of error, Type I or Type II?

b. To proceed with more expensive testing of a drug that is in fact useless is called a *false positive*. A false positive corresponds to which type of error, Type I or Type II?

c. Which of the two errors is more serious? Explain.

8.3 Hypothesis-Testing Methodology: Test Statistics and Rejection Regions

Once we have formulated the null and alternative hypotheses and selected the significance level α, we are ready to carry out the test. For example, suppose we want to test the null hypothesis

$$H_0: \mu = \mu_0$$

where μ_0 is some fixed value of the population mean. The next step is to obtain a random sample from the population of interest. The information provided by this sample, in the form of a sample statistic, will help us decide whether to reject the null hypothesis. The sample statistic upon which we base our decision is called the *test statistic.* For this example, we are hypothesizing about the value of the population mean μ. Since our best guess about the value of μ is the sample mean $\bar{x}$ (see Section 7.2), it seems reasonable to use $\bar{x}$ as a test statistic.

In general, when the hypothesis test involves a specific population parameter, the test statistic to be used is the conventional point estimate of that parameter in standardized form. For example, standardizing $\bar{x}$ (i.e., converting $\bar{x}$ to its z score) leads to the test statistic:

$$z = \frac{\bar{x} - \mu_0}{\sigma_{\bar{x}}} = \frac{\bar{x} - \mu_0}{\sigma/\sqrt{n}} \tag{8.1}$$

Definition 8.9

The **test statistic** is a sample statistic, computed from the information provided by the sample, upon which the decision concerning the null and alternative hypotheses is based.

Now we need to specify the range of possible computed values of the test statistic for which the null hypothesis will be rejected. That is, what specific values of the test statistic will lead us to reject the null hypothesis in favor of the alternative hypothesis? These specific values are known collectively as the *rejection region* for the test. We learn (in the next example) that the rejection region depends on the value of α selected by the researcher.

Definition 8.10

The **rejection region** is the set of possible computed values of the test statistic for which the null hypothesis will be rejected.

To complete the test, we make our decision by observing whether the computed value of the test statistic lies within the rejection region. If the computed value falls within the rejection region, we will reject the null hypothesis; otherwise, we do not reject the null hypothesis.

A summary of the **hypothesis-testing procedure** we have developed is given in the box.

Outline for Testing a Hypothesis

Step 1 Specify the null and alternative hypotheses, H_0 and H_a, and the significance level, α.

Step 2 Obtain a random sample from the population(s) of interest.

Step 3 Determine an appropriate test statistic and compute its value using the sample data.

Step 4 Specify the rejection region. (This will depend on the value of α selected.)

Step 5 Make the appropriate conclusion by observing whether the computed value of the test statistic lies within the rejection region. If so, reject the null hypothesis; otherwise, do not reject the null hypothesis.

EXAMPLE 8.7

REJECTION REGION FOR AN UPPER-TAILED TEST

Specify the form of the rejection region for a test of

$$H_0\text{: } \mu = 72$$

$$H_a\text{: } \mu > 72$$

at a significance level of $\alpha = .05$.

Solution

The hypothesized value of μ in H_0 is 72. Substituting this value in equation 8.1, we obtain the test statistic

$$z = \frac{\bar{x} - \mu_0}{\sigma_{\bar{x}}} = \frac{\bar{x} - 72}{\sigma/\sqrt{n}}$$

This z score gives us a measure of how many standard deviations the observed $\bar{x}$ is from what we would expect to observe if H_0 were true. If z is "large," i.e., if $\bar{x}$ is "sufficiently greater" than 72, we will reject H_0.

Now examine Figure 8.1a and observe that the chance of obtaining a value of $\bar{x}$ more than 1.645 standard deviations above 72 is only .05, *when the true value of μ is 72.* (We are assuming that the sample size is large enough to ensure that the sampling distribution of $\bar{x}$ is approximately normal.)

Thus, if we observe a sample mean located more than 1.645 standard deviations above 72, then either H_0 is true and a relatively rare (with probability .05 or less) event has occurred, or H_a is true and the population mean exceeds 72. We would favor the latter explanation for obtaining such a large value of $\bar{x}$, and would then reject H_0.

In summary, our rejection region consists of all values of z that are greater than 1.645 (i.e., all values of $\bar{x}$ that are more than 1.645 standard deviations above 72). The *critical value* 1.645 is shown in Figure 8.1b. In this situation, the probability

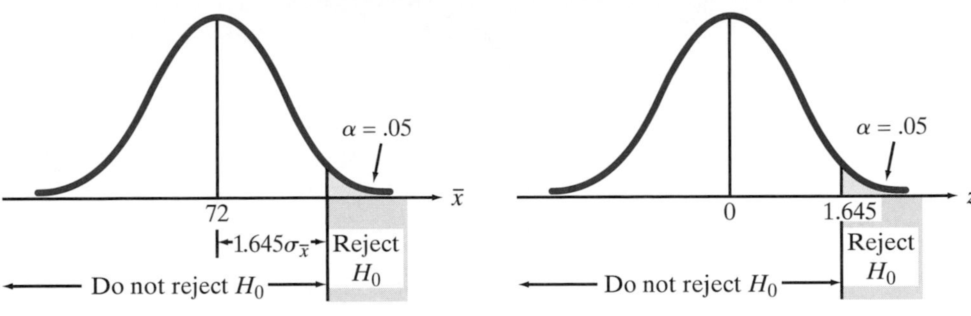

a. Rejection region in terms of $\bar{x}$

b. Rejection region in terms of z

Figure 8.1 Location of Rejection Region for Example 8.7

of a Type I error—that is, deciding in favor of H_a if in fact H_0 is true—is equal to our selected significance level, $\alpha = .05$.

> **Definition 8.11**
>
> In specifying the rejection region for a particular test of hypothesis, the value at the boundary of the rejection region is called the **critical value.**

EXAMPLE 8.8

REJECTION REGION FOR A LOWER-TAILED TEST

Specify the form of the rejection region for a test of

$$H_0\colon \mu = 72$$
$$H_a\colon \mu < 72$$

at significance level $\alpha = .01$.

Solution

Here, we want to be able to detect the directional alternative that μ is *less than* 72; in this case, it is "sufficiently small" values of the test statistic $\bar{x}$ that would cast doubt on the null hypothesis. As in Example 8.7, we use equation 8.1 to standardize the value of the test statistic and obtain a measure of the distance between $\bar{x}$ and the null hypothesized value of 72:

$$z = \frac{\bar{x} - \mu_0}{\sigma_{\bar{x}}} = \frac{\bar{x} - 72}{\sigma/\sqrt{n}}$$

This z value tells us how many standard deviations the observed $\bar{x}$ is from what would be expected *if H_0 were true.* (Again, we have assumed that the sample size n is large so that the sampling distribution of $\bar{x}$ will be approximately normal. The appropriate modifications for small samples will be discussed in Section 8.5.)

Figure 8.2a (p. 400) shows us that *when the true value of μ is 72,* the chance of observing a value of $\bar{x}$ more than 2.33 standard deviations below 72 is only .01. Thus, at a significance level (probability of Type I error) of $\alpha = .01$, we would reject the null hypothesis for all values of z that are less than -2.33 (see Figure 8.2b), i.e., for all values of $\bar{x}$ that lie more than 2.33 standard deviations below 72.

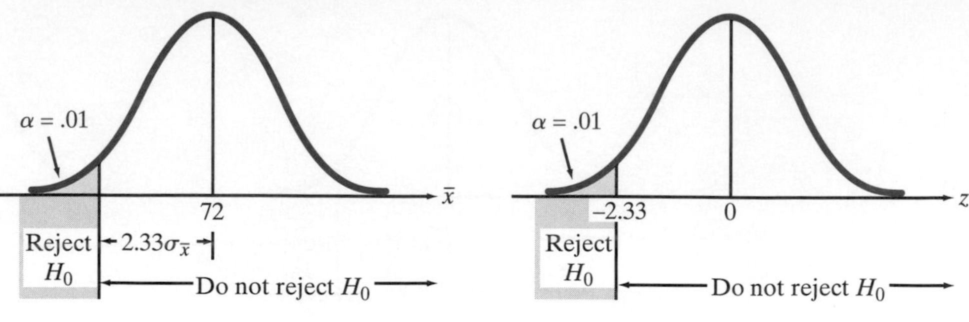

a. Rejection region in terms of $\bar{x}$ b. Rejection region in terms of z

Figure 8.2 Location of Rejection Region for Example 8.8

EXAMPLE 8.9

REJECTION REGION FOR TWO-TAILED TEST

Specify the form of the rejection region for a test of

$$H_0: \mu = 72$$

$$H_a: \mu \neq 72$$

where we are willing to tolerate a .05 chance of making a Type I error.

Solution

For this two-sided (nondirectional) alternative, $\alpha = .05$. We will reject the null hypothesis for "sufficiently small" or "sufficiently large" values of the standardized test statistic

$$z = \frac{\bar{x} - \mu_0}{\sigma_{\bar{x}}} = \frac{\bar{x} - 72}{\sigma/\sqrt{n}}$$

Now, from Figure 8.3a, we note that the chance of observing a sample mean $\bar{x}$ more than 1.96 standard deviations below 72 *or* more than 1.96 standard deviations above 72, *when H_0 is true,* is only $\alpha = .05$. Thus, the rejection region consists of two sets of values: We will reject H_0 if z is either less than -1.96 or greater than 1.96 (see Figure 8.3b). For this rejection rule, the probability of a Type I error is $\alpha = .05$.

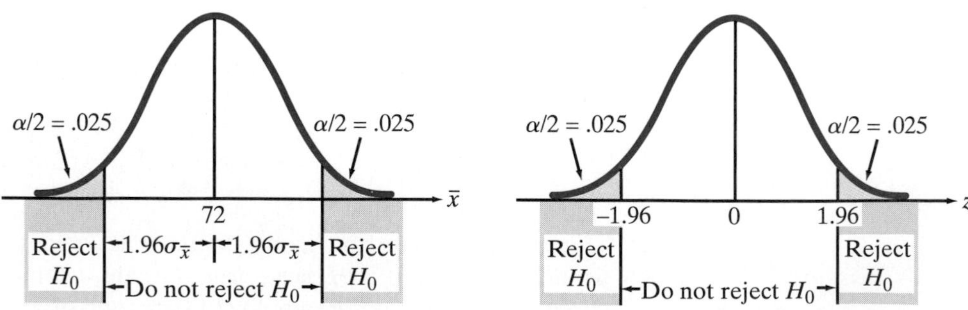

a. Rejection region in terms of $\bar{x}$ b. Rejection region in terms of z

Figure 8.3 Location of Rejection Region for Example 8.9

The three previous examples all exhibit certain common characteristics regarding the rejection region, as indicated in the box.

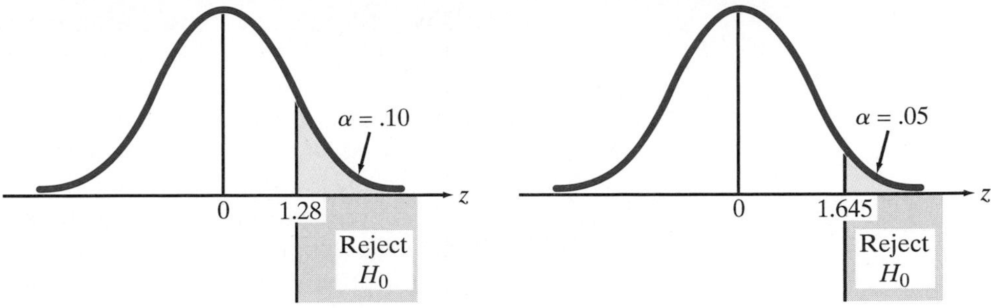

Figure 8.4 Size of the Upper-Tailed Rejection Region for Different Values of α

GUIDELINES FOR SPECIFYING THE REJECTION REGION OF A TEST

1. The value of α, the probability of a Type I error, is specified in advance by the researcher. It can be made as small or as large as desired; typical values are $\alpha = .01, .05$, and $.10$. For a fixed sample size, the size of the rejection region decreases as the value of α decreases (see Figure 8.4). That is, for smaller values of α, more extreme departures of the test statistic from the null hypothesized parameter value are required to permit rejection of H_0.
2. The test statistic (i.e., the point estimate of the target parameter) is standardized to provide a measure of how great its departure is from the null hypothesized value of the parameter. The standardization is based on the sampling distribution of the point estimate, assuming H_0 is true. For means and proportions, the general formula is:

$$\text{Standardized test statistic} = \frac{\text{Point estimate} - \text{Hypothesized value in } H_0}{\text{Standard error of point estimate}}$$

3. The location of the rejection region depends on whether the test is one-tailed or two-tailed, and on the prespecified significance level, α.
   a. For a one-tailed test in which the symbol ">" occurs in H_a (an **upper-tailed test**), the rejection region consists of values in the upper tail of the sampling distribution of the standardized test statistic. The critical value is selected so that the area to its right is equal to α. (See Figure 8.5a.)
   b. For a one-tailed test in which the symbol "<" occurs in H_a (a **lower-tailed test**), the rejection region consists of values in the lower tail of the sampling distribution of the standardized test statistic. The critical value is selected so that the area to its left is equal to α. (See Figure 8.5b.)
   c. For a two-tailed test, in which the symbol "≠" occurs in H_a, the rejection region consists of two sets of values. The critical values are selected so that the area in each tail of the sampling distribution of the standardized test statistic is equal to $\alpha/2$. (See Figure 8.5c.)

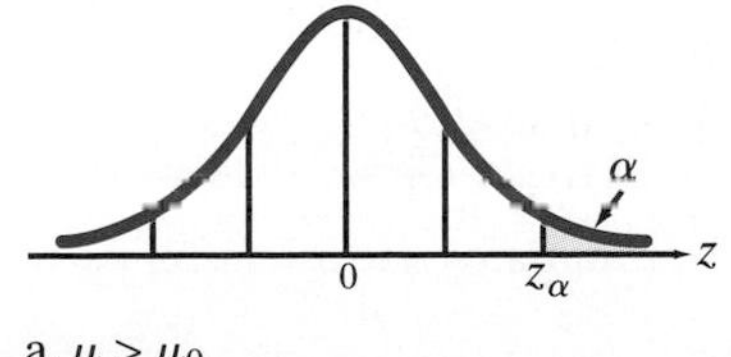

a. $\mu > \mu_0$

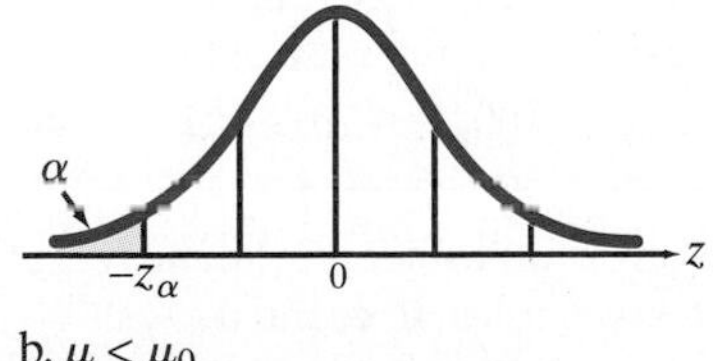

b. $\mu < \mu_0$

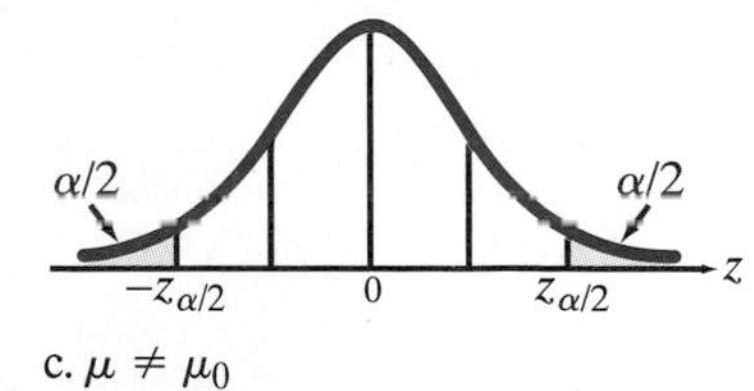

c. $\mu \neq \mu_0$

Figure 8.5 The Rejection Region for Various Alternative Hypotheses

Once we have chosen a test statistic from the sample information and computed its value (step 3, p. 398), we determine if its standardized value lies within the rejection region so we can decide whether to reject the null hypothesis (step 5).

Throughout this section, we have assumed that the sample size n is sufficiently large (i.e., $n \geq 30$). A summary of this large-sample test of hypothesis is provided in the box. Note that the test statistic will depend on the parameter of interest and its point estimate. In the remainder of this chapter, we cover the test statistics for a population mean μ (Section 8.5), a population proportion p (Section 8.7), a population variance σ^2 (optional Section 8.8), and population category probabilities $p_1, p_2, \ldots, p_k$ (optional Section 8.9). First, we provide some guidelines on selecting the parameter of interest in the next section.

LARGE-SAMPLE TEST OF HYPOTHESIS ABOUT A POPULATION PARAMETER

H_0: Parameter = Hypothesized value

Upper-Tailed Test	**Two-Tailed Test**	**Lower-Tailed Test**
H_a: Parameter > Hypothesized value	H_a: Parameter ≠ Hypothesized value	H_a: Parameter < Hypothesized value

$$\textit{Test statistic:} = \frac{\text{Point estimate} - \text{Hypothesized value}}{\text{Standard error of point estimate}} \quad \textbf{(8.2)}$$

Rejection region:	*Rejection region:*	*Rejection region:*
$z > z_{\alpha}$	$z < -z_{\alpha/2}$ or $z > z_{\alpha/2}$	$z < -z_{\alpha}$

Self-Test 8.3

At $\alpha = .10$, find the rejection region for a large-sample test of

$$H_0: \mu = 150$$

$$H_a: \mu < 150$$

PROBLEMS FOR SECTION 8.3

Using the Tools

8.11 Suppose we want to test H_0: $\mu = 65$. Specify the form of the rejection region for each of the following (assume that the sample size will be sufficient to guarantee the approximate normality of the sampling distribution of $\bar{x}$):

a. H_a: $\mu \neq 65$, $\alpha = .05$ **b.** H_a: $\mu > 65$, $\alpha = .05$

c. H_a: $\mu < 65$, $\alpha = .01$ **d.** H_a: $\mu < 65$, $\alpha = .10$

8.12 Refer to Problem 8.11. Calculate the test statistic for each of the following sample results (assume s is a good approximation of σ):

a. $n = 100, \bar{x} = 70, s = 15$ **b.** $n = 50, \bar{x} = 70, s = 15$

c. $n = 50, \bar{x} = 60, s = 15$ **d.** $n = 50, \bar{x} = 60, s = 30$

8.13 Refer to Problems 8.11 and 8.12. Give the appropriate conclusions for each of the tests, parts **a**, **b**, **c**, and **d**.

8.14 For each of the following rejection regions, determine the value of α, the probability of a Type I error:

a. $z < -1.96$ **b.** $z > 1.645$ **c.** $z < -2.58$ or $z > 2.58$

Applying the Concepts

8.15 Suppose you are interested in testing whether μ, the true mean score on an aptitude test used for determining admission to graduate study in your field, differs from 500. To conduct the test, you randomly select 75 applicants to graduate study and record the aptitude score, x, for each.

a. Set up the null and alternative hypotheses for the test.

b. Set up the rejection region for the test if $\alpha = .10$.

8.16 The Computer-Assisted Hypnosis Scale (CAHS) is designed to measure a person's susceptibility to hypnosis. CAHS scores range from 0 (no susceptibility to hypnosis) to 12 (extremely high susceptibility to hypnosis). *Psychological Assessment* (Mar. 1995) reported that University of Tennessee undergraduates had a mean CAHS score of $\mu = 4.6$. Suppose you want to test whether undergraduates at your college or university are more susceptible to hypnosis than University of Tennessee undergraduates.

a. Set up the null and alternative hypotheses for the test.

b. Select a value of α and interpret it.

c. Find the rejection region for the test if you plan on sampling $n = 100$ undergraduates at your college.

8.17 The purchase of a coin-operated laundry is being considered by a potential entrepreneur. The present owner claims that over the past five years the average daily revenue was \$675. The entrepreneur wants to test whether the true average daily revenue of the laundry is less than \$675.

a. Set up the null and alternative hypotheses for the test.

b. Find the rejection region for the test if $\alpha = .05$. Assume a large sample of days will be selected from the past five years and the revenue recorded for each day.

8.18 According to the *Journal of Psychology and Aging* (May 1992), older workers (i.e., workers 45 years or older) have a mean job satisfaction rating of 4.3 on a 5-point scale. (Higher scores indicate higher levels of job satisfaction.) In a random sample of 100 older workers, suppose the sample mean job satisfaction rating is $\bar{x} = 4.1$.

a. Set up the null and alternative hypotheses for testing whether the true mean job satisfaction rating of older workers differs from 4.3.

b. At $\alpha = .05$, locate the rejection region for the test.

c. Compute the test statistic if $\sigma = .5$. What is your conclusion?

d. Repeat part **c**, but assume $\sigma = 1.5$.

8.19 *Psychological Assessment* (Mar. 1995) published the results of a study of World War II aviators who were shot down and became prisoners of war (POWs). A post-traumatic stress disorder (PTSD) score was determined for each in a sample of 33 aviators. The results, where higher scores indicate higher PTSD levels, are $\bar{x} = 9.00$, $s = 9.32$.

a. Psychologists have established that the mean PTSD score for Vietnam veterans is 16. Set up the null and alternative hypotheses for determining whether the true mean PTSD score μ of all World War II aviator POWs is less than 16.

b. Calculate the test statistic (assume $s \approx \sigma$).

c. Specify the form of the rejection region if the level of significance is $\alpha = .01$. Locate the rejection region, α, and the critical value on a sketch of the standard normal curve.

d. Use the results of parts **b** and **c** to make the proper conclusion in terms of the problem.

8.20 A certain species of beetle produces offspring with either blue eyes or black eyes. Suppose a biologist wants to determine which, if either, of the two eye colors is dominant for this species of beetle. Let p represent the true proportion of offspring that possess blue eyes. If the beetles produce blue-eyed and black-eyed offspring at an equal rate, then $p = .5$. Thus, the biologist desires to test

$$H_0: p = .5$$
$$H_a: p \neq .5$$

Give the form of the rejection region if the biologist is willing to tolerate a Type I error probability of $\alpha = .05$. Locate the rejection region, α, and the critical value(s) on a sketch of the normal curve. Assume that the sample size n is large. [*Hint:* The test statistic is the large-sample z statistic.]

8.4 Guidelines for Determining the Target Parameter

The key to correctly diagnosing a hypothesis test is to first determine the parameter of interest.

The following three-step process is useful for converting the words of the problem into the symbols needed for the parameter of interest:

(1) Identify the experimental unit (i.e., the objects upon which the measurements are taken).

(2) Identify the type of variable, quantitative or qualitative, measured on each experimental unit.

(3) Determine the target parameter based on the phenomenon of interest and the variable measured. For quantitative data, the target parameter will be either a population mean or variance; for qualitative data the parameter will be a population proportion.

Often, there are one or more key words in the statement of the problem that indicate the appropriate population parameter. In this section, we will present several examples illustrating how to determine the parameter of interest. First, we state in the next box the key words to look for when conducting a hypothesis test about a single population parameter.

DIAGNOSING A HYPOTHESIS TEST: DETERMINING THE TARGET PARAMETER

Parameter	Key Words or Phrases
μ	Mean; average
p	Proportion; percentage; fraction; rate
σ^2	Variance; variation; spread; precision

EXAMPLE 8.10 **CHOOSING THE TARGET PARAMETER**

The "Pepsi Challenge" was a marketing strategy used by Pepsi-Cola. A consumer is presented with two cups of cola and asked to select the one that tastes better. Unknown to the consumer, one cup is filled with Pepsi, the other with Coke. Marketers of Pepsi claim that the true fraction of consumers who select their product will exceed .50.

a. What is the parameter of interest to the Pepsi marketers?

b. Set up H_0 and H_a.

Solution

a. In this problem, the experimental units are the consumers and the variable measured is *qualitative*—the consumer chooses either Pepsi or Coke. The key word in the statement of the problem is "fraction." Thus, the parameter of

interest is p, where

p = True fraction of consumers who favor Pepsi over Coke in the taste test

b. To support their claim that the true fraction who favor Pepsi exceeds .5, the marketers will test

$$H_0: p = .5$$
$$H_a: p > .5$$

EXAMPLE 8.11

CHOOSING THE TARGET PARAMETER

The administrator at a large hospital wants to know if the average length of stay of the hospital's patients is less than five days. To check this, lengths of stay were recorded for 100 randomly selected hospital patients.

a. What is the parameter of interest to the administrator?

b. Set up H_0 and H_a.

Solution

a. The experimental units for this problem are the hospital patients. For each patient, the variable measured is length of stay (in days)—a *quantitative* variable. Since the administrator wants to make a decision about the average length of stay, the target parameter is

$$\mu = \text{average length of stay of hospital patients}$$

b. To determine whether the average length of stay is less than five days, the administrator will test

$$H_0: \mu = 5$$
$$H_a: \mu < 5$$

EXAMPLE 8.12

CHOOSING THE TARGET PARAMETER

According to specifications, the variation in diameters of 2-inch bolts produced on an assembly line should be .001. A quality control engineer wants to determine whether the variation in bolt diameters differs from these specifications. What is the parameter of interest?

Solution

For the quality control engineer, the experimental units are the bolts produced by the assembly line, and the variable measured is *quantitative*—the diameter of the bolt. The key word in the statement of the problem is "variation." Therefore, the parameter of interest is σ^2.

Self-Test 8.4

According to *Chance* (Fall 1994), the average number of faxes transmitted in the United States each minute is 88,000. Suppose we believe the population average exceeds 88,000.

a. What is the parameter of interest?

b. Set up the null and alternative hypotheses.

In the remaining sections of this chapter, we present a summary of the hypothesis-testing procedures for each of the parameters listed in the box on p. 404. In the problems that follow each section, the target parameter can be identified by simply noting the title of the section. However, to properly diagnose a hypothesis test, it is essential to search for the key words in the statement of the problem.

Statistics in the Real World Revisited

Identifying the Parameter of Interest for the KLEENEX® Survey

In Kimberly-Clark Corporation's survey of people with colds, 250 customers were asked to keep count of their use of KLEENEX® tissues in diaries. One goal of the company was to determine how many tissues to package in a cold-care box of KLEENEX®; consequently, the total number of tissues used was recorded for each person surveyed. Since number of tissues is a quantitative variable, the parameter of interest is μ, the mean number of tissues used by all customers with colds.

8.5 Testing a Population Mean

Large Samples

When testing a hypothesis about a population mean μ, the procedure that we use will depend on whether the sample size n is large (say, $n \geq 30$) or small. The accompanying box contains the elements of a large-sample hypothesis test for μ based on the z statistic. Note that for this case, the only assumption required for the validity of the procedure is that the random sample is, in fact, large so that the sampling distribution of $\bar{x}$ is normal. Technically, the true population standard deviation σ must be known in order to use the z statistic, and this is rarely, if ever, the case. However, we established in Chapter 7 that when n is large, the sample standard deviation s provides a good approximation to σ and the z statistic can be approximated as shown in the box.

LARGE-SAMPLE (z) TEST OF HYPOTHESIS ABOUT A POPULATION MEAN μ

Upper-Tailed Test	**Two-Tailed Test**	**Lower-Tailed Test**
$H_0: \mu = \mu_0$	$H_0: \mu = \mu_0$	$H_0: \mu = \mu_0$
$H_a: \mu > \mu_0$	$H_a: \mu \neq \mu_0$	$H_a: \mu < \mu_0$

$$\textit{Test statistic: } = \frac{\bar{x} - \mu_0}{\sigma_{\bar{x}}} \approx \frac{\bar{x} - \mu_0}{s/\sqrt{n}} \quad \textbf{(8.3)}$$

Rejection region:	*Rejection region:*	*Rejection region:*
$z > z_\alpha$	$\lvert z \rvert > z_{\alpha/2}$	$z < -z_\alpha$

where z_α is the z value such that $P(z > z_\alpha) = \alpha$; $-z_\alpha$ is the z value such that $P(z < -z_\alpha) = \alpha$; and $z_{\alpha/2}$ is the z value such that $P(z > z_{\alpha/2}) = \alpha/2$.

Assumption: The random sample must be sufficiently large (say, $n \geq 30$) so that the sampling distribution of $\bar{x}$ is approximately normal and so that s provides a good approximation to σ.

EXAMPLE 8.13

LARGE-SAMPLE TEST OF μ

Humerus bones from the same species of animal tend to have approximately the same length-to-width ratios. When fossils of humerus bones are discovered, archaeologists can often determine the species of animal by examining the length-to-width ratios of the bones. It is known that species A has a mean ratio of 8.5. Suppose 41 fossils of humerus bones were unearthed at an archaeological site in East Africa, which species A is believed to have inhabited. (Assume that the unearthed bones were all from the same unknown species.) The length-to-width ratios of the bones were measured and are listed in Table 8.3.

BONES

TABLE 8.3 Length-to-Width Ratios of a Sample of Humerus Bones

10.73	9.59	8.37	9.35	9.39	8.38
8.89	8.48	6.85	8.86	9.17	11.67
9.07	8.71	8.52	9.93	9.89	8.30
9.20	9.57	8.87	8.91	8.17	9.17
10.33	9.29	6.23	11.77	8.93	12.00
9.98	9.94	9.41	10.48	8.80	9.38
9.84	8.07	6.66	10.39	10.02	

We wish to test the hypothesis that μ, the population mean ratio of all bones of this particular species, is equal to 8.5 against the alternative that it is different from 8.5, i.e., we wish to test whether the unearthed bones are from species A.

a. Suppose we want a very small chance of rejecting H_0 if μ is equal to 8.5. That is, it is important that we avoid making a Type I error. Select an appropriate value of the significance level α.

b. Test whether μ, the population mean length-to-width ratio, is different from 8.5 using the significance level selected in part **a**.

c. What are the practical implications of the result, part **b**?

Solution

a. The hypothesis-testing procedure that we have developed gives us the advantage of being able to choose any significance level that we desire. Since the significance level α is also the probability of a Type I error, we will choose α to be very small. In general, researchers who consider a Type I error to have very serious practical consequences should perform the test at a very low α value—say, $\alpha = .01$. Other researchers may be willing to tolerate an α value as high as .10 if a Type I error is not deemed a serious error to make in practice. For this example, we will test at $\alpha = .01$.

b. We formulate the following hypotheses:

$$H_0\text{: } \mu = 8.5$$
$$H_a\text{: } \mu \neq 8.5$$

The sample size is large ($n = 41$); thus, we may proceed with the large-sample (z) test about μ. The data (Table 8.3) were entered into an Excel spreadsheet; the output obtained from PHStat and Excel are displayed in Figure 8.6. From Figure 8.6a, $\bar{x} = 9.26$ and $s = 1.20$. Substituting these values into equation 8.3, we obtain

$$z = \frac{\bar{x} - 8.5}{s/\sqrt{n}} = \frac{9.26 - 8.5}{1.20/\sqrt{41}} = 4.055$$

E Figure 8.6 Large Sample Hypothesis Test for μ Using Excel and PHStat, Example 8.13

	A	B
1	*Length to width ratio*	
2		
3	Mean	9.257561
4	Standard Error	0.187965
5	Median	9.2
6	Mode	9.17
7	Standard Deviation	1.203565
8	Sample Variance	1.448569
9	Kurtosis	1.066261
10	Skewness	-0.08495
11	Range	5.77
12	Minimum	6.23
13	Maximum	12
14	Sum	379.56
15	Count	41
16	Largest(1)	12
17	Smallest(1)	6.23

a.

	A		B
1	**Test of Hypothesis for the Mean**		
2			
3	**Data**		
4	Null Hypothesis	μ=	8.5
5	Level of Significance		0.01
6	Population Standard Deviation		1.2
7	Sample Size		41
8	Sample Mean		9.26
9			
10	Intermediate Calculations		
11	Standard Error of the Mean		0.187408514
12	Z Test Statistic		4.055312017
13			
14	**Two-Tailed Test**		
15	Lower Critical Value		-2.575834515
16	Upper Critical Value		2.575834515
17	p-Value		5.00929E-05
18	**Reject the null hypothesis**		

b.

This value is shown in Figure 8.6b. At $\alpha = .01$, the rejection region for this two-tailed test is

$$|z| > z_{.005} = 2.58$$

This rejection region is shown in Figure 8.7. Since $z = 4.055$ falls in the rejection region, we reject H_0 and conclude that the mean length-to-width ratio of all humerus bones of this particular species is significantly different from 8.5. If the null hypothesis is in fact true (i.e., if $\mu = 8.5$), then the probability that we have incorrectly rejected it is equal to $\alpha = .01$.

Figure 8.7 Rejection Region for Example 8.13

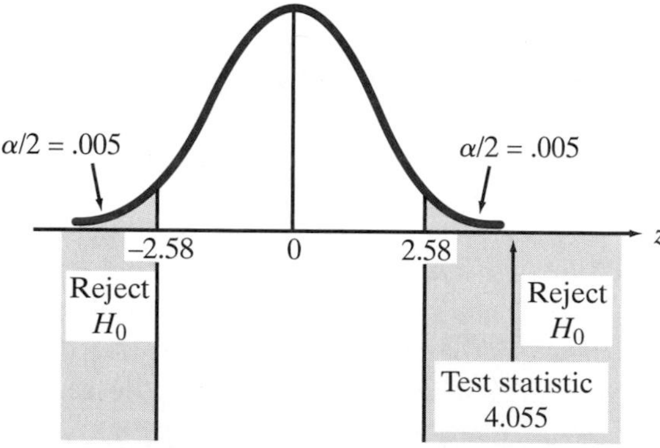

c. The *practical* implications of the result remain to be studied further. Perhaps the animal discovered at the archaeological site is of some species other than A. Alternatively, the unearthed humerus bones may have larger than normal length-to-width ratios because of unusual feeding habits of species A. **It is not always the case that a statistically significant result implies a practically significant result.** The researcher must retain his or her objectivity and judge the practical significance using, among other criteria, knowledge of the subject matter and the phenomenon under investigation.

Self-Test 8.5

Prior to the institution of a new safety program, the average number of on-the-job accidents per day at a factory was 4.5. To determine whether the safety program was effective, a factory foreman will conduct a test of

H_0: $\mu = 4.5$ (no change in average number of on-the-job accidents per day)

H_a: $\mu < 4.5$ (average number of on-the-job accidents per day has decreased)

where μ represents the mean number of on-the-job accidents per day at the factory after institution of the new safety program. Summary statistics for the daily number of accidents for a random sample of 120 days after institution of the new program are: $\bar{x} = 3.7, s = 2.6$. Conduct the test using $\alpha = .01$.

Small Samples

Time and cost considerations sometimes limit the sample size. In this case, the assumption required for a large-sample test of hypothesis about μ will be violated and s will not provide a reliable estimate of σ. We need, then, a procedure that is appropriate for use with small samples.

A hypothesis test about a population mean μ for small samples ($n < 30$) is based on a t statistic. The elements of the test are listed in the next box.

SMALL-SAMPLE (t) TEST OF HYPOTHESIS ABOUT A POPULATION MEAN μ

Upper-Tailed Test	Two-Tailed Test	Lower-Tailed Test
H_0: $\mu = \mu_0$	H_0: $\mu = \mu_0$	H_0: $\mu = \mu_0$
H_a: $\mu > \mu_0$	H_a: $\mu \neq \mu_0$	H_a: $\mu < \mu_0$

$$\textit{Test statistic: } t = \frac{\bar{x} - \mu_0}{s/\sqrt{n}} \quad \textbf{(8.4)}$$

Rejection region:	*Rejection region:*	*Rejection region:*
$t > t_\alpha$	$\lvert t \rvert > t_{\alpha/2}$	$t < -t_\alpha$

where the distribution of t is based on $(n - 1)$ degrees of freedom, t_α is the t value such that $P(t > t_\alpha) = \alpha$; $-t_\alpha$ is the t value such that $P(t < -t_\alpha) = \alpha$; and $t_{\alpha/2}$ is the t value such that $P(t > t_{\alpha/2}) = \alpha/2$.

Assumption: The relative frequency distribution of the population from which the random sample was selected is approximately normal.

As we noticed in the development of confidence intervals (Chapter 7), when making inferences based on small samples more restrictive assumptions are required than when making inferences from large samples. In particular, this hypothesis test requires the assumption that the population from which the sample is selected is approximately normal.

Notice that the test statistic given in the box is a t statistic and is calculated exactly as our approximation to the large-sample test statistic z given earlier in this section. Therefore, just like z, the computed value of t indicates the direction and approximate distance (in units of standard deviations) that the sample mean $\bar{x}$ is from the hypothesized population mean μ_0.

EXAMPLE 8.14 SMALL-SAMPLE TEST OF μ

The building specifications in a certain city require that the sewer pipe used in residential areas have a mean breaking strength of more than 2,500 pounds per lineal foot. A manufacturer who would like to supply the city with sewer pipe has submitted a bid and provided the following additional information: An independent contractor randomly selected seven sections of the manufacturer's pipe and tested each for breaking strength. The results (pounds per lineal foot) are shown here:

PIPES

2,610	2,750	2,420	2,510	2,540	2,490	2,680

a. Compute $\bar{x}$ and s for the sample.

b. Is there sufficient evidence to conclude that the manufacturer's sewer pipe meets the required specifications? Use a significance level of $\alpha = .10$.

Solution

a. The data were entered into an Excel spreadsheet and descriptive statistics were generated. The Excel printout is shown in Figure 8.8a. The values of the sample mean breaking strength and standard deviation of breaking strengths (highlighted) are $\bar{x} = 2{,}571.4$ and $s = 115.1$.

Figure 8.8 Small-Sample Hypothesis Test for μ Using Excel and PHStat, Example 8.14

	A	B
1	*Breaking Strength of sewer pipe*	
2		
3	Mean	2571.429
4	Standard Error	43.50307
5	Median	2540
6	Mode	#N/A
7	Standard Deviation	115.0983
8	Sample Variance	13247.62
9	Kurtosis	-0.76526
10	Skewness	0.421794
11	Range	330
12	Minimum	2420
13	Maximum	2750
14	Sum	18000
15	Count	7
16	Largest(1)	2750
17	Smallest(1)	2420

a.

	A		B
1	t Test of Hypothesis for the mean		
2			
3	Data		
4	Null Hypothesis	μ=	2500
5	Level of Significance		0.1
6	Sample Size		7
7	Sample Mean		2571.4
8	Sample Standard Deviation		115.1
9			
10	Intermediate Calculations		
11	Standard Error of the Mean		43.50371084
12	Degrees of Freedom		6
13	t Test Statistic		1.641239302
14			
15	Upper-Tail Test		
16	Upper Critical Value		1.439755124
17	p-Value		0.075928304
18	Reject the null hypothesis		

b.

b. The relevant hypothesis test has the following elements:

H_0: $\mu = 2{,}500$ (the manufacturer's pipe does not meet the city's specifications)

H_a: $\mu > 2{,}500$ (the pipe meets the specifications)

where μ represents the true mean breaking strength (in pounds per lineal foot) for all sewer pipe produced by the manufacturer.

This small-sample ($n = 7$) test requires the assumption that the relative frequency distribution of the population values of breaking strength for the manufacturer's pipe is approximately normal. Then the test will be based on a t distribution with $(n - 1) = 6$ degrees of freedom. We will thus reject H_0 if

$$t > t_{.10} = 1.440 \quad \text{(see Figure 8.9)}$$

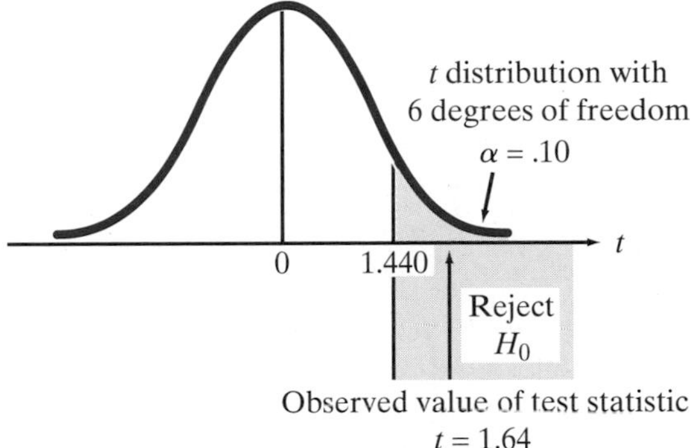

Figure 8.9 Rejection Region for Example 8.14

Substituting the values $\bar{x} = 2{,}571.4$ and $s = 115.1$ into equation 8.4 yields the test statistic

$$t = \frac{\bar{x} - \mu_0}{s/\sqrt{n}} = \frac{2{,}571.4 - 2{,}500}{115.1/\sqrt{7}} = 1.64$$

This value is also shown on the Excel and PHStat printout of the test results, Figure 8.8b. Since this value of t is larger than the critical value of 1.440, we reject H_0. There is sufficient evidence (at significance level $\alpha = .10$) that the manufacturer's pipe meets the city's building specifications.

Self-Test 8.6

Refer to Self-Test 8.5 (p. 409). Conduct the test H_0: $\mu = 4.5$ against H_a: $\mu < 4.5$, but assume the sample size is $n = 12$.

Remember that the small-sample test of Example 8.14 requires the assumption that the sampled population has a distribution that is approximately normal. If you know that the population is highly skewed (based on, for example, a stem-and-leaf plot of the sample data), then any inferences derived from the t test are suspect.

CAUTION

When the sampled population is decidedly nonnormal (e.g., highly skewed), any inferences derived from the small-sample t test for μ are suspect. In this case, one alternative is to use a nonparametric test discussed in Daniel (1990).

PROBLEMS FOR SECTION 8.5

Using the Tools

8.21 Compute the value of the test statistic z for each of the following situations:

a. H_0: $\mu = 9.8$, H_a: $\mu > 9.8$, $\bar{x} = 10.0$, $s = 4.3$, $n = 50$

b. H_0: $\mu = 80$, H_a: $\mu < 80$, $\bar{x} = 75$, $s^2 = 19$, $n = 86$

c. H_0: $\mu = 8.3$, H_a: $\mu \neq 8.3$, $\bar{x} = 8.2$, $s^2 = .79$, $n = 175$

8.22 A random sample of n observations is selected from a population with unknown mean μ and variance σ^2. For each of the following situations, specify the test statistic and rejection region.

a. H_0: $\mu = 50$, H_a: $\mu > 50$, $n = 36$, $\bar{x} = 60$, $s = 8$, $\alpha = .05$

b. H_0: $\mu = 140$, H_a: $\mu \neq 140$, $n = 40$, $\bar{x} = 143.2$, $s = 9.4$, $\alpha = .01$

c. H_0: $\mu = 10$, H_a: $\mu < 10$, $n = 50$, $\bar{x} = 9.5$, $s = .35$, $\alpha = .10$

8.23 To test the null hypothesis H_0: $\mu = 10$, a random sample of n observations is selected from a normal population. Specify the rejection region for each of the following combinations of H_a, n, and α:

a. H_a: $\mu \neq 10$, $n = 15$, $\alpha = .05$

b. H_a: $\mu \neq 10$, $n = 15$, $\alpha = .01$

c. H_a: $\mu < 10$, $n = 15$, $\alpha = .05$

d. H_a: $\mu > 10$, $n = 5$, $\alpha = .10$

e. H_a: $\mu > 10$, $n = 25$, $\alpha = .10$

8.24 A random sample of n observations is selected from a normal population. For each of the following situations, specify the rejection region, test statistic, and conclusion.

a. H_0: $\mu = 3{,}000$, H_a: $\mu \neq 3{,}000$, $\bar{x} = 2{,}958$, $s = 39$, $n = 8$, $\alpha = .05$

b. H_0: $\mu = 6$, H_a: $\mu > 6$, $\bar{x} = 6.3$, $s = .3$, $n = 7$, $\alpha = .01$

c. H_0: $\mu = 22$, H_a: $\mu < 22$, $\bar{x} = 13.0$, $s = 6$, $n = 17$, $\alpha = .05$

8.25 A random sample of five measurements from a normally distributed population yielded the following data: 12, 4, 3, 5, 5.

a. Test the null hypothesis that $\mu = 4$ against the alternative hypothesis that $\mu \neq 4$. Use $\alpha = .01$.

b. Test the null hypothesis that $\mu = 4$ against the alternative hypothesis that $\mu > 4$. Use $\alpha = .01$.

Applying the Concepts

8.26 The room and board costs for a sample of 20 American universities and colleges are listed in the table on p. 413. Test the hypothesis that the true mean room and board costs of all American colleges and universities differ from \$6,000. Use $\alpha = .05$.

8.27 Television commercials most often employ females, or "feminized" males, to pitch a company's product. Research published in *Nature* (Aug. 27, 1998) revealed that people are, in fact, more attracted to "feminized" faces, regardless of gender. In one experiment, 50 human subjects viewed both a Japanese female and a Caucasian male face on a computer. Using special computer graphics, each subject could morph the faces (by making them more feminine or more masculine) until they attained the "most attractive" face. The level of feminization x (measured as a percentage) was measured.

a. For the Caucasian male face, $\bar{x} = 15.0\%$ and $s = 25.1$. The researchers used this sample information to test the null hypothesis of a mean level of feminization equal to 0%. Verify that the test statistic is equal to 4.22.

b. Give the rejection region at $\alpha = .10$ if the alternative hypothesis is that the mean level of feminization exceeds 0%.

c. Give the appropriate conclusion for the test.

8.28 One of the most feared predators in the ocean is the great white shark. Although it is known that the great white shark grows to a mean length of 21 feet, a marine

ROOMBOARD

College/University	Room and Board Costs
Arizona State University	$4,300
Babson College	7,600
Boston University	7,000
California State University–Fresno	5,400
Case Western Reserve University	5,000
Emory University	6,500
George Washington University	6,900
Harvard University	7,000
Lafayette College	6,300
LaSalle University	6,700
Lehigh University	6,000
Montclair State University	5,300
Niagara University	5,400
New York University	7,800
Northwestern University	6,100
Notre Dame	4,800
Purdue University	4,500
Rochester Institute of Technology	6,100
Stanford University	7,300
Sienna College	5,400

Sources: Individual 1996–1997 institutional catalogues.

Problem 8.26

biologist believes that great white sharks off the Bermuda coast grow much longer because of unusual feeding habits. To test this claim, researchers plan to capture a number of full-grown great white sharks off the Bermuda coast, measure them, then set them free. However, because capturing sharks is difficult, costly, and very dangerous, only three are sampled. Their lengths are 24, 20, and 22 feet.

a. Do the data provide sufficient evidence to support the marine biologist's claim? Test at significance level $\alpha = .05$.

b. What assumptions are required for the hypothesis test of part **a** to be valid? Do you think these assumptions are likely to be satisfied in this particular sampling situation?

8.29 The symbol pH represents the hydrogen ion concentration in a liter of a solution. The higher the pH value, the less acidic the solution. The table below lists the pH values for a sample of foods as determined by the Food and Drug Administration (FDA). Suppose an FDA spokesperson claims that the mean pH level of all food items is $\mu = 4.0$. Test this claim at $\alpha = .10$. Interpret the results.

PH

Food	pH	Food	pH	Food	pH
Lima beans	6.5	Orange juice	3.9	Corn syrup	5.0
Cauliflower	5.6	Buttermilk	4.5	Cider	3.1
Figs	4.6	Crabs	7.0	Caviar	5.4
Lemons	2.3	Fresh fish	6.7	Mayonnaise	4.35
Potatoes	6.1	Blueberries	3.7	Celery	5.85
Peppers	5.15	Cheddar cheese	5.9	Artichokes	5.6
Apples	3.4	Egg whites	8.0	Parsnips	5.3
Sauerkraut	3.5	Bread	5.55	Dill pickles	3.35
Lettuce	5.9	Eclairs	4.45	Tomatoes	4.55
Sweet corn	7.3	Raisins	3.9	Bananas	4.85
Ham	6.0	Honey	3.9	Ground beef	5.55
Nectarines	3.9	Cocoa	6.3	Veal	6.0

8.30 *Supremism* is defined as an attitude of superiority based on race. Psychologists at Wake Forest University investigated the degree of supremism exhibited by future teachers (*Journal of Black Psychology,* Nov. 1997). Ten European-American college students who were enrolled in a 4th-year teacher education course participated in the study. Each student was given a picture of a 7-year-old African-American child and instructed to estimate the child's IQ. The IQ estimates are summarized as follows: $\bar{x} = 94.1, s = 10.3$. Is there sufficient evidence to indicate that the true mean IQ estimated by all European-American future teachers is less than 100? Test using $\alpha = .05$.

8.31 A group of researchers at the University of Texas–Houston conducted a comprehensive study of pregnant cocaine-dependent women (*Journal of Drug Issues,* Summer 1997). All the women in the study used cocaine on a regular basis (at least three times a week) for more than a year. One of the many variables measured was birthweight (in grams) of the baby delivered. For a sample of 16 cocaine-dependent women, the mean birthweight was 2,971 grams and the standard deviation was 410 grams. Test the hypothesis that the true mean birthweight of babies delivered by cocaine-dependent women is less than 3,500 grams. Use $\alpha = .05$.

8.32 A consumers' advocate group would like to evaluate the average energy efficiency rating (EER) of window-mounted, large-capacity (i.e., in excess of 7,000 Btu) air-conditioning units. A random sample of 36 such air-conditioning units is selected and tested for a fixed period of time with their EER recorded as follows:

EERAC

8.9	9.1	9.2	9.1	8.4	9.5	9.0	9.6	9.3
9.3	8.9	9.7	8.7	9.4	8.5	8.9	8.4	9.5
9.3	9.3	8.8	9.4	8.9	9.3	9.0	9.2	9.1
9.8	9.6	9.3	9.2	9.1	9.6	9.8	9.5	10.0

a. Using the .05 level of significance, is there evidence that the average EER is different from 9.0?

b. What assumptions are being made in order to perform this test?

c. What will your answer in part **a** be if the last data value is 8.0 instead of 10.0?

8.33 During the National Football League (NFL) season, Las Vegas oddsmakers establish a point spread on each game for betting purposes. For example, the Baltimore Ravens were established as 3-point favorites over the New York Giants in the 2001 Super Bowl. The final scores of NFL games were compared against the final point spreads established by the oddsmakers in *Chance* (Fall 1998). The difference between the game outcome and point spread (called a point-spread error) was calculated for 240 NFL games. The mean and standard deviation of the point-spread errors are $\bar{x} = -1.6$ and $s = 13.3$. Use this information to test the hypothesis that the true mean point-spread error for all NFL games is 0. Conduct the test at $\alpha = .01$ and interpret the result.

8.6 Reporting Test Results: *p*-Values

The statistical hypothesis-testing technique that we have developed requires us to choose the significance level α (i.e., the maximum probability of a Type I error that we are willing to tolerate) before obtaining the data and computing the test statistic. By choosing α a priori, we fix the rejection region for the test. Thus, no matter how large or how small the observed value of the test statistic, our decision regarding H_0 is clear-cut: Reject H_0 (i.e., conclude that the test results are statistically significant) if the observed value of the test statistic falls into the rejection region, and do not reject H_0 (i.e., conclude that the test results are insignificant) otherwise. This **"fixed" significance level** α then serves as a measure of the reliability of our inference. However, there is one drawback to a test conducted in this manner—namely, a measure of the *degree* of significance of the test results is not

readily available. That is, if the value of the test statistic falls into the rejection region, we have no measure of the extent to which the data disagree with the null hypothesis.

EXAMPLE 8.15

THE DEGREE OF DISAGREEMENT BETWEEN THE SAMPLE DATA AND H_0

A large-sample test of H_0: $\mu = 72$ against H_a: $\mu > 72$ is to be conducted at a fixed significance level of $\alpha = .05$. Consider the following possible values of the computed test statistic:

$$z = 1.82 \quad \text{and} \quad z = 5.66$$

a. Which of the above values of the test statistic provides stronger evidence for the rejection of H_0?

b. How can we measure the extent of disagreement between the sample data and H_0 for each of the computed values?

Solution

a. The appropriate rejection region for this upper-tailed test, at $\alpha = .05$, is given by

$$z > z_{.05} = 1.645$$

Clearly, for either of the test statistic values, $z = 1.82$ or $z = 5.66$, we will reject H_0; hence, the result in each case is statistically significant. Recall, however, that the appropriate test statistic for a large-sample test concerning μ is simply the z score for the observed sample mean $\bar{x}$, calculated by using the hypothesized value of μ in H_0 (in this case, $\mu = 72$). The larger the z score, the greater the distance (in units of standard deviations) that $\bar{x}$ is from the hypothesized value of $\mu = 72$. Thus, a z score of 5.66 would present stronger evidence that the true mean is larger than 72 than would a z score of 1.82. This reasoning stems from our knowledge of the sampling distribution of $\bar{x}$; if in fact $\mu = 72$, we would certainly not expect to observe an $\bar{x}$ with a z score as large as 5.66.

b. One way of measuring the amount of disagreement between the observed data and the value of μ in the null hypothesis is to calculate the probability of observing a value of the test statistic equal to or greater than the actual computed value of z, if H_0 were true. That is, if z_c is the computed value of the test statistic, calculate

$$P(z \geq z_c)$$

assuming that the null hypothesis is true. This "disagreement" probability, or ***p*-value,** is calculated here for each of the computed test statistics $z = 1.82$ and $z = 5.66$, using Table B.3 in Appendix B:

$$P(z \geq 1.82) = .5 - .4656 = .0344$$

$$P(z \geq 5.66) \approx .5 - .5 = 0$$

From the discussion in part **a**, you can see that the smaller the p-value, the greater the extent of disagreement between the data and the null hypothesis—that is, the more significant the result.

In general, p-values for tests based on large samples are computed as shown in the next box. (p-Values for small-sample tests are obtained by replacing the z statistic with the t statistic.)

MEASURING THE DISAGREEMENT BETWEEN THE DATA AND H_0: p-VALUES

Upper-tailed test: p-value $= P(z \geq z_c)$
Lower-tailed test: p-value $= P(z \leq z_c)$
Two-tailed test: p-value $= 2P(z \geq |z_c|)$

where z_c is the computed value of the test statistic and $|z_c|$ denotes the absolute value of z_c (which will always be positive).

Notice that the p-value for a two-tailed test is twice the probability for the one-tailed test. This is because the disagreement between the data and H_0 can be in two directions.

Reporting Test Results as p-Values: How to Decide Whether to Reject H_0

1. Choose the maximum value of α that you are willing to tolerate.
2. If the observed significance level (p-value) of the test is less than α, then reject the null hypothesis.

When publishing the results of a statistical test of hypothesis in journals, case studies, reports, etc., most researchers use p-values. Instead of selecting α a priori and then conducting a test using a rejection region, the researcher will compute and report the value of the appropriate test statistic and its associated p-value. The use of p-values has been facilitated by the routine inclusion of p-values as part of the output of statistical software and spreadsheet packages, such as Excel and MINITAB. Given the p-value, it is left to the reader of the report to judge the significance of the result. The reader must determine whether to reject the null hypothesis in favor of the alternative, based on the reported p-value. This p-value is often referred to as the **observed significance level** of the test. The null hypothesis will be rejected if the observed significance level is *less* than the fixed significance level α chosen by the reader (see Figure 8.10).

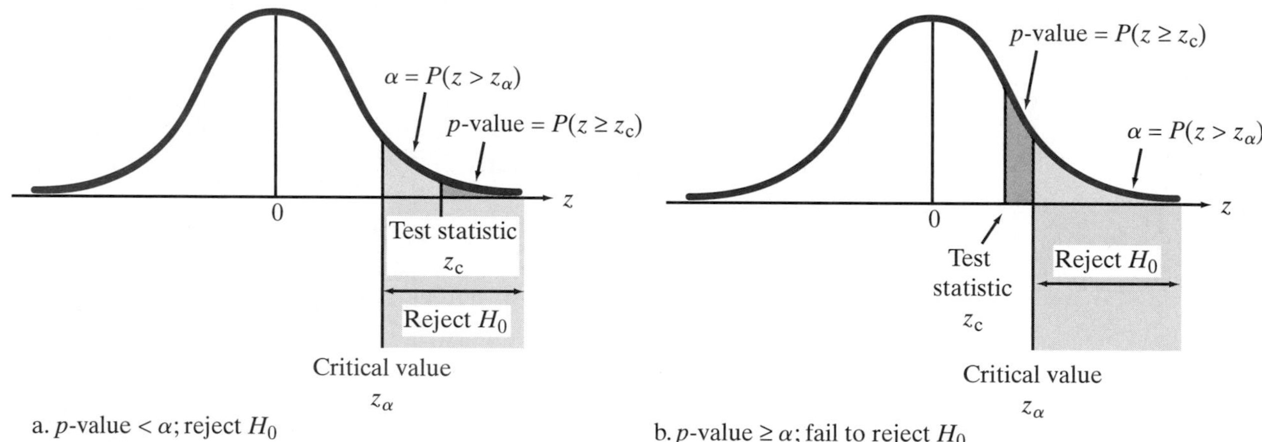

Figure 8.10 Using p-Values to Make Conclusions

There are two inherent advantages of reporting test results using p-values: (1) Readers are permitted to select the maximum value of α that they would be willing to tolerate in carrying out a standard test of hypothesis in the manner outlined in this chapter; and (2) it is an easy way to present the results of test calculations performed by a statistical software or spreadsheet package.

EXAMPLE 8.16 **p-VALUE FOR A ONE-TAILED TEST**

Refer to Example 8.15 and the test H_0: $\mu = 72$ versus H_a: $\mu > 72$. Suppose the test statistic is $z = .42$. Compute the observed significance level of the test and interpret its value.

Solution

In this large-sample test concerning a population mean μ, the computed value of the test statistic is $z_c = .42$. Since the test is upper-tailed, the associated p-value is given by

$$P(z \geq z_c) = P(z \geq .42)$$

From Table B.3, we obtain

$$P(z \geq .42) = .5 - .1628 = .3372 \quad \text{(see Figure 8.11)}$$

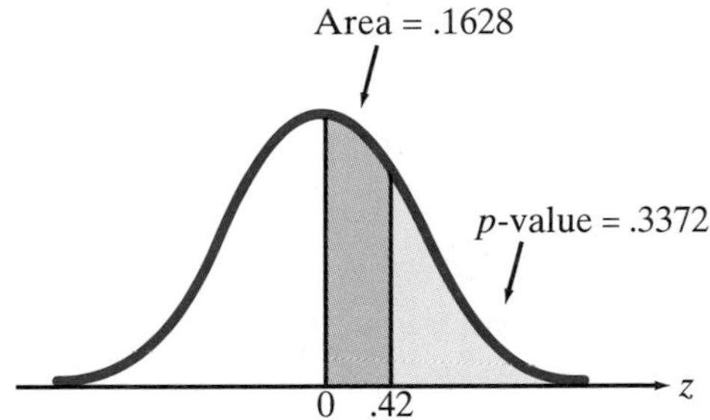

Figure 8.11 p-Value for Example 8.16

Thus, the observed significance level of the test is $p = .3372$. To reject the null hypothesis H_0: $\mu = 72$, we would have to be willing to risk a Type I error probability α of at least .3372. Most researchers would not be willing to take this risk and would deem the result insignificant (i.e., conclude that there is insufficient evidence to reject H_0).

EXAMPLE 8.17

p-VALUE FOR A TWO-TAILED TEST

Refer to Example 8.13 (p. 407) and the large-sample test of H_0: $\mu = 8.5$ versus H_a: $\mu \neq 8.5$. The Excel printout for this test was given in Figure 8.6 (p. 408).

a. Compute the observed significance level of the test. Compare to the value shown on the Excel printout.

b. Make the appropriate conclusion if you are willing to tolerate a Type I error probability of $\alpha = .05$.

Solution

a. The computed test statistic for this large-sample test about μ was given as $z_c = 4.055$. Since the test is two-tailed, the associated p-value is

$$2P(z \geq |z_c|) = 2P(z \geq |4.055|) = 2P(z \geq 4.055) \quad \text{(see Figure 8.12)}$$

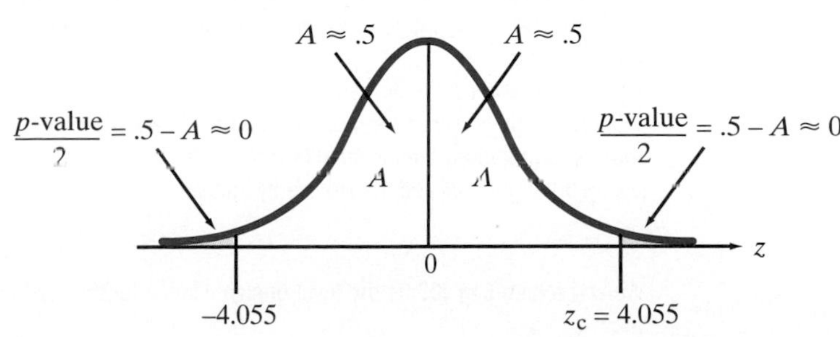

Figure 8.12 p-Value for Example 8.17

Since $P(z \geq 4.055)$ is very near 0, the observed significance level of the test is approximately 0. Note that our computed value agrees with the Excel value, $p = .00005$ (5.00929E-05, in scientific notation), highlighted on Figure 8.6b.

b. Since the p-value is less than the maximum tolerable Type I error probability of $\alpha = .01$, we will reject H_0 and conclude that the mean length-to-width ratio of humerus bones is significantly different from 8.5. In fact, we could choose an even smaller Type I error probability (e.g., $\alpha = .001$) and still have sufficient evidence to reject H_0. Thus, the result is highly significant.

Self-Test 8.7

Consider testing the null hypothesis H_0: $\mu = 2$. For each situation below, make the appropriate conclusion about H_0.

a. $\alpha = .01$, p-value $= .15$
b. $\alpha = .05$, p-value $= .02$
c. $\alpha = .10$, p-value $= .05$

Whether we conduct a test using p-values or the rejection region approach, our choice of a maximum tolerable Type I error probability becomes critical to the decision concerning H_0 and should not be hastily made. In either case, care should be taken to weigh the seriousness of committing a Type I error in the context of the problem.

Note: In this section, we demonstrated the procedure for computing p-values for a large-sample test of hypothesis based on the standard normal (z) statistic. For small samples, p-values are computed using the t distribution. However, the t table (Table B.4 in Appendix B) is not suited for this task. Consequently, it is critical to use statistical or spreadsheet software to find p-values for small-sample tests of hypotheses.

Statistics in the Real World Revisited

Testing a Population Mean in the KLEENEX® Survey

Refer to Kimberly-Clark Corporation's survey of 250 people who kept a count of their use of KLEENEX® tissues in diaries (p. 389). According to a *Wall Street Journal* report, there was "little doubt that the company should put 60 tissues" in a cold-care box of KLEENEX® tissues. This statement was based on a claim made by marketing experts that 60 is the average number of times a person will blow his/her nose during a cold.

Let μ represent the mean number of times a person actually blows his/her nose during a cold for the population of all people with colds. Suppose we disbelieve the claim that $\mu = 60$, believing instead that the population mean is smaller than 60 tissues. In order to test the claim against our belief, we set up the following null and alternative hypotheses:

$$H_0: \mu = 60 \qquad H_a: \mu < 60$$

We will select $\alpha = .05$ as the level of significance for the test.

(Continued)

The survey results for the 250 sampled KLEENEX® users are stored in the **TISSUES** data file. A MINITAB analysis of the data yielded the printout displayed in Figure 8.13.

```
Test of mu = 60.00 vs mu < 60.00

Variable        N      Mean    StDev   SE Mean       T       P
Number Used   250     56.68    25.03      1.58   -2.10   0.018
```

Figure 8.13 MINITAB Test of $\mu = 60$ for KLEENEX® Survey

The observed significance level of the test, highlighted on the printout, is p-value $= .018$. Since this p-value is less than $\alpha = .05$, we have sufficient evidence to reject H_0; therefore, we conclude that the mean number of times a person actually blows his/her nose during a cold is less than 60.

[*Note:* If we conduct the same test using $\alpha = .01$ as the level of significance, we would have insufficient evidence to reject H_0 since p-value $= .018$ is greater than $\alpha = .01$. Thus, at $\alpha = .01$, there is no evidence to support our alternative that the population mean is less than 60.]

PROBLEMS FOR SECTION 8.6

Using the Tools

8.34 For a large-sample test of

$$H_0: \mu = 0$$
$$H_a: \mu > 0$$

compute the p-value associated with each of the following computed test statistic values:

a. $z_c = 1.96$ **b.** $z_c = 1.645$
c. $z_c = 2.67$ **d.** $z_c = 1.25$

8.35 For a large-sample test of

$$H_0: p = .75$$
$$H_a: p \neq .75$$

compute the p-value associated with each of the following computed test statistic values:

a. $z_c = -1.01$ **b.** $z_c = -2.37$
c. $z_c = 4.66$ **d.** $z_c = -1.45$

8.36 Give the approximate observed significance level of the test H_0: $\mu = 16$ for each of the following combinations of test statistic value and H_a:

a. $z_c = 3.05$, H_a: $\mu \neq 16$ **b.** $z_c = -1.58$, H_a: $\mu < 16$
c. $z_c = 2.20$, H_a: $\mu > 16$ **d.** $z_c = -2.97$, H_a: $\mu \neq 16$

8.37 The p-value and the value of α for a test of H_0: $\mu = 150$ are provided for each part. Make the appropriate conclusion regarding H_0.

a. p-value $= .217$, $\alpha = .10$ **b.** p-value $= .033$, $\alpha = .05$
c. p-value $= .001$, $\alpha = .05$ **d.** p-value $= .866$, $\alpha = .01$
e. p-value $= .025$, $\alpha = .01$

Applying the Concepts

8.38 Refer to the *Psychological Assessment* study of World War II aviator POWs, Problem 8.19 (p. 403). You tested whether the true mean post-traumatic stress disorder score of World War II aviator POWs is less than 16. Recall that $\bar{x} = 9.00$ and $s = 9.32$ for a sample of $n = 33$ POWs. Find the p-value of the test and interpret the result.

8.39 Refer to the *Nature* (Aug. 27, 1998) study of whether people are more attracted to "feminized" faces, Problem 8.27 (p. 412). Recall that 50 human subjects viewed both a Japanese female and a Caucasian male face on a computer and morphed the faces until they attained the "most attractive" face. The level of feminization x (measured as a percentage) was measured. For the Japanese female face, $\bar{x} = 10.2\%$ and $s = 31.3\%$. The researchers reported the test statistic for testing the null hypothesis of a mean level of feminization equal to 0% as 2.3 with an observed significance level of p-value $= .027$. Verify and interpret these results.

8.40 Refer to Problem 8.29 (p. 413) and the test to determine whether the mean pH level of food items differs from 4.0. Locate the observed significance level (p-value) of the test on the PHStat and Excel printout shown below. Interpret the result.

	A	B
1	**t Test for Hypothesis of the Mean**	
2		
3	**Data**	
4	**Null Hypothesis** μ=	**4**
5	**Level of Significance**	**0.1**
6	**Sample Size**	**36**
7	**Sample Mean**	**5.081944444**
8	**Sample Standard Deviation**	**1.291187315**
9		
10	Intermediate Calculations	
11	Standard Error of the Mean	0.215197886
12	Degrees of Freedom	35
13	*t* Test Statistic	5.027672276
14		
15	**Two-Tailed Test**	
16	**Lower Critical Value**	**-1.689572855**
17	**Upper Critical Value**	**1.689572855**
18	***p*-Value**	**1.47561E-05**
19	**Reject the null hypothesis**	

E **Problem 8.40**

8.41 Refer to Problem 8.32 (p. 414) and the test to determine whether the average energy efficiency rating is different from 9.0. Find the approximate observed significance level (p-value) of the test. What is your decision regarding H_0 if you are willing to risk a maximum Type I error probability of only $\alpha = .01$?

8.42 Marine scientists at the University of South Florida investigated the feeding habits of midwater fish inhabiting the eastern Gulf of Mexico (*Prog. Oceanog.,* Vol. 38, 1996). Most of the fish captured fed heavily on copepods, a type of zooplankton. The table gives the percent of shallow-living copepods in the diets of each of 35 species of myctophid fish. Use the p-value approach to test the hypothesis (at $\alpha = .05$) that the true mean percentage of shallow-living copepods in the diets of all myctophid fish inhabiting the Gulf of Mexico differs from 90%.

COPEPODS

79	100	100	98	95	96	56	51	90	95	94	93
92	71	100	100	99	100	100	100	99	80	54	56
59	92	88	81	88	59	100	85	82	74	66	

Source: Hopkins, T. L., Sutton, T. T., and Lancraft, T. M. "The tropic structure and predation impact of low latitude midwater fish assemblage." *Prog. Oceanog.,* Vol. 38, No. 3, 1996, p. 223 (Table 6).

8.43 According to *USA Today* (Dec. 30, 1999), the average age of viewers of MSNBC cable television news programming is 50 years. A random sample of 50 U.S. households that receive cable television programming yielded the following additional

information about the ages of MSNBC news viewers: $\bar{x} = 51.3$ years and $s = 7.1$ years. Do the sample data provide sufficient evidence to conclude that the average age of MSNBC's viewers is greater than 50 years?

a. Conduct the appropriate hypothesis test at $\alpha = .01$.

b. Find the observed significance level (p-value) of the test and interpret its value in the context of the problem.

c. Would the p-value have been larger or smaller if $\bar{x}$ had been larger? Explain.

8.44 During the summer of 1993, Iowa received an exceptional amount of precipitation. The abundance of standing water led to an unusually high mosquito population. Entomologists at Iowa State University collected data on the mosquito population at six Iowa sites during 1993 (*Journal of the American Mosquito Control Association,* June 1995). Using light traps, mosquitos were collected daily and their species identified and counted. A light trap index (LTI), measured as the number of mosquitos per light trap per night, was measured at each location. At one site (Des Moines), the sample mean LTI in 1993 was $\bar{x} = 108.15$. The entomologists want to know whether the true 1993 mean LTI at Des Moines exceeds the mean LTI at Des Moines for the previous 10 years, $\mu = 47.82$.

a. Set up the null and alternative hypotheses for the appropriate test.

b. The p-value for the test was reported at $p = .04$. Make the appropriate conclusion at $\alpha = .05$.

c. Repeat part **b**, but use $\alpha = .01$.

8.7 Testing a Population Proportion

The procedure described in the box is used to test a hypothesis about a population proportion p, based on a *large* sample from the target population. (Recall that p represents the probability of success in a binomial experiment.)

LARGE-SAMPLE (z) TEST OF HYPOTHESIS ABOUT A POPULATION PROPORTION p

Upper-Tailed Test	**Two-Tailed Test**	**Lower-Tailed Test**
$H_0: p = p_0$	$H_0: p = p_0$	$H_0: p = p_0$
$H_a: p > p_0$	$H_a: p \neq p_0$	$H_a: p < p_0$

$$\text{Test statistic: } z = \frac{\hat{p} - p_0}{\sqrt{p_0(1 - p_0)/n}} \qquad (8.5)$$

where $\hat{p} = \dfrac{x}{n}$ = sample proportion of "successes"

Rejection region:	*Rejection region:*	*Rejection region:*
$z > z_\alpha$	$\lvert z \rvert > z_{\alpha/2}$	$z < -z_\alpha$

Assumption: The sample size n is large.

EXAMPLE 8.18 **LARGE-SAMPLE TEST OF p**

An *American Demographics* study conducted in 1980 found that 40% of new car buyers were women. Suppose that in a random sample of $n = 120$ new car buyers in 2000, 57 were women. Does this evidence indicate that the true proportion of new car buyers in 2000 who were women is significantly larger than .40, the

1980 proportion? Use the rejection region approach, testing at significance level $\alpha = .10$.

Solution

We wish to perform a large-sample test about a population proportion p.

$$H_0: p = .40 \text{ (no change from 1980 to 2000)}$$

$$H_a: p > .40 \text{ (proportion of new car buyers who were women was greater in 2000)}$$

where p represents the true proportion of all new car buyers in 2000 who were women.

At significance level $\alpha = .10$, the rejection region for this one-tailed test consists of all values of z for which

$$z > z_{.10} = 1.28 \quad \text{(see Figure 8.14)}$$

Figure 8.14 Rejection Region for Example 8.18

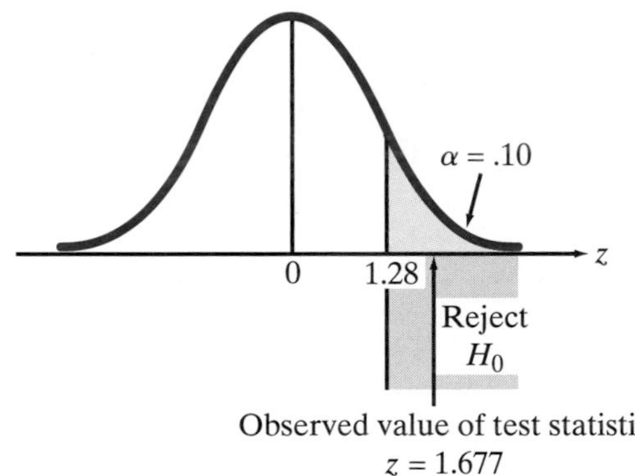

The test statistic requires the calculation of the sample proportion $\hat{p}$ of new car buyers who were women:

$$\hat{p} = \frac{\text{Number of sampled new car buyers who were women}}{\text{Number of new car buyers sampled}}$$

$$= \frac{57}{120} = .475$$

Substituting into equation 8.5, we obtain the following value of the test statistic:

$$z = \frac{\hat{p} - p_0}{\sqrt{p_0(1 - p_0)/n}} = \frac{.475 - .40}{\sqrt{(.40)(.60)/120}} = 1.677$$

This value of z lies within the rejection region; we thus conclude that the proportion of new car buyers in 2000 who were women increased significantly from .40. The probability of our having made a Type I error (rejecting H_0 when, in fact, it is true) is $\alpha = .10$.

EXAMPLE 8.19 ***p*-VALUE FOR A PROPORTION TEST**

Refer to Example 8.18. Find the observed significance level (p-value) of the test and interpret its value.

Solution

For this large-sample, upper-tailed test, the computed value of the test statistic is $z_c \approx 1.68$. Thus, the p-value is approximated by finding

$$P(z > z_c) = P(z > 1.68)$$

From Table B.3, this probability is $.5 - .4535 = .0465$. (The exact p-value for this test, .0468, is shown on the PHStat and Excel printout in Figure 8.15.

Figure 8.15 PHStat and Excel Output for Example 8.19

	A	B
1	**Z Test of Hypothesis for the Proportion**	
2		
3	**Data**	
4	**Null Hypothesis** $p=$	**0.4**
5	**Level of Significance**	**0.1**
6	**Number of Successes**	**57**
7	**Sample Size**	**120**
8		
9	Intermediate Calculations	
10	Sample Proportion	0.475
11	Standard Error	0.04472136
12	Z Test Statistic	1.677050983
13		
14	**Upper-Tail Test**	
15	**Upper Critical Value**	**1.281550794**
16	***p*-Value**	**0.046766232**
17	**Reject the null hypothesis**	

Remember, p-value $= .0465$ is the minimum α value that leads to a rejection of the null hypothesis. Consequently, we will reject H_0 for any α value exceeding .0465 (as in Example 8.18) but will fail to reject H_0 for α values below .0465. Therefore, a researcher who desires a Type I error rate of only $\alpha = .01$ or $\alpha = .02$ will have insufficient evidence to say that the proportion of new car buyers in 2000 who were women exceeds .40. However, $\alpha = .10$ (as in Example 8.18) or $\alpha = .05$ will lead the researcher to reject H_0 and claim that the proportion of new car buyers who are women has increased.

Self-Test 8.8

In a nationwide telephone poll, 1,532 adults were asked: "Do you believe that people who become victims of muggings or holdups have a right to take matters into their own hands?" Forty-seven percent answered "yes." Test the hypothesis that fewer than 50% of American adults believe that crime victims have a right to take matters into their own hands. Use a significance level of $\alpha = .10$.

Although small-sample procedures are available for testing hypotheses about a population proportion, the details are omitted from our discussion. It is our experience that they are of limited utility, since most surveys of binomial populations (for example, opinion polls) performed in the real world use samples that are large enough to employ the techniques of this section.

Statistics in the Real World Revisited

Testing a Population Proportion in the KLEENEX® Survey

In the previous Statistics in the Real World Revisited section (p. 418), we investigated Kimberly-Clark Corporation's assertion that the company should put 60 tissues in a cold-care box of KLEENEX® tissues. We did this by testing the claim that the mean number of times a person blows his/her nose during a cold is $\mu = 60$ using data collected from a survey of 250 KLEENEX® users. Another approach to the problem is to consider the proportion of KLEENEX® users who use fewer than 60 tissues when they have a cold. Now, the population parameter of interest is p, the proportion of all KLEENEX® users who use fewer than 60 tissues when they have a cold.

Kimberly-Clark Corporation's belief that the company should put 60 tissues in a cold-care box will be supported if half of the KLEENEX® users surveyed use less than 60 tissues and half use more than 60 tissues, i.e., if $p = .5$. Is there evidence to indicate that the population proportion differs from .5? To answer this question, we set up the following null and alternative hypotheses:

$$H_0: p = .5 \qquad H_a: p \neq .5$$

Recall that the survey results for the 250 sampled KLEENEX® users are stored in the **TISSUES** data file. In addition to the number of tissues used by each person, the file contains a qualitative variable representing whether or not the person used fewer than 60 tissues. A MINITAB analysis of this variable yielded the printout displayed in Figure 8.16.

```
Test of p = 0.5 vs p not = 0.5

Success = BELOW

Variable        X     N  Sample p         95.0 % CI          Z-Value  P-Value
USE60         143   250  0.572000  (0.510666, 0.633334)        2.28    0.023
```

Figure 8.16 MINITAB Test of $p = .5$ for KLEENEX® Survey

On the MINITAB printout, X represents the number of the 250 people with colds that used less than 60 tissues. Note that X = 143. This value is used to compute the test statistic, $z = 2.28$, highlighted on the printout. The p-value of the test, also highlighted on the printout, is p-value = .023. Since this value is less than $\alpha = .05$, there is sufficient evidence (at $\alpha = .05$) to reject H_0; we conclude that the proportion of all KLEENEX® users who use fewer than 60 tissues when they have a cold differs from .5. However, if we test at $\alpha = .01$, there is insufficient evidence to reject H_0. Consequently, our choice of α (as in the previous Statistics in the Real World Revisited section) is critical to our decision.

PROBLEMS FOR SECTION 8.7

Using the Tools

8.45 A random sample of n observations is selected from a binomial population to test the null hypothesis that $p = .40$. Specify the rejection region for each of the following combinations of H_a and α:

a. $H_a: p \neq .40, \alpha = .05$ **b.** $H_a: p < .40, \alpha = .05$
c. $H_a: p > .40, \alpha = .10$ **d.** $H_a: p < .40, \alpha = .01$
e. $H_a: p \neq .40, \alpha = .01$

8.46 A random sample of n observations is selected from a binomial population. For each of the following situations, specify the rejection region, test statistic value, and conclusion:

a. $H_0: p = .10, H_a: p > .10, p = .13, n = 200, \alpha = .10$
b. $H_0: p = .05, H_a: p < .05, p = .04, n = 1{,}124, \alpha = .05$
c. $H_0: p = .90, H_a: p \neq .90, p = .73, n = 125, \alpha = .01$

8.47 A random sample of 100 observations from a binomial population resulted in 45 successes.

a. Test the null hypothesis that the true proportion of successes in the population is .5 against the alternative hypothesis that $p < .5$. Use $\alpha = .05$.

b. Test H_0: $p = .35$ against H_a: $p \neq .35$ at $\alpha = .05$.

c. Test H_0: $p = .4$ against H_a: $p > .4$ at $\alpha = .05$.

8.48 Consider a test of H_0: $p = .75$ and let x = number of successes in a sample of size n. Compute the test statistic for each of the following:

a. $x = 100, n = 500$

b. $x = 125, n = 200$

c. $x = 1{,}122, n = 1{,}500$

d. $x = 31, n = 50$

Applying the Concepts

8.49 An article in *The Wall Street Journal* (July 25, 2000) implies that more than half of all Americans would prefer being given \$100 than be given a day off work. This statement is made based on a survey conducted by American Express Incentive Services, where 593 of 1,040 Americans indicated that they would rather have the \$100.

a. At $\alpha = .05$, is there evidence based on the survey data that more than half of all Americans would rather have \$100 than a day off?

b. Find the p-value and interpret its meaning.

8.50 According to a survey sponsored by the American Society for Microbiology, 50% of the people using the restrooms in New York City's Penn Station do not wash their hands after going to the bathroom (Associated Press, Sept. 19, 2000). Suppose you observe 427 people who fail to wash their hands in a random sample of 1,000 people who use the Penn Station restrooms. Use this information to test the claim made by the American Society for Microbiology. Test using $\alpha = .05$.

8.51 *Jeopardy!* is a popular television game show in which three contestants answer general knowledge questions on a wide variety of topics and earn money for each correct answer. The contestant with the highest total amount wins, thereby earning the right to defend his/her championship the next day. *Chance* (Spring 1994) collected data for a sample of 218 *Jeopardy!* games. In these 218 games, the defending champion won 126 times. Does this imply that the defending *Jeopardy!* champion has no better or no worse than a 50-50 chance of winning the next day? Test using $\alpha = .01$.

8.52 The *New England Journal of Medicine* (Oct. 6, 1994) published a study on the link between guns in the home and homicides. Studying 420 homicides that occurred in the homes of victims, the researchers found that 322 were killed by a spouse, family member, or acquaintance. Is there sufficient evidence (at $\alpha = .05$) to claim that a spouse, family member, or acquaintance is the killer in more than 70% of all homicides that occur in the victim's home?

8.53 According to a survey reported in the *Journal of the American Medical Association* (Apr. 22, 1998), only 608 of 1,066 randomly selected Americans recognized the warning signs of a stroke.

a. Test the hypothesis that the true percentage of Americans who recognize the warning signs of a stroke exceeds 50%. Use $\alpha = .05$.

b. Test the hypothesis that the true percentage of Americans who recognize the warning signs of a stroke exceeds 55%. Use the p-value method and $\alpha = .05$.

8.54 What percentage of Web users are addicted to the Internet? Researchers at ABC News modified a questionnaire developed for gambling addiction and made it available online at *ABCNEWS.com*. Of a total of 17,251 responses to the questionnaire, they determined that 990 users were addicted to the Internet (Associated Press, Aug. 23, 1999).

a. Identify the target population and parameter of interest for this study.

b. Assuming the sample was randomly selected from the population, conduct a test to determine if the population percentage of users addicted to the Internet exceeds 5%. Test using $\alpha = .01$.

c. Comment on the validity of the test based on the method of collecting the data.

8.55 Archaeologists have found a paucity of antique grinding implements (e.g., grindstones, seed grinders) in Australia. Past excavations have indicated that 2.5% of all artifacts found in Australia are grinding implements. A recent archaeological dig at Wanmara in Central Australia found a total of 1,436 artifacts. Of these, 102 were identified as grinding implements (*Archaeol. Oceania,* July 1997). Conduct a test to determine whether the true percentage of artifacts found at Wanmara that are grinding implements differs from 2.5%. Use $\alpha = .01$.

8.56 According to a study in *Nature* (Dec. 1995), 1-year-old babies are more likely to resemble their dads in physical appearance. The researchers asked a sample of 122 people to compare photos of 1-year-old babies with photos of three possible parents (the dad, the mom, and a stranger) and to select the "parent" that most closely resembles the baby. For one particular 1-year-old baby, suppose 70 of the 122 people chose the dad's photo as most similar in appearance.

a. Conduct a test to determine whether the true proportion of people who believe the dad most resembles the baby exceeds .3333. Use $\alpha = .01$.

b. Conduct a test to determine whether the true proportion of people who believe the dad most resembles the baby exceeds .5. Use $\alpha = .01$.

8.8 Testing a Population Variance (Optional)

Hypothesis tests about a population variance σ^2 are conducted using the chi-square (χ^2) distribution introduced in optional Section 7.9. The test is outlined in the box. Note that the assumption of a normal population is required regardless of whether the sample size n is large or small.

TEST OF HYPOTHESIS ABOUT A POPULATION VARIANCE σ^2

Upper-Tailed Test	Two-Tailed Test	Lower-Tailed Test
$H_0: \sigma^2 = \sigma_0^2$	$H_0: \sigma^2 = \sigma_0^2$	$H_0: \sigma^2 = \sigma_0^2$
$H_a: \sigma^2 > \sigma_0^2$	$H_a: \sigma^2 \neq \sigma_0^2$	$H_a: \sigma^2 < \sigma_0^2$

$$\textit{Test statistic: } \chi^2 = \frac{(n-1)s^2}{\sigma_0^2} \qquad \textbf{(8.6)}$$

Rejection region:	*Rejection region:*	*Rejection region:*
$\chi^2 > \chi^2_\alpha$	$\chi^2 < \chi^2_{(1-\alpha/2)}$ or $\chi^2 > \chi^2_{\alpha/2}$	$\chi^2 < \chi^2_{(1-\alpha)}$

where χ^2_α and $\chi^2_{(1-\alpha)}$ are values of χ^2 that locate an area of α to the right and α to the left, respectively, of a chi-square distribution based on $(n-1)$ degrees of freedom.

Assumption: The population from which the random sample is selected has an approximate normal distribution.

EXAMPLE 8.20 **TEST FOR σ^2**

Regulatory agencies specify that the standard deviation of the amount of fill for 8-ounce cans of chicken noodle soup produced in a cannery should be less than .1 ounce. To check that specifications are being met, the quality control supervisor

at the cannery sampled ten 8-ounce cans and measured the amount of fill in each. The data (in ounces) are shown here.

CANS

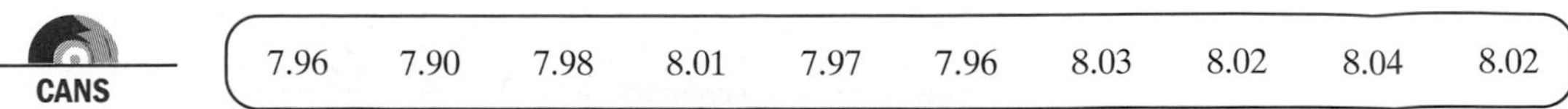

7.96	7.90	7.98	8.01	7.97	7.96	8.03	8.02	8.04	8.02

Is there sufficient evidence to indicate that the standard deviation σ of the fill measurements is less than .1 ounce?

a. Conduct the test using the rejection region approach.
b. Conduct the test using the p-value approach.

Solution

a. Since the null and alternative hypotheses are stated in terms of σ^2 (rather than σ), we will want to test the null hypothesis that $\sigma^2 = .01$ against the alternative that $\sigma^2 < .01$. Therefore, the elements of the test are

$$H_0\text{: } \sigma^2 = .01 \; (\sigma = .1)$$
$$H_a\text{: } \sigma^2 < .01 \; (\sigma < .1)$$

Assumption: The population of amounts of fill of the cans is approximately normal.

Test statistic:

$$\chi^2 = \frac{(n-1)s^2}{\sigma_0^2}$$

Rejection region:

The smaller the value of s^2 we observe, the stronger the evidence in favor of H_a. Thus, we reject H_0 for "small values" of the test statistic. With $\alpha = .05$ and $(n-1) = 9$ df, the critical χ^2 value is found in Table B.5 in Appendix B and illustrated in Figure 8.17. We will reject H_0 if $\chi^2 < 3.32511$. (Remember that the area given in Table B.5 is the area to the *right* of the numerical value in the table. Thus, to determine the lower-tail value that has $\alpha = .05$ to its *left*, we use the $\chi^2_{.95}$ column in Table B.5.)

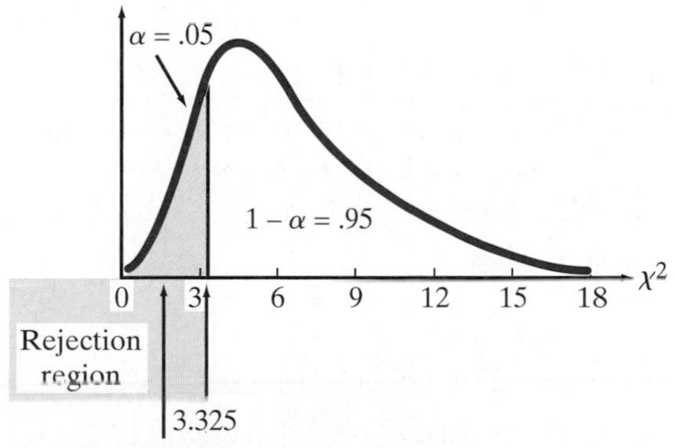

Figure 8.17 Rejection Region for Example 8.20

E Figure 8.18 Excel and PHStat Output for Example 8.20

	A	B
1	*Amount of fill in ounces*	
2		
3	Mean	7.989
4	Standard Error	0.013618
5	Median	7.995
6	Mode	7.96
7	Standard Deviation	0.043063
8	Sample Variance	0.001854
9	Kurtosis	0.479371
10	Skewness	-0.8538
11	Range	0.14
12	Minimum	7.9
13	Maximum	8.04
14	Sum	79.89
15	Count	10
16	Largest(1)	8.04
17	Smallest(1)	7.9

a.

	A	B
1	Chi-Square Test of Hypothesis for the Variance	
2		
3	Null Hypothesis σ^2=	0.01
4	Level of significance	0.05
5	Sample size	10
6	Sample standard deviation	0.043
7	Degrees of freedom	9
8	Chi-Square Test Statistic	1.6641
9		
10	Lower tailed test	
11	LowerCritical Value	3.32511514
12	p-value	0.004263521
13	Reject the null hypothesis	

b.

To compute the test statistic, we need to find the sample standard deviation s. Numerical descriptive statistics for the sample data are provided in the Excel printout, Figure 8.18a. The value of s, highlighted in Figure 8.18a, is $s = .043$. Substituting $s = .043$, $n = 10$, and $\sigma_0^2 = .01$ into equation 8.6, we obtain the test statistic

$$\chi^2 = \frac{(10 - 1)(.043)^2}{.01} = 1.66$$

Conclusion: Since the test statistic $\chi^2 = 1.66$ is less than 3.32511, the supervisor can reject H_0 and conclude that the standard deviation σ of the population of all amounts of fill is less than .1 at $\sigma = .05$. Thus, the quality control supervisor is confident in the decision that the cannery is operating within the desired limits of variability.

b. The p-value for the test is shown on the PHStat and Excel printout, Figure 8.18b. Its value (highlighted) is $p \approx .0043$. Since the p-value is less than $\alpha = .05$, we reject H_0 and conclude (as in part **a**) that $\sigma < .1$.

Self-Test 8.9

A random sample of $n = 10$ observations yielded $\bar{x} = 231.7$ and $s^2 = 15.5$. Test the null hypothesis H_0: $\sigma^2 = 20$ against the alternative hypothesis H_a: $\sigma^2 < 20$. Use $\alpha = .05$. What assumptions are necessary for the test to be valid?

PROBLEMS FOR SECTION 8.8

Using the Tools

8.57 Calculate the test statistic for H_0: $\sigma^2 = \sigma_0^2$ in a normal population for each of the following:

a. $n = 25, \bar{x} = 17, s^2 = 8, \sigma_0^2 = 10$

b. $n = 12, \bar{x} = 6.2, s^2 = 1.7, \sigma_0^2 = 1$

c. $n = 50, \bar{x} = 106, s^2 = 31, \sigma_0^2 = 50$

d. $n = 100, \bar{x} = .35, s^2 = .04, \sigma_0^2 = .01$

8.58 A random sample of n observations, selected from a normal population, is used to test the null hypothesis H_0: $\sigma^2 = 9$. Specify the appropriate rejection region for each of the following:

a. H_a: $\sigma^2 > 9, n = 20, \alpha = .01$ **b.** H_a: $\sigma^2 \neq 9, n = 20, \alpha = .01$

c. H_a: $\sigma^2 < 9, n = 12, \alpha = .05$ **d.** H_a: $\sigma^2 < 9, n = 12, \alpha = .10$

8.59 The following measurements represent a random sample of $n = 5$ observations from a normal population: 11, 7, 2, 9, 13. Is this sufficient evidence to conclude that $\sigma^2 \neq 2$? Test using $\alpha = .10$.

8.60 A random sample of $n = 35$ observations yielded $\bar{x} = .107$ and $s = .022$. Test the null hypothesis H_0: $\sigma = .04$ against the alternative hypothesis H_a: $\sigma \neq .04$. Use $\alpha = .10$. What assumptions are necessary for the test to be valid?

Applying the Concepts

8.61 By law, the levels of toxic organic compounds in fish are constantly monitored. A study of a new method for extracting toxic organic compounds was published in *Chromatographia* (Mar. 1995). Uncontaminated fish fillets were injected with a known amount of a toxic substance and the method was used to extract the toxicant. To test the precision of the new method, seven measurements (percentage of toxic compound recovered) were obtained on a single fish fillet injected with the toxin. Summary statistics for the recovery percentages are as follows: $\bar{x} = 99$, $s^2 = 81$. Determine whether the variance of the recovery percentages for the new method is less than 100. (Use $\alpha = .10$.)

8.62 Polychlorinated biphenyls (PCBs) are extremely hazardous contaminants when released into the environment. The Environmental Protection Agency (EPA) is experimenting with a new device for measuring PCB concentration in fish. To check the precision of the new instrument, seven PCB readings were taken on the same fish sample. The data are recorded here (in parts per million):

PCB

6.2	5.8	5.7	6.3	5.9	5.8	6.0

Suppose the EPA requires an instrument that yields PCB readings with a variance of less than .1. Does the new instrument meet the EPA's specifications? Test using $\alpha = .05$.

8.63 A candy manufacturer must monitor the temperature at which candies are baked. Too much variation will cause inconsistency in the taste of the candy. When the process is operating correctly, the standard deviation of the baking temperature is 1.3°F. A random sample of 50 batches of candy is selected and the sample standard deviation of the baking temperature is $s = 2.1$°F. Is there evidence that the true standard deviation of the baking temperature has increased above 1.3°F? Test using $\alpha = .05$.

8.64 Refer to the *Journal of Drug Issues* (Summer 1997) study of pregnant cocaine-dependent women, Problem 8.31 (p. 414). Test (at $\alpha = .01$) to determine whether the variance in the birthweights of babies delivered by cocaine-dependent women is less than 200,000 grams^2. Recall that for a sample of $n = 16$ cocaine-dependent women, the babies delivered had a mean birthweight of $\bar{x} = 2{,}971$ grams and a standard deviation of $s = 410$ grams.

8.65 Refer to the *Chance* (Fall 1998) study of point-spread errors in NFL games, Problem 8.33 (p. 414). Recall that the difference between the actual game outcome and the point spread established by oddsmakers—the point-spread error—was calculated for 240 NFL games. The results are summarized as follows: $\bar{x} = -1.6$, $s = 13.3$. Suppose the researcher wants to know whether the true standard deviation of the point-spread errors differs from 15.

a. Set up H_0 and H_a for the researcher.

b. Compute the value of the test statistic.

c. Conduct the test at $\alpha = .10$. Interpret the results in the words of the problem.

8.66 A company that produces a fast-drying rubber cement in 32-ounce aluminum cans is interested in testing whether the variance of the amount of rubber cement dispensed into the cans is more than .3. If so, the dispensing machine needs adjustment. Since inspection of the canning process requires that the dispensing machines be shut down, and shutdowns for any lengthy period of time cost the company thousands of dollars in lost revenue, the company is able to obtain a random sample of only 10 cans for testing. After measuring the weights of their contents, the following summary statistics are obtained:

$$\bar{x} = 31.55 \text{ ounces} \qquad s = .48 \text{ ounce}$$

a. Does the sample evidence indicate that the dispensing machines are in need of adjustment? Test at significance level $\alpha = .05$.

b. What assumption is necessary for the hypothesis test of part **a** to be valid?

8.67 Following the Persian Gulf War of 1991, the Pentagon changed its delivery system to a "just-in-time" system. Deliveries from factories to foxholes are now expedited using bar codes, laser cards, radio tags, and databases to track supplies (*Business Week,* Dec. 11, 1995). During the Persian Gulf War, the Pentagon found the standard deviation of the order-to-delivery times for all shipments to be $\sigma = 10.5$ days. To gauge whether the standard deviation has decreased under the new system, the Pentagon randomly sampled nine shipments made during the 1995 Bosnian War. The order-to-delivery times (in days) for these nine shipments are listed below:

BOSNIA

15.1	6.4	5.0	11.4	6.5	6.5	3.0	7.0	5.5

Conduct a test to determine whether the standard deviation of the order-to-delivery times during the Bosnian War is less than 10.5 days. Use $\alpha = .05$.

8.68 To improve the signal-to-noise ratio (SNR) in the electrical activity of the brain, neurologists repeatedly stimulate subjects and average the responses—a procedure that assumes that single responses are homogeneous. A study was conducted to test the homogeneous signal theory (*IEEE Engineering in Medicine and Biology Magazine,* Mar. 1990). The null hypothesis is that the variance of the SNR readings of subjects equals the "expected" level under the homogeneous signal theory. For this study, the "expected" level was assumed to be .54. If the SNR variance exceeds this level, the researchers will conclude that the signals are nonhomogeneous.

a. Set up the null and alternative hypotheses for the researchers.

b. SNRs recorded for a sample of 41 normal children ranged from .03 to 3.0. Use this information to obtain an estimate of the sample standard deviation. [*Hint:* Assume that the distribution of SNRs is normal, and that most SNRs in the population will fall within $\mu \pm 2\sigma$, i.e., from $\mu - 2\sigma$ to $\mu + 2\sigma$. Note that the range of the interval equals 4σ.]

c. Use the estimate of s in part **b** to conduct the test of part **a**. Test using $\alpha = .10$.

8.9 Testing Categorical Probabilities for a Qualitative Variable (Optional)

In Section 8.7, we developed a hypothesis test for p, the population proportion of successes in a binomial experiment. Recall from Chapter 5 that a binomial experiment involves a single qualitative variable that results in one of two categorical outcomes (success or failure). In this optional section, we consider data collected on a qualitative variable with more than 2 (say, k) possible outcomes. Typically, we want to make inferences about the true proportions that occur in the k categories based on sample information.

For example, when you fill out your income tax return for the Internal Revenue Service, you must select your Filing Status. Filing Status is a qualitative variable with five possible outcomes: Single, Married filing joint return, Married filing separate return, Head of household with qualifying person, and Qualifying widow(er). Suppose we record the filing status for each in a random sample of 50 taxpayers and count how many taxpayers fall in each category. Such an experiment satisfies the properties of a **multinomial experiment** (as opposed to a binomial experiment).

PROPERTIES OF THE MULTINOMIAL EXPERIMENT

1. The experiment consists of n identical trials.
2. There are k possible outcomes to each trial.
3. The probabilities of the k outcomes, denoted by $p_1, p_2, \ldots, p_k$, remain the same from trial to trial, where $p_1 + p_2 + \cdots + p_k = 1$.
4. The trials are independent.
5. The random variables of interest are the counts $n_1, n_2, \ldots, n_k$ in each of the k categories.

In this section we demonstrate how to conduct a test of hypothesis about the category proportions, $p_1, p_2, \ldots, p_k$, in a multinomial experiment. The appropriate test is based on the deviations of the **observed category counts**, $n_1, n_2, \ldots, n_k$, from their **expected** (*or mean*) **category counts**.

EXAMPLE 8.21

FINDING EXPECTED CATEGORY COUNTS

Consider a study of aphasia published in the *Journal of Communication Disorders* (Mar. 1995). Aphasia is the "impairment or loss of the faculty of using or understanding spoken or written language." Three types of aphasia have been identified by researchers: Broca's, conduction, and anomic. The researchers want to determine whether one type of aphasia occurs more often than any other. Consequently, they measured aphasia type for a sample of 22 adult aphasiacs. If, in fact, there is no difference in the proportions of aphasiacs in the three categories, how many of the 22 sampled patients would you expect of each type?

Solution

For this study, the qualitative variable of interest is Aphasia type with three possible outcomes (categories): Broca's, conduction, and anomic. Define p_1, p_2, and p_3 as the population proportions of aphasiacs falling in each of the three categories, respectively. If the proportions are no different (and since they must sum to 1), then $p_1 = p_2 = p_3 = 1/3$. To determine the number of sampled aphasiacs we expect to fall in each type, we employ the formula given in the box on p. 432. Substituting, we have

Expected number in Broca's:	$np_1 = 22(1/3) = 7.33$
Expected number in Conduction:	$np_2 = 22(1/3) = 7.33$
Expected number in Anomic:	$np_3 = 22(1/3) = 7.33$

Therefore, we expect to observe about 7.33 patients of each aphasia type in the sample of 22 aphasiacs, if in fact, the population proportions are equal.

CALCULATING EXPECTED CATEGORY COUNTS IN A MULTINOMIAL EXPERIMENT

The expected number of observations falling in a particular category is obtained by multiplying the total sample size by the hypothesized probability that an observation will fall in that category, i.e.,

$$E_i = np_i \quad \textbf{(8.7)}$$

where

E_i = Expected count for category i
n = Sample size
p_i = Hypothesized probability that an observation will fall in category i

Self-Test 8.10

There are four candidates for an elected political postion. If the voters have no overall preference for any one of the candidates, how many votes do you expect each candidate to get in a sample of 200 voters?

EXAMPLE 8.22

TESTING CATEGORY PROBABILITIES

Refer to Example 8.21. The results of the researchers' study are summarized in Table 8.4. Note that the table gives the total number of aphasiacs found of each type in the sample of 22 adult aphasiacs. Use the sample information to test the hypothesis that the types of aphasia have the same probability of occurring. Test using $\alpha = .05$.

TABLE 8.4 Summary of Data on 22 Adult Aphasiacs

Type of Aphasia	Number of Aphasiacs
Broca's	5
Conduction	7
Anomic	10
Total	22

Source: Li, E. C., Williams, S. E., and Volpe, A. D., "The effects of topic and listener familiarity of discourse variables in procedural and narrative discourse tasks." *Journal of Communication Disorders,* Vol. 28, No. 1, Mar. 1995, p. 44 (Table 1).

Solution

We want to test the null hypothesis that the category probabilities (or population proportions) are equal (that is, $p_1 = p_2 = p_3 = 1/3$) against the alternative hypothesis that one type of aphasia occurs more often than another type (that is, at least one of the probabilities p_1, p_2, and p_3 exceeds 1/3). Thus, we want to test

$$H_0\text{: } p_1 = p_2 = p_3 = 1/3$$

$$H_a\text{: At least one of the proportions exceeds } 1/3$$

The details for carrying out this test are shown in the box on p. 433. Note that the test statistic compares the observed category counts (listed in Table 8.4) to their respective expected counts (calculated in Example 8.21). Large deviations between the observed and expected category counts would provide evidence to indicate

that the category probabilities specified in the null hypothesis are incorrect. This test, often called a **goodness-of-fit test,** is based on the chi-square distribution. We calculate the test statistic and specify the rejection region as follows:

$$\textit{Test statistic: } \chi^2 = \frac{(5 - 7.33)^2}{7.33} + \frac{(7 - 7.33)^2}{7.33} + \frac{(10 - 7.33)^2}{7.33} = 1.73$$

Rejection region: At $\alpha = .05$, reject H_0 if $\chi^2 > \chi^2_{.05} = 5.99147$ (where $\chi^2_{.05}$ is based on $k - 1 = 2$ degrees of freedom and obtained from Table B.5).

Since the value of the test statistic does not fall in the rejection region, we cannot reject H_0; the sample data do not provide sufficient evidence (at $\alpha = .05$) to say that at least one type of aphasia has a population proportion that exceeds 1/3. Thus, the researchers cannot conclude that one type of aphasia occurs more often than any other.

[*Note:* The chi-square test for a multinomial experiment can be conducted using Microsoft Excel. The Excel printout for the data of Table 8.4 is shown in Figure 8.19. Both the test statistic value and p-value of the test are highlighted on the printout. Since $\alpha = .05$ is less than p-value $= .422$, there is insufficient evidence to reject H_0.]

E Figure 8.19 Excel Output for Example 8.22

	A	B	C	D	E
1	Test of Hypothesis in a Multinomial Experiment with k Outcomes				
2					
3	Category	Observed	Expected		
4	Broca's	5	7.333333		
5	Conduction	7	7.333333		
6	Anomic	10	7.333333		
7	Total	22			
8					
9	Level of Significance	0.05			
10	Number of categories	3			
11	Degrees of freedom	2			
12	Critical Value	5.991476			
13	Chi-Square test statistic	1.727272			
14	p-value	0.421626			
15	Do not reject the null hypothesis				

A TEST OF HYPOTHESIS ABOUT CATEGORY PROBABILITIES IN A MULTINOMIAL EXPERIMENT WITH k OUTCOMES

$$H_0: p_1 = p_{1,0}, p_2 = p_{2,0}, \ldots, p_k = p_{k,0}$$

where $p_{1,0}, p_{2,0}, \ldots, p_{k,0}$ represent hypothesized probabilities that sum to 1

H_a: At least 1 of the category probabilities does not equal its hypothesized value

$$\textit{Test statistic: } \chi^2 = \sum_{i=1}^{k} \frac{(n_i - E_i)^2}{E_i} \quad \textbf{(8.8)}$$

where n_i is the observed count and $E_i = np_{i,0}$ is the expected count for category i, assuming H_0 is true

Rejection region: $\chi^2 > \chi^2_\alpha$ (where χ^2_α is based on $(k - 1)$ df)

Assumptions:

1. A multinomial experiment has been conducted. This is generally satisfied by taking a random sample from the population of interest.
2. The sample size n will be large enough so that for every category, the expected category count E_i will be equal to 5 or more.*

*The assumption that all expected category counts are at least 5 is necessary in order to ensure that the χ^2 approximation is appropriate. Exact methods for conducting the test of a hypothesis exist and may be used for small expected cell counts, but these methods are beyond the scope of this text.

PROBLEMS FOR SECTION 8.9

Using the Tools

8.69 Find the rejection region for a χ^2 test of category probabilities in a multinomial experiment with k outcomes if:

a. $k = 3; \alpha = .05$

b. $k = 5; \alpha = .10$

c. $k = 4; \alpha = .01$

d. $k = 8; \alpha = .01$

8.70 Find the expected category counts for a χ^2 test of category probabilities in a multinomial experiment with k outcomes if:

a. $k = 5; H_0: p_1 = p_2 = p_3 = p_4 = p_5; n = 200$

b. $k = 4; H_0: p_1 = .2, p_2 = .3, p_3 = .1, p_4 = .1, p_5 = .3; n = 200$

c. $k = 3; H_0: p_1 = .4, p_2 = .3, p_3 = .3; n = 1{,}000$

d. $k = 6; H_0: p_1 = p_2 = p_3 = p_4 = p_5 = p_6; n = 300$

8.71 A random sample of $n = 100$ observations were classified into $k = 4$ categories as shown in the accompanying table. Consider a test of the null hypothesis $H_0: p_1 = .1$, $p_2 = .3, p_3 = .3, p_4 = .3$.

Category	1	2	3	4	Total
Number	14	35	27	24	100

a. Calculate the expected category counts.

b. Find the rejection region for $\alpha = .10$.

c. State the alternative hypothesis for the test.

d. Calculate the test statistic.

e. Make the appropriate conclusion.

8.72 A random sample of $n = 500$ observations were classified into $k = 5$ categories as shown in the accompanying table. Consider a test of the null hypothesis $H_0: p_1 = p_2 = p_3 = p_4 = p_5$.

Category	1	2	3	4	5
Number	27	62	241	69	101

a. Calculate the expected category counts.

b. Find the rejection region for $\alpha = .05$

c. State the alternative hypothesis for the test.

d. Calculate the test statistic.

e. Make the appropriate conclusion.

Applying the Concepts

8.73 *Bon Appetit* magazine polled 200 of its readers concerning which of four vegetables—brussel sprouts, okra, lima beans, and cauliflower—is their least favorite. The results (adapted from *Adweek,* Feb. 21, 2000) are presented in the table. Let p_1, p_2, p_3, and p_4 represent the proportions of all *Bon Appetit* readers who indicate brussel sprouts, okra, lima beans, and cauliflower, respectively, as their least favorite vegetable.

Brussel Sprouts	Okra	Lima Beans	Cauliflower
46	76	44	34

a. If, in general, *Bon Appetit* readers do not have a preference for their least favorite vegetable, what are the values of p_1, p_2, p_3, and p_4?

b. Specify the null and alternative hypotheses to determine whether *Bon Appetit* readers have a preference for one of the vegetables as "least favorite."

c. Conduct the test you described in part **b** using $\alpha = .05$. Report your conclusion in the context of the problem.

d. What assumptions must hold to ensure the validity of the test you conducted in part **c**? Which, if any, of these assumptions may be a concern in this application?

8.74 M&M's plain chocolate candies come in six different colors: brown, yellow, red, orange, green, and blue. According to the manufacturer (Mars, Inc.), the color ratio in each large production batch is 30% brown, 20% yellow, 20% red, 10% orange, 10% green, and 10% blue. To test this claim, a professor at Carleton College (Minnesota) had students count the colors of M&M's found in "fun size" bags of the candy (*Teaching Statistics,* Spring 1993). The results for 370 M&M's are displayed in the table.

Color	brown	yellow	red	orange	green	blue
# M&M's	84	79	75	49	36	47

Source: Johnson, R. W. "Testing color proportions of M&M's." *Teaching Statistics,* Vol. 15, No. 1, Spring 1993, p. 2 (adapted from Table 1).

a. Assuming the manufacturer's stated percentages are accurate, calculate the expected numbers falling into the six categories.

b. Calculate the value of χ^2.

c. Conduct a test to determine whether the true percentages of the colors produced differ from the manufacturer's stated percentages. Use $\alpha = .05$.

8.75 The Center for Economic and Development Research (CEDR) at the University of South Florida periodically conducts a "Business Retention Monitor." The monitor is based on a survey of industries in the Tampa Bay (Florida) area. In one survey, a sample of 83 industries responded to a question concerning their energy consumption as compared to a year ago. The responses are summarized in the table. Determine whether the percentages of industries falling into the three response categories are significantly different. Test using $\alpha = .10$.

Change in Energy Consumption	Number
Slight expansion	12
Same as last year	65
Slight contraction	6
Total	83

8.76 Interferons are proteins produced naturally by the human body that help fight viral infections and regulate the immune system. A drug developed from interferons is available for treating patients with multiple sclerosis (MS). In a clinical study, 85 MS patients received weekly injections of the drug over a 2-year period (Biogen, Inc., 1997). The number of exacerbations (i.e., flare-ups of symptoms) was recorded for each patient and is summarized in the table on p. 436. For MS patients who take a placebo (no drug) over a similar 2-year period, it is known from previous studies that 26% will experience no exacerbations, 30% one exacerbation, 11% two exacerbations, 14% three exacerbations, and 19% four or more exacerbations. Conduct a test to determine whether the exacerbation percentages for MS patients who take the interferon drug differ from the percentages reported for placebo patients. Test using $\alpha = .05$.

Exacerbations	none	1	2	3	4 or more
# Patients	32	26	15	6	6

Problem 8.76

8.77 Refer to the *Quality Engineering* (Vol. 11, 1999) study of a process improvement program implemented at a large injection-molding company, Problem 2.11 (p. 66). The accompanying table summarizes the causes of 6,324 defective plastic molds found in computer keyboards produced during a 3-month period by the company. Conduct a test to determine if the defective plastic molds are more likely to occur in one cause category than another. Use $\alpha = .01$.

Cause	Frequency	Percentage
Black spot	413	6.53
Damage	1,039	16.43
Jetting	258	4.08
Pin mark	834	13.19
Scratches	442	6.99
Shot mold	275	4.35
Silver streak	413	6.53
Sink mark	371	5.87
Spray mark	292	4.62
Warpage	1,987	31.42
Total	6,324	100.00

Source: Acharya, U. H., and Mahesh, C. "Winning back the customer's confidence: A case study on the application of design of experiments to an injection-molding process," *Quality Engineering,* 11, 1999, 357–363.

8.10 Potential Hypothesis-Testing Pitfalls and Ethical Issues

When planning to carry out a test of hypothesis based on some designed experiment or research study under investigation, several questions need to be raised to ensure that proper methodology is used:

1. What is the goal of the experiment or research? Can it be translated into null and alternative hypotheses?
2. Is the hypothesis test going to be two-tailed or one-tailed?
3. Can a random sample be drawn from the population of interest?
4. What kind of measurements will be obtained from the sample? Is the variable measured quantitative or qualitative?
5. At what significance level, or risk of committing a Type I error, should the hypothesis test be conducted?
6. When testing means, is the intended sample size large enough to use the large-sample (z) test, or should the small-sample (t) test be applied?
7. What kind of conclusions and interpretations can be drawn from the results of the hypothesis test?

Questions like these need to be raised and answered in the planning stage of a survey or designed experiment, so a person with substantial statistical training should be consulted and involved early in the process. All too often such an individual is consulted far too late in the process, after the data have been collected. Typically in such a situation, all that can be done at such a late stage is to choose the statistical test procedure that would be best for the obtained data. We are forced to assume that certain biases that have been built into the study (because

of poor planning) are negligible. But this is a large assumption. Good research involves good planning. To avoid biases, adequate controls must be built in from the beginning.

In this section, we want to distinguish between what is poor research methodology and what is unethical behavior. Ethical considerations arise when a researcher is manipulative of the hypothesis-testing process. Some of the ethical issues that arise when dealing with hypothesis-testing methodology are listed below, followed by a brief description of each problem.

- Data collection method—randomization
- Informed consent from human subjects being "treated"
- Type of test—two-tailed or one-tailed
- Choice of level of significance α
- Data snooping
- Discarding data outliers
- Reporting findings

Data Collection Method—Randomization

To eliminate the possibility of potential biases in the results, we must use proper data collection methods. To be able to draw meaningful conclusions, the data we obtain must be the outcomes of a random sample from the target population or the outcomes from some experiment in which a **randomization** process was employed. For example, potential subjects should not be permitted to self-select for a study. In a similar manner, a researcher should not be permitted to purposely select the subjects for the study. Aside from the potential ethical issues that may be raised, such a lack of randomization can result in selection biases and destroy the value of the study.

Informed Consent from Human Subjects Being "Treated"

Ethical considerations require that any individual who is to be subjected to some "treatment" in an experiment be apprised of the research endeavor and any potential behavioral or physical side effects and provide informed consent with respect to participation. A researcher is not permitted to dupe or manipulate the subjects in a study.

Type of Test—Two-Tailed or One-Tailed

If we have prior information that leads us to test the null hypothesis against a specifically directed alternative, then a one-tailed test will be more powerful than a two-tailed test. On the other hand, we should realize that if we are interested only in *differences* from the null hypothesis, not in the *direction* of the difference, the two-tailed test is the appropriate procedure to utilize. This is an important point. For example, if previous research and statistical testing have already established the difference in a particular direction, or if an established scientific theory states that it is only possible for results to occur in one direction, then a one-tailed or directional test may be employed. However, these conditions are not often satisfied in practice, and it is recommended that one-tailed tests be used cautiously.

Choice of Level of Significance α

In a well-designed experiment or study, the level of significance α is selected in advance of data collection. One cannot be permitted to alter the level of significance, after the fact, to achieve a specific result. This would be **data snooping.** One answer to this issue of level of significance is to always report the p-value, not just the results of the test.

Data Snooping

Data snooping is never permissible. It would be unethical to perform a hypothesis test on a set of data, look at the results, and then select whether it should be two-tailed or one-tailed and/or choose the level of significance. These steps must be done first, as part of the planned experiment or study, before the data are collected, for the conclusions drawn to have meaning. In those situations in which a statistician is consulted by a researcher late in the process, with data already available, it is imperative that the null and alternative hypotheses be established and the level of significance chosen prior to carrying out the hypothesis test.

Discarding Data Outliers

After editing, coding, and transcribing the data, one should always look for outliers—i.e., any observations whose measurements seem to be extreme or unusual. Stem-and-leaf displays and box-and-whisker plots aid in this *exploratory data analysis* stage. In addition, the exploratory data analysis enables us to examine the data graphically with respect to the assumptions underlying a particular hypothesis test procedure.

The process of outlier detection raises a major ethical question. Should an observation be removed from a study? The answer is a qualified "yes." If it can be determined that a measurement is incomplete or grossly in error owing to some equipment problem or unusual behavioral occurrence unrelated to the study, a decision to discard the observation may be made. Sometimes there is no choice—an individual may decide to quit a particular study he has been participating in before a final measurement can be made. In a well-designed experiment or study, the researcher would plan, in advance, decision rules regarding the possible discarding of data.

Reporting Findings

When conducting research, it is vitally important to document both good and bad results so that individuals who follow up on such research do not have to "reinvent the wheel." It would be inappropriate to report the results of hypothesis tests that show statistical significance but not those for which there was insufficient evidence in the findings.

Ethical Considerations: A Summary

Again, when discussing ethical issues concerning the hypothesis-testing methodology, the key is *intent.* We must distinguish between poor data analysis and unethical practice. Unethical behavior occurs when a researcher willfully causes a selection bias in data collection, manipulates the treatment of human subjects without informed consent, uses data snooping to select the type of test (two-tailed or one-tailed) and/or level of significance to his or her advantage, hides the facts by discarding observations that do not support a stated hypothesis, or fails to report pertinent findings.

KEY TERMS

[Note: Starred (*) terms are from the optional sections of this chapter.]

Alternative hypothesis 391
Critical value 399
Data snooping 437
*Expected category count 431
Fixed significance level 414
*Goodness-of-fit test 433
Hypothesis-testing procedure 398
Level of significance 394
Lower-tailed test 401
*Multinomial experiment 431
Null hypothesis 391
*Observed category count 431
Observed significance level 416
One-tailed test 392
p-value 415
Randomization 437
Rejection region 397
Statistical hypothesis 391
Test statistic 397
Two-tailed test 392
Type I error 393
Type II error 393
Upper-tailed test 401

KEY FORMULAS

[Note: Starred (*) terms are from the optional sections of this chapter.]

Large-sample test statistic: $z = \dfrac{\text{Estimator} - \text{Hypothesized } (H_0) \text{ value}}{\text{Standardized error}}$ **(8.3)**, 406 **(8.5)**, 421

Small-sample test statistic: $t = \dfrac{\text{Estimator} - \text{Hypothesized } (H_0) \text{ value}}{\text{Standardized error}}$ **(8.4)**, 409

Note: The respective estimators and standard errors for the population parameters μ and p are provided in Table 7.8 of Chapter 7 (p. 372). [The test statistics for σ^2 and for multinomial category probabilities from the optional sections of this chapter are given below.]

Test statistic for H_0: $\sigma^2 = \sigma_0^2$:* $\chi^2 = \dfrac{(n-1)s^2}{\sigma_0^2}$ **(8.6), 426

Test statistic for H_0: $p_1 = p_{1,0}, p_2 = p_{2,0}, \ldots, p_k = p_{k,0}$:* $\chi^2 = \sum_{i=1}^{k} \dfrac{(n_i - E_i)^2}{E_i}$ **(8.8), 433

where $E_i = np_{i,0}$ **(8.7)**, 432

KEY SYMBOLS

SYMBOL	DESCRIPTION
H_0	Null hypothesis
H_a	Alternative hypothesis
α	Probability of a Type I error (reject H_0 when H_0 is true)
β	Probability of a Type II error (accept H_0 when H_0 is false)
μ_0	Hypothesized value of μ in H_0
p_0	Hypothesized value of p in H_0
σ_0^2	Hypothesized value of σ^2 in H_0
E_i	Expected category count

CHECKING YOUR UNDERSTANDING

1. What is the difference between a null hypothesis and an alternative hypothesis?
2. What is the difference between a Type I error and a Type II error?
3. What do the probabilities α and β represent?
4. What is the difference between a one-tailed test and a two-tailed test?
5. What key words help you identify whether the parameter of interest is a mean, proportion, or variance?
6. Explain when a confidence interval for a population parameter provides the same information as a hypothesis test for the parameter.
7. What is meant by a p-value?
8. Give an outline of the steps involved in hypothesis testing.
9. What are some of the ethical issues to be concerned with in performing a hypothesis test?
10. What are the characteristics of a multinomial experiment?

SUPPLEMENTARY PROBLEMS

Starred (*) problems refer to the optional sections of this chapter.

8.78 Specify the form of the rejection region for a two-tailed test of hypothesis conducted at each of the following significance levels:

a. $\alpha = .01$ **b.** $\alpha = .02$ **c.** $\alpha = .04$

Locate the rejection region, α, and the critical values on a sketch of the standard normal curve for each part of the exercise. (Assume that the sampling distribution of the test statistic is approximately normal.)

8.79 For each of the following rejection regions, determine the value of α, the probability of a Type I error:

a. $z > 2.58$ **b.** $z < -1.29$ **c.** $z < -1.645$ or $z > 1.645$

8.80 Suppose the observed significance level (p-value) of a test is .07.

a. For what values of α would you reject H_0?

b. For what values of α would you fail to reject H_0?

8.81 "Deep hole" drilling is a family of drilling processes used when the ratio of hole depth to hole diameter exceeds 10. Successful deep hole drilling depends on the satisfactory discharge of the drill chip. An experiment was conducted to investigate the performance of deep hole drilling when chip congestion exists (*Journal of Engineering for Industry,* May 1993). The length, in millimeters (mm), of 50 drill chips resulted in the following summary statistics: $\bar{x} = 81.2$ mm, $s = 50.2$ mm. Conduct a test to determine whether the true mean drill chip length μ differs from 75 mm. Use a significance level of $\alpha = .01$.

8.82 Each March the 64 best college basketball teams in the U.S. play in a single-elimination tournament to determine the National Collegiate Athletic Association (NCAA) champion. The NCAA groups the 64 teams into four regions (East, South, Midwest, and West) of 16 teams each. The teams in each region are ranked (seeded) from 1 to 16 based on their performance during the regular season. Statisticians Hal Stern and Barbara Mock analyzed data from 13 past NCAA tournaments and published their results in *Chance* (Winter 1998). The results of first-round games are summarized in the table.

Summary of First-Round NCAA Tournament Games, 1985–1997

Match-up (Seeds)	Number of Games	Number Won by Favorite (Higher Seed)	Margin of Victory (Points)	
			Mean	Standard Deviation
1 vs 16	52	52	22.9	12.4
2 vs 15	52	49	17.2	11.4
3 vs 14	52	41	10.6	12.0
4 vs 13	52	42	10.0	12.5
5 vs 12	52	37	5.3	10.4
6 vs 11	52	36	4.3	10.7
7 vs 10	52	35	3.2	10.5
8 vs 9	52	22	−2.1	11.0

Source: Stern, H. S., and Mock, B. "College basketball upsets: Will a 16-seed ever beat a 1-seed?" *Chance,* Vol. 11, No. 1, Winter 1998, p. 29 (Table 3).

a. A common perception among fans, media, and gamblers is that the higher seeded team has a better than 50-50 chance of winning a first-round game. Is there evidence to support this perception? Conduct the appropriate test for each matchup. What trends do you observe?

b. Is there evidence to support the claim that a 1, 2, 3, or 4-seeded team will win by an average of more than 10 points in first-round games? Conduct the appropriate test for each matchup.

c. Is there evidence to support the claim that a 5, 6, 7, or 8-seeded team will win by an average of less than five points in first-round games? Conduct the appropriate test for each matchup.

***d.** For each matchup, test the null hypothesis that the standard deviation of the victory margin is 11 points.

8.83 A telephone survey of adult literacy was administered to a large sample of San Diego (California) residents (*Journal of Literacy Research,* Dec. 1996). Each person interviewed was presented with a list of magazine names (e.g., *New Yorker, Psychology Today*) and asked, "Do you recognize the magazine to be real?" For one of the fictitious magazines given (*American Journal Review*), 170 of the 486 respondents believed the magazine to be real. Suppose you want to know whether more than $\frac{1}{3}$ of the respondents believe *American Journal Review* to be a real magazine.

a. Compute the test statistic appropriate for conducting the test.

b. Set up the rejection region for the test if you are willing to tolerate a Type I error probability of $\alpha = .10$. Locate the pertinent quantities on a sketch of the standard normal curve.

c. Give a full conclusion in terms of the problem.

d. What are the consequences of a Type I error?

e. Find the p-value of the test and interpret its value.

***8.84** In order to study consumer preferences for health care reform in the U.S., researchers from the University of Michigan surveyed 500 U.S. households (*Journal of Consumer Affairs,* Winter 1999). Heads of household were asked whether they are in favor of, neutral about, or opposed to a national health insurance program in which all Americans are covered and costs are paid by tax dollars. The 434 useable responses are summarized in the following table. Is there sufficient evidence to conclude that opinions are not evenly divided on the issue of national health insurance? Conduct the appropriate test using $\alpha = .01$.

Preferences for National Health Insurance (Number of Respondents)

Favor	Neutral	Oppose
234	119	81

Source: Hong, G., and White-Means, S. "Consumer Preferences for Health Care Reform", *Journal of Consumer Affairs,* Vol. 33, No. 2 (Winter 1999), pp. 237–253.

8.85 How does lack of sleep have an impact on one's creative ability? A British study found that loss of sleep sabotages creative faculties and the ability to deal with unfamiliar situations (*Sleep,* Jan. 1989). In the study, 12 healthy college students, deprived of one night's sleep, received an array of tests intended to measure thinking time, fluency, flexibility, and originality of thought. The overall test scores of the sleep-deprived students were compared to the average score one would expect from students who received their accustomed sleep. Suppose the overall scores of the 12 sleep-deprived students had a mean of $\bar{x} = 63$ and a standard deviation of 17. (Lower scores are associated with a decreased ability to think creatively.)

a. Test the hypothesis that the true mean score of sleep-deprived subjects is less than 80, the mean score of subjects who received sleep prior to taking the test. Use $\alpha = .05$.

b. What assumption is required for the hypothesis test of part **a** to be valid?

***c.** Some sleep experts believe that although most people lose their ability to think creatively when deprived of sleep, others are not nearly as affected. Consequently, they theorize that the variation in test scores for sleep-deprived subjects is larger than 225, the estimated variance in scores of those who receive their usual amount of sleep. Test the hypothesis that σ^2, the variance in test scores of sleep-deprived subjects, is larger than 225. Use $\alpha = .01$.

8.86 Although shoplifting in the U.S. costs retailers about $15 billion a year, Shoplifters Alternative of Jericho, New York, claims that only 50% of all shoplifters are

turned over to police (*Athens Daily News*, Dec. 12, 1999). A random sample of 40 U.S. retailers were questioned concerning the disposition of the most recent shoplifter they apprehended. Only 24 were turned over to police. Do these data provide sufficient evidence to contradict Shoplifters Alternative? Test using using $\alpha = .05$.

8.87 Any sentence that contains an animate noun as a direct object and another noun as a second object (e.g., "Sue offered Ann a cookie.") is termed a double object dative (DOD). The connection between certain verbs and a DOD was investigated in *Applied Psycholinguistics* (June 1998). The subjects were 35 native English speakers who were enrolled in an introductory English composition course at a Hawaiian community college. After viewing pictures of a family, each subject was asked to write a sentence about the family using a specified verb. Of the 35 sentences using the verb "buy," 10 had a DOD structure. Conduct a test to determine if the true fraction of sentences with the verb "buy" that are DODs is less than 1/3. Use $\alpha = .05$.

Vision	Number
normal	15
farsighted	61
nearsighted	34

***8.88** According to the American Optometric Association, 10% of the U.S. population have normal visual acuity, 60% suffer from farsightedness, and 30% suffer from nearsightedness. Each in a random sample of 100 Americans was examined by an optometrist and classified as shown in the table at left. Use the sample information to test the American Optometric Association's claim at $\alpha = .05$.

8.89 Radium-226 is a naturally occurring radioactive gas. Elevated levels of radium-226 in metro-Dade County (Florida) were recently investigated (*Florida Scientist*, Summer/Autumn 1991). The data in the table are radium-226 levels (measured in pCi/L) for 26 soil specimens collected in southern Dade County. The Environmental Protection Agency (EPA) has set maximum exposure levels of radium-226 at 4.0 pCi/L. Use the information in the PHStat and Excel printout below to determine whether the mean radium-226 level of soil specimens collected in southern Dade County is less than the EPA limit of 4.0 pCi/L. Use $\alpha = .10$.

RADIUM

1.46	0.58	4.31	1.02	0.17	2.92	0.91	0.43	0.91
1.30	8.24	3.51	6.87	1.43	1.44	4.49	4.21	1.84
5.92	1.86	1.41	1.70	2.02	1.65	1.40	0.75	

Source: Moore, H. E., and Gussow, D. G. "Radium and radon in Dade County ground water and soil samples." *Florida Scientist*, Vol. 54, No. 3/4, Summer/Autumn 1991, p. 155 (portion of Table 3).

	A	B
1	**t Test for Hypothesis of the Mean**	
2		
3	**Data**	
4	**Null Hypothesis** $\mu=$	**4**
5	**Level of Significance**	**0.1**
6	**Sample Size**	**26**
7	**Sample Mean**	**2.413461538**
8	**Sample Standard Deviation**	**2.080755521**
9		
10	Intermediate Calculations	
11	Standard Error of the Mean	0.408069731
12	Degrees of Freedom	25
13	t Test Statistic	-3.887910181
14		
15	**Lower-Tail Test**	
16	**Lower Critical Value**	**-1.316345788**
17	**p-Value**	**0.000330179**
18	**Reject the null hypothesis**	

E Problem 8.89

8.90 *Marine Technology* (Jan. 1995) reported on the spillage amount and cause of puncture for 50 recent major oil spills from tankers and carriers in the United States. The data are reproduced in the table on p. 444.

a. Test the hypothesis that the true mean spillage of all major oil spills in the U.S. exceeds 50 thousand metric tons. Use $\alpha = .10$.

b. Test the hypothesis that the true proportion of major oil spills in the U.S. that are caused by hull failure differs from .25. Use $\alpha = .10$.

***c.** Test the hypothesis that the true spillage variance of all major oil spills in the U.S. differs from 3,000.

d. Does it appear that the spillage amounts are normally distributed? Explain.

e. How does your answer to part **d** impact the validity of the test, part **a**?

***f.** How does your answer to part **d** impact the validity of the test, part **c**?

8.91 How confident are you in your ability to do well? Researchers at Bowling Green University attempted to determine whether confidence is gender-related (*Psychological Reports,* Aug. 1997). In one part of the study, 87 female psychology students were given a 26-item test covering general social psychological knowledge and were asked to indicate (privately) the number of questions they were confident they had answered correctly. The number of questions the female students indicated that they were confident they had answered correctly had a mean of $\bar{x} = 9.7$ and a standard deviation of $s = 5.3$. Is there sufficient evidence to say that, on average, female students are confident on less than half (13) of the questions on the 26-item test? Use $\alpha = .10$.

CELLPHONE

8.92 Is there a link between cellular telephone usage while driving and the risk of a motor vehicle collision? Researchers at the University of Toronto identified 699 drivers with cell phones who had been involved in a collision with significant damage, but no personal injury (*Chance,* Spring 1997). For each sampled case, the researchers recorded the values of two qualitative variables. The first variable was whether or not the driver made a cell phone call during the 10-minute interval immediately before the collision. (This 10-minute interval is called the *hazard* interval). The second variable was whether or not the driver made a cell phone call in the car during a similar interval on the day prior to the collision. (This prior day interval is called the *control* interval.) The data on these two qualitative variables for all 699 sampled cases is stored in the **CELLPHONE** data file.

a. Is there evidence (at $\alpha = .01$) to indicate that the likelihood of a driver making a cell phone call during the hazard interval exceeds 5%?

b. Is there evidence (at $\alpha = .01$) to indicate that the likelihood of a driver making a cell phone call during the control interval exceeds 5%?

REFERENCES

Daniel, W. W. *Applied Nonparametric Statistics,* 2nd ed. Boston: PWS-Kent, 1990.

Snedecor, G. W., and Cochran, W. G. *Statistical Methods,* 7th ed. Ames, Iowa: Iowa State University Press, 1980.

Wackerly, D., Mendenhall, W., and Scheaffer, R. *Mathematical Statistics with Applications,* 5th ed. Belmont, Calif.: Duxbury Press, 1996.

Tanker	Spillage (metric tons, thousands)	Cause of Spillage: Collision	Grounding	Fire/ Explosion	Hull Failure	Unknown
Atlantic Empress	257	X				
Castillo De Bellver	239			X		
Amoco Cadiz	221				X	
Odyssey	132			X		
Torrey Canyon	124		X			
Sea Star	123	X				
Hawaiian Patriot	101				X	
Independento	95	X				
Urquiola	91		X			
Irenes Serenade	82			X		
Khark 5	76			X		
Nova	68	X				
Wafra	62		X			
Epic Colocotronis	58		X			
Sinclair Petrolore	57			X		
Yuyo Maru No 10	42	X				
Assimi	50			X		
Andros Patria	48			X		
World Glory	46				X	
British Ambassador	46				X	
Metula	45		X			
Pericles G.C.	44			X		
Mandoil II	41	X				
Jacob Maersk	41		X			
Burmah Agate	41	X				
J. Antonio Lavalleja	38		X			
Napier	37		X			
Exxon Valdez	36		X			
Corinthos	36	X				
Trader	36				X	
St. Peter	33			X		
Gino	32	X				
Golden Drake	32			X		
Ionnis Angelicoussis	32			X		
Chryssi	32				X	
Irenes Challenge	31				X	
Argo Merchant	28		X			
Heimvard	31	X				
Pegasus	25					X
Pacocean	31				X	
Texaco Oklahoma	29				X	
Scorpio	31		X			
Ellen Conway	31		X			
Caribbean Sea	30				X	
Cretan Star	27					X
Grand Zenith	26				X	
Athenian Venture	26			X		
Venoil	26	X				
Aragon	24				X	
Ocean Eagle	21		X			

Source: Daidola, J. C. "Tanker structure behavior during collision and grounding." *Marine Technology,* Vol. 32, No. 1, Jan. 1995, p. 22 (Table 1). Reprinted with permission of The Society of Naval Architects and Marine Engineers (SNAME), 601 Pavonia Ave., Jersey City, NJ 07306, USA, (201) 798-4800. Material appearing in The Society of Naval Architect and Marine Engineers (SNAME) publications cannot be reprinted without obtaining written permission.

Using Microsoft® Excel

8.E.1 Testing the Mean with the Z Test (σ Known)

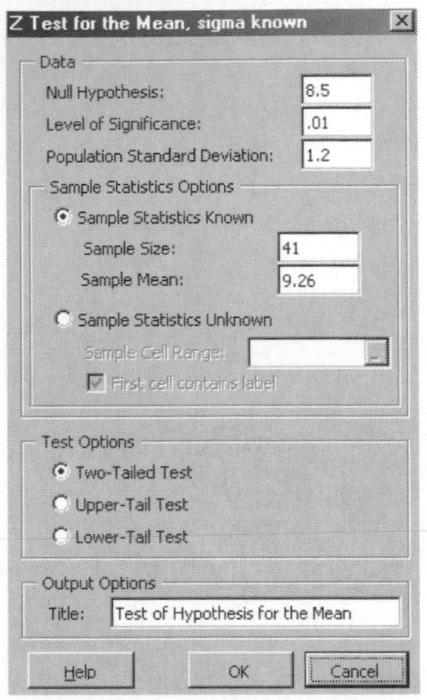

E Figure 8.E.1 PHStat Z Test for the Mean, Sigma Known Dialog Box

Use the PHStat **One-Sample Tests | Z Test for the Mean, sigma known** procedure to perform a Z test of the hypothesis for the mean (σ known). For example, to perform the Z test of the hypothesis for the mean shown in Figure 8.6 on p. 408 for the length-to-width ratio of humerus bones

1. Select PHStat | One-Sample Tests | Z Test for the Mean, sigma known.
2. In the Z Test for the Mean, sigma known dialog box (see Figure 8.E.1):
   a. Enter 8.5 in the Null Hypothesis edit box.
   b. Enter .01 in the Level of Significance edit box.
   c. Enter 1.2 in the Population Standard Deviation edit box.
   d. Select the Sample Statistics Known option button and enter 41 in the Sample Size edit box and 9.26 in the Sample Mean edit box.
   e. Select the Two-Tailed Test option button.
   f. Enter a title in the Title edit box.
   g. Click the OK button.

For similar problems in which the sample mean is not known and needs to be calculated, select the Sample Statistics Unknown option button in step 2d and enter the cell range of the sample data in the Sample Cell Range edit box, shown dimmed in Figure 8.E.1.

For one-tailed tests, select the Upper-Tail Test or Lower-Tail Test option button instead of the Two-Tailed Test option button in step 2e.

8.E.2 Testing the Mean with the t Test (σ Unknown)

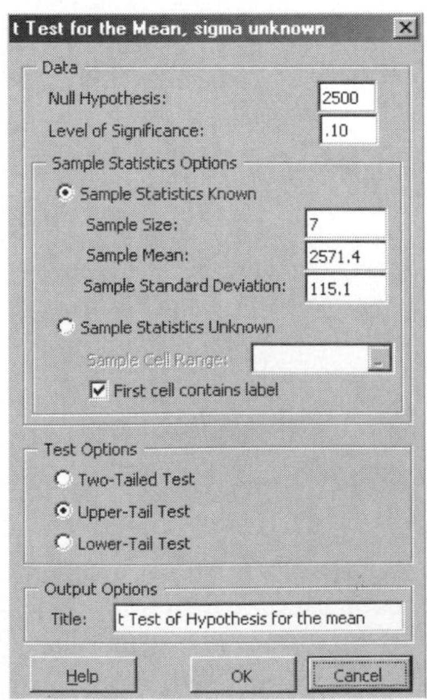

E Figure 8.E.2 t Test for the Mean, Sigma Unknown Dialog Box

Use the PHStat **One-Sample Tests | t Test for the Mean, sigma unknown** procedure to perform a t test of the hypothesis for the mean (σ unknown). For example, to perform the t test of the hypothesis for the mean for the breaking strength of pipes, shown in Example 8.14 on p. 410, open the **PIPES.XLS** workbook to the Data worksheet and

1. Select PHStat | One-Sample Tests | t Test for the Mean, sigma unknown.
2. In the t Test for the Mean, sigma unknown dialog box (see Figure 8.E.2):
   a. Enter 2500 in the Null Hypothesis edit box.
   b. Enter .10 in the Level of Significance edit box.
   c. Select the Sample Statistics Known option button and enter 7 in the Sample Size edit box, 2571.4 in the Sample Mean edit box, and 115.1 in the Sample Standard Deviation edit box.
   d. Select the Upper-Tail Test option button.
   e. Enter a title in the Title edit box.
   f. Click the OK button.

For similar problems in which the sample size, sample mean, and sample standard deviation are not known, select the Sample Statistics Unknown option button in step 2c and enter the cell range in the appropriate edit box, shown dimmed in Figure 8.E.2. For two-tailed tests, select the Two-Tailed Test button instead of the Upper-Tail Test option button in step 2d.

8.E.3 Testing a Proportion with the Z Test

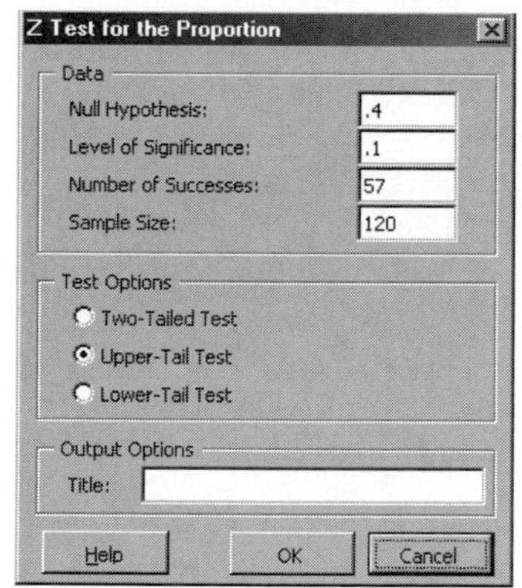

E Figure 8.E.3 Z Test for the Proportion Dialog Box

Use the PHStat **One-Sample Tests | Z Test for the Proportion** procedure to perform a Z test of the hypothesis for the proportion. For example, to perform the Z test of the hypothesis for the proportion shown in Figure 8.15 on p. 423 for the proportion of new car buyers who are women, open to an empty worksheet and

1. Select PHStat | One-Sample Tests | Z Test for the Proportion.
2. In the Z Test for the Proportion dialog box (see Figure 8.E.3):
   a. Enter .4 in the Null Hypothesis edit box.
   b. Enter .1 in the Level of Significance edit box.
   c. Enter 57 in the Number of Successes edit box.
   d. Enter 120 in the Sample Size edit box.
   e. Select the Upper-Tail Test option button.
   f. Enter a title in the Title edit box.
   g. Click the OK button.

For two-tailed tests, select the Two-Tailed Test option button instead of the Upper-Tail Test option button in step 2e.

8.E.4 Testing the Variance with the χ^2 Test

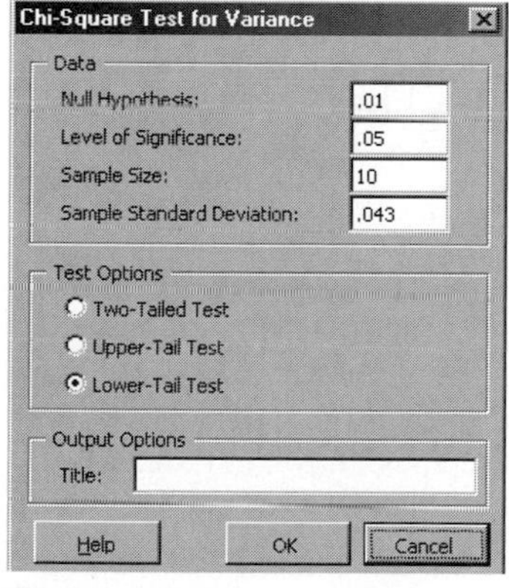

E Figure 8.E.4 Chi-Square Test for Variance Dialog Box

Use the PHStat **One-Sample Tests | χ^2 Test for Variance** procedure to perform a χ^2 test of hypothesis for the variance. For example, to perform the χ^2 test for the variance shown in Figure 8.18 on p. 428 for the cereal filling process problem, open to an empty worksheet and

1. Select PHStat | One-Sample Tests | χ^2 Test for Variance.
2. In the χ^2 Test for Variance dialog box (see Figure 8.E.4):
   a. Enter .01 (the variance) in the Null Hypothesis edit box.
   b. Enter .05 in the Level of Significance edit box.
   c. Enter 10 in the Sample Size edit box.
   d. Enter .043 in the Sample Standard Deviation edit box.
   e. Select the Lower-Tail Test option button.
   f. Enter a title in the Title edit box.
   g. Click the OK button.

For two-tailed tests, select the Two-Tailed Test option button instead of the Lower-Tail Test option button in step 2e.

8.E.5 Testing Categorical Probabilities with the χ^2 Test

The CHIINV and CHITEST worksheet functions can be used to perform a test of hypothesis for categorical probabilities. The CHITEST function uses the format

CHITEST(*observed frequencies cell range*, *expected frequencies cell range*)

to return the probability of the χ^2 test statistic. The CHIINV function uses the format

CHIINV(*upper tail probability*, *degrees of freedom*)

to return the value of the χ^2 statistic. Figure 8.E.5 represents the worksheet design for Example 8.22 on p. 432 concerning the study of aphasia.

E Figure 8.E.5 Worksheet Design for a Chi-Square Test of Hypothesis for Categorical Probabilities

	A	B	C
1	**Test of Hypothesis in a Multinomial Experiment with k Outcomes**		
2			
3	Category	Observed	Expected
4	Broca's	5	=$B$7/$B$10
5	Conduction	7	=$B$7/$B$10
6	Anomic	10	=$B$7/$B$10
7	Total	=SUM(B4:B6)	
8			
9	Level of Significance	0.05	
10	Number of categories	=COUNT(B4:B6)	
11	Degrees of freedom	=B10-1	
12	Critical Value	=CHIINV(B9,B11)	
13	Chi-Square test statistic	=CHIINV(B14,B11)	
14	*p*-value	=CHITEST(B4:B6,C4:C6)	
15	=IF(B14<B9,"Reject the null hypothesis","Do not reject the null hypothesis")		

Note that this design is for an example that has three categories that are hypothesized to have an equal probability of occurrence. If you are solving a problem in which there are more than three categories, you need to insert additional rows after row 6 of the worksheet and copy the formulas in column C down through the added rows. If the null hypothesis specifies a test for unequal probabilities in the categories, the formulas in column C for the expected frequencies need to be changed to reflect these probabilities.

Be aware that when the *p*-value approaches zero, due to a fault in the Excel worksheet function, the CHIINV function may return the error message #NUM. This does not affect the decision reported on the worksheet or the *p*-value.

Using MINITAB®

8.M.1 Testing the Mean with the t Test (σ Unknown)

MINITAB can be used for a test of the mean when σ is unknown by using Stat | Basic Statistics | 1-Sample t from the Menu bar.

To illustrate the test of hypothesis for the mean when σ is unknown, return to the KLEENEX® tissue example of Figure 8.13 on p. 419. Open the **TISSUE.MTW** worksheet, and select **Stat | Basic Statistics | 1-Sample t.** Enter **C1** or **'Number Used'** in the Variables edit box. Select the **Test Mean** option button, and enter **60** in the Test Mean edit box (see Figure 8.M.1). In the Alternative drop-down list box, select less than or greater than for one-tailed tests or not equal for a two-tailed test. Click the **OK** button.

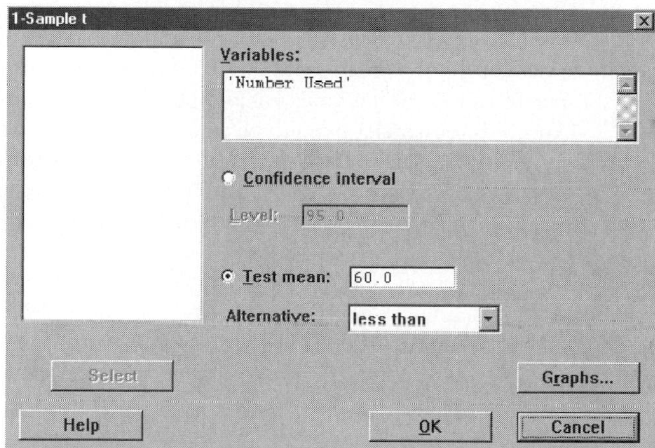

Figure 8.M.1 MINITAB 1-Sample t Dialog Box

8.M.2 Testing a Proportion with the Z Test

MINITAB can be used to obtain a test of hypothesis for the proportion. For example, to perform the Z test of the hypothesis for the proportion for the KLEENEX® tissue example (see Figure 8.16 on p. 424), select **Stat | Basic Statistics | 1 Proportion.** In the 1 Proportion dialog box (see Figure 8.M.2), select the **Samples in columns** option button. Enter **C2** or **'USE60'** in the edit box. Click the **Options** button. In the 1 Proportion—Options dialog box (see Figure 8.M.3), enter **95.0** in the Confidence level edit box and **0.5** in the Test proportion edit box. Select the **Use test and interval based on the normal distribution** check box. In the Alternative drop-down list box, select less than or greater than for one-tailed tests or not equal for a two-tailed test. Click the **OK** button to return to the 1 Proportion dialog box. Click the **OK** button.

If you were using summarized data, instead of selecting the Samples in columns option button, you would select the Summarized data option button and enter the sample size and number of successes in their appropriate edit boxes.

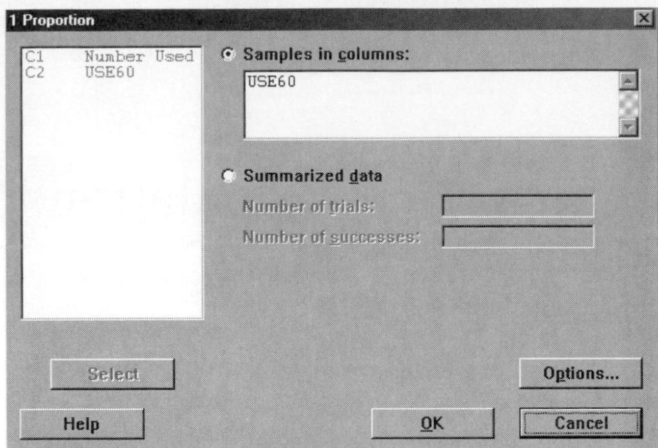

Figure 8.M.2 MINITAB 1 Proportion Dialog Box

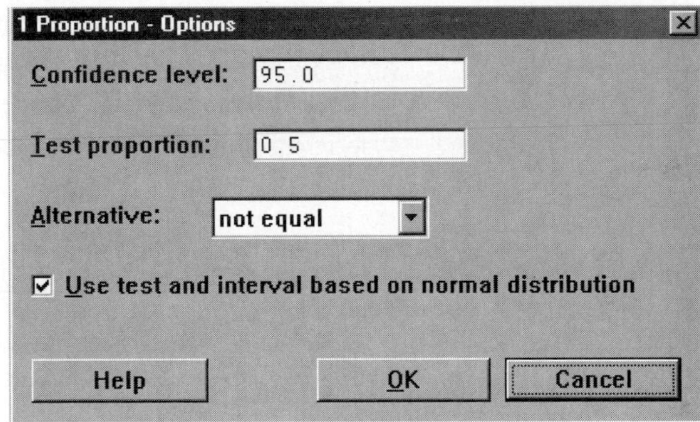

Figure 8.M.3 MINITAB 1 Proportion—Options Dialog Box

CHAPTER 9

Inferences about Population Parameters: Two Samples

OBJECTIVES

1. To extend the basic principles of hypothesis testing and confidence intervals to two populations
2. To present hypothesis tests and confidence intervals for comparing two populations based on independent samples
3. To present hypothesis tests and confidence intervals for comparing two populations based on paired samples
4. To introduce nonparametric tests as alternative tests of hypothesis

CONTENTS

EXCEL TUTORIAL

MINITAB TUTORIAL

Statistics in the Real World

An IQ Comparison of Identical Twins Reared Apart

How much of our personality, our likes and dislikes, our individuality, is predetermined by our genes? And which of our traits are shaped and changed by our environment? Identical twins, because they share an identical genotype, make ideal subjects for investigating the degree to which various environmental conditions may instigate change. The classical method of studying this phenomenon is the study of identical twins separated early in life and reared apart. The book *Twins: And What They Tell Us About Who We Are* (Lawrence Wright, Wiley: 1998) and ABC-TV's *20/20* news program "Separated at Birth" (May 30, 1999) have publicized the research of psychologists who study identical twins reared apart.

Identical twins are formed when a single egg, fertilized by a single sperm, splits into two parts and each develops into a separate embryo. In contrast, fraternal twins develop when two eggs are released from one or both ovaries and each is fertilized by a different sperm. Although they have been the subjects in many studies of twins, fraternal twins are no more genetically similar than two siblings. Therefore, a study of fraternal twins would leave unanswered the question of whether a given individual trait is due to hereditary or environmental differences. Likewise, identical twins reared in the same family share almost identical environments and make it difficult to separate the two factors. Thus, many psychologists theorize that the clearest demarcation of heredity and environment is found when identical twins have been separated early in life and reared apart in different homes, by different parents, and often in widely varying socioeconomic and geographic circumstances.

Over the past twenty years, several studies of identical twins have been conducted. The most publicized study was begun in 1979 by University of Minnesota psychologist Thomas Bouchard and continues today. Bouchard and his colleagues at the Minnesota Center for Twin and Adoption Research have published over 129 scientific papers on the subject. In this real-world application, we consider a similar study carried out by psychologist Susan Farber and published in her book, *Identical Twins Reared Apart* (Basic Books, 1981).

Farber chronicles and analyzes data for 95 pairs of identical twins reared apart. Much of her discussion focuses on a comparison of IQ scores. The question of concern is, "Are there significant differences between the IQ scores of identical twins, where one member of the pair is reared by the natural parents and the other member of the pair is not?" IQ scores for 32 of Farber's 95 twins are stored in the data file named **TWINS.** One member (A) of each of the $n = 32$ pairs of twins was reared by a natural parent, whereas the other member (B) was reared by a relative or some other person. In the following Statistics in the Real World Revisited sections, we apply the statistical methodology presented in this chapter to this data set.

Statistics in the Real World Revisited

- Determining the Target Parameter for the Twins-IQ Study (p. 453)
- Identifying the Appropriate Method for Analyzing the Twins-IQ Data (p. 461)
- Comparing the Mean IQ Scores of Identical Twins Reared Apart (p. 470)

9.1 Determining the Target Parameter

In the previous two chapters, we covered confidence intervals and hypothesis tests applied to a single sample. In this chapter, we present confidence intervals and hypothesis tests for comparing two samples. The population parameters covered include the **difference between two population means $\mu_1 - \mu_2$, the difference between two population proportions $p_1 - p_2$,** and (optionally) **the ratio of two population variances σ_1^2/σ_2^2.** The key words or phrases that help you identify the target parameter are reported in the next box. Remember, the type of data (quantitative or qualitative) collected will aid in your decision. With quantitative data, you're interested in comparing means or variances. With qualitative data, a comparison of proportions is appropriate.

DETERMINING THE TARGET PARAMETER: TWO SAMPLES

Parameter	Key Words or Phrases
$(\mu_1 - \mu_2)$	Difference in means or averages; mean difference; unpaired comparisons of means or averages
μ_d	Mean difference; average difference; paired comparison of means or averages
$(p_1 - p_2)$	Difference in proportions, percentages, fractions, or rates; comparison of proportions, percentages, fractions, or rates
σ_1^2/σ_2^2	Ratio of variances; difference in variation; comparison of variances

EXAMPLE 9.1

CHOOSING THE TARGET PARAMETER

A study at the University of Minnesota's Heart Disease Prevention Clinic found that "participants who added 3 cups daily of Cheerios (cereal) to a diet low in saturated fat and cholesterol dropped their cholesterol levels, on average, more than those who followed the low fat diet alone" (*Minneapolis Star Tribune,* Jan. 14, 1998). Identify the parameter of interest in the study.

Solution

In this problem, the experimental units are people who participated in the study and the variable measured is cholesterol level—a *quantitative* variable. The key words in the statement of this problem are "average" and "more than;" thus it is clear that there are two means to be compared:

μ_1 = Mean cholesterol level of people on a low fat diet

and

μ_2 = Mean cholesterol level of people on a low fat diet supplemented with Cheerios

Consequently, the parameter of interest is $(\mu_1 - \mu_2)$.

EXAMPLE 9.2

CHOOSING THE TARGET PARAMETER

A biologist wants to compare the proportions of fish infected with a tapeworm parasite in two locations—the Mediterranean Sea and the Atlantic Ocean. Fish at both locations are captured and dissected, and the number found to be infected with the tapeworm parasite determined. What is the parameter of interest?

Solution

For this study, the experimental units are the fish captured and dissected, and the variable measured is *qualitative*—either a fish is infected with a tapeworm parasite or it is not. The two key words in the statement of the problem are "compare" and "proportions." Therefore, the parameter of interest is $(p_1 - p_2)$, where

p_1 = Proportion of Mediterranean Sea fish infected with a tapeworm parasite

p_2 = Proportion of Atlantic Ocean fish infected with a tapeworm parasite

Self-Test 9.1

Determine the parameter of interest for each of the following:

a. The manager of a pet supply store wants to determine whether there is a difference in the average amount of money spent by owners of cats and owners of dogs.

b. The pet store manager also wants to know if the variation in amount spent by cat owners differs from the variation in amount spent by dog owners.

Statistics in the Real World Revisited

Determining the Target Parameter for the Twins-IQ Study

In her study, psychologist Susan Farber collected data on IQ scores for pairs of identical twins, where one member of the pair (denoted A) is reared by the natural parents and the other member of the pair (denoted B) is not (see p. 451). The research question of interest is, "Are there significant differences between the mean IQ scores of identical twins reared apart?" Since IQ score is a quantitative variable, and we want to compare the scores for the two groups of twins, the parameter of interest is $(\mu_A - \mu_B)$, where

μ_A = Mean IQ score of the population of twins who are reared by their natural parents

and

μ_B = Mean IQ score of the population of twins who are reared by a relative or some other person

9.2 Comparing Two Population Means: Independent Samples

Suppose you want to use the information in two samples to compare two population means. For example, you may want to compare the mean starting salaries of college graduates with engineering and journalism degrees; or the mean gasoline consumptions that may be expected this year for drivers in two areas of the country; or the mean reaction times of men and women to a visual stimulus. The methods we present in this section are extensions of those used for estimating or testing a single population mean in Chapter 8.

EXAMPLE 9.3 **SELECTING A POINT ESTIMATE**

Researchers at Rochester Institute of Technology investigated the use of isolation timeout as a behavioral management technique (*Exceptional Children,* Feb. 1995). Subjects for the study were 155 emotionally disturbed students enrolled in a special education facility. The students were randomly assigned to one of two types of classroom—Option I classrooms (one teacher, one paraprofessional, and a maximum of 12 students) and Option II classrooms (one teacher, one paraprofessional, and a maximum of six students). Over the academic year the number of behavioral incidents resulting in an isolation timeout was recorded for each student. Summary statistics for the two groups of students are shown in Table 9.1 on p. 454. Calculate a point estimate for the difference between the mean number of timeout incidents for the two groups of students.

TABLE 9.1 Summary of Information for Example 9.3

	Option I	Option II
Number of students	100	55
Mean number of timeout incidents	78.67	102.87
Standard deviation	59.08	69.33

Source: Costenbader, V., and Reading-Brown, M. "Isolation timeout used with students with emotional disturbance." *Exceptional Children,* Vol. 61, No. 4, Feb 1995, p. 359 (Table 3).

Solution

Let the subscript "1" refer to the Option I classrooms and the subscript "2" to the Option II classrooms; then also define the following notation:

μ_1 = Population mean number of timeout incidents for Option I students

μ_2 = Population mean number of timeout incidents for Option II students

Similarly, let $\bar{x}_1$ and $\bar{x}_2$ denote the respective sample means; s_1 and s_2, the respective sample deviations; and n_1 and n_2, the respective sample sizes.

Now, to estimate $(\mu_1 - \mu_2)$, it seems logical to use the difference between the sample means

$$(\bar{x}_1 - \bar{x}_2) = 78.67 - 102.87 = -24.20$$

as our point estimate of the difference between the population means.

Large Samples

Confidence intervals and hypothesis tests for $(\mu_1 - \mu_2)$ are based on the sampling distribution of the point estimate $(\bar{x}_1 - \bar{x}_2)$. The properties of the sampling distribution of $(\bar{x}_1 - \bar{x}_2)$ are shown in the next box (see also Figure 9.1).

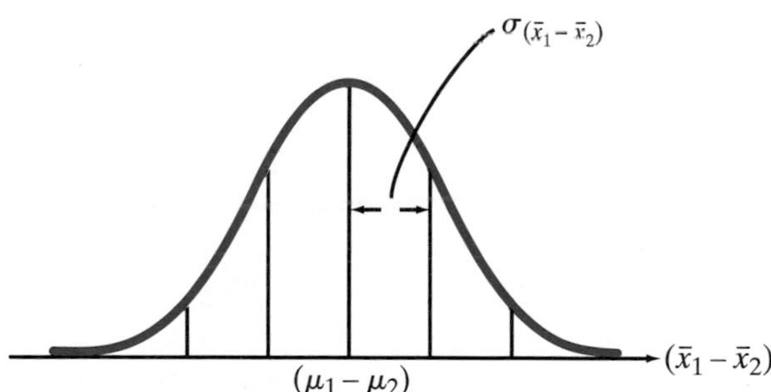

Figure 9.1 Sampling Distribution of $(x_1 - \bar{x}_2)$

SAMPLING DISTRIBUTION OF $(\bar{x}_1 - \bar{x}_2)$

For sufficiently large sample sizes (say, $n_1 \geq 30$ and $n_2 \geq 30$), the sampling distribution of $(\bar{x}_1 - \bar{x}_2)$, based on independent random samples from two populations, is approximately normal with mean equal to the difference between the two population means and standard deviation equal to the square root of $(\sigma_1^2/n_1 + \sigma_2^2/n_2)$, i.e.,

$$\text{Mean: } \mu_{(\bar{x}_1-\bar{x}_2)} = (\mu_1 - \mu_2) \quad \textbf{(9.1)}$$

$$\text{Standard deviation: } \sigma_{(\bar{x}_1-\bar{x}_2)} = \sqrt{\frac{\sigma_1^2}{n_1} + \frac{\sigma_2^2}{n_2}} \quad \textbf{(9.2)}$$

where σ_1^2 and σ_2^2 are the variances of the two populations from which the samples were selected.

As was the case with large-sample inferences about a single population mean, the requirements of "large" sample sizes enable us to apply the Central Limit Theorem to obtain the sampling distribution of $(\bar{x}_1 - \bar{x}_2)$; they also justify the use of s_1^2 and s_2^2 as approximations to the respective population variances σ_1^2 and σ_2^2.

The procedures for testing and estimating $(\mu_1 - \mu_2)$ with large samples appear in the accompanying boxes.

LARGE-SAMPLE TEST OF HYPOTHESIS ABOUT $\mu_1 - \mu_2 = 0$

Upper-Tailed Test	Two-Tailed Test	Lower-Tailed Test
H_0: $(\mu_1 - \mu_2) = 0$	H_0: $(\mu_1 - \mu_2) = 0$	H_0: $(\mu_1 - \mu_2) = 0$
H_a: $(\mu_1 - \mu_2) > 0$	H_a: $(\mu_1 - \mu_2) \neq 0$	H_a: $(\mu_1 - \mu_2) < 0$

$$\text{Test statistic: } z = \frac{(\bar{x}_1 - \bar{x}_2) - 0}{\sigma_{(\bar{x}_1 - \bar{x}_2)}} = \frac{(\bar{x}_1 - \bar{x}_2) - 0}{\sqrt{\frac{\sigma_1^2}{n_1} + \frac{\sigma_2^2}{n_2}}} \approx \frac{(\bar{x}_1 - \bar{x}_2) - 0}{\sqrt{\frac{s_1^2}{n_1} + \frac{s_2^2}{n_2}}} \quad \textbf{(9.3)}$$

Rejection region:	*Rejection region:*	*Rejection region:*
$z > z_\alpha$	$\lvert z \rvert > z_{\alpha/2}$	$z < -z_\alpha$

LARGE-SAMPLE $(1 - \alpha)100\%$ CONFIDENCE INTERVAL FOR $(\mu_1 - \mu_2)$

$$(\bar{x}_1 - \bar{x}_2) \pm z_{\alpha/2}\sigma_{(\bar{x}_1 - \bar{x}_2)} = (\bar{x}_1 - \bar{x}_2) \pm z_{\alpha/2}\sqrt{\frac{\sigma_1^2}{n_1} + \frac{\sigma_2^2}{n_2}}$$

$$\approx (\bar{x}_1 - \bar{x}_2) \pm z_{\alpha/2}\sqrt{\frac{s_1^2}{n_1} + \frac{s_2^2}{n_2}} \quad \textbf{(9.4)}$$

[*Note:* We have used the sample variances s_1^2 and s_2^2 as approximations to the corresponding population parameters.]

ASSUMPTIONS FOR LARGE-SAMPLE INFERENCES ABOUT $(\mu_1 - \mu_2)$

1. The two random samples are selected in an independent manner from the target populations. That is, the choice of elements in one sample does not affect, and is not affected by, the choice of elements in the other sample.
2. The sample sizes n_1 and n_2 are sufficiently large. (We recommend $n_1 \geq 30$ and $n_2 \geq 30$.)

EXAMPLE 9.4 **LARGE-SAMPLE TEST OF $(\mu_1 - \mu_2)$**

Refer to Example 9.3 (p. 453). Is there evidence that the mean number of time-out incidents for special education students assigned to Option I classrooms is less than the corresponding mean for special education students assigned to Option II classrooms? Test using $\alpha = .05$.

Solution

For this problem, we want to test the hypotheses

H_0: $(\mu_1 - \mu_2) = 0$ (no difference between mean number of timeout incidents for Option I and Option II students)

H_a: $(\mu_1 - \mu_2) < 0$ (mean number of timeout incidents for Option I students is less than the mean for Option II students)

This lower-tailed, large-sample (since both n_1 and n_2 exceed 30) test is based on a z statistic. Thus, we will reject H_0 if $z < -z_{.05} = -1.645$ (see Figure 9.2). We compute the test statistic by substituting the summary information in Table 9.1 into equation 9.3:

$$z \approx \frac{(\bar{x}_1 - \bar{x}_2) - 0}{\sqrt{\frac{s_1^2}{n_1} + \frac{s_2^2}{n_2}}} = \frac{(78.67 - 102.87) - 0}{\sqrt{\frac{(59.08)^2}{100} + \frac{(69.33)^2}{55}}} = -2.19$$

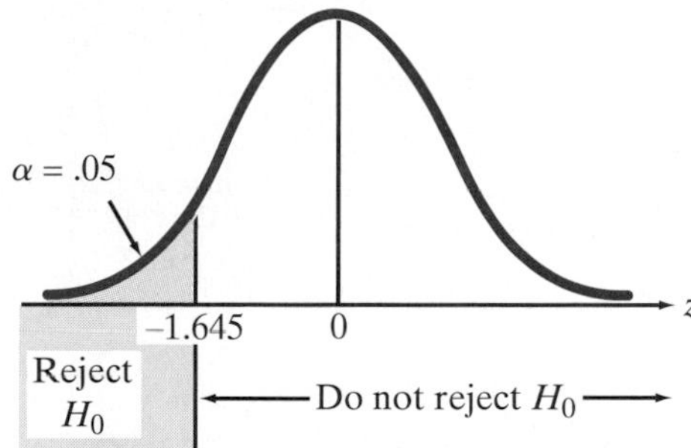

Figure 9.2 Rejection Region for Example 9.4

Since this computed value of $z = -2.19$ lies in the rejection region, there is sufficient evidence (at $\alpha = .05$) to conclude that the mean number of timeout incidents for Option I students is significantly less than the mean number of timeout incidents for Option II students.

The same conclusion can be reached by computing the observed significance level (p-value) of the test, described in Section 8.6. For this lower-tailed test,

$$p\text{-value} = P(z < -2.19)$$

From the standard normal table in Table B.3 in Appendix B, we find

$$p\text{-value} = P(z < -2.19) = .5 - P(-2.19 < z < 0) = .5 - .4857 = .0143$$

Since p-value $= .0143$ is less than $\alpha = .05$, we again reject H_0 in favor of H_a.

EXAMPLE 9.5

LARGE-SAMPLE CONFIDENCE INTERVAL FOR $(\mu_1 - \mu_2)$

Refer to Examples 9.3 and 9.4.

a. Construct a 95% confidence interval for $(\mu_1 - \mu_2)$, the difference between the mean number of timeout incidents for the two groups of special education students.

b. Interpret the interval. Compare your inference here to the results of Example 9.4.

Solution

a. The general form of a 95% confidence interval for $(\mu_1 - \mu_2)$, based on large samples from the target populations, is given by equation 9.4. Substituting $z_{.025} = 1.96$

and the summary information in Table 9.1 into the formula, we obtain:

$$(78.67 - 102.87) \pm 1.96\sqrt{\frac{\sigma_1^2}{100} + \frac{\sigma_2^2}{55}}$$

$$\approx (78.67 - 102.87) \pm 1.96\sqrt{\frac{(59.08)^2}{100} + \frac{(69.33)^2}{55}}$$

$$= -24.20 \pm 21.68$$

or $(-45.88, -2.52)$.

b. Since the lower and upper limits in our interval are negative, we are 95% confident that the mean number of timeout incidents for Option I students is between 2.52 and 45.88 less than the mean number of timeout incidents for Option II students. Option I students have, on average, fewer timeout incidents than Option II students.

In Example 9.4, the hypothesis test also detected a lower mean for Option I students at $\alpha = .05$. However, our confidence interval provides additional information on the magnitude of the difference.

Self-Test 9.2

Independent random samples of $n_1 = 45$ blue-collar workers and $n_2 = 38$ white-collar workers yielded the following summary statistics on number of sick days taken in a recent year:

$$\bar{x}_1 = 11.5 \quad s_1 = 10.2 \quad \bar{x}_2 = 9.0 \quad s_2 = 5.6$$

a. Construct a 95% confidence interval for $(\mu_1 - \mu_2)$.
b. Test H_0: $(\mu_1 - \mu_2) = 0$ against H_a: $(\mu_1 - \mu_2) \neq 0$ at $\alpha = .05$.
c. Compare the results, parts **a** and **b**.

Small Samples

When estimating or testing the difference between two population means based on small samples from each population, we must make specific assumptions about the distributions of the two populations, as indicated in the box.

ASSUMPTIONS REQUIRED FOR SMALL-SAMPLE INFERENCES ABOUT $(\mu_1 - \mu_2)$

1. Both of the populations from which the samples are selected have distributions that are approximately normal.
2. The variances σ_1^2 and σ_2^2 of the two populations are equal.
3. The random samples are selected in an independent manner from the two populations.

Figure 9.3 on p. 458 illustrates the form of the population distributions implied by assumptions 1 and 2. Observe that both populations have distributions that are approximately normal. Although the means of the two populations may differ, we require the variances σ_1^2 and σ_2^2, which measure the spread of the two distributions, to be equal. When these assumptions are satisfied, we use the Student's t distribution (specified in the next box) to make inferences about $(\mu_1 - \mu_2)$ based

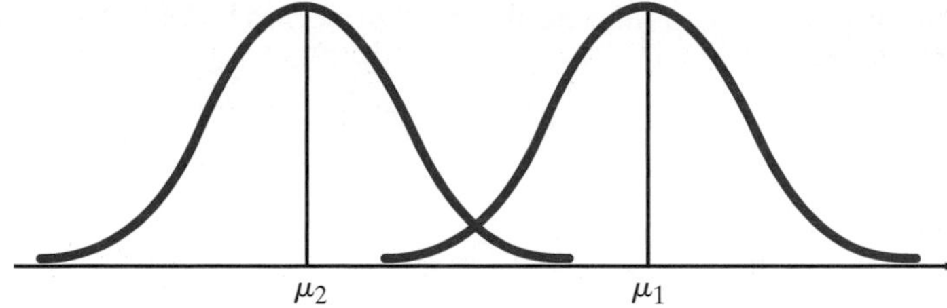

Figure 9.3 Assumptions Required for Small-Sample Estimation of $(\mu_1 - \mu_2)$: Normal Distributions with Equal Variances

on small samples (say, $n_1 < 30$ or $n_2 < 30$) from the respective populations. Since we assume that the two populations have equal variances (i.e., $\sigma_1^2 = \sigma_2^2 = \sigma^2$), we construct a **pooled estimate of σ^2** (denoted s_p^2) based on the information contained in *both* samples.

SMALL-SAMPLE TEST OF HYPOTHESIS ABOUT $(\mu_1 - \mu_2) = 0$

Upper-Tailed Test	**Two-Tailed Test**	**Lower-Tailed Test**
H_0: $(\mu_1 - \mu_2) = 0$	H_0: $(\mu_1 - \mu_2) = 0$	H_0: $(\mu_1 - \mu_2) = 0$
H_a: $(\mu_1 - \mu_2) > 0$	H_a: $(\mu_1 - \mu_2) \neq 0$	H_a: $(\mu_1 - \mu_2) < 0$

$$\textit{Test statistic: } t = \frac{(\bar{x}_1 - \bar{x}_2) - 0}{\sqrt{s_p^2\left(\frac{1}{n_1} + \frac{1}{n_2}\right)}} \quad \textbf{(9.5)}$$

where s_p^2 is given in equation 9.7 in the next box.

Rejection region:	*Rejection region:*	*Rejection region:*
$t > t_\alpha$	$\lvert t \rvert > t_{\alpha/2}$	$t < -t_\alpha$

where the distribution of t is based on $(n_1 + n_2 - 2)$ degrees of freedom.

SMALL-SAMPLE $(1 - \alpha)100\%$ CONFIDENCE INTERVAL FOR $(\mu_1 - \mu_2)$

$$(\bar{x}_1 - \bar{x}_2) \pm t_{\alpha/2}\sqrt{s_p^2\left(\frac{1}{n_1} + \frac{1}{n_2}\right)} \quad \textbf{(9.6)}$$

where

$$s_p^2 = \frac{(n_1 - 1)s_1^2 + (n_2 - 1)s_2^2}{n_1 + n_2 - 2} \quad \textbf{(9.7)}$$

and the value of $t_{\alpha/2}$ is based on $(n_1 + n_2 - 2)$ degrees of freedom.

TESTING FOR A NONZERO DIFFERENCE

Equations 9.3 and 9.5 give the test statistics for testing the null hypothesis of no difference in the means, i.e., H_0: $\mu_1 - \mu_2 = 0$. To test for a nonzero difference, i.e., H_0: $\mu_1 - \mu_2 = D$ where $D \neq 0$, use the test statistic

$$\text{Large sample: } z = \frac{(\bar{x}_1 - \bar{x}_2) - D}{\sqrt{\frac{\sigma_1^2}{n_1} + \frac{\sigma_2^2}{n_2}}} \quad \textbf{(9.8)}$$

$$\text{Small sample: } t = \frac{(\bar{x}_1 - \bar{x}_2) - D}{\sqrt{s_p^2\left(\frac{1}{n_1} + \frac{1}{n_2}\right)}} \quad \textbf{(9.9)}$$

EXAMPLE 9.6

SMALL-SAMPLE TEST FOR $(\mu_1 - \mu_2)$

To study the effectiveness of a new type of dental anesthetic, a dentist conducted an experiment with 10 randomly selected patients. Five patients were randomly assigned to receive the standard anesthetic (Novocaine), whereas the remaining five patients received the proposed new anesthetic. While being treated, each patient was asked to give a measure of his or her discomfort, on a scale from 0 to 100. (Higher scores indicate greater discomfort.) The discomfort scores for the 10 patients are shown in Table 9.2.

DENTAL

TABLE 9.2 Discomfort Scores for Example 9.6

Novocaine	62	71	44	50	42
New Anesthetic	38	45	27	42	30

If the new anesthetic is more effective, then the mean discomfort level for Novocaine will exceed the mean discomfort level for the new treatment. Use a test of hypothesis to make an inference about the effectiveness of the new anesthetic. Test at $\alpha = .01$.

Solution

Let μ_1 and μ_2 represent the true mean discomfort levels of patients receiving Novocaine and the new anesthetic, respectively. Since the samples selected for the study are small ($n_1 = n_2 = 5$), the following assumptions are required:

1. The populations of discomfort levels of dental patients receiving Novocaine and the new anesthetic both have approximately normal distributions.
2. The variance σ^2 in the discomfort levels is the same for both groups of patients.
3. The samples were independently and randomly selected from the two target populations.

The dentist wants to test

H_0: $\mu_1 - \mu_2 = 0$ (no difference in mean discomfort levels of the two treatments)

H_a: $\mu_1 - \mu_2 > 0$ (the mean discomfort level of patients using Novocaine [μ_1] is greater than the mean discomfort level of patients on the new anesthetic [μ_2])

If these assumptions are valid, the test statistic will have a t distribution with $(n_1 + n_2 - 2) = (5 + 5 - 2) = 8$ degrees of freedom. With a significance level of $\alpha = .01$, we require $t_{.01}$ for this one-tailed test. From Table B.4 in Appendix B, $t_{.01} = 2.896$. Thus, the rejection region for this upper-tailed test is:

$$t > 2.896 \quad \text{(see Figure 9.4)}$$

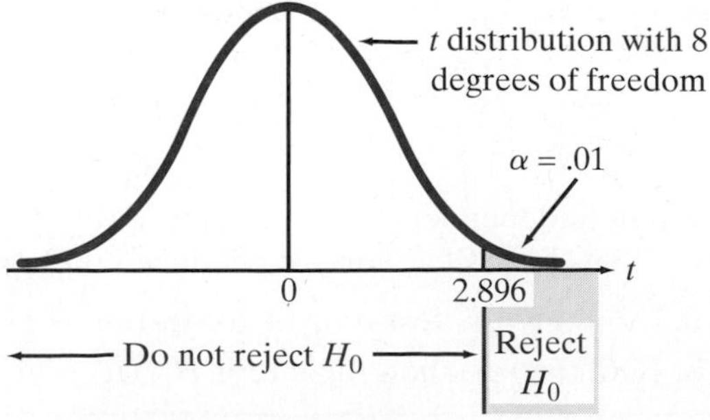

Figure 9.4 Rejection Region for Example 9.6

	A	B	C
1		Novocaine	New
2			
3	Mean	53.8	36.4
4	Standard Error	5.535341	3.443835
5	Median	50	38
6	Mode	#N/A	#N/A
7	Standard Deviation	12.3774	7.700649
8	Sample Variance	153.2	59.3
9	Kurtosis	-1.56299	-2.33196
10	Skewness	0.659365	-0.25271
11	Range	29	18
12	Minimum	42	27
13	Maximum	71	45
14	Sum	269	182
15	Count	5	5

a.

	A	B	C
1	t-Test: Two-Sample Assuming Equal Variances		
2			
3		Novocaine	New
4	Mean	53.8	36.4
5	Variance	153.2	59.3
6	Observations	5	5
7	Pooled Variance	106.25	
8	Hypothesized Mean Difference	0	
9	df	8	
10	t Stat	2.669038	
11	P(T<=t) one-tail	0.014202	
12	t Critical one-tail	2.896468	
13	P(T<=t) two-tail	0.028405	
14	t Critical two-tail	3.355381	

t statistic (t Stat); *p*-value (P(T<=t) one-tail)

b.

Figure 9.5 Excel Output for Example 9.6

To compute the test statistic, we need to find $\bar{x}_1, \bar{x}_2, s_1$, and s_2. These summary statistics are highlighted in the Excel printout, Figure 9.5a, as $\bar{x}_1 = 53.8$, $\bar{x}_2 = 36.4$, $s_1 = 12.38$, and $s_2 = 7.70$. Since we have assumed that the two populations have equal variances (i.e., $\sigma_1^2 = \sigma_2^2 = \sigma^2$), the next step is to compute an estimate of this common variance. Our pooled estimate is given by equation 9.7:

$$s_p^2 = \frac{(n_1 - 1)s_1^2 + (n_2 - 1)s_2^2}{n_1 + n_2 - 2} = \frac{(5 - 1)(12.38)^2 + (5 - 1)(7.70)^2}{5 + 5 - 2}$$

$$= \frac{849.98}{8} = 106.25$$

We now substitute the appropriate quantities into equation 9.5 to obtain the test statistic:

$$t = \frac{(\bar{x}_1 - \bar{x}_2) - 0}{\sqrt{s_p^2\left(\frac{1}{n_1} + \frac{1}{n_2}\right)}} = \frac{(53.8 - 36.4) - 0}{\sqrt{106.25\left(\frac{1}{5} + \frac{1}{5}\right)}} = 2.67$$

Since the test statistic $t = 2.67$ does not fall in the rejection region, we fail to reject H_0 in favor of H_a. That is, there is insufficient evidence (at $\alpha = .01$) to infer that μ_2, the mean discomfort level for patients receiving the new anesthetic, is less than μ_1, the corresponding mean for patients treated with Novocaine. Therefore, there is no evidence that the new anesthetic is more effective than the standard one.

The test could also be performed using the *p*-value approach. The *p*-value of this upper-tailed test, highlighted on the Excel printout, Figure 9.5b, is .0142. Since *p*-value = .0142 is greater than $\alpha = .01$, we cannot reject H_0.

Self-Test 9.3

The summary data for Self-Test 9.2 are repeated below:

$$\bar{x}_1 = 11.5 \quad s_1 = 10.2 \quad \bar{x}_2 = 9.0 \quad s_2 = 5.6$$

Assume independent random samples of $n_1 = 15$ blue-collar workers and $n_2 = 13$ white-collar workers produced these statistics.

a. Compute the test statistic for testing H_0: $(\mu_1 - \mu_2) = 0$

b. Find the rejection region for H_a: $(\mu_1 - \mu_2) > 0$ at $\alpha = .05$.

Note: The assumptions required for estimating and testing $(\mu_1 - \mu_2)$ with small samples do not have to be satisfied exactly for the procedures to be useful in practice. Slight departures from these assumptions do not seriously affect the level of confidence or α value. For situations where the variances σ_1^2 and σ_2^2 of the sampled populations are unequal, a modified t test is available (see Satterthwaite, 1946). When the sampled populations depart greatly from normality, the t distribution is invalid and any inferences derived from the procedure are suspect.

CAUTION

When the sampled populations are decidedly nonnormal (e.g., highly skewed), any inferences derived from the small-sample t test or confidence interval for $(\mu_1 - \mu_2)$ are suspect. In this case, one alternative is to use the nonparametric Wilcoxon rank sum test of optional Section 9.8.

Statistics in the Real World Revisited

Identifying the Appropriate Method for Analyzing the Twins-IQ Data

TWINS

Refer to the study of identical twins reared apart (p. 451). The psychologist wants to compare the mean IQ scores of two populations of twins, where one population (A) consists of all twins who were raised by their natural parents and the other population (B) consists of twins who are reared by a relative or some other person. One way to compare the two population IQ means, μ_A and μ_B, is to apply the independent samples t test procedure to the data stored in the **TWINS** file. A MINITAB printout of the analysis is displayed in Figure 9.6.

```
Two sample T for Twin A IQ score vs Twin B IQ score

            N       Mean      StDev   SE Mean
Twin A I   32       93.2       15.2       2.7
Twin B I   32       96.1       13.9       2.4

90% CI for mu Twin A I - mu Twin B I: ( -9.0,  3.2)
T-Test mu Twin A I = mu Twin B I (vs not =): T = -0.80  P = 0.43  DF = 62
Both use Pooled StDev = 14.5
```

p-value

t statistic

M **Figure 9.6** MINITAB Two-Sample *t* Test Output for Twins IQ Data

The p-value for testing the null hypothesis, H_0: $(\mu_A - \mu_B) = 0$, against the alternative hypothesis, H_a: $(\mu_A - \mu_B) \neq 0$, is highlighted on the printout as well as a 90% confidence interval for $(\mu_A - \mu_B)$. Since the p-value (.43) is greater than $\alpha = .10$ and the confidence interval, (9.0, 3.2), contains 0, there is insufficient evidence to claim that the population mean IQ scores are different.

Based on this analysis, a researcher would likely conclude that there is no significant difference between the mean IQ scores of identical twins reared apart. However, the statistical method applied here requires that the two samples of twins be collected *independently*. In fact, the two samples are dependent; one member of sample A is the identical twin of a member of sample B! Consequently, the validity of any inferences derived from the analysis is suspect. To properly analyze the data, we need a method that accounts for the dependency of the measurements in the two samples. This is the subject of the next section.

PROBLEMS FOR SECTION 9.2

Using the Tools

9.1 Independent random samples are selected from two populations with means μ_1 and μ_2, respectively. Determine approximate values of

$$\mu_{(\bar{x}_1 - \bar{x}_2)} \quad \text{and} \quad \sigma_{(\bar{x}_1 - \bar{x}_2)}$$

for each of the following situations. (Assume $\sigma_1^2 = \sigma_2^2$ in each case.)

a. $\bar{x}_1 = 150, s_1^2 = 36; \bar{x}_2 = 140, s_2^2 = 24; n_1 = n_2 = 35$

b. $\bar{x}_1 = 125, s_1^2 = 225; n_1 = 90; \bar{x}_2 = 112, s_2^2 = 90; n_2 = 60$

9.2 Consider two independent random samples with 30 observations selected from population 1 and 40 from population 2. The resulting sample means and variances are shown in the accompanying table. (Assume $\sigma_1^2 = \sigma_2^2$.) Construct a confidence interval for $(\mu_1 - \mu_2)$ using a confidence coefficient of:

a. .10 **b.** .05 **c.** .01

Sample from Population 1	Sample from Population 2
$\bar{x}_1 = 15$	$\bar{x}_2 = 23$
$s_1^2 = 16$	$s_2^2 = 20$
$n_1 = 30$	$n_2 = 40$

Problem 9.2

9.3 Refer to Problem 9.2. Find the test statistic for testing H_0: $\mu_1 - \mu_2 = 0$.

9.4 Two independent random samples, both large ($n > 30$), are selected from populations with unknown means μ_1 and μ_2. Specify the rejection region for each of the following combinations of H_0, H_a, and α.

a. H_0: $(\mu_1 - \mu_2) = 0$, H_a: $(\mu_1 - \mu_2) \neq 0$, $\alpha = .01$

b. H_0: $(\mu_1 - \mu_2) = 0$, H_a: $(\mu_1 - \mu_2) \neq 0$, $\alpha = .10$

c. H_0: $(\mu_1 - \mu_2) = 0$, H_a: $(\mu_1 - \mu_2) > 0$, $\alpha = .05$

d. H_0: $(\mu_1 - \mu_2) = 0$, H_a: $(\mu_1 - \mu_2) < 0$, $\alpha = .05$

9.5 Two independent random samples are selected from normal populations with unknown means μ_1 and μ_2 and variances $\sigma_1^2 = \sigma_2^2$. Specify the rejection region for each of the following combinations of H_a, n_1, n_2, and α when testing the null hypothesis H_0: $(\mu_1 - \mu_2) = 0$.

a. H_a: $(\mu_1 - \mu_2) \neq 0$, $n_1 = 10$, $n_2 = 10$, $\alpha = .05$

b. H_a: $(\mu_1 - \mu_2) > 0$, $n_1 = 8$, $n_2 = 4$, $\alpha = .10$

c. H_a: $(\mu_1 - \mu_2) < 0$, $n_1 = 6$, $n_2 = 5$, $\alpha = .01$

9.6 The following tables show summary statistics for random samples selected from two normal populations that are assumed to have the same variance. In each case, find s_p^2, the pooled estimate of the common variance.

a.

Sample from Population 1	Sample from Population 2
$\bar{x}_1 = 552$	$\bar{x}_2 = 369$
$s_1^2 = 4{,}400$	$s_2^2 = 7{,}481$
$n_1 = 6$	$n_2 = 7$

b.

Sample from Population 1	Sample from Population 2
$\bar{x}_1 = 10.8$	$\bar{x}_2 = 8.4$
$s_1^2 = .313$	$s_2^2 = .499$
$n_1 = 8$	$n_2 = 8$

9.7 Independent random samples from two normal populations with equal variances produced the sample means and sample variances listed in the accompanying table. Construct a confidence interval for $(\mu_1 - \mu_2)$ using a confidence coefficient of:

a. .10 **b.** .05 **c.** .01

Sample from Population 1	Sample from Population 2
$n_1 = 14$	$n_2 = 7$
$\bar{x}_1 = 53.2$	$\bar{x}_2 = 43.4$
$s_1^2 = 96.8$	$s_2^2 = 102.0$

Problem 9.7

9.8 Refer to Problem 9.7. Find the test statistic for testing H_0: $\mu_1 - \mu_2 = 0$.

9.9 To use the t statistic in making small-sample inferences about the difference between the means of two populations, what assumptions must be made about the two populations? About the two samples?

Applying the Concepts

9.10 A Johns Hopkins University medical researcher compared the health preferences of 14 healthy female volunteers to those of 13 healthy male volunteers (*Women &*

Therapy, Vol. 16, 1995). The volunteers rated several health conditions on a 5-point scale (1 = very desirable, . . . , 5 = very undesirable). In response to the condition "body pains," the women had an average rating of 4.64 while the men had an average rating of 4.07.

a. Based on only this information, can a reliable inference be made about the true difference in the mean "body pain" ratings of men and women? Explain.

b. A test of the null hypothesis H_0: $\mu_{\text{women}} = \mu_{\text{men}}$ resulted in an observed significance level of .006. Interpret this result.

9.11 When grazing, sheep tend to cluster together. An experiment was conducted to determine if spacing distance has an impact on sheep grazing behavior (*Applied Animal Behaviour Science,* Nov. 2000). Twenty female Scottish Blackface sheep were randomly divided into two equal groups. One group of 10 sheep was restricted during grazing to a 50-square-meter area; the other group of 10 sheep grazed in an unrestricted area. One variable of interest to the researchers was grazing time (measured in number of minutes per day).

a. Set up H_0 and H_a for testing whether the mean grazing times of the two groups of sheep differ.

b. The observed significance level of the test, part **a**, was reported as p-value = .001. Make the appropriate conclusion.

c. The sample means for the two groups of sheep were 622 (restricted) and 720 (unrestricted) minutes per day. The estimated standard error of the difference between the sample means was 18 minutes per day. Use this information to find a 95% confidence interval for the difference between the population means. Interpret the result.

d. What assumptions, if any, are necessary for the validity of the inferences in parts **b** and **c**?

9.12 Many psychologists believe that knowledge of a college student's relationships with his or her parents can be useful in predicting the student's future interpersonal relationships both on the job and in private life. Researchers at the University of South Alabama compared the attitudes of male and female students toward their fathers (*Journal of Genetic Psychology,* Mar. 1998). Using a five-point Likert-type scale, they asked each group to complete the following statement: My relationship with my father can best be described as: (1) Awful, (2) Poor, (3) Average, (4) Good, or (5) Great. The data (adapted from the article) is listed in the table.

FATHER

Father Ratings for Males

4 4 4 3 3 5 4 5 4 5 5 3 4 4 5 2 3 2 5 4 5 3 5 5 4 3 5 4 4 3
5 4 2 4 3 4 4 5 3 4 5 3 2 5

Father Ratings for Females

5 2 5 5 3 3 2 5 2 1 5 4 2 5 3 5 4 2 5 5 2 3 4 3 5 4 4 4 2 1
5 3 5 4 4 5 5 5 3 5 5 4 5 4 5 1 5 4 4 5 4 4 4 2 1 1 4 2 3 4
2 4 2 3 5 2 4 5 5 5 5 5 5 1 2 2 5 4 1 2 5 4 3 5 5 3 5 5 5 2

Source: Adapted from Vitulli, William F., and Richardson, Deanna K. "College students' attitudes toward relationships with parents: A five-year comparative analysis," *Journal of Genetic Psychology,* Vol. 159, No. 1, (Mar. 1998), pp. 45–52.

a. Do male college students tend to have better relationships, on average, with their fathers than female students? Conduct the appropriate hypothesis test using $\alpha = .01$.

b. What assumptions, if any, about the samples did you have to make to ensure the validity of the hypothesis test you conducted in part **a**?

c. Refer to part **b**. If you made assumptions, check to see if they are reasonably satisfied. If no assumptions were necessary, explain why.

9.13 Students enrolled in music classes at the University of Texas (Austin) participated in a study to compare the teacher evaluations of music education majors and non-music education majors (*Journal of Research in Music Education,* Winter 1991). Independent random samples of 100 music majors and 100 nonmusic majors rated the overall performance of their teacher using a 6-point scale, where 1 = lowest rating and 6 = highest rating.

	Music Majors	Nonmusic Majors
Sample size	100	100
Mean "overall" rating	4.26	4.59
Standard deviation	.81	.78

Source: Duke, R. A., and Blackman, M. D. "The relationship between observers' recorded teacher behavior and evaluation of music instruction." *Journal of Research in Music Education,* Vol. 39, No. 4, Winter 1991 (Table 2).

a. Conduct a test of hypothesis to compare the mean teacher ratings of the two groups of music students. Use $\alpha = .05$.

b. Repeat part **a**, but use a 95% confidence interval. Interpret the result.

9.14 Refer to the *Journal of Agricultural, Biological, and Environmental Statistics* (Sept. 2000) study of the impact of the *Exxon Valdez* tanker oil spill on the seabird population in Alaska, Problem 2.62 (p. 107). The CD file called **EVOS** contains data on the number of seabirds found at each of 96 shoreline transects and whether or not the transect was in an oiled area. Compare the mean number of seabirds found in oiled and unoiled areas. Is there sufficient evidence of a difference between the means at $\alpha = .01$? Use the accompanying MINITAB printout to make your decision.

```
Two sample T for Count

Oil          N       Mean      StDev   SE Mean
no          36       18.7       46.3       7.7
yes         60       20.1       40.8       5.3

95% CI for mu (no ) - mu (yes): ( -19.3,  16.6)
T-Test mu (no ) = mu (yes) (vs not =): T = -0.15  P = 0.88  DF = 94
Both use Pooled StDev = 42.9
```

M Problem 9.14

9.15 As a result of advances in educational telecommunications, many colleges and universities are utilizing instruction by interactive television for "distance" education. For example, each semester Ball State University televises six graduate business courses to students at remote off-campus sites (*Journal of Education for Business,* Jan./Feb. 1991). To compare the performance of the off-campus MBA students at Ball State (who take the televised classes) to the on-campus MBA students (who have a "live" professor), a comprehensive test was administered to a sample of both groups of students. The test scores (50 points maximum) are summarized in the table. Based on these results, the researchers report that "there was no significant difference between the two groups of students."

	Mean	Standard Deviation
On-campus students	41.93	2.86
Off-campus TV students	44.56	1.42

Source: Arndt, T. L., and LaFollette, W. R. "Interactive television and the nontraditional student," *Journal of Education for Business,* Jan./Feb. 1991, p. 184.

a. Note that the sample sizes were not given in the journal article. Assuming 50 students are sampled from each group, perform the desired analysis. Do you agree with the researchers' findings?

b. Repeat part **a**, but assume 15 students are sampled from each group.

9.16 Intensive supervision (ISP) officers are parole officers who receive extensive training on the principles of effective intervention in order to help the criminal offender on parole or probation. The *Prison Journal* (Sept. 1997) published a study comparing the attitudes of ISP officers to regular parole officers who had not received any intervention training. Independent random samples of 11 ISP parole officers and 61 regular parole officers were recruited for the study. The officers' responses to their subjective role (7–42 point scale) and strategy (4–24 point scale) are summarized in the table.

			Subjective Role		Strategy	
		Sample Size	**Mean**	**Standard Deviation**	**Mean**	**Standard Deviation**
Group	**Regular**	61	26.15	5.51	12.69	2.94
	ISP	11	19.50	4.35	9.00	2.00

Source: Fulton, B., et al. "Moderating probation and parole officer attitudes to achieve desired outcomes." *The Prison Journal,* Vol. 77, No. 3, Sept. 1997, p. 307 (Table 3).

a. At $\alpha = .05$, conduct a test to detect a difference between the mean subjective role responses for the two groups of parole officers.

b. Construct and interpret a 95% confidence interval for the difference between the mean strategy responses for the two groups of parole officers.

c. What assumptions are necessary for the validity of the inferences in parts **a** and **b**?

9.17 *Psychological Reports* (Aug. 1997) published a study on whether confidence in your ability to do well is gender-related. Undergraduate students in a social psychology class were given a 26-item pretest early in the semester. Each student was asked to indicate the number of questions they were confident that they had answered correctly. A summary of the results, by gender, is given in the table.

			Number of "Confident" Answers		Actual Number of Correct Answers	
		Sample Size	**Mean**	**Standard Deviation**	**Mean**	**Standard Deviation**
Gender	**Men**	29	14.1	6.0	12.1	2.5
	Women	87	9.7	5.3	12.1	3.2

Source: Yarab, P. E., Sensibaugh, C. C., and Allgeier, E. R. "Over-confidence and underperformance: Men's perceived accuracy and actual performance in a course." *Psychological Reports,* Vol. 81, No. 1, Aug. 1997, p. 77 (Table 1).

a. Compare the mean number of "confident" answers for men to the corresponding mean for women with a test of hypothesis. Use $\alpha = .01$.

b. Compare the mean number of actual correct answers for men to the corresponding mean for women with a test of hypothesis. Use $\alpha = .01$.

9.18 Refer to the Harris Corporation/University of Florida study to determine whether a manufacturing process performed at a remote location can be established locally, Problem 3.38 (p. 154). Test devices (pilots) were set up at both the old and new locations and voltage readings on 30 production runs at each location were obtained. The data are reproduced in the table on p. 466. Summary results are displayed in the accompanying Excel and MINITAB printouts. [*Note:* Smaller voltage readings are better than larger voltage readings.]

VOLTAGE

Old Location			New Location		
9.98	10.12	9.84	9.19	10.01	8.82
10.26	10.05	10.15	9.63	8.82	8.65
10.05	9.80	10.02	10.10	9.43	8.51
10.29	10.15	9.80	9.70	10.03	9.14
10.03	10.00	9.73	10.09	9.85	9.75
8.05	9.87	10.01	9.60	9.27	8.78
10.55	9.55	9.98	10.05	8.83	9.35
10.26	9.95	8.72	10.12	9.39	9.54
9.97	9.70	8.80	9.49	9.48	9.36
9.87	8.72	9.84	9.37	9.64	8.68

Source: Harris Corporation, Melbourne, Fla.

Problem 9.18

	A	B	C
1	t-Test: Two-Sample Assuming Equal Variances		
2			
3		*Old Location*	*New Location*
4	Mean	9.803666667	9.422333333
5	Variance	0.29258954	0.229321954
6	Observations	30	30
7	Pooled Variance	0.260955747	
8	Hypothesized Mean Difference	0	
9	df	58	
10	t Stat	2.891125647	
11	P(T<=t) one-tail	0.002697103	
12	t Critical one-tail	1.671553491	
13	P(T<=t) two-tail	0.005394206	
14	t Critical two-tail	2.001715984	

E Problem 9.18

```
Two sample T for Old Location vs New Location

           N      Mean     StDev   SE Mean
Old Loca  30     9.804     0.541     0.099
New Loca  30     9.422     0.479     0.087

95% CI for mu Old Loca - mu New Loca: ( 0.117,  0.645)
T-Test mu Old Loca = mu New Loca (vs not =): T = 2.89  P = 0.0054  DF = 58
Both use Pooled StDev = 0.511
```

M Problem 9.18

a. Compare the mean voltage readings at the two locations using a test of hypothesis ($\alpha = .05$).

b. Based on the test results, does it appear that the manufacturing process can be established locally?

9.19 Refer to the *Applied Psycholinguistics* (June 1998) study comparing the language skills of children from low income and middle income families, Problem 3.40 (p. 155). Descriptive statistics for the variable "sentence complexity score" for each of the two groups of children are reproduced in the table below. Compare the population mean scores for the two groups using a 99% confidence interval. Which income group appears to have the highest mean sentence complexity score?

	Low Income	Middle Income
Sample size	65	195
Mean	7.62	15.55
Standard deviation	8.91	12.24

Source: Arriaga, R. I., et al. "Scores on the MacArthur Communicative Development Inventory of children from low-income and middle-income families." *Applied Psycholinguistics,* Vol. 19, No. 2, June 1998, p. 217 (Table 7).

9.3 Comparing Two Population Means: Matched Pairs

The procedures for comparing two population means presented in Section 9.2 were based on the assumption that the samples were randomly and independently selected from the target populations. Sometimes we can obtain more information about the difference between population means ($\mu_1 - \mu_2$) by selecting paired observations or repeated measurements on the same experimental unit.

EXAMPLE 9.7

DRAWBACK TO USING INDEPENDENT SAMPLES

An elementary education teacher wants to compare two methods for teaching reading skills to first graders. One way to design the experiment is to randomly select 20 students from all available first graders, and then randomly assign 10 students to method 1 and 10 students to method 2. The reading achievement test scores obtained after completion of the experiment would represent independent random samples of scores attained by students taught reading skills by the two different methods. Consequently, the procedures described in Section 9.2 could be used to make inferences about ($\mu_1 - \mu_2$), the difference between the mean achievement test scores of the two methods.

a. Comment on the potential drawbacks of using independent random samples to make inferences about ($\mu_1 - \mu_2$).

b. Propose a better method of sampling, one that will yield more information on the parameter of interest.

Solution

a. Assume that method 1 is truly more effective than method 2 in teaching reading skills to first graders. A potential drawback to the independent sampling plan is that the differences in the reading skills of first graders due to IQ, learning ability, socioeconomic status, and other factors are not taken into account. For example, by chance the sampling plan may assign the 10 "worst" students to method 1 and the 10 "best" students to method 2. This unbalanced assignment may mask the fact that method 1 is more effective than method 2, i.e., the resulting test of hypothesis or confidence interval on ($\mu_1 - \mu_2$) may fail to show that μ_1 exceeds μ_2.

b. A better method of sampling is one that attempts to remove the variation in achievement test scores due to extraneous factors such as IQ, learning ability, and socioeconomic status. One way to do this is to match the first graders in pairs, where the students in each pair have similar IQ, socioeconomic status, etc. From each pair, one member would be randomly selected to be taught by method 1; the other member would be assigned to the class taught by method 2 (see Figure 9.7). The differences between the **matched pairs** of achievement test scores should provide a clearer picture of the true difference in achievement for the two reading methods because the matching would tend to cancel the effects of the extraneous factors that formed the basis of the matching.

Figure 9.7 Matched Pairs Experiment, Example 9.7 (Pair Members Identified as Student A and Student B)

First-Grader Pair	Assignment	
	Method 1	Method 2
1	A	B
2	B	A
⋮	⋮	⋮
10	A	B

The sampling plan shown in Figure 9.7 is commonly known as a *matched-pairs experiment.* In the next box we give the procedures for estimating and testing the difference between two population means based on matched-pairs data. You can see that once the differences in the paired observations are obtained, the analysis proceeds as a one-sample problem. That is, a confidence interval or test for a single mean (the mean of the difference μ_d) is computed.

MATCHED PAIRS TEST OF HYPOTHESIS ABOUT $\mu_d = (\mu_1 - \mu_2) = 0$

Let $d_1, d_2, \ldots, d_n$ represent the differences between the pairwise observations in a *random sample* of n matched pairs, $\bar{d}$ = mean of the n sample differences, and s_d = standard deviation of the n sample differences.

Upper-Tailed Test	Two-Tailed Test	Lower-Tailed Test
H_0: $\mu_d = 0$	H_0: $\mu_d = 0$	H_0: $\mu_d = 0$
H_a: $\mu_d > 0$	H_a: $\mu_d \neq 0$	H_a: $\mu_d < 0$

Large Sample:

$$\textit{Test statistic: } z = \frac{\bar{d} - 0}{\sigma_d/\sqrt{n}} \approx \frac{\bar{d} - 0}{s_d/\sqrt{n}} \qquad \textbf{(9.10)}$$

Rejection region:	*Rejection region:*	*Rejection region:*
$z > z_\alpha$	$\lvert z \rvert > z_{\alpha/2}$	$z < -z_\alpha$

Small Sample:

$$\textit{Test statistic: } t = \frac{\bar{d} - 0}{s_d/\sqrt{n}} \qquad \textbf{(9.11)}$$

Rejection region:	*Rejection region:*	*Rejection region:*
$t > t_\alpha$	$\lvert t \rvert > t_{\alpha/2}$	$t < -t_\alpha$

where t_α and $t_{\alpha/2}$ are based on $(n - 1)$ degrees of freedom.

$(1 - \alpha)100\%$ CONFIDENCE INTERVAL FOR $\mu_d = (\mu_1 - \mu_2)$ MATCHED PAIRS

Large Sample:

$$\bar{d} \pm z_{\alpha/2}\left(\frac{\sigma_d}{\sqrt{n}}\right) \qquad \textbf{(9.12)}$$

where σ_d is the population standard deviation of differences.

Small Sample:

$$\bar{d} \pm t_{\alpha/2}\left(\frac{s_d}{\sqrt{n}}\right) \qquad \textbf{(9.13)}$$

where $t_{\alpha/2}$ is based on $(n - 1)$ degrees of freedom.

MATCHED PAIRS ASSUMPTIONS WITH A SMALL SAMPLE ($n < 30$)

1. The relative frequency distribution of the population of differences is approximately normal.
2. The paired differences are randomly selected from the population of differences.

EXAMPLE 9.8 **HYPOTHESIS TEST FOR μ_d**

In the comparison of two methods for teaching reading discussed in Example 9.7, suppose that the $n = 10$ pairs of achievement test scores were as shown in Table 9.3. At $\alpha = .05$, test for a difference between the mean achievement scores of students taught by the two methods.

READING

TABLE 9.3 Reading Achievement Test Scores for Example 9.8

	Student Pair									
	1	2	3	4	5	6	7	8	9	10
Method 1 score	78	63	72	89	91	49	68	76	85	55
Method 2 score	71	44	61	84	74	51	55	60	77	39
Paired difference	7	19	11	5	17	−2	13	16	8	16

Solution

We want to test H_0: $\mu_d = 0$ against H_a: $\mu_d \neq 0$, where $\mu_d = (\mu_1 - \mu_2)$. The differences between the $n = 10$ matched pairs of reading achievement test scores are computed as

$$d = (\text{Method 1 score}) - (\text{Method 2 score})$$

and are shown in the third row of Table 9.3. Since $n = 10$ (i.e., a small sample), we must assume that these differences are from an approximately normal population in order to conduct the small-sample test. The mean and standard deviation of these sample differences are $\bar{d} = 11.0$ and $s_d = 6.53$. Substituting these values into the formula for the small-sample test statistic (equation 9.11), we obtain

$$t = \frac{\bar{d} - 0}{s_d/\sqrt{n}} = \frac{11.0 - 0}{6.53/\sqrt{10}} = 5.33$$

This test statistic is highlighted on the Excel printout, Figure 9.8, as well as the two-tailed p-value of the test. Since p-value $= .000478$ is less than $\alpha = .05$, there is sufficient evidence to reject H_0 in favor of H_a. Our conclusion is that method 1 produces a mean achievement test score that is statistically different than the mean score for method 2.

E Figure 9.8 Excel Printout for Example 9.8

	A	B	C
1	t-Test: Paired Two Sample for Means		
2			
3		*Method 1*	*Method 2*
4	Mean	72.6	61.6
5	Variance	198.0444	217.8222
6	Observations	10	10
7	Pearson Correlation	0.09842	
8	Hypothesized Mean	0	
9	df	9	
10	t Stat	5.325352	
11	P(T<=t) one-tail	0.000239	
12	t Critical one-tail	1.833114	
13	P(T<=t) two-tail	0.000478	
14	t Critical two-tail	2.262159	

t statistic

p-value

Self-Test 9.4

A random sample of $n = 200$ paired observations yielded the following summary information: $\bar{d} = 115$, $s_d = 250$.

a. Is there evidence that μ_1 differs from μ_2? Test at $\alpha = .05$.

b. Find a 95% confidence interval for $\mu_d = \mu_1 - \mu_2$.

Often, the pairs of observations in a matched-pairs experiment arise naturally by recording two measurements on the same experimental unit at two different

points in time. For example, a professor at Brandeis University investigated the effectiveness of a prescription drug in improving the SAT scores of nervous test takers (*Newsweek*, Nov. 16, 1987). The experimental units (the objects upon which the measurements are taken) were 22 high school juniors who took the SAT twice, once without and once with the drug. Thus, two measurements were taken on each junior: (1) SAT score with no drug and (2) SAT score with the drug. These two observations, taken for all students in the study, formed the "matched pairs" of the experiment.

In an analysis of matched-pairs observations, it is important to stress that the pairing of the experimental units must be performed *before* the data are collected. Recall that the objective is to compare two methods of "treating" the experimental units. By using matched pairs of units that have similar characteristics, we are able to cancel out the effects of the variables used to match the pairs.

As in the previous section, we close with a caution.

CAUTION

It is inappropriate to apply the small-sample matched-pairs *t* procedure when the population of differences is decidedly nonnormal (e.g., highly skewed). In this case, an alternative procedure is the nonparametric Wilcoxon signed ranks test of optional Section 9.9.

Statistics in the Real World Revisited

Comparing the Mean IQ Scores of Identical Twins Reared Apart

TWINS

In the study of $n = 32$ pairs of identical twins reared apart (p. 451), one member (A) of each pair was reared by a natural parent and one member (B) was reared by a relative or some other person. Consequently, the psychologist has collected matched pairs of experimental units. To properly compare the mean IQ scores of the two populations of twins, an analysis for a matched-pairs experiment should be conducted on the data in the **TWINS** file. Here, the population parameter of interest is $\mu_d = (\mu_A - \mu_B)$, where μ_A and μ_B are the IQ means for populations A and B, respectively.

A MINITAB printout of the matched-pairs analysis is displayed in Figure 9.9. The *p*-value for testing the null hypothesis, H_0: $\mu_d = 0$, against the alternative hypothesis, H_a: $\mu_d \neq 0$, is highlighted on the printout as well as a 90% confidence interval for μ_d.

```
Paired T for Twin A IQ score - Twin B IQ score

              N      Mean     StDev   SE Mean
Twin A I     32     93.22     15.17      2.68
Twin B I     32     96.12     13.86      2.45
Difference   32     -2.91      8.89      1.57

90% CI for mean difference: (-5.57, -0.24)
T-Test of mean difference = 0 (vs not = 0): T-Value = -1.85  P-Value = 0.074
```

t statistic

M Figure 9.9 MINITAB Matched-Pairs Output for Twins IQ Data

If we select a Type I error rate of $\alpha = .10$, then *p*-value $= .074$ is less than $\alpha = .10$ and there is sufficient evidence to reject H_0. Our conclusion is that there is a statistically significant difference between the mean IQ scores of identical twins reared apart. (Note that this inference differs from the inference derived from applying the invalid independent samples *t* test, p. 461.) However, the 90% confidence interval, $(-5.57, -.24)$, implies that this difference is very small; we are 90% confident that the difference in mean IQ scores is, at maximum, only 5.57 points and could be as small as .24 point. From a practical viewpoint, most psychologists would agree that the detected difference is negligible. (This is another example of a "statistically" significant result that is not "practically" significant.)

PROBLEMS FOR SECTION 9.3

Using the Tools

9.20 The data for a random sample of four paired observations are shown in the accompanying table.

Pair	Observation from Population A	Observation from Population B
1	2	0
2	5	7
3	10	6
4	8	5

a. Calculate the difference within each pair, subtracting observation B from observation A. Use the differences to calculate $\bar{d}$ and s_d.

b. If μ_1 and μ_2 are the means of populations A and B, respectively, express μ_d in terms of μ_1 and μ_2.

c. Compute the test statistic for testing H_0: $\mu_d = 0$ against H_a: $\mu_d > 0$.

d. Find the rejection region ($\alpha = .05$) for the test, part **c**.

e. What is the conclusion of the test, parts **c** and **d**?

f. Construct a 95% confidence interval for μ_d.

9.21 A matched-pairs experiment is used to test the hypothesis H_0: $(\mu_1 - \mu_2) = 0$. For each of the following situations, specify the rejection region, test statistic, and conclusion.

a. H_a: $(\mu_1 - \mu_2) \neq 0$, $\bar{d} = 400$, $s_d = 435$, $n = 100$, $\alpha = .01$

b. H_a: $(\mu_1 - \mu_2) > 0$, $\bar{d} = .48$, $s_d = .08$, $n = 5$, $\alpha = .05$

c. H_a: $(\mu_1 - \mu_2) < 0$, $\bar{d} = -1.3$, $s_d^2 = .95$, $n = 5$, $\alpha = .10$

9.22 A random sample of 10 paired observations yielded the following summary information: $\bar{d} = 2.3$, $s_d = 2.67$.

a. Find a 90% confidence interval for μ_d.

b. Find a 95% confidence interval for μ_d.

c. Find a 99% confidence interval for μ_d.

9.23 A random sample of $n = 50$ paired observations yielded the following summary statistics: $\bar{d} = 19.3$, $s_d = 5.2$.

a. Compute the test statistic for testing H_0: $\mu_d = 0$ against H_a: $\mu_d \neq 0$.

b. Find the p-value for the test, part **a**.

c. What is the conclusion of the test, parts **a** and **b**?

d. Construct a 95% confidence interval for μ_d.

9.24 The data for a random sample of seven paired observations are shown in the accompanying table. Let $\mu_d = (\mu_A - \mu_B)$.

Pair	Observation from Population A	Observation from Population B
1	48	54
2	50	56
3	47	50
4	50	55
5	63	64
6	65	65
7	55	61

a. Test H_0: $\mu_d = 0$ against H_a: $\mu_d < 0$ at $\alpha = .05$.

b. Test H_0: $\mu_d = 0$ against H_a: $\mu_d \neq 0$ at $\alpha = .05$.

c. Find a 90% confidence interval for μ_d.

9.25 List the assumptions required to make valid inferences about μ_d based on:

a. a large sample of matched pairs

b. a small sample of matched pairs

Applying the Concepts

9.26 When searching for an item (e.g., a roadside traffic sign, a lost earring, or a tumor in a mammogram), common sense dictates that you will not re-examine items previously rejected. However, researchers at Harvard Medical School found that a visual search has no memory (*Nature,* Aug. 6, 1998). In their experiment, nine subjects searched for the letter "T" among several letters "L." Each subject conducted the search under two conditions: random and static. In the random condition, the locations of the letters were changed every 111 milliseconds; in the static condition, the locations of the letters remained unchanged. In each trial, the reaction time (i.e., the amount of time it took the subject to locate the target letter) was recorded in milliseconds.

a. One goal of the research is to compare the mean reaction times of subjects in the two experimental conditions. Explain why the data should be analyzed as a paired-difference experiment.

b. If a visual search had no memory, then the mean reaction times in the two conditions will not differ. Specify the null and alternative hypotheses appropriate for testing the "no memory" theory.

c. The test statistic for comparing the two means was calculated as $t = 1.52$ with an associated p-value of .15. What conclusion can you make?

9.27 Internet search engines, which allow users to simply enter key words and click the "Search" button, are the most heavily used on-line services, with millions of searches performed each day. The journal *Online* (May/June 1999) published the results of a test of five major search engines—AltaVista, Excite, HotBot, InfoSeek, and Lycos. Based on the engine's ability to find sites that match the key words, the researchers computed a "matching" index for each engine. The lower the index, the fewer instances where the selected site (i.e., the "hit") did not satisfy the matching criterion. The table below lists the matching indexes for each search engine based on the top 20 "hits" and the top 100 "hits." Assuming the five search engines represent a random sample of all Internet search engines, find a 90% confidence interval for the difference between the mean matching indexes for top 20 and top 100 hits. Interpret the result.

SEARCHENG

Search Engine	Top 20 Hits	Top 100 Hits
AltaVista	.120	.320
Excite	.050	.190
HotBot	.123	.195
InfoSeek	.170	.240
Lycos	.054	.084

9.28 According to *Webster's New World Dictionary,* a tongue-twister is "a phrase that is hard to speak rapidly." Do tongue-twisters have an effect on the length of time it takes to read silently? To answer this question, 42 undergraduate psychology students participated in a reading experiment (*Memory & Cognition,* Sept. 1997). Two lists, each comprised of 600 words, were constructed. One list contained a series of tongue-twisters and the other list (called the *control*) did not contain any tongue-twisters. Each student read both lists and the length of time (in minutes) required to complete the lists was recorded. The researchers used a test of hypothesis to compare the mean reading response times for the tongue-twister and control lists.

a. Set up the null hypothesis for the test.

b. Use the information in the accompanying table to find the test statistic and p-value of the test.

c. Give the appropriate conclusion. Use $\alpha = .05$.

		Response Time (Minutes)	
		Mean	**Standard Deviation**
List Type	**Tongue-twister**	6.59	1.94
	Control	6.34	1.92
	Difference	.25	.78

Source: Robinson, D. H., and Katayama, A. D. "At-lexical, articulatory interference in silent reading: The 'upstream' tongue-twister effect." *Memory & Cognition,* Vol. 25, No. 5, Sept. 1997, p. 663.

Problem 9.28

9.29 Refer to the *Chemosphere* study of Vietnam veterans' exposure to Agent Orange, Problem 7.28 (p. 359). In addition to the amount of TCDD (measured in parts per million) in blood plasma, the TCDD in fat tissue drawn from 20 exposed Vietnam veterans was recorded. The data are shown in the table. Compare the mean TCDD level in plasma to the mean TCDD level in fat tissue for Vietnam veterans exposed to Agent Orange using a test of hypothesis at $\alpha = .05$.

TCDDFAT

Veteran	TCDD Levels in Plasma	TCDD Levels in Fat Tissue	Veteran	TCDD Levels in Plasma	TCDD Levels in Fat Tissue
1	2.5	4.9	11	6.9	7.0
2	3.1	5.9	12	3.3	2.9
3	2.1	4.4	13	4.6	4.6
4	3.5	6.9	14	1.6	1.4
5	3.1	7.0	15	7.2	7.7
6	1.8	4.2	16	1.8	1.1
7	6.0	10.0	17	20.0	11.0
8	3.0	5.5	18	2.0	2.5
9	36.0	41.0	19	2.5	2.3
10	4.7	4.4	20	4.1	2.5

Source: Schecter, A., et al. "Partitioning of 2,3,7,8-chlorinated dibenzo-*p*-dioxins and dibenzofurans between adipose tissue and plasma lipid of 20 Massachusetts Vietnam veterans." *Chemosphere,* Vol. 20, Nos. 7–9, 1990, pp. 954–955 (Tables I & II).

9.30 Refer to the *Journal of Genetic Psychology* (Mar. 1998) comparison of male and female college students' attitudes toward their fathers, Problem 9.12 (p. 463). In this problem, data adapted from the same study are used to compare male students' attitudes toward their fathers with their attitudes toward their mothers. Each of a

FATHMOTH

Student	Attitude toward Father	Attitude toward Mother
1	2	3
2	5	5
3	4	3
4	4	5
5	3	4
6	5	4
7	4	5
8	2	4
9	4	5
10	5	4
11	4	5
12	5	4
13	3	3

Source: Adapted from Vitulli, William F., and Richardson, Deanna K. "College students' attitudes toward relationships with parents: A five-year comparative analysis," *Journal of Genetic Psychology,* Vol. 159, No. 1, (March 1998), pp. 45–52.

sample of 13 males from the original sample of 44 males was asked to complete the following statement about each of their parents: My relationship with my father (mother) can best be described as: (1) Awful, (2) Poor, (3) Average, (4) Good, or (5) Great. The data are shown on p. 473:

a. Specify the appropriate hypotheses for testing whether male students' attitudes toward their fathers differ from their attitudes toward their mothers, on average.

b. Conduct the test of part **a** at $\alpha = .05$. Interpret the results in the context of the problem.

9.31 The drug clonidine has been shown to be beneficial in the treatment of Tourette's Syndrome. Since patients with Tourette's disorder often stutter, clonidine may be useful in treating stuttering. The *American Journal of Psychiatry* (July 1995) reported on a study of 25 stuttering children who were treated with clonidine. Prior to treatment, a speech sample was recorded for each child and the frequency of speech repetitions was measured. The experiment was repeated eight weeks later after the children were treated with a daily dose of clonidine. The difference in stuttering frequency before and after treatment was recorded for each child. These differences are summarized here:

$$\bar{d} = -2.33 \qquad s_d = 16.98$$

a. The journal reported a 95% confidence interval for the true mean difference in stuttering frequencies as (−9.3, 4.7). Do you agree with the result?

b. The journal reported that "clonidine was not found to have a positive effect on stuttering." Do you agree? Explain.

c. Conduct a test of hypothesis (at $\alpha = .05$) to confirm the result, part **b**.

9.32 Merck Research Labs conducted an experiment to evaluate the effect of a new drug using the Single-T swim maze. Nineteen impregnated dam rats were captured and allocated a dosage of 12.5 milligrams of the drug. One male and one female pup were randomly selected from each resulting litter to perform in the swim maze. Each rat pup is placed in the water at one end of the maze and allowed to swim until it successfully escapes at the opposite end. If the rat pup fails to escape after a certain period of time, it is placed at the beginning of the maze and given another attempt to escape. The experiment is repeated until three successful escapes are accomplished by each rat pup. The number of swims required by each pup to perform three successful escapes is reported in the accompanying table. Is there sufficient evidence of a difference between the mean number of swims required by male and female rat pups? Test using $\alpha = .10$.

SWIMMAZE

Litter	Male	Female	Litter	Male	Female
1	8	5	11	6	5
2	8	4	12	6	3
3	6	7	13	12	5
4	6	3	14	3	8
5	6	5	15	3	4
6	6	3	16	8	12
7	3	8	17	3	6
8	5	10	18	6	4
9	4	4	19	9	5
10	4	4			

Source: Thomas E. Bradstreet, Merck Research Labs, BL 3-2, West Point, Penn. 19486

9.33 Refer to the *Journal of Geography* (May/June 1997) study of the effectiveness of computer-based interactive maps in the classroom, Problem 2.27 (p. 78). Each of 45 seventh grade geography students was given two assignments. For the first assignment, the students used traditional library resources (books, atlases, and encyclopedias) to research a particular African region. For the second assignment,

the students used computer resources (electronic atlases and encyclopedias, interactive maps) to research a different African region. At the end of each assignment, a test is administered to the students. The test scores are listed in the accompanying table.

GEOSCORE

Student	Test #1	Test #2	Student	Test #1	Test #2	Student	Test #1	Test #2
1	9	9	16	56	59	31	38	47
2	16	9	17	25	31	32	94	91
3	9	3	18	59	75	33	94	100
4	56	56	19	81	66	34	75	75
5	38	47	20	59	47	35	50	66
6	31	63	21	81	63	36	81	91
7	50	47	22	41	56	37	75	78
8	53	59	23	91	75	38	88	63
9	69	72	24	53	69	39	78	78
10	63	56	25	56	63	40	44	63
11	41	63	26	38	56	41	72	69
12	56	47	27	25	72	42	59	63
13	50	47	28	63	72	43	69	81
14	59	53	29	47	69	44	84	88
15	53	63	30	50	63	45	69	91

Source: Linn, S. E. "The effectiveness of interactive maps in the classroom: A selected example in studying Africa." *Journal of Geography,* Vol. 96, No. 3, May/June 1997, p. 167 (Table 1).

Problem 9.33

a. Is there evidence of a difference between the two mean test scores at $\alpha = .10$?

b. What assumptions (if any) are necessary for the inference in part **a** to be valid? Explain.

9.4 Comparing Two Population Proportions: Independent Samples

In this section we present methods for making inferences about the difference between two population proportions when the samples are collected independently. For example, one may be interested in comparing the proportions of married and unmarried persons who are overweight, or the proportions of homes in two states that are heated by natural gas, or the proportions of teenagers today and 20 years ago who smoke, etc.

EXAMPLE 9.9

SELECTING A POINT ESTIMATE

In a *Journal of Business Ethics* study of ethical management decision making, 48 of 50 female managers responded that concealing one's on-the-job errors was very unethical. In contrast, only 30 of 50 male managers responded in a similar manner. Construct a point estimate for the difference between the proportions of female and male managers who believe that concealing one's errors is very unethical.

Solution

For this example, define

p_1 = Population proportion of female managers who believe that concealing on-the-job errors is unethical

p_2 = Population proportion of male managers who believe that concealing on-the-job errors is unethical

x_1 = Number of females in the sample who believe that concealing on-the-job errors is unethical

x_2 = Number of males in the sample who believe that concealing on-the-job errors is unethical

As a point estimate of $(p_1 - p_2)$, we will use the difference between the corresponding sample proportions $(\hat{p}_1 - \hat{p}_2)$, where

$$\hat{p}_1 = \frac{x_1}{n_1} = \frac{48}{50} = .96$$

and

$$\hat{p}_2 = \frac{x_2}{n_2} = \frac{30}{50} = .60$$

Thus, the point estimate of $(p_1 - p_2)$ is

$$(\hat{p}_1 - \hat{p}_2) = .96 - .60 = .36$$

To judge the reliability of the point estimate $(\hat{p}_1 - \hat{p}_2)$, we need to know the characteristics of its performance in repeated independent sampling from two binomial populations. This information is provided by the sampling distribution of $(\hat{p}_1 - \hat{p}_2)$, illustrated in Figure 9.10 and described in the accompanying box.

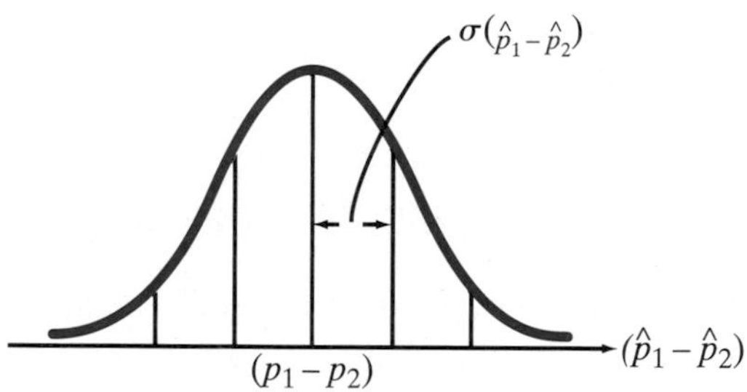

Figure 9.10 Sampling Distribution of $\hat{p}_1 - \hat{p}_2$

SAMPLING DISTRIBUTION OF $\hat{p}_1 - \hat{p}_2$

For sufficiently large sample sizes n_1 and n_2, the sampling distribution of $(\hat{p}_1 - \hat{p}_2)$, based on independent random samples from two binomial populations, is approximately normal with

$$\text{Mean: } \mu_{(\hat{p}_1 - \hat{p}_2)} = (p_1 - p_2) \tag{9.14}$$

and

$$\text{Standard deviation: } \sigma_{(\hat{p}_1 - \hat{p}_2)} = \sqrt{\frac{p_1(1 - p_1)}{n_1} + \frac{p_2(1 - p_2)}{n_2}} \tag{9.15}$$

Large-sample confidence intervals and tests of hypothesis for $(p_1 - p_2)$ are derived from the sampling distribution of $(\hat{p}_1 - \hat{p}_2)$. Details are presented in the next three boxes.

LARGE-SAMPLE TEST OF HYPOTHESIS ABOUT $(p_1 - p_2) = 0$

Upper-Tailed Test	**Two-Tailed Test**	**Lower-Tailed Test**
$H_0: (p_1 - p_2) = 0$	$H_0: (p_1 - p_2) = 0$	$H_0: (p_1 - p_2) = 0$
$H_a: (p_1 - p_2) > 0$	$H_a: (p_1 - p_2) \neq 0$	$H_a: (p_1 - p_2) < 0$

$$\text{Test stastistic: } z = \frac{(\hat{p}_1 - \hat{p}_2) - 0}{\sigma_{(\hat{p}_1 - \hat{p}_2)}} \approx \frac{(\hat{p}_1 - \hat{p}_2) - 0}{\sqrt{\hat{p}(1-\hat{p})\left(\frac{1}{n_1} + \frac{1}{n_2}\right)}} \quad (9.16)$$

where

$$\hat{p}_1 = \frac{x_1}{n_1} = \frac{\text{Number of successes in sample 1}}{n_1}$$

$$\hat{p}_2 = \frac{x_2}{n_2} = \frac{\text{Number of successes in sample 2}}{n_2}$$

$$\hat{p} = \frac{x_1 + x_2}{n_1 + n_2} = \frac{\text{Total number of successes}}{\text{Total sample size}}$$

Rejection region:	*Rejection region:*	*Rejection region:*
$z > z_\alpha$	$\lvert z \rvert > z_{\alpha/2}$	$z < -z_\alpha$

LARGE-SAMPLE $(1 - \alpha)100\%$ CONFIDENCE INTERVAL FOR $(p_1 - p_2)$

$$(\hat{p}_1 - \hat{p}_2) \pm z_{\alpha/2}\sigma_{(\hat{p}_1 - \hat{p}_2)} \approx (\hat{p}_1 - \hat{p}_2) \pm z_{\alpha/2}\sqrt{\frac{\hat{p}_1(1-\hat{p}_1)}{n_1} + \frac{\hat{p}_2(1-\hat{p}_2)}{n_2}} \quad (9.17)$$

where $\hat{p}_1$ and $\hat{p}_2$ are the sample proportions of observations with the characteristic of interest.

ASSUMPTIONS FOR MAKING INFERENCES ABOUT $(p_1 - p_2)$

1. Independent random samples are collected.
2. The samples are sufficiently large (i.e., $n_1 \geq 30$ and $n_2 \geq 30$).

EXAMPLE 9.10

HYPOTHESIS TEST FOR $(p_1 - p_2)$

Refer to Example 9.9 on p. 475. Conduct a test to determine whether the proportion of female managers who believe that concealing one's errors is very unethical exceeds the corresponding proportion for male managers. Use $\alpha = .05$.

Solution

Since we want to know whether p_1, the proportion for female managers, exceeds p_2, the proportion for male managers, we will test

$$H_0: p_1 - p_2 = 0$$
$$H_a: p_1 - p_2 > 0$$

For this upper-tailed test, the null hypothesis will be rejected at $\alpha = .05$ if

$$z > z_{.05} = 1.645 \quad \text{(see Figure 9.11 on p. 478)}$$

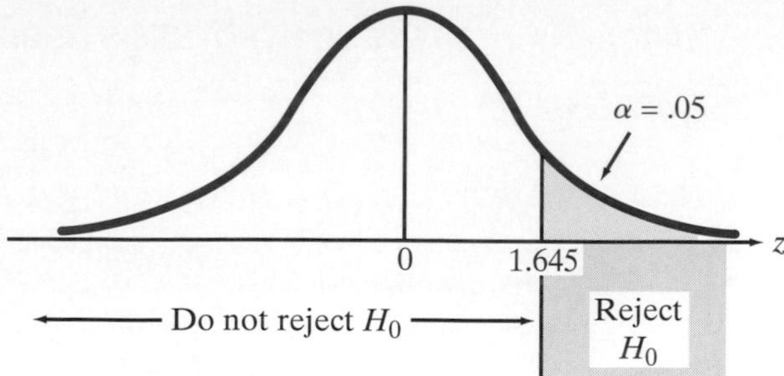

Figure 9.11 Rejection Region for Example 9.10

From Example 9.9, we have $n_1 = 50$, $n_2 = 50$, $\hat{p}_1 = .96$, and $\hat{p}_2 = .60$. Since we are testing whether the two proportions are equal, the best estimate of this unknown p is obtained by "pooling" the information from both samples into a single estimator $\hat{p}$:

$$\hat{p} = \frac{x_1 + x_2}{n_1 + n_2} = \frac{48 + 30}{50 + 50} = \frac{78}{100} = .78$$

Substituting the values of $\hat{p}_1$, $\hat{p}_2$, and $\hat{p}$ into equation 9.16, we obtain the test statistic:

$$z = \frac{(\hat{p}_1 - \hat{p}_2) - 0}{\sqrt{\hat{p}(1 - \hat{p})\left(\frac{1}{n_1} + \frac{1}{n_2}\right)}} = \frac{(.96 - .60) - 0}{\sqrt{(.78)(.22)\left(\frac{1}{50} + \frac{1}{50}\right)}} = 4.35$$

Since the test statistic $z = 4.35$ is greater than 1.645, it falls in the rejection region. Consequently, there is sufficient evidence (at $\alpha = .05$) to conclude that the proportion of female managers who believe that concealing on-the-job errors is very unethical exceeds the corresponding proportion of male managers.

Using the p-value approach, we will reach the same conclusion. The p-value of this upper-tailed test is, by definition,

$$p\text{-value} = P(z > 4.35) \approx 0$$

Since this p-value of 0 is less than $\alpha = .05$, we reject H_0.

Self-Test 9.5

Independent random samples of size $n_1 = n_2 = 500$ from populations 1 and 2 produced sample successes $x_1 = 220$ and $x_2 = 260$.

a. Test H_0: $(p_1 - p_2) = 0$ against H_a: $(p_1 - p_2) \neq 0$ using $\alpha = .05$.
b. Find a 95% confidence interval for $(p_1 - p_2)$.

The procedures presented in this section require large, independent random samples from the populations. When the samples are not "sufficiently large," we must resort to another statistical technique, as stated in the box.

CAUTION

The z test for comparing p_1 and p_2 is inappropriate when the sample sizes are not "sufficiently large." In this case, p_1 and p_2 can be compared using a statistical technique to be discussed in Section 9.5.

PROBLEMS FOR SECTION 9.4

Using the Tools

9.34 Suppose you want to estimate $p_1 - p_2$, the difference between the proportions of two populations. Based on independent random samples, you can compute the sample proportions of successes, $\hat{p}_1$ and $\hat{p}_2$, respectively. Estimate $\mu_{(\hat{p}_1 - \hat{p}_2)}$ and $\sigma_{(\hat{p}_1 - \hat{p}_2)}$ for each of the following results:

a. $n_1 = 150, \hat{p}_1 = .3; n_2 = 130, \hat{p}_2 = .4$

b. $n_1 = 100, \hat{p}_1 = .10; n_2 = 100, \hat{p}_2 = .05$

c. $n_1 = 200, \hat{p}_1 = .76; n_2 = 200, \hat{p}_2 = .96$

9.35 Independent random samples are taken from two populations. The accompanying table shows the sample sizes and the number of observations with the characteristic of interest.

Sample from Population 1	Sample from Population 2
$n_1 = 400$	$n_2 = 350$
$x_1 = 200$	$x_2 = 210$

a. Calculate the test statistic for testing $H_0: (p_1 - p_2) = 0$ against $H_a: (p_1 - p_2) \neq 0$.

b. Find the rejection region for the test, part **a**, at $\alpha = .05$.

c. Make the proper conclusion for the test, parts **a** and **b**.

d. Find a 90% confidence interval for $(p_1 - p_2)$.

e. Find a 95% confidence interval for $(p_1 - p_2)$.

f. Find a 99% confidence interval for $(p_1 - p_2)$.

9.36 Independent random samples of size n_1 and n_2 are selected from two binomial populations to test the null hypothesis $H_0: (p_1 - p_2) = 0$. For each of the following situations, specify the rejection region, test statistic, and conclusion.

a. $H_a: (p_1 - p_2) \neq 0, x_1 = 600, x_2 = 600, n_1 = 1{,}500, n_2 = 2{,}000, \alpha = .01$

b. $H_a: (p_1 - p_2) > 0, x_1 = 50, x_2 = 20, n_1 = n_2 = 1{,}000, \alpha = .05$

c. $H_a: (p_1 - p_2) < 0, x_1 = 72, x_2 = 78, n_1 = n_2 = 120, \alpha = .01$

9.37 Consider independent random samples from populations 1 and 2. Construct a confidence interval for $(p_1 - p_2)$ if:

a. $\alpha = .10, \hat{p}_1 = .20, \hat{p}_2 = .15, n_1 = 70, n_2 = 95$

b. $\alpha = .05, \hat{p}_1 = .73, \hat{p}_2 = .68, n_1 = 1{,}025, n_2 = 640$

c. $\alpha = .01, \hat{p}_1 = .51, \hat{p}_2 = .23, n_1 = 500, n_2 = 500$

9.38 Independent random samples of 250 observations each are selected from two populations. The samples from populations 1 and 2 produced, respectively, 100 and 75 observations possessing the characteristic of interest.

a. Test $H_0: (p_1 - p_2) = 0$ against $H_a: (p_1 - p_2) > 0$ at $\alpha = .05$.

b. Test $H_0: (p_1 - p_2) = 0$ against $H_a: (p_1 - p_2) \neq 0$ at $\alpha = .01$.

c. Construct a 90% confidence interval for $(p_1 - p_2)$.

d. Construct a 99% confidence interval for $(p_1 - p_2)$.

Applying the Concepts

9.39 *Oncology Nursing Forum* (July 2000) published a study of the effects of advanced nursing care on quality of life of women with breast cancer. Breast cancer patients were randomly assigned to either an intervention group or a control group. All patients received standard medical care, but those in the intervention group received additional care from an advanced practice nurse (APN). The next table describes

the number of patients who needed to make at least one emergency room visit during the study period. Compare the percentage of breast cancer patients who make emergency room visits for the two groups with a 90% confidence interval. What do you conclude?

	Control Group	**Intervention Group**
Sample size	74	78
Number who made emergency room visits	71	58

Problem 9.39

9.40 A University of South Florida biologist conducted an experiment to determine if increased levels of carbon dioxide kill leaf-eating moths (*USF Magazine,* Winter 1999). Moth larvae were placed in open containers filled with oak leaves. Half the containers had normal carbon dioxide levels while the other half had double the normal level of carbon dioxide. Ten percent of the larvae in the containers with high carbon dioxide levels died, compared to 5 percent in the containers with normal levels. Assume that 80 moth larvae were placed, at random, in each of the two types of containers. Do the experimental results demonstrate that an increased level of carbon dioxide is effective in killing a higher percentage of leaf-eating moth larvae? Test using $\alpha = .01$.

9.41 *Sports Illustrated* (Dec. 8, 1997) asked a sample of middle school and high school students whether they agreed or disagreed with the statement: "African-American players have become so dominant in sports such as football and basketball that many whites feel they can't compete at the same level as blacks." Forty-five percent of the white males in the sample and 27% of the black males in the sample disagreed with the statement. Assume that 500 white males and 500 black males participated in the study.

a. Using a 95% confidence interval, compare the percentages of white and black males who disagreed with the statement.

b. Make the comparison, part **a**, using a test of hypothesis. Use $\alpha = .05$.

9.42 The herbal medicine EGb, made from the extract of the Ginkgo biloba tree, may improve the condition of Alzheimer's patients (*Journal of the American Medical Association,* Oct. 21, 1997). Patients with mild to moderately severe dementia were randomly divided into two groups. One group received a daily dosage of EGb while the other group received a placebo. (The experiment was "double blind"—neither the patients nor the researchers knew who received which treatment.) Assume there were 30 patients in each group and that 21 of the EGb patients demonstrated improved cognitive ability compared to eight of the placebo patients. Use a test of hypothesis to compare the proportion of patients with dementia who showed improved cognitive ability for the two treatments. Use $\alpha = .10$.

9.43 The National Institute of Justice (Oct. 1998) reported on a research project that examined the criminal behavior of youth gang members and at-risk youths in Cleveland, Ohio. Independent random samples of 50 gang members and 50 nongang members were interviewed about any crimes they have committed. The table on p. 481 gives the number of youths in each group who engaged in auto theft, concealed weapons, police assault, and shoplifting. For each of the four crimes, compare the proportion of gang members who have committed the crime to the corresponding proportion of nongang members using either a confidence interval or a test of hypothesis. (Use $\alpha = .05$ for each comparison.) Report your findings in the words of the problem.

Crime	Gang Members ($n = 50$)	Nongang Members ($n = 50$)
Auto theft	22	2
Concealed weapons	39	11
Police assault	5	7
Shoplifting	15	8

Problem 9.43

9.44 Refer to the *Psychological Science* (Mar. 1995) experiment in which 120 subjects were given the "water-level task," Problem 2.39 (p. 88). Recall that the subjects are shown a drawing of a glass tilted at a 45° angle and asked to draw a line representing the surface of the water. The researchers recorded whether or not the line drawn is within 5° from the true line. (The degree deviations for all 120 subjects are contained in the data file **WATERTASK.**) Below is a MINITAB printout that compares the proportion of males who are within 5° of the true line to the corresponding proportion of females. Locate a 95% confidence interval for the difference between the population proportions on the printout and interpret the result.

Test and Confidence Interval for Two Proportions

```
Sample       X      N  Sample p
F           24     60  0.400000
M           39     60  0.650000

Estimate for p(1) - p(2):  -0.25
95% CI for p(1) - p(2):  (-0.423007, -0.0769932)
```

M **Problem 9.44**

9.45 A new mood stabilizer drug, called Depakote, has been developed for treating mania. In advertisements for Depakote published in medical journals (e.g., the *American Journal of Psychiatry,* Jan. 1998), the effectiveness of the new drug is described. In a double-blind clinical trial, 120 manic patients were randomly assigned to one of two treatment groups for 21 days. (Assume 60 patients were in each group.) The first group received a daily dosage of Depakote and the second group received a placebo (i.e., no drug) every day. Each patient was administered the Manic Rating Scale (MRS) prior to and following the clinical trial. (MRS is designed to measure a patient's mania level.) The results are shown in the table. Conduct a test of hypothesis to compare the proportions of manic patients with significant improvement in MRS score for the Depakote and placebo treatment groups. Use $\alpha = .05$.

Drug	Proportion of Patients with Significant Improvement in MRS Score
Depakote	.58
Placebo	.29

Source: Abbott Laboratories, Inc., North Chicago, IL, Nov. 1997.

9.46 A recent survey of investors with Internet access divided them into two groups, those who trade online and those who do not (traditional traders). The survey found that 48% of the traditional investors were bullish on the market, while 69% of the online investors were bullish on the market (*New York Times,* June 7, 2000). Suppose that the survey was based on 500 traditional investors and 500 online investors.

a. Find a 90% confidence interval for the difference between the population proportions of traditional investors and online investors who were bullish on the market at the time of the survey.

b. Interpret the interval, part **a**, in the words of the problem.

c. Are you confident that online investors were more bullish on the market than traditional investors? Explain.

9.47 Sports fans often argue that the outcomes of National Basketball Association (NBA) games are not decided until the fourth (and last) quarter. In contrast, the conventional "wisdom" in Major League Baseball (MLB) suggests that most nine-inning games are "over" by the seventh inning. To test these theories, University of Missouri researchers collected and analyzed data for 200 NBA games and 100 MLB games played during the 1990 regular season (*Chance,* Vol. 5, 1992). The accompanying table reports the number and percentage of games in which the team leading after three quarters or seven innings lost. [*Note:* Tied games after three quarters or seven innings were removed from the data set.] Use a test of hypothesis to compare the loss percentage of NBA teams leading after three quarters to the loss percentage of MLB teams leading after seven innings. Test using $\alpha = .05$.

Sports	Games	Leader Wins	Leader Loses	% Reversals
NBA	189	150	39	20.6
MLB	92	86	6	6.5

Source: Cooper, H., DeNeve, K. M., and Mosteller, F. "Predicting professional sports game outcomes from intermediate game scores." *Chance,* Vol. 5, Nos. 3–4, 1992, pp. 18–22 (Table 1).

9.5 Comparing Population Proportions: Contingency Tables

Data are often categorized according to two qualitative variables for the purpose of determining whether the two variables are related. For example, we may want to determine whether religious affiliation and marital status are related, or whether socioeconomic status is related to race. Recall from Section 2.4 that such data can be summarized in the form of a cross-classification table. In this section, we present a method for making inferences about the proportions in a cross-classification table.

A cross-classification table is often called a **contingency table** because the objective of the study is to investigate whether the proportions associated with the categories of one of the qualitative variables *depend* (or are *contingent*) on the proportions associated with the categories of the other variable. We illustrate with an example.

EXAMPLE 9.11 INDEPENDENCE OF TWO QUALITATIVE VARIABLES

Refer to the *American Sociological Review* study of whether men and women interrupt a speaker equally often, Example 2.5 (p. 79). Undergraduate students were organized into discussion groups and their interactions recorded on video. Recall that the researchers recorded two variables each time the group speaker was interrupted: (1) the gender of the speaker and (2) the gender of the interrupter. A cross-classification (or contingency) table for a sample of $n = 40$ interruptions is shown in Table 9.4. In terms of proportions, what does it mean to say that speaker gender and interrupter gender are independent?

TABLE 9.4 Contingency Table for Example 9.11

		Interrupter		
		Male	Female	Totals
Speaker	Male	15	5	20
	Female	10	10	20
	Totals	25	15	40

Solution

From our discussion of independent events in Chapter 4, we know that the probability of one event occurring (e.g., a male interrupter) does not depend on the other event (e.g., a female speaker) if the events are independent. Consequently, if the two variables, speaker gender and interrupter gender, are independent, then the proportion of times a male interrupts will be the same for both male and female speakers.

If we let p_1 represent the true proportion of male speakers who are interrupted by a male and p_2 represent the true proportion of female speakers who are interrupted by a male, then independence implies $p_1 = p_2$.

A test of the null hypothesis that two qualitative variables are independent uses a test statistic that compares the actual number of observations in each cell of the table to the number expected if H_0 is true. The following examples illustrate the procedure for conducting the test.

EXAMPLE 9.12

COMPUTING EXPECTED NUMBERS IN A CONTINGENCY TABLE

Refer to Example 9.11. Based on the assumption that the two qualitative variables, speaker gender and interrupter gender, are independent, how many observations would you expect to fall in each of the cells of Table 9.4?

Solution

Note that 20 of the 40 interruptions sampled are in row 1 of Table 9.4 and 20 are in row 2. If speaker gender is independent of interrupter gender, we would expect $20/40 = \frac{1}{2}$ of the 25 male interruptions to fall in the first row of Table 9.4 and $\frac{1}{2}$ to fall in the second row. Therefore, the expected number of observations falling in row 1, column 1, is

$$E_{11} = \left(\frac{20}{40}\right)(25) = \left(\frac{\text{Row 1 total}}{n}\right)(\text{Column 1 total}) = 12.5$$

and the expected number of observations falling in row 2, column 1, is

$$E_{21} = \left(\frac{20}{40}\right)(25) = \left(\frac{\text{Row 2 total}}{n}\right)(\text{Column 1 total}) = 12.5$$

We now move to column 2 and note that the column total is 15, i.e., there were 15 total female interruptions. Again, we expect $20/40 = \frac{1}{2}$ of these 15 observations to fall in the first row and $\frac{1}{2}$ to fall in the second row. Thus, the expected number for row 1, column 2, is

$$E_{12} = \left(\frac{20}{40}\right)(15) = \left(\frac{\text{Row 1 total}}{n}\right)(\text{Column 2 total}) = 7.5$$

and the expected number for row 2, column 2, is

$$E_{22} = \left(\frac{20}{40}\right)(15) = \left(\frac{\text{Row 2 total}}{n}\right)(\text{Column 2 total}) = 7.5$$

These expected numbers are shown in Table 9.5.

TABLE 9.5 Expected Numbers if Speaker Gender and Interrupter Gender Are Independent

		Interrupter		
		Male	**Female**	**Totals**
Speaker	**Male**	12.5	7.5	20
	Female	12.5	7.5	20
	Totals	25	15	40

The formula for calculating any expected value in a contingency table can be deduced from the values previously calculated. Each **expected cell count** is equal to the product of its respective row and column totals divided by the total sample size n. The general formula for calculating the expected cell count for a cell in any row and column of a contingency table is given in the box.

FORMULA FOR COMPUTING EXPECTED CELL COUNTS IN A CONTINGENCY TABLE

$$E_{ij} = \frac{(R_i)(C_j)}{n} \qquad \textbf{(9.18)}$$

where

E_{ij} = expected cell count for the cell in row i and column j
R_i = row total corresponding to row i
C_j = column total corresponding to column j
n = sample size

Self-Test 9.6

Calculate the expected cell counts for the following contingency table:

		Columns (Variable 2)	
		1	**2**
Rows (Variable 1)	**1**	20	10
	2	80	50

EXAMPLE 9.13

COMPUTING THE TEST STATISTIC

Refer to Examples 9.11 and 9.12. To test the null hypothesis that the two qualitative variables, speaker gender and interrupter gender, are independent, we compare the observed number and expected number of each cell in the contingency table in the following manner:

$$\frac{(\text{Observed cell count} - \text{Expected cell count})^2}{\text{Expected cell count}} \tag{9.19}$$

The sum of these quantities over all cells, called χ^2 (chi-squared), represents the test statistic for the test of independence. Calculate χ^2 using the information in Tables 9.4 and 9.5.

Solution

We substitute the observed cell count from Table 9.4 and the expected cell count from Table 9.5 into equation 9.19 for each cell to obtain the following quantities:

Row 1, Column 1: $\frac{(15 - 12.5)^2}{12.5} = .5$

Row 1, Column 2: $\frac{(5 - 7.5)^2}{7.5} = .833$

Row 2, Column 1: $\frac{(10 - 12.5)^2}{12.5} = .5$

Row 2, Column 2: $\frac{(10 - 7.5)^2}{7.5} = .833$

Summing these values, we obtain the test statistic:

$$\chi^2 = .5 + .833 + .5 + .833 = 2.67$$

EXAMPLE 9.14

FINDING THE REJECTION REGION

Specify the rejection region for the test described in Example 9.13. Use $\alpha = .05$. Then test to determine whether speaker gender and interrupter gender are independent.

Solution

If the two qualitative variables are, in fact, independent, then the observed cell counts should be nearly equal to the expected cell counts. Consequently, the greater the difference between the observed and expected cell counts, the greater is the evidence to indicate a lack of independence. Since the value of chi-square increases as the differences between the observed and expected cell counts increase, we will reject

H_0: Speaker gender and interrupter gender are independent

for values of χ^2 *larger* than the critical value $\chi^2_{.05}$.

The appropriate degrees of freedom for this χ^2 test will always be $(r - 1)(c - 1)$, where r is the number of rows and c is the number of columns in the contingency

table. For these data, we have $r = 2$ rows and $c = 2$ columns; hence, the appropriate number of degrees of freedom for χ^2 is

$$\text{df} = (r - 1)(c - 1) = (1)(1) = 1$$

The tabulated value of $\chi^2_{.05}$ corresponding to 1 degree of freedom, obtained from Table B.5 in Appendix B, is 3.84146. Therefore, the rejection region (highlighted in Figure 9.12) is

$$\textit{Rejection region: } \chi^2 > 3.84146$$

Since the computed value $\chi^2 = 2.67$ is less than the critical value 3.84146, we fail to reject the null hypothesis of independence at $\alpha = .05$. There is not sufficient evidence to claim that the proportion of times the speaker is interrupted by a male depends on the gender of the speaker.

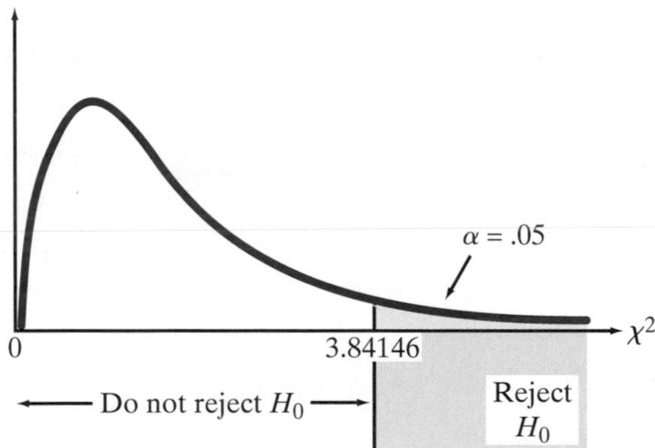

Figure 9.12 Rejection Region for the Contingency Table, Example 9.14

The elements of a χ^2 test for a contingency table are summarized in the box.

A TEST OF HYPOTHESIS FOR INDEPENDENCE OF TWO QUALITATIVE VARIABLES IN A CONTINGENCY TABLE

H_0: The two qualitative variables comprising the contingency table are independent

H_a: The two qualitative variables comprising the contingency table are dependent

$$\textit{Test statistic: } \chi^2 = \sum_{i=1}^{r}\sum_{j=1}^{c}\frac{(O_{ij} - E_{ij})^2}{E_{ij}} \qquad \textbf{(9.20)}$$

where

r = number of rows in the table

c = number of columns in the table

O_{ij} = observed number of responses in the cell in row i and column j

E_{ij} = expected number of responses in cell $ij = \dfrac{(R_i)(C_j)}{n}$

$$\textit{Rejection region: } \chi^2 > \chi^2_\alpha$$

where χ^2_α is the tabulated value of the chi-square distribution based on $(r - 1)(c - 1)$ degrees of freedom such that $P(\chi^2 > \chi^2_\alpha) = \alpha$.

Assumption: The sample size n is "large." This will be satisfied if the expected count for each of the $r \times c$ cells is at least 1. (See the next box.)

"LARGE" n IN CONTINGENCY TABLES

For the χ^2 test to give accurate results, the expected counts of the $r \times c$ cells must be large. There has been much debate among statisticians as to the definition of "large." Some statisticians adopt a conservative approach and require each expected count to be at least 5. Others have found that the test is accurate if no more than 20% of the cells contain expected frequencies less than 5 and no cells have expected frequencies less than 1. In the case of a $2 \times c$ table, statisticians have found the test gives accurate results as long as all expected cell counts equal at least 0.5. *A reasonable compromise between these points of view is to be sure that all expected counts are at least 1.*

Self-Test 9.7

The contingency table for Self-Test 9.6 is reproduced below:

		Columns (Variable 2)	
		1	**2**
Rows (Variable 1)	**1**	20	10
	2	80	50

a. Calculate the χ^2 statistic for testing whether the two qualitative variables are independent.

b. Give the rejection region when $\alpha = .01$.

c. Conduct the test.

The contingency table of Examples 9.11–9.14 is a 2×2 table—that is, there are two rows (representing the two outcomes for the first qualitative variable) and two columns (representing the two outcomes for the second qualitative variable). For this special case, we can assume the rows represent two different populations. In the examples above, the two populations would be (1) male speakers and (2) female speakers. Now, consider the proportion of male interrupters for each population and define

p_1 = Proportion of male interrupters for the population of male speakers

p_2 = Proportion of male interrupters for the population of female speakers

As we learned in Example 9.14, if the two variables are independent, $p_1 = p_2$. Consequently, for the 2×2 table, a test of independence is equivalent to testing

$$H_0\colon (p_1 - p_2) = 0$$

This null hypothesis is equivalent to that tested when comparing two population proportions with a large-sample z test (Section 9.4). Therefore, both the z test and the χ^2 test for a 2×2 contingency table can be used to test for the difference between two population proportions. The z test, however, requires "large" samples (i.e., $n_1 \geq 30$ and $n_2 \geq 30$). In contrast, the χ^2 test requires only that n be large

enough so that the expected number in each cell is at least 1. Thus, the χ^2 test may be used as an alternative to the z test when the samples are small, as long as the expected cell counts are all at least 1.

For contingency tables with more than two rows or two columns, the χ^2 test is usually performed using computer software because of the numerous calculations involved, as the next example illustrates.

EXAMPLE 9.15

2 × 3 CONTINGENCY TABLE ANALYSIS

A study published in the *Journal of Education for Business* used responses from a mail survey to classify a sample of 215 personnel directors according to two qualitative variables: foreign language hiring preference (yes, no, or neutral) and type of business (foreign or domestic). The response data are summarized in the contingency table, Table 9.6. Is there sufficient evidence to claim that foreign language hiring preference depends on type of business? Test using $\alpha = .01$.

TABLE 9.6 Contingency Table for Example 9.15

		Foreign Language Hiring Preference			
		Yes	**Neutral**	**No**	**Totals**
Firm Type	**U.S.**	50	57	19	126
	Foreign	60	22	7	89
	Totals	110	79	26	215

Source: Cornick, M. F., et al. "The value of foreign language skills for accounting and business majors." *Journal of Education for Business,* Jan./Feb. 1991, p. 162 (Table 2).

Solution

We want to test

H_0: Foreign language hiring preference and type of business are independent

H_a: Foreign language hiring preference and type of business are dependent

The results of the test are shown in the Excel printout, Figure 9.13. Note that the expected counts are given in the cells of the contingency table shown on the printout, and that each expected count is at least 1. Thus, it is appropriate to apply the χ^2 test.

The test statistic and corresponding p-value are

$$\chi^2 = 16.062 \qquad p\text{-value} = .00033$$

Since the p-value is less than $\alpha = .01$, there is sufficient evidence to reject H_0 and claim that foreign language hiring preference does depend on type of business. In fact, the estimated proportion of U.S. firms that favor foreign language hiring is $50/126 = .397$, while the estimated proportion of foreign firms that give foreign language hiring preference is $60/89 = .674$—almost twice as high.

E Figure 9.13 PHStat and Excel Output, Example 9.15

	A	B	C	D	E
3	Observed Frequencies				
4		Foreign Language Hiring Preference			
5	Firm Type	Yes	Neutral	No	Total
6	U.S.	50	57	19	126
7	Foreign	60	22	7	89
8	Total	110	79	26	215
9					
10		Expected Frequencies			
11		Foreign Language Hiring Preference			
12	Firm Type	Yes	Neutral	No	Total
13	U.S.	64.46511628	46.29767	15.23721	126
14	Foreign	45.53488372	32.70233	10.76279	89
15	Total	110	79	26	215
16					
17	Data				
18	Level of Significance	0.01			
19	Number of Rows	2			
20	Number of Columns	3			
21	Degrees of Freedom	2			
22					
23	Results				
24	Critical Value	9.210351036			
25	Chi-Square Test Statistic	16.06213606			
26	p-Value	0.000325201			
27	Reject the null hypothesis				

CAUTION

Inferences derived from χ^2 tests that do not satisfy the sample size requirement outlined earlier (i.e., expected cell counts of at least 1) are suspect. To make sure the sample size requirement is satisfied, it may be necessary to collapse two or more low-frequency categories into one category in the contingency table prior to performing the test. Alternative procedures are available (see Marascuilo and McSweeney, 1977) if the combining or pooling of categories is undesirable or nonsensical.

PROBLEMS FOR SECTION 9.5

Using the Tools

9.48 Refer to the 2×2 contingency table shown here.

		Columns 1	Columns 2	Totals
Rows	1	133	219	352
	2	201	247	448
	Totals	334	466	800

a. Calculate the expected cell counts for the contingency table.

b. Calculate the chi-square statistic for the table.

c. At $\alpha = .05$, are the rows and columns independent?

9.49 Refer to the 2×3 (two rows and three columns) contingency table on p. 490.

a. Calculate the expected cell counts for the contingency table.

b. Calculate the chi-square statistic for the table.

c. At $\alpha = .05$, are the rows and columns independent?

		Columns			
		1	2	3	Totals
Rows	1	14	37	23	74
	2	21	32	38	91
	Totals	35	69	61	165

Problem 9.49

9.50 Give the degrees of freedom for a test of independence of the two qualitative variables comprising a contingency table with:

a. $r = 2$ rows and $c = 2$ columns **b.** $r = 4$ rows and $c = 2$ columns

c. $r = 3$ rows and $c = 3$ columns **d.** $r = 3$ rows and $c = 4$ columns

9.51 Test the null hypothesis of independence of the two qualitative variables for the 2×2 contingency table shown here. Use $\alpha = .05$.

		Variable 2	
		0	1
Variable 1	0	6	24
	1	16	36

9.52 Refer to the 3×4 contingency table given here.

		Color				
		Red	White	Blue	Yellow	Totals
Length	Short	18	12	21	37	88
	Medium	7	10	15	31	63
	Long	9	6	14	30	59
	Totals	34	28	50	98	210

a. Calculate the expected cell counts for the contingency table.

b. Calculate the chi-square statistic for the table.

c. Test the hypothesis of independence of the two qualitative variables for the 3×4 table. Use $\alpha = .01$.

Applying the Concepts

9.53 Many Internet users shopped on-line during the holiday season. In a survey of 1,500 customers who did holiday shopping on-line, 270 indicated that they were not satisfied with their experience. Of the customers that were not satisfied, 143 indicated that they did not receive their product in time for the holidays (*The Wall Street Journal,* April 17, 2000). Suppose the following complete set of results were reported. Conduct a test at $\alpha = .01$ to determine whether the percentage of satisfied on-line shoppers who did not receive their product in time for the holidays differs from the corresponding percentage of unsatisfied on-line shoppers.

		Received Products in Time for Holidays		
		Yes	No	Total
Satisfied with Experience	Yes	1,197	33	1,230
	No	127	143	270
	Totals	1,324	176	1,500

9.54 Refer to the *Oncology Nursing Forum* (July 2000) study of the effects of advanced nursing care on quality of life of women with breast cancer, Problem 9.39 (p. 479). The cancer patients either received standard medical care (control group) or standard care plus care from an advanced practice nurse (intervention group). The researchers wanted to know whether the two groups of patients differed with respect to their medical insurance coverage. Insurance was classified as Health Maintenance Organization (HMO), non-HMO, or Medicare. (If the patients were truly randomly assigned to the two study groups, the proportion of patients with each type of coverage should be about the same for the two groups.) The table gives the number of patients with each type of coverage in each group. Do the data indicate that the two study groups differ with respect to type of medical coverage? Test using $\alpha = .10$.

Medical Coverage	Intervention Group	Control Group
HMO	60	53
Non-HMO	22	26
Medicare	24	25

9.55 A University of Florida sociologist once claimed that, "If you kill a white person (in Florida), the chance of getting the death penalty is greater than if you kill a black person." Concentrating only on crimes against strangers, the sociologist classified the data of 326 murder cases in Florida according to race of the victim and death sentence, as shown in the accompanying table.

		Death Sentence		
		Yes	No	Totals
Race of Victim	White	30	184	214
	Black	6	106	112
	Totals	36	290	326

a. Conduct a test to determine whether the victim's race and death sentence are independent. Use $\alpha = .05$.

b. Do the results, part **a**, support the statement made by the sociologist?

c. Construct a 95% confidence interval for $(p_1 - p_2)$, where p_1 is the percentage of murderers of whites who received the death penalty and p_2 is the percentage of murderers of blacks who received the death penalty. Interpret the interval. Do the results agree with the test, part **a**?

9.56 Refer to the *Sociology of Sport Journal* (Vol. 14, 1997) study of "stacking" positions in professional basketball, Problem 2.36 (p. 87). The contingency table, reproduced below, summarizes the race and positions of 368 National Basketball Association (NBA) players in a recent year. The researchers are interested in whether black and white NBA players are distributed proportionately across positions. Conduct the appropriate test of hypothesis to answer the researchers' question. Test using $\alpha = .05$.

		Position			
		Guard	Forward	Center	Totals
Race	White	26	30	28	84
	Black	128	122	34	284
	Totals	154	152	62	368

9.57 Refer to the *Industrial Marketing Management* (Feb. 1993) study of humor in trade magazine advertisements, Problem 2.35 (p. 87). Each in a sample of 665 ads was classified according to nationality (British, German, American) of the trade magazine in which it appeared and whether or not it was considered humorous by a panel of judges. The numbers of ads falling into each of the categories are provided in the accompanying table. Conduct an analysis to determine whether nationality of trade magazine and humor in advertisements are independent. Use $\alpha = .01$.

		Humorous	
		Yes	**No**
Nationality	**British**	52	151
	German	44	148
	American	56	214

Source: McCullough, L. S., and Taylor, R. K. "Humor in American, British, and German ads." *Industrial Marketing Management,* Vol. 22, No. 1, Feb. 1993, p. 22 (Table 3).

9.58 Refer to the *Psychological Science* (Mar. 1995) "water-level task" experiment, Problem 2.39 (p. 88). Recall that groups of subjects completed a task in which they were asked to judge the location of the water line in a glass tilted at a 45° angle. The group and the angle of the judged line were recorded for each subject. A contingency table for the study is provided here, along with an Excel printout.

		Group					
		Students	**Waitresses**	**Housewives**	**Bartenders**	**Bus Drivers**	**Totals**
Judge Line	**More than 5° above or below surface**	11	15	14	12	5	57
	Within 5° of surface	29	5	6	8	15	63
	TOTALS	40	20	20	20	20	120

	A	B	C	D	E	F	G
3		**Observed Frequencies**					
4				**Group**			
5	**Judge Line**	**Students**	**Waitresses**	**Housewives**	**Bartenders**	**Bus Drivers**	**Total**
6	**More than 5 degrees above or below surface**	**11**	**15**	**14**	**12**	**5**	**57**
7	**Within 5 degrees of surface**	**29**	**5**	**6**	**8**	**15**	**63**
8	**Total**	**40**	**20**	**20**	**20**	**20**	**120**
9							
10		Expected Frequencies					
11				Group			
12	Judge Line	Students	Waitresses	Housewives	Bartenders	Bus Drivers	Total
13	More than 5 degrees above or below surface	19	9.5	9.5	9.5	9.5	57
14	Within 5 degrees of surface	21	10.5	10.5	10.5	10.5	63
15	Total	40	20	20	20	20	120
16							
17	**Data**						
18	**Level of Significance**	**0.05**					
19	Number of Rows	2					
20	Number of Columns	5					
21	Degrees of Freedom	4					
22							
23	**Results**						
24	**Critical Value**	9.487728					
25	**Chi-Square Test Statistic**	21.85464					
26	***p*-Value**	0.000214					
27	**Reject the null hypothesis**						

E Problem 9.58

a. Test the hypothesis that the judged location of the water line depends on group. Use $\alpha = .05$.

b. Repeat part **a**, but only use the data for waitresses and bartenders.

9.59 According to *Nature* (Sept. 1993), biologists define a "hotspot" as a species-rich, 10-kilometer-square geographical area. Do rare flying insects tend to congregate in hotspots of their own species? To answer this question, 105 rare British flying insect species were captured and the following two variables measured for each: (1) type of species captured and (2) type of hotspot where they were found. The data are summarized in the accompanying table. Is there evidence to indicate that hotspot type depends on rare species type? Use $\alpha = .10$.

		Hotspot Type			
		Butterfly	Dragonfly	Bird	Totals
Rare Species Type	Butterfly	31	17	20	68
	Dragonfly	11	13	13	37
	Totals	42	30	33	105

9.60 Refer to the *Journal of Gang Research* (Winter, 1997) study of approximately 10,000 confined offenders, Problem 2.40 (p. 89). The offenders were classified according to gang classification score and whether or not the offender had ever carried a homemade weapon (e.g., knife) while in custody. The data are summarized in the following contingency table. Conduct a test to determine if carrying a homemade weapon in custody depends on gang classification score. Use $\alpha = .01$.

		Homemade Weapon Carried	
		Yes	No
Gang Classification Score	0 (Never joined a gang, no close friends in a gang)	255	2,551
	1 (Never joined a gang, 1–4 close friends in a gang)	110	560
	2 (Never joined a gang, 5 or more friends in a gang)	151	636
	3 (Inactive gang member)	271	959
	4 (Active gang member, no position of rank)	175	513
	5 (Active gang member, holds position of rank)	476	831

Source: Knox, G. W., et al. "A gang classification system for corrections," *Journal of Gang Research,* Vol. 4, No. 2, Winter 1997, p. 54 (Table 4).

9.6 Comparing Two Population Variances (Optional)

In this optional section, we present a test of hypothesis for comparing two population variances, σ_1^2 and σ_2^2. Variance tests have broad applications. For example, a production manager may be interested in comparing the variation in the length of eye-screws produced on each of two assembly lines. A line with a large variation produces too many individual eye-screws that do not meet specifications (either too long or too short), even though the mean length may be satisfactory. Similarly, an investor might want to compare the variation in the monthly rates of return for two different stocks that have the same mean rate of return. In this case, the stock with the smaller variance may be preferred because it is less risky—that is, it is less likely to have many very low and very high monthly return rates.

Variance tests can also be applied before conducting a small-sample t test for $(\mu_1 - \mu_2)$, discussed in Section 9.2. Recall that the t test requires the assumption that the variances of the two sampled populations are equal. If the two population variances are greatly different, any inferences derived from the t test are suspect. Consequently, it is important that we detect a significant difference between the two variances, if it exists, before applying the small-sample t test.

EXAMPLE 9.16 **SETTING UP THE HYPOTHESIS TEST**

Suppose you want to conduct a test of hypothesis to compare two population variances. Let σ_1^2 represent the variance of population 1 and σ_2^2 represent the variance of population 2. Assume independent random samples of size $n_1 = 13$ and $n_2 = 18$ are selected and that the two populations are both normal.

a. Set up the null and alternative hypotheses for testing whether σ_1^2 exceeds σ_2^2.

b. Give the form of the test statistic.

c. Specify the rejection region for $\alpha = .05$.

Solution

a. To detect whether σ_1^2 exceeds σ_2^2, we test

$$H_0\colon \sigma_1^2 = \sigma_2^2$$

$$H_a\colon \sigma_1^2 > \sigma_2^2$$

However, the common statistical procedure for comparing the two variances makes an inference about the ratio σ_1^2/σ_2^2. This is because the sampling distribution of the estimator of σ_1^2/σ_2^2 is well known when the *samples are randomly and independently selected from two normal populations*. Consequently, we typically write the null and alternative hypotheses as follows:

$$H_0\colon \sigma_1^2/\sigma_2^2 = 1$$

$$H_a\colon \sigma_1^2/\sigma_2^2 > 1$$

b. Since s_1^2 and s_2^2 are the best estimates of σ_1^2 and σ_2^2, respectively, the best estimate of σ_1^2/σ_2^2 is s_1^2/s_2^2. Consequently, the test statistic, called an F statistic, is

$$F = \frac{s_1^2}{s_2^2} \qquad \textbf{(9.21)}$$

c. Under the assumption that both samples come from normal populations, the F statistic, $F = s_1^2/s_2^2$, possesses an *F distribution* with $\nu_1 = (n_1 - 1)$ numerator degrees of freedom and $\nu_2 = (n_2 - 1)$ denominator degrees of freedom. An F distribution can be symmetric about its mean, skewed to the left, or skewed to the right; its exact shape depends on the values of ν_1 and ν_2. In this example, $\nu_1 = (n_1 - 1) = 12$ and $\nu_2 = (n_2 - 1) = 17$. An F distribution with $\nu_1 = 12$ numerator df and $\nu_2 = 17$ denominator df is shown in Figure 9.14. You can see that this particular F distribution is skewed to the right.

Upper-tail critical values of F are found in Table B.6 of Appendix B. Table 9.7, a partial reproduction of Table B.6, gives F values that correspond to $\alpha = .05$ upper-tail areas for different pairs of degrees of freedom. The columns of the table correspond to various numerator degrees of freedom (ν_1), while the rows correspond to various denominator degrees of freedom (ν_2).

Thus, if the numerator degrees of freedom are $\nu_1 = 12$ and the denominator degrees of freedom are $\nu_2 = 17$, we find the F value,

$$F_{.05} = 2.38$$

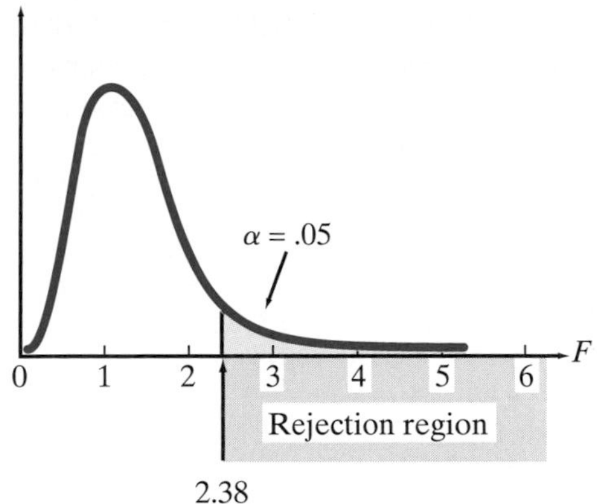

Figure 9.14 Rejection Region for the Example 9.16

TABLE 9.7 Reproduction of Part of Table B.6 of Appendix B ($\alpha = .05$)

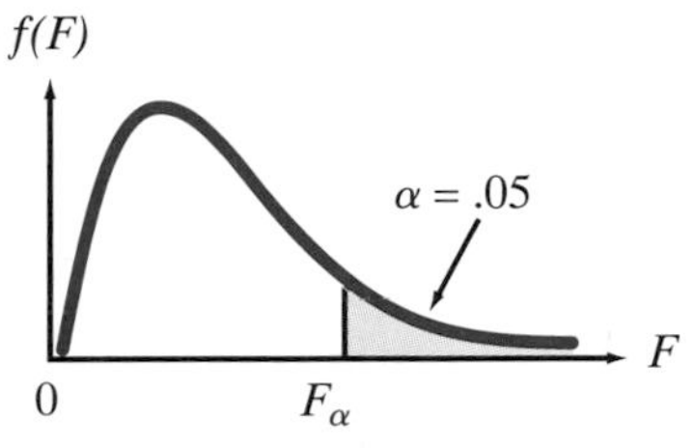

Numerator Degrees of Freedom

Denominator Degrees of Freedom $\nu_2 \backslash \nu_1$	10	12	15	20	24	30	40	60	120	∞
1	241.9	243.9	245.9	248.0	249.1	250.1	251.1	252.2	253.3	254.3
2	19.40	19.41	19.43	19.45	19.45	19.46	19.47	19.48	19.49	19.50
3	8.79	8.74	8.70	8.66	8.64	8.62	8.59	8.57	8.55	8.53
4	5.96	5.91	5.86	5.80	5.77	5.75	5.72	5.69	5.66	5.63
5	4.74	4.68	4.62	4.56	4.53	4.50	4.46	4.43	4.40	4.36
6	4.06	4.00	3.94	3.87	3.84	3.81	3.77	3.74	3.70	3.67
7	3.64	3.57	3.51	3.44	3.41	3.38	3.34	3.30	3.27	3.23
8	3.35	3.28	3.22	3.15	3.12	3.08	3.04	3.01	2.97	2.93
9	3.14	3.07	3.01	2.94	2.90	2.86	2.83	2.79	2.75	2.71
10	2.98	2.91	2.85	2.77	2.74	2.70	2.66	2.62	2.58	2.54
11	2.85	2.79	2.72	2.65	2.61	2.57	2.53	2.49	2.45	2.40
12	2.75	2.69	2.62	2.54	2.51	2.47	2.43	2.38	2.34	2.30
13	2.67	2.60	2.53	2.46	2.42	2.38	2.34	2.30	2.25	2.21
14	2.60	2.53	2.46	2.39	2.35	2.31	2.27	2.22	2.18	2.13
15	2.54	2.48	2.40	2.33	2.29	2.25	2.20	2.16	2.11	2.07
16	2.49	2.42	2.35	2.28	2.24	2.19	2.15	2.11	2.06	2.01
17	2.45	2.38	2.31	2.23	2.19	2.15	2.10	2.06	2.01	1.96

Source: From M. Merrington and C. M. Thompson, "Tables of Percentage Points of the Inverted Beta (*F*)-Distribution." *Biometrika,* 1943, 33, 73–88. Reproduced by permission of the *Biometrika* Trustees and Oxford University Press.

As shown in Figure 9.14, $\alpha = .05$ is the tail area to the right of 2.38 in the F distribution with 12 numerator df and 17 denominator df. Thus, the probability that the F statistic will exceed 2.38 is $\alpha = .05$.

Given this information on the F distribution, we are now able to find the rejection region for this test. Logically, large values of $F = s_1^2/s_2^2$ will lead us to

reject H_0 in favor of H_a. (If H_a is, in fact, true, then $\sigma_1^2 > \sigma_2^2$. Consequently, s_1^2 will very likely exceed s_2^2 and F will be large.) How large? The critical F value depends on our choice of α. The general form of the rejection region is $F > F_\alpha$. For $\alpha = .05$, we have $F_{.05} = 2.38$ (based on $\nu_1 = 12$ and $\nu_2 = 17$ df). Thus, the rejection region is

$$\text{Reject } H_0 \text{ if } F > 2.38$$

Self-Test 9.8

Consider independent random samples of size $n_1 = 10$ and $n_2 = 25$ selected from normal populations. Give the rejection region for testing H_0: $\sigma_1^2/\sigma_2^2 = 1$ against H_a: $\sigma_1^2/\sigma_2^2 > 1$ at $\alpha = .10$.

The elements of both one-tailed and two-tailed hypothesis tests for the ratio of two population variances σ_1^2/σ_2^2 are summarized in the box.

TEST OF HYPOTHESIS FOR THE RATIO OF TWO POPULATION VARIANCES σ_1^2/σ_2^2

Upper-Tailed Test	**Two-Tailed Test**	**Lower-Tailed Test**
H_0: $\sigma_1^2/\sigma_2^2 = 1$ $(\sigma_1^2 = \sigma_2^2)$	H_0: $\sigma_1^2/\sigma_2^2 = 1$ $(\sigma_1^2 = \sigma_2^2)$	H_0: $\sigma_1^2/\sigma_2^2 = 1$ $(\sigma_1^2 = \sigma_2^2)$
H_a: $\sigma_1^2/\sigma_2^2 > 1$ $(\sigma_1^2 > \sigma_2^2)$	H_a: $\sigma_1^2/\sigma_2^2 \neq 1$ $(\sigma_1^2 \neq \sigma_2^2)$	H_a: $\sigma_1^2/\sigma_2^2 < 1$ $(\sigma_1^2 < \sigma_2^2)$
Test statistic:	*Test statistic:*	*Test statistic:*
$F = s_1^2/s_2^2$	$F = \dfrac{\text{Larger sample variance}}{\text{Smaller sample variance}}$	$F = s_2^2/s_1^2$
Rejection region:	*Rejection region:*	*Rejection region:*
$F > F_\alpha$	$F > F_{\alpha/2}$	$F > F_\alpha$
where	where	where
$\nu_1 = n_1 - 1$	$\nu_1 = n - 1$ for sample variance in numerator	$\nu_1 = n_2 - 1$
$\nu_2 = n_2 - 1$	$\nu_2 = n - 1$ for sample variance in denominator	$\nu_2 = n_1 - 1$

Assumptions:

1. Both the populations from which the samples are selected have relative frequency distributions that are approximately normal.
2. The random samples are selected in an independent manner from the two populations.

EXAMPLE 9.17 **CONDUCTING THE TEST FOR COMPARING VARIANCES**

Ethylene oxide (ETO) is used quite frequently in sterilizing hospital supplies. A study was conducted to investigate the effect of ETO on hospital personnel involved with the sterilization process. Thirty subjects were randomly selected and randomly assigned to one of two tasks: 19 subjects were assigned the task of opening the sterilization package that contains ETO (task 1); the remaining 11 subjects

were assigned the task of opening and unloading the sterilizer gun filled with ETO (task 2). After the tasks were performed, researchers measured the amount of ETO (in milligrams) present in the bloodstream of each subject. A summary of the results appears in Table 9.8. Do the data provide sufficient evidence to indicate a difference in the variability of the ETO levels in subjects assigned to the two tasks? Test using $\alpha = .10$.

TABLE 9.8 Data on ETO Levels, Example 9.17

	Task 1	Task 2
Sample size	19	11
Mean	5.90	5.60
Standard deviation	1.93	4.10

Solution

Let

$$\sigma_1^2 = \text{Population variance of ETO levels in subjects assigned task 1}$$

$$\sigma_2^2 = \text{Population variance of ETO levels in subjects assigned task 2}$$

For this test to yield valid results, we must assume that both samples of ETO levels come from normal populations and that the samples are independent.

Now, following the guidelines in the box, the elements of this two-tailed hypothesis test follow:

$$H_0: \sigma_1^2/\sigma_2^2 = 1 \qquad (\sigma_1^2 = \sigma_2^2)$$

$$H_a: \sigma_1^2/\sigma_2^2 \neq 1 \qquad (\sigma_1^2 \neq \sigma_2^2)$$

$$\textit{Test statistic: } F = \frac{\text{Larger } s^2}{\text{Smaller } s^2} = \frac{s_2^2}{s_1^2} = \frac{(4.10)^2}{(1.93)^2} = 4.51$$

Rejection region: For this two-tailed test, $\alpha = .10$ and $\alpha/2 = .05$. Thus, the rejection region is

$$F > F_{.05} = 2.41 \quad \text{(from Table B.6 in Appendix B)}$$

where the distribution of F is based on $\nu_1 = (n_2 - 1) = 10$ and $\nu_2 = (n_1 - 1) = 18$ degrees of freedom.

Conclusion: Since the test statistic $F = 4.51$ falls in the rejection region, we reject H_0. Therefore, the data provide sufficient evidence at $\alpha = .10$ to indicate that the population variance of ETO levels for hospital personnel involved with opening the sterilization package (task 1) differs from the population variance of ETO levels for those involved with opening and unloading the sterilizer gun (task 2).

Example 9.17 illustrates the technique for calculating the test statistic and rejection region for a two-tailed F test. The reason we place the larger sample variance in the numerator of the test statistic is that only upper-tail values of F are shown in Table B.6 of Appendix B—no lower-tail values are given. By placing the larger sample variance in the numerator, we make certain that only the upper tail of the rejection region is used. The fact that the upper-tail area is $\alpha/2$ reminds us that the test is two-tailed.

The problem of not being able to locate an F value in the lower tail of the F distribution is avoided in a one-tailed test because we can control how we specify the ratio of the population variances in H_0 and H_a. That is, we can always make a one-tailed test an upper-tailed test. For example, if we want to test whether σ_1^2 is greater than σ_2^2, then we write the alternative hypothesis as

$$H_a: \frac{\sigma_1^2}{\sigma_2^2} > 1 \qquad (\sigma_1^2 > \sigma_2^2)$$

and the appropriate test statistic is $F = s_1^2/s_2^2$. Conversely, if we want to test whether σ_1^2 is less than σ_2^2 (i.e., whether σ_2^2 is greater than σ_1^2), we write

$$H_a: \frac{\sigma_2^2}{\sigma_1^2} > 1 \qquad (\sigma_2^2 > \sigma_1^2)$$

and the corresponding test statistic is $F = s_2^2/s_1^2$, as shown in the box.

Self-Test 9.9

Consider independent random samples of size $n_1 = 6$ and $n_2 = 10$ selected from normal populations. Find the rejection region for testing H_0: $\sigma_1^2 = \sigma_2^2$ against H_a: $\sigma_1^2 < \sigma_2^2$ at $\alpha = .05$.

CAUTION

The F test is much less robust (i.e., much more sensitive) to departures from normality than the t test for comparing population means (Section 9.2). If you have doubts about the normality of the two populations, avoid using the F test for comparing the two variances. Rather, use a nonparametric test (see Hollander and Wolfe, 1973).

PROBLEMS FOR SECTION 9.6

Using the Tools

9.61 Find F_α for an F distribution with 15 numerator df and 12 denominator df for the following values of α:

a. $\alpha = .025$ **b.** $\alpha = .05$ **c.** $\alpha = .10$

9.62 Find $F_{.05}$ for an F distribution with:

a. Numerator df = 7, denominator df = 25
b. Numerator df = 10, denominator df = 8
c. Numerator df = 30, denominator df = 60
d. Numerator df = 15, denominator df = 4

9.63 Calculate the value of the test statistic for testing H_0: $\sigma_1^2/\sigma_2^2 = 1$ in each of the following cases:

a. H_a: $\sigma_1^2/\sigma_2^2 > 1$; $s_1^2 = 1.75$; $s_2^2 = 1.23$
b. H_a: $\sigma_1^2/\sigma_2^2 < 1$; $s_1^2 = 1.52$; $s_2^2 = 5.90$
c. H_a: $\sigma_1^2/\sigma_2^2 \neq 1$; $s_1^2 = 2{,}264$; $s_2^2 = 4{,}009$

9.64 Under what conditions does the sampling distribution of s_1^2/s_2^2 have an F distribution?

Applying the Concepts

9.65 *Ingratiation* is defined as a class of strategic behaviors designed to make others believe in the attractiveness of one's personal qualities. An index that measures ingratiatory behavior, called the Measure of Ingratiatory Behaviors in Organization Settings (MIBOS), was applied independently to a sample of managers employed by four manufacturing companies in the southeastern U.S. and to clerical personnel from a large university in the northwestern U.S. (*Journal of Applied Psychology,* Dec. 1998.) Scores are reported on a five-point scale with higher scores indicating more extensive ingratiatory behavior. Summary statistics are shown in the table.

Managers	Clerical Personnel
$n = 288$	$n = 110$
$\bar{x} = 2.41$	$\bar{x} = 1.90$
$s = .74$	$s = .59$

Source: Harrison, A. W., Hochwarter, W. A., Perrewe, P. L., and Ralston, D. A. "The ingratiation construct: An assessment of the validity of the Measure of Ingratiatory Behaviors in Organization Settings (MIBOS)." *Journal of Applied Psychology,* Vol. 86, No. 6, (Dec. 1998), pp. 932–943.

a. Specify the null and alternative hypotheses you would use to test for a difference in the variability of MIBOS index scores between managers and clerical personnel.

b. Conduct the test of part **a** using $\alpha = .05$. Interpret the results of the test in the context of the problem.

9.66 An *Environmental Science & Technology* (Oct. 1993) study used the data in the table to compare the mean oxon/thion ratios at a California orchard under two weather conditions: foggy and clear/cloudy. Test the assumption of equal variances required for the comparison of means to be valid. Use $\alpha = .05$.

OTRATIO

Date	Condition	Thion	Oxon	Oxon/Thion Ratio
Jan. 15	Fog	38.2	10.3	.270
17	Fog	28.6	6.9	.241
18	Fog	30.2	6.2	.205
19	Fog	23.7	12.4	.523
20	Clear	74.1	45.8	.618
21	Fog	88.2	9.9	.112
21	Clear	46.4	27.4	.591
22	Fog	135.9	44.8	.330
23	Fog	102.9	27.8	.270
23	Cloudy	28.9	6.5	.225
25	Fog	46.9	11.2	.239
25	Clear	44.3	16.6	.375

Source: Selber, J. N., et al. "Air and fog deposition residues of four organophosphate insecticides used on dormant orchards in the San Joaquin Valley, California." *Environmental Science & Technology,* Vol. 27, No. 10, Oct. 1993, p. 2240 (Table V).

9.67 Refer to the *Prison Journal* (Sept. 1997) study of two groups of parole officers, Problem 9.16 (p. 465). Summary statistics on subjective role and strategy for regular officers and for intensive supervision (ISP) officers are reproduced in the table on the next page.

			Subjective Role		Strategy	
		Sample Size	Mean	Standard Deviation	Mean	Standard Deviation
Group	**Regular**	61	26.15	5.51	12.69	2.94
	ISP	11	19.50	4.35	9.00	2.00

Source: Fulton, B., et al. "Moderating probation and parole officer attitudes to achieve desired outcomes." *The Prison Journal,* Vol. 77, No. 3, Sept. 1997, p. 307 (Table 3).

Problem 9.67

a. In Problem 9.16**a**, you made an inference about the difference between the mean subjective role response for the two groups of parole officers, assuming the variances for the two groups were equal. Conduct a test to determine if this assumption is violated. Use $\alpha = .05$.

b. In Problem 9.16**b**, you made an inference about the difference between the mean strategy responses for the two groups of parole officers, assuming the variances for the two groups were equal. Conduct a test to determine if this assumption is violated. Use $\alpha = .05$.

9.68 Hermaphrodites are animals that possess the reproductive organs of both sexes. *Genetical Research* (June 1995) published a study of the mating systems of hermaphroditic snail species. The mating habits of the snails were classified into two groups: (1) self-fertilizing (selfing) snails that mate with snails of the same sex and (2) cross-fertilizing (outcrossing) snails that mate with snails of the opposite sex. One variable of interest in the study was the effective population size of the snail species. The means and standard deviations of the effective population size for independent random samples of 17 outcrossing snail species and five selfing snail species are given in the accompanying table. Geneticists are often more interested in comparing the variation in population size of the two types of mating systems. Conduct this analysis for the researcher using $\alpha = .10$. Interpret the result.

			Effective Population Size	
		Sample Size	Mean	Standard Deviation
Snail Mating System	**Outcrossing**	17	4,894	1,932
	Selfing	5	4,133	1,890

Source: Jarne, P. "Mating system, bottlenecks, and generic polymorphism in hermaphroditic animals." *Genetical Research,* Vol. 65, No. 3, June 1995, p. 197 (Table 4).

9.69 A study in the *Journal of Occupational and Organizational Psychology* (Dec. 1992) investigated the relationship of employment status and mental health. A sample of working and unemployed people was selected, and each person was given a mental health examination using the General Health Questionnaire (GHQ), a widely recognized measure of mental health. Although the article focused on comparing the mean GHQ levels, a comparison of the variability of GHQ scores for employed and unemployed men and women is of interest as well.

a. In general terms, what does the amount of variability in GHQ scores tell us about the group?

b. What are the appropriate null and alternative hypotheses to compare the variability of the mental health scores of the employed and unemployed groups? Define any symbols you use.

c. The standard deviation for a sample of 142 employed men was 3.26, while the standard deviation for 49 unemployed men was 5.10. Conduct the test you set up in part **b** using $\alpha = .05$. Interpret the results.

d. What assumptions are necessary to assure the validity of the test?

AUTO2000

9.70 Refer to the *Consumer Reports* (April 2000) data on new (year 2000) automobiles, stored in the data file **AUTO2000.** Suppose you want to compare the variation in turning circle requirements of front-wheel drive cars and rear-wheel drive cars.

a. Conduct a two-tailed test of hypothesis of the difference in variances at $\alpha = .05$. Interpret the results.

b. What assumptions are required for the inference made in part **a** to be valid? Check to see if these assumptions are reasonably satisfied.

9.71 An experiment was conducted to compare the variability in the sentence perception of normal-hearing individuals with no prior experience in speechreading to those with experience in speechreading (*Journal of the Acoustical Society of America,* Feb. 1986). The sample consisted of 24 inexperienced and 12 experienced subjects. All subjects were asked to verbally reproduce sentences while speechreading. A summary of the results (percentage of correct syllables) for the two groups is given in the table. Conduct a test to determine whether the variance in the percentage of correctly reproduced syllables differs between the two groups of speechreaders. Test using $\alpha = .10$.

Inexperienced Speechreaders	Experienced Speechreaders
$n_1 = 24$	$n_2 = 12$
$\bar{x}_1 = 87.1$	$\bar{x}_2 = 86.1$
$s_1 = 8.7$	$s_2 = 12.4$

Source: Breeuwer, M., and Plomp, R. "Speechreading supplemented with auditorily presented speech parameters." *Journal of the Acoustical Society of America,* Vol. 79, No. 2, Feb. 1986, p. 487.

9.7 Distribution-Free Tests (Optional)

The z tests and t tests developed in Sections 9.1 and 9.2 involve making inferences about population parameters. Consequently, they are often referred to as *parametric statistical tests.* For small samples, these parametric methods rely on the assumption that the data are sampled from normally distributed populations. When the data are normal, these tests are *most powerful.* That is, the use of these parametric tests maximizes the chance the researcher correctly rejects the null hypothesis.

Power of a Test
The probability of rejecting H_0 when H_0 is false is called the *power* of the test.

Consider populations of data that are clearly nonnormal. For example, the distributions might be very flat, peaked, or strongly skewed to the right or left (see Figure 9.15). Applying the two-sample t test to such a data set may result in serious consequences. Since the normality assumption is clearly violated, the results of the t test may be unreliable: (1) the probability of a Type I error (i.e., rejecting H_0 when it is true) may be larger than the value of α selected; and (2) the power of the test (see box at left) may be different from what we expect.

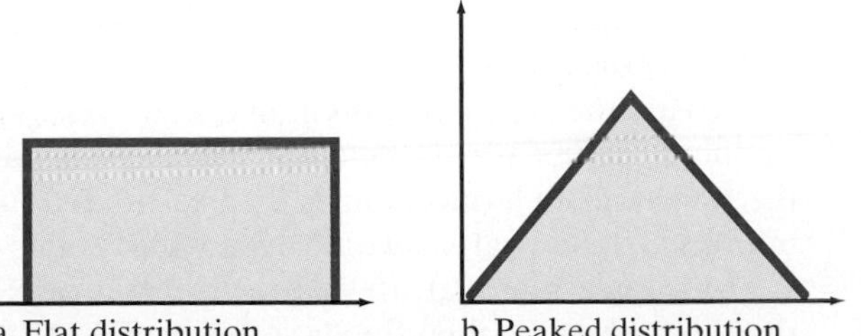

Figure 9.15 Some Nonnormal Distributions

A variety of *nonparametric techniques* are available for analyzing data that do not follow a normal distribution. Nonparametric tests do not depend on the distribution of the sampled population; thus, they are called *distribution-free tests.* Also, nonparametric methods focus on the location of the probability distribution of the population, rather than specific parameters of the population, such as the mean (hence, the name "nonparametrics").

Definition 9.1

Distribution-free tests are statistical tests that do not rely on any underlying assumptions about the probability distribution of the sampled population.

Definition 9.2

The branch of inferential statistics devoted to distribution-free tests is called **nonparametrics.**

Nonparametric tests are also appropriate when the data are nonnumerical in nature, but can be ranked. For example, when taste-testing foods or in other types of consumer product evaluations, we can say we like product A better than product B, and B better than C, but we cannot obtain exact quantitative values for the respective measurements. Nonparametric tests based on the ranks of measurements are called *rank tests*.

Definition 9.3

Nonparametric statistics (or tests) based on the ranks of measurements are called **rank statistics** (or **rank tests**).

In the optional sections that follow, we present two useful nonparametric methods. Keep in mind that these nonparametric tests are more powerful than their corresponding parametric counterparts in those situations where either the data are nonnormal or the data are ranked.

9.8 A Nonparametric Test for Comparing Two Populations: Independent Random Samples (Optional)

In Section 9.1 we presented *parametric* tests (tests about population parameters based on the z and t statistics) to test for a difference between two population means μ_1 and μ_2. Recall that both the mean and median of a population measure the *location* of the population distribution. If the two population distributions have the same shape, the equivalent nonparametric test is a test about the difference between population medians η_1 and η_2. The test, based on independent random samples of n_1 and n_2 observations from the respective populations, is known as the *Wilcoxon rank sum test.*

To use the Wilcoxon rank sum test, we first rank all $(n_1 + n_2)$ observations, assigning a rank of 1 to the smallest, 2 to the second smallest, and so on. The sum of the ranks, called a **rank sum,** is then calculated for each sample. If the two distributions are identical, we would expect the sample rank sums, designated as T_1 and T_2, to be nearly equal. In contrast, if one rank sum—say, T_1—is much larger than the other, T_2, then the data suggest that the distribution for population 1 is

shifted to the right of the distribution for population 2. The procedure for conducting a **Wilcoxon rank sum test** is summarized in the box and illustrated in Example 9.18.

*Another statistic used for comparing two populations based on independent random samples is the *Mann-Whitney U statistic.* The U statistic is a simple function of the rank sums, T_1 and T_2. It can be shown that the Wilcoxon test and the U test are equivalent.

WILCOXON RANK SUM TEST FOR A DIFFERENCE BETWEEN POPULATION MEDIANS: INDEPENDENT RANDOM SAMPLES*

Let η_1 and η_2 represent the medians for populations 1 and 2, respectively.

Upper-Tailed Test	Two-Tailed Test	Lower-Tailed Test
H_0: $\eta_1 = \eta_2$	H_0: $\eta_1 = \eta_2$	H_0: $\eta_1 = \eta_2$
H_a: $\eta_1 > \eta_2$	H_a: $\eta_1 \neq \eta_2$	H_a: $\eta_1 < \eta_2$

Test statistic: Either rank sum T_1 or T_2

Either rank sum can be used if $n_1 = n_2$ (denote this rank sum as T for the two-tailed test)

Rejection region:	*Rejection region:*	*Rejection region:*
If T_1: $T_1 \geq T_U$	Either $T \leq T_L$ or $T \geq T_U$	If T_1: $T_1 \leq T_L$
If T_2: $T_2 \leq T_L$		If T_2: $T_2 \geq T_U$

where T_L and T_U are the lower- and upper-tailed critical values obtained from Table B.7 of Appendix B. [*Note:* Tied observations are assigned ranks equal to the average of the ranks that would have been assigned to the observations had they not been tied.]

Assumption: The shapes of the two population distributions are approximately the same.

EXAMPLE 9.18

APPLYING THE RANK SUM TEST

In a comparison of visual acuity of deaf and hearing children, eye movement rates are taken on 10 deaf and 10 hearing children. (See Table 9.9.) A clinical psychologist believes that deaf children have greater visual acuity than hearing children. Test the psychologist's claim by using the data in Table 9.9. (The larger a child's eye movement rate, the more visual acuity the child possesses.) Use $\alpha = .05$.

DEAF

TABLE 9.9 Visual Acuity of Children

Deaf Children		Hearing Children	
2.75	1.95	1.15	1.23
3.14	2.17	1.65	2.03
3.23	2.45	1.43	1.64
2.30	1.83	1.83	1.96
2.64	2.23	1.75	1.37

Solution

The psychologist believes that the visual acuity of deaf children is greater than that of hearing children. If so, the median visual acuity of deaf children, η_1, will exceed the median for hearing children, η_2. Therefore, we will test

$$H_0: \eta_1 = \eta_2$$
$$H_a: \eta_1 > \eta_2$$

To conduct the test, we need to rank all 20 visual acuity observations in the combined samples. The ranks are shown in Table 9.10. Tied ranks receive the average of the ranks the scores would have received if they had not been tied.

TABLE 9.10 Calculation of Rank Sums for Example 9.18

Deaf Children		Hearing Children	
Visual Acuity	Rank	Visual Acuity	Rank
2.75	18	1.15	1
3.14	19	1.65	6
3.23	20	1.43	4
2.30	15	1.83	8.5
2.64	17	1.75	7
1.95	10	1.23	2
2.17	13	2.03	12
2.45	16	1.64	5
1.83	8.5	1.96	11
2.23	14	1.37	3
	$T_1 = 150.5$		$T_2 = 59.5$

Since the sample sizes n_1 and n_2 are equal, the Wilcoxon rank sum test statistic is (according to the box) either T_1 (the sum of the ranks for deaf children) or T_2 (the sum of the ranks for hearing children). Arbitrarily, we choose T_1 and our test statistic is $T_1 = 150.5$.

Table B.7 of Appendix B gives lower- and upper-tailed critical values of the rank sum distribution, denoted T_L and T_U, respectively, for values $n_1 \leq 10$ and $n_2 \leq 10$. The portion of Table B.7 for a one-tailed test with $\alpha = .05$ and for a two-tailed test with $\alpha = .10$ is reproduced in Table 9.11. Values of n_1 are given across the top of the table; values of n_2 are given at the left.

Examining Table 9.11, you will find that the critical values (highlighted) corresponding to $n_1 = n_2 = 10$ are $T_L = 83$ and $T_U = 127$. Therefore, for a one-tailed (upper-tailed) test at $\alpha = .05$, we will reject H_0 if $T_1 \geq T_U$, i.e.,

$$\text{Reject } H_0 \text{ if } T_1 \geq 127$$

Since the observed value of the test statistic, $T_1 = 150.5$, is greater than 127, we reject H_0 and conclude (at $\alpha = .05$) that the median of the visual acuity levels of

TABLE 9.11 A Portion of the Wilcoxon Rank Sum Table, Table B.7

$n_2 \backslash n_1$	3		4		5		6		7		8		9		10	
	T_L	T_U	T_L	T_U	T_L	T_U	T_L	T_U	T_L	T_U	T_L	T_U	T_L	T_U	T_L	T_U
3	6	15	7	17	7	20	8	22	9	24	9	27	10	29	11	31
4	7	17	12	24	13	27	14	30	15	33	16	36	17	39	18	42
5	7	20	13	27	19	36	20	40	22	43	24	46	25	50	26	54
6	8	22	14	30	20	40	28	50	30	54	32	58	33	63	35	67
7	9	24	15	33	22	43	30	54	39	66	41	71	43	76	46	80
8	9	27	16	36	24	46	32	58	41	71	52	84	54	90	57	95
9	10	29	17	39	25	50	33	63	43	76	54	90	66	105	69	111
10	11	31	18	42	26	54	35	67	46	80	57	95	69	111	83	127

	A	B
1	**Wilcoxon Rank Sum Test**	
2		
3	**Data**	
4	**Level of Significance**	**0.05**
5		
6	Population 1 Sample	
7	Sample Size	10
8	Sum of Ranks	150.5
9	Population 2 Sample	
10	Sample Size	10
11	Sum of Ranks	59.5
12		
13	Intermediate Calculations	
14	Total Sample Size n	20
15	T1 Test Statistic	150.5
16	T1 Mean	105
17	Standard Error of T1	13.22876
18	Z Test Statistic	3.439477
19		
20	**Upper-Tail Test**	
21	**Upper Critical Value**	**1.644853**
22	***p*-value**	**0.000291**
23	**Reject the null hypothesis**	

E Figure 9.16 PHStat and Excel Output of Wilcoxon Rank Sum Test, Example 9.18

deaf children (η_1) exceeds the median of the visual acuity levels of hearing children (η_2).

The same conclusion can be obtained by using statistical software. The PHStat and Excel output of the Wilcoxon rank sum analysis is displayed in Figure 9.16. Both the test statistic (150.5) and p-value of the test (≈ 0) are indicated on the printout. Since the p-value is less than $\alpha = .05$, we again have evidence to reject H_0 and to support the psychologist's belief.

Self-Test 9.10

The following sample data represent independent random samples selected from two populations.

Sample 1:	31	40	27	22	25
Sample 2:	50	32	20	48	45

a. Rank the data for the combined samples.

b. Compute the rank sums T_1 and T_2.

c. Find the critical value for testing whether the median of population 1 is less than the median of population 2. Use $\alpha = .05$.

d. Give the conclusion of the test, part **c**.

Many nonparametric test statistics have sampling distributions that are approximately normal when n_1 and n_2 are large. For these situations we can test hypotheses using the large-sample z test of Chapter 8. The procedure is outlined in the following box.

THE WILCOXON RANK SUM TEST FOR LARGE SAMPLES

Let η_1 and η_2 represent the medians for populations 1 and 2, respectively.

Upper-Tailed Test	**Two-Tailed Test**	**Lower-Tailed Test**
H_0: $\eta_1 = \eta_2$	H_0: $\eta_1 = \eta_2$	H_0: $\eta_1 = \eta_2$
H_a: $\eta_1 > \eta_2$	H_a: $\eta_1 \neq \eta_2$	H_a: $\eta_1 < \eta_2$

$$\text{Test statistic: } z = \frac{T_1 - \left[\dfrac{n_1 n_2 + n_1(n_1+1)}{2}\right]}{\sqrt{\dfrac{n_1 n_2(n_1+n_2+1)}{12}}} \qquad \textbf{(9.22)}$$

Rejection region:	*Rejection region:*	*Rejection region:*
$z > z_\alpha$	$\lvert z\rvert > z_{\alpha/2}$	$z < -z_\alpha$

[*Note:* The sample sizes n_1 and n_2 must both be at least 10.]

EXAMPLE 9.19

LARGE-SAMPLE RANK SUM TEST

Refer to Example 9.18 and the Wilcoxon rank sum test statistic $T_1 = 150.5$. Show that the large-sample Wilcoxon rank sum z test gives the same result as the exact test performed in Example 9.18.

Solution

According to equation 9.22 in the box, the value of the large-sample Wilcoxon rank sum z test statistic is

$$z = \frac{T_1 - \left[\frac{n_1 n_2 + n_1(n_1 + 1)}{2}\right]}{\sqrt{\frac{n_1 n_2(n_1 + n_2 + 1)}{12}}} = \frac{150.5 - \left[\frac{(10)(10) + (10)(11)}{2}\right]}{\sqrt{\frac{(10)(10)(10 + 10 + 1)}{12}}}$$

$$= \frac{150.5 - 105}{\sqrt{2{,}100/12}}$$

$$= 3.44$$

(This value is shown on the Excel printout, Figure 9.16, as "Z Test Statistic.")

Since we want to detect if the median for population 1 (deaf children) is greater than the median for population 2 (hearing children), we will reject H_0 for values of z in the upper tail of the z distribution. For $\alpha = .05$, the rejection region is $z > z_{.05}$, or

$$\text{Reject } H_0 \text{ if } z > 1.645 \quad \text{(see Figure 9.17)}$$

Figure 9.17 Rejection Region for Example 9.19

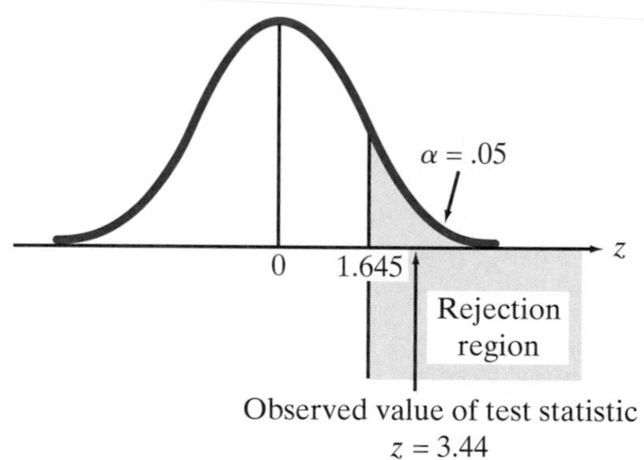

Also, the p-value of the large-sample test is

$$p\text{-value} = P(z \geq 3.44) \approx 0$$

Since $z = 3.44$ is greater than 1.645 or p-value $= 0$ is less than $\alpha = .05$, our conclusion is the same as that for Example 9.18—reject H_0 and conclude that the median visual acuity level of deaf children (population 1) is greater than the median visual acuity level of hearing children (population 2).

PROBLEMS FOR SECTION 9.8

Using the Tools

9.72 Specify the rejection region for the Wilcoxon rank sum test for independent samples in each of the following situations. Assume the test statistic is T_1.

a. H_0: $\eta_1 = \eta_2$, H_a: $\eta_1 \neq \eta_2$, $n_1 = 5$, $n_2 = 10$, $\alpha = .05$

b. H_0: $\eta_1 = \eta_2$, H_a: $\eta_1 > \eta_2$, $n_1 = 8$, $n_2 = 8$, $\alpha = .05$

c. H_0: $\eta_1 = \eta_2$, H_a: $\eta_1 < \eta_2$, $n_1 = 5$, $n_2 = 7$, $\alpha = .05$

9.73 Repeat Problem 9.72, but assume the test statistic is T_2.

9.74 For each of the following situations, find the large-sample Wilcoxon rank sum test statistic and corresponding rejection region:

a. Lower-tailed test, $T_1 = 71, n_1 = 10, n_2 = 14, \alpha = .05$

b. Upper-tailed test, $T_1 = 750, n_1 = 25, n_2 = 25, \alpha = .10$

c. Two-tailed test, $T_1 = 430, n_1 = 20, n_2 = 15, \alpha = .05$

9.75 Independent random samples were selected from two populations. The data are shown in the table. Suppose you want to determine whether the median of population 1 is greater than the median of population 2.

Sample from Population 1		Sample from Population 2	
15	12	5	7
17	16	9	4
12		13	5
14		10	10

a. Compute the rank sums T_1 and T_2.

b. Give the rejection region for the test using $\alpha = .05$. (Use T_1 as a test statistic.)

c. State the appropriate conclusion.

	A	B
1	**Wilcoxon Rank Sum Test**	
2		
3	**Data**	
4	**Level of Significance**	**0.1**
5		
6	Population 1 Sample	
7	Sample Size	20
8	Sum of Ranks	600.5
9	Population 2 Sample	
10	Sample Size	20
11	Sum of Ranks	219.5
12		
13	Intermediate Calculations	
14	Total Sample Size n	40
15	T1 Test Statistic	600.5
16	T1 Mean	410
17	Standard Error of T1	36.96846
18	Z Test Statistic	5.153042
19		
20	**Upper-Tail Test**	
21	**Upper Critical Value**	**1.281551**
22	***p*-value**	**1.28E-07**
23	**Reject the null hypothesis**	

E Problem 9.76

Applying the Concepts

9.76 A study was conducted to compare the self-perceived assertiveness of women with traditional and nontraditional (i.e., less stereotypical) gender attitudes (*Small Group Research,* Nov. 1997). Forty female college undergraduates (20 with traditional and 20 with nontraditional attitudes) were selected to be group leaders. Each group was asked to complete a card sort task in 25 minutes. After completing the task, the assertiveness of each leader was measured using the Rathus Assertiveness Schedule (RAS). The RAS scores (simulated, based on information provided in the journal article) are listed in the table below. (Higher scores indicate higher levels of self-perceived assertiveness.) Use the Excel printout at left to compare the medians of the RAS scores of the two groups of leaders. Do nontraditional leaders appear to be more self-assertive than traditional leaders? Test using $\alpha = .10$.

RAS

Nontraditional Leader			Traditional Leader		
17	28	25	6	9	7
28	24	37	4	8	11
27	23	29	10	8	2
20	26	30	10	9	6
9	23	26	7	13	7
43	42	21	4	13	8
40	28		6	21	

9.77 Type II collagen is a candidate drug for suppressing the symptoms of rheumatoid arthritis. Medical researchers at Harvard University conducted a clinical trial to test the ability of collagen to reduce swollen joints in rheumatoid arthritis sufferers (*Science,* Sept. 24, 1993.) Each of 59 patients with severe, active rheumatoid arthritis was randomly assigned to receive either a daily dose of collagen or an indistinguishable placebo over a 30-day period. (Twenty eight received collagen, 31 received the placebo.) The variable of interest was change in the number of swollen joints (after treatment minus before treatment) in each patient. The means and standard deviations for the two samples are reported in the table on p. 508 .

a. Although no information on the distribution of the variable of interest for the two patient groups was provided in the *Science* article, the researchers used the

	Change in Number of Swollen Joints (After–Before)	
	Collagen	Placebo
Sample size	28	31
Mean	−2.7	2.0
Std. deviation	.5	1.4

Source: Trentham, D. E., et al. "Effects of oral administration of Type II collagen on rheumatoid arthritis." *Science,* Vol. 261, No. 5129, Sept. 24, 1993, p. 1727 (Table 1).

Wilcoxon rank sum test rather than a t test to compare the two groups. Give two reasons why a nonparametric test may be more appropriate in this study.

b. Give the null hypothesis tested by the Wilcoxon rank sum procedure in this study.

c. The observed significance level of the test was reported to be smaller than .05 (i.e., p-value $< .05$). Make the appropriate conclusion.

9.78 Refer to the *Journal of Geography* (May/June 1997) study of the influence of computer software on learning geography, Problem 2.27 (p. 78). Recall that two groups of seventh graders were given a test on the regions of Africa. One group used computer resources (e.g., electronic atlases and interactive maps) to research the region, while the other group used conventional methods (e.g., notes, worksheets). The test scores are reproduced in the accompanying tables. Is there evidence that the median test score of the computer resource group is larger than the median of the conventional group? Conduct the appropriate nonparametric test using $\alpha = .01$.

GEO

Group 1 (Computer Resources) Student Scores											
41	53	44	41	66	91	69	44	31	75	66	69
75	53	44	78	91	28	69	78	72	53	72	63
66	75	97	84	63	91	59	84	75	78	88	59
97	69	75	69	91	91	84					

Group 2 (No Computer) Student Scores											
56	59	9	59	25	66	31	44	47	50	63	19
66	53	44	66	50	66	19	53	78	78		

Source: Linn, S. E. "The effectiveness of interactive maps in the classroom: A selected example in studying Africa." *Journal of Geography,* Vol. 96, No. 3, May/June 1997, p. 167 (Table 1).

9.79 The Harris Corporation/University of Florida data on the voltage readings of production runs at two locations, Problem 9.18 (p. 465) are reproduced on p. 509. If the median voltage reading at the new location is significantly higher (at $\alpha = .05$) than the median at the old (remote) location, the production process will be permanently shifted to the new location. What decision should Harris Corporation make?

9.80 In an epidemiological study, 55 female patients were diagnosed with full syndrome *bulimia nervosa* (an eating disorder). Each of these patients was asked to rate her satisfaction with life in general on a 7-point scale, where 1 = extremely dissatisfied and 7 = extremely satisfied. Their responses were compared to those of an independent sample of 4,208 normal female subjects (*American Journal of Psychiatry,* July 1995). The researchers hypothesized that female bulimia nervosa patients will have lower satisfaction levels than normal females.

a. The two groups' satisfaction levels were compared using the large-sample Wilcoxon rank sum test. State the null hypothesis of this test.

Old Location			New Location		
9.98	10.12	9.84	9.19	10.01	8.82
10.26	10.05	10.15	9.63	8.82	8.65
10.05	9.80	10.02	10.10	9.43	8.51
10.29	10.15	9.80	9.70	10.03	9.14
10.03	10.00	9.73	10.09	9.85	9.75
8.05	9.87	10.01	9.60	9.27	8.78
10.55	9.55	9.98	10.05	8.83	9.35
10.26	9.95	8.72	10.12	9.39	9.54
9.97	9.70	8.80	9.49	9.48	9.36
9.87	8.72	9.84	9.37	9.64	8.68

Source: Harris Corporation, Melbourne, Fla.

Problem 9.79

b. State the alternative hypothesis of interest to the researchers.

c. The large-sample test statistic was calculated to be $z = 5.48$. Use this result to carry out the test. What is your conclusion if $\alpha = .01$?

9.81 *Environmental Science & Technology* (Oct. 1993) reported on a study of insecticides used on dormant orchards in the San Joaquin Valley, California. Ambient air samples were collected and analyzed daily at an orchard site during the most intensive period of spraying. The oxon/thion ratios (in ng/m^3) in the air samples are recorded in the table. Compare the median oxon/thion ratio on foggy days to the median ratio on clear/cloudy days at the orchard using a nonparametric test. Use $\alpha = .05$.

Date	Condition	Oxon/Thion Ratio
Jan. 15	Fog	.270
17	Fog	.241
18	Fog	.205
19	Fog	.523
20	Clear	.618
21	Fog	.112
21	Clear	.591
22	Fog	.330
23	Fog	.270
23	Cloudy	.225
25	Fog	.239
25	Clear	.375

Source: Selber, J. N., et al. "Air and fog deposition residues of four organophosphate insecticides used on dormant orchards in the San Joaquin Valley, California." *Environmental Science & Technology,* Vol. 27, No. 10, Oct. 1993, p. 2240 (Table V).

9.9 A Nonparametric Test for Comparing Two Populations: Matched-Pairs Design (Optional)

Recall from Section 9.2 that the parametric matched-pairs t test for a difference in population means is based on the differences within the matched pairs of observations. The nonparametric alternative is the *Wilcoxon signed ranks test* for the median difference. To perform the test, we assign ranks to the absolute values of the differences and then base the comparison on the rank sum of the positive (T^+) differences. Differences equal to 0 are eliminated, and the number n of differences is reduced accordingly. Tied absolute differences receive ranks equal to the average of the ranks they would have received had they not been tied. The **Wilcoxon signed ranks test** is summarized in the box and is illustrated in Example 9.20.

THE WILCOXON SIGNED RANKS TEST FOR MEDIAN DIFFERENCE: MATCHED PAIRS

Let η_D represent the median of the differences of paired observations from populations 1 and 2, where the difference is population 1's measurement minus population 2's measurement.

Upper-Tailed Test	Two-Tailed Test	Lower-Tailed Test
H_0: $\eta_D = 0$	H_0: $\eta_D = 0$	H_0: $\eta_D = 0$
H_a: $\eta_D > 0$	H_a: $\eta_D \neq 0$	H_a: $\eta_D < 0$

Calculate the difference within each of the n matched pairs of observations. Then rank the absolute values of the n differences from the smallest (rank 1) to the highest (rank n) and calculate the rank sum T^+ of the positive differences.

Test statistic: T^+

Rejection region:	*Rejection region:*	*Rejection region:*
$T^+ \geq T_U$	$T^+ \leq T_L$ or $T^+ \geq T_U$	$T^+ \leq T_L$

where T_L and T_U are the lower and upper critical values, respectively, given in Table B.8 of Appendix B.

[*Note:* Differences equal to 0 are eliminated and the number n of differences is reduced accordingly. Tied absolute differences receive ranks equal to the average of the ranks they would have received had they not been tied.]

EXAMPLE 9.20

APPLYING THE SIGNED RANKS TEST

In Example 9.8 (p. 468), we used the parametric t test to determine whether the mean reading achievement test scores for first graders taught by two methods are different. First graders were matched according to IQ, socioeconomic status, and learning ability and one member of each pair was taught by method 1 and the other by method 2. The data for the $n = 10$ matched pairs are reproduced in Table 9.12. Conduct a test of hypothesis to determine if the median reading achievement test score differs for the two teaching methods. Use the Wilcoxon signed ranks test with $\alpha = .05$.

TABLE 9.12 Reading Test Scores and Differences, Example 9.20

	Teaching Method				
First-Grader Pair	1	2	Difference	Absolute Value of Difference	Rank of Absolute Value
1	78	71	7	7	3
2	63	44	19	19	10
3	72	61	11	11	5
4	89	84	5	5	2
5	91	74	17	17	9
6	49	51	−2	2	1
7	68	55	13	13	6
8	76	60	16	16	7.5
9	85	77	8	8	4
10	55	39	16	16	7.5

Solution

Let η_D represent the population median of the differences in test scores, where each difference is calculated as test score for method 1 minus test score for method 2.

Then we want to test

$$H_0\colon \eta_D = 0 \qquad H_a\colon \eta_D \neq 0$$

To apply the Wilcoxon signed ranks test, we need to rank the absolute values of the differences between the matched pairs of observations. The differences and ranks are shown in Table 9.12, with the ranks of the positive differences highlighted. The rank sum of the positive differences is

$$\begin{aligned} T^+ &= 3 + 10 + 5 + 2 + 9 + 6 + 7.5 + 4 + 7.5 \\ &= 54 \end{aligned}$$

Since we want to conduct a two-tailed test, we need to find the lower and upper critical values, T_L and T_U, respectively.

The critical values of the Wilcoxon signed ranks statistic for one-tailed and two-tailed tests and different values of α are provided in Table B.8 of Appendix B. A portion of Table B.8 is reproduced in Table 9.13. For this two-tailed test at $\alpha = .05$, locate the column labeled "Two-Tailed: $\alpha = .05$" and the row $n = 10$ in Table 9.13. The values of T_L and T_U (highlighted in Table 9.13) are $T_L = 8$ and $T_U = 47$. Thus, our rejection region is:

$$\text{Reject } H_0 \text{ if} \quad T^+ \leq 8 \quad \text{or} \quad T^+ \geq 47$$

TABLE 9.13 A Portion of the Wilcoxon Signed Ranks Table, Table B.8

n	One-Tailed: $\alpha = .05$ Two-Tailed: $\alpha = .10$	$\alpha = .025$ $\alpha = .05$	$\alpha = .01$ $\alpha = .02$	$\alpha = .005$ $\alpha = .01$
		(Lower, Upper)		
5	0, 15	—, —	—, —	—, —
6	2, 19	0, 21	—, —	—, —
7	3, 25	2, 26	0, 28	—, —
8	5, 31	3, 33	1, 35	0, 36
9	8, 37	5, 40	3, 42	1, 44
10	10, 45	8, 47	5, 50	3, 52
11	13, 53	10, 56	7, 59	5, 61
12	17, 61	13, 65	10, 68	7, 71
13	21, 70	17, 74	12, 79	10, 81
14	25, 80	21, 84	16, 89	13, 92
15	30, 90	25, 95	19, 101	16, 104

Clearly, our test statistic, $T^+ = 54$, is greater than $T_U = 47$. Therefore, we reject H_0 and conclude that there is sufficient evidence of a median difference in the reading achievement test scores for first graders taught by the two methods. (This is the same conclusion we obtained by using the t test in Example 9.8.)

A MINITAB printout of this nonparametric test is illustrated in Figure 9.18. Both the test statistic (called "Wilcoxon statistic") and the p-value are highlighted

Wilcoxon Signed Rank Test

```
Test of median = 0.000000 versus median not = 0.000000

                  N for   Wilcoxon             Estimated
             N    Test   Statistic       P       Median
Differen    10      10       54.0    0.008        11.50
                             T+      p-value
```

M **Figure 9.18** MINITAB Output for Wilcoxon Signed Ranks Test, Example 9.20

on the printout. Since p-value $= .008$ is less than $\alpha = .05$, our conclusion to reject H_0 is verified.

Self-Test 9.11

A random sample of $n = 5$ pairs of measurements is shown below.

	Pair Number				
	1	2	3	4	5
Measurement 1	600	510	622	331	308
Measurement 2	721	615	580	422	295

a. Compute and rank the difference between the two measurements.

b. Find the rank sum T^+.

c. Find the critical value for testing for a difference in the locations of the distributions for the two measurements. Use $\alpha = .10$.

d. Give the conclusion of the test, part **c.**

The Wilcoxon signed ranks statistic also has a sampling distribution that is approximately normal when the number n of pairs is large—say, $n \geq 20$. This large-sample nonparametric matched-pairs test is summarized in the following box.

WILCOXON SIGNED RANKS TEST FOR LARGE SAMPLES

Let η_D represent the population median difference.

Upper-Tailed Test	**Two-Tailed Test**	**Lower-Tailed Test**
H_0: $\eta_D = 0$	H_0: $\eta_D = 0$	H_0: $\eta_D = 0$
H_a: $\eta_D > 0$	H_a: $\eta_D \neq 0$	H_a: $\eta_D < 0$

$$\text{Test statistic: } z = \frac{T^+ - [n(n+1)/4]}{\sqrt{[n(n+1)(2n+1)]/24}} \qquad \textbf{(9.23)}$$

Rejection region:	*Rejection region:*	*Rejection region:*
$z > z_\alpha$	$|z| > z_{\alpha/2}$	$z < -z_\alpha$

Assumptions: The sample size n is greater than or equal to 20. Differences equal to 0 are eliminated and the number n of differences is reduced accordingly. Tied absolute differences receive ranks equal to the average of the ranks they would have received had they not been tied.

PROBLEMS FOR SECTION 9.9

Using the Tools

9.82 Specify the test statistic and rejection region for the Wilcoxon signed ranks test for the matched-pairs design in each of the following situations:

a. H_0: $\eta_D = 0$, H_a: $\eta_D \neq 0$, $n = 19$, $\alpha = .05$

b. H_0: $\eta_D = 0$, H_a: $\eta_D > 0$, $n = 36$, $\alpha = .01$

c. H_0: $\eta_D = 0$, H_a: $\eta_D < 0$, $n = 50$, $\alpha = .005$

9.83 Suppose you want to test the hypothesis that two treatments, A and B, are equivalent against the alternative that the responses for A tend to be larger than those for B.

a. If $n = 8$ and $\alpha = .01$, give the rejection region for a Wilcoxon signed ranks test.

b. Suppose you want to determine if the population median difference differs from 0. If $n = 7$ and $\alpha = .10$, give the rejection region for the Wilcoxon signed ranks test.

9.84 For each of the following sample differences in a matched-pairs experiment, compute T^+.

a. −4, −1, 16, −2, 3, 1, 2, −5

b. 21.2, 1.7, −4.0, 3.3, 11.5, 6.6, −15.7, 20.0, 31.5

c. 1,700, 1,334, −45, 911, 726, −1,271

9.85 A random sample of nine pairs of measurements is shown in the table.

Pair	Sample Data from Population 1	Sample Data from Population 2
1	8	7
2	10	1
3	6	4
4	10	10
5	7	4
6	8	3
7	4	6
8	9	2
9	8	4

a. Use the Wilcoxon signed ranks test to determine whether observations from population 1 tend to be larger than those from population 2. Test using $\alpha = .05$.

b. Use the Wilcoxon signed ranks test to determine whether a median difference exists. Test using $\alpha = .05$.

Applying the Concepts

9.86 One of the most critical aspects of a new atlas design is its thematic content. In a survey of atlas users (*Journal of Geography,* May/June 1995), a large sample of high school teachers in British Columbia ranked 12 thematic atlas topics for usefulness. The consensus rankings of the teachers (based on the percentage of teachers who responded they "would definitely use" the topic) are given in the table. These

ATLAS

	Rankings	
Theme	**High School Teachers**	**Geography Alumni**
Tourism	10	2
Physical	2	1
Transportation	7	3
People	1	6
History	2	5
Climate	6	4
Forestry	5	8
Agriculture	7	10
Fishing	9	7
Energy	2	8
Mining	10	11
Manufacturing	12	12

Source: Keller, C. P., et al. "Planning the next generation of regional atlases: Input from educators." *Journal of Geography,* Vol. 94, No. 3, May/June 1995, p. 413 (Table 1).

teacher rankings were compared to the rankings made by a group of university geography alumni. Compare the theme rankings for the two groups with an appropriate nonparametric test. Use $\alpha = .05$. Interpret the results.

9.87 Eleven prisoners of the war in Croatia were evaluated for neurological impairment after their release from a Serbian detention camp (*Collegium Antropologicum,* June 1997). All eleven released POWs received blows to the head and neck and/or loss of consciousness during imprisonment. Neurological impairment was assessed by measuring the amplitude of the visual evoked potential (VEP) in both eyes at two points in time: 157 days and 379 days after release from prison. (The higher the VEP value, the greater the neurological impairment.) The data for the 11 POWs are shown in the table. Determine whether the VEP measurements of POWs 157 days after their release tend to be less than the VEP measurements of POWs 379 days after their release. Test using $\alpha = .05$.

POWVEP

POW	157 Days after Release	379 Days after Release
1	2.46	3.73
2	4.11	5.46
3	3.93	7.04
4	4.51	4.73
5	4.96	4.71
6	4.42	6.19
7	1.02	1.42
8	4.30	8.70
9	7.56	7.37
10	7.07	8.46
11	8.00	7.16

Source: Vrca, A., et al. "The use of visual evoked potentials to follow up prisoners of war after release from detention camps." *Collegium Antropologicum,* Vol. 21, No. 1, June 1997, p. 232. (Data simulated from information provided in Table 3.)

9.88 Researchers at Purdue University compared human real-time scheduling in a processing environment to an automated approach that utilizes computerized robots and sensing devices (*IEEE Transactions,* Mar. 1993). The experiment consisted of eight simulated scheduling problems. Each task was performed by a human scheduler and by the automated system. Performance was measured by the *throughput* rate, defined as the number of good jobs produced weighted by product quality. The resulting throughput rates are shown in the accompanying table. Compare the throughput rates of tasks scheduled by a human and by the automated method using a nonparametric test. Use $\alpha = .01$.

THRUPUT

Task	Human Scheduler	Automated Method	Task	Human Scheduler	Automated Method
1	185.4	180.4	5	240.0	269.3
2	146.3	248.5	6	253.8	249.6
3	174.4	185.5	7	238.8	282.0
4	184.9	216.4	8	263.5	315.9

Source: Yih, Y., Liang, T., and Moskowitz, H. "Robot scheduling in a circuit board production line: A hybrid OR/ANN approach." *IEEE Transactions,* Vol. 25, No. 2, Mar. 1993, p. 31 (Table 1).

9.89 Dental researchers have developed a new material for preventing cavities—a plastic sealant that is applied to the chewing surfaces of teeth. To determine whether the sealant is effective, it was applied to half of the teeth of each of 12 school-age children. After five years, the numbers of cavities in the sealant-coated teeth and untreated teeth were counted. The results are given in the table on p. 515. Is there sufficient evidence to indicate that sealant-coated teeth are less prone to cavities than are untreated teeth? Test using $\alpha = .05$.

CAVITIES

Child	Sealant-Coated	Untreated	Child	Sealant-Coated	Untreated
1	3	3	7	1	5
2	1	3	8	2	0
3	0	2	9	1	6
4	4	5	10	0	0
5	1	0	11	0	3
6	0	1	12	4	3

Problem 9.89

9.90 Refer to the Merck Research Labs experiment to evaluate a new drug using rats in a swim maze, Problem 9.32 (p. 474). Recall that 19 impregnated dam rats were given a dosage of the drug, then one male and one female rat pup from each resulting litter were selected to perform in the swim maze. The number of swims required by each rat pup to escape three times is reproduced in the table.

SWIMMAZE

Litter	Male	Female	Litter	Male	Female
1	8	5	11	6	5
2	8	4	12	6	3
3	6	7	13	12	5
4	6	3	14	3	8
5	6	5	15	3	4
6	6	3	16	8	12
7	3	8	17	3	6
8	5	10	18	6	4
9	4	4	19	9	5
10	4	4			

Source: Thomas E. Bradstreet, Merck Research Labs, BL 3-2, West Point, Penn. 19486.

a. Compare the number of swim attempts for male and female pups using the Wilcoxon signed ranks test. Use $\alpha = .10$.

b. Compare the results of the nonparametric test, part **a**, to the results of the parametric test you conducted in Problem 9.32.

9.91 Refer to the *Environmental Science & Technology* (Oct. 1993) study of air deposition residues of the insecticide diazinon used on dormant orchards, Problem 9.81 (p. 509). Ambient air samples were collected and analyzed at an orchard site for each of 11 days during the most intensive period of spraying. The levels of diazinon residue (in ng/m^3) during the day and at night are recorded in the table. The researchers want to know whether the diazinon residue levels tend to differ from day to night. Perform the appropriate nonparametric analysis at $\alpha = .05$ and interpret the results.

DIAZINON

	Diazinon Residue			Diazinon Residue	
Date	Day	Night	Date	Day	Night
Jan. 11	5.4	24.3	Jan. 17	6.1	104.3
12	2.7	16.5	18	7.7	96.9
13	34.2	47.2	19	18.4	105.3
14	19.9	12.4	20	27.1	78.7
15	2.4	24.0	21	16.9	44.6
16	7.0	21.6			

Source: Selber, J. N., et al. "Air and fog deposition residues for organophosphate insecticides used on dormant orchards in the San Joaquin Valley, California." *Environmental Science & Technology,* Vol. 27, No. 10, Oct. 1993, p. 2240 (Table IV).

KEY TERMS

Starred (*) terms are from the optional sections of this chapter.

Contingency table 482
Difference between means 451
Difference between proportions 451
*Distribution-free tests 502
Expected cell counts 484
Independence of two qualitative variables 482
Independent samples 453
Matched pairs 467
*Nonparametrics 502
Pooled estimate of variance 458
*Rank statistics 502
*Rank sum 502
Ratio of two variances 451
*Wilcoxon rank sum test 503
*Wilcoxon signed ranks test 509

KEY FORMULAS

Starred (*) formulas are from the optional sections of this chapter.

Confidence Interval for Comparing Means or Proportions

Large Sample:

$$\text{Estimator} \pm (z_{\alpha/2})(\text{Standard error})$$ **(9.4)**, 455; **(9.12)**, 468; **(9.17)**, 477

Small Sample:

$$\text{Estimator} \pm (t_{\alpha/2})(\text{Standard error})$$ **(9.6)**, 458; **(9.13)**, 468

Test Statistic for Comparing Means or Proportions

$$z = \frac{\text{Estimator} - \text{Hypothesized value}}{\text{Standard error}}$$ **(9.3)**, 455; **(9.10)**, 468; **(9.16)**, 477

$$t = \frac{\text{Estimator} - \text{Hypothesized value}}{\text{Standard error}}$$ **(9.5)**, 458; **(9.11)**, 468

[*Note:* The respective point estimator and standard error for each parameter are provided in Table 9.14.]

TABLE 9.14 Estimators and Standard Errors for Population Parameters

Parameter	Point Estimate	Standard Error	Estimated Standard Error
$(\mu_1 - \mu_2)$	$(\bar{x}_1 - \bar{x}_2)$	$\sqrt{\frac{\sigma_1^2}{n_1} + \frac{\sigma_2^2}{n_2}}$	$\sqrt{\frac{s_1^2}{n_1} + \frac{s_2^2}{n_2}}$
μ_d	$\bar{d}$	$\sigma_d/\sqrt{n}$	$s_d/\sqrt{n}$
$(p_1 - p_2)$	$(\hat{p}_1 - \hat{p}_2)$	$\sqrt{\frac{p_1(1-p_1)}{n_1} + \frac{p_2(1-p_2)}{n_2}}$	$\sqrt{\frac{\hat{p}_1(1-\hat{p}_1)}{n_1} + \frac{\hat{p}_2(1-\hat{p}_2)}{n_2}}$
$*\sigma_1^2/\sigma_2^2$	s_1^2/s_2^2	(not necessary)	(not necessary)

Test Statistic for χ^2 Test of Independence

$$\chi^2 = \sum\sum\frac{(O_{ij} - E_{ij})^2}{E_{ij}}$$ **(9.20)**, 486

$$\text{where } E_{ij} = \frac{R_iC_j}{n}$$ **(9.18)**, 484

*Test Statistic for Comparing Variances

$$F = s_1^2/s_2^2$$ **(9.21)**, 494

*Wilcoxon Large-Sample Rank Sum Test Statistic

$$z = \frac{T_1 - \left[\frac{n_1(n_1 + n_2 + 1)}{2}\right]}{\sqrt{\frac{n_1n_2(n_1 + n_2 + 1)}{12}}}$$ **(9.22)**, 505

*Wilcoxon Large-Sample Signed Ranks Test Statistic

$$z = \frac{T^+ - \left[\frac{n(n + 1)}{4}\right]}{\sqrt{\frac{n(n + 1)(2n + 1)}{24}}}$$ **(9.23)**, 512

KEY SYMBOLS

Starred (*) symbols are from the optional sections of this chapter.

SYMBOL	DEFINITION
$\mu_1 - \mu_2$	Difference between two population means (independent samples)
$p_1 - p_2$	Difference between two population proportions (independent samples)
μ_d	Population mean difference (matched pairs)
σ_d	Population standard deviation of differences (matched pairs)
$\bar{x}_1 - \bar{x}_2$	Difference between two sample means (estimates $\mu_1 - \mu_2$)
$\hat{p}_1 - \hat{p}_2$	Difference between two sample proportions (estimates $p_1 - p_2$)
$\bar{d}$	Sample mean difference, matched pairs (estimates μ_d)
s_p^2	Pooled sample variance (estimates $\sigma_1^2 = \sigma_2^2$)
s_d	Sample standard deviation of differences, matched pairs (estimates σ_d)
σ_1^2/σ_2^2	*Ratio of two population variances
s_1^2/s_2^2	*Ratio of two sample variances (estimates σ_1^2/σ_2^2)
F_α	*Critical F value used to compare variances
R_i	Total for row i
C_j	Total for column j
E_{ij}	Expected count for cell in row i and column j
η	*Population median
T_i	*Sum of ranks in sample i
T^+	*Sum of ranks of positive differences

CHECKING YOUR UNDERSTANDING

1. What are the key words for determining the population parameter $(\mu_1 - \mu_2)$?
2. What are the key words for determining the population parameter $(p_1 - p_2)$?
3. What are the key words for determining the population parameter σ_1^2/σ_2^2?
4. Under what conditions is the two-sample t test appropriate for comparing two population means with small samples?
5. What is the difference between a matched-pairs design and an independent sampling design?
6. Under what conditions is it appropriate to apply the F test for comparing population variances?
7. What is a distribution-free test?
8. When is it more appropriate to apply a nonparametric test over a parametric test?
9. What is the difference between the Wilcoxon rank sum test and the Wilcoxon signed ranks test?

SUPPLEMENTARY PROBLEMS

Starred (*) problems refer to the optional sections of this chapter.

9.92 In developing countries such as India, working women are often the target of vio lence. Thus, in theory, these women are under greater stress. *Collegium Antropologicum* (June 1997) reported on a study to compare the anxiety levels of working and nonworking mothers in India. A random sample of 94 working mothers had a mean anxiety level of 33.44 (on a 50-point scale) with a standard deviation of 12.15. An independent random sample of 94 nonworking mothers had a mean anxiety level of 35.14 with a standard deviation of 11.09.

a. Is there sufficient evidence of a difference in mean anxiety levels of working mothers and nonworking mothers in India? Use $\alpha = .10$.

b. Estimate the difference between the true mean anxiety levels of working and nonworking mothers in India with a 95% confidence interval.

9.93 In evaluating the usefulness and validity of a questionnaire, researchers often pretest the questionnaire on different independently selected samples of respondents. Knowledge of the differences and similarities of the samples and their respective populations is important for interpreting the questionnaire's validity. *Educational and Psychological Measurement* (Feb. 1998) reported on a newly developed questionnaire for measuring the career success expectations of employees. The instrument was tested on the two independent samples described below.

	Managers/Professionals	Part-Time MBA Students
Sample size	162	109
Gender (% males)	95.0	68.9
Marital status (% married)	91.2	53.4

Source: Stephens, G. K., Szajna, B., and Broome, K. M. "The Career Success Expectation Scale: An exploratory and confirmatory factor analysis." *Educational and Psychological Measurement,* Vol. 58, No. 1, (Feb. 1998) pp. 129–141.

a. Does the population of managers/professionals from which the sample was drawn consist of a higher percentage of males than the part-time MBA population? Conduct the appropriate test using $\alpha = .05$.

b. Does the population of managers/professionals consist of a higher percentage of married individuals than the part-time MBA population? Conduct the appropriate hypothesis test using $\alpha = .01$.

9.94 A study compared a traditional approach to teaching basic nursing skills with an innovative approach (*Journal of Nursing Education,* Jan. 1992). Forty-two students enrolled in an upper-division nursing course participated in the study. Half (21) were randomly assigned to labs that utilized the innovative approach. After completing the course, all students were given short-answer questions about scientific principles underlying each of 10 nursing skills. The objective of the research is to compare the mean scores of the two groups of students.

a. What is the appropriate test to use to compare the two groups?

b. Are any assumptions required for the test?

c. One question dealt with the use of clean/sterile gloves. The mean scores for this question were 3.28 (traditional) and 3.40 (innovative). Is there sufficient information to perform the test?

d. Refer to part **c.** The p-value for the test was reported as .79. Interpret this result.

e. Another question concerned the choice of a stethoscope. The mean scores of the two groups were 2.55 (traditional) and 3.60 (innovative) with an associated p-value of .02. Interpret these results.

9.95 Did you know that the use of aspirin to alleviate the symptoms of viral infections in children may lead to serious complications (Reyes' syndrome)? A random sample of 500 children with viral infections received no aspirin to alleviate symptoms, and 12 developed Reyes' syndrome. In a random sample of 450 children with viral infections who were given aspirin, 23 developed Reyes' syndrome.

a. Is there sufficient evidence to indicate that the proportion of children with viral infections who develop Reyes' syndrome is greater for those who take aspirin than for those who do not? Test at $\alpha = .05$.

b. Construct a 95% confidence interval for the difference in the proportions of children who develop Reyes' syndrome between those who receive no aspirin and those who receive aspirin during a viral infection. Interpret the interval.

9.96 Propranolol, a class of heart drugs called *beta blockers,* have been used in an attempt to reduce anxiety in students taking the Scholastic Aptitude Test (SAT). In one

study, 22 high school juniors who had not performed as well as expected on the SAT were administered a beta blocker one hour prior to retaking the test in their senior year. The junior and senior year SAT scores for these 22 students are shown in the table. Nationally, students who retake the test without special preparations will increase their score by an average of 38 points (reported in *Newsweek,* Nov. 16, 1987). Test the hypothesis that the true mean increase in SAT scores for those students who take a beta blocker prior to the exam exceeds the national average increase of 38. Use $\alpha = .05$.

SATSCORES

	SAT Scores			SAT Scores	
Student	**Junior Year**	**Senior Year**	**Student**	**Junior Year**	**Senior Year**
1	810	984	12	933	945
2	965	1015	13	811	897
3	707	1006	14	774	875
4	652	995	15	780	923
5	983	997	16	913	884
6	822	963	17	655	931
7	874	860	18	906	1136
8	900	915	19	737	854
9	693	847	20	788	927
10	1115	1202	21	878	872
11	749	910	22	912	1054

Note: SAT scores are simulated based on information provided in *Newsweek.*

Problem 9.96

***9.97** Refer to Problem 9.96. Apply the appropriate nonparametric test to the data. Interpret the results.

9.98 Refer to the *Journal of Travel Research* (May 1999) survey of 5,026 pleasure travelers in the Tampa Bay region, Problem 2.34 (p. 86). The travelers were classified according to education level and their use of the Internet to seek travel information. The table summarizes the results of the interviews. The researchers theorized that travelers who use the Internet to search for travel information are more likely to be college educated. Test this theory at $\alpha = .05$.

		Use of Internet to Seek Travel Information	
		Yes	**No**
Education	**College degree or more**	1,072	1,287
	Less than a college degree	640	2,027

Source: Bonn, M., Furr, L., and Susskind, A. "Predicting a behavioral profile for pleasure travelers on the basis of Internet use segmentation." *Journal of Travel Research,* Vol. 37, (May 1999), pp. 333–340.

9.99 Do carbohydrate-electrolyte drinks (e.g., Gatorade, Powerade, All Sport) improve performance in sports? To answer this question, researchers at Springfield College (Mass.) recruited eight trained male competitive cyclists to participate in a study (*International Journal of Sport Nutrition,* June 1995). During one cycling trial, each cyclist was given a carbohydrate-electrolyte drink. During a second trial, the cyclists were given a placebo drink (Crystal Light). At the end of each trial, several physiological variables were measured. These variables were analyzed using matched pairs. A summary of the results is shown in the table on p. 520.

	Placebo		Carbohydrate-Electrolyte		Matched Pairs
Variable	Mean	Std. Dev.	Mean	Std. Dev.	*p*-Value
1. Perceived exertion rating	5.60	1.01	4.99	1.11	.003
2. Heart rate (beats/minute)	161.90	6.80	160.10	10.00	>.050
3. Peak power (WAT)	747.60	97.30	794.00	83.50	.020
4. Fatigue index (%)	21.30	12.40	20.90	10.40	>.050

Source: Ball, T. C., et al. "Periodic carbohydrate replacement during 50 min. of high intensity cycling improves subsequent sprint performance." *International Journal of Sport Medicine,* Vol. 5, No. 2, June 1995, p. 155 (Table 3).

Problem 9.99

a. Interpret the *p*-values shown in the table. Make a practical conclusion for each variable.

b. Sufficient information is provided in the table to analyze the data as independent random samples. Perform these tests and compare the results to those of part **a**.

c. Explain why the matched pairs tests are more appropriate than the independent samples tests.

9.100 Refer to the *Journal of Genetic Psychology* (Mar. 1998) study of the attitudes of male and female students toward their fathers, Problem 9.12 (p. 463). A sample of University of South Alabama students were asked whether their relationship with their father was (1) Awful, (2) Poor, (3) Average, (4) Good, or (5) Great. Use a 99% confidence interval to compare the proportion of male students who have a "great" relationship with their father to the corresponding proportion of female students. Can you infer that one group of students has a higher proportion with a "great" father relationship than the other group? Explain.

9.101 Epidemiologists have theorized that the risk of coronary heart disease can be reduced by an increased consumption of fish. One study monitored the diet and health of a random sample of middle-age men who live in The Netherlands. The men were divided into groups according to the number of grams of fish consumed per day. Twenty years later, the level of dietary cholesterol (one of the risk factors for coronary disease) present in each was recorded. The results for two groups of subjects, the "no fish consumption" group (0 grams per day) and the "high fish consumption" group (greater than 45 grams per day), are summarized in the table. (Dietary cholesterol is measured in milligrams per 1,000 calories.)

	No Fish Consumption (0 Grams/day)	High Fish Consumption (>45 Grams/day)
Sample size	159	79
Mean	146	158
Standard deviation	66	75

Source: Kromhout, D., Bosschieter, E. B., and Coulander, C. L. "The inverse relationship between fish consumption and 20-year mortality from coronary heart disease." *New England Journal of Medicine,* May 9, 1985, Vol. 312, No. 19, pp. 1205–1209.

a. Calculate a 99% confidence interval for the difference between the mean levels of dietary cholesterol present in the two groups.

b. Based on the interval constructed in part **a**, what can you infer about the true difference? Explain.

***c.** Is there evidence of a difference between the variances of the cholesterol levels for the two groups of men? Test using $\alpha = .05$.

d. How does the result, part **c**, impact the validity of the inference made in part **b**? Explain.

9.102 Refer to the *Dalton Transactions* (Dec. 1997) study of bond lengths of new metal compounds, Problem 2.73 (p. 110). The data, reproduced in the accompanying table, are bond lengths for samples of 36 copper–selenium compounds and 7 copper–phosphorus compounds.

Copper–Selenium					
256.1	259.0	256.0	264.7	262.5	239.0
238.9	237.4	265.9	240.2	239.7	238.3
273.8	246.4	246.0	247.4	290.6	260.9
255.6	253.1	283.7	261.4	271.9	267.9
238.7	237.6	240.0	241.1	240.3	240.8
246.5	248.3	246.0	250.3	250.1	254.6
Copper–Phosphorus					
223.8	222.7	223.3	226.1	226.3	226.0
225.0					

Source: Deveson, A., et al. "Syntheses and structures of four new copper (I)–selenium clusters: Size dependence of the cluster on the reaction conditions." *Dalton Transactions,* No. 23, Dec. 1997, p. 4492 (Table 1).

a. Find a point estimate for the true difference between the mean bond lengths of copper–selenium and copper–phosphorus metal compounds.

b. Form a 99% confidence interval around the point estimate, part **a**. Interpret the result.

c. Compare the mean bond lengths of the two metal compounds with a two-tailed test of hypothesis. Use $\alpha = .01$. Does your inference agree with that of part **b**?

***d.** Test the assumption required for the inferences, parts **b** and **c**, to be valid.

***e.** Conduct an appropriate nonparametric test on the data. Compare the results to part **c**.

9.103 Geneticists at Duke University Medical Center have identified the E2F1 transcription factor as an important component of cell proliferation control (*Nature,* Sept. 23, 1993). The researchers induced DNA synthesis in two batches of serum-starved cells. Each cell in one batch was micro-injected with the E2F1 gene, while the cells in the second batch (the controls) were not exposed to E2F1. After 30 hours, the number of cells in each batch that exhibited altered growth was determined. The results of the experiment are summarized in the table.

	Control	E2F1-Treated Cells
Total number of cells	158	92
Number of growth-altered cells	15	41

Source: Johnson, D. G., et al. "Expression of transcription factor E2F1 induces quiescent to enter S phase." *Nature,* Vol. 365, No. 6444, Sept. 23, 1993, p. 351 (Table 1).

a. Compare the percentages of cells exhibiting altered growth in the two batches with a test of hypothesis. Use $\alpha = .10$.

b. Use the results, part **a**, to make an inference about the ability of the E2F1 transcription factor to alter cell growth.

***9.104** In the early 1960s, the air conditioning systems of a fleet of Boeing 720 jet airplanes came under investigation. The table on p. 522 presents the lifelengths (in hours) of the air-conditioning systems in two different Boeing 720 planes. Assuming the data represent random samples from the respective populations, is there evidence of a difference between the median lifelengths of the air-conditioning systems for the two Boeing 720 planes? Test using $\alpha = .05$.

BOEING720

Plane 1			Plane 2		
23	156	76	59	66	67
118	49	62	32	230	34
90	10		14	54	
29	310		102	152	

Source: Hollander, M., Park, D. H., and Proschan, F. "Testing whether F is 'more NBU' than is G." *Microelectronics and Reliability,* Vol. 26, No. 1, 1986, p. 43, Table 1, Pergamon Press, Ltd.

Problem 9.104

9.105 A group of University of Florida psychologists investigated the effects of age and gender on the short-term memory of adults (*Cognitive Aging Conference,* Apr. 1996). Each person in a sample of 152 adults was asked to place 20 common household items (e.g., eyeglasses, keys, hat, hammer) into the rooms of a computer-image house. After performing some unrelated activities, each subject was asked to recall the locations of the objects they had placed. The number of correct responses (out of 20) was recorded.

a. The researchers theorized that women will have a higher mean recall score than men. Set up the null and alternative hypotheses to test this theory.

b. Refer to part **a**. The 43 men in the study had a mean recall score of 13.5, while the 109 women had a mean recall score of 14.4. The p-value for comparing these two means was found to be .0001. Interpret this value.

c. The researchers also hypothesized that younger adults would have a higher mean recall score than older adults. Set up H_0 and H_a to test this theory.

d. The p-value for the test of part **c** was reported as .0001. Interpret this result.

9.106 Researchers at Mount Sinai Medical Center in New York believe that ALS, a neurological disorder (also known as "Lou Gehrig's" disease) that slowly paralyzes and kills its victims, may be linked to household pets, especially small dogs. A five-member medical team found that 72% of the afflicted patients studied had small household dogs at least 20 years before contracting ALS. In contrast, only 33% of a healthy control group had pet dogs early in life. (Assume that 100 afflicted patients and 100 healthy controls were studied.) Is there sufficient evidence to indicate that the percentage of all patients with ALS who had pet dogs early in life is larger than the corresponding percentage for nonafflicted people? Test using $\alpha = .01$.

9.107 Magnetotherapy involves the use of electromagnetic fields in the healing process of muscles. Rehabilitation researchers in Croatia experimented with magnetotherapy in the treatment of the quadriceps in the leg muscle (*Collegium Antropologicum,* June 1997). Two groups of 30 patients each were studied. Both groups received the standard exercise treatment; however, magnetotherapy was added to the treatment of the second group. Treatment ended when therapeutic results were achieved. The length of time (in days) each patient spent in therapy was recorded. The data are summarized in the table at left. Is there sufficient evidence to conclude that the average time spent in therapy for patients in the exercise group is greater than the average time for patients treated with magnetotherapy? Test at $\alpha = .05$.

Exercise Group	Magnetotherapy Group
$n_1 = 30$	$n_2 = 30$
$\bar{x}_1 = 39.83$	$\bar{x}_2 = 33.50$
$s_1 = 7.01$	$s_2 = 6.87$

Problem 9.107

REACTIME

9.108 A federal traffic safety researcher was hired to ascertain the effect of wearing safety devices (shoulder harnesses, seat belts) on reaction times to peripheral stimuli. To investigate this question, he randomly selected 15 subjects from the students enrolled in a driver education program. Each subject performed a simulated driving task that allowed reaction times to be recorded under two conditions, wearing a safety device (restrained condition) and no safety device (unrestrained condition). Thus, each subject received two reaction-time scores, one for the restrained condition and one for the unrestrained condition. The data (in hundredths of a second) are shown in the table on p. 523.

a. Is there evidence of a difference between mean reaction-time scores for the restrained and unrestrained drivers? Test using $\alpha = .05$.

	Driver														
	1	2	3	4	5	6	7	8	9	10	11	12	13	14	15
Restrained	36.7	37.5	39.3	44.0	38.4	43.1	36.2	40.6	34.9	31.7	37.5	42.8	32.6	36.8	38.0
Unrestrained	36.1	35.8	38.4	41.7	38.3	42.6	33.6	40.9	32.5	30.7	37.4	40.2	33.1	33.6	37.5

Problem 9.108

b. What assumptions are necessary for the validity of the test of part **a**?

c. Based on the test of part **a**, what would you infer about the mean reaction times for the driving conditions?

***d.** Use a nonparametric test to compare the median reaction times for restrained and unrestrained drivers. (Test at $\alpha = .05$.)

***e.** Compare the results of the tests, parts **a** and **d**.

9.109 Suppose you are investigating allegations of gender discrimination in the hiring practice of a particular firm. An equal-rights group claims that females are less likely to be hired than males with the same background, experience, and other qualifications. Data on hiring status and gender were collected on 28 former applicants and are shown in the table.

DISCRIM

Hiring Status	Gender	Hiring Status	Gender
Not hired	Female	Hired	Male
Not hired	Male	Not hired	Female
Hired	Male	Not hired	Male
Hired	Male	Hired	Male
Not hired	Female	Not hired	Female
Hired	Female	Not hired	Male
Not hired	Male	Not hired	Female
Not hired	Female	Not hired	Male
Not hired	Female	Hired	Male
Hired	Female	Not hired	Female
Not hired	Male	Not hired	Female
Not hired	Female	Not hired	Male
Not hired	Female	Hired	Male
Not hired	Female	Hired	Female

a. Summarize the data in a contingency table.

b. Test the hypothesis that hiring status and gender are independent. Use $\alpha = .01$.

REFERENCES

Cochran, W. G. "The χ^2 test of goodness of fit." *Annals of Mathematical Statistics,* 1952, p. 23.

Conover, W. J. *Practical Nonparametric Statistics,* 2nd ed. New York: Wiley, 1980.

Daniel, W. W. *Applied Nonparametric Statistics,* 2nd ed. Boston: PWS-Kent, 1990.

Farber, Susan. *Identical Twins Reared Apart.* New York: Basic Books, 1981.

Friedman, M. "The use of ranks to avoid the assumption of normality implicit in the analysis of variance." *Journal of the American Statistical Association,* Vol. 32, 1937.

Hollander, M., and Wolfe, D. A. *Nonparametric Statistical Methods.* New York: Wiley, 1973.

Lehmann, E. L. *Nonparametrics: Statistical Methods Based on Ranks.* San Francisco: Holden-Day, 1975.

Lewontin, R. C., and Felsenstein, J. "Robustness of homogeneity tests in 2 × *n* tables." *Biometrics,* Vol. 21, Mar. 1965, pp. 19–33.

Marascuilo, L. A., and McSweeney, M. *Nonparametric and Distribution-Free Methods for the Social Sciences.* Monterey, Calif.: Brooks/Cole, 1977.

Satterthwaite, F. E. "An approximate distribution of estimates of variance components." *Biometrics Bulletin,* Vol. 2, 1946, pp. 110–114.

Snedecor, G. W., and Cochran, W. *Statistical Methods,* 7th ed. Ames: Iowa State University Press, 1980.

Wilcoxon, F., and Wilcox, R. A. "Some rapid approximate statistical procedures." The American Cyanamid Co., 1964.

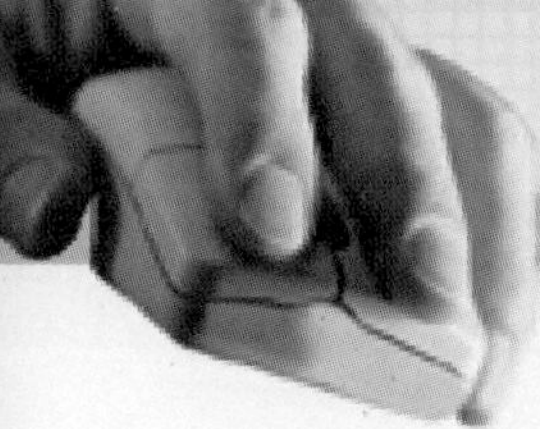

Using Microsoft® Excel

9.E.1 Comparing Two Population Means: Independent Samples

Performing the Pooled-Variance *t* Test for Differences in Two Means Using Sample Data

Use the Data Analysis **t-Test: Two-Sample Assuming Equal Variances** procedure to perform the pooled-variance t test for differences in two means using sample data. For example, to perform this test for the dental discomfort data of Table 9.2 on p. 459, open the **DENTAL.XLS** workbook to the Data sheet and

1. Select Tools | Data Analysis.
2. Select t-Test: Two-Sample Assuming Equal Variances from the Analysis Tools list box in the Data Analysis dialog box and click the OK button.
3. In the t-Test: Two-Sample Assuming Equal Variances dialog box (see Figure 9.E.1):

E Figure 9.E.1 Microsoft Excel t-Test: Two-Sample Assuming Equal Variances Dialog Box

a. Enter A1:A6 in the Variable 1 Range edit box.
b. Enter B1:B6 in the Variable 2 Range edit box.
c. Enter 0 (zero) in the Hypothesized Mean Difference edit box.
d. Select the Labels check box.
e. Enter 0.01 in the Alpha edit box.
f. Select the New Worksheet Ply option button and enter a name for the new sheet.
g. Click the OK button.

This procedure requires that the data for each group be in separate columns, an arrangement known as unstacked data. To use this procedure with worksheet data that are stacked, first use the PHStat **Data Preparation | Unstack Data** command.

The procedure generates a worksheet that does *not* dynamically change. Therefore, you will need to rerun the procedure if you make any changes to the underlying data in order to update the results of the test.

Performing the Pooled-Variance *t* Test for Differences in Two Means Using Summary Data

Use the PHStat **Two-Sample Tests | t Test for Differences in Two Means** procedure to perform the pooled-variance t test for differences in two means using summary data.

For example, to perform this test for the dental discomfort data of Table 9.2, using the summary statistics shown in Figure 9.5 on p. 460, open to an empty worksheet and

1. Select PHStat | Two-Sample Tests | t Test for Differences in Two Means.
2. In the t Test for Differences in Two Means dialog box (see Figure 9.E.2):

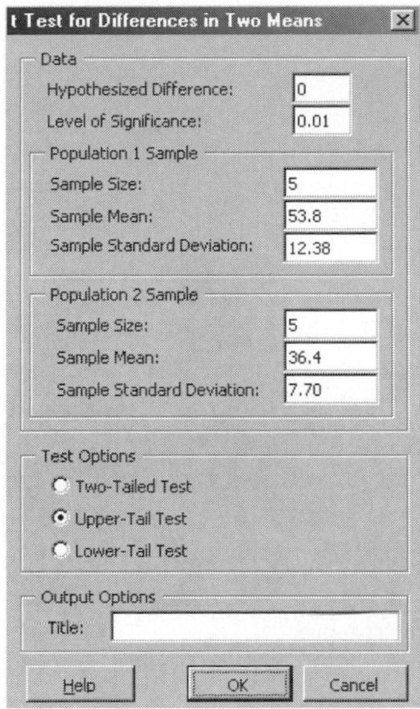

E Figure 9.E.2 PHStat t Test for Differences in Two Means Dialog Box

a. Enter 0 (zero) in the Hypothesized Difference edit box.
b. Enter 0.01 in the Level of Significance edit box.
c. Enter the summary data of Figure 9.5 for both samples. Enter 5, 53.8, and 12.38 as the sample size, sample mean, and sample standard deviation for the population 1 sample. Enter 5, 36.4, and 7.70 as the sample size, sample mean, and sample standard deviation for the population 2 sample.
d. Select the Upper-Tail Test option button.
e. Enter a title in the Title edit box.
f. Click the OK button.

9.E.2 Comparing Two Population Means: Matched Pairs

Use the Data Analysis **t-Test: Paired Two Sample for Means** procedure to perform the *t* test for matched pairs. For example, to perform this test for the reading scores data of Table 9.3 on p. 469, open the **READING.XLS** workbook to the Data sheet and

1. Select Tools | Data Analysis.
2. Select t-Test: Paired Two Sample for Means from the Analysis Tools list box in the Data Analysis dialog box and click the OK button.
3. In the t-Test: Paired Two Sample for Means dialog box (see Figure 9.E.3):

a. Enter A1:A11 in the Variable 1 Range edit box.
b. Enter B1:B11 in the Variable 2 Range edit box.
c. Enter 0 (zero) in the Hypothesized Mean Difference edit box.
d. Select the Labels check box.
e. Enter 0.05 in the Alpha edit box.
f. Select the New Worksheet Ply option button and enter a name for the new sheet.
g. Click the OK button.

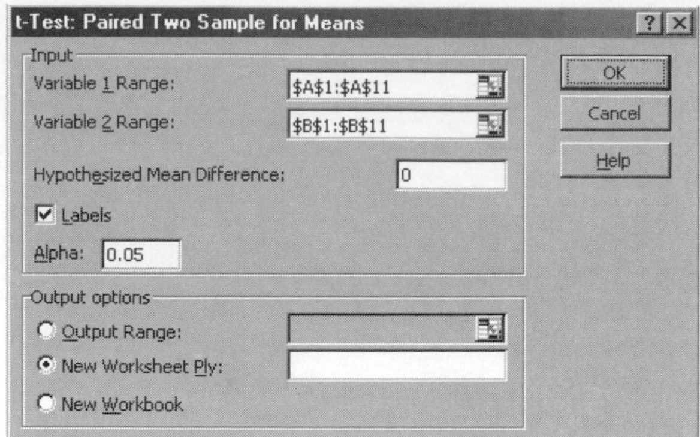

E **Figure 9.E.3** Microsoft Excel t-Test: Paired Two Sample for Means Dialog Box

The worksheet generated by this procedure will *not* dynamically change and if changes are made to the underlying data, the procedure must be repeated to update the results of the test.

This Data Analysis procedure requires that the data for each group be in separate columns, an arrangement known as unstacked data. To use this procedure with worksheet data that are stacked, first select the PHStat **Data Preparation | Unstack Data** procedure to properly unstack the data.

9.E.3 Comparing Two Population Proportions

Use the PHStat **Two-Sample Tests | Z Test for the Difference in Two Proportions** procedure to perform the z test for differences in two proportions. For example, to perform this test for the ethical decision making data of Example 9.10 on p. 477, open to an empty worksheet and

1. Select PHStat | Two-Sample Tests | Z Test for the Difference in Two Proportions.
2. In the Z Test for the Difference in Two Proportions dialog box (see Figure 9.E.4):

E **Figure 9.E.4** PHStat Z Test for the Difference in Two Proportions Dialog Box

   a. Enter 0 in the Hypothesized Difference edit box.
   b. Enter 0.05 in the Level of Significance edit box.
   c. Enter the number of successes and sample size for both samples. Enter 48 in the number of successes and 50 as the sample size for the population 1 sample. Enter 30 as the number of successes and 50 as the sample size for the population 2 sample.

d. Select the Upper-Tail Test option button.

e. Enter a title in the Title edit box.

f. Click the OK button.

9.E.4 Analyzing Contingency Tables

Performing the χ^2 Test for Differences in Two Proportions

Use the PHStat **Two-Sample Tests | Chi-Square Test for Differences in Two Proportions** procedure to perform the χ^2 test for differences in two proportions. For example, to perform this test for the speaker interruption data of Table 9.4 on p. 483, open to an empty worksheet and

1. Select PHStat | Two-Sample Tests | Chi-Square Test for Differences in Two Proportions.
2. In the Chi-Square Test for Differences in Two Proportions dialog box (see Figure 9.E.5):

a. Enter 0.05 in the Level of Significance edit box.

E **Figure 9.E.5** PHStat Chi-Square Test for Differences in Two Proportions Dialog Box

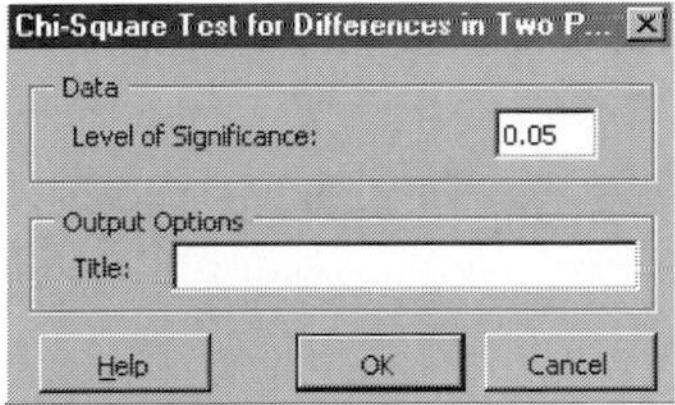

b. Enter a title in the Title edit box.

c. Click the OK button.

3. In the newly inserted worksheet:

a. Enter replacement labels for the row and column variables. Enter Speaker in cell A5 and Interrupter in cell B4.

b. Enter replacement labels for the row and column categories. Enter Male in cell A6 and Female in cell A7. Enter Male in cell B5 and Female in cell C5, widening the columns as necessary (see p. 13) to display the entire label.

c. Enter the Table 9.4 data (see p. 483) for the Male and Female counts for the Speaker and Interrupter in the cell range B6:C7. Enter 15 in cell B6, 10 in cell B7, 5 in cell C6, and 10 in cell C7.

Note: before you enter the Table 9.4 data, many cells display the error message #DIV/0!. These messages disappear once you enter all of the observed frequencies.

Performing the χ^2 Test for Differences in *c* Proportions

Use the PHStat **Multiple-Sample Tests | Chi-Square Test** procedure to perform the χ^2 test for differences in *c* proportions. For example, to perform this test for the foreign language hiring preference data of Table 9.6 on p. 488, open to an empty worksheet and

1. Select PHStat | Multiple-Sample Tests | Chi-Square Test.
2. In the Chi-Square Test dialog box (see Figure 9.E.6):

a. Enter 0.01 in the Level of Significance edit box.

b. Enter 2 in the Number of Rows edit box.

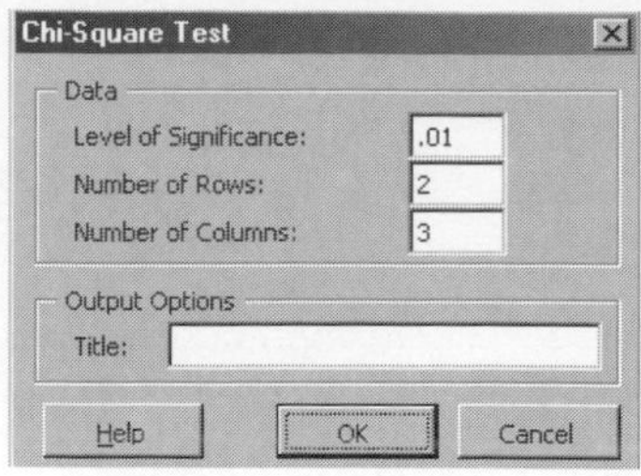

E **Figure 9.E.6** PHStat Chi-Square Test Dialog Box

c. Enter 3 in the Number of Columns edit box.

d. Enter a title in the Title edit box.

e. Click the OK button.

3. In the newly inserted worksheet:

a. Enter replacement labels for the row and column variables. Enter Firm Type in cell A5 and Foreign Language Hiring Preference in cell B4.

b. Enter replacement labels for the row and column categories. Enter U.S. in cell A6 and Foreign in cell A7. Enter Yes in cell B5, Neutral in cell C5, and No in cell D5, widening columns as necessary (see p. 13) to display entire label.

c. Enter the Table 9.6 data (see p. 488) for the U.S. and Foreign counts for the three foreign language hiring preferences in the cell range B6:D7. Enter 50 in cell B6, 60 in cell B7, 57 in cell C6, 22 in cell C7, 19 in cell D6, and 7 in cell D7.

Performing the χ^2 Test for the $R \times C$ Table

Use the procedures first discussed in the section Performing the χ^2 Test for Differences in c Proportions to perform a χ^2 test of independence. If using the PHStat **Multiple-Sample Tests | Chi-Square Test** procedure, adjust the number or rows and columns and observed frequencies as necessary.

9.E.5 Comparing Two Population Variances

Use the Data Analysis **F-Test: Two-Sample for Variances** procedure to perform the F test for differences in the variances of two populations using sample data. For example, to perform this test for the dental discomfort data of Table 9.2 on p. 459, open the **DENTAL.XLS** workbook to the Data sheet and

1. Select Tools | Data Analysis.

2. Select F-Test Two-Sample for Variances from the Analysis Tools list box in the Data Analysis dialog box and click the OK button.

3. In the F-Test Two-Sample for Variances dialog box (see Figure 9.E.7):

E **Figure 9.E.7** Microsoft Excel F-Test: Two-Sample for Variances Dialog Box

F-Test Two-Sample for Variances
Input
Variable 1 Range: A1:A6
Variable 2 Range: B1:B6
Labels
Alpha: 0.05
Output options
Output Range:
New Worksheet Ply:
New Workbook
OK
Cancel
Help

a. Enter A1:A6 in the Variable 1 Range edit box.
b. Enter B1:B6 in the Variable 2 Range edit box.
c. Select the Labels check box.
d. Enter 0.05 in the Alpha edit box.
e. Select the New Worksheet Ply option button and enter a name for the new sheet.
f. Click the OK button.

The worksheet generated by this procedure will *not* dynamically change and if changes are made to the underlying data, the procedure must be repeated in order to update the results of the test. This Data Analysis procedure requires that the data for each group be in separate columns, an arrangement known as unstacked data. To use this procedure with worksheet data that are stacked, first select the PHStat **Data Preparation | Unstack Data** procedure to properly unstack the data.

Performing the *F* Test for Differences in Two Variances Using Summary Data

Use the PHStat **Two-Sample Tests | F Test for Differences in Two Variances** procedure to perform the *F* test for differences in the variances of two populations using summary data. For example, to perform this test for the hospital sterilization data of Table 9.8 on p. 497, open to an empty worksheet and

1. Select PHStat | Two-Sample Tests | F Test for Differences in Two Variancces.

2. In the F Test for Differences in Two Variances dialog box (see Figure 9.E.8):

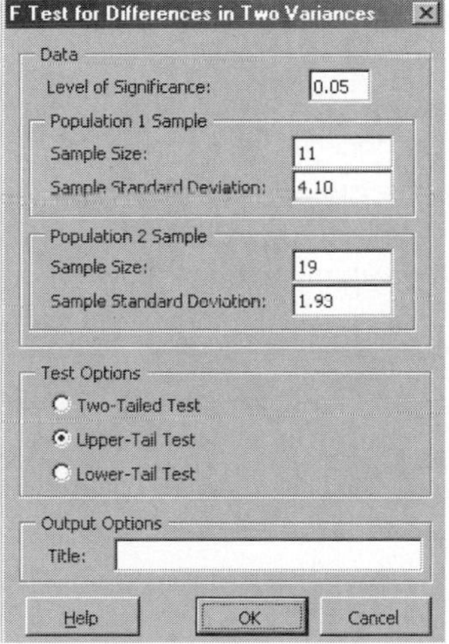

E Figure 9.E.8 PHStat F Test for Differences in Two Variances Dialog Box

a. Enter 0.05 in the Level of Significance edit box.
b. Enter the summary data of Table 9.8 for both samples. Enter 11 and 4.10 as the sample size and sample standard deviation for the population 1 sample. Enter 19 and 1.93 as the sample size and sample standard deviation for the population 2 sample.
c. Select the Upper-Tail Test option button.
d. Enter a title in the Title edit box.
e. Click the OK button.

9.E.6 Comparing Two Populations: Wilcoxon Rank Sum Test

Use the PHStat **Two-Sample Tests | Wilcoxon Rank Sum Test** procedure to perform the Wilcoxon rank sum test for differences in two medians. For example, to perform this test

for the visual acuity data of Table 9.9 on p. 503, open the **DEAF.XLS** workbook to the Data sheet and

1. Select PHStat | Two-Sample Tests | Wilcoxon Rank Sum Test
2. In the Wilcoxon Rank Sum Test dialog box (see Figure 9.E.9):

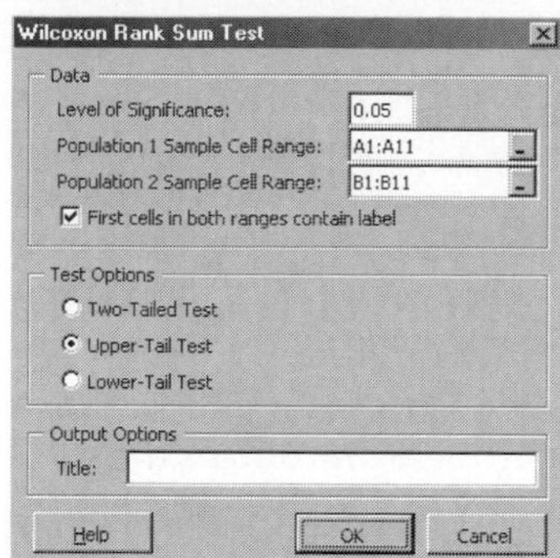

E Figure 9.E.9 PHStat Wilcoxon Rank Sum Test Dialog Box

a. Enter 0.05 in the Level of Significance edit box.
b. Enter A1:A11 in the Population 1 Sample Cell Range edit box.
c. Enter B1:B11 in the Population 2 Sample Cell Range edit box.
d. Select the First cells in both ranges contain label edit box.
e. Select the Upper-Tail Test option button.
f. Enter a title in the Title edit box.
g. Click the OK button.

This procedure requires that the data for each group be in separate columns, an arrangement known as unstacked data. To use this procedure with data that are stacked, first select the PHStat **Data Preparation | Unstack Data** procedure to properly unstack the data.

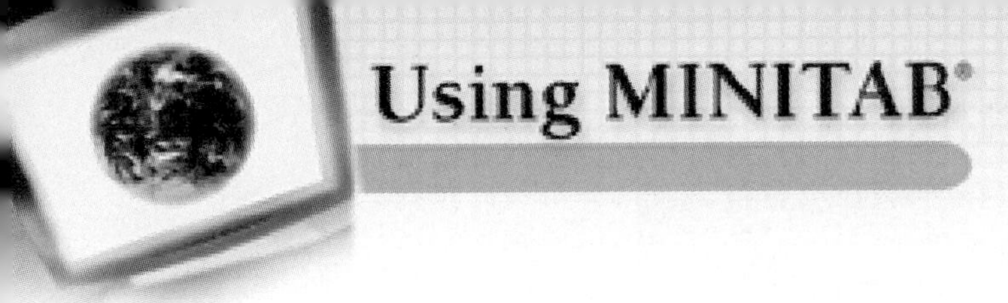

Using MINITAB®

9.M.1 Comparing Two Population Means: Independent Samples

To illustrate the use of MINITAB for the t test of the difference in two means, open the **DENTAL.MTW** worksheet. Select **Stat | Basic Statistics | 2-Sample t.** If the data are unstacked as they are in this worksheet, with the samples in different columns, select the **Samples in different columns** option button. (If the values are in one column and the categories in a second column, select the **Samples in one column** option button, enter the variable name in the Samples edit box and the category name in the Subscripts edit box.) In the 2-Sample t dialog box (see Figure 9.M.1), in the First: edit box, enter **C1** or **'Novocaine'.** In the Second: edit box, enter **C2** or **'New'.** If you wish to assume equal variances, select the Assume equal variances check box. In the Alternative drop-down list box, select less than or greater than for a one-tailed test or not equal for a two-tailed test. Click on the **Graphs** options button, and select the **Boxplots** of data check box. Click the **OK** button to return to the 2-Sample t dialog box. Click the **OK** button.

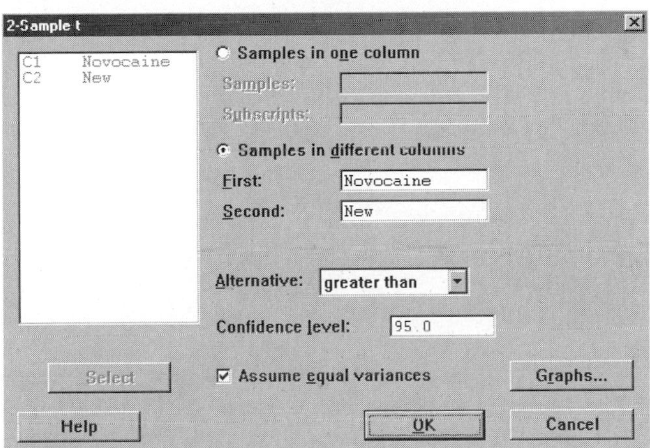

M Figure 9.M.1 MINITAB 2-Sample t Dialog Box

9.M.2 Comparing Two Population Means: Matched Pairs

To illustrate the use of MINITAB for the paired t test, open the **READING.MTW** worksheet. Select **Stat | Basic Statistics | Paired t.** In the Paired t dialog box (see Figure 9.M.2 on p. 532), in the First sample edit box, enter **C1** or **'Method 1'.** In the Second sample edit box, enter **C2** or **'Method 2'.** Click the **Options** button. In the **Alternative** drop-down list box, select less than or greater than for a one-tailed test or not equal for a two-tailed test. Click the **OK** button to return to the Paired t dialog box. Click the **Graphs** button, and select the **Boxplot of Differences** check box. Click the **OK** button to return to the Paired t dialog box. Click the **OK** button.

9.M.3 Comparing Two Population Proportions

To illustrate the use of MINITAB to compare two population proportions, examine Example 9.10 on p. 477 concerning ethical decision making. Select **Stat | Basic Statistics | 2 Proportions.** In the 2 Proportions dialog box (see Figure 9.M.3 on p. 532), select the **Summarized data** option button. In the First sample row, enter **50** in the Trials edit box and **48**

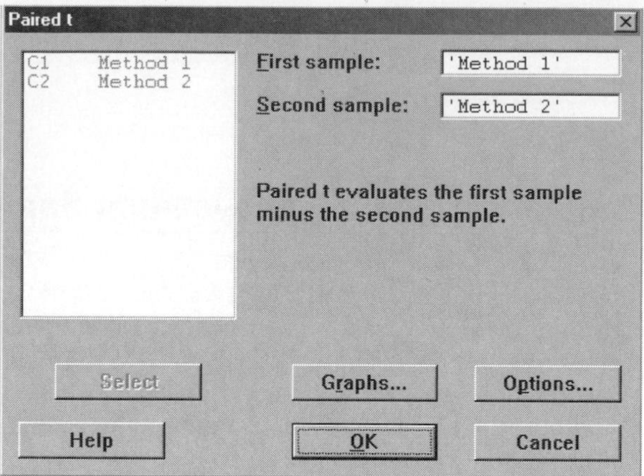

Figure 9.M.2 MINITAB Paired t Dialog Box

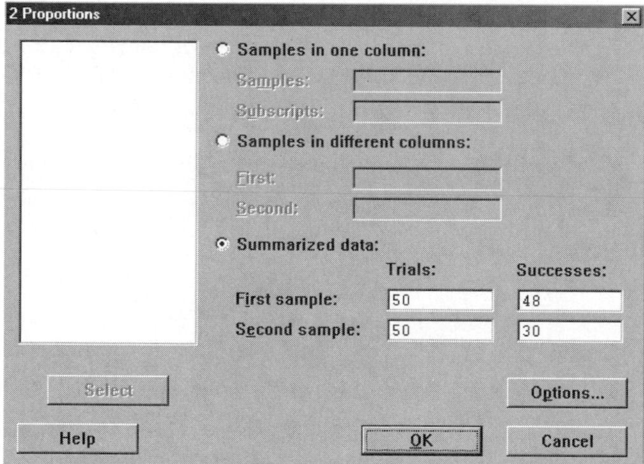

Figure 9.M.3 MINITAB 2 Proportions Dialog Box

in the Successes edit box. In the Second sample row, enter **50** in the Trials edit box and **30** in the Successes edit box. Click the **Options** button. Select the **Use pooled estimate of p for test** check box. Click **OK** to return to the 2 proportions dialog box. Click the **OK** button.

To compare population proportions when raw data are available, if the data are stacked, select the Samples in one column option button. Enter the name or column number of the variable to be analyzed in the Samples edit box and the variable containing the subscripts in the Subscripts edit box. If the data are unstacked, select the Samples in different columns option button. Enter the first variable column number or name in the First edit box and the second variable column number or name in the Second edit box.

9.M.4 Analyzing Contingency Tables

To obtain a two-way contingency table from raw data, open the file of interest. Select **Stat | Tables | Cross Tabulation.** In the Cross Tabulation dialog box (see Figure 9.M.4), in the Classification variables edit box, enter the two variables to be cross-classified. Select the **Counts, Row percents, Column percents,** and **Total percents** check boxes. Select the **Chi-Square analysis** check box. Select the **Above and expected count** option button. Click the **OK** button.

If the cell frequencies are available as in the speaker interruption data of Table 9.4 on p. 483, enter them in columns in a MINITAB worksheet. For the speaker interruption 2×2 table, enter **15** and **10** in column C1 and **5** and **10** in column C2. Select **Stat | Tables | Chi-Square Test.** In the Chi-Square Test dialog box (see Figure 9.M.5), in the Columns containing the table edit box, enter **C1** and **C2.** Click the **OK** button.

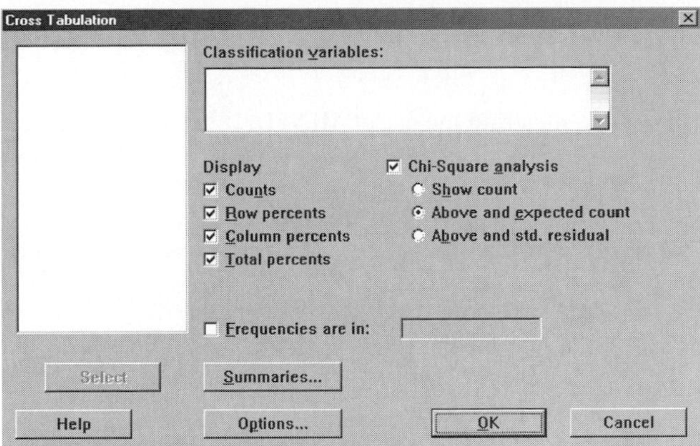

Figure 9.M.4 MINITAB Cross Tabulation Dialog Box

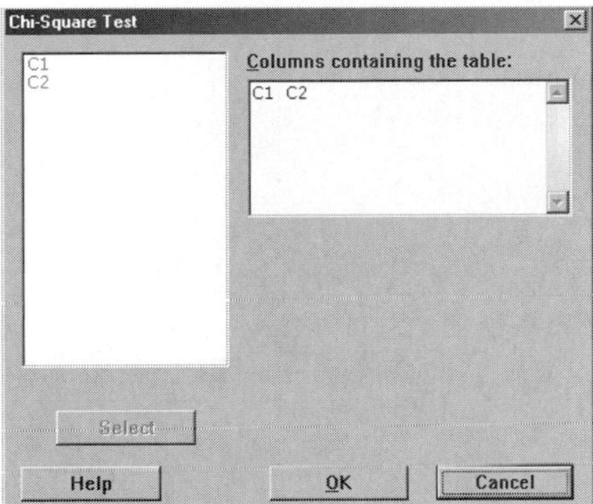

Figure 9.M.5 MINITAB Chi-Square Test Dialog Box

9.M.5 Comparing Two Population Variances

To illustrate the use of MINITAB for the *F* test of the difference in two variances, open the **DENTAL.MTW** worksheet. The data need to be stacked with the values in one column and the categories in a second column. Select **Manip | Stack/Unstack | Stack Columns.** In the Stack the following columns edit box, enter **C1 or 'Novocaine'** and **C2** or **'New'.** In the Store the stacked data in edit box, enter **C3.** In the Store subscripts in edit box, enter **C4.** Click the **OK** button. Enter Discomfort Score as the label for C3 and Anesthetic as the label for C4.

Select **Stat | ANOVA | Homogeneity of Variances.** In the Homogeneity of Variance Test dialog box (see Figure 9.M.6), in the Response edit box, enter **C3** or **'Discomfort Score'.** In the Factors edit box, enter **C4** or **'Anesthetic'.** Click the **OK** button.

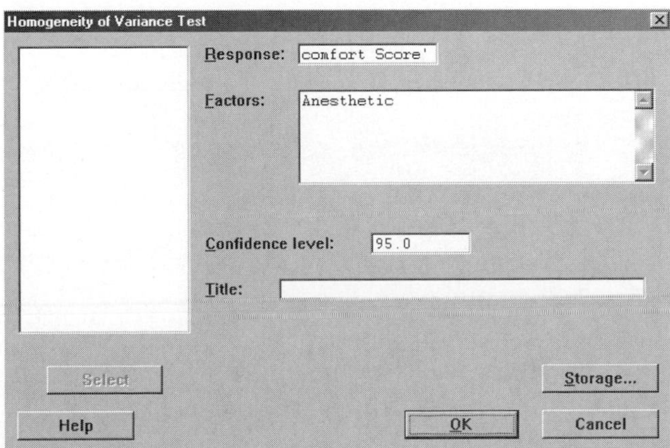

Figure 9.M.6 MINITAB Homogeneity of Variance Test Dialog Box

9.M.6 Comparing Two Populations: Wilcoxon Rank Sum Test

To illustrate the use of MINITAB for the Wilcoxon rank sum test, open the **DEAF.MTW** worksheet. Select **Stat | Nonparametrics | Mann-Whitney.** In the Mann-Whitney dialog box (see Figure 9.M.7) in the First Sample edit box, enter **C1** or **'Deaf'.** In the Second Sample edit box, enter **C2** or **'Hearing'.** In the Alternative drop-down list box, select less than or greater than for a one-tailed test or not equal for a two-tailed test. Click the **OK** button.

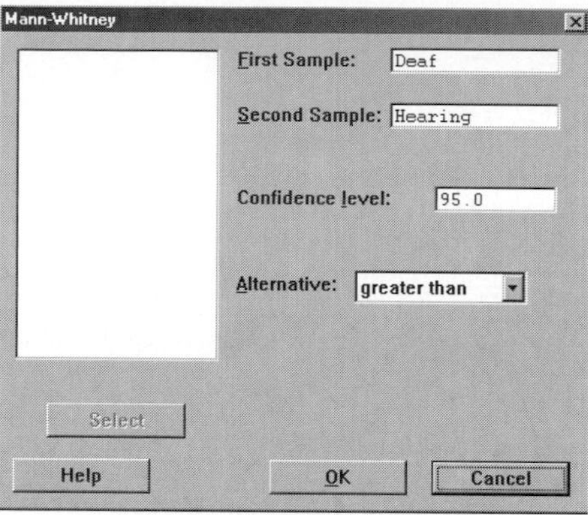

M **Figure 9.M.7** MINITAB Mann-Whitney Dialog Box

9.M.7 Comparing Two Populations: Wilcoxon Signed Ranks Test

To illustrate the use of MINITAB for the Wilcoxon signed ranks test, open the **READING.MTW** worksheet. To compute the differences, select **Calc | Calculator.** In the Store Result in Variable edit box, enter **C3.** In the Expression edit box, enter **C1–C2.** Click the **OK** button. Enter the label Difference for C3.

Select **Stat | Nonparametrics | 1-Sample Wilcoxon** test. In the 1-Sample Wilcoxon dialog box (see Figure 9.M.8), in the Variables edit box, enter **C3** or **'Difference'.** Select the Test median option button, and enter **0.0** in the edit box. In the Alternative drop-down list box, select less than or greater than for a one-tailed test or not equal for a two-tailed test. Click the **OK** button.

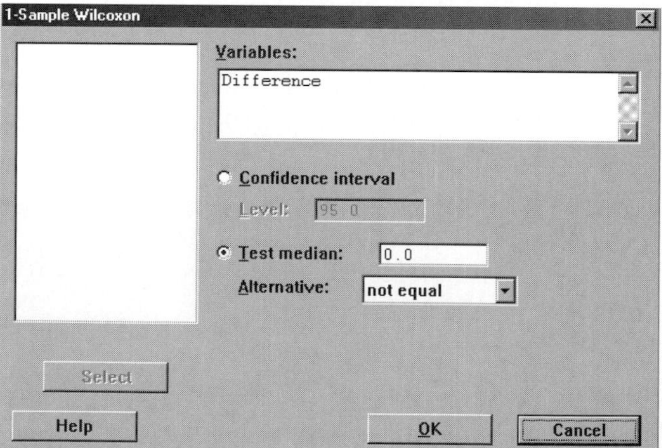

M **Figure 9.M.8** MINITAB 1-Sample Wilcoxon Dialog Box

CHAPTER 10

Regression Analysis

OBJECTIVES

1. To develop the simple linear regression model as a means of using one quantitative variable to predict another quantitative variable
2. To assess the fit of a simple linear regression model
3. To make inferences about the coefficient of correlation for two variables
4. To develop a multiple regression model for predicting a quantitative variable using more than one independent variable
5. To study the pitfalls involved in using regression models

CONTENTS

EXCEL TUTORIAL

MINITAB TUTORIAL

Statistics in the Real World

Relating Residential Property Sales to Appraisals

TAMPALMS

A property appraiser is an elected officer responsible for determining the fair and just value of all real and tangible property in a specified geographical area (city or county). Factors that are taken into account by appraisers when evaluating property are property use, the size and conditions of improvements on the site, and the local real estate market. Appraisals need to be accurate since various taxing authorities use the property value to determine the amount of real estate taxes to be paid on an individual property and as the basis for setting the tax rate in the region.

Appraised property values do not necessarily reflect, however, what the properties would sell for on the open market. Nevertheless, real estate investors, home buyers, and home owners often use the appraised value as a basis for predicting sale price. Does such a relationship exist? That is, can the appraised value of a property be used to obtain accurate predictions of sale price?

To investigate this relationship, we have obtained data on appraisals and sales for a large sample of residential properties that sold in 1999 in Hillsborough County (Tampa), Florida. For each residential property, the appraised land value, appraised value of improvements (e.g., the value of the home built on the property), total appraised value, and sale price were recorded. In addition, the geographic location (i.e., neighborhood) in which the property is located was identified.

In the following Statistics in the Real World Revisited sections, we apply the statistical methodology presented in this chapter to a subset of this data. The examples analyze the sales-appraisal data for a particular Tampa neighborhood, called Tampa Palms. Tampa Palms is an upscale subdivision of over 1,000 homes located near the University of South Florida. The subdivision, built around an 18-hole golf course, contains amenities such as an elementary school, 10 lighted clay tennis courts, an Olympic-size swimming pool, and a state-of-the-art fitness center. The data for 92 Tampa Palms homes sold in 1999 are saved in the file named **TAMPALMS.**

Statistics in the Real World Revisited

10.1 Introduction to Regression Models

Suppose a university administrator wants to predict the annual merit raise (in dollars) awarded to professors employed by the university. To do this, the administrator will build a mathematical equation, i.e., a **model,** for the annual merit raise for any particular professor. The process of finding a mathematical model to predict the value of a variable is part of a statistical method known as **regression analysis.**

In regression analysis, the variable y to be predicted is called the *dependent* (or *response*) *variable*. In this example, y = annual merit raise for a professor.

Definition 10.1

The variable to be predicted (or modeled), y, is called the **dependent** (or **response**) **variable.**

The administrator knows that the actual value of y will vary from professor to professor depending on the professor's rank, teaching rating, tenure status, and numerous other factors. These factors used to predict y are called *independent* (or *explanatory*) *variables*.

Definition 10.2

The variables used to predict (or model) y are called **independent** (or **explanatory) variables** and are denoted by the symbols x_1, x_2, x_3, etc.

A **general regression model** takes the form shown in the next box.

GENERAL REGRESSION MODEL FOR *y*

$$y = \beta_0 + \beta_1 x_1 + \beta_2 x_2 + \cdots + \beta_k x_k + \varepsilon \quad (10.1)$$

where $x_1, x_2, \ldots, x_k$ are the independent variables, $\beta_0, \beta_1, \ldots, \beta_k$ are the unknown model parameters, and ε is unexplainable (or random) error.

Suppose the administrator wants to use teaching rating x as the only predictor of merit raise y. When a single independent variable is used to predict y, the model simplifies to

$$y = \beta_0 + \beta_1 x + \varepsilon$$

The values β_0 and β_1 are unknown population parameters that need to be estimated from the sample data and ε represents unexplainable (or random) error. The process involves obtaining a sample of professors and recording teaching rating x as well as merit raise y. Subjecting this sample data to a regression analysis will yield estimates of the model parameters and enable the administrator to predict the merit raise y for a particular professor. The prediction equation takes the form

$$\hat{y} = \mathrm{b}_0 + \mathrm{b}_1 x$$

where $\hat{y}$ is the predicted value of y and b_0 and b_1 are estimates of the model parameters β_0 and β_1, respectively.

EXAMPLE 10.1

IDENTIFYING THE REGRESSION VARIABLES

The National Collegiate Athletic Association (NCAA) attempts to predict the academic success of college athletes using a regression model. The NCAA wants to predict an athlete's GPA as a function of several variables including high school GPA, Scholastic Aptitude Test (SAT) score, and number of hours tutored during an academic year.

a. Identify the dependent variable y of interest to the NCAA.

b. Identify the independent variables.

c. Give the equation of the regression model for y if the NCAA uses only high school GPA as a predictor.

d. Give the equation of the regression model for y if the NCAA uses all the independent variables as predictors.

Solution

a. The dependent variable is the variable to be predicted. Since the NCAA wants to predict the GPA of an athlete, the dependent variable is

$$y = \text{GPA of a college athlete}$$

b. The independent variables the NCAA wants to use to predict y are:

x_1 = High school GPA of a college athlete
x_2 = SAT score of a college athlete
x_3 = Number of hours the athlete is tutored during the year

c. Using only x_1 as a predictor variable, the regression model is

$$y = \beta_0 + \beta_1 x_1 + \varepsilon$$

where ε represents random error.

d. Applying equation 10.1 shown in the box (p. 537), the full model takes the form

$$y = \beta_0 + \beta_1 x_1 + \beta_2 x_2 + \beta_3 x_3 + \varepsilon$$

Self-Test 10.1

A real estate appraiser wants to use square footage of living area, assessed value, and number of rooms to predict the sale price of a home.

a. Identify the dependent and independent variables.

b. Write the equation of a regression model that can be used to make the prediction.

An overview of the steps involved in a regression analysis is given in the box.

STEPS IN A REGRESSION ANALYSIS

1. Propose a model for y as a function of the independent variables $x_1, x_2, \ldots, x_k$.
2. Collect the sample data.
3. Use the sample data to estimate unknown parameters in the model.
4. Make specific assumptions about the probability distribution of the random error term ε and estimate any unknown parameters of this distribution.
5. Statistically check the usefulness of the model.
6. Check whether the assumptions on the random error term (step 4) are satisfied. If not, make model modifications.
7. When satisfied that the (modified) model is useful, use it for prediction, for estimation, and for making inferences about the model parameters.

We present the simplest of all regression models—called the straight-line model—in Sections 10.2–10.11. More complex models are discussed in optional Section 10.12.

10.2 The Straight-Line Model: Simple Linear Regression

Recall that the Federal Trade Commission collects data on the levels of tar, nicotine, and carbon monoxide in smoke emitted from domestic cigarettes. (These data are stored in the **FTC** data file.) Suppose you are interested in predicting a cigarette brand's carbon monoxide (CO) measurement from its nicotine content.

Then, the dependent variable is

y = Carbon monoxide (CO) content (measured in milligrams)

and the only independent variable is

x = Nicotine content (also measured in milligrams)

In the following examples, we develop a model for y called the **straight-line model.**

EXAMPLE 10.2

INTERPRETING A SCATTERPLOT

Table 10.1 is a list of the nicotine and CO measurements for a hypothetical sample of 18 different cigarette brands. Examine the relationship between CO content y and nicotine content x. What type of regression model is suggested by this relationship?

CONICOTINE

TABLE 10.1 Carbon Monoxide–Nicotine Data for Example 10.2

CO *y*	Nicotine *x*	CO *y*	Nicotine *x*	CO *y*	Nicotine *x*
6	0.4	9	0.8	12	1.2
8	0.4	15	0.8	18	1.2
6	0.5	11	0.9	13	1.3
9	0.5	15	0.9	17	1.3
9	0.7	13	1.1	14	1.4
11	0.7	16	1.1	22	1.4

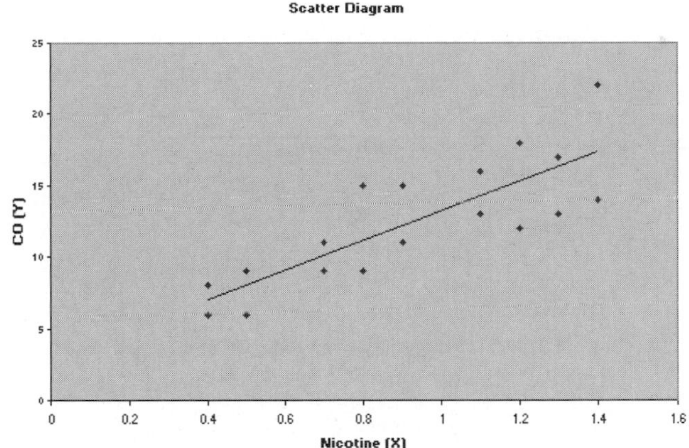

E Figure 10.1 Excel Scatterplot of CO Content versus Nicotine Content, Example 10.2

Solution

Recall (Section 3.8) that we can gain insight into the relationship between two quantitative variables by constructing a scatterplot for the sample data. Figure 10.1 shows a scatterplot for the sample data in Table 10.1 generated with Excel.

You can see that the level of CO (y) increases as nicotine content (x) increases. Although the points do not fall in a perfect pattern (e.g., a straight line or a curve), such a result is expected with regression models. The **random error component ε** in a regression model allows for random fluctuation of the points around some relationship between x and y. The points in Figure 10.1 appear to fluctuate around a straight line (illustrated on the graph). Therefore, it is reasonable to propose a straight-line relationship between y and x.

EXAMPLE 10.3 EQUATION OF A STRAIGHT-LINE MODEL

Refer to Example 10.2. Write the straight-line equation relating CO content y to nicotine content x.

Solution

A straight-line equation involves two parameters, the y-intercept and the slope of the line. In regression, the Greek symbols β_0 and β_1 represent the y-intercept and slope, respectively, since they are population parameters that will be known only if we have access to the entire population of (x, y) measurements. Therefore, the equation of the model is:

$$y = \beta_0 + \beta_1 x + \varepsilon$$

Note that this is the simple, one-variable model discussed in Section 10.1. A graph of this model is shown in Figure 10.2. β_0 (the **y-intercept**) represents the value of y when $x = 0$, and β_1 (the **slope**) is the amount of change in y for every 1-unit increase in x.

Figure 10.2 The Straight-Line Model

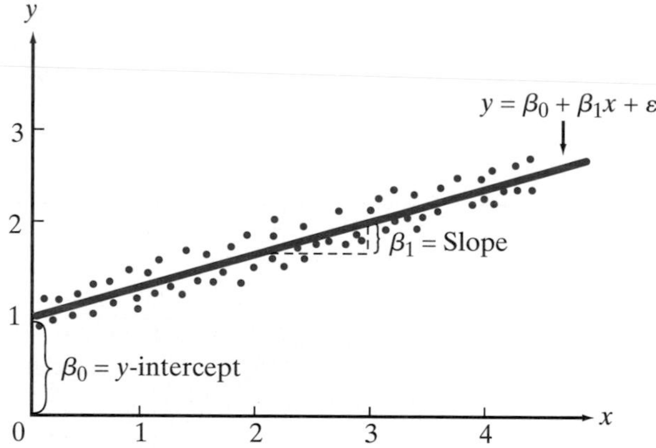

In this chapter, we will primarily consider the simplest of regression models—the straight-line model. The analysis of the straight-line model is commonly known as **simple linear regression** analysis. The elements of the simple linear regression model are summarized in the box.

THE SIMPLE LINEAR REGRESSION (STRAIGHT-LINE) MODEL

$$y = \beta_0 + \beta_1 x + \varepsilon \qquad \textbf{(10.2)}$$

where

y = quantitative dependent variable
x = quantitative independent variable
ε = random error component
β_0 (beta zero) = y-intercept of the line (the value of y for $x = 0$—see Figure 10.2)
β_1 (beta one) = slope of the line (amount of increase, or decrease, in y for every 1-unit increase in x—see Figure 10.2)

EXAMPLE 10.4

IDENTIFYING THE MODEL PARAMETERS

Suppose it is known that y is related to x by the straight-line model

$$y = 17 + 4x + \varepsilon$$

Give the y-intercept and slope of the line.

Solution

From equation 10.2 in the box on p. 540, the slope β_1 is the number multiplied by x in the model. Thus, $\beta_1 = 4$. The y-intercept β_0 is the number that is not multiplied by x. Therefore, $\beta_0 = 17$.

Self-Test 10.2

Give the y-intercept and slope of the straight-line model

$$y = -100 + 67x + \varepsilon$$

Statistics in the Real World Revisited

Graphing the Sales-Appraisal Data

TAMPALMS

Consider the data on sale prices and appraised values of residential properties sold in 1999 in the upscale Tampa, Florida, neighborhood named Tampa Palms (p. 536). Recall that the data for a sample of 92 Tampa Palms homes are saved in the file named **TAMPALMS.** To gain insight into whether the appraised property value can be used to obtain accurate predictions of sale price for residential properties in this neighborhood, we construct a scatterplot of the data.

A MINITAB scatterplot is shown in Figure 10.3. Sale price (recorded in thousands of dollars) is plotted on the vertical axis and total appraised value (also recorded in thousands of dollars) is plotted on the horizontal axis. The graph reveals a clearly increasing trend between the two variables—as appraised value of a property increases, the sale price also increases. Also, it appears that the increasing trend can be approximated by a straight line. We will estimate this line in the Statistics in the Real World Revisited section on p. 578.

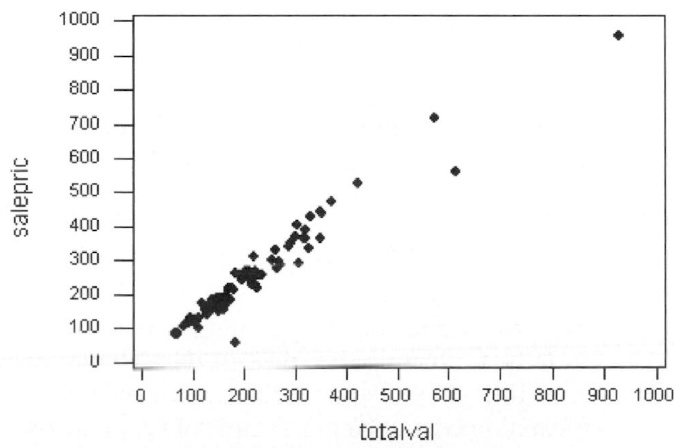

M **Figure 10.3** MINITAB Scatterplot of Sales-Appraisal Data for Tampa Palms

PROBLEMS FOR SECTION 10.2

Using the Tools

10.1 Suppose that y is related to x by the straight-line equation $y = 1.5 + 2x$.

a. Find the y-intercept for the line.

b. Find the slope of the line.

c. If you increase x by one unit, how much will y increase or decrease?

d. If you decrease x by one unit, how much will y increase or decrease?

e. What is the value of y when $x = 0$?

10.2 Suppose that y is related to x by the straight-line equation $y = 1.5 - 2x$.

a. Find the y-intercept for the line.

b. Find the slope of the line.

c. If you increase x by one unit, how much will y increase or decrease?

d. What is the value of y when $x = 0$?

e. What does this line have in common with the line in Problem 10.1? How do the two lines differ?

10.3 Graph the lines corresponding to each of the following equations:

a. $y = 1 + 3x$ **b.** $y = 1 - 3x$

c. $y = -1 + .5x$ **d.** $y = -1 - 3x$

e. $y = 2 - .5x$ **f.** $y = -1.5 + x$

g. $y = 3x$ **h.** $y = -2x$

10.4 Give the values of β_0 and β_1 corresponding to each of the lines in Problem 10.3.

Applying the Concepts

10.5 In a straight-line model, what do the values β_0 and β_1 represent?

10.6 If a straight-line model accurately represents the true relationship between y and an independent variable x, does it imply that every value of y will always fall exactly on the line? Explain.

10.3 Estimating and Interpreting the Model Parameters

The following example illustrates the technique we will use to *fit the straight-line model to the data,* i.e., to estimate the slope and y-intercept of the line using information provided by the sample data.

EXAMPLE 10.5

HOW TO ESTIMATE β_0 AND β_1

Suppose a psychologist wants to model the relationship between the creativity score y and the flexibility score x of a mentally retarded child. Based on practical experience, the psychologist hypothesizes the straight-line model

$$y = \beta_0 + \beta_1 x + \varepsilon$$

If the psychologist were able to obtain the flexibility and creativity scores of *all* mentally retarded children, i.e., the entire population of (x, y) measurements, then the values of the population parameters β_0 and β_1 could be determined exactly. Collecting this mass of data would, of course, be impossible. The problem, then, is to estimate the unknown population parameters based on the information contained in a sample of (x, y) measurements. Suppose the psychologist tests 10 randomly selected mentally retarded children. The creativity and flexibility

scores (both measured on a scale of 1 to 20) are given in Table 10.2. How can we best use the sample information to estimate the unknown y-intercept β_0 and the slope β_1?

CREATIVITY

TABLE 10.2 Creativity–Flexibility Scores for Example 10.5

Child	Flexibility Score x	Creativity Score y
1	2	2
2	6	11
3	4	7
4	5	10
5	3	5
6	5	8
7	3	4
8	3	3
9	7	11
10	9	13

Solution

Estimates of the unknown parameters β_0 and β_1 are obtained by finding the "best-fitting" straight line through the sample data points of Table 10.2. (These points are graphed in the Excel scatterplot, Figure 10.4.) We denote the estimates as b_0 and b_1, respectively. Then the "best-fitting" line can be written

$$\hat{y} = b_0 + b_1 x$$

where $\hat{y}$ is the predicted value of the creativity score y.

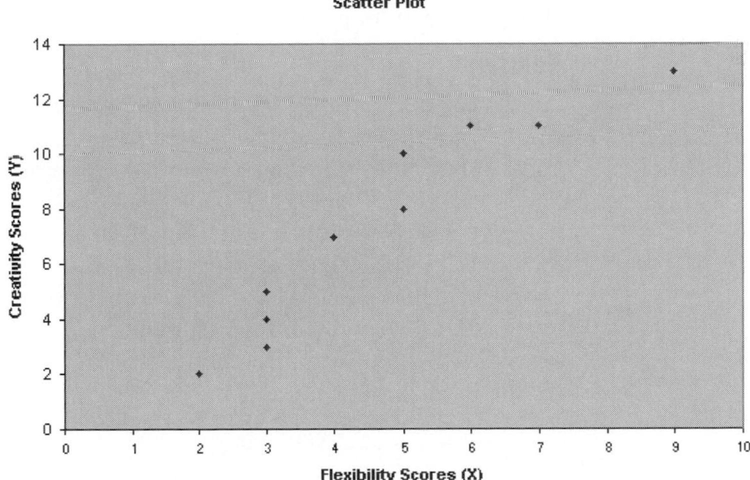

E **Figure 10.4** Excel Scatterplot of Creativity–Flexibility Score Data, Table 10.2

Although "best fit" can be defined in a variety of ways, the simplest approach involves making the difference between the actual y and predicted y, i.e., $y - \hat{y}$, as small as possible. We call the difference, $y - \hat{y}$, the error of prediction. Because the error of prediction is positive for some observations and negative for others, it is better, mathematically, to minimize the sum of the squared errors, denoted SSE. A mathematical technique that finds the estimate of β_0 and β_1 and minimizes SSE is known as the **method of least squares.**

LEAST SQUARES CRITERION FOR FINDING THE "BEST-FITTING" LINE

Choose the line that minimizes the sum of squared errors,

$$\text{SSE} = \sum (y - \hat{y})^2 \qquad \textbf{(10.3)}$$

This is called the **least squares line,** or the **least squares prediction equation.**

With the least squares method, we find b_0 and b_1 using the formulas given below.

The sum of the errors for the least squares line will always equal 0, i.e., $\Sigma(y - \hat{y}) = 0$. Since there are many other fitted lines that also have this property, we do not use this as the only criterion for choosing the "best-fitting" line.

FORMULAS FOR THE LEAST SQUARES ESTIMATES

$$\textit{Slope: } b_1 = \frac{\text{SSxy}}{\text{SSxx}} \qquad \textbf{(10.4)}$$

$$\textit{y-Intercept: } b_0 = \bar{y} - b_1\bar{x} \qquad \textbf{(10.5)}$$

where

$$\text{SSxy} = \sum xy - \frac{(\sum x)(\sum y)}{n} \qquad \textbf{(10.6)}$$

$$\text{SSxx} = \sum x^2 - \frac{(\sum x)^2}{n} \qquad \textbf{(10.7)}$$

n = sample size

EXAMPLE 10.6

FINDING b_0 AND b_1

Refer to Example 10.5. Use the least squares method to estimate β_0 and β_1 in the straight-line model.

Solution

All spreadsheet and statistical software perform the least squares computations in simple linear regression. Here, we use Microsoft Excel to find the estimates of β_0 and β_1. (For those who are interested, these computations are illustrated in optional Section 10.10.)

Figure 10.5 is an Excel printout of the simple linear regression analysis of the data in Table 10.2. The estimates of β_0 and β_1 are indicated in Figure 10.5 in the column labeled **Coefficients.** The estimate of β_0, the y-intercept, is given in the row labeled **Intercept** and the estimate of β_1, the slope, is given in the row labeled **Flexibility Score.** (*Note:* The estimate of the slope will always be in the row named for the independent variable in the model.) These estimates are:

$$b_0 = -.437 \qquad b_1 = 1.667$$

Consequently, the "best-fitting" line—also known as the least squares line—is

$$\hat{y} = -.437 + 1.667x$$

This line is plotted in Figure 10.6 using Excel in order for you to see how well the least squares line fits the data.

EXAMPLE 10.7

INTERPRETING b_0 AND b_1

Refer to Example 10.6. Interpret the values of the estimated y-intercept and estimated slope.

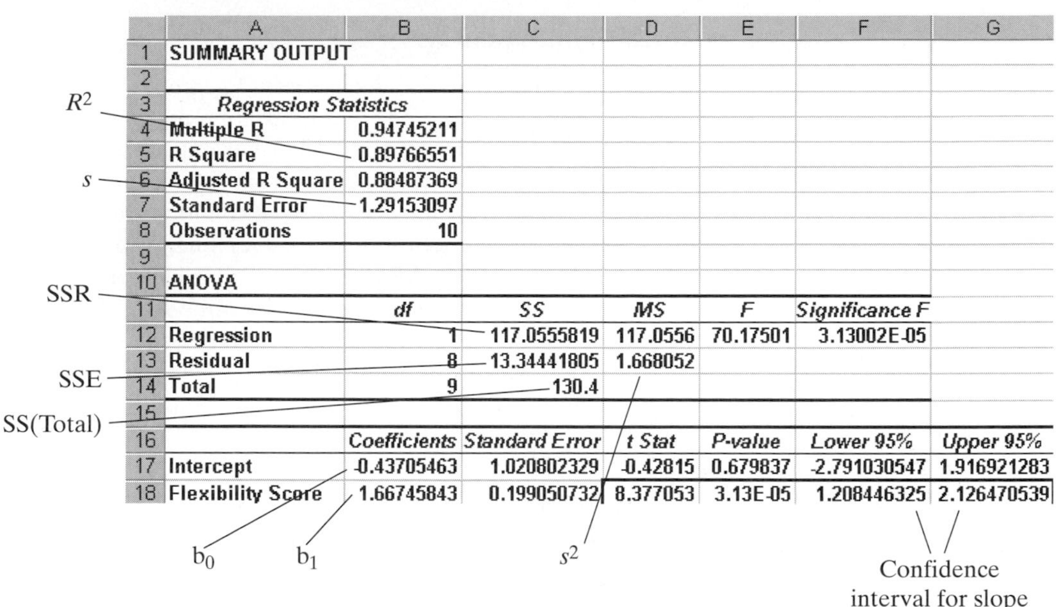

	A	B	C	D	E	F	G
1	SUMMARY OUTPUT						
2							
3	*Regression Statistics*						
4	Multiple R	0.94745211					
5	R Square	0.89766551					
6	Adjusted R Square	0.88487369					
7	Standard Error	1.29153097					
8	Observations	10					
9							
10	ANOVA						
11		*df*	*SS*	*MS*	*F*	*Significance F*	
12	Regression	1	117.0555819	117.0556	70.17501	3.13002E-05	
13	Residual	8	13.34441805	1.668052			
14	Total	9	130.4				
15							
16		*Coefficients*	*Standard Error*	*t Stat*	*P-value*	*Lower 95%*	*Upper 95%*
17	Intercept	-0.43705463	1.020802329	-0.42815	0.679837	-2.791030547	1.916921283
18	Flexibility Score	1.66745843	0.199050732	8.377053	3.13E-05	1.208446325	2.126470539

E Figure 10.5 Excel Simple Linear Regression Analysis of Data in Table 10.2

E Figure 10.6 Excel Plot of Least Squares Line, Example 10.6

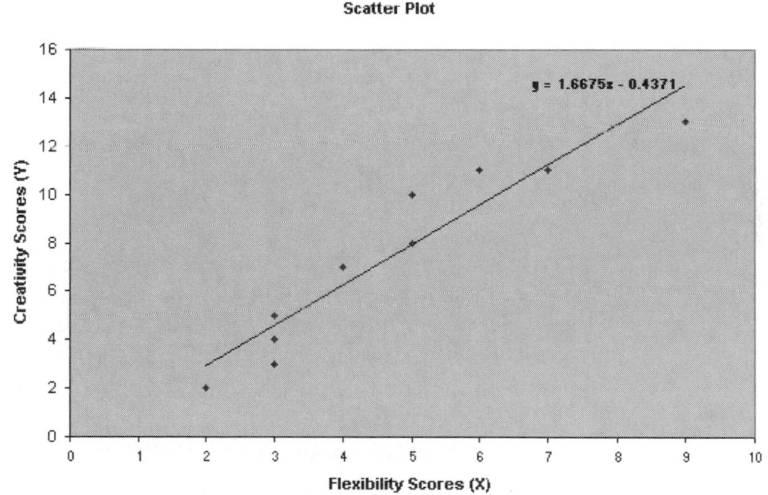

In simple linear regression, the estimated y-intercept will often not have a practical interpretation. It will, however, be practical if the value $x = 0$ is meaningful and within the range of the x-values in the sample.

Solution

First, we will interpret the estimated slope b_1. Recall from Section 10.2 that the slope represents the change in y for every 1-unit increase in x. Therefore, $b_1 = 1.667$ is the estimated change in creativity score y for every 1-point increase in flexibility score x. Since b_1 is positive, we estimate creativity score to *increase* 1.667 points for every 1-point increase in flexibility score.

The y-intercept is the value of y when $x = 0$. Consequently, the estimated y-intercept b_0 can be interpreted as the predicted creativity score y when flexibility score x is 0. However, in this application b_0 is meaningless since it is impractical for a mentally retarded child to have a flexibility score of 0. In fact, none of the children in the sample had flexibility scores this low.

EXAMPLE 10.8 INTERPRETING b_0 AND b_1

CRASH

Refer to the National Car Assessment Program (NCAP) crash test data stored in the file **CRASH.** Two of the many variables measured are severity of driver's head injury (DRIVHEAD) and overall driver injury rating (DRIVSTAR). Recall that overall driver injury rating ranges from 1 to 5; the higher the rating, the less the chance of an injury. Suppose we want to model the overall driver injury rating y as a straight-line function of the driver's head injury score x. The straight-line model is fit to the NCAP data collected for 98 crash-tested cars. A MINITAB printout of the simple linear regression analysis is displayed in Figure 10.7. Interpret the estimates of β_0 and β_1 shown on the printout.

M **Figure 10.7** Simple Linear Regression Output Obtained from MINITAB, Example 10.8

```
The regression equation is
DriverStar = 5.67 - 0.00289 DriveHead

Predictor        Coef        StDev          T        P
Constant       5.6718       0.1682      33.72    0.000
DriveHea   -0.0028873    0.0002665     -10.84    0.000

S = 0.4864        R-Sq = 55.0%       R-Sq(adj) = 54.5%

Analysis of Variance

Source            DF         SS         MS         F        P
Regression         1     27.784     27.784    117.42    0.000
Residual Error    96     22.716      0.237
Total             97     50.500
```

(Annotations: b_0 points to 5.6718; b_1 points to -0.0028873.)

Solution

The estimates of β_0 and β_1, indicated on Figure 10.7, are

$$b_0 = 5.6718 \qquad b_1 = -.0028873$$

The estimated y-intercept, $b_0 = 5.6718$, will have a practical interpretation only if the value $x = 0$ is meaningful. In this problem, a driver's head injury score of $x = 0$ is very unlikely to occur when a car is crashed into a wall at 35 mph. Even if $x = 0$ was theoretically possible, none of the 98 cars crash-tested in the sample had a driver's head injury score of 0. In fact, the lowest x-value in the sample data is 238! Therefore, the value $b_0 = 5.6718$ has no practical interpretation.

The estimated slope, $b_1 = -.00289$, represents the change in overall driver's injury rating y for every 1-point increase in driver's head injury score x. The negative value implies that the overall injury rating *decreases* by .00289 point as the head injury score increases by 1 point.

EXAMPLE 10.9 PREDICTING y

Refer to Example 10.8. Use the least squares line to predict the overall driver's injury rating y when the severity of head injury is $x = 1{,}000$.

Solution

The least squares prediction equation, obtained in Example 10.8, is $\hat{y} = 5.6718 - .00289x$. To obtain the predicted value of y, we substitute $x = 1{,}000$ into the least squares prediction equation:

$$\hat{y} = 5.6718 - .00289(1{,}000) = 2.78$$

Consequently, we predict a driver involved in a head-on crash at 35 mph to have an overall injury rating of 2.78 (a fairly low rating) when the severity of his or her head injury rating is 1,000. We show how to provide a measure of reliability for this inference in Section 10.10.

Self-Test 10.3

Determining the nutritive quality of a new food product is often accomplished by feeding the food product to animals whose metabolic processes are very similar to our own (e.g., rats). This technique was used to evaluate the protein efficiency of two forms of a new product—one solid and the other liquid. Ten rats were randomly assigned a solid diet and 10 a liquid diet. At the end of the feeding period, the total protein intake x (in grams) and the weight gain y (in grams) were recorded for each of the rats. A straight line was fit to the 10 (x, y) data points for each diet by the method of least squares. The least squares prediction equations follow:

$$\text{Liquid: } \hat{y} = 110 - .75x$$

$$\text{Solid: } \hat{y} = 84 + 3.66x$$

a. Interpret the estimated y-intercept and slope for each diet.

b. For each diet, find the predicted weight gain of a rat with a protein intake of 10 grams.

PROBLEMS FOR SECTION 10.3

Applying the Concepts

10.7 The data on retail price and energy cost for 10 large side-by-side refrigerators, Problems 2.45 (p. 94) and 3.76 (p. 176), are reproduced in the table.

REFRIGERATOR

Brand	Price ($)	Energy Cost per Year ($)
Maytag MTB2156B	850	48
Amana TR21V2	760	54
Kenmore(Sears) 7820	900	58
Whirlpool Gold GT22DXXG	870	66
GEProfile TBX22PRY	1100	77
KitchenAid Prestige KTRP22KG	800	66
Whirlpool ET21PKXG	650	70
GE TBX22ZIB	750	81
FrigidaireGallery FRT20NGC	750	72
Hotpoint CTX21DIB	570	78

Data Source: "Cold Storage," *Consumer Reports,* Feb. 1999, 49. Although these data sets originally appeared in *Consumer Reports,* the selective adaptation and resulting conclusions presented are those of the authors and are not sanctioned or endorsed in any way by Consumers Union, the publisher of *Consumer Reports.*

a. Give the equation of a straight-line model relating retail price y to energy cost x.

b. The model, part **a**, was fit to the data using Excel. The Excel printout is shown on p. 548. Locate the estimated slope and y-intercept of the line on the printout.

c. Write the equation of the least squares line.

	A	B	C	D	E	F	G
1	Regression Analysis						
2							
3	*Regression Statistics*						
4	Multiple R	0.164132702					
5	R Square	0.026939544					
6	Adjusted R Square	-0.094693013					
7	Standard Error	151.7801248					
8	Observations	10					
9							
10	ANOVA						
11		*df*	*SS*	*MS*	*F*	*Significance F*	
12	Regression	1	5102.349624	5102.349624	0.221483003	0.650476756	
13	Residual	8	184297.6504	23037.2063			
14	Total	9	189400				
15							
16		*Coefficients*	*Standard Error*	*t Stat*	*P-value*	*Lower 95%*	*Upper 95%*
17	Intercept	946.7199248	315.4320015	3.001343936	0.017036751	219.3319546	1674.107895
18	Energy Cost	-2.189849624	4.653118275	-0.470619807	0.650476756	-12.91996655	8.540267298

E Problem 10.7

d. Why is the line, part **c**, the "best" fitting line through the data?

e. Interpret the values of b_0 and b_1.

f. Use the least squares line to predict the retail price of a refrigerator with an energy cost of $80 per year.

10.8 Refer to *The American Statistician* (Feb. 1991) study of a process for drilling rock, Problems 2.46 (p. 000) and 3.77 (p. 176). Recall that experimenters wanted to determine whether the time y it takes to dry drill a distance of five feet in rock increases with depth x at which drilling begins. The data are reproduced in the table.

DRILLROCK

Depth at which Drilling Begins x (feet)	Time to Drill 5 Feet y (minutes)
0	4.90
25	7.41
50	6.19
75	5.57
100	5.17
125	6.89
150	7.05
175	7.11
200	6.19
225	8.28
250	4.84
275	8.29
300	8.91
325	8.54
350	11.79
375	12.12
395	11.02

Source: Penner, R., and Watts, D. G. "Mining information." *The American Statistician,* Vol. 45, No. 1, Feb. 1991, p. 6 (Table 1).

a. Write a straight-line model relating drill time y to depth x.

b. Use Excel or MINITAB to fit the simple linear regression model, part **a**, to the data.

c. Refer to part **b**. Interpret the estimated slope of the line.

d. Refer to part **b**. Interpret the estimated y-intercept of the line.

10.9 Carbon dioxide–baited traps are typically used by entomologists to monitor mosquito populations. An article in the *Journal of the American Mosquito Control Association* (Mar. 1995) investigated whether temperature influences the number of mosquitos caught in a trap. Six mosquito samples were collected on each of nine consecutive days. For each day, two variables were measured: average temperature (degrees Centigrade) and mosquito catch ratio (the number of mosquitos caught in each sample divided by the largest sample caught).

MOSQUITO

Date	Average Temperature	Catch Ratio
July 24	16.8	.66
25	15.0	.30
26	16.5	.46
27	17.7	.44
28	20.6	.67
29	22.6	.99
30	23.3	.75
31	18.2	.24
Aug. 1	18.6	.51

Source: Petric, D., et al. "Dependence of CO_2-baited suction trap captures on temperature variations." *Journal of the Mosquito Control Association,* Vol. 11, No. 1, Mar. 1995, p. 8.

a. Construct a scatterplot relating catch ratio to average temperature. Is there a linear trend in the data?

b. Use Excel or MINITAB to fit a straight-line model relating catch ratio y to average temperature x. Interpret the β estimates.

10.10 Refer to the *American Journal of Psychiatry* (July 1995) study relating verbal memory retention y and right hippocampal volume x of 21 Vietnam veterans, Problems 2.48 (p. 95) and 3.80 (p. 177). The data, reproduced in the table, were subjected to a simple linear regression analysis using MINITAB. The MINITAB printout is shown on p. 550.

BRAIN

Veteran	Verbal Memory Retention y	Right Hippocampal Volume x
1	26	960
2	22	1,090
3	30	1,180
4	65	1,000
5	70	1,010
6	60	1,070
7	60	1,080
8	58	1,040
9	65	1,040
10	83	1,030
11	69	1,045
12	60	1,210
13	62	1,220
14	66	1,215
15	76	1,220
16	65	1,300
17	66	1,350
18	84	1,350
19	90	1,370
20	102	1,400
21	102	1,460

```
The regression equation is
Verbal Memory retention = - 29.7 + 0.0813 Right Hippocampal Volume

Predictor        Coef       StDev          T        P
Constant       -29.65       30.72      -0.97    0.347
Right Hi      0.08132     0.02598       3.13    0.006

S = 17.61      R-Sq = 34.0%      R-Sq(adj) = 30.5%

Analysis of Variance

Source            DF          SS          MS        F        P
Regression         1      3037.4      3037.4     9.80    0.006
Residual Error    19      5890.4       310.0
Total             20      8927.8
```

M Problem 10.10

a. Locate the estimated slope and y-intercept on the printout.

b. Write the equation of the least squares line.

c. Interpret the value of the estimated slope.

d. Interpret the value of the estimated y-intercept.

e. Predict the verbal memory retention of a Vietnam veteran with a right hippocampal volume of 1,000.

10.11 Is it possible to predict the number of wins of a professional basketball team during the regular National Basketball Association (NBA) season? Charles Fisk, a California data analyst, collected performance data on NBA teams from 1979 to 1996. A regression model was developed to predict y, the number of wins for a team in a particular year, standardized using a z score calculation. Independent variables

NBA

Team	Wins y	2-Point Field Goals x_1	Rebounds x_2	Turnovers x_3
Chicago	2.21	0.77	2.14	2.02
Seattle	1.64	1.70	0.29	0.67
Orlando	1.35	1.28	−0.34	0.69
San Antonio	1.28	1.02	−0.23	0.54
Utah	1.00	1.49	1.11	1.61
LA Lakers	0.85	1.10	−0.65	1.50
Indiana	0.78	0.73	0.86	−0.67
Houston	0.50	0.50	−0.99	−0.18
New York	0.43	0.81	−0.56	0.18
Cleveland	0.43	0.18	−0.47	1.84
Detroit	0.36	0.29	0.36	−0.54
Atlanta	0.36	−0.89	−0.06	1.56
Portland	0.21	0.58	1.93	−1.63
Miami	0.07	0.92	0.78	−0.93
Phoenix	0.00	−0.05	0.49	−0.14
Charlotte	0.00	−0.60	−0.27	−0.23
Washington	−0.14	0.54	−0.86	0.31
Sacramento	−0.14	−0.29	0.36	−0.76
Golden State	−0.36	−0.90	0.21	0.28
Denver	−0.43	−0.23	0.89	−1.19
Boston	−0.57	−0.93	−0.60	0.11
New Jersey	−0.78	−1.14	1.86	−0.93
LA Clippers	−0.85	−0.10	−0.90	0.02
Dallas	−1.07	−2.26	−0.10	0.69
Minnesota	−1.07	−0.44	−0.38	−0.73
Milwaukee	−1.14	−0.31	−0.66	−0.73
Toronto	−1.42	−0.44	−0.35	−1.92
Philadelphia	−1.64	−1.66	−1.78	−1.11
Vancouver	−1.85	−1.65	−2.07	0.67

(all standardized using z scores) that were found to be the best predictors included:

x_1 = The difference between a team's 2-point field goal percentage and its opponents' percentage during the year

x_2 = The difference between a team's rebounds and its opponents' rebounds during the year

x_3 = The difference between opponents' turnovers and a team's turnovers during the year

The values of these variables for each of the 29 NBA teams during the 1995–96 season are listed in the table on p. 550.

a. Plot each of the independent variables against the dependent variable y. Which independent variable appears to have the strongest relationship with standardized number of wins?

b. Use Excel or MINITAB to fit a straight-line model relating standardized wins y to x_1.

c. Interpret the values of b_0 and b_1 obtained in part **b**.

d. Repeat parts **b** and **c** for independent variable x_2.

e. Repeat parts **b** and **c** for independent variable x_3.

10.12 Civil engineers often use a straight-line equation to model the relationship between the shear strength y of masonry joints and precompression stress x. To test this theory, a series of stress tests was performed on solid bricks arranged in triplets and joined with mortar (*Proceedings of the Institute of Civil Engineers*, Mar. 1990). The precompression stress was varied for each triplet; the ultimate shear load just before failure (called the shear strength) was recorded. The stress results for seven triplets (measured in N/mm^2) are shown in the accompanying table.

TRIPLETS

	Triplet Test						
	1	**2**	**3**	**4**	**5**	**6**	**7**
Shear strength y	1.00	2.18	2.21	2.41	2.59	2.82	3.06
Precompression stress x	0	.60	1.20	1.33	1.43	1.75	1.75

Source: Riddington, J. R., and Ghazali, M. Z. "Hypothesis for shear failure in masonry joints." *Proceedings of the Institute of Civil Engineers, Part 2,* Mar. 1990, Vol. 89, p. 96 (Figure 7).

a. Construct a scatterplot for the seven data points. Does the relationship between shear strength and precompression stress appear to be linear?

b. Use the method of least squares (Excel or MINITAB) to estimate the parameters of the linear model.

c. Interpret the values of b_0 and b_1.

d. Predict the shear strength for a triplet with a precompression stress of 1 N/mm^2.

10.13 The Sasakawa Sports Foundation recently conducted a national survey to assess the physical activity patterns of Japanese adults. The table on p. 552 lists the frequency (average number of days in the past year) and duration of time (average number of minutes) Japanese adults spent participating in a sample of 11 sports activities.

a. Construct a scatterplot for the data. Do you observe a trend?

b. Write the equation of a straight-line model relating duration (y) to frequency (x).

c. Find the least squares prediction equation using Excel or MINITAB.

d. Graph the least squares line on the scatterplot.

e. Interpret the estimated y-intercept and slope of the line.

f. Use the least squares line to predict the duration of time Japanese adults participate in a sport that they play 25 times a year.

10.14 The Mechanics Baseline and the Force Concept Inventory are two standardized tests designed to measure the competency of high school students in physics. The *American Journal of Physics* (July 1995) reported the inventory and baseline

JAPANSPORTS

Activity	Frequency x (days/year)	Duration y (minutes)
Jogging	135	43
Cycling	68	99
Aerobics	44	61
Swimming	39	60
Volleyball	30	80
Tennis	21	100
Softball	16	91
Baseball	19	127
Skating	7	115
Skiing	10	249
Golf	5	262

Source: Bennett, J., ed., *Statistics in Sport,* London: Arnold, 1998 (adapted from Figure 11.6).

Problem 10.13

PHYSICS

Student	Inventory Score y (%)	Baseline Score x (%)	Student	Inventory Score y (%)	Baseline Score x (%)
1	97	100	15	83	68
2	97	93	16	83	52
3	90	84	17	76	52
4	90	80	18	70	52
5	90	75	19	70	58
6	97	72	20	80	58
7	86	75	21	62	58
8	86	65	22	76	61
9	86	61	23	62	38
10	83	65	24	42	38
11	76	65	25	67	41
12	72	65	26	52	45
13	95	68	27	45	33
14	95	52			

Source: Wells, M., Hestenes, D., and Swackhamer, G. "A modeling method for high school physics instruction." *American Journal of Physics,* Vol. 63, No. 7, July 1995, p. 612 (adapted from Fig. 3), p. 162.

Problem 10.14

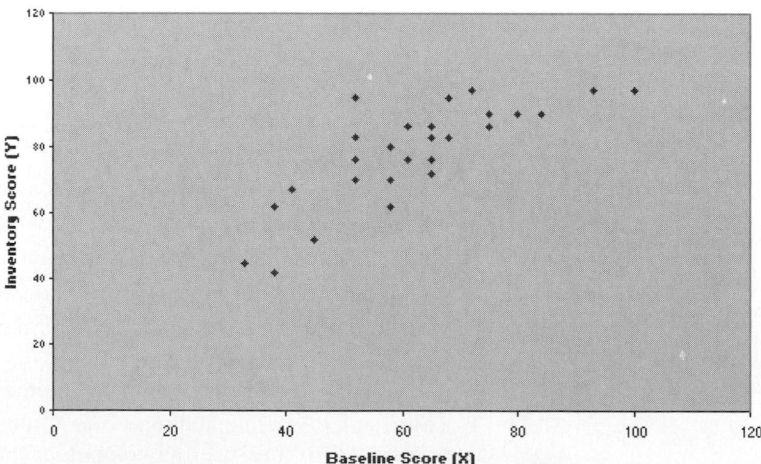

E **Problem 10.14**

scores for 27 students in an honors physics course taught by one of the developers of the tests. These scores are given in the second table on p. 552. Suppose you are interested in predicting a student's memory score y from his or her baseline score x.

a. An Excel plot of the data is shown on p. 552. Does baseline score appear to be a good predictor of inventory score?

b. Write the equation of a straight-line model relating the two variables.

c. An Excel simple linear regression printout for the data is shown below. Locate and interpret the estimates of β_0 and β_1 on the printout.

d. What is the predicted inventory score of a physics student with a baseline score of 50%?

	A	B	C	D	E	F	G
1	SUMMARY OUTPUT						
2							
3	*Regression Statistics*						
4	Multiple R	0.79529465					
5	R Square	0.63249359					
6	Adjusted R Squar	0.61779333					
7	Standard Error	9.57283768					
8	Observations	27					
9							
10	ANOVA						
11		*df*	*SS*	*MS*	*F*	*Significance F*	
12	Regression	1	3942.871321	3942.871	43.02602	7.16126E-07	
13	Residual	25	2290.98053	91.63922			
14	Total	26	6233.851852				
15							
16		*Coefficients*	*Standard Error*	*t Stat*	*P-value*	*Lower 95%*	*Upper 95%*
17	Intercept	31.3146973	7.362792722	4.2531	0.000258	16.15075243	46.47864211
18	Baseline Score	0.7541835	0.114977118	6.559423	7.16E-07	0.517383856	0.990983138

E Problem 10.14

10.4 Model Assumptions

In the following sections, we describe the statistical methods (e.g., tests of hypotheses and confidence intervals) appropriate for making inferences from a simple linear regression analysis. As with most statistical procedures, the validity of the inferences depends on certain assumptions being satisfied. These assumptions, made about the random error term in the straight-line probabilistic model, are summarized in the box.

ASSUMPTIONS ABOUT THE RANDOM ERROR ε REQUIRED FOR A SIMPLE LINEAR REGRESSION ANALYSIS

1. The mean of the probability distribution of ε is 0.*
2. The variance of the probability distribution of ε is constant for all values of the independent variable x and is equal to σ^2.
3. The probability distribution of ε is normal.
4. The errors associated with any two observations are independent. That is, the error associated with one value of y has no effect on the errors associated with other y values.

*The assumption that mean error is 0 is *not* guaranteed by the method of least squares. The method of least squares yields a *sample* mean error of 0, i.e., $[\Sigma(y - \hat{y})]/n = 0$, for all models regardless of whether the model is the correct one. The zero mean assumption in simple linear regression is equivalent to assuming that the form of the correct model is, in fact, a straight line.

Figure 10.8 on p. 554 shows a pictorial representation of the assumptions given in the box. For each value of x shown in the figure, the relative frequency

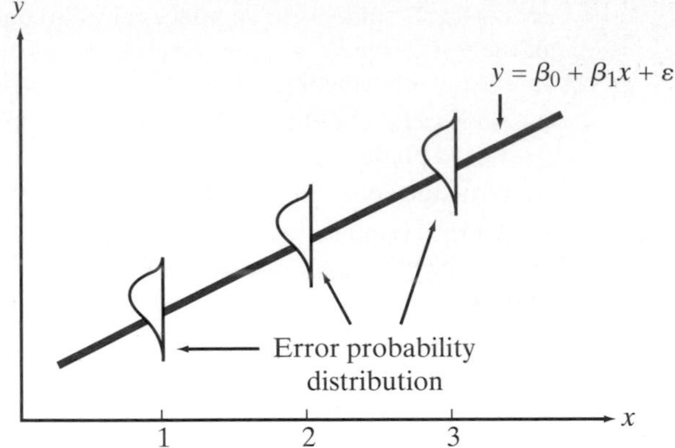

Figure 10.8 The Probability Distribution of the Random Error Component ε

distribution of the errors is normal with mean 0 and with a constant variance (all the distributions shown have the same amount of spread or variability) equal to σ^2.

Statistical techniques are available for detecting when one or more of the assumptions are grossly violated. These methods are based on an analysis of the least squares errors of prediction, or *residuals*. (We present an analysis of residuals in optional Section 10.11.) In practice, however, you will never know whether the data exactly satisfy the four assumptions. Fortunately, the estimators and test statistics used in a simple linear regression have sampling distributions that remain relatively stable for minor departures from the assumptions.

In regression analysis, the symbol μ_y is used in place of μ to represent the true mean (or *expected value*) of y.

There is an additional assumption that is implied in a simple linear regression analysis, but often forgotten—namely, the assumption that the relationship between the mean value of y, denoted μ_y, and the independent variable x is correctly modeled by a *straight line*. In a real application, the relationship between μ_y and x may possess some curvature. Therefore, when we conduct a simple linear regression analysis, we are assuming that this curvature is minimal over the set of values for which x is measured. The implications of this assumption can be seen in Figure 10.9. If x is measured over the interval between two points, say x_L and x_U, a simple linear regression analysis may produce a very good prediction equation for predicting y for values of x between x_L and x_U, but very poor estimates and predictions for values of x outside this range. The rule, then, is *never to attempt to predict values of y for values of x outside the range of values used in the regression analysis.*

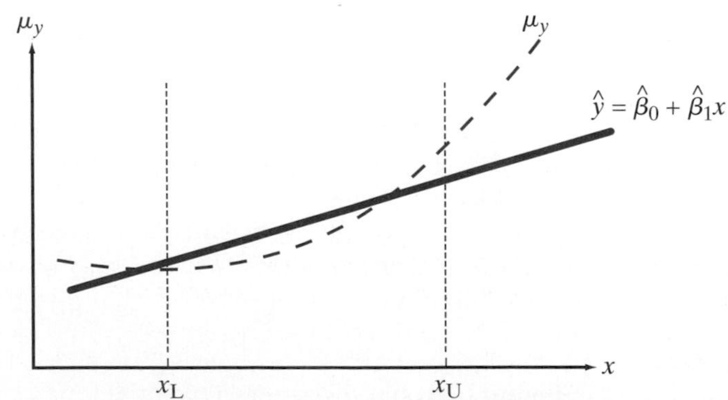

Figure 10.9 Hypothetical Comparison of the True Relationship between μ_y and x with the Simple Linear Regression Model

10.5 Measuring Variability around the Least Squares Line

In Example 10.5, is the creativity score y really related to the flexibility score x, or is the linear relation that we seem to see a result of chance? That is, could it be the case that x and y are completely unrelated, and that the apparent linear configuration of the data points in the scatterplot of Figure 10.4 is due to random variation? The statistical method that will answer this question requires that we know how much y will vary for a given value of x. That is, we must know the value of the quantity, called σ^2, that measures the variability of the y values about the least squares line. This value is the variance σ^2 of the random error identified in Section 10.4.

Figure 10.10a shows a hypothetical situation where the value of σ^2 is small. Note that there is little variation in the y values about the least squares line. In contrast, Figure 10.10b illustrates a situation where σ^2 is large; the y values deviate greatly about the least squares line.

Figure 10.10 Illustration of Error Variance σ^2

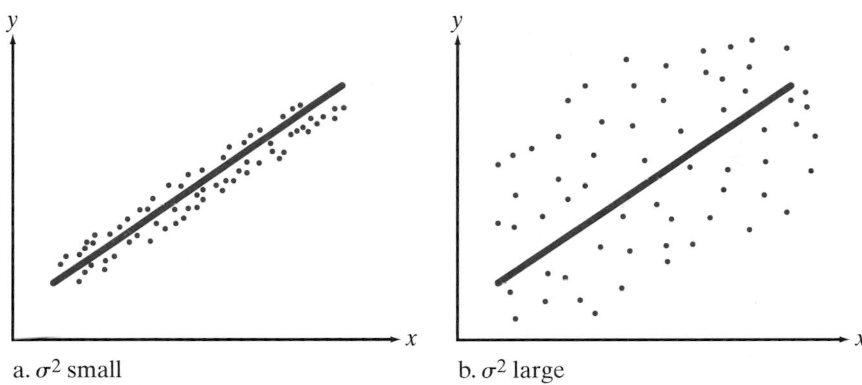

Since the variance σ^2 will rarely be known, we estimate its value using the sum of squared errors, SSE, and the formulas shown in the box. However, we can rely on computer software such as Excel or MINITAB to compute an estimate of σ^2.

Degrees of Freedom for Estimating σ^2
The divisor of s^2, $n - 2$, represents the *number of degrees of freedom available for estimating* σ^2. The value results from the fact that we "lose" two degrees of freedom for estimating the model parameters β_0 and β_1.

ESTIMATION OF σ^2, A MEASURE OF THE VARIABILITY OF THE y VALUES ABOUT THE LEAST SQUARES LINE

An estimate of σ^2 is given by

$$s^2 = \frac{\text{SSE}}{n - 2} \quad \textbf{(10.8)}$$

where

$$\text{SSE} = \sum (y - \hat{y})^2 = \text{SSyy} - b_1\text{SSxy} \quad \textbf{(10.9)}$$

$$\text{SSyy} = \sum (y - \bar{y})^2 = \sum y^2 - \frac{(\sum y)^2}{n} \quad \textbf{(10.10)}$$

$$\text{SSxy} = \sum (x - \bar{x})(y - \bar{y}) = \sum xy - \frac{(\sum x)(\sum y)}{n} \quad \textbf{(10.6)}$$

EXAMPLE 10.10

ESTIMATING σ^2 AND σ

Refer to the simple linear regression relating creativity score to flexibility score, Example 10.5.

a. Estimate the value of the error variance σ^2.

b. Estimate and interpret the value of σ.

Solution

a. Refer to the Excel output for the simple linear regression relating creativity score y to flexibility score x, Figure 10.5 (p. 545). The value of SSE (indicated on the printout) is 13.3444 and the estimate of σ^2 (also indicated on the printout) is 1.66805.

b. An estimate of the standard deviation σ of the random error is obtained by taking the square root of s^2. Therefore, our estimate is

$$s = \sqrt{s^2} = \sqrt{1.66805} = 1.29153$$

This value is also indicated on Figure 10.5. Since s measures the spread of the distribution of the y values about the least squares line, we should not be surprised to find that most of the observations lie within $2s = 2(1.29) = 2.58$ of the least squares line. From Figure 10.11 we see that, for this example, all 10 data points have y values that lie within $2s$ of $\hat{y}$, the least squares predicted value. Since the distribution of errors is assumed to be normal, we expect about 95% of the y values to lie within $2s$ of the least squares line.

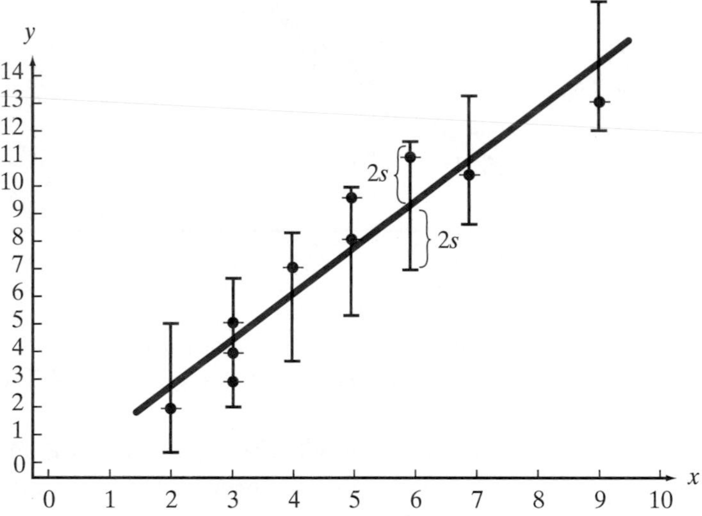

Figure 10.11 Observations within $2s$ of the Least Squares Line

PRACTICAL INTERPRETATION OF s, THE ESTIMATED STANDARD DEVIATION OF σ

Given the assumptions outlined in the box in Section 10.4, we expect most (about 95%) of the observed y values to lie within $2s$ of their respective least squares predicted values y.

Self-Test 10.4

Refer to Self-Test 10.3 (p. 547). Recall that simple linear regression was used to fit the model $y = \beta_0 + \beta_1 x + \varepsilon$ to $n = 10$ data points, where y is the weight gain for a rat on a diet of a new food product and x is the total protein intake of the rat. For rats on a solid diet, SSE = 300.

a. Compute an estimate of σ.

b. Interpret the estimate in part **a**.

PROBLEMS FOR SECTION 10.5

Using the Tools

10.15 Suppose you fit a least squares line to 10 data points and find SSE = .22.

a. Find s^2, the estimate of σ^2.

b. Find s, the estimate of σ.

10.16 Repeat Problem 10.15, but assume $n = 100$.

10.17 Repeat Problem 10.15, but assume SSE = 22.

10.18 Suppose you fit a straight line to 25 data points. Find s^2 and s for each of the following values of SSE.

a. 100 **b.** 6.20 **c.** 8,865

Applying the Concepts

10.19 Locate and interpret the values of SSE, s^2, and s on the Excel printout, Problem 10.7 (p. 548).

10.20 Locate and interpret the values of SSE, s^2, and s on the MINITAB printout, Problem 10.10 (p. 550).

10.21 Refer to the three straight-line models for number of NBA wins y, Problem 10.11 (p. 550). For each model, find the estimate of σ on the Excel or MINITAB printout and interpret it. Which of the three independent variables, x_1, x_2, or x_3, yields the linear prediction equation with the smallest error variance?

10.22 Refer to Problem 10.13 (p. 551).

a. Find SSE and s^2.

b. Find s, the estimate of σ.

c. Interpret the value of s.

10.23 Locate and interpret the values of SSE, s^2, and s on the Excel printout, Problem 10.14 (p. 553).

10.6 Inferences about the Slope

After fitting the model to the data and computing an estimate of σ^2, we can statistically check the usefulness of the model. That is, we can use a statistical inferential procedure (a test of hypothesis or confidence interval) to determine whether the least squares straight-line (linear) model is a reliable tool for predicting y for a given value of x.

EXAMPLE 10.11

HOW TO TEST THE MODEL

Consider the straight-line model

$$y = \beta_0 + \beta_1 x + \varepsilon$$

How do we determine statistically whether this model is useful for prediction purposes? In other words, how could we test whether x provides useful information for the prediction of y?

Solution

Suppose x is *completely unrelated* to y. What could we say about the values of β_0 and β_1 in the probabilistic model, if in fact x contributes no information for the prediction of y? For y to be independent of x, the true slope β_1 of the line must be equal to 0. Therefore, to test the null hypothesis that x contributes no information for the prediction of y against the alternative that these variables are linearly

related with a slope differing from 0, we test

$$H_0: \beta_1 = 0$$
$$H_a: \beta_1 \neq 0$$

If the data support the alternative hypothesis, we will conclude that x does contribute information for the prediction of y using the straight-line model, although the true relationship between μ_y and x could be more complex than a straight line.

The sampling distribution of b_1, the estimator of the slope β_1, has a t distribution with standard error

$$s_{b_1} = \frac{s}{\sqrt{SSxx}} \qquad \textbf{(10.11)}$$

Using the hypothesis-testing technique developed in Chapter 8, we set up the test as shown in the next box.

TEST OF HYPOTHESIS FOR THE SLOPE OF THE STRAIGHT-LINE MODEL

Upper-Tailed Test	**Two-Tailed Test**	**Lower-Tailed Test**
$H_0: \beta_1 = 0$	$H_0: \beta_1 = 0$	$H_0: \beta_1 = 0$
$H_a: \beta_1 > 0$	$H_a: \beta_1 \neq 0$	$H_a: \beta_1 < 0$

$$\textit{Test statistic: } t = \frac{b_1}{s_{b_1}} \approx \frac{b_1}{s/\sqrt{SSxx}} \qquad \textbf{(10.12)}$$

Rejection region:	*Rejection region:*	*Rejection region:*
$t > t_\alpha$	$|t| > t_{\alpha/2}$	$t < -t_\alpha$

where the distribution of t is based on $(n - 2)$ degrees of freedom, t_α is the t value such that $P(t > t_\alpha) = \alpha$, and $t_{\alpha/2}$ is the t value such that $P(t > t_{\alpha/2}) = \alpha/2$.

Assumptions: See Section 10.4.

Inferences based on this hypothesis test require the standard least squares assumptions about the random error term listed in the box in Section 10.4. However, the test statistic has a sampling distribution that remains relatively stable for minor departures from the assumptions. That is, our inferences remain valid for practical cases in which the assumptions are nearly, but not completely, satisfied.

EXAMPLE 10.12 **TESTING THE MODEL**

Refer to the simple linear regression relating creativity score y to flexibility score x, Example 10.5 (p. 542). At a significance level $\alpha = .05$, test whether the straight-line model is useful for predicting y.

Solution

Testing the usefulness of the model requires testing the hypotheses

$$H_0: \beta_1 = 0$$
$$H_a: \beta_1 \neq 0$$

With $n = 10$ and $\alpha = .05$, the critical value based on $(10 - 2) = 8$ degrees of

freedom is obtained from Table B.4 in Appendix B:

$$t_{\alpha/2} = t_{.025} = 2.306$$

Thus, we will reject H_0 if $|t| > 2.306$ (see Figure 10.12)

Figure 10.12 Rejection Region for Example 10.12

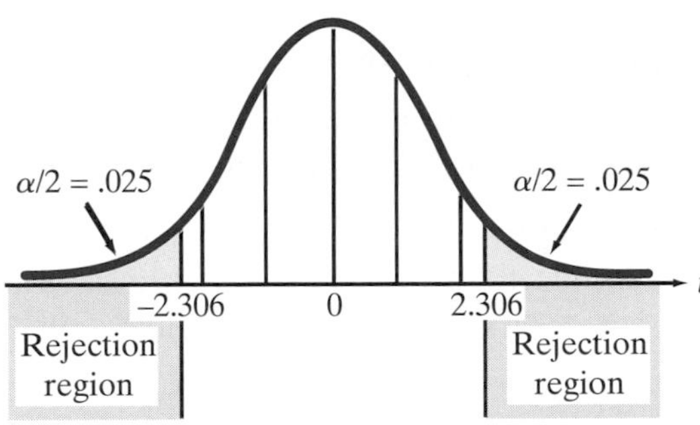

Rather than compute the test statistic using equation 10.12, we will use the results obtained from Excel. A portion of the Excel simple linear regression analysis is reproduced in Figure 10.13.

E **Figure 10.13** Excel Simple Linear Regression Output, Example 10.12

16		*Coefficients*	*Standard Error*	*t Stat*	*P-value*	*Lower 95%*	*Upper 95%*
17	**Intercept**	-0.43705463	1.020802329	-0.42815	0.679837	-2.791030547	1.916921283
18	**Flexibility Score**	1.66745843	0.199050732	8.377053	3.13E-05	1.208446325	2.126470539

b_1 s_{b_1} t p-value   95% CI for β_1

The estimate $b_1 = 1.667$ and its standard error $s_{b_1} = .199$ are indicated in the printout as well as the value of the test statistic $t = 8.38$. Since this calculated t value falls in the upper tail of the rejection region (see Figure 10.12), we reject the null hypothesis and conclude that the slope β_1 is not 0.

At the $\alpha = .05$ level of significance, the sample data provide sufficient evidence to conclude that flexibility score *does* contribute useful information for the prediction of creativity score using the straight-line model.

The same conclusion can be reached by using the p-value approach. The p-value for this two-tailed test (indicated on Figure 10.13) is near 0.* Since this value is less than $\alpha = .05$, we reject H_0 and conclude that the slope β_1 differs from 0.

**Scientific Notation:* The p-value on the Excel printout is expressed in scientific notation as 3.13E-05. This means that the decimal point should be moved 5 places to the left. Therefore, the p-value is .0000313.

If the test in Example 10.12 had resulted in "fail to reject H_0," would we have concluded that $\beta_1 = 0$? The answer to this question is "no" (recall the discussion on Type II errors in Chapter 8). Rather, we acknowledge that additional data might indicate that β_1 differs from 0, or that a more complex relationship (other than a straight line) may exist between y and x.

Self-Test 10.5

A consumer investigator obtained the following least squares straight-line model (based on a sample of $n = 100$ families) relating the yearly food cost y (thousands of dollars) for a family of four to annual income x:

$$\hat{y} = 467 + .26x$$

In addition, the investigator computed the standard error $s_{b_1} = .22$. Conduct a test to determine whether mean yearly food cost y increases as annual income x increases, i.e., whether the slope β_1 of the line is positive. Use $\alpha = .05$.

In addition to testing whether the slope β_1 is 0, we may also be interested in estimating its value with a confidence interval. The procedure is illustrated in the following example.

EXAMPLE 10.13 CONFIDENCE INTERVAL FOR THE SLOPE

Refer to Example 10.12. Construct a 95% confidence interval for the slope β_1 in the straight-line model relating creativity score y to flexibility score x.

Solution

The method of Chapter 7 can be used to construct a confidence interval for β_1. The interval, derived from the sampling distribution of b_1, is given in the next box.

A $(1 - \alpha)100\%$ CONFIDENCE INTERVAL FOR THE SLOPE β_1

$$b_1 \pm (t_{\alpha/2})s_{b_1} = b_1 \pm t_{\alpha/2}\left(\frac{s}{\sqrt{SSxx}}\right) \quad (10.13)$$

where the distribution of t is based on $(n - 2)$ degrees of freedom and $t_{\alpha/2}$ is the value of t such that $P(t > t_{\alpha/2}) = \alpha/2$.

For a 95% confidence interval, $\alpha = .05$. Therefore, we need to find the value of $t_{.025}$ based on $(10 - 2) = 8$ df. In Example 10.12, we found that $t_{.025} = 2.306$. From the Excel printout, Figure 10.13 (p. 559), we have

$$b_1 = 1.667 \quad \text{and} \quad s_{b_1} = .199$$

Now we apply equation 10.13 to obtain a 95% confidence interval for the slope in the model relating creativity score to flexibility score:

$$b_1 \pm (t_{.025})s_{b_1} = 1.667 \pm (2.306)(.199)$$
$$= 1.667 \pm .459$$

Our interval estimate of the slope parameter β_1 is then 1.208 to 2.126. This confidence interval also appears on the Excel printout, Figure 10.13, as indicated.

EXAMPLE 10.14 INTERPRETING THE CONFIDENCE INTERVAL FOR β_1

Interpret the interval estimate of β_1 derived in Example 10.13.

Solution

Since all the values in the interval (1.208, 2.126) are positive, we say that we are 95% confident that the slope β_1 is positive. That is, we are 95% confident that the mean creativity score μ_y increases as flexibility score x increases. In addition, we can say that for every 1-point increase in flexibility score x, the increase in mean creativity score μ_y could be as small as 1.208 or as large as 2.126 points. However, the relatively large width of the interval reflects the small number of data points (and, consequently, a lack of information) in the experiment. We could expect a narrower interval if the sample size were increased.

> **Self-Test 10.6**
>
> Use the information in Self-Test 10.5 (p. 559) to find a 90% confidence interval for β_1. Interpret the results.

PROBLEMS FOR SECTION 10.6

Using the Tools

10.24 Suppose you want to test H_0: $\beta_1 = 0$ versus H_a: $\beta_1 \neq 0$ in a simple linear regression model. Give the degrees of freedom associated with the value of the test statistic for each of the following sample sizes.

a. $n = 6$ **b.** $n = 10$

c. $n = 25$ **d.** $n = 50$

10.25 For each of the following, specify the rejection region for testing the null hypothesis H_0: $\beta_1 = 0$ in simple linear regression.

a. H_a: $\beta_1 \neq 0, n = 5, \alpha = .05$

b. H_a: $\beta_1 > 0, n = 5, \alpha = .10$

c. H_a: $\beta_1 < 0, n = 7, \alpha = .01$

10.26 A simple linear regression analysis of $n = 10$ data points yielded the following results:

$$b_1 = 56 \qquad s_{b_1} = 16$$

a. Test the null hypothesis that the slope β_1 of the line equals 0 against the alternative hypothesis that β_1 is not equal to 0. Use $\alpha = .10$.

b. Find a 90% confidence interval for the slope β_1.

10.27 A simple linear regression analysis of $n = 45$ data points yielded the following results:

$$b_1 = -300.7 \qquad s_{b_1} = 193.4$$

a. Test the null hypothesis that the slope β_1 of the line equals 0 against the alternative hypothesis that β_1 is negative. Use $\alpha = .10$.

b. Find a 95% confidence interval for the slope β_1.

Applying the Concepts

10.28 The data on temperature at flight time and O-ring damage index for 23 launches of the space shuttle *Challenger,* first presented in Problem 2.50 (p. 97), are reproduced in the table on p. 562.

a. Hypothesize a straight-line model relating O-ring damage y to flight time temperature x.

b. Fit the model, part **a**, to the data using Excel or MINITAB.

c. Theoretically, O-ring damage will decrease as flight time temperatures increase. Test this theory at $\alpha = .01$.

DRILLROCK

10.29 Refer to *The American Statistician* investigation of dry drilling in rock, Problem 10.8 (p. 548). Is there evidence to indicate that dry drill time y increases with depth x? Test using $\alpha = .10$.

MOSQUITO

10.30 Refer to the *Journal of the American Mosquito Control Association* study of the influence of temperature on number of mosquitos caught in carbon dioxide–baited traps, Problem 10.9 (p. 549).

a. Construct a 95% confidence interval for β_1, the slope of the line relating catch ratio y to temperature x.

b. Interpret the result, part **a**.

ORING

Flight Number	Temperature (°F)	O-Ring Damage Index
1	66	0
2	70	4
3	69	0
5	68	0
6	67	0
7	72	0
8	73	0
9	70	0
41-B	57	4
41-C	63	2
41-D	70	4
41-G	78	0
51-A	67	0
51-B	75	0
51-C	53	11
51-D	67	0
51-F	81	0
51-G	70	0
51-I	67	0
51-J	79	0
61-A	75	4
61-B	76	0
61-C	58	4

Note: Data for flight number 4 are omitted due to an unknown O-ring condition.

Primary Sources: Report of the Presidential Commission on the Space Shuttle Challenger *Accident,* Washington, D.C., 1986, Vol. II (pp. H1–H3) and Volume IV (p. 664). *Post-*Challenger *Evaluation of Space Shuttle Risk Assessment and Management,* Washington, D.C., 1988, pp. 135–136.

Secondary Source: Tufte, E. R. *Visual and Statistical Thinking: Displays of Evidence for Making Decisions.* Graphics Press, 1997.

Problem 10.28

JAPANSPORTS

10.31 Refer to the Sasakawa Sports Foundation study of the physical activities of Japanese adults, Problem 10.13 (p. 551). Do the data provide sufficient evidence to indicate that frequency of participation in a sport x contributes information for the prediction of average duration of play y? Test using $\alpha = .05$.

PHYSICS

10.32 Refer to the *American Journal of Physics* study of two standardized physics competency tests, Problem 10.14 (p. 551). Is there evidence of a positive linear relationship between the Mechanics Baseline score x and Force Concept Inventory score y? Use $\alpha = .01$.

10.33 Refer to the *Scientific American* (Dec. 1993) study of fertility rates in developing countries, Problem 3.81 (p. 177). The data on fertility rate y and percentage of married women who use contraception x for each of 27 developing countries are reproduced in the table on p. 563. The simple linear regression model was fit to the data using Excel. The Excel printout is also shown on p. 563.

a. Give the least squares prediction equation.

b. Conduct a test to determine whether fertility rate y and contraceptive prevalence x are linearly related. Use $\alpha = .05$.

c. Construct a 95% confidence interval for β_1. Interpret the interval.

d. Does the confidence interval, part **c**, support the results of the test, part **b**? Explain.

10.34 *Organizational Science* (Mar.–April 2000) published a study on the use of communication media by mid-level managers. One question of interest to the researchers was whether the use of electronic mail (e-mail) is influenced by the user's subjective

FERTIL

Country	Contraceptive Prevalence x	Fertility Rate y
Mauritius	76	2.2
Thailand	69	2.3
Colombia	66	2.9
Costa Rica	71	3.5
Sri Lanka	63	2.7
Turkey	62	3.4
Peru	60	3.5
Mexico	55	4.0
Jamaica	55	2.9
Indonesia	50	3.1
Tunisia	51	4.3
El Salvador	48	4.5
Morocco	42	4.0
Zimbabwe	46	5.4
Egypt	40	4.5
Bangladesh	40	5.5
Botswana	35	4.8
Jordan	35	5.5
Kenya	28	6.5
Guatemala	24	5.5
Cameroon	16	5.8
Ghana	14	6.0
Pakistan	13	5.0
Senegal	13	6.5
Sudan	10	4.8
Yemen	9	7.0
Nigeria	7	5.7

Source: Robey, B., et al. "The fertility decline in developing countries." *Scientific American,* Dec. 1993, p. 62. [*Note:* The data values are estimated from a scatterplot.]

Problem 10.33

	A	B	C	D	E	F	G
1	SUMMARY OUTPUT						
2							
3	*Regression Statistics*						
4	Multiple R	0.86501984					
5	R Square	0.74825933					
6	Adjusted R Square	0.7381897					
7	Standard Error	0.69571072					
8	Observations	27					
9							
10	ANOVA						
11		*df*	*SS*	*MS*	*F*	*Significance F*	
12	Regression	1	35.96633167	35.96633	74.30855	5.86482E-09	
13	Residual	25	12.10033499	0.484013			
14	Total	26	48.06666667				
15							
16		*Coefficients*	*Standard Error*	*t Stat*	*P-value*	*Lower 95%*	*Upper 95%*
17	Intercept	6.73192924	0.290342518	23.18616	2.02E-18	6.133958053	7.32990043
18	Contraceptive Prevalence	-0.0546103	0.006335123	-8.62024	5.86E-09	-0.067657702	-0.04156286

E Problem 10.33

perception of "flow," where flow is defined as playful and exploratory interaction with the medium. For each in a sample of 426 managers, the researchers recorded the number of e-mail messages y sent during a typical week and flow x (measured on an 11-point scale, where higher values represent stronger beliefs that flow exists). Consider a simple linear regression analysis conducted on the data.

a. The estimate of the slope in the regression model is $b_1 = .08$. Interpret this value.

b. A test of hypothesis on the slope resulted in a p-value of .475. Interpret this result.

10.35 A Florida real estate appraisal company regularly employs a paired sales technique to evaluate land sales. When two parcels of land are sold, the ratio of the acreage of the larger parcel to the smaller parcel is calculated as well as the difference in sale prices (measured as a percentage of the higher sale price and called a downward adjustment). The variables land sale ratio x and downward price adjustment y were measured for a random sample of 14 pairs of land sales in Seminole County, Florida. The data (provided by a confidential source) are shown in the accompanying table.

LANDSALE

Pair	Land Sale Ratio x	Downward Price Adjustment y (%)
1	1.7	8
2	2.4	7
3	3.3	8
4	6.8	23
5	5.7	30
6	18.2	52
7	26.2	51
8	33.6	52
9	43.2	57
10	48.4	51
11	59.2	55
12	80.0	57
13	85.3	55
14	140.9	61

A MINITAB simple linear regression analysis of the data is shown in the printout.

```
The regression equation is
Downward Price Adjustment = 25.8 + 0.371 Land Sale Ratio

Predictor        Coef       StDev        T        P
Constant       25.796       5.556     4.64    0.001
Land Sal      0.37099     0.09969     3.72    0.003

S = 14.62       R-Sq = 53.6%      R-Sq(adj) = 49.7%

Analysis of Variance

Source            DF        SS          MS        F        P
Regression         1    2958.2      2958.2    13.85    0.003
Residual Error    12    2563.3       213.6
Total             13    5521.5
```

M Problem 10.35

a. Give the least squares prediction equation.

b. Conduct a test (at $\alpha = .05$) to determine whether downward price adjustment y is linearly related to land sale ratio x.

10.7 Inferences about the Correlation Coefficient (Optional)

In the previous section, we discovered that b_1, the least squares slope, provides useful information on the linear relationship between two variables y and x. You may recall that another way to measure this association is to compute the **correlation coefficient r.** The correlation coefficient, defined in Section 3.8 (p. 171), provides a quantitative measure of the strength of the linear relationship between x and y in the sample, just as does the least squares slope b_1. The values of r and b_1 will always have the same sign (either both positive or both negative) in simple

linear regression. However, unlike the slope, the correlation coefficient r is not expressed in units along a scale. (Recall that r is always between -1 and $+1$, no matter what the units of x and y are.)

Just as b_1 estimates the population parameter β_1 (the true slope of the line), r estimates the parameter that represents the coefficient of correlation for the population of all (x, y) data points. The **population correlation coefficient** is denoted by the Greek letter ρ (rho). A test of the null hypothesis that there is no linear association between x and y in the population is described in the next box.

TEST OF HYPOTHESIS FOR LINEAR CORRELATION

Upper-Tailed Test	Two-Tailed Test	Lower-Tailed Test
$H_0: \rho = 0$	$H_0: \rho = 0$	$H_0: \rho = 0$
$H_a: \rho > 0$	$H_a: \rho \neq 0$	$H_a: \rho < 0$

$$\textit{Test statistic: } t = \frac{r}{\sqrt{\dfrac{1 - r^2}{n - 2}}} \qquad \textbf{(10.14)}$$

where

$$r = \frac{\text{SSxy}}{\sqrt{(\text{SSxx})(\text{SSyy})}} \qquad \textbf{(10.15)}$$

Rejection region:	*Rejection region:*	*Rejection region:*
$t > t_\alpha$	$\|t\| > t_{\alpha/2}$	$t < -t_\alpha$

where the distribution of t depends on $(n - 2)$ df, and t_α, $t_{\alpha/2}$, and $-t_\alpha$ are the critical values obtained from Table B.4 in Appendix B.

Assumptions: The sample of (x, y) values is randomly selected from a (bivariate) normal population.*

*A bivariate normal population implies that the probability distributions of both x and y are normal.

The next example illustrates the test for correlation. In addition, it demonstrates how the correlation coefficient r may be a misleading measure of the strength of the association between x and y in situations where the true relationship is nonlinear.

EXAMPLE 10.15 TEST FOR ZERO CORRELATION

Underinflated or overinflated tires can increase tire wear and decrease gas mileage. A manufacturer of a new tire tested the tire for wear at different pressures with the results shown in Table 10.3.

TIRES

TABLE 10.3 Data for Example 10.15

Pressure x (pounds per sq. inch)	Mileage y (thousands)	Pressure x (pounds per sq. inch)	Mileage y (thousands)
30	29.5	33	37.6
30	30.2	34	37.7
31	32.1	34	36.1
31	34.5	35	33.6
32	36.3	35	34.2
32	35.0	36	26.8
33	38.2	36	27.4

a. Test the hypothesis of no correlation between mileage y and tire pressure x. Use $\alpha = .05$.

b. Interpret the result.

Solution

a. A MINITAB printout of the correlation analysis is shown in Figure 10.14. The value of r is $r = -.114$. This relatively small value for r describes a weak linear relationship between pressure x and mileage y.

M **Figure 10.14** MINITAB Printout of Correlation Analysis of Data in Table 10.3

```
Correlation of Pressure and Mileage = -0.114, P-Value = 0.699
```

The p-value for conducting the test H_0: $\rho = 0$ against H_a: $\rho \neq 0$, shown on Figure 10.14, is .699. Since this p-value exceeds $\alpha = .05$, there is insufficient evidence to reject H_0 and conclude that pressure x and mileage y are linearly related.

b. Although the test of part **a** was nonsignificant, the manufacturer would be remiss in concluding that tire pressure has little or no impact on wear of the tire. On the contrary, the relationship between pressure and wear is fairly strong, as the Excel scatterplot in Figure 10.15 illustrates. The relationship shown is curvilinear; both underinflated tires (low pressure values) and overinflated tires (high pressure values) lead to low mileages.

E **Figure 10.15** Scatterplot of Data in Table 10.3

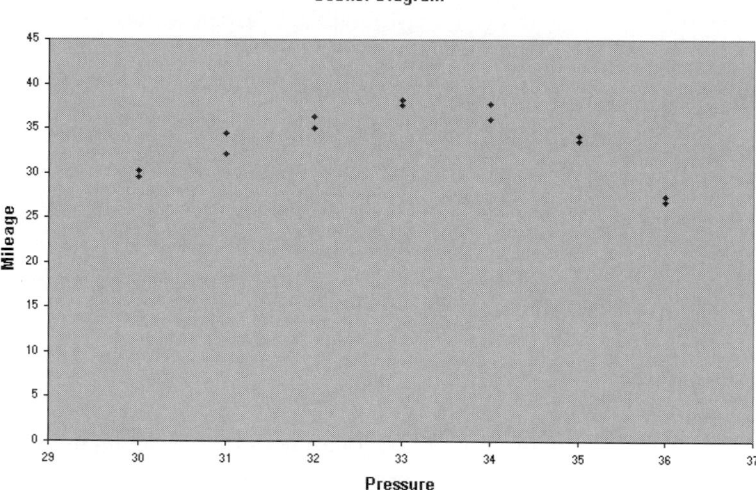

Example 10.15 points out the danger of using r to determine how well x predicts y: The correlation coefficient r describes only the *linear* relationship between x and y. For nonlinear relationships, the value of r may be misleading and we must resort to other methods for describing and testing such a relationship. Regression models for curvilinear relationships are presented in optional Section 10.12.

We conclude this section by pointing out that the null hypothesis H_0: $\rho = 0$ is equivalent to the null hypothesis H_0: $\beta_1 = 0$. Consequently, a test for zero correlation in the population is equivalent to testing for a zero slope in a straight-line model. For example, when we tested the null hypothesis H_0: $\beta_1 = 0$ in the straight-line model relating creativity score y to flexibility score x (Example 10.12), the data led to a rejection of the hypothesis for $\alpha = .05$. This implies that the null hypothesis of a zero linear correlation between the two variables (creativity score and

flexibility score) can also be rejected at $\alpha = .05$. The only real difference between the least squares slope b_1 and the coefficient of correlation r is the measurement scale. Therefore, the information they provide about the utility of the least squares model is redundant and either test can be used. If, however, you want to know how much y increases (or decreases) for every 1-unit increase in x, you need to construct a confidence interval for the slope β_1; the coefficient of correlation r does not provide this information.

Self-Test 10.7

A simple linear regression analysis on $n = 40$ (x, y) data points yielded the following results:

$$b_1 = 35.5 \qquad s_{b_1} = 10.0 \qquad r = .50$$

Show that the test statistic used to test H_0: $\rho = 0$ is equivalent to the test statistic used to test H_0: $\beta_1 = 0$.

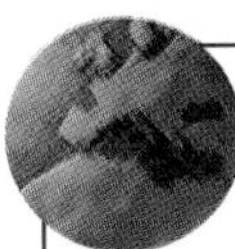

Statistics in the Real World Revisited

The Correlation between Sale Price and Appraised Value

TAMPALMS

Consider, again, the data on sale prices and appraised values for a sample of 92 Tampa Palms homes. Recall that the data is saved in the file named **TAMPALMS.** Are appraised property value and sale price significantly correlated for residential properties sold in this neighborhood? That is, is there sufficient evidence (at $\alpha = .01$) to indicate that ρ, the correlation between appraised value and sale price for the population of all Tampa Palms residential properties, is nonzero?

A test of H_0: $\rho = 0$ for the data is illustrated in the MINITAB printout, Figure 10.16. Both the sample correlation coefficient, r, and the p-value of the test are highlighted on the printout. Since p-value $= 0$ is less than $\alpha = .01$, we reject H_0 and conclude that there is a significant correlation between appraised value and sale price. The fact that $r = .972$ is close to positive 1 implies that the correlation between the two variables is strong and positive. However, this result does not guarantee that appraised value will be an excellent predictor of sale price. In the next Statistics in the Real World Revisited section (p. 578) we use a simple linear regression model to assess the predictive ability of appraised value.

```
Correlation of salepric and totalval = 0.972, P-Value = 0.000
```

M **Figure 10.16** MINITAB Test of Zero Correlation for Sales-Appraisal Data

PROBLEMS FOR SECTION 10.7

Using the Tools

10.36 What value does r assume if all the sample points fall on the same straight line and if the line has:

a. A positive slope? **b.** A negative slope?

10.37 For each of the following situations, specify the rejection region and state your conclusion for testing the null hypothesis H_0: There is no linear correlation between x and y:

a. H_a: The variables x and y are positively correlated; $r = .68$, $n = 10$, $\alpha = .01$

b. H_a: The variables x and y are negatively correlated; $r = -.68$, $n = 52$, $\alpha = .01$

c. H_a: The variables x and y are linearly correlated; $r = -.84$, $n = 10$, $\alpha = .10$

Applying the Concepts

10.38 Can teaching effectiveness be gauged from very brief observations of behavior? In an attempt to answer this question, Harvard psychologists N. Ambady and R. Rosenthal collected data on a random sample of 13 graduate student teachers (*Chance,* Fall 1997). Two variables were measured for each: (1) the teacher's end-of-semester average student evaluation rating (on a scale of 1 to 5); and (2) the average nonverbal behavior rating of the teacher (on a scale of 1 to 9) as judged by nine undergraduates who viewed short video clips of each teacher's behavior in the classroom.

a. The coefficient of correlation relating the two variables is $r = .76$. Interpret this result.

b. Use the result, part **a**, to calculate the test statistic for testing for a positive correlation between teacher effectiveness and nonverbal behavior.

c. Is there evidence (at $\alpha = .05$) of a positive linear relationship between teacher effectiveness and nonverbal behavior?

Object	Correlation r
Piano	.447
Bench	−.057
Motorbike	.619
Armchair	.294
Teapot	.949

Problem 10.39

10.39 Refer to the *Perception & Psychophysics* (July 1998) study of how people view three-dimensional objects projected onto a rotating two-dimensional image, Problem 3.79 (p. 177). Recall that each in a sample of 25 university students viewed a depth-rotated object (e.g., a piano) until they recognized the object. The recognition exposure time y (in milliseconds) and the "goodness of view" x of the object (measured on a numerical scale) were recorded for each student. The experiment was repeated for several different rotated objects. The table at left gives the correlation coefficient, r, between recognition exposure time and goodness of view for each of five objects. For each object, conduct a test to determine if x and y are significantly correlated. Conduct each test at $\alpha = .05$.

10.40 The booming economy of the 1990s created many new billionaires. The 1999 *Forbes 400* ranks the 400 wealthiest people in the United States. The top 15 billionaires on this list are described in the following table.

FORBES400

Name	Age	Net Worth ($ millions)
Gates, William H. III	43	85,000
Allen, Paul Gardner	40	40,000
Buffett, Warren Edward	69	31,000
Ballmer, Steven Anthony	43	23,000
Dell, Michael	34	20,000
Walton, Jim C.	51	17,300
Walton, Helen R.	80	17,000
Walton, Alice L.	50	16,900
Walton, John T.	53	16,800
Walton, S. Robson	55	16,600
Moore, Gordon Earl	70	15,000
Ellison, Lawrence Joseph	55	13,000
Anschutz, Philip F.	59	11,000
Kluge, John Werner	85	11,000
Anthony, Barbara Cox	76	9,700

Source: Forbes, Oct. 11, 1999, p. 414.

a. Construct a scatterplot for these data. What does the plot suggest about the relationship between the age and net worth of billionaires?

b. Find the coefficient of correlation and explain what it tells you about the relationship between age and net worth.

c. Conduct a test of hypothesis to determine if age and net worth are correlated. Test using $\alpha = .10$.

10.41 Researchers at Boston College investigated the tactics used by sociology professors to learn students' names (*Teaching Sociology,* July 1995). Questionnaire data were obtained from a sample of 138 sociology professors. Two of the many variables measured were (1) the number of years the professor had been employed as a full-time faculty member and (2) the total number of name-learning tactics used by the professor. [*Note:* 18 tactics were listed on the questionnaire (e.g., name tags, seating chart) and the respondent circled those he or she used in the past two years.]

a. Set up the null and alternative hypotheses for testing whether years of full-time teaching are correlated with total number of name-learning tactics used.

b. The researchers reported that the test, part **a**, was nonsignificant at $\alpha = .10$, but did not report the value of the correlation coefficient, r. Show that r, for this problem, was less than .14.

10.42 Passive exposure to environmental tobacco has been associated with growth suppression and an increased frequency of respiratory tract infections in normal children. Is this association more pronounced in children with cystic fibrosis? To answer this question, 43 children (18 girls and 25 boys) attending a 2-week summer camp for cystic fibrosis patients were studied (*New England Journal of Medicine,* Sept. 20, 1990). Among several variables measured were the child's weight percentile y and the number of cigarettes smoked per day in the child's home x.

a. For the 18 girls, the coefficient of correlation between y and x was reported as $r = -.50$. Interpret this result.

b. Refer to part **a.** The p-value for testing H_0: $\rho = 0$ against H_a: $\rho \neq 0$ was reported as $p = .03$. Interpret this result.

c. For the 25 boys, the coefficient of correlation between y and x was reported as $r = -.12$. Interpret this result.

d. Refer to part **c.** The p-value for testing H_0: $\rho = 0$ against H_a: $\rho \neq 0$ was reported as $p = .57$. Interpret this result.

10.43 Refer to the *Journal of Literacy Research* (Dec. 1996) study of adult literacy, Problem 8.83 (p. 441). Total literacy for each of the over 500 adults surveyed was measured by summing the scores of four separate literacy tests—Author Recognition, Magazine Recognition, Cultural Literacy, and Vocabulary Literacy Tests—administered over the telephone. Several other variables were measured in addition to total literacy, including years of education.

a. For the fathers who participated in the study, the correlation between total literacy and years of education was $r = .10$. A test for positive correlation resulted in a significance level of $p = .025$. Interpret the results.

b. For the mothers who participated in the study, the correlation between total literacy and years of education was found to be nonsignificant at $\alpha = .10$. Interpret this result.

c. Refer to part **b**. For mothers, the correlation between years of education and score on the Magazine Recognition test was $r = .18$ (with p-value $= .002$). Interpret this result.

10.44 A Global Positioning System (GPS) is a satellite system used to locate wetland boundaries for regulatory purposes. Although the GPS can locate boundaries much faster than a traditional land survey, its accuracy varies with the number of satellites used in the system. Correlation analysis and simple linear regression was used to examine the relationship between the accuracy of a GPS and the number of satellites used (*Wetlands,* Mar. 1995). Multiple GPS readings were collected for each of sixteen points located on a wetland boundary in South Carolina. The dependent variable is the distance y between the average GPS reading and the actual location (in meters). The independent variable is the percentage x of readings obtained when three satellites were in use. The data and output of the correlation analysis and simple linear regression, generated by Excel, are shown on p. 570.

Survey Location	Distance Error y	Percent of Readings with Three Satellites x
1	7.8	94
2	4.2	83
3	4.5	23
4	2.8	58
5	1.5	78
6	3.0	100
7	1.4	100
8	14.9	100
9	0.4	28
10	18.9	100
11	5.8	75
12	0.6	2
13	30.9	100
14	30.3	100
15	31.4	100
16	7.5	53

Source: Hook, D. D., et al. "Locating delineated wetland boundaries in coastal South Carolina using Global Positioning Systems." *Wetlands,* Vol. 15, No. 1, Mar. 1995, p. 34 (Table 2).

Problem 10.44

	A	B
1	t test for the Existence of Correlation	
2		
3	Level of Significance	0.1
4	Sample Size	16
5	Sample Correlation Coefficier	0.544849
6	Degrees of Freedom	14
7	t Test Statistic	2.431193
8		
9	Two-Tailed Test	
10	Lower Critical value	-1.76131
11	Upper Critical Value	1.761309
12	p-Value	0.029076
13	Reject the null hypothesis	

E Problem 10.44

	A	B	C	D	E	F	G
1	SUMMARY OUTPUT						
2							
3	*Regression Statistics*						
4	Multiple R	0.544848659					
5	R Square	0.296860061					
6	Adjusted R Square	0.24663578					
7	Standard Error	9.840587094					
8	Observations	16					
9							
10	ANOVA						
11		*df*	*SS*	*MS*	*F*	*Significance F*	
12	Regression	1	572.3742141	572.3742	5.910688	0.029076377	
13	Residual	14	1355.720161	96.83715			
14	Total	15	1928.094375				
15							
16		*Coefficients*	*Standard Error*	*t Stat*	*P-value*	*Lower 95%*	*Upper 95%*
17	Intercept	-3.82499881	6.335358253	-0.60375	0.555664	-17.41300295	9.7630053
18	Percent of Readings	0.190200989	0.078233678	2.431191	0.029076	0.022406288	0.3579957

E Problem 10.44

a. Locate the coefficient of correlation r on the printout and interpret it.

b. Locate the p-value for a test of correlation between distance error y and percent x of readings with three satellites. Is there evidence of a correlation between y and x in the population at $\alpha = .10$?

c. Locate the p-value for a test of H_0: $\beta_1 = 0$ against H_a: $\beta_1 \neq 0$. Conduct the test at $\alpha = .10$.

d. Do the results, parts **b** and **c**, agree?

10.45 In cotherapy two or more therapists lead a group. An article in the *American Journal of Dance Therapy* (Spring/Summer 1995) examined the use of cotherapy in dance/movement therapy. Two of several variables measured on each of a sample of 136 professional dance/movement therapists were years of formal training x and reported success rate y (measured as a percentage) of coleading dance/movement therapy groups.

a. Propose a linear model relating y to x.

b. The researcher hypothesized that dance/movement therapists with more years in formal dance training will report higher perceived success rates in cotherapy relationships. State the hypothesis in terms of the parameter of the model, part **a**.

c. The correlation coefficient for the sample data was reported as $r = -.26$. Interpret this result.

d. Does the value of r in part **c** support the hypothesis in part **b**? Test using $\alpha = .05$.

10.8 The Coefficient of Determination

In this section, we define a numerical descriptive measure of how well the least squares line fits the sample data. This measure, called the *coefficient of determination,* is very useful for assessing how much the errors of prediction of y can be reduced by using the information provided by x. We develop this descriptive measure of model adequacy in the next three examples.

EXAMPLE 10.16

BEST PREDICTOR OF *y* WITHOUT *x*

Refer to the creativity–flexibility data, Table 10.2 (p. 543). Suppose you do not use x, the flexibility score, to predict y, the creativity score.

a. If you have access to the sample of 10 mentally retarded children's creativity scores only, what measure would you use as the best predictor for any y value?

b. Use this predictor to compute the sum of squared errors.

Solution

a. If we have no information on the distribution of the y values other than that provided by the sample, then $\bar{y}$, the sample average creativity score, would be the best predictor for *any* y value. From Table 10.2, we find $\bar{y} = 7.4$.

b. The errors of prediction using $\bar{y}$ as the predictor of y for the data of Table 10.2 are illustrated in Figure 10.17 on p. 572. With $\bar{y}$ as our predictor, the sum of squared errors [denoted SS(Total)] is:

$$\text{SS(Total)} = \sum(\text{Actual } y - \text{Predicted } y)^2 = \sum(y - \bar{y})^2$$
$$= (2 - 7.4)^2 + (11 - 7.4)^2 + \cdots + (13 - 7.4)^2 = 130.4$$

The magnitude of SS(Total) is an indicator of how well $\bar{y}$ behaves as a predictor of y. It is also commonly described as the total variation of the sample y values around their mean.

SS(Total) is equivalent to SSyy. See equation 10.10 on p. 555.

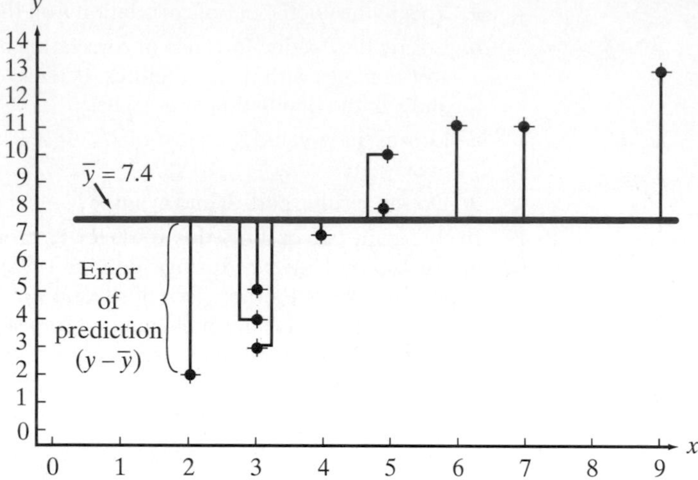

Figure 10.17 Errors of Prediction: Using $\bar{y}$ to Predict Creativity Score y

EXAMPLE 10.17

BEST PREDICTOR OF *y* USING *x*

Refer to Example 10.16. Suppose now that you use a straight-line model with flexibility score x to predict creativity score y. How do we measure the additional information provided by using the value of x in the least squares prediction equation, rather than $\bar{y}$, to predict y?

Solution

If we use the model with x to predict y, then $\hat{y}$ is our predicted value, where $\hat{y} = b_0 + b_1x$. The errors of prediction using $\hat{y}$ for the data of Table 10.2 are illustrated in Figure 10.18. Notice how much smaller these errors are relative to the errors shown in Figure 10.17.

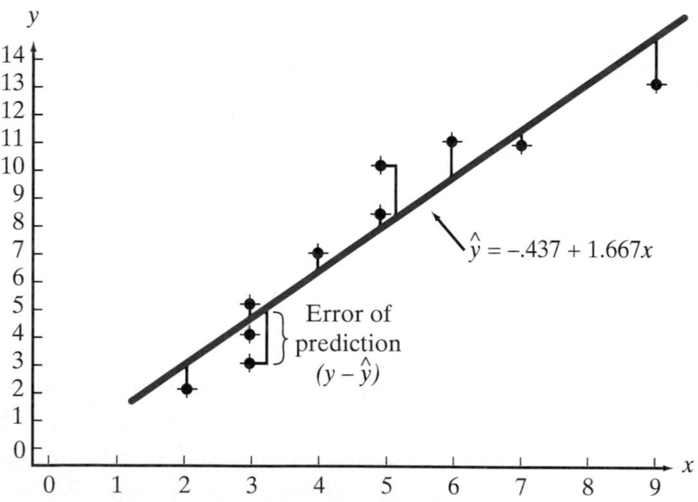

Figure 10.18 Errors of Prediction: Using the Least Squares Model

We already know (from Example 10.10) that the sum of squared errors for the least squares line is

$$SSE = \sum (y - \hat{y})^2 = 13.3444$$

A convenient way of measuring how well the least squares equation performs as a predictor of y is to compare the sum of squared errors using $\bar{y}$ as a predictor [i.e., the total sample variation, SS(Total)] to the sum of squared errors using $\hat{y}$ as a predictor (i.e., SSE). The difference between these, SS(Total) − SSE, represents the sum of squares *explained* by the linear regression between x and y, and is called

SSR. The ratio of SSR to SS(Total) is called the *coefficient of determination,* and is denoted by the symbol R^2.

Definition 10.3

The **coefficient of determination** is

$$R^2 = \frac{\text{SS(Total)} - \text{SSE}}{\text{SS(Total)}} = \frac{\text{SSR}}{\text{SS(Total)}} \qquad \textbf{(10.16)}$$

where

$$\text{SS(Total)} = \sum (y - \bar{y})^2$$
$$\text{SSE} = \sum (y - \hat{y})^2$$
$$\text{SSR} = \text{SS(Total)} - \text{SSE}$$

and $0 \le R^2 \le 1$. R^2 represents the proportion of the sum of squares of deviations of the y values about their mean that can be attributed to a linear relationship between y and x. [*Note:* In simple linear regression, it may also be computed as the square of the coefficient of correlation, i.e., $R^2 = (r)^2$.]

Since r is between -1 and $+1$, the coefficient of determination R^2 will always be between 0 and 1. Thus, $R^2 = .75$ means that 75% of the sum of squares of deviations of the y values about their mean is attributable to the linear relationship between y and x. In other words, the error of prediction can be reduced by 75% when the least squares equation, rather than $\bar{y}$, is used to predict y.

EXAMPLE 10.18

COMPUTING R^2

Find the coefficient of determination for the flexibility–creativity score data in Table 10.2 (p. 543) and interpret its value.

Solution

From Examples 10.16 and 10.17, we have SS(Total) = 130.4 and SSE = 13.3444. Applying equation 10.16, we obtain

$$\frac{\text{SSR}}{\text{SS(Total)}} = \frac{(130.4 - 13.3444)}{130.4} = \frac{117.0556}{130.4} = .8977$$

This value is also indicated on the Excel printout, Figure 10.5 (p. 545), next to the cell labeled **R Square**. We interpret this value as follows: The use of flexibility score x to predict creativity score y with the least squares line

$$\hat{y} = -.44 + 1.667x$$

accounts for 89.8% of the total variation of the 10 sample creativity scores about their mean. That is, we can reduce the sum of squares errors by nearly 90% by using the least squares equation $\hat{y} = -.44 + 1.667x$, instead of $\bar{y}$, to predict y.

PRACTICAL INTERPRETATION OF THE COEFFICIENT OF DETERMINATION R^2

About $100(R^2)$% of the total variation in the sample y values about their mean $\bar{y}$ can be explained by (or attributed to) using x to predict y in the straight-line model.

Self-Test 10.8

The coefficient of determination for a straight-line model relating annual salary y to years of experience x for factory workers is $R^2 = .62$. Interpret this value.

Obviously, the coefficient of correlation r and the coefficient of determination R^2 are very closely related. Hence, there may be some confusion as to when each should be used. Our recommendations are as follows: If you are interested only in measuring the strength of the linear relationship between two variables x and y, use the coefficient of correlation r. However, if you want to determine how well the least squares straight-line model fits the data, use the coefficient of determination R^2.

PROBLEMS FOR SECTION 10.8

Using the Tools

10.46 For a set of $n = 20$ data points, SS(Total) $= 210$ and SSE $= 31$. Find R^2.

10.47 For a set of $n = 30$ data points, $\Sigma(y - \bar{y})^2 = 12$ and $\Sigma(y - \hat{y})^2 = 2.74$. Find R^2.

10.48 The coefficient of correlation between x and y is $r = .72$. Calculate the coefficient of determination R^2.

10.49 The coefficient of correlation between x and y is $r = -.35$. Calculate the coefficient of determination R^2.

Applying the Concepts

DRILLROCK

10.50 Refer to the simple linear regression relating drill time y and depth at which drilling begins x, Problem 10.8 (p. 548). Find and interpret R^2.

JAPANSPORTS

10.51 Refer to the simple linear regression relating frequency of sports participation x and duration of activity y, Problem 10.13 (p. 551). Find and interpret the coefficient of determination R^2.

PHYSICS

10.52 Refer to the simple linear regression relating Force Concept Inventory score y to Mechanics Baseline score x, Problem 10.14 (p. 551). Find and interpret R^2.

ORING

10.53 Refer to the simple linear regression relating O-ring damage y to flight time temperature x, Problem 10.28 (p. 561). Find and interpret R^2.

FERTIL

10.54 Refer to the simple linear regression relating fertility rate y to contraceptive prevalence x, Problem 10.33 (p. 562). Find and interpret R^2.

10.55 In Problem 10.38 (p. 568), the coefficient of correlation between teacher effectiveness and nonverbal behavior rating was $r = .76$. Find the coefficient of determination R^2 for a straight-line model relating teacher effectiveness y to nonverbal behavior rating x. Interpret the result.

10.56 In Problem 10.41 (p. 569), assume the coefficient of correlation between total number of name-learning techniques and years of full-time teaching was $r = .12$. Find the coefficient of determination R^2 for a straight-line model relating number of name-learning techniques y to years of teaching x. Interpret the result.

10.9 Using the Model for Estimation and Prediction

After we have tested our straight-line model and are satisfied that x is a statistically useful linear predictor of y, we are ready to accomplish our original objective—using the model for prediction and estimation.

The most common uses of a regression model for making inferences can be divided into two categories, which are listed in the box.

USES OF THE REGRESSION MODEL FOR MAKING INFERENCES

1. Use the model for estimating μ_y, the mean value of y, for a specific value of x.
2. Use the model for predicting a particular y value for a given value of x.

In the first case, we want to estimate the mean value of y for a very large number of experiments at a given x value. For example, a psychologist may want to estimate the mean creativity score for *all* mentally retarded children with flexibility scores of 3. In the second case, we wish to predict the outcome of a single experiment (predict an individual value of y) at the given x value. For example, the psychologist may want to predict the creativity score of a particular mentally retarded child who scored 3 on the flexibility test.

In Section 10.3, we showed how to use the least squares line

$$\hat{y} = b_0 + b_1x$$

to predict a particular value of y for a given x. The least squares line is also used to obtain μ_y, an estimate of the mean value of y, as demonstrated in the next example.

EXAMPLE 10.19

POINT ESTIMATOR OF μ_Y

Refer to Example 10.6 (p. 544). We found the least squares model relating creativity score y to flexibility score x to be

$$\hat{y} = -.437 + 1.667x$$

Give a point estimate for the mean creativity score of all mentally retarded children who have a flexibility score of 5.

Solution

Just as we used $\hat{y}$ to predict an individual value of y (see Example 10.9, p. 546), we use $\hat{y}$ to estimate μ_y for a particular x value. When $x = 5$, we have

$$\hat{y} = -.437 + 1.667(5) = 7.90$$

Thus, the estimated mean creativity score for all mentally retarded children with flexibility score 5 is 7.90.

Self-Test 10.9

The least squares line relating annual salary y to years of experience x for factory workers is $\hat{y} = 22{,}000 + 1{,}000x$.

a. Find a point estimate of y for a worker who has 10 years of experience.

b. Find a point estimate of μ_y for all workers who have eight years of experience.

Since the least squares model is used to obtain both the estimator of μ_y and the predictor of y, how do the two inferences differ? The difference lies in the accuracies with which the estimate and the prediction are made. These accuracies are best measured by the repeated *sampling errors* of the least squares line when it is used as an estimator and predictor, respectively. These sampling errors are used in estimation and prediction intervals as shown below.

ESTIMATION AND PREDICTION INTERVALS IN SIMPLE LINEAR REGRESSION

$(1 - \alpha)100\%$ Confidence Interval for μ_y at a Fixed x:

$$\hat{y} \pm (t_{\alpha/2})s\sqrt{\frac{1}{n} + \frac{(x - \bar{x})^2}{\Sigma(x - \bar{x})^2}} \quad \textbf{(10.17)}$$

$(1 - \alpha)100\%$ Prediction Interval for an Individual y at a Fixed x:

$$\hat{y} \pm (t_{\alpha/2})s\sqrt{1 + \frac{1}{n} + \frac{(x - \bar{x})^2}{\Sigma(x - \bar{x})^2}} \quad \textbf{(10.18)}$$

EXAMPLE 10.20 **CONFIDENCE INTERVAL FOR μ_Y**

Refer to the simple linear regression relating creativity score y to flexibility score x.

a. Find and interpret a 95% confidence interval for the mean creativity score of all mentally retarded children who have a flexibility score of 5.

b. Using a 95% prediction interval, predict the creativity score of a particular retarded child if his flexibility score is 5. Interpret the interval.

Solution

a. We used PHStat and Excel to obtain the 95% confidence interval. (See Section 10.10 for the calculations required to obtain the interval.) The 95% confidence interval for μ_y when $x = 5$ is indicated on the printout, Figure 10.19. The interval is (6.95, 8.85). Hence, we are 95% confident that the mean

E Figure 10.19 Output from PHStat and Excel 95% Confidence Interval for μ_y

	A	B	
1	Regression Analysis for Creativity Scores		
2			
3	X Value	5	
4	Confidence Level	95%	
5	Sample Size	10	
6	Degrees of Freedom	8	
7	*t* Value	2.306006	
8	Sample Mean	4.7	
9	Sum of Squared Difference	42.10	
10	Standard Error of the Estimate	1.291531	
11	*h* Statistic	0.102138	
12	Average Predicted Y (YHat)	7.900238	
13			
14			
15	For Average Predicted Y (YHat)		
16	Interval Half Width	0.951828	
17	Confidence Interval Lower Limit	6.94841	95% CI for μ_y
18	Confidence Interval Upper Limit	8.852065	
19			
20	For Individual Response Y		
21	Interval Half Width	3.126678	
22	Prediction Interval Lower Limit	4.77356	95% PI for y
23	Prediction Interval Upper Limit	11.02692	

creativity score μ_y for all mentally retarded children with a flexibility score of $x = 5$ ranges from 6.95 to 8.85.

b. The 95% prediction interval is also indicated on the Excel printout, Figure 10.19. (Again, consult Section 10.10 for details on the calculations.) The interval is (4.77, 11.03). Thus, with 95% confidence, we predict that the creativity score for a retarded child with a flexibility score of 5 will fall between 4.77 and 11.03.

In comparing the results in Example 10.20, it is important to note that the prediction interval for an individual creativity score y is wider than the corresponding confidence interval for the mean creativity score μ_y. By examining the formulas for the two intervals, you can see that this will always be true.

Additionally, over the range of the sample data, the widths of both intervals increase as the value of x gets farther from $\bar{x}$. (See Figure 10.20.) Thus, the more x deviates from $\bar{x}$, the less useful the interval will be in practice. In fact, when x is selected far enough away from $\bar{x}$ so that it falls outside the range of the sample data, it is dangerous to make any inferences about μ_y or y, as the following box explains.

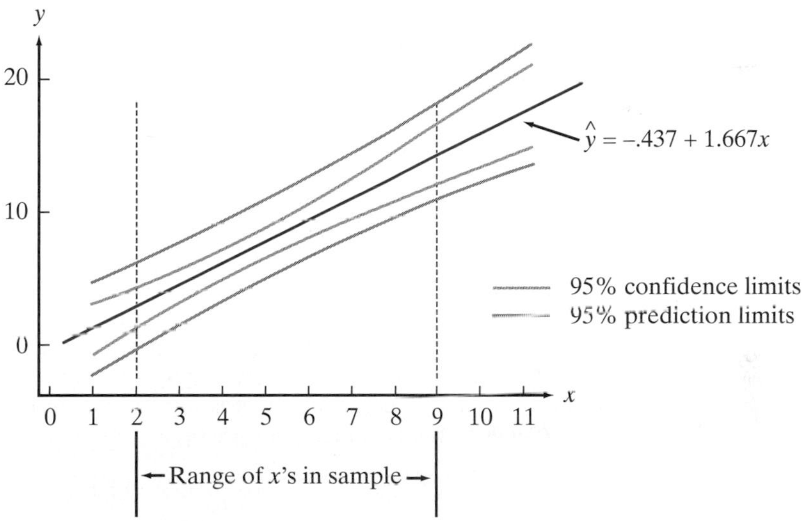

Figure 10.20 Comparison of Widths of 95% Confidence Interval and Prediction Interval

CAUTION

Using the least squares prediction equation to estimate the mean value of y or to predict a particular value of y for values of x that fall *outside the range* of the values of x contained in your sample data may lead to errors of estimation or prediction that are much larger than expected. Although the least squares model may provide a very good fit to the data over the range of x values contained in the sample, *it could give a poor representation of the true model for values of x outside this region.*

Self-Test 10.10

Refer to Self-Test 10.9 (p. 575) and the line relating annual salary y to years of experience x.

a. A 95% confidence interval for μ_y when $x = 8$ is (27,500, 32,500). Interpret this interval.

b. A 95% prediction interval for y when $x = 8$ is (23,000, 37,000). Interpret this interval.

Statistics in the Real World Revisited

Applying Simple Linear Regression to the Sales-Appraisal Data

TAMPALMS

In the two previous Statistics in the Real World Revisited examples in this chapter (pp. 541, 567), we discovered that appraised property value and sale price are positively related for homes sold in the Tampa Palms subdivision. To determine if appraised value (x) will yield accurate predictions of sale price (y), we fit a straight-line model to the data stored in the **TAMPALMS** file using MINITAB's simple linear regression program. The MINITAB output is shown in Figure 10.21.

```
The regression equation is
salepric = 20.9 + 1.07 totalval

Predictor        Coef       StDev         T        P
Constant       20.942       6.446      3.25    0.002
totalval      1.06873     0.02709     39.45    0.000

S = 32.79       R-Sq = 94.5%      R-Sq(adj) = 94.5%

Analysis of Variance

Source            DF          SS          MS         F         P
Regression         1     1673142     1673142   1556.48     0.000
Residual Error    90       96746        1075
Total             91     1769888


Predicted Values

     Fit  StDev Fit          95.0% CI               95.0% PI
  341.56       4.33   (  332.95,   350.17)   (  275.86,   407.26)
```

Figure 10.21 MINITAB Simple Linear Regression Output for Sales-Appraisal Data

Several key statistics are highlighted on the printout, including the least squares line, $\hat{y} = 20.9 + 1.07x$; p-value = 0 for testing H_0: $\beta_1 = 0$; the coefficient of determination, $R^2 = .945$; and, the estimated model standard deviation, $s = 32.79$. A graph of the least squares line is shown in the MINITAB printout, Figure 10.22. You can see that the line appears to provide a good fit to the data.

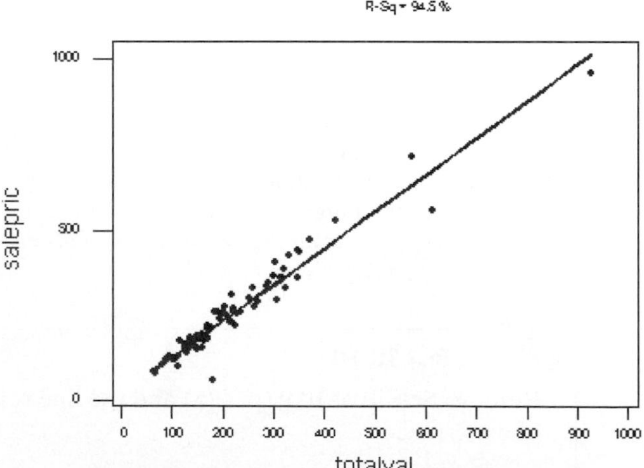

Figure 10.22 MINITAB Plot of Least Squares Regression Line for Sales-Appraisal Model

Our interpretation of the other model statistics follows: Since *p*-value = 0 is less than $\alpha = .05$, the test for zero slope implies that the straight-line model is statistically useful for predicting sale price. From the coefficient of determination, we know that the model can explain 94.5% of the sample variation in sale prices; from the standard deviation, we learn that the model can predict sale price to within about $2s = 2(32.79) = 65.6$ thousand dollars of its true value. The significant slope test, high value of R^2, and relatively low value of *s* all indicate that the model will be useful for predicting sale prices of residential properties in the neighborhood.

The bottom of the MINITAB printout gives both a 95% confidence interval for mean sale price, μ_y, and a 95% prediction interval for sale price, *y*, for properties that have an appraised value of $300,000 (i.e., $x = 300$). The 95% confidence interval (highlighted) is (332.95, 350.17) and the 95% prediction interval (also highlighted) is (275.86, 407.26). We interpret these results as follows:

95% confidence interval for μ_y: With 95% confidence, the mean sale price for all Tampa Palms residential properties with an appraised value of $300,000 will fall between $332,950 and $350,170.

95% prediction interval for y: With 95% confidence, the sale price for a single Tampa Palms residential property with an appraised value of $300,000 will fall between $275,860 and $407,260.

[*Note:* Some real estate analysts could consider the prediction interval to be of limited use in practice due to its relatively large width. If so, these analysts should seek to improve the prediction accuracy of the model by considering additional variables (e.g., home size) as predictors of sale price.]

PROBLEMS FOR SECTION 10.9

Using the Tools

10.57 A simple linear regression analysis based on $n = 20$ data points produced the following results:

$$\hat{y} = 2.1 + 3.4x \qquad s = .475 \qquad \bar{x} = 2.5 \qquad \Sigma(x - \bar{x})^2 = 4.77$$

a. Obtain $\hat{y}$ when $x = 2.5$. What two quantities does $\hat{y}$ estimate?

b. Find a 95% confidence interval for μ_y when $x = 2.5$. Interpret the interval.

c. Find a 95% confidence interval for μ_y when $x = 2.0$. Interpret the interval.

d. Find a 95% confidence interval for μ_y when $x = 3.0$. Interpret the interval.

e. Examine the widths of the confidence intervals obtained in parts **b**, **c**, and **d**. What happens to the width of the confidence interval for μ_y as the value of *x* moves away from the value of $\bar{x}$?

f. Find a 95% prediction interval for a value of *y* to be observed in the future when $x = 2.5$. Interpret the interval.

g. Compare the width of the confidence interval from part **b** with the width of the prediction interval from part **f**. Explain why for a given value of *x*, a prediction interval for *y* is always wider than a confidence interval for μ_y.

10.58 In fitting a least squares line to $n = 22$ data points, the following results were obtained:

$$\hat{y} = 1.4 + .8x \qquad s = .22 \qquad \bar{x} = 2 \qquad \Sigma(x - \bar{x})^2 = 25$$

a. Obtain $\hat{y}$ when $x = 1$. What two quantities does $\hat{y}$ estimate?

b. Find a 95% confidence interval for μ_y when $x = 1$.

c. Find a 95% prediction interval for *y* when $x = 1$.

10.59 A simple linear regression analysis on $n = 85$ data points yielded the following results:

$$\hat{y} = 67.6 - 14.3x \qquad s_{\hat{y}} = 3.5 \qquad s_{(y-\hat{y})} = 4.7$$

a. Estimate the mean value of *y* when $x = 2$, using a 90% confidence interval. Interpret the interval.

	A	B
1	**Confidence Interval Estimate**	
2		
3	**Data**	
4	**X Value**	70
5	**Confidence Level**	95%
6		
7	Intermediate Calculations	
8	Sample Size	10
9	Degrees of Freedom	8
10	t Value	2.306006
11	Sample Mean	67
12	Sum of Squared Difference	1064
13	Standard Error of the Estimate	151.7801
14	h Statistic	0.108459
15	Average Predicted Y (YHat)	793.4305
16		
17	**For Average Predicted Y (YHat)**	
18	Interval Half Width	115.2676
19	**Confidence Interval Lower Limit**	**678.1628**
20	**Confidence Interval Upper Limit**	**908.6981**
21		
22	**For Individual Response Y**	
23	Interval Half Width	368.4979
24	**Predicition Interval Lower Limit**	**424.9326**
25	**Prediction Interval Upper Limit**	**1161.928**

E Problem 10.61

b. Suppose you plan to observe the value of y for a particular experimental unit with $x = 2$. Find a 90% prediction interval for the value of y that you will observe. Interpret the interval.

c. Which of the two intervals constructed in parts **a** and **b** is wider? Why?

10.60 A simple linear regression analysis on $n = 40$ data points yielded the following results:

$$\hat{y} = -1{,}867 + 45x \qquad s_{\hat{y}} = 104 \qquad s_{(y-\hat{y})} = 198$$

a. Estimate the mean value of y when $x = 100$, using a 95% confidence interval. Interpret the interval.

b. Suppose you plan to observe the value of y for a particular experimental unit with $x = 100$. Find a 95% prediction interval for the value of y that you will observe. Interpret the interval.

c. Which of the two intervals constructed in parts **a** and **b** is wider? Why?

Applying the Concepts

10.61 The PHStat and Excel printout for Problem 10.7 (p. 547) is given at left. Locate a 95% prediction interval for the retail price of a large side-by-side refrigerator that has an energy cost of $x = \$70$ per year. Interpret the interval.

DRILLROCK

10.62 Refer to Problem 10.8 (p. 548). Use PHStat and Excel or MINITAB to find a 95% prediction interval for drill time y when drilling begins at a depth of $x = 300$ feet. Interpret your result.

JAPANSPORTS

10.63 Refer to Problem 10.13 (p. 551) and the simple linear regression relating average duration of activity y to frequency of sports participation x.

a. Use PHStat and Excel or MINITAB to find a 95% confidence interval for μ_y when $x = 25$ and interpret it.

b. Use PHStat and Excel or MINITAB to find a 95% prediction interval for y when $x = 25$ and interpret it.

10.64 In forestry, the diameter of a tree at breast height (which is fairly easy to measure) is used to predict the height of the tree (a difficult measurement to obtain). Silviculturalists working in British Columbia's boreal forest attempted to predict the heights of several species of trees. The data in the accompanying table are the breast height diameters (in centimeters) and heights (in meters) for a sample of 36 white spruce trees.

FOREST1

Breast Height Diameter x (cm)	Height y (m)	Breast Height Diameter x (cm)	Height y (m)
18.9	20.0	16.6	18.8
15.5	16.8	15.5	16.9
19.4	20.2	13.7	16.3
20.0	20.0	27.5	21.4
29.8	20.2	20.3	19.2
19.8	18.0	22.9	19.8
20.3	17.8	14.1	18.5
20.0	19.2	10.1	12.1
22.0	22.3	5.8	8.0
23.6	18.9	20.7	17.4
14.8	13.3	17.8	18.4
22.7	20.6	11.4	17.3
18.5	19.0	14.4	16.6
21.5	19.2	13.4	12.9
14.8	16.1	17.8	17.5
17.7	19.9	20.7	19.4
21.0	20.4	13.3	15.5
15.9	17.6	22.9	19.2

Source: Scholz, H., Northern Lights College, British Columbia.

a. Construct a scatterplot for the data. Do you detect a trend?

b. A printout of a simple linear regression analysis of the data using MINITAB is shown below. Find the least squares line.

c. Plot the least squares line on your graph, part **a**.

d. Do the data provide sufficient evidence to indicate that the breast height diameter x contributes information for the prediction of tree height y? Test using $\alpha = .05$.

e. A 95% confidence interval for the average height of white spruce trees with a breast height diameter of 20 cm is shown on the printout. Interpret the interval.

```
The regression equation is
Height = 9.15 + 0.481 Breast Height Diameter

Predictor        Coef        StDev          T        P
Constant        9.147        1.121       8.16    0.000
Breast H      0.48147      0.05967       8.07    0.000

S = 1.678        R-Sq = 65.7%       R-Sq(adj) = 64.7%

Analysis of Variance

Source            DF          SS          MS          F         P
Regression         1      183.24      183.24      65.10     0.000
Residual Error    34       95.70        2.81
Total             35      278.95

Predicted Values

     Fit  StDev Fit          95.0% CI                95.0% PI
  18.776      0.300    ( 18.167,  19.385)    ( 15.313,  22.240)
```

M **Problem 10.64**

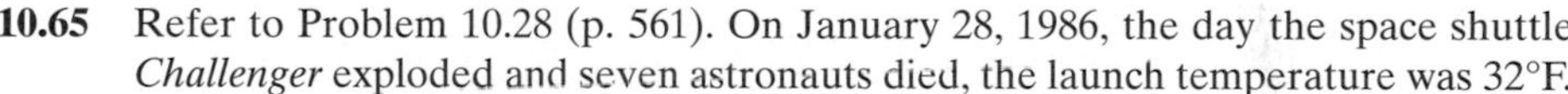

10.65 Refer to Problem 10.28 (p. 561). On January 28, 1986, the day the space shuttle *Challenger* exploded and seven astronauts died, the launch temperature was 32°F.

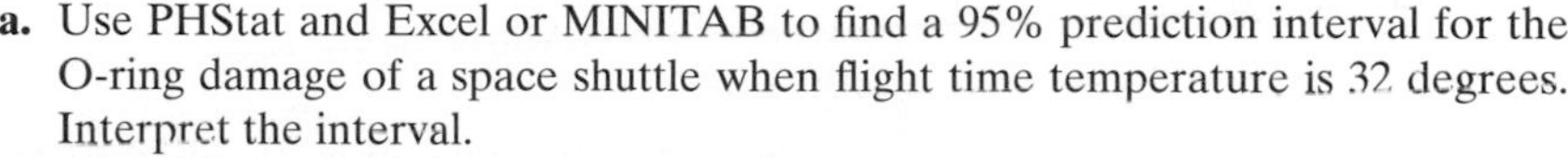

ORING

a. Use PHStat and Excel or MINITAB to find a 95% prediction interval for the O-ring damage of a space shuttle when flight time temperature is 32 degrees. Interpret the interval.

b. Comment on the validity of the prediction interval, part **a**. (Consider the range of temperatures in the sample data used in the simple linear regression analysis.)

10.66 Refer to Problem 10.33 (p. 562). Use PHStat or MINITAB to find a 95% confidence interval for the mean fertility rate of all developing countries with a contraceptive prevalence of 30%. Interpret the interval.

10.67 Refer to Problem 10.44 (p. 569). Use PHStat or MINITAB to find a 95% prediction interval for GPS distance error of a wetland when 100% of the readings are obtained with three satellites. Interpret the interval.

10.10 Computations in Simple Linear Regression (Optional)

In our development of the simple linear (straight-line) model, we have primarily focused on using the output of software such as Microsoft Excel or MINITAB. In this optional section, we demonstrate the computations that were involved in obtaining the numerical results used in a simple linear regression analysis. Although many of the computing formulas were given earlier in the chapter, they are repeated here for convenience.

Consider, again, the simple linear regression analysis of the creativity–flexibility data, Table 10.2 (p. 543). The computations involved in each step of the analysis are illustrated in the examples that follow. Recall that we want to model creativity score y as a straight-line function of flexibility score x. The initial step is to hypothesize the model

$$y = \beta_0 + \beta_1 x + \varepsilon$$

and the second step is to collect the sample data. We begin with the third step in the analysis.

EXAMPLE 10.21

ESTIMATING β_0 AND β_1 (STEP 3)

The formulas for computing estimates of the y-intercept β_0 and the slope β_1 were given in equations 10.4–10.7. These are repeated in the box. Apply these formulas to the creativity–flexibility data of Table 10.2.

FORMULAS FOR THE LEAST SQUARES ESTIMATES

$$\textit{Slope: } b_1 = \frac{SSxy}{SSxx} \quad (10.4)$$

$$\textit{y-Intercept: } b_0 = \bar{y} - b_1\bar{x} \quad (10.5)$$

where

$$SSxy = \sum xy - \frac{(\sum x)(\sum y)}{n} \quad (10.6)$$

$$SSxx = \sum x^2 - \frac{(\sum x)^2}{n} \quad (10.7)$$

Solution

According to the formulas, we need to compute the sum of the x values ($\sum x$), the sum of the y values ($\sum y$), the sums of squares of the x values ($\sum x^2$), and the sums of squares of the cross-products of the x and y values ($\sum xy$). A table of these sums is shown in Table 10.4. (We have also included $\sum y^2$ in the table. This quantity will be used in step 4.)

TABLE 10.4 Sums of Squares for Data from Table 10.2

	x	y	x^2	xy	y^2
	2	2	4	4	4
	6	11	36	66	121
	4	7	16	28	49
	5	10	25	50	100
	3	5	9	15	25
	5	8	25	40	64
	3	4	9	12	16
	3	3	9	9	9
	7	11	49	77	121
	9	13	81	117	169
TOTALS	$\sum x = 47$	$\sum y = 74$	$\sum x^2 = 263$	$\sum xy = 418$	$\sum y^2 = 678$

Substituting the values of Σx, Σx^2, and Σxy into the formulas for SSxy (equation 10.6) and SSxx (equation 10.7), we obtain

$$\text{SSxy} = \sum xy - \frac{(\Sigma x)(\Sigma y)}{n} = 418 - \frac{(47)(74)}{10} = 70.2$$

$$\text{SSxx} = \sum x^2 - \frac{(\Sigma x)^2}{n} = 263 - \frac{(47)^2}{10} = 42.1$$

Now we can compute b_1 using equation 10.4:

$$b_1 = \frac{\text{SSxy}}{\text{SSxx}} = \frac{70.2}{42.1} = 1.667$$

To obtain b_0, we substitute the values of b_1, Σx, and Σy into equation 10.5:

$$b_0 = \bar{y} - b_1\bar{x} = \left(\frac{\Sigma y}{n}\right) - b_1\left(\frac{\Sigma x}{n}\right)$$

$$= \frac{74}{10} - (1.667)\left(\frac{47}{10}\right) = -.437$$

Note that these estimates, $b_0 = -.437$ and $b_1 = 1.667$, agree (except for rounding) with the values shown on the Excel printout, Figure 10.5 (p. 545).

Self-Test 10.11

Consider the data listed in the table.

x	−1	0	1	2	3
y	−1	1	2	4	5

Find the least squares line relating y to x.

EXAMPLE 10.22

COMPUTING SSE AND *s* (STEP 4)

The formulas for computing SSE and an estimate of σ^2 are repeated in the next box. Apply these formulas to the creativity–flexibility data of Table 10.2. Also, compute an estimate of σ.

ESTIMATION OF σ^2, A MEASURE OF THE VARIABILITY OF THE *y* VALUES ABOUT THE LEAST SQUARES LINE

$$s^2 = \frac{\text{SSE}}{n-2} \qquad \textbf{(10.8)}$$

where

$$\text{SSE} = \sum (y - \hat{y})^2 = \text{SSyy} - b_1(\text{SSxy}) \qquad \textbf{(10.9)}$$

Solution

The formula for computing SSE is given in equation 10.9. Note that it requires the values of SSxy and b_1 (calculated in Example 10.21) and SSyy. To obtain SSyy, we use equation 10.10 and the values of Σy^2 and Σy calculated in Table 10.4:

$$\text{SSyy} = \Sigma y^2 - \frac{(\Sigma y)^2}{n} = 678 - \frac{(74)^2}{10} = 130.4$$

Substituting SSyy, SSxy, and b_1 into equation 10.9, we obtain the value of SSE:

$$SSE = SSyy - b_1 SSxy = 130.4 - (1.667)(70.2) = 13.38$$

Except for rounding, this value agrees with the value of SSE shown in the Excel printout, Figure 10.5 (p. 545).

To obtain s^2, the estimate of σ^2, we use equation 10.8:

$$s^2 = \frac{SSE}{n-2} = \frac{13.38}{8} = 1.67$$

Then s, the estimate of σ, is

$$s = \sqrt{s^2} = \sqrt{1.67} = 1.29$$

Again, except for rounding, both $s^2 = 1.67$ and $s = 1.29$ match the values shown in the Excel printout, Figure 10.5 (p. 545).

Self-Test 10.12

Refer to the data, Self-Test 10.11. Compute SSE and s for the simple linear regression.

EXAMPLE 10.23

TESTING THE SLOPE (STEP 5)

Refer to the creativity–flexibility data. Apply equation 10.12 (repeated below) to obtain the test statistic for testing H_0: $\beta_1 = 0$.

COMPUTING THE TEST STATISTIC FOR TESTING H_0: $\beta_1 = 0$ IN SIMPLE LINEAR REGRESSION

$$t = \frac{b_1}{\left(\frac{s}{\sqrt{SSxx}}\right)} \qquad \textbf{(10.12)}$$

Solution

To compute the test statistic t we require the values of b_1, s, and SSxx, computed previously. Substitution into equation 10.12 yields

$$t = \frac{b_1}{\frac{s}{\sqrt{SSxx}}} = \frac{1.667}{\frac{1.29}{\sqrt{42.1}}} = \frac{1.667}{.199} = 8.38$$

This value agrees (except for rounding) with that computed by Excel in Figure 10.5 (p. 545).

Self-Test 10.13

Refer to Self-Tests 10.11 and 10.12. Compute the test statistic for testing H_0: $\beta_1 = 0$.

EXAMPLE 10.24

CONFIDENCE INTERVAL FOR THE SLOPE (STEP 5)

Construct a 95% confidence interval for the slope of the line relating creativity score y to flexibility score x.

Solution

The formula for computing the confidence interval is repeated in the next box.

COMPUTING A CONFIDENCE INTERVAL FOR β_1 IN SIMPLE LINEAR REGRESSION

$$b_1 \pm (t_{\alpha/2})\frac{s}{\sqrt{\text{SSxx}}} \qquad \textbf{(10.13)}$$

where $t_{\alpha/2}$ is based on $(n - 2)$ degrees of freedom.

From previous examples, we found

$$b_1 = 1.667 \qquad s = 1.29 \qquad \text{SSxx} = 42.1$$

For a 95% confidence interval, $\alpha = .05$ and $\alpha/2 = .025$. We need to find the value of $t_{.025}$ in Table B.4 in Appendix B. For $n - 2 = 8$ df, $t_{.025} = 2.306$. Substituting these values into equation 10.13, we obtain

$$b_1 \pm (t_{\alpha/2})\frac{s}{\sqrt{\text{SSxx}}} = 1.667 \pm (2.306)\left(\frac{1.29}{\sqrt{42.1}}\right)$$

$$= 1.667 \pm .459$$

$$= (1.208, 2.126)$$

Self-Test 10.14

Refer to Self-Tests 10.11–10.13. Compute a 95% confidence interval for β_1.

EXAMPLE 10.25

COMPUTING R^2 (STEP 5)

Compute the coefficient of determination R^2 for the straight-line model relating creativity score y to flexibility score x.

Solution

The formula for computing R^2 is repeated in the next box.

COMPUTING R^2 IN SIMPLE LINEAR REGRESSION

$$R^2 = \frac{\text{SSyy} - \text{SSE}}{\text{SSyy}} = \frac{\text{SSR}}{\text{SSyy}} \qquad \textbf{(10.16)}^*$$

where SSyy $= \Sigma(y - \bar{y})^2$

*SSyy = SS(Total)

From previous examples, we found SSE = 13.38 and SSyy = 130.4. Substituting these values into equation 10.16, we obtain

$$R^2 = \frac{130.4 - 13.38}{130.4} = \frac{117.02}{130.4} = .897$$

Self-Test 10.15

Refer to Self-Tests 10.11–10.14. Compute R^2.

Step 6 (check model assumptions) is covered in the next section. We continue with Step 7 in the next example.

EXAMPLE 10.26

ESTIMATION AND PREDICTION COMPUTATIONS (STEP 7)

Refer to the straight-line model relating creativity score y to flexibility score x. Use the data in Table 10.2 (p. 543) to compute

a. a 95% confidence interval for μ_y when $x = 5$

b. a 95% prediction interval for y when $x = 5$

Solution

As we demonstrated in Example 10.19 (p. 575), the point estimate for both y and μ_y is $\hat{y}$, where $\hat{y}$ is obtained by substituting $x = 5$ into the prediction equation:

$$\hat{y} = \mathrm{b}_0 + \mathrm{b}_1 x = -.437 + 1.667(5) = 7.90$$

The formulas for computing the confidence interval for μ_y and the prediction interval for y are repeated in the box below.

COMPUTING A CONFIDENCE INTERVAL FOR μ_y OR A PREDICTION INTERVAL FOR y FOR $x = x_p$

$(1 - \alpha)100\%$ Confidence Interval for μ_y:

$$\hat{y} \pm (t_{\alpha/2})s\sqrt{\frac{1}{n} + \frac{(x_p - \bar{x})^2}{\mathrm{SSxx}}} \quad \textbf{(10.17)}^\dagger$$

$(1 - \alpha)100\%$ Prediction Interval for an Individual y:

$$\hat{y} \pm (t_{\alpha/2})s\sqrt{1 + \frac{1}{n} + \frac{(x_p - \bar{x})^2}{\mathrm{SSxx}}} \quad \textbf{(10.18)}^\dagger$$

where x_p is the value of x used to predict y or estimate μ_y, and $t_{\alpha/2}$ is based on $(n - 2)$ degrees of freedom.

$^\dagger\mathrm{SSxx} = \Sigma(x - \bar{x})^2$

From previous examples, we found $n = 10$, $s = 1.29$, $\bar{x} = 4.7$, SSxx = 42.1, and $t_{\alpha/2} = t_{.025} = 2.306$. The value of x used to predict y, denoted x_p, is $x_p = 5$. Now we substitute these values into the formulas:

a. 95% confidence interval for μ_y:

$$\hat{y} \pm (t_{\alpha/2})s\sqrt{\frac{1}{n} + \frac{(x_p - \bar{x})^2}{\mathrm{SSxx}}}$$

$$= 7.90 \pm (2.306)(1.29)\sqrt{\frac{1}{10} + \frac{(5 - 4.7)^2}{42.1}}$$

$$= 7.90 \pm .951$$

$$= (6.95, 8.85)$$

b. 95% prediction interval for y:

$$\hat{y} \pm (t_{\alpha/2})s\sqrt{1 + \frac{1}{n} + \frac{(x_p - \bar{x})^2}{SSxx}}$$

$$= 7.90 \pm (2.306)(1.29)\sqrt{1 + \frac{1}{10} + \frac{(5 - 4.7)^2}{42.1}}$$

$$= 7.90 \pm 3.12$$

$$= (4.78, 11.02)$$

Except for rounding, these intervals agree with those obtained from the PHStat add-in for Excel in Example 10.20 (p. 576).

Self-Test 10.16

Refer to Self-Tests 10.11–10.15.

a. Find a 95% prediction interval for y when $x = 0$.
b. Find a 95% confidence interval for μ_y when $x = 2$.

PROBLEMS FOR SECTION 10.10

Using the Tools

x	−2	0	1	4	3
y	−1	3	5	9	5

Problem 10.68

10.68 Consider the data listed in the table at left. Compute each of the following.

a. SSxy **b.** SSxx **c.** $\bar{y}$ **d.** $\bar{x}$ **e.** b_1
f. b_0 **g.** SSyy **h.** SSE **i.** s **j.** s_{b_1}
k. t for testing H_0: $\beta_1 = 0$ **l.** a 95% confidence interval for β_1
m. R^2 **n.** $\hat{y}$ when $x = 2$
o. a 95% prediction interval for y when $x = 2$

x	−1	0	1	2	3
y	−1	1	1	2.5	3.5

Problem 10.69

10.69 Consider the five data points shown in the table at left.

a. Construct a scatterplot for the data.
b. Use the computing formulas to find the least squares prediction equation.
c. Graph the least squares line on the scatterplot and visually confirm that it provides a good fit to the data points.
d. Compute the test statistic for testing H_0: $\beta_1 = 0$ against H_a: $\beta_1 > 0$.
e. Give the conclusion for the test of part **d**.
f. Compute and interpret R^2.
g. Compute a 95% confidence interval for μ_y when $x = 0$. Interpret the result.

x	1	1.5	1.9	2.5
y	3.1	2.2	1.0	.3

Problem 10.70

10.70 Consider the four data points shown in the table at left.

a. Construct a scatterplot for the data.
b. Use the computing formulas to find the least squares prediction equation.
c. Graph the least squares line on the scatterplot and visually confirm that it provides a good fit to the data points.

d. Compute the test statistic for testing H_0: $\beta_1 = 0$ against H_a: $\beta_1 < 0$.
e. Give the conclusion for the test of part **d**.
f. Compute and interpret R^2.
g. Compute a 95% confidence interval for μ_y when $x = 1.5$. Interpret the result.

10.71 Consider the seven data points given in the table below.

x	−5	−3	−1	0	1	3	5
y	.8	1.1	2.5	3.1	5.0	4.7	6.2

a. Construct a scatterplot for the data.
b. Using the computing formulas, find the least squares prediction equation.
c. Graph the least squares line on the scatterplot and visually confirm that it provides a good fit to the data points.
d. Compute the test statistic for testing H_0: $\beta_1 = 0$ against H_a: $\beta_1 \neq 0$.
e. Give the conclusion for the test of part **d**.
f. Compute and interpret R^2.
g. Compute a 90% prediction interval for y when $x = -4$. Interpret the result.

10.72 Consider the four data points in the table at left.

x	−3.0	2.4	−1.1	2.0
y	2.7	.4	1.3	.5

Problem 10.72

a. Construct a scatterplot for the data.
b. Using the computing formulas, find the least squares prediction equation.
c. Graph the least squares line on the scatterplot and visually confirm that it provides a good fit to the data points.
d. Compute the test statistic for testing H_0: $\beta_1 = 0$ against H_a: $\beta_1 \neq 0$.
e. Give the conclusion for the test of part **d**.
f. Compute and interpret R^2.
g. Compute a 90% confidence interval for μ_y when $x = 1.0$. Interpret the result.

10.11 Residual Analysis: Checking the Assumptions (Optional)

In Section 10.4 we listed four assumptions about the random error ε in the model required in a simple linear regression analysis. For convenience, they are repeated here.

ASSUMPTIONS REQUIRED IN REGRESSION

1. Mean of ε is 0 (i.e., $\mu_\varepsilon = 0$)
2. Variance of ε is constant for all values of x (i.e., $\text{Var}(\varepsilon) = \sigma^2$)
3. Probability distribution of ε is normal
4. ε's are independent

It is unlikely that these assumptions are ever satisfied exactly in a practical application of regression analysis. Fortunately, experience has shown that least squares regression produces reliable statistical tests, confidence intervals, and prediction intervals as long as departures from the assumptions are not too great. In this section, we present some methods for determining whether the data indicate significant departures from the assumptions.

The methods of this section are based on an analysis of *residuals.*

Definition 10.4

A regression **residual** is defined as the difference between an observed y value and its corresponding predicted value:

$$\text{Residual} = (y - \hat{y}) \tag{10.19}$$

For each assumption, we show how a graph of the residuals can be used to detect violations and we provide some guidance on how to remedy the problem.

Mean Error of Zero

The assumption that the average of the errors is 0 will be violated if an inappropriate regression model is used. For example, if you hypothesize a straight-line relationship between y and x, when the true relationship is more complex (e.g., curvilinear), then you have misspecified the model and this assumption will be violated.

One way to detect model misspecification is to plot the value of each residual versus the corresponding value of the independent variable, x. (If the model contains more than one independent variable, a plot would be constructed for each of the independent variables.) If the plot shows a random pattern of residuals, the assumption is likely to be satisfied. However, if the plot reveals a strong pattern, the assumption is violated. The following example illustrates this method.

EXAMPLE 10.27

DETECTING MODEL MISSPECIFICATION

Fit the straight-line model $y = \beta_0 + \beta_1 x + \varepsilon$ to the data shown in Table 10.5. Then calculate the residuals, plot them versus x, and analyze the plot.

TABLE 10.5 Data for Example 10.27

x	y	x	y
0	1	4	9
1	4	5	10
2	6	6	10
3	8	7	8

Solution

We obtained the least squares equation for the data using Excel. The Excel printout for the simple linear regression is shown in Figure 10.23a. You can see that the resulting prediction equation is

$$\hat{y} = 3.167 + 1.0952x$$

Substituting each value of x into this prediction equation, we can calculate $\hat{y}$ and the corresponding residual $(y - \hat{y})$. The predicted value $\hat{y}$ and the residual $(y - \hat{y})$ are shown in Figure 10.23b (p. 590) for each of the data points.

E **Figure 10.23** Excel Printout for the Simple Linear Model, Example 10.27

	A	B	C	D	E	F	G
1	Regression Analysis						
2							
3	*Regression Statistics*						
4	Multiple R	0.848367781					
5	R Square	0.719727891					
6	Adjusted R Square	0.673015873					
7	Standard Error	1.808270242					
8	Observations	8					
9							
10	ANOVA						
11		*df*	*SS*	*MS*	*F*	*Significance F*	
12	Regression	1	50.38095238	50.38095238	15.40776699	0.007754791	
13	Residual	6	19.61904762	3.26984127			
14	Total	7	70				
15							
16		*Coefficients*	*Standard Error*	*t Stat*	*P-value*	*Lower 95%*	*Upper 95%*
17	Intercept	3.166666667	1.167233422	2.71296778	0.034967891	0.310547284	6.02278605
18	X	1.095238095	0.279022156	3.925272856	0.007754791	0.412494976	1.777981215

a.

22	RESIDUAL OUTPUT		
23			
24	*Observation*	*Predicted Y*	*Residuals*
25	1	3.166666667	-2.166666667
26	2	4.261904762	-0.261904762
27	3	5.357142857	0.642857143
28	4	6.452380952	1.547619048
29	5	7.547619048	1.452380952
30	6	8.642857143	1.357142857
31	7	9.738095238	0.261904762
32	8	10.83333333	-2.833333333

b.

A plot of these residuals versus the independent variable x is shown in Figure 10.24. Instead of varying in a random pattern as x increases, the values of the residuals are curvilinear. This cyclical behavior is caused because we have fit a straight-line model to data for which a more complex relationship is appropriate—that is, we have misspecified the model.

E **Figure 10.24** Excel Residual Plot for Example 10.27

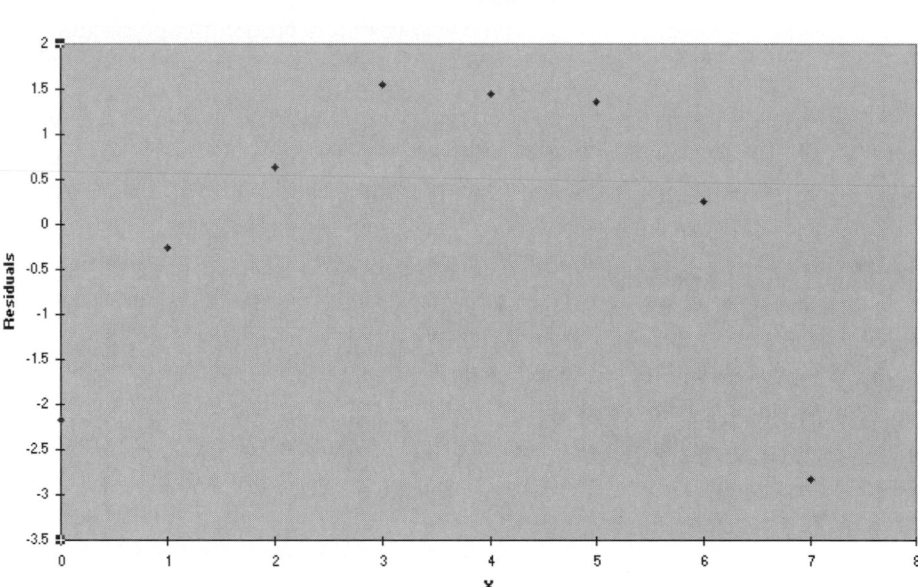

Figure 10.24 shows why fitting the wrong model to a set of data can produce patterns in the residuals when they are plotted versus an independent variable. For this example, the nonrandom (in this case, curvilinear) behavior of the residuals can be eliminated by fitting the curvilinear model

$$y = \beta_0 + \beta_1 x + \beta_2 x^2 + \varepsilon$$

to the data. In general, certain patterns in the values of the residuals may suggest a need to modify the regression model, but the exact change that is needed may not always be obvious. We consider more complex models (such as the curvilinear model above) in optional Section 10.12.

Constant Error Variance

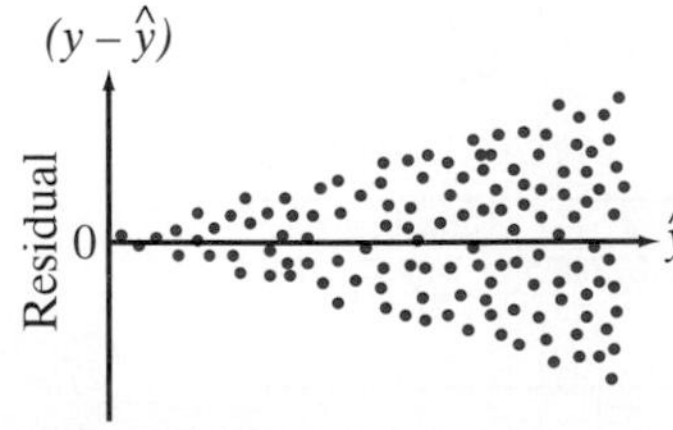

Figure 10.25 Residual Plot Showing Changes in the Variance of ε

To detect a violation of assumption #2, that the variation in the error is constant for all x, we plot the regression residuals against $\hat{y}$, the predicted value of y. For example, a plot of the residuals versus the predicted value $\hat{y}$ may display a pattern as shown in Figure 10.25. In the figure, the range in values of the residuals increases as $\hat{y}$ increases, indicating that the variance of the random error ε becomes larger as $\hat{y}$ increases in value. Since according to the model, increasing x

will increase (or decrease) $\hat{y}$, this implies that the variance of ε is *not* constant for all x—a violation of assumption #2.

Residual plots of the type shown in Figure 10.25 are not uncommon because the variance of y often depends on the mean value of y. Dependent variables that represent counts per unit of area, volume, time, etc. (e.g., Poisson random variables—see Section 5.4) are cases in point. Other types of dependent variables that tend to violate this assumption are business and economic data (e.g., prices, salaries) and proportions or percentages.

EXAMPLE 10.28 DETECTING A NONCONSTANT ERROR VARIANCE

The data in Table 10.6 are the salaries y and years of experience x for a sample of 50 auditors. The straight-line model was fit to the data using Excel. The Excel printout is shown in Figure 10.26, followed by an Excel plot of the residuals versus $\hat{y}$ in Figure 10.27. Interpret the results.

AUDITOR

TABLE 10.6 Salary Data for Example 10.28

Years of Experience x	Salary y	Years of Experience x	Salary y	Years of Experience x	Salary y
7	$26,075	21	$43,628	28	$99,139
28	79,370	4	16,105	23	52,624
23	65,726	24	65,644	17	50,594
18	41,983	20	63,022	25	53,272
19	62,308	20	47,780	26	65,343
15	41,154	15	38,853	19	46,216
24	53,610	25	66,537	16	54,288
13	33,697	25	67,447	3	20,844
2	22,444	28	64,785	12	32,586
8	32,562	26	61,581	23	71,235
20	43,076	27	70,678	20	36,530
21	56,000	20	51,301	19	52,745
18	58,667	18	39,346	27	67,282
7	22,210	1	24,833	25	80,931
2	20,521	26	65,929	12	32,303
18	49,727	20	41,721	11	38,371
11	33,233	26	82,641		

E Figure 10.26 Excel Output for Example 10.28

	A	B	C	D	E	F	G
1	SUMMARY OUTPUT						
2							
3	*Regression Statistics*						
4	Multiple R	0.8870796					
5	R Square	0.78691022					
6	Adjusted R Square	0.78247085					
7	Standard Error	8642.4414					
8	Observations	50					
9							
10	ANOVA						
11		*df*	*SS*	*MS*	*F*	*Significance F*	
12	Regression	1	13239655469	1.32E+10	177.2572	9.86906E-18	
13	Residual	48	3585206077	74691793			
14	Total	49	16824861546				
15							
16		*Coefficients*	*Standard Error*	*t Stat*	*P-value*	*Lower 95%*	*Upper 95%*
17	Intercept	11368.7211	3160.316978	3.597336	0.000758	5014.481697	17722.96057
18	Years	2141.38073	160.8392325	13.3138	9.87E-18	1817.99197	2464.769494

Figure 10.27 Excel Residual Plot for the Data of Example 10.28

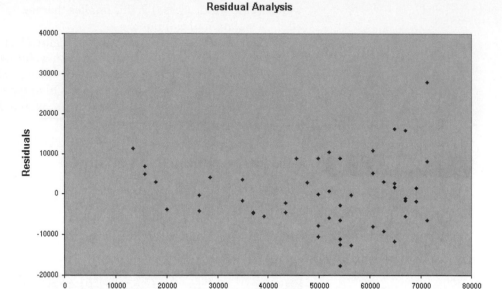

Solution

The Excel printout, Figure 10.26, suggests that the straight-line model provides an adequate fit to the data. The R^2 value indicates that the model explains 78.7% of the sample variation in salaries. The t value for testing β_1, 13.31, is highly significant (p-value ≈ 0) and indicates that the model contributes information for the prediction of y. However, an examination of the residuals plotted against $\hat{y}$ (Figure 10.27) reveals a potential problem. Note the "cone" shape of the residual variability; the size of the residuals increases as the estimated mean salary increases.

This residual plot strongly suggests that a nonconstant error variance exists. Thus, assumption #2 appears to be violated.

Statisticians have found that certain transformations on the dependent variable y can stabilize the error variance. For example, when the data are prices or salaries (as in Example 10.28), the error variance will remain constant if the dependent variable in the model is expressed as the natural logarithm of y [i.e., $\log(y)$]. Similarly, with Poisson data, the error variance can be stabilized by using the square root of y (i.e., $\sqrt{y}$) as the dependent variable. The transformations $\log(y)$ and $\sqrt{y}$ are called **variance-stabilizing transformations** on the dependent variable y because they lead to regression models that satisfy the assumption of a constant error variance.

Normal Errors

Of the four regression assumptions, the assumption that the random error is normally distributed is the least restrictive when we apply regression analysis in practice. That is, moderate departures from the assumption of normality have very little effect on the validity of the statistical tests, confidence intervals, and prediction intervals. In this case we say that regression is **robust** with respect to nonnormality. However, great departures from normality cast doubt on any inferences derived from the regression analysis.

The simplest way to determine whether the data grossly violate the assumption of normality is to construct either a histogram, a stem-and-leaf display, or a normal probability plot of the residuals, as illustrated in the next example.

EXAMPLE 10.29

CHECKING FOR NORMAL ERRORS

Refer to the straight-line model relating salary to experience, fit in Example 10.28. A histogram of the residuals of the model, generated by Excel, is shown in Figure 10.28. Interpret the plot.

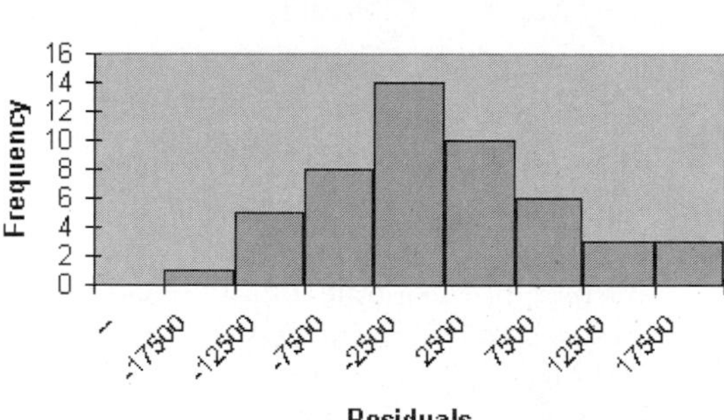

E Figure 10.28 Excel Histogram of the Regression Residuals, Example 10.29

Solution

You can see from Figure 10.28 that the distribution of the residuals is bell-shaped and reasonably symmetric about 0. Consequently, it is unlikely that the normality assumption would be violated using these data.

When gross nonnormality of the random error term is detected, it can often be rectified by applying a transformation on the dependent variable. For example, if the relative frequency distribution of the residuals is highly skewed to the right (as it is for Poisson data), the square-root transformation on y will stabilize (approximately) the variance and, at the same time, will reduce skewness in the distribution of residuals. Nonnormality may also be due to outliers, discussed next.

Outliers

Residual plots can also be used to detect *outliers,* values of y that appear to be in disagreement with the model. Since almost all values of y should lie within 3σ of μ_y, the mean value of y, we would expect most of them to lie within $3s$ of $\hat{y}$. If a residual is larger than $3s$ (in absolute value), we consider it an outlier and seek background information that might explain the reason for its large value.

Definition 10.5

A residual that is larger than $3s$ (in absolute value) is considered to be an **outlier.**

To detect outliers we can construct horizontal lines located a distance of $3s$ above and below 0 (see Figure 10.29 on p. 594) on a residual plot. Any residual falling outside the band formed by these lines would be considered an outlier. We would then initiate an investigation to seek the cause of the departure of such observations from expected behavior.

Although some analysts advocate eliminating outliers, regardless of whether cause can be assigned, others encourage correcting only those outliers that can be traced to specific causes. The best philosophy is probably a compromise between

Figure 10.29 3s Lines Used to Locate Outliers

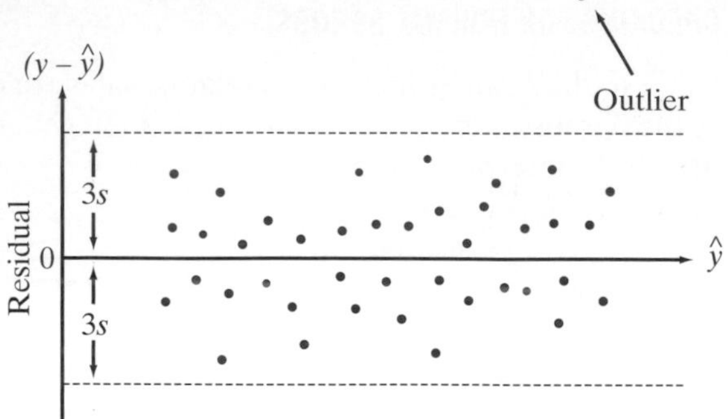

these extremes. For example, before deciding the fate of an outlier, you may want to determine how much influence it has on the regression analysis. When an accurate outlier (i.e., an outlier that is not due to recording or measurement error) is found to have a dramatic effect on the regression analysis, it may be the model and not the outlier that is suspect. Omission of important variables could be the reason why the model is not predicting well for the outlying observation. Several sophisticated numerical techniques are available for identifying outlying influential observations. Consult the references given at the end of this chapter (e.g., Belsley, Kuh, and Welsch [1980], Berenson, Levine, and Krehbiel [2002], Mendenhall and Sincich [1996]) for a discussion of these techniques.

Self-Test 10.17

A simple linear regression analysis yielded $\hat{y} = 7 + 15x$ and $s = 5$.

a. Find the residual for an observation with $y = 14$ and $x = 2$.

b. Is the observation, part **a**, an outlier?

Independent Errors

The assumption that the random errors are independent (uncorrelated) is most often violated when the data employed in a regression analysis are a **time series.** With time series data, the experimental units in the sample are time periods (e.g., years, months, or days) in consecutive time order.

For most business and economic time series, there is a tendency for the regression residuals to have positive and negative runs over time. For example, consider fitting a straight-line regression model to yearly time series data. The model takes the form

$$y = \beta_0 + \beta_1 t + \varepsilon$$

where y is the value of the time series in year t. A plot of the yearly residuals may appear as shown in Figure 10.30. Note that if the residual for year t is positive (or negative), there is a tendency for the residual for year $(t + 1)$ to be positive (or negative). That is, neighboring residuals tend to have the same sign and appear to be correlated. Thus, the assumption of independent errors is likely to be violated, and any inferences derived from the model are suspect.

Consult the references at the end of this chapter (e.g., Neter et al. [1996]) to learn how to fit and apply time series models.

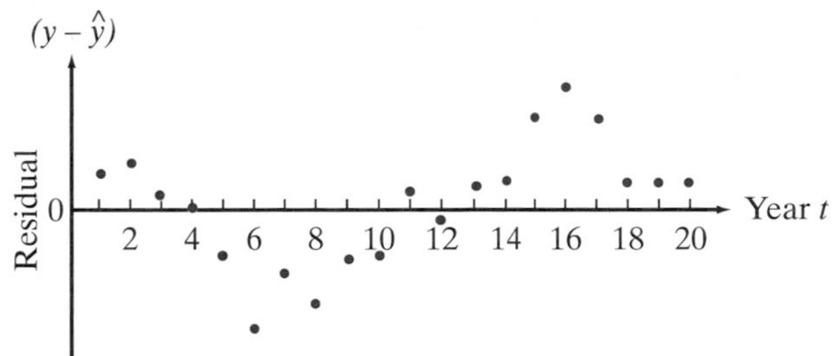

Figure 10.30 Residual Plot for Yearly Time Series Model

A SUMMARY OF STEPS TO FOLLOW IN A RESIDUAL ANALYSIS

1. Check for a *misspecified model* by plotting the residuals $(y - \hat{y})$ against each independent variable in the model. A curvilinear trend detected in a plot implies that a quadratic term for that particular x variable will probably improve model adequacy.
2. Check for *unequal variances* by plotting the residuals against the predicted values $\hat{y}$. If you detect a cone-shape pattern, refit the model using an appropriate variance-stabilizing transformation on y.
3. Check for *nonnormal errors* by constructing a stem-and-leaf display, histogram, or normal probability plot of the residuals. If you detect extreme skewness in the data, look for one or more outliers.
4. Check for *outliers* by locating residuals that lie a distance of $3s$ or more above or below 0 on a residual plot versus $\hat{y}$. Before eliminating an outlier from the analysis, you should conduct an investigation to determine its cause. If the outlier is found to be the result of a coding or recording error, fix it or remove it. Otherwise, you may want to determine how influential the outlier is before deciding its fate.
5. Check for *correlated errors* by plotting the residuals in time order. If you detect runs of positive and negative residuals, propose a time series model to account for the residual correlation.

Statistics in the Real World Revisited

An Analysis of the Residuals from the Sales-Appraisal Regression

Refer to the previous Statistics in the Real World Revisited (p. 578), where we conducted a simple linear regression on the data stored in the **TAMPALMS** file to determine if appraised value (x) will yield accurate predictions of sale price (y). To complete the analysis, we need to examine the residuals from the model to check for violations of the standard regression assumptions and to check for outliers. Figure 10.31 (p. 596) is a MINITAB printout of the residual analysis.

Figure 10.31a is a portion of the MINITAB regression printout (not shown previously) that lists the residuals for observations determined to be "unusual." In addition to the residuals, MINITAB also prints the values of **standardized residuals,** i.e., residual values divided by s, the model standard deviation. Any standardized residual that is larger than 3 in absolute value is considered an outlier (sometimes called a y-outlier). You can see from Figure 10.31a that observation #43 (with a standardized residual of -4.81) and observation #73 (with a standardized residual of -3.88) are both y-outliers. In addition, MINITAB lists observations that have a strong influence on the analysis due to unusual x values. (The statistic calculated to determine x-outliers, called *leverage,* is beyond the scope of this text.) Two more observations, #74 and #82, are x-outliers.

(Continued)

Figure 10.31a
MINITAB Residual Analysis of Sales-Appraisal Model: List of Unusual Observations

```
Unusual Observations
Obs   totalval    salepric          Fit   StDev Fit     Residual    St Resid
 43        182       59.00       215.74        3.46      -156.74      -4.81R
 73        616      560.00       678.83       11.72      -118.83     -3.88RX
 74        573      715.00       632.81       10.61        82.19      2.65RX
 82        929      957.50      1014.21       20.01       -56.71     -2.18RX

R denotes an observation with a large standardized residual
X denotes an observation whose X value gives it large influence.
```

A histogram of the standardized residuals is shown in Figure 10.31b. Although the distribution is bell-shaped, you can see that it is skewed to the left, casting doubt on the validity of the assumption of normal errors. This skewness is due to the two outliers with large negative residuals (y-outliers) identified above. However, since least squares regression is robust to moderate departures from normality, the results of the analysis remain valid.

Figure 10.31b
MINITAB Residual Analysis of Sales-Appraisal Model: Histogram of Standardized Residuals

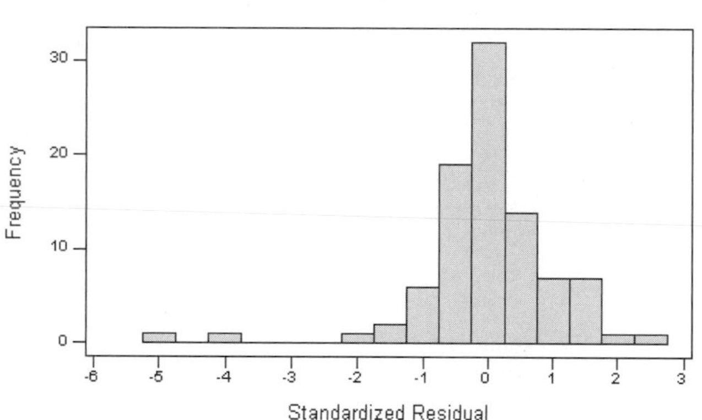

To check the assumption of a constant error variance, the residuals of the straight-line model are plotted against the predicted sale price in Figure 10.31c. The funnel pattern indicates a possible violation of this assumption. However, the pattern seems to be due to the four data points highlighted on the plot. These points correspond to the four observations identified as either y-outliers or x-outliers by MINITAB in Figure 10.31a. When these observations are removed from the simple linear regression analysis, the funnel shaped pattern disappears.

Figure 10.31c
MINITAB Residual Analysis of Sales-Appraisal Model: Plot of Residuals vs. Predicted Values

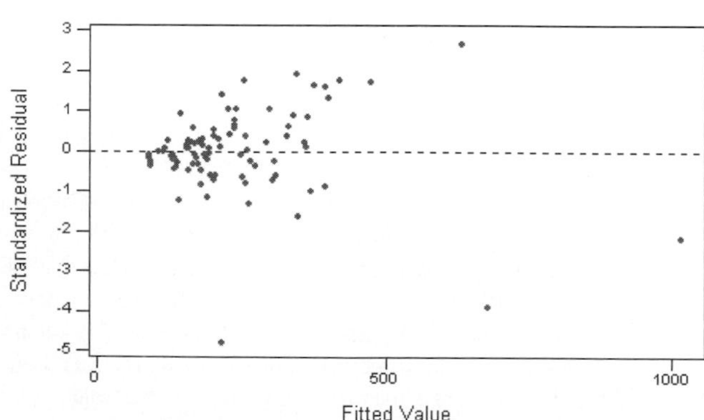

Caution: Removing influential observations (outliers) from a data set and running a regression analysis on the remaining sample observations often leads to dramatic model improvements. However, the outliers could be an indication that the form of the model is incorrect or that other independent variables need to be added to the model.

PROBLEMS FOR SECTION 10.11

Using the Tools

10.73 Identify the problem(s) with each of the following residual plots:

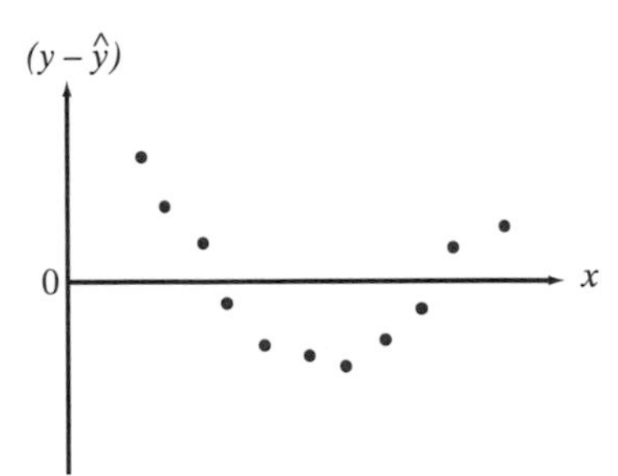

a.

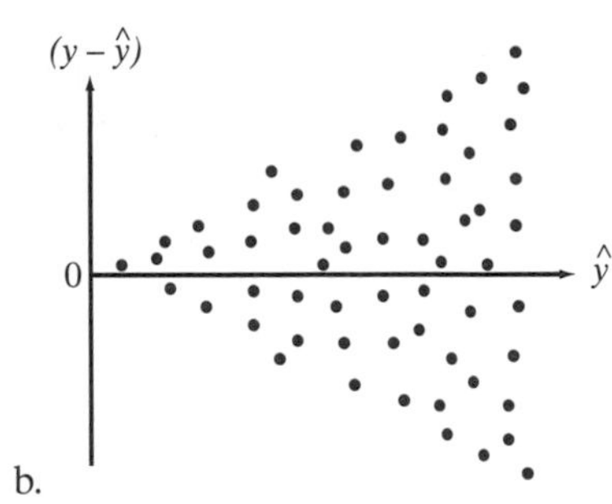

b.

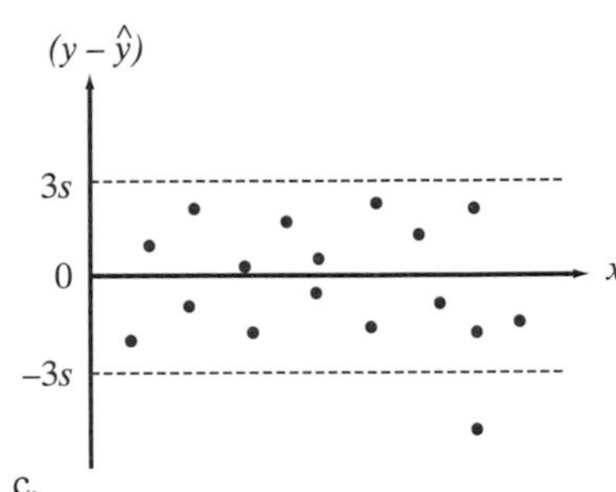

c.

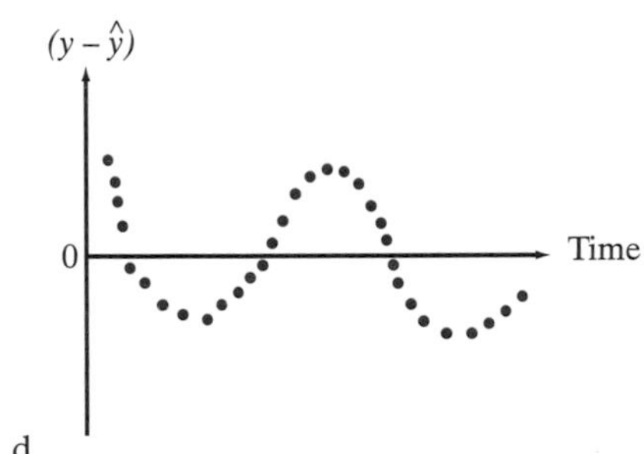

d.

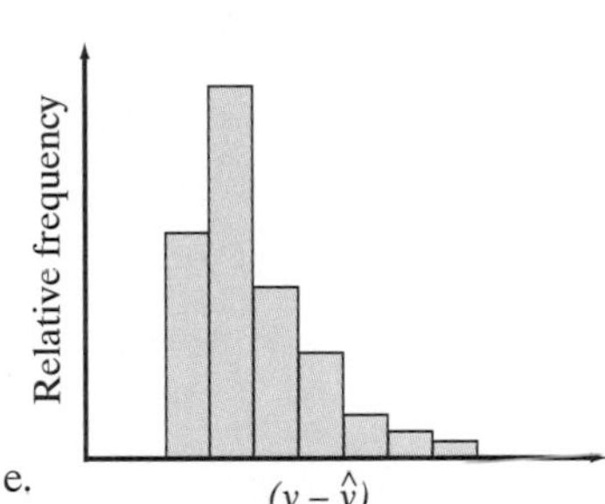

e.

10.74 A straight-line model is fit to the data shown in the table below, with the following results: $\hat{y} = 2.588 + .541x$, $s = .356$.

x	−2	−2	−1	−1	0	0	1	1	2	2	3	3
y	1.1	1.3	2.0	2.1	2.7	2.8	3.4	3.6	4.0	3.9	3.8	3.6

a. Calculate the residuals for the model.
b. Plot the residuals versus x. Do you detect any trends? If so, what does the pattern suggest about the model?
c. Plot the residuals versus $\hat{y}$. Identify any outliers on the plot.
d. Refer to the residual plot constructed in part **c**. Do you detect any trends? If so, what does the pattern suggest about the model?

10.75 A straight-line model is fit to the data shown in the table below, with the following results: $\hat{y} = -3.179 + 2.491x$, $s = 4.154$.

x	2	4	7	10	12	15	18	20	21	25
y	5	10	12	22	25	27	39	50	47	65

a. Calculate the residuals for the model.
b. Plot the residuals versus x. Do you detect any trends? If so, what does the pattern suggest about the model?
c. Plot the residuals versus $\hat{y}$. Identify any outliers on the plot.
d. Refer to the residual plot constructed in part **c**. Do you detect any trends? If so, what does the pattern suggest about the model?

Applying the Concepts

10.76 Refer to the *American Journal of Psychiatry* simple linear regression relating verbal memory retention y to right hippocampal volume of the brain x, Problem 10.10 (p. 549). Several MINITAB residual plots are shown here. Interpret these plots. Do the least squares assumptions appear to be satisfied?

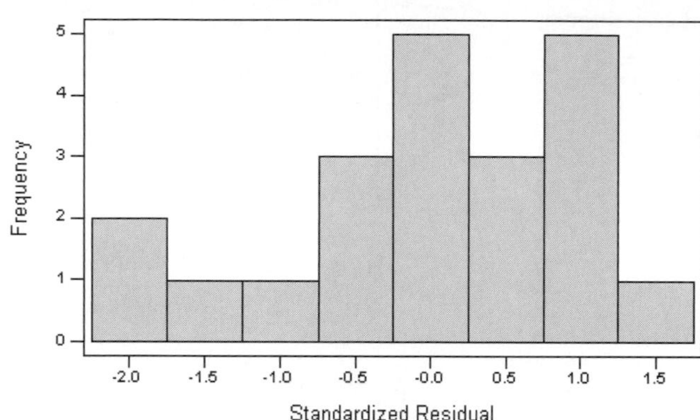

M **Problem 10.76a**

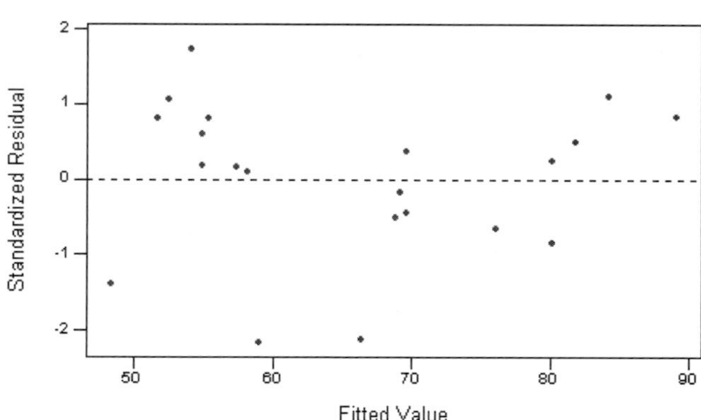

M **Problem 10.76b**

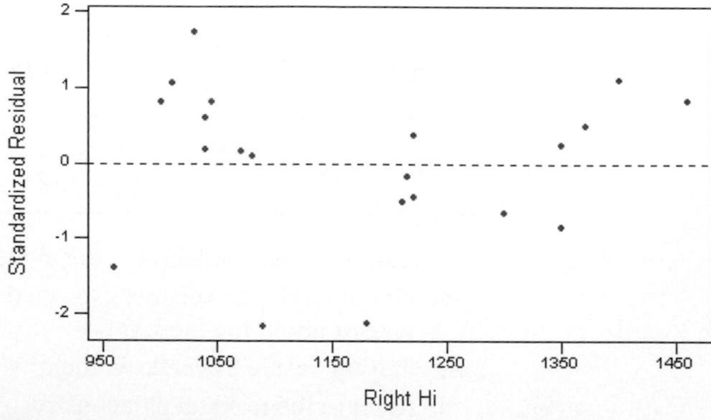

M **Problem 10.76c**

PHYSICS

10.77 Refer to the *American Journal of Physics* simple linear regression relating inventory score y to baseline score x for 27 physics students, Problem 10.14 (p. 551). Use Excel or MINITAB to conduct a residual analysis.

LANDSALE

10.78 Refer to Problem 10.35 (p. 564) and the simple linear regression relating downward price adjustment y and land sale ratio x. The real estate appraisal company has been advised that a curvilinear relationship exists between y and x and, therefore, that the straight-line model is inappropriate for predicting price adjustment. Plot the residuals of the straight-line model against x. Does the plot support this claim?

10.79 Refer to *The New England Journal of Medicine* study of passive exposure to environmental tobacco smoke in children with cystic fibrosis, Problem 10.42 (p. 569). Recall that the researchers investigated the correlation between a child's weight percentile y and the number of cigarettes smoked per day in the child's home x. The accompanying table lists the data for the 25 boys.

SMOKE

Weight Percentile y	Number of Cigarettes Smoked per Day x	Weight Percentile y	Number of Cigarettes Smoked per Day x
6	0	43	0
6	15	49	0
2	40	50	0
8	23	49	22
11	20	46	30
17	7	54	0
24	3	58	0
25	0	62	0
17	25	66	0
25	20	66	23
25	15	83	0
31	23	87	44
35	10		

Source: Rubin, B. K. "Exposure of children with cystic fibrosis to environmental tobacco smoke." *The New England Journal of Medicine,* Sept. 20, 1990, Vol. 323, No. 12, p. 785 (data extracted from Figure 3).

a. Use Excel or MINITAB to fit the straight-line model relating y to x and obtain the residuals.

b. Verify that the sum of the residuals is 0.

c. Plot the residuals against the number of cigarettes smoked per day x.

d. Do you detect any patterns, trends, or unusual observations on the plot, part **c**?

e. Suggest a remedy for any problems identified in part **d**.

10.80 Breakdowns of machines that produce steel cans are very costly. The more breakdowns, the fewer cans produced, and the smaller the company's profits. To help anticipate profit loss, the owners of a can company would like to find a model that will predict the number of breakdowns on the assembly line. One model proposed by the company's statisticians is

$$y = \beta_0 + \beta_1 x + \varepsilon$$

where y is the number of breakdowns per 8-hour shift and x is the number of inexperienced personnel working on the assembly line. After the model is fit using the least squares procedure, the residuals are plotted against $\hat{y}$, as shown in the figure.

a. Do you detect a pattern in the residual plot? What does this suggest about the least squares assumptions?

b. Given the nature of the response variable y and the pattern detected in part **a**, what model adjustments would you recommend?

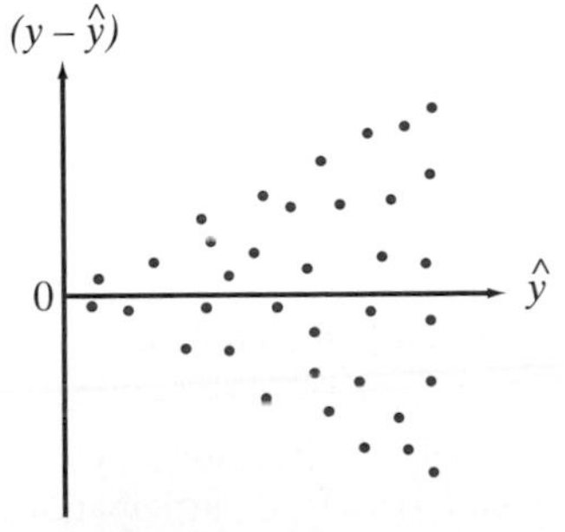

Problem 10.80

10.81 A certain type of rare gem serves as a status symbol for many of its owners. In theory, as the price of the gem increases, the demand will decrease at low prices, level off at moderate prices, and increase at high prices as a result of the status the owners believe they gain by obtaining the gem. Although a nonlinear model would seem to match the theory, the model proposed to explain the demand for the gem by its price is the straight-line model

$$y = \beta_0 + \beta_1 x + \varepsilon$$

where y is the demand (in thousands) and x is the retail price per carat (dollars). Consider the following data.

GEM

x	100	700	450	150	500	800	70	50	300	350	750	700
y	130	150	60	120	50	200	150	160	50	40	180	130

a. Use Excel or MINITAB to fit the model to the 12 data points given in the table. Obtain the residuals.

b. Plot the residuals against retail price per carat x.

c. Can you detect any trends in the residual plot? What does this imply?

10.82 In 1919, the Cincinnati Reds defeated the heavily favored Chicago White Sox in the World Series, five games to three. A year later, eight Chicago players (known as the "Black Sox") were banned from major league baseball for conspiring to lose the 1919 World Series. The data in the table, published in *The American Statistician* (Nov. 1993), are performance statistics for the 16 players who played in every game of the 1919 World Series. Batting player game percentage (PGP) y represents the player's ability to hit in the "clutch" when the outcome of the game is at stake. (The higher the number, the better the player's performance in the clutch.) Slugging average x is total bases accumulated by a player's hits divided by number of at bats. The players marked by an asterisk in the table are "Black Sox" players.

BLACKSOX

Player	Batting PGP y	Slugging Average x
1	−1.6	.05
*2	−3.6	.17
3	−0.8	.20
*4	−1.8	.23
5	−1.4	.24
6	−0.8	.26
*7	−0.5	.30
8	1.2	.30
9	−1.0	.31
10	−0.8	.32
11	−0.2	.35
12	0.4	.36
23	1.0	.38
14	−0.1	.46
*15	−0.3	.50
*16	2.1	.56

Source: Bennett, J. "Did Shoeless Joe Jackson throw the 1919 World Series?" *The American Statistician,* Vol. 47, No. 4, Nov. 1993, p. 246 (adapted from Figure 2).

Consider the straight-line model $y = \beta_0 + \beta_1 x + \varepsilon$.

a. Use Excel or MINITAB to obtain the least squares prediction equation.

b. Use Excel or MINITAB to obtain the regression residuals.

c. Do you detect any outliers in the data? If so, are these outliers Black Sox players? [*Note:* Player #16 in the table is Shoeless Joe Jackson. His alleged participation in the "fix" is the only factor preventing his election into baseball's Hall of Fame.]

10.83 PCBs make up a family of hazardous chemicals that are often dumped, illegally, by industrial plants into the surrounding streams, rivers, or bays. The table reports the concentrations of PCBs (measured in parts per billion) in water samples collected from 37 U.S. bays and estuaries for two different years. An official from the Environmental Protection Agency (EPA) wants to model the PCB concentration y of a bay in year 2 as a function of the PCB concentration x in year 1.

PCBWATER

Bay	State	PCB Concentration	
		Year 1	**Year 2**
Casco Bay	ME	95.28	77.55
Merrimack River	MA	52.97	29.23
Salem Harbor	MA	533.58	403.1
Boston Harbor	MA	17104.86	736
Buzzards Bay	MA	308.46	192.15
Narragansett Bay	RI	159.96	220.6
East Long Island Sound	NY	10	8.62
West Long Island Sound	NY	234.43	174.31
Raritan Bay	NJ	443.89	529.28
Delaware Bay	DE	2.5	130.67
Lower Chesapeake Bay	VA	51	39.74
Pamlico Sound	NC	0	0
Charleston Harbor	SC	9.1	8.43
Sapelo Sound	GA	0	0
St. Johns River	FL	140	120.04
Tampa Bay	FL	0	0
Apalachicola Bay	FL	12	11.93
Mobile Bay	AL	0	0
Round Island	MS	0	0
Mississippi River Delta	LA	34	30.14
Barataria Bay	LA	0	0
San Antonio Bay	TX	0	0
Corpus Christi Bay	TX	0	0
San Diego Harbor	CA	422.1	531.67
San Diego Bay	CA	6.74	9.3
Dana Point	CA	7.06	5.74
Seal Beach	CA	46.71	46.47
San Pedro Canyon	CA	159.56	176.9
Santa Monica Bay	CA	14	13.69
Bodega Bay	CA	4.18	4.89
Coos Bay	OR	3.19	6.6
Columbia River Mouth	OR	8.77	6.73
Nisqually Beach	WA	4.23	4.28
Commencement Bay	WA	20.6	20.5
Elliott Bay	WA	329.97	414.5
Lutak Inlet	AK	5.5	5.8
Nahku Bay	AK	6.6	5.08

a. Use Excel or MINITAB to fit a straight-line model to the data.

b. Is the model adequate for predicting y? Explain.

c. Construct a residual plot for the data. Do you detect any outliers? If so, identify them.

d. Refer to part **c**. Although the residual for Boston Harbor is not, by definition, an outlier, the EPA believes that it strongly influences the regression because of its large y value. Remove the observation for Boston Harbor from the data and refit the model using Excel or MINITAB. Has model adequacy improved?

10.12 Multiple Regression Models (Optional)

Many practical applications of regression analysis require models that are more complex than the simple straight-line model. For example, a realistic probabilistic model for the carbon monoxide rating y of an American-made cigarette would include more variables than the nicotine content x of the cigarette (discussed in Example 10.2). Additional variables such as tar content, length, filter type, and menthol flavor might also be related to carbon monoxide ranking. Thus, we would want to incorporate these and other potentially important independent variables into the model if we needed to make accurate predictions of the carbon monoxide ranking y.

A more complex model relating y to various independent variables, say x_1, x_2, $x_3, \ldots,$ is called a **multiple regression model.** The general form of a multiple regression model is shown in the box.

GENERAL MULTIPLE REGRESSION MODEL

$$y = \beta_0 + \beta_1 x_1 + \beta_2 x_2 + \cdots + \beta_k x_k + \varepsilon$$

where

y is the dependent variable (the variable to be predicted)

$x_1, x_2, \ldots, x_k$ are the independent variables

β_i determines the contribution of the independent variable x_i

ε is the random error component of the model

[*Note:* The symbols $x_1, x_2, \ldots, x_k$ may represent functions of other x's. For example, $x_3 = x_1x_2$ and $x_4 = x_1^2$.]

The steps to follow in a multiple regression analysis are similar to those in a simple linear regression, as outlined in the box.

STEPS TO FOLLOW IN A MULTIPLE REGRESSION ANALYSIS

1. Hypothesize the form of the model.
2. State assumptions about the random error. (These are the same assumptions as those listed in Section 10.4.)
3. Estimate the unknown parameters $\beta_0, \beta_1, \beta_2, \ldots, \beta_k$ using the method of least squares.
4. Check whether the fitted model is useful for predicting y.
5. Check that the assumptions of step 2 are satisfied using a residual analysis. Make modifications to the model, if necessary.
6. If we decide that the model is useful and the assumptions are satisfied, use it to estimate the mean value of y or to predict a particular value of y for given values of the independent variables.

In addition to those outlined in the box, there are other similarities. A t test on an individual β parameter in multiple regression is performed identically to the t test on β_1 in simple linear regression. Interpretations of s, the estimate of σ, and R^2, the coefficient of determination, are also identical. There are some key differences, however, between analyzing a multiple and simple linear regression model. For example, the interpretations of the β estimates in the multiple regression model are not always the same; they will depend on the form of the model.

Also, the methodology for testing the overall adequacy of the multiple regression model differs from that in simple linear regression. We illustrate these concepts in the next several examples.

The Basic Multiple Regression Model

We will start by considering the simplest of all multiple regression models, called a **first-order model.** This model is described in the box for the two independent variables case.

A FIRST-ORDER MULTIPLE REGRESSION MODEL WITH TWO INDEPENDENT VARIABLES

$$y = \beta_0 + \beta_1 x_1 + \beta_2 x_2 + \varepsilon$$

where β_i represents the slope of the line relating y to x_i when the other x is held fixed (i.e., β_i measures the change in y for every 1-unit increase in x_i, holding the other x fixed).

[*Note:* Both independent variables in the model are quantitative.]

EXAMPLE 10.30

BIDRIG

INTERPRETING β ESTIMATES IN THE FIRST-ORDER MODEL

Refer to the Florida Department of Transportation (DOT) study of bid-rigging on road construction contracts in the state (p. 129). Recall that the DOT collected data for 279 contracts. These data are stored in the **BIDRIG** file. Suppose the DOT wants to predict y, the price bid by the lowest bidder (thousands of dollars), using the following two independent variables: x_1 = the DOT engineer's estimate of fair price (thousands of dollars) and x_2 = the number of bidders on the contract. Consider the first-order multiple regression model

$$\mu_y = \beta_0 + \beta_1 x_1 + \beta_2 x_2$$

a. Use the method of least squares to estimate the unknown parameters β_0, β_1, and β_2 in the model.

b. Interpret the estimates of the β parameters.

Solution

a. The model hypothesized in this example is fit to the data in the **BIDRIG** file using MINITAB. The MINITAB printout is reproduced in Figure 10.32.

M **Figure 10.32** MINITAB Output for Low-Bid Price Model, Example 10.30

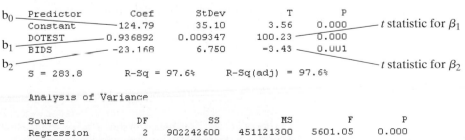

```
The regression equation is
LOWBID = 125 + 0.937 DOTEST - 23.2 BIDS

Predictor        Coef      StDev        T       P
Constant       124.79      35.10     3.56   0.000
DOTEST       0.936892   0.009347   100.23   0.000
BIDS          -23.168      6.750    -3.43   0.001

S = 283.8       R-Sq = 97.6%      R-Sq(adj) = 97.6%

Analysis of Variance

Source            DF         SS          MS         F       P
Regression         2  902242600   451121300   5601.05   0.000
Residual Error   276   22229671       80542
Total            278  924472271
```

The least squares estimates of the parameters appear in the column labeled **Coef.** You can see that $b_0 = 124.79$, $b_1 = .937$, and $b_2 = -23.17$. Therefore, the equation that minimizes SSE for this data set (i.e., the least squares prediction equation) is

$$\hat{y} = 124.79 + .937x_1 - 23.17x_2$$

b. Recall that in the straight-line model, β_0 represents the *y*-intercept of the line and β_1 represents the slope of the line. From our earlier discussion, β_1 has a practical interpretation—it represents the mean change in *y* for every 1-unit increase in *x*. When the independent variables are *quantitative,* the β parameters in the first-order model have similar interpretations. The difference is that when we interpret the β that multiplies one of the variables (e.g., x_1), we must be certain to hold the values of the remaining independent variables (e.g., x_2) fixed. Therefore, β_1 measures the change in *y* for every 1-unit increase in x_1 when the other variable in the model, x_2, is held fixed. A similar statement can be made about β_2: β_2 measures the change in *y* for every 1-unit increase in x_2 when the other variable in the model, x_1, is held fixed. Consequently, we obtain the following interpretations:

$b_1 = .937$: We estimate the low-bid price of a road contract to increase .937 thousand dollars (i.e., \$937) for every 1-unit (i.e., \$1,000) increase in DOT estimated price (x_1) when number of bidders (x_2) is held fixed.

$b_2 = -23.17$: We estimate the low-bid price of a road contract to decrease 23.17 thousand dollars (i.e., \$23,170) for each additional bidder (x_2) on the contract when DOT estimated price (x_1) is held fixed.

The value $b_0 = 124.79$ does not have a meaningful interpretation in this example. To see this, note that $\hat{y} = b_0$ when $x_1 = x_2 = 0$. Thus, $b_0 = 124.79$ represents the predicted low-bid price when the values of both independent variables are set equal to 0. Since a road construction contract with these characteristics—DOT estimated price of \$0 and 0 bidders on the contract—is not practical, the value of b_0 has no meaningful interpretation. In general, b_0 will not have a practical interpretation unless it makes sense to set the values of the *x*'s simultaneously equal to 0.

Self-Test 10.18

The least squares prediction equation for a multiple regression model is:

$$\hat{y} = -2 + 5x_1 + 700x_2$$

a. What is the estimated slope of the line relating *y* to x_1, holding x_2 fixed?

b. What is the estimated increase in *y* for every 1-unit increase in x_2, holding x_1 fixed?

c. Interpret the estimated *y*-intercept. When is the interpretation meaningful?

CAUTION

The interpretation of the β parameters in multiple regression will depend on the terms specified in the model. The interpretations given in Example 10.30 are for a first-order model only. In practice, you should be sure that a first-order model is the correct model for *y* before making these β interpretations.

Test of Overall Model Adequacy

One of the goals in a multiple regression analysis is to conduct a test of the adequacy of the model—that is, test to determine whether the model is really useful for predicting y (step 4). In simple linear regression we performed this step by conducting a t test on the slope parameter β_1. In multiple regression, conducting t tests on each β parameter in a model is generally *not* an appropriate way to determine whether a model is contributing information for the prediction of y. This is due to the fact that a t test for, say, β_1 in a multiple regression model tests the contribution of x_1, holding all the other x's in the model fixed. To test the overall model, we need to conduct a test involving *all* the β parameters (except β_0) simultaneously. The null and alternative hypotheses for this overall test of model utility are given in the box.

HYPOTHESES FOR TESTING WHETHER A MULTIPLE REGRESSION MODEL IS USEFUL FOR PREDICTING *y*

H_0: $\beta_1 = \beta_2 = \cdots = \beta_k = 0$

H_a: At least one of the β parameters in H_0 is nonzero

EXAMPLE 10.31

CONDUCTING THE *F* TEST

Refer to Example 10.30. Test (using $\alpha = .05$) whether the first-order model in the two quantitative independent variables is useful for predicting low-bid price y by testing the null hypothesis

$$H_0: \beta_1 = \beta_2 = 0$$

against the alternative hypothesis

H_a: At least one of the model parameters, β_1 and/or β_2, differs from 0

Solution

The test statistic used in the test for model utility is an F statistic, as described in the next box.

PROCEDURE FOR TESTING WHETHER THE OVERALL MULTIPLE REGRESSION MODEL IS USEFUL FOR PREDICTING *y*

H_0: $\beta_1 = \beta_2 = \cdots = \beta_k = 0$

H_a: At least one of the parameters, $\beta_1, \beta_2, \ldots, \beta_k$, differs from 0

$$\textit{Test statistic: } F = \frac{\text{Mean square for model}}{\text{Mean square for error}} = \frac{\text{SS(Model)}/k}{\text{SSE}/[n - (k + 1)]} \quad \textbf{(10.20)}$$

$$\textit{Rejection region: } F > F_\alpha$$

where

n = number of observations

k = number of parameters in the model (excluding β_0)

and the distribution of F depends on k numerator degrees of freedom and $n - (k + 1)$ denominator degrees of freedom.

The F value for the low-bid price model, indicated on the MINITAB printout shown in Figure 10.32 (p. 603), is $F = 5601.05$. The p-value of the test, also indicated on the printout, is approximately 0. Since the p-value is less than $\alpha = .05$, there is sufficient evidence to reject H_0 and to conclude that at least one of the model coefficients, β_1 and/or β_2, is nonzero. This test result implies that the first-order model is useful for predicting the low-bid price of road construction contracts.

After we have determined that the overall multiple regression model is useful for predicting y using the F test, we may elect to conduct one or more t tests on the individual β parameters. However, when the model includes a large number of independent variables, we should limit the number of t tests conducted to avoid a potential problem of making too many Type I errors. In addition to these tests, we should examine the value of s (the estimate of σ) and R^2. Ideally, we want s to be small and R^2 to be near 1. Once we are satisfied with the results of the F test (and possible t tests) and the values of R^2 and s, we will be comfortable using the model for estimation and prediction (step 6 on p. 602).

GUIDELINES FOR CHECKING THE UTILITY OF A MULTIPLE REGRESSION MODEL

1. Conduct a test of overall model adequacy using the F test, i.e., test

$$H_0: \beta_1 = \beta_2 = \cdots = \beta_k$$

If the model is deemed adequate (i.e., if you reject H_0), then proceed to step 2. Otherwise, you should hypothesize and fit another model. The new model may include more independent variables or more complex terms.

2. Conduct t tests on those β parameters that you are particularly interested in (i.e., the "most important" β's). However, it is a safe practice to limit the number of β's tested. Conducting a series of t tests leads to a high probability of making at least one Type I error.
3. Examine the values of R^2 and s. Models with high R^2 values and low s values are preferred.

EXAMPLE 10.32

TESTING INDIVIDUAL β PARAMETERS

Refer to the first-order model relating low-bid price y to DOT estimated price x_1 and number of bidders x_2, Examples 10.30 and 10.31. Conduct the tests $H_0: \beta_1 = 0$ and $H_0: \beta_2 = 0$, each at $\alpha = .05$. Interpret the results.

Solution

The test statistics and corresponding p-values of the two tests are indicated on the MINITAB printout, Figure 10.32. The results are shown below.

$$H_0: \beta_1 = 0, t = 100.23, p\text{-value} = .000$$

$$H_0: \beta_2 = 0, t = -3.43, p\text{-value} = .001$$

Note that both p-values are less than $\alpha = .05$; hence, we reject H_0 in both cases.

The test for β_1 implies that there is sufficient evidence that the DOT estimated price x_1 contributes to the prediction of y when the number of bidders x_2 is held fixed. Similarly, the test for β_2 implies that number of bidders x_2 contributes to the prediction of y when the DOT estimated price x_1 is held fixed. Both variables appear to be contributing to the overall model's ability to predict low-bid price.

EXAMPLE 10.33

INTERPRETING *s* AND R^2

Refer to the multiple regression model of Examples 10.30–10.32, and the MINITAB printout, Figure 10.32 (p. 603).

a. Locate s, the estimate of σ, on the MINITAB printout and interpret its value.

b. Locate R^2, the coefficient of determination, on the MINITAB printout and interpret its value.

Solution

a. The value of s, shown in the middle of Figure 10.32, is $s = 283.8$. Thus, we expect to predict low-bid price y to within $2s = 2(283.8) \approx 567.6$ thousand dollars of its true value using the first-order multiple regression model.

b. The value of R^2 is also shown in the middle of Figure 10.32. The value $R^2 = .976$ implies that about 97.6% of the sample variation in low-bid price y can be explained by the multiple regression model with the DOT estimate x_1 and the number of bidders x_2 as independent variables.

Adjusted Coefficient of Determination

The value of R^2 in Example 10.33 is high (almost 98%), indicating the model fits the data well. We already know (from the F test in Example 10.31) that the model is statistically useful for predicting low-bid price y. Consequently, one might assume that a model with a high R^2 will always be useful for predicting y. Unfortunately, this is not always the case.

For example, consider a first-order model with 10 parameters (i.e., nine independent variables):

$$y = \beta_0 + \beta_1 x_1 + \beta_2 x_2 + \cdots + \beta_9 x_9 + \varepsilon$$

Now suppose (unknown to us) that none of the nine x's are useful predictors of y. If we fit the model to a data set with $n = 10$ data points (equal to the number of β parameters in the model), the value of R^2 will equal 1. That is, we will obtain a "perfect fit" using a model with independent variables that are unrelated to y! The problem stems from the fact that for a fixed sample size n, R^2 will increase artificially as independent variables (no matter how useful) are added to the model.

CAUTION

In a multiple regression analysis, the value of R^2 will increase as more independent variables are added to the model. Consequently, use the value of R^2 as a measure of model utility only if the sample contains substantially more data points than the number of β parameters in the model.

As an alternative to using R^2 as a measure of model adequacy, the *adjusted coefficient of determination,* denoted R_a^2, is often reported. The formula for R_a^2 is shown in the box.

THE ADJUSTED COEFFICIENT OF DETERMINATION

The **adjusted coefficient of determination** is given by

$$R_a^2 = 1 - \left[\frac{n-1}{n-(k+1)}\right](1 - R^2) \qquad \textbf{(10.21)}$$

Unlike R^2, R_a^2 takes into account ("adjusts" for) both the sample size n and the number of β parameters in the model. R_a^2 will always be less than or equal to R^2, and more important, R_a^2 cannot be "forced" to 1 by simply adding more and more indepen-dent variables to the model. Consequently, analysts prefer the more conservative R_a^2 when choosing a measure of model adequacy in multiple regression.

EXAMPLE 10.34 **INTERPRETING R_a^2**

Refer to Examples 10.30–10.33. Locate the value of R_a^2 on the MINITAB printout, Figure 10.32 (p. 603). Interpret its value.

Solution

The value of R_a^2 is shown on the MINITAB printout next to the value of R^2. Note that $R_a^2 = .976$, a value (after rounding) equal to R^2. Our interpretation is that after adjusting for sample size and the number of parameters in the model, approximately 98% of the sample variation in low-bid price can be "explained" by the first-order model.

As stated in the beginning of this section, multiple regression models can be more complex than the first-order model. These models may include **quadratic** (squared) terms (e.g., x_1^2); *cross-product* terms (e.g., x_1x_2), commonly known as **interaction** terms; and terms for a qualitative independent variable (called **dummy variables**). Some of these other multiple regression models are listed below. Details are beyond the scope of this text. Consult the references (e.g., Mendenhall and Sincich [1996], Berenson, Levine, and Krehbiel [2002], Neter, et. al [1996]) for Chapter 10 to learn more about these models.

QUADRATIC (SECOND-ORDER) MODEL FOR A SINGLE QUANTITATIVE INDEPENDENT VARIABLE

$$y = \beta_0 + \beta_1 x + \beta_2 x^2 + \varepsilon \quad \textbf{(10.22)}$$

DUMMY VARIABLE MODEL FOR A SINGLE QUALITATIVE INDEPENDENT VARIABLE AT TWO LEVELS

$$y = \beta_0 + \beta_1 x + \varepsilon, \text{ where } x = \begin{cases} 1 & \text{if level A} \\ 0 & \text{if level B} \end{cases} \quad \textbf{(10.23)}$$

INTERACTION MODEL FOR TWO QUANTITATIVE INDEPENDENT VARIABLES

$$y = \beta_0 + \beta_1 x_1 + \beta_2 x_2 + \beta_3 x_1 x_2 + \varepsilon \quad \textbf{(10.24)}$$

PROBLEMS FOR SECTION 10.12

Using the Tools

10.84 Give the equation of a first-order model relating y to three quantitative indepen-dent variables.

10.85 Give the equation of a first-order model relating y to five quantitative independent variables.

10.86 The least squares prediction equation for a first-order multiple regression model in two quantitative independent variables is: $\hat{y} = 655 + 47x_1 - 1.6x_2$.

a. What is the estimated slope of the line relating y to x_1, holding x_2 fixed?

b. What is the estimated increase in y for every 1-unit increase in x_2, holding x_1 fixed?

10.87 Suppose y is related to four quantitative independent variables x_1, x_2, x_3, and x_4 by the first-order model

$$y = \beta_0 + \beta_1 x_1 + \beta_2 x_2 + \beta_3 x_3 + \beta_4 x_4 + \varepsilon$$

You fit this model to a set of $n = 15$ data points and find $R^2 = .74$, SS(Total) $= 1.690$, and SSE $= .439$.

a. Calculate s^2, the estimate of the variance of the random error.

b. Calculate the F statistic for testing H_0: $\beta_1 = \beta_2 = \beta_3 = \beta_4 = 0$.

c. Do the data provide sufficient evidence to indicate that the model contributes information for predicting y? Test using $\alpha = .05$.

10.88 Suppose you fit the first-order multiple regression model

$$y = \beta_0 + \beta_1 x_1 + \beta_2 x_2 + \varepsilon$$

to $n = 25$ data points and obtain the prediction equation

$$\hat{y} = 6.4 + 3.1x_1 + .92x_2$$

The estimated standard deviations of the sampling distributions of b_1 and b_2 are 2.3 and .27, respectively.

a. Test H_0: $\beta_1 = 0$ against H_a: $\beta_1 > 0$. Use $\alpha = .05$.

b. Test H_0: $\beta_2 = 0$ against H_a: $\beta_2 \neq 0$. Use $\alpha = .05$.

c. Find a 90% confidence interval for β_1. Interpret the interval.

d. Find a 99% confidence interval for β_2. Interpret the interval.

10.89 Suppose you fit the first-order multiple regression model

$$y = \beta_0 + \beta_1 x_1 + \beta_2 x_2 + \beta_3 x_3 + \varepsilon$$

to $n = 20$ data points and obtain $R^2 = .2623$. Test the null hypothesis H_0: $\beta_1 = \beta_2 = \beta_3 = 0$ against the alternative hypothesis that at least one of the β parameters is nonzero. Use $\alpha = .05$.

Applying the Concepts

10.90 Backpacks are commonly seen in many places, especially college campuses, shopping malls, airplanes, and hiking trails. In the Aug. 1997 issue of *Consumer Reports,* information was provided concerning different features of backpacks—including their price (dollars), volume (cubic inches), and number of 5-inch by 7-inch books that the backpack can hold. The results are listed in the table.

BACKPACKS

Price	Volume	Books	Price	Volume	Books	Price	Volume	Books
48	2,200	59	35	1,950	49	35	1,519	43
45	1,670	49	32	1,385	45	95	1,102	73
50	2,200	48	40	1,700	38	40	1,316	55
42	1,700	52	35	2,000	51	40	1,760	43
29	1,875	52	28	1,500	46	25	1,844	42
50	1,500	49	40	1,950	46	50	2,150	52
48	1,874	50	40	1,810	44	35	1,810	50
38	1,586	47	45	1,910	48	50	2,180	46
33	1,910	53	27	1,875	42	35	1,635	40
40	1,500	49	25	1,450	42	15	1,245	47

Data Source: "Packs for town and country," *Consumer Reports,* August 1997, pp. 20–21. Although these data sets originally appeared in *Consumer Reports,* the selective adaptation and resulting conclusions presented are those of the authors and are not sanctioned or endorsed in any way by Consumers Union, the publisher of *Consumer Reports.*

Suppose that we want to develop a first-order multiple regression model to predict the price of a backpack y based on the volume x_1 and the number of books it can hold x_2.

a. Write the equation of the model.

b. Use Excel or MINITAB to obtain the multiple regression results and then give the least squares prediction equation.

c. Interpret the meaning of the estimated model parameters.

d. Predict the price of a backpack that has a volume of 2,000 cubic inches and can hold 50 5" × 7" books.

e. Determine whether there is a significant relationship between price and the two explanatory variables (volume and number of books) at $\alpha = .05$.

f. Find and interpret s.

g. Find and interpret adjusted R^2.

h. At $\alpha = .05$, determine whether each explanatory variable makes a significant contribution to the regression model.

i. Do the results, parts **d–h**, support the use of this model in predicting backpack price? Explain.

10.91 UCLA political science professor A. F. Simon investigated the predictors of the amount of U.S. media coverage given to foreign earthquakes (*Journal of Communication,* Summer 1997). Three independent variables were used to predict y, the total number of seconds devoted to the earthquake on the national news television broadcasts of ABC, CBS, and NBC:

x_1 = Number of people killed by the earthquake

x_2 = Number of people affected by the earthquake

x_3 = Distance (in thousands of miles) from the earthquake to New York City

The first-order model $y = \beta_0 + \beta_1 x_1 + \beta_2 x_2 + \beta_3 x_3 + \varepsilon$ was fit to data collected for $n = 22$ recent foreign earthquakes.

a. The adjusted R^2 for the model is .4924. Interpret this value.

b. The F value for testing H_0: $\beta_1 = \beta_2 = \beta_3 = 0$ is $F = 7.79$. Use this value to make an inference about the utility of the model at $\alpha = .05$.

c. The t values (and associated p-values) for each independent variable in the model are shown in the table below. Interpret these results.

Independent Variable	*t* Value	*p*-Value
x_1	3.73	.002
x_2	0.53	.603
x_3	−3.00	.008

10.92 Perfectionists are persons who set themselves standards and goals that cannot be reasonably met or accomplished. One theory suggests that those individuals who are depressed have a tendency toward perfectionism. To study this phenomenon, 76 members of an introductory psychology class completed four questionnaires: (1) the ASO scale, designed to measure self-acceptance; (2) the Burns scale, designed to measure perfectionism; (3) the Zung scale, designed to measure depression; and (4) the Rotter scale, designed to measure perceptions between actions and reinforcement (*The Journal of Adlerian Theory, Research, and Practice,* Mar. 1986).

a. Give the equation of a first-order model relating depression (Zung scale) to self-acceptance (ASO scale), perfectionism (Burns scale), and reinforcement (Rotter scale).

b. The model, part **a**, was fit to the $n = 76$ points and resulted in a coefficient of determination of $R^2 = .70$. Interpret this value.

c. The model resulted in $F = 56$. Is there sufficient evidence to indicate that the model is useful for predicting depression (Zung scale) score? Test using $\alpha = .05$.

d. A t test for the perfectionism (Burns scale) variable resulted in a (two-tailed) p-value of .87. Interpret this value.

10.93 Refer to the *Prison Journal* (Sept. 1997) study of parole officer attitudes, Problem 9.16 (p. 465). Multiple regression analysis was used to model the subjective role y of a parole officer (measured on a scale from 7 to 42). Data collected on 72 parole officers were used to fit the model

$$y = \beta_0 + \beta_1 x_1 + \beta_2 x_2 + \beta_3 x_3 + \beta_4 x_4 + \beta_5 x_5 + \varepsilon$$

where

$$x_1 = \begin{cases} 1 & \text{if female} \\ 0 & \text{if male} \end{cases}$$

$$x_2 = \begin{cases} 1 & \text{if intensive supervision officer} \\ 0 & \text{if regular officer} \end{cases}$$

$$x_3 = \begin{cases} 1 & \text{if work in midwestern agency} \\ 0 & \text{if work in northeastern agency} \end{cases}$$

x_4 = length of time (in months) worked as a probation officer

x_5 = age (in years)

The results are displayed in the accompanying table.

Independent Variable	β Estimate	Standard Error	t Value	p-Value
x_1	−.540	1.432	−.377	.7087
x_2	−7.410	1.930	−3.839	.0003
x_3	−1.060	1.434	.739	.4630
x_4	−.0017	.010	−.170	.8639
x_5	−.136	.117	−1.162	.2501

$R^2 = .212 \quad R_a^2 = .144 \quad F = 3.12$ (p-value = .0146)

a. Identify the type (quantitative or qualitative) of the independent variables.

b. At $\alpha = .05$, is the model adequate for predicting subjective role y?

c. Interpret the estimate of β_4.

d. At $\alpha = .05$, is there sufficient evidence that age x_5 is a useful predictor of subjective role y?

e. Interpret R^2 and R_a^2. Why do these values differ?

10.94 A manufacturer of boiler drums wants to use regression to predict the number of man-hours needed to erect the drums in future projects. To accomplish this, data for 35 boilers were collected. In addition to man-hours y, the variables measured were boiler capacity (x_1 = pounds per hour), boiler design pressure (x_2 = pounds per square inch), boiler type (x_3 = 1 if industry field erected, 0 if utility field erected), and drum type (x_4 = 1 if steam, 0 if mud). The data are provided in the table on p. 612 followed by an Excel printout for the model $y = \beta_0 + \beta_1 x_1 + \beta_2 x_2 + \beta_3 x_3 + \beta_4 x_4 + \varepsilon$.

a. Conduct a test of the overall adequacy of the model. Use $\alpha = .05$.

b. Test the hypothesis that boiler capacity x_1 is positively linearly related to man-hours y. Use $\alpha = .05$.

Man-Hours y	Boiler Capacity x_1	Design Pressure x_2	Boiler Type x_3	Drum Type x_4
3,137	120,000	375	Industrial	Steam
3,590	65,000	750	Industrial	Steam
4,526	150,000	500	Industrial	Steam
10,825	1,073,877	2,170	Utility	Steam
4,023	150,000	325	Industrial	Steam
7,606	610,000	1,500	Utility	Steam
3,748	88,200	399	Industrial	Steam
2,972	88,200	399	Industrial	Steam
3,163	88,200	399	Industrial	Steam
4,065	90,000	1,140	Industrial	Steam
2,048	30,000	325	Industrial	Steam
6,500	441,000	410	Industrial	Steam
5,651	441,000	410	Industrial	Steam
6,565	441,000	410	Industrial	Steam
6,387	441,000	410	Industrial	Steam
6,454	627,000	1,525	Utility	Steam
6,928	610,000	1,500	Utility	Steam
4,268	150,000	500	Industrial	Steam
14,791	1,089,490	2,170	Utility	Steam
2,680	125,000	750	Industrial	Steam
2,974	120,000	375	Industrial	Mud
1,965	65,000	750	Industrial	Mud
2,566	150,000	500	Industrial	Mud
1,515	150,000	250	Industrial	Mud
2,000	150,000	500	Industrial	Mud
2,735	150,000	325	Industrial	Mud
3,698	610,000	1,500	Utility	Mud
2,635	90,000	1,140	Industrial	Mud
1,206	30,000	325	Industrial	Mud
3,775	441,000	410	Industrial	Mud
3,120	441,000	410	Industrial	Mud
4,206	441,000	410	Industrial	Mud
4,006	441,000	410	Industrial	Mud
3,728	627,000	1,525	Utility	Mud
3,211	610,000	1,500	Utility	Mud
1,200	30,000	325	Industrial	Mud

Source: Dr. Kelly Uscategui, University of Connecticut.

Problem 10.94

	A	B	C	D	E	F	G
3	*Regression Statistics*						
4	Multiple R	0.950241938					
5	R Square	0.90295974					
6	Adjusted R Square	0.890438416					
7	Standard Error	894.6031853					
8	Observations	36					
9							
10	ANOVA						
11		*df*	*SS*	*MS*	*F*	*Significance F*	
12	Regression	4	230854854.1	57713714	72.11376	2.97665E-15	
13	Residual	31	24809760.63	800314.9			
14	Total	35	255664614.8				
15							
16		*Coefficients*	*Standard Error*	*t Stat*	*P-value*	*Lower 95%*	*Upper 95%*
17	Intercept	-3783.43295	1205.489975	-3.1385	0.003711	-6242.04734	-1324.819
18	Boiler Capacity	0.008749011	0.000903468	9.683809	6.86E-11	0.006906375	0.0105916
19	Design Pressure	1.926477177	0.648906909	2.968804	0.005723	0.603022072	3.2499323
20	Boiler Type	3444.254644	911.7282884	3.77772	0.000675	1584.771503	5303.7378
21	Drum Type	2093.353564	305.6336847	6.849224	1.12E-07	1470.009206	2716.6979

E Problem 10.94

c. Test the hypothesis that boiler pressure x_2 is positively linearly related to man-hours y. Use $\alpha = .05$.

d. Construct a 95% confidence interval for β_3.

e. It can be shown that β_3 represents the difference between the mean number of man-hours required for industrial and utility field erection boilers. Use this information to interpret the confidence interval of part **d**.

f. Construct a 95% confidence interval for β_4 and interpret the result. [*Hint:* $\beta_4 = \mu_{\text{Steam}} - \mu_{\text{Mud}}$, where μ_i represents the mean number of man-hours required for drum type i.]

10.95 Furman University researchers used multiple regression to determine the most important indicators of educational achievement in South Carolina (*Furman Studies,* June 1996). The following multiple regression model was fit to data collected for $n = 91$ South Carolina school districts during the 1992–1993 academic year:

$$y = \beta_0 + \beta_1 x_1 + \beta_2 x_2 + \beta_3 x_3 + \beta_4 x_4 + \beta_5 x_5 + \beta_6 x_6 + \beta_7 x_7 + \beta_8 x_8 + \beta_9 x_9 + \varepsilon$$

where

y = percentage of eleventh grade students who performed above the 50th national percentile on the Stanford-8 Achievement Test

x_1 = average number of years of total education of teachers

x_2 = average number of years of total education of the administrative staff

x_3 = operation expenditures per pupil

x_4 = ratio of students to teaching staff

x_5 = percentage of students classified as nonwhite

x_6 = percentage of population 20 years or over with less than a twelfth-grade education

x_7 = per capita personal income

x_8 = percentage of total school membership enrolled in private schools

x_9 = percentage of population meeting all reading and math standards

a. The analysis yielded $R^2 = .82$. Interpret this result.

b. The global F value for the model is $F = 47.07$ with an associated p-value of $p \approx 0$. Interpret these results.

c. The estimate of β_1 in the model is $b_1 = 1.98$. Interpret this result.

d. The test statistic for testing H_0: $\beta_1 = 0$ is $t = 4.644$ with an associated p-value of $p \approx 0$. Interpret these results.

e. The test statistic for testing H_0: $\beta_8 = 0$ is $t = .88$ with an associated p-value of $p = .381$. Interpret these results.

10.96 Is it possible to predict the number of wins achieved by a Major League Baseball (MLB) team during a season? Crazy Dave has collected various team statistics for the 2000 Major League Baseball season for this purpose. The data is stored in the file **BB2000.** He hypothesizes that total number of wins are related to two quantitative variables: earned run average (ERA) and runs scored.

BB2000

a. Propose a first-order multiple regression model relating number of wins to ERA and runs scored.

b. Use Excel or MINITAB to fit the model to the data. Interpret the meaning of the slopes in the prediction equation.

c. Predict the average number of wins for a team that has an ERA of 4.00 and scored 750 runs.

d. Perform a test of the overall adequacy of the model at $\alpha = .05$.

e. Interpret the meaning of the adjusted coefficient of determination for this model.

f. Perform a residual analysis on your regression results. Are the assumptions reasonably satisfied?

g. Find a 95% confidence interval for the population slope between wins and ERA. Interpret the interval.

10.97 Most educators agree that effective parent involvement in school programs can be facilitated by the actions and attitudes of the school principal. Principals usually employ three strategies when interacting with parents: (1) cooperation (attempting to achieve a common goal), (2) socialization (molding parents' attitudes to those of the school), and (3) formalization (adopting formal measures that weaken parental demands). A study was conducted to investigate the influence of these three strategies on the degree of parent involvement in public elementary schools in Taipei, China (*Proceedings of the National Science Council, Republic of China,* July 1997). Using survey data collected for 172 elementary school principals, the following variables were measured:

$$y = \text{Overall parent involvement score}$$
$$x_1 = \text{Level of principal's cooperation strategy}$$
$$x_2 = \text{Level of principal's socialization strategy}$$
$$x_3 = \text{Level of principal's formalization strategy}$$

a. A first-order model was fit to the data. The resulting F statistic was $F = 11.09$ with a corresponding p-value of $p \approx 0$. Interpret these results.

b. The coefficient of determination for the model, part **a**, was $R^2 = .165$. Interpret this result.

c. The β estimates and corresponding t values (and p-values) are listed in the table. Give a practical interpretation of each β estimate.

Variable	β Estimate	t Value	p-Value
x_1	.341	4.77	.000
x_2	.070	1.00	.319
x_3	.050	1.26	.209

d. Refer to part **c**. Conduct a test to determine whether principal's cooperation x_1 is linearly related to parent involvement y. Use $\alpha = .05$.

e. Refer to part **c**. Conduct a test to determine whether principal's formalization x_3 is linearly related to parent involvement y. Use $\alpha = .05$.

10.98 Because the coefficient of determination R^2 always increases when a new independent variable is added to the model, it may be tempting to include many variables in a model to force R^2 to be near 1. However, doing so reduces the degrees of freedom available for estimating σ^2, which adversely affects our ability to make reliable inferences. As an example, suppose you want to predict the selling price of a used car using 18 independent variables (such as make, model, year, and odometer reading). You fit the model

$$y = \beta_0 + \beta_1 x_1 + \beta_2 x_2 + \cdots + \beta_{17} x_{17} + \beta_{18} x_{18} + \varepsilon$$

where y = selling price and $x_1, x_2, \ldots, x_{18}$ are the predictor variables. Using the relevant information on $n = 20$ used cars to fit the model, you obtain $R^2 = .95$ and $F = 1.056$. Test to determine whether this value of R^2 is large enough for you to infer that this model is useful—i.e., that at least one term in the model is important for predicting used car selling price. Use $\alpha = .05$.

10.99 According to the 1990 census, the number of homeless people in the U.S. is more than a quarter of a million. Yet, little is known about what causes homelessness.

Economists at the City University of New York used multiple regression to assist in determining the factors that cause homelessness in American cities (*American Economic Review,* Mar. 1993). Data on the number y of homeless per 100,000 population in $n = 50$ metropolitan areas were obtained from the Department of Housing and Urban Development. In addition, the 16 independent variables listed in the accompanying table were measured for each city and a multiple regression analysis performed by fitting the first-order model

$$y = \beta_0 + \beta_1 x_1 + \beta_2 x_2 + \cdots + \beta_{16} x_{16} + \varepsilon$$

Independent Variable	β Estimate	t Value
Intercept	307.54	(−)
Rental price (10th percentile)	2.87	3.93
Vacancy rate (10th percentile)	−872.9	−1.58
Rent-control law (yes or no)	−15.50	−.23
Employment growth	−859.09	−1.58
Share of employment in service industries	−347.69	−1.33
Size of low-skill labor market	−1,003.87	−.38
Households (per 100,000) below poverty level	.013	1.22
Public welfare expenditures	.11	.59
AFDC benefits	−.95	−2.58
SSI benefits	1.07	2.14
Percent reduction in AFDC (nonpoor percents)	146.62	1.49
AFDC accuracy rate	98.15	.13
Mental health in-patients (per 100,000)	−.83	−1.50
Fraction of births to teenage mothers	−1,173.00	−1.39
Blacks (per 100,000)	.004	1.78
1984 population (100,000s)	1.22	1.44

Source: Honig, M., and Filer, R. K. "Cause of intercity variation in homelessness." *American Economic Review,* Vol. 83, No. 1, Mar. 1993, p. 251 (Table 2).

a. Interpret the β estimate for the independent variable rental price.

b. Test the hypothesis that the incidence of homelessness decreases as employment growth increases. Use $\alpha = .05$.

c. What is the danger in performing t tests for all 16 independent variables to determine model adequacy?

d. For this model, $R_a^2 = .83$. Interpret this result.

10.13 A Nonparametric Test for Rank Correlation (Optional)

We learned in Sections 10.6 and 10.7 how to use a parametric t test to conduct a test for the slope β_1 of a simple linear regression model or the population coefficient of correlation ρ. (Recall that the tests are equivalent.) These tests are suspect, however, if the assumptions about the random error term ε in the model are violated. In this situation, an alternative procedure is to conduct a nonparametric test for linear correlation ρ in the population.

The nonparametric test is based on **Spearman's rank correlation coefficient,** denoted r_S. The formula for computing r_S and the nonparametric test of hypothesis for rank correlation are shown in the box. Example 10.35 illustrates the procedure.

SPEARMAN'S NONPARAMETRIC TEST FOR RANK CORRELATION

Upper-Tailed Test	Two-Tailed Test	Lower-Tailed Test
H_0: $\rho = 0$	H_0: $\rho = 0$	H_0: $\rho = 0$
H_a: $\rho > 0$	H_a: $\rho \neq 0$	H_a: $\rho < 0$

$$\textit{Test statistic: } r_S = 1 - \frac{6\sum_{i=1}^{n} d_i^2}{n(n^2 - 1)} \qquad \textbf{(10.25)}$$

where d_i is the difference between the y rank and x rank for the ith observation.

Rejection region:	*Rejection region:*	*Rejection region:*
$r_S > r_\alpha$	$\lvert r_S \rvert > r_{\alpha/2}$	$r_S < -r_\alpha$

where the values of r_α and $r_{\alpha/2}$ are given in Table B.9 of Appendix B.

[*Note:* In the case of ties, calculate r_S by substituting the ranks of the y's and the ranks of the x's for the actual y values and x values in the formula for r given in Section 3.9 (p. 171).]

EXAMPLE 10.35

APPLYING SPEARMAN'S TEST FOR RANK CORRELATION

An experiment was designed to learn whether eye pupil size is related to a person's attempt at deception. Fifteen college students were asked to respond verbally to a series of questions. Before questioning began, the pupil size of each student was measured and the students were instructed to answer some of the questions dishonestly. During questioning, the percentage increase in pupil size was recorded. After answering the questions, each student was given a deception score based on the proportion answered dishonestly. (High scores indicate a large number of deceptive responses.) The data are shown in Table 10.7.

DECEPT

TABLE 10.7 Data for Example 10.35

Student	Deception Score y	Percent Increase in Pupil Size x
1	87	10.5
2	63	6.2
3	95	1.1
4	50	7.4
5	43	0.8
6	89	15.2
7	33	4.4
8	55	5.9
9	80	9.3
10	59	7.5
11	45	8.8
12	72	10.0
13	23	0.5
14	91	6.5
15	85	13.4

a. Calculate Spearman's rank correlation coefficient as a measure of the strength of the relationship between deception score and percentage increase in pupil size.

b. Is there sufficient evidence to indicate that percentage increase in pupil size is positively correlated with deception score? Test using $\alpha = .01$.

Solution

a. Spearman's rank correlation coefficient is found by first ranking the values of each variable separately. (Ties are treated by averaging the tied ranks.) Then r_S is computed in exactly the same way as the Pearson correlation coefficient r; the only difference is that the values of x and y that appear in the formula for r are replaced by their ranks. That is, the *ranks* of the raw data are used to compute r_S rather than the raw data themselves. When there are no (or few) ties in the ranks, this formula reduces to the simple expression given in equation 10.25:

$$r_S = 1 - \frac{6\sum_{i=1}^{n} d_i^2}{n(n^2 - 1)}$$

where d_i is the difference between the rank of y and x for the ith observation. The ranks of y and x, the differences between the ranks, and the squared differences for each of the 15 students are shown in Table 10.8. Note that the sum of the squared differences is $\sum d_i^2 = 276$. Substituting this value into equation 10.25, we obtain

$$r_S = 1 - \frac{6\sum_{i=1}^{n} d_i^2}{n(n^2 - 1)} = 1 - \frac{6(276)}{15(224)} = .507$$

This positive value of r_S implies that a moderate positive linear correlation exists between deception score y and percent increase in pupil size x in the sample.

TABLE 10.8 Calculation Table for Example 10.35

Student	Deception Score y	Rank	Percent Increase in Pupil Size x	Rank	d_i	d_i^2
1	87	12	10.5	13	−1	1
2	63	8	6.2	6	2	4
3	95	15	1.1	3	12	144
4	50	5	7.4	8	−3	9
5	43	3	0.8	2	1	1
6	89	13	15.2	15	−2	4
7	33	2	4.4	4	−2	4
8	55	6	5.9	5	1	1
9	80	10	9.3	11	−1	1
10	59	7	7.5	9	−2	4
11	45	4	8.8	10	−6	36
12	72	9	10.0	12	−3	9
13	23	1	0.5	1	0	0
14	91	14	6.5	7	7	49
15	85	11	13.4	14	−3	9
					$\sum d_i^2 = 276$	

b. To determine whether a positive correlation exists in the population, we test

$$H_0: \rho = 0$$

$$H_a: \rho > 0$$

using r_S as a test statistic. As you would expect, we reject H_0 for large values of r_S. Upper-tailed critical values of Spearman's r_S are provided in Table B.9

TABLE 10.9 A Portion of the Spearman's r_S Table, Table B.9

	$\alpha = .10$	$\alpha = .05$	$\alpha = .02$	$\alpha = .01$	← Two-Tailed
n	$\alpha = .05$	$\alpha = .025$	$\alpha = .01$	$\alpha = .005$	← One-Tailed
15	.441	.525	.623	.689	
16	.425	.507	.601	.666	
17	.412	.490	.582	.645	
18	.399	.476	.564	.625	
19	.388	.462	.549	.608	
20	.377	.450	.534	.591	
21	.368	.438	.521	.576	
22	.359	.428	.508	.562	
23	.351	.418	.496	.549	
24	.343	.409	.485	.537	
25	.336	.400	.475	.526	

of Appendix B. This table is partially reproduced in Table 10.9. For $\alpha = .01$ and $n = 15$, the critical value (highlighted in Table 10.9) is $r_{.01} = .623$. Thus, the rejection region for the test is

$$\text{Reject } H_0 \text{ if } r_S > .623$$

Since the test statistic, $r_S = .507$, does not fall in the rejection region, there is insufficient evidence (at $\alpha = .01$) of positive linear correlation between deception score y and percent increase in pupil size x in the population.

Self-Test 10.19

Data for a random sample of $n = 5$ (x, y) values are listed below:

x	.2	.4	.8	1.1	1.5
y	30	25	20	21	13

a. Rank the five x values.
b. Rank the five y values.
c. Compute the differences between the ranks.
d. Find Spearman's rank correlation coefficient r_S.

Spearman's rank correlation coefficient r_S has a sampling distribution that is approximately normal when the sample size n is large—say, $n > 30$. For large samples, the critical values of r_S are obtained using the formulas shown in the box. The test is then conducted in the usual manner.

CRITICAL VALUES OF r_S FOR LARGE SAMPLES

$$\text{One-tailed test: } r_\alpha = \frac{z_\alpha}{\sqrt{n-1}} \quad \textbf{(10.26)}$$

$$\text{Two-tailed test: } r_{\alpha/2} = \frac{z_{\alpha/2}}{\sqrt{n-1}} \quad \textbf{(10.27)}$$

where z_α and $z_{\alpha/2}$ are obtained from the standard normal table, Table B.3 in Appendix B.

EXAMPLE 10.36

SPEARMAN'S LARGE-SAMPLE TEST FOR RANK CORRELATION

For a large sample of $n = 40$ (x, y) pairs, suppose Spearman's rank correlation is calculated to be $r_S = -.38$. Test the null hypothesis $H_0: \rho = 0$ against the alternative hypothesis $H_a: \rho \neq 0$ using $\alpha = .05$.

Solution

The test statistic is $r_S = -.38$. For this two-tailed, large-sample test, we use equation 10.27 to find the critical value. For $\alpha = .05$, we have $\alpha/2 = .025$, and $z_{.025} = 1.96$. Substituting these values into equation 10.27, we obtain

$$r_{\alpha/2} = r_{.025} = \frac{z_{.025}}{\sqrt{n-1}} = \frac{1.96}{\sqrt{40-1}} = .31385$$

Consequently, we reject H_0 if $|r_S| > r_{.025} = .31385$. Since $|r_S| = |-.38| = .38$ exceeds the critical value, we reject H_0 and conclude (at $\alpha = .05$) that the population rank correlation ρ differs from 0.

Nonparametric tests are also available for the general multiple regression model. These tests are very sophisticated, however, and require the use of specialized statistical software. Consult the Chapter 10 references if you want to learn more about these nonparametric techniques.

PROBLEMS FOR SECTION 10.13

Using the Tools

10.100 Specify the rejection region for Spearman's nonparametric test for rank correlation in each of the following situations:

a. $H_a: \rho > 0, n = 15, \alpha = .05$ **b.** $H_a: \rho < 0, n = 20, \alpha = .01$

c. $H_a: \rho \neq 0, n = 50, \alpha = .05$ **d.** $H_a: \rho \neq 0, n = 10, \alpha = .05$

e. $H_a: \rho > 0, n = 100, \alpha = .05$

10.101 A random sample of seven pairs of observations is recorded on two variables, x and y. The data are shown in the table.

Pair	x	y
1	65	59
2	57	61
3	55	58
4	38	23
5	29	34
6	43	38
7	49	37

a. Rank the values of each variable, x and y. (Note that there are no tied ranks.)

b. Compute the test statistic r_S.

c. Do the data provide sufficient evidence to conclude that the ranked pairs are correlated? Test using $\alpha = .05$.

10.102 A random sample of five pairs of observations is recorded on two variables, x and y. The data are shown in the table.

a. Rank the values of each variable, x and y. (Note that there are no tied ranks.)

b. Compute the test statistic, r_S.

c. Do the data provide sufficient evidence to conclude that the ranked pairs are correlated? Test using $\alpha = .05$.

Pair	x	y
1	185	16
2	188	20
3	210	13
4	243	10
5	277	7

Problem 10.102

Applying the Concepts

10.103 Professional tennis players are ranked based on the number of tournament points earned during the playing season. (The most points are earned by winning a major tennis tournament such as Wimbledon or the U.S. Open.) *Tennis* magazine (Feb. 2000) claims that "tennis players who tie the knot often see their games unravel." The table below lists a sample of players and their performance rankings on their wedding days and on their first anniversaries.

TENNISWED

Player	Wedding Day Rank	First Anniversary Rank
Arthur Ashe	12	130
Jonathan Stark	67	165
Richey Reneberg	28	97
Paul Haarhuis	28	73
Richard Fromberg	40	79
Byron Black	44	77
Sabine Appelmans	16	49
Petr Korda	7	11
Dominique Van Roost	43	46
Ivan Lendl	1	3
John McEnroe	7	9
Stefan Edberg	2	3
Chris Evert	4	4
Mats Wilander	3	3
Sandrine Testud	14	12
Zina Garrison	6	4
Yevgeny Kafelnikov	8	4
Boris Becker	11	3
Michael Stich	15	6
Julie Halard-Decugis	32	15
Todd Woodbridge	71	27
Jason Stoltenberg	82	31

Source: Tennis, Feb. 2000, p. 14.

a. Construct a scatterplot for these data. Does it tend to support or refute the magazine's claim? Justify your answer.

b. Rank the data on wedding day ranking (x) and first anniversary ranking (y).

c. Find the rank correlation between wedding day ranking (x) and first anniversary ranking (y) based on the ranks in part **b**.

d. Is there a significant negative rank correlation between x and y (at $\alpha = .05$)?

e. Why is a rank correlation test preferred over a parametric test for correlation?

10.104 The *Journal of Financial Planning* (Jan. 1993) offered advice on selecting stocks and bonds. One of the many exhibits presented in the article involved a comparison of 5-year net returns and expense ratios (operating expenses as a percentage of net assets) for 18 insured AAA-rated municipal bond funds. The 18 funds were

ranked according to both of these variables; the rankings are displayed in the accompanying table.

AAABONDS

Municipal Bond Fund	Expense Ratio Rank	Net Return Rank	Municipal Bond Fund	Expense Ratio Rank	Net Return Rank
1	1	1	10	10	18
2	2	6	11	11	15
3	3	3	12	12	17
4	4	10	13	13	2
5	5	11	14	14	7
6	6	9	15	15	14
7	7	8	16	16	13
8	8	5	17	17	16
9	9	4	18	18	12

Source: Bogle, J. C. "A crystal ball look at U.S. markets in the 1990s." *Journal of Financial Planning,* Jan. 1993, p. 19 (Exhibit 15).

a. Give a reason why nonparametric statistics should be used to analyze the data.

b. Use Spearman's rank correlation method to analyze the data. Interpret the results.

10.105 Refer to the *Journal of the American Mosquito Control Association* (Mar. 1995) study of the relationship between temperature and mosquito catch ratio, Problem 10.9 (p. 549). The data are reproduced in the accompanying table. Use Spearman's method to test for a positive rank correlation between the two variables. Use $\alpha = .05$.

MOSQUITO

Date	Average Temperature	Catch Ratio
July 24	16.8	.66
25	15.0	.30
26	16.5	.46
27	17.7	.44
28	20.6	.67
29	22.6	.99
30	23.3	.75
31	18.2	.24
Aug. 1	18.6	.51

Source: Petric, D., et al. "Dependence of CO_2-baited suction trap captures on temperature variations." *Journal of the Mosquito Control Association,* Vol. 11, No. 1, Mar. 1995, p. 8.

10.106 Refer to the study of the variables related to the number of wins by a professional basketball (NBA) team, Problem 10.11 (p. 550). The accompanying table gives the value of Spearman's rank correlation coefficient relating standardized wins y to each of the three independent variables, 2-point field goals, x_1, rebounds, x_2, and turnovers, x_3.

	r_S
Two-point field goals x_1	.8399
Rebounds x_2	.3798
Turnovers x_3	.5218

a. Interpret each of the three values of r_S.

b. Is there evidence of rank correlation between standardized wins y and standardized 2-point field goals x_1? Test using $\alpha = .01$.

c. Is there evidence of rank correlation between standardized wins y and standardized rebounds x_2? Test using $\alpha = .01$.

d. Is there evidence of rank correlation between standardized wins y and standardized turnovers x_3? Test using $\alpha = .01$.

10.107 Refer to the *Wetlands* (Mar. 1995) study of the relationship between the accuracy of a land survey determined by a Global Positioning System and the percentage of readings obtained when three satellites are used, Problem 10.44 (p. 570). Spearman's rank correlation relating distance error y to percent of readings with three satellites x is $r_S = .5956$. Give a full interpretation of this result (using $\alpha = .10$).

10.108 The *Journal of Archaeological Science* (Oct. 1997) reported on two indexes developed for classifying whale bones. The architectural utility index (AUI) is based on bone length, shape, and weight. The meat utility index (MUI) is based on estimated flesh weight per skeletal portion. Thirteen different skeletal body parts of a prehistoric bowhead whale were ranked from low (1) to high (9) according to the AUI. In addition, the MUI (measured as a percentage) was determined for each part. The data are listed in the table.

WHALES

Whale Part	AUI Ranking	MUI (%)
Cranium	9	3.80
Maxillae/premaxillae	8	7.55
Mandible	9	7.55
Hyoid	2	88.80
Cervical vert.	8	4.30
Thoracic vert.	6	49.20
Lumbar vert.	7	100.00
Caudal vert.	3	91.20
Rib	7	39.70
Sternum	1	2.10
Scapula	6	4.80
Humerus	5	1.90
Radius/ulna	4	3.90

Source: Savelle, J. M. "The role of architectural utility in the formation of zooarchaeological whale bone assemblages." *Journal of Archaeological Science,* Vol. 24, No. 10, Oct. 1997, pp. 872–873 (Tables 2 and 3).

a. Find the rank correlation between AUI rank and MUI percent. Interpret the result.

b. Is there evidence (at $\alpha = .10$) of a significant rank correlation between AUI and MUI?

c. The research archaeologist also ranked the whale body parts according to how they were distributed among whaling crews. (The portion of the whale retained by the boat's captain—the edible material—is rated the highest, while the portion distributed to others who did not participate in the kill—the skin or blubber—is rated the lowest.) The rank correlation between MUI and whaling crew distribution rating is $r_S = .9118$ for the 13 whole body parts. Interpret this value.

10.14 Pitfalls in Regression and Ethical Issues

Regression analysis is perhaps the most widely used and, unfortunately, the most widely misused statistical technique applied to data. A few of the most frequently encountered difficulties are discussed below.

Pitfalls

1. Assumptions: Many users of regression lack an awareness of the standard least squares regression assumptions. (This problem has been magnified by the ease with which computer software can be employed to fit a regression model. Nowhere on the printout are the four assumptions listed as a reminder to the user!) Of those users who are aware of the assumptions, many do not know how to evaluate them or what to do if, in fact, any of the assumptions are violated. Consequently, the user with this lack of knowledge risks drawing inferences and making decisions from an invalid model.

To illustrate, consider the data in Table 10.10. In a classical pedagogical piece of statistical literature that deals with the importance of observation through scatterplots and residual analysis, Anscombe (1973) showed that the following simple linear regression results are obtained for each of the four artificial data sets given in Table 10.10: $y = 3 + .5x$, $s = 1.237$, and $R^2 = .667$.

TABLE 10.10 Four Sets of Artificial Data

Data Set A		Data Set B		Data Set C		Data Set D	
x	y	x	y	x	y	x	y
10	8.04	10	9.14	10	7.46	8	6.58
14	9.96	14	8.10	14	8.84	8	5.76
5	5.68	5	4.74	5	5.73	8	7.71
8	6.95	8	8.14	8	6.77	8	8.84
9	8.81	9	8.77	9	7.11	8	8.47
12	10.84	12	9.13	12	8.15	8	7.04
4	4.26	4	3.10	4	5.39	8	5.25
7	4.82	7	7.26	7	6.42	19	12.50
11	8.33	11	9.26	11	7.81	8	5.56
13	7.58	13	8.74	13	12.74	8	7.91
6	7.24	6	6.13	6	6.08	8	6.89

Source: Anscombe, F. J. "Graphs in Statistical Analysis," *American Statistician,* Vol. 27 (1973), pp. 17–21.

Thus, with respect to these simple linear regression statistics, the four data sets are identical. Had the analysis stopped at this point, valuable information in the data would be lost. This can be observed by examining Figure 10.33 (p. 624), which presents scatterplots for each of the four data sets, and Figure 10.34 (p. 625), which shows the corresponding residual plots.

From the scatterplots of Figure 10.33 and residual plots of Figure 10.34, you see how different the data sets are. The only data set that seems to follow an approximate straight line is data set A. The residual plot for data set A does not show any obvious patterns or outlying residuals. However, this is certainly not the case for data sets B, C, and D. The scatterplot for data set B shows that a curvilinear regression model should be considered—a conclusion reinforced by the clear parabolic trend on the corresponding residual plot. Both plots for data set C clearly depict what is an outlying observation. If this is the case, you may want to remove the outlier and re-estimate the basic model. Similarly, the plots for data set D show a highly unusual data point ($x_8 = 19$ and $y_8 = 12.50$) on which the fitted model is heavily dependent. Any regression model would have to be evaluated cautiously if its regression coefficients are heavily dependent on a single observation.

Figure 10.33 Scatterplots for the Four Data Sets of Table 10.10

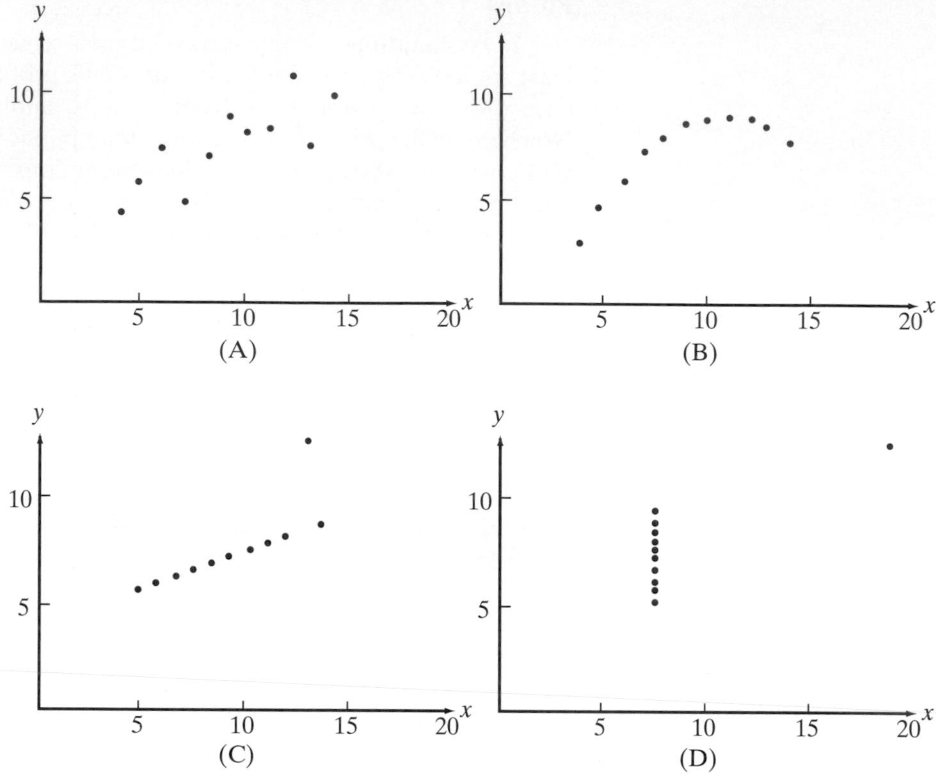

To avoid pitfalls such as those illustrated in Figure 10.34, residual plots are of vital importance to a regression analysis. Always check the assumptions with a residual analysis before using the model statistics to make inferences.

2. Parameter Interpretations: In any regression analysis, it is important to interpret the estimates of the β parameters correctly. A typical misconception is that β_i *always* measures the effect of x_i on μ_y, *independent* of the other x variables in the model. This may be true for some models, but it is not true in general. Generally, the interpretation of an individual β parameter becomes increasingly more difficult as the model becomes more complex. (See the boxes, pp. 602, 606.)

Another misconception about parameter estimates is that a statistically significant b_i value establishes a *cause-and-effect* relationship between μ_y and x_i. That is, if b_i is found to be significantly greater than 0, then some practitioners would infer that an increase in x_i *causes* an increase in the mean response μ_y. Unfortunately, it is dangerous to infer a causal relationship between two variables based on a regression analysis. There may be many other independent variables (some of which we may have included in our model, some of which we may have omitted) that affect the mean response. Unless we can control the values of these other variables, we are uncertain about what is actually causing the observed increase in y. In Chapter 11, we introduce the notion of *designed experiments,* where the values of the independent variables are set in advance before the value of y is observed. Only with such an experiment can a cause-and-effect relationship be established.

3. Multicollinearity: Often, two or more of the independent variables used in the model for μ_y will contribute redundant information. That is, the independent variables will be correlated with each other. For example, suppose we want to construct a model to predict the gasoline mileage rating y of a truck as a function of the weight of its load x_1 and the horsepower x_2 of its engine. In general, you would expect heavier loads to require greater horsepower and to result in lower mileage

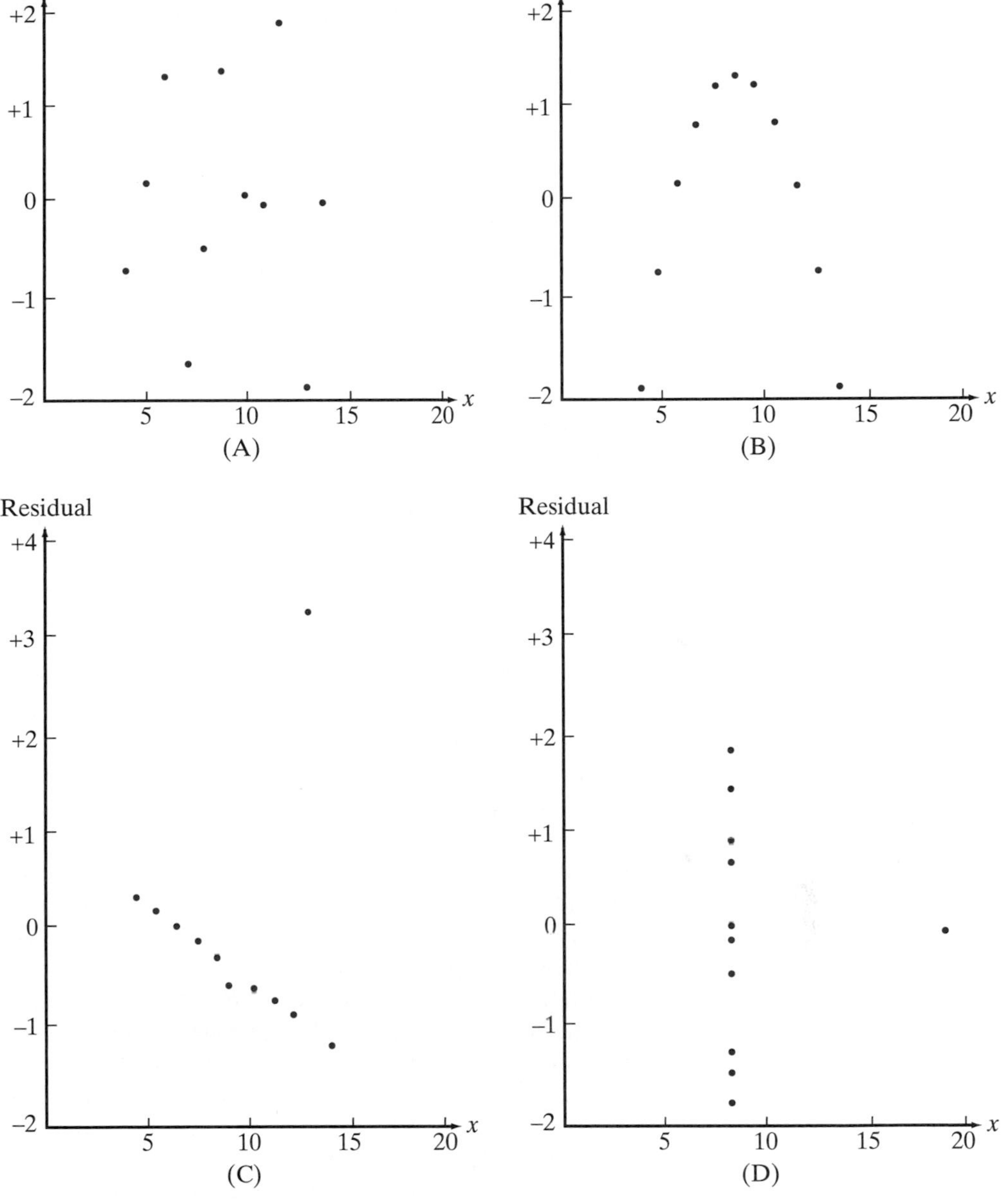

Figure 10.34 Residual Plots for the Four Data Sets of Table 10.10

ratings. Thus, although both x_1 and x_2 contribute information for the prediction of mileage rating, some of the information is overlapping because x_1 and x_2 are correlated. When the independent variables are correlated, we say that *multicollinearity* exists. In practice, it is not uncommon to observe correlations among the independent variables. However, a few problems arise when serious multicollinearity is present in the regression analysis.

Definition 10.6

When the independent variables in a multiple regression analysis exhibit a high degree of correlation, **multicollinearity** exists.

First, high correlations among the independent variables increase the likelihood of rounding errors in the calculations of the β estimates, standard errors, and so forth. Second, the regression results may be confusing and misleading.

To illustrate, if the gasoline mileage rating model

$$y = \beta_0 + \beta_1 x_1 + \beta_2 x_2 + \varepsilon$$

were fit to a set of data, we might find that the t values for both b_1 and b_2 (the least squares estimates) are nonsignificant. However, the F test for H_0: $\beta_1 = \beta_2 = 0$ would probably be highly significant. The tests may seem to be contradictory, but really they are not. The t tests indicate that the contribution of one variable, say x_1 = load, is not significant after the effect of x_2 = horsepower has been discounted (because x_2 is also in the model). The significant F test, on the other hand, tells us that at least one of the two variables is making a contribution to the prediction of y (i.e., either β_1, β_2, or both differ from 0). In fact, both are probably contributing, but the contribution of one overlaps with that of the other.

Multicollinearity can also have an effect on the signs of the parameter estimates. More specifically, a value of b_i may have the opposite sign from what is expected. For example, we expect the signs of both of the parameter estimates for the gasoline mileage rating model to be negative, yet the regression analysis for the model might yield the estimates $b_1 = .2$ and $b_2 = -.7$. The positive value of b_1 seems to contradict our expectation that heavy loads will result in lower mileage ratings. However, it is dangerous to interpret a β coefficient when the independent variables are correlated. Because the variables contribute redundant information, the effect of load weight x_1 on mileage rating is measured only partially by b_1.

Several methods are available for detecting multicollinearity in regression. A simple technique is to examine the coefficient of correlation r between each pair of independent variables in the model. If one or more of the r values is near 1 (in absolute value), the variables in question are highly correlated and a severe multicollinearity problem may exist. Other indications of the presence of multicollinearity include those mentioned above—namely, nonsignificant t tests for the individual β parameters when the F test for overall model adequacy is significant, and parameter estimates with opposite signs from what is expected. More formal methods for detecting multicollinearity (such as variance-inflation factors) are available. Consult the references given at the end of the chapter (e.g., Mendenhall and Sincich [1996], Berenson, Levine, and Krehbiel [2002], Neter, et al. [1996]) if you want to learn more about these methods.

One way to avoid the problems of multicollinearity in regression is to conduct a designed experiment (Chapter 11) so that the levels of the x variables are uncorrelated. Unfortunately, time and cost constraints may prevent you from collecting data in this manner. Most analysts, when confronted with highly correlated independent variables, choose to include only one of the correlated variables in the model.

4. Predicting Outside the Experimental Region: The fitted regression model enables us to construct a confidence interval for μ_y and a prediction interval for y for values of the independent variable only within the region of experimentation, i.e., within the range of values of the independent variables used in the experiment. For example, suppose that you conduct experiments on the mean strength of plastic molded at several different temperatures in the interval 200°F to 400°F. The regression model that you fit to the data is valid for estimating μ_y or for predicting values of y for values of temperature x in the range 200°F to

400°F. However, if you attempt to extrapolate beyond the experimental region, you risk the possibility that the fitted model is no longer a good approximation to the mean strength of the plastic (see Figure 10.35). For example, the plastic may become too brittle when formed at 500°F and possess no strength at all. Estimating and predicting outside of the experimental region are sometimes necessary. If you do so, keep in mind the possibility of a large extrapolation error.

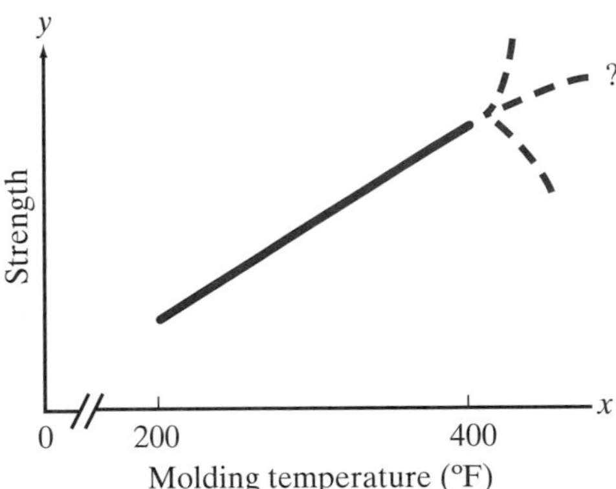

Figure 10.35 Using a Regression Model Outside the Experimental Region

5. Model Building: In many regression applications, only a single model (usually a first-order model) is fit and evaluated. Even if the model is deemed adequate (based on statistics such as adjusted R^2, s, and the F statistic), the model is not guaranteed to be the best predictor of y. A more complex model—one with squared terms, interactions, or other independent variables—may be a significantly better predictor of y in practice. A user unaware of this alternative model will risk making erroneous conclusions about the relationship between y and the independent variables.

In addition to the initial model fit, the most successful regression analysts always consider one or more alternative models. Model comparisons are made and the model judged "best" is used to make inferences and predictions.

Ethical Issues

Ethical considerations arise when a user of regression manipulates the process of developing the regression model. The key here is intent. Some examples of unethical behavior on the part of the regression user are listed below.

1. Predicting the dependent variable of interest with the willful intent of possibly excluding certain independent variables from consideration in the model.
2. Deleting observations from the model to obtain a better model without giving reasons for deleting these observations.
3. Making inferences about the model without providing an evaluation of the assumptions when he or she knows that the assumptions of least squares regression are violated.
4. Willfully failing to remove independent variables from the model that exhibit a high degree of multicollinearity.

All of these situations emphasize the importance of following the steps of regression (Sections 10.1 and 10.12).

KEY TERMS

Starred (*) terms are from the optional sections of this chapter.

*Adjusted coefficient of determination 607
Coefficient of determination 573
*Correlation coefficient 564
Dependent (response) variable 536
*Dummy variable 608
*First-order model 603
General regression model 537
Independent (explanatory) variable 537
*Interaction model 608
Least squares line 544
Least squares prediction equation 544
Method of least squares 543
Model 536
Multicollinearity 625
Multiple regression model* 602
*Outlier 593
*Population correlation coefficient 565
*Quadratic model 608
Random error component 539
Regression analysis 536
*Residual 589
*Robust 592
Simple linear regression 540
Slope 540
*Spearman's rank correlation coefficient 615
Standardized residuals 595
Straight-line model 539
*Time series 594
*Variance-stabilizing transformation 592
y-intercept 540

KEY FORMULAS

Starred (*) formulas are from the optional sections of this chapter.

Simple Linear Regression

Least squares estimates of the β's:

$$\mathrm{b}_1 = \frac{\mathrm{SSxy}}{\mathrm{SSxx}} \qquad \textbf{(10.4)}, 544$$

$$\mathrm{b}_0 = \bar{y} - \mathrm{b}_1\bar{x} \qquad \textbf{(10.5)}, 544$$

where

$$\mathrm{SSxy} = \sum xy - \frac{(\sum x)(\sum y)}{n} \qquad \textbf{(10.6)}, 544$$

$$\mathrm{SSxx} = \sum x^2 - \frac{(\sum x)^2}{n} \qquad \textbf{(10.7)}, 544$$

$$\mathrm{SSyy} = \sum y^2 - \frac{(\sum y)^2}{n} \qquad \textbf{(10.10)}, 555$$

Least squares line:

$$\hat{y} = \mathrm{b}_0 + \mathrm{b}_1 x$$

Sum of squared errors:

$$\mathrm{SSE} = \sum (y - \hat{y})^2 = \mathrm{SSyy} - \mathrm{b}_1\mathrm{SSxy} \qquad \textbf{(10.9)}, 555$$

Estimated variance of ε:

$$s^2 = \frac{\mathrm{SSE}}{n-2} = \mathrm{MSE} \qquad \textbf{(10.8)}, 555$$

Estimated standard error of b_1:

$$s_{\mathrm{b}_1} = s/\sqrt{\mathrm{SSxx}} \qquad \textbf{(10.11)}, 558$$

Test statistic for testing H_0: $\beta_1 = 0$:

$$t = \frac{\mathrm{b}_1}{s_{\mathrm{b}_1}} \qquad \textbf{(10.12)}, 558$$

$(1-\alpha)100\%$ confidence interval for β_1:

$$\mathrm{b}_1 \pm (t_{\alpha/2})s_{\mathrm{b}_1} \qquad \textbf{(10.13)}, 560$$

Coefficient of determination:

$$R^2 = \frac{\mathrm{SSR}}{\mathrm{SSyy}} = \frac{\mathrm{SSyy} - \mathrm{SSE}}{\mathrm{SSyy}} \qquad \textbf{(10.16)}, 573$$

$(1-\alpha)100\%$ confidence interval for μ_y when $x = x_p$:

$$\hat{y} \pm (t_{\alpha/2})s\sqrt{\frac{1}{n} + \frac{(x_p - \bar{x})^2}{\mathrm{SSxx}}} \qquad \textbf{(10.17)}, 576$$

$(1-\alpha)100\%$ prediction interval for y when $x = x_p$:

$$\hat{y} \pm (t_{\alpha/2})s\sqrt{1 + \frac{1}{n} + \frac{(x_p - \bar{x})^2}{\mathrm{SSxx}}} \qquad \textbf{(10.18)}, 576$$

Multiple Regression

*Estimated variance of ε:

$$s^2 = \frac{\mathrm{SSE}}{n - (k+1)} = \mathrm{MSE}$$

*Test statistic for testing H_0: $\beta_i = 0$:

$$t = \frac{\mathrm{b}_i}{s_{\mathrm{b}_i}}$$

*$(1 - \alpha)100\%$ confidence interval for β_i: $b_i \pm (t_{\alpha/2})s_{b_i}$

*Test statistic for testing H_0: $\beta_1 = \beta_2 = \cdots = \beta_k = 0$: $F = \frac{\text{MS(Model)}}{\text{MSE}}$ **(10.20)**, 605

*Residual: $y - \hat{y}$ **(10.19)**, 589

Spearman's rank correlation coefficient $r_S = 1 - \frac{6\sum d_i^2}{n(n^2 - 1)}$ **(10.25)**, 616

*(shortcut formula):

KEY SYMBOLS

Symbol	Description
y	Dependent variable (variable to be predicted or modeled)
x	Independent (predicator) variable
μ_y	Mean value of y
β_0	y-intercept of true line
β_1	Slope of true line
b_0	Least squares estimate of y-intercept
b_1	Least squares estimate of slope
ε	Random error
$\hat{y}$	Predicted value of y
$(y - \hat{y})$	Error of prediction (or residual)
SSE	Sum of squared errors (will be smallest for least squares line)
SSxx	Sum of squares of x values
SSyy	Sum of squares of y values
SSxy	Sum of squares of cross products, $x \cdot y$
r	Coefficient of correlation
R^2	Coefficient of determination
R_a^2	Adjusted coefficient of determination
x_p	Value of x used to predict y
x_1^2	Quadratic term that allows for curvature in the relationship between y and x
x_1x_2	Interaction term
β_i	Coefficient of x_i in the multiple regression model
b_i	Least squares estimate of β_i
s_{b_i}	Estimated standard error of b_i
F	Test statistic for testing global usefulness of model
r_S	Spearman's rank correlation coefficient

CHECKING YOUR UNDERSTANDING

1. What is the objective of a regression analysis?
2. What is the interpretation of the y-intercept in a simple linear regression model?
3. What is the interpretation of the slope in a simple linear regression model?
4. What are the properties of the least squares line?
5. What is the difference between the coefficient of correlation and the coefficient of determination?
6. What are the assumptions of regression analysis and how can they be evaluated?
7. What are regression residuals and how are they used?
8. What is the interpretation of the model standard deviation?

9. How do we test the statistical utility of a regression model?

10. When would you desire a confidence interval for the mean of y for a given x?

11. When would you desire a prediction interval for y for a given x?

12. What is an outlier in regression?

13. When is it appropriate to conduct a multiple regression analysis?

14. How do you test the overall utility of a multiple regression model?

15. What is multicollinearity and how does it impact a regression analysis?

SUPPLEMENTARY PROBLEMS

Starred (*) problems refer to one of the optional sections of this chapter.

10.109 A medical item used to administer to a hospital patient is called a factor. For example, factors can be intravenous (I.V.) tubing, I.V. fluid, needles, shave kits, bedpans, diapers, dressings, medications, and even code carts. The coronary care unit at Bayonet Point Hospital (St. Petersburg, Florida) investigated the relationship between the number of factors per patient x and the patient's length of stay (in days) y. The data for a random sample of 50 coronary care patients are given in the table.

FACTORS

Number of Factors x	Length of Stay y (days)	Number of Factors x	Length of Stay y (days)
231	9	354	11
323	7	142	7
113	8	286	9
208	5	341	10
162	4	201	5
117	4	158	11
159	6	243	6
169	9	156	6
55	6	184	7
77	3	115	4
103	4	202	6
147	6	206	5
230	6	360	6
78	3	84	3
525	9	331	9
121	7	302	7
248	5	60	2
233	8	110	2
260	4	131	5
224	7	364	4
472	12	180	7
220	8	134	6
383	6	401	15
301	9	155	4
262	7	338	8

Source: Bayonet Point Hospital, Coronary Care Unit.

a. Construct a scatterplot of the data.

b. Use Excel or MINITAB to find the least squares line for the data and show it on your scatterplot.

c. Define β_1 in the context of this problem.

d. Test the hypothesis that the number of factors per patient x contributes no information for the prediction of the patient's length of stay y when a linear model is used (use $\alpha = .05$). Draw the appropriate conclusions.

e. Find and interpret a 95% confidence interval for β_1.

f. Find and interpret the coefficient of correlation for the data.

g. Find and interpret the coefficient of determination for the model.

h. Find a 95% prediction interval for the length of stay of a coronary care patient who is administered a total of $x = 200$ factors.

i. Explain why the prediction interval obtained in part **h** is so wide. How could you reduce the width of the interval?

j. Conduct a residual analysis of the data. What model modifications (if any) do you recommend?

10.110 A study was conducted to model the thermal performance of tubes used in the refrigeration and process industries (*Journal of Heat Transfer,* Aug. 1990). Twenty-four specially manufactured copper tubes were used in the experiment. Vapor was released downward into each tube, and the heat transfer ratio was measured. Theoretically, heat transfer will be related to the area at the top of the tube that is "unflooded" by condensation of the vapor. The data in the table are the unflooded area ratio x and heat transfer ratio y values recorded for the 24 copper tubes. Use Excel or MINITAB to perform a complete simple linear regression analysis of the data. Be sure to check the regression assumptions with a residual analysis.

HEAT

Unflooded Area Ratio x	Heat Transfer Ratio y	Unflooded Area Ratio x	Heat Transfer Ratio y
1.93	4.4	2.00	5.2
1.95	5.3	1.77	4.7
1.78	4.5	1.62	4.2
1.64	4.5	2.77	6.0
1.54	3.7	2.47	5.8
1.32	2.8	2.24	5.2
2.12	6.1	1.32	3.5
1.88	4.9	1.26	3.2
1.70	4.9	1.21	2.9
1.58	4.1	2.26	5.3
2.47	7.0	2.04	5.1
2.37	6.7	1.88	4.6

Source: Marto, P. J., et al. "An experimental study of R-113 film condensation on horizontal integral-fin tubes." *Journal of Heat Transfer,* Vol. 112, Aug. 1990, p. 763 (Table 2).

10.111 In the long jump, the absolute distance between the front edge of the takeoff board and where the toe actually lands on the board prior to jumping is called "takeoff error." Is takeoff error in the long jump linearly related to best jumping distance? To answer this question, kinesiology researchers videotaped the performances of 18 novice long jumpers at a high school track meet (*Journal of Applied Biomechanics,* May 1995). The average takeoff error x and best jumping distance (out of three jumps) y for each jumper are recorded in the table on p. 632. Use Excel or MINITAB to conduct a complete simple linear regression analysis of the data. Assume one of the researchers' goals is to predict the best jumping distance for a jumper with an average takeoff error of .05 meter.

Variable	Coefficient of Correlation r
x_1 and x_2	.33
x_1 and x_3	.08
x_2 and x_3	.16

Problem 10.112

10.112 Refer to the *Proceedings of the National Science Council, Republic of China* (July 1997) study of the variables related to y, parent involvement in public elementary schools, Problem 10.97. Three independent variables were included in the regression model: level of principal's cooperation x_1, level of principal's socialization x_2, and level of principal's formalization x_3. The intercorrelations among these variables are listed in the table. Do you detect extreme multicollinearity in the data? What impact does this have, if any, on the analysis of the multiple regression model, Problem 10.97 (p. 614)?

HELIUM

10.113 At temperatures approaching absolute zero (273 degrees below zero Celsius), helium exhibits traits that defy many laws of conventional physics. An experiment has been conducted with helium in solid form at various temperatures near absolute zero. The solid helium is placed in a dilution refrigerator along with a solid impure substance, and the proportion (by weight) of the impurity passing through the solid

LONGJUMP

Jumper	Best Jumping Distance y (meters)	Average Takeoff Error x (meters)
1	5.30	.09
2	5.55	.17
3	5.47	.19
4	5.45	.24
5	5.07	.16
6	5.32	.22
7	6.15	.09
8	4.70	.12
9	5.22	.09
10	5.77	.09
11	5.12	.13
12	5.77	.16
13	6.22	.03
14	5.82	.50
15	5.15	.13
16	4.92	.04
17	5.20	.07
18	5.42	.04

Source: W. P. Berg and N. L. Greer, "A kinematic profile of the approach run of novice long jumpers." *Journal of Applied Biomechanics,* Vol. 11, No. 2, May 1995, p. 147 (Table 1).

Problem 10.111

Proportion of Impurity Passing through Helium y	Temperature x (°C)
.315	−262
.202	−265
.204	−256
.620	−267
.715	−270
.935	−272
.957	−272
.906	−272
.985	−273
.987	−273

Problem 10.113

helium is recorded. (This phenomenon of solids passing directly through solids is known as *quantum tunneling*.) The data are given in the table at left.

a. Use Excel or MINITAB to conduct a complete simple linear regression analysis of the data. Be sure to check the assumptions and make any model modifications, if necessary.

b. Would you recommend using the fitted model, part **a**, to predict the proportion of impurity passing through helium at a temperature of −250°C? Explain.

10.114 A company that has the distribution rights to home video sales of previously released movies would like to be able to predict the number of units that it can expect to sell (based on the box office gross sales of the movie). Data are available for 30 movies that indicate the box office gross (in millions of dollars) and the number of units sold (in thousands) of home videos. The results are listed in table on p. 633.

Using Excel or MINITAB, conduct a complete simple linear regression analysis of the data. One of your goals is to predict the video unit sales for a movie that had a box office gross of $20 million.

*__10.115__ Refer to the *Journal of Communications* (Summer 1997) study of homicide newsworthiness, Problem 1.7 (p. 28). The researchers fit the following multiple regression model to data collected for $n = 100$ Milwaukee homicides:

$$y = \beta_0 + \beta_1 x_1 + \beta_2 x_2 + \beta_3 x_3 + \beta_4 x_4 + \beta_5 x_5 + \beta_6 x_6 + \beta_7 x_7 + \beta_8 x_8 + \beta_9 x_1 x_2 + \varepsilon$$

where

y = average length of published news stories on a homicide

$$x_1 = \begin{cases} 1 & \text{if white victim or suspect} \\ 0 & \text{if not} \end{cases}$$

$$x_2 = \begin{cases} 1 & \text{if female suspect} \\ 0 & \text{if not} \end{cases}$$

$$x_3 = \begin{cases} 1 & \text{if female victim} \\ 0 & \text{if not} \end{cases}$$

$$x_4 = \begin{cases} 1 & \text{if white child or senior citizen} \\ 0 & \text{if not} \end{cases}$$

VIDEOS

Movie	Box Office Gross ($ millions)	Home Video Units Sold (thousands)
1	1.10	57.18
2	1.13	26.17
3	1.18	92.79
4	1.25	61.60
5	1.44	46.50
6	1.53	85.06
7	1.53	103.52
8	1.69	30.88
9	1.74	49.29
10	1.77	24.14
11	2.42	115.31
12	5.34	87.04
13	5.70	128.45
14	6.43	126.64
15	8.59	107.28
16	9.36	190.80
17	9.89	121.57
18	12.66	183.30
19	15.35	204.72
20	17.55	112.47
21	17.91	162.95
22	18.25	109.20
23	23.13	280.79
24	27.62	229.51
25	37.09	277.68
26	40.73	226.73
27	45.55	365.14
28	46.62	218.64
29	54.70	286.31
30	58.51	254.58

Problem 10.114

x_5 = per capita income of census tract in which crime occurred

$$x_6 = \begin{cases} 1 & \text{if suspect and victim had any type of prior relationship} \\ 0 & \text{if not} \end{cases}$$

$$x_7 = \begin{cases} 1 & \text{if victim was involved with drugs, gambling, gangs, or prostitution} \\ 0 & \text{if not} \end{cases}$$

$$x_8 = \begin{cases} 1 & \text{if police information ban} \\ 0 & \text{if not} \end{cases}$$

The regression results are summarized in the accompanying table.

Variable	β Estimate	p-Value
Race (x_1)	.42	$p < .001$
Gender of suspect (x_2)	−.33	$.001 < p < .01$
Gender of victim (x_3)	.26	$.01 < p < .05$
Age of victim (x_4)	.33	$p < .001$
Per capita income (x_5)	−.02	$p > .10$
Relationship (x_6)	−.11	$p > .10$
Risky behavior (x_7)	−.06	$p > .10$
Information ban (x_8)	−.15	$p > .10$
Race × Gender (x_1x_2)	−.25	$.01 < p < .05$
Overall model: $R^2 = .29, p < .001$		

Source: Pritchard, D., and Hughes, K. D. "Patterns of deviance in crime news." *Journal of Communication,* Vol. 47, No. 3, Summer 1997, p. 59 (Table 2).

a. Identify the qualitative independent variables used in the model.

b. Give the null hypothesis for testing whether the overall model is useful for predicting the average length of published news stories on a homicide.

c. Conduct the test, part **b**. Use $\alpha = .01$.

d. Interpret the value of R^2.

e. Interpret the estimate of β_3.

f. Interpret the estimate of β_5.

g. Is there evidence of interaction between race x_1 and gender of suspect x_2? Test using $\alpha = .05$.

***10.116** A study of women's comprehension of quantitative-based messages about mammography was published in the *Annals of Internal Medicine* (Vol. 127, 1997). A random sample of women was selected from a registry of female veterans in Vermont. Each veteran was mailed one of four questionnaires that differed only in how the same information on average risk reduction with mammography was presented. Three basic probability questions were asked and the number of correct answers (0, 1, 2, or 3)—called *numeracy score*—was recorded for each respondent. In addition, the researchers measured the *perceived absolute risk reduction* to determine how well each respondent applied the information on breast cancer given in the survey questionnaire. For each survey group, the rank correlation between numeracy score and perceived absolute risk reduction is shown below. Which groups (if any) show a significant correlation between the two variables? Test using $\alpha = .05$.

Group	Sample Size	Rank Correlation
1	59	−.1899
2	52	−.3904
3	58	−.2576
4	61	−.1956

***10.117** Refer to the *Journal of Agricultural, Biological, and Environmental Statistics* (Sept. 2000) study of the impact of the *Exxon Valdez* tanker oil spill on the seabird population in Alaska, Problem 2.62 (p. 107). Data collected on 96 shoreline locations (called transects) are stored in the file **EVOS.** For each transect, let y = the number of seabirds found, x_1 = the length (in kilometers) of the transect, and $x_2 = 1$ if the transect was in an oiled area and 0 if not. Conduct a complete multiple regression analysis for the first-order model relating number of seabirds found to transect length and whether or not the transect was in an oiled area. Interpret the results of your analysis.

***10.118** Does extensive media coverage of a military crisis influence public opinion on how to respond to the crisis? Political scientists at UCLA researched this question and reported their results in *Communication Research* (June 1993). The military crisis of interest was the 1990 Persian Gulf War, precipitated by Iraqi leader Saddam Hussein's invasion of Kuwait. The researchers used multiple regression analysis to model the level y of support Americans had for a military (rather than a diplomatic) response to the crisis. Values of y ranged from 0 (preference for a diplomatic response) to 4 (preference for a military response). The following independent variables were used in the model:

x_1 = Level of TV news exposure in a selected week (number of days)

x_2 = Knowledge of 7 political figures (1 point for each correct answer)

x_3 = Gender (1 if male, 0 if female)

x_4 = Race (1 if nonwhite, 0 if white)

x_5 = Partisanship (0–6 scale, where 0 = strong Democrat and 6 = strong Republican)

x_6 = Defense spending attitude (1–7 scale, where 1 = greatly decrease spending and 7 = greatly increase spending)

x_7 = Education level (1–7 scale, where 1 = less than eight grades and 7 = college)

Data from a survey of 1,763 Americans were used to fit the model

$$y = \beta_0 + \beta_1 x_1 + \beta_2 x_2 + \beta_3 x_3 + \beta_4 x_4 + \beta_5 x_5 + \beta_6 x_6 + \beta_7 x_7 + \beta_8 x_2 x_3 + \beta_9 x_2 x_4 + \varepsilon$$

The regression results are shown in the accompanying table.

Variable	β Estimate	Standard Error	Two-Tailed p-Value
TV news exposure (x_1)	.02	.01	.03
Political knowledge (x_2)	.07	.03	.03
Gender (x_3)	.67	.11	<.001
Race (x_4)	−.76	.13	<.001
Partisanship (x_5)	.07	.01	<.001
Defense spending (x_6)	.20	.02	<.001
Education (x_7)	.07	.02	<.001
Knowledge × Gender ($x_2 x_3$)	−.09	.04	.02
Knowledge × Race ($x_2 x_4$)	.10	.06	.08

Source: Iyengar, S., and Simon, A. "News coverage of the Gulf Crisis and public opinion." *Communication Research,* Vol. 20, No. 3, June 1993, p. 380 (Table 2).

a. Interpret the β estimate for the variable x_1, TV news exposure.

b. Conduct a test to determine whether an increase in TV news exposure is associated with an increase in support for a military resolution of the crisis. Use $\alpha = .05$.

c. Is there sufficient evidence to indicate that the relationship between support for a military resolution y and gender x_3 depends on political knowledge x_2? Test using $\alpha = .05$.

d. Is there sufficient evidence to indicate that the relationship between support for a military resolution y and race x_4 depends on political knowledge x_2? Test using $\alpha = .05$.

e. The coefficient of determination for the model was $R^2 = .194$. Interpret this value.

f. For this model, $F = 46.88$. Conduct a global test for model utility. Use $\alpha = .05$.

10.119 Is there a link between the loneliness of parents and their offspring? Research psychologists examined this question in the *Journal of Marriage and the Family* (Aug. 1986). The participants in the study were 130 female college undergraduates and their parents. Each triad of daughter, mother, and father completed the UCLA Loneliness Scale, a 20-item questionnaire designed to assess loneliness and several variables theoretically related to loneliness, such as social accessibility to others, difficulty in making friends, and depression. Correlations relating daughter's loneliness to parent's loneliness score as well as the other variables were calculated. The results are summarized in the table on p. 636.

a. The researchers concluded that "mother and daughter loneliness scores were (positively) significantly correlated at $\alpha = .01$." Do you agree?

b. Determine which, if any, of the other sample correlations are large enough to indicate (at $\alpha = .01$) that linear correlation exists between daughter's loneliness score and the variable measured.

c. Explain why it would be dangerous to conclude that a causal relationship exists between mother's loneliness and daughter's loneliness.

d. Explain why it would be dangerous to conclude that the variables with non-significant correlations in the table are unrelated.

Variable	Correlation r between Daughter's Loneliness and Parental Variables	
	Mother	Father
Loneliness	.26	.19
Depression	.11	.06
Self-esteem	−.14	−.06
Assertiveness	−.05	.01
Number of friends	−.21	−.10
Quality of friendships	−.17	.01

Source: Lobdell, J., and Perlman, D. "The intergenerational transmission of loneliness. A study of college females and their parents." *Journal of Marriage and the Family,* Vol. 48, No. 8, Aug. 1986, p. 592.

Problem 10.119

***10.120** Amorphous alloys have been found to have superior corrosion resistance. *Corrosion Science* (Sept. 1993) reported on the resistivity of an amorphous iron-boron-silicon alloy after crystallization. Five alloy specimens were annealed at 700°C, each for a different length of time. The passivation potential—a measure of resistivity of the crystallized alloy—was then measured for each specimen. The experimental data are shown in the table below.

Annealing Time x (minutes)	Passivation Potential y (mV)
10	−408
20	−400
45	−392
90	−379
120	−385

Source: Chattoraj, I., et al. "Polarization and resistivity measurements of post-crystallization changes in amorphous Fe–B–Si alloys." *Corrosion Science,* Vol. 49, No. 9, Sept. 1993, p. 712 (Table 1).

a. Calculate Spearman's correlation coefficient between annealing time x and passivation potential y. Interpret the result.

b. Use the result, part **a**, to test for a significant correlation between annealing time and passivation potential. Use $\alpha = .05$.

***10.121** Refer to the FTC cigarette data stored in the **FTC** file. Select a random sample of $n = 50$ cigarette brands from the data set. Suppose you want to use the data to fit the model $y = \beta_0 + \beta_1 x_1 + \beta_2 x_2 + \varepsilon$, where y = carbon monoxide content, x_1 = tar content, and x_2 = nicotine content.

a. Use Excel or MINITAB to obtain the correlation between x_1 and x_2. What does this imply?

b. Using Excel or MINITAB, fit the model $y = \beta_0 + \beta_1 x_1 + \beta_2 x_2 + \varepsilon$. Do you detect any signs of multicollinearity?

***10.122** An article published in *Geography* (July 1980) used multiple regression to predict annual rainfall levels in California. Data on the average annual precipitation y, altitude x_1, latitude x_2, and distance from the Pacific coast x_3 for 30 meteorological stations scattered throughout California are listed in the table on p. 637. Consider the first-order model $y = \beta_0 + \beta_1 x_1 + \beta_2 x_2 + \beta_3 x_3 + \varepsilon$.

Station	Average Annual Precipitation y (inches)	Altitude x_1 (feet)	Latitude x_2 (degrees)	Distance from Coast x_3 (miles)
1. Eureka (W)	39.57	43	40.8	1
2. Red Bluff (L)	23.27	341	40.2	97
3. Thermal (L)	18.20	4,152	33.8	70
4. Fort Bragg (W)	37.48	74	39.4	1
5. Soda Springs (W)	49.26	6,752	39.3	150
6. San Francisco (W)	21.82	52	37.8	5
7. Sacramento (L)	18.07	25	38.5	80
8. San Jose (L)	14.17	95	37.4	28
9. Giant Forest (W)	42.63	6,360	36.6	145
10. Salinas (L)	13.85	74	36.7	12
11. Fresno (L)	9.44	331	36.7	114
12. Pt. Piedras (W)	19.33	57	35.7	1
13. Pasa Robles (L)	15.67	740	35.7	31
14. Bakersfield (L)	6.00	489	35.4	75
15. Bishop (L)	5.73	4,108	37.3	198
16. Mineral (W)	47.82	4,850	40.4	142
17. Santa Barbara (W)	17.95	120	34.4	1
18. Susanville (L)	18.20	4,152	40.3	198
19. Tule Lake (L)	10.03	4,036	41.9	140
20. Needles (L)	4.63	913	34.8	192
21. Burbank (W)	14.74	699	34.2	47
22. Los Angeles (W)	15.02	312	34.1	16
23. Long Beach (W)	12.36	50	33.8	12
24. Los Banos (L)	8.26	125	37.8	74
25. Blythe (L)	4.05	268	33.6	155
26. San Diego (W)	9.94	19	32.7	5
27. Daggett (L)	4.25	2,105	34.09	85
28. Death Valley (L)	1.66	−178	36.5	194
29. Crescent City (W)	74.87	35	41.7	1
30. Colusa (L)	15.95	60	39.2	91

Source: Taylor, P. J. "A pedagogic application of multiple regression analysis." *Geography,* July 1980, Vol. 65, pp. 203–212.

Problem 10.122

a. Conduct a complete analysis of the multiple regression model. Is the model useful for predicting precipitation level y?

b. Conduct an analysis of the regression residuals. Are model modifications necessary? Explain.

c. Refer to your plot of the residuals against $\hat{y}$ in part **b**. The *Geography* researcher noted that stations located on the westward-facing slopes of the California mountains (identified by the symbol "W" in the table) invariably had positive residuals (i.e., the least squares model underpredicted the level of precipitation) whereas stations on the leeward side of the mountains (identified by the symbol "L" in the table) had negative residuals (i.e., the least squares model overpredicted the level of precipitation). In the researcher's words, "This suggests a very clear shadow effect of the mountains, for which California is known." Verify this observation on your residual plot.

d. The shadow effect detected in part **c** suggests that a dummy variable for leeward/westward will improve the fit of the model. Let $x_4 = 1$ if leeward side and $x_4 = 0$ if westward side and then fit the modified model $y = \beta_0 + \beta_1 x_1 + \beta_2 x_2 + \beta_3 x_3 + \beta_4 x_4 + \varepsilon$ to the data. Does the addition of the β_4 term improve the fit of the model? Explain.

AUTO

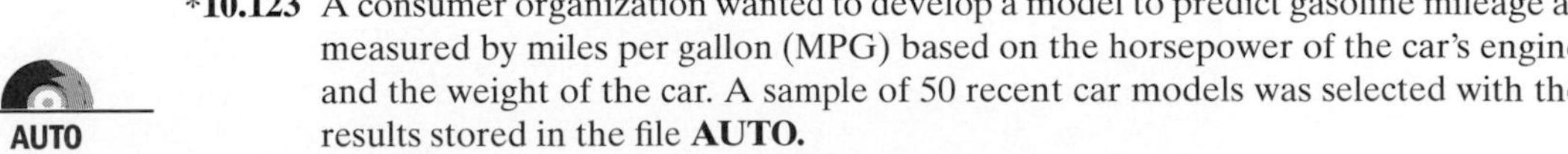

***10.123** A consumer organization wanted to develop a model to predict gasoline mileage as measured by miles per gallon (MPG) based on the horsepower of the car's engine and the weight of the car. A sample of 50 recent car models was selected with the results stored in the file **AUTO.**

a. Give the equation of the first-order multiple regression model.
b. Fit the model to the data using Excel or MINITAB.
c. Interpret the estimated slopes in this least squares prediction equation.
d. Predict the average miles per gallon for a car that has 60 horsepower and weighs 2,000 pounds.
e. Find and interpret a 95% confidence interval for the average miles per gallon for all cars with 60 horsepower and weight of 2,000 pounds.
f. Find a 95% prediction interval for the miles per gallon for an individual car that has 60 horsepower and weighs 2,000 pounds.
g. Find the adjusted R^2 and interpret its value.
h. Determine whether the first-order model with horsepower and weight is statistically useful for predicting gasoline mileage at $\alpha = .05$.

REFERENCES

Anscombe, F. J. "Graphs in Statistical Analysis," *American Statistician,* Vol. 27, 1973.

Barnett, V., and Lewis, T. *Outliers in Statistical Data.* New York: Wiley, 1978.

Belsley, D. A., Kuh, E., and Welsch, R. E. *Regression Diagnostics: Identifying Influential Data and Sources of Collinearity.* New York: Wiley, 1980.

Berenson, M., Levine, D., and Krehbiel, T. C. *Basic Business Statistics,* 8th ed. Upper Saddle River, N.J.: Prentice Hall, 2002.

Chatterjee, S., Hadi, A. S., and Price, B. *Regression Analysis by Example,* 3rd ed. New York: Wiley, 1999.

Conover, W. J. *Practical Nonparametric Statistics,* 2nd ed. New York: Wiley, 1980.

Daniel, W. W. *Applied Nonparametric Statistics,* 2nd ed. Boston: PWS-Kent, 1990.

Draper, N., and Smith, H. *Applied Regression Analysis,* 3rd ed. New York: Wiley, 1998.

Hollander, M., and Wolfe, D. A. *Nonparametric Statistical Methods.* New York: Wiley, 1973.

Mendenhall, W., and Sincich, T. *A Second Course in Statistics: Regression Analysis,* 5th ed. Upper Saddle River, N. J.: Prentice Hall, 1996.

Mosteller, F., and Tukey, J. W. *Data Analysis and Regression: A Second Course in Statistics.* Reading, Mass.: Addison-Wesley, 1977.

Neter, J., Kutner, M., Nachtsheim, C. J., and Wasserman, W. *Applied Linear Statistical Models,* 4th ed. Homewood, Ill.: Richard Irwin, 1996.

Rousseeuw, P. J., and Leroy, A. M. *Robust Regression and Outlier Detection.* New York: Wiley, 1987.

Using Microsoft® Excel

10.E.1 Generating Scatterplots and a Regression Line

Use the Microsoft Excel Chart Wizard to generate a scatter diagram (see p. 119–120) and then modify the scatter diagram by adding a chart trend line to generate a line of regression. For example, to generate the scatter diagram with line of regression for the data of Table 10.2 on p. 543, open the **CREATIVITY.XLS** workbook to the Data worksheet, select **Insert | Chart** from the Excel menu bar, and do the following to first produce the scatter diagram:

1. In the Step 1 dialog box:

a. Select the Standard Types tab and then select XY (Scatter) from the Chart type list box.

b. Select the first (top) Chart sub-type choice, described as "Scatter. Compares pairs of values." when selected, and click the Next button.

2. In the Step 2 dialog box:

a. Select the Data Range tab. Enter Data!B1:C11 in the Data range edit box.

b. Select the Columns option button in the Series in group and click Next.

3. In the Step 3 dialog box:

a. Select the Titles tab. Enter Scatter Diagram as the Chart title, Flexibility Score in the Value (X) axis edit box, and Creativity Score in the Value (Y) axis edit box.

b. Select, in turn, the Axes, Gridlines, Legend, and Data Label tabs and adjust settings as discussed in the Excel Primer on p. 14.

c. Click Next.

4. In the Step 4 dialog box, select the As new sheet option button and click the Finish button to generate the chart.

When generating scatter diagrams, the Chart Wizard always assumes that the first column (or row) of data of the data range entered in the Step 2 dialog box contains values for the x variable (as it does in the preceding example). Had the x variable data been located in the second column of the data range, it would have been necessary to select the Series tab of the Step 2 dialog box and change the cell ranges in the X Values and Y Values edit boxes. Furthermore, due to a quirk in this wizard, the revised cell ranges must be entered into those edit boxes as formulas that include sheet names, e.g., =Data!B1:B11. Using the simplified cell range form, e.g., B1:B11, would cause Microsoft Excel to display the misleading "The formula you typed contains an error" error message.

To add a line of regression to the new chart produced in steps 1–4 above, select the new chart sheet and

1. Select Chart | Add Trendline. (The Chart choice appears on the Microsoft Excel menu bar only when a chart or chart sheet is selected.)

2. In the Add Trendline dialog box:

a. Select the Type tab and select the Linear choice in the Trend/Regression type group (see Figure 10.E.1).

b. Select the Options tab and select the Automatic option button and the Display equation on chart and Display R-squared value on chart check boxes (see Figure 10.E.2).

c. Click the OK button.

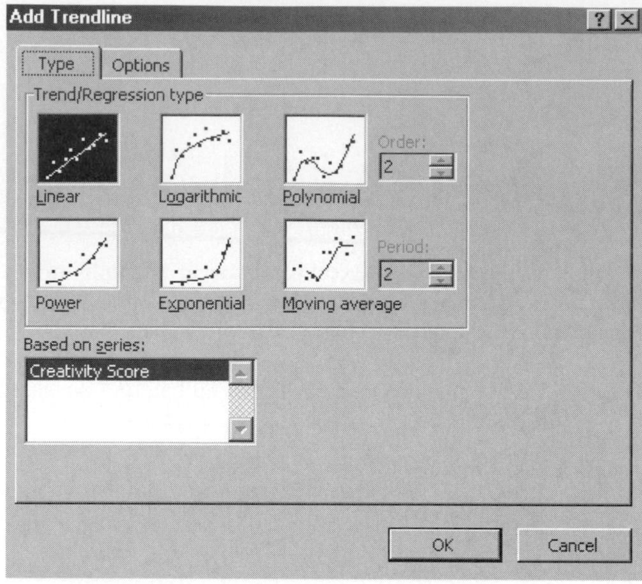

E **Figure 10.E.1** Microsoft Excel Chart Wizard Add Trendline Dialog Box

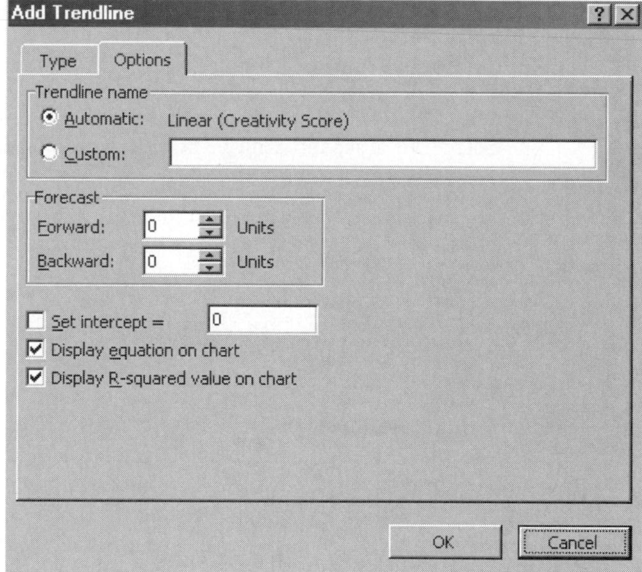

E **Figure 10.E.2** Microsoft Excel Chart Wizard Options Tab Dialog Box

10.E.2 Fitting Simple Linear Regression Models

Use the PHStat **Regression | Simple Linear Regression** procedure to calculate the regression coefficients and other statistics for a simple linear regression analysis. For example, to calculate the regression coefficients and other statistics for the simple linear regression analysis for the creativity–flexibility scores data of Table 10.2 on p. 543, open the **CREATIVITY.XLS** workbook to the Data worksheet and

1. Select PHStat | Regression | Simple Linear Regression.
2. In the Simple Linear Regression dialog box (see Figure 10.E.3):

   a. Enter C1:C11 in the Y Variable Cell Range edit box.

   b. Enter B1:B11 in the X Variable Cell Range edit box.

E Figure 10.E.3 PHStat Simple Linear Regression Dialog Box

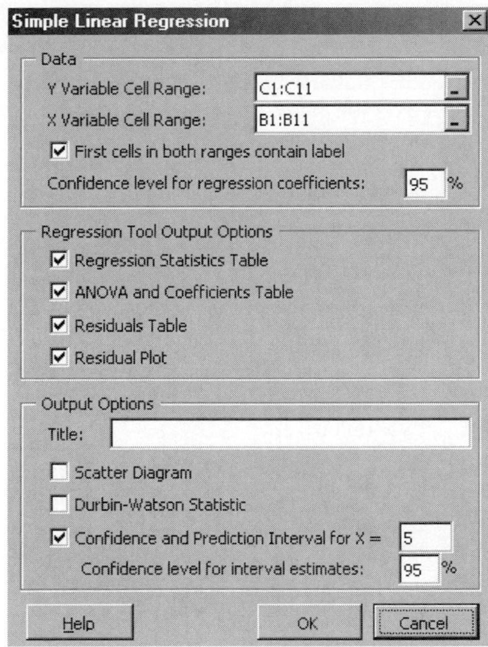

c. Select the First cells in both ranges contain label check box.

d. Enter 95 in the Confidence level for regression coefficients edit box.

e. Select the Regression Statistics Table, ANOVA and Coefficients Table, Residuals Table, and Residual Plot check boxes.

f. Select the Confidence and Prediction Interval for X = check box and enter 5 in its edit box. Enter 95 in the Confidence level for interval estimates edit box.

g. Enter a title in the Title edit box.

h. Click the OK button.

10.E.3 Testing for Correlation

To perform a *t* test of the hypothesis of no correlation, implement a worksheet based on the design shown in Table 10.E.1. This design uses the TINV function and the TDIST function. The format of these functions is

TINV(*level of significance*, *degrees of freedom*)

TDIST(*absolute value of the t test statistic*, *degrees of freedom*, *tails*)

where *tails* = 1 for a one-tailed test and 2 for a two-tailed test.

TINV returns the critical value of the *t* statistic. TDIST returns the probability of exceeding a given *t* value for a given number of degrees of freedom and type of test.

As an example, to test the null hypothesis that the correlation coefficient is zero for the tire pressure problem of Example 10.15 on p. 565, implement rows 1 through 13 of the Table 10.E.1 design. Only those rows need to be implemented, as only the two-tailed test is needed for this example. To implement these rows, do the following:

1. Select File | Open to open a new workbook (or open the existing workbook into which the hypothesis testing worksheet is to be inserted).

2. Select an unused worksheet (or select Insert | Worksheet if there are none) and rename the sheet hypothesis.

TABLE 10.E.1 Worksheet Design for Testing for the Existence of Correlation

	A	B
1	*t* Test for the Existence of Correlation	
2		
3	Level of Significance	.XX
4	Sample Size	XXX
5	Sample Correlation Coefficient	XXX
6	Degrees of Freedom	=B4-2
7	*t* Test Statistic	=B5/SQRT(1-B5^2)
8		
9	Two-Tailed Test	
10	Lower Critical Value	=-TINV(B3,B6)
11	Upper Critical Value	=TINV(B3,B6)
12	*p*-Value	=TDIST(ABS(B7),B6,2)
13	=IF(B12<$B$3, "Reject the null hypothesis", Do not reject the null hypothesis")	
14		
15	Lower-Tail Test	
16	Lower Critical Value	=-TINV(2*B3,B6)
17	*p*-Value	=IF(B7<0,E19,E20)
18	=IF(B17<$B$3, "Reject the null hypothesis", Do not reject the null hypothesis")	
19		
20	Upper-Tail Test	
21	Upper Critical Value	=TINV(2*B3,B6)
22	*p*-Value	=IF(B7<0,E20,E19)
23	=IF (B22<$B$3, "Reject the null hypothesis", Do not reject the null hypothesis")	

	D	E
17	Calculations Area	
18	For one-tail tests	
19	TDIST value	=TDIST(ABS(B7),B6,1)
20	1- TDIST value	1 – E19

3. Enter the title, labels, and formulas for column A as shown in Table 10.E.1. Enter the formulas in cells A13, A18, and A23, which have been typeset as two lines, as one continuous line.

4. Enter the level of significance, sample size, and sample correlation coefficient in the cell range B3:B5. If the correlation coefficient needs to be computed, use the CORREL function explained in Section 3.E.4. For the tire pressure example, enter 0.05 in cell B3, 14 in cell B4, and −0.114 in cell B5.

5. Enter the formulas for cells B6, B7, and B10:B12.

10.E.4 Fitting Multiple Regression Models (Optional)

Use the PHStat **Regression | Multiple Regression** procedure to calculate the regression coefficients and other statistics for a multiple regression analysis. Note that all x variables must be in contiguous columns. For example, to calculate the regression coefficients and other statistics for the multiple regression analysis for the boiler drum data of Problem 10.94 on p. 611, open the **BOILER.XLS** workbook to the Data worksheet and

1. Select PHStat | Regression | Multiple Regression.
2. In the Multiple Regression dialog box (see Figure 10.E.4):

E Figure 10.E.4 PHStat Multiple Regression Dialog Box

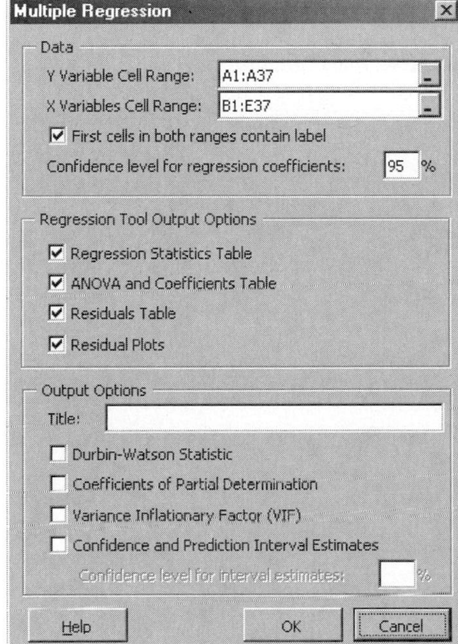

a. Enter A1:A37 in the Y Variable Cell Range edit box.
b. Enter B1:E37 in the X Variables Cell Range edit box.
c. Select the First cells in both ranges contain label check box.
d. Enter 95 in the Confidence level for regression coefficients edit box.
e. Select the Regression Statistics Table, the ANOVA and Coefficients Table, the Residuals Table, and Residual Plots check boxes.
f. Enter a title in the Title edit box.
g. Click the OK button.

If a second-order (quadratic) regression model is to be used, we can use simple formulas to create the square of an explanatory variable. For example, if the cell range for an explanatory variable is B2:B20, a variable that is the square of this x variable can be created by entering the formula = B2^2 in cell C2, and copying this formula through cell C20.

If a regression model with dummy variables is to be used, and the category values for a dummy variable are provided rather than numerical codes, the Excel Find and Replace procedure can be used to code the category names into 0 and 1 codes. Suppose that the category codes Yes and No were contained in the cell range C1:C51 To change these categorical labels into the numerical codes of 0 for No and 1 for Yes, do

the following:

1. Select the range C1:C51 containing the categorical responses Yes and No, then select Edit | Replace.
2. In the Replace dialog box:
   a. Enter Yes in the Find what edit box.
   b. Enter 1 in the Replace with edit box.
   c. Click the Replace All button.
3. With the cell range C1:C51 still selected, select Edit | Replace a second time.
4. In the Replace dialog box:
   a. Enter No in the Find what edit box.
   b. Enter 0 in the Replace with edit box.
5. Click the Replace All button.

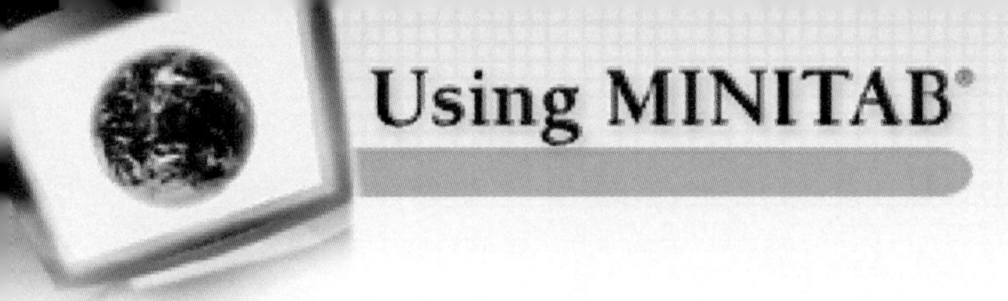

10.M.1 Generating Scatterplots and a Regression Line

Use the MINITAB **Stat | Regression | Fitted Line Plot** procedure to obtain a scatter diagram with a fitted regression line. To illustrate the use of MINITAB to obtain a scatter diagram with a fitted regression line with the creativity–flexibility data of Example 10.5 on p. 542, open the **CREATIVITY.MTW** worksheet and

1. Select Stat | Regression | Fitted Line Plot.
2. In the Fitted Line Plot dialog box (see Figure 10.M.1), enter **C3** or **'Creativity Score'** in the Response edit box and **C2** or **'Flexibility Score'** in the Predictor edit box.
3. Click the **OK** button.

Figure 10.M.1 MINITAB Fitted Line Plot Dialog Box

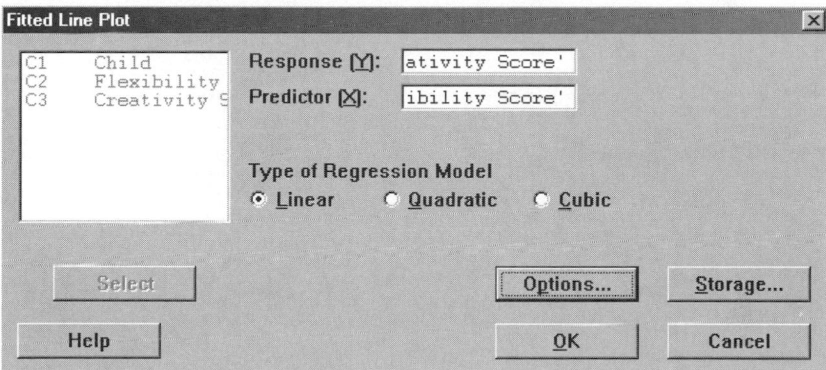

10.M.2 Fitting Simple Linear Regression Models

MINITAB can be used for simple linear regression by using **Stat | Regression | Regression.** To illustrate the use of MINITAB for simple linear regression with the creativity–flexibility data of Example 10.5 on p. 542, open the **CREATIVITY.MTW** worksheet and

1. Select Stat | Regression | Regression.
2. In the Regression dialog box (see Figure 10.M.2 on p. 646), enter **C3** or **'Creativity Score'** in the Response edit box and **C2** or **'Flexibility Score'** in the Predictors edit box. Click the **Graphs** option button.
3. In the Graphs dialog box (see Figure 10.M.3 on p. 646), select the **Standardized** option button under Residuals for Plots. For Residual Plots, select the **Histogram of residuals, Normal plot of residuals,** and **Residuals versus fits** check boxes. In the Residuals versus the variables edit box, enter **C2** or **'Flexibility Score'**. Click the **OK** button to return to the Regression dialog box.
4. Click the **Results** option button. In the Results dialog box (see Figure 10.M.4 on p. 646), select the **Regression equation, table of coefficients, s, R-squared, and basic analysis of variance** option button. Click the **OK** button to return to the Regression dialog box. Click the **OK** button.

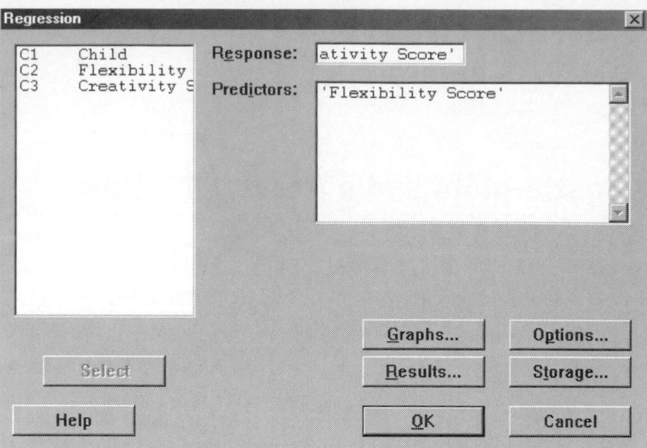

Figure 10.M.2 MINITAB Regression Dialog Box

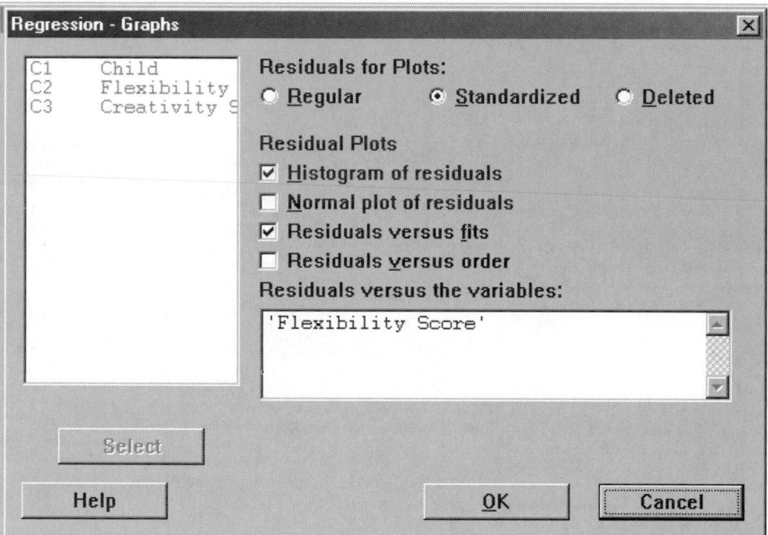

Figure 10.M.3 MINITAB Regression—Graphs Dialog Box

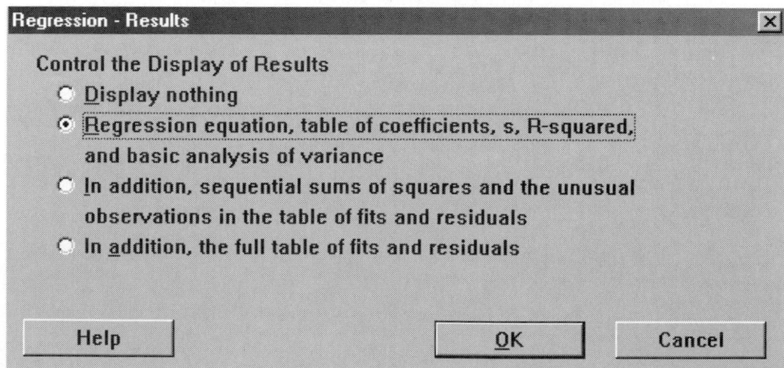

Figure 10.M.4 MINITAB Regression—Results Dialog Box

10.M.3 Testing for Correlation

Use the MINITAB **Stat | Basic Statistics | Correlation** procedure to obtain a test for a correlation coefficient. To illustrate the use of MINITAB to obtain a test for a correlation coefficient with the tire data of Example 10.15 on p. 565, open the **TIRES.MTW**

worksheet and

1. Select Stat | Basic Statistics | Correlation.
2. In the Correlation dialog box (see Figure 10.M.5), enter **C1** or **'Pressure'** and **C2** or **'Mileage'** in the Variables edit box
3. Select the Display p-values check box.
4. Click the **OK** button.

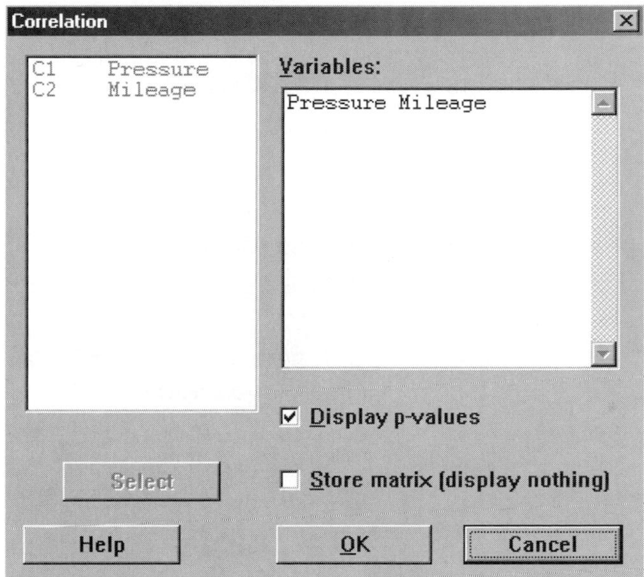

M Figure 10.M.5 MINITAB Correlation Dialog Box

10.M.4 Fitting Multiple Regression Models (Optional)

In Section 10.M.2, instructions were provided for using MINITAB for simple linear regression. The same set of instructions are valid in using MINITAB for multiple regression. To obtain a regression analysis for the bid-rigging data of Example 10.30 on p. 603, open the **BIDRIG.MTW** worksheet, and select **Stat | Regression | Regression.** Enter **C1** or **'LOW-BID'** in the Response edit box and **C2** or **'DOTEST'** and **C6** or **'BIDS'** in the Predictors edit box. Click the **Graphs** button. In the Residuals for Plots edit box, select the **Standardized** option button. For Residual Plots, select the **Histogram of residuals** check box. In the Residuals versus the variables edit box, select **C2** or **'DOTEST'** and **C6** or **'BIDS.'** Click the **OK** button to return to the Regression dialog box. Click the **Results** option button. In the Regression Results dialog box, click the **In addition, the full table of fits and residuals** option button. Click the **OK** button to return to the Regression dialog box. Click the **OK** button.

CHAPTER 11

Analysis of Variance

OBJECTIVES

1. To introduce the concepts of experimental design
2. To develop the analysis of variance for data collected from a completely randomized design
3. To develop a procedure for making multiple comparisons of means
4. To develop the analysis of variance for data collected from a factorial design
5. To present a nonparametric alternative to the one-way analysis of variance procedure

CONTENTS

EXCEL TUTORIAL

MINITAB TUTORIAL

Statistics in the Real World

On the Trail of the Cockroach

Entomologists have long established that insects such as ants, bees, caterpillars, and termites use chemical or "odor" trails for navigation. These trails are used as highways between sources of food and the insect nest. Until recently, however, "bug" researchers believed that the navigational behavior of cockroaches scavenging for food was random and not linked to a chemical trail.

One of the first researchers to challenge the "random-walk" theory for cockroaches was professor and entomologist Dini Miller of Virginia Tech University. According to Miller, "the idea that roaches forage randomly means that they would have to come out of their hiding places every night and bump into food and water by accident. But roaches never seem to go hungry." Since cockroaches had never before been evaluated for trail-following behavior, Miller designed an experiment to test a cockroach's ability to follow a trail of their fecal material (*Explore*, Research at the University of Florida, Fall 1998).

First, Dr. Miller developed a methanol extract from roach feces—called a pheromone. She theorized that "pheromones are communication devices between cockroaches. If you have an infestation and have a lot of fecal material around, it advertises, 'Hey, this is a good cockroach place.'" Then, she created a chemical trail with the pheromone on a strip of white chromatography paper and placed the paper at the bottom of a plastic, V-shaped container, 122 square centimeters in size. German cockroaches were released into the container at the beginning of the trail, one at a time, and a video surveillance camera was used to monitor the roach's movements.

In addition to the trail containing the fecal extract (the treatment), a trail using methanol only was created. This second trail served as a "control" to compare back against the treated trail. Since Dr. Miller also wanted to determine if trail-following ability differed among cockroaches of different age, sex, and reproductive status, four roach groups were utilized in the experiment: adult males, adult females, gravid (pregnant) females, and nymphs (immatures). Twenty roaches of each type were randomly assigned to the treatment trail and ten of each type were randomly assigned to the control trail. Thus, a total of 120 roaches were used in the experiment.

ROACH

The movement pattern of each cockroach tested was translated into (x, y) coordinates every one-tenth of a second by the Dynamic Animal Movement Analyzer (DAMA) program. Miller measured the perpendicular distance of each (x, y) coordinate from the trail and then averaged these distances, or deviations, for each cockroach. The average trail deviations (measured in "pixels", where 1 pixel equals approximately 2 centimeters) for each of the 120 cockroaches in the study are stored in the data file named **ROACH.**

In the following Statistics in the Real World Revisited sections, we apply the statistical methodology presented in this chapter to the cockroach data.

Statistics in the Real World Revisited

- A One-Way Analysis of the Cockroach Data (p. 660)
- Ranking the Means of the Cockroach Groups (p. 670)
- Checking the Assumptions for the Cockroach ANOVA (p. 689)
- A Nonparametric Analysis of the Cockroach Data (p. 702)

11.1 Experimental Design

As we have seen in Chapters 7, 8, and 9, the solutions to many statistical problems are based on inferences about population means. In Chapters 7 and 8 we considered inferences about a single population mean, and in Chapter 9 we developed inferences for the difference between two population means. In this chapter we expand our discussion to a comparison of three or more means. The procedure for selecting sample data for a comparison of several population means is called the *design of the experiment,* and the statistical procedure for comparing the means is called an *analysis of variance.* The objective of this chapter is to introduce some aspects of experimental design and the analysis of data from such experiments using an analysis of variance.

The study of experimental design originated in England and, in its early years, was associated solely with agricultural experimentation. The need for experimental design in agriculture was very clear: It takes a full year to obtain a single observation on the yield of a new variety of wheat. Consequently, the need to save time and money led to a study of ways to obtain more information using smaller samples. Similar motivation led to its subsequent acceptance and wide use in fields such as biology, engineering, psychology, and marketing. Despite this fact, the terminology associated with experimental design clearly indicates its early association with the biological sciences.

We call the process of collecting sample data an *experiment* and the variable to be measured the *response.* (In this chapter we consider only experiments in which the response is a quantitative variable. Experiments with qualitative responses are beyond the scope of this text.) The planning of the sampling procedure is called the *design* of the experiment. The object upon which the response measurement is taken is called an *experimental* (or *sampling*) unit.

Definition 11.1

The process of collecting sample data is called an **experiment.**

Definition 11.2

The plan for collecting the sample is called the **design** of the experiment.

Definition 11.3

The variable measured in the experiment is called the **response variable.** (In this chapter, all response variables will be quantitative variables.) The response will be denoted with the symbol y.

Definition 11.4

The object upon which the response variable is measured is called an **experimental** (or **sampling**) **unit.**

Variables that may be related to a response variable are called *factors*. The value assumed by a factor in an experiment is called a *level.* The combinations of levels of the factors for which the response will be observed are called *treatments.*

Definition 11.5

The variables, quantitative or qualitative, that are related to a response variable are called **factors.**

Definition 11.6

The intensity setting of a factor (i.e., the value assumed by a factor in an experiment) is called a **factor level.**

Definition 11.7

A **treatment** is a particular combination of levels of the factors involved in an experiment.

EXAMPLE 11.1

A DESIGNED EXPERIMENT

A marketing study is conducted to investigate the effects of brand and shelf location on mean weekly coffee sales. Coffee sales are recorded for each of the two brands (brand A and brand B) and three shelf locations (bottom, middle, and top). The $2 \times 3 = 6$ combinations of brand and shelf location were varied each week for a period of 18 weeks. A layout of the design is displayed in Figure 11.1. For this experiment, identify

a. the experimental unit
b. the response y
c. the factors
d. the factor levels
e. the treatments
f. the objective of the analysis

Figure 11.1 Layout for Designed Experiment of Example 11.1

		Shelf Location		
		Bottom	Middle	Top
Brand	A	Week 1 9 14	Week 2 7 16	Week 4 12 17
	B	Week 5 10 13	Week 3 8 18	Week 6 11 15

Solution

a. Since our data will be collected each week for a period of 18 weeks, the experimental unit is a week.

b. The variable of interest, i.e., the response, is y = weekly coffee sales. Note that weekly coffee sales is a quantitative variable.

c. Since we are interested in investigating the effect of brand and shelf location on mean sales, *brand* and *shelf location* are the factors. Note that both factors are qualitative variables, although, in general, they may be quantitative or qualitative.

d. For this experiment, brand is measured at two levels (A and B) and shelf location at three levels (bottom, middle, and top).

e. Since coffee sales are recorded for each of the six brand-shelf location combinations (brand A, bottom), (brand A, middle), (brand A, top), (brand B, bottom), (brand B, middle), and (brand B, top), the experiment involves six treatments. The term *treatments* is used to describe the factor-level combinations to be included in an experiment because many experiments involve "treating" or doing something to alter the nature of the experimental unit. Thus, we might view the six brand-shelf location combinations as treatments on the experimental units in the marketing study involving coffee sales.

f. The objective of the analysis is to compare the mean weekly coffee sales associated with the six treatments.

Now that you understand some of the terminology, it is helpful to think of the design of an experiment in four steps.

DESIGNING AN EXPERIMENT

1. Select the factors to be included in the experiment and identify the parameters that are the object of the study. Usually, the target parameters are the population means associated with the factor-level combinations (i.e., treatments).
2. Choose the treatments (the factor-level combinations) to be included in the experiment.
3. Determine the number of observations (sample size) to be made for each treatment. (This will usually depend on the standard error(s) that you desire. See Section 7.5.)
4. Plan how the treatments will be assigned to the experimental units. That is, decide on which design to employ.

Self-Test 11.1

The United States Golf Association (USGA) regularly tests golf equipment to ensure that it conforms to USGA standards. Suppose it wishes to compare the mean distance traveled by four different brands of golf balls when struck by a driver (the club used to maximize distance). The following experiment is conducted: 10 balls of each brand are randomly selected. Each is struck by "Iron Byron" (the USGA's golf robot named for the famous golfer Byron Nelson) using a driver, and the distance traveled is recorded. Identify each of the following elements in this experiment: response, factors, levels, treatments, and experimental units.

Entire texts are devoted to properly executing these steps for various experimental designs (e.g., Cochran and Cox [1957]). The main objective of this chapter, however, is to show how to analyze the data that are collected in a designed experiment.

In Sections 11.3 and 11.5, we consider two widely used experimental designs and demonstrate how to analyze the data for each design. First, in Section 11.2, we present a short discussion of the logic behind the analysis of data collected from such experiments.

11.2 ANOVA Fundamentals

Once the data for a designed experiment have been collected, we use the sample information to make inferences about the population means associated with the various treatments. The method used to compare the treatment means is known as **analysis of variance,** or **ANOVA.** The concept behind an analysis of variance can be explained using the following simple example.

EXAMPLE 11.2 **CONCEPT OF ANOVA**

Suppose we want to compare the means of two populations (μ_1 and μ_2) using independent random samples of size $n_1 = n_2 = 5$ from each of the populations. The sample observations and the sample means are listed in Table 11.1 and shown on a dot plot in Figure 11.2.

a. Do you think these data provide sufficient evidence to indicate a difference between the population means μ_1 and μ_2?

TABLE 11.1 Data for Example 11.2a

Sample from Population 1	Sample from Population 2
6	8
−1	1
0	3
4	7
1	6
$\bar{y}_1 = 2$	$\bar{y}_2 = 5$

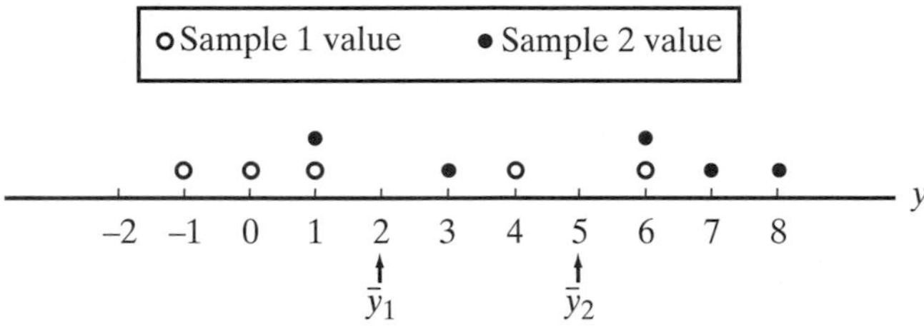

Figure 11.2 Dot Plot of Data in Table 11.1

b. Now look at two more samples of $n_1 = n_2 = 5$ measurements from the populations, as listed in Table 11.2 and plotted in Figure 11.3. Do these data appear to provide evidence of a difference between μ_1 and μ_2?

TABLE 11.2 Data for Example 11.2b

Sample from Population 1	Sample from Population 2
2	5
3	5
2	5
2	4
1	6
$\bar{y}_1 = 2$	$\bar{y}_2 = 5$

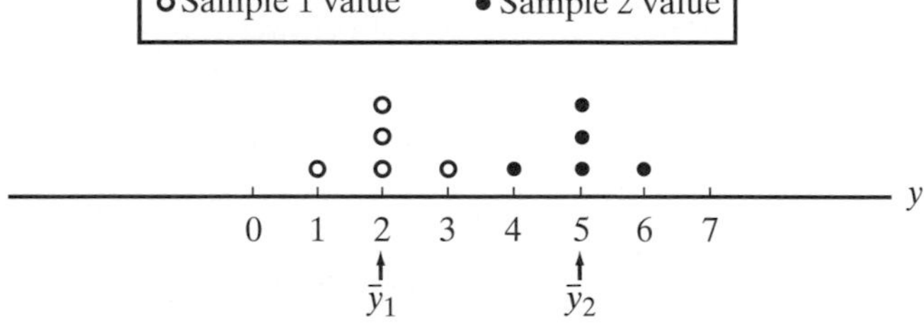

Figure 11.3 Dot Plot of Data in Table 11.2

Solution

a. One way to determine whether a difference exists between the population means μ_1 and μ_2 is to examine the difference *between* the sample means, $(\bar{y}_1 - \bar{y}_2)$, and to compare it to a measure of variability *within* the samples. The greater the disparity in these two measures, the greater will be the evidence to indicate a difference between μ_1 and μ_2.

For the data of Table 11.1, you can see in Figure 11.2 that *the difference between the sample means is small relative to the variability within the sample observations.* Thus, we think you will agree that the difference between $\bar{y}_1$ and $\bar{y}_2$ is probably not large enough to indicate a difference between μ_1 and μ_2.

b. Notice that the difference between the sample means for the data of Table 11.2 is identical to the difference shown in Table 11.1. However, since there is now very little variability within the sample observations (see Figure 11.3), *the difference between the sample means is large compared to the variability within the sample observations*. Thus, the data appear to give evidence of a difference between μ_1 and μ_2.

To conduct a formal test of hypothesis, we need to quantify these two measures of variation. The **within-sample variation** is measured by the pooled s^2 that we computed for the independent random samples t test of Section 9.2, namely

$$\text{Within-sample variation: } s^2 = \frac{\sum_{i=1}^{n_1}(y_{i1} - \bar{y}_1)^2 + \sum_{i=1}^{n_2}(y_{i2} - \bar{y}_2)^2}{n_1 + n_2 - 2} \tag{11.1}$$

$$= \frac{\text{SSE}}{n_1 + n_2 - 2} = \frac{\text{SSE}}{n - 2} = \text{MSE}$$

where y_{i1} is the ith observation in sample 1, y_{i2} is the ith observation in sample 2, and $n = (n_1 + n_2)$ is the total sample size. The quantity in the numerator of s^2 is often denoted SSE, the sum of squared errors. SSE measures unexplained variability—that is, it measures variability *unexplained* by the differences between the sample means. The ratio $\text{SSE}/(n - 2)$ is also called **mean square for error,** or **MSE.** Thus, MSE is another name for s^2.

A measure of the **between-sample variation** is given by the weighted sum of squares of deviations of the individual sample means about the mean for all n observations, $\bar{y}$, divided by the number of samples minus 1, i.e.,

$$\text{Between-sample variation: } s^2 = \frac{n_1(\bar{y}_1 - \bar{y})^2 + n_2(\bar{y}_2 - \bar{y})^2}{2 - 1} \tag{11.2}$$

$$= \frac{\text{SST}}{1} = \text{MST}$$

The quantity in the numerator is often denoted SST, the **sum of squares for treatments,** because it measures the variability *explained* by the differences between the sample means of the two treatments. The ratio $\text{SST}/(2 - 1)$ is also called **mean square for treatments,** or **MST.**

For this experimental design, SSE and SST sum to a known total, namely,

$$\text{SS(Total)} = \sum (y_i - \bar{y})^2$$

In the next section we will see that

$$F = \frac{\text{Between-sample variation}}{\text{Within-sample variation}} = \frac{\text{MST}}{\text{MSE}} \tag{11.3}$$

has an F distribution with $\nu_1 = 1$ and $\nu_2 = n_1 + n_2 - 2$ degrees of freedom (df) and therefore can be used to test the null hypothesis of no difference between the treatment means. Large values of F lead us to conclude that a difference between the treatment (population) means exists.

The additivity property of the sums of squares led early researchers to view this analysis as a *partitioning* of $\text{SS(Total)} = \sum(y_i - \bar{y})^2$ into sources corresponding to the factors included in the experiment and to SSE. The formulas for computing the sums of squares, the additivity property, and the form of the test statistic made it natural for this procedure to be called an *analysis of variance.* We demonstrate

the analysis of variance procedures for the general problem of comparing k population means for two special types of experimental designs in Sections 11.3 and 11.5.

11.3 Completely Randomized Designs: One-Way ANOVA

The simplest experimental design employed in practice is called a *completely randomized design.* This experiment involves a comparison of the means for a number, say k, of treatments, based on independent random samples of n_1, $n_2, \ldots, n_k$ observations, drawn from populations associated with treatments 1, 2, . . . , k, respectively. A layout for a completely randomized design is shown in Figure 11.4

Figure 11.4 Layout for a Completely Randomized Design

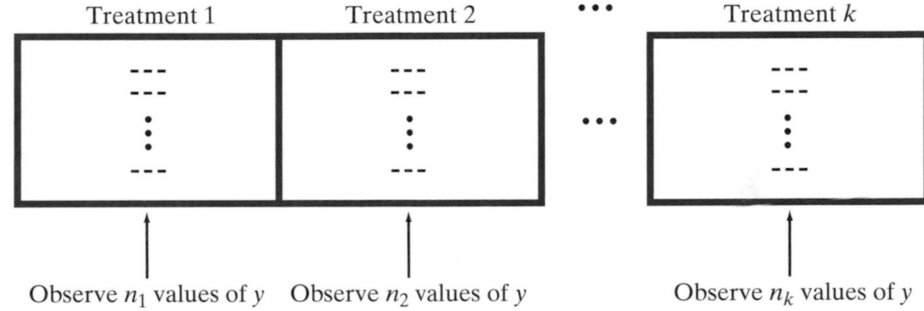

Definition 11.8

A **completely randomized design** to compare k treatment (population) means is one in which the treatments are randomly assigned to the experimental units, or in which independent random samples are drawn from each of the k target populations.

A key feature of a completely randomized design is that there is only a single factor of interest. (For this reason, the analysis of the data is often called a *one-way analysis of variance.)* The levels of this factor represent the k treatments (or populations) to be compared. If we define μ_i as the mean of the population of measurements associated with treatment i, for $i = 1, 2, \ldots, k$, then the null hypothesis to be tested is that the k treatment means are equal, i.e.,

$$H_0\colon \mu_1 = \mu_2 = \cdots = \mu_k$$

and the alternative hypothesis is that at least two of the treatment means differ.

EXAMPLE 11.3 **SETTING UP THE ONE-WAY ANOVA**

Sociologists often conduct experiments to investigate the relationship between socioeconomic status and college performance. Socioeconomic status is generally partitioned into three groups: lower class, middle class, and upper class. Consider the problem of comparing the mean grade point averages of those college freshmen associated with the lower class, those associated with the middle class, and those associated with the upper class. The grade point averages for random samples of seven college freshmen associated with each class—a total of 21 freshmen—

were selected from a university's files at the end of the academic year. The data are recorded in Table 11.3.

GPACLASS

TABLE 11.3 Grade Point Averages for Three Socioeconomic Groups, Example 11.3

	(1) Lower Class	(2) Middle Class	(3) Upper Class
	2.87	3.23	2.25
	2.16	3.45	3.13
	3.14	2.78	2.44
	2.51	3.77	3.27
	1.80	2.97	2.81
	3.01	3.53	1.36
	2.16	3.01	2.53
Sample Means:	$\bar{y}_1 = 2.52$	$\bar{y}_2 = 3.25$	$\bar{y}_3 = 2.54$

a. Verify that the experimental design is completely randomized, and identify the treatments.

b. State the null and alternative hypotheses for a test to determine whether there is a difference among the true mean grade point averages (GPAs) for the three socioeconomic classes.

Solution

a. Since the data are collected from independent random samples of college freshmen from each population (socioeconomic class), this is a completely randomized design. The three socioeconomic classes—lower class, middle class, and upper class—represent the three treatments (or populations).

b. Our objective is to determine whether differences exist among the three population (socioeconomic class) GPA means. Consequently, we will test the null hypothesis "All three population means—μ_1, μ_2, and μ_3—are equal." That is,

$$H_0\colon \mu_1 = \mu_2 = \mu_3$$

where μ_1 is the mean GPA for those college freshmen associated with the lower socioeconomic class, μ_2 is the mean GPA for those associated with the middle class, and μ_3 is the mean GPA for those associated with the upper class. The alternative hypothesis of interest is

$$H_a\text{: At least two of the population means differ}$$

If an analysis of variance of the data in Table 11.3 shows that the differences among the sample means are large enough to indicate differences among the corresponding population means, we will reject the null hypothesis H_0 in favor of the alternative hypothesis H_a; otherwise, we will not reject H_0 and will conclude that there is insufficient evidence to indicate differences among the population means.

An analysis of variance for a completely randomized design partitions SS(Total) into two components, SSE and SST (see Figure 11.5). Recall that the quantity SST denotes the sum of squares for treatments and measures the variation explained by

Figure 11.5 The Partitioning of SS(Total) for a Completely Randomized Design

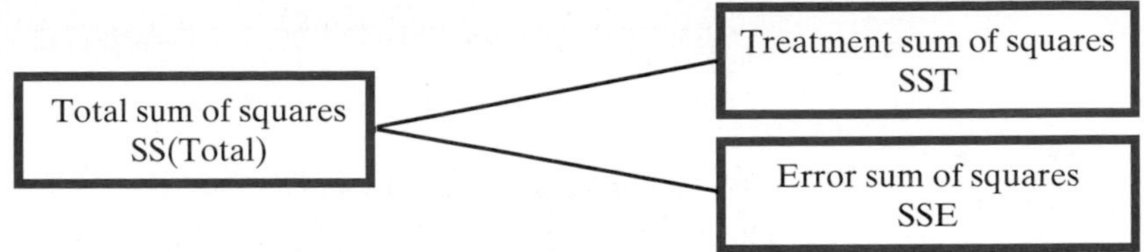

the differences between the treatment means. The sum of squares for error, SSE, is a measure of the unexplained variability, obtained by calculating a pooled measure of the variability *within* the k samples. If the treatment means truly differ, then SSE should be substantially smaller than SST. We compare the two sources of variability by forming an F statistic:

$$F = \frac{\text{SST}/(k - 1)}{\text{SSE}/(n - k)} = \frac{\text{MST}}{\text{MSE}} \quad \textbf{(11.4)}$$

where n is the total number of measurements.

Under certain conditions, the F statistic has an F *distribution* that depends on ν_1 numerator degrees of freedom and ν_2 denominator degrees of freedom. For the completely randomized design, F is based on $\nu_1 = (k - 1)$ and $\nu_2 = (n - k)$ degrees of freedom. If the computed value of F is greater than the upper critical value F_α, we reject H_0 and conclude that at least two of the treatment means differ.

ANOVA *F* TEST TO COMPARE *k* POPULATION MEANS FOR A COMPLETELY RANDOMIZED DESIGN

H_0: $\mu_1 = \mu_2 = \cdots = \mu_k$ (i.e., no difference in the treatment [population] means)

H_a: At least two treatment means differ

Test statistic: $F = \frac{\text{MST}}{\text{MSE}}$ **(11.4)**

Rejection region: $F > F_\alpha$

where the distribution of F is based on $(k - 1)$ numerator df and $(n - k)$ denominator df, and F_α is the F value found in Table B.6 of Appendix B such that $P(F > F_\alpha) = \alpha$.

Assumptions:

1. All k population probability distributions are normal.
2. The k population variances are equal.
3. The samples from each population are random and independent.

The results of an analysis of variance are usually summarized and presented in an **analysis of variance (ANOVA) table.** Such a table shows the sources of variation, their respective degrees of freedom, sums of squares, mean squares, and computed F statistic. The general form of the ANOVA table for a completely randomized design is shown in the box on p. 658. In practice, the entries in the ANOVA table are obtained using spreadsheet or statistical software. (For those who are interested, we provide the calculation formulas in optional Section 11.7.)

GENERAL ANOVA TABLE FOR A COMPLETELY RANDOMIZED DESIGN

Source	df	SS	MS	F
Treatments	$k - 1$	SST	$\text{MST} = \dfrac{\text{SST}}{k-1}$	$F = \dfrac{\text{MST}}{\text{MSE}}$
Error	$n - k$	SSE	$\text{MSE} = \dfrac{\text{SSE}}{n-k}$	
TOTAL	$n - 1$	SS(Total)		

EXAMPLE 11.4

CONDUCTING THE ONE-WAY ANOVA

Refer to Example 11.3 and the data presented in Table 11.3. Conduct the test of the hypotheses

$$H_0: \mu_1 = \mu_2 = \mu_3$$

H_a: At least two population means are different

where μ_1 is the true mean GPA for lower-class college freshmen, μ_2 is the true mean GPA for middle-class college freshmen, and μ_3 is the true mean GPA for upper-class college freshmen. Use $\alpha = .05$.

Solution

An Excel printout of the analysis is shown in Figure 11.6. The sums of squares (SST and SSE) and mean squares (MST and MSE) in the resulting ANOVA table are indicated on the printout. The test statistic, $F = 4.59$, is highlighted on the printout. To carry out the test, we can use either the rejection region (at $\alpha = .05$) or the p-value of the test. Both the critical F value and p-value of the test are also highlighted in Figure 11.6.

E Figure 11.6 Excel Output for One-Way ANOVA, Example 11.4

	A	B	C	D	E	F	G
1	Anova: Single Factor						
2							
3	SUMMARY						
4	*Groups*	*Count*	*Sum*	*Average*	*Variance*		
5	Lower Class	7	17.65	2.521429	0.254114		
6	Middle Class	7	22.74	3.248571	0.124348		
7	Upper Class	7	17.79	2.541429	0.406748		
8							
9							
10	ANOVA						
11	*Source of Variation*	*SS*	*df*	*MS*	*F*	*P-value*	*F crit*
12	Between Groups (Treatments)	2.401438095 (SST)	2	1.200719 (MST)	4.587511	0.024543	3.554561
13	Within Groups (Error)	4.711257143 (SSE)	18	0.261737 (MSE)			
14	Total	7.112695238	20				

The critical F value for the $\alpha = .05$ rejection region, given as **F crit** in Figure 11.6, is $F_{.05} = 3.55$. Thus, we will reject the null hypothesis that the three means are equal if

$$F > F_{.05} = 3.55$$

Since the computed value of the test statistic, $F = 4.59$, is greater than the critical value, we have sufficient evidence (at $\alpha = .05$) to conclude that the true mean freshmen GPAs differ for at least two of the three socioeconomic classes.

The same conclusion is obtained using the p-value of the test. Since $p = .0245$ is less than $\alpha = .05$, we have sufficient evidence to reject H_0.

Self-Test 11.2

Suppose the USGA wants to compare the mean distances associated with four different brands of golf balls when struck with a driver. A completely randomized design is employed, with Iron Byron, the USGA's robotic golfer, using a driver to hit a random sample of 10 balls of each brand in a random sequence. The ANOVA table is shown below.

Source	df	SS	MS	F	p-value
Brand	3	2,794	931.5	43.99	.0001
Error	36	762	21.2		
TOTAL	39	3,556			

a. State the null hypothesis.

b. What is the value of the test statistic?

c. Make the appropriate conclusion at $\alpha = .05$.

An analysis of variance for a completely randomized design may include the construction of confidence intervals for a single mean or for the difference between two means. Because the independent sampling design involves the selection of independent random samples, we can find a confidence interval for a single mean using the method of Section 7.3 and for the difference between two population means using the method of Section 9.2.*

However, the researcher is usually interested in making all possible pairwise comparisons of the means. Consequently, several intervals need to be constructed, inflating the overall probability of a Type I error. We discuss a better, alternative approach to constructing confidence intervals for means in ANOVA in Section 11.4.

Before ending our discussion of completely randomized designs, we make the following comment. The proper application of the ANOVA procedure requires that certain assumptions be satisfied, i.e., all k populations are approximately normal with equal variances. In Section 11.6, we provide details on how to use the data to determine if these assumptions are satisfied to a reasonable degree. If the analysis, for example, reveals that one or more of the populations are nonnormal (e.g., highly skewed), then any inferences derived from the ANOVA of the data are suspect. In this case, we can apply a nonparametric technique.

*The only modification we will make in these two procedures is that we will use an estimate of σ^2 based on the information contained in all k samples—namely, the pooled measure of variability within the k samples:

$$\text{MSE} = s^2 = \frac{\text{SSE}}{(n-k)}.$$

CAUTION

When the assumptions for analyzing data collected from a completely randomized design are violated, any inferences derived from the ANOVA are suspect. An alternative technique to use for the completely randomized design is the nonparametric Kruskal-Wallis H test (see optional Section 11.8).

Statistics in the Real World Revisited

A One-Way Analysis of the Cockroach Data

Consider the experiment designed to investigate the trail-following ability of German cockroaches (p. 649). Recall that an entomologist created a chemical trail with either a methanol extract from roach feces or just methanol (control). Cockroaches were then released into a container at the beginning of the trail, one at a time, and a video surveillance camera was used to monitor the roach's movements. The movement pattern of each cockroach was measured by its average trail deviation (in pixels) and the data stored in the **ROACH** file.

For this application, consider only the cockroaches assigned to the fecal extract trail. Four roach groups were utilized in the experiment—adult males, adult females, gravid females, and nymphs—with 20 roaches of each type independently and randomly selected. Is there sufficient evidence to say that the ability to follow the extract trail differs among cockroaches of different age, sex, and reproductive status? In other words, is there evidence to suggest that the mean trail deviation μ differs for the four roach groups?

ROACH2

To answer this question, we conduct a one-way analysis of variance on a subset of the data in the **ROACH** data file. The data file **ROACH2** contains only those cockroaches assigned to the fecal extract trail. The dependent (response) variable of interest is extract trail deviation, while the treatments are the four different roach groups. Thus, we want to test the null hypothesis:

$$H_0: \mu_{\text{Male}} = \mu_{\text{Female}} = \mu_{\text{Gravid}} = \mu_{\text{Nymph}}$$

A MINITAB printout of the ANOVA is displayed in Figure 11.7. The p-value of the test (highlighted on the printout) is 0. Since this value is less than, say, $\alpha = .05$, we reject the null hypothesis and conclude (at the .05 level of significance) that the mean extract trail deviation differs among the populations of adult male, adult female, gravid, and nymph cockroaches.

M Figure 11.7 MINITAB One-Way ANOVA for Extract Trail Deviation

One-way Analysis of Variance

```
Analysis of Variance for Deviate
Source     DF        SS        MS        F        P
Group       3     14164      4721    11.61    0.000
Error      76     30918       407
Total      79     45083
                                    Individual 95% CIs For Mean
                                    Based on Pooled StDev
Level       N      Mean     StDev   --+---------+---------+---------+----
Female     20     21.07     26.13            (-----*-----)
Gravid     20     44.03     24.84                            (-----*-----)
Male       20      7.37      8.61    (-----*-----)
Nymph      20     18.73     15.92           (-----*-----)
                                    --+---------+---------+---------+----
Pooled StDev =    20.17               0        15        30        45
```

The sample means for the four cockroach groups are also highlighted on Figure 11.7. Note that adult males have the smallest sample mean deviation (7.37) while gravids have the largest sample mean deviation (44.03). In the next Statistics in the Real World Revisited section (p. 670), we demonstrate how to rank, statistically, the four population means based on their respective sample means.

PROBLEMS FOR SECTION 11.3

Using the Tools

11.1 Independent random samples were selected from two populations, with the results shown in the table on p. 661.

a. Construct a dot plot of the data similar to Figures 11.2 and 11.3. Do you think the data provide evidence of a difference between the population means?

b. Calculate the within-sample variation (MSE) for the data using equation 11.1 (p. 654).

Sample 1	Sample 2
10	12
7	8
8	13
11	10
10	10
9	11
9	

Problem 11.1

c. Calculate the between-sample variation (MST) for the data using equation 11.2 (p. 654).

d. Using equation 11.3 (p. 654) compute the test statistic appropriate for testing H_0: $\mu_1 = \mu_2$ against the alternative hypothesis that the two means differ.

e. Summarize the results of parts **b–d** in an ANOVA table.

f. Specify the rejection region, using a significance level of $\alpha = .05$.

g. Make the proper conclusion. How does this compare to your answer to part **a**?

11.2 Problem 11.1 involves a test of the null hypothesis H_0: $\mu_1 = \mu_2$ based on independent random sampling. Recall that a test of this hypothesis was conducted in Section 9.2 using a t statistic.

a. Use the t test to test H_0: $\mu_1 = \mu_2$ against the alternative hypothesis H_a: $\mu_1 \neq \mu_2$. Test using $\alpha = .05$.

b. It can be shown (proof omitted) that an F statistic with $\nu_1 = 1$ numerator degree of freedom and ν_2 denominator degrees of freedom is equal to t^2, where t is a t statistic based on ν_2 degrees of freedom. Square the value of t that you calculated in part **a** and show that it is equal to the value of F calculated in Problem 11.1.

c. Is the analysis of variance F test for comparing two population means a one- or a two-tailed test of H_0: $\mu_1 = \mu_2$? [*Hint:* Although the t test can be used to test for either H_a: $\mu_1 > \mu_2$ or H_a: $\mu_1 < \mu_2$, the alternative hypothesis for the F test is H_a: The two means are different.]

11.3 Independent random samples were selected from three populations. The data are shown in the accompanying table, followed by an Excel printout of the analysis of variance.

Sample 1	Sample 2	Sample 3
2.1	4.4	1.1
3.3	2.6	.2
.2	3.0	2.0
	1.9	

	A	B	C	D	E	F	G
1	Anova: Single Factor						
2							
3	SUMMARY						
4	*Groups*	*Count*	*Sum*	*Average*	*Variance*		
5	Sample 1	3	5.6	1.866667	2.443333		
6	Sample 2	4	11.9	2.975	1.109167		
7	Sample 3	3	3.3	1.1	0.81		
8							
9							
10	ANOVA						
11	*Source of Variatio*	*SS*	*df*	*MS*	*F*	*P-value*	*F crit*
12	Between Groups	6.221833	2	3.110917	2.214363	0.179825	4.737416
13	Within Groups	9.834167	7	1.404881			
14							
15	Total	16.056	9				

E Problem 11.3

a. Locate the value of MST. What type of variability is measured by this quantity?

b. Locate the value of MSE. What type of variability is measured by this quantity?

c. How many degrees of freedom are associated with MST?

d. How many degrees of freedom are associated with MSE?

e. Locate the value of the test statistic for testing H_0: $\mu_1 = \mu_2 = \mu_3$ against the alternative hypothesis that at least one population mean is different from the other two.

f. Summarize the results of parts **a–e** in an ANOVA table.

g. Specify the rejection region, using a significance level of $\alpha = .05$.

h. State the proper conclusion.

i. Locate and interpret the p-value for the test of part **e**. Does this agree with your answer to part **h**?

11.4 A partially completed ANOVA table for a completely randomized design is shown here.

Source	df	SS	MS	F
Treatment	4	24.7	—	—
Error	—	—	—	
TOTAL	34	62.4		

a. Complete the ANOVA table.

b. How many treatments are involved in the experiment?

c. Do the data provide sufficient evidence to indicate a difference among the population means? Test using $\alpha = .10$.

Applying the Concepts

11.5 *Bulimia nervosa* is an eating disorder characterized by episodes of binge eating, extreme efforts to counteract ingested calories, and a morbid fear of becoming fat. In a study published in the *American Journal of Psychiatry* (July 1995), a random sample of 75 subjects was selected from females clinically diagnosed with bulimia nervosa. The subjects were divided into three groups based on when they were born: (1) before 1950, (2) between 1950 and 1959, and (3) after 1959. For one part of the study, the variable of interest was age (in years) at onset of the eating disorder. The goal was to compare the mean ages at onset of the three groups.

a. Identify the response variable in the experiment.

b. Identify the treatments in the experiment.

c. State the null and alternative hypotheses to be tested.

d. The observed significance level of the ANOVA F test was less than .01. Interpret this result.

11.6 In robotics, researchers are attempting to create machines that interact with an unpredictable environment. One possible solution is to train robots to behave like ants in an ant colony. A study published in *Nature* (Aug. 2000) investigated whether ant-derived computer algorithms allow robots to achieve tasks in a cooperative manner. Robots were trained and randomly assigned to "colonies" (i.e., groups) consisting of 3, 6, 9, or 12 robots. The robots were assigned the task of foraging for "food" and to recruit another robot when they identified a resource-rich area. One goal of the experiment was to compare the mean energy expended (per robot) of the four different colony sizes.

a. What type of experimental design was employed?

b. Identify the treatments and the dependent variable.

c. Set up the null and alternative hypotheses of the test.

d. The following ANOVA results were reported: $F = 7.70$, numerator df $= 3$, denominator df $= 56$, p-value $< .001$. Conduct the test at a significance level of $\alpha = .05$ and interpret the result.

11.7 *Fortune* (Oct. 25, 1999) published a list of the 50 most powerful women in America. The data on age (in years) and title of each of these 50 women are stored

in the **WOMENPOWER** file. Suppose you want to compare the average ages of powerful women in three groups based on their position (title) within the firm: Group 1 (CEO, CFO, CIO, or COO); Group 2 (Chairman, President, or Director); and Group 3 (Vice President, Vice Chairman, or General Manager). A MINITAB analysis of variance is conducted on the data, with the results shown in the accompanying printout. Fully interpret the results. Give the null and alternative hypotheses tested.

```
Analysis of Variance for Age
Source     DF        SS        MS        F        P
Group       2     114.3      57.1     1.62    0.209
Error      47    1658.5      35.3
Total      49    1772.7
                                    Individual 95% CIs For Mean
                                    Based on Pooled StDev
Level       N      Mean     StDev  ---------+---------+---------+-------
1          21    48.952     7.159                 (--------*--------)
2          16    49.188     5.648                 (---------*---------)
3          13    45.615     3.595  (----------*----------)
                                   ---------+---------+---------+-------
Pooled StDev =    5.940                  45.0      48.0      51.0
```

M Problem 11.7

11.8 Refer to the Harris Corporation/University of Florida study to compare the mean voltage emitted by a manufacturing process established at two locations, Problems 3.38 (p. 154) and 9.18 (p. 465). The voltage readings of 30 production runs at the old location and the new location are reproduced in the accompanying table. A one-way analysis of variance was performed on the data using Excel. The Excel printout is also shown.

Old Location			New Location		
9.98	10.12	9.84	9.19	10.01	8.82
10.26	10.05	10.15	9.63	8.82	8.65
10.05	9.80	10.02	10.10	9.43	8.51
10.29	10.15	9.80	9.70	10.03	9.14
10.03	10.00	9.73	10.09	9.85	9.75
8.05	9.87	10.01	9.60	9.27	8.78
10.55	9.55	9.98	10.05	8.83	9.35
10.26	9.95	8.72	10.12	9.39	9.54
9.97	9.70	8.80	9.49	9.48	9.36
9.87	8.72	9.84	9.37	9.64	8.68

Source: Harris Corporation, Melbourne, Fla.

	A	B	C	D	E	F	G
1	Anova: Single Factor						
2							
3	SUMMARY						
4	*Groups*	*Count*	*Sum*	*Average*	*Variance*		
5	Old Location	30	294.11	9.803667	0.29259		
6	New Location	30	282.67	9.422333	0.229322		
7							
8							
9	ANOVA						
10	*Source of Variation*	*SS*	*df*	*MS*	*F*	*P-value*	*F crit*
11	Between Groups	2.181227	1	2.181227	8.358608	0.005394	4.006864
12	Within Groups	15.13543	58	0.260956			
13							
14	Total	17.31666	59				

E Problem 11.8

a. Locate on the printout the F statistic for testing whether the mean voltage readings at the two locations are equal.

b. Is there evidence (at $\alpha = .05$) to indicate that the mean voltage readings at the two locations are different?

c. In Problem 9.18**a** you compared the two voltage means using a two-sample t test. Compare the t statistic to the value of F obtained in part **a.** Verify that $F = t^2$ and that the two tests yield the same conclusion at $\alpha = .05$.

FISH

11.9 Consider the data on DDT measurements of contaminated fish in the Tennessee River (Alabama) stored in the **FISH** file. Recall that the qualitative variable, SPECIES, stored in the workbook represents the species of fish captured—channel catfish, smallmouth buffalo, or largemouth bass. According to the U.S. Army Corps of Engineers, who collected the data, it is important to know whether the mean DDT levels of the three species of fish differ and, if so, which species has the highest mean level. Using Excel or MINITAB, carry out the proper analysis at $\alpha = .05$. Summarize the results in an ANOVA table.

11.10 A statistics professor wants to study four different strategies of playing the game of Blackjack (Twenty-One). The four strategies are

1. Dealer's strategy
2. Five-count strategy
3. Basic ten-count strategy
4. Advanced ten-count strategy

A calculator that is programmed to play Blackjack is utilized and data from five sessions of each strategy are collected. The profits (or losses) from each session are given in the accompanying table. The professor wants to know whether there is evidence of a difference among the four Blackjack strategies. Use Excel or MINITAB to conduct the analysis at $\alpha = .01$.

BLCKJACK

Strategy			
Dealer's	**Five Count**	**Basic Ten Count**	**Advanced Ten Count**
−$56	−$26	+$16	+$60
−$78	−$12	+$20	+$40
−$20	+$18	−$14	−$16
−$46	−$8	+$6	+$12
−$60	−$16	−$25	+$4

11.11 Do Hispanic visitors to a national forest in the U.S. have the same environmental concerns as Anglo visitors? To answer this question, 334 people touring the Angeles National Forest and the San Bernardino National Forest were surveyed on several environmental issues (e.g., oil spills, car emissions, and wildfires). Each visitor was classified into one of four ethnic categories: (1) U.S.-born Anglos, (2) U.S.-born Hispanics, (3) Mexican-born Hispanics, and (4) Central American-born Hispanics. Environmental concern for each issue was measured on a 5-point scale where 1 = not harmful, 2 = somewhat harmful, 3 = neutral, 4 = harmful, and 5 = very harmful. A summary of the results on the issues of car emissions and wildfires, published in the *Journal of Environmental Education* (Spring 1995), is provided below.

		Mean Response	
Ethnic Group	**Sample Size**	**Car Emissions**	**Wildfires**
U.S.-born Anglo	39	3.61	3.61
U.S.-born Hispanic	79	3.32	4.00
Mexican-born Hispanic	173	3.13	3.55
Central American-born Hispanic	43	2.61	3.35

a. State the null and alternative hypotheses appropriate for determining whether the four ethnic groups differ regarding their mean responses on the issue of car emissions.

b. At $\alpha = .05$, the test of part **a** was statistically significant. Interpret this result.

c. State the null and alternative hypotheses appropriate for determining whether the four ethnic groups differ regarding their mean responses on the issue of wildfires.

d. At $\alpha = .05$, the test of part **c** was not statistically significant. Interpret this result.

11.12 What do people infer from facial expressions of emotion? This was the research question of interest in an article published in the *Journal of Nonverbal Behavior* (Fall 1996). A sample of 36 introductory psychology students were randomly divided into 6 groups. Each group was assigned to view one of six slides showing a person making a facial expression.* The six expressions were (1) angry, (2) disgusted, (3) fearful, (4) happy, (5) sad, and (6) neutral faces. After viewing the slides, the students rated the degree of dominance they inferred from the facial expression (on a scale ranging from −15 to +15). The data (simulated from summary information provided in the article) are listed in the table. Conduct an analysis of variance to determine whether the mean dominance ratings differ among the six facial expressions. Use $\alpha = .10$.

*In the actual experiment, each group viewed all six facial expression slides and the design employed was a Latin Square (beyond the scope of this text).

FACES

Angry	Disgusted	Fearful	Happy	Sad	Neutral
2.10	.40	.82	1.71	.74	1.69
.64	.73	−2.93	−.04	−1.26	−.60
.47	−.07	−.74	1.04	−2.27	−.55
.37	−.25	.79	1.44	−.39	.27
1.62	.89	−.77	1.37	−2.65	−.57
−.08	1.93	−1.60	.59	−.44	−2.16

11.4 Follow-Up Analysis: Multiple Comparisons of Means

Many designed experiments are conducted with the ultimate goal of determining the largest (or the smallest) treatment mean. For example, suppose a drugstore is considering five floor displays for a new product. After conducting an ANOVA on the data and discovering differences among the mean weekly sales associated with the five displays, the drugstore wants to determine which display yields the greatest mean weekly sales of the product. Similarly, a psychologist might want to determine which among six behavior types achieves the highest mean score on an anxiety exam. An entomologist might want to choose one species of ant, from among several, that produces the highest mean number of offspring, and so on.

Once differences among treatment means have been detected in an ANOVA, choosing the treatment with the largest population mean μ might appear to be a simple matter. For example, if there were three treatment means, $\bar{y}_1$, $\bar{y}_2$, and $\bar{y}_3$, we could compare them by constructing a $(1 - \alpha)100\%$ confidence interval for the difference between each pair of means. In general, if there are k treatment means, there are

$$c = k(k - 1)/2$$

pairs of means to be compared. In the case of $k = 3$ means, there are $c = 3(3 \quad 1)/2 - 3$ confidence intervals to construct, one for each of the following differences: $(\mu_1 - \mu_2)$, $(\mu_1 - \mu_3)$, and $(\mu_2 - \mu_3)$.

Number of Pairwise Comparisons c When Comparing k Treatment Means

$$c = k(k - 1)/2$$

However, there is a problem associated with this procedure. If you construct a series of $(1 - \alpha)100\%$ confidence intervals, the risk of making *at least one* Type I

error in the series of inferences—called the *experimentwise error rate*—will be larger than the value of α specified for a single interval. For example, if 10 95% confidence intervals are constructed, the probability of at least one Type I error is greater than .05 and may be as high as .40! In this case, there is a good chance that we will conclude that a difference between some pair of means exists, when, in fact, the means are equal.

Definition 11.9

The **experimentwise error rate (EER)** in a designed experiment to compare k means is the probability of making at least one Type I error in a series of inferences about the population means, based on $(1 - \alpha)100\%$ confidence intervals.

$$\text{EER} = P(\text{at least one Type I error})$$

where $P(\text{Type I error}) = P(\text{Reject } H_0\text{: } \mu_i = \mu_j \mid \mu_i = \mu_j) = \alpha$.

There are a number of alternative procedures for comparing and ranking a group of treatment means as part of a *follow-up analysis* to the ANOVA. One method that is useful for making pairwise comparisons in many experimental designs is called the **Bonferroni multiple comparisons procedure.*** Bonferroni's method is designed to control the experimentwise error rate EER. The user selects a small value of EER (say, EER $\leq$ .10) so that inferences derived from Bonferroni's method can be made with an overall confidence level of (1 − EER). (See Miller [1981] for proof of this result.)

*Other procedures, such as Tukey, Tukey-Kramer, Duncan, and Scheffé, may be more appropriate in certain sampling situations. Consult the Chapter 11 references for details on how to use these multiple comparisons methods.

The formula for constructing confidence intervals for the differences between treatment means using Bonferroni's procedure is provided in the next box. Note that confidence intervals for the difference between two treatment means for all possible pairs of treatments are produced based on the EER selected by the analyst.

BONFERRONI'S MULTIPLE COMPARISONS PROCEDURE FOR ALL PAIRWISE COMPARISONS OF *k* TREATMENT MEANS

1. Select the experimentwise error rate EER.
2. Determine c, the number of pairwise comparisons (i.e., confidence intervals), where

$$c = k(k - 1)/2 \tag{11.5}$$

3. Compute

$$\alpha = \frac{\text{EER}}{c} \tag{11.6}$$

Each interval constructed will have a confidence level of $(1 - \alpha)$.

4. For each treatment pair (i, j) calculate the critical difference:

$$B_{ij} = (t_{\alpha/2})\sqrt{\text{MSE}\left(\frac{1}{n_i} + \frac{1}{n_j}\right)} \tag{11.7}$$

where

$t_{\alpha/2}$ = the critical t value with tail area $\alpha/2$ based on ν degrees of freedom

ν = df(Error) in the ANOVA table

(Continued)

MSE = mean square for error in the ANOVA table
n_i = sample size for treatment i
n_j = sample size for treatment j

5. For each treatment pair (i, j), compute the confidence interval

$$(\bar{y}_i - \bar{y}_j) \pm \mathrm{B}_{ij} \qquad \textbf{(11.8)}$$

6. a. If the interval contains 0, then conclude that the two treatment means are *not* significantly different.
   b. If the interval contains all positive numbers, then conclude that $\mu_i > \mu_j$.
   c. If the interval contains all negative numbers, then conclude that $\mu_j > \mu_i$.

[*Note:* The level of confidence associated with all inferences drawn from the analysis simultaneously is at least (1 − EER).]

EXAMPLE 11.5 COMPARING MEANS IN A ONE-WAY TABLE

Refer to the one-way ANOVA for the completely randomized design discussed in Examples 11.3 and 11.4 (pp. 665–656). Recall that we rejected the null hypothesis of no differences among the mean GPAs of freshmen in three socioeconomic classes. Use Bonferroni's method to perform pairwise comparisons of the means for the three groups of freshmen. Use an experimentwise error rate of .09.

Solution

Step 1 We selected EER = .09. This value represents the probability of making at least one Type I error in the analysis. Also, the overall level of confidence associated with all the inferences derived from the analysis is (1 − .09) = .91.

Step 2 Since there are three treatments (i.e., socioeconomic classes) in this experiment, $k = 3$. Then the number of possible pairwise comparisons is (from equation 11.5) $c = 3(3 - 1)/2 = 3$. In other words, we will construct $c = 3$ confidence intervals, one for each of the following treatment pairs: (lower class, middle class), (lower class, upper class), and (middle class, upper class).

Step 3 The value of α used to construct each confidence interval is (from equation 11.6)

$$\alpha = \frac{\mathrm{EER}}{c} = \frac{.09}{3} = .03$$

Since $1 - \alpha = .97$, we will construct three 97% confidence intervals in order to obtain an experimentwise error rate of .09.

Step 4 To find the critical difference for each pair of treatment means, we require the quantities $t_{\alpha/2}$, df(Error), and MSE. From the ANOVA table on the Excel printout, Figure 11.6 (p. 658), ν = df(Error) = 18 and MSE = .262. We know from step 3 that $\alpha = .03$ and therefore $\alpha/2 = .015$. We need to find $t_{\alpha/2} = t_{.015}$ where the distribution of t is based on $\nu = 18$ df. If you examine Table B.4 of Appendix B, you will not find a column for $t_{.015}$. However, Excel's TINV function can provide this information. Using Excel, we obtain $t_{.015} = 2.356$.

Since all three treatments have seven observations, i.e., $n_1 = n_2 = n_3 = 7$, the Bonferroni critical difference B_{ij} will be the same for any pair of treatment means. In other words, $\mathrm{B}_{12} = \mathrm{B}_{13} = \mathrm{B}_{23}$. Applying

equation 11.7, we find the critical difference

$$B_{ij} = t_{\alpha/2}\sqrt{\text{MSE}\left(\frac{1}{n_i} + \frac{1}{n_j}\right)}$$

$$= 2.356\sqrt{.262\left(\frac{1}{7} + \frac{1}{7}\right)} = .64$$

Step 5 From the Excel printout, Figure 11.6 (p. 658), we obtain the following sample means:

$$\bar{y}_1 = 2.52 \qquad \bar{y}_2 = 3.25 \qquad \bar{y}_3 = 2.54$$

Substituting these values and B_{ij} into equation 11.8 produces the following three confidence intervals:

Confidence interval for $(\mu_1 - \mu_2)$: $(\bar{y}_1 - \bar{y}_2) \pm B_{12}$

$$= (2.52 - 3.25) \pm .64$$
$$= -.73 \pm .64$$
$$= (-1.37, -.09)$$

Confidence interval for $(\mu_1 - \mu_3)$: $(\bar{y}_1 - \bar{y}_3) \pm B_{13}$

$$= (2.52 - 2.54) \pm .64$$
$$= -.02 \pm .64$$
$$= (-.66, .62)$$

Confidence interval for $(\mu_2 - \mu_3)$: $(\bar{y}_2 - \bar{y}_3) \pm B_{23}$

$$= (3.25 - 2.54) \pm .64$$
$$= .71 \pm .64$$
$$= (.07, 1.35)$$

Overall, we are 91% confident that the intervals *collectively* contain the differences between the true mean GPAs.

Step 6 We demonstrated how to interpret confidence intervals for the difference between two means in Section 9.2. If zero is included in the interval, then there is no evidence of a difference between the means compared. Note that the confidence interval for $(\mu_1 - \mu_3)$ contains zero. Consequently, there is no significant difference between the lower class and upper class GPA means μ_1 and μ_3.

The confidence intervals for $(\mu_1 - \mu_2)$ and $(\mu_2 - \mu_3)$ do not contain zero. Since the first interval, $(\mu_1 - \mu_2)$, contains all negative numbers, we infer that the mean GPA for the middle class (μ_2) exceeds the mean for the lower class (μ_1). Similarly, the third interval, $(\mu_2 - \mu_3)$, contains all positive numbers, implying the middle class mean (μ_2) exceeds the upper class mean (μ_3).

These inferences can also be derived by comparing the absolute difference between the treatment means, $|\bar{y}_i - \bar{y}_j|$, to the Bonferroni critical difference B_{ij}. These results are shown on an Excel printout of the Bonferroni analysis, Figure 11.8.

The absolute differences and Bonferroni critical differences are highlighted on the printout for each pair of treatment means. If

E **Figure 11.8** Excel Output for the Bonferroni Analysis, Example 11.5

	A	B	C	D
1	Bonferroni Procedure for GPA			
2				
3	Group 1		Group 1 to Group 2 Comparison	
4	Sample Mean	2.52	Absolute Difference	0.73
5	Sample Size	7	Standard Error of Difference	0.2736003
6	Group 2		Bonferroni Critical Difference	0.6446523
7	Sample Mean	3.25	Means are different	
8	Sample Size	7		
9	Group 3		Group 1 to Group 3 Comparison	
10	Sample Mean	2.54	Absolute Difference	0.02
11	Sample Size	7	Standard Error of Difference	0.2736003
12	MSE	0.262	Bonferroni Critical Difference	0.6446523
13	Error DF	18	Means are not different	
14	Experimentwise Error Rate	0.09		
15	Number of Groups	3	Group 2 to Group 3 Comparison	
16	Treatment pairs	3	Absolute Difference	0.71
17	α value for comparisons	0.03	Standard Error of Difference	0.2736003
18	t statistic for comparisons	2.356182	Bonferroni Critical Difference	0.6446523
19			Means are different	

$|\bar{y}_i - \bar{y}_j|$ exceeds B_{ij}, then Excel indicates that the treatment means are significantly different. If $|\bar{y}_i - \bar{y}_j|$ is less than B_{ij}, then Excel states that the treatments are not significantly different. You can see from Figure 11.8 that Excel found no difference between the GPA means for the lower (1) and upper (3) classes.

A convenient summary of the results of the Bonferroni multiple comparisons procedure is a listing of the treatment means from lowest to highest, with a solid line connecting those means that are *not* significantly different. This summary is shown in Figure 11.9. Our interpretation is that the mean GPA for the middle-class freshmen students exceeds that of either of the other two mean GPAs; but the means of the lower and upper classes are not significantly different. Again, all these inferences are made with an overall confidence level of 91% since our experimentwise error rate is .09.

Figure 11.9 Summary of Bonferroni's Multiple Comparisons, Example 11.5

Mean GPA	$\overline{2.52}$	$\overline{2.54}$	3.25
Treatment (Class)	Lower	Upper	Middle

Self-Test 11.3

Refer to the one-way ANOVA, Self-Test 11.2 (p. 659). Bonferroni's procedure was used to rank the four golf ball brand mean distances with EER = .06. The results are summarized below.

Mean	$\overline{249.3}$	$\overline{250.8}$	261.1	270.0
Brand	D	A	B	C

a. How many pairwise comparisons of the four brand means are included in the analysis?

b. Find the value of α used to form the confidence intervals.

c. Find the value of the Bonferroni critical difference used to rank the means.

d. Which brand has the highest ranked mean? The smallest?

e. Interpret the meaning of EER = .06.

In addition to a completely randomized design, Bonferroni's multiple comparisons of means procedure can be applied to a wide variety of experimental designs in ANOVA (including the factorial design of optional Section 11.5). Keep in mind, however, that many other methods of making multiple comparisons are available and one or more of these techniques may be more appropriate to use in your particular application. Consult the Chapter 11 references for details on other techniques (e.g., Hsu [1996]).

In closing, we remind you that multiple comparisons of treatment means should be performed only as a follow-up analysis to the ANOVA, i.e., only after you have conducted the appropriate analysis of variance *F* test(s) and determined that sufficient evidence exists of differences among the treatment means. Be wary of conducting multiple comparisons when the ANOVA *F* test indicates no evidence of a difference among a small number of treatment means—this may lead to confusing and contradictory results.*

*When a large number of treatments are to be compared, a borderline, nonsignificant *F* value (e.g., $.05 < p$-value $< .10$) may mask differences between some of the means. In this situation, it is better to ignore the *F* test and proceed directly to a multiple comparisons procedure.

CAUTION

In practice, it is advisable to avoid conducting multiple comparisons of a small number of treatment means when the corresponding ANOVA *F* test is nonsignificant; otherwise, confusing and contradictory results may occur.

Statistics in the Real World Revisited

Ranking the Means of the Cockroach Groups

ROACH2

Refer to the experiment designed to investigate the trail-following ability of German cockroaches. In the previous Statistics in the Real World Revisited section (p. 660), we applied a one-way ANOVA to the **ROACH2** data and discovered statistically significant differences among the mean extract trail deviations for the four groups of cockroaches–adult males, adult females, gravid females, and nymphs. In order to determine which group has the highest degree of trail-following ability, we want to rank the population means from largest to smallest. That is, we want to follow up the ANOVA by conducting multiple comparisons of the four treatment means.

Figure 11.10 is a MINITAB printout of the ANOVA and the multiple comparisons results. MINITAB uses **Tukey's multiple comparisons procedure,** a method very similar to the Bonferroni procedure presented in this section. The experimentwise error rate (EER) of .05 is highlighted on the printout as well as the Tukey confidence intervals for all possible pairs of means (at the bottom of the printout). The information contained in these confidence intervals will enable us to rank the treatment (population) means.

For example, the confidence interval for $(\mu_{\text{Female}} - \mu_{\text{Gravid}})$ is $(-39.74, -6.18)$. Since the endpoints of the interval are both negative, the difference between the means is negative. This implies that the population mean trail deviation for adult female cockroaches is less than the population mean for gravids, i.e., $\mu_{\text{Female}} < \mu_{\text{Gravid}}$.

Now consider the confidence interval for $(\mu_{\text{Female}} - \mu_{\text{Male}})$. The interval shown on the printout is $(-3.08, 30.47)$. Since the value 0 is included in the interval, there is no evidence of a significant difference between the two treatment means. Similar interpretations are made for the confidence intervals for $(\mu_{\text{Female}} - \mu_{\text{Nymph}})$ and $(\mu_{\text{Male}} - \mu_{\text{Nymph}})$ since both these intervals contain 0.

Finally, note that the confidence intervals for $(\mu_{\text{Gravid}} - \mu_{\text{Male}})$ and $(\mu_{\text{Gravid}} - \mu_{\text{Nymph}})$ both have positive endpoints, implying that the differences between the means is positive. Thus, the population mean trail deviation for gravids is greater than either the population mean for adult males ($\mu_{\text{Gravid}} > \mu_{\text{Male}}$) or the population mean for nymphs ($\mu_{\text{Gravid}} > \mu_{\text{Nymph}}$).

M Figure 11.10
MINITAB ANOVA with Multiple Comparisons of Extract Trail Deviation Means

One-way Analysis of Variance

```
Analysis of Variance for Deviate
Source     DF        SS        MS        F        P
Group       3     14164      4721    11.61    0.000
Error      76     30918       407
Total      79     45083
                                   Individual 95% CIs For Mean
                                   Based on Pooled StDev
Level       N      Mean     StDev  --+---------+---------+---------+----
Female     20     21.07     26.13            (-----*-----)
Gravid     20     44.03     24.84                           (-----*-----)
Male       20      7.37      8.61   (-----*-----)
Nymph      20     18.73     15.92          (-----*-----)
                                   --+---------+---------+---------+----
Pooled StDev =    20.17              0        15        30        45

Tukey's pairwise comparisons

    Family error rate = 0.0500
Individual error rate = 0.0103

Critical value = 3.72

Intervals for (column level mean) - (row level mean)

            Female      Gravid      Male

Gravid      -39.74
             -6.18

Male         -3.08       19.88
             30.47       53.43

Nymph       -14.44        8.52     -28.13
             19.12       42.08       5.42
```

The results of these multiple comparisons are summarized in Table 11.4. With an overall significance level of .05, we conclude that gravid cockroaches have a mean extract trail deviation larger than any of the other three groups; there are no significant differences among adult males, adult females, or nymphs.

TABLE 11.4 Ranking of the Cockroach Group Means

Treatment Mean	7.38	18.73	21.07	44.03
Cockroach Group	Male	Nymph	Female	Gravid

PROBLEMS FOR SECTION 11.4

Using the Tools

11.13 Consider a completely randomized design with k treatments. For each of the following values of k, find the number of pairwise comparisons of treatments to be made in a Bonferroni analysis.

a. $k = 3$ **b.** $k = 5$ **c.** $k = 4$ **d.** $k = 8$

11.14 Refer to Problem 11.13. For each part, find the value of α used to construct the Bonferroni confidence intervals if EER = .10.

11.15 Find the Bonferroni critical difference B_{ij} for comparing k treatment means in each of the following situations:

a. $n_i = 10, n_j = 10$, EER $= .06, k = 3$, MSE $= 25$

b. $n_i = 8, n_j = 12$, EER $= .05, k = 5$, MSE $= 100$

c. $n_i = 25, n_j = 20$, EER $= .10, k = 5$, MSE $= 4.8$

11.16 Bonferroni's multiple comparisons procedure was applied to data from a completely randomized design with five treatments (A, B, C, D, and E). The findings are summarized as follows:

Mean	10.2	11.3	14.7	14.9	20.6
Treatment	B	C	A	E	D

a. Identify the treatment(s) in the group with the statistically largest mean(s).

b. Identify the treatment(s) in the group with the statistically smallest mean(s).

11.17 In a design with four treatments (A, B, C, and D), Bonferroni's multiple comparisons procedure found that (1) treatment A had a statistically smaller mean than the other three treatments, (2) the means for treatments C and D are not statistically different, and (3) the mean for treatment B is statistically larger than those for treatments C and D. Summarize these results by using lines to connect treatment means that are not statistically different.

Applying the Concepts

11.18 *Science Education* (Jan. 1995) investigated whether the level of chemistry education impacts performance on a test about conservation of matter. A sample of 120 students was divided into six groups of 20 subjects each according to age and level of chemistry instruction. The six experimental groups were (1) seventh graders, (2) ninth graders, (3) eleventh grade science students, (4) eleventh grade nonscience students, (5) college psychology majors, and (6) college chemistry majors. The mean proportion of correct answers on the exam for each group is listed in the accompanying table.

Group	Mean Score
College chemistry majors	.891
College psychology majors	.674
11th grade nonscience students	.660
11th grade science students	.588
9th graders	.519
7th graders	.384

a. An ANOVA F test for group differences resulted in a p-value less than .001. Interpret this result.

b. A multiple comparisons of treatment means was performed using an experimentwise error rate of .05. The vertical lines in the table summarize the results of the multiple comparisons. Which group(s) has the largest mean proportion of correct answers? The smallest?

c. Interpret the value of EER = .05.

11.19 Refer to the *Organizational Science* (Mar.–April 2000) study on the use of communication media by mid-level managers, Problem 10.34 (p. 562). The researchers utilized a designed experiment in which managers were randomly assigned to one of four "media" groups: electronic mail (e-mail), facsimile (fax), meetings, or letters. Each manager was asked to remember the last time he/she chose the assigned medium to send a message. One of the dependent variables of interest was "perceived richness" of the message (measured on a 40-point scale). The researchers compared the means of the richness variable across the four media groups with a one-way analysis of variance.

a. Set up the null and alternative hypotheses of the test.

b. The following results were reported: $F = 50.39$, $p\text{-value} < .05$. Conduct the test at a significance level of $\alpha = .05$.

Sample mean	23.98	26.50	28.11	31.99
Media	Letters	E-mail	Fax	Meetings

oblem 11.19

e results of a multiple comparisons of the means (at EER = .05) are shown the table above. Interpret these results.

to the *American Journal of Psychiatry* study of bulimia nervosa, Problem 11.5 2). Recall that females diagnosed with bulimia nervosa were divided into three oups based on year of birth. A multiple comparisons procedure was used to he mean age (in years) at onset of the eating disorder for the three groups, an experimentwise error rate of .01. The results are summarized below. What isions can you draw from this analysis?

.6	18.8	17.5
fore 950	Between 1950 and 1959	After 1959

to the *Nature* (Aug. 2000) study of whether ant-derived computer algorithms robots to achieve tasks in a cooperative manner, Problem 11.6 (p. 662). Mul-omparisons of mean energy expended for the four colony sizes were conducted an experimentwise error rate of .05. The results are summarized in the table.

nple mean	.97	.95	.93	.80
up size	3	6	9	12

ow many pairwise comparisons are conducted in this analysis?

terpret the results shown in the table.

to the analysis of variance on the DDT data stored in **FISH,** Problem 11.9 4). Excel was used to make a Bonferroni multiple comparisons of the mean levels of the three species of fish using an experimentwise error rate of .05. pret the results shown on the Excel printout.

	A	B	C	D
1	Bonferroni Procedure for DDT			
2				
3	Group 1		Group 1 to Group 2 Comparison	
4	Sample Mean	33.29938	Absolute Difference	25.137713
5	Sample Size	96	Standard Error of Difference	19.19763115
6	Group 2		Bonferroni Critical Difference	46.51347392
7	Sample Mean	8.161667	Means are not different	
8	Sample Size	36		
9	Group 3		Group 1 to Group 3 Comparison	
10	Sample Mean	1.38	Absolute Difference	31.91938
11	Sample Size	12	Standard Error of Difference	30.07690767
12	MSE	9649.284	Bonferroni Critical Difference	72.87260859
13	Error DF	141	Means are not different	
14	Experimentwise Error Rate	0.05		
15	Number of Groups	3	Group 2 to Group 3 Comparison	
16	Treatment pairs	3	Absolute Difference	6.781667
17	α value for comparisons	0.01666667	Standard Error of Difference	32.7435897
18	t statistic for comparisons	2.4228757	Bonferroni Critical Difference	79.33364768
19			Means are not different	

E Problem 11.22

BLCKJACK

11.23 Refer to the one-way ANOVA on the Blackjack profit data, Problem 11.10 (p. 664). Apply the Bonferroni method (using Excel) to rank the four Blackjack strategies. Use EER = .10.

11.24 Refer to the *Journal of Environmental Education* study of environmental concerns, Problem 11.11 (p. 664). On the issue of oil spills, the ANOVA *F* test comparing the means of the four ethnic groups was significant at $\alpha = .05$. A Bonferroni follow-up analysis yielded the following results at an experimentwise error rate of .05. What conclusions can you draw from the analysis?

Mean response	4.74	4.53	4.04	3.79
Ethnic group	U.S.-born Anglos	U.S.-born Hispanics	Mexican-born Hispanics	Central American-born Hispanics

FACES

11.25 Refer to the *Journal of Nonverbal Behavior* study of students' evaluations of facial expressions, Problem 11.12 (p. 665). Use a multiple comparisons procedure in Excel or MINITAB to rank the dominance rating means of the six facial expressions. (Use an experimentwise error rate of .10.)

11.5 Factorial Designs: Two-Way ANOVA (Optional)

In Section 11.3 we presented a one-way ANOVA for analyzing data from an experiment on a single factor. Suppose we want to investigate the effect of two factors on the mean value of a response variable. That is, we want to conduct a two-way ANOVA. The design appropriate for a two-way ANOVA is a **factorial design.**

In a factorial design, experimental units are measured for various combinations of the factor levels. For example, suppose an experiment involves two factors, one at three levels and the other at two levels. If the response is measured for each of the $2 \times 3 = 6$ factor-level combinations, the design is called a *complete* 2×3 *factorial design* (since all $2 \times 3 = 6$ possible treatments are included in the experiment).

Definition 11.10

A **complete factorial design** is an experiment that includes *all* possible factor-level combinations (treatments).

If one observation on the response variable y is taken for each of the six factor-level combinations, we say that we conducted one replication of a 2×3 factorial experiment. Stating that we conducted one replication of the experiment means that we obtained one measurement on y for each of the six factor-level combinations. As we will see, most factorial designs involve two or more replications.

Definition 11.11

A single **replication** of a factorial experiment is one in which the response variable is observed once for every possible factor-level combination. [*Note:* In practice, the number of replications r for a factorial experiment is almost always chosen to be two or larger.]

The data for a two-factor factorial experiment are presented in a two-way table, with rows corresponding to the levels of one factor and columns corresponding to the levels of the other factor. For each combination of factor levels,

the data fall in one of the row-column cells of the table. To illustrate, consider the following example.

EXAMPLE 11.6 2 × 3 FACTORIAL DESIGN

A company that stamps gaskets out of sheets of rubber, plastic, and cork wants to compare the mean number of gaskets produced per hour for two different types of stamping machines. Practically, the manufacturer wants to determine whether one machine is more productive than the other and, even more important, whether one machine is more productive in producing rubber gaskets while the other is more productive in producing plastic or cork gaskets. To answer these questions, the manufacturer decides to conduct an experiment using the three types of gasket material, cork (C), rubber (R), and plastic (P), with each of the two stamping machines, (1) and (2). Each machine is operated for three 1-hour time periods for each of the gasket materials, with the 18 1-hour time periods assigned to the six machine-material combinations in a random order. (The purpose of the randomization is to eliminate the possibility that uncontrolled environmental factors might bias the results.) The data for the experiment, the number of gaskets (in thousands) produced per hour, are shown in Table 11.5.

GASKET

TABLE 11.5 Data for the 2 × 3 Factorial Experiment of Example 11.6

		Gasket Material		
		Cork (C)	Rubber (R)	Plastic (P)
Stamping Machine	1	4.31 4.27 4.40	3.36 3.42 3.48	4.01 3.94 3.89
	2	3.94 3.81 3.99	3.91 3.80 3.85	3.48 3.53 3.42

a. Identify the response y, factors, and treatments (factor-level combinations) in the experiment.

b. Is the design selected a complete factorial design?

c. How many replications are in the experiment?

Solution

a. Since the company will produce gaskets using three types of material with each of two types of stamping machine, the experiment involves two factors, machine and material, with machine at two levels (1, 2) and material at three levels (C, R, P). Each of the $2 \times 3 = 6$ combinations of machine and material—(1, C), (1, R), (1, P), (2, C), (2, R), and (2, P)—represents the treatments in the experiment. For this reason the experiment is referred to as a 2×3 factorial design.

For each treatment, the response (i.e., quantitative variable) measured is y = number of gaskets (in thousands) produced per hour. One goal of the experiment will be to compare the mean responses of the six treatments.

b. Since all possible treatments are included in the experiment, a complete factorial design is employed.

c. Note that there are three observations on y recorded for each treatment in Table 11.5. Hence, the experiment consists of $r = 3$ replications.

Self-Test 11.4

Refer to Self-Test 11.2 (p. 659). Suppose the USGA tests four brands of golf balls (A, B, C, D) and two different clubs (driver, 5-iron) using a complete factorial design. Iron Byron (the robotic golfer) will hit four balls of each brand with the driver, then four balls of each brand with the 5-iron—a total of 32 hits (four hits for each of the $4 \times 2 = 8$ brand-club combinations). The goal is to compare the mean distances of the eight brand-club combinations.

a. Identify the response y, factors, and treatments in this experiment.

b. How many replications are in the experiment?

As stated in Example 11.6a, one objective of the factorial experiment is to compare the means for the six treatments. One way to accomplish this is to investigate how the means for one factor (material) differ when the levels of the other factor (machine) are varied. The next example illustrates this idea.

EXAMPLE 11.7

FACTOR INTERACTION

Refer to the data of Table 11.5. Suppose that we have calculated the six cell (treatment) means. Consider two possible outcomes. Figure 11.11 shows two hypothetical plots of the six means. For both plots, the three means for stamping machine 1 are connected by dark-color line segments. The corresponding three means for machine 2 are connected by lighter-color line segments. What do these plots imply?

Figure 11.11 Hypothetical Plots of the Means for the Six Machine-Material Combinations

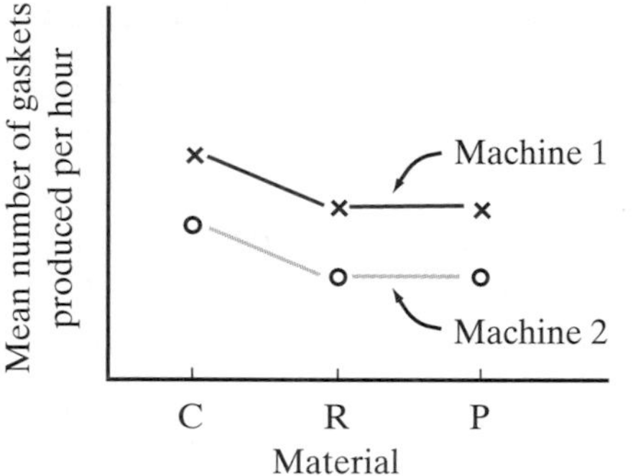

a. No interaction

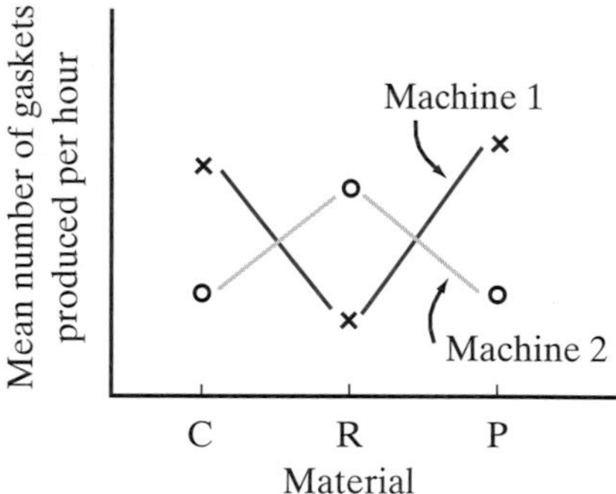

b. Interaction

Solution

Figure 11.11a suggests that machine 1 produces a larger number of gaskets per hour, regardless of the gasket material, and is therefore superior to machine 2. On the average, machine 1 stamps more cork (C) gaskets per hour than rubber or plastic, but the *difference* in the mean numbers of gaskets produced by the two machines remains approximately the same, regardless of the gasket material. Thus, the *difference* between the mean numbers of gaskets produced by the two machines is *independent* of the gasket material used in the stamping process.

In contrast, Figure 11.11b shows the productivity of machine 1 to be larger than for machine 2 when the gasket material is cork (C) or plastic (P). But the means are reversed for rubber (R) gasket material. For this material, machine 2 produces, on the average, more gaskets per hour than machine 1. Thus, Figure 11.11b illustrates a situation where the mean value of the response variable *depends* on the combination of the factor levels. When this situation occurs, we say that the factors

interact. Thus, one of the most important objectives of a factorial experiment is to detect factor interaction if it exists.

Definition 11.12

In a factorial design with two factors A and B, when the difference between the mean levels of factor A depends on the different levels of factor B, we say there is *interaction* between factors A and B. If the difference is independent of the levels of B, then there is *no interaction* between factors A and B.

The analysis of variance for a two-factor factorial design (i.e., a two-way ANOVA) involves a partitioning of SS(Total) into four parts as shown in Figure 11.12. These sums of squares allow us to investigate and test for factor interaction as well as for **factor main effects;** i.e., an overall effect of factor A (called the main effect for A) and an overall effect of factor B (called the main effect for B). (The formulas for the sums of squares and the corresponding mean squares for a factorial experiment are given in optional Section 11.7.) As in the previous section, we will use Excel or MINITAB to perform the analysis of the data and to summarize the ANOVA results in a table. In general, the ANOVA table for a two-factor factorial design, with factor A at a levels, factor B at b levels, and with r replications, appears as shown in the box. A summary of the ANOVA F tests for a factorial design follows.

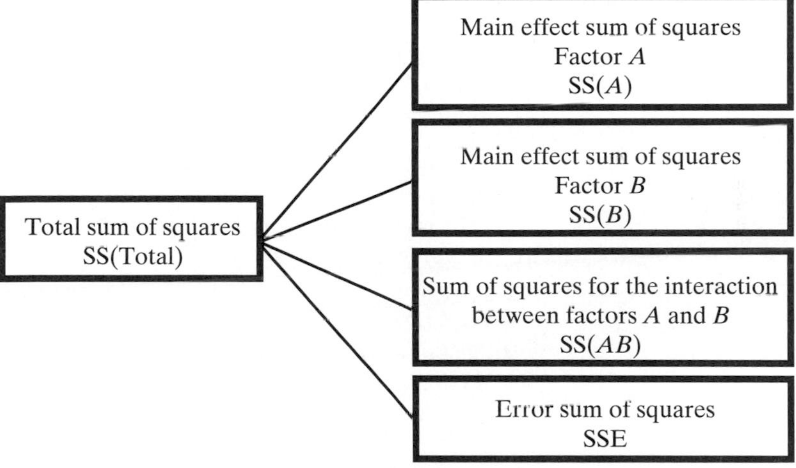

Figure 11.12 Partitioning of the Total Sum of Squares for a Two-Factor Factorial Design

GENERAL ANOVA TABLE FOR A TWO-FACTOR FACTORIAL DESIGN

Source	df	SS	MS	F
Main effect, A	$a-1$	SS(A)	$\text{MS}(A) = \dfrac{\text{SS}(A)}{a-1}$	$F = \dfrac{\text{MS}(A)}{\text{MSE}}$
Main effect, B	$b-1$	SS(B)	$\text{MS}(B) = \dfrac{\text{SS}(B)}{b-1}$	$F = \dfrac{\text{MS}(B)}{\text{MSE}}$
AB interaction	$(a-1) \times (b-1)$	SS(AB)	$\text{MS}(AB) = \dfrac{\text{SS}(AB)}{(a-1)(b-1)}$	$F = \dfrac{\text{MS}(AB)}{\text{MSE}}$
Error	$ab(r-1)$	SSE	$\text{MSE} = \dfrac{\text{SSE}}{ab(r-1)}$	
TOTAL	$abr-1$	SS(Total)		

ANOVA *F* TESTS FOR A TWO-FACTOR FACTORIAL DESIGN

Test for Factor Interaction

H_0: No interaction between factors A and B

H_a: Factors A and B interact

$$\textit{Test statistic: } F = \frac{\text{MS}(AB)}{\text{MSE}} = \frac{\text{MS}(AB)}{s^2} \tag{11.9}$$

Rejection region: $F > F_\alpha$, where F is based on $\nu_1 = (a-1)(b-1)$ and $\nu_2 = ab(r-1)$ df

Test for Main Effects for Factor A

H_0: There are no differences among the means for main effect A

H_a: At least two of the main effect A means differ

$$\textit{Test statistic: } F = \frac{\text{MS}(A)}{\text{MSE}} = \frac{\text{MS}(A)}{s^2} \tag{11.10}$$

Rejection region: $F > F_\alpha$, where F is based on $\nu_1 = (a-1)$ and $\nu_2 = ab(r-1)$ df

Test for Main Effects for Factor B

H_0: There are no differences among the means for main effect B

H_a: At least two of the main effect B means differ

$$\textit{Test statistic: } F = \frac{\text{MS}(B)}{\text{MSE}} = \frac{\text{MS}(B)}{s^2} \tag{11.11}$$

Rejection region: $F > F_\alpha$, where F is based on $\nu_1 = (b-1)$ and $\nu_2 = ab(r-1)$ df

Assumptions:

1. The population distribution of the observations for any factor-level combination is approximately normal.
2. The variance of the probability distribution is constant and the same for all factor-level combinations.
3. The treatments (factor-level combinations) are randomly assigned to the experimental units.
4. The observations for each factor-level combination represent independent random samples.

CAUTION

When the assumptions for analyzing data collected from a factorial experiment are violated, any inferences derived from the ANOVA are suspect. Nonparametric methods are available for analyzing factorial experiments, but they are beyond the scope of this text. Consult the references given at the end of this chapter if you want to learn about such techniques.

EXAMPLE 11.8 TWO-WAY ANOVA

Refer to Example 11.6. Excel printouts for the 2×3 factorial ANOVA of the data in Table 11.5 (p. 675) are shown in Figure 11.13a and b. Summarize the results in an ANOVA table.

E Figure 11.13 Excel Output for Factorial ANOVA, Example 11.8

	A	B	C	D	E
1	Anova: Two-Factor With Replication				
2					
3	SUMMARY	C	R	P	Total
4	*Machine 1*				
5	Count	3	3	3	9
6	Sum	12.98	10.26	11.84	35.08
7	Average	4.326667	3.42	3.946667	3.897778
8	Variance	0.004433	0.0036	0.003633	0.158394
9					
10	*Machine 2*				
11	Count	3	3	3	9
12	Sum	11.74	11.56	10.43	33.73
13	Average	3.913333	3.853333	3.476667	3.747778
14	Variance	0.008633	0.003033	0.003033	0.045694
15					
16	*Total*				
17	Count	6	6	6	
18	Sum	24.72	21.82	22.27	
19	Average	4.12	3.636667	3.711667	
20	Variance	0.05648	0.058987	0.068937	

a.

	23	ANOVA						
	24	*Source of Variation*	*SS*	*df*	*MS*	*F*	*P-value*	*F crit*
Machines	25	Sample	0.10125	1	0.10125	23.04046	0.000434	4.747221
Materials	26	Columns	0.811944	2	0.405972	92.38306	5.15E-08	3.88529
	27	Interaction	0.768033	2	0.384017	87.38685	7.03E-08	3.88529
Error	28	Within	0.052733	12	0.004394			
	29							
	30	Total	1.733961	17				

b.

Solution

Figure 11.13a gives the sample means for each of the $2 \times 3 = 6$ treatments. The sums of squares and mean squares for the sources of variation, machine main effect, material main effect, machine $\times$ material interaction, and error, are indicated on the Excel printout, Figure 11.13b. These values make up the ANOVA table for the experiment.

EXAMPLE 11.9 *F* TEST FOR INTERACTIONS

Refer to Example 11.8. A plot of the six means corresponding to the six machine-material combinations is shown in Figure 11.14.

E Figure 11.14 Plot of the Means for the Six Machine-Material Combinations

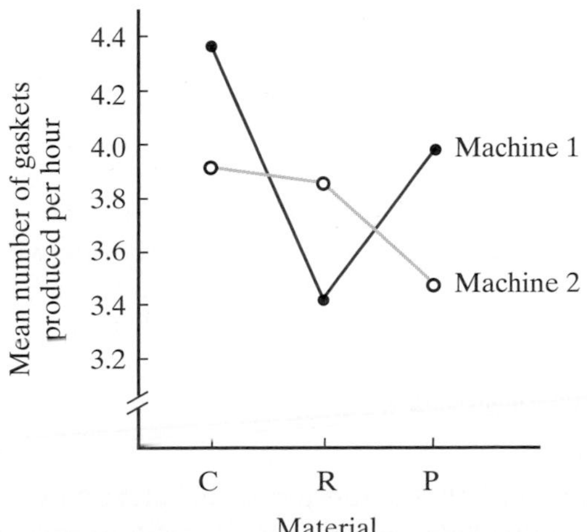

a. Based on the plot, do you think that the factors machine and material interact?

b. Perform a test of hypothesis to determine whether the data provide sufficient evidence to indicate that the more productive stamping machine depends on the gasket material. Test using $\alpha = .05$.

Solution

a. Figure 11.14 shows that the difference between the means for machine 1 and machine 2 changes depending on the material stamped. For example, when cork is stamped, machine 1 is more productive (i.e., has the higher mean). But when rubber is stamped, machine 2 is more productive. Consequently, it appears that the two factors interact. Remember, however, that the plot is based on *sample* means. A statistical test is required to determine whether the pattern of sample means shown in Figure 11.14 implies interaction in the population.

b. The test statistic, critical F value, and p-value for testing the null hypothesis of no interaction between machines and materials, highlighted in the Excel printout, Figure 11.13b, are

$$F = 87.387 \qquad F_{.05} = 3.885 \qquad p\text{-value} \approx 0$$

Since the p-value is less than $\alpha = .05$ (or the test statistic F is greater than the critical value), we conclude that there is an interaction between machines and materials. Therefore, there is sufficient evidence to indicate that neither machine is the more productive for all three materials. If the differences in mean productivity are large enough, the manufacturer should use both machines, selecting the machine that gives the greater productivity for a specific material.

Tests for differences in the mean levels of the main effects in a factorial experiment are relevant *only when the factors do not interact.* When there is no factor interaction, the differences in the mean levels of factor A are the same for all levels of factor B. The test for main effect A tests the significance of these differences. In the presence of interaction, however, the main effect test is irrelevant since the differences in the mean levels of factor A are not the same at each level of factor B. The following example demonstrates the tests for main effects and discusses their implications.

EXAMPLE 11.10 TESTING FOR MAIN EFFECTS

A 2×2 factorial design was employed to determine the effects of work scheduling (factor A) and method of payment (factor B) on attitude toward the job (y). Two types of scheduling were employed, the standard 8:00 A.M.–5:00 P.M. work day and a modification whereby the worker was permitted to vary the starting time and the length of the lunch hour. The two methods of payment were a standard hourly rate and a reduced hourly rate with an added piece rate based on worker production. Four workers were randomly assigned to each of the four scheduling-payment combinations; each completed an attitude test after one month on the job. An ANOVA table for the test scores data is shown in Table 11.6. Interpret the results using $\alpha = .05$.

Solution

First notice that the test for factor interaction is not statistically significant (i.e., there is no evidence of factor interaction) since the p-value for the test, .8899, is

TABLE 11.6 ANOVA Table for the 2 × 2 Factorial Design, Example 11.10

Source	df	SS	MS	F	p-value
Schedule (A)	1	361	361	7.37	.0189
Payment (B)	1	1,444	1,444	29.47	.0002
Schedule × Payment interaction (AB)	1	1	1	.02	.8899
Error	12	588	49		
TOTAL	15	2,394			

greater than $\alpha = .05$. Therefore, we focus on the main effects for the two factors, schedule (A) and payment (B).

Note that the p-values for the tests of main effect A (schedule) and main effect B (payment) are .0189 and .0002, respectively, and both are less than $\alpha = .05$. Thus, there is evidence (at $\alpha = .05$) of differences in the mean levels of the respective main effects. Practically, this implies that the mean worker attitude scores differ for the two schedules, but that this difference does not depend on method of payment (due to lack of interaction). Similarly, the mean worker attitude scores for the two methods of payment differ, but the difference does not depend on the work schedule.

An analysis of variance for a factorial experiment is based on the assumption that the observations in the cells of the two-way table represent independent random samples. Under this assumption, formulas for constructing confidence intervals for the difference between two cell means are exactly the same as for the completely randomized design.

CAUTION

In a factorial experiment, the F test for factor interaction should always be conducted first because the F tests for factor main effects are usually relevant only when factor interaction is *not* significant. If interaction is detected, *do not* perform the F tests for main effects.

Self-Test 11.5

Refer to the factorial experiment described in Self-Test 11.4 (p. 676). The p-values for the ANOVA F tests are listed below.

Factor	p-Value
Club main effect	.0001
Brand main effect	.3420
Club × Brand interaction	.2603

a. Which ANOVA F test should be conducted first?

b. Is there evidence (at $\alpha = .01$) of interaction between club and brand?

c. Based on the results of the test, part **b**, should tests for main effects be conducted? If so, perform these tests using $\alpha = .01$.

PROBLEMS FOR SECTION 11.5

Using the Tools

11.26 Suppose you conduct a 3 × 4 factorial experiment with two observations per treatment. Construct an analysis of variance table for the experiment showing the sources of variation and degrees of freedom for each.

11.27 The analysis of variance for a 3 × 2 factorial design with four observations per treatment produced the ANOVA table entries shown here.

Source	df	SS	MS	F
A	—	100	—	—
B	1	—	—	—
AB interaction	2	—	2.5	—
Error	—	—	2.0	
TOTAL	—	700		

a. Complete the ANOVA table.

b. Test for interaction between factor A and factor B. Use $\alpha = .05$.

c. Test for differences in main effect means for factor A. Use $\alpha = .05$.

d. Test for differences in main effect means for factor B. Use $\alpha = .05$.

11.28 Consider a two-factor factorial design with factor A at a levels and factor B at b levels. For each of the following, find the number of pairwise comparisons of treatments to be made in a Bonferroni analysis.

a. $a = 2, b = 3$ **b.** $a = 2, b = 2$

c. $a = 4, b = 2$ **d.** $a = 3, b = 2$

11.29 Refer to Problem 11.28. For each part, find the value of α used to construct the Bonferroni confidence intervals if EER = .05.

11.30 The data for a 3 × 4 factorial design with two observations per treatment are shown in the accompanying table. An analysis of variance was conducted on the data using Excel. The results are reported in the accompanying Excel printout.

PROB11_30

		Factor B			
		1	2	3	4
Factor A	1	5 4	7 9	6 5	5 7
	2	6 4	10 9	5 8	9 7
	3	8 10	7 6	5 8	6 5

29	ANOVA						
30	*Source of Variation*	*SS*	*df*	*MS*	*F*	*P-value*	*F crit*
31	Sample (A)	6.583333	2	3.291667	1.837209	0.201344	3.88529
32	Columns (B)	13.79167	3	4.597222	2.565891	0.103331	3.4903
33	Interaction	35.08333	6	5.847222	3.263566	0.038502	2.996117
34	Within	21.5	12	1.791667			
35							
36	Total	76.95833	23				

E Problem 11.30

a. Do the data provide sufficient evidence of interaction between factor *A* and factor *B*? Test using $\alpha = .05$.

b. Given the results of the test in part **a**, would you recommend that tests for main effects be conducted? Explain.

Applying the Concepts

11.31 Parapsychologists define "lucky" people as individuals who report that seemingly chance events consistently tend to work out in their favor. A team of British psychologists designed a study to examine the effects of luckiness and competition on performance in a guessing task (*The Journal of Parapsychology,* Mar. 1997). Each in a sample of 56 college students was classified as lucky, unlucky, or uncertain based on their responses to a Luckiness Questionnaire. In addition, the participants were randomly assigned to either a competitive or noncompetitive condition. All students were then asked to guess the outcomes of 50 flips of a coin. The response variable measured was percentage of coin-flips correctly guessed.

a. An ANOVA for a 2×3 factorial design was conducted on the data. Identify the factors and their levels for this design.

b. The results of the ANOVA are summarized in the table. Fully interpret the results using $\alpha = .05$.

Source	df	*F*	*p*-value
Luckiness (L)	2	1.39	.26
Competition (C)	1	2.84	.10
L × C	2	0.72	.72
Error	50		
TOTAL	55		

11.32 A videocassette recorder (VCR) repair service wished to study the effect of VCR brand and service center on the repair time measured in minutes. Three VCR brands (A, B, C) and three service centers were specifically selected for analysis. Each service center repaired two VCRs of each brand. The results are shown in the table.

VCR

		VCR Brands		
		A	**B**	**C**
	1	52 57	48 39	59 67
Service Centers	**2**	51 43	61 52	58 64
	3	37 46	44 50	65 69

a. Use Excel or MINITAB to obtain an ANOVA table for this 3×3 factorial.

b. Conduct the test for factor interaction at $\alpha = .05$. Interpret the results.

c. Is there a main effect due to service centers? Test at $\alpha = .05$.

d. Is there a main effect due to VCR brand? Test at $\alpha = .05$.

11.33 In the *Journal of Nutrition* (July 1995), University of Georgia researchers examined the impact of a vitamin-B supplement (nicotinamide) on the kidney. The experimental "subjects" were 28 Zucker rats—a species that tends to develop kidney problems. Half of the rats were classified as obese and half as lean. Within each group, half were randomly assigned to receive a vitamin-B-supplemented diet and half were not. Thus, a 2×2 factorial experiment was conducted with seven rats assigned to each of the four combinations of size (lean or obese) and diet

(supplement or not). One of the response variables measured was weight (in grams) of the rat's kidney at the end of a 20-week feeding period. The data (simulated from summary information provided in the journal article) are shown in the table.

KIDNEY

		Diet			
		Regular		**Vitamin-B Supplement**	
Rat Size	**Lean**	1.62	1.47	1.51	1.63
		1.80	1.37	1.65	1.35
		1.71	1.71	1.45	1.66
		1.81		1.44	
	Obese	2.35	2.84	2.93	2.63
		2.97	2.05	2.72	2.61
		2.54	2.82	2.99	2.64
		2.93		2.19	

a. Use Excel or MINITAB to conduct an analysis of variance on the data. Summarize the results in an ANOVA table.

b. Conduct the appropriate ANOVA F tests at $\alpha = .01$. Interpret the results.

11.34 Do women enjoy the thrill of a close basketball game as much as men? To answer this question, male and female undergraduate students were recruited to participate in an experiment (*Journal of Sport & Social Issues,* Feb. 1997). The students watched one of eight live televised games of a recent NCAA basketball tournament. (None of the games involved a home team to which the students could be considered emotionally committed.) The "suspense" of each game was classified into one of four categories according to the closeness of scores at the game's conclusion: minimal (15 point or greater differential), moderate (10–14 point differential), substantial (5–9 point differential), and extreme (1–4 point differential). After the game, each student rated his or her enjoyment on an 11-point scale ranging from 0 (not at all) to 10 (extremely). The enjoyment rating data were analyzed as a 4×2 factorial design, with suspense (four levels) and gender (two levels) as the two factors. The $4 \times 2 = 8$ treatment means are shown in the table at left.

	Gender	
Suspense	**Male**	**Female**
Minimal	1.77	2.73
Moderate	5.38	4.34
Substantial	7.16	7.52
Extreme	7.59	4.92

Source: Gan, Su-lin, et al. "The thrill of a close game: Who enjoys it and who doesn't?" *Journal of Sport & Social Issues,* Vol. 21, No. 1, Feb. 1997, pp. 59–60.

Problem 11.34

a. Plot the treatment means in a graph similar to Figure 11.11 (p. 676). Does the pattern of means suggest interaction between suspense and gender? Explain.

b. The ANOVA F test for interaction yielded the following results: numerator df = 3, denominator df = 68, $F = 4.42$, p-value = .007. What can you infer from these results?

c. Based on the test, part **b**, is the difference between the mean enjoyment levels of males and females the same, regardless of the suspense level of the game?

d. Multiple comparisons (using EER = .05) of the four suspense means for male viewers are summarized in the table below. Interpret the results.

1.77	5.38	7.16	7.59
Minimal	Moderate	Substantial	Extreme

e. Multiple comparisons (using EER = .05) of the four suspense means for female viewers are summarized in the table below. Interpret the results.

2.73	4.34	4.92	7.52
Minimal	Moderate	Extreme	Substantial

11.35 The *Accounting Review* (Jan. 1991) reported on a study of the effect of two factors, confirmation of accounts receivable and verification of sales transactions, on account misstatement risk by auditors. Both factors were held at the same two levels: completed or not completed. Thus, the experimental design is a 2×2 factorial design.

a. Identify the factors, factor levels, and treatments for this experiment.
b. Explain what factor interaction means for this experiment.
c. A graph of the hypothetical mean misstatement risks for each of the $2 \times 2 = 4$ treatments is displayed below. In this hypothetical case, does it appear that interaction exists?

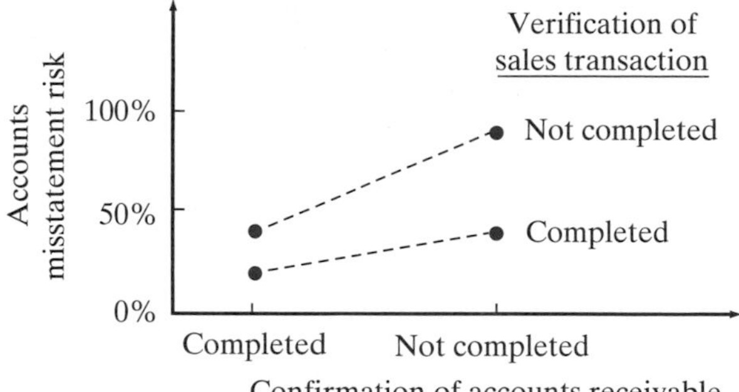

Source: Brown, C. E., and Solomon, I. "Configural information processing in auditing: The role of domain-specific knowledge." *The Accounting Review,* Vol. 66, No. 1, Jan. 1991, p. 105 (Figure 1).

Problem 11.35

11.36 The citrus mealybug is an insect that feeds on citrus fruit, especially grapefruit. An experiment was conducted to determine whether the density of mealybugs in an orchard depends on time of year (*Environmental Entomology,* June 1995). Thirty-five grapefruits were randomly sampled from each of two orchards (A and B) in each of four time periods (June, July, September, and October), and the number of mealybugs per fruit was recorded. The data were subjected to a 2×4 factorial ANOVA, with the results shown in the ANOVA table. Using $\alpha = .05$, conduct the appropriate ANOVA F tests. What conclusions can you draw from the ANOVA?

Source	df	SS	MS	F	p-value
Orchard	1	1.1535	1.1535	3.44	.06
Period	3	20.6278	6.8759	20.52	.001
Orchard × Period	3	2.9707	.9902	2.96	.03
Error	272	91.1437	.3351		

11.37 A hospital administrator wished to examine postsurgical hospitalization periods following knee surgery. A random sample of 30 patients was selected, five for each combination of age group and type of surgery. The results, in number of postsurgical hospitalization days, are listed in the accompanying table. Conduct a complete analysis of the data using Excel or MINITAB. Interpret the results.

KNEESURG

		Age Group		
		Under 30	**30 to 50**	**Over 50**
Type of Knee Surgery	**Arthroscopy**	1	4	3
		3	3	5
		2	2	2
		6	3	3
		2	2	3
	Arthrotomy	3	4	4
		10	5	8
		6	11	12
		7	5	10
		8	6	3

11.38 The impact of a fictional story on real-world beliefs was investigated in *Psychonomic Bulletin & Review* (Sept. 1997). The researchers theorized that readers unfamiliar with the setting of a fictional story would be vulnerable to its assertion, while readers familiar with the setting would not. In a designed experiment, each in a group of Yale University students was assigned to read a story that took place either at Yale (familiar setting) or at Princeton University (unfamiliar setting). Also, the student read either a story about a real-world truth (e.g., sunlight is bad for your skin) or a story that contained a false assertion (e.g., sunlight is good for your skin). After reading the story, the student was asked to rate the believability of a true statement about the story item (e.g., sunlight's effect on skin). Thus, there were $2 \times 2 = 4$ experimental conditions, one for each of the combinations of setting (Yale or Princeton) and story (assertion true or false). The students' standardized agreement ratings were analyzed as a 2×2 factorial design ANOVA.

a. Identify the factors, factor levels, and treatments in this experiment.

b. Identify the response variable y.

c. The ANOVA F test for interaction resulted in a p-value of less than .01. Interpret this result.

d. Based on the result, part **c**, would you recommend conducting F tests on the main effects? Explain.

e. The sample means for the four experimental conditions are shown in the table. [*Note:* Positive standardized agreement ratings reflect the tendency for the students to agree with the statement. Negative ratings reflect the tendency for the students to disagree with the statement.] Do the sample means support the researchers' theory? Explain.

		Story Assertion	
		True	**False**
Setting	**Yale (familiar)**	−.16	−.26
	Princeton (unfamiliar)	.69	−.56

11.39 Those who care for a spouse with Alzheimer's disease often experience both physical and mental health problems. Four strategies caregivers use to cope with the stress are *wishfulness* ("wished you were a stronger person"), *acceptance* ("accepted the situation"), *intrapsychic* ("fantasies about how things will turn out"), and *instrumental* ("followed a plan of action"). In one experiment, 78 caregivers were classified into one of the four coping strategies and as having a high or low level of stress. All subjects then completed a standardized test designed to measure level of coping (*Journal of Applied Gerontology,* Mar. 1997). The mean coping scores were calculated for subjects in each of the $4 \times 2 = 8$ strategy $\times$ stress classes. A graph of these eight means is shown in the figure on p. 687. An ANOVA for a 4×2 factorial design was conducted on the data.

a. Based on the graph, do you believe that strategy and stress interact?

b. In practical terms, what does it mean to say that strategy and stress interact?

11.40 A study published in *Teaching Psychology* (May 1998) examined how external clues influence student performance. Introductory psychology students were randomly assigned to one of four different midterm examinations. Form 1 was printed on blue paper and contained difficult questions, while form 2 was also printed on blue paper but contained simple questions. Form 3 was printed on red paper, with difficult questions; form 4 was printed on red paper with simple questions. The researchers

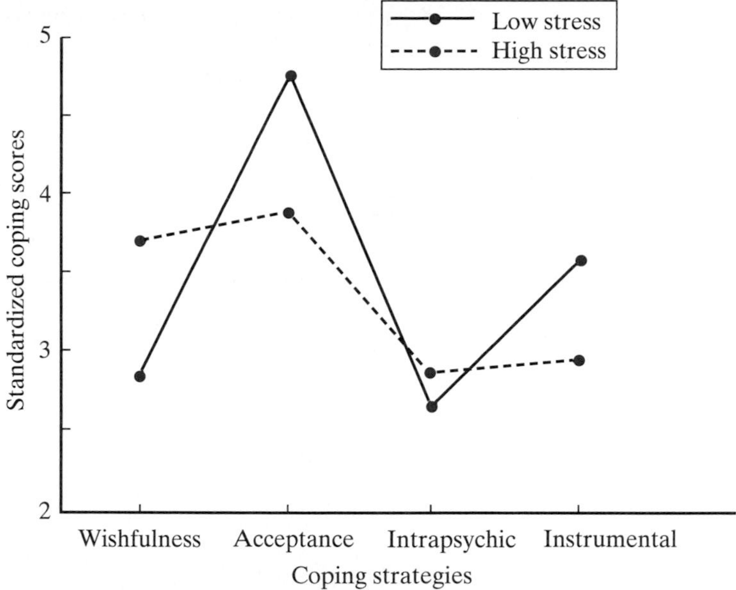

Source: Rose, S. K., et al. "The relationship of self-restraint and distress to coping among spouses caring for persons with Alzheimer's disease." *Journal of Applied Gerontology,* Vol. 16, No. 1, Mar. 1997, p. 97 (Figure 1).

Problem 11.39

were interested in the impact that Color (red or blue) and Question (simple or difficult) had on mean exam score.

a. What experimental design was employed in this study? Identify the factors and treatments.

b. The researchers conducted an ANOVA and found a significant interaction between Color and Question (p-value $< .03$). Interpret this result.

c. The sample mean scores (percentage correct) for the four exam forms are listed below. Plot the four means on a graph to illustrate the Color × Question interaction.

Form	Color	Question	Mean Score
1	Blue	Difficult	53.3
2	Blue	Simple	80.0
3	Red	Difficult	39.3
4	Red	Simple	73.6

11.6 Checking ANOVA Assumptions

For both of the experimental designs discussed in this chapter, we listed the assumptions underlying the analysis in the relevant boxes. The assumptions for a completely randomized design are as follows: (1) the k probability distributions of the response y corresponding to the k treatments are normal; and (2) the population variances of the k treatments are equal. Similarly, for factorial designs, the data for the treatments must come from normal probability distributions with equal variances.

Checks on these ANOVA assumptions can be performed using graphs of the response variable y. A brief overview of these techniques is given in the box.

CHECKING ANOVA ASSUMPTIONS

Detecting Nonnormal Populations

1. For each treatment, construct a histogram, stem-and-leaf display, box-and-whisker plot, or normal probability plot of the response (or dependent) variable y. Look for highly skewed distributions. [*Note:* ANOVA, like regression, is robust with respect to the normality assumption. That is, slight departures from normality will have little impact on the validity of the inferences derived from the analysis. However, if the sample size for each treatment is small, then these graphs will probably be of limited use.]
2. Formal statistical tests of normality are also available. The null hypothesis is that the probability distribution of the response y is normal. These tests, however, are sensitive to slight departures from normality. Since in most scientific applications the normality assumption will not be satisfied exactly, these tests will likely result in a rejection of the null hypothesis and, consequently, are of limited use in practice. Consult the references for more information on these formal tests.
3. If the distribution of the response y departs greatly from normality, a *normalizing transformation* may be necessary. For example, for highly skewed distributions, transformations such as $\log(y)$ or $\sqrt{y}$ tend to "normalize" the data since these functions "pull" the observations in the tail of the distribution back toward the mean.

Detecting Unequal Variances

1. For each treatment, construct a dot plot for the values of the response y and look for differences in the spread (variability) shown in the plots. (See Figure 11.15 [p. 689] for an example.)
2. When the sample sizes are small for each treatment, only a few points are plotted on the frequency plots, making it difficult to detect differences in variation. In this situation, you may want to use one of several formal statistical tests of homogeneity of variances that are available. Consult the chapter references for information on these tests (e.g., Gujarati [1995]).
3. When unequal variances are detected, use a *variance stabilizing transformation* such as $\log(y)$ or $\sqrt{y}$.

EXAMPLE 11.11 **CHECKING ASSUMPTIONS**

Refer to the completely randomized ANOVA, Examples 11.3 and 11.4. Check the ANOVA assumptions for this experiment.

Solution

For this completely randomized design, the three treatments are the three socioeconomic classes (lower, middle, and upper) and the response y is grade point average (GPA). Thus, we require the GPA distribution for each class to be normally distributed with equal variances. Recall that the sample size for each treatment was seven.

With such a small number of observations per treatment, a stem-and-leaf display to check for normality will not be very revealing. Not knowing whether the

GPAs are normally distributed is not a major concern, since ANOVA, like regression, is robust with respect to nonnormal data. That is, the procedure yields valid inferences even when the distribution of the response y deviates from normality.

To check for equal variances, we plotted the GPAs for each socioeconomic class in Figure 11.15. The dot plot shows the spread of the GPAs for each class. Except for one low GPA (1.36) in the upper class, the variability of the response is about the same for each treatment; thus, the assumption of equal variances appears to be reasonably satisfied.

Figure 11.15 Dot Plot for Example 11.11

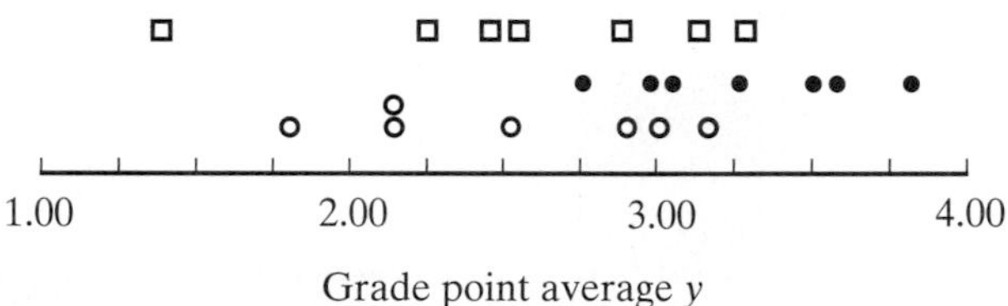

Self-Test 11.6

Refer to the one-way ANOVA, Self-Test 11.2 (p. 659). Dot plots of the distances for each brand of golf ball are illustrated in the accompanying figure. Does it appear that the brands have approximately the same variation in distance?

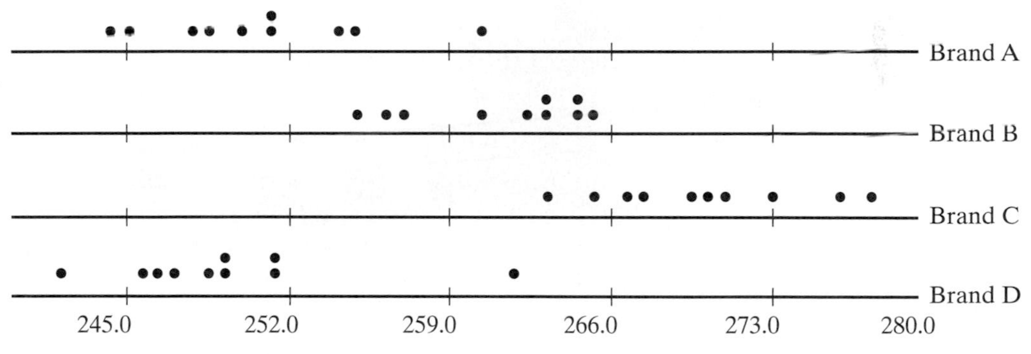

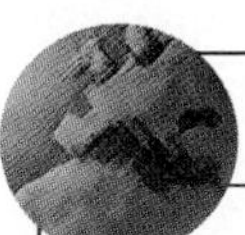

Statistics in the Real World Revisited

Checking the Assumptions for the Cockroach ANOVA

In the previous Statistics in the Real World Revisited sections (p. 660 and p. 670), we used ANOVA to conclude that the population mean trail deviations for four groups of cockroaches are significantly different. In particular, we found the mean for gravid females was greater than the means for the other three groups (adult males, adult females, and nymphs). For these inferences to be valid, the assumptions required for an ANOVA need to be reasonably satisfied. Stated in terms of the application, the two assumptions are:

(Continued)

1. For each cockroach group, the trail deviations are normally distributed
2. The four cockroach groups have equal trail deviation variances

To check these assumptions, we graphed the data in the **ROACH2** file using MINITAB; the MINITAB graphs are shown in Figure 11.16.

Figures 11.16a–d show histograms of the trail deviation data by group, Figure 11.16e shows dot plots for the groups, and Figure 11.16f shows box plots for the groups. Except possibly for gravid females (Figure 11.16c), the histograms in Figures 11.16a–d yield distributions that appear to be nonnormal. Consequently, the assumption of normal data for each group is likely to be violated. The dot plots and box plots in Figures 11.16e–f show the spread of the trail deviations for the four groups to be about the same. Thus, it appears that the assumption of equal variances is reasonably satisfied.

M Figure 11.16a
MINITAB Histogram of Trail Deviation for Adult Males

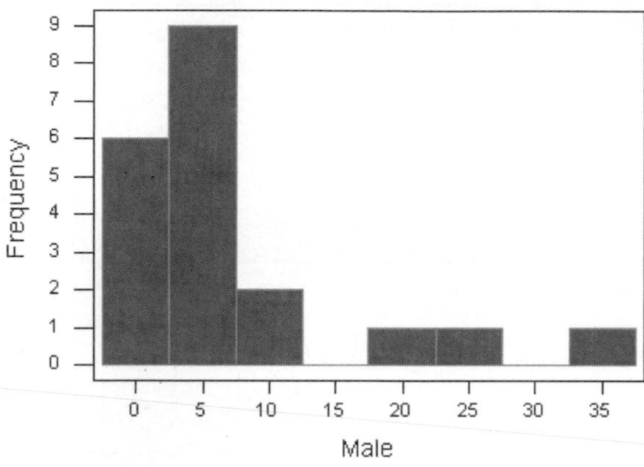

M Figure 11.16b
MINITAB Histogram of Trail Deviation for Adult Females

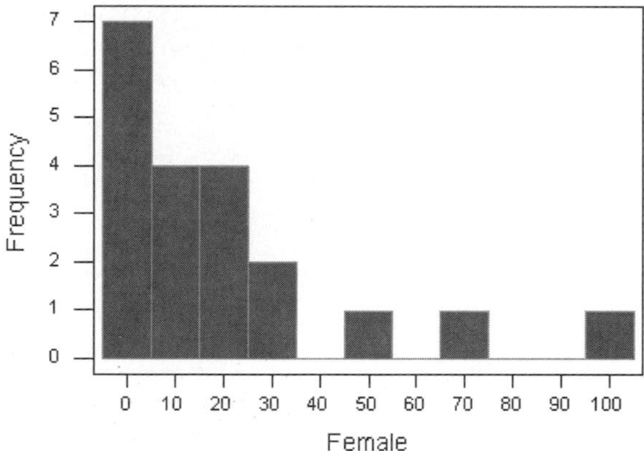

M Figure 11.16c
MINITAB Histogram of Trail Deviation for Gravid Females

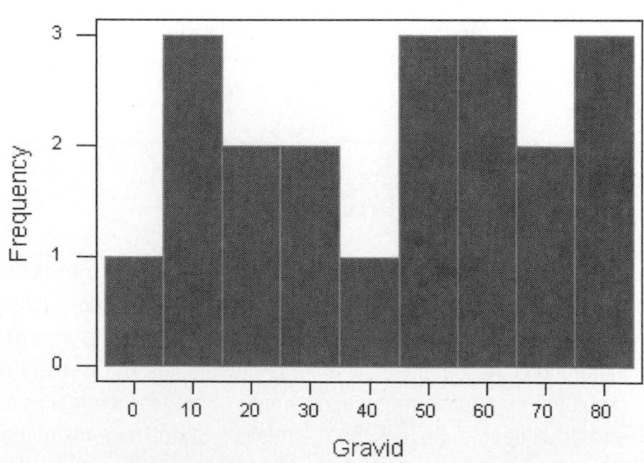

M Figure 11.16d MINITAB Histogram of Trail Deviation for Nymphs

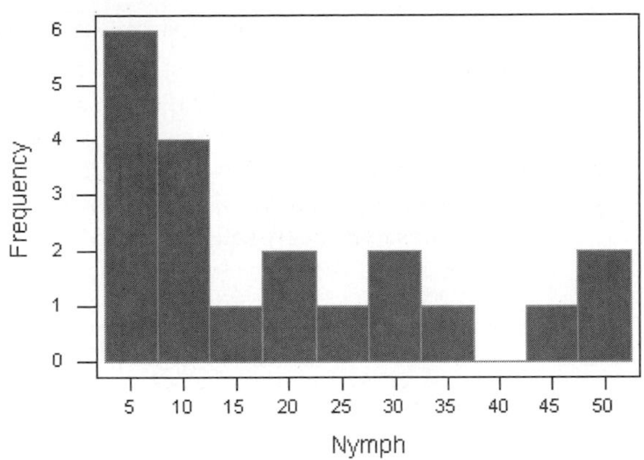

M Figure 11.16e MINITAB Dot Plots for Extract Trail Deviation

Dot Plot for Deviate by Group

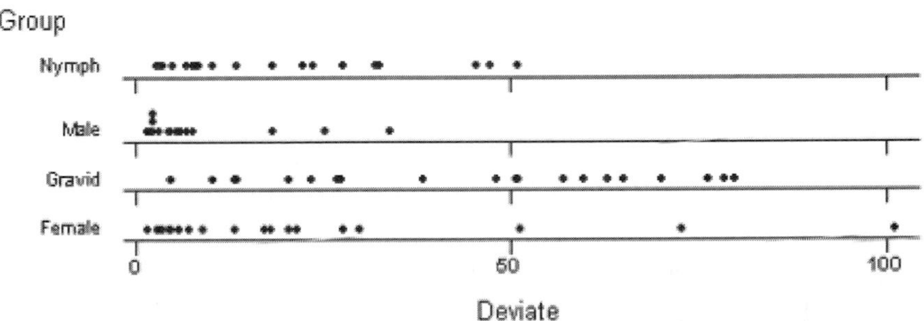

M Figure 11.16f MINITAB Box Plots for Extract Trail Deviation

Boxplots of Deviate by Group
(means are indicated by solid circles)

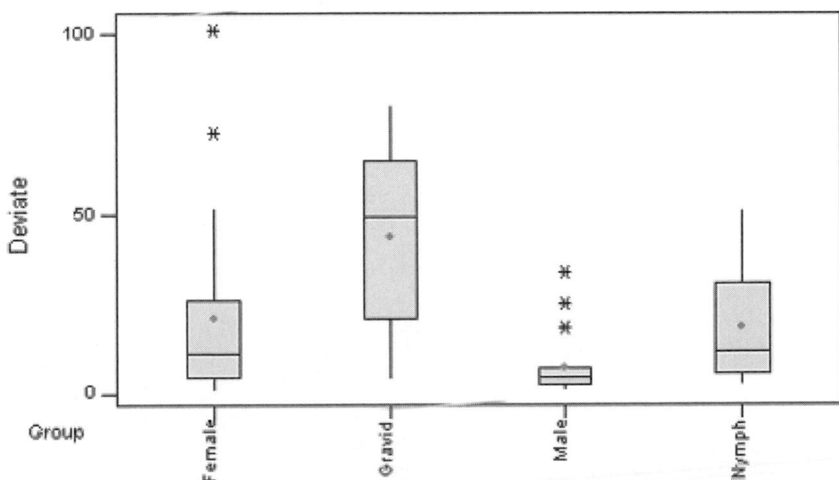

Although ANOVA is robust with respect to departures from normality, the highly skewed histograms for three of the four cockroach groups may cause concern about the validity of our ANOVA inferences. In the next Statistics in the Real World Revisited section, we apply a nonparametric method to the data.

PROBLEMS FOR SECTION 11.6

Applying the Concepts

11.41 Refer to the *Fortune* (Oct. 25, 1999) study of the 50 most powerful women in America, Problem 11.7 (p. 662). You conducted an ANOVA to compare the average ages of powerful women in three groups based on their position within the firm. Use the accompanying MINITAB plots to check the ANOVA assumptions. Are the assumptions reasonably satisfied?

```
Stem-and-leaf of Age        Group = 1        N  = 21
Leaf Unit = 1.0


    1     3 6
    1     3
    3     4 11
    4     4 3
    8     4 4555
    9     4 6
   (2)    4 88
   10     5 011
    7     5 2222
    3     5
    3     5
    3     5 8
    2     6 0
    1     6
    1     6
    1     6
    1     6 8


Stem-and-leaf of Age        Group = 2        N  = 16
Leaf Unit = 1.0


    1     4 0
    1     4
    3     4 55
    8     4 66677
    8     4 889
    5     5
    5     5 33
    3     5 55
    1     5
    1     5
    1     6
    1     6
    1     6 4


Stem-and-leaf of Age        Group = 3        N  = 13
Leaf Unit = 1.0


    1     3 7
    1     3
    1     4
    3     4 23
    5     4 45
   (4)    4 6667
    4     4 89
    2     5 00
```

11.42 Check the assumptions for the completely randomized design ANOVA of Problem 11.8 (p. 663).

11.43 Check the assumptions for the completely randomized design ANOVA of Problem 11.9 (p. 664).

11.44 Check the assumptions for the completely randomized design ANOVA of Problem 11.10 (p. 664).

11.45 Check the assumptions for the completely randomized design ANOVA of Problem 11.12 (p. 665).

11.46 (Optional) Check the assumptions for the factorial design ANOVA of Problem 11.33 (p. 683).

11.7 Calculation Formulas for ANOVA (Optional)

In this optional section we demonstrate the calculation formulas for constructing an ANOVA table for both experimental designs discussed in this chapter. Except for rounding, these formulas will produce results identical to those obtained from statistical such as MINITAB or spreadsheet software such as Excel.

ANOVA CALCULATION FORMULAS FOR A COMPLETELY RANDOMIZED DESIGN

1. Let k = number of means (treatments) to be compared

n_i = number of observations for treatment i

n = total sample size $= n_1 + n_2 + \cdots + n_k$

CM = Correction for the mean

$$= \frac{(\text{Total of all observations})^2}{\text{Total number of observations}} = \frac{(\sum y)^2}{n} \quad \textbf{(11.12)}$$

2. SS(Total) = Total sum of squares

= (Sum of squares of all observations) − CM

$$= \sum y^2 - \text{CM} \quad \textbf{(11.13)}$$

3. SST = Sum of squares for treatments

$$= \left(\begin{array}{c}\text{Sum of squares of treatment totals}\\ \text{with each square divided by the}\\ \text{number of observations in that treatment}\end{array}\right) - \text{CM}$$

$$= \frac{T_1^2}{n_1} + \frac{T_2^2}{n_2} + \cdots + \frac{T_k^2}{n_k} - \text{CM} \quad \textbf{(11.14)}$$

where T_i is the total of all observations for treatment i, $i = 1, 2, \ldots, k$.

4. SSE = Sum of squares for error

$$= \text{SS(Total)} - \text{SST} \quad \textbf{(11.15)}$$

5. $$\text{MST} = \text{Mean square for treatment} = \frac{\text{SST}}{k-1} \quad \textbf{(11.16)}$$

6. $$\text{MSE} = \text{Mean square for error} = \frac{\text{SSE}}{n-k} \quad \textbf{(11.17)}$$

7. $$F = \text{Test statistic} = \frac{\text{MST}}{\text{MSE}} \quad \textbf{(11.18)}$$

ANOVA CALCULATION FORMULAS FOR A TWO-FACTOR FACTORIAL DESIGN

1. Let

$n = abr$

a = number of levels of factor A

b = number of levels of factor B

r = number of replications of the factorial experiment

CM = Correction for the mean

$$= \frac{(\text{Total of all } n \text{ observations})^2}{n} = \frac{(\sum y)^2}{n} \quad \textbf{(11.19)}$$

2. SS(Total) = (Sum of squares of all observations) − CM

$$= \sum y^2 - \text{CM} \quad \textbf{(11.20)}$$

3. Main effect sum of squares for factor A:

$$\text{SS}(A) = \frac{A_1^2 + A_2^2 + \cdots + A_a^2}{br} - \text{CM} \quad \textbf{(11.21)}$$

where A_i is the total of all observations at level i for factor A.

4. Main effect sum of squares for factor B:

$$\text{SS}(B) = \frac{B_1^2 + B_2^2 + \cdots + B_b^2}{ar} - \text{CM} \quad \textbf{(11.22)}$$

where B_j is the total of all observations at level j for factor B.

5. Interaction sum of squares:

$$\text{SS}(AB) = \sum_{i,j} \frac{(AB_{ij})^2}{r} - \text{SS}(A) - \text{SS}(B) - \text{CM} \quad \textbf{(11.23)}$$

where AB_{ij} is the sum of all observations in the cell corresponding to the ith level of factor A and the jth level of factor B. [*Note:* To find SS(AB), square each cell total, sum the squares for all cell totals, divide by r, and then subtract SS(A), SS(B), and CM.]

6.

$$\text{SSE} = \text{SS(Total)} - \text{SS}(A) - \text{SS}(B) - \text{SS}(AB) \quad \textbf{(11.24)}$$

7.

$$\text{MS}(A) = \frac{\text{SS}(A)}{a - 1} \quad \textbf{(11.25)}$$

$$\text{MS}(B) = \frac{\text{SS}(B)}{b - 1} \quad \textbf{(11.26)}$$

$$\text{MS}(AB) = \frac{\text{SS}(AB)}{(a - 1)(b - 1)} \quad \textbf{(11.27)}$$

$$\text{MSE} = s^2 = \frac{\text{SSE}}{ab(r - 1)} \quad \textbf{(11.28)}$$

8.

$$F \text{ statistic for testing } AB \text{ interaction: } F = \frac{\text{MS}(AB)}{\text{MSE}} \quad \textbf{(11.29)}$$

9.

$$F \text{ statistic for testing } A \text{ main effect: } F = \frac{\text{MS}(A)}{\text{MSE}} \quad \textbf{(11.30)}$$

10.

$$F \text{ statistic for testing } B \text{ main effect: } F = \frac{\text{MS}(B)}{\text{MSE}} \quad \textbf{(11.31)}$$

EXAMPLE 11.12

USING CALCULATION FORMULAS FOR A ONE-WAY ANOVA

The data for the completely randomized design of Example 11.3 are reproduced in Table 11.7. Use the calculation formulas shown in the box to construct an ANOVA table.

TABLE 11.7 Data (GPAs) and Sums for the Completely Randomized ANOVA, Example 11.12

	Lower Class	Middle Class	Upper Class
	2.87	3.23	2.25
	2.16	3.45	3.13
	3.14	2.78	2.44
	2.51	3.77	3.27
	1.80	2.97	2.81
	3.01	3.53	1.36
	2.16	3.01	2.53
Sums:	$T_1 = 17.65$	$T_2 = 22.74$	$T_3 = 17.79$

Solution

Step 1 The first step is to find the sum of all $n = 21$ sample observations, or $\sum y$. The sums of the GPAs for each class (treatment) are shown at the bottom of Table 11.7. Adding these three sums yields

$$\sum y = T_1 + T_2 + T_3 = 17.65 + 22.74 + 17.79 = 58.18$$

Then we compute CM—called the correction for the mean—using equation 11.12:

$$\text{CM} = \frac{(\sum y)^2}{n} = \frac{(58.18)^2}{21} = 161.1863$$

Step 2 SS(Total) is obtained using equation 11.13. (Note that this requires that we sum the squares of all $n = 21$ sample GPA values.)

$$\begin{aligned}\text{SS(Total)} &= \sum y^2 - \text{CM} \\ &= (2.87)^2 + (2.16)^2 + (3.14)^2 + \cdots + (2.53)^2 - 161.1863 \\ &= 168.299 - 161.1863 = 7.1127\end{aligned}$$

Step 3 Applying equation 11.14, the sum of squares for treatments, SST, is

$$\begin{aligned}\text{SST} &= \frac{T_1^2}{n_1} + \frac{T_2^2}{n_2} + \frac{T_3^2}{n_3} - \text{CM} \\ &= \frac{(17.65)^2}{7} + \frac{(22.74)^2}{7} + \frac{(17.79)^2}{7} \\ &= 163.5877 - 161.1863 = 2.4014\end{aligned}$$

Step 4 The sum of squared errors, SSE, is obtained using equation 11.15:

$$\begin{aligned}\text{SSE} &= \text{SS(Total)} - \text{SST} \\ &= 7.1127 - 2.4014 = 4.7113\end{aligned}$$

Step 5 Applying equation 11.16, we obtain the mean square for treatments:

$$\text{MST} = \frac{\text{SST}}{k - 1} = \frac{2.4014}{3 - 1} = 1.2007$$

Step 6 Applying equation 11.17, we obtain the mean square for error:

$$\text{MSE} = \frac{\text{SSE}}{n - k} = \frac{4.7113}{21 - 3} = .2617$$

Step 7 Finally, from equation 11.18, the test statistic is

$$F = \frac{\text{MST}}{\text{MSE}} = \frac{1.2007}{.2617} = 4.59$$

These computations are summarized in the ANOVA table shown in Table 11.8. Except for rounding, these numbers agree with those shown in the Excel printout, Figure 11.6 (p. 658).

TABLE 11.8 ANOVA Table for Example 11.12

Source	df	SS	MS	F
Class (treatments)	2	2.4014	1.2007	4.59
Error	18	4.7113	.2617	
TOTAL	20	7.1127		

EXAMPLE 11.13

USING CALCULATION FORMULAS FOR A TWO-WAY ANOVA

The data for the factorial design of Example 11.6 are reproduced in Table 11.9. Use the calculation formulas shown in the box to construct an ANOVA table for the data.

TABLE 11.9 Data (Thousands of Gaskets) and Sums for the Factorial ANOVA, Example 11.13

		Material			
		C	R	P	Sums
Machine	1	4.31 4.27 4.40 Sum = 12.98	3.36 3.42 3.48 Sum = 10.26	4.01 3.94 3.89 Sum = 11.84	$A_1 = 35.08$
	2	3.94 3.81 3.99 Sum = 11.74	3.91 3.80 3.85 Sum = 11.56	3.48 3.53 3.42 Sum = 10.43	$A_2 = 33.73$
Sums		$B_1 = 24.72$	$B_2 = 21.82$	$B_3 = 22.27$	TOTAL = 68.81

Solution

For this 2×3 factorial design, let machine represent factor A and material represent factor B. Then the number of levels of factors A and B, respectively, are $a = 2$ and $b = 3$. Also, there are $r = 3$ replications in this experiment.

Step 1 First, we calculate the correction factor for the mean, CM. From Table 11.9, the sum of all $n = 18$ sample observations is 68.81. Substituting into equation 11.19, we find:

$$\text{CM} = \frac{(\Sigma y)^2}{n} = \frac{(68.81)^2}{18} = 263.04534$$

Step 2 Applying equation 11.20, we obtain SS(Total):

$$\begin{aligned}\text{SS(Total)} &= \sum y^2 - \text{CM} \\ &= (4.31)^2 + (4.27)^2 + (4.40)^2 + \cdots + (3.42)^2 - 263.04534 \\ &= 264.7793 - 263.04534 = 1.73396\end{aligned}$$

Step 3 The main effect sum of squares for factor A (machine) requires that we first find the totals for each level of A. These sums, $A_1 = 35.08$ and $A_2 = 33.73$, are shown in Table 11.9. Substituting these values into equation 11.21, we obtain:

$$\begin{aligned}\text{SS}(A) &= \frac{A_1^2 + A_2^2}{br} - \text{CM} \\ &= \frac{(35.08)^2 + (33.73)^2}{(3)(3)} - 263.04534 \\ &= 263.14659 - 263.04534 = .10125\end{aligned}$$

Step 4 To find the main effect sum of squares for factor B (material), we need the totals for each level of B. These sums, $B_1 = 24.72$, $B_2 = 21.82$, and $B_3 = 22.27$, are also shown in Table 11.9. Substituting into equation 11.22, we find

$$\begin{aligned}\text{SS}(B) &= \frac{B_1^2 + B_2^2 + B_3^2}{ar} - \text{CM} \\ &= \frac{(24.72)^2 + (21.82)^2 + (22.27)^2}{(2)(3)} - 263.04534 \\ &= 263.85728 - 263.04534 = .81194\end{aligned}$$

Step 5 To calculate the interaction sum of squares, we need the sum of responses in each combination of the levels of the two factors. These sums are shown in the six cells of Table 11.9. Applying equation 11.23 yields

$$\begin{aligned}\text{SS}(AB) &= \frac{(A_1B_1)^2 + (A_1B_2)^2 + (A_1B_3)^2 + (A_2B_1)^2 + (A_2B_2)^2 + (A_2B_3)^2}{r} - \text{SS}(A) - \text{SS}(B) - \text{CM} \\ &= \frac{(12.98)^2 + (10.26)^2 + (11.84)^2 + (11.74)^2 + (11.56)^2 + (10.43)^2}{3} - .10125 - .81194 - 263.04534 \\ &= 264.72657 - .10125 - .81194 - 263.04534 \\ &= .76803\end{aligned}$$

Step 6 The sum of squares for error, SSE, is obtained using equation 11.24:

$$\begin{aligned}\text{SSE} &= \text{SS(Total)} - \text{SS}(A) - \text{SS}(B) - \text{SS}(AB) \\ &= 1.73396 - .10125 - .81194 - .76803 = .05274\end{aligned}$$

Step 7 Using equations 11.25–11.28, we obtain the following mean squares:

$$\text{MS}(A) = \frac{\text{SS}(A)}{a-1} = \frac{.10125}{2-1} = .10125$$

$$\text{MS}(B) = \frac{\text{SS}(B)}{b-1} = \frac{.81194}{3-1} = .40597$$

$$\text{MS}(AB) = \frac{\text{SS}(AB)}{(a-1)(b-1)} = \frac{.76803}{(2-1)(3-1)} = .384015$$

$$\text{MSE} = \frac{\text{SSE}}{ab(r-1)} = \frac{.05274}{(2)(3)(3-1)} = .004395$$

Steps 8–10 The ANOVA F statistics are calculated using equations 11.29–11.31:

$$\text{Interaction: } F = \frac{\text{MS}(AB)}{\text{MSE}} = \frac{.384015}{.004395} = 87.38$$

$$\text{Main Effect } A \text{ (machine): } F = \frac{\text{MS}(A)}{\text{MSE}} = \frac{.10125}{.004395} = 23.04$$

$$\text{Main Effect } B \text{ (machine): } F = \frac{\text{MS}(B)}{\text{MSE}} = \frac{.40597}{.004395} = 92.37$$

These computations are summarized in the ANOVA table shown in Table 11.10. Except for rounding, these numbers agree with those in the Excel printout, Figure 11.13 (p. 679).

TABLE 11.10 ANOVA Table for Example 11.13

Source	df	SS	MS	F
Machine (A)	1	.10125	.10125	23.04
Material (B)	2	.81194	.40597	92.37
$A \times B$	2	.76803	.384015	87.38
Error	12	.05274	.004395	
TOTAL	17	1.73396		

PROBLEMS FOR SECTION 11.7

Using the Tools

11.47 Use the ANOVA calculation formulas for a completely randomized design to construct the ANOVA summary table for Problem 11.3 (p. 661).

11.48 Use the ANOVA calculation formulas for a completely randomized design to construct the ANOVA summary table for Problem 11.8 (p. 663).

11.49 Use the ANOVA calculation formulas for a completely randomized design to construct the ANOVA summary table for Problem 11.12 (p. 665).

11.50 Use the ANOVA calculation formulas for a factorial design to construct the ANOVA summary table for Problem 11.30 (p. 682).

11.51 Use the ANOVA calculation formulas for a factorial design to construct the ANOVA summary table for Problem 11.33 (p. 683).

11.8 A Nonparametric One-Way ANOVA (Optional)

In Section 11.3, we mentioned that a nonparametric test can be used in place of the ANOVA F test for comparing population means. The advantage of the nonparametric test is that we do not need to make restrictive assumptions (e.g., normality, equal variances) about the sampled populations.

The *Kruskal-Wallis H test* provides a nonparametric alternative to the one-way analysis of variance F test based on independent random samples (i.e., the completely randomized design). The test is designed to compare the medians (rather than the means) of three or more populations. As with most nonparametric tests, no assumptions regarding the normality or variances of the sampled populations are required.

The sample observations are ranked from the smallest to the largest and the rank sums are calculated for each sample. For example, if you had three samples with $n_1 = 8$, $n_2 = 6$, and $n_3 = 7$, you would rank the $n_1 + n_2 + n_3 = 21$ observations from the smallest (rank 1) to the largest (rank 21) and then calculate the rank sums, T_1, T_2, and T_3, for the three samples. The Kruskal-Wallis H test uses these rank sums to calculate an H statistic that possesses an approximate *chi-square* sampling distribution. The elements of the **Kruskal-Wallis *H* test** are summarized in the box and illustrated in Example 11.14.

THE KRUSKAL-WALLIS *H* TEST FOR COMPARING *k* POPULATION MEDIANS: COMPLETELY RANDOMIZED DESIGN

Let η_i represent the median for population i.

H_0: $\eta_1 = \eta_2 = \cdots = \eta_k$

H_a: At least two of the population medians are different

$$\text{Test statistic: } H = \frac{12}{n(n+1)} \sum_{i=1}^{k} \frac{T_i^2}{n_i} - 3(n+1) \qquad \textbf{(11.32)}$$

where

n_i = number of observations in sample i

T_i = rank sum of sample i

n = total sample size = $n_1 + n_2 + \cdots + n_k$

[*Note:* Tied observations are assigned ranks equal to the average of the ranks that would have been assigned had they not been tied.]

Rejection region: $H > \chi^2_\alpha$ where χ^2_α is based on $(k-1)$ degrees of freedom

Assumptions:

1. The k samples have been independently and randomly selected from their respective populations.
2. For the chi-square approximation to be adequate, there should be five or more observations in each sample.

EXAMPLE 11.14 **APPLYING THE KRUSKAL-WALLIS *H* TEST**

In Examples 11.3 and 11.4 (pp. 655 and 658), we used an ANOVA F test to compare the mean grade point averages of college freshmen from three different socioeconomic backgrounds. Independent random samples of $n_1 = n_2 = n_3 = 7$ freshmen were selected from each of the three populations. The data are reproduced in Table 11.11. Use the Kruskal-Wallis H test to determine whether the data provide sufficient evidence to indicate that the median grade point average of freshmen depends on the students' socioeconomic backgrounds. Test using $\alpha = .05$.

GPACLASS

TABLE 11.11 Grade Point Averages for Three Socioeconomic Groups

Lower Class	Middle Class	Upper Class
2.87	3.23	2.25
2.16	3.45	3.13
3.14	2.78	2.44
2.51	3.77	3.27
1.80	2.97	2.81
3.01	3.53	1.36
2.16	3.01	2.53

Solution

Let $\eta_1, \eta_2,$ and η_3 represent the population median GPAs of lower class, middle class, and upper class college freshmen, respectively. For this problem, we want to test:

$$H_0: \eta_1 = \eta_2 = \eta_3$$

H_a: At least two population medians differ

The first step in finding the Kruskal-Wallis H test statistic is to rank the $n_1 + n_2 + n_3 = 7 + 7 + 7 = 21$ observations from the smallest to the largest. Thus, we give the smallest observation (1.36) a rank of 1, the next smallest (1.80) a rank of 2, . . . , and the largest observation (3.77) a rank of 21. The original data, their associated ranks, and the sample rank sums T_1, T_2, and T_3 are shown in Table 11.12.

TABLE 11.12 Rank Sums for Grade Point Average Data

Lower Class		Middle Class		Upper Class	
GPA	Rank	GPA	Rank	GPA	Rank
2.87	11	3.23	17	2.25	5
2.16	3.5	3.45	19	3.13	15
3.14	16	2.78	9	2.44	6
2.51	7	3.77	21	3.27	18
1.80	2	2.97	12	2.81	10
3.01	13.5	3.53	20	1.36	1
2.16	3.5	3.01	13.5	2.53	8
	$T_1 = 56.5$		$T_2 = 111.5$		$T_3 = 63$

From equation 11.32,

$$H = \frac{12}{n(n+1)} \sum_{i=1}^{3} \frac{T_i^2}{n_i} - 3(n+1)$$

where $n_1 = n_2 = n_3 = 7$, $n = 21$, $T_1 = 56.5$, $T_2 = 111.5$, and $T_3 = 63$. Substituting these values into the formula for H, we obtain

$$H = \frac{12}{21(21+1)}\left[\frac{(56.5)^2}{7} + \frac{(111.5)^2}{7} + \frac{(63)^2}{7}\right] - 3(21+1)$$

$$= \frac{12}{462}\left(\frac{3192.25}{7} + \frac{12{,}432.25}{7} + \frac{3969}{7}\right) - 66$$

$$= 6.70$$

The rejection region for the H test is $H > \chi^2_\alpha$, where χ^2_α is based on $(k - 1)$ degrees of freedom. Since we have selected $\alpha = .05$ and we want to compare $k = 3$ population relative frequency distributions, we need the value of $\chi^2_{.05}$ for $(k - 1) = (3 - 1) = 2$ degrees of freedom. This value is given in Table B.5 as $\chi^2_{.05} = 5.99147$. Therefore, the rejection region for the test includes all values of H greater than 5.99147 (see Figure 11.17).

Figure 11.17 Rejection Region for the Kruskal-Wallis *H* Test of Example 11.14

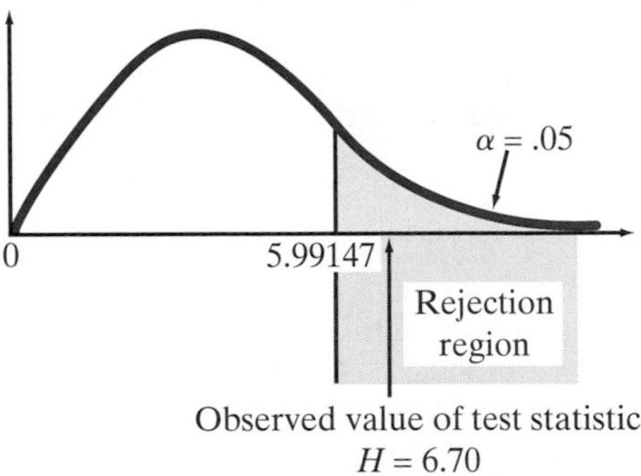

The final step in conducting the test is to determine whether the value of the test statistic falls in the rejection region. Since the value computed for the grade point data, $H = 6.70$, is greater than $\chi^2_{.05} = 5.99147$, we reject the null hypothesis, i.e., there is sufficient evidence to indicate differences in median freshmen grade point averages among the three socioeconomic populations.

The same conclusion can be reached using PHStat and Excel or MINITAB. The PHStat and Excel printout of the analysis is shown in Figure 11.18. The test statistic (rounded) is indicated, as is the p-value of the test. Note that the p-value (.035) is less than $\alpha = .05$, resulting in our conclusion of "reject H_0."

E Figure 11.18 PHStat and Excel Output, Example 11.14

	A	B
1	**GPA Comparison**	
2		
3	**Level of Significance**	**0.05**
4	**Group 1**	
5	**Sum of Ranks**	**56.5**
6	**Sample Size**	**7**
7	**Group 2**	
8	**Sum of Ranks**	**111.5**
9	**Sample Size**	**7**
10	**Group 3**	
11	**Sum of Ranks**	**63**
12	**Sample Size**	**7**
13	Sum of Squared Ranks/Sample Size	2799.071
14	Sum of Sample Sizes	21
15	Number of groups	3
16	*H* Test Statistic	6.703154
17	**Critical Value**	**5.991476**
18	***p*-Value**	**0.035029**
19	**Reject the null hypothesis**	

Self-Test 11.7

The rank sums for samples of five measurements selected from each of four populations (a total of 20 measurements) are $T_1 = 84$, $T_2 = 15$, $T_3 = 50$, and $T_4 = 61$.

a. For each sample, find T_i^2/n_i.

b. Compute the Kruskal-Wallis H test statistic for the data.

c. Find the critical value (at $\alpha = .10$) for testing whether the distributions of the four populations have the same median.

d. Give the conclusion for the test, part **c**.

Statistics in the Real World Revisited

A Nonparametric Analysis of the Cockroach Data

ROACH2

Refer to the experiment designed to investigate the trail-following ability of German cockroaches (p. 649). In previous Statistics in the Real World Revisited sections, we discovered that the assumptions required for conducting a one-way ANOVA on the data in the **ROACH2** file may be violated. In particular, the trail deviation measurements do not appear to be normally distributed for each of the cockroach groups. As an alternative method of analysis, we conducted the Kruskal-Wallis nonparametric test on the data. The null hypothesis is that the population median trail deviations for the four groups are equal, i.e.,

$$H_0\text{: } \eta_{\text{Male}} = \eta_{\text{Female}} = \eta_{\text{Gravid}} = \eta_{\text{Nymph}}$$

A MINITAB printout of the nonparametric ANOVA is displayed in Figure 11.19. The *p*-value of the test (highlighted on the printout) is 0. Since this value is less than $\alpha = .05$, we reject the null hypothesis and conclude that the median extract trail deviation differs among the populations of adult male, adult female, gravid, and nymph cockroaches. Note that this conclusion agrees with the inference derived from the ANOVA *F* test on p. 671. (This result is another example of the robustness of the ANOVA *F* test when the data are nonnormal.)

Figure 11.19 MINITAB Nonparametric One-Way ANOVA for Extract Trail Deviation

```
Kruskal-Wallis Test on Deviate

Group       N   Median   Ave Rank       Z
Female     20   11.150       38.4   -0.46
Gravid     20   49.600       61.0    4.56
Male       20    4.500       21.7   -4.18
Nymph      20   11.950       40.9    0.08
Overall    80                40.5

H = 28.82  DF = 3  P = 0.000
H = 28.82  DF = 3  P = 0.000  (adjusted for ties)
```

The sample median trail deviations of the four cockroach groups (also highlighted on Figure 11.19) are ranked as follows:

Males: 4.50

Females: 11.15

Nymphs: 11.95

Gravids: 49.60

In order to determine which population medians are significantly different, we should employ a procedure similar to the Bonferroni method for ranking treatment means. A nonparametric multiple comparisons of medians that controls the experimentwise error rate is beyond the scope of this introductory text. Consult the chapter references [e.g., Daniel (1990) and Dunn (1964)] if you want to learn how to conduct this analysis.

PROBLEMS FOR SECTION 11.8

Using the Tools

11.52 For each part, compute the rank sums of the three samples.

a. Sample 1: 6, 18, 20, 2; Sample 2: 7, 11, 6, 5; Sample 3: 19, 40, 3, 3

b. Sample 1: 100, 123, 194; Sample 2: 215, 208, 137; Sample 3: 204, 256, 245

c. Sample 1: −16, 41, 57, 8, 0; Sample 2: 11, 32, 24, 16, 15; Sample 3: 20, 14, 10, 8; Sample 4: −3, −12, 15, 4

11.53 Consider a test to determine whether three distributions differ in location. Independent random samples of size $n = 5$ were collected from each distribution and all 15 observations ranked. For each set of rank sums, compute the Kruskal-Wallis

H test statistic.

a. $T_1 = 60, T_2 = 40, T_3 = 20$ **b.** $T_1 = 55, T_2 = 45, T_3 = 20$
c. $T_1 = 70, T_2 = 15, T_3 = 35$ **d.** $T_1 = 30, T_2 = 30, T_3 = 60$

11.54 Find the rejection region for comparing *k* medians with the Kruskal-Wallis *H* test if:

a. $k = 3, n = 50, \alpha = .01$ **b.** $k = 5, n = 30, \alpha = .10$
c. $k = 4, n = 100, \alpha = .05$ **d.** $k = 10, n = 75, \alpha = .05$

11.55 Suppose you want to use the Kruskal-Wallis *H* test to compare the medians of three populations. The following are independent random samples selected from the three populations:

PROB11_55

I:	45	33	55	88	58	
II:	22	31	16	25	30	33
III:	91	96	102	75	88	

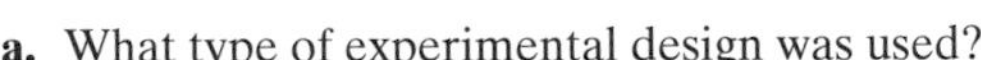

a. What type of experimental design was used?
b. Specify the null and alternative hypotheses you would test.
c. Specify the rejection region that would be used for your hypothesis test at $\alpha = .05$.
d. Calculate the rank sums for the three samples.
e. Calculate the test statistic.
f. State the appropriate conclusion of the test.

Applying the Concepts

11.56 The *Journal of the American Mosquito Control Association* (Mar. 1995) published a study of biting flies. The effect of wind speeds in kilometers per hour (kph) on the biting rate of the fly on Stanbury Island, Utah, was investigated by exposing samples of volunteers to one of six wind speed conditions. The distributions of the biting rates for the six wind speeds were compared using the Kruskal-Wallis *H* test. The rank sums of the biting rates for the six conditions are shown in the accompanying table.

Wind Speed (kph)	Number of Volunteers (n_i)	Rank Sum of Biting Rates (T_i)
<1	11	1,804
1–2.9	49	6,398
3–4.9	62	7,328
5–6.9	39	4,075
7–8.9	35	2,660
9–20	21	1,388
TOTALS	217	23,653

Source: Strickman, D., et al. "Meteorological effects on the biting activity of *Leptoconops americanus* (Diptera: Ceratopogonidae)." *Journal of the American Mosquito Control Association,* Vol. II, No. 1, Mar. 1995, p. 17 (Table 1).

a. The researchers reported the test statistic as $H = 35.2$. Verify this value.
b. Find the rejection region for the test using $\alpha = .01$.
c. Make the proper conclusions.
d. The researchers reported the *p*-value of the test as $p < .01$. Does this value support your inference in part **c**? Explain.

11.57 In disagreements between the Internal Revenue Service (IRS) and the taxpayer that end up in litigation, taxpayers are permitted by law to choose the court forum. Three trial courts are available: (1) U.S. Tax Court, (2) Federal District Court, and (3) U.S. Claims Court. Each court possesses different requirements and restrictions that make the choice an important one for the taxpayer. A study of taxpayers' choice of forum in litigating tax issues was published in the *Journal of Applied Business*

Research (Fall 1996). In a random sample of 161 litigated tax disputes, the researchers measured the taxpayer's choice of forum (Tax, District, or Claims Court) and tax deficiency, DEF (i.e., the disputed amount, in dollars). One of the objectives of the study was to determine if the mean DEF values for the three tax courts are significantly different.

a. The researchers applied the nonparametric Kruskal-Wallis H test, rather than a parametric test, to compare the median DEF values of the three tax litigation forums. Give a plausible reason for their choice.

b. The table below summarizes the data analyzed by the researchers. Use the information in the table to compute the appropriate test statistic.

Court Selected by Taxpayer	Sample Size	Sample Mean DEF	Rank Sum of DEF Values
Tax	67	$ 80,357	5,335
District	57	74,213	3,937
Claims	37	185,648	3,769

Source: Billings, B. A., Green, B. P., and Volz, W. H. "Selection of forum for litigated tax issues." *Journal of Applied Business Research,* Vol. 12, No. 4, 1996, p. 38 (Table 2).

c. The observed significance level (p-value) of the test was reported as $p = .0037$. Fully interpret this result.

NEMATODE

January	May	August
35.0	224.5	217.4
149.8	104.9	202.3
71.0	216.9	203.0
105.0	312.0	10.1
1.2	268.3	155.9

Problem 11.58

11.58 A field experiment was conducted to determine if the prevalence of nematodes (roundworms) in sedimentation in the Baltic Sea depends on season (*Estuarine, Coastal and Shelf Science,* Aug. 1997). Five sediment samples were randomly selected during one month in each of three seasons—January, May, and August—and the biomass (dry weight) of nematodes in each sediment sample was determined. The data (recorded in grams) for a particular nematode species are provided in the table at left. (The data are simulated based on summary statistics from the journal article.)

a. The researchers used the nonparametric Kruskal-Wallis H test to compare the median dry weights of the nematode species across the three seasons. Explain why this test is more appropriate than the ANOVA F test.

b. Using Excel or MINITAB, analyze the data using the Kruskal-Wallis H test. Interpret the results at $\alpha = .05$.

FISH

11.59 Refer to the data on contaminated fish captured in the Tennessee River, Problem 11.9 (p. 664). Recall that three species of fish were investigated: channel catfish, smallmouth buffalo, and largemouth bass. Suppose you want to compare the median levels of the pollutant DDT in the three species of fish.

a. Use Excel or MINITAB to graphically check whether the DDT distribution for each species is approximately normal.

b. Explain why the graphs, part **a**, suggest that a nonparametric procedure is more appropriate for analyzing the data.

c. Use Excel or MINITAB to perform the appropriate nonparametric test at $\alpha = .05$. Interpret the results.

11.60 In a study to evaluate the effectiveness of performance appraisal training, each in a sample of middle-level managers was randomly assigned to one of three training conditions: standard (in-class) training, Internet-assisted training, or Internet-assisted training plus a behavior modeling workshop. After the formal training, the managers were administered a 25-question multiple-choice test of managerial knowledge and the number of correct answers was recorded for each. The data are provided in the table. Is there sufficient evidence of differences in the median scores for the three types of performance appraisal training? Test using $\alpha = .01$.

Standard Training	Internet-Assisted Training	Internet Training Plus Workshop
16	19	12
18	22	19
11	13	18
14	15	22
23	20	16
	18	25
	21	

Problem 11.60

11.61 Organic chemical solvents are used for cleaning fabricated metal parts in industries such as aerospace, electronics, and automobiles. These solvents, when disposed of, have the potential to become hazardous waste. The *Journal of Hazardous Materials* (July 1995) published the results of a study of the chemical properties of three different types of hazardous organic solvents used to clean metal parts: aromatics, chloroalkanes, and esters. One variable studied was sorption rate, measured as mole percentage. Independent samples of solvents from each type were tested and their sorption rates recorded in the table. What do the results imply about the median sorption rates for the three solvents? Use $\alpha = .01$.

Aromatics		Chloroalkanes		Esters		
1.06	.95	1.58	1.12	.29	.43	.06
.79	.65	1.45	.91	.06	.51	.09
.82	1.15	.57	.83	.44	.10	.17
.89	1.12	1.16	.43	.61	.34	.60
1.05				.55	.53	.17

Source: Reprinted from *Journal of Hazardous Materials,* Vol. 42, No. 2, J. D. Ortego et al., "A review of polymeric geosynthetics used in hazardous waste facilities," p. 142 (Table 9), July 1995, Elsevier Science-NL, Sara Burgerhartstraat 25, 1055 KV Amsterdam, The Netherlands.

11.62 Phosphoric acid is chemically produced by reacting phosphate rock with sulfuric acid. An important consideration in the chemical process is the length of time required for the chemical reaction to reach a specified temperature. An experiment was conducted to compare the reactivity of phosphate rock mined in north, central, south, and the panhandle of Florida. Rock samples were collected from each location and placed in vacuum bottles with a sulfuric acid solution. The time (in seconds) for the chemical reaction to reach 200°F was recorded for each sample. Do the data provide sufficient evidence to indicate differences among the median reaction times of phosphoric rock mined at the four locations? Test using $\alpha = .05$.

South	Central	North	Panhandle
40	41	25	20
42	38	36	22
37	40	28	51
38	33	31	37
41	35	29	
		22	
		27	

KEY TERMS

Starred (*) terms are from the optional sections in this chapter.

KEY FORMULAS

Starred (*) formulas are from the optional sections in this chapter.

Completely randomized design:

Testing treatments	$F = \frac{\text{MST}}{\text{MSE}}$	**(11.4)**, 657
Number of pairwise comparisons for k treatment means	$c = \frac{k(k-1)}{2}$	**(11.5)**, 666
Bonferroni critical difference	$B_{ij} = (t_{\alpha/2})\sqrt{\text{MSE}\left(\frac{1}{n_i} + \frac{1}{n_j}\right)}$	**(11.7)**, 666
*Factorial design with 2 factors:		
*Testing $A \times B$ interaction	$F = \frac{\text{MS}(AB)}{\text{MSE}}$	**(11.9)**, 678
*Testing main effect A	$F = \frac{\text{MS}(A)}{\text{MSE}}$	**(11.10)**, 678
*Testing main effect B	$F = \frac{\text{MS}(B)}{\text{MSE}}$	**(11.11)**, 678
*Kruskal-Wallis H test statistic:	$H = \frac{12}{n(n+1)} \sum \frac{T_i^2}{n_i} - 3(n+1)$	**(11.32)**, 699

KEY SYMBOLS

SYMBOL	DESCRIPTION
ANOVA	Analysis of variance
SS(Total)	Total sum of squares of deviations
SST	Sum of squares for treatments
SSE	Sum of squared errors
MST	Mean square for treatments
MSE	Mean square for errors
F	Ratio of mean squares
SS(A)	Sum of squares for main effect A
SS(AB)	Sum of squares for interaction between factors A and B
MS(A)	Mean square for main effect A
MS(AB)	Mean square for AB interaction
H	Kruskal-Wallis test statistic

CHECKING YOUR UNDERSTANDING

1. What is a designed experiment?
2. How are factors, factor levels, and treatments related?

3. What are the features of a completely randomized design?
4. What are the features of a complete factorial design?
5. When is it appropriate to analyze the data using a one-way ANOVA?
6. When is it appropriate to analyze the data using a two-way ANOVA?
7. When and how should multiple comparisons procedures for evaluating pairwise combinations of treatment means be used?
8. What is an experimentwise error rate?
9. What are the assumptions of an ANOVA?
10. What do we mean by the concept of interaction in a factorial design?
11. When is it appropriate to apply the nonparametric Kruskal-Wallis H test in ANOVA?
12. When is it appropriate to conduct tests for factor main effects in a factorial design?

SUPPLEMENTARY PROBLEMS

Starred (*) problems refer to one of the optional sections in this chapter.

11.63 Bell Communications Research (Bellcore) conducted a study of four automated information filtering methods (*Communications of the Association for Computing Machinery,* Dec. 1992). These methods are (1) keyword match-word profile, (2) latent semantic indexing (LSI)-word profile, (3) keyword match-document profile, and (4) LSI-document profile. A sample of technical memos was filtered by each method and the relevance of the filtered information was rated on a 7-point scale by a panel of Bellcore employees. In addition, a subset of memos was randomly selected and rated by the panel. The mean relevance ratings of the five filtering methods (four automated methods plus the random selection method) were compared using an analysis of variance for a completely randomized design.

a. Identify the treatments in the experiment.

b. Identify the response variable.

c. The ANOVA resulted in a test statistic of $F = 117.5$, based on 4 numerator degrees of freedom and 132 denominator degrees of freedom. Interpret this result at a significance level of $\alpha = .05$.

***11.64** A study of the work-related attitudes of truck drivers was conducted (*Transportation Journal,* Fall 1993). The two factors considered in the study were career stage and time spent on the road. Career stage was set at three levels: early (less than 2 years), mid-career (between 2 and 10 years), and late (more than 10 years). Road time was dichotomized as short (gone for one weekend or less) and long (gone for longer than one weekend). Data were collected on job satisfaction for drivers sampled in each of the $3 \times 2 = 6$ combinations of career stage and road time. (Job satisfaction was measured on a 5-point scale, where 1 = really dislike and 5 = really like.)

a. Identify the response variable for this experiment.

b. Identify the factors for this experiment.

c. Identify the treatments for this experiment.

d. The ANOVA table for the analysis is provided below. Fully interpret the results.

Source	*F* Value	*p*-Value
Career stage (CS)	26.67	$p \le .001$
Road time (RT)	.19	$p > .05$
CS × RT	1.59	$p > .05$

Source: McElroy, J. C., et al. "Career stage, time spent on the road, and truckload driver attitudes." *Transportation Journal,* Vol. 33, No. 1, Fall 1993, p. 10 (Table 2).

e. The researchers theorized that the impact of road time on job satisfaction may be different depending on the career stage of the driver. Do the results support this theory?

f. The researchers also theorized that career stage impacts the job satisfaction of truck drivers. Do the results support this theory?

11.65 Studies conducted at the University of Melbourne (Australia) indicate that there may be a difference in the pain thresholds of blonds and brunettes. Men and women of various ages were divided into four categories according to hair color: light blond, dark blond, light brunette, and dark brunette. The purpose of the experiment was to determine whether hair color is related to the amount of pain produced by common types of mishaps and assorted types of trauma. Each person in the experiment was given a pain threshold score based on his or her performance in a pain sensitivity test (the higher the score, the lower the person's pain tolerance). The data are shown in the table.

HAIRPAIN

Light Blond	Dark Blond	Light Brunette	Dark Brunette
62	63	42	32
60	57	50	39
71	52	41	51
55		37	30
48			35

a. Use Excel or MINITAB to determine if there is evidence of a difference among the mean pain thresholds for people with the four hair color types. Use $\alpha = .05$.

b. Which of the four hair-color categories shows the lowest mean pain threshold? Use an EER of .05.

***c.** Conduct a nonparametric test to determine whether the median pain thresholds differ for the four hair colors. (Use $\alpha = .05$). Compare the results to those of part **a**.

11.66 The Minnesota Multiphasic Personality Inventory (MMPI) is a questionnaire used to gauge personality type. *Psychological Assessment* (Mar. 1995) published a study that investigated the effectiveness of the MMPI in detecting deliberately distorted responses. A completely randomized design with four treatments was employed. The treatments consisted of independent random samples of females in the following four groups: NFP—nonforensic psychiatric patients ($n_1 = 65$); FP—forensic psychiatric patients ($n_2 = 28$); CSH—college students who were requested to respond honestly ($n_3 = 140$); and CSFB—college students who were instructed to provide "fake bad" responses ($n_4 = 45$). All 278 participants were given the MMPI and the scores were recorded for each. The score was treated as a response variable and an analysis of variance conducted. The ANOVA F value was found to be $F = 155.8$.

a. Determine whether the mean scores of the four groups completing the MMPI differ significantly. Use $\alpha = .05$ for each test.

b. The table gives the results of the Bonferroni multiple comparison of means (EER = .05). If the MMPI is effective in detecting distorted responses, then the mean score for the "fake bad" treatment group will be largest. Based on the information provided, can the researchers make an inference about the effectiveness of the MMPI? Explain.

Mean score	7.1	11.3	14.6	33.6
Group	CHS	NFP	FP	CSFB

*11.67 The reluctance for a person to transmit bad news to peers is termed the "MUM effect" by psychologists. The *Journal of Experimental Social Psychology* (Vol. 23, 1987) published the results of a designed experiment to investigate the factors that influence the MUM effect. Each of 40 undergraduates at Duke University was asked to administer an IQ test to another student and then provide the test taker with his or her score. Unknown to the subject, the test taker was working with the researchers. The experiment manipulated two factors, *subject visibility* and *test taker success*, each at two levels. Subject visibility was either visible (i.e., the subject was visible to the test taker) or not visible. Test taker success was either success (i.e., the test taker supplied a set of answers that placed him in the top 20% of all Duke undergraduates) or failure (i.e., the bottom 20%). Ten subjects were randomly assigned to each of the 2 × 2 = 4 experiment conditions; thus, a 2 × 2 factorial design with 10 replications was employed. The dependent variable measured during the experiment was *latency to feedback*, defined as the time (in seconds) between the end of the test and the delivery of feedback (i.e., the test score) from the subject to the test taker. The experimental data, shown in the table below, are stored in the **MUM** file.

MUM

		Test Taker Success			
		Success		**Failure**	
Subject Visibility	**Visible**	67.6	46.3	118.7	169.4
		83.8	75.9	153.6	171.2
		75.7	51.4	127.1	156.2
		84.9	87.0	141.3	137.6
		54.8	104.3	153.9	143.2
	Not Visible	79.1	65.7	76.2	62.0
		106.2	99.6	90.3	85.0
		87.2	105.3	83.3	64.8
		104.5	93.3	46.2	37.9
		100.5	55.0	83.4	96.4

Note: Data are simulated based on summary statistics reported in the journal article.

a. Conduct a complete analysis of variance for the data, including a multiple comparisons of treatment means (if necessary). Use $\alpha = .05$.

b. The researchers concluded that "subjects appear reluctant to transmit bad news—but only when they are visible to the news recipient." Do you agree?

*11.68 The *Journal of Personal Selling & Sales Management* (Fall 1990) reported on a study to examine the effects of salesperson credibility on buyer persuasion. The experiment involved two factors, each at two levels: brand quality (high versus low) and salesperson credibility (high versus low). Each of 64 undergraduate students was randomly assigned to one of the 2 × 2 = 4 experimental treatments (16 students per treatment). After viewing a presentation on laptop computers, the students' intentions to buy were measured with a questionnaire. The ANOVA table for this experiment is shown below.

Source of Variation	df	SS	MS	F	p-Value
Brand quality (A)	1	59.30	59.30	39.74	.001
Salesperson credibility (B)	1	3.51	3.51	2.35	.130
$A \times B$	1	15.13	15.13	10.14	.002

Source: Sharma, A. "The persuasive effect of salesperson credibility: conceptual and empirical examination." *Journal of Personal Selling & Sales Management*, Fall 1990, Vol. 10, pp. 71–80 (Table 2).

a. Interpret the results of the ANOVA.

b. Bonferroni's multiple comparisons procedure (with EER = .01) was used to compare the mean buyer intentions for the $2 \times 2 = 4$ experimental conditions. Since the two factors (brand quality and salesperson credibility) interact, the means for high and low salesperson credibility were compared for each of the two levels of brand quality. The results for high brand quality are shown in the table below. Interpret the results.

Mean buyer intention	4.88	5.27
Sales credibility	High	Low

Problem 11.68

c. Refer to part **b**. The results for low brand quality are shown in the table below. Interpret the results.

Mean buyer intention	2.35	3.92
Sales credibility	High	Low

Problem 11.68

***11.69** The *American Journal of Psychology* (Winter 1991) reported on a study designed to investigate the way in which adolescents with low reading ability comprehend simple text-based problems. Fourteen-year-old students were divided into four groups: (1) learning disabled, low socioeconomic status; (2) nondisabled, low socioeconomic status; (3) learning disabled, high socioeconomic status; and (4) nondisabled, high socioeconomic status. Each student was asked to read and retell a "story problem"; however, the way in which the problem was presented was varied. Some students read a "no-priority" problem (i.e., a problem with no clear goals and/or objectives), while others read a "priority" problem (i.e., a problem with a clear statement of the character's priority). The experiment was designed as a 4×2 factorial, with group at four levels and problem type (priority or no-priority) at two levels. One of the dependent variables measured was proportion of ideas recalled correctly.

a. The test for group by problem type interaction was nonsignificant at $\alpha = .01$. Interpret this result.

b. The test for problem type main effects was nonsignificant at $\alpha = .01$. Interpret this result.

c. The test for group main effects was statistically significant at $\alpha = .01$. Interpret this result.

d. The mean proportions of ideas recalled correctly for the four groups were compared with a multiple comparisons procedure (EER = .05); the results are shown below. Interpret the results.

Mean	.252	.361	.379	.589
Group	LD-Low SES	ND-Low SES	LD-High SES	ND-High SES

11.70 In business, the prevailing theory is that companies can be categorized into one of four types based on their strategic profile: reactors (marginal competitors, unstable, victims of industry forces); defenders (specialize in established products, lower costs while maintaining quality); prospectors (develop new/improved products); and analyzers (operate in two product areas—one stable, one dynamic). The *American Business Review* (Jan. 1990) reported on a study that proposes a fifth organization type, balancers, who operate in three product spheres—one stable and two dynamic. Each firm in a sample of 78 glassware firms was categorized into one of these five types; the level of performance (process research and development ratio) of each was measured.

a. A completely randomized design ANOVA of the data resulted in a significant (at $\alpha = .05$) F value for treatments (organization types). Interpret this result.

b. Bonferroni multiple comparisons of the five performance levels (using EER = .05) are summarized in the following table. Interpret the results.

Mean	.138	.235	.820	.826	.911
Type	Reactor	Prospector	Defender	Analyzer	Balancer

Source: Wright, P., et al. "Business performance and conduct of organization types: A study of select special-purpose and laboratory glassware firms." *American Business Review,* Jan. 1990, p. 95 (Table 4).

***11.71** The cattle raised on the Biological Reserve of Doñana (Spain) live under free-range conditions, with virtually no human interference. The cattle population is organized into four herds (LGN, MTZ, PLC, and QMD). The *Journal of Zoology* (July 1995) investigated the ranging behavior of the four herds across the four seasons. Thus, a 4×4 factorial experiment was employed, with herd and season representing the two factors. Three animals from each herd during each season were sampled and the home range of each individual was measured (in square kilometers). The data were subjected to an ANOVA, with the results shown in the table.

Source	df	F	p-Value
Herd (H)	3	17.2	$p < .001$
Season (S)	3	3.0	$p > .05$
$H \times S$	9	1.2	$p > .05$
Error	32		
TOTAL	47		

a. Conduct the appropriate ANOVA F tests and interpret the results.

b. The researcher ranked the four herd means independently of season. Do you agree with this strategy? Explain.

c. Refer to part **b**. The Bonferroni rankings of the four herd means (EER = .05) are shown below. Interpret the results.

Mean home range (km^2)	.75	1.0	2.7	3.8
Herd	PLC	LGN	QMD	MTZ

11.72 Operators of sonar display consoles must learn to decode abbreviated words quickly and accurately. In one experiment, 20 Navy and civilian personnel participated in a study of abbreviations on a sonar console. Five of these subjects were highly familiar with the sonar system. The 15 subjects unfamiliar with the system were randomly divided into three groups of five. Thus, the study consisted of a total of four groups (one experienced and three inexperienced groups), with five subjects per group. The experienced group and one inexperienced group (denoted TE and TI, respectively) were assigned to learn the simple method of abbreviation. One of the remaining inexperienced groups was assigned the conventional single abbreviation method (denoted CS), whereas the other was assigned the conventional multiple abbreviation method (denoted CM). Each subject was then given a list of 75 abbreviations to learn, one at a time, through the display console of a minicomputer. The number of trials until the subject accurately decoded at least 90% of the words on the list was recorded as shown in the table on p. 712. Use Excel or MINITAB to determine (at $\alpha = .05$) if differences exist among the mean numbers of trials required for the four groups. If so, which group performed the best? Be sure to evaluate the ANOVA assumptions.

CM	CS	TE	TI
4	6	5	8
7	9	5	4
5	5	7	8
6	7	8	10
8	6	7	3

Source: Data are simulated values based on the group means reported in *Human Factors,* Feb. 1984. Copyright 1984 by the Human Factors Society, Inc.

Problem 11.72

11.73 Vanadium (V) is an essential trace element. An experiment was conducted to compare the concentrations of V in biological materials using isotope dilution mass spectrometry. The accompanying table gives the quantities of V (measured in nanograms per gram) in dried samples of oyster tissue, citrus leaves, bovine liver, and human serum.

VANADIUM

Oyster Tissue	Citrus Leaves	Bovine Liver	Human Serum
2.35	2.32	.39	.10
1.30	3.07	.54	.17
.34	4.09	.30	.14

Source: Fassett, J. D., and Kingston, H. M. "Determination of nanogram quantities of vanadium in biological material by isotope dilution thermal ionization mass spectrometry with ion counting detection." *Analytical Chemistry,* Vol. 57, No. 13, Nov. 1985, p. 2474 (Table II).

a. Use Excel or MINITAB to analyze the data. Summarize the results in an ANOVA table.

b. Is there sufficient evidence (at $\alpha = .05$) to indicate that the mean V concentrations differ among the four biological materials?

c. Use a multiple comparisons procedure to rank the mean V concentrations of the four biological materials. Use an experimentwise error rate of .05.

***11.74** What is the optimal method of directing newcomers to a specific location in a complex building? Researchers at Ball State University (Indiana) investigated this "wayfinding" problem and reported their results in *Human Factors* (Mar. 1993). Subjects met in a starting room on a multilevel building and were asked to locate the "goal" room as quickly as possible. (Some of the subjects were provided directional aids, while others were not.) Upon reaching their destination, the subjects returned to the starting room and were given a second room to locate. (One of the goal rooms was located in the east end of the building, the other in the west end.) The experimentally controlled variables in the study were aid type at three levels (signs, map, no aid) and room order at two levels (east/west, west/east). Subjects were randomly assigned to each of the $3 \times 2 = 6$ experimental conditions and the travel time (in seconds) recorded. The results of the analysis of the east room data for this 3×2 factorial design are provided in the table below. Interpret the results.

Source	df	MS	*F*	*p*-Value
Aid type	2	511,323.06	76.67	$<.001$
Room order	1	13,005.08	1.95	$>.10$
Aid $\times$ Order	2	8,573.13	1.29	$>.10$
Error	46	6,668.94		

Source: Butler, D. L., et al. "Wayfinding by newcomers in a complex building." *Human Factors,* Vol. 35, No. 1, Mar. 1993, p. 163 (Table 2).

*11.75 A major defense contractor examined the effects of varying inspection levels and incoming test times in detecting early transformer part failure or fatigue. The levels of inspection selected were full military inspection (A), reduced military specification level (B), and commercial grade (C). Operational burn-in test times chosen for this study were at 1-hour increments from 1 hour to 9 hours. The response was failures per 1,000 pieces obtained from samples taken from lot sizes inspected to a specified level and burned in over a prescribed time length. Three replications were randomly sequenced under each condition making this a complete 3 × 9 factorial experiment (a total of 81 observations). The data for the study are shown in the following table. Conduct a complete two-way ANOVA on the data. Be sure to check that the ANOVA assumptions are satisfied.

TRANSFORM

	Inspection Levels								
Burn-in (hours)	**Full Mil. Spec. A**			**Reduced Mil. Spec B**			**Commercial C**		
1	7.60	7.50	7.67	7.70	7.10	7.20	6.16	6.13	6.21
2	6.54	7.46	6.84	5.85	6.15	6.15	6.21	5.50	5.64
3	6.53	5.85	6.38	5.30	5.60	5.80	5.41	5.45	5.35
4	5.66	5.98	5.37	5.38	5.27	5.29	5.68	5.47	5.84
5	5.00	5.27	5.39	4.85	4.99	4.98	5.65	6.00	6.15
6	4.20	3.60	4.20	4.50	4.56	4.50	6.70	6.72	6.54
7	3.66	3.92	4.22	3.97	3.90	3.84	7.90	7.47	7.70
8	3.76	3.68	3.80	4.37	3.86	4.46	8.40	8.60	7.90
9	3.46	3.55	3.45	5.25	5.63	5.25	8.82	9.76	8.52

Source: Danny La Nuez, College of Business Administration, graduate student, University of South Florida, 1989–1990.

REFERENCES

Box, G. E. P., Hunter, W. G., and Hunter, J. S. *Statistics for Experimenters*. New York: Wiley, 1978.

Cochran, W. G., and Cox, G. M. *Experimental Designs,* 2nd ed. New York: Wiley, 1957.

Daniel, W. W. *Applied Nonparametric Statistics,* 2nd ed Boston: PWS-Kent, 1990.

Dunn, O. J. "Multiple comparisons using rank sums." *Technometrics,* Vol. 6, 1964.

Gujarati, D. N. *Basic Econometrics,* 3rd ed. New York: McGraw-Hill, 1995 (Chapter 11).

Hsu, J. C. *Multiple Comparisons: Theory and Methods.* London: Chapman & Hall, 1996.

Johnson, N., and Leone, F. *Statistics and Experimental Design in Engineering and the Physical Sciences,* Vol. II, 2nd ed. New York: Wiley, 1977.

Kirk, R. E. *Experimental Design: Procedures for the Behavioral Sciences.* Monterey, Ca.: Brooks/Cole, 1968.

Kramer, C. Y. "Extension of multiple range tests to group means with unequal number of replications." *Biometrics,* Vol. 12, 1956, pp. 307–310.

Kruskal, W. H., and Wallis, W. A. "Use of ranks in one-criterion variance analysis." *Journal of the American Statistical Association,* Vol. 47, 1952.

Mason, R. L., Gunst, R. F., and Hess, J. L. *Statistical Design and Analysis of Experiments.* New York: Wiley, 1989.

Mendenhall, W. *Introduction to Linear Models and the Design and Analysis of Experiments.* Belmont, Ca.: Wadsworth, 1968.

Miller, R. G., Jr. *Simultaneous Statistical Inference.* New York: Springer-Verlag, 1981.

Neter, J., Kutner, M., Nachtsheim, C., and Wasserman, W. *Applied Linear Statistical Models,* 4th ed. Homewood, Ill.: Richard Irwin, 1996.

Scheffé, H. *The Analysis of Variance.* New York: Wiley, 1959.

Snedecor, G. W., and Cochran, W. G. *Statistical Methods,* 7th ed. Ames, Iowa: Iowa State University Press, 1980.

Steel, R. G. D., and Torrie, J. H. *Principles and Procedures of Statistics: A Biometrical Approach,* 2nd ed. New York: McGraw-Hill, 1980.

Tukey, J. W. "Comparing individual means in the analysis of variance." *Biometrics,* Vol. 5, 1949, pp. 99–114.

Wilcoxon, F., and Wilcox, R. A. *Some Rapid Approximate Statistical Procedures.* The American Cyanamid Co., 1964.

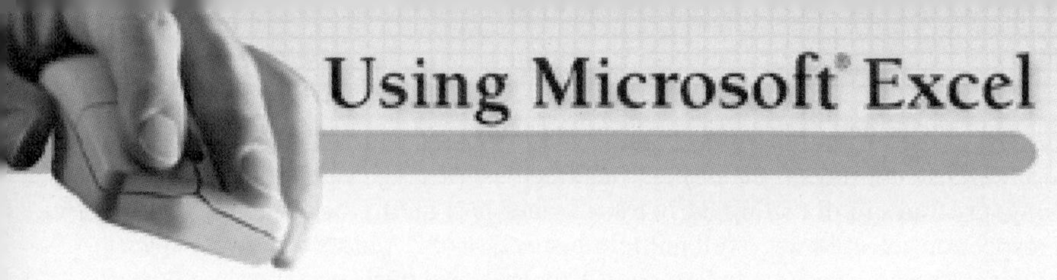

11.E.1 Conducting a One-Way ANOVA

Use the Data Analysis **Anova: Single Factor** procedure to perform the one-factor analysis of variance. For example, to perform the one-factor analysis of variance for the grade point average–socioeconomic class data of Figure 11.6 on p. 658, open the **GPACLASS.XLS** workbook to the Data sheet and

1. Select Tools | Data Analysis.
2. Select Anova: Single Factor from the Analysis Tools list box in the Data Analysis dialog box. Click the OK button.
3. In the Anova: Single Factor dialog box (see Figure 11.E.1):

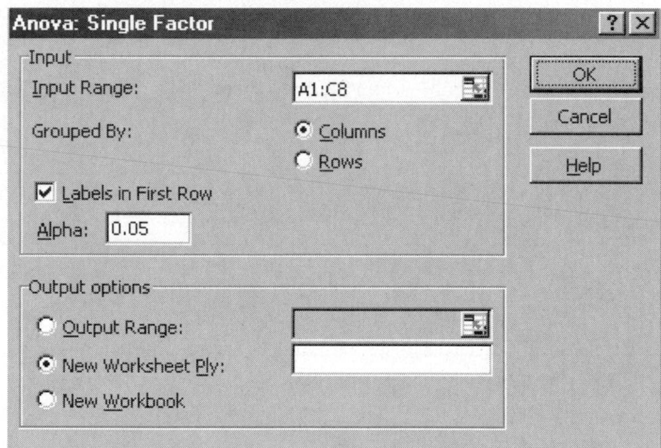

E **Figure 11.E.1** Anova: Single Factor Dialog Box

a. Enter A1:C8 in the Input Range edit box.
b. Select the Columns option button.
c. Select the Labels in the First Row check box.
d. Enter 0.05 in the Alpha edit box.
e. Select the New Worksheet Ply option button and enter a name for the new sheet.
f. Click the OK button.

The worksheet generated by this procedure will *not* dynamically change and if changes are made to the underlying data, the procedure must be repeated in order to update the results of the test.

This Data Analysis procedure requires that the data for each group be in separate columns, an arrangement known as unstacked data. To use this procedure with worksheet data that are stacked, first select the PHStat **Data Preparation | Unstack Data** procedure to properly unstack the data.

11.E.2 Conducting Multiple Comparisons of Means

Use the Bonferroni procedure to conduct multiple comparisons and obtain the sample means and sample sizes of each group, the degrees of freedom within groups, and the MSE value from the Data Analysis Anova: Single Factor output.

To perform the Bonferroni procedure for making multiple comparisons between all possible pairs of k means, first obtain the necessary statistics for the procedure. As an example, consider the grade point average examples of Section 11.3. To generate multiple comparisons between all possible pairs of the means of grade point average data for the three socioeconomic classes, do the following:

1. Generate a worksheet containing analysis of variance statistics using the instructions of Section 11.E.1.

2. From the worksheet generated by the Data Analysis tool in step 1:

a. Obtain the sample sizes and means from Count and Average columns, respectively, of the SUMMARY table that begins in row 4.

b. Obtain the degrees of freedom within groups from the cell in the "df" column and "Within Groups" row of the ANOVA table.

c. Obtain the MSE value from the cell in the "MS" column and "Within Groups" row of the ANOVA table.

BONFERRONI

Implementing a Worksheet to Perform the Bonferroni Procedure

We can use arithmetic and logical formulas and the values obtained in steps 1 and 2 as the basis for performing the Bonferroni procedure for making multiple comparisons between all possible pairs of k means. Tables 11.E.1 and 11.E.2 present a Bonferroni sheet design that performs the Bonferroni procedure using the grade point averages data of Table 11.3. Similar to the worksheet designs presented in Section 10.E.3, this design also includes a formula that uses the IF function to display a message informing the user whether or not a pair of means significantly differs.

To implement the Tables 11.E.1 and 11.E.2 design, do the following:

TABLE 11.E.1 Bonferroni Procedure Sheet Design for Columns A and B for the Grade Point Average Examples of Section 11.3

	A	B
1	Bonferroni Procedure for Grade Point Average	
2		
3	Group 1	
4	Sample Mean	xx
5	Sample Size	xx
6	Group 2	
7	Sample Mean	xx
8	Sample Size	xx
9	Group 3	
10	Sample Mean	xx
11	Sample Size	xx
12	MSE	xx
13	Error DF	=(B5-1)+(B8-1)+(B11-1)
14	Experiment Error Rate	0.xx
15	Number of Groups	3
16	Treatment pairs	=B15*(B15-1)/2
17	α value for comparisons	=B14/B16
18	t statistic for comparisons	=TINV(B17,B13)

TABLE 11.E.2 Bonferroni Procedure Sheet Design for Columns C and D for the Grade Point Average Examples of Section 11.3

	C	D
1		
2		
3	Group 1 to Group 2 Comparison	
4	Absolute Difference	=ABS(B4-B7)
5	Standard Error of Difference	=SQRT(B12*(1/B5+1/B8))
6	Bonferroni Critical Difference	=B18*D5
7	=IF(D4>D6, "Means are different", "Means are not different")	
8		
9	Group 1 to Group 3 Comparison	
10	Absolute Difference	=ABS(B4-B10)
11	Standard Error of Difference	=SQRT(B12*(1/B5+1/B11))
12	Bonferroni Critical Difference	=B18*D11
13	=IF(D10>D12, "Means are different", "Means are not different")	
14		
15	Group 2 to Group 3 Comparison	
16	Absolute Difference	=ABS(B7-B10)
17	Standard Error of Difference	=SQRT(B12*(1/B8+1/B11))
18	Bonferroni Critical Difference	=B18*D17
19	=IF(D16>D18, "Means are different", "Means are not different")	

1. Select File | New to open a new workbook (or open the existing workbook into which the Bonferroni worksheet is to be inserted).
2. Select an unused worksheet (or select Insert | Worksheet if there are none) and rename the sheet Bonferroni.
3. Enter the title, headings, and labels for column A as shown in Table 11.E.1.
4. Enter the sample size and sample mean for each group, the MSE statistic value, the experimentwise error rate, and the number of groups in column B as shown in Table 11.E.1.
5. Enter the headings, labels, and formulas for columns C and D as shown in Table 11.E.2. Formulas for cells C7, C13, and C19, shown as two lines, should be entered as a single continuous line
6. Select the cell range A3:B3 and click the Merge and Center button on the formatting toolbar (see Section EP.2.9). Repeat this step for the cell ranges A6:B6, A9:B9, and for the two-cell ranges in columns C and D in rows 3, 7, 9, 13, 15, and 19.

The completed worksheet will be similar to the one shown in Figure 11.8 on p. 669.

11.E.3 Conducting a Two-Way ANOVA

Use the Data Analysis **Anova: Two-Factor with Replication** procedure to perform the two-factor analysis of variance. For example, to perform the two-factor analysis of variance for

the machine-material data of Figure 11.13 on p. 679, open the **GASKET.XLS** workbook to the Data sheet and

1. Select Tools | Data Analysis.
2. Select Anova: Two-Factor With Replication from the Analysis Tools list box in the Data Analysis dialog box and click the OK button.
3. In the Anova: Two-Factor With Replication dialog box (see Figure 11.E.2):

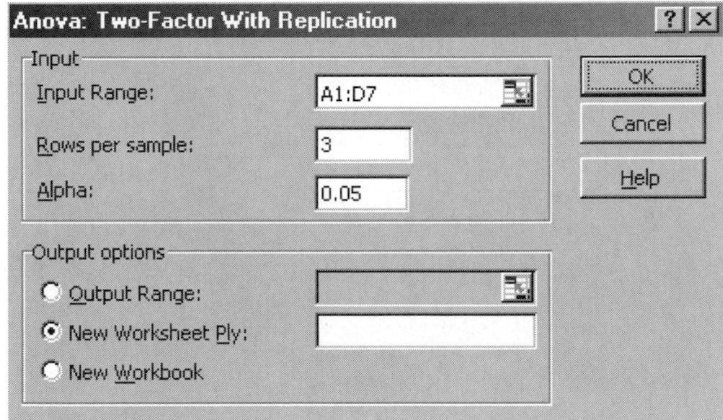

E **Figure 11.E.2** Anova: Two-Factor with Replication Dialog Box

a. Enter A1:D7 in the Input Range edit box.
b. Enter 3 in the Rows per sample edit box.
c. Enter 0.05 in the Alpha edit box.
d. Select the New Worksheet Ply option button and enter a name for the new sheet.
e. Click the OK button.

The worksheet generated by this procedure will *not* dynamically change and if changes are made to the underlying data, the procedure must be repeated in order to update the results of the test. On the Data sheet, note that the data have been set up in a format that is similar to Table 11.5 on p. 675, except that column A provides a label corresponding to each level of factor A, and cell A1 is blank.

11.E.4 Conducting the Kruskal-Wallis *H* Test

Use the PHStat **Multiple-Sample Tests | Kruskal-Wallis Rank Sum Test** procedure to perform the Kruskal-Wallis rank test (*H* test) for differences in *k* medians. For example, to perform this test for the grade point average–socioeconomic class data of Figure 11.18 on p. 701, open the **GPACLASS.XLS** workbook to the Data sheet and

1. Select PHStat | Multiple-Sample Tests | Kruskal-Wallis Rank Test
2. In the Kruskal-Wallis Rank Test dialog box (see Figure 11.E.3):

a. Enter .05 in the Level of Significance edit box.
b. Enter A1:C8 in the Sample Data Cell Range edit box.
c. Select the First cells contain label check box.
d. Enter a title in the Title edit box.
e. Click the OK button.

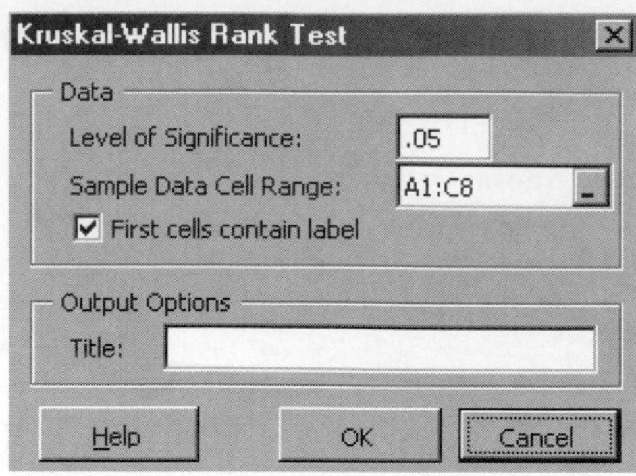

E **Figure 11.E.3** PHStat Kruskal-Wallis Rank Test Dialog Box

This procedure requires that the data for each group be in separate columns, an arrangement known as unstacked data. To use this procedure with data that are stacked, first select the PHStat **Data Preparation | Unstack Data** procedure to properly unstack the data.

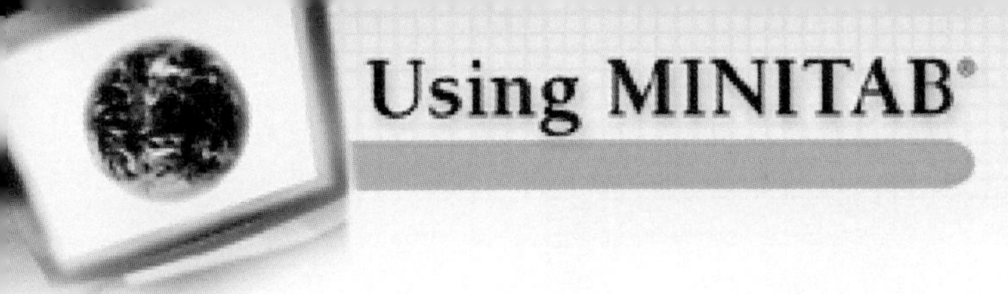

11.M.1 Conducting a One-Way ANOVA

To illustrate the use of MINITAB for the one-factor ANOVA, open the **GPACLASS.MTW** worksheet. Note that the data have been stored in an unstacked format with each level in a separate column. Select **Stat | ANOVA | OneWay(Unstacked).** In the OneWay(Unstacked) dialog box (see Figure 11.M.1), in the Responses (in separate columns) edit box, enter **C1 C2 C3.** Click the **Graphs** option button and select the **Boxplot of data** check box. Click the **OK** button to return to the OneWay(Unstacked) dialog box. Click the **OK** button.

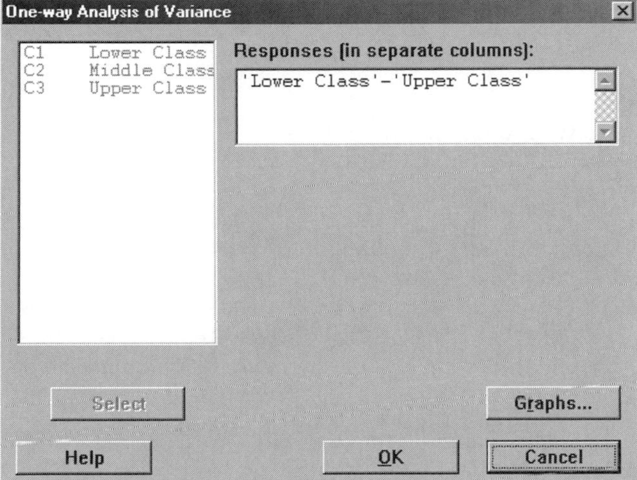

Figure 11.M.1 MINITAB One-way (Unstacked) Dialog Box

If the data are stored in a stacked format, select **Stat | ANOVA | One-way.** Enter the column in which the response variable is stored in the Response edit box, and the column in which the factor is stored in the Factor edit box.

11.M.2 Conducting Multiple Comparisons of Means

To illustrate the use of MINITAB for multiple comparisons of means, open the **GPACLASS.MTW** worksheet. Note that the data have been stored in an unstacked format with each level in a separate column. In order to use the Tukey procedure, you need to stack the data. Select **Manip | Stack/Unstack | Stack Columns.** Enter **C1–C3** in the Stack the following Columns edit box, **C4** in the Store the Stacked data in edit box, and **C5** in the Subscripts in edit box. Click the **OK** button. Enter labels for GPA in C4 and Class in C5. Select **Stat | ANOVA | One-way.** In the One-way dialog box, enter **C4** or **GPA** in the Response edit box, and **C5** or **Class** in the Factor edit box. Select the **Comparisons** box. In the One-way Multiple Comparisons dialog box (see Figure 11.M.2), select the Tukey's, family error rate check box, and enter **.05** in the edit box. Click the **OK** button to return to the One-way dialog box. Click the **OK** button.

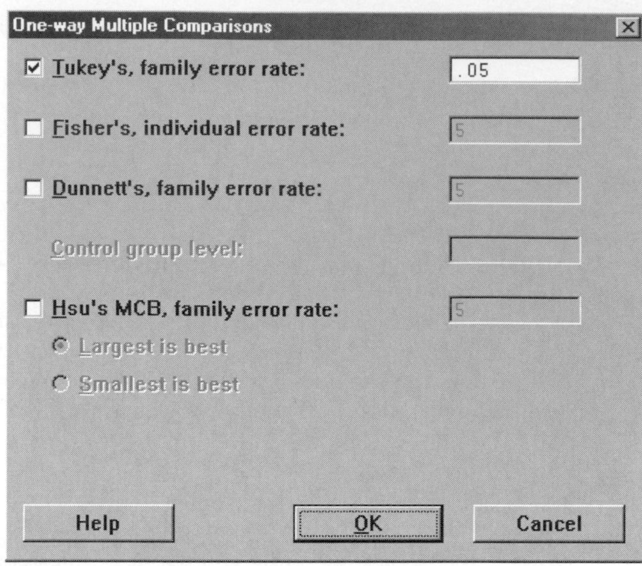

Figure 11.M.2 MINITAB One-way Multiple Comparisons Dialog Box

11.M.3 Conducting a Two-Way ANOVA

To illustrate the use of MINITAB for the two-factor design, open the **GASKET.MTW** worksheet. Note that Machine is stored in C1, Material in C2, and Gaskets Produced in C3. Select **Stat | ANOVA | Two-way.** In the Two-way Analysis of Variance dialog box (see Figure 11.M.3), enter **C3** or **'Gaskets Produced'** in the Response edit box, **C1** or **Machine** in the Row factor edit box, and **C2** or **Material** in the Column factor edit box. Select the Display means check box for the row and column factors. Click the **OK** button.

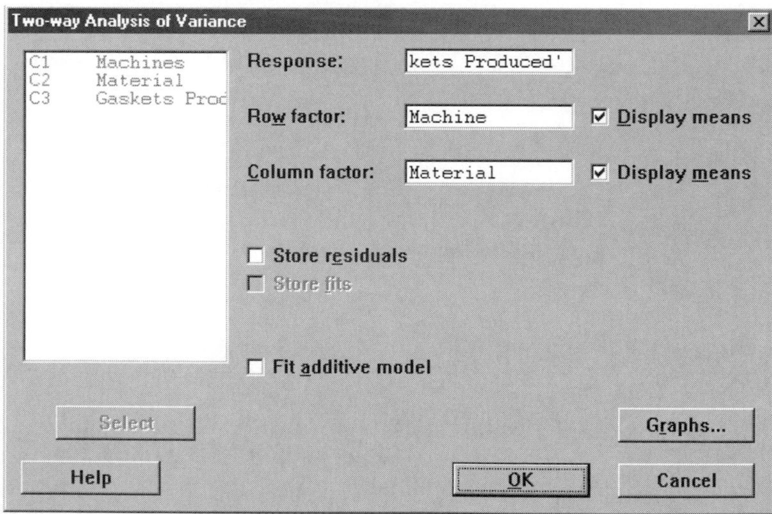

Figure 11.M.3 MINITAB Two-Way Analysis of Variance Dialog Box

11.M.4 Conducting the Kruskal-Wallis *H* Test

To illustrate the use of MINITAB for the Kruskal-Wallis *H* test, open the **GPACLASS.MTW** worksheet. Note that the data have been stored in an unstacked format with each level in a separate column. In order to perform the Kruskal-Wallis *H* test, we need to stack the

data. Select **Manip | Stack/Unstack | Stack Columns.** Enter **C1–C3** in the Stack the following Columns edit box, **C4** in the Store the Stacked data in edit box, and **C5** in the Subscripts in edit box. Click the **OK** button. Enter labels for GPA in C4 and Class in C5.

Select **Stat | Nonparametrics | Kruskal-Wallis.** In the Kruskal-Wallis dialog box (see Figure 11.M.4), enter **C4** in the Response edit box, and **C5** in the Factor edit box. Click the **OK** button.

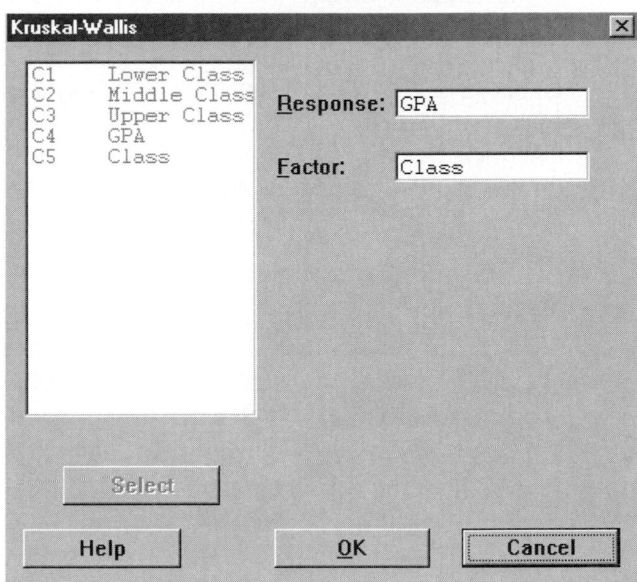

Figure 11.M.4 MINITAB Kruskal-Wallis Dialog Box

Appendix A

Review of Arithmetic and Algebra

In writing this text, we realize that there are wide differences in the mathematical background of students taking a basic statistics course. Some students may have taken various courses in calculus and matrix algebra, while other students may not have taken any mathematics courses since high school. Since the emphasis in this text is on the concepts of statistical methods and the interpretation of output from Microsoft Excel and MINITAB, no mathematical prerequisite beyond elementary algebra is needed, since no formal mathematical proofs are derived. However, a proper foundation in basic arithmetic and algebraic skills will enable the student to focus on understanding the concepts of statistics rather than the mechanics of computing results.

With this goal in mind, we should make it clear that the object of this appendix is to review arithmetic and algebra for those students whose basic skills have become rusty over a period of time. If students need to actually learn arithmetic and algebra, then this review will probably not provide enough depth.

In order to assess their arithmetic and algebraic skills, and increase their confidence, we suggest that students take the following quiz before studying the text.

Part I: Fill in the correct answer.

1. $\dfrac{1}{\frac{2}{3}} =$
2. $.4^2 =$
3. $1 + \frac{2}{3} =$
4. $(\frac{1}{3})^4 =$
5. $1/5 =$ (in decimals)
6. $1 - (-.3) =$
7. $4 \times .2 \times (-8) =$
8. $(\frac{1}{4} \times \frac{2}{3}) =$
9. $(1/100) + (1/200) =$
10. $\sqrt{16} =$

Part II: Select the correct answer.

1. If $a = bc$, then $c =$

   a. ab

   b. b/a

   c. a/b

   d. None of the above

2. If $x + y = z$, then $y =$

   a. z/x

   b. $z + x$

   c. $z - x$

   d. None of the above

3. $(x^3)(x^2) =$

   a. x^5

   b. x^6

   c. x^1

   d. None of the above

4. $x^0 =$

   a. x

   b. 1

   c. 0

   d. None of the above

5. $x(y - z) =$

   a. $xy - xz$

   b. $xy - z$

   c. $(y - z)/x$

   d. None of the above

6. $(x + y)/z =$

   a. $(x/z) + y$

   b. $(x/z) + (y/z)$

   c. $x + (y/z)$

   d. None of the above

7. $x/(y + z) =$

a. $(x/y) + (1/z)$

b. $(x/y) + (x/z)$

c. $(y + z)/x$

d. None of the above

8. If $x = 10$, $y = 5$, $z = 2$, and $w = 20$, then $(xy - z^2)/w =$

a. 5

b. 2.3

c. 46

d. None of the above

9. $(8x^4)/(4x^2) =$

a. $2x^2$

b. 2

c. $2x$

d. None of the above

10. $\sqrt{x/y}$

a. $\sqrt{y}/\sqrt{x}$

b. $\sqrt{1}/\sqrt{xy}$

c. $\sqrt{x}/\sqrt{y}$

d. None of the above

The answers to both parts of this quiz appear at the end of this appendix. Most students will find the review of arithmetic and algebraic operations presented in the following sections to be helpful. A student who gets fewer than ten of the twenty items correct might need a more extensive review of basic algebra.

Symbols

Each of the four basic arithmetic operations—addition, subtraction, multiplication, and division—is indicated by an appropriate symbol.

$+$	add	$\times$ or $\cdot$	multiply
$-$	subtract	$\div$ or $/$	divide

In addition to these operations, the following symbols are used to indicate equality or inequality:

$=$ equals $\quad \neq$ not equal $\quad \approx$ approximately equal to $\quad >$ greater than

$<$ less than $\quad \geq$ greater than or equal to $\quad \leq$ less than or equal to

Addition

The process of addition refers to the summation or accumulation of a set of numbers. In adding numbers together, there are two basic principles or laws, the commutative law and the associative law.

The *commutative law* states that the order in which numbers are added is irrelevant. This can be seen in the following two examples:

$$1 + 2 = 3 \qquad 2 + 1 = 3$$

$$x + y = z \qquad y + x = z$$

In each example, it did not matter which number was added first and which was added second, the result was the same.

The *associative law* of addition states that in adding several numbers, any subgrouping of the numbers can be added first, last, or in the middle. This can be seen in the following examples:

1. $2 + 3 + 6 + 7 + 4 + 1 = 23$
2. $(5) + (6 + 7) + 4 + 1 = 23$
3. $5 + 13 + 5 = 23$
4. $5 + 6 + 7 + 4 + 1 = 23$

In each of these cases, the order in which the numbers have been added has no effect on the result.

Subtraction

The process of subtraction is the opposite or inverse of addition. The operation of subtracting 1 from 2 (i.e., $2 - 1$) means that one unit is to be taken away from two units, leaving a remainder of one unit. In contrast to addition, the commutative and associative laws do not hold for subtraction. Therefore, as indicated in the following examples, we have

$$8 - 4 = 4 \quad \text{but} \quad 4 - 8 = -4$$
$$3 - 6 = -3 \quad \text{but} \quad 6 - 3 = 3$$
$$8 - 3 - 2 = 3 \quad \text{but} \quad 3 - 2 - 8 = -7$$
$$9 - 4 - 2 = 3 \quad \text{but} \quad 2 - 4 - 9 = -11$$

When subtracting negative numbers, we must remember that the same result occurs when subtracting a negative number as adding a positive number. Thus we would have

$$4 - (-3) = +7 \qquad 4 + 3 = 7$$
$$8 - (-10) = +18 \qquad 8 + 10 = 18$$

Multiplication

The operation of multiplication is actually a shortcut method of addition when the same number is to be added several times. For example, if 7 is to be added three times ($7 + 7 + 7$), we could just multiply 7 by 3 to obtain the product of 21.

In multiplication as in addition, the commutative laws and associative laws are in operation so that:

$$a \times b = b \times a$$
$$4 \times 5 = 5 \times 4 = 20$$
$$(2 \times 5) \times 6 = 10 \times 6 = 60$$

A third law of multiplication, the *distributive law,* applies to the multiplication of one number by the sum of several other numbers. Thus,

$$a(b + c) = ab + ac$$
$$2(3 + 4) = 2(7) = 2(3) + 2(4) = 14$$

Here the resulting product is the same regardless of whether b and c are summed and multiplied by a, or a is multiplied by b and by c and the two products are added together. We also need to remember when multiplying negative numbers that a negative number multiplied by a negative number equals a positive number. Thus,

$$(-a) \times (-b) = ab$$
$$(-5) \times (-4) = +20$$

Division

Just as subtraction is the opposite of addition, division is the opposite or inverse of multiplication. When we discussed multiplication, we viewed it as a shortcut to addition in certain situations. Similarly, division can be viewed as a shortcut to subtraction. When we divide 20 by 4, we are actually determining the number of times that 4 can be subtracted from 20. In general, however, the number of times one number can be divided by another does not have to be an exact integer value,

since there could be a remainder. For example, if 21 rather than 20 were divided by 4, the answer would be 5 with a remainder of 1, or $5\frac{1}{4}$.

As in the case of subtraction, neither the commutative nor associative law of addition and multiplication holds for division. Thus

$$a \div b \neq b \div a$$
$$9 \div 3 \neq 3 \div 9$$
$$6 \div (3 \div 2) = 4$$
$$(6 \div 3) \div 2 = 1$$

The distributive law will only hold when the numbers to be added are contained in the numerator, not the denominator. Thus,

$$\frac{a+b}{c} = \frac{a}{c} + \frac{b}{c} \quad \text{but} \quad \frac{a}{b+c} \neq \frac{a}{b} + \frac{a}{c}$$

For example,

$$\frac{6+9}{3} = \frac{6}{3} + \frac{9}{3} = 2 + 3 = 5$$

$$\frac{1}{2+3} = \frac{1}{5} \quad \text{but} \quad \frac{1}{2+3} \neq \frac{1}{2} + \frac{1}{3}$$

The last important property of division is that if the numerator and the denominator are each multiplied or divided by the same number, the resulting quotient will not be affected. Therefore, if we have

$$\frac{80}{40} = 2$$

then

$$\frac{5(80)}{5(40)} = \frac{400}{200} = 2$$

and

$$\frac{80 \div 5}{40 \div 5} = \frac{16}{8} = 2$$

Fractions

A fraction is a number that consists of a combination of whole numbers and/or parts of whole numbers. For instance, the fraction $\frac{1}{6}$ consists of only a portion of a number, while the fraction $\frac{7}{6}$ consists of the whole number 1 plus the fraction $\frac{1}{6}$. Each of the operations of addition, subtraction, multiplication, and division can be applied to fractions. When adding and subtracting fractions, one must obtain the lowest common denominator for each fraction prior to adding or subtracting them. Thus, in adding $\frac{1}{3} + \frac{1}{5}$, the lowest common denominator is 15 so that we have

$$\frac{5}{15} + \frac{3}{15} = \frac{8}{15}$$

In subtracting $\frac{1}{4} - \frac{1}{6}$, the same principle can be applied so that we would have a lowest common denominator of 12, producing a result of

$$\frac{3}{12} - \frac{2}{12} = \frac{1}{12}$$

The operations of multiplication of fractions and division of fractions do not have the lowest common denominator requirement associated with the addition

and subtraction of fractions. Thus, if a/b is multiplied by c/d we obtain

$$\frac{ac}{bd}$$

The resulting numerator, ac, is the product of the numerators a and c, while the denominator, bd, is the product of the two denominators b and d. The resulting fraction can sometimes be reduced to a lower term by dividing numerator and denominator by a common factor. For example, taking

$$\frac{2}{3} \times \frac{6}{7} = \frac{12}{21}$$

and dividing numerator and denominator by 3 produces a result of $\frac{4}{7}$.

Division of fractions can be thought of as the inverse of multiplication, so that the divisor can be inverted and multiplied by the original fraction. Thus,

$$\frac{9}{5} \div \frac{1}{4} = \frac{9}{5} \times \frac{4}{1} = \frac{36}{5}$$

Finally, the division of a fraction can be thought of as a way of converting the fraction to a decimal number. For example, the fraction $\frac{2}{5}$ can be converted to a decimal number by dividing its numerator, 2, by its denominator, 5, to produce the decimal number 0.40.

Exponents and Square Roots

The process of exponentiation provides a shortcut in writing out numerous multiplications. For example, if we have $2 \times 2 \times 2 \times 2 \times 2$, then we may also write this as $2^5 = 32$. The 5 represents the exponent of the number 2, telling us that 2 is to be multiplied by itself five times.

There are several rules that can be applied for multiplying or dividing numbers that contain exponents.

Rule 1

$$x^a \cdot x^b = x^{(a+b)}$$

If two numbers involving powers of the same number are multiplied, the product is that same number raised to the sum of the powers. Thus

$$4^2 \cdot 4^3 = (4 \cdot 4)(4 \cdot 4 \cdot 4) = 4^5$$

Rule 2

$$(x^a)^b = x^{ab}$$

If we take the power of a number that is already taken to a power, the result will be a number that is raised to the product of the two powers. For example,

$$(4^2)^3 = (4^2)(4^2)(4^2) = 4^6$$

Rule 3

$$\frac{x^a}{x^b} = x^{(a-b)}$$

If a number raised to a power is divided by the same number raised to a power, the quotient will be the number raised to the difference of the powers. Thus

$$\frac{3^5}{3^3} = \frac{3 \cdot 3 \cdot 3 \cdot 3 \cdot 3}{3 \cdot 3 \cdot 3} = 3^2$$

If the denominator has a higher power than the numerator, the resulting quotient will be a negative power. Thus

$$\frac{3^3}{3^5} = \frac{3 \cdot 3 \cdot 3}{3 \cdot 3 \cdot 3 \cdot 3 \cdot 3} = \frac{1}{3^2} = 3^{-2}$$

If the difference between the powers of the numerator and denominator is 1, the result will be the actual number itself, so that $x^1 = x$. For example,

$$\frac{3^3}{3^2} = \frac{3 \cdot 3 \cdot 3}{3 \cdot 3} = 3^1 = 3$$

If, however, there is no difference in the power of the numbers in the numerator and denominator, the result will be 1. Thus,

$$\frac{x^a}{x^a} = x^{a-a} = x^0 = 1$$

Therefore, any number raised to the zero power will equal 1. For example,

$$\frac{3^3}{3^3} = \frac{3 \cdot 3 \cdot 3}{3 \cdot 3 \cdot 3} = 3^0 = 1$$

The square root represents a special power of a number, the $\frac{1}{2}$ power. It indicates the value that when multiplied by itself will produce the original number. It is represented by the symbol $\sqrt{\ }$.

Equations

In statistics, many formulas are expressed as equations where one unknown value is a function of some other value. Therefore, it is extremely useful that we know how to manipulate equations into various forms. The rules of addition, subtraction, multiplication, and division can be used to work with equations. For example, if we have the equation $x - 2 = 5$, we can solve for x by adding 2 to each side of the equation. Thus, we would have $x - 2 + 2 = 5 + 2$ and, therefore, $x = 7$. If we had $x + y = z$, we could solve for x by subtracting y from both sides of the equation so that $x = z - y$.

If we have the product of two variables equal to the third, such as $x \cdot y = z$, we can solve for x by dividing both sides of the equation by y. Thus, $x = z/y$. On the other hand, if $x/y = z$, we can solve for x by multiplying both sides of the equation by y. Thus, x would equal yz. Therefore, the various operations of addition, subtraction, multiplication, and division can be applied to equations as long as the same operation is performed on each side of the equation, thereby maintaining the equality.

Answers to Quiz

Part I

1. $\frac{3}{2}$
2. 0.16
3. $\frac{5}{3}$
4. $\frac{1}{81}$
5. 0.20
6. 1.30
7. −6.4
8. $+\frac{1}{6}$
9. 3/200
10. 4

Part II

1. c
2. c
3. a
4. b
5. a
6. b
7. d
8. b
9. a
10. c

Appendix B

Statistical Tables

Contents

Table B.1 Random Numbers

Row \ Column	1	2	3	4	5	6	7	8	9	10	11	12	13	14
1	10480	15011	01536	02011	81647	91646	69179	14194	62590	36207	20969	99570	91291	90700
2	22368	46573	25595	85393	30995	89198	27982	53402	93965	34095	52666	19174	39615	99505
3	24130	48360	22527	97265	76393	64809	15179	24830	49340	32081	30680	19655	63348	58629
4	42167	93093	06243	61680	07856	16376	39440	53537	71341	57004	00849	74917	97758	16379
5	37570	39975	81837	16656	06121	91782	60468	81305	49684	60672	14110	06927	01263	54613
6	77921	06907	11008	42751	27756	53498	18602	70659	90655	15053	21916	81825	44394	42880
7	99562	72905	56420	69994	98872	31016	71194	18738	44013	48840	63213	21069	10634	12952
8	96301	91977	05463	07972	18876	20922	94595	56869	69014	60045	18425	84903	42508	32307
9	89579	14342	63661	10281	17453	18103	57740	84378	25331	12566	58678	44947	05585	56941
10	85475	36857	53342	53988	53060	59533	38867	62300	08158	17983	16439	11458	18593	64952
11	28918	69578	88231	33276	70997	79936	56865	05859	90106	31595	01547	85590	91610	78188
12	63553	40961	48235	03427	49626	69445	18663	72695	52180	20847	12234	90511	33703	90322
13	09429	93969	52636	92737	88974	33488	36320	17617	30015	08272	84115	27156	30613	74952
14	10365	61129	87529	85689	48237	52267	67689	93394	01511	26358	85104	20285	29975	89868
15	07119	97336	71048	08178	77233	13916	47564	81056	97735	85977	29372	74461	28551	90707
16	51085	12765	51821	51259	77452	16308	60756	92144	49442	53900	70960	63990	75601	40719
17	02368	21382	52404	60268	89368	19885	55322	44819	01188	65255	64835	44919	05944	55157
18	01011	54092	33362	94904	31273	04146	18594	29852	71585	85030	51132	01915	92747	64951
19	52162	53916	46369	58586	23216	14513	83149	98736	23495	64350	94738	17752	35156	35749
20	07056	97628	33787	09998	42698	06691	76988	13602	51851	46104	88916	19509	25625	58104
21	48663	91245	85828	14346	09172	30168	90229	04734	59193	22178	30421	61666	99904	32812
22	54164	58492	22421	74103	47070	25306	76468	26384	58151	06646	21524	15227	96909	44592
23	32639	32363	05597	24200	13363	38005	94342	28728	35806	06912	17012	64161	18296	22851
24	29334	27001	87637	87308	58731	00256	45834	15398	46557	41135	10367	07684	36188	18510
25	02488	33062	28834	07351	19731	92420	60952	61280	50001	67658	32586	86679	50720	94953
26	81525	72295	04839	96423	24878	82651	66566	14778	76797	14780	13300	87074	79666	95725
27	29676	20591	68086	26432	46901	20849	89768	81536	86645	12659	92259	57102	80428	25280
28	00742	57392	39064	66432	84673	40027	32832	61362	98947	96067	64760	64584	96096	98253
29	05366	04213	25669	26422	44407	44048	37937	63904	45766	66134	75470	66520	34693	90449
30	91921	26418	64117	94305	26766	25940	39972	22209	71500	64568	91402	42416	07844	69618
31	00582	04711	87917	77341	42206	35126	74087	99547	81817	42607	43808	76655	62028	76630
32	00725	69884	62797	56170	86324	88072	76222	36086	84637	93161	76038	65855	77919	88006
33	69011	65795	95876	55293	18988	27354	26575	08625	40801	59920	29841	80150	12777	48501
34	25976	57948	29888	88604	67917	48708	18912	82271	65424	69774	33611	54262	85963	03547
35	09763	83473	73577	12908	30883	18317	28290	35797	05998	41688	34952	37888	38917	88050

continued

Table B.1 Continued

Row \ Column	1	2	3	4	5	6	7	8	9	10	11	12	13	14
36	91576	42595	27958	30134	04024	86385	29880	99730	55536	84855	29080	09250	79656	73211
37	17955	56349	90999	49127	20044	59931	06115	20542	18059	02008	73708	83517	36103	42791
38	46503	18584	18845	49618	02304	51038	20655	58727	28168	15475	56942	53389	20562	87338
39	92157	89634	94824	78171	84610	82834	09922	25417	44137	48413	25555	21246	35509	20468
40	14577	62765	35605	81263	39667	47358	56873	56307	61607	49518	89656	20103	77490	18062
41	98427	07523	33362	64270	01638	92477	66969	98420	04880	45585	46565	04102	46880	45709
42	34914	63976	88720	82765	34476	17032	87589	40836	32427	70002	70663	88863	77775	69348
43	70060	28277	39475	46473	23219	53416	94970	25832	69975	94884	19661	72828	00102	66794
44	53976	54914	06990	67245	68350	82948	11398	42878	80287	88267	47363	46634	06541	97809
45	76072	29515	40980	07391	58745	25774	22987	80059	39911	96189	41151	14222	60697	59583
46	90725	52210	83974	29992	65831	38857	50490	83765	55657	14361	31720	57375	56228	41546
47	64364	67412	33339	31926	14883	24413	59744	92351	97473	89286	35931	04110	23726	51900
48	08962	00358	31662	25388	61642	34072	81249	35648	56891	69352	48373	45578	78547	81788
49	95012	68379	93526	70765	10592	04542	76463	54328	02349	17247	28865	14777	62730	92277
50	15664	10493	20492	38391	91132	21999	59516	81652	27195	48223	46751	22923	32261	85653
51	16408	81899	04153	53381	79401	21438	83035	92350	36693	31238	59649	91754	72772	02338
52	18629	81953	05520	91962	04739	13092	97662	24822	94730	06496	35090	04822	86774	98289
53	73115	35101	47498	87637	99016	71060	88824	71013	18735	20286	23153	72924	35165	43040
54	57491	16703	23167	49323	45021	33132	12544	41035	80780	45393	44812	12512	98931	91202
55	30405	83946	23792	14422	15059	45799	22716	19792	09983	74353	68668	30429	70735	25499
56	16631	35006	85900	98275	32388	52390	16815	69290	82732	38480	73817	32523	41961	44437
57	96773	20206	42559	78985	05300	22164	24369	54224	35083	19687	11052	91491	60383	19746
58	38935	64202	14349	82674	66523	44133	00697	35552	35970	19124	63318	29686	03387	59846
59	31624	76384	17403	53363	44167	64486	64758	75366	76554	31601	12614	33072	60332	92325
60	78919	19474	23632	27889	47914	02584	37680	20801	72152	39339	34806	08930	85001	87820
61	03931	33309	57047	74211	63445	17361	62825	39908	05607	91284	68833	25570	38818	46920
62	74426	33278	43972	10110	89917	15665	52872	73823	73144	88662	88970	74492	51805	99378
63	09066	00903	20795	95452	92648	45454	09552	88815	16553	51125	79375	97596	16296	66092
64	42238	12426	87025	14267	20979	04508	64535	31355	86064	29472	47689	05974	52468	16834
65	16153	08002	26504	41744	81959	65642	74240	56302	00033	67107	77510	70625	28725	34191
66	21457	40742	29820	96783	29400	21840	15035	34537	33310	06116	95240	15957	16572	06004
67	21581	57802	02050	89728	17937	37621	47075	42080	97403	48626	68995	43805	33386	21597
68	55612	78095	83197	33732	05810	24813	86902	60397	16489	03264	88525	42786	05269	92532
69	44657	66999	99324	51281	84463	60563	79312	93454	68876	25471	93911	25650	12682	73572
70	91340	84979	46949	81973	37949	61023	43997	15263	80644	43942	89203	71795	99533	50501

Table B.1 Continued

Row \ Column	1	2	3	4	5	6	7	8	9	10	11	12	13	14
71	91227	21199	31935	27022	84067	05462	35216	14486	29891	68607	41867	14951	91696	85065
72	50001	38140	66321	19924	72163	09538	12151	06878	91903	18749	34405	56087	82790	70925
73	65390	05224	72958	28609	81406	39147	25549	48542	42627	45233	57202	94617	23772	07896
74	27504	96131	83944	41575	10573	08619	64482	73923	36152	05184	94142	25299	84387	34925
75	37169	94851	39117	89632	00959	16487	65536	49071	39782	17095	02330	74301	00275	48280
76	11508	70225	51111	38351	19444	66499	71945	05422	13442	78675	84081	66938	93654	59894
77	37449	30362	06694	54690	04052	53115	62757	95348	78662	11163	81651	50245	34971	52924
78	46515	70331	85922	38329	57015	15765	97161	17869	45349	61796	66345	81073	49106	79860
79	30986	81223	42416	58353	21532	30502	32305	86482	05174	07901	54339	58861	74818	46942
80	63798	64995	46583	09785	44160	78128	83991	42865	92520	83531	80377	35909	81250	54238
81	82486	84846	99254	67632	43218	50076	21361	64816	51202	88124	41870	52689	51275	83556
82	21885	32906	92431	09060	64297	51674	64126	62570	26123	05155	59194	52799	28225	85762
83	60336	98782	07408	53458	13564	59089	26445	29789	85205	41001	12535	12133	14645	23541
84	43937	46891	24010	25560	86355	33941	25786	54990	71899	15475	95434	98227	21824	19585
85	97656	63175	89303	16275	07100	92063	21942	18611	47348	20203	18534	03862	78095	50136
86	03299	01221	05418	38982	55758	92237	26759	86367	21216	98442	08303	56613	91511	75928
87	79626	06486	03574	17668	07785	76020	79924	25651	83325	88428	85076	72811	22717	50585
88	85636	68335	47539	03129	65651	11977	02510	26113	99447	68645	34327	15152	55230	93448
89	18039	14367	64337	06177	12143	46609	32989	74014	64708	00533	35398	58408	13261	47908
90	08362	15656	60627	36478	65648	16764	53412	09013	07832	41574	17639	82163	60859	75567
91	79556	29068	04142	16268	15387	12856	66227	38358	22478	73373	88732	09443	82558	05250
92	92608	82674	27072	32534	17075	27698	98204	63863	11951	34648	88022	56148	34925	57031
93	23982	25835	40055	67006	12293	02753	14827	23235	35071	99704	37543	11601	35503	85171
94	09915	96306	05908	97901	28395	14186	00821	80703	70426	75647	76310	88717	37890	40129
95	59037	33300	26695	62247	69927	76123	50842	43834	86654	70959	79725	93872	28117	19233
96	42488	78077	69882	61657	34136	79180	97526	43092	04098	73571	80799	76536	71255	64239
97	46764	86273	63003	93017	31204	36692	40202	35275	57306	55543	53203	18098	47625	88684
98	03237	45430	55417	63282	90816	17349	88298	90183	36600	78406	06216	95787	42579	90730
99	86591	81482	52667	61582	14972	90053	89534	76036	49199	43716	97548	04379	46370	28672
100	38534	01715	94964	87288	65680	43772	39560	12918	86537	62738	19636	51132	25739	56947

Source: Abridged from W. H. Beyer (ed.). *CRC Standard Mathematical Tables,* 24th edition. (Cleveland: The Chemical Rubber Company), 1976.

Table B.2 Cumulative Binomial Probabilities

a. $n = 5$

k \ p	.01	.05	.1	.2	.3	.4	.5	.6	.7	.8	.9	.95	.99
0	.9510	.7738	.5905	.3277	.1681	.0778	.0313	.0102	.0024	.0003	.0000	.0000	.0000
1	.9990	.9774	.9185	.7373	.5282	.3370	.1875	.0870	.0308	.0067	.0005	.0000	.0000
2	1.0000	.9988	.9914	.9421	.8369	.6826	.5000	.3174	.1631	.0579	.0086	.0012	.0000
3	1.0000	1.0000	.9995	.9933	.9692	.9130	.8125	.6630	.4718	.2627	.0815	.0226	.0010
4	1.0000	1.0000	1.0000	.9997	.9976	.9898	.9687	.9222	.8319	.6723	.4095	.2262	.0490

b. $n = 6$

k \ p	.01	.05	.1	.2	.3	.4	.5	.6	.7	.8	.9	.95	.99
0	.9415	.7351	.5314	.2621	.1176	.0467	.0156	.0041	.0007	.0001	.0000	.0000	.0000
1	.9985	.9672	.8857	.6554	.4202	.2333	.1094	.0410	.0109	.0016	.0001	.0000	.0000
2	1.0000	.9978	.9841	.9011	.7443	.5443	.3437	.1792	.0705	.0170	.0013	.0001	.0000
3	1.0000	.9999	.9987	.9830	.9295	.8208	.6562	.4557	.2557	.0989	.0158	.0022	.0000
4	1.0000	1.0000	.9999	.9984	.9891	.9590	.8906	.7667	.5798	.3446	.1143	.0328	.0015
5	1.0000	1.0000	1.0000	.9999	.9993	.9959	.9844	.9533	.8824	.7379	.4686	.2649	.0585

c. $n = 7$

k \ p	.01	.05	.1	.2	.3	.4	.5	.6	.7	.8	.9	.95	.99
0	.9321	.6983	.4783	.2097	.0824	.0280	.0078	.0016	.0002	.0000	.0000	.0000	.0000
1	.9980	.9556	.8503	.5767	.3294	.1586	.0625	.0188	.0038	.0004	.0000	.0000	.0000
2	1.0000	.9962	.9743	.8520	.6471	.4199	.2266	.0963	.0288	.0047	.0002	.0000	.0000
3	1.0000	.9998	.9973	.9667	.8740	.7102	.5000	.2898	.1260	.0333	.0027	.0002	.0000
4	1.0000	1.0000	.9998	.9953	.9712	.9037	.7734	.5801	.3529	.1480	.0257	.0038	.0000
5	1.0000	1.0000	1.0000	.9996	.9962	.9812	.9375	.8414	.6706	.4233	.1497	.0444	.0020
6	1.0000	1.0000	1.0000	1.0000	.9998	.9984	.9922	.9720	.9176	.7903	.5217	.3017	.0679

d. $n = 8$

k \ p	.01	.05	.1	.2	.3	.4	.5	.6	.7	.8	.9	.95	.99
0	.9227	.6634	.4305	.1678	.0576	.0168	.0039	.0007	.0001	.0000	.0000	.0000	.0000
1	.9973	.9423	.8131	.5033	.2553	.1064	.0352	.0085	.0013	.0001	.0000	.0000	.0000
2	.9999	.9942	.9619	.7969	.5518	.3154	.1445	.0498	.0113	.0012	.0000	.0000	.0000
3	1.0000	.9996	.9950	.9437	.8059	.5941	.3633	.1737	.0580	.0104	.0004	.0000	.0000
4	1.0000	1.0000	.9996	.9896	.9420	.8263	.6367	.4059	.1941	.0563	.0050	.0004	.0000
5	1.0000	1.0000	1.0000	.9988	.9887	.9502	.8555	.6346	.4482	.2031	.0381	.0058	.0001
6	1.0000	1.0000	1.0000	.9999	.9987	.9915	.9648	.8936	.7447	.4967	.1869	.0572	.0027
7	1.0000	1.0000	1.0000	1.0000	.9999	.9993	.9961	.9832	.9424	.8322	.5695	.3366	.0773

e. $n = 9$

k \ p	.01	.05	.1	.2	.3	.4	.5	.6	.7	.8	.9	.95	.99
0	.9135	.6302	.3874	.1342	.0404	.0101	.0020	.0003	.0000	.0000	.0000	.0000	.0000
1	.9966	.9288	.7748	.4362	.1960	.0705	.0195	.0038	.0004	.0000	.0000	.0000	.0000
2	.9999	.9916	.9470	.7382	.4623	.2318	.0898	.0250	.0043	.0003	.0000	.0000	.0000
3	1.0000	.9994	.9917	.9144	.7297	.4826	.2539	.0994	.0253	.0031	.0001	.0000	.0000
4	1.0000	1.0000	.9991	.9804	.9012	.7334	.5000	.2666	.0988	.0196	.0009	.0000	.0000
5	1.0000	1.0000	.9999	.9969	.9747	.9006	.7461	.5174	.2703	.0856	.0083	.0006	.0000
6	1.0000	1.0000	1.0000	.9997	.9957	.9750	.9102	.7682	.5372	.2618	.0530	.0084	.0001
7	1.0000	1.0000	1.0000	1.0000	.9996	.9962	.9805	.9295	.8040	.5638	.2252	.0712	.0034
8	1.0000	1.0000	1.0000	1.0000	1.0000	.9997	.9980	.9899	.9596	.8658	.6126	.3698	.0865

Table B.2 Continued

f. $n = 10$

k \ p	.01	.05	.1	.2	.3	.4	.5	.6	.7	.8	.9	.95	.99
0	.9044	.5987	.3487	.1074	.0282	.0060	.0010	.0001	.0000	.0000	.0000	.0000	.0000
1	.9957	.9139	.7361	.3758	.1493	.0464	.0107	.0017	.0001	.0000	.0000	.0000	.0000
2	.9999	.9885	.9298	.6778	.3828	.1673	.0547	.0123	.0016	.0001	.0000	.0000	.0000
3	1.0000	.9990	.9872	.8791	.6496	.3823	.1719	.0548	.0106	.0009	.0000	.0000	.0000
4	1.0000	.9999	.9984	.9672	.8497	.6331	.3770	.1662	.0473	.0064	.0001	.0000	.0000
5	1.0000	1.0000	.9999	.9936	.9527	.8338	.6230	.3669	.1503	.0328	.0016	.0001	.0000
6	1.0000	1.0000	1.0000	.9991	.9894	.9452	.8281	.6177	.3504	.1209	.0128	.0010	.0000
7	1.0000	1.0000	1.0000	.9999	.9984	.9877	.9453	.8327	.6172	.3222	.0702	.0115	.0001
8	1.0000	1.0000	1.0000	1.0000	.9999	.9983	.9893	.9536	.8507	.6242	.2639	.0861	.0043
9	1.0000	1.0000	1.0000	1.0000	1.0000	.9999	.9990	.9940	.9718	.8926	.6513	.4013	.0956

g. $n = 15$

k \ p	.01	.05	.1	.2	.3	.4	.5	.6	.7	.8	.9	.95	.99
0	.8601	.4633	.2059	.0352	.0047	.0005	.0000	.0000	.0000	.0000	.0000	.0000	.0000
1	.9904	.8290	.5490	.1671	.0353	.0052	.0005	.0000	.0000	.0000	.0000	.0000	.0000
2	.9996	.9638	.8159	.3980	.1268	.0271	.0037	.0003	.0000	.0000	.0000	.0000	.0000
3	1.0000	.9945	.9444	.6482	.2969	.0905	.0176	.0019	.0001	.0000	.0000	.0000	.0000
4	1.0000	.9994	.9873	.8358	.5155	.2173	.0592	.0093	.0007	.0000	.0000	.0000	.0000
5	1.0000	.9999	.9978	.9389	.7216	.4032	.1509	.0338	.0037	.0001	.0000	.0000	.0000
6	1.0000	1.0000	.9997	.9819	.8689	.6098	.3036	.0950	.0152	.0008	.0000	.0000	.0000
7	1.0000	1.0000	1.0000	.9958	.9500	.7869	.5000	.2131	.0500	.0042	.0000	.0000	.0000
8	1.0000	1.0000	1.0000	.9992	.9848	.9050	.6964	.3902	.1311	.0181	.0003	.0000	.0000
9	1.0000	1.0000	1.0000	.9999	.9963	.9662	.8491	.5968	.2784	.0611	.0022	.0001	.0000
10	1.0000	1.0000	1.0000	1.0000	.9993	.9907	.9408	.7827	.4845	.1642	.0127	.0006	.0000
11	1.0000	1.0000	1.0000	1.0000	.9999	.9981	.9824	.9095	.7031	.3518	.0556	.0055	.0000
12	1.0000	1.0000	1.0000	1.0000	1.0000	.9997	.9963	.9729	.8732	.6020	.1841	.0362	.0004
13	1.0000	1.0000	1.0000	1.0000	1.0000	1.0000	.9995	.9948	.9647	.8329	.4510	.1710	.0096
14	1.0000	1.0000	1.0000	1.0000	1.0000	1.0000	1.0000	.9995	.9953	.9648	.7941	.5367	.1399

h. $n = 20$

k \ p	.01	.05	.1	.2	.3	.4	.5	.6	.7	.8	.9	.95	.99
0	.8179	.3585	.1216	.0115	.0008	.0000	.0000	.0000	.0000	.0000	.0000	.0000	.0000
1	.9831	.7358	.3917	.0692	.0076	.0005	.0000	.0000	.0000	.0000	.0000	.0000	.0000
2	.9990	.9245	.6769	.2061	.0355	.0036	.0002	.0000	.0000	.0000	.0000	.0000	.0000
3	1.0000	.9841	.8670	.4114	.1071	.0160	.0013	.0000	.0000	.0000	.0000	.0000	.0000
4	1.0000	.9974	.9568	.6296	.2375	.0510	.0059	.0003	.0000	.0000	.0000	.0000	.0000
5	1.0000	.9997	.9887	.8042	.4164	.1256	.0207	.0016	.0000	.0000	.0000	.0000	.0000
6	1.0000	1.0000	.9976	.9133	.6080	.2500	.0577	.0065	.0003	.0000	.0000	.0000	.0000
7	1.0000	1.0000	.9996	.9679	.7723	.4159	.1316	.0210	.0013	.0000	.0000	.0000	.0000
8	1.0000	1.0000	.9999	.9900	.8867	.5956	.2517	.0565	.0051	.0001	.0000	.0000	.0000
9	1.0000	1.0000	1.0000	.9974	.9520	.7553	.4119	.1275	.0171	.0006	.0000	.0000	.0000
10	1.0000	1.0000	1.0000	.9994	.9829	.8725	.5881	.2447	.0480	.0026	.0000	.0000	.0000
11	1.0000	1.0000	1.0000	.9999	.9949	.9435	.7483	.4044	.1133	.0100	.0001	.0000	.0000
12	1.0000	1.0000	1.0000	1.0000	.9987	.9790	.8684	.5841	.2277	.0321	.0004	.0000	.0000
13	1.0000	1.0000	1.0000	1.0000	.9997	.9935	.9423	.7500	.3920	.0867	.0024	.0000	.0000
14	1.0000	1.0000	1.0000	1.0000	1.0000	.9984	.9793	.8744	.5836	.1958	.0113	.0003	.0000
15	1.0000	1.0000	1.0000	1.0000	1.0000	.9997	.9941	.9490	.7625	.3704	.0432	.0026	.0000
16	1.0000	1.0000	1.0000	1.0000	1.0000	1.0000	.9987	.9840	.8929	.5886	.1330	.0159	.0000
17	1.0000	1.0000	1.0000	1.0000	1.0000	1.0000	.9998	.9964	.9645	.7939	.3231	.0755	.0010
18	1.0000	1.0000	1.0000	1.0000	1.0000	1.0000	1.0000	.9995	.9924	.9308	.6083	.2642	.0169
19	1.0000	1.0000	1.0000	1.0000	1.0000	1.0000	1.0000	1.0000	.9992	.9885	.8784	.6415	.1821

continued

Table B.2 Continued

i. $n = 25$

k \ p	.01	.05	.1	.2	.3	.4	.5	.6	.7	.8	.9	.95	.99
0	.7778	.2774	.0718	.0038	.0001	.0000	.0000	.0000	.0000	.0000	.0000	.0000	.0000
1	.9742	.6424	.2712	.0274	.0016	.0001	.0000	.0000	.0000	.0000	.0000	.0000	.0000
2	.9980	.8729	.5371	.0982	.0090	.0004	.0000	.0000	.0000	.0000	.0000	.0000	.0000
3	.9999	.9659	.7636	.2340	.0332	.0024	.0001	.0000	.0000	.0000	.0000	.0000	.0000
4	1.0000	.9928	.9020	.4207	.0905	.0095	.0005	.0000	.0000	.0000	.0000	.0000	.0000
5	1.0000	.9988	.9666	.6167	.1935	.0294	.0020	.0001	.0000	.0000	.0000	.0000	.0000
6	1.0000	.9998	.9905	.7800	.3407	.0736	.0073	.0003	.0000	.0000	.0000	.0000	.0000
7	1.0000	1.0000	.9977	.8909	.5118	.1536	.0216	.0012	.0000	.0000	.0000	.0000	.0000
8	1.0000	1.0000	.9995	.9532	.6769	.2735	.0539	.0043	.0001	.0000	.0000	.0000	.0000
9	1.0000	1.0000	.9999	.9827	.8106	.4246	.1148	.0132	.0005	.0000	.0000	.0000	.0000
10	1.0000	1.0000	1.0000	.9944	.9022	.5858	.2122	.0344	.0018	.0000	.0000	.0000	.0000
11	1.0000	1.0000	1.0000	.9985	.9558	.7323	.3450	.0778	.0060	.0001	.0000	.0000	.0000
12	1.0000	1.0000	1.0000	.9996	.9825	.8462	.5000	.1538	.0175	.0004	.0000	.0000	.0000
13	1.0000	1.0000	1.0000	.9999	.9940	.9222	.6550	.2677	.0442	.0015	.0000	.0000	.0000
14	1.0000	1.0000	1.0000	1.0000	.9982	.9656	.7878	.4142	.0978	.0056	.0000	.0000	.0000
15	1.0000	1.0000	1.0000	1.0000	.9995	.9868	.8852	.5754	.1894	.0173	.0001	.0000	.0000
16	1.0000	1.0000	1.0000	1.0000	.9999	.9957	.9461	.7265	.3231	.0468	.0005	.0000	.0000
17	1.0000	1.0000	1.0000	1.0000	1.0000	.9988	.9784	.8464	.4882	.1091	.0023	.0000	.0000
18	1.0000	1.0000	1.0000	1.0000	1.0000	.9997	.9927	.9264	.6593	.2200	.0095	.0002	.0000
19	1.0000	1.0000	1.0000	1.0000	1.0000	.9999	.9980	.9706	.8065	.3833	.0334	.0012	.0000
20	1.0000	1.0000	1.0000	1.0000	1.0000	1.0000	.9995	.9905	.9095	.5793	.0980	.0072	.0000
21	1.0000	1.0000	1.0000	1.0000	1.0000	1.0000	.9999	.9976	.9668	.7660	.2364	.0341	.0001
22	1.0000	1.0000	1.0000	1.0000	1.0000	1.0000	1.0000	.9996	.9910	.9018	.4629	.1271	.0020
23	1.0000	1.0000	1.0000	1.0000	1.0000	1.0000	1.0000	.9999	.9984	.9726	.7288	.3576	.0258
24	1.0000	1.0000	1.0000	1.0000	1.0000	1.0000	1.0000	1.0000	.9999	.9962	.9282	.7226	.2222

Table B.3 Normal Curve Areas

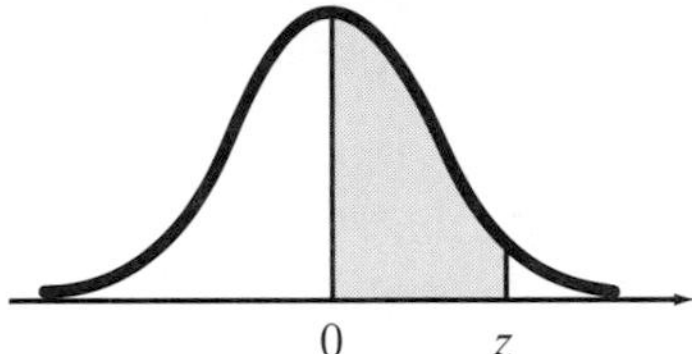

z	.00	.01	.02	.03	.04	.05	.06	.07	.08	.09
.0	.0000	.0040	.0080	.0120	.0160	.0199	.0239	.0279	.0319	.0359
.1	.0398	.0438	.0478	.0517	.0557	.0596	.0636	.0675	.0714	.0753
.2	.0793	.0832	.0871	.0910	.0948	.0987	.1026	.1064	.1103	.1141
.3	.1179	.1217	.1255	.1293	.1331	.1368	.1406	.1443	.1480	.1517
.4	.1554	.1591	.1628	.1664	.1700	.1736	.1772	.1808	.1844	.1879
.5	.1915	.1950	.1985	.2019	.2054	.2088	.2123	.2157	.2190	.2224
.6	.2257	.2291	.2324	.2357	.2389	.2422	.2454	.2486	.2517	.2549
.7	.2580	.2611	.2642	.2673	.2704	.2734	.2764	.2794	.2823	.2852
.8	.2881	.2910	.2939	.2967	.2995	.3023	.3051	.3078	.3106	.3133
.9	.3159	.3186	.3212	.3238	.3264	.3289	.3315	.3340	.3365	.3389
1.0	.3413	.3438	.3461	.3485	.3508	.3531	.3554	.3577	.3599	.3621
1.1	.3643	.3665	.3686	.3708	.3729	.3749	.3770	.3790	.3810	.3830
1.2	.3849	.3869	.3888	.3907	.3925	.3944	.3962	.3980	.3997	.4015
1.3	.4032	.4049	.4066	.4082	.4099	.4115	.4131	.4147	.4162	.4177
1.4	.4192	.4207	.4222	.4236	.4251	.4265	.4279	.4292	.4306	.4319
1.5	.4332	.4345	.4357	.4370	.4382	.4394	.4406	.4418	.4429	.4441
1.6	.4452	.4463	.4474	.4484	.4495	.4505	.4515	.4525	.4535	.4545
1.7	.4554	.4564	.4573	.4582	.4591	.4599	.4608	.4616	.4625	.4633
1.8	.4641	.4649	.4656	.4664	.4671	.4678	.4686	.4693	.4699	.4706
1.9	.4713	.4719	.4726	.4732	.4738	.4744	.4750	.4756	.4761	.4767
2.0	.4772	.4778	.4783	.4788	.4793	.4798	.4803	.4808	.4812	.4817
2.1	.4821	.4826	.4830	.4834	.4838	.4842	.4846	.4850	.4854	.4857
2.2	.4861	.4864	.4868	.4871	.4875	.4878	.4881	.4884	.4887	.4890
2.3	.4893	.4896	.4898	.4901	.4904	.4906	.4909	.4911	.4913	.4916
2.4	.4918	.4920	.4922	.4925	.4927	.4929	.4931	.4932	.4934	.4936
2.5	.4938	.4940	.4941	.4943	.4945	.4946	.4948	.4949	.4951	.4952
2.6	.4953	.4955	.4956	.4957	.4959	.4960	.4961	.4962	.4963	.4964
2.7	.4965	.4966	.4967	.4968	.4969	.4970	.4971	.4972	.4973	.4974
2.8	.4974	.4975	.4976	.4977	.4977	.4978	.4979	.4979	.4980	.4981
2.9	.4981	.4982	.4982	.4983	.4984	.4984	.4985	.4985	.4986	.4986
3.0	.4987	.4987	.4987	.4988	.4988	.4989	.4989	.4989	.4990	.4990

Source: Abridged from Table I of A. Hald, *Statistical Tables and Formulas* (New York: John Wiley & Sons, Inc.), 1952. Reproduced by permission of A. Hald and the publisher.

Table B.4 Critical Values for Student's *t*

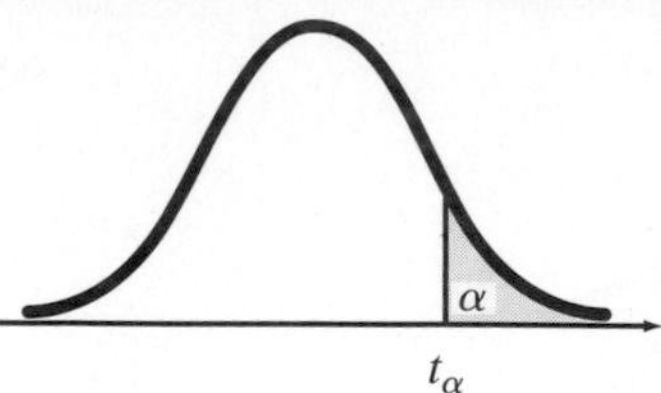

Degrees of Freedom	$t_{.100}$	$t_{.050}$	$t_{.025}$	$t_{.010}$	$t_{.005}$	$t_{.001}$	$t_{.0005}$
1	3.078	6.314	12.706	31.821	63.657	318.31	636.62
2	1.886	2.920	4.303	6.965	9.925	22.326	31.598
3	1.638	2.353	3.182	4.541	5.841	10.213	12.924
4	1.533	2.132	2.776	3.747	4.604	7.173	8.610
5	1.476	2.015	2.571	3.365	4.032	5.893	6.869
6	1.440	1.943	2.447	3.143	3.707	5.208	5.959
7	1.415	1.895	2.365	2.998	3.499	4.785	5.408
8	1.397	1.860	2.306	2.896	3.355	4.501	5.041
9	1.383	1.833	2.262	2.821	3.250	4.297	4.781
10	1.372	1.812	2.228	2.764	3.169	4.144	4.587
11	1.363	1.796	2.201	2.718	3.106	4.025	4.437
12	1.356	1.782	2.179	2.681	3.055	3.930	4.318
13	1.350	1.771	2.160	2.650	3.012	3.852	4.221
14	1.345	1.761	2.145	2.624	2.977	3.787	4.140
15	1.341	1.753	2.131	2.602	2.947	3.733	4.073
16	1.337	1.746	2.120	2.583	2.921	3.686	4.015
17	1.333	1.740	2.110	2.567	2.898	3.646	3.965
18	1.330	1.734	2.101	2.552	2.878	3.610	3.922
19	1.328	1.729	2.093	2.539	2.861	3.579	3.883
20	1.325	1.725	2.086	2.528	2.845	3.552	3.850
21	1.323	1.721	2.080	2.518	2.831	3.527	3.819
22	1.321	1.717	2.074	2.508	2.819	3.505	3.792
23	1.319	1.714	2.069	2.500	2.807	3.485	3.767
24	1.318	1.711	2.064	2.492	2.797	3.467	3.745
25	1.316	1.708	2.060	2.485	2.787	3.450	3.725
26	1.315	1.706	2.056	2.479	2.779	3.435	3.707
27	1.314	1.703	2.052	2.473	2.771	3.421	3.690
28	1.313	1.701	2.048	2.467	2.763	3.408	3.674
29	1.311	1.699	2.045	2.462	2.756	3.396	3.659
30	1.310	1.697	2.042	2.457	2.750	3.385	3.646
40	1.303	1.684	2.021	2.423	2.704	3.307	3.551
60	1.296	1.671	2.000	2.390	2.660	3.232	3.460
120	1.289	1.658	1.980	2.358	2.617	3.160	3.373
∞	1.282	1.645	1.960	2.326	2.576	3.090	3.291

Source: This table is reproduced with the kind permission of the Trustees of Biometrika and Oxford University Press from E. S. Pearson and H. O. Hartley (eds.), *The Biometrika Tables for Statisticians,* Vol. 1, 3rd ed., *Biometrika,* 1966.

Table B.5 Critical Values for the χ^2 Statistic

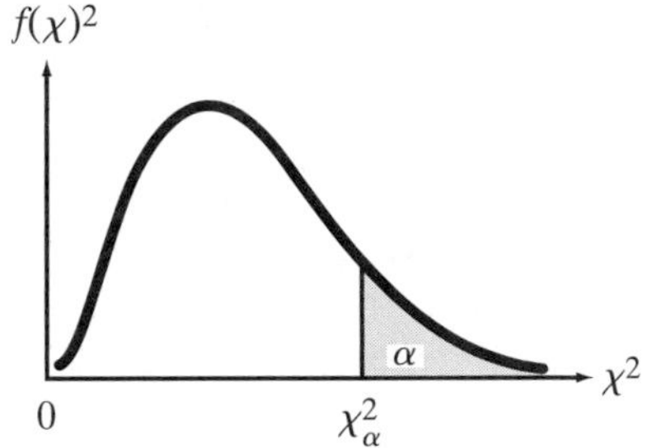

Degrees of Freedom	$\chi^2_{.995}$	$\chi^2_{.990}$	$\chi^2_{.975}$	$\chi^2_{.950}$	$\chi^2_{.900}$
1	.0000393	.0001571	.0009821	.0039321	.0157908
2	.0100251	.0201007	.0506356	.102587	.210720
3	.0717212	.114832	.215795	.351846	.584375
4	.206990	.297110	.484419	.710721	1.063623
5	.411740	.554300	.831211	1.145476	1.61031
6	.675727	.872085	1.237347	1.63539	2.20413
7	.989265	1.239043	1.68987	2.16735	2.83311
8	1.344419	1.646482	2.17973	2.73264	3.48954
9	1.734926	2.087912	2.70039	3.32511	4.16816
10	2.15585	2.55821	3.24697	3.94030	4.86518
11	2.60321	3.05347	3.81575	4.57481	5.57779
12	3.07382	3.57056	4.40379	5.22603	6.30380
13	3.56503	4.10691	5.00874	5.89186	7.04150
14	4.07468	4.66043	5.62872	6.57063	7.78953
15	4.60094	5.22935	6.26214	7.26094	8.54675
16	5.14224	5.81221	6.90766	7.96164	9.31223
17	5.69724	6.40776	7.56418	8.67176	10.0852
18	6.26481	7.01491	8.23075	9.39046	10.8649
19	6.84398	7.63273	8.90655	10.1170	11.6509
20	7.43386	8.26040	9.59083	10.8508	12.4426
21	8.03366	8.89720	10.28293	11.5913	13.2396
22	8.64272	9.54249	10.9823	12.3380	14.0415
23	9.26042	10.19567	11.6885	13.0905	14.8479
24	9.88623	10.8564	12.4011	13.8484	15.6587
25	10.5197	11.5240	13.1197	14.6114	16.4734
26	11.1603	12.1981	13.8439	15.3791	17.2919
27	11.8076	12.8786	14.5733	16.1513	18.1138
28	12.4613	13.5648	15.3079	16.9279	18.9392
29	13.1211	14.2565	16.0471	17.7083	19.7677
30	13.7867	14.9535	16.7908	18.4926	20.5992
40	20.7065	22.1643	24.4331	26.5093	29.0505
50	27.9907	29.7067	32.3574	34.7642	37.6886
60	35.5346	37.4848	40.4817	43.1879	46.4589
70	43.2752	45.4418	48.7576	51.7393	55.3290
80	51.1720	53.5400	57.1532	60.3915	64.2778
90	59.1963	61.7541	65.6466	69.1260	73.2912
100	67.3276	70.0648	74.2219	77.9295	82.3581
150	109.142	112.668	117.985	122.692	128.275
200	152.241	156.432	162.728	168.279	174.835
300	240.663	245.972	253.912	260.878	269.068
400	330.903	337.155	346.482	354.641	364.207
500	422.303	429.388	439.936	449.147	459.926

Source: From C. M. Thompson, "Tables of the Percentage Points of the χ^2-Distribution," *Biometrika,* 1941, Vol. 32, pp. 188–189. Reproduced by permission of the *Biometrika* Trustees and Oxford University Press.

Table B.5 Continued

Degrees of Freedom	$\chi^2_{.100}$	$\chi^2_{.050}$	$\chi^2_{.025}$	$\chi^2_{.010}$	$\chi^2_{.005}$
1	2.70554	3.84146	5.02389	6.63490	7.87944
2	4.60517	5.99147	7.37776	9.21034	10.5966
3	6.25139	7.81473	9.34840	11.3449	12.8381
4	7.77944	9.48773	11.1433	13.2767	14.8602
5	9.23635	11.0705	12.8325	15.0863	16.7496
6	10.6446	12.5916	14.4494	16.8119	18.5476
7	12.0170	14.0671	16.0128	18.4753	20.2777
8	13.3616	15.5073	17.5346	20.0902	21.9550
9	14.6837	16.9190	19.0228	21.6660	23.5893
10	15.9871	18.3070	20.4831	23.2093	25.1882
11	17.2750	19.6751	21.9200	24.7250	26.7569
12	18.5494	21.0261	23.3367	26.2170	28.2995
13	19.8119	22.3621	24.7356	27.6883	29.8194
14	21.0642	23.6848	26.1190	29.1413	31.3193
15	22.3072	24.9958	27.4884	30.5779	32.8013
16	23.5418	26.2962	28.8454	31.9999	34.2672
17	24.7690	27.5871	30.1910	33.4087	35.7185
18	25.9894	28.8693	31.5264	34.8053	37.1564
19	27.2036	30.1435	32.8523	36.1908	38.5822
20	28.4120	31.4104	34.1696	37.5662	39.9968
21	29.6151	32.6705	35.4789	38.9321	41.4010
22	30.8133	33.9244	36.7807	40.2894	42.7956
23	32.0069	35.1725	38.0757	41.6384	44.1813
24	33.1963	36.4151	39.3641	42.9798	45.5585
25	34.3816	37.6525	40.6465	44.3141	46.9278
26	35.5631	38.8852	41.9232	45.6417	48.2899
27	36.7412	40.1133	43.1944	46.9630	49.6449
28	37.9159	41.3372	44.4607	48.2782	50.9933
29	39.0875	42.5569	45.7222	49.5879	52.3356
30	40.2560	43.7729	46.9792	50.8922	53.6720
40	51.8050	55.7585	59.3417	63.6907	66.7659
50	63.1671	67.5048	71.4202	76.1539	79.4900
60	74.3970	79.0819	83.2976	88.3794	91.9517
70	85.5271	90.5312	95.0231	100.425	104.215
80	96.5782	101.879	106.629	112.329	116.321
90	107.565	113.145	118.136	124.116	128.299
100	118.498	124.342	129.561	135.807	140.169
150	172.581	179.581	185.800	193.208	198.360
200	226.021	233.994	241.058	249.445	255.264
300	331.789	341.395	349.874	359.906	366.844
400	436.649	447.632	457.305	468.724	476.606
500	540.930	553.127	563.852	576.493	585.207

Table B.6a Critical Values of the *F* Statistic, $\alpha = .10$

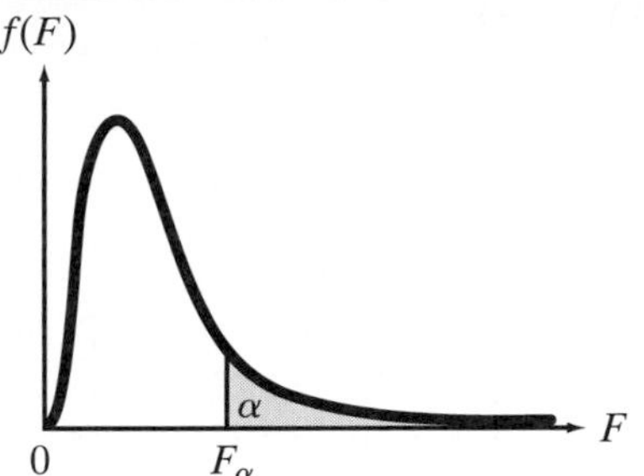

ν_2 \ ν_1	NUMERATOR DEGREES OF FREEDOM 1	2	3	4	5	6	7	8	9
1	39.86	49.50	53.59	55.83	57.24	58.20	58.91	59.44	59.86
2	8.53	9.00	9.16	9.24	9.29	9.33	9.35	9.37	9.38
3	5.54	5.46	5.39	5.34	5.31	5.28	5.27	5.25	5.24
4	4.54	4.32	4.19	4.11	4.05	4.01	3.98	3.95	3.94
5	4.06	3.78	3.62	3.52	3.45	3.40	3.37	3.34	3.32
6	3.78	3.46	3.29	3.18	3.11	3.05	3.01	2.98	2.96
7	3.59	3.26	3.07	2.96	2.88	2.83	2.78	2.75	2.72
8	3.46	3.11	2.92	2.81	2.73	2.67	2.62	2.59	2.56
9	3.36	3.01	2.81	2.69	2.61	2.55	2.51	2.47	2.44
DENOMINATOR DEGREES OF FREEDOM 10	3.29	2.92	2.73	2.61	2.52	2.46	2.41	2.38	2.35
11	3.23	2.86	2.66	2.54	2.45	2.39	2.34	2.30	2.27
12	3.18	2.81	2.61	2.48	2.39	2.33	2.28	2.24	2.21
13	3.14	2.76	2.56	2.43	2.35	2.28	2.23	2.20	2.16
14	3.10	2.73	2.52	2.39	2.31	2.24	2.19	2.15	2.12
15	3.07	2.70	2.49	2.36	2.27	2.21	2.16	2.12	2.09
16	3.05	2.67	2.46	2.33	2.24	2.18	2.13	2.09	2.06
17	3.03	2.64	2.44	2.31	2.22	2.15	2.10	2.06	2.03
18	3.01	2.62	2.42	2.29	2.20	2.13	2.08	2.04	2.00
19	2.99	2.61	2.40	2.27	2.18	2.11	2.06	2.02	1.98
20	2.97	2.59	2.38	2.25	2.16	2.09	2.04	2.00	1.96
21	2.96	2.57	2.36	2.23	2.14	2.08	2.02	1.98	1.95
22	2.95	2.56	2.35	2.22	2.13	2.06	2.01	1.97	1.93
23	2.94	2.55	2.34	2.21	2.11	2.05	1.99	1.95	1.92
24	2.93	2.54	2.33	2.19	2.10	2.04	1.98	1.94	1.91
25	2.92	2.53	2.32	2.18	2.09	2.02	1.97	1.93	1.89
26	2.91	2.52	2.31	2.17	2.08	2.01	1.96	1.92	1.88
27	2.90	2.51	2.30	2.17	2.07	2.00	1.95	1.91	1.87
28	2.89	2.50	2.29	2.16	2.06	2.00	1.94	1.90	1.87
29	2.89	2.50	2.28	2.15	2.06	1.99	1.93	1.89	1.86
30	2.88	2.49	2.28	2.14	2.05	1.98	1.93	1.88	1.85
40	2.84	2.44	2.23	2.09	2.00	1.93	1.87	1.83	1.79
60	2.79	2.39	2.18	2.04	1.95	1.87	1.82	1.77	1.74
120	2.75	2.35	2.13	1.99	1.90	1.82	1.77	1.72	1.68
∞	2.71	2.30	2.08	1.94	1.85	1.77	1.72	1.67	1.63

Source: From M. Merrington and C. M. Thompson, "Tables of Percentage Points of the Inverted Beta (*F*)-Distribution," *Biometrika*, 1943, 33, 73–88. Reproduced by permission of the *Biometrika* Trustees and Oxford University Press.

Table B.6a Continued

ν_2 \ ν_1	NUMERATOR DEGREES OF FREEDOM									
DENOMINATOR DEGREES OF FREEDOM	**10**	**12**	**15**	**20**	**24**	**30**	**40**	**60**	**120**	∞
1	60.19	60.71	61.22	61.74	62.00	62.26	62.53	62.79	63.06	63.33
2	9.39	9.41	9.42	9.44	9.45	9.46	9.47	9.47	9.48	9.49
3	5.23	5.22	5.20	5.18	5.18	5.17	5.16	5.15	5.14	5.13
4	3.92	3.90	3.87	3.84	3.83	3.82	3.80	3.79	3.78	3.76
5	3.30	3.27	3.24	3.21	3.19	3.17	3.16	3.14	3.12	3.10
6	2.94	2.90	2.87	2.84	2.82	2.80	2.78	2.76	2.74	2.72
7	2.70	2.67	2.63	2.59	2.58	2.56	2.54	2.51	2.49	2.47
8	2.54	2.50	2.46	2.42	2.40	2.38	2.36	2.34	2.32	2.29
9	2.42	2.38	2.34	2.30	2.28	2.25	2.23	2.21	2.18	2.16
10	2.32	2.28	2.24	2.20	2.18	2.16	2.13	2.11	2.08	2.06
11	2.25	2.21	2.17	2.12	2.10	2.08	2.05	2.03	2.00	1.97
12	2.19	2.15	2.10	2.06	2.04	2.01	1.99	1.96	1.93	1.90
13	2.14	2.10	2.05	2.01	1.98	1.96	1.93	1.90	1.88	1.85
14	2.10	2.05	2.01	1.96	1.94	1.91	1.89	1.86	1.83	1.80
15	2.06	2.02	1.97	1.92	1.90	1.87	1.85	1.82	1.79	1.76
16	2.03	1.99	1.94	1.89	1.87	1.84	1.81	1.78	1.75	1.72
17	2.00	1.96	1.91	1.86	1.84	1.81	1.78	1.75	1.72	1.69
18	1.98	1.93	1.89	1.84	1.81	1.78	1.75	1.72	1.69	1.66
19	1.96	1.91	1.86	1.81	1.79	1.76	1.73	1.70	1.67	1.63
20	1.94	1.89	1.84	1.79	1.77	1.74	1.71	1.68	1.64	1.61
21	1.92	1.87	1.83	1.78	1.75	1.72	1.69	1.66	1.62	1.59
22	1.90	1.86	1.81	1.76	1.73	1.70	1.67	1.64	1.60	1.57
23	1.89	1.84	1.80	1.74	1.72	1.69	1.66	1.62	1.59	1.55
24	1.88	1.83	1.78	1.73	1.70	1.67	1.64	1.61	1.57	1.53
25	1.87	1.82	1.77	1.72	1.69	1.66	1.63	1.59	1.56	1.52
26	1.86	1.81	1.76	1.71	1.68	1.65	1.61	1.58	1.54	1.50
27	1.85	1.80	1.75	1.70	1.67	1.64	1.60	1.57	1.53	1.49
28	1.84	1.79	1.74	1.69	1.66	1.63	1.59	1.56	1.52	1.48
29	1.83	1.78	1.73	1.68	1.65	1.62	1.58	1.55	1.51	1.47
30	1.82	1.77	1.72	1.67	1.64	1.61	1.57	1.54	1.50	1.46
40	1.76	1.71	1.66	1.61	1.57	1.54	1.51	1.47	1.42	1.38
60	1.71	1.66	1.60	1.54	1.51	1.48	1.44	1.40	1.35	1.29
120	1.65	1.60	1.55	1.48	1.45	1.41	1.37	1.32	1.26	1.19
∞	1.60	1.55	1.49	1.42	1.38	1.34	1.30	1.24	1.17	1.00

Table B.6b Critical Values of the F Statistic, $\alpha = .05$

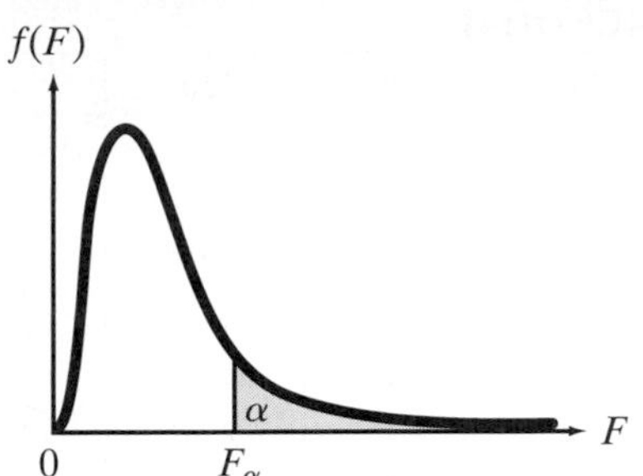

ν_1 / ν_2	NUMERATOR DEGREES OF FREEDOM								
DENOMINATOR DEGREES OF FREEDOM	**1**	**2**	**3**	**4**	**5**	**6**	**7**	**8**	**9**
1	161.4	199.5	215.7	224.6	230.2	234.0	236.8	238.9	240.5
2	18.51	19.00	19.16	19.25	19.30	19.33	19.35	19.37	19.38
3	10.13	9.55	9.28	9.12	9.01	8.94	8.89	8.85	8.81
4	7.71	6.94	6.59	6.39	6.26	6.16	6.09	6.04	6.00
5	6.61	5.79	5.41	5.19	5.05	4.95	4.88	4.82	4.77
6	5.99	5.14	4.76	4.53	4.39	4.28	4.21	4.15	4.10
7	5.59	4.74	4.35	4.12	3.97	3.87	3.79	3.73	3.68
8	5.32	4.46	4.07	3.84	3.69	3.58	3.50	3.44	3.39
9	5.12	4.26	3.86	3.63	3.48	3.37	3.29	3.23	3.18
10	4.96	4.10	3.71	3.48	3.33	3.22	3.14	3.07	3.02
11	4.84	3.98	3.59	3.36	3.20	3.09	3.01	2.95	2.90
12	4.75	3.89	3.49	3.26	3.11	3.00	2.91	2.85	2.80
13	4.67	3.81	3.41	3.18	3.03	2.92	2.83	2.77	2.71
14	4.60	3.74	3.34	3.11	2.96	2.85	2.76	2.70	2.65
15	4.54	3.68	3.29	3.06	2.90	2.79	2.71	2.64	2.59
16	4.49	3.63	3.24	3.01	2.85	2.74	2.66	2.59	2.54
17	4.45	3.59	3.20	2.96	2.81	2.70	2.61	2.55	2.49
18	4.41	3.55	3.16	2.93	2.77	2.66	2.58	2.51	2.46
19	4.38	3.52	3.13	2.90	2.74	2.63	2.54	2.48	2.42
20	4.35	3.49	3.10	2.87	2.71	2.60	2.51	2.45	2.39
21	4.32	3.47	3.07	2.84	2.68	2.57	2.49	2.42	2.37
22	4.30	3.44	3.05	2.82	2.66	2.55	2.46	2.40	2.34
23	4.28	3.42	3.03	2.80	2.64	2.53	2.44	2.37	2.32
24	4.26	3.40	3.01	2.78	2.62	2.51	2.42	2.36	2.30
25	4.24	3.39	2.99	2.76	2.60	2.49	2.40	2.34	2.28
26	4.23	3.37	2.98	2.74	2.59	2.47	2.39	2.32	2.77
27	4.21	3.35	2.96	2.73	2.57	2.46	2.37	2.31	2.25
28	4.20	3.34	2.95	2.71	2.56	2.45	2.36	2.29	2.24
29	4.18	3.33	2.93	2.70	2.55	2.43	2.35	2.28	2.22
30	4.17	3.32	2.92	2.69	2.53	2.42	2.33	2.27	2.21
40	4.08	3.23	2.84	2.61	2.45	2.34	2.25	2.18	2.12
60	4.00	3.15	2.76	2.53	2.37	2.25	2.17	2.10	2.04
120	3.92	3.07	2.68	2.45	2.29	2.17	2.09	2.02	1.96
∞	3.84	3.00	2.60	2.37	2.21	2.10	2.01	1.94	1.88

Source: From M. Merrington and C. M. Thompson, "Tables of Percentage Points of the Inverted Beta (F)-Distribution." *Biometrika*, 1943, 33, 73–88. Reproduced by permission of the *Biometrika* Trustees and Oxford University Press.

Table B.6b Continued

ν_2 \ ν_1	NUMERATOR DEGREES OF FREEDOM									
DENOMINATOR DEGREES OF FREEDOM	**10**	**12**	**15**	**20**	**24**	**30**	**40**	**60**	**120**	∞
1	241.9	243.9	245.9	248.0	249.1	250.1	251.1	252.2	253.3	254.3
2	19.40	19.41	19.43	19.45	19.45	19.46	19.47	19.48	19.49	19.50
3	8.79	8.74	8.70	8.66	8.64	8.62	8.59	8.57	8.55	8.53
4	5.96	5.91	5.86	5.80	5.77	5.75	5.72	5.69	5.66	5.63
5	4.74	4.68	4.62	4.56	4.53	4.50	4.46	4.43	4.40	4.36
6	4.06	4.00	3.94	3.87	3.84	3.81	3.77	3.74	3.70	3.67
7	3.64	3.57	3.51	3.44	3.41	3.38	3.34	3.30	3.27	3.23
8	3.35	3.28	3.22	3.15	3.12	3.08	3.04	3.01	2.97	2.93
9	3.14	3.07	3.01	2.94	2.90	2.86	2.83	2.79	2.75	2.71
10	2.98	2.91	2.85	2.77	2.74	2.70	2.66	2.62	2.58	2.54
11	2.85	2.79	2.72	2.65	2.61	2.57	2.53	2.49	2.45	2.40
12	2.75	2.69	2.62	2.54	2.51	2.47	2.43	2.38	2.34	2.30
13	2.67	2.60	2.53	2.46	2.42	2.38	2.34	2.30	2.25	2.21
14	2.60	2.53	2.46	2.39	2.35	2.31	2.27	2.22	2.18	2.13
15	2.54	2.48	2.40	2.33	2.29	2.25	2.20	2.16	2.11	2.07
16	2.49	2.42	2.35	2.28	2.24	2.19	2.15	2.11	2.06	2.01
17	2.45	2.38	2.31	2.23	2.19	2.15	2.10	2.06	2.01	1.96
18	2.41	2.34	2.27	2.19	2.15	2.11	2.06	2.02	1.97	1.92
19	2.38	2.31	2.23	2.16	2.11	2.07	2.03	1.98	1.93	1.88
20	2.35	2.28	2.20	2.12	2.08	2.04	1.99	1.95	1.90	1.84
21	2.32	2.25	2.18	2.10	2.05	2.01	1.96	1.92	1.87	1.81
22	2.30	2.23	2.15	2.07	2.03	1.98	1.94	1.89	1.84	1.78
23	2.27	2.20	2.13	2.05	2.01	1.96	1.91	1.86	1.81	1.76
24	2.25	2.18	2.11	2.03	1.98	1.94	1.89	1.84	1.79	1.73
25	2.24	2.16	2.09	2.01	1.96	1.92	1.87	1.82	1.77	1.71
26	2.22	2.15	2.07	1.99	1.95	1.90	1.85	1.80	1.75	1.69
27	2.20	2.13	2.06	1.97	1.93	1.88	1.84	1.79	1.73	1.67
28	2.19	2.12	2.04	1.96	1.91	1.87	1.82	1.77	1.71	1.65
29	2.18	2.10	2.03	1.94	1.90	1.85	1.81	1.75	1.70	1.64
30	2.16	2.09	2.01	1.93	1.89	1.84	1.79	1.74	1.68	1.62
40	2.08	2.00	1.92	1.84	1.79	1.74	1.69	1.64	1.58	1.51
60	1.99	1.92	1.84	1.75	1.70	1.65	1.59	1.53	1.47	1.39
120	1.91	1.83	1.75	1.66	1.61	1.55	1.50	1.43	1.35	1.25
∞	1.83	1.75	1.67	1.57	1.52	1.46	1.39	1.32	1.22	1.00

Table B.6c Critical Values of the *F* Statistic, $\alpha = .025$

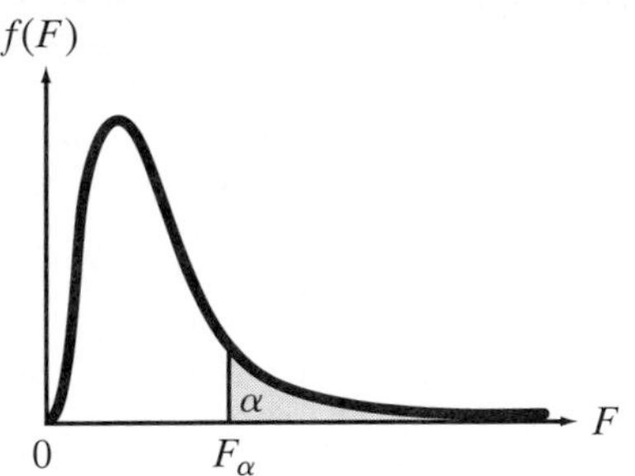

ν_2 \ ν_1	NUMERATOR DEGREES OF FREEDOM 1	2	3	4	5	6	7	8	9
DENOMINATOR DEGREES OF FREEDOM 1	647.8	799.5	864.2	899.6	921.8	937.1	948.2	956.7	963.3
2	38.51	39.00	39.17	39.25	39.30	39.33	39.36	39.37	39.39
3	17.44	16.04	15.44	15.10	14.88	14.73	14.62	14.54	14.47
4	12.22	10.65	9.98	9.60	9.36	9.20	9.07	8.98	8.90
5	10.01	8.43	7.76	7.39	7.15	6.98	6.85	6.76	6.68
6	8.81	7.26	6.60	6.23	5.99	5.82	5.70	5.60	5.52
7	8.07	6.54	5.89	5.52	5.29	5.12	4.99	4.90	4.82
8	7.57	6.06	5.42	5.05	4.82	4.65	4.53	4.43	4.36
9	7.21	5.71	5.08	4.72	4.48	4.32	4.20	4.10	4.03
10	6.94	5.46	4.83	4.47	4.24	4.07	3.95	3.85	3.78
11	6.72	5.26	4.63	4.28	4.04	3.88	3.76	3.66	3.59
12	6.55	5.10	4.47	4.12	3.89	3.73	3.61	3.51	3.44
13	6.41	4.97	4.35	4.00	3.77	3.60	3.48	3.39	3.31
14	6.30	4.86	4.24	3.89	3.66	3.50	3.38	3.29	3.21
15	6.20	4.77	4.15	3.80	3.58	3.41	3.29	3.20	3.12
16	6.12	4.69	4.08	3.73	3.50	3.34	3.22	3.12	3.05
17	6.04	4.62	4.01	3.66	3.44	3.28	3.16	3.06	2.98
18	5.98	4.56	3.95	3.61	3.38	3.22	3.10	3.01	2.93
19	5.92	4.51	3.90	3.56	3.33	3.17	3.05	2.96	2.88
20	5.87	4.46	3.86	3.51	3.29	3.13	3.01	2.91	2.84
21	5.83	4.42	3.82	3.48	3.25	3.09	2.97	2.87	2.80
22	5.79	4.38	3.78	3.44	3.22	3.05	2.93	2.84	2.76
23	5.75	4.35	3.75	3.41	3.18	3.02	2.90	2.81	2.73
24	5.72	4.32	3.72	3.38	3.15	2.99	2.87	2.78	2.70
25	5.69	4.29	3.69	3.35	3.13	2.97	2.85	2.75	2.68
26	5.66	4.27	3.67	3.33	3.10	2.94	2.82	2.73	2.65
27	5.63	4.24	3.65	3.31	3.08	2.92	2.80	2.71	2.63
28	5.61	4.22	3.63	3.29	3.06	2.90	2.78	2.69	2.61
29	5.59	4.20	3.61	3.27	3.04	2.88	2.76	2.67	2.59
30	5.57	4.18	3.59	3.25	3.03	2.87	2.75	2.65	2.57
40	5.42	4.05	3.46	3.13	2.90	2.74	2.62	2.53	2.45
60	5.29	3.93	3.34	3.01	2.79	2.63	2.51	2.41	2.33
120	5.15	3.80	3.23	2.89	2.67	2.52	2.39	2.30	2.22
∞	5.02	3.69	3.12	2.79	2.57	2.41	2.29	2.19	2.11

Source: From M. Merrington and C. M. Thompson, "Tables of Percentage Points of the Inverted Beta (*F*)-Distribution," *Biometrika*, 1943, 33, 73–88. Reproduced by permission of the *Biometrika* Trustees and Oxford University Press.

Table B.6c Continued

ν_2 \ ν_1	NUMERATOR DEGREES OF FREEDOM									
DENOMINATOR DEGREES OF FREEDOM	**10**	**12**	**15**	**20**	**24**	**30**	**40**	**60**	**120**	∞
1	968.6	976.7	984.9	993.1	997.2	1,001	1,006	1,010	1,014	1,018
2	39.40	39.41	39.43	39.45	39.46	39.46	39.47	39.48	39.49	39.50
3	14.42	14.34	14.25	14.17	14.12	14.08	14.04	13.99	13.95	13.90
4	8.84	8.75	8.66	8.56	8.51	8.46	8.41	8.36	8.31	8.26
5	6.62	6.52	6.43	6.33	6.28	6.23	6.18	6.12	6.07	6.02
6	5.46	5.37	5.27	5.17	5.12	5.07	5.01	4.96	4.90	4.85
7	4.76	4.67	4.57	4.47	4.42	4.36	4.31	4.25	4.20	4.14
8	4.30	4.20	4.10	4.00	3.95	3.89	3.84	3.78	3.73	3.67
9	3.96	3.87	3.77	3.67	3.61	3.56	3.51	3.45	3.39	3.33
10	3.72	3.62	3.52	3.42	3.37	3.31	3.26	3.20	3.14	3.08
11	3.53	3.43	3.33	3.23	3.17	3.12	3.06	3.00	2.94	2.88
12	3.37	3.28	3.18	3.07	3.02	2.96	2.91	2.85	2.79	2.72
13	3.25	3.15	3.05	2.95	2.89	2.84	2.78	2.72	2.66	2.60
14	3.15	3.05	2.95	2.84	2.79	2.73	2.67	2.61	2.55	2.49
15	3.06	2.96	2.86	2.76	2.70	2.64	2.59	2.52	2.46	2.40
16	2.99	2.89	2.79	2.68	2.63	2.57	2.51	2.45	2.38	2.32
17	2.92	2.82	2.72	2.62	2.56	2.50	2.44	2.38	2.32	2.25
18	2.87	2.77	2.67	2.56	2.50	2.44	2.38	2.32	2.26	2.19
19	2.82	2.72	2.62	2.51	2.45	2.39	2.33	2.27	2.20	2.13
20	2.77	2.68	2.57	2.46	2.41	2.35	2.29	2.22	2.16	2.09
21	2.73	2.64	2.53	2.42	2.37	2.31	2.25	2.18	2.11	2.04
22	2.70	2.60	2.50	2.39	2.33	2.27	2.21	2.14	2.08	2.00
23	2.67	2.57	2.47	2.36	2.30	2.24	2.18	2.11	2.04	1.97
24	2.64	2.54	2.44	2.33	2.27	2.21	2.15	2.08	2.01	1.94
25	2.61	2.51	2.41	2.30	2.24	2.18	2.12	2.05	1.98	1.91
26	2.59	2.49	2.39	2.28	2.22	2.16	2.09	2.03	1.95	1.88
27	2.57	2.47	2.36	2.25	2.19	2.13	2.07	2.00	1.93	1.85
28	2.55	2.45	2.34	2.23	2.17	2.11	2.05	1.98	1.91	1.83
29	2.53	2.43	2.32	2.21	2.15	2.09	2.03	1.96	1.89	1.81
30	2.51	2.41	2.31	2.20	2.14	2.07	2.01	1.94	1.87	1.79
40	2.39	2.29	2.18	2.07	2.01	1.94	1.88	1.80	1.72	1.64
60	2.27	2.17	2.06	1.94	1.88	1.82	1.74	1.67	1.58	1.48
120	2.16	2.05	1.94	1.82	1.76	1.69	1.61	1.53	1.43	1.31
∞	2.05	1.94	1.83	1.71	1.64	1.57	1.48	1.39	1.27	1.00

Table B.6d Critical Values of the F Statistic, $\alpha = .01$

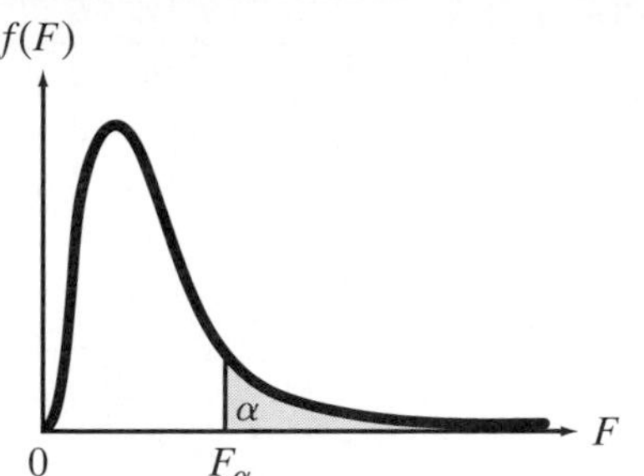

ν_2 \ ν_1	NUMERATOR DEGREES OF FREEDOM								
DENOMINATOR DEGREES OF FREEDOM	1	2	3	4	5	6	7	8	9
1	4,052	4,999.5	5,403	5,625	5,764	5,859	5,928	5,982	6,022
2	98.50	99.00	99.17	99.25	99.30	99.33	99.36	99.37	99.39
3	34.12	30.82	29.46	28.71	28.24	27.91	27.67	27.49	27.35
4	21.20	18.00	16.69	15.98	15.52	15.21	14.98	14.80	14.66
5	16.26	13.27	12.06	11.39	10.97	10.67	10.46	10.29	10.16
6	13.75	10.92	9.78	9.15	8.75	8.47	8.26	8.10	7.98
7	12.25	9.55	8.45	7.85	7.46	7.19	6.99	6.84	6.72
8	11.26	8.65	7.59	7.01	6.63	6.37	6.18	6.03	5.91
9	10.56	8.02	6.99	6.42	6.06	5.80	5.61	5.47	5.35
10	10.04	7.56	6.55	5.99	5.64	5.39	5.20	5.06	4.94
11	9.65	7.21	6.22	5.67	5.32	5.07	4.89	4.74	4.63
12	9.33	6.93	5.95	5.41	5.06	4.82	4.64	4.50	4.39
13	9.07	6.70	5.74	5.21	4.86	4.62	4.44	4.30	4.19
14	8.86	6.51	5.56	5.04	4.69	4.46	4.28	4.14	4.03
15	8.68	6.36	5.42	4.89	4.56	4.32	4.14	4.00	3.89
16	8.53	6.23	5.29	4.77	4.44	4.20	4.03	3.89	3.78
17	8.40	6.11	5.18	4.67	4.34	4.10	3.93	3.79	3.68
18	8.29	6.01	5.09	4.58	4.25	4.01	3.84	3.71	3.60
19	8.18	5.93	5.01	4.50	4.17	3.94	3.77	3.63	3.52
20	8.10	5.85	4.94	4.43	4.10	3.87	3.70	3.56	3.46
21	8.02	5.78	4.87	4.37	4.04	3.81	3.64	3.51	3.40
22	7.95	5.72	4.82	4.31	3.99	3.76	3.59	3.45	3.35
23	7.88	5.66	4.76	4.26	3.94	3.71	3.54	3.41	3.30
24	7.82	5.61	4.72	4.22	3.90	3.67	3.50	3.36	3.26
25	7.77	5.57	4.68	4.18	3.85	3.63	3.46	3.32	3.22
26	7.72	5.53	4.64	4.14	3.82	3.59	3.42	3.29	3.18
27	7.68	5.49	4.60	4.11	3.78	3.56	3.39	3.26	3.15
28	7.64	5.45	4.57	4.07	3.75	3.53	3.36	3.23	3.12
29	7.60	5.42	4.54	4.04	3.73	3.50	3.33	3.20	3.09
30	7.56	5.39	4.51	4.02	3.70	3.47	3.30	3.17	3.07
40	7.31	5.18	4.31	3.83	3.51	3.29	3.12	2.99	2.89
60	7.08	4.98	4.13	3.65	3.34	3.12	2.95	2.82	2.72
120	6.85	4.79	3.95	3.48	3.17	2.96	2.79	2.66	2.56
∞	6.63	4.61	3.78	3.32	3.02	2.80	2.64	2.51	2.41

Source: From M. Merrington and C. M. Thompson, "Tables of Percentage Points of the Inverted Beta (F)-Distribution," *Biometrika,* 1943, 33, 73–88. Reproduced by permission of the *Biometrika* Trustees and Oxford University Press.

Table B.6d Continued

ν_2 \ ν_1	NUMERATOR DEGREES OF FREEDOM									
	10	12	15	20	24	30	40	60	120	∞
1	6,056	6,106	6,157	6,209	6,235	6,261	6,287	6,313	6,339	6,366
2	99.40	99.42	99.43	99.45	99.46	99.47	99.47	99.48	99.49	99.50
3	27.23	27.05	26.87	26.69	26.60	26.50	26.41	26.32	26.22	26.13
4	14.55	14.37	14.20	14.02	13.93	13.84	13.75	13.65	13.56	13.46
5	10.05	9.89	9.72	9.55	9.47	9.38	9.29	9.20	9.11	9.02
6	7.87	7.72	7.56	7.40	7.31	7.23	7.14	7.06	6.97	6.88
7	6.62	6.47	6.31	6.16	6.07	5.99	5.91	5.82	5.74	5.65
8	5.81	5.67	5.52	5.36	5.28	5.20	5.12	5.03	4.95	4.86
9	5.26	5.11	4.96	4.81	4.73	4.65	4.57	4.48	4.40	4.31
10	4.85	4.71	4.56	4.41	4.33	4.25	4.17	4.08	4.00	3.91
11	4.54	4.40	4.25	4.10	4.02	3.94	3.86	3.78	3.69	3.60
12	4.30	4.16	4.01	3.86	3.78	3.70	3.62	3.54	3.45	3.36
13	4.10	3.96	3.82	3.66	3.59	3.51	3.43	3.34	3.25	3.17
14	3.94	3.80	3.66	3.51	3.43	3.35	3.27	3.18	3.09	3.00
15	3.80	3.67	3.52	3.37	3.29	3.21	3.13	3.05	2.96	2.87
16	3.69	3.55	3.41	3.26	3.18	3.10	3.02	2.93	2.84	2.75
17	3.59	3.46	3.31	3.16	3.08	3.00	2.92	2.83	2.75	2.65
18	3.51	3.37	3.23	3.08	3.00	2.92	2.84	2.75	2.66	2.57
19	3.43	3.30	3.15	3.00	2.92	2.84	2.76	2.67	2.58	2.49
20	3.37	3.23	3.09	2.94	2.86	2.78	2.69	2.61	2.52	2.42
21	3.31	3.17	3.03	2.88	2.80	2.72	2.64	2.55	2.46	2.36
22	3.26	3.12	2.98	2.83	2.75	2.67	2.58	2.50	2.40	2.31
23	3.21	3.07	2.93	2.78	2.70	2.62	2.54	2.45	2.35	2.26
24	3.17	3.03	2.89	2.74	2.66	2.58	2.49	2.40	2.31	2.21
25	3.13	2.99	2.85	2.70	2.62	2.54	2.45	2.36	2.27	2.17
26	3.09	2.96	2.81	2.66	2.58	2.50	2.42	2.33	2.23	2.13
27	3.06	2.93	2.78	2.63	2.55	2.47	2.38	2.29	2.20	2.10
28	3.03	2.90	2.75	2.60	2.52	2.44	2.35	2.26	2.17	2.06
29	3.00	2.87	2.73	2.57	2.49	2.41	2.33	2.23	2.14	2.03
30	2.98	2.84	2.70	2.55	2.47	2.39	2.30	2.21	2.11	2.01
40	2.80	2.66	2.52	2.37	2.29	2.20	2.11	2.02	1.92	1.80
60	2.63	2.50	2.35	2.20	2.12	2.03	1.94	1.84	1.73	1.60
120	2.47	2.34	2.19	2.03	1.95	1.86	1.76	1.66	1.53	1.38
∞	2.32	2.18	2.04	1.88	1.79	1.70	1.59	1.47	1.32	1.00

(Row label: DENOMINATOR DEGREES OF FREEDOM)

Table B.7 Critical Values for the Wilcoxon Rank Sum Test

Test statistic is the rank sum associated with the smaller sample (if equal sample sizes, either rank sum can be used).

a. α = .025 one-tailed; α = .05 two-tailed

n_2 \ n_1	3		4		5		6		7		8		9		10	
	T_L	T_U	T_L	T_U	T_L	T_U	T_L	T_U	T_L	T_U	T_L	T_U	T_L	T_U	T_L	T_U
3	5	16	6	18	6	21	7	23	7	26	8	28	8	31	9	33
4	6	18	11	25	12	28	12	32	13	35	14	38	15	41	16	44
5	6	21	12	28	18	37	19	41	20	45	21	49	22	53	24	56
6	7	23	12	32	19	41	26	52	28	56	29	61	31	65	32	70
7	7	26	13	35	20	45	28	56	37	68	39	73	41	78	43	83
8	8	28	14	38	21	49	29	61	39	73	49	87	51	93	54	98
9	8	31	15	41	22	53	31	65	41	78	51	93	63	108	66	114
10	9	33	16	44	24	56	32	70	43	83	54	98	66	114	79	131

b. α = .05 one-tailed; α = .10 two-tailed

n_2 \ n_1	3		4		5		6		7		8		9		10	
	T_L	T_U	T_L	T_U	T_L	T_U	T_L	T_U	T_L	T_U	T_L	T_U	T_L	T_U	T_L	T_U
3	6	15	7	17	7	20	8	22	9	24	9	27	10	29	11	31
4	7	17	12	24	13	27	14	30	15	33	16	36	17	39	18	42
5	7	20	13	27	19	36	20	40	22	43	24	46	25	50	26	54
6	8	22	14	30	20	40	28	50	30	54	32	58	33	63	35	67
7	9	24	15	33	22	43	30	54	39	66	41	71	43	76	46	80
8	9	27	16	36	24	46	32	58	41	71	52	84	54	90	57	95
9	10	29	17	39	25	50	33	63	43	76	54	90	66	105	69	111
10	11	31	18	42	26	54	35	67	46	80	57	95	69	111	83	127

Source: From F. Wilcoxon and R. A. Wilcox, "Some Rapid Approximate Statistical Procedures," 1964, 20–23. Courtesy of Lederle Laboratories Division of American Cyanamid Company, Madison, NJ.

Table B.8 Critical Values for the Wilcoxon Signed Ranks Test

n	One-Tailed: $\alpha = .05$ Two-Tailed: $\alpha = .10$	$\alpha = .025$ $\alpha = .05$	$\alpha = .01$ $\alpha = .02$	$\alpha = .005$ $\alpha = .01$
	(Lower, Upper)			
5	0,15	—,—	—,—	—,—
6	2,19	0,21	—,—	—,—
7	3,25	2,26	0,28	—,—
8	5,31	3,33	1,35	0,36
9	8,37	5,40	3,42	1,44
10	10,45	8,47	5,50	3,52
11	13,53	10,56	7,59	5,61
12	17,61	13,65	10,68	7,71
13	21,70	17,74	12,79	10,81
14	25,80	21,84	16,89	13,92
15	30,90	25,95	19,101	16,104
16	35,101	29,107	23,113	19,117
17	41,112	34,119	27,126	23,130
18	47,124	40,131	32,139	27,144
19	53,137	46,144	37,153	32,158
20	60,150	52,158	43,167	37,173

Source: Adapted from Table 2 of F. Wilcoxon and R. A. Wilcox, *Some Rapid Approximate Statistical Procedures* (Pearl River, NY: Lederle Laboratories, 1964), with permission of the American Cyanamid Company.

Table B.9 Critical Values of Spearman's Rank Correlation Coefficient

The α values correspond to a one-tailed test of H_0: $\rho = 0$. The value should be doubled for two-tailed tests.

n	$\alpha = .05$	$\alpha = .025$	$\alpha = .01$	$\alpha = .005$	n	$\alpha = .05$	$\alpha = .025$	$\alpha = .01$	$\alpha = .005$
5	.900	—	—	—	**18**	.399	.476	.564	.625
6	.829	.886	.943	—	**19**	.388	.462	.549	.608
7	.714	.786	.893	—	**20**	.377	.450	.534	.591
8	.643	.738	.833	.881	**21**	.368	.438	.521	.576
9	.600	.683	.783	.833	**22**	.359	.428	.508	.562
10	.564	.648	.745	.794	**23**	.351	.418	.496	.549
11	.523	.623	.736	.818	**24**	.343	.409	.485	.537
12	.497	.591	.703	.780	**25**	.336	.400	.475	.526
13	.475	.566	.673	.745	**26**	.329	.392	.465	.515
14	.457	.545	.646	.716	**27**	.323	.385	.456	.505
15	.441	.525	.623	.689	**28**	.317	.377	.448	.496
16	.425	.507	.601	.666	**29**	.311	.370	.440	.487
17	.412	.490	.582	.645	**30**	.305	.364	.432	.478

Source: From E. G. Olds, "Distribution of Sums of Squares of Rank Differences for Small Samples," *Annals of Mathematical Statistics,* 1938, 9. Reproduced with the permission of the Institute of Mathematical Statistics.

Appendix C

Documentation for CD-ROM Data Files

C.1 CD-ROM Overview

The CD-ROM that is packaged with this textbook contains software resources that support your learning of statistics. The CD-ROM contains the following folders, or directories:

PHStat2: Contains the PHStat2 setup program and related files. This setup program must be run to properly set up the PHStat2 statistical software that is used throughout the text. (See Appendix E for more information about setting up and using PHStat2.)

Excel: Contains the Microsoft Excel workbook data files (with the extension .XLS) used in the textbook. A detailed list of the files found in this folder appears on pp. 754–759. A searchable version of this list also appears in the File Finder workbook file (File Finder.xls) located in this folder.

MINITAB: Contains the MINITAB worksheet files (with the extension .MTW) used in the textbook. A detailed list of the files found in this folder appears on pp. 754–759.

> You should also review the CD-ROM readme file in the root folder, if present, for any updated information about the contents of the CD-ROM.

Most likely, you will be using files in the Excel or MINITAB folder on a regular basis. You can retrieve those files directly from the CD-ROM or copy them first to a hard disk folder. If you plan to copy the files to a hard disk folder, first make sure that your hard disk has 10MB of free space. When you copy the files, the files will likely have read-only status. While this status is not a problem in Microsoft Excel—read-only files can always be saved under a different name by selecting the File | Save As command—you can change the status of the files to read-write by doing the following:

1. Open Windows Explorer to the hard disk folder containing the files.
2. Select the files to make read-write. (Select Edit | Select All to select every file in list.)
3. Select Files | Properties and in the Properties dialog box that appears, deselect (uncheck) the Read-only attribute check box and click the OK button.

(Due to the nature of the CD-ROM medium, CD-ROM files are always read-only and cannot be changed to read-write status.)

C.2 Excel and MINITAB Folder File Contents

The following is an alphabetical listing and description of all data files found the Excel and MINITAB folders. The CD icons that appear throughout the text identify these files. When you see this icon in the text, you should normally retrieve the similarly named file in the Excel or MINITAB directory.

File Name	Description of Variables
AAABONDS	Expense ratio rank and net return rank (Chapter 10)
ALLOY	Annealing time and passivation potential (Chapter 10)
AMMONIA	Daily ammonia concentrations (Chapter 3)
APPRTRAIN	Training and performance score (Chapter 11)
ATLAS	Theme, high school teacher ranking, geography alumni ranking (Chapter 9)
ATTENT	Attention times (Chapter 7)
AUDITOR	Years of experience and salary (Chapter 10)
AUTO	Car model, gas mileage, engine horsepower, and weight (Chapter 10)
AUTO2000	Car model, gas mileage, fuel tank capacity, length, wheelbase, width, turning circle requirement, weight, luggage capacity, front shoulder room, front leg room, front head room, rear shoulder room, rear leg room, rear head room (Chapters 3, 9)
BACKPACKS	Price of backpack, volume, number of books (Chapter 10)
BANKCOST	Bounced check fees (Chapters 2, 3)
BB2000	League (American = 0, National = 1), wins, ERA, runs scored, hits allowed, walks allowed, saves, and errors (Chapter 10)
BEERCAL	Beer brand and number of calories per 12-ounce serving (Chapters 2, 3, 7)
BIDRIG	Low-bid price, DOT estimate, low-bid/DOT-estimate ratio, bid status, district, number of bidders, number of days to complete project, road length, percent of cost allocated to asphalt, subcontractor use (Chapters 3, 10)
BILE	Belief in lunar effects scale scores (Chapter 2)
BLACKSOX	Member of Black Sox (0 = No, 1 = Yes), Chicago White Sox player number, game percentage, slugging average (Chapter 10)
BLCKJACK	Profit and blackjack strategy (Chapter 11)
BOEING720	Plane and life length of air-conditioning system (Chapter 9)
BOILER	Man-hours, boiler capacity, design pressure, boiler type, drum type (Chapter 10)
BOND	Length and whether the metal compound is copper-phosphorus (Chapters 2, 3, 9)
BONES	Length-to-width ratios of human bones (Chapter 8)
BONFERRONI	Excel workbook for obtaining Bonferroni multiple comparisons (Chapter 11)
BOSNIA	Order-to-delivery times (Chapter 8)
BOY16	Number of destructive responses for 16-year-old boys (Chapter 3)
BRAIN	Veteran, verbal memory retention, right hippocampal volume (Chapters 3, 10)

CANS	Amount of fill in ounces (Chapter 8)
CAVITIES	Child and cavities for sealant-coated teeth and untreated teeth (Chapter 9)
CEODEGREES	Highest degree obtained (Chapter 2)
CELLPHONE	Case number, hazard interval, control interval (Chapter 8)
CHAPTER5-E(X)	Excel worksheet for expected value (Chapter 5)
CHARITY	Tax exempt charitable (Chapter 7)
CHKOUT	Supermarket checkout times in seconds (Chapters 1, 2, 6)
COLOGNE	Men's cologne brand and intensity rating (Chapter 2)
COMMIT	Employee number, age, gender, organizational tenure, job tenure, continuance commitment scale, affective commitment scale, moral commitment scale, quitting scale, search scale, intent to leave scale, leave status (Chapter 5)
CO2	Year, carbon dioxide level (Chapter 2)
CONICOTINE	Nicotine level and carbon monoxide level (Chapter 10)
COPEPOD	Percent of shallow-living copepods in diet of myctophid fish (Chapters 3, 8)
CORRTEST	Excel workbook for a test for a correlation (Chapter 10)
CRACK	Crack size and presence/absence of flaw in specimens (Chapter 2)
CRASH	Class, make, car model, number of doors, weight, driver star rating, passenger star rating, driver air bag status, passenger air bag status, driver head injury rating, passenger head injury rating, driver chest injury rating, passenger chest injury rating, driver left femur injury rating, driver right femur injury rating, passenger left femur injury rating, passenger right femur injury rating (Chapters 2, 3, 6)
CREATIVITY	Child, flexibility score, creativity score (Chapter 10)
CROPWT	Crop weights (Chapter 3)
DARTS	Distance from target and frequency (Chapter 6)
DAYSABSENT	Number of days absent (Chapter 6)
DBHL	Signal intensity score (Chapter 6)
DEAF	Hearing status, visual acuity score (Chapter 9)
DECAY	Decay rate of particles (Chapter 7)
DECEPT	Student, deception score, percent increase in pupil size (Chapter 10)
DECODE	Group and number of trials required for decoding (Chapter 11)
DENTAL	Discomfort level of novocaine and new anesthetic (Chapter 9)
DESTRUCT	Number of destructive responses (Chapter 7)
DIAZINON	Day, level of diazinon residue during day, level of diazinon residue at night (Chapter 9)

DISCRIM	Hiring status and gender (Chapter 9)
DRILLROCK	Drilling depth and time to drill 5 feet (Chapters 2, 3, 10)
EERAC	Energy efficiency ratings (Chapter 8)
ENTRAP	Points awarded in entrapment study (Chapter 3)
EVOS	Transect number, number of seabirds found, length, and oil spill status (Chapters 2, 9, 10)
FACES	Dominance degree, facial expression (Chapter 11)
FACTORS	Number of medical factors and length of hospital stay (Chapter 10)
FASTFOOD	City and weekly sales at stores (Chapter 3)
FATHER	Attitude score towards father, gender (Chapter 9)
FATHMOTH	Student, attitude toward father, attitude toward mother (Chapter 9)
FERTIL	Country, contraceptive prevalence, fertility rate (Chapters 3, 10)
FIRE	Deaths from hotel fires (Chapters 3, 7)
FIREFRAUD	Month of sale, invoice number, selling price, profit, profit margin (Chapter 6)
FISH	Location, fish species, length, weight, amount of DDT (Chapters 1, 2, 3, 6, 7, 11)
FORBES400	Billionaire, age, net worth (millions of dollars) (Chapter 10)
FOREST	Breast-height diameters of trembling aspen trees (Chapter 6)
FOREST1	Breast-height diameter and height of white spruce trees (Chapter 10)
FTC	Cigarette brand, length, menthol, filter, light, pack, tar content, nicotine content, carbon monoxide level (Chapters 3, 6, 7, 10)
GASKET	Machine number, material stamped, number of gaskets (Chapter 11)
GEM	Retail price per carat, demand for gem (Chapter 10)
GEO	Group number, geography score (Chapters 2, 3, 9)
GEOSCORE	Student, score on test 1, score on test 2 (Chapter 9)
GOLD	Gold production levels (Chapter 3)
GPACLASS	Socioeconomic class, grade point average (Chapter 11)
GPS	Location, distance error, percent of readings with 3 satellites (Chapter 10)
HAIRPAIN	Hair color, pain threshold score (Chapter 11)
HEAT	Unflooded area ratio, heat transfer ratio (Chapter 10)
HELIUM	Proportion of impurity passing through helium, temperature (Chapter 10)
HPLC	Specimen number, percent label strength added, percent label strength found, recovery percentage (Chapter 2)
ICECREAM	Number of calories, percentage of fat per half-cup serving (Chapters 2, 3)

IDTWIN	Twin A IQ, twin B IQ (Chapter 2)
INTERRUPT	Interruption number, speaker gender, interrupter gender (Chapter 2)
IODINE	Iodine concentration measurements (Chapter 7)
IOUNITS	Input-output units utilized (Chapter 6)
JAPANSPORTS	Sport, frequency of play, duration of participation (Chapter 10)
KIDNEY	Rat number, weight, size, diet (Chapter 11)
KNEESURG	Knee surgery type, age group, number of post-surgical hospitalization days (Chapter 11)
LANDSALE	Pair, land sale ratio, downward price adjustment (Chapter 10)
LIQCO2	Amounts of liquid carbon monoxide (Chapter 7)
LONGJUMP	Jumper, best jumping distance, average takeoff error (Chapter 10)
MANHRS	Man-hours required (Chapter 3)
MATHCPU	Time to solve programming problems (Chapter 7)
MONRENTS	Monthly rents for 2-bedroom apartments (Chapter 3)
MOSQUITO	Date, average temperature, mosquito catch ratio (Chapter 10)
MOUSING PRACTICE	Excel workbook for practicing mouse operations (Chapter 1)
MUCK	Depth of muck (Chapter 11)
MUM	Subject visibility, time to deliver bad news, test-taker success and failure (Chapter 11)
MULTINOMIAL TEST	Excel workbook for multinomial test (Chapter 8)
NBA	Team, standardized wins, standardized difference between team and opponent 2-point field goal percentage, standardized difference between team and opponent rebounds, standardized difference between team and opponent turnovers (Chapter 10)
NEMATODE	Dry weight of nematode, month (Chapter 11)
NOSHOW	Number of hotel cancellations (Chapter 3)
OILSPILL	Spillage amount, cause of puncture (Chapter 8)
ORING	Space flight number, temperature, O-ring damage index (Chapters 2, 3, 10)
OTRATIO	Day, condition, thion amount, oxon amount, oxon/thion ratio (Chapter 9)
OXONTHION	Day, weather condition, oxon/thion ratio (Chapter 9)
PCB	Levels of PCB (Chapter 8)
PCBWATER	Year 1 PCB amount, year 2 PCB amount, name of bay (Chapter 10)
PH	Food and pH values (Chapter 8)
PHOSROCK	Location and time for chemical reaction (Chapter 11)
PHYSICS	Student, inventory score, baseline score (Chapter 10)
PIPES	Breaking strength (Chapter 8)
PMI	Postmortem intervals (Chapter 2)

POWVEP	POW number, VEP measurement 157 days after release, VEP measurement 379 days after release (Chapter 9)
PROB11_30	Factor A, factor B, value (Chapter 11)
PROB11_55	Group and value (Chapter 11)
PROTEIN	Food, calories, protein, fat, saturated fat, cholesterol (Chapters 2, 3)
QUIZ	Scores on introductory statistics quiz (Chapter 3)
RADIUM	Radium-226 levels in soil samples (Chapter 8)
RAIN	Station number, location, direction facing, average annual precipitation, altitude, latitude, distance from coast, dummy variable for direction facing (Chapter 10)
RAS	Gender attitude, RAS score (Chapter 9)
REACTIME	Driver, restrained reaction time, unrestrained reaction time (Chapter 9)
READING	Method 1 reading test score, method 2 reading test score (Chapter 9)
REFRIGERATOR	Refrigerator brand, price, energy cost (Chapters 2, 3, 10)
RESTRATE	Zagat ratings for food, decor, service, price per person, city location (Chapter 3)
ROACH	Average trail deviations, group, trail (Chapter 11)
ROACH2	(Only for those with fecal extract trail) Deviate, group, deviate for female, gravid, male, and nymph (Chapter 11)
ROOMBOARD	College, room cost, board cost (Chapter 8)
RUBIDIUM	Trace amounts of rubidium (Chapter 6)
SAL50	Starting salaries (Chapter 2)
SAMPLE50	Sample number, 5 measurements (Chapter 6)
SATSCORES	Student, junior year SAT score, senior year SAT score (Chapter 9)
SCALLOPS	Scallop weights (Chapter 7)
SEARCHENG	Internet search engine, match index for top 20 hits, match index for top 100 hits (Chapter 9)
SHIPSANIT	Cruise ship, sanitation inspection score (Chapters 2, 3, 7)
SILICA	Silica determinations (parts per million) (Chapter 7)
SMALLGRP	Group, grade (Chapter 6)
SMOKE	Weight percentile, number of cigarettes smoked in home (Chapter 10)
SMOKING	Brand, CO, Nicotine (Chapter 2)
SOCIOLOGY	Student, verbal ability test score, final sociology grade (Chapters 2, 3)
SORPRATE	Sorption rate, hazardous organic solvent (Chapter 11)
SPL	Sound pressure levels (Chapter 3)
SUICIDE	Days in jail before committing suicide for all inmates (Chapter 3)

SUICIDE14	Days in jail before committing suicide for 14 murderers (Chapter 7)
SWIMMAZE	Litter, number of swims for males, number of swims for females (Chapter 9)
TAMPALMS	Appraised land value, appraised improvements, total appraised value, sale price, sales-to-appraisal ratio (Chapter 10)
TAXRATE	Software program, satisfaction score (Chapters 3, 10)
TCDD	TCDD level in plasma (Chapters 2, 3, 7)
TCDDFAT	Veteran, TCDD level in plasma, TCDD level in fat (Chapters 3, 9)
TENNISWED	Tennis player, rank on wedding day, rank on first anniversary (Chapter 11)
THRUPUT	Task number, throughput rate for human scheduler, throughput rate for automated system (Chapter 9)
TIRES	Tire pressure, mileage (Chapter 10)
TISSUES	Number of tissues during a cold, below/above 60 status (Chapter 8)
TRANSFORM	Hours, inspection level, failures per 1,000 pieces (Chapter 11)
TRIPLETS	Shear strength, pre-compression stress (Chapter 10)
TWINS	Pair number, twin A IQ score, twin B IQ score (Chapter 9)
VANADIUM	Tissue type, vanadium trace amount (Chapter 11)
VCR	Service center, VCR brand, repair time (Chapter 11)
VIDEOS	Movie, box office gross, home video units sold (Chapter 10)
VOLTAGE	Location, voltage reading (Chapters 3, 6, 9, 11)
WAREHOUS	Number of vehicles, congestion time (Chapter 2)
WATERTASK	Deviation of judged line from true line (Chapters 2, 3, 9)
WHALES	Whale part, AUI ranking, MUI percent (Chapter 10)
WOMENPOWER	Name, age, company, title, group (Chapter 11)

Appendix D

Microsoft Excel Configuration and Customization

Use this appendix to adjust the settings of your copy of Excel to match the settings used in this text. If you share your copy of Excel with other users, be aware that the instructions in this appendix will affect what those users see when they use your copy of Excel.

D.1 Configuring the Excel Application Window

All of the textbook illustrations of the Excel application window were produced using the same configuration. To configure your copy of the Excel application so that it matches the look of illustrations such as Figure EP.3 on p. 2, first, adjust the display of the various bars on the window:

a. Select **View** and if the Formula Bar choice is not checked, select it.

b. Select **View** and if the Status Bar choice is not checked, select it.

c. Select **View | Toolbars** and if the Standard choice is not checked, select it.

d. Select **View | Toolbars** and if the Formatting choice is not checked, select it.

If any bar is not aligned to the top of the Excel application window, drag the bar toward the top of the screen until the border of the bar changes to a slender band of small dots. Then release the mouse button. The bar will then snap into place.

Once these bars are properly displayed, configure the worksheet area by selecting **Tools | Options.** In the Options dialog box that appears, select the View tab and select all of the Windows options check boxes *except* Page breaks and Formulas and then click the OK button. Figure D.1 shows a properly configured View tab for a copy of Excel 2000 (other versions will differ slightly).

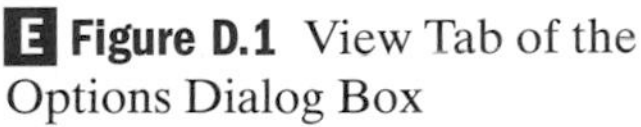

E Figure D.1 View Tab of the Options Dialog Box

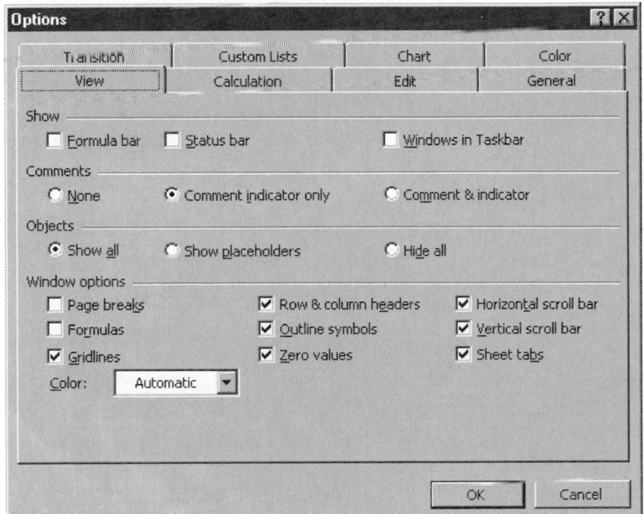

D.2 Special Configuration Notes for Excel 2000 Users

In addition to the instructions of section D.1, if you are using Excel 2000, you will need to further configure the display of the Excel menu bar and toolbars. To do

so, select **Tools | Customize.** In the Customize dialog box, deselect (uncheck) the Standard and Formatting toolbars share one row and the Menus show recently used commands first check boxes and click the Close button.

If you are using Excel 2000 version SR-1 or later, you also will need to configure your security settings if you plan on using PHStat. To configure these settings for PHStat, select **Tools | Macro | Security** and in the Security Level tab of the Security dialog box select the Medium option button and click the OK button. (If you have a version of Excel earlier than 2000 SR-1, your Tools | Macro submenu will not contain the Security choice and you should ignore the instructions in this paragraph.)

D.3 Configuring Excel General Settings

The Excel-related descriptions and procedures of this text assume a particular set of (standard) configuration settings. To confirm that your copy of Excel is similarly configured, do the following:

To verify calculation, edit, and general display settings, select **Tools | Options** and in the Options dialog box that appears:

a. Select the Calculation tab and verify that the Automatic option button of the Calculation group has been selected.

b. Select the Edit tab and verify that all check boxes *except* the Fixed decimal, Provide feedback with animation, and Enable automatic percent entry have been selected. (Note: The Automatic percent entry check box is not present in Excel 97.)

c. Select the General tab. Verify that the R1C1 reference style check box is deselected (unchecked) and that the Macro virus protection check box (Excel 97 only) is selected. Also enter 3 in the Sheets in new workbook edit box, select Arial (or similar font) from the Standard font list box, select 10 from the Size drop-down list box, and change the values in the Default file location and User name edit boxes, if desired.

d. Click the OK button.

To verify the installation of the Excel Data Analysis ToolPak add-ins, select **Tools** and verify that Data Analysis is one of the choices on this menu. If Data Analysis does not appear as a choice on the Tools menu, select **Tools | Add-Ins** from the menu and in the Add-Ins dialog box that appears (see Figure D.2), select the Analysis ToolPak and Analysis ToolPak - VBA check boxes from the Add-Ins available list and click the OK button.

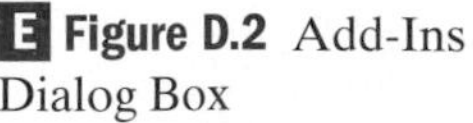
Figure D.2 Add-Ins Dialog Box

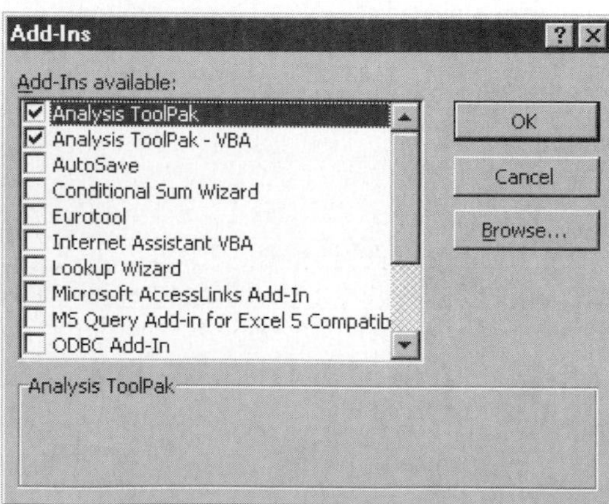

If you cannot find either the Analysis ToolPak or Analysis ToolPak - VBA check box in the Add-Ins available list, these Excel add-in files were not included when your copy of Excel was installed using the Microsoft Excel (or Office) setup program. You will need to rerun the Excel setup program and add these components of Excel before continuing. (See the Microsoft Excel help system or the Add-In Installation Notes file in the Additional Information folder of the companion CD for further information.)

To verify the installation of the latest version of the custom CD program PHStat, select **PHStat | About PHStat Add-in** and note the version number in the About dialog box. You can then visit the Prentice Hall Web site for PHStat, located through www.prenhall.com/phstat, and verify that you have the latest version of PHStat installed. If you have a version less than the most current version, you will be able to download a free update from the Web site.

If PHStat does not appear on your Excel menu bar, PHStat either was not loaded when Excel was opened or was never installed with the PHStat setup program that is found on the companion CD. If you are certain that PHStat has been installed, you can manually **File | Open** the PHStat.xla file from the folder to which it was installed. If you have never run the PHStat setup program, you will need to do so before you can use PHStat. See the PHStat Readme file on the companion CD for complete instructions.

Running the PHStat setup program requires that you have write access to the Windows Registry. Users in a networked environment, especially users in a computing lab, will not normally have such access and will need to have a network administrator run the setup program. For more information, including all of the technical requirements and program issues, see the PHStat Readme file in the Additional Information folder on the companion CD.

D.4 Customizing Excel Printouts

Excel allows you to customize your printouts by changing settings in the Page Setup dialog box (see Figure D.3). To change settings, select **File | Page Setup** from the

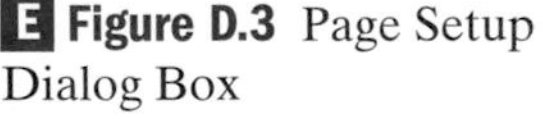
Figure D.3 Page Setup Dialog Box

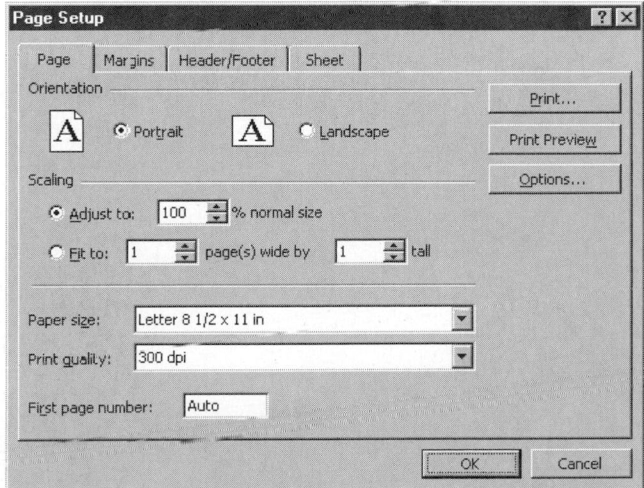

Excel menu bar or click the Setup button in the Print Preview screen. Some useful settings from each of the four tabs of this dialog box are presented below:

Page tab. Select the Landscape orientation option button to fit more columns on a single printed sheet or if your worksheet has a greater width than length.

Header/Footer tab. You can choose a predefined header or footer from the Header and Footer drop-down lists or click the Custom Header or Custom Footer buttons to define your format.

Sheet tab. Select the Gridlines and Row and Column headings check boxes of the Print group to produce a printout that will most look like the worksheet as displayed onscreen.

Margins tab. Change the values in the Left, Right, Top, and Bottom edit boxes to adjust your printout as needed.

Remember to click the OK button of the dialog box to have Excel accept your settings.

Appendix E

More about PHStat2

PHStat2 is Windows software that assists you in learning statistics by allowing you to perform many types of statistical analyses in Excel. You use PHStat2 by first selecting commands from the PHStat menu and then entering the values and/or cell ranges necessary to perform a particular analysis. PHStat2 generates new worksheets and chart sheets to hold the results of an analysis. Many of these sheets contain formulas and are interactive, meaning that you can change the underlying data and see new results without having to rerun a procedure. Although the use of PHStat2 is optional with this text, but for a few analyses that would be too cumbersome or too complex to construct manually in Excel, you are encouraged to use PHStat2, at least initially, as it will allow you to avoid the distractions of operating Excel.

E.1 Setting Up PHStat2

In order to use PHStat2, you first must set up the software on your computer system by running the PHStat2 setup program. Setting up PHStat2 copies program files to your hard disk; updates Windows system files, if necessary; and creates entries in the Windows Registry for use by PHStat2. If you are a student using a computer system in a campus lab, or a faculty member using a computer that is part of a campus network, your user account may not allow you to set up any Windows software including PHStat2. If this is the case, ask your network or lab technician for assistance in setting up PHStat2.

To begin setting up PHStat2, first review Appendix D and configure your copy of Excel as necessary. Then run the PHStat2 setup program (**Setup.exe**) that can be found in the PHStat2 folder on the CD-ROM that accompanies this textbook. The setup program begins with an opening dialog box that greets you and invites you to continue. Step through the dialog boxes, reading all informational messages and clicking the Next button. During this time, you will have the opportunity to specify a target directory for the PHStat2 files other than the default\Program Files\PHStat2 directory in a dialog box similar to the one shown in Figure E.1 on p. 766.

After the setup program completes successfully, desktop icons and a Start menu program group are both created for PHStat2. With the exception of certain older systems that may require a reboot after the setup program updates certain Windows system files, PHStat2 is ready to be used.

E.2 Using PHStat2

You use PHStat2 by double-clicking the Desktop PHStat2 icon or selecting PHStat2 from the Start menu PHStat2 program group. If you prefer, you can also open the **PHStat2.xla** file, containing the component of PHStat2 that works inside Excel, in the Excel (File) Open dialog box (see the Excel Primer). Once properly loaded, PHStat2 will add a new PHStat menu to the Excel menu bar. You are then ready to begin work with PHStat2.

The **PHStat2.xla** component is technically an Excel *add-in*, and some users prefer to have Excel automatically load their add-ins when Excel itself opens. Appendix D discusses instructions for making sure that two add-ins that are supplied

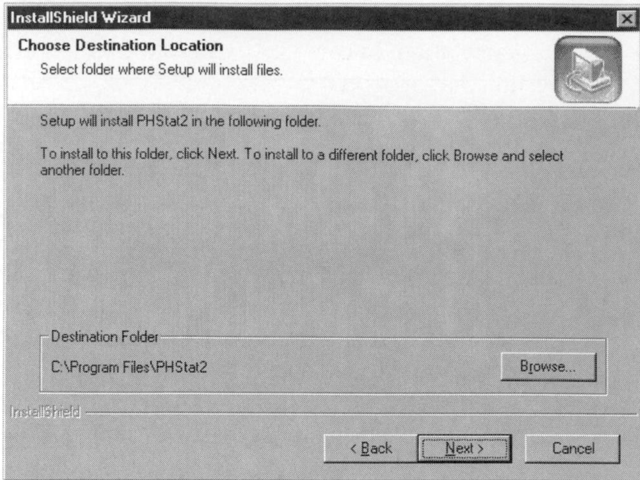

E Figure E.1 Choose Destination Location Dialog Box

with Excel and are necessary for the operation of PHStat2 are always automatically loaded. These instructions can be adapted by you if you are an experienced Excel user and would like to "install," as Excel calls it, the PHStat2 add-in. However, if you install the PHStat2 add-in in this manner it will need to be uninstalled before any future update is applied to your system (see the section Updating PHStat2 below).

E.3 Preparing Data for Analysis by PHStat2

Many PHStat2 procedures require you to supply worksheet data. For such procedures, prepare your data for analysis by PHStat2 by placing it in columns. If you are selecting only part of a column, copy that partial column range to a new column or worksheet. This good practice will allow you to easily verify your data later as well as avoid some common types of user errors.

Due to the limitations of Excel, select the worksheet containing the data to be analyzed before selecting the PHStat2 procedure that will use the data. If you forget to do this, you may encounter an error message that refers to an unexpected error or an error in handling a cell range, although some PHStat2 procedures are more forgiving of this than others. For procedures that require two or more cell ranges, such as all of the regression procedures, make sure that all your cell ranges are from the same worksheet.

E.4 Updating PHStat2

From time to time, enhanced versions of PHStat2 that add new capabilities or clarify user issues may be published on the Prentice Hall Web site. If you plan to use PHStat2, you should regularly check the PHStat Web site at www.prenhall.com/phstat for updates. Such updates will be freely downloadable by owners of this textbook who possess the original CD-ROM packaged with the textbook. Instructions for applying future updates and other general information about PHStat2 will also be found on the PHStat Web site.

E.5 What PHStat2 Is Not

PHStat2 is not designed as a commercial statistical package and should not be used as a replacement for packages such as SAS, SPSS, or MINITAB. Because PHStat2 emphasizes learning, PHStat2 tries to use the methods of calculation that

a learner could follow and would find presented in a textbook that discusses manual calculation. In other cases, PHStat2 uses pre-existing Excel methods to produce results that are too complex to manually calculate or present to the learner. Both of these choices will occasionally cause PHStat2 and Excel to produce results not as precise as those produced by commercial packages that use methods of calculation fine-tuned to the limitations of computing technology. While these differences are not significant for most sets of data, real-world data sets with unusual numerical properties or with extreme values could produce significant anomalies when used with PHStat2 and Excel. If you are a student, use PHStat2 only with the data discussed in your text or required by your instructor. For such sets of data, the results produced by PHStat2 will allow you to form the proper statistical conclusion about your data.

Answers to Self-Test Problems

Chapter 1

1.1 **a.** qualitative **b.** quantitative **c.** quantitative **d.** qualitative

1.2 **a.** set of all U.S. school children
b. 1,000 children that were included in the survey
c. number of cavities; quantitative

1.3 quantitative; quantitative; qualitative; observational study; population

1.5 selection bias error

Chapter 2

2.1 potential growth deficiency

2.2 **a.** 1, 2, 3, 4, and 5
c. .15, .35, .35, .10, .05

2.3 **a.** 20
b. .2667

2.4 **a.** 22% of the males sampled indicated that they experienced early puberty

Chapter 3

3.1 **a.** 12 **b.** 40 **c.** 7 **d.** 21

3.2 5.7, 4.5, 4

3.3 8, 7.556, 2.749

3.4 **a.** approximately 70% **b.** approximately 95%
c. approximately 100%

3.5 **a.** 80%, 20% **b.** 50%, 50% **c.** 14%, 86%
d. 75%, 25% **e.** 25%, 75% **f.** 90%, 10%

3.6 IQR = 10

3.7 **a.** −.667, not an outlier **b.** 6, outlier

3.8 −.9163

Chapter 4

4.1 **b.** one eighth of all families with three children have 3 boys

4.2 (First does not use coupons and second does not use coupons), (First does use coupons and second does use coupons)

4.3 $P(A) = 1/10, P(B) = 6/10, P(C) = 3/10$

4.4 9/10

4.5 120

4.6 1/3

4.7 1/36

4.8 **a.** .7 **b.** .75

Chapter 5

5.1 **a.** continuous **b.** discrete **c.** discrete **d.** continuous

5.2 **a.** .10 **b.** 1.3 **c.** 1.345

5.3 **a.** .0090 **b.** .001488 **c.** .00145

5.4 yes, $P(x < 14) = .0023861$

5.5 $\mu = 200, \sigma = 6.325$

5.6 **a.** .25102 **b.** .55783 **c.** 1.5

5.7 **a.** 4/7 **b.** 4/35

Chapter 6

6.1 **a.** .2734 **b.** .4873 **c.** .0735 **d.** .84

6.2 .0080

6.3 **a.** no **b.** yes **c.** no **d.** no

6.5 **a.** 5 **b.** .1

6.6 .2266

Chapter 7

7.2 485 ± 15.56

7.3 1.943

7.4 485 ± 150.21

7.5 **a.** .4 **b.** $.4 \pm .033$

7.6 68

7.7 26.1190

7.8 (64.86, 300.96)

Chapter 8

8.1 $H_0: p = .65, H_a: p < .65$

8.2 **a.** claim $\mu < 16$ when, in fact, $\mu = 16$
b. claim $\mu = 16$ when, in fact, $\mu < 16$
c. Type II
d. Type I

8.3 $z < -1.282$

8.4 **a.** average number of faxes transmitted
b. $H_0: \mu = 88{,}000, H_a: \mu > 88{,}000$

8.5 $z = -3.37$, reject H_0

8.6 $t = -1.07$, do not reject H_0

8.7 **a.** do not reject H_0
b. reject H_0 **c.** reject H_0

8.8 $z = -2.35$, reject H_0

8.9 $\chi^2 = 6.975$, do not reject H_0

8.10 50

Chapter 9

9.1 **a.** $\mu_1 - \mu_2$ **b.** σ_1^2/σ_2^2

9.2 **a.** 2.5 ± 3.47 **b.** $z = 1.41$, do not reject H_0

9.3 **a.** $t = .786$ **b.** $t > 1.706$

9.4 **a.** yes, $z = 6.51$ **b.** 115 ± 34.6

9.5 **a.** $z = -2.53$, reject H_0 **b.** $-.08 \pm .062$

9.6 $e_{11} = 18.75, e_{12} = 11.25, e_{21} = 81.25, e_{22} = 48.75$

9.7 **a.** $\chi^2 = .2735$ **b.** $\chi^2 > 6.63490$ **c.** do not reject H_0

9.8 $F > 1.91$

9.9 $F > 4.77$

9.10 **b.** $T_1 = 21, T_2 = 34$
c. $T_1 \leq 19, T_2 \geq 36$ **d.** do not reject H_0

9.11 **b.** 3 **c.** $T^+ \leq 0$ or $T^+ \geq 15$
d. do not reject H_0

Chapter 10

10.1 **a.** y = sales price; x_1 = square footage of living area, x_2 = assessed value, x_3 = number of rooms
b. $y = \beta_0 + \beta_1 x_1 + \beta_2 x_2 + \beta_3 x_3 + \varepsilon$

10.2 y-intercept $= -100$, slope $= 67$

10.3 **b.** Liquid: 102.5 grams; Solid: 120.6 grams

10.4 **a.** 6.12

10.5 $t = 1.18$, do not reject H_0

10.6 $.26 \pm .365$

10.7 $t = 3.55$

10.8 62% of total variation of annual salary values about their mean is explained by linear relationship with the years of experience.

10.9 **a.** \$32,000 **b.** \$30,000

10.10 **a.** 95% confidant that mean salary of all factory workers with 8 years of experience falls between \$27,500 and \$32,500
b. 95% confidant that salary of a factory worker with 8 years of experience falls between \$23,000, and \$37,000

10.11 $\hat{y} = .7 + 1.5x$

10.12 SSE $= .3$, $s = .3162$

10.13 $t = 15$

10.14 $1.5 \pm .3182$

10.15 .9868

10.16 **a.** $.7 \pm 1.147$ **b.** $3.7 \pm .551$

10.17 **a.** -23 **b.** yes

10.18 **a.** 5 **b.** 700

10.19 **a.** 1, 2, 3, 4, 5 **b.** 5, 4, 2, 3, 1
c. $-4, -2, 1, 1, 4$ **d.** $-.9$

Chapter 11

11.1 Response: distance traveled by golf ball; Factor: golf ball brand; Levels: the four brands tested; Treatment: the four golf ball brands; Experimental unit: a golf ball

11.2 **a.** H_0: $\mu_1 = \mu_2 = \mu_3 = \mu_4$ **b.** 43.99 **c.** reject H_0

11.3 **a.** 6 **b.** .01 **c.** 5.60 **d.** Brand C; Brands A and D

11.4 **a.** Response: distance traveled by golf ball; Factor 1: golf ball brand (levels A, B, C, D); Factor 2: club (levels driver, 5-iron); Treatments: $(A, D), (A, 5), (B, D), (B, 5), (C, D), (C, 5), (D, D), (D, 5)$

11.5 **a.** interaction test **b.** no **c.** yes; reject H_0: $\mu_D = \mu_5$; do not reject H_0: $\mu_A = \mu_B = \mu_C = \mu_D$

11.6 yes

11.7 **a.** 1: 1,411.2; 2: 45; 3: 500; 4: 744.2
b. 14.154 **c.** $H > 6.25139$ **d.** reject H_0

Answers to Selected Problems

Chapter 1

1.1 **a.** quantitative **b.** qualitative **c.** qualitative **d.** quantitative **e.** quantitative

1.3 **a.** individual property
b. leased fee value: quantitative; lot size: quantitative; neighborhood: qualitative; location of lot: qualitative

1.5 **a.** qualitative **b.** quantitative

1.7 **a.** quantitative **b.** quantitative **c.** qualitative **d.** qualitative **e.** qualitative **f.** quantitative **g.** qualitative **h.** quantitative

1.11 **a.** sample or population **b.** sample **c.** population **d.** sample

1.13 **a.** survey **b.** set of all Web users **c.** probably not

1.15 **a.** 1; type of question **b.** individual question **c.** qualitative **d.** sample

1.17 **a.** all light cigarette brands; all regular cigarette brands
b. data from the 25 light cigarette brands collected; data from the 25 regular cigarette brands collected

1.19 **a.** diameters of stones found in the delta region of the Amazon
b. diameters of the 50 stones collected by the geographer
c. extremely unlikely

1.21 observational

1.23 survey

1.25 **a.** population: all people; sample: 12 volunteers
b. designed experiment

1.27 **a.** 001 **b.** 020 **c.** 750

1.33 method B; method A

1.35 students probably not representative of business employees

1.37 nonresponse bias

1.43 **a.** quantitative **b.** quantitative **c.** quantitative **d.** quantitative **e.** quantitative **f.** qualitative **g.** qualitative **h.** qualitative

1.45 **a.** set of directions that all sea turtle hatchlings travel under the three conditions
b. set of directions that the 60 sea turtle hatchlings travel under the three conditions
c. qualitative
d. designed experiment

1.47 **a.** high school juniors who had not performed as well on the SAT as they should have
b. change in the SAT score
c. differences in SAT scores of all students taking a beta blocker
d. designed experiment
e. 22 SAT differences of the students administered the beta blocker
f. beta blockers are effective in increasing SAT scores

Chapter 2

2.1 **a.** .26%; .56%; .18%

2.7 **c.** 46%

2.9 **a.** artifact found; type of artifact found
c. drillpoints; 57%

2.11 **b.** warpage

2.15 **a.** 3, 4, and 5

2.17 **a.** 21, 22, 23, 24, 25, 26, 27, 28, 29, 30, 31, and 32

2.19 **a.** 10.5–15.5, 15.5–20.5, 20.5–25.5, 25.5–30.5, 30.5–35.5, 35.5–40.5
b. .05, .10, .20, .25, .20, .20

2.23 **c.** 113/121 = .9339

2.29 **b.** yes

2.31 **a.** (X, A), (X, B), (Y, A), (Y, B), (Z, A), and (Z, B)
b. 1; 3; 3; 2; 1; 2
c. A: .2, .6, .2; B: .4286, .2857, .2857
d. X: .25, .75; Y: .6, .4; Z: .333, .667

2.33 **a.** 400 **b.** 100 **c.** 25
d. A: .1053, .2105, .4211, .2632; B: .4000, .3200, .0800, .2000; C: .3529, .2353, .1176, .2941
e. North: .2, .5, .3; East: .4, .4, .2; South: .8, .1, .1; West: .5, .25, .25

2.35 **a.** nationality (British, German, or American); ad humorous (yes or no)
b. British: .2562, .7438; German: .2292, .7708; American: .2074, .7926

2.37 **a.** complication type (redo surgery, post-op infection, both, or none); drug (yes or no)
b. yes: .1228, .1228, .0526, .7018; no: .0877, .0702, .0175, .8246

2.41 **a.** positive linear relationship
b. no linear relationship
c. negative linear relationship

2.43 **b.** slight positive relationship

2.45 **b.** negative **c.** yes

2.47 **b.** positive linear **c.** yes

2.49 **b.** yes, strong positive

2.51 **a.** strong positive **b.** no

2.57 **a.** clear and concise **b.** contains chart junk

2.59 **a.** African black: .1914; African white: .6231; Asian Sumatran: .0294; Asian Javan: .0052; Asian Indian: .1509

c. .8145; .1855

2.61 **b.** .726

2.67 **b.** positively related

2.71 **a.** quantitative **b.** relative frequency histogram, 64%

2.73 **b.** skewed to the right

Chapter 3

3.1 **a.** 58 **b.** 690 **c.** 3,364 **d.** −7 **e.** 27

3.3 **a.** 17.2 **b.** 30 **c.** 680

3.5 **a.** 6 **b.** 50 **c.** 42.8

3.7 **a.** 6 **b.** 12.2 **c.** .25

3.9 mean = 4.6; median = 4

3.11 **a.** 5; 5; 5 **b.** 12; 5; 5 **c.** 0; 0; 0 **d.** 0; 4.5; 9

3.13 **a.** 9 **b.** 7.6 **c.** 8

3.15 **a.** 174.12 **b.** 160 **c.** 150 **d.** median

e. mean = 9.53%, median = 9%, mode = 9%; median

3.17 mean = 90.34, median = 92, mode = 93

3.19 mean = 4.246; median = 4.2; mode = 4.4

3.21 mean = 75.575; median = 75.15; no mode

3.23 **a.** mean = 68.86; median = 69

b. mean = 50.045; median = 53

c. group 1: approximately symmetric; group 2: skewed to the left

d. yes

3.25 **a.** 10, 14, 3.742 **b.** 10, 10.4, 3.225 **c.** 0, 0, 0

3.27 $\bar{x} = 4.64$, $s = 1.912$

3.29 $\bar{x} = 12.75$, $s = 3.51$

3.31 **a.** 15 **b.** 21.742 **c.** 4.663 **d.** 22.864 **e.** (13.54, 32.19)

3.33 **a.** 4.5 **b.** 1.317 **c.** (0.566, 5.834)

3.35 **a.** (−10.13, 16.67) **b.** (−8.441, 15.431) **c.** unoiled

3.37 **c.** red

3.39 **a.** $\bar{x} = 84.91$; $s^2 = 261.26$; $s = 16.16$

b. (52.59, 117.23)

c. 34/35 = .9714

3.41 **a.** $\bar{x} = 6.03$; $s^2 = 66.10$; $s = 8.13$

b. approximately 95%

c. 19/20 = .95 **d.** decrease; decrease

3.43 **a.** −1.6 **b.** .6 **c.** 1.6

3.45 **a.** $M = 6.1$, $Q_1 = 3.3$, $Q_3 = 8.2$ **b.** 1.2

3.47 **a.** 40% **b.** 5% **c.** 67% **d.** 85%

3.51 $Q_1 = 88$, $M = 92$, $Q_3 = 94$

3.53 **a.** $z = -.76$ **b.** $z = 1.87$

3.55 **a.** $Q_1 = 74$, $M = 92$, $Q_3 = 99$

b. 100 **c.** $z = -0.242$

3.57 **a.** 98 **b.** 122 **c.** 24 **d.** 75 **e.** 143 **g.** no skewness

3.59 $x_{\text{smallest}} = 213$, $Q_1 = 228$, $M = 266.5$, $Q_3 = 291$, $x_{\text{largest}} = 320$

3.61 **a.** .5 **b.** 7.5 **c.** −3.9 **d.** −2

3.63 **b.** skewed right

c. five-number summary: 1, 4, 15, 41.5, 309; five largest values are outliers

d. the largest value is an outlier; no

3.65 **a.** no skewness

b. yes, 4 nicotine values fall above 1.7

3.67 **a.** 2.8, 3.95, 5.65, 8.2; 82

b. yes, 82 in city C

c. $\bar{x} = 9.07$; $s = 15.69$; $z = 4.65$

d. $\bar{x} = 6.00$; $s = 2.28$; five-number summary: 2.8, 3.95, 5.65, 8.05, 10.6; no outliers

3.69 **a.** .20 **b.** −.70 **c.** .625 **d.** −1.00

3.71 **a.** .993 **b.** .174 **c.** −.989

3.73 .2509

3.75 positive

3.77 .7937

3.79 piano: moderate positive; bench: little or none; motor bike: moderate positive; arm chair: weak positive; teapot: strong positive

3.81 −.865

3.83 decreasing order of skewness: Avila, Tampa Palms, Carrolwood Village, Northdale, Ybor City, and Town & Country

3.87 **a.** calories: 193.4, 190, 98, 299, 4,799.5, 69.28, 157, 239, 82; protein: 26.52, 27, 27, 17, 14.09, 3.75, 24, 29, 5; fat: 36.56, 37, 37, 71, 407.4, 20.18, 18, 51, 33; saturated fat: 12.72, 14, several modes, 27, 70.71, 8.41, 0, 20, 15; cholesterol: 102.2, 89, 89, 443, 6,992.67, 83.62, 71, 94, 23

c. calories and cholesterol: skewed to the right; protein: skewed to the left; fat and saturated fat: symmetric

d. calories: (124.12, 262.68) 64%, (54.84, 331.96) 96%, (−14.44, 401.24) 100%; protein: (22.77, 30.27) 80%, (19.01, 34.03) 96%, (15.26, 37.78) 100%; fat: (16.38, 56.74) 60%, (−3.81, 76.93) 96%, (−23.99, 97.11) 100%; saturated fat: (4.31, 21.13) 60%, (−4.10, 29.54) 100%, (−12.51, 37.95) 100%; cholesterol: (18.58, 185.82) 96%, (−65.04, 269.44) 96%, (−148.67, 353.07) 96%

e. 482 mg for cholesterol

f. calorie, protein: .4644; calorie, fat: .8509, calorie, saturated fat: .8888; calorie, cholesterol: .1777; protein, fat: .4606; protein, saturated fat: .4984, protein, cholesterol: .1417; fat, saturated fat: .9299; fat, cholesterol: .1088; saturated fat, cholesterol: .1083

3.89 **a.** .04563 **b.** .53514 **c.** −.74174

3.91 **a.** ants: (−1.9, 7.7); no ants: (−1.2, 16.4) **b.** approximately 95%

3.93 **a.** mean = 117.82, median = 117.5, modes = 97, 112, 124, 128, 131

b. range = 62, variance = 225.3343, standard deviation = 15.01114

c. (102.809, 132.831), 31, .62; (87.798, 147.842), 49, .98; (72.787, 162.853), 50, 1.00

3.95 **a.** $\bar{x} = 29.45$, $s = 9.33$

b. 7, 23.25, 30, 35, 57

c. (20.12–38.78), 28, .70; (10.79–48.11), 38, .95; (1.46–57.44): 40, 1.00

d. 18

f. 57 inches

Chapter 4

4.1 **a.** 1/6

4.3 **a.** .01

b. yes

4.5 500

4.7 **a.** $P(A) = .60$, $P(B) = .60$, $P(C) = .75$

b. $A' = \{S_3, S_5\}$, $B' = \{S_1, S_4\}$, $C' = \{S_4\}$

c. $P(A') = .40$, $P(B') = .40$, $P(C') = .25$

4.9 **a.** 1/4 **b.** 1/4 **c.** 3/4 **d.** 3/4

4.13 3/650 = .0046154

4.15 **a.** Huggies, Pampers, Private label, Luvs, and Other brands

b. .317, .265, .210, .142, .066 **c.** .407 **d.** .683

4.17 **a.** .1739 **b.** 0 **c.** Yes **d.** .6522

4.19 **a.** .2685 **b.** .1128 **c.** .8794

4.21 **a.** .5694 **b.** .3873 **c.** .0433 **d.** .6127 **e.** .6127

4.23 **a.** .1914 **b.** .1855 **c.** .8145

4.25 **a.** 32 **b.** .25

4.27 **a.** 1 **b.** 50 **c.** 50 **d.** 1

4.29 **a.** 10 **b.** $ABC, ABD, ABE, ACD, ACE, ADE, BCD, BCE, BDE, CDE$

4.31 28

4.33 1,820

4.35 **a.** 2,598,960 **b.** 4 **c.** .0000015

4.37 .18

4.39 **a.** no **b.** no **c.** no

4.41 **a.** .40, .60 **b.** .16

4.43 **a.** .0057 **b.** .0007 **c.** .0010 **d.** .017925 **e.** .006435

4.45 **a.** $P(S) = .34$, $P(I) = .32$, $P(S|I) = .58$, $P(S|I') = .23$

b. no

4.47 **b.** .0009766 **c.** .9980469

4.49 **a.** .2283 **b.** .1685 **c.** .3478 **d.** .3333 **e.** .1197

4.51 .6

4.53 .06

4.55 **a.** 6/36 **b.** 11/36 **c.** 2/36 **d.** 15/36

4.57 .5111

4.59 **b.** .147 **c.** .85 **d.** .65 **e.** yes

4.61 **a.** .82 **b.** .8827 **c.** .7980 **d.** .0220

e. .9047 **f.** .2020 **g.** .9041

4.63 **a.** .53 **b.** .05 **c.** .89 **d.** .41 **e.** .0205 **f.** B and C

4.65 **b.** .035

4.67 **a.** .275 **b.** .170 **c.** .3818 **d.** no

4.69 **a.** .625

4.71 **a.** P(win Publishers Clearing House) = 1/181,795,000, P(win American Family Publishers) = 1/200,000,000, P(win Reader's Digest) = 1/84,000,000

b. 3.27×10^{-25} **c.** .9999999776 **d.** .0000000224

4.73 **a.** all are mutually exclusive

4.75 **a.** .46 **b.** .68 **c.** .20

4.77 **a.** 142,506 **b.** .000007

4.79 **a.** .0612 **b.** .1400 **c.** .0065 **d.** .0902 **e.** .0397 **f.** .9671

4.81 **a.** .0256 **b.** .4182 **c.** independence

4.83 635,013,559,600

4.85 **a.** .4927 **b.** .5073 **c.** .1169

4.87 **b.** no, the man has a better chance of having two boys than does the woman

Chapter 5

5.1 **a.** −5, 0, 2, 5 **b.** 2 **c.** .5 **d.** .2 **f.** $\mu = .3$, $\sigma = 3.0017$

5.3 **a.** $p(1) = .8$, $p(2) = .16$, $p(3) = .032$, $p(4) = .0064$, $p(5) = .00128$

b. no, $P(x > 5) = .00032$

c. .96

5.5 **b.** .24 **c.** .39 **d.** 1.80

5.7 **b.** .1875 **c.** .5875 **d.** 12.425 **e.** 2.235
f. (7.955, 16.895) **g.** .95

5.9 $p(0) = .09, p(1) = .42, p(2) = .49$

5.11 **a.** $p(0) = .0909, p(1) = .48485, p(2) = .42424$
b. $\mu = 1.333, \sigma = .6356$

5.13 **a.** 24 **b.** 4 **c.** 10 **d.** .064 **e.** .2304

5.15 no

5.17 **a.** .8192 **b.** .1808 **c.** complements **d.** sum to 1

5.19 **a.** .9983223 **b.** .9877054 **c.** .0016777

5.21 **a.** $\mu = 297, \sigma = 1.723$
b. $\mu = 160, \sigma = 5.657$
c. $\mu = 50, \sigma = 5$
d. $\mu = 20, \sigma = 4$
e. $\mu = 10, \sigma = 3.1464$

5.23 **a.** yes **b.** .03 **c.** .1328 **d.** .1413

5.25 **b.** .40 **c.** $\mu = 4.00, \sigma = 1.549$ **d.** .9536

5.27 **a.** 14.586 **b.** 3.75 **c.** $z = -3.353$

5.29 **a.** $p(0) = .125, p(1) = .375, p(2) = .375, p(3) = .125$
b. .875 **c.** no

5.31 no

5.33 **a.** 24
b. support

5.35 **a.** .001245 **b.** .≈ .015 **c.** 41.6 **d.** 3.816 **e.** effective

5.37 **a.** .082085, .2052, .2565, .2138, .1336, .0668, .0278, .0099, .0031, .00086, .00022
b. $\mu = 2.5, \sigma = 1.58$ **c.** .958

5.39 **a.** .981 **b.** .8571 **c.** .5367 **d.** .265 **e.** .05915

5.41 **a.** .368 **b.** .264 **c.** .920

5.43 **a.** .205212 **b.** .543813 **c.** no

5.45 **a.** .349084 **b.** .030282 **c.** $\sigma = 4.243$; (9.514, 26.486)

5.47 **a.** 350/792 **b.** 246/792 **c.** 21/792
d. 196/792 **e.** 2.92 **f.** .88

5.49 **a.** 0 **b.** 21/252 **c.** 1/3 **d.** 1/6

5.51 **a.** .0000000387 **b.** .00001115 **c.** .000655124
d. .01339365 **e.** .475139722
f. .0000002605, .00005315, .0021923, .0311798, .3503832

5.53 **a.** .3 **b.** 1

5.55 **a.** 17.4 **b.** 3.176 **c.** $z = -3.906$

5.57 **a.** $\mu = 15.14, \sigma = 1.114$
b. (12.912, 17.368)
c. .29

5.59 **a.** .5 **b.** .25 **c.** .125 **d.** .0625

5.61 **a.** 38,000 **b.** 184.93 **c.** $z = -16.22$

5.63 .291021672

5.65 no

5.69 **a.** .0282475 **b.** .9984096 **c.** .0001437

Chapter 6

6.1 **a.** .3849 **b.** .4319 **c.** .1844 **d.** .4147 **e.** .0918

6.3 **a.** .25 **b.** .92 **c.** 1.28 **d.** 1.65 **e.** 1.96

6.5 **a.** .75 **b.** −1 **c.** −1.625 **d.** 2 **e.** −2

6.7 **a.** −1.28 **b.** −1.04 **c.** −.52 **d.** 0

6.9 **a.** .0735 **b.** .3651 **c.** 7.29

6.11 **a.** .2033 **b.** no **c.** 0 **d.** no **e.** 23.24 meters

6.13 **a.** 9% **b.** 11.01% **c.** 13.92%

6.15 **a.** .0228 **b.** .9544 **c.** 38.58 seconds

6.17 **a.** .2621 **b.** .0143

6.19 **a.** .9406 **b.** .0068

6.21 **a.** .6826 **b.** .9544 **c.** .9974

6.23 c

6.25 sentence complexity scores cannot be negative, distribution is not symmetric

6.27 no

6.29 immature group is approximately normal; mature group is not normal

6.31 no

6.33 **a.** no **b.** yes **c.** no **d.** yes **e.** yes **f.** yes

6.35 **a.** .2500, .2483 **b.** .0065, .0060 **c.** .8925, .8904

6.37 **a.** .9922 **b.** 0 **c.** no

6.39 .2451

6.41 **a.** $p(1)=.04, p(1.5)=.12, p(2)=.17, p(2.5)=.20, p(3)=.20, p(3.5)=.14, p(4)=.08, p(4.5)=.04, p(5)=.01$
c. .05

6.43 **c.** $\bar{x} = 4.74, s = 1.275$

6.47 **d.** $\mu_{\bar{x}} = 70, \sigma_{\bar{x}} = 4.96$

6.51 **a.** $\mu_{\bar{x}} = 60, \sigma_{\bar{x}} = 1$
b. $\mu_{\bar{x}} = 60, \sigma_{\bar{x}} = .63$
c. $\mu_{\bar{x}} = 60, \sigma_{\bar{x}} = .45$
d. $\mu_{\bar{x}} = 60, \sigma_{\bar{x}} = .37$
e. $\mu_{\bar{x}} = 60, \sigma_{\bar{x}} = .32$
f. $\mu_{\bar{x}} = 60, \sigma_{\bar{x}} = .14$
g. $\mu_{\bar{x}} = 60, \sigma_{\bar{x}} = .10$

6.53 **a.** .9932 **b.** .2061 **c.** .0159

6.55 **a.** $\mu_{\bar{x}} = \mu, \sigma_{\bar{x}} = \sigma/\sqrt{344}$ **b.** .0322
c. .8924 **d.** 19.1 **e.** less than

6.57 **b.** Central Limit Theorem **c.** .2119

6.59 **a.** .0166 **b.** evidence that $\mu > 3$

6.61 **a.** 5.5, .783 **b.** no **c.** yes, $P(\bar{x} < 5.3) = .3974$

6.63 **a.** .0668 **b.** .5881 **c.** .0212 **d.** unlikely **e.** 19.44 cycles

6.65 **a.** approximately normal, $\mu_{\bar{x}} = 400$, $\sigma_{\bar{x}} = 10$
b. .1587 **c.** .3023

6.67 **a.** .0548 **b.** 97 days

6.69 **a.** .7642 **b.** .2037 **c.** 54,175 **d.** 1

6.71 **a.** $\mu = 6{,}000$, $\sigma^2 = 2{,}400$ **b.** .0202 **c.** no

6.73 all variables are approximately normal

6.75 **a.** .5199 **b.** 4.3934

6.77 .6915

6.79 **a.** .0179 **b.** yes

6.81 **a.** $\mu_{\bar{x}} = 121.74$, $\sigma_{\bar{x}} = 4.86$, approximately normal
b. .7348

Chapter 7

7.3 **a.** 81 ± 1.97 **b.** 81 ± 2.35 **c.** 81 ± 3.09

7.5 **a.** $5.7 \pm .66$ **b.** 5.7 ± 1.03

7.9 **a.** $2.8 \pm .11$ **b.** $.1 \pm .04$

7.11 **a.** (89.3, 91.4)

7.13 **a.** $1.13 \pm .671$ **b.** yes

7.15 **a.** late gabbro: $3.04 \pm .036$; massive gabbro: $2.83 \pm .015$; cumberlandite: $3.05 \pm .044$

7.19 **a.** 5 ± 2.02 **b.** 5 ± 4.37

7.23 **a.** 37.3 ± 7.70
c. true mean of sleepers 1 hour before waking is higher

7.25 (2.26, 4.14)

7.27 **a.** (128.50, 142.59)

7.29 **a.** .031 **b.** .016 **c.** .043

7.31 **a.** $.2 \pm .078$ **b.** $.2 \pm .035$

7.33 **a.** .26 **b.** $.26 \pm .070$

7.35 **c.** yes **d.** $.1438 \pm .0163$

7.37 $.2785 \pm .0830$

7.39 **a.** $.10 \pm .013$ **b.** $.06 \pm .032$

7.41 **a.** .833 **b.** $.833 \pm .133$

7.43 **a.** 153 **b.** 423 **c.** 6,766

7.45 **a.** 385 **b.** 139

7.47 60

7.49 1,068

7.51 $\approx$14,735

7.53 **a.** (4.54, 8.81)
b. (.00024, .00085)
c. (641.86, 1,809.09)
d. (.95, 12.66)

7.55 (4.274, 65.988)

7.57 (.0028, .0105)

7.59 (195.9, 1,397.0)

7.61 **a.** (4.727, 9.445) **b.** no

7.65 **a.** $.6805 \pm .0284$ **b.** 2,089 **c.** 8,353

7.67 **a.** $3.39 \pm .047$

7.69 $.43 \pm .012$

7.71 **a.** (178.88, 299.52)
b. (231.0, 16,584.4)

7.73 **a.** $.044 \pm .104$

7.75 **a.** (.163, 29.142)
b. 85.65 ± 1.33

7.77 **a.** .0129 **b.** yes

Chapter 8

8.5 H_0: $\mu = 9$, H_a: $\mu \neq 9$

8.7 **a.** H_0: $\mu = 22$, H_a: $\mu < 22$
b. conclude mean breaking strength is less than 22 pounds when it really is not
c. conclude mean breaking strength equals 22 pounds when it is really less

8.9 **a.** H_0: $\mu = 4$, H_a: $\mu > 4$
b. conclude that the mean amount of radium-226 exceeds 4 when it does not
c. conclude that the mean amount of radium-226 equals 4 when it is really higher

8.11 **a.** $|z| > 1.96$ **b.** $z > 1.645$
c. $z < -2.326$ **d.** $z < -1.282$

8.13 **a.** reject H_0 **b.** reject H_0
c. reject H_0 **d.** do not reject H_0

8.15 **a.** H_0: $\mu = 500$, H_a: $\mu \neq 500$
b. $|z| > 1.645$

8.17 **a.** H_0: $\mu = 675$, H_a: $\mu < 675$ **b.** $z < -1.645$

8.19 **a.** H_0: $\mu = 16$, H_a: $\mu < 16$
b. $z = -4.31$
c. $z < -2.326$
d. reject H_0

8.21 **a.** $z = .329$ **b.** $z = -10.64$ **c.** $z = -1.49$

8.23 **a.** $|t| > 2.145$ **b.** $|t| > 2.977$ **c.** $t < -1.761$
d. $t > 1.533$ **e.** $t > 1.318$

8.25 **a.** $t = 1.13$, do not reject H_0 **b.** $t = 1.13$, do not reject H_0

8.27 **b.** $z > 1.282$
c. reject H_0

8.29 $z = 5.03$, reject H_0

8.31 $t = -5.16$, reject H_0

8.33 $z = -1.86$, do not reject H_0

8.35 **a.** .3124 **b.** .0178 **c.** 0 **d.** .1470

8.37 **a.** do not reject H_0 **b.** reject H_0
c. reject H_0 **d.** do not reject H_0
e. do not reject H_0

8.39 reject H_0 at $\alpha = .05$

8.41 reject H_0

8.43 **a.** $z = 1.29$, do not reject H_0
b. .0985 **c.** smaller

8.45 **a.** $|z| > 1.96$
b. $z < -1.645$
c. $z > 1.28$
d. $z < -2.33$
e. $|z| > 2.58$

8.47 **a.** $z = -1$, do not reject H_0
b. $z = 2.097$, reject H_0
c. $z = 1.02$, do not reject H_0

8.49 **a.** yes, $z = 4.53$
b. 0

8.51 $z = 2.30$, do not reject H_0

8.53 **a.** $z = 4.59$, reject H_0
b. $z = 1.34$, do not reject H_0

8.55 $z = 11.17$, reject H_0

8.57 **a.** 19.2 **b.** 18.7 **c.** 30.38 **d.** 396

8.59 yes, $\chi^2 = 35.6$

8.61 $\chi^2 = 4.86$, do not reject H_0

8.63 yes, $\chi^2 = 127.86$

8.65 **a.** H_0: $\sigma^2 = 225$, H_a: $\sigma^2 \neq 225$
b. $\chi^2 = 187.8965$
c. do not reject H_0

8.67 $\chi^2 = .969$, reject H_0

8.69 **a.** $\chi^2 > 5.99147$ **b.** $\chi^2 > 7.77944$
c. $\chi^2 > 11.3449$ **d.** $\chi^2 > 18.4753$

8.71 **a.** 10, 30, 30, 30
b. $\chi^2 > 6.25139$
c. at least one of the category probabilities does not equal its hypothesized value
d. $\chi^2 = 3.933$
e. do not reject H_0

8.73 **a.** $p_1 = p_2 = p_3 = p_4 = .25$
b. H_0: $p_1 = p_2 = p_3 = p_4 = .25$
c. $\chi^2 = 19.68$, reject H_0

8.75 $\chi^2 = 76.217$, reject H_0

8.77 $\chi^2 = 4{,}151.715$, reject H_0

8.79 **a.** .0050 **b.** .0985 **c.** .1000

8.81 $z = .87$, do not reject H_0

8.83 **a.** $z = .77$ **b.** $z > 1.282$
c. do not reject H_0 **e.** .2206

8.85 **a.** $t = -3.46$, reject H_0
b. population of test scores is normally distributed
c. $\chi^2 = 14.13$, do not reject H_0

8.87 $z = -.59$, do not reject H_0

8.89 $t = -3.898$, p-value $= .0003$, reject H_0

8.91 $z = -5.81$, reject H_0

Chapter 9

9.1 **a.** 10; 1.31
b. 13; 2.181

9.3 $z = -7.87$

9.5 **a.** $|t| > 2.101$ **b.** $t > 1.372$ **c.** $t < -2.821$

9.7 **a.** 9.8 ± 7.94 **b.** 9.8 ± 9.61 **c.** 9.8 ± 13.14

9.11 **a.** H_0: $\mu_1 - \mu_2 = 0$, H_a: $\mu_1 - \mu_2 \neq 0$
b. reject H_0
c. -98 ± 37.818

9.13 **a.** $z = -2.93$, reject H_0
b. $-.33 \pm .22$

9.15 **a.** $z = 5.82$, reject H_0
b. $t = 3.19$, reject H_0

9.17 **a.** $z = 3.52$, reject H_0
b. $z = 0$, do not reject H_0

9.19 -7.93 ± 3.63; middle

9.21 **a.** $|z| > 2.575$; $z = 9.195$; reject H_0
b. $t > 2.132$; $t = 13.42$; reject H_0
c. $t < -1.533$; $t = -2.98$; reject H_0

9.23 **a.** $z = 26.24$ **b.** 0 **c.** reject H_0 **d.** 19.3 ± 1.44

9.27 $-.1024 \pm .0642$

9.29 $t = -1.08$, p-value $= .3033$, do not reject H_0

9.31 **a.** yes **b.** yes **c.** $t = -.686$, do not reject H_0

9.33 **a.** yes, $t = -2.49$, p-value $= .0167$

9.35 **a.** $z = -2.74$ **b.** $|z| > 1.96$
c. reject H_0 **d.** $-.10 \pm .060$
e. $-.10 \pm .071$ **f.** $-.10 \pm .093$

9.37 **a.** .05 ± .099

b. .05 ± .045

c. .28 ± .075

9.39 (0.12623, 0.30551)

9.41 **a.** .18 ± .058 **b.** $z = 5.93$, reject H_0

9.43 Auto Theft: $z = 4.68$, reject H_0; Concealed Weapon: $z = 5.6$, reject H_0; Police Assault: $z = -0.615$, do not reject H_0; Shoplifting: $z = 1.66$, do not reject H_0

9.45 $z = 3.20$, reject H_0

9.47 $z = 3.03$, reject H_0

9.49 **a.** $e_{11} = 15.697, e_{12} = 30.945, e_{13} = 27.358,$ $e_{21} = 19.303, e_{22} = 38.055, e_{23} = 33.642$

b. $\chi^2 = 3.7396$

c. yes

9.51 $\chi^2 = 1.124$, do not reject H_0

9.53 $\chi^2 = 540.44$, reject H_0

9.55 **a.** $\chi^2 = 5.61$, reject H_0 **b.** yes **c.** .09 ± .062, yes

9.57 $\chi^2 = 1.56$, do not reject H_0

9.59 no, $\chi^2 = 2.6177$

9.61 **a.** 3.18 **b.** 2.62 **c.** 2.10

9.63 **a.** 1.42 **b.** 3.88 **c.** 1.77

9.65 **a.** H_0: $\sigma_1^2/\sigma_2^2 = 1$, H_a: $\sigma_1^2/\sigma_2^2 \neq 1$

b. $F = 1.57$, reject H_0

9.67 **a.** $F = 1.60$, do not reject H_0

b. $F = 2.1609$, do not reject H_0

9.69 **b.** H_0: $\sigma_E^2/\sigma_U^2 = 1$, H_a: $\sigma_E^2/\sigma_U^2 \neq 1$

c. $F = 2.447$, reject H_0

9.71 $F = 2.03$, do not reject H_0

9.73 **a.** $T_2 \leq 24$ or $T_2 \geq 56$ **b.** $T_2 \leq 52$ **c.** $T_2 \geq 43$

9.75 **a.** $T_1 = 67, T_2 = 38$ **b.** $T_1 \geq 58$ **c.** reject H_0

9.77 **b.** H_0: The distributions of the change in the number of swollen joints are identical for the placebo and collagen groups

c. reject H_0

9.79 $z = 3.238$, reject H_0

9.81 $T = 35$, do not reject H_0

9.83 **a.** $T^+ \geq 35$

b. $T^+ \leq 3$ or $T^+ \geq 25$

9.85 **a.** $T^+ = 33.5$, reject H_0

b. $T^+ = 33.5$, reject H_0

9.87 $T^+ = 9$, reject H_0

9.89 $T^+ = 11$, reject H_0

9.91 $T^+ = 1$, reject H_0

9.93 **a.** yes, $z = 5.82$ **b.** yes, $z = 7.14$

9.95 **a.** yes, $z = -2.21$

b. −.027 ± .024

9.97 $z = -3.05$, reject H_0

9.99 **a.** reject H_0 at $\alpha = .05$ for variable 1 and 3

9.101 **a.** −12 ± 25.62 **b.** no evidence of a difference

c. no, $F = 1.29$

9.103 **a.** $z = -6.41$, reject H_0

9.105 **a.** H_0: $\mu_w - \mu_m = 0$, H_a: $\mu_w - \mu_m > 0$

b. reject H_0

c. H_0: $\mu_y - \mu_o = 0$, H_a: $\mu_y - \mu_o > 0$

d. reject H_0

9.107 yes, $z = -3.53$

9.109 **a.** Hired/Male: 6; Hired/Female: 3; Not Hired/Male: 7; Not Hired/Female: 12

b. $\chi^2 = 2.184$, do not reject H_0

Chapter 10

10.1 **a.** 1.5 **b.** 2 **c.** increase by 2

d. decrease by 2 **e.** 1.5

10.5 β_0 = y-intercept; β_1 = slope

10.7 **a.** $y = \beta_0 + \beta_1 x + \varepsilon$

b. $b_0 = 946.72$, $b_1 = -2.19$

c. $\hat{y} = 946.72 - 2.19x$

f. 771.53

10.9 **a.** yes

b. $b_0 = -.643$, $b_1 = .0638$

10.11 **a.** two-point field goals

b. $\hat{y} = -.00021 + .8116x$

d. $b_0 = .0002$, $b_1 = .467$

e. $b_0 = -.0174$, $b_1 = .515$

10.13 **a.** yes, negative

b. $y = \beta_0 + \beta_1 x + \varepsilon$

c. $\hat{y} = 155.91 - 1.09x$

f. 128.753 minutes

10.15 **a.** .0275 **b.** .16583

10.17 **a.** 2.75 **b.** 1.6583

10.19 SSE = 184,297.7; $s^2 = 23{,}037.21$, $s = 151.78$

10.21 x_1: $s = .596$; x_2: $s = .901$; x_3: $s = .861$; x_1

10.23 SSE = 2,290.98; $s^2 = 91.639$; $s = 9.573$

10.25 **a.** $|t| > 3.182$ **b.** $t > 1.638$ **c.** $t < -3.365$

10.27 **a.** $t = -1.55$; reject H_0

b. −300.7 ± 390.86

10.29 yes, $t = 5.053$

10.31 no, $t = -2.05$, p-value = .0705

10.33 **a.** $\hat{y} = 6.732 - .0546x$
b. $t = -8.62$, reject H_0
c. $-.0546 \pm .01305$
d. yes

10.35 **a.** $\hat{y} = 25.796 + .37099x$ **b.** $t = 3.72$, reject H_0

10.37 **a.** $t = 2.623$, $t > 2.896$, do not reject H_0
b. $t = -6.558$; $t < -2.403$, reject H_0
c. $t = -4.379$, $|t| > 1.860$; reject H_0

10.39 piano: $t = 2.40$, reject H_0; bench: $t = -.27$, do not reject H_0; motorbike: $t = 3.78$, reject H_0; armchair: $t = 1.48$, do not reject H_0; teapot: $t = 14.44$, reject H_0

10.41 **a.** H_0: $\rho = 0$, H_a: $\rho \neq 0$ **b.** if $r = .14$, $t = 1.649$

10.43 **a.** reject H_0
b. do not reject H_0
c. reject H_0

10.45 **a.** $y = \beta_0 + \beta_1 x + \varepsilon$ **b.** H_0: $\beta_1 = 0$, H_a: $\beta_1 > 0$
c. slight negative relationship **d.** no

10.47 .7717

10.49 .1225

10.51 .3185

10.53 .386

10.55 .5776

10.57 **a.** 10.6 **b.** $10.6 \pm .223$ **c.** $8.9 \pm .319$
d. $12.3 \pm .319$ **e.** width increases **f.** 10.6 ± 1.023

10.59 **a.** 39 ± 5.85 **b.** 39 ± 7.85 **c.** prediction interval

10.61 ($424.90, $1,161.90)

10.63 **a.** (83.61, 173.89) **b.** (0, 279.09)

10.65 **a.** (3.51, 17.20)
b. 32° falls outside the range of the sampled temperatures

10.67 (−6.97, 37.36)

10.69 **b.** $\hat{y} = .35 + 1.05x$ **d.** $t = 7$ **e.** reject H_0
f. .942 **g.** $.35 \pm .827$

10.71 **b.** $\hat{y} = 3.343 + .576x$ **d.** $t = 7.993$
e. reject H_0 **f.** .927 **g.** 1.04 ± 1.42

10.73 **a.** misspecified model
b. unequal error variances **c.** outlier
d. correlated errors **e.** nonnormal errors

10.75 **a.** 3.197, 3.215, −2.258, .269, −1.713, −7.186, −2.659, 3.359, −2.132, 5.904
b. curvilinear; add quadratic term
c. no outliers

10.77 no problems

10.79 **d.** outlier

10.81 **d.** yes; add curvature term

10.83 **a.** $\hat{y} = 85.0138 + .04045x$
b. yes, $t = 4.666$
c. no **d.** yes

10.85 $y = \beta_0 + \beta_1 x_1 + \beta_2 x_2 + \beta_3 x_3 + \beta_4 x_4 + \beta_5 x_5 + \varepsilon$

10.87 **a.** .0439 **b.** 1.251 **c.** yes, reject H_0

10.89 $F = 1.896$, do not reject H_0

10.91 **b.** reject H_0
c. x_2 does not seem to be a useful linear predictor

10.93 **a.** Qualitative: x_1, x_2, and x_3; Quantitative: x_4 and x_5
b. yes, $F = 3.12$
c. For every 1 month increase in length in time, we estimate that subjective role will decrease by .0017 points, holding x_1, x_2, x_3, and x_5 constant.
d. no, $t = -1.162$

10.95 **a.** model explains 82% of sample variation in y
b. model is useful
c. For every 1 additional average number of years of total education of teachers, we estimate y to increase by 1.98%, holding x_2, x_3, x_4, x_5, x_6, x_7, x_8, and x_9 fixed.
d. x_1 is a useful predictor of y
e. do not reject H_0

10.97 **a.** model is useful
b. About 16.5% of the total variation of the overall parent involvement scores about their mean can be explained by the model.
d. $t = 4.77$, reject H_0
e. $t = 1.26$, do not reject H_0

10.99 **a.** For every percentile increase in rental price, number of homeless increases by 2.87.
b. Reject H_0; $\beta_4 = 0$ in favor of H_a; $\beta_4 < 0$
c. Inflated overall Type 1 error rate.
d. Model explains 83% of sample variation in number of homeless.

10.101 **b.** .893 **c.** yes

10.103 **c.** .7635
d. no

10.105 $r_s = .7167$, reject H_0

10.107 reject H_0

10.109 **b.** $\hat{y} = 3.306 + .014755x$
d. $t = 5.356$, reject H_0: $\beta_1 = 0$
e. $.014755 \pm .0055679$
f. .6116

g. .3740

h. (1.9898, 10.5242)

j. no model modifications

10.111 $\hat{y} = 5.348 + .530x$; $t = .57$, do not reject H_0; $R^2 = .02$

10.113 **a.** $\hat{y} = -13.622 - .0533x$; $t = -6.80$, reject H_0; $R^2 = .852$; add x^2 term to model

b. no

10.115 **a.** $x_1, x_2, x_3, x_4, x_6, x_7$, and x_8

b. H_0: $\beta_1 = \beta_2 = \cdots = \beta_9 = 0$

c. reject H_0, $p < .001$

d. 29% of total variation in sample average lengths about their mean is explained by the model.

e. Estimated difference in average length of the story between female suspects and male suspects, holding all other variables constant, is .26.

f. For every $1 increase in per capita income, estimate the length of the story to decrease by $.02, holding all other variables constant.

g. no

10.117 $\hat{y} = -3.02 + 4.172x_1 - .073x_2$; $F = 2.158$, do not reject H_0: $\beta_1 = \beta_2 = 0$

10.119 **a.** yes, $t = 3.05$

b. $r = -.21$

10.121 **a.** .922 **b.** no

Chapter 11

11.1 **b.** 2.381 **c.** 7.502 **d.** 3.15

e.

SOURCE	df	SS	MS	F
Treatments	1	7.502	7.502	3.15
Error	11	26.190	2.381	
Total	12	33.692		

f. $F > 4.84$

g. do not reject H_0

11.3 **a.** 3.111 **b.** 1.405 **c.** 2

d. 7 **e.** $F = 2.214$

f.

SOURCE	df	SS	MS	F
Treatment	2	6.222	3.111	2.214
Error	7	9.834	1.405	
Total	9	16.056		

g. $F > 4.74$

h. do not reject H_0

i. .1798; yes

11.5 **a.** age at onset of the eating disorder

b. three groups of birth dates: (1) before 1950, (2) between 1950 and 1959, and (3) after 1960

c. H_0: $\mu_1 = \mu_2 = \mu_3$

d. reject H_0

11.7 H_0: $\mu_1 = \mu_2 = \mu_3$, $F = 1.62$, do not reject H_0

11.9 $F = 1.22$, reject H_0

11.11 **a.** H_0: $\mu_1 = \mu_2 = \mu_3 = \mu_4$

b. reject H_0

c. H_0: $\mu_1 = \mu_2 = \mu_3 = \mu_4$

d. do not reject H_0

11.13 **a.** 3 **b.** 10 **c.** 6 **d.** 28

11.15 **a.** 5.557 **b.** 14.619 **c.** 1.748

11.17 A $\overline{\text{C D}}$ B

11.19 **a.** H_0: $\mu_1 = \mu_2 = \mu_3 = \mu_4$

b. reject H_0

c. $\mu_L < (\mu_E, \mu_F) < \mu_M$

11.21 **a.** 6 **b.** $(\mu_3, \mu_6, \mu_9) > \mu_{12}$

11.23 $\mu_1 < (\mu_2, \mu_3, \mu_4)$

11.25 $(\mu_{\text{happy}}, \mu_{\text{angry}}) > \mu_{\text{sad}}$

11.27 **a.**

SOURCE	df	SS	MS	F
A	2	100	50	25
B	1	559	559	279.5
AB Interaction	2	5	2.5	1.25
Error	18	36	2	
Total	23	700		

b. $F = 1.25$, do not reject H_0

c. $F = 25$, reject H_0

d. $F = 279.5$, reject H_0

11.29 **a.** .00333 **b.** .00833 **c.** .00179 **d.** .00333

11.31 **a.** luckiness: lucky, unlucky, uncertain; competition: competitive, noncompetitive

b. L × C: $F = .72$, do not reject H_0; L: $F = 1.39$, do not reject H_0; C: $F = 2.84$, do not reject H_0

11.33 **b.** Interaction: $F = .6375$, do not reject H_0; Diet: $F = .2175$, do not reject H_0; Weight: $F = 141.18$, reject H_0

11.35 **a.** treatments: (completed$_A$, completed$_B$), (completed$_A$, not completed$_B$), (not completed$_A$, completed$_B$), (not completed$_A$, not completed$_B$),

c. yes

11.37 Interaction: $F = .075$, do not reject H_0; Age: $F = .282$, do not reject H_0; Surgery: $F = 19.33$, reject H_0

11.39 **a.** yes

11.41 no, unequal variances

11.43 normal and equal variance assumptions appear to be violated

11.47 see answer to 11.3f

11.49

SOURCE	df	SS	MS	F
Treatment	5	23.0852	4.6170	3.9589
Error	30	34.987	1.1662	
Total	35	58.0722		

11.51

SOURCE	df	SS	MS	F
A	1	8.0679	8.0679	141.18
B	1	.01243	.01243	0.218
$A \times B$	1	.03643	.03643	0.638
Error	24	1.3715	.05715	
Total	27	9.4883		

11.53 **a.** 8 **b.** 6.5 **c.** 15.5 **d.** 6

11.55 **a.** independent samples

b. H_0: $\eta_{I} = \eta_{II} = \eta_{III}$ **c.** $H > 5.99147$

d. $T_1 = 46$, $T_2 = 21.5$, $T_3 = 68.5$

e. $H = 12.47$ **f.** reject H_0

11.57 **b.** $H = 11.20$ **c.** reject H_0

11.59 **c.** $H = 37.65$, reject H_0

11.61 $H = 20.21$, reject H_0

11.63 **a.** five filtering methods

b. relevance rating **c.** reject H_0

11.65 **a.** $F = 9.46$, reject H_0 **b.** $\mu_{\text{light blond}}, \mu_{\text{dark blond}}$

11.67 **a.** $F = 63.37$, reject H_0; $\mu_{\text{VF}} > \mu_{\text{NV-F}}, \mu_{\text{VS}}, \mu_{\text{NV-S}}$

11.71 **a.** interaction: do not reject H_0; herds: reject H_0; seasons: do not reject H_0

b. yes **c.** $\mu_{\text{PLC}}, \mu_{\text{LGN}} < \mu_{\text{QMD}}, \mu_{\text{MTZ}}$

11.73 **b.** $F = 12.33$, reject H_0 **c.** $\mu_{\text{C}} > (\mu_{\text{H}}, \mu_{\text{B}})$

11.75

SOURCE	df	SS	MS	F
Burn-In	8	27.9744	3.497	54.63
Inspection	2	43.0841	21.542	336.54
Interaction	16	97.5535	6.097	95.25
Within	54	3.4565	0.064	
Total	80	172.0685		

Microsoft Excel and MINITAB Index

Subject Index

T

U

V

LICENSE AGREEMENT

YOU SHOULD CAREFULLY READ THE FOLLOWING TERMS AND CONDITIONS BEFORE BREAKING THE SEAL ON THE PACKAGE. AMONG OTHER THINGS, THIS AGREEMENT LICENSES THE ENCLOSED SOFTWARE TO YOU AND CONTAINS WARRANTY AND LIABILITY DISCLAIMERS. BY BREAKING THE SEAL ON THE PACKAGE, YOU ARE ACCEPTING AND AGREEING TO THE TERMS AND CONDITIONS OF THIS AGREEMENT. IF YOU DO NOT AGREE TO THE TERMS OF THIS AGREEMENT, DO NOT BREAK THE SEAL. YOU SHOULD PROMPTLY RETURN THE PACKAGE UNOPENED.

LICENSE.

Subject to the provisions contained herein, Prentice-Hall, Inc. ("PH") hereby grants to you a non-exclusive, non-transferable license to use the object code version of the computer software product ("Software") contained in the package on a single computer of the type identified on the package.

SOFTWARE AND DOCUMENTATION.

PH shall furnish the Software to you on media in machine-readable object code form and may also provide the standard documentation ("Documentation") containing instructions for operation and use of the Software.

LICENSE TERM AND CHARGES.

The term of this license commences upon delivery of the Software to you and is perpetual unless earlier terminated upon default or as otherwise set forth herein.

TITLE.

Title, and ownership right, and intellectual property rights in and to the Software and Documentation shall remain in PH and/or in suppliers to PH of programs contained in the Software. The Software is provided for your own internal use under this license. This license does not include the right to sublicense and is personal to you and therefore may not be assigned (by operation of law or otherwise) or transferred without the prior written consent of PH. You acknowledge that the Software in source code form remains a confidential trade secret of PH and/or its suppliers and therefore you agree not to attempt to decipher or decompile, modify, disassemble, reverse engineer or prepare derivative works of the Software or develop source code for the Software or knowingly allow others to do so. Further, you may not copy the Documentation or other written materials accompanying the Software.

UPDATES.

This license does not grant you any right, license, or interest in and to any improvements, modifications, enhancements, or updates to the Software and Documentation. Updates, if available, may be obtained by you at PH's then current standard pricing, terms, and conditions.

LIMITED WARRANTY AND DISCLAIMER.

PH warrants that the media containing the Software, if provided by PH, is free from defects in material and workmanship under normal use for a period of sixty (60) days from the date you purchased a license to it.

THIS IS A LIMITED WARRANTY AND IT IS THE ONLY WARRANTY MADE BY PH. THE SOFTWARE IS PROVIDED 'AS IS' AND PH SPECIFICALLY DISCLAIMS ALL WARRANTIES OF ANY KIND, EITHER EXPRESS OR IMPLIED, INCLUDING, BUT NOT LIMITED TO, THE IMPLIED WARRANTY OF MERCHANTABILITY AND FITNESS FOR A PARTICULAR PURPOSE. FURTHER, COMPANY DOES NOT WARRANT, GUARANTY OR MAKE ANY REPRESENTATIONS REGARDING THE USE, OR THE RESULTS OF THE USE, OF THE SOFTWARE IN TERMS OF CORRECTNESS, ACCURACY, RELIABILITY, CURRENTNESS, OR OTHERWISE AND DOES NOT WARRANT THAT THE OPERATION OF ANY SOFTWARE WILL BE UNINTERRUPTED OR ERROR FREE. COMPANY EXPRESSLY DISCLAIMS ANY WARRANTIES NOT STATED HEREIN. NO ORAL OR WRITTEN INFORMATION OR ADVICE GIVEN BY PH, OR ANY PH DEALER, AGENT, EMPLOYEE OR OTHERS SHALL CREATE, MODIFY OR EXTEND A WARRANTY OR IN ANY WAY INCREASE THE SCOPE OF THE FOREGOING WARRANTY, AND NEITHER SUBLICENSEE OR PURCHASER MAY RELY ON ANY SUCH INFORMATION OR ADVICE. If the media is subjected to accident, abuse, or improper use; or if you violate the terms of this Agreement, then this warranty shall immediately be terminated. This warranty shall not apply if the Software is used on or in conjunction with hardware or programs other than the unmodified version of hardware and programs with which the Software was designed to be used as described in the Documentation.

LIMITATION OF LIABILITY.

Your sole and exclusive remedies for any damage or loss in any way connected with the Software are set forth below. UNDER NO CIRCUMSTANCES AND UNDER NO LEGAL THEORY, TORT, CONTRACT, OR OTHERWISE, SHALL PH BE LIABLE TO YOU OR ANY OTHER PERSON FOR ANY INDIRECT, SPECIAL, INCIDENTAL, OR CONSEQUENTIAL DAMAGES OF ANY CHARACTER INCLUDING, WITHOUT LIMITATION, DAMAGES FOR LOSS OF GOODWILL, LOSS OF PROFIT, WORK STOPPAGE, COMPUTER FAILURE OR MALFUNCTION, OR ANY AND ALL OTHER COMMERCIAL DAMAGES OR LOSSES, OR FOR ANY OTHER DAMAGES EVEN IF PH SHALL HAVE BEEN INFORMED OF THE POSSIBILITY OF SUCH DAMAGES, OR FOR ANY CLAIM BY ANY OTHER PARTY. PH'S THIRD PARTY PROGRAM SUPPLIERS MAKE NO WARRANTY, AND HAVE NO LIABILITY WHATSOEVER, TO YOU. PH's sole and exclusive obligation and liability and your exclusive remedy shall be: upon PH's election, (i) the replacement of your defective media; or (ii) the repair or correction of your defective media if PH is able, so that it will conform to the above warranty; or (iii) if PH is unable to replace or repair, you may terminate this license by returning the Software. Only if you inform PH of your problem during the applicable warranty period will PH be obligated to honor this warranty. You may contact PH to inform PH of the problem as follows:

SOME STATES OR JURISDICTIONS DO NOT ALLOW THE EXCLUSION OF IMPLIED WARRANTIES OR LIMITATION OR EXCLUSION OF CONSEQUENTIAL DAMAGES, SO THE ABOVE LIMITATIONS OR EXCLUSIONS MAY NOT APPLY TO YOU. THIS WARRANTY GIVES YOU SPECIFIC LEGAL RIGHTS AND YOU MAY ALSO HAVE OTHER RIGHTS WHICH VARY BY STATE OR JURISDICTION.

MISCELLANEOUS.

If any provision of this Agreement is held to be ineffective, unenforceable, or illegal under certain circumstances for any reason, such decision shall not affect the validity or enforceability (i) of such provision under other circumstances or (ii) of the remaining provisions hereof under all circumstances and such provision shall be reformed to and only to the extent necessary to make it effective, enforceable, and legal under such circumstances. All headings are solely for convenience and shall not be considered in interpreting this Agreement. This Agreement shall be governed by and construed under New York law as such law applies to agreements between New York residents entered into and to be performed entirely within New York, except as required by U.S. Government rules and regulations to be governed by Federal law.

YOU ACKNOWLEDGE THAT YOU HAVE READ THIS AGREEMENT, UNDERSTAND IT, AND AGREE TO BE BOUND BY ITS TERMS AND CONDITIONS. YOU FURTHER AGREE THAT IT IS THE COMPLETE AND EXCLUSIVE STATEMENT OF THE AGREEMENT BETWEEN US THAT SUPERSEDES ANY PROPOSAL OR PRIOR AGREEMENT, ORAL OR WRITTEN, AND ANY OTHER COMMUNICATIONS BETWEEN US RELATING TO THE SUBJECT MATTER OF THIS AGREEMENT.

U.S. GOVERNMENT RESTRICTED RIGHTS.

Use, duplication or disclosure by the Government is subject to restrictions set forth in subparagraphs (a) through (d) of the Commercial Computer-Restricted Rights clause at FAR 52.227-19 when applicable, or in subparagraph (c) (1) (ii) of the Rights in Technical Data and Computer Software clause at DFARS 252.227-7013, and in similar clauses in the NASA FAR Supplement.